Genetics and Evolution of Infectious Diseases

Genetics and Evolution of Infectious Diseases

Edited by

Michel Tibayrenc

AMSTERDAM • BOSTON • HEIDELBERG • LONDON • NEW YORK • OXFORD
PARIS • SAN DIEGO • SAN FRANCISCO • SINGAPORE • SYDNEY • TOKYO

Elsevier
32 Jamestown Road London NW1 7BY
30 Corporate Drive, Suite 400, Burlington, MA 01803, USA

First edition 2011

Copyright © 2011 Elsevier Inc. All rights reserved

No part of this publication may be reproduced or transmitted in any form or by any means, electronic or mechanical, including photocopying, recording, or any information storage and retrieval system, without permission in writing from the publisher. Details on how to seek permission, further information about the Publisher's permissions policies and our arrangement with organizations such as the Copyright Clearance Center and the Copyright Licensing Agency, can be found at our website: www.elsevier.com/permissions

This book and the individual contributions contained in it are protected under copyright by the Publisher (other than as may be noted herein).

Notices

Knowledge and best practice in this field are constantly changing. As new research and experience broaden our understanding, changes in research methods, professional practices, or medical treatment may become necessary.

Practitioners and researchers must always rely on their own experience and knowledge in evaluating and using any information, methods, compounds, or experiments described herein. In using such information or methods they should be mindful of their own safety and the safety of others, including parties for whom they have a professional responsibility.

To the fullest extent of the law, neither the Publisher nor the authors, contributors, or editors, assume any liability for any injury and/or damage to persons or property as a matter of products liability, negligence or otherwise, or from any use or operation of any methods, products, instructions, or ideas contained in the material herein.

British Library Cataloguing in Publication Data
A catalogue record for this book is available from the British Library

Library of Congress Cataloging-in-Publication Data
A catalog record for this book is available from the Library of Congress

ISBN: 978-0-12-384890-1

For information on all Elsevier publications
visit our website at www.elsevierdirect.com

This book has been manufactured using Print On Demand technology. Each copy is produced to order and is limited to black ink. The online version of this book will show color figures where appropriate.

Working together to grow
libraries in developing countries

www.elsevier.com | www.bookaid.org | www.sabre.org

ELSEVIER   BOOK AID International   Sabre Foundation

Cover image: Jenny Telleria

Contents

List of Contributors	xiii
Preface	xix
Introduction	xxi
Michel Tibayrenc	

Part 1   Methodological/Generalist Chapters — 1

1   Virus Species — 3
Marc H.V. Van Regenmortel
- 1.1   The Species Problem in Biology — 3
- 1.2   Classes Versus Individuals — 4
- 1.3   Species Taxa as Cluster Classes — 4
- 1.4   Species: Concrete Versus Abstract Entities — 5
- 1.5   The Nature of Viruses — 6
- 1.6   Viruses Should not be Reduced to Virions — 6
- 1.7   What is a Virus Species? — 8
- 1.8   Species and Quasi-Species — 9
- 1.9   Definition of Virus Species Versus Identification of Viruses — 9
- 1.10 Using Viral Genomes to Define Virus Species — 10
- 1.11 Names and Typography of Virus Species — 12

2   A Theory-Based Pragmatism for Discovering and Classifying Newly Divergent Bacterial Species — 21
Sarah Kopac and Frederick M. Cohan
- 2.1   Introduction — 21
- 2.2   Ecological Breadth of Recognized Species — 24
- 2.3   Models of Bacterial Speciation — 27
- 2.4   Algorithms for Identifying Ecotypes — 31
- 2.5   Confirming the Ecological Distinctness of Ecotypes — 32
- 2.6   Ecological Homogeneity within Ecotypes — 34
- 2.7   Are Bacterial Ecotypes Cohesive? — 35
- 2.8   Incorporating Ecology into Bacterial Systematics — 36

3   Population Structure of Pathogenic Bacteria — 43
D. Ashley Robinson, Jonathan C. Thomas and William P. Hanage
- 3.1   Introduction — 43
- 3.2   Bacterial Population Structure by Other Means — 44

	3.3	What, if Anything, is a Bacterial Population?	48
	3.4	Conclusions	52
4	**Epidemiology and Evolution of Fungal Pathogens in Plants and Animals**		**59**
	Pierre Gladieux, Edmond J. Byrnes III, Gabriela Aguileta, Matthew C. Fisher, Joseph Heitman and Tatiana Giraud		
	4.1	Introduction	59
	4.2	Major Human and Animal Pathogenic Fungi	61
	4.3	New and Emerging Mycoses	64
	4.4	Plant Pathogenic Fungi	70
	4.5	New and Emerging Plant Diseases	72
	4.6	Modern Molecular Epidemiologic Tools for Investigating Fungal Diseases	74
	4.7	Population Genetics of Pathogenic Fungi	78
	4.8	Species and Speciation in Pathogenic Fungi	94
	4.9	Mating and Pathogenesis	99
	4.10	Genomics of Fungi: What makes a Fungus Pathogenic?	102
	4.11	Conclusion	106
5	**Clonal Evolution**		**133**
	Thierry de Meeûs and Franck Prugnolle		
	5.1	Introduction	133
	5.2	Definitions	134
	5.3	The Origin of Life, the Origin of Propagation and Recombination	135
	5.4	Clonal Modes	136
	5.5	Quantifying the Importance of Asexuality in the Biosphere	137
	5.6	Genetic Consequences of Asexuality	137
	5.7	Evolution and the Paradox of Sex	138
	5.8	Clonal Microevolution	140
	5.9	Conclusions	143
6	**Coevolution of Host and Pathogen**		**147**
	Andrew D. Morgan and Britt Koskella		
	6.1	Coevolution of Host and Pathogen	147
	6.2	The Process of Antagonistic Coevolution	153
	6.3	Testing for Host—Pathogen Coevolution	156
	6.4	Implications of Coevolution	162
	6.5	Summary/Future Outlook	163
7	**Elucidating Human Migrations by Means of their Pathogens**		**173**
	Aude Gilabert and Thierry Wirth		
	7.1	Introduction	173
	7.2	Using Pathogens as Genetic Tracers for Host History	174

	7.3	Candidates	176
	7.4	Conclusion	193

8 Phylogenetic Analysis of Pathogens   203
David A. Morrison

8.1	Introduction	203
8.2	The Uses of Phylogenies	204
8.3	The Logic of Phylogeny Reconstruction	206
8.4	Characters and Samples	208
8.5	The Practice of Phylogeny Reconstruction	212
8.6	Choosing a Method	217
8.7	Phylogenetic Trees	219
8.8	Phylogenetic Networks	223

9 Evolutionary Responses to Infectious Disease   233
Gregory Cochran and Henry Harpending

9.1	Introduction	233
9.2	Parasites as Our Friends	234
9.3	Demography and Parasites	235
9.4	Agriculture	236
9.5	Some Lessons from Malaria	237
9.6	Disease and Standard of Living in Preindustrial Societies: A Simple Model	240
9.7	Population Limitation	241
9.8	Disease, Mating, and Reproductive Strategy	244
9.9	Prosperity and the Postindustrial Era Mortality Decline	245

10 Infectious Disease Genomics   249
Yu-Tsueng Liu

10.1	Introduction	249
10.2	Vaccine Target	251
10.3	Drug Target	253
10.4	Vector Control	253
10.5	Diagnostic Target and Pathogen Discovery	254
10.6	Conclusion	255

11 Proteomics and Host–Pathogen Interactions: A Bright Future?   263
David G. Biron, Dobrin Nedelkov, Dorothée Missé and Philippe Holzmuller

11.1	Introduction	263
11.2	Interest of Proteomics to Study Host–Pathogen Interactions	264
11.3	Retrospective Analysis of Previous Proteomics Studies	265
11.4	Pitfalls of the Current Conceptual Approach in many "Parasito-Proteomics" Studies	271

	11.5	Toward New Conceptual Approaches to Decipher the Host–Parasite Interactions for Parasites with Short or Complex Life Cycle	272
	11.6	Population Proteomics, an Emerging Discipline, to Study Host–Parasite Interactions	278
	11.7	5-Year View	286
	11.8	Conclusion	289
	Glossary		290
12	**The Evolution of Antibiotic Resistance**		**305**
	Fernando González-Candelas, Iñaki Comas, José Luis Martínez, Juan Carlos Galán and Fernando Baquero		
	12.1	Introduction	305
	12.2	Mechanisms and Sources of Antibiotic Resistance	309
	12.3	Evolution of Antibiotic-Resistance Genes	314
	12.4	Limitations to Adaptation and the Cost of Resistance	320
	12.5	Can the Evolution of Antibiotic Resistance be Predicted?	326
	12.6	Conclusions and Perspectives	327
	Glossary		328
13	**General Mechanisms of Antiviral Resistance**		**339**
	Anthony Vere Hodge and Hugh J. Field		
	13.1	Introduction	339
	13.2	Evolutionary Outcomes that have Enabled Viruses to Resist Control	343
	13.3	One Principle Mechanism for Development of Resistance	346
	13.4	Viruses with Segmented Genomes: Additional Resistance Mechanism	347
	13.5	Evolution of Resistant Mutants	347
	13.6	Illustrative Examples of Resistance to Specific Antiviral Drugs	348
	13.7	Optimizing Drug Combinations to Avoid Resistance	353
	13.8	Unexpected Consequences of Resistance Mutations	355
	13.9	A Role for Compounds Targeting Host Proteins for Antiviral Therapy	357
	13.10	Conclusion	358
14	**Evolution of Resistance to Insecticide in Disease Vectors**		**363**
	Pierrick Labbé, Haoues Alout, Luc Djogbénou, Nicole Pasteur and Mylène Weill		
	14.1	Introduction	363
	14.2	Part 1: Insecticide Resistance: Definition and History	365
	14.3	Part 2: Mechanisms of Resistance	371
	14.4	Part 3: Treatment Practices and Resistance Management Strategies	386
	14.5	Conclusion	393

| 15 | Genetics of Major Insect Vectors | 411 |

Patricia L. Dorn, François Noireau, Elliot S. Krafsur,
Gregory C. Lanzaro and Anthony J. Cornel

	15.1	Introduction	411
	15.2	Genetics of Tsetse Flies, *Glossina* spp. (Diptera: Glossinidae), and African Trypanosomiasis	412
	15.3	Genetics of the Triatominae (Hemiptera, Reduviidae) and Chagas Disease	426
	15.4	The *Anopheles gambiae* Complex	446
	Glossary		459

| 16 | Modern Morphometrics of Medically Important Insects | 473 |

Jean-Pierre Dujardin

	16.1	Introduction	473
	16.2	Landmark-Based Geometric Morphometry	474
	16.3	Nonenvironmental Sources of Metric Change	477
	16.4	Environmental Sources of Metric Changes	480
	16.5	The Regulation of Phenotype	483
	16.6	Applications in Medical Entomology	485
	16.7	Conclusion	494

| 17 | Multilocus Sequence Typing of Pathogens | 503 |

Marcos Pérez-Losada, Megan L. Porter, Raphael P. Viscidi and Keith A. Crandall

	17.1	Introduction	503
	17.2	Molecular Design and Development of MLST	504
	17.3	MLST Databases	508
	17.4	Advantages and Disadvantages of MLST	508
	17.5	Analytical Approaches	510
	17.6	Applications of MLST	514
	17.7	Conclusions and Prospects	517

| 18 | Omics, Bioinformatics, and Infectious Disease Research | 523 |

Konrad H. Paszkiewicz and Mark van der Giezen

	18.1	The Need for Bioinformatics	523
	18.2	Metagenomics	524
	18.3	Comparative Genomics	525
	18.4	Pan-Genomics	528
	18.5	Transcriptomics	529
	18.6	Proteomics	531
	18.7	Structural Genomics/Proteomics	532
	18.8	A "How-To" of Second-Generation Sequencing	533
	18.9	Alignment or Assembly of Second-generation Sequences	534
	18.10	Concluding Remarks	536

19	**Genomics of Infectious Diseases and Private Industry**	**541**
	Guy Vernet	
	19.1 Introduction	541
	19.2 Customers and Their Needs	542
	19.3 Technologies and Companies	545
	19.4 Conclusion and Perspectives	551
20	**Current Progress in the Pharmacogenetics of Infectious Disease Therapy**	**555**
	Tabitha Mahungu and Andrew Owen	
	20.1 Introduction	555
	20.2 The Role of Pharmacogenetics in Communicable Diseases	556
	20.3 Strategies for Investigating New Pharmacogenetic Associations	556
	20.4 Implementation of Pharmacogenetics	558
	20.5 Pharmacogenetics of HIV Therapy	560
	20.6 Pharmacogenetics of Antimalarial Therapy	566
	20.7 Pharmacogenetics of Antituberculosis Therapy	568
	20.8 Summary and Perspective	568

Part 2 Specialized Chapters 579

21	**Genetic Exchange in Trypanosomatids and Its Relevance to Epidemiology**	**581**
	Wendy Gibson, Michael D. Lewis and Michael Miles	
	21.1 Introduction	581
	21.2 *Trypanosoma brucei*	582
	21.3 *Trypanosoma cruzi*	588
	21.4 *Leishmania*	596
22	**Genomic Insights into the Past, Current and Future Evolution of Human Parasites of the Genus *Plasmodium***	**607**
	Colin J. Sutherland and Spencer D. Polley	
	22.1 Introduction	607
	22.2 Evolution of *Plasmodium*: The Past 10 Million Years	614
	22.3 Evolution of *Plasmodium*: The Twenty-First Century in Three Courses	621
	22.4 Evolution of *Plasmodium* and the Eradication Agenda	627
23	**Integrated Genetic Epidemiology of Chagas Disease**	**637**
	Michel Tibayrenc, Jenny Telleria, Patricio Diosque, Juan Carlos Dib and Christian Barnabé	
	23.1 What Is Integrated Genetic Epidemiology?	637
	23.2 Chagas Disease: A Major Health Problem in Latin America and Other Countries	638
	23.3 The Chagas Disease Cycle	638

23.4	Host Genetic Susceptibility to Chagas Disease	639
23.5	Vector Genetic Diversity	643
23.6	Parasite Genetic Diversity	643
23.7	Concluding Remarks	645
Glossary		645

24 The Rise and Fall of the *Mycobacterium tuberculosis* Complex — 651
Marcel A. Behr and S. Sebastien Gagneux

24.1	Overview	651
24.2	*Mycobacterium tuberculosis* Complex: Definitions and Epidemiology	652
24.3	The Mycobacterium Genus	653
24.4	*Mycobacterium tuberculosis* Complex Evolution	657
24.5	Pending Questions and Concluding Thoughts	661
Glossary		663

25 The Evolution and Dynamics of Methicillin-Resistant *Staphylococcus aureus* — 669
Patrick Basset, Edward J. Feil, Giorgio Zanetti and Dominique S. Blanc

25.1	Introduction	669
25.2	The Staphylococcal Cassette Chromosome *mec*	670
25.3	Molecular Epidemiology of MRSA	672
25.4	Evolution of *Staphylococcus aureus* and MRSA	677
25.5	Conclusion	681

26 Origin and Emergence of HIV/AIDS — 689
Lucie Etienne, Eric Delaporte and Martine Peeters

26.1	History of AIDS	689
26.2	HIV Is Closely Related to Simian Immunodeficiency Viruses (SIV) from Nonhuman Primates	691
26.3	HIV-1 Is Derived from SIVs Circulating among African Apes	695
26.4	Origin of HIV-2: An Other Emergence, an Other Epidemic	700
26.5	Ongoing Exposure of Humans to a Large Diversity of SIVs: Risk for a Novel HIV?	701
26.6	Conclusion	704

27 Evolution of SARS Coronavirus and the Relevance of Modern Molecular Epidemiology — 711
Zhengli Shi and Lin-Fa Wang

27.1	A Brief History of SARS	711
27.2	SARS Coronavirus	713
27.3	The Animal Link	716
27.4	Natural Reservoirs of SARS-CoV	717
27.5	Molecular Evolution of SARS-CoV in Humans and Animals	719

		27.6	Virus Surveillance in Wild Animals	**723**
		27.7	Concluding Remarks	**724**
28	**Ecology and Evolution of Avian Influenza Viruses**			**729**
	Josanne H. Verhagen, Vincent J. Munster and Ron A.M. Fouchier			
		28.1	Introduction to Influenza A Virus	**729**
		28.2	Influenza Viruses in Birds	**732**
		28.3	Evolutionary Genetics of Avian Influenza Viruses	**738**
		28.4	Future Perspective	**743**

List of Contributors

Gabriela Aguileta
Ecologie, Systématique et Evolution,
UMR 8079, CNRS
Bâtiment 360, Univ. Paris-Sud,
F-91405 Orsay,
France

Haoues Alout
Institut des Sciences de l'Evolution
(CNRS UMR 5554) Université
Montpellier II,
34095 Montpellier,
France

Fernando Baquero
CIBER en Epidemiología y Salud
Pública,
Spain; IRYCIS,
Servicio de Microbiología,
Hospital Ramón y Cajal,
Madrid,
Spain

Christian Barnabé
Unité Mixte de Recherche n° 2724:
Institut de Recherche pour le
Développement (IRD)/Centre National
de la Recherche Scientifique (CNRS):
Génétique et Evolution des Maladies
Infectieuses,
Centre IRD, BP 64501,
34394 Montpellier Cedex 5,
France

Patrick Basset
Hospital Preventive Medicine,
Centre Hospitalier Universitaire
Vaudois and University Hospital
of Lausanne, Lausanne,
Switzerland

Marcel A. Behr
McGill University, Montreal
Canada

David G. Biron
HPMCT Proteome Analysis,
St-Elphège, QC,
Canada

Dominique S. Blanc
Hospital Preventive Medicine,
Centre Hospitalier Universitaire
Vaudois and University Hospital
of Lausanne, Lausanne,
Switzerland

Edmond J. Byrnes III
Department of Molecular Genetics and
Microbiology,
Duke University Medical Center,
Durham,
North Carolina,
USA

Gregory Cochran
Department of Anthropology
University of Utah,
Salt Lake City, UT,
USA

Frederick M. Cohan
Department of Biology,
Wesleyan University,
Middletown, CT,
USA

Iñaki Comas
National Institute for Medical
Research,
London, UK

Anthony J. Cornel
University of California at Davis,
Davis, CA,
USA and Mosquito Control
Research Lab,
Parlier, CA,
USA

Keith A. Crandall
Department of Biology,
Brigham Young University,
Provo, UT,
USA

Eric Delaporte
UMR145,
Institut de Recherche pour le
Développement (IRD),
University of Montpellier,
Montpellier,
France

Juan Carlos Dib
University of Antioquia,
Antioquia,
Colombia

Patricio Diosque
CONICET,
University of Salta,
Salta,
Argentina

Luc Djogbénou
Institut de Recherche en Santé
Publique (IRSP),
Université Abomey-Calavi,
Ouidah,
Bénin

Patricia L. Dorn
Loyola University New Orleans,
New Orleans, LA,
USA

Jean-Pierre Dujardin
GEMI,
IRD,
Montpellier, France

Lucie Etienne
UMR145,
Institut de Recherche pour le
Développement (IRD),
University of Montpellier,
Montpellier, France

Edward J. Feil
Department of Biology and
Biochemistry,
University of Bath, Bath, UK

Hugh J. Field
Department of Veterinary Medicine,
Cambridge University,
Madingley Road,
Cambridge, CB3 0ES, UK

Matthew C. Fisher
Department of Infectious Disease
Epidemiology,
School of Public Health,
Imperial College London, St. Mary's
Hospital, Norfolk Place,
London, W2 1PG, UK

Ron A.M. Fouchier
Department of Virology and National
Influenza Center,
Erasmus Medical Center,
Rotterdam,
The Netherlands

Sébastien Gagneux
Swiss Tropical and Public Health
Institute, Basel, Switzerland;
University of Basel, Switzerland;
MRC National Institute for Medical
Research, London,
UK

List of Contributors

Juan Carlos Galán
CIBER en Epidemiología y Salud Pública,
Spain;
IRYCIS,
Servicio de Microbiología,
Hospital Ramón y Cajal,
Madrid,
Spain

Wendy Gibson
School of Biological Sciences,
University of Bristol,
Bristol, UK

Mark van der Giezen
Biosciences,
College of Life and Environmental Sciences,
University of Exeter,
Exeter, UK

Aude Gilabert
Muséum National d'Histoire Naturelle,
Laboratoire Origine Structure Evolution de la Biodiversité,
Paris, France;
Muséum National d'Histoire Naturelle,
Laboratoire de Biologie intégrative des populations,
Ecole Pratique des Hautes Etudes,
Paris,
France

Tatiana Giraud
Ecologie, Systématique et Evolution,
UMR 8079, CNRS,
Bâtiment 360, Univ. Paris-Sud,
F-91405 Orsay, France

Pierre Gladieux
Ecologie, Systématique et Evolution,
UMR 8079, CNRS,
Bâtiment 360, Univ. Paris-Sud,
F-91405 Orsay,
France

Fernando González-Candelas
Unidad Mixta de Investigación
"Genómica y Salud" CSISP-UV/
Instituto Cavanilles,
Valencia, Spain;
CIBER en Epidemiología y Salud Pública, Spain

William P. Hanage
Department of Epidemiology,
Harvard School of Public Health,
Boston, MA, USA

Henry Harpending
Department of Anthropology
University of Utah,
Salt Lake City, UT,
USA

Joseph Heitman
Department of Molecular Genetics and Microbiology,
Duke University Medical Center,
Durham, North Carolina,
USA

Philippe Holzmuller
CIRAD UMR 17 Trypanosomes,
Campus International de Baillarguet,
Montpellier, France

Sarah Kopac
Department of Biology,
Wesleyan University,
Middletown, CT,
USA

Britt Koskella
Department of Zoology,
University of Oxford, Oxford, UK

Elliot S. Krafsur
Iowa State University,
Ames, IA,
USA

Pierrick Labbé
Institut des Sciences de l'Evolution
(CNRS UMR 5554) Université
Montpellier II,
34095 Montpellier, France

Gregory C. Lanzaro
University of California at Davis,
Davis, CA, USA

Michael D. Lewis
Pathogen Molecular Biology Unit,
Department of Infectious and Tropical
Diseases,
London School of Hygiene and
Tropical Medicine (LSHTM),
London, UK

Yu-Tsueng Liu
Division of Infectious Diseases,
Department of Medicine,
Moores Cancer Center,
University of California San Diego,
La Jolla, CA, USA

José Luis Martínez
CIBER en Epidemiología y Salud
Pública,
Spain; Centro Nacional de
Biotecnología (CNB),
CSIC, Madrid, Spain

Tabitha Mahungu
Hospital for Tropical Diseases,
London, UK

Thierry de Meeûs
Interactions Hôtes-Vecteurs-Parasites
dans les infections par des
Trypanosomatidae (TRYPANOSOM),
Centre International de Recherche-
Développement sur l'Elevage en zone
Subhumide (CIRDES),
Bobo-Dioulasso, Burkina-Faso;
CNRS, Délégation Languedoc-
Roussillon, Montpellier,
France

Michael Miles
Pathogen Molecular Biology Unit,
Department of Infectious and Tropical
Diseases,
London School of Hygiene and
Tropical Medicine (LSHTM),
London, UK

Dorothée Missé
GEMI,
Centre IRD de Montpellier,
Montpellier,
France

Andrew D. Morgan
Institute for Evolutionary Biology,
University of Edinburgh, Edinburgh,
UK

David A. Morrison
Section for Parasitology (SWEPAR),
Department of Biomedical Sciences
and Veterinary Public Health,
Swedish University of Agricultural
Sciences, Uppsala,
Sweden

Vincent J. Munster
Laboratory of Virology,
Rocky Mountain Laboratories,
National Institute of Allergy and
Infectious Diseases,
National Institutes of Health,
Hamilton, MT,
USA

Dobrin Nedelkov
Institute for Population Proteomics and
Intrinsic Bioprobes, Inc.,
Tempe, AZ,
USA

François Noireau
IRD,
Montpellier,
France

List of Contributors

Andrew Owen
Department of Molecular and Clinical Pharmacology,
Institute of Translational Medicine,
University of Liverpool,
Liverpool, UK

Nicole Pasteur
Institut des Sciences de l'Evolution
(CNRS UMR 5554) Université Montpellier II,
34095 Montpellier, France

Konrad H. Paszkiewicz
Biosciences,
College of Life and Environmental Sciences,
University of Exeter,
Exeter, UK

Martine Peeters
UMR145,
Institut de Recherche pour le Développement (IRD),
University of Montpellier,
Montpellier, France

Marcos Pérez-Losada
CIBIO,
Centro de Investigação em Biodiversidade e Recursos Genéticos,
Universidade do Porto,
Campus Agrário de Vairão,
Vairão, Portugal

Spencer D. Polley
Department of Clinical Parasitology,
Hospital for Tropical Diseases,
London, UK

Megan L. Porter
Department of Biological Sciences,
University of Maryland,
Baltimore County,
Baltimore, MD,
USA

Franck Prugnolle
Génétique et Evolution des Maladies Infectieuses (GEMI),
UMR CNRS/IRD 2724,
Centre IRD de Montpellier,
Montpellier,
France

D. Ashley Robinson
Department of Microbiology,
University of Mississippi Medical Center,
Jackson, MS,
USA

Zhengli Shi
State Key Laboratory of Virology,
Wuhan Institute of Virology,
Chinese Academy of Sciences (CAS),
Wuhan, China

Colin J. Sutherland
HPA Malaria Reference Laboratory,
London School of Hygiene and Tropical Medicine,
London, UK

Jenny Telleria
Unité Mixte de Recherche n° 2724:
Institut de Recherche pour le Développement (IRD)/Centre National de la Recherche Scientifique (CNRS):
Génétique et Evolution des Maladies Infectieuses,
Centre IRD, BP 64501,
34394 Montpellier Cedex 5,
France

Jonathan C. Thomas
Department of Microbiology,
University of Mississippi Medical Center,
Jackson, MS,
USA

Michel Tibayrenc
Unité Mixte de Recherche n° 2724:
Institut de Recherche pour le
Développement (IRD)/Centre National
de la Recherche Scientifique (CNRS):
Génétique et Evolution des Maladies
Infectieuses,
Centre IRD, BP 64501,
34394 Montpellier Cedex 5,
France

Marc H.V. Van Regenmortel
Institut de Recherche de l'Ecole de
Biotechnologie de Strasbourg, CNRS/
Université de Strasbourg,
Strasbourg,
France

Anthony Vere Hodge
Vere Hodge Antivirals Ltd,
Old Denshott,
Reigate,
Surrey, RH2 8RD, UK

Josanne H. Verhagen
Department of Virology and National
Influenza Center,
Erasmus Medical Center,
Rotterdam,
The Netherlands

Guy Vernet
Fondation Mérieux,
Lyon,
France

Raphael P. Viscidi
Department of Pediatrics,
Johns Hopkins University School of
Medicine,
Baltimore, MD,
USA

Lin-Fa Wang
CSIRO Livestock Industries,
Australian Animal Health Laboratory,
Geelong,
Australia

Mylène Weill
Institut des Sciences de l'Evolution
(CNRS UMR 5554) Université
Montpellier II, 34095 Montpellier,
France

Thierry Wirth
Muséum National d'Histoire Naturelle,
Laboratoire Origine Structure
Evolution de la Biodiversité,
Paris, France;
Muséum National d'Histoire Naturelle,
Laboratoire de Biologie intégrative des
populations,
Ecole Pratique des Hautes Etudes,
Paris,
France

Giorgio Zanetti
Hospital Preventive Medicine,
Centre Hospitalier Universitaire
Vaudois and University Hospital of
Lausanne, Lausanne, Switzerland

Preface

Research and scholarship are characterized by a deep division between C.P. Snow's two cultures, the humanities and the sciences. A colleague recently pointed out that within the sciences there is a division nearly as deep between, roughly, those whose language is mathematics and those whose language is organic chemistry. For much of the twentieth century, genetics and epidemiology spoke mathematics in the sense that there was faith that models, equations, and numbers would lead us to answers to our questions. In the latter part of the century, speakers of organic chemistry came to prevail as much of the focus was on DNA, its function, and the mechanics of sequencing. A great milestone was the completion of the human sequence along with subsequent surveys of genome-wide sequence diversity in humans and in other species.

There was an implicit faith that with enough data we could simply look and find answers to our questions. This faith thus far has not been rewarded. We can download floods of genetic data from the Internet, but we cannot make much sense of it. It seems to be time for "speakers of mathematics" to again take a place in the discipline. Michel Tibayrenc has understood this very well for years, and he has been at the forefront of bringing the disciplines of ecology and evolutionary biology, genetics, infectious disease epidemiology, and immunology together to generate useful understandings of how to manipulate the relations between ourselves and our parasites.

This volume brings together a diverse set of authors who share an appreciation of the complexities of context and history in the study of human disease.

Henry Harpending
Member of the National Academy of Sciences USA
Department of Anthropology
University of Utah
Salt Lake City, UT, USA

Introduction

Michel Tibayrenc

Institut de Recherche pour le Développement (IRD)/Centre National de la
Recherche Scientifique (CNRS), Montpellier, France

The present multidisciplinary book is at the crossroads between two major scientific fields of the twenty-first century: evolutionary biology on one hand and infectious diseases on the other (Tibayrenc, 2001). Evolutionary biology is taken here in a broad sense and includes genetics, genomics, postgenomics, bioinformatics, and population biology. The genomic revolution has upset modern biology and has revolutionized our approach to ancient disciplines such as evolutionary studies. In particular, this revolution is profoundly changing our view on genetically driven human phenotypic diversity, and this is especially true in disease genetic susceptibility.

On the other hand, infectious diseases are indisputably the major challenge of medicine at the dawn of the twenty-first century. When considering a global view, they are the number one killers of humans, and therefore the main selective pressure exerted on our species. Even in industrial countries, infectious diseases are now far less under control than 20 years ago. In New York City, 25% of the strains of *Mycobacterium tuberculosis* are resistant to the main antituberculosis antibiotics. In France, nosocomial infections kill twice as many people as road accidents. The industrial world is now threatened by bird flu, which has reached West European poultry farms. The threat of a major pandemic is becoming increasingly probable (Ragnar Norrby, 2004; Fedson, 2005).

This book is composed of two parts:

1. A set of generalist chapters exposes the main features and applications of modern technologies in the study of infectious diseases. Rather than limiting itself to technological aspects, this part of the book will insist on the theoretical means needed for interpreting the data. Indeed, the major challenge for modern biology is not the development of even more powerful technologies (even if this is always welcome), but rather the search for new theoretical tools able to sort out and interpret the flood of data generated by mega technologies (automatic sequencing, micro arrays, proteomics, mega computers). In genetics and evolution, we are probably facing a conceptual revolution comparable to the transition between Newtonian physics and particle physics (Tibayrenc, 2001). Today technology is ahead of theory and this gap needs to be filled. This is one of the main goals of this book.
2. More specialized chapters delve into today's major stars of infection (malaria, SARS, avian flu, HIV, tuberculosis, nosocomial infections) and a few other pathogens that will

be taken as examples to illustrate the power of modern technologies and the value of evolutionary approaches.

This book will consider the three links of the epidemiological chain (host, pathogen, and vector in the case of vector-borne diseases), as well as the coevolution phenomena among them. This concept of an integrated approach to infectious diseases is also the basis for the MEEGID (Molecular Epidemiology and Evolutionary Genetics of Infectious Diseases) congresses, successfully held since 1996 (next session: Atlanta, Georgia, USA, 31st October−2nd November 2012; see http://www.meegidconference.com/), and of the new journal *Infection, Genetics and Evolution* (http://www.elsevier.com/locate/meegid), started in 2001 (2009 official impact factor: 3.223). This shows that the concept is meeting resounding success.

The readers of this book will include not only specialized scientists, but also medical doctors, health professionals, professors, students, and the educated public. The authors will make every attempt to avoid overly specialized jargon, to make their chapters accessible to a broad public, while exposing the most updated science.

The double goal of the book is hopefully clear: (i) to emphasize the value of evolutionary approaches to all professionals working in the field of infectious diseases and (ii) to demonstrate the high potentiality of infectious models to all people interested in genetics and evolution.

References

Tibayrenc, M., 2001. The golden age of genetics and the dark age of infectious diseases. Infect. Genet. Evol. 1 (1), 1−2.

Fedson, D.S., 2005. Preparing for pandemic vaccination: an international policy agenda for vaccine development. J. Public Health Policy 26, 4−29.

Ragnar Norrby, S, 2004. Alert to a European epidemic. Nature 431, 507−508.

Part 1

Methodological/Generalist Chapters

1 Virus Species

*Marc H.V. Van Regenmortel**

Institut de Recherche de l'Ecole de Biotechnologie de Strasbourg, CNRS/ Université de Strasbourg, Strasbourg, France

1.1 The Species Problem in Biology

Before dealing with the species concept used in virology, it will help to first review the many different ways the term species has been used in biology. It may be surprising to learn that, 150 years after the publication of Darwin's On the Origin of Species, biologists still do not agree about what constitutes a plant, animal, or microbial species. The term "species" is universally used for the lowest grouping of organisms in a hierarchical biological classification, but what a species actually is (i.e., its ontology) remains a hotly debated issue in the philosophy of biology (Ghiselin, 1997; Wheeler and Meier, 2000; Stamos, 2003; Ereshefsky, 2009; Claridge, 2010). In the proceedings of a conference published in 1997, no less than 22 different species concepts were described (Mayden, 1997).

An important distinction between two meanings of the word species should be kept in mind. On the one hand, the term is used to refer to the innumerable individual species taxa that have organisms as their members. The species *Canis familiaris*, for instance, is the species taxon that has dogs as its members. The term species, however, is also used to refer to the lowest category in a hierarchical biological classification, below the categories genus and family. The difference between these two meanings of species is somewhat analogous to the difference between the element gold (the class of all gold atoms) and the concept of element (the class of all the elements), which is an abstract category that does not refer to any particular type of atom. This analogy, of course, is not perfect since all gold atoms are identical whereas the dog species refers to a heterogeneous grouping of related but dissimilar organisms. It is important to note that the category species is an abstract, taxonomic concept (the class of all species taxa) that does not refer to any particular organism. Species taxa, on the other hand, are classes that have real organisms as their members in the same way that higher taxa such as genera and families are classes with organisms as their members.

*E-mail: vanregen@esbs.u-strasbg.fr

All classification schemes are human, conceptual constructions made up of classes. Class membership allows a bridge to be established between two different logical categories—the class as an abstract entity and the concrete members of the class (Buck and Hull, 1966; Mahner and Bunge, 1997, p. 230). Class membership must be distinguished from part–whole relationships that exist only between concrete objects, one being a part of the other, in the way cells are parts of an organism. An organism can be a member of a species class but not a part of it. Similarly, a thought or concept cannot be part of a concrete object. Finally, the relationship between two classes in the classification hierarchy, such as a species and a genus, is known as "class inclusion" rather than class membership. Since all classes are abstract constructions, it is indeed not possible for one class to be a concrete member of another class (Buck and Hull, 1966; Mahner and Bunge, 1997, p. 20). The mixing of logical categories has been a fertile source of confusion in taxonomy (Stamos, 2003; Van Regenmortel, 2003).

1.2 Classes Versus Individuals

Many philosophers have been reluctant to accept that species are classes because they only considered so-called Aristotelian, universal classes, which have constant properties that do not change with time. Since species change during evolution, giving rise to new species, they considered that species were historical entities with a definite localization in space and a beginning and an end in time and could not be abstract classes, immutable and timeless. This viewpoint gave rise to the bionominalist school of thought, which regards species as concrete individuals rather than abstract classes (Ghiselin, 1974, 1997; Hull, 1976, 1988). Although the species-as-individuals (SAI) thesis has many followers, it still remains controversial (Stamos, 2003, pp. 181–283; Mahner and Bunge, 1997, pp. 253–270), one reason being that the concept of "individual" is as fuzzy (Hull, 1992) as the concept of species (Van Regenmortel, 1998). The main shortcoming of the SAI thesis is that it claims that asexual organisms cannot form species because species integration is impossible in the absence of gene pools (Ghiselin, 1997). In effect, this makes the species concept inapplicable to a large portion of the biological realm. If species appeared during evolution only after sexual reproduction had become widespread, most of life's history on Earth could not be described in terms of evolving species, a conclusion most biologists find unacceptable.

1.3 Species Taxa as Cluster Classes

Instead of viewing species as universal classes defined by a single property or combination of properties necessarily present in every member of the class, many biologists consider species to be cluster classes, sometimes also called polytypic or polythetic classes (Beckner, 1959; Stamos, 2003). Each class is defined by a variable combination of several properties of its members, with no single property being necessary or sufficient for membership in the class. A cluster class is similar to the family resemblance concept that Wittgenstein (1953) introduced in philosophy to describe a concept such as "game." Human activities as disparate as board games, card games, ball games, and sports are all referred to as games although they lack a single, common defining property (Pigliucci, 2003). Using the category "species" as a family resemblance concept in taxonomy has the same advantage as using the category "game" for referring to a human activity, since in both cases it does away with the need for a definition in terms of necessary and sufficient properties. The inherent vagueness of the membership conditions of a cluster class also makes the concept particularly useful for dealing with the usual absence of clear discontinuities between species that result from the continuous nature of biological variation (Van Regenmortel, 1998; Schaefer and Wilson, 2002).

1.4 Species: Concrete Versus Abstract Entities

Philosophers often claim that the term species should be reserved either for an abstract taxonomic class that requires a definition, or for its concrete referents and members (Stamos, 2003). Such a restriction tends to be counterproductive (Hull, 1988, p. 214) if it obscures the need to distinguish a concrete entity from its conceptual representation. When the term species is defined as an intellectual construction that is not physically real, this does not prevent the concept to have referents in the form of concrete, physical organisms that are members of the species (Van Regenmortel, 2010a). Furthermore, when species taxa are defined as fuzzy cluster classes, the boundaries between the actual members of such species will also be fuzzy.

Many definitions of the species category have been proposed over the years such as the biological species concept based on gene pools and reproductive isolation (Mayr, 1970) and the ecological and phylogenetic species concepts (Baum and Donoghue, 1995; Mayden, 1997; Stamos, 2003). None of these concepts have gained general acceptance because they are based on defining properties that cannot be applied to all types of organisms. The general lineage species concept proposed by de Queiroz (2007) is also unsatisfactory because it does not allow a clear differentiation between species and genera (Ereshefsky, 2010).

It seems that only the family resemblance concept (Pigliucci, 2003) is able to unify various types of species taxa defined by genealogy, reproductive isolation, phylogeny, or ecology into a single cluster class category. However, many biologists share the viewpoint of Darwin, who regarded the species category to be no more real or definable than the categories genus and family. The unwillingness of Darwin to argue over the definition of species has been called a modern solution to the species problem (Ereshefsky, 2009).

The species problem is very much a philosophical question concerning the nature of reality: it all depends on whether we are willing to accept that a concept such as the species category "exists" as a "real" category in nature, irrespective of any human conceptualization, or whether we consider it merely as an artifact of human thinking (Ereshefsky, 2010). However, the question whether the category species is "real" and has an existence outside of human minds is rarely debated by biologists, because they tend to be more concerned with identifying organisms as members of a species taxon than with finding an appropriate definition of the species category.

1.5 The Nature of Viruses

Viruses have been defined as molecular genetic parasites that use cellular systems for their own replication (Villarreal, 2005). Viruses do not replicate or self-replicate themselves, but are being replicated in a passive rather than active manner through the metabolic activities of the cells they have infected. The replication of viruses occurs through a process of copying carried out by the cellular machinery of the host cell, whereas cells reproduce themselves by a process of fission. Viruses are considered to be biological entities because they possess a genome and are able to adapt to particular hosts and biotic habitats. However, viruses lack many of the essential attributes of living organisms, such as the ability to capture and store free energy, and they do not possess the characteristic autonomy and self-repairing capacity of organisms (Van Regenmortel, 2010b). Most biologists accept that the simplest system that can be said to be alive is a cell, and that subcellular constituents such as organelles, macromolecules, or genes are not themselves alive. The recent discovery of the large Mimivirus containing a 1.2-Mbp genome and several genes involved in transcription and translation led some authors to suggest that this virus represented a new branch in the tree of life that predated the divergence of the three domains of cellular life (Raoult et al., 2004; Raoult and Forterre, 2008). However, phylogenetic analyses have demonstrated that the bacterial and eukaryotic genes present in Mimivirus were most probably acquired by horizontal gene transfer from organisms (Moreira and Brochier-Armanet, 2008). There is therefore no compelling reason for including viruses in the tree of life (Doolittle and Bapteste, 2007) nor for considering them as living organisms (Wolkowicz and Schaechter, 2008; Lopez-Garcia and Moreira, 2009; Moreira and Lopez-Garcia, 2009; Van Regenmortel, 2010a).

1.6 Viruses Should not be Reduced to Virions

A virus cannot be reduced to a virus particle or virion, which is only one stage of the viral replication cycle. The function of the virion is to protect the viral genome from degradation by nucleases and other environmental attack and to act as a vehicle to deliver the genome to new host cells. Virions of many sizes and morphologies exist, which are among the most useful characteristics for recognizing and identifying different viruses. A virion can be fully described by its intrinsic and chemical properties, and the composition of a single particle of poliovirus has been described by Wimmer (2006) as $C_{332,652}$ $H_{492,288}$ $N_{98,245}$ $O_{131,196}$ P_{7501} S_{2340}. However, such a reductionist description of a biological entity in chemical terms is not an adequate characterization of the entity poliovirus (Van Regenmortel, 2010a). It leaves out essential features such as nucleotide and amino sequences, as well as the conformation of viral proteins. The protein conformation gives rise to the binding site that is recognized by host cell receptors and is required for initiating the infection process. The viral receptor binding site is a relational structure, emerging from a specific relationship with a complementary receptor arising during biological evolution, which allows the virus to infect certain host cells. To describe a virus fully, it is thus necessary to include a number of relational properties involving hosts and vectors that are required for virus transmission, infection, and replication and which make the virus a genetic parasite as well as a biological entity. Confusing "virus" with "virion" is somewhat similar to confusing the entity "insect," which includes several different life stages, with a single one of these stages, such a pupa, a caterpillar, or a butterfly (Van Regenmortel, 2010b).

It is also important to differentiate between the genome and genotype and between the phenome and phenotype of a virus, which are differences based on the type-token distinction. The terms "genotype" and "phenotype" refer to classes of viruses that satisfy some genetic or phenetic criteria, whereas the individual members of these classes are tokens of those types. The type-token distinction is exemplified by the statement that every particular dollar bill is a token or occurrence of the "one dollar" type (Bunge, 2003). The genome of a virus is the actual genetic material found in virions, and it is the description of this genome in terms of particular genetic properties that determine the genotype of which the individual genome is a token (Lewontin, 1992). Since viral genomes have mutation rates that are orders of magnitudes higher than those of cellular genomes, a viral clone will always contain many mutated genetic tokens that are not exactly identical copies of the replicated genotype.

The DNA or RNA genome sequence found in a virion is actually part of the phenome of the virus since it corresponds to a part of the virion's chemical structure. The phenome of a virus includes all its observable physical manifestations, such as virion morphology, molecular composition, biochemical activities, and relational interactions with hosts and vectors. The properties of the phenome determine the phenotype of which the individual virus is a token and its description has a temporal dimension since the phenome changes over time, for instance, during viral replication when multiple tokens of the same phenotype are produced.

A virus classification based on nucleotide sequences is actually a phenotypic classification with the discriminating characters being molecular rather than morphological, physiological, or relational (Mahner and Bunge, 1997, p. 287). A phylogenetic virus classification, on the other hand, must be based on inferred or hypothesized phylogeny rather than simply on nucleotide sequences or motifs, which by themselves have no greater significance than other phenotypic traits

(Van Regenmortel, 2006). A classification based only on genotypic and phylogenetic characteristics of viruses would amount to a classification of genomes and would be of limited value to laboratory virologists (Calisher et al., 1995).

1.7 What is a Virus Species?

Although virus classification always included the classical hierarchical categories of family and genus, it took many years before the species category became accepted by the virological community. Plant virologists, in particular, were opposed to the idea that the species category could be applied to viruses because they assumed that the only legitimate species definition was the biological species concept developed for sexually reproducing organisms and not applicable to viruses that replicate as clones (Milne, 1984). Some virologists were also opposed to the use of the species category in virology because they feared that the introduction of virus species would inevitably lead to Latinized names, which they strongly opposed (Matthews, 1983).

The absence of a satisfactory definition of the concept of virus species was an additional reason why many virologists were reluctant to use the species concept. Various species definitions had been proposed over the years but none had gained general acceptance (Van Regenmortel, 2007).

In 1991, the International Committee on Taxonomy of Viruses (ICTV), which is the body empowered by the International Union of Microbiological Societies to make decisions on matters of virus classification and nomenclature, endorsed the following definition: a virus species is a polythetic class of viruses that constitute a replicating lineage and occupy a particular ecological niche (Van Regenmortel, 1989; Pringle, 1991). It should be emphasized that this is only a definition of the species category (i.e., the class of all the individual species taxa in virology) and that it does not provide guidelines on how individual virus species should be demarcated (Van Regenmortel et al., 1997). However, once the virus species category was accepted, ICTV Study Groups composed of virologists with in-depth knowledge of particular areas of virology were able to embark on the task of establishing and delineating hundreds of virus species taxa.

The concept of polythetic class used in the definition of virus species was introduced in biology by Beckner (1959), who called this type of class "polytypic" to differentiate it from monotypic, universal classes that can be defined by properties that are both necessary and sufficient for establishing membership in the class. In a polythetic class, each member of the class shares a large but unspecified number of properties, each property is present in a large but unspecified number of members of the class, and no property is necessarily present in all the members of the class (Van Regenmortel, 2000a).

It must be stressed that the definition of the category species as a polythetic class cannot provide guidelines for deciding whether particular viruses are members of one or other species taxon. One reason is that a species taxon has first to be established by taxonomists before it becomes possible to ascertain if a certain property is present in a sufficient number of the taxon's members. Another reason is that many of the properties of viruses, especially those used at the level of species demarcation such as host range, pathogenicity, and degree of genome relatedness, are easily modified by mutation and can only reveal blurred discontinuities between species (Van Regenmortel et al., 1997). Furthermore, the notion of character or trait is itself problematic and is often confused with parts (Inglis, 1991; Fristrup, 1992), although a part of a thing is a thing and not a property (Mahner and Bunge, 1997, p. 11). It is unclear, for instance, what should count as a single character: a complete genome map, a particular nucleotide motif, or the presence of a certain nucleotide at a given position in the viral genome sequence (Van Regenmortel, 2006). In fact, a major advantage of defining virus species as polythetic classes is that it makes it possible to accommodate within a species taxon members that lack one or other characteristic that would usually be considered typical of the species. The concept of polythetic species is thus particularly useful for dealing with replicating entities like viruses that possess extensive variability (Van Regenmortel, 2000a).

1.8 Species and Quasi-Species

The term "quasi-species" is often used to refer to populations of RNA viruses that contain large numbers of mutants (Holland et al., 1992), but this term is unrelated to the concept of species discussed here. The term quasi-species was introduced by Eigen (1996) to describe the self-replicating RNA molecules believed to be the first genes on Earth and which, because of mutations, do not consist of a unique molecular species. In this context, species refers to a chemical species, which always consists of identical molecules, and the term quasi-species is used to indicate that the RNA molecules do not have a unique sequence (Sanjuan, 2010).

When a population of RNA viruses is called a quasi-species (implying some sort of imperfect species), this does not mean that there could be such a thing as a "true" virus species with its members all possessing exactly the same genome sequence (Van Regenmortel et al., 1997). Quasi-species theory is a complex field of population genetics but it has no taxonomic implications (Domingo and Holland, 1997; Smith et al., 1997; Holmes and Moya, 2002; Sanjuan, 2010) since a taxonomic species is always necessarily a quasi-species in the chemical sense.

1.9 Definition of Virus Species Versus Identification of Viruses

Viruses are classified by allocating them to the hierarchical classes of order, family, subfamily, genus, and species. For instance, the species *Measles virus* is included in the genus *Morbillivirus*, which in turn is included in the family *Paramyxoviridae*, itself included in the order *Mononegavirales* (Pringle, 1997). The last three categories correspond to universal classes defined by a small number of properties that are both necessary and sufficient for establishing membership in the class. For instance, all the members of the family *Adenoviridae* are nonenveloped viruses that have an icosahedral virion and double-stranded DNA, with protecting fibers at the vertices of the protein shell. This means that allocating a virus to a genus or a family is an easy task since it is sufficient to show that the virus possesses a small number of invariant, stable properties that are present, without exception, in every member of the class.

The situation is completely different in the case of virus species since they are defined by a variable combination of phenotypic and genotypic properties, none of which are necessarily present in every member of the species. When they establish individual species, ICTV Study Groups must rely on properties that are not present in all the members of the genus to which the species belongs since such properties obviously would not allow individual species to be differentiated. Stable properties such as virion morphology, genome organization, or method of replication shared by all the members of a genus or family are therefore not useful for demarcating individual species. Properties useful for discriminating between virus species within a genus are the viral host range, pathogenicity, host cell and tissue tropism, small differences in genome sequence, and other traits that often vary in different members of the same species. When ICTV Study Groups create new virus species, they may have to draw boundaries across a continuous range of genomic and phenotypic variability, and such taxonomic decisions often involve a strong subjective element (Van Regenmortel, 2007, 2010a).

The 8th ICTV Report published in 2005 lists 3 orders, 73 families, 287 genera, and 1950 virus species (Fauquet et al., 2005). For each genus, a type species is designated, although this does not imply that the properties of the type species are typical and representative of the properties of all species in the genus (Mayo et al., 2002). The demarcation criteria that are used to delineate individual species within each genus, such as differences in host range, vectors, pathogenicity, antigenicity, and degree of genome similarity are listed for most species. In agreement with the polythetic definition of virus species, no single property is supposed to be used as the defining property of a virus species (Ball, 2005).

1.10 Using Viral Genomes to Define Virus Species

As more viral genome sequences become available, there is an increasing tendency to rely mainly on genome data to establish new viral species. Genome sequence similarities in the members of a virus family are usually visualized by plotting the frequency distribution of pairwise identity percentages using all available genome sequences in the family (Bao et al., 2010). These pairwise sequence comparisons (PASC) produce multimodal distributions where the peaks are attributed somewhat arbitrarily to clusters of sequences corresponding to strains, species, genera, or even sometimes ill-defined variants and isolates (Fauquet et al., 2008). Attributing a peak to "isolates" is actually meaningless since the term "virus isolate" refers to any virus culture that is being studied, and an isolate could therefore be a member of a strain, a species, or a genus (Van Regenmortel, 2007). The distribution peaks represent the average degree of sequence identity between pairs of individual virus isolates allocated to one or other taxonomic category. The percentage demarcation thresholds that are chosen to allocate peaks to the different categories determine how many separate species will be recognized in each genus (Van Regenmortel, 2007). For instance, in the family *Flexiviridae* it has been accepted that members of different species have less than about 72% identical nucleotides between their entire coat protein genes (Adams et al., 2004). In the genus *Begomovirus* of the family *Geminiviridae* of single-stranded DNA plant viruses, different species were established on the basis of a pairwise sequence identity of less than 89% in the DNA-A genome component. This led to the recognition of more than 100 *Begomovirus* species (Fauquet and Stanley, 2005) and subsequently to an additional 85 species (Carstens and Ball, 2009). Many members of what are labeled as different species infect the same host, such as cotton or tomato, and produce very similar disease symptoms but are given different names by including the geographical location of the first isolation of the virus. This leads to dozens of species names like *Tomato leaf curl Comoros virus*, *Tomato leaf curl Guangxi virus*, *Tomato leaf curl Hsinchu virus*, *Tomato leaf curl Joydebpur virus*, *Tomato leaf curl New Delhi virus*, *Tomato leaf curl Pune virus*, and so on.

Many of these species could be considered strains of the same virus if a different threshold demarcation percentage for differentiating strains from species had been chosen. The allocation of different begomovirus isolates to the category strain or variant is equally arbitrary (Fauquet et al., 2008). In the genus *Mastrevirus* of the family *Geminiviridae*, a cutoff figure of 75% sequence identity in DNA-A nucleotide sequences was used to establish different species (Fauquet and Stanley, 2005), and this led to the recognition of only 12 separate species. Establishing valid demarcation criteria in the family *Geminiviridae* is also made difficult by the frequent recombination events that occur between different geminiviruses (Padidam et al., 1999). The inordinate increase in the number of species in the genus *Begomovirus* is due to excessive reliance on a single arbitrary PASC criterion for demarcating species, which is at odds with the polythetic definition of virus species. It is doubtful if all these species really correspond to biologically distinguishable stable entities justifying the label species, and it seems that the result is more a classification of genomes than of viruses. It may be relevant to mention that the ICTV now approves such taxonomic decisions on the creation of species by an electronic ratification process that involves 11 ICTV life members, 74 ICTV subcommittee members, and 53 ICTV national representatives, without actually reporting the number of individuals who responded and voted in favor (Carstens and Ball, 2009).

Recently it has been proposed that the term "polythetic" should be removed from the ICTV definition of virus species and that a species should instead be defined as a monothetic or universal class, on the basis that a unique nucleotide motif is necessarily present in all the members of a species (Gibbs and Gibbs,

2006). This proposal overlooks the fact that if a taxonomist were to create a new monothetic species by relying on a single defining property such as a nucleotide motif, he or she would have to know beforehand that this motif is present in all the members of the species and absent in other species (Inglis, 1991; Van Regenmortel, 2006). In actual fact, the motif can only be discovered when the sequences of numerous members of a preestablished species have been compared with other species. This means that the nucleotide motif is not a property useful for initially defining and creating a species, but is only a diagnostic marker used for identifying viruses that are members of a species created beforehand as a polythetic class.

Species are created by taxonomists who stipulate which covariant sets of shared properties must be present in most members of a species, and a subsequently discovered diagnostic marker is not one of the original species-defining properties. It is the frequent, combined occurrence of several covariant properties in individual members of a species that makes it possible to predict many of the properties of a newly discovered virus once it has been identified as a member of a particular species. If a species could be defined monothetically by a single diagnostic marker, identifying a virus as a member of a species would not be very informative (Van Regenmortel, 2007). The erroneous claim that virus species are monothetic universal classes instead of polythetic cluster classes (Gibbs and Gibbs, 2006) arises from the failure to distinguish between the creation of species by taxonomists on the basis of defining characters and the identification of members of an established species by means of diagnostic markers (Van Regenmortel, 2003, 2007).

In recent years, short sequences of nucleotides located in an appropriate part of an organism's genome have become increasingly used for identifying members of species. This approach, known as DNA barcoding (Hebert and Gregory, 2005), is often presented as providing an additional character for creating and recognizing new species. In reality, DNA barcoding is useful only for identifying members of an already established species and not for defining or creating the millions of species that have not yet been recognized by taxonomists on the basis of phenotypic and other criteria (Ebach and Holdrege, 2005; Waugh, 2007; Wheeler and Valdecasas, 2007). Unwarranted claims regarding the species-defining potential of barcoding again illustrates the need to distinguish the task of defining new species from the task of identifying members of established species (Van Regenmortel, 2010a).

1.11 Names and Typography of Virus Species

For many years the names of viral orders, families, subfamilies, and genera have been written in italics with a capital initial letter (Murphy et al., 1995), which is a different typography from that advocated by the Biological Nomenclature Code (Van Regenmortel, 2001). The International Code of Virus Classification and Nomenclature was modified in 1998, and the names of virus species were now also italicized and capitalized, providing a visible sign that species were taxonomic classes just like genera and families (Mayo and Horzinek, 1998). The ICTV Executive Committee (EC) accepted the proposal that the common English names of viruses should become the species names (written in italics instead of Roman) but rejected a second proposal that new non-Latinized binomial names (NLBNs) should be introduced for each species (Van Regenmortel, 2007). Such NLBNs had been used for many years in the indices of ICTV Reports (Fenner, 1976; Matthews, 1979, 1982) and in plant virology publications and books (Albouy and Devergne, 1998; Bos, 1999b). These binomials are obtained by replacing the word "virus" appearing at the end of all English common names of viruses by the genus name of which the virus is a member, leading to a species name such as *Measles morbillivirus*. Since the species names of animals, plants, and microorganisms are always binomials that include a genus designation, such NLBNs should be easily recognized as virus species names whereas the vernacular virus names used in different languages would obviously be the names of the viruses. Since all such binomial names for virus species end with the suffix −*virus*, they clearly indicate that the names refer to viral entities. This is an advantage compared to the Latin species names used in biology, which do not indicate to the uninitiated whether the organism referred to is an animal, a plant, or a microorganism. Although NLBNs could have been introduced immediately for more than 90% of the 1550 species recognized at the time (Van Regenmortel and Fauquet, 2002) the ICTV decided instead to adopt *Measles virus* as the species name and measles virus as the virus name. The decision to use the same name to refer to the virus (in Roman) and to the species (in italics) resulted in considerable confusion among virologists since they now had to differentiate between a virus and a species only on the basis of typography (Drebot et al., 2000; Calisher and Mahy, 2004; Van Regenmortel, 2003, 2007).

Kuhn and Jahrling (2010) have argued that many virologists do not see the point of differentiating between a virus (a concrete object) and a virus species (a concept) because many species have only one virus as a member. In spite of arguments to the contrary (Bos, 2007), it is important to make the distinction because a virus can be isolated and manipulated experimentally and always exists in the form of many mutants, variants, or strains with different genome sequences, whereas a species, being a taxonomic construct, does not have a sequence and cannot be injected, isolated, or otherwise manipulated. It is therefore incorrect to write, as is often done, that the species *Measles virus* (rather than measles virus) has been isolated, transmitted to a host, or sequenced. Species classes are concepts that do not possess "physical" properties and they can only be defined by listing certain properties of their members (Van Regenmortel, 2007; Kuhn and Jahrling, 2010).

In biology, the majority of animals, plants, and microorganisms do not have vernacular names in English or other languages, and scientists will therefore write that *Escherichia coli* (i.e., the species) has been infected by a bacterial virus, implying that a taxonomic concept can be infected. In virology, such incorrect statements can always be avoided since all viruses have vernacular names, and the name of the virus can therefore be used to refer to the infectious agent rather than to the species into which it has been placed. Unfortunately virologists continue to use

Table 1.1 Latin Names Given in the Past to Tobacco Mosaic Virus

Names	References (See Francki, 1981)
Stangyloplasma iwanowskii	B.T. Palm (1922)
Marmor tabaci	F.O. Holmes (1939)
Musivum tabaci	W.D. Valleau (1940)
Phytovirus nicomosaicum	H.H. Thornberry (1941)
Nicotianavir communae	H.S. Fawcett (1942)
Minchorda nicotianae	H.P. Hansen (1957)
Protovirus tabaci	A. Lwoff and P. Tournier (1966)
Vironicotum maculans	H.H. Thornberry (1968)
Virothrix iwanowskii	A.E. Procenko (1970)

wrong terminology and go on writing that a virus species (using italicized typography) can infect a certain host or can be sequenced (Calisher and Mahy, 2004; Van Regenmortel, 2006; Calisher and Van Regenmortel, 2009; Kuhn and Jahrling, 2010).

Virologists would certainly find it easier to differentiate between a virus and a species if the ICTV introduced species names that were clearly different from virus names. Many proposals for new species names have been made in the past, including the suggestion that binomial Latin names should be introduced (Bos, 1999a; Agut, 2002; Gibbs, 2003; Van Regenmortel, 2007). However, virologists have mostly been opposed to the introduction of Latin names (Matthews, 1983, 1985; Milne, 1984; Van Regenmortel, 2007) and the bewildering variety of Latin names that have been given to tobacco mosaic virus in the past (Table 1.1) certainly suggests that reaching agreement on Latin names for thousands of virus species would be no easy task. It is also unlikely that virologists would be in favor of introducing thousands of totally new species names. As an alternative, adopting the NLBNs as species names would seem to be a simple and elegant solution, since binomial names are always associated with taxonomic entities and the inclusion of the genus name gives useful additional information on the properties of members of the species. For the majority of virus species, it would require the creation of no new names, since it combines well-known English virus names with accepted genus names (Table 1.2). Only in a small number of cases would it be necessary to introduce changes to existing names (Van Regenmortel, 2001), but this would certainly be preferable to the introduction of completely new names for more than 2000 virus species. The introduction of binomial species names could sometimes lead to odd names such as *Marburg marburgvirus*, but these are no worse than many accepted names in biology, such as *Rattus rattus* for the rat and *Bombina bombina* for the toad. Most of the problems posed by NLBNs in a limited number of genera are due to the fact that these genera have names that do not follow ICTV rules (Van Regenmortel and Fauquet, 2002) but these difficulties could be resolved by the relevant ICTV Study Groups. Using binomial species names implies that name

Table 1.2 Examples of Binomial Species Names for Vertebrate Viruses

Virus Names	Binomial Species Names
California encephalitis virus	*California encephalitis orthobunyavirus*
hepatitis A virus	*hepatitis A hepatovirus*
hepatitis B virus	*Hepatitis B orthohepadnavirus*
hepatitis C virus	*Hepatitis C hepacivirus*
hepatitis E virus	*Hepatitis E hepevirus*
Lassa virus	*Lassa arenavirus*
louping ill virus	*Louping ill flavivirus*
measles virus	*Measles morbillivirus*
mumps virus	*Mumps rubulavirus*
rabies virus	*Rabies lyssavirus*
rubella virus	*Rubella rubivirus*
Sendai virus	*Sendai respirovirus*
Sindbis virus	*Sindbis alphavirus*
Sin Nombre virus	*Sin Nombre Hantavirus*
West Nile virus	*West Nile flavivirus*

changes would occur when species are moved from one genus to another but this could be perceived as clarifying advantage rather than an alleged disadvantage (Mayo and Haenni, 2006). It has also been pointed out that species in a family but not yet assigned to a genus could not be given a binomial name (Mayo and Haenni, 2006) but such exceptional cases hardly seem a valid reason for not adopting NLBNs.

In 1998, most members of the ICTV-EC were opposed to the introduction of NLBNs, but by 2004, 50% of them no longer objected to the system. Surveys conducted in 2002 among laboratory virologists had shown that more than 80% of those who responded were in favor of NLBNs (Mayo, 2002; Van Regenmortel and Fauquet, 2002; Mayo and Haenni, 2006; Van Regenmortel, 2007). Democratic decision making in matters of viral taxonomy has always been difficult to achieve since few virologists express an opinion in ballot polls (Matthews, 1982; Van Regenmortel, 2000b) with the result that the minority view is heard disproportionally. Even within the ICTV, only a minority of the 80 Study Groups responded when asked about their opinion on NLBNs (Ball and Mayo, 2004). In view of such inertia, the ICTV often has ratified decisions in the past by accepting that a no answer vote was a vote in favor (Van Regenmortel, 2007). In conclusion, there is no doubt that virologists would find it easier to use correct terminology if the ICTV introduced new names for virus species (Kuhn and Jahrling, 2010) and the option of using binomials based on existing genus and virus names should therefore receive serious consideration.

References

Adams, M.J., Antoniw, J.F., Bar-Joseph, M., 2004. The new plant virus family *Flexiviridae* and assessment of molecular criteria for species demarcation. Arch. Virol. 149, 1045–1060.
Agut, H., 2002. Back to Latin and tradition: a proposal for an official nomenclature of virus species. Arch. Virol. 147, 1465–1470.
Albouy, J., Devergne, J.C., 1998. Maladies à Virus des Plantes Ornementales. Editions INRA, Paris.
Ball, L.A., 2005. The universal taxonomy of viruses in theory and practice. In: Fauquet, C.M., et al. (Eds.), Eighth ICTV Report. Elsevier, pp. 11–16.
Ball, L.A., Mayo, M.A., 2004. Report from the 33rd Meeting of the ICTV executive committee. Arch. Virol. 149, 1259–1263.
Bao, Y., Kapustin, Y., Tatusova, T., 2010. Virus classification by Pairwise Sequence Comparison (PASC). In: Van Regenmortel, M.H.V., Mahy, B. (Eds.), Desk Encyclopedia of General Virology. Academic Press, Elsevier, Oxford, pp. 95–100.
Baum, D., Donoghue, M., 1995. Choosing among alternative "phylogenetic" species concepts. Syst. Biol. 20, 560–573.
Beckner, M., 1959. The Biological Way of Thought. Columbia University Press, New York, NY.
Bos, L., 1999a. The naming of viruses: an urgent call to order. Arch. Virol. 144, 631–636.
Bos, L., 1999b. Plant Viruses, Unique and Intriguing Pathogens—A Textbook of Plant Virology. Backhuys Publishers, Leiden.
Bos, L., 2007. Coming to grips with the naming of viruses; continuing discord, or a way out. Arch. Virol. 152, 649–653.
Buck, R.C., Hull, D.L., 1966. The logical structure of the Linnaean hierarchy. Syst. Zool. 15, 97–111.
Bunge, M., 2003. Philosophical Dictionary. Prometheus Books, Amherst, NY.
Calisher, C.H., Van Regenmortel, M.H.V., 2009. Should all other biologists follow the lead of virologists and stop italicizing the names of living organisms? A proposal. Zootaxa 2113, 63–68.
Calisher, C.H., Horzinek, M.C., Mayo, M.A., Ackermann, H.W., Maniloff, J., 1995. Sequence analyses and a unifying system of virus taxonomy: consensus via consent. Arch. Virol. 140, 2093–2099.
Calisher, C.H., Mahy, B.M.J., 2003. Taxonomy: get it right or leave it alone. Am. J. Trop. Med. Hyg. 68, 505–506.
Carstens, E.B., Ball, L.A., 2009. Ratification vote on taxonomic proposals to the International Committee on Taxonomy of Viruses (2008). Arch. Virol. 154, 1181–1188.
Claridge, M.F., 2010. Species are real biological entities. In: Ayala, F.J., Arp, R. (Eds.), Contemporary Debates in Philosophy of Biology. Wiley-Blackwell, Chichester, pp. 91–109.
de Queiroz, K., 2007. Species concepts and species delimitation. Syst. Biol. 56, 879–886.
Domingo, E., Holland, J.J., 1997. RNA virus mutations and fitness for survival. Ann. Rev. Microbiol. 51, 151–178.
Doolittle, W.F., Bapteste, E., 2007. Pattern pluralism and the Tree of Life hypothesis. Proc. Natl. Acad. Sci. (USA) 104, 2043–2049.
Drebot, M.A., Henchal, E., Hjelle, B., LeDuc, J.W., Repik, P.M., Roehrig, J.T., Schmaljohn, C.S., Shope, R.E., Tesch, R.B., Weaver, S.C., Calisher, C.H., 2002. Improved clarity of

meaning from the use of both formal species names and common (vernacular) virus names in virological literature. Arch. Virol. 147, 2465–2471.
Ebach, M.C., Holdrege, C., 2005. More taxonomy, not DNA barcoding. Bioscience 55, 822–823.
Eigen, M., 1996. On the nature of viral quasispecies. Trends Microbiol. 4, 216–218.
Ereshefsky, M., 2009. Darwin's solution to the species problem. Synthese 10.1007/s11229-009-9538-4.
Ereshefsky, M., 2010. Mystery of mysteries: Darwin and the species problem. Cladistics 26, 1–13.
Fauquet, C.M., Stanley, J., 2005. Revising the way we conceive and name viruses below the species level: a review of geminivirus taxonomy calls for new standardized isolate descriptors. Arch. Virol. 150, 2151–2179.
Fauquet, C.M., Mayo, M.A., Maniloff, J., Desselberger, U., Ball, L.A., 2005. Virus Taxonomy, Classification and Nomenclature of Viruses, Eighth Report of the ICTV. Elsevier, San Diego, CA.
Fauquet, C.M., Briddo, R.W., Brown, J.K., Moriones, E., Stanley, J., Zerbini, M., et al., 2008. Geminivirus strain demarcation and nomenclature. Arch. Virol. 153, 783–821.
Fenner, F., 1976. The classification and nomenclature of viruses. Second report of the International Committee on Taxonomy of Viruses. Intervirology 7, 1–115.
Francki, R.I.B., 1981. Plant virus taxonomy. In: Kurstak, E. (Ed.), Handbook of Plant Virus Infections and Comparative Diagnosis. Elsevier, Amsterdam, pp. 3–16.
Fristrup, K., 1992. Character: current usages. In: Keller, E.F., Lloyd, E.A. (Eds.), Keywords in Evolutionary Biology. Harvard University Press, Cambridge, MA, pp. 45–51.
Ghiselin, M.T., 1974. A radical solution to the species problem. Syst. Zool. 23, 536–544.
Ghiselin, M.T., 1997. Metaphysics and the Origin of Species. New York State University Press, New York, NY.
Gibbs, A.J., 2003. Virus nomenclature, what next? Arch. Virol. 148, 1645–1653.
Gibbs, A., Gibbs, M.J., 2006. A broader definition of the "virus species". Arch. Virol. 148, 1645–1653.
Hebert, P.D.N., Gregory, T.R., 2005. The promise of DNA barcoding for taxonomy. Syst. Biol. 54, 852–859.
Holland, J.J., De la Torre, J.C., Steinhauer, D.A., 1992. RNA virus populations as quasispecies. In: Holland, J.J. (Ed.), Genetic Diversity of RNAViruses. Springer-Verlag, New York, NY.
Holmes, E.C., Moya, A., 2002. Is the quasi-species concept relevant to RNA viruses? J. Virol. 76, 460–462.
Hull, D.L., 1976. Are species really individuals? Syst. Zool 25, 174–191.
Hull, D.L., 1988. Science as a Process. The University of Chicago Press, Chicago, IL.
Hull, D., 1992. Individual. In: Fox-Keller, E., Lloyd, E. (Eds.), Keywords in Evolutionary Biology. Harvard University Press, Cambridge, MA, pp. 181–187.
Inglis, W.G., 1991. Characters: the central mystery of taxonomy and systematics. Biol. J. Linn. Soc. 44, 121–139.
Kuhn, J.H., Jahrling, P.B., 2010. Clarification and guidance on the proper usage of virus and virus species names. Arch. Virol. 155, 445–453.
Lewontin, R.C., 1992. Genotype and phenotype. In: Fox Keller, E., Lloyd, E.A. (Eds.), Keywords in Evolutionary Biology. Harvard University Press, Cambridge, MA, pp. 137–144.
Lopez-Garcia, P., Moreira, D., 2009. Yet viruses cannot be included in the tree of life. Nat. Rev. Microbiol. 7, 615–617 (August 2009) doi: 10.1038/nrmicro2108-c7.

Mahner, M., Bunge, M., 1997. Foundations of Biophilosophy. Springer-Verlag, Berlin.
Matthews, R.E.F., 1982. Classification and nomenclature of viruses. Fourth report of the International Committee on Taxonomy of Viruses. Intervirology 17, 1–200.
Matthews, R.E.F., 1983. The history of viral taxonomy. In: Matthews, R.E.F. (Ed.), A Critical Appraisal of Viral Taxonomy. CRC Press, Boca Raton, FL, pp. 1–35.
Matthews, R.E.F., 1985. Viral taxonomy for the non-virologist. Ann. Rev. Microbiol. 39, 451–474.
Mayden, R.L., 1997. A hierarchy of species concepts: the denouement in the saga of the species problem. In: Claridge, M.F, Dawah, H.A., Wilson, M.R. (Eds.), Species: The Units of Biodiversity. Chapman and Hall, London, pp. 381–424.
Mayo, M.A., 2002. ICTV at the Paris ICV: results of the plenary session and the binomial ballot. Arch. Virol. 147, 2254–2260.
Mayo, M.A., Haenni, A.-L., 2006. Report from the 36th and 37th meetings of the Executive Committee of the ICTV. Arch. Virol. 151, 1031–1037.
Mayo, M.A., Horzinek, M.C., 1998. A revised version of the International Code of Virus Classification and Nomenclature. Arch. Virol. 143, 1645–1654.
Mayo, M.A., Maniloff, J., Van Regenmortel, M.H.V., Fauquet, C.M., 2002. The type species in virus taxonomy. Arch. Virol. 147, 1271–1274.
Mayr, E., 1970. Populations, Species and Evolution. Harvard University Press, Cambridge, MA.
Milne, R.G., 1984. The species problem in plant virology. Microbiol. Sci. 1, 113–122.
Moreira, D., Brochier-Armanet, C., 2008. Giant viruses, giant chimeras: the multiple evolutionary histories of Mimivirus genes. BMC Evol. Biol. 8, e12.
Moreira, D., Lopez-Garcia, P., 2009. Ten reasons to exclude viruses from the tree of life. Nat. Rev. Microbiol. 7, 306–311.
Murphy, F.A., Fauquet, C.M., Bishop, D.H.L, Ghabrial, S.A., Jarvis, A.W., Martelli, G.P., et al. (Eds.), 1995. Virus Taxonomy. Sixth Report of the ICTV. Springer, Vienna.
Padidam, M., Sawyer, S., Fauquet, C.M., 1999. Possible emergence of new geminiviruses by frequent recombination. Virology 265, 218–225.
Pigliucci, M., 2003. Species as family resemblance concepts: the (dis-)solution of the species problem? Bioessays 25, 596–602.
Pringle, C.R., 1991. The 20th meeting of the executive committee of the ICTV. Virus species, higher taxa, a universal database and other matters. Arch. Virol. 119, 303–304.
Pringle, C.R., 1997. The order *Mononegavirales*. Arch. Virol. 142, 2321–2326.
Raoult, D., Forterre, P., 2008. Redefining viruses: lessons from Mimivirus. Nat. Rev. Microbiol. 6, 315–319.
Raoult, D., Audic, S., Robert, C., Abergel, C., Renesto, P., Ogata, H., et al., 2004. The 1.2-megabase genome sequence of Mimivirus. Science 306, 1344–1350.
Sanjuan, R., 2010. Quasispecies. In: Mahy, B.W.J., Van Regenmortel, M.H.V. (Eds.), Desk Encyclopedia of General Virology. Academic Press, Elsevier, Oxford, pp. 42–48.
Schaefer, J.A., Wilson, C.C., 2002. The fuzzy structure of populations. Can. J. Zool. 80, 2235–2241.
Smith, D.B., McAllister, J., Casino, C., Simmonds, P., 1997. Virus "quasi-species": making a mountain out of a molehill? J. Gen. Virol. 78, 1511–1519.
Stamos, D.N., 2003. The Species Problem. Biological Species, Ontology and the Metaphysics of Biology. Lexington Books, Oxford.
Van Regenmortel, M.H.V., 1989. Applying the species concept to plant viruses. Arch. Virol. 104, 1–17.
Van Regenmortel, M.H.V., 1998. From absolute to exquisite specificity. Reflections on the fuzzy nature of species, specificity and antigenic sites. J. Immunol. Meth. 216, 37–48.

Van Regenmortel, M.H.V., 2000a. Introduction to the species concept in virus taxonomy. In: Van Regenmortel, M.H.V, Fauquet, C., Bishop, D.H.L., Carstens, E.B., Estes, M.K., Lemon, S.M., et al., Virus Taxonomy. Seventh Report of the International Committee on Taxonomy of Viruses. Academic Press, San Diego, CA, pp. 3–16.

Van Regenmortel, M.H.V., 2000b. On the relative merits of italics, Latin and binomial nomenclature in virus taxonomy. Arch. Virol. 145, 433–441.

Van Regenmortel, M.H.V., 2001. Perspectives on binomial names of virus species. Arch. Virol. 146, 1637–1640.

Van Regenmortel, M.H.V., 2003. Viruses are real, virus species are man-made taxonomic constructions. Arch. Virol. 148, 2481–2488.

Van Regenmortel, M.H.V., 2006. Virologists, taxonomy and the demand of logic. Arch. Virol. 151, 1251–1255.

Van Regenmortel, M.H.V., 2007. Virus species and virus identification: past and current controversies. Infect. Gen. Evol. 7, 133–144.

Van Regenmortel, M.H.V., 2010a. Logical puzzles and scientific controversies: the nature of species, viruses and living organisms. Syst. Appl. Microbiol. 33, 1–6.

Van Regenmortel, M.H.V., 2010b. The nature of viruses. In: Mahy, B.W.J., Van Regenmortel, M.H.V. (Eds.), Desk Encyclopedia of General Virology. Academic Press, Elsevier, Oxford, pp. 23–29.

Van Regenmortel, M.H.V., Fauquet, C.M., 2002. Only italicised species names of viruses have a taxonomic meaning. Arch. Virol. 147, 2247–2250.

Van Regenmortel, M.H.V., Bishop, D.H.L., Fauquet, C.M., Mayo, M.A., Maniloff, J., Calisher, C.H., 1997. Guidelines to the demarcation of virus species. Arch. Virol. 142, 1505–1518.

Villarreal, L.P., 2005. Viruses and The Evolution of Life. ASM Press, Washington, DC.

Waugh, J., 2007. DNA barcoding in animal species: progress, potential and pitfalls. Bioessays 29, 188–197.

Wheeler, Q., Meier, R. (Eds.), 2000. Species Concepts and Phylogenetic Theory: A Debate. Columbia University Press, New York, NY.

Wheeler, Q.D., Valdecasas, A.G., 2007. Taxonomy: myths and misconceptions. An. Jard. Bot. Madr. 64, 237–241.

Wimmer, E., 2006. The test-tube synthesis of a chemical called poliovirus: the simple synthesis of a virus has far-reaching societal implications. EMBO Rep. 7, S3–S9.

Wittgenstein, L., 1953. Philosophical Investigations. Macmillan, New York, NY.

Wolkowicz, R., Schaechter, M., 2008. What makes a virus a virus? Nat. Rev. Microbiol. 6, 643.

2 A Theory-Based Pragmatism for Discovering and Classifying Newly Divergent Bacterial Species

*Sarah Kopac and Frederick M. Cohan**

Department of Biology, Wesleyan University, Middletown, CT, USA

2.1 Introduction

Are bacterial species real? They are real enough to the systematists who classify them, and to the practitioners of microbiology who depend on bacterial classification. Bacterial systematists have routinely identified species as closely related groups that differ in their disease-causing properties, in their ecological roles in biological communities, and in their physiological capacities (Rosselló-Mora and Amann, 2001). To provide this service, systematists have taken a simple and pragmatic approach—to define species as groups (or clusters) of close relatives separated by large gaps in phenotypic and molecular characters (Vandamme et al., 1996; Rosselló-Mora and Amann, 2001). This practical approach has the cachet of approval from no less an evolutionary biologist than Charles Darwin (Darwin, 1859; Mallet, 2008b). Darwin proposed that animal and plant species should be defined as closely related groups that can coexist as phenotypically distinct clusters (Darwin, 1859; Stamos, 2007; Mallet, 2008b), and this is largely the approach taken by bacterial systematists. This cluster-based approach has proved to be remarkably robust, even as the criteria for defining bacterial species have changed over the decades, from being based on phenotype (usually metabolism) to whole-genome similarity (as measured through genome hybridization) to sequence identity (Rosselló-Mora and Amann, 2001). Bacterial systematists have argued about whether the species they recognize are too narrowly or too broadly defined, and whether they are using the best criteria for demarcating species, but they have agreed that species should hold the essential property of being clusters of close relatives with gaps between them (Gevers et al., 2005).

*E-mail: fcohan@wesleyan.edu

However, many microbiologists and most systematists outside of microbiology have understood species to be more than closely related groups separated by gaps (Mayr, 1982; de Queiroz, 2005). They have viewed the species level of taxonomy as having a reality beyond human attempts at classification. Largely under the influence of Ernst Mayr, the property of cohesion has come to be understood to be a quintessential aspect of species (Mayr, 1963, 1982; Templeton, 1989; de Queiroz, 2005). In this view, species are real because they are the largest groups whose diversity is constrained by a force of cohesion. In the case of the highly sexual animals and plants, the force constraining diversity within species is understood to be genetic exchange. In Mayr's Biological Species Concept, speciation requires certain unusual circumstances that allow newly divergent populations to break free of cohesion by recurrent, high-frequency genetic exchange; speciation is therefore understood to be rare (Mayr, 1963). Recently, zoologists have questioned whether animal species are really cohesive across their geographic ranges, and whether cohesion by genetic exchange actually prevents speciation (Mallet, 2008a). This controversy has raised our doubts as to whether bacterial species are cohesive (Cohan, 2010), an issue to which we shall return.

Many concepts of species have been developed since Mayr's Biological Species Concept, and most have in common certain quintessential features, all related to cohesion (de Queiroz, 2005; Cohan, 2010): the diversity within a species is limited by a force of cohesion, species are invented only once, and different species are ecologically distinct and irreversibly separate. In what we might consider Mayrian concepts of species, these essential properties have been extended to other groups where genetic exchange is rare or absent, such as the bacteria (Templeton, 1989; Cohan, 1994). With our colleagues, we have developed the "ecotype" theory of bacterial species, in which the diversity within an ecotype is constrained by recurrent forces of cohesion such as periodic selection or genetic drift (Cohan, 1994, 2010; Ward, 1998; Cohan and Perry, 2007).

Models of species cohesion depend on homogeneity among members of a species. In the case of animals and plants, cohesion across populations by genetic exchange is widely thought to require homogeneity of reproductive features, such that genes can be successfully exchanged (Mayr, 1963; Templeton, 1989). Likewise, the ecotype model is premised on the existence of ecotypes whose members are *ecologically* homogeneous and interchangeable, such that one competitively superior adaptive mutant can replace all other members of the ecotype (Cohan, 1994).

There is an important pragmatic reason for bacterial species to be demarcated as cohesive and ecologically homogeneous units. The animal ecologist Evelyn Hutchinson saw species as groups that should be homogeneous in their physiological, biochemical, morphological, and ecological characteristics (Hutchinson, 1968). He noted that species so defined have the useful property that the characteristics of any individual classified to a species could be easily predicted. While we believe Hutchinson was overly optimistic about the homogeneity of animal and plant species, microbiologists could probably agree that a taxonomy based on homogeneity, if possible, would be extremely beneficial (Cohan and Perry, 2007). For example, under this approach, the total membership of a pathogenic species would have the same disease-causing properties, the same tissue tropisms, the same transmission properties, and the same host range, while organisms with significantly different properties would be recognized as different species.

It is widely understood that the species recognized by bacterial systematics are far from satisfying the Hutchinsonian property of homogeneity. The named species have long been known to be metabolically and ecologically diverse (Barrett et al., 1986; Cohan et al., 2006; Smith et al., 2006; Walk et al., 2007; Dykhuizen et al., 2008; Hunt et al., 2008; Koeppel et al., 2008; Manning et al., 2008; Walk et al., 2009; Connor et al., 2010). One goal of this chapter is to propose a method to identify and classify ecologically homogeneous groups we may define as Hutchinsonian species, if they exist.

There are reasons to suspect that Hutchinsonian species may be extremely limited in their phylogenetic breadth, that they are perhaps limited to containing as phylogenetically narrow a group as a single cell and its immediate descendants (Doolittle and Zhaxybayeva, 2009). Recent genome comparisons suggest the possibility that, at least in some taxa, extremely close relatives are distinct in their genome content (Welch et al., 2002; Whittam and Bumbaugh, 2002; Tettelin et al., 2005; Lefébure and Stanhope, 2007; Rasko et al., 2008; Touchon et al., 2009; Paul et al., 2010). That is, bacteria may acquire genes by horizontal genetic transfer at such a high rate that the set of homogeneous organisms may be too small to be worth the trouble to recognize as a species entity. The second goal of this chapter is to lay out a protocol for determining the phylogenetic extent of ecological homogeneity.

Approaches to discovering homogeneous, cohesive species of bacteria are handicapped by various features of bacterial ecology and evolution, which make it difficult to recognize the ecological dimensions by which species diverge or the physiological adaptations that underlie the ecological divergence of new species (Cohan and Perry, 2007). This is because we cannot just look at bacteria and infer how they are different ecologically, as we can with birds of different beak size or shape. Also, horizontal genetic transfer is thought to be one mechanism responsible for the formation of new ecotypes (Gogarten et al., 2002; Cohan and Koeppel, 2008; Palenik et al., 2009), and we cannot predict the genes transferred or their donor source. We therefore cannot always anticipate the dimensions of ecological and physiological divergence among new bacterial species, even in groups that are well characterized (Cohan and Perry, 2007).

The discovery of newly divergent bacterial species requires a universal method that is not based on a priori knowledge or intuition about the ecological dimensions of speciation. The approach we outline for discovering the homogeneous, cohesive species of the bacterial world is ecology-blind, where we aim to hypothesize ecotype demarcations from sequence data, confirm the ecological distinctness of ecotypes, and then test for their homogeneity and cohesion, all without a priori knowledge of the ecological dimensions of ecotype distinctness (Cohan and Koeppel, 2008). We will lay out a process to identify groups of organisms that fit into species that are real, in the sense that they are homogeneous and cohesive; we also allow for the possibility that some (perhaps many) groups of bacteria fit only into reified units of close relatives that are neither homogeneous nor cohesive. Finally, we will present a new pragmatism for bacterial systematics, which will recognize the real, ecologically homogeneous units of bacterial diversity, where practical, and will recognize reified, heterogeneous units of close relatives where necessary.

2.2 Ecological Breadth of Recognized Species

The classification scheme of bacterial systematics focuses on finding species that are significantly different from one another in DNA sequence identity, genome content, and physiology (Rosselló-Mora and Amann, 2001), but places almost no emphasis on ensuring that each individual species is homogeneous in any characteristic (Cohan, 2002; Staley, 2006; Ward et al., 2006). Under this system, two individuals may be in the same species if they show a critical (previously 97%, now 99%) sequence identity in their 16S rRNA genes (Stackebrandt and Ebers, 2006). This degree of genetic diversity allows for enormous ecological (and disease-causing) differences within a species, as illustrated by *Escherichia coli*. Members of *E. coli* may be specialized as pathogens or commensals, and may be specialized to colonize the large intestine or other parts of the body (Touchon et al., 2009; Walk et al., 2009). These are vastly different environments where the bacteria encounter different extracellular secretions, pH, and notable differences in the extracellular matrix, which they must attach to. Moreover, different *E. coli* populations may be specialized to different hosts (Gordon and Lee, 1999) and different outside environments (Walk et al., 2007). The profound ecological and physiological differences among *E. coli* populations are reflected by huge genomic differences, with three divergent populations sharing only 39% of their genes (Welch et al., 2002). Other named species have also been found to contain a high diversity of ecologically, physiologically, and genomically distinct members (Marri et al., 2006; Kettler et al., 2007; Lefébure and Stanhope, 2007; Vernikos et al., 2007; Paul et al., 2010).

How did systematists come to agree to house such a huge amalgam of diversity within the species they recognize? In the case of the animals and plants, humans have developed an "umwelt," a foundation for demarcating natural groups of consequence for survival, through natural selection and cultural evolution (Yoon, 2009). However, the bacteria until recently escaped the attention of human interest in biodiversity, and so systematists of bacteria had to develop a way of seeing and classifying the diversity of bacteria from scratch (Cohan, 2010). Moreover, as we have mentioned, bacterial systematists have not had the advantage of being able to anticipate either the ecological differences between close relatives of bacteria or the physiological differences underlying their ecological divergence.

Bacteriologists were successful from the mid-century on in developing an objectively based umwelt for species demarcation. While limited at the time to metabolic and other phenotypic characteristics, "numerical taxonomy" allowed bacterial systematists to develop standard levels of phenotypic diversity within and between species (Sneath and Sokal, 1973; Yoon, 2009) (Figure 2.1A). In principle, species

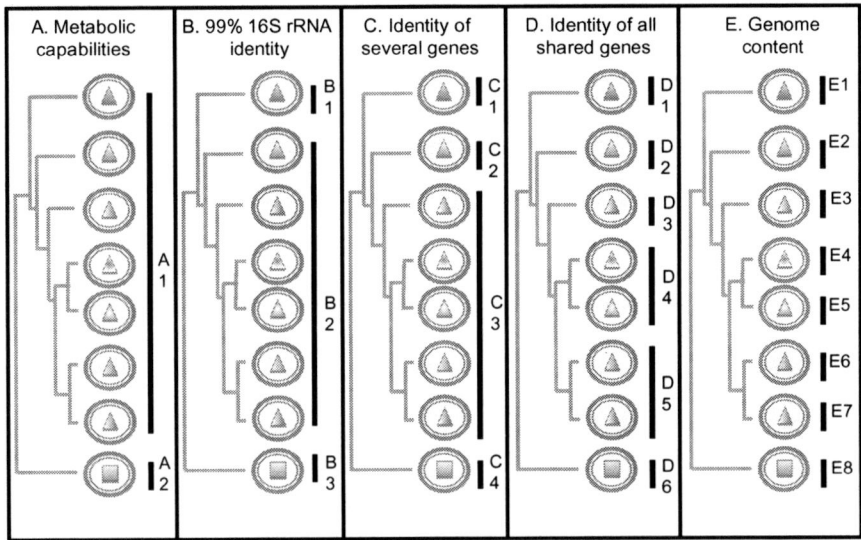

Figure 2.1 Species demarcations under different criteria. Each oval represents a set of closely related cells with identical characteristics of metabolism and ecology, sequences of shared genes, and genome content. Different shapes within the ovals (triangle versus square) represent extremely divergent metabolic capabilities (correlated with ecological function), and variations in shading within a particular shape represent more subtle divergence in metabolism and ecology. The species demarcations under each criterion are indicated by a black vertical bar and a species label (e.g., A1). A. Species were originally defined as groups that differ to a large extent in metabolic capability (indicated by triangle versus square), frequently with much metabolic diversity within each species (indicated here by shading differences within the triangles). B. Defining a species as a group of organisms sharing at least 99% 16S rRNA identity can split the metabolically defined species in the previous panel, as seen here by the splitting of species A1 into B1 and B2. C. Defining species as clusters based on several protein-coding gene sequences can split a 16S-defined species into groups that are each more ecologically homogeneous. This is seen here by the splitting of species B2 into C2 and C3. D. Defining species as clusters based on sequence identity for all shared genes can divide species even further with, for example, species C3 being split into D3, D4, and D5. This may be the most highly resolving method for identifying species based on sequences of shared genes. Within species D4, we can see the possibility that even with this level of resolution for species demarcation, there may still be ecological heterogeneity (indicated by the difference in shading between cells in species D4). Species D5 shows an alternative model where this high level of resolution finds clusters that are ecologically homogeneous, as noted by the same shading patterns among members of D5. E. Defining species by identity of genome content could spuriously split close relatives that are ecologically identical into different species. Note that the two organisms within D5, with the same ecology, are split on the basis of genome content into different species. In this case, E6 and E7 would most likely be different for phage or insertion sequence genes that do not specify ecological niche.

could have been narrowly defined on metabolic grounds, but systematists made a pragmatic, but fateful decision early on to include strains within a species that were heterogeneous in the presence versus absence of many metabolic capabilities. Bacterial species were from the start defined to be extremely diverse in their physiological and, hence, ecological characteristics.

Subsequent incorporation of molecular technology has improved species identification in some important ways (Rosselló-Mora and Amann, 2001; Cohan, 2010). Molecular approaches have provided universal and readily available methods and criteria for species demarcation to all systematists. Using sequence-based criteria, systematists have been able to avoid recognition of polyphyletic groups. Also, because systematics has been based to some extent on whole-genome assays, such as DNA–DNA hybridization, classification has not been deeply affected by recombination across species. Finally, universally applying molecular criteria has led to a pragmatic demarcation scheme that most systematists can agree on (Rosselló-Mora and Amann, 2001) (Figure 2.1B).

Nevertheless, molecular technology has not brought about a refinement in the breadth of diversity subsumed within a recognized taxon. Rather, as each new technology has been embraced, including DNA–DNA hybridization (Wayne et al., 1987), 16S rRNA sequence (Stackebrandt and Ebers, 2006), multilocus sequence analysis (Gevers et al., 2005), and genome-wide average nucleotide identity (ANI) (Konstantinidis and Tiedje, 2005), systematists have attempted to calibrate every new method to yield the existing species taxa (Cohan, 2002; Cohan and Perry, 2007).

Thus, while the approaches of systematics have brought pragmatic solutions for the practice of systematists, we might ask whether these approaches have been pragmatic for microbiologists outside of systematics. The problem is that when systematists reify an amalgam of ecological and functional diversity into a species taxon, other microbiologists tend to assume that each such species constitutes a natural and fundamental unit of biodiversity. This has led to numerous unfortunate consequences for microbiologists outside of systematics. One such consequence is the classification of genes within a recognized species as essential, "core" genes, shared by all "species" members, versus the nonessential, "dispensable" (Tettelin et al., 2005) or "flexible" (Kettler et al., 2007) genes that are shared only by a subset of species members. This dichotomy is false because it is based on the reification of the named species. Also, this gives the impression that those genes held only by one subclade are somehow not essential to the ecology or physiology of that group.

This reification of the core genome may have a real, negative impact on vaccine development. Vaccine development can be based on choosing a target protein that is a member of the core genome (Tettelin et al., 2005). However, if the pathogenic strains of concern constitute only a single ecotype within the species diversity, the choice of vaccine target is unnecessarily restricted to the small core genome of the entire named species, rather than the larger set of genes shared among the pathogen ecotype.

The broad definition of recognized species has led to innumerable errors in population genetic estimation of the critical parameters of evolution. For example, attributing the name *Escherichia coli* to the huge diversity of ecological specialists within the species gives population geneticists the impression that they are dealing with a group of ecologically interchangeable organisms, such as the members of the fruit fly species *Drosophila pseudoobscura* within a particular habitat. This has led to incorrect application of various algorithms for estimating effective population size and recombination rates (Gordon and Lee, 1999), which assume that the organisms sampled are interchangeable (McVean et al., 2002). In the case of estimating effective population size from sequence diversity, ecological heterogeneity artificially increases the sequence diversity, and thereby the estimate of population size, by conflating the divergence between populations (which is not affected by population size) with divergence within them. Sequence-based estimations of migration rates have also erred by pooling within a taxon a number of ecologically distinct groups (Roberts and Cohan, 1995).

In addition to the errors caused by species reification, the broad brush of systematics has also incurred an opportunity cost for different subfields of microbiology, starting with systematics itself. When a systematist discovers a new species and sees that it can be squeezed into one species taxon, there is no further motivation from systematics to further explore the ecologically distinct clades within the species. Hence, the research in systematics is impoverished by a standard of detail that leaves much of a clade's diversity uncharacterized.

The broad brush also incurs an opportunity cost on epidemiology. In preparation for the next epidemic, epidemiologists might find it useful to identify all the ecologically distinct populations that already exist within a named pathogenic species. We could then prepare for a future epidemic by characterizing, in advance, the disease-causing properties of each population (Cohan and Perry, 2007). Biotechnologists could also take advantage of a more fine-grained systematics of species. After discovering a strain with a valuable enzyme, one could then search for homologs of the enzyme across closely related, ecologically distinct populations, if they were highlighted by taxonomic recognition (Cohan and Perry, 2007; Jensen, 2010).

The molecular revolution has taken us far beyond the early days of systematics, when species demarcation was based entirely on metabolism and other phenotypic traits. Sequencing has now revealed ecologically distinct populations within the recognized species, yet we do not take advantage of this information to refine the demarcations of species. The time has come to incorporate the high resolution of molecular technology into our taxonomy, so that the physiological and ecological diversity we know to exist within the named species can be officially recognized. An important challenge is to develop universal algorithms to analyze sequence data to identify populations that are each ecologically homogeneous and ecologically distinct from one another.

2.3 Models of Bacterial Speciation

In order to integrate ecology into taxonomic classification, we need to take into account the various ways in which bacterial species form and diversity within species is constrained. In the Stable Ecotype model, ecotypes are long-lived, giving each ecotype ample opportunity to acquire a unique set of neutral mutations in each gene in the genome; also, cohesion results from recurrent periodic selection events within each ecotype lineage (Cohan and Perry, 2007). Ecotypes are founded when a single individual acquires a mutation (or a recombination event) that changes its ecology, through utilizing a new set of resources, thriving under a new set of environmental conditions, or adopting some other change in lifestyle. Since new ecotypes are each founded by a single individual, they start out with zero diversity. A new ecotype is not in direct competition with the members of the parental ecotype because it lives in a different place or uses at least somewhat different resources. For example, a member of a primarily impetigo-causing (skin-infecting) ecotype of *Streptococcus pyogenes* might mutate or acquire a gene that allows it to primarily infect the throat (Bessen, 2009), thus founding a new ecotype. Although the new ecotype may share the same host as the parental ecotype, it is utilizing different host resources, and so the two ecotypes may not experience the same periodic selection events.

Each ecotype remains homogeneous via periodic selection events (Cohan and Perry, 2007). In periodic selection, an individual acquires a genetic change that allows it to be more efficient in its present niche. The individual and its nearly clonal descendants then outcompete all other members of the ecotype and constitute the only lineage that persists into the future (with the exception of a very small level of diversity that survives because of rare recombination) (Cohan, 2005). The result is that the ecotype (surviving through a single lineage) now has extremely little genetic or ecological variation.

How do we know that periodic selection occurs in nature? Periodic selection events are difficult to observe directly in nature. One could try to infer the existence of a history of periodic selection events from an ecotype's low level of sequence diversity (e.g., an ANI of 99%). If one could rule out genetic drift as responsible for limiting the divergence, one might suspect that periodic selection was responsible (since it can purge an ecotype of its sequence diversity regardless of the population size). However, it is difficult to rule out the possibility that an ecotype is simply too young to have accumulated substantial diversity (Koeppel et al., 2008).

It is possible to document periodic selection events when the fitness advantage of an adaptive mutation has transcended beyond its original ecotype, to provide adaptation for another ecotype (the Adapt Globally, Act Locally model) (Majewski and Cohan, 1999). In this case, an adaptive mutation rises to 100% frequency in its original ecotype, becomes transferred to another ecotype by genetic exchange, and then sets in motion a periodic selection event within the recipient ecotype. It is important to keep in mind here that while the adaptive value of the mutation (the allele) transcends beyond its ecotype, the fitness of the mutant (the organism) is limited to the ecological niche of its ecotype; thus, each ecotype has its own periodic selection event, although based on the same mutation (Cohan and Perry, 2007). The genome-wide result of such Adapt Globally, Act Locally periodic selection events is that ecotypes remain divergent throughout their genomes, except that the ecotypes become nearly identical in the gene(s) driving the periodic selection, as well as in linked genes transferred between ecotypes (Koeppel et al., 2008).

There are numerous examples of such anomalous localized identity between otherwise divergent ecotypes. For example, we found that two hot spring *Synechococcus* ecotypes, which are adapted to different temperatures and are 13% divergent throughout the shared parts of their genomes, are nearly identical, within and between ecotypes, in their nitrogen-fixing operon (Bhaya et al., 2007). We interpret this as evidence for a periodic selection caused by acquisition of the nitrogen-fixing genes within each ecotype (Koeppel et al., 2008).

One consequence of many recurrent periodic selection events in each of several long-standing ecotypes is that ecotypes are expected to correspond to sequence clusters for any gene in the genome (Palys et al., 1997). This is because, while diversity in each gene in the genome is recurrently purged within an ecotype, different long-standing ecotypes can accumulate unique mutations in every gene. Assuming the Stable Ecotype model, various algorithms have been developed to find the sequence clusters most likely to correspond to ecotypes (Corander et al., 2008; Hunt et al., 2008; Koeppel et al., 2008; Barraclough et al., 2009).

However, we need to consider some alternative models where ecotypes do not correspond to sequence clusters. In some models, more than one ecotype is subsumed within a sequence cluster; in other models, more than one sequence cluster is contained within a single ecotype (Cohan and Perry, 2007). We believe that all of these models apply under certain circumstances, and that the idealized Stable Ecotype model might occur in only a minority of cases (Cohan, 2010).

In the Speedy Speciation model, cohesion occurs through periodic selection (and/or genetic drift), just as in the Stable Ecotype model (Cohan and Perry, 2007). The difference is that speciation is greatly accelerated as in an adaptive radiation, with the practical consequence that there are many newly divergent species that cannot be distinguished by neutral divergence in a small number of randomly chosen genes (i.e., genes not involved in the adaptive divergence between species, such as those used in most multilocus analyses). Depending on the rate of speciation, species could perhaps be distinguished by neutral sequence variation if the whole genome were sequenced in many isolates. Moreover, sequencing of the whole genome may reveal the genes responsible for ecological divergence.

The Species-Less model is profoundly different from the Stable Ecotype model and all models that assume cohesion within species (Cohan and Perry, 2007). This is a model of rapid speciation, as well as rapid extinction, leading to a high turnover of species. In this case, a species might not persist long enough, from its time of origin to its extinction, to undergo any periodic selection events. In the Species-Less model, each ecotype, while ecologically homogeneous, could not be considered a cohesive unit. Like the case of the Speedy Speciation model, where species are cohesive, the Species-Less model will lead to a diversity of ecotypes that cannot be easily distinguished as sequence clusters.

In the Species-Less model, ecotypes evolve not by becoming more efficient in utilizing their current ecological niche, but instead by evolving to invade a new ecological niche. The Species-Less model may apply to the case for pathogens, where immune-escape mutations may each constitute a new ecotype (Achtman and Wagner, 2008; Cohan, 2010). Also, the Species-Less model may apply in cases where an environment undergoes a succession process, where organisms at a site must adapt to rapidly changing conditions, for example the successions that occur on mine tailings, with pH and oxidation levels changing predictably and quickly (Remonsellez et al., 2009).

The Nano-Niche model also assumes a high rate of speciation, but here cohesion can occur across ecotypes (Cohan and Perry, 2007). In the Nano-Niche model, closely related ecotypes are subtly and only quantitatively different in their ecology. These "nano-niche ecotypes" use the same set of resources and conditions, but they coexist by using their shared resources and conditions *in different proportions*. Not having any unique resources that might constitute a haven from competition from other ecotypes, each ecotype is vulnerable to extinction from competition with other ecotypes. For a time, the various nano-niche ecotypes may coexist while each has its own private periodic selection events. At some point, however, an extremely competitive adaptive mutant (which bears what we call a speciation-quashing mutation) from one ecotype may extinguish not only the other members of its own ecotype, but also other closely related ecotypes (Cohan, 2005). In the Nano-Niche model, divergence among very closely related ecotypes is limited by these speciation-quashing mutations. Many closely related ecotypes might not last long enough to appear as separate sequence clusters, as based on niche-neutral genes not involved in ecological divergence.

The Nano-Niche model may apply to bacterial ecotypes that adapt over a long time to a given host individual (e.g., a commensal or chronic pathogen). The course of evolutionary adaptation to one human body may bring about multiple periodic selection events in that nano-niche population, but the individual hosts might not be different enough to support unique ecotypes into the indefinite future. Any speciation-quashing mutation that makes an individual not just superior in its own host but also in other hosts would put an end to the speciation among the various nano-niche ecotypes. In the Nano-Niche model, extinction of the nano-niche ecotypes might be too rapid for us to discern them as sequence clusters; also, in this case cohesion (the limit to diversification) occurs across ecologically distinct groups. In our quest to identify ecologically homogeneous groups, we should perhaps be satisfied with finding sets of nano-niche ecotypes whose continued coexistence and divergence we predict will end with a speciation-quashing mutation.

In the Recurrent Niche Invasion model, mobile genetic elements such as plasmids or phage may determine bacterial niches (Cohan and Perry, 2007). For example, in the case of *Rhizobium*, a bacterial lineage may acquire a symbiotic plasmid that adapts it as an endosymbiotic mutualist for a particular set of legume hosts; the lineage may then lose that plasmid and gain another, which adapts it to another set of legume hosts. In the Recurrent Niche Invasion model, the dynamics are particularly interesting when there is no specialization among bacteria to different mutualistic plasmids. In this case, plasmids conferring different ecological niches may come and go among a large group of host bacteria, and any adaptive mutant in the bacterial host population will purge the diversity in the entire bacterial host population. Thus, sequence clusters will correspond to a set of interchangeable bacterial hosts that can each accommodate a diverse set of plasmids. Here, any single sequence cluster we recognize would actually be adapted to a diversity of plasmid-provided ecological niches. Cohesion extends beyond the individual, plasmid-encoded ecotypes to the whole suite of host organisms that can profit from the set of plasmids (Cohan, 2010).

The Cohesive Recombination model provides another mechanism by which ecologically distinct populations will fail to be recognizable as sequence clusters (Cohan and Perry, 2007). As analyzed quantitatively by Hanage et al. (2006) and Buckley (personal communication), bacteria with the highest rates of recombination may exchange genes so frequently that ecotypes do not accumulate sequence divergence in niche-neutral genes, and so we will not be able to discern the ecotypes as distinct sequence clusters. We note, however, that the rates of genetic exchange in bacteria are never sufficient to hinder or reverse adaptive divergence in niche-specifying genes, as we have previously discussed (Cohan and Koeppel, 2008). Thus, while genetic exchange in rapidly recombining bacteria will not prevent ecotype formation, it may prevent our ability to discover ecotypes using niche-neutral sequence diversity.

In some cases, a single ecotype from a given community may fall into several distinct sequence clusters, as seen in the Geotype plus Boeing model (Cohan and Perry, 2007). Provided that a given taxon has not dispersed frequently, geographically isolated members of a single ecotype may diverge into different clusters (geotypes), even while remaining ecologically interchangeable (Papke and Ward, 2004). This would yield a different sequence cluster in each geographical region, a common phenomenon in the systematics of all organisms that do not readily disperse (Whitaker et al., 2003). In the case of bacteria, geotypes may be a source of confusion for systematists if geotypes have historically been isolated but now with modern human transport, their dispersal has been recently accelerated. In this case (the Geotype plus Boeing model), members of one ecotype isolated from a single site may contain multiple clusters representing the ecotype's various, formerly isolated geotypes from all over the world. In addition, patterns of genetic drift can yield multiple, ecologically interchangeable sequence clusters within a single ecotype.

2.4 Algorithms for Identifying Ecotypes

There are multiple tools available for discerning ecotypes from sequence data, including AdaptML (Hunt et al., 2008), Ecotype Simulation (Koeppel et al., 2008), BAPS (Corander et al., 2008), and GYMC (Barraclough et al., 2009). In contrast to the approaches of bacterial systematics, none of these algorithms assumes a universal criterion for demarcation. Rather, each algorithm uses sequence data from the taxon of focus to identify the appropriate sequence divergence criterion for distinguishing ecotypes.

AdaptML differs from the rest in requiring the habitat of isolation as input data, while the others are blind to ecology (i.e., no information about the ecology or habitat of the strains is taken into account in the analysis) (Hunt et al., 2008). Both approaches have their advantages (Cohan and Koeppel, 2008). AdaptML is useful when associations with certain habitats are suspected, as this algorithm can simultaneously discover ecotypes and confirm their preferences to habitats specified by the investigator. In contrast, the ecology-blind algorithms do not require the researcher to know anything about the potential environmental differences being analyzed. As a result, multiple ecotypes can be found even in environments which were a priori thought to be homogeneous. However, because Ecotype Simulation does not incorporate ecological data, it cannot confirm the ecological distinctness of ecotypes, and so additional tests must be performed to independently confirm that the clusters are ecologically distinct. Thus far, the clusters found by Ecotype Simulation have consistently been shown to be significantly distinct from one another in their habitat associations; in many cases, the algorithm has hypothesized multiple ecotypes within recognized species (Cohan et al., 2006; Ward et al., 2006; Koeppel et al., 2008; Connor et al., 2010). Likewise, AdaptML has identified ecologically distinct populations within recognized species of *Vibrio* (Hunt et al., 2008). The ecotypes identified by Ecotype Simulation and AdaptML have largely been the same, although in some cases Ecotype Simulation hypothesizes a diversity of ecotypes within one ecotype found by AdaptML (Connor et al., 2010; Melendrez et al., unpublished data). We believe this is because AdaptML can only detect those ecotypes that are different in preferences for habitats anticipated by the investigator. Less frequently, AdaptML hypothesizes multiple ecotypes within one ecotype demarcated by Ecotype Simulation (Connor et al., 2010).

Ecotype Simulation is the only algorithm of its kind that incorporates both periodic selection and speciation. The algorithm is thereby designed to find ecotypes that are each subject to periodic selection, and it estimates how often diversity is purged and how often new ecotypes are formed (Koeppel et al., 2008).

For all of the aforementioned models, the resolution of the analysis depends upon the rates of evolution of the genes analyzed. We have observed that some newly divergent ecotypes may be discerned with protein-coding gene diversity but not with 16S rRNA sequences (Ward et al., 2006; Koeppel et al., 2008). This is because 16S rRNA sequences offer fewer informative sites than protein-coding genes, especially a concatenation of multiple protein-coding genes (Palys et al., 2000) (Figure 2.1B,C). Analysis of a concatenation of all the shared genes (orthologs) among genomes, made possible by whole-genome sequencing, may be the most discerning approach to finding newly divergent ecotypes (Figure 2.1D).

2.5 Confirming the Ecological Distinctness of Ecotypes

If we knew the Stable Ecotype model to apply universally, we could identify ecotypes with confidence as sequence clusters for any gene in the genome (Cohan and Perry, 2007). However, under some models, particularly the Geotype plus Boeing model and models with strong genetic drift, one ecotype may contain multiple sequence clusters. Therefore, each sequence cluster deemed to be an ecotype must be confirmed to be ecologically distinct.

Ecotypes can be confirmed as ecologically distinct when they are significantly different in their habitat associations (Ward et al., 2006; Koeppel et al., 2008). It is important to note that ecotypes need not be absolutely specialized to different habitats (Hunt et al., 2008). Closely related species frequently are at least somewhat ecologically generalized, and coexist by quantitative differences in habitat distinction (Hunt et al., 2008). Various methods are available for confirming quantitative habitat preferences, such as contingency tests (such as Fisher's Exact Test) and AdaptML, designed for bacterial ecotypes (Hunt et al., 2008). We have confirmed the ecological distinctness of putative ecotypes by their habitat associations, including differences in solar exposure, soil texture, rhizospheres, and elevation for soil *Bacillus* (Koeppel et al., 2008; Connor et al., 2010; Kopac et al., unpublished data), temperature and depth in the photic zone in hot spring *Synechococcus* (Ward et al., 2006; Becraft et al., unpublished data), and host range in *Legionella* (Cohan et al., 2006); differences in associations with particles of different size and with season among ecotypes in *Vibrio splendidus* have been identified and confirmed by AdaptML (Hunt et al., 2008). In addition, many ecologists have noted that very closely related sequence clusters (demarcated by eye, rather than by a computer algorithm) are different in their habitat associations (Brisson and Dykhuizen, 2004; Smith et al., 2006; Walk et al., 2007; Manning et al., 2008; Walk et al., 2009).

A disadvantage of the habitat association approach for confirming ecotypes is that it requires the investigator to anticipate the ecological dimensions by which ecotypes diverge. We have previously suggested a different approach where the habitat differences need not be initially known (Cohan and Koeppel, 2008). Each environment from which the focus taxon is isolated may be biotically characterized by identifying the other microbes that live there, using high throughput sequencing of 16S rRNA from each site. Then, environments may be clustered by community type, and one may test whether ecotypes of the focus group differ in their associations with different community types. One may even infer the preferred physical habitats of each ecotype from previous ecological knowledge about the dominant microbial players in an ecotype's preferred community.

The ecological distinctness of hypothesized ecotypes may be confirmed by showing that the groups respond to an environmental perturbation in different ways. In some cases, the natural distribution of putative ecotypes may suggest the direction of response of putative ecotypes to a perturbation. For example, when investigators shaded a Yellowstone hot spring mat, this increased the frequencies of ecotypes normally found in the most-shaded depths of the photic zone (Becraft et al., unpublished data). In other cases, putative ecotypes may respond distinctly to a given perturbation, but with no prediction available for the direction of response. For example, disturbing the thermocline of a lake was shown to affect the vertical distribution of different putative ecotypes in distinct ways (Youngblut and Whitaker, unpublished data). When putative ecotypes respond in distinct ways to an environmental perturbation, whether the directions of response were predicted or not, this indicates the ecological distinctness of the hypothesized ecotypes.

The ecological distinctness of ecotypes may also be confirmed by finding physiological differences between ecotypes that adapt them to their preferred habitats.

For example, bacteria may evolve adjustments of their temperature optima by changing the rigidity of their cell membranes, and such changes may be assayed by fatty acid content (Sikorski and Nevo, 2007; Connor et al., 2010). In other cases, investigators may demonstrate that ecotypes grow fastest under laboratory conditions simulating one component of the preferred habitat. For example, ecotypes associated with hotter regions of a Yellowstone hot spring mat were found to have a higher temperature optimum for growth in the laboratory (Allewalt et al., 2006).

The genomic revolution has allowed identification of ecologically distinguishing features even when we do not have any previous knowledge or intuition about them. One approach involves comparing genome content. For example, differences in genome content showed two ecotypes of hot spring *Synechococcus* to differ in nitrogen storage and phosphonate uptake capacity, revealing that a population more distant from the hot spring source is not simply adapted to cooler temperature; it is also adapted to a lifestyle of scavenging elements (N and P) that are less available downstream (Bhaya et al., 2007). Also, genome-wide comparisons of orthologous genes shared across ecotypes can potentially reveal ecological divergence. Shared genes that have participated in the ecological divergence among ecotypes may show an acceleration of divergence, and Vos has argued that detection of such differences can identify ecotypes (as well as the ecological differences among them) (unpublished data).

2.6 Ecological Homogeneity within Ecotypes

Demonstrating an ecological difference among suspected ecotypes appears to be easy, given the success of the approaches we have discussed. It is quite another matter to confirm the ecological homogeneity within a putative ecotype. When one does not know ahead of time which organisms within a suspected ecotype are likely to be different, attempts to identify ecological differences can be expensive and time-consuming.

Doolittle and colleagues have suggested that ecological homogeneity does not even exist beyond the closest of relatives (in the extreme form of this argument, one cell and its immediate offspring) (Doolittle and Zhaxybayeva, 2009). Suspicion that lineages rapidly change their ecology emerged out of a study of genome size variation among closely related isolates of marine *Vibrio* (Thompson et al., 2004). Even bacteria that were indistinguishable by sequence in a marker gene had diverged in genome size. More recently, owing to the ease and economy of fully sequencing genomes, multiple organisms within various named species have been fully sequenced. Members of a species have thus been shown to differ in genomic content that may change the ecology (Rasko et al., 2008; Touchon et al., 2009; Tenaillon et al., 2010).

An important limitation of these studies is that investigators have rarely chosen to sequence extremely close relatives within a named species. Therefore, discovering differences in genome content of ecological import demonstrates only that there is ecological diversity within a named species, something that was already well known. What is needed is a survey of extremely closely related organisms that have been hypothesized to constitute a single ecotype, as based on sequence analysis. If hypothesized ecotypes are found to be heterogeneous in genome content of ecological significance, this would indicate that the phylogenetic breadth of ecological homogeneity may indeed be extremely limited (Figure 2.1E).

We emphasize that such tests must be careful to show that genome content differences among close relatives are niche specifying. This is because genome content differences between close relatives appear to reflect mostly the coming and going of phage and insertion sequences, and not the acquisition of new ecological traits (Touchon et al., 2009), demonstrating that heterogeneity of genome content does not necessarily imply heterogeneity of ecology.

Full sequencing of close relatives may be the most effective way to both identify and confirm the ecological distinctness of ecotypes. The concatenation of all shared genes will grant extremely high resolution to algorithms such as Ecotype Simulation. Moreover, as we have noted, genome comparisons can also identify the ecological differences among ecotypes. It will be particularly interesting to find whether this most finely resolving approach will find ecologically homogeneous ecotypes, or whether even among closest relatives, there is ecological heterogeneity.

2.7 Are Bacterial Ecotypes Cohesive?

We began with the issue of the reality of bacterial species, whether there is something biologically unique about the level of species. The concept of cohesion has been argued to provide a key dynamic property of species throughout the biological world—that diversity within a species is limited by certain forces but that divergence above the species level is not (Templeton, 1989; de Queiroz, 2005). The ecotype concept (and particularly the Stable Ecotype model) assumes cohesion within ecotypes, in that diversity within an ecotype is limited by periodic selection and genetic drift, but that divergence between them is not. This cohesion requires ecological homogeneity within an ecotype (Cohan and Perry, 2007).

However, ecological homogeneity is not sufficient to ensure cohesion by forces such as periodic selection and drift, which act recurrently over the lifetime of an ecotype. As we have seen in the Species-Less model, it is possible that new, ecologically homogeneous populations may not persist long enough to encounter a periodic selection event before it goes extinct. In a world with a high turnover of bacterial species, with rapid invention and extinction of ecotypes, the only force limiting the diversity within an ecotype would be its short lifetime before extinction. We have previously proposed that some, perhaps most, bacterial ecotypes within a taxon may not represent species-like cohesive groups, while others may be long-lasting and cohesive, and may even extend over broad geographical areas (Cohan, 2010). We have proposed a phylogenetic test to determine whether newly formed ecotypes are cohesive groups (Cohan, 2010). For the purpose of building systematics that might aim to identify and classify all the ecological diversity within a taxon, it is probably sufficient to focus on finding the ecologically homogeneous clades, without concern for their cohesiveness.

2.8 Incorporating Ecology into Bacterial Systematics

We suggest that systematics should recognize ecologically homogeneous ecotypes rather than the broadly defined, ecologically heterogeneous amalgams currently recognized. To this end, we lay out a protocol for selecting ecotypes that systematists might have the confidence and motivation to recognize. First, we suggest using sequence data to demarcate ecotypes that appear to represent phylogenetic groups with a history of coexistence as ecologically distinct lineages. Ecotypes could be hypothesized by any of the various universal, sequence-based methods, including AdaptML, Ecotype Simulation, GMYC, and BAPS. If such analyses were to be based on many genes, in the extreme the entire set of shared genes in the genome, more newly divergent ecotypes could be resolved.

Second, the most closely related ecotypes should be confirmed to be ecologically distinct from one another by differences in habitat association or in physiology.

Third, in keeping with an important tradition of bacterial systematics, ecotypes should be confirmed to be phenotypically distinct (Rosselló-Mora and Amann, 2001), and we add that ideally the phenotypic differences should confer the ecological niche specificity of the ecotypes.

Fourth, if possible, an ecotype should be confirmed to be ecologically homogeneous, although as we have pointed out, this may be difficult short of sequencing the full genomes of many members of the ecotype.

Fifth, we suggest that we should not be compelled to recognize every ecotype—only those of interest or consequence. This is because some focus taxa may contain multiple, extremely young, ecologically distinct populations that are unlikely to persist into the future (as in the case of the Nano-Niche model). Here we see that there is a conflict between ensuring homogeneity of ecotypes and recognizing only those of potential interest. Thus, the reform we suggest aims to identify the real, ecologically homogeneous groups where possible, but when impractical, we suggest classifying an ecologically heterogeneous clade as an ecotype, provided that it has been identified by sequence-based algorithms as a putative ecotype and has shown to be ecologically distinct from other closely related ecotypes.

We first consider those cases where a recognized, legacy species is found to contain multiple ecotypes, such as is the case for *Bacillus simplex* (Koeppel et al., 2008), *Vibrio splendidus* (Hunt et al., 2008), and probably many cases where sequence clusters within a pathogenic species are known to differ in host range and/or tissue tropism (Gordon and Cowling, 2003; Smith et al., 2006; Walk et al., 2007; Walk et al., 2009). In these cases, we suggest keeping the existing species binomial in order to maintain stability of the taxonomy, but suggest adding a trinomial "ecovar" epithet to describe the ecotype taxon. For example, an oak-forest-associated and a grassland-associated ecotype within *Bacillus simplex*, from a canyon near Haifa, Israel (Koeppel et al., 2008), might be named *B. simplex* ecovar Alon and *B. simplex* ecovar Esev (based on the Hebrew words for oak and grass). For ecotypes that are found to be outside the phylogenetic range of existing, recognized species, we suggest naming each ecotype as a species.

We believe that the approach we have laid out is pragmatic both for systematists and for those whose work would benefit from a full accounting of the ecological diversity among close relatives. The proposed system is pragmatic because it identifies the likely ecotypes through universally available and applicable techniques of genomics and DNA sequencing, as well as computer algorithms to recognize the ecotypes from sequence diversity patterns. It also does not reify heterogeneous groups by attempting to apply a universal molecular criterion to all bacterial species. Microbiologists outside of systematists would benefit from systematics that would recognize the most recent products of bacterial speciation. Perhaps most importantly, we will more effectively come to know the unique ecological roles played by each member of a vast and diverse microbial community.

Acknowledgment

This work was sponsored by NSF FIBR grant (EF-0328698) and by research funds from Wesleyan University.

References

Achtman, M., Wagner, M., 2008. Microbial diversity and the genetic nature of microbial species. Nat. Rev. Microbiol. 6, 431–440.

Allewalt, J.P., Bateson, M.M., Revsbech, N.P., Slack, K., Ward, D.M., 2006. Effect of temperature and light on growth of and photosynthesis by *Synechococcus* isolates typical of those predominating in the octopus spring microbial mat community of Yellowstone National Park. Appl. Environ. Microbiol. 72, 544–550.

Barraclough, T.G., Hughes, M., Ashford-Hodges, N., Fujisawa, T., 2009. Inferring evolutionarily significant units of bacterial diversity from broad environmental surveys of single-locus data. Biol. Lett. 5, 425–428.

Barrett, E.L., Solanes, R.E., Tang, J.S., Palleroni, N.J., 1986. *Pseudomonas fluorescens* biovar V: its resolution into distinct component groups and the relationship of these groups to other *P. fluorescens* biovars, to *P. putida*, and to psychrotrophic pseudomonads associated with food spoilage. J. Gen. Microbiol. 132, 2709–2721.

Bessen, D.E., 2009. Population biology of the human restricted pathogen, *Streptococcus pyogenes*. Infect. Genet. Evol. 9, 581–593.

Bhaya, D., Grossman, A.R., Steunou, A.S., Khuri, N., Cohan, F.M., Hamamura, N., et al., 2007. Population level functional diversity in a microbial community revealed by comparative genomic and metagenomic analyses. ISME J. 1, 703–713.

Brisson, D., Dykhuizen, D.E., 2004. *ospC* diversity in *Borrelia burgdorferi*: different hosts are different niches. Genetics 168, 713–722.
Cohan, F.M., 1994. The effects of rare but promiscuous genetic exchange on evolutionary divergence in prokaryotes. Am. Nat. 143, 965–986.
Cohan, F.M., 2002. What are bacterial species? Annu. Rev. Microbiol. 56, 457–487.
Cohan, F.M., 2005. Periodic selection and ecological diversity in bacteria. In: Nurminsky, D. (Ed.), Selective Sweep. Landes Bioscience, Georgetown, TX, pp. 78–93.
Cohan, F.M., 2010. Are species cohesive? A view from bacteriology. In: Walk, S., Feng, P. (Eds.), Bacterial Population Genetics: A Tribute to Thomas S. Whittam. ASM Press, Washington, DC.
Cohan, F.M., Koeppel, A.F., 2008. The origins of ecological diversity in prokaryotes. Curr. Biol. 18, R1024–R1034.
Cohan, F.M., Perry, E.B., 2007. A systematics for discovering the fundamental units of bacterial diversity. Curr. Biol. 17, R373–R386.
Cohan, F.M., Koeppel, A., Krizanc, D., 2006. Sequence-based discovery of ecological diversity within *Legionella*. In: Cianciotto, N.P., Abu Kwaik, Y., Edelstein, P.H., Fields, B.S., Geary, D.F., Harrison, T.G., et al., (Eds.), *Legionella*: State of the Art 30 Years after Its Recognition. ASM Press, Washington, DC, pp. 367–376.
Connor, N., Sikorski, J., Rooney, A.P., Kopac, S., Koeppel, A.F., Burger, A., et al., 2010. The ecology of speciation in *Bacillus*. Appl. Environ. Microbiol. 76, 1349–1358.
Corander, J., Marttinen, P., Siren, J., Tang, J., 2008. Enhanced Bayesian modelling in BAPS software for learning genetic structures of populations. BMC Bioinformatics 9, 539.
Darwin, C., 1859. On the Origin of Species by Means of Natural Selection, or the Preservation of Favoured Races in the Struggle for Life. John Murray, London.
de Queiroz, K., 2005. Ernst Mayr and the modern concept of species. Proc. Natl. Acad. Sci. U.S.A. 102 (Suppl. 1), 6600–6607.
Doolittle, W.F., Zhaxybayeva, O., 2009. On the origin of prokaryotic species. Genome Res. 19, 744–756.
Dykhuizen, D.E., Brisson, D., Sandigursky, S., Wormser, G.P., Nowakowski, J., Nadelman, R.B., et al., 2008. The propensity of different *Borrelia burgdorferi* sensu stricto genotypes to cause disseminated infections in humans. Am. J. Trop. Med. Hyg. 78, 806–810.
Gevers, D., Cohan, F.M., Lawrence, J.G., Spratt, B.G., Coenye, T., Feil, E.J., et al., 2005. Opinion: re-evaluating prokaryotic species. Nat. Rev. Microbiol. 3, 733–739.
Gogarten, J.P., Doolittle, W.F., Lawrence, J.G., 2002. Prokaryotic evolution in light of gene transfer. Mol. Biol. Evol. 19, 2226–2238.
Gordon, D.M., Cowling, A., 2003. The distribution and genetic structure of *Escherichia coli* in Australian vertebrates: host and geographic effects. Microbiology 149, 3575–3586.
Gordon, D.M., Lee, J., 1999. The genetic structure of enteric bacteria from Australian mammals. Microbiology 145 (Pt 10), 2673–2682.
Hanage, W.P., Fraser, C., Spratt, B.G., 2006. Sequences, sequence clusters and bacterial species. Philos. Trans. R. Soc. Lond. B. Biol. Sci. 361, 1917–1927.
Hunt, D.E., David, L.A., Gevers, D., Preheim, S.P., Alm, E.J., Polz, M.F., 2008. Resource partitioning and sympatric differentiation among closely related bacterioplankton. Science 320, 1081–1085.
Hutchinson, G.E., 1968. When are species necessary? In: Lewontin, R.C. (Ed.), Population Biology and Evolution. Syracuse University Press, Syracuse, NY, pp. 177–186.
Jensen, P.R., 2010. Linking species concepts to natural product discovery in the postgenomic era. J. Ind. Microbiol. Biotechnol. 37, 219–224.

Kettler, G.C., Martiny, A.C., Huang, K., Zucker, J., Coleman, M.L., Rodrigue, S., et al., 2007. Patterns and implications of gene gain and loss in the evolution of *Prochlorococcus*. PLoS Genet. 3, e231.

Koeppel, A., Perry, E.B., Sikorski, J., Krizanc, D., Warner, W.A., Ward, D.M., et al., 2008. Identifying the fundamental units of bacterial diversity: a paradigm shift to incorporate ecology into bacterial systematics. Proc. Natl. Acad. Sci. U.S.A. 105, 2504–2509.

Konstantinidis, K.T., Tiedje, J.M., 2005. Genomic insights that advance the species definition for prokaryotes. Proc. Natl. Acad. Sci. U.S.A. 102, 2567–2572.

Lefébure, T., Stanhope, M.J., 2007. Evolution of the core and pan-genome of *Streptococcus*: positive selection, recombination, and genome composition. Genome Biol. 8, R71.

Majewski, J., Cohan, F.M., 1999. Adapt globally, act locally: the effect of selective sweeps on bacterial sequence diversity. Genetics 152, 1459–1474.

Mallet, J., 2008a. Hybridization, ecological races and the nature of species: empirical evidence for the ease of speciation. Philos. Trans. R. Soc. Lond. B. Biol. Sci. 363, 2971–2986.

Mallet, J., 2008b. Mayr's view of Darwin: was Darwin wrong about speciation? Biol. J. Linn. Soc. 95, 3–16.

Manning, S.D., Motiwala, A.S., Springman, A.C., Qi, W., Lacher, D.W., Ouellette, L.M., et al., 2008. Variation in virulence among clades of *Escherichia coli* O157:H7 associated with disease outbreaks. Proc. Natl. Acad. Sci. U.S.A. 105, 4868–4873.

Marri, P.R., Hao, W., Golding, G.B., 2006. Gene gain and gene loss in *Streptococcus*: is it driven by habitat? Mol. Biol. Evol. 23, 2379–2391.

Mayr, E., 1963. Animal Species and Evolution. Belknap Press of Harvard University Press, Cambridge, MA.

Mayr, E., 1982. The Growth of Biological Thought: Diversity, Evolution, and Inheritance. Harvard University Press, Cambridge, MA, Chapter 6.

McVean, G., Awadalla, P., Fearnhead, P., 2002. A coalescent-based method for detecting and estimating recombination from gene sequences. Genetics 160, 1231–1241.

Palenik, B., Ren, Q., Tai, V., Paulsen, I.T., 2009. Coastal *Synechococcus* metagenome reveals major roles for horizontal gene transfer and plasmids in population diversity. Environ. Microbiol. 11, 349–359.

Palys, T., Nakamura, L.K., Cohan, F.M., 1997. Discovery and classification of ecological diversity in the bacterial world: the role of DNA sequence data. Int. J. Syst. Bacteriol. 47, 1145–1156.

Palys, T., Berger, E., Mitrica, I., Nakamura, L.K., Cohan, F.M., 2000. Protein-coding genes as molecular markers for ecologically distinct populations: the case of two *Bacillus* species. Int. J. Syst. Evol. Microbiol. 50 (Pt 3), 1021–1028.

Papke, P.T., Ward, D.M., 2004. The importance of physical isolation in microbial evolution. FEMS Microbiol. Ecol. 48, 293–303.

Paul, S., Dutta, A., Bag, S.K., Das, S., Dutta, C., 2010. Distinct, ecotype-specific genome and proteome signatures in the marine cyanobacteria *Prochlorococcus*. BMC Genomics 11, 103.

Rasko, D.A., Rosovitz, M.J., Myers, G.S., Mongodin, E.F., Fricke, W.F., Gajer, P., et al., 2008. The pangenome structure of *Escherichia coli*: comparative genomic analysis of *E. coli* commensal and pathogenic isolates. J. Bacteriol. 190, 6881–6893.

Remonsellez, F., Galleguillos, F., Moreno-Paz, M., Parro, V., Acosta, M., Demergasso, C., 2009. Dynamic of active microorganisms inhabiting a bioleaching industrial heap of low-grade copper sulfide ore monitored by real-time PCR and oligonucleotide prokaryotic acidophile microarray. Microb. Biotechnol. 2, 613–624.

Roberts, M.S., Cohan, F.M., 1995. Recombination and migration rates in natural populations of *Bacillus subtilis* and *Bacillus mojavensis*. Evolution 49, 1081–1094.
Rosselló-Mora, R., Amann, R., 2001. The species concept for prokaryotes. FEMS Microbiol. Rev. 25, 39–67.
Sikorski, J., Nevo, E., 2007. Patterns of thermal adaptation of *Bacillus simplex* to the microclimatically contrasting slopes of "Evolution Canyons" I and II, Israel. Environ. Microbiol. 9, 716–726.
Smith, N.H., Gordon, S.V., de la Rua-Domenech, R., Clifton-Hadley, R.S., Hewinson, R.G., 2006. Bottlenecks and broomsticks: the molecular evolution of *Mycobacterium bovis*. Nat. Rev. Microbiol. 4, 670–681.
Sneath, P., Sokal, R., 1973. Numerical Taxonomy: the Principles and Practice of Numerical Classification. W.H. Freeman, San Francisco, CA.
Stackebrandt, E., Ebers, J., 2006. Taxonomic parameters revisited: tarnished gold standards. Microbiol. Today 33, 152–155.
Staley, J.T., 2006. The bacterial species dilemma and the genomic-phylogenetic species concept. Philos. Trans. R. Soc. Lond. B. Biol. Sci. 361, 1899–1909.
Stamos, D.N., 2007. Darwin and the Nature of Species. State University of New York Press, Albany, NY.
Templeton, A., 1989. The meaning of species and speciation: a genetic perspective. In: Otte, D., Endler, J. (Eds.), Speciation and Its Consequences. Sinauer Associates, Sunderland, MA, pp. 3–27.
Tenaillon, O., Skurnik, D., Picard, B., Denamur, E., 2010. The population genetics of commensal *Escherichia coli*. Nat. Rev. Microbiol. 8, 207–217.
Tettelin, H., Masignani, V., Cieslewicz, M.J., Donati, C., Medini, D., Ward, N.L., et al., 2005. Genome analysis of multiple pathogenic isolates of *Streptococcus agalactiae*: implications for the microbial "pan-genome". Proc. Natl. Acad. Sci. U.S.A. 102, 13950–13955.
Thompson, J.R., Pacocha, S., Pharino, C., Klepac-Ceraj, V., Hunt, D.E., Benoit, J., et al., 2004. Genotypic diversity within a natural coastal bacterioplankton population. Science 307, 1311–1313.
Touchon, M., Hoede, C., Tenaillon, O., Barbe, V., Baeriswyl, S., Bidet, P., et al., 2009. Organised genome dynamics in the *Escherichia coli* species results in highly diverse adaptive paths. PLoS Genet. 5, e1000344.
Vandamme, P., Pot, B., Gillis, M., de Vos, P., Kersters, K., Swings, J., 1996. Polyphasic taxonomy, a consensus approach to bacterial systematics. Microbiol. Rev. 60, 407–438.
Vernikos, G.S., Thomson, N.R., Parkhill, J., 2007. Genetic flux over time in the *Salmonella* lineage. Genome. Biol. 8, R100.
Walk, S.T., Alm, E.W., Calhoun, L.M., Mladonicky, J.M., Whittam, T.S., 2007. Genetic diversity and population structure of *Escherichia coli* isolated from freshwater beaches. Environ. Microbiol. 9, 2274–2288.
Walk, S.T., Alm, E.W., Gordon, D.M., Ram, J.L., Toranzos, G.A., Tiedje, J.M., et al., 2009. Cryptic lineages of the genus *Escherichia*. Appl. Environ. Microbiol. 75, 6534–6544.
Ward, D.M., 1998. A natural species concept for prokaryotes. Curr. Opin. Microbiol. 1, 271–277.
Ward, D.M., Bateson, M.M., Ferris, M.J., Kühl, M., Wieland, A., Koeppel, A., et al., 2006. Cyanobacterial ecotypes in the microbial mat community of Mushroom Spring (Yellowstone National Park, Wyoming) as species-like units linking microbial community composition, structure and function. Philos. Trans. R. Soc. Lond. B. Biol. Sci. 361, 1997–2008.

Wayne, L.G., Brenner, D.J., Colwell, R.R., Grimont, P.A.D., Kandler, O., Krichevsky, M.I., et al., 1987. Report of the ad hoc committee on reconciliation of approaches to bacterial systematics. Int. J. Syst. Bacteriol. 37, 463–464.

Welch, R.A., Burland, V., Plunkett 3rd, G., Redford, P., Roesch, P., Rasko, D., et al., 2002. Extensive mosaic structure revealed by the complete genome sequence of uropathogenic *Escherichia coli*. Proc. Natl. Acad. Sci. U.S.A. 99, 17020–17024.

Whitaker, R.J., Grogan, D.W., Taylor, J.W., 2003. Geographic barriers isolate endemic populations of hyperthermophilic archaea. Science 301, 976–978.

Whittam, T.S., Bumbaugh, A.C., 2002. Inferences from whole-genome sequences of bacterial pathogens. Curr. Opin. Genet. Dev. 12, 719–725.

Yoon, C.K., 2009. Naming Nature: The Clash Between Instinct and Science. Norton, New York, NY.

3 Population Structure of Pathogenic Bacteria

D. Ashley Robinson[1,]*, Jonathan C. Thomas[1] and William P. Hanage[2]

[1]Department of Microbiology, University of Mississippi Medical Center, Jackson, MS, USA, [2]Department of Epidemiology, Harvard School of Public Health, Boston, MA, USA

3.1 Introduction

The population structure of pathogenic bacteria is a manifestation of the evolutionary and ecological history that these bacteria have experienced. All known bacterial pathogens of humans belong to the eubacterial domain of life (Cavicchioli et al., 2003), but the diversity represented by these species is immense. For example, the differences in 16S-rRNA sequence between *Bacillus* and *Escherichia*, genera which each include important human pathogens, is of similar magnitude to the differences between corn and frogs (Woese, 1987). The role of recombination in shuffling this diversity and the niches inhabited by pathogenic bacteria also differ immensely. In fact, some environments mastered by these species qualify as extreme even by archaeal standards; *Helicobacter pylori* can survive stomach acid with pH less than 2.0, and *Staphylococcus aureus* can survive salt solutions greater than 3.5 M (Chapman, 1945; Chen et al., 1997).

To attempt to infer the processes that give rise to this diversity may seem a daunting task, but it is a worthwhile task. Population genetics inference can provide knowledge about such important factors as the dynamics of antimicrobial resistance and vaccine escape, and the ability to robustly identify the impact of natural selection on bacterial populations and separate it from other processes, such as drift or demographic factors. The enterprise essentially consists of measuring genetic variation, partitioning it into various components and, along the way, attempting to understand how the variation arose. The population structure of pathogenic bacteria is a balance between many different processes, including those that produce genetic variation (i.e., polymorphisms) and those that modulate the frequency of polymorphisms in a population. In this chapter, we call attention to the underemphasized processes at work in bacterial populations, to biases that can impact our

*E-mail: darobinson@umc.edu

inferences, and we discuss various definitions for what a bacterial population is and methods for identifying and studying them.

3.2 Bacterial Population Structure by Other Means

3.2.1 Beyond I_a

Although it is known that bacteria reproduce asexually through binary fission, the three parasexual processes of conjugation, transduction, and transformation provide mechanisms for gene exchange and subsequent recombination between bacteria. The relentless emergence of antimicrobial-resistant bacteria during the twentieth century led researchers to ask questions about the spread of resistance and the role of recombination on bacterial population structure (Smith et al., 2000). Multilocus enzyme electrophoresis (MLEE) (Selander et al., 1986a) provided the first data to address these questions. Briefly, MLEE uses the electrophoretic mobility of expressed water-soluble enzymes to indirectly assess genetic variation. Rates of migration during electrophoresis are affected by the amino acid sequences of the enzymes. Different electrophoretic mobility variants are equated with different alleles at the loci that encode the enzymes.

MLEE data are well-suited for detecting linkage disequilibrium, which is a nonrandom association of alleles at different loci (Brown et al., 1980). Such associations are not expected in the presence of high rates of recombination. The index of association (I_a) captured linkage disequilibrium in a single statistic and became very popular. I_a makes use of the variance in the distribution of pairwise allelic mismatches among bacteria; recombination reduces the variance of this distribution and may be easily simulated for hypothesis testing purposes. Smith et al. (1993) used I_a to demonstrate that linkage disequilibrium and recombination were compatible: recombination needed to be about 20 times more common than mutation for I_a to reflect a random association of alleles at different loci (Smith, 1994), so a population could be very far from purely clonal yet still exhibit significant linkage disequilibrium. It was also thought that some multilocus genotypes were temporarily overrepresented due to "epidemic" spread in an otherwise recombinant population. This phenomenon seemed to explain data for a number of bacterial pathogens such as *Neisseria meningitidis* and *Streptococcus pneumoniae*, where recombination appeared prevalent but some clones appeared stable. It was subsequently found that I_a also showed that some species, such as *N. gonorrhoeae*, were essentially panmictic while others, such as some serovars of *Salmonella enterica*, were essentially clonal.

Multilocus sequence typing (MLST) (Maiden et al., 1998) has succeeded MLEE as the tool of choice for bacterial population genetics. Briefly, in MLST the DNA sequences of multiple housekeeping gene fragments are determined, and alleles at these loci are assigned based on their unique nucleotide sequences. These data are ideally suited for studying the relative roles of mutation and recombination on bacterial population structure; several such analysis tools have been developed (Feil et al., 2000; Fraser et al., 2005; Didelot and Falush, 2007). The results of these analyses are broadly concordant with those of I_a (Hanage et al., 2006). Some salient points have emerged from a large body of work in this area: (1) recombination has a far-reaching impact on bacterial population structure (Spratt and Maiden, 1999); (2) recombination affects most bacterial species to some extent, including the prototype of the clonal paradigm, *E. coli* (Touchon et al., 2009); and (3) recombination can vary widely even within the confines of named species (Didelot and Maiden, 2010).

3.2.2 The Fate of Bacterial Genetic Variation

The population structure of pathogenic bacteria is not solely the result of the mutational and recombinational processes that generate genetic variation. It is also a result of population-level processes that determine the fate of that variation. However, little progress has been made in developing analysis tools that assess the relative roles of neutral genetic drift and natural selection on shaping bacterial genetic variation. It is ironic that the birth of bacterial population genetics occurred when Milkman (1973) used *E. coli* to test Kimura's (1968) neutral theory of molecular evolution, which proposed that drift accounted for most genetic variation. Milkman found a high proportion of polymorphic MLEE loci but concluded that the effective number of alleles per locus was too few to be due to drift. Whittam et al. (1983) subsequently suggested a role for both drift and selection in shaping the population structure of *E. coli*. While we may assert that all genetic variation passes through nature's sieves of drift and selection, data quantifying their relative roles in bacterial populations are limited. Drift contributes a stochastic element to bacterial evolution that is predictable only in the magnitude by which variants change over time, not in the direction of change. Selection, on the other hand, provides a deterministic element to bacterial evolution as the direction of change is clear; the most-fit variant in a given environment should eventually prevail. The size of bacterial populations is crucial in determining the rate of drift and the efficiency of selection. For those reasons, it is important to discuss what is meant by "population size" and what is meant by a "population."

The foundational work of Wright and Fisher (WF) produced a simple model that permits drift and selection to be studied (Fisher, 1930; Wright, 1931). They proposed a population of constant size, where individuals are selected at random with replacement to make up the next generation. Although this may not be the most appropriate model of bacterial reproduction (Schierup and Wiuf, 2010), it is illustrative of how drift and selection can impact a population. In the WF population, the fate of a new variant (be it a point mutation, recombination, or insertion−deletion) that has no impact on fitness is determined strictly by N, the number of individuals in the population (Kimura, 1962). The probability that the variant will eventually reach fixation where all individuals possess the variant is simply $1/N$. Likewise, the probability that the variant will eventually be lost is $1 - 1/N$. Examining different values of N immediately makes two points: the new variant is more likely to be lost than to be fixed and fixation is more likely in a smaller population than in a larger population.

Selection is introduced by adding a coefficient, s, to describe a fitness effect such that the new fixation probability is $(1 + s)/N$. The variant is under the influence of drift when $s < 1/N$ and under the influence of selection when $s > 1/N$ (Kimura and Takahata, 1983). Examining different values of N for a given s illustrates that the same probability of fixation can be produced by drift in a small population and by selection in a large population. In a small population, $1/N$ can be a tall stochastic hurdle to clear for a variant to be maintained regardless of how big s may be. On the other hand, in a large population, $1/N$ can be a barrier for removing variants that are only slightly deleterious. Thus, not all favorable variations will be fixed and, especially for less-recombinant species, large numbers of slightly deleterious variations may take a long time to be purified from the population (Rocha et al., 2006).

As previously indicated, the vast majority of newly minted polymorphisms are destined to be lost from the WF population. In each generation, allele frequencies are expected to change by $p(1-p)/N$, where p is the initial allele frequency. This formula represents the standard binomial sampling variance. In the WF population, this variance increases each generation as alleles are lost. Moreover, diversity (defined by the parameter $1 - \Sigma p^2$) is expected to decrease by $1/N$ each generation. One approach to estimating drift in bacterial populations is to measure changes in allele frequency and diversity over time. A so-called temporal method (Jorde and Ryman, 2007) based on allele frequency change has been developed for sexual diploids and may be useful for bacteria (unpublished data). The potential for samples collected over multiple time points to capture evolutionary processes in action has not gone unnoticed (Drummond et al., 2003). At this point, temporal samples have been exploited only for estimating mutation rates (Perez-Losada et al., 2007; Wilson et al., 2009; Smyth et al., 2010), but their use for disentangling drift and selection remains promising.

In real bacterial populations, N, which is the census population size, is not the best measure of drift because only a fraction of the total number of individuals will contribute genetic material to the next generation. This reduced N is called the effective population size, N_e. Wright (1969) indicated that N_e was whatever must be substituted into these formulas to describe the actual change in allele frequency and diversity over time. At this point, we have a limited number of N_e estimates for different bacterial populations (Ochman and Wilson, 1987). The census size of bacterial populations is incredibly large in comparison to that of sexual diploids. Consider that the number of individual bacteria on the skin of a single human adult is of similar magnitude to the number of humans in the USA, $\sim 3 \times 10^8$ (Whitman et al., 1998; Census Bureau, 2010). Bacterial population size in the human gut is even larger. However, for many pathogenic bacteria, there are a number of dynamics that might make N_e extremely low relative to this large N, which would indicate a major role for drift in those populations. For example, N_e may be substantially reduced by the recent origin of some species or their clonal lineages (Achtman, 2008; Feng et al., 2008), population bottlenecks arising from between-host transmission events or adaptation to a new niche, and fluctuating population sizes (Fraser et al., 2009).

It is fair to say that studies of natural selection in bacterial populations have received much more attention than studies of drift, not considering the usual procedures of using "neutrality" as a null model for detecting selection. This situation may be due to the perception that everything interesting in biology must be the consequence of selection. However, we have seen the potential for a selected variant to be stochastically lost from a small population, and so the observed variation may have been the product of a more complicated and nuanced history than assumed in naïve selective scenarios. There is a rich literature on the evolution of antimicrobial resistance and on individual cases of positive selection on proteins that are exposed on the bacterial cell surface. There are also new results that report positive selection on housekeeping proteins (Buckee et al., 2008; Chattopadhyay et al., 2009), long presumed to evolve through negative (or purifying) selection. It is worth pointing out that both drift and selection can be involved in shaping bacterial population structure, much like mutation and recombination are both involved in producing genetic variation. In a small population, a new variation will be impacted by drift even if selection operates on this variation later. This discussion is applicable to describe the fate of any bacterial genetic variation, from a single point mutation to the architecture of an entire bacterial genome. With the genomic era in full bloom, a dialogue may once again be revived about the relative roles of drift and selection in shaping bacterial genetic variation (Lynch and Conery, 2003; Daubin and Moran, 2004).

3.2.3 Biases and the Inference of Bacterial Population Structure

One sample bias of special relevance to infectious diseases is the impact of transmission and repeated sampling of the same transmission chain or host network. An ideal sample of a bacterial population would be a cross-section of independent, randomly selected individuals in the host population. However, such an ideal sample is rarely achieved. The potential of sampling to distort findings is obvious in the example of a study design in which serial samples are taken from the same host. Subsequent samples from an individual host are not independent: if the duration of infection (or colonization) is long, the same type may be found more than once. If the organism in question produces strong immunity, subsequent samples from the same host are more likely to be of a different antigenic type than samples from the population at random. In the same way, hosts within a contact network are more likely to be epidemiologically linked, and such individuals are more likely to be infected (or colonized) by the same strain than hosts picked at random from the community at large. In other words, the samples are not independent. The existence and importance of such host networks have been appreciated for some time, and in theoretical epidemiology a distinction is made between the individual and household reproductive numbers (Pellis et al., 2009). Important host networks include but are far from limited to households (Hope Simpson, 1952), day care centers (Leino et al., 2008; Hoti et al., 2009), and core groups in sexually transmitted diseases (Wohlfeiler and Potterat, 2005). The concept may also apply to health care settings (e.g., hospitals; Donker et al., 2010) or particular services within them.

The predicted consequence of repeated sampling of short transmission chains is an excess of indistinguishable strains. One study that examined the potential of this phenomenon to distort analyses of population structure found evidence for such "microepidemics" in samples of three important pathogenic bacteria (Fraser et al., 2005). Strikingly, once this excess of indistinguishable strains was accounted for, the population structure (assessed using the diversity of MLST data [Maiden et al., 1998] and eBURST [Turner et al., 2007]) was consistent with that produced solely by drift. That is to say, there was no evidence of particular lineages being favored by selection. It is important to realize that the predicted consequences of selection are not merely an increased frequency of the selected strain, but also other closely related strains that share the selected feature. However, in these samples it was only an excess of indistinguishable strains that were detected, leading to the conclusion that their population structure was not compatible with any simple selective scenario. Of numerous modifications to the basic neutral model, none reproduced the distinctive excess of indistinguishable strains leading to the suggestion of microepidemic structure (Fraser et al., 2005).

Another bias of relevance to bacterial population structure is the genetic markers that we choose to study. Acquiring genetic data for large numbers of bacterial isolates can be laborious and expensive. One may conduct a study with a discovery phase that screens a set of genetic markers for variation using a small number of isolates (the discovery isolates), followed by a typing phase that uses only the variable markers to type a much larger sample. While this procedure might ensure that the more expensive effort to type the larger sample would yield some genetic variation, it could also introduce an insidious bias referred to as ascertainment bias or phylogenetic discovery bias. It can also occur when a small number of genome sequences are used to discover variations (e.g., single-nucleotide polymorphisms [SNPs]), for subsequent typing in a larger sample, even if there are hundreds of such variations. This bias has consequences for both phylogenetic and population genetic analysis. It causes phylogenetic trees to become linear, with the collapse of branches along the paths that lie between the discovery isolates (Keim et al., 2004). It also causes an excess of higher frequency polymorphisms and a scarcity of lower frequency polymorphisms in population genetic analyses (Nielsen et al., 2004); this can result in rejection of neutral models. It is possible that rapidly evolving markers, such as short sequence repeats or variable number of tandem repeats (e.g., microsatellites, minisatellites), might be able to "outevolve" this bias (unpublished data). Nonetheless, bias is most likely to occur when the discovery isolates represent a limited portion of the species diversity (Rosenblum and Novembre, 2007), so it is recommended to make the discovery isolates as diverse as possible based on preexisting data.

3.3 What, if Anything, is a Bacterial Population?

3.3.1 Definitions and A Priori Approaches

There is no universally accepted definition for a bacterial population. This deficiency might be viewed as a consequence of the long-standing problems in defining bacterial species (Spratt et al., 2006); for if we do not know what a species is, how can we know about groups that subtend species? On the other hand, working up the biological hierarchy from individuals has yielded infraspecific taxa with some common usage (e.g., isolates, strains, clones) (Van Belkum et al., 2007). Even with sexual diploids, where the definition of species seems less ad hoc and much population genetic theory has been tested, there still exist fundamental questions about the nature of a population. Waples and Gaggiotti (2006) listed a dozen definitions of a "population" from the literature, though none involved bacteria. These definitions come from separate evolutionary and ecological paradigms where populations are defined based on either the genetic or demographic cohesion of the individuals, respectively. As discussed later, the definition of a bacterial population may differ depending on the role of recombination in the species.

Long-term stability of clonal lineages may produce cohesion within less-recombinant bacterial species. Rannala et al. (2000) studied the Lyme disease spirochete, *Borrelia burgdorferi*, and suggested that clonal lineages rather than individual isolates should define bacterial populations. By this definition, we note that different clonal (i.e., low recombination) species can be structured differently. For example, the methicillin-resistant *S. aureus* lineage known as ST239-MRSA-III shows strong geographic structuring by continent (Smyth et al., 2010), whereas the *S. enterica* lineage known as serotype Typhi shows no geographic structuring (Roumagnac et al., 2006). Both lineages have human reservoirs, demonstrating that relative clonality and host associations alone do not predict all aspects of bacterial population structure.

Within more recombinant species, recombination itself may define a cohesion. However, the cohesive influence of recombination depends on the relatedness of the donor and recipient. Among closely related isolates, recombination homogenizes genetic variation, but among more distantly related isolates, recombination promotes divergence (Schierup and Hein, 2000). Balloux (2010) suggested that the basic level of bacterial population subdivision should be where equilibrium occurs in terms of both random mating and demography. Inferences could then be made based on a model of this population structure. Though a number of models of population structure might be applied to bacteria, the metapopulation concept is appealing (Achtman and Wagner, 2008).

It is a common approach to define bacterial populations a priori with respect to a characteristic of interest (e.g., geography or host). A G_{ST}-like statistic (Nei, 1977) can then be used with MLST data to determine whether the between-population differentiation in allele frequencies is greater than zero. This approach was first recommended for bacterial populations by Selander et al. (1986b). It has been used to study bacterial population structure in a variety of contexts. For example, Whittam et al. (1989) used MLEE to study *E. coli* from humans, cats, and dogs. He found no evidence that *E. coli* was structured by host species; G_{ST} was equal to 0.01, meaning that only 1% of the genetic variation could be accounted for by between-host species differences. As another example, Filliol et al. (2006) used SNPs, determined from a small number of *Mycobacterium* genomes, to study the global genetic structure of *M. tuberculosis*. After defining a number of lineages using the SNP data, they used G_{ST} to determine how well a new panel of genetic markers (short sequence repeats) could differentiate among the SNP-defined lineages. They found three potentially diagnostic markers where >50% of the short sequence repeat variation was accounted for by between-lineage differences.

While G_{ST}-like statistics can provide new insights into bacterial population structure, some issues should be noted. First, with large enough samples and powerful enough markers, all populations will be differentiated from each other. Recall that neutral genetic drift alone is expected to generate allele frequency differences. Perhaps it is more informative to consider the range of differentiation values possible for a given sample, as reflected by confidence intervals, rather than a P-value that tests a null hypothesis of zero differentiation. Second, G_{ST}-like statistics do not work properly with highly variable markers. With the notable exceptions of some bacterial pathogens of low diversity (e.g., *M. tuberculosis*), many bacterial samples will be characterized with markers that present multiple alleles and reveal diverse populations. It is known that in these situations G_{ST}-like statistics need to be corrected, otherwise an artifact of low differentiation may be found (Hedrick, 2005). An investigation of this phenomenon by Jost (2008) led to the surprising conclusion that the standard procedures for partitioning genetic variation are erroneous, and that measures based on the effective number of alleles are more appropriate mathematically. Jost's D provides a differentiation statistic that may be applied to a wide variety of bacterial typing tools, so long as the results are interpreted as a differentiation in allele frequencies.

3.3.2 Model-Based Approaches for the Identification of Populations

Over the past decade, interest has grown in the use of MLST data to identify populations in a relatively "unsupervised" fashion. Researchers studying sexual diploids are spoilt for choice when it comes to algorithms and programs to address population structure. Each of these analysis tools makes slightly different assumptions, and is expected to give broadly concordant yet slightly different results as a consequence. The results also depend on the definition, touched on above, of what a population is. The application of such tools to bacteria is complicated by their variably recombinant lifestyles. All approaches to the unsupervised identification of populations work by identifying groups with more genetic features in common than would be expected by chance. As a result, in a clonal or near clonal species the populations identified will, to a large extent, represent the phylogeny, with subpopulations representing well-supported clonal lineages. As noted previously, most bacteria undergo some recombination, which complicates phylogenetic and population assignments: two isolates may share traits because of common ancestry or because of recombination. Population analysis estimates the probability that any polymorphic site is derived from a particular ancestral population, even if its history has since involved recombination.

Helicobacter pylori is noted among bacteria for its very high mutation and recombination rates, and is largely vertically transmitted. In a startling demonstration of the potential of population analysis for bacteria, the program STRUCTURE

(Pritchard et al., 2000; Falush et al., 2003a, 2007) was used to show that the population structure of *H. pylori* recapitulated that of its human host (Falush et al., 2003b). A further finding of this work was the existence of admixed variants of *H. pylori* that contained sequence typical of more than one population, which could be related back to contacts between the distinct human populations bearing them. As STRUCTURE assumes random mating and performs better with strong genetic differentiation (Waples and Gaggiotti, 2006), it was well-suited to *H. pylori* data.

As another example, STRUCTURE was used to examine lineages of *E. coli* typed by MLST, finding some to have a far more extensive history of recombination than others (Wirth et al., 2006). As noted above, to an extent these results largely reflect the phylogeny, and the admixture analysis suggested a gradient of recombination between populations corresponding roughly to the *E. coli* lineages of A, B1, B2, and D (themselves defined by MLEE data) (Ochman and Selander, 1984). In this sample, strains considered to be pathogenic (EPEC, EHEC, EIEC, and *Shigella*) were more likely to contain polymorphisms at the sequenced loci that were characteristic of more than one population inferred by STRUCTURE. This suggested an intriguing link between homologous recombination and virulence. However, these results, and all others of this sort, should be treated with some caution because of the sample studied: it is easier to assign polymorphisms accurately to groups if you have examples of those groups. If there is a portion of the population for which you have little or no data, the estimates of admixture may reflect this rather than genuine mosaic ancestry. In other words, STRUCTURE does not model the underlying phylogeny of the bacteria so it might confuse common ancestry with recombination if the sample is incomplete. Recently, it has been proposed that more than four lineages exist within *E. coli*, and that the new lineages (C, E, and F) as well as some *Shigella* lineages may define the lineages previously characterized to be recombinant (Denamur et al., 2010).

Bayesian Analysis of Population Structure (BAPS; Corander et al., 2003, 2004; Corander and Marttinen, 2006) is another program that performs population analysis, and has been used to study populations within *S. pneumoniae* typed by MLST (Hanage et al., 2009). In this case a group of genotypes with a greater history of recombination than the rest of the species was detected, and this group was found to be significantly more likely to be resistant to antibiotics of multiple classes. It has been known for some time that *S. pneumoniae* can acquire resistance by recombination, occasionally with other species (Dowson et al., 1993), and it is interesting to note that this recombining group identified by the BAPS analysis also contained almost all cases of MLST loci transferred from other oral streptococci. It was suggested that this was the result of a history of hyper-recombination, in which a lineage that is recombining at a higher rate was more likely to acquire both resistance determinants and anomalous DNA at the MLST loci. This interpretation should be subject to the same caveats as the previously mentioned *E. coli* case, though in this case the sample was much larger. In fact, attempts to use STRUCTURE to analyze this dataset failed to separate the *S. pneumoniae* sequences from those of other species in the data. However, when both programs were applied to a wider *E. coli* sample, they were broadly concordant (Gordon et al., 2008). BAPS differs from

STRUCTURE in the algorithm it uses to fit the model to the data, employing a nonreversible stochastic search operator (Corander et al., 2006), rather than the more commonly used Gibbs sampler that can have trouble converging with very complex datasets (Robert and Casella, 2005).

The nature of "recombinant" groups deserves some attention. Are they genuine monophyletic lineages distinguished by a higher recombination rate, or a polyphyletic grouping of strains brought together by the fact that they all contain divergent sequences uncharacteristic of all the other populations captured in the sample? This sort of confusion is explicitly discussed in Marttinen et al. (2008), which introduces Bayesian Recombination Tracker (BRAT), a new means of detecting recombinant sequences in a sample. Simulations illustrate that divergent sequences tend to be grouped together as a single heterogeneous population. This does not necessarily mean that they share a recent common ancestor. The phenomenon can be intuitively understood as being similar to long branch attraction in phylogenetic analysis (Huelsenbeck, 1997).

In each of these examples, it is not clear exactly what the identified populations are in any intuitive sense. In *H. pylori*, they derive their interest and relevance from their relation to human population structure. In *E. coli*, they appear to be related to deep branching clades. In *S. pneumoniae*, three populations were identified, but only the population apparently associated with recombination could be tied to a distinct phenotype (resistance). Finally, we note that STRUCTURE as well as BAPS may be inappropriate for less-recombinant species, as aptly demonstrated with STRUCTURE for *Salmonella* MLST data (Falush et al., 2006). With such species, phylogenetic approaches are preferred.

As we collect larger databases, even extending to those from whole genomes, it will be increasingly important that the computational demands of our analysis tools scale in a reasonable fashion with the complexity and volume of data. The ideal tools for future work in this field will be parallelizable. BAPS has already been applied to a dataset of more than 100 bacterial genomes, producing results consistent with a phylogeny based on nonrecombinant sites (Corander, personal communication), suggesting it has potential, but the wider application of these tools remains a subject for research.

3.4 Conclusions

There are some things we know: the vast majority of newly arising polymorphisms have no future. Even a selected polymorphism may be stochastically lost. Though the census population size of bacteria is almost unimaginably large, several dynamics may make the effective size much smaller. Then there are many things we do not know: how much of the variation that we can observe is produced by drift and how much by selection? What exactly are the populations detected by BAPS, STRUCTURE, and other relatively unsupervised approaches—deep clades, lineages that preferentially recombine with themselves, or some artifact of our assumptions?

And given the power of selection to wipe out variation in a selective sweep (Majewski and Cohan, 1999), why are so many bacterial species so very diverse?

Any claim that the population structures of pathogenic bacteria follow simple, easily understood rules is manifestly false. We have not directed enough effort to collecting truly unbiased samples. We also should not overlook the sheer protean diversity of the bacteria and their complicated sex lives. With that said, we already have extraordinary ability to produce whole genome sequences from many bacteria, and yet more technological advances await in the near future. We must work to ensure that our theory does not lag behind our technology, because if it does so, we may find ourselves with extraordinary data, but little idea of how to interpret them.

Population analyses have great potential in the study of pathogenic bacteria, especially those in which there is a high rate of homologous recombination rendering straightforward phylogenetic analysis inappropriate. It is wise, however, to remember that just as a phylogeny is one possible hypothesis about the history of the sequences making up the sample, so too are the results of analyses of population structure. These tools detect signal, but the processes that have given rise to that signal may not be immediately clear and should be interpreted within a biological context.

References

Achtman, M., 2008. Evolution, population structure, and phylogeography of genetically monomorphic bacterial pathogens. Annu. Rev. Microbiol. 62, 53−70.
Achtman, M., Wagner, M., 2008. Microbial diversity and the genetic nature of microbial species. Nat. Rev. Microbiol. 6, 431−440.
Balloux, F., 2010. Demographic influences on bacterial population structure. In: Robinson, D.A., Falush, D., Feil, E.J. (Eds.), Bacterial Population Genetics in Infectious Disease. Wiley-Blackwell, Hoboken, NJ, pp. 103−120.
Brown, A.H., Feldman, M.W., Nevo, E., 1980. Multilocus structure of natural populations of *Hordeum spontaneum*. Genetics 96, 523−536.
Buckee, C.O., Jolley, K.A., Recker, M., Penman, B., Kriz, P., Gupta, S., et al., 2008. Role of selection in the emergence of lineages and the evolution of virulence in *Neisseria meningitidis*. Proc. Natl. Acad. Sci. U.S.A. 105, 15082−15087.
Cavicchioli, R., Curmi, P.M., Saunders, N., Thomas, T., 2003. Pathogenic archaea: do they exist? Bioessays 25, 1119−1128.
Census Bureau U.S., 2010. http://factfinder.census.gov.
Chapman, G.H., 1945. The significance of sodium chloride in studies of Staphylococci. J. Bacteriol. 50, 201−203.
Chattopadhyay, S., Weissman, S.J., Minin, V.N., Russo, T.A., Dykhuizen, D.E., Sokurenko, E.V., 2009. High frequency of hotspot mutations in core genes of *Escherichia coli* due to short-term positive selection. Proc. Natl. Acad. Sci. U.S.A. 106, 12412−12417.
Chen, G., Fournier, R.L., Varanasi, S., Mahama-Relue, P.A., 1997. *Helicobacter pylori* survival in gastric mucosa by generation of a pH gradient. Biophys. J. 73, 1081−1088.
Corander, J., Marttinen, P., 2006. Bayesian identification of admixture events using multilocus molecular markers. Mol. Ecol. 15, 2833−2843.

Corander, J., Waldmann, P., Sillanpaa, M.J., 2003. Bayesian analysis of genetic differentiation between populations. Genetics 163, 367–374.
Corander, J., Waldmann, P., Marttinen, P., Sillanpaa, M.J., 2004. BAPS 2: enhanced possibilities for the analysis of genetic population structure. Bioinformatics 20, 2363–2369.
Corander, J., Gyllenberg, M., Koski, T., 2006. Bayesian model learning based on a parallel MCMC strategy. Stat. Comput. 16, 355–362.
Daubin, V., Moran, N.A., 2004. Comment on "The origins of genome complexity". Science 306, 978, author reply 978.
Denamur, E., Picard, B., Tenaillon, O., 2010. Population genetics of pathogenic *Escherichia coli*. In: Robinson, D.A., Falush, D., Feil, E.J. (Eds.), Bacterial Population Genetics in Infectious Disease. Wiley-Blackwell, Hoboken, NJ, pp. 269–286.
Didelot, X., Falush, D., 2007. Inference of bacterial microevolution using multilocus sequence data. Genetics 175, 1251–1266.
Didelot, X., Maiden, M.C., 2010. Impact of recombination on bacterial evolution. Trends Microbiol. 18, 315–322.
Donker, T., Wallinga, J., Grundmann, H., 2010. Patient referral patterns and the spread of hospital-acquired infections through national health care networks. PLoS Comput. Biol. 6, e1000715.
Dowson, C.G., Coffey, T.J., Kell, C., Whiley, R.A., 1993. Evolution of penicillin resistance in *Streptococcus pneumoniae*; the role of *Streptococcus mitis* in the formation of low affinity PBP2B in *S. pneumoniae*. Mol. Microbiol. 9, 635–643.
Drummond, A., Pybus, O.G., Rambaut, A., 2003. Inference of viral evolutionary rates from molecular sequences. Adv. Parasitol. 54, 331–358.
Falush, D., Stephens, M., Pritchard, J.K., 2003a. Inference of population structure using multilocus genotype data: linked loci and correlated allele frequencies. Genetics 164, 1567–1587.
Falush, D., Wirth, T., Linz, B., Pritchard, J.K., Stephens, M., Kidd, M., et al., 2003b. Traces of human migrations in *Helicobacter pylori* populations. Science 299, 1582–1585.
Falush, D., Torpdahl, M., Didelot, X., Conrad, D.F., Wilson, D.J., Achtman, M., 2006. Mismatch induced speciation in *Salmonella*: model and data. Philos. Trans. R. Soc. Lond. B. Biol. Sci. 361, 2045–2053.
Falush, D., Stephens, M., Pritchard, J.K., 2007. Inference of population structure using multilocus genotype data: dominant markers and null alleles. Mol. Ecol. Notes 7, 574–578.
Feil, E.J., Smith, J.M., Enright, M.C., Spratt, B.G., 2000. Estimating recombinational parameters in *Streptococcus pneumoniae* from multilocus sequence typing data. Genetics 154, 1439–1450.
Feng, L., Reeves, P.R., Lan, R., Ren, Y., Gao, C., Zhou, Z., et al., 2008. A recalibrated molecular clock and independent origins for the cholera pandemic clones. PLoS One 3, e4053.
Filliol, I., Motiwala, A.S., Cavatore, M., Qi, W., Hazbon, M.H., Bobadilla del Valle, M., et al., 2006. Global phylogeny of *Mycobacterium tuberculosis* based on single nucleotide polymorphism (SNP) analysis: insights into tuberculosis evolution, phylogenetic accuracy of other DNA fingerprinting systems, and recommendations for a minimal standard SNP set. J. Bacteriol. 188, 759–772.
Fisher, R., 1930. The Genetical Theory of Natural Selection, Clarendon Press, Oxford.
Fraser, C., Hanage, W.P., Spratt, B.G., 2005. Neutral microepidemic evolution of bacterial pathogens. Proc. Natl. Acad. Sci. U.S.A. 102, 1968–1973.
Fraser, C., Alm, E.J., Polz, M.F., Spratt, B.G., Hanage, W.P., 2009. The bacterial species challenge: making sense of genetic and ecological diversity. Science 323, 741–746.

Gordon, D.M., Clermont, O., Tolley, H., Denamur, E., 2008. Assigning *Escherichia coli* strains to phylogenetic groups: multi-locus sequence typing versus the PCR triplex method. Environ. Microbiol. 10, 2484–2496.
Hanage, W.P., Fraser, C., Spratt, B.G., 2006. The impact of homologous recombination on the generation of diversity in bacteria. J. Theor. Biol. 239, 210–219.
Hanage, W.P., Fraser, C., Tang, J., Connor, T.R., Corander, J., 2009. Hyper-recombination, diversity, and antibiotic resistance in pneumococcus. Science 324, 1454–1457.
Hedrick, P.W., 2005. A standardized genetic differentiation measure. Evolution 59, 1633–1638.
Hope Simpson, R.E., 1952. Infectiousness of communicable diseases in the household (measles, chickenpox, and mumps). Lancet 2, 549–554.
Hoti, F., Erasto, P., Leino, T., Auranen, K., 2009. Outbreaks of *Streptococcus pneumoniae* carriage in day care cohorts in Finland—implications for elimination of transmission. BMC Infect. Dis. 9, 102.
Huelsenbeck, J.P., 1997. Is the Felsenstein zone a fly trap? Syst. Biol. 46, 69–74.
Jorde, P.E., Ryman, N., 2007. Unbiased estimator for genetic drift and effective population size. Genetics 177, 927–935.
Jost, L., 2008. G(ST) and its relatives do not measure differentiation. Mol. Ecol. 17, 4015–4026.
Keim, P., Van Ert, M.N., Pearson, T., Vogler, A.J., Huynh, L.Y., Wagner, D.M., 2004. Anthrax molecular epidemiology and forensics: using the appropriate marker for different evolutionary scales. Infect. Genet. Evol. 4, 205–213.
Kimura, M., 1962. On the probability of fixation of mutant genes in a population. Genetics 47, 713–719.
Kimura, M., 1968. Evolutionary rate at the molecular level. Nature 217, 624–626.
Kimura, M., Takahata, N., 1983. Selective constraint in protein polymorphism: study of the effectively neutral mutation model by using an improved pseudosampling method. Proc. Natl. Acad. Sci. U.S.A. 80, 1048–1052.
Leino, T., Hoti, F., Syrjanen, R., Tanskanen, A., Auranen, K., 2008. Clustering of serotypes in a longitudinal study of *Streptococcus pneumoniae* carriage in three day care centres. BMC Infect. Dis. 8, 173.
Lynch, M., Conery, J.S., 2003. The origins of genome complexity. Science 302, 1401–1404.
Maiden, M.C., Bygraves, J.A., Feil, E., Morelli, G., Russell, J.E., Urwin, R., 1998. Multilocus sequence typing: a portable approach to the identification of clones within populations of pathogenic microorganisms. Proc. Natl. Acad. Sci. U.S.A. 95, 3140–3145.
Majewski, J., Cohan, F.M., 1999. Adapt globally, act locally: the effect of selective sweeps on bacterial sequence diversity. Genetics 152, 1459–1474.
Marttinen, P., Baldwin, A., Hanage, W.P., Dowson, C., Mahenthiralingam, E., Corander, J., 2008. Bayesian modeling of recombination events in bacterial populations. BMC Bioinformatics 9, 421.
Milkman, R., 1973. Electrophoretic variation in Escherichia coli from natural sources. Science 182, 1024–1026.
Nei, M., 1977. F-statistics and analysis of gene diversity in subdivided populations. Ann. Hum. Genet. 41, 225–233.
Nielsen, R., Hubisz, M.J., Clark, A.G., 2004. Reconstituting the frequency spectrum of ascertained single-nucleotide polymorphism data. Genetics 168, 2373–2382.
Ochman, H., Selander, R.K., 1984. Standard reference strains of *Escherichia coli* from natural populations. J. Bacteriol. 157, 690–693.

Ochman, H., Wilson, A.C., 1987. Evolutionary history of enteric bacteria. In: Neidhart, F.C. (Ed.), *Escherichia coli* and *Salmonella typhimurium*: Cellular and Molecular Biology. ASM Press, Washington, DC, pp. 1649–1654.
Pellis, L., Ferguson, N.M., Fraser, C., 2009. Threshold parameters for a model of epidemic spread among households and workplaces. J. R. Soc. Interface 6, 979–987.
Perez-Losada, M., Crandall, K.A., Zenilman, J., Viscidi, R.P., 2007. Temporal trends in gonococcal population genetics in a high prevalence urban community. Infect. Genet. Evol. 7, 271–278.
Pritchard, J., Stephens, M., Donnelly, P., 2000. Inference of population structure using multilocus genotype data. Genetics 155, 945–959.
Rannala, B., Qiu, W.G., Dykhuizen, D.E., 2000. Methods for estimating gene frequencies and detecting selection in bacterial populations. Genetics 155, 499–508.
Robert, C.P., Casella, G., 2005. Monte Carlo Statistical Methods, second ed. Springer, New York, NY.
Rocha, E.P., Smith, J.M., Hurst, L.D., Holden, M.T., Cooper, J.E., Smith, N.H., et al., 2006. Comparisons of dN/dS are time dependent for closely related bacterial genomes. J. Theor. Biol. 239, 226–235.
Rosenblum, E.B., Novembre, J., 2007. Ascertainment bias in spatially structured populations: a case study in the eastern fence lizard. J. Hered. 98, 331–336.
Roumagnac, P., Weill, F.X., Dolecek, C., Baker, S., Brisse, S., Chinh, N.T., et al., 2006. Evolutionary history of *Salmonella typhi*. Science 314, 1301–1304.
Schierup, M.H., Hein, J., 2000. Consequences of recombination on traditional phylogenetic analysis. Genetics 156, 879–891.
Schierup, M.H., Wiuf, C., 2010. The coalescent of bacterial populations. In: Robinson, D.A., Falush, D., Feil, E.J. (Eds.), Bacterial Population Genetics in Infectious Disease. Wiley-Blackwell, Hoboken, NJ, pp. 3–18.
Selander, R.K., Caugant, D.A., Ochman, H., Musser, J.M., Gilmour, M.N., Whittam, T.S., 1986a. Methods of multilocus enzyme electrophoresis for bacterial population genetics and systematics. Appl. Environ. Microbiol. 51, 873–884.
Selander, R.K., Korhonen, T.K., Vaisanen-Rhen, V., Williams, P.H., Pattison, P.E., Caugant, D.A., 1986b. Genetic relationships and clonal structure of strains of *Escherichia coli* causing neonatal septicemia and meningitis. Infect. Immun. 52, 213–222.
Smith, J.M., 1994. Estimating the minimum rate of genetic transformation in bacteria. J. Evol. Biol. 7, 525–534.
Smith, J.M., Smith, N.H., O'Rourke, M., Spratt, B.G., 1993. How clonal are bacteria? Proc. Natl. Acad. Sci. U.S.A. 90, 4384–4388.
Smith, J.M., Feil, E.J., Smith, N.H., 2000. Population structure and evolutionary dynamics of pathogenic bacteria. Bioessays 22, 1115–1122.
Smyth, D.S., McDougal, L.K., Gran, F.W., Manoharan, A., Enright, M.C., Song, J.H., et al., 2010. Population structure of a hybrid clonal group of methicillin-resistant *Staphylococcus aureus*, ST239-MRSA-III. PLoS One 5, e8582.
Spratt, B.G., Maiden, M.C., 1999. Bacterial population genetics, evolution and epidemiology. Philos. Trans. R. Soc. Lond. B. Biol. Sci. 354, 701–710.
Spratt, B.G., Staley, J.T., Fisher, M.C., 2006. Introduction: species and speciation in microorganisms. Philos. Trans. R. Soc. Lond. B. Biol. Sci. 361, 1897–1898.
Touchon, M., Hoede, C., Tenaillon, O., Barbe, V., Baeriswyl, S., Bidet, P., et al., 2009. Organised genome dynamics in the *Escherichia coli* species results in highly diverse adaptive paths. PLoS Genet. 5, e1000344.

Turner, K.M., Hanage, W.P., Fraser, C., Connor, T.R., Spratt, B.G., 2007. Assessing the reliability of eBURST using simulated populations with known ancestry. BMC Microbiol. 7, 30.

Van Belkum, A., Tassios, P.T., Dijkshoorn, L., Haeggman, S., Cookson, B., Fry, N.K., et al., 2007. Guidelines for the validation and application of typing methods for use in bacterial epidemiology. Clin. Microbiol. Infect. 13, 1–46.

Waples, R.S., Gaggiotti, O., 2006. What is a population? An empirical evaluation of some genetic methods for identifying the number of gene pools and their degree of connectivity. Mol. Ecol. 15, 1419–1439.

Whittam, T.S., Ochman, H., Selander, R.K., 1983. Multilocus genetic structure in natural populations of *Escherichia coli*. Proc. Natl. Acad. Sci. U.S.A. 80, 1751–1755.

Whittam, T.S., Wolfe, M.L., Wilson, R.A., 1989. Genetic relationships among *Escherichia coli* isolates causing urinary tract infections in humans and animals. Epidemiol. Infect. 102, 37–46.

Whitman, W.B., Coleman, D.C., Wiebe, W.J., 1998. Prokaryotes: the unseen majority. Proc. Natl. Acad. Sci. U.S.A. 95, 6578–6583.

Wilson, D.J., Gabriel, E., Leatherbarrow, A.J., Cheesbrough, J., Gee, S., Bolton, E., et al., 2009. Rapid evolution and the importance of recombination to the gastroenteric pathogen *Campylobacter jejuni*. Mol. Biol. Evol. 26, 385–397.

Wirth, T., Falush, D., Lan, R., Colles, F., Mensa, P., Wieler, L.H., et al., 2006. Sex and virulence in *Escherichia coli*: an evolutionary perspective. Mol. Microbiol. 60, 1136–1151.

Woese, C.R., 1987. Bacterial evolution. Microbiol. Rev. 51, 221–271.

Wohlfeiler, D., Potterat, J.J., 2005. Using gay men's sexual networks to reduce sexually transmitted disease (STD)/human immunodeficiency virus (HIV) transmission. Sex Transm. Dis. 32, S48–S52.

Wright, S., 1931. Evolution in Mendelian populations. Genetics 16, 97–159.

Wright, S., 1969. Evolution and the Genetics of Populations, vol. 2. The Theory of Gene Frequencies, University of Chicago Press, Chicago, IL.

4 Epidemiology and Evolution of Fungal Pathogens in Plants and Animals

Pierre Gladieux[1], Edmond J. Byrnes III[2], Gabriela Aguileta[1], Matthew C. Fisher[3], Joseph Heitman[2] and Tatiana Giraud[1,]*

[1]Ecologie, Systématique et Evolution, UMR 8079, CNRS, Bâtiment 360, Univ. Paris-Sud, F-91405 Orsay, France, [2]Department of Molecular Genetics and Microbiology, Duke University Medical Center, Durham, North Carolina, USA, [3]Department of Infectious Disease Epidemiology, School of Public Health, Imperial College London, St. Mary's Hospital, Norfolk Place, London, W2 1PG, UK

4.1 Introduction

Although parasitism is one of the most common lifestyles among eukaryotes, population genetics on pathogens lag far behind those on free-living organisms, probably because they are rarely conspicuous in the environment, do not possess the visible morphologic or behavioral variation used in the early studies of population genetics, and are less charismatic than the macrofauna. However, the advent of molecular markers offers great tools for studying key processes of pathogen biology, such as dispersal, mating systems, host adaptation, and patterns of speciation. Population genetics studies have also valuable practical applications, for instance for studying the evolution of drug resistance or new virulence. Another reason to study epidemiology and evolution in pathogens is that they display a huge diversity of life cycles and lifestyles, thus providing great opportunity for comparative studies to test pathogen-specific questions or general issues about evolution. Nevertheless, the field of parasitology has yet to attract more evolutionary biologists. This is especially true for fungal pathogens, despite their importance in crop diseases, and even in animal and human diseases. Furthermore, despite their obvious common interests there are few connections so far between scientists working on fungal pathogens versus other pathogens.

* E-mail: tatiana.giraud@u-psud.fr

Approximately 100,000 species of fungi have been described so far (1.5 million fungal species are estimated to exist; Hawksworth, 1991), of which a high percentage obtain nutrients by living in close association with other organisms, mainly plants. Many fungi are pathogenic and can have important impact on human health or lead to severe economic losses due to infected crops or to animal diseases. Fungal species parasitizing animals and plants are found interspersed with saprophytes and mutualists in fungal phylogenies (Berbee, 2001; James et al., 2006), suggesting that transitions between these life-history strategies have occurred repeatedly within the fungal kingdom.

True fungi belong to the opisthokont clade, as do animals. The two major groups that have been traditionally recognized among the true fungi are the Ascomycota, including the yeasts and filamentous fungi, with several important model species (e.g., *Saccharomyces cerevisiae*, *Neurospora crassa*), and the Basidiomycota, including the conspicuous mushrooms, the rusts and the smuts. Ascomycota and Basidiomycota have been resolved as sister taxa (Lutzoni et al., 2004; James et al., 2006) and they have been called the Dikaryomycota (Schaffer, 1975). The Dikaryomycota contain the majority (ca. 98%) of the fungal species, including most of the human and plant pathogens. Basal to the Dikaryomycota branch there are several other fungal groups. The Glomeromycota, mycorrhizal mutualists, are united within a clade with the Dikaryomycota (James et al., 2006). The Zygomycota are common in terrestrial and aquatic ecosystems, but they are rarely noticed by humans because they are of microscopic size. Some fungi among the Zygomycota are pathogens of animals (including humans), plants, amoebae, and other fungi (mycoparasites). Zygomycota branch at the base of the clade containing the Dikaryomycota and the Glomeromycota. The Chytridiomycota are defined as fungi with flagellated cells and were long thought to be the sister group of all the other true fungi, nonflagellated. However, recent phylogenies suggested that the chytrids may in fact be polyphyletic, representing early diverging lineages having retained the ancestral flagellum (James et al., 2006). Chytrids also encompass plant and animal pathogens. Microsporidia are obligate endoparasitic, protist-like organisms with highly reduced morphology and genomes; they have recently been proposed to belong to the fungi, as the most basal group (James et al., 2006). Oomycetes have long been considered as fungi but were recently recognized to belong to the distant Stramenopiles (Keeling et al., 2005). These filamentous organisms however share many morphologic and physiologic characteristics with fungi and continue to be studied by mycologists. They also contain plant pathogens, such as *Plasmopara viticola*, responsible for the grape mildew, and *Phytophthora* species, causing devastating emerging diseases, in particular on trees. Some oomycetes are pathogens on fishes or amphibians. We will therefore also consider oomycetes in this chapter.

Most fungi have been dependent on other organisms for their resources through much of their evolutionary history, in particular, fungal pathogens. During the past century, however, many new fungal diseases have emerged. This is probably due to human activities that have completely modified the ecosystems on earth at a global scale (e.g., climate warming, widespread deforestation, habitat fragmentation and urbanization, changes in agricultural practices, global trade) (Kareiva et al., 2007).

Of these, the intensification and globalization of agriculture as well as the increase in international trade and travel have broken down many natural barriers to dispersal causing an unprecedented redistribution of many organisms (Kolar and Lodge, 2001). Concomitantly, there is growing evidence that these global changes play a key role in the emergence of infectious diseases in humans (Tatem et al., 2006), wildlife (Daszak et al., 2000), domestic animals (Cleaveland et al., 2001), and plants (Anderson et al., 2004).

To understand how new diseases emerge, and more generally to understand the spread and maintenance of diseases, it is essential to study dispersal, mating systems, host adaptation, and mechanisms of speciation. The advent of molecular markers offers great tools for studying these key processes of pathogen biology (Criscione et al., 2005; Giraud et al., 2008a). Molecular markers, together with mathematical modeling and experiments, have also been instrumental to unravel the mechanisms of fungal speciation (Giraud et al., 2008b). The recent development of full genome sequencing, especially among fungi because they have small genomes (Galagan et al., 2005), has allowed comparative genomics to begin drawing inference on the mechanisms of pathogenicity (Aguileta et al., 2009).

In this chapter, we will thus describe the main pathogenic fungi, parasitizing humans, animals, and plants, and having important consequences on human health or human activities. We will focus on some examples of recent emerging fungal diseases on humans, animals, and plants. We will then review (1) the modern molecular tools used for epidemiology and population genetics of fungal pathogens, the types of markers most useful, and the different types of analyses that can be performed to unravel their mating systems and dispersal; (2) the criteria used for species delimitation in fungi and the mechanisms of fungal speciation that have been elucidated to date; (3) the recent advances in fungal genomics, in particular the insights that have been gained so far regarding the pathogenic lifestyles; and (4) the relationship between mating and pathogenesis in fungi.

4.2 Major Human and Animal Pathogenic Fungi

Each of the four major fungal phyla has representatives that cause serious disease in both humans and a vast range of other animals. Although less prevalent than plant pathogens, the animal pathogens pose serious threats to entire animal populations and continue to cause serious morbidity and mortality among immunocompromised patients and otherwise healthy individuals worldwide. In many cases, the incidence of disease is increasing due to a rise in susceptible hosts, while at the same time the treatment options have remained limited in comparison to other classes of pathogens. A major factor influencing treatment is that, unlike bacteria and viruses, the fungi are eukaryotic siblings to the animals. These issues cause major obstacles in the search and development of new antimicrobials that target fungi without causing major toxic side effects against animal metabolism. Here we summarize the morbidity and mortality associated with several of the major classes of human and animal pathogenic fungi.

4.2.1 Ascomycetes: The Candida Species Complex, Aspergillus fumigatus, Pneumocystis, the Dimorphic Fungi, and Others

Within the fungal kingdom, the ascomycetes harbor the majority of fungal pathogens that afflict humans. Among these, *Candida* species are the most common causes of invasive fungal infections in humans. Infections can range from readily treatable mucocutaneous disorders, although these may be acute in AIDS-infected patients, to severe invasive disease that can result in significant morbidity and mortality, most often occurring in patients with immune system suppression (Pappas et al., 2003, 2009). Candidemia is the fourth most common cause of nosocomial bloodstream infections in the USA (Wisplinghoff et al., 2004), with similar levels in many other developed countries. It has been estimated that the attributable mortality of invasive candidiasis is approximately 15—25% for adults and 10—15% for the pediatric population (Morgan et al., 2005; Zaoutis et al., 2005), with one study reporting mortality rates reaching levels >45% (Gudlaugsson et al., 2003).

Another of the major causes of human fungal infections is the filamentous pathogen, *A. fumigatus* and other closely related *Aspergillus* species. Aspergillosis, primarily invasive aspergillosis, is an emerging disease in the immunocompromised population (Walsh et al., 2008). The spores are widely prevalent in all environments, and are readily inhaled, causing both respiratory and disseminated disease in immunocompromised patients. There is a particularly high incidence of aspergillosis among stem cell and solid organ transplant recipients (Paterson and Singh, 1999; Marr et al., 2002a,b). Additionally, infected patients often have long and costly hospitals visits (Dasbach et al., 2000), making this disease a major concern, particularly in hospital settings.

A group of pathogenic fungi in humans that cause serious disease in both healthy and immunocompromised individuals are the dimorphic fungi. The name dimorphic stems from the common feature that all of these pathogens grow in a filamentous mold form in the environment, and based on changes in temperature grow as yeast at mammalian host temperatures and in the infected host (Rappleye and Goldman, 2006). This class of fungi includes the primary pathogens *Histoplasma capsulatum*, *Coccidioides immitis*, and *Coccidioides posadasii*, as well as species that more often infect immunocompromised individuals, including *Blastomyces dermatitidis*, *Sporothrix schenckii*, *Paracoccidioides brasiliensis*, and *Penicillium marneffei* (Fraser et al., 2007; Reis et al., 2009; Sharpton et al., 2009). All of these species (except *S. schenckii* and *H. capsulatum*) are known as "endemic mycoses". These are pathogenic fungi that have restricted ranges and tend to be associated with specific ecologic niches. For instance, *C. immitis* and *C. posadasii* are associated with the Lower Sonoran Life Zone—low hot deserts found only in northern, central, and southern America. Within these regions, the fungus exists as mycelia in sandy soils, as well as infections in small desert mammals such as kangaroo rats (*Dipodomys*) (Rippon, 1988). Human infection occurs as a consequence of the inhalation of arthroconidia, which can then undergo a temperature-determined dimorphic transition into septate spherules that disseminate throughout the body via hematogenous spread, causing a severe and life-threatening disease in a variable proportion of individuals. In common with *Coccidioides*, other endemic dimorphic mycoses inhabit recognizable ecologic niches that are often associated with animals. For instance, *P. marneffei* is the only pathogenic species of *Penicillium* from the highly speciose biverticilliate clade, and exhibits a highly constrained distribution to the wet tropics of Southeast Asia (Vanittanakom et al., 2006). Within this region, the pathogen is found infecting a high proportion of the tropical bamboo rat species *Rhizomys* and *Cannomys*, and is a key HIV-associated mycosis across the region. Although the route of human infection has yet to be confirmed, it is likely to stem from the inhalation of airborne conidia whereupon a thermally regulated dimorphic transition to a fission arthroconidium form occurs. Dissemination throughout the body causes significant pathologic effects with involvement documented for most of the major body organs (Vanittanakom et al., 2006).

Although the incidence of HIV-associated mycoses has decreased since the establishment of highly active antiretroviral therapy (HAART), pneumocystis infections are another of the major human pathogens infecting immunosuppressed hosts, with high incidences observed in the AIDS-infected and stem cell transplant recipient populations (Cushion, 2004). Infections result in a severe pneumonia, and are predominantly caused by *Pneumocystis jirovecii*, a widely prevalent species known to principally infect mammalian lung cells (Calderon, 2010). Research in this group of pathogens is difficult because axenic in vitro cultivation remains elusive. Culturing techniques are currently limited to growth in mammalian tissue culture cell lines (Cushion and Walzer, 1984a,b).

4.2.2 Basidiomycetes: The Pathogenic Cryptococcus Species Complex

Cryptococcus neoformans and *Cryptococcus gattii* comprise the pathogenic *Cryptococcus* species complex. They are related basidiomycete yeast species that are common fungal pathogens of both humans and animals. The two species are distinguished in that *C. neoformans* is prevalent, ubiquitous worldwide, largely associated with pigeon guano, and a common cause of meningitis in immunocompromised hosts (Perfect, 1989; Casadevall and Perfect, 1998; Carlile et al., 2001). *C. gattii* is generally geographically restricted to tropical and subtropical regions, associated with trees, and commonly infects immunocompetent hosts, although cases in immunocompromised patients also occur (Kwon-Chung and Bennett, 1984a,b; Sorrell, 2001). It is estimated that the two sibling species diverged ~ 37.5 million years ago, which may explain the observed differences in ecology and host range (Kwon-Chung et al., 2002). Additionally, the "tropical" status of *C. gattii* has been recently challenged by the occurrence of an outbreak that began in 1999, initially on Vancouver Island, Canada at latitude 49.28°. This emerging infection has since expanded into mainland British Columbia and the Pacific Northwest region of the USA (Kidd et al., 2004; Fraser et al., 2005; MacDougall et al., 2007; Upton et al., 2007; Byrnes et al., 2009b; Datta et al., 2009).

C. neoformans can be further subdivided into two serotypes (A and D) based on unique antigenic profiles and sequence divergence (Kwon-Chung and Varma,

2006; Bovers et al., 2008). This distinction is clinically relevant, as serotype A strains cause the vast majority of infections globally, with high incidences in the AIDS and transplant populations (Casadevall, 1998; Blankenship et al., 2005; Singh et al., 2008). Overall, >99% of AIDS-related infections and >95% of overall cases are attributable to serotype A (Casadevall, 1998). The global burden of disease is significant, with a recent report documenting almost 1 million annual cases with over 620,000 attributable mortalities, resulting in approximately one-third of all deaths in AIDS patients (Park et al., 2009). While less prevalent globally, *C. gattii* has also been a significant cause of morbidity and mortality, with high incidences in humans and animals reported in North America, Australia, Southeast Asia, and South America (Sukroongreung et al., 1996; Chen et al., 2000; Sorrell, 2001; Lizarazo et al., 2007; MacDougall et al., 2007; Galanis and MacDougall, 2010). Thus, the *Cryptococcus* species complex remains a global health concern for both humans and a wide range of domestic, agrarian, and wild mammals.

4.2.3 Globally Emerging Fungal Infections in Wildlife Species

While fungi are recognized as serious pathogens to their human hosts, it is also becoming clear that fungal pathogens have the capacity to cause severe disease in wildlife species. Notably, several of the fungi that are currently causing impacts on biodiversity were not previously recorded as pathogens. This illustrates not only the vast pool of undescribed fungal taxa, but also the capability of any branch of the fungal tree to give rise to serious pathogens. For instance, globally spreading chytridiomycosis in amphibians stems from a basal fungal lineage that was never before found to infect vertebrates (Fisher and Garner, 2007; Voyles et al., 2009). Similarly, *Nosema ceranae*, another microsporidian basal fungal lineage, has been discovered as a contributing agent for the currently mysterious declines in honeybee colonies (Klee et al., 2007). More recently, a new fungal infection of bats called white-nose syndrome (WNS) has swept though the northeastern USA since 2008, causing the deaths of >1 million bats and extirpating some well-known cave roosts (Blehert et al., 2008). The etiologic agent has been described as an ascomycete fungus *Geomyces destructans*, related to the human skin-infecting fungus *Geomyces pannorum* (Meteyer et al., 2009). These fungi exemplify the wide range of disease syndromes that are attributable to fungi, and the breadth of hosts that they are able to infect.

4.3 New and Emerging Mycoses

4.3.1 Evolution and Emergence of Pathogenic C. gattii Genotypes in the Pacific Northwest

As of 1999, *C. gattii* emerged as a primary pathogen in northwestern North America, including both Canada and the USA (Kidd et al., 2004, 2005; Fraser et al., 2005; MacDougall et al., 2007; Bartlett et al., 2008; Byrnes et al., 2009b;

Byrnes and Heitman, 2009; Galanis and MacDougall, 2010). This outbreak now spans a large geographic range, with levels of infection as high or higher than anywhere else globally, with an annual incidence on Vancouver Island of approximately 25 cases/million (Galanis and MacDougall, 2010). The only two reports with higher overall levels are one examination of native Aboriginals in the Northern Territory of Australia, and a study conducted in the central province of Papua New Guinea (Fisher et al., 1993; Seaton, 1996; Galanis and MacDougall, 2010). Specifically, *C. gattii* is classified into four discrete molecular types (VGI–VGIV), with molecular types VGI and VGII as the two most frequent causes of illness in otherwise healthy individuals (Byrnes and Heitman, 2009). Infections due to VGI have been reported at high rates among populations in Australia, while the levels of VGII infection are high in the Pacific Northwest, where ~95% of all cases are attributable to this molecular type (Sorrell, 2001; Fraser et al., 2003, 2005; Bovers et al., 2008; Byrnes and Heitman, 2009). The appearance of *C. gattii* in North America is startling because this is the first major emergence in a temperate climate (MacDougall and Fyfe, 2006; Kidd et al., 2007b). To examine the evolutionary aspects of this unprecedented emergence, efforts were undertaken to study the molecular epidemiology and characteristics of isolates collected from humans, animals, and the environment. These efforts have and will continue to shed light onto several key features of this outbreak, while other contributing factors remain elusive.

The first efforts to elucidate the molecular types of the isolates collected in the Vancouver Island area revealed that two genotypes, now known as VGIIa/major and VGIIb/minor, are responsible for the vast majority of cases (Kidd et al., 2004; Fraser et al., 2005). *C. gattii* was identified in the number of environments including several tree species, the air, soil, seawater, and freshwater (Kidd et al., 2007a,b; Bartlett et al., 2008). These studies then led to questions surrounding the properties of the common genotypes in the region. The VGIIa/major genotype was found to be highly virulent in a murine model of infection (Fraser et al., 2005). In addition, the examination of the isolates, particularly the discovery of a homozygous VGIIa/major diploid and the molecular characterization of the genome and mating-type locus, led to the hypothesis that same-sex mating was involved in this α only outbreak (Fraser et al., 2005). Together, these efforts showed that *C. gattii* VGII was now endemic in much of the region, and that the genotype responsible for the majority of cases is highly virulent in animal models of infection and also possibly in humans.

The next question in the field became focused on a possible expansion of the outbreak zone and the molecular and phenotypic characterizations of virulent isolates.

In 2007 and 2008, the first reports of *C. gattii* in the Pacific Northwest of the USA were published. The report of Upton and colleagues (2007) illustrated the first confirmed case of the Vancouver Island outbreak VGIIa/major in the USA (2006) from a patient in Puget Sound, Washington. Additionally, in 2005, MacDougall and colleagues discovered an increased number of outbreak-related cases on the mainland of British Columbia and related *C. gattii* VGII genotypes in the USA, including one later recognized as a VGIIc/novel isolate. Shortly thereafter, studies by Byrnes et al. documented a large cohort of clinical and veterinary cases from the VGIIa/major outbreak genotype in both Washington and Oregon (Byrnes et al., 2009a,b). These studies also reported VGIIb/minor in the USA, and importantly, defined a novel VGIIc genotype that was unique to Oregon and observed in both human and animal cases (Byrnes et al., 2009b).

Recent phenotypic examinations have also begun to address several key aspects of the outbreak genotypes. Studies in the mouse model revealed that *C. gattii* isolates from the outbreak induced less protective inflammation than *C. neoformans*, indicating that *C. gattii* may thrive in immunocompetent hosts by evading or suppressing the protective immune responses that normally limit *C. neoformans* disease progression (Cheng et al., 2009). Another unique feature of the outbreak VGIIa/major genotype is its ability to proliferate at high levels within macrophages as well as the ability to form highly tubular mitochondria after intracellular parasitism (Ma et al., 2009). These unique features were also shown to be positively correlated with murine virulence (Ma et al., 2009). Recently, it had also been shown that the VGIIc/novel genotype shares similar intracellular proliferation rates, mitochondrial morphology, and murine virulence characteristics with the VGIIa/major genotype, further supporting the hypothesis that the genotypes seen in the region are uncharacteristically enhanced for virulence (Byrnes, Lewit, Li, et al., PLoS Pathogens 2010 in press).

Over the past decade, we have witnessed the emergence and expansion of a tropical/subtropical pathogen into a temperate climate, leading to the formation of a multidisciplinary *C. gattii* working group established to address the epidemiology, clinical features, and basic science questions surrounding this outbreak (Datta et al., 2009). Although the overall incidence remains low, little is currently known about how or why specific humans and animals become infected and may involve unique host factors, including possible genetic predispositions. In addition, the origins of the VGIIa/major and VGIIc/novel genotypes remain elusive. Substantial progress has been achieved in addressing the molecular epidemiology and expansion of the outbreak, and also the phenotypic characteristics that make these genotypes unique. However, many critical questions remain to be addressed in the future to understand the evolutionary dynamics of this unprecedented *C. gattii* emergence in the region of the world, including expanded environmental sampling, further phenotypic characterizations of associations with host animals and plants, and genome sequencing of more representative *C. gattii* mitochondrial and nuclear genomes.

The Global Emergence of the Amphibian Pathogen Batrachochytrium dendrobatidis

The ability of fungi to cause severe disease in nonhuman vertebrate species has been dramatically illustrated by global declines in amphibian biodiversity caused by the fungus *Batrachochytrium dendrobatidis* (*Bd*). Only discovered in 1997 (Berger et al., 1998) and named in 1999 (Longcore et al., 1999), *Bd* is a basal fungal lineage in the Chytridiomycota; these fungi are characteristically aquatic and unique from other fungi in that they have a motile, flagellate zoospore (James et al., 2006). Many species of chytrid have been described in aquatic environments and soils, as free-living or commensal organisms, and as pathogens of algae, invertebrates, and fungi (Gleason et al., 2008). Of these, *Bd* is unique in that it is the only chytrid known to parasitize vertebrates, by infecting and developing within the keratinized epidermal cells of living amphibian skin (Pessier et al., 1999; Piotrowski et al., 2004). *Bd* is now known to be widespread in all continents except Antarctica (where amphibian hosts do not occur). A global-mapping project for this pathogen has shown that *Bd* infects over 350 species of amphibian, and has been implicated in driving the declines and extinctions of over 200 of these (http://www.spatialepidemiology.net/bd-maps/; Fisher et al., 2009).

Following the discovery that *Bd* was a driver of declines in amphibian species in Australia, the Americas, and Europe, much attention has been focused on finding out how *Bd* was being spread, and from where. In eastern Australia, prospective and retrospective sampling of amphibians has shown that populations were initially *Bd*-negative prior to 1978, followed by an expansion north and south from a center in southern Queensland; western Australia was *Bd*-negative until mid-1985, whereupon the spread of disease was detected and documented (Berger et al., 1998). Mesoamerica has witnessed a rapid wave-like front of expansion from an apparent origin in Monteverde, Costa Rica, southward at estimated rates of 17–43 km/year, and has recently jumped the Panama Canal (Lips et al., 2008). The epidemic front of chytridiomycosis along the North-South transect of Central America has been predictable to the extent that researchers have been able to anticipate the arrival of *Bd* in uninfected regions, such as El Copé in Panama, and to document the collapse of the amphibian community upon arrival of the pathogen and the onset of chytridiomycosis (Lips et al., 2006).

Given these patterns of declines, where is the original source of *Bd*? Answers to this question have been sought by attempting to identify geographic regions where *Bd* has had a long and stable association with host species, indicative of coevolution, as well as substantially increased levels of genetic diversity when compared against the various regional epizootics. One such study by Weldon et al. (2004) has identified Africa as a potential source of the panzootic. Histology on historical museum specimens showed that *Bd* has infected amphibians in Southern Africa since at least 1938, and the "*Bd* Out of Africa" hypothesis was coined to suggest that *Bd* was spread around the world via the extensive trade in the African clawed frog *Xenopus laevis* from the 1930s onward. However, the recently published molecular analysis by James et al. (2009) on global strains of the pathogen failed to find evidence that Africa contains more diversity than occurs in other regions, and in fact found that North American isolates of *Bd* were more highly diverse than elsewhere and that a single globalized lineage is causing the current panzootic. However, recent discovery that genotypes of *Bd* occur in the Japanese archipelago that appear basal to the panzootic lineage suggests that there may yet be other potential sources of *Bd* diversity (Goka et al., 2009). Therefore, the overarching question on the origin of *Bd* remains unanswered to date. What is clear, however, is that the global trade in amphibians is a potent force in spreading *Bd* into naïve populations and species. This statement is especially true for the so-called Typhoid Mary species such as *X. laevis* and the North American bullfrog *Rana catesbeiana*; these species carry *Bd* infections but rarely exhibit the disease, chytridiomycosis. They are also widely traded and are often highly invasive when introduced by accident or purpose into new environments (Fisher and Garner, 2007). Therefore, these two species constitute ideal vectors for introducing *Bd* into uninfected regions of the globe (Garner et al., 2006) and are likely a major source of new *Bd* infections when released into naïve environments.

Currently, it is not known whether the genome of *Bd* harbors the genes for mating and meiosis, although such genes have been found in the related Zygomycete lineage *Phycomyces blakesleeanus* (Idnurm et al., 2008). Importantly, all population genetic studies thus far have shown that *Bd* exhibits levels of heterozygosity that are consistent with a predominately asexual mode of reproduction. Of James' 17 sequenced polymorphic loci, 8 of these exhibited heterozygote excess. By anchoring the 17 sequenced loci to the genome-scaffolds, James et al. (2009) showed that levels of heterozygosity were not uniformly distributed across the genome, but were significantly reduced on the largest inferred chromosome where loss of heterozygosity (LOH) had occurred. This pattern of LOH is not consistent with sexual reproduction and segregation, but rather with a model of chromosome-specific variation in mitotic (somatic) recombination, a process that is well documented in other fungi including the diploid pathogenic fungus *Candida albicans* that exhibits vegetative diploidy (Odds et al., 2007).

This model of asexual LOH driving the diversity of *Bd* isolates is not, however, consistent across all studied populations. For instance, *Bd* sampled from Sierra Nevada populations of the mountain yellow-legged frogs, *Rana muscosa*, showed that, while allelic diversity was still found to be low throughout the dataset, within some local populations genotypic diversity was high. In these "high diversity" populations no new alleles appeared to have been introduced, and no genotypes were shared between different infected populations. Thus, it was suggested that local recombination had occurred within introduced lineages infecting particular lakes (Morgan et al., 2007). These findings have two interpretations: either *Bd* has the potential for sex that is largely unrealized due to population bottlenecks causing the loss of complementary mating types or that LOH can occur at variable rates in different populations, generating a spurious "signal" of genetic recombination. Recent efforts to sample more global isolates of *Bd* coupled to next-generation sequencing techniques are likely to reveal with greater clarity the mechanisms by which the *Bd* genome evolves.

Despite the apparent rapid spread of *Bd* and the high degree of genetic similarity between isolates, data is accumulating showing that genotypes differ significantly in their virulence. Fisher et al. (2010) showed that the sporangia of five isolates of *Bd* from the Balearic Island of Mallorca, all with identical genotypes, were similar in size, but differed significantly from those isolates recovered from amphibians in mainland Spain and the UK. When the virulence of a Mallorcan isolate of *Bd* (TF5a1) and a UK isolate of *Bd* (UKTvB) was assayed in *Bufo bufo* (Fisher et al., 2010), the Mallorcan strain of *Bd* was avirulent in comparison against the UK strain of the pathogen. Proteomic profiling of a global set of isolates showed that there was significant interisolate variation in patterns of protein expression. The amount of differentiation among isolates at neutral genetic markers and biologic (morphologic and proteomic) characters was greatest for morphologic traits, suggesting that these characters are under selection, and that this is possibly related to local environmental conditions (Fisher et al., 2009). These data suggest that, if *Bd* is able to generate functional diversity from a genetically depauperate genetic background via recombination, and thus increase its rate of adaptation to new environments, then the pathogen likely has the capacity to adapt to new climates and/or host species. This raises the possibility that *Bd* may increase its fitness to new environments and/or species combinations, and in this way change future patterns of disease in ways that parallel to those seen in other species of pathogenic fungi such as *C. gattii*.

4.3.2 Origin of Human Pathogens: Cryptococcus and Candida from Saprobes Associated with Insects

The origin and evolution of pathogens remain central questions in studies of both plant and animal diseases. One method to examine the likely origins of pathogens is to phylogenetically place the species into the context of closely related saprobic relatives. As mentioned earlier, *Cryptococcus* and *Candida* represent major classes of mammalian fungal pathogens, and in both cases their closest related species are associated with insects. Although these sibling species are less often studied than their medically relevant counterparts, they offer important insights into the evolution of the animal pathogens and how these pathogenic species might have arisen from insect-associated saprophytes.

Phylogenetic analyses indicate that the *Cryptococcus* species complex likely arose from the *Tremella* lineage and that it clusters closely with the Tremellales, Trichosporonales, Filobasidiales, and Cystofilobasidiales (Ergin et al., 2004; Rimek et al., 2004; Sampaio et al., 2004). Several of the species within these lineages are saprophytes that are commonly associated with insect debris, leading to the hypothesis that the pathogens emerged from an association within this environmental niche (Findley et al., 2009). In support of this hypothesis, *C. gattii* has been isolated from both insect frass and wasp nests, and *C. neoformans* has been isolated from honeybee hives, indicating that these animal pathogens may still in some cases act as an insect-associated saprophyte in the environment (Gezuele et al., 1993; Kidd et al., 2003; Ergin et al., 2004). While the evolutionary factors influencing the emergence of a mammalian pathogen from saprobes are still unclear, the support for this hypothesis of emergence based on phylogenetic and ecologic studies gives insights into the emergence of the pathogenic *Cryptococcus* species complex.

In addition to the studies mentioned earlier, interactions between *Cryptococcus* and insects have been further supported by the development of a well-validated insect model of pathogenesis in the model insect *Galleria mellonella* (Mylonakis et al., 2005; Fuchs and Mylonakis, 2006; London et al., 2006). Results from this system have been shown to correlate with the murine model of infection, and studies examining *C. neoformans* and *C. gattii* both show successful survival assays (Byrnes et al., 2009c; Findley et al., 2009; Velagapudi et al., 2009). Additionally, the pathogenic *Cryptococcus* species were shown to be more virulent than their nonpathogenic insect-associated relatives (Findley et al., 2009). This invertebrate model of infection is low-cost and poses fewer ethical issues, allowing for more facile, high-throughput analyses of virulence.

Studies in the *Candida* clade have shown several species to be associated with insects, particularly plant-associated beetles (Suh et al., 2004b, 2005, 2006, 2008). In most cases, these yeast species are xylose-fermenting, closely related to the pathogenic *Candida* species, and isolated from the insect gastrointestinal (GI) tract, leading to the hypothesis that this is a symbiotic relationship between the insects and fungi (Suh et al., 2003, 2004a,b, 2005, 2006; Suh and Blackwell, 2004, 2005; Nguyen et al., 2007). The large number of *Candida* taxa that are associated with a diverse range of host beetles also suggests that this phylogenetic lineage has a strong possibility of being involved in symbiotic insect associations (Suh et al., 2006). Recently, four novel species from the *C. albicans/Lodderomyces* clade were isolated from insect guts, with one, *Candida blackwellae*, the closest relative to *C. albicans* yet to be reported (Ji et al., 2009). Furthermore, *Candida dubliniensis*, a human pathogen, was isolated from seabird-associated ticks suggesting a possible reservoir and ecologic niche for this pathogen (Nunn et al., 2007).

Although the phylogeny and ecology of the clade are being uncovered, little is known about how several representative species from this lineage may have emerged and expanded into mammals. One commonality between the pathogenic mammalian and insect-associated species is that they are associated with the gut (Nguyen et al., 2006, 2007; Suh et al., 2006; Schulze and Sonnenborn, 2009; Tampakakis et al., 2009). While the environments are largely different, there may be common links between the insect and mammalian GI tracts that are important for survival and proliferation. An increase in number of studies examining the role of fungi in the human GI tract will also enhance the understanding of what roles the fungi, particularly the *Candida* clade, play in animal microbiomes.

4.4 Plant Pathogenic Fungi

Although several important fungal pathogens attack animals, land plants have probably been the main nutrient source of fungi through much of their evolutionary history, given the predominance of plant saprophytes, pathogens, and mycorrhizal species in fungi (Berbee, 2001; Berbee and Taylor, 2001; James et al., 2006). Collectively, fungi cause more plant diseases than any other group of plant pests (such as viruses or bacteria), with over 8000 species shown to cause disease. The life cycles of many of these are complex and involve two or more host plants. Plant diseases caused by fungi exhibit a huge diversity of symptoms. Pathogenic fungi can indeed be responsible for lesions on leaves or on flowers, stem cankers, root and fruit rot, or for sterilizing plants.

Fungal pathogens are therefore a serious concern for agriculture, as they reduce crop yield and lower product quality by attacking cultivated plants and their products (seeds, fruits, grains). Nearly all crops have their pathogenic fungi, and often several of them, from cereals to corn, rice, potatoes, beans, peas, soybean, fruit trees (including coffee and cacao), and ornamental plants and trees. Some of the world's great famines and human suffering can be blamed on plant pathogenic fungi. Wheat crops of the Middle Ages were commonly destroyed when the grains became infected with a dark, dusty powder now known to be the spores of the fungus called bunt or stinking smut (*Tilletia* spp.). Epidemics caused by rust fungi have also been noted for millennia. These epidemics were recognized in ancient Greece and described in the writings of Aristotle and Theophrastus. The Romans held a religious ceremony/festival, the Robigalia, to appease the gods Robigo and Robigus, whom they believed responsible for the rust epidemics. Potato late blight, caused by the oomycete *Phytophthora infestans*, is the most important biotic constraint to potato production worldwide. It caused epidemics during the 1840s, because of which more than 1 million people died from starvation or famine-related diseases, and more than 1.5 million emigrated from Ireland. A more recent epidemic that resulted in large-scale famine was caused by a fungus responsible for brown spot of rice; 2 million people died of starvation during the great Bengal famine of 1942. A related fungus, which attacks corn and causes southern leaf blight, resulted in a widespread epidemic in the USA in 1970; ca. 15% of the total corn crop was lost, with yields in some states reduced to 50%. In the USA alone, hundreds of millions of bushels of wheat have been lost in epidemic years due to stem rust (*Puccinia graminis tritici*). Rice blast, caused by the fungus *Magnaporthe oryzae*, is an important disease on rice, among many other diseases. It is found wherever rice is grown, it is always important, and it is always a threat. Coffee rust, caused by *Hemileia vastatrix*, caused epidemics on cultivated coffee in Ceylon (Sri Lanka, which was British at the time) in the nineteenth century. All exports of coffee from Ceylon had to be stopped; planters turned to tea in place of Arabian coffee, and tea became the social drink of the British (Staples, 2000). Scarcity of wheat caused by epidemics of wheat stem rust is the historical reason that the bread of central Europe is often made of rye and that cornbread is so popular in the southern US (Horsfall, 1983; Palm, 2001). The *Botrytis* gray mold is a common disease of greenhouse floral crops and all ornamental plants can be infected by powdery mildews. These are only few examples of the many pathogenic fungi devastating crops.

In addition to being agents of preharvest and postharvest diseases and rots, fungi produce highly toxic, hallucinogenic, and carcinogenic chemicals that not only affected the lives of millions of people historically, but also continue to be of problems today. In 2006, dozens of dogs perished in the USA from food tainted with aflatoxin, a chemical produced by several *Aspergillus* species. These fungi can grow on corn and fill the seed with the toxin that not only attacks the liver, but is one of the most carcinogenic substances known. Another example comes from the genus *Fusarium*, which contains numerous phytopathogenic species, *F. culmorum* and *F. graminearum* being particularly important pathogens of cereal crops in many areas of the world. They are responsible for head and seedling blight of small grains such as wheat and barley, ear and stalk rot of corn, and stem rot of carnation. Besides causing yield reduction, these *Fusarium* diseases come with the production of mycotoxins, which are highly toxic to both plants and animals, including humans (Desjardins et al., 1993).

Several methods are used in modern agroecosystems to control fungal pathogens, including spraying fungicides, creating resistant varieties, crop rotations, and a variety of cultural practices aimed at reducing plant infections. Fungicides and resistance can be very efficient at first, but are expensive, polluting, and often do not last. Fungi have indeed huge effective sizes, with millions of spores produced by a single diseased plant, great dispersal abilities, and several generations per year, enabling rapid adaptation. Fungicide breakdown often occurs within a few years, as does resistance resistance (Brown, 1994).

Pathogenic fungi are also widespread in natural ecosystems, with great impacts on the compositions of natural communities. Forest trees, for instance, are attacked by many pathogenic fungi. *Armillaria* root disease, causing branch dieback and crown thinning, is often one of the most important diseases of trees in temperate regions of the world, especially in native forests. The most infamous tree diseases include Dutch elm disease caused by *Ophiostoma* species, chestnut blight caused by *Cryphonectria parasitica*, and sudden oak death, ramorum leaf blight, and ramorum shoot blight all caused by the oomycete *Phytophthora ramorum*. These diseases have dramatic consequences on forest composition and their associated biota, with some tree species even disappearing from continents. For instance, the chestnut blight fungus caused the death of 80% of the native American chestnut trees throughout eastern forests from Maine to Georgia during the first half of the twentieth century. The Dutch elm disease fungus, *Ophiostoma ulmi*, has led to the destruction of American elm trees and has altered urban landscapes by killing ornamental elms across the country. It has been estimated that more than 77 million elms have died. Not only trees, but virtually all natural plants have their own pathogenic fungi. Examples include choke disease on grasses caused by *Epichloë* species, anther smut disease on Caryophyllaceae caused by *Microbotryum violaceum*, and powdery mildew on many natural plants.

4.5 New and Emerging Plant Diseases

Fungi are also responsible for about 30% of emerging diseases in plants (sensu lato, i.e., including oomycetes), which is 3 times more than for emerging diseases in humans or wildlife (Anderson et al., 2004). These patterns of fungal disease emergence in plants have elicited great concern for several reasons. First, epidemics caused by invasive pathogens have been repeatedly reported to alter natural ecosystems (Anderson et al., 2004; Desprez-Loustau et al., 2007). Well-documented examples of emergent diseases in natural plant communities include some of the ones mentioned earlier, such as the spread of *Cryphonectria parasitica* that eliminated the dominant chestnut forests throughout eastern North America at the end of the nineteenth century. The Dutch elm disease caused by *Ophiostoma ulmi* and *O. novo-ulmi* appeared in Europe around 1919, and the fungus was described in Holland in 1921; it was first found in the USA in Ohio in 1930 (Anagnostakis, 2001). *Phytophthora cinnamomi* that threatens native forests throughout Australia is also an emerging disease (Anderson et al., 2004; Desprez-Loustau et al., 2007). Many powdery mildews appear as invasive fungi (Kiss, 2005). Such dramatic diseases not only affect the host plants, but also the whole associated fauna, including insects, birds, and mammals.

Second, our primary food production is at risk due to emerging crop diseases; the most dramatic historical example being the Irish Potato Famine caused by *P. infestans* on cultivated potato in the beginning of the twentieth century (Birch and Whisson, 2001). Other examples of invasive fungi parasitizing crops include *Plasmopara viticola*, an oomycete causing the grapevine downy mildew, that has been introduced from North America to Europe in the past two centuries, *Plasmopara halstedii*, another oomycete causing sunflower downy mildew (Delmotte et al., 2008), the soybean rust in North America and the coffee rust in Asia and South America (Staples, 2000). Crop plants are in fact particularly susceptible to the emergence of new diseases because of the large-scale planting of genetically uniform varieties.

Third, epidemics on crop plants generate a huge production of potentially infectious spores (Brown and Hovmoller, 2002). In addition to spreading disease over agricultural areas, this high propagule pressure on surrounding areas may contribute to disease emergence in natural plant communities. Indeed, though nonhost resistance is quite durable, host shifts are known to occur (Stukenbrock and McDonald, 2008; Tellier et al., 2010), so epidemics of crop species may pose an undetected and poorly recognized danger to natural plant communities (Power and Mitchell, 2004).

There has been an increasing focus on identifying the factors that drive the emergence of new fungal diseases (Anderson et al., 2004; Thrall et al., 2006; Desprez-Loustau et al., 2007; Stukenbrock and McDonald, 2008). As mentioned earlier, introduction of pathogens in a new area is one of the most obvious causes. It has been estimated that between 65% and 85% of plant pathogens worldwide are alien in the location where they were recorded (Pimentel et al., 2001). The emergence of a new fungal disease following an introduction may be due to the reunification of a pathogen and a crop that had been introduced to new continents. This has been the case for instance for the potato late blight, rubber leaf blight, and coffee rust (Staples, 2000; Birch and Whisson, 2001; Desprez-Loustau et al., 2007). Fungal pathogens introduced into new continents have also been responsible for disease emergence in natural plant communities. Examples include the sudden oak death in North America caused by *Phytophthora ramorum* (Rizzo et al., 2005), the oak decline in Europe and Jarrah decline in Australia caused by *Phytophthora cinnamomi* (Hardham, 2005), the Dutch elm disease caused by *Ophiostoma ulmi* and *O. novo-ulmi* (Brasier, 2001), and the chestnut blight caused by *Cryphonectria parasitica* (Anagnostakis, 2001).

Horizontal gene transfer (HGT) (i.e., the exchange of specific genes between species that are normally reproductively isolated) has also been invoked to explain the emergence of new fungal diseases (Stukenbrock and McDonald, 2008). The most convincing example is a HGT between *Phaeosphaeria nodorum* and *Pyrenophora tritici-repentis*, both fungal pathogens of wheat with similar foliar symptoms worldwide. *P. tritici-repentis* is a very recently emerged pathogen, which was suggested to be due to the acquisition of a host-specific toxin gene by horizontal transfer from *P. nodorum* (Friesen et al., 2006).

Interspecific hybridization involves whole genomes, in contrast to HGTs. Hybridization is quite common in fungi, and can also lead to the emergence of new hybrid species (see Section 4.8). Transgressive traits in hybrids could even lead to the emergence of a disease on a new host that the hybrid would be able to parasitize while its parent species would not (Brasier, 2001; Olson and Stenlid, 2002; Stukenbrock and McDonald, 2008). In fact, this scenario has been suggested for some emerging fungal diseases, in particular the rusts caused by *Melampsora* spp. on poplar (Newcombe et al., 2000) and the diseased caused by a complex of *Phytophthora* species on alder in Europe (Ioos et al., 2006).

4.6 Modern Molecular Epidemiologic Tools for Investigating Fungal Diseases

To understand the dynamics of fungal diseases and the dynamics of emergence of new diseases, epidemiology is a necessary step. Epidemiology is indeed a discipline concerned with understanding the factors affecting the dynamics of disease in space and in time, with an emphasis on being quantitative and predictive. During the past decade, the integration of molecular biology into traditional epidemiologic research has revolutionized the discipline (Tibayrenc, 1998; Taylor et al., 1999a). This led to the development of a new field, *molecular epidemiology*, which addresses epidemiologic problems using "the various molecular methods that aim to identify the relevant units of analysis of pathogens involved in transmissible disease" (Tibayrenc, 2005). Increasingly sophisticated, sensitive, and reproducible detection methods have made studies of spatial patterns of disease, nosocomial infections, or disease outbreaks much more convincing and have provided unprecedented opportunities to track pathogen populations with particularly harmful characteristics in pathogenicity, virulence, or resistance to chemicals. Early adoption of molecular typing techniques to address applied questions in fungi of medical, veterinary, and agronomical relevance was soon cross-fertilized by the related field of molecular evolutionary genetics (Milgroom, 1996; Taylor et al., 1999b). However, in common with other microbial fields, a plethora of novel platforms and typing techniques was developed to study the molecular epidemiology of fungi during the 1990s. This multiplication may have, to some extent, confused rather than elucidated fungal epidemiology (Achtman, 1996; Urwin and Maiden, 2003). Fortunately, the list of esoteric acronyms for typing techniques (or YATMs, for yet another typing method; Achtman, 1996) eventually ceased to expand as research coalesced on those platforms that yielded unambiguous and portable data. Specifically, two methods in particular are now predominant in molecular epidemiologic studies of fungal pathogens: MLST, for multilocus strain typing (Maiden et al., 1998) and MLMT, for multilocus microsatellite typing (Fisher et al., 2001). The advantages of these two techniques as typing approaches are their portability and reproducibility (typing can be done in any molecular biology laboratory and generates identical results for identical DNA samples), their archiveability (results can be compiled by multiple contributors in publicly available web databases) and their amenability to high-throughput automation. The development of these methods, designed to be enduring, would not have been possible without the technologic advances that eased and reduced the cost of generating, determining, storing, and interpreting genetic data. Currently, new advances in next-generation high-throughput sequencing techniques mean that MLST and MLMT typing schemes are on the brink of being absorbed into whole-genome single nucleotide polymorphism (SNP)-typing platforms. However, the nature of MLST/MLMT typing schemes is such that older data will be incorporated into new schemes rather than discarded, as was previously the case.

MLST schemes compare nucleotide polymorphisms within regions of ca. 500 nucleotides from five to seven genes. Traditionally, the regions represent coding sequences of housekeeping genes, which are under purifying selection to retain function and persist in genomes. MLST was originally developed to facilitate studies of epidemiology in bacterial populations (Maiden et al., 1998). Mycologists were using a similar technique since the mid-1990s championed by J.W. Taylor and his associates, but it was lacking an acronym and more oriented toward the study of evolutionary features of fungal species than issues of epidemiology (Taylor et al., 1999a; Taylor and Fisher, 2003). The principle of an MLST scheme is simple: allelic variants, resulting from SNPs, are recorded as series of integers that together constitute a barcode called a strain sequence type (ST). The MLMT scheme is a direct extension of the MLST approach to microsatellite loci, which are short tandem repeats of 2–6 nucleotides. Allelic variants at ca. 5–15 microsatellite loci, here resulting from variation in the number of repeats of the microsatellite motif, are recorded as series of integers called microsatellite types (MT). Both MLST and MLMT have been greatly enhanced by the availability of complete fungal genomes that render the steps of isolation of new markers less cumbersome and more likely to succeed. This is particularly true for MLMT schemes, whose development is often hindered in fungi by the weak representation of microsatellite loci in fungal genomes, the low abundance of motifs with potentially high mutability (i.e., with high number of repeats), and the small size of genomes (Dutech et al., 2007).

The main use of fungal MLST is for diagnosis and species identification (Taylor and Fisher, 2003). Issues such as quarantine, selection of appropriate treatments and disease control measures, identification of significant sources of inoculum for epidemics, or nosocomial infections depend on a proper taxonomic assignment (Palm, 2001). Traditional methods based on phenotype or mating tests, if applicable, were laborious and compromised in their accuracy. Phylogenetic methods using genealogical concordance phylogenetic recognition (GCPSR) are progressively superseding previous approaches (Taylor et al., 2000; Giraud et al., 2008b). Gene-sequence data from studies using GCPSR can easily be converted to STs and stored in web databases accessible to the whole community of mycologists. Species recognition studies are also an elementary prerequisite to the design of an MLST scheme, as molecular epidemiology tools are based on the idea that pathogen populations form discrete entities that are stable enough to be identifiable. These entities, and their biologic relevance, are conditioned by evolutionary forces; hence, the usefulness of a preliminary exploration of species limits and population structure by evolutionary biologists (Tibayrenc, 2005). The delimitation of species, and therefore implicitly the design of a new MLST scheme, requires assembling a diverse collection of isolates on the basis of current knowledge on the ecology of the fungus and existing typing information (Urwin and Maiden, 2003). Clinical or field isolates can be quite readily collected, but even a modest sampling of environmental individuals of pathogenic fungi can turn out to be a very difficult task (Greene et al., 2000; Taylor and Fisher, 2003). Where fungi have been thoroughly sampled across their range and habitats and have provided a sufficient diversity of genotypes, MLST schemes can assist in dissecting the factors behind a disease outbreak. For instance, a recent MLST study of the chytrid fungus *Batrachochytrium dendrobatidis* (James et al., 2009), suggested as a principal cause for the worldwide decline of amphibians, found the global epidemic owes to the global dispersal of a single genotype. This data was used to argue that the observed low allelic diversity and high heterozygosity provide strong support that the fungus is a novel pathogen introduced into naïve host populations, over the alternative hypothesis that the species is an endemic pathogen whose emergence is due to recent changes in the environment. By contrast, an MLST study of *Coccidioides immitis*, the etiologic agent of coccidiomycosis revealed that the epidemic observed in California in the early 1990s was not due to the emergence of a virulent genotype but rather governed by the synchrony of environmental factors (Fisher et al., 2000). In this study, analyses of clinical isolates with MLST data revealed extensive genetic, genotypic diversity and a lack of significant association across loci, rejecting the hypothesis of aggressive clonal spread, and in conjunction with these elements, multivariate statistical treatment of environmental data showing that the number of cases of disease was best explained by the interaction between two climatic factors.

MLMT-based techniques are more useful in discriminating genotypes within species and inbred populations than among species, which make their use complementary to MLST. A primary reason is that the cross-species transferability of microsatellite appears lower in fungi than in other organisms (Dutech et al., 2007), which limits their utility to discriminate among species (with some exceptions, e.g,. Fisher et al., 2002; Matute et al., 2006). A second reason is that microsatellites typically have high mutation rates in fungi, even though they appear less polymorphic in this kingdom than in others (Dutech et al., 2007). Factors such as recent speciation, demographic bottlenecks, or selective sweeps associated with a lack of recombination can dramatically reduce variation at housekeeping genes and thus seriously hamper the discriminatory power of MLST schemes at the intraspecific level (Morehouse et al., 2003; Couch et al., 2005; Bain et al., 2007). In such cases where nucleotide variation is not sufficient to address questions at the intraspecific level, MLMT provide a powerful alternative (Fisher et al., 2004; Morgan et al., 2007). In a recent study Ivors et al. (2004) sequenced three housekeeping genes in outbreak isolates of *Phytophthora ramorum*, the etiologic agent of the devastating "sudden oak death" disease. This study showed that all sequences were identical among all isolates and therefore completely uninformative on the nature of epidemic. MLMT tools developed later (Prospero et al., 2004, 2007; Ivors et al., 2006) proved very useful in tracking the pathogen as it spread in the USA (Cooke, 2007; Prospero et al., 2007). Analyses of MLMT data provided evidence of a historical link between nursery and wild populations of the pathogen, and identified three common genotypes as the likely founders of the Californian epidemics (Mascheretti et al., 2008). The potential of MLMT as a tool to resolve interstrain differences even at fine-scale can also be useful to study basic features of the biology of fungi. For instance, in plant pathology it is critical to know the relative importance of ascospores, mycelium, and conidia in dispersal of fungi and whether the source of primary inoculum is soil, plant debris, infected seeds, other plants, etc. The most obvious application of this knowledge in disease management is to eliminate the source of inoculum by sanitation, debris, and weed management, or crop rotation (Milgroom and Peever, 2003). MLMT also provides a useful tool to infer the source and type of primary pathogen inoculum, which are often impossible to identify by direct observation or using the traditional epidemiologic approach of studying the distribution of disease foci (Douhan et al., 2002; Peever et al., 2004). A last original application of MLMT tools in fungi is in plant breeding efforts. Here, the basic idea is the use of a better representative of existing variation in pathogen populations to screen for resistant germplasms in order to assist in breeding plant varieties with more durable and effective resistance. Based on the assumption that variation at microsatellite markers can be used as a proxy for variation in pathogenicity traits, MLMT schemes have found use in selecting a core collection of pathogen genotypes that are more representative of extant genetic diversity than are a random sample of the local inoculum found in the immediate neighborhood of the nursery (Peever et al., 2000).

What biostatistical methods can be used to track genotypes and species using MLST and MLMT data? Many fungal species exhibit limited recombination in nature and form complexes of genetically related haplotypes (Taylor et al., 1999b). For these taxa, split decomposition (Huson and Bryant, 2006), statistical parsimony (Clement et al., 2000), or eBURST (Feil et al., 2004; Spratt et al., 2004) can provide a graphical representation of the relationships among genotypes and their prevalence (Morehouse et al., 2003; Urwin and Maiden, 2003; Couch et al., 2005; Fisher et al., 2005; Morgan et al., 2007). These network approaches can be used to infer the origin of particular isolates, provided that the species have sufficient host-specific or geographical population structure. Where population structures are highly recombining, however, network approaches are improper for epidemiologic tracking because recombination mixes genotypes at individual loci and renders multilocus genotypes unstable (Tibayrenc, 2005). Therefore, in recombining species, determining source populations requires comparing individual isolates in statistical settings that explicitly model the associations amongst loci. A number of Bayesian methods have been developed to produce robust assignments for organisms with extensive genetic recombination (Rannala and Mountain, 1997; Pritchard et al., 2000; Falush et al., 2003; Piry et al., 2004). An example of the utility of this approach is provided by Fisher et al., 2001 to identify the source populations for

Coccidioides isolates recovered from patient treated outside the endemic area of *Coccidioides* sp. These methods can also be used to exclude possible sources. In one study, Bayesian tests were used to demonstrate that populations of *Venturia inaequalis*, the apple scab fungus, spreading in France on apple varieties harboring the *Vf* major resistance gene were not derived from local apple scab populations, but rather introduced from another region (Guérin et al., 2007).

Recent technologic developments have increased the rate of data generation and concomitantly greatly enhanced our understanding of fungal species and populations. The next step will be to increase the level of data sharing and to promote the development of curated Internet databases that can accommodate the incoming avalanche of SNP-diversity that is generated by next-generation sequencing schemes. Web-based community tools are still underdeveloped in fungi in comparison with other microbial pathogens, and an extensive sharing of data could allow more sophisticated spatio-temporal surveys of epidemics. Recent progress is illustrated by the development of an MLST-typing scheme for the fungal pathogen *C. neoformans* that integrates MLST-approaches with new scalable mapping technologies to ascertain regional and global patterns of spread (http://cneoformans.mlst.net/earth/maps/) (Meyer et al., 2009). Such informatic technologies will in the future be integrated with next-generation sequencing and combined with the development of predictive models of disease spread to relate strain typing data with phenotypic traits, environmental data, and disease risk-assessment decision platforms.

4.7 Population Genetics of Pathogenic Fungi

Population genetics is also needed to understand fungal diseases. The genetic structure of a species refers to the amount and distribution of genetic variation within and among populations. Population genetics aims to understand the evolutionary processes that shape the genetic structures of species. For pathogenic fungi, population genetics questions are not simply of academic interest, as these questions have genuinely practical applications. Issues such as breakdown of plant resistance, resistance to fungicides, emergence and spread of virulent strains, or the design of tools for identification are related to the genetic structure of fungal populations. By providing an understanding of the processes that shaped the structure of a pathogen species in the past, population genetics offers the opportunity to forecast the emergence of genotypes, populations, or species with detrimental characteristics for human affairs (McDonald and Linde, 2002; Giraud et al., 2010), and also to inform practical attempts to bring fungal pathogens into durably effective human control (Williams, 2009). Currently, a great part of the research effort of population geneticists on fungal pathogens is devoted to fuel the development of risk-assessment models. Herein, we provide an overview of the tools available to understand three important components of population structure of fungal pathogens that are related to their evolutionary potential: the reproductive system, gene flow, and population subdivision. We also provide examples of case studies where the methods have been successfully applied to elucidate genetic structures in fungi.

4.7.1 Reproductive System

Fungi present a striking diversity of life cycles, and studying their reproductive biology is a challenging task. However, this information is critical to assess the risk posed by pathogens and for the design of disease management strategies (McDonald and Linde, 2002). For instance, outcrossing promotes genetic exchange and can accelerate the spread of new mutations in combination with other beneficial alleles, which is critical in the context of an arms race between hosts (or the humans that breed or grow them) and pathogens. By contrast, selfing or asexual reproduction provides insurance of reproduction for species having a low probability of finding a mate, and these species can therefore invade distant territories more easily and/or more rapidly (Taylor et al., 2006). Asexual reproduction is also an expeditious way of multiplying rapidly favorable combinations of genes built by past selection (Otto, 2009) and a more efficient strategy of transmitting genes to the next generation. Indeed, an asexual parent transmits 100% of its genes to the next generation, against only 50% for a sexual parent, which is called "the twofold cost of sex" in anisogamous species (Bell, 1982). In the following, we briefly define the terminology used to qualify different aspects of fungal reproductive systems, and then we provide an overview of the methods available for their analysis, with some case studies among the fungal pathogens.

Terminology

Inconsistent use of key terms might be a cause of the slow integration of fungi in the field of evolution, and more generally unclear definitions of concepts are often an obstacle in the progress of science (Neal and Anderson, 2005). A proper identification of the key features of the reproductive system of fungal pathogens is also fundamental for the correct selections of appropriate models to study population structure (Giraud et al., 2008a). Three aspects of the fungal reproductive system can be distinguished: the reproductive mode, the breeding system, and the mating system.

Sexual reproduction is the process by which progeny is formed through the combination of two parental nuclei, generally involving syngamy and meiotic recombination (Schurko et al., 2009). In fungi, genes can be transmitted across generations through asexual, sexual, or mixed modes of reproduction ("mixed" referring to the alternation of sexual and asexual reproduction during the life cycle). Approximately one-fifth of described fungal species have been thought to be asexual, but population genetic studies have revealed that most show footprints of recombination, which is incompatible with strictly asexual reproduction (Taylor et al., 1999b). The difficulty in determining the reproductive mode mostly stems from the failure of morphologic observations of sexual structures. There is also the complication that fungal species often participate in both sexual and asexual reproduction, and therefore the same fungal species can display different reproductive modes in different places or at different times. It is often asserted that mitotic recombination via parasexuality can mix parental genomes and mimic the effect of sexual reproduction in fungi (Taylor et al., 1999b), but the importance of parasexuality in nature remains to be determined (but see Milgroom et al., 2009).

The term "breeding system" refers to the physiologic determinants of compatibility among individuals (Neal and Anderson, 2005). Mating compatibility in fungi is regulated strictly in the haploid stage by mating-type loci. For most species, the successful fusion of gametes can occur only between haploids carrying functionally different mating-type alleles, a phenomenon called heterothallism. Compatibility can be determined by alleles at a single locus (a condition termed bipolar heterothallism) or by alleles at two unlinked loci (a condition termed tetrapolar heterothallism). Some fungi are homothallic, meaning that they do not require genetic differences for mating compatibility.

The term "mating system" refers to the degree of genetic relatedness between mates. Outcrossing corresponds to the mating between cells derived from meioses in two different unrelated diploid genotypes, whereas inbreeding corresponds to the mating between related individuals. Inbreeding can be caused by selfing, the mating between meiotic products of the same diploid genotype. Three subcategories of selfing can be distinguished: intertetrad, intratetrad, and intrahaploid mating. Intertetrad mating refers to the union of cells derived from the same diploid individual but from different meiotic tetrads. Intratetrad mating refers to the union of cells derived from the same meiotic tetrad. Intrahaploid mating is allowed by homothallism, where genetic differences between pairing individuals are not required, permitting union between haploid mitotic descendants of the same meiotic product. We invite the reader to note that, contrary to persistent misconceptions in the fungal literature, the breeding system has little influence on the mating system. For instance, heterothallism does not prevent selfing, because any diploid individual is necessarily heterozygous at the mating-type locus (see Giraud et al., 2008c, for more details), and homothallism may have been selected for more efficient outcrossing rather than for allowing intrahaploid mating, the latter having little advantage over asexuality while retaining some of the costs of sex (Billiard et al., 2010). Tetrapolarity is often suggested to promote outcrossing. If one considers biallelic breeding systems, it is true that tetrapolarity is less favorable to intratetrad mating than bipolarity, since the chance that any two siblings are compatible is 50% in a bipolar cross compared to only 25% in a tetrapolar cross (Hsueh et al., 2008). However, if gametes disperse before mating, these odds of compatibility within a progeny will be of little relevance. Whereas the breeding system cannot be determined without laboratory experiments, the mating system and reproductive mode of fungi cannot be inferred without analyzing patterns of genetic variation in natural populations.

Analysis of the Reproductive System

Following Milgroom (1996), we can distinguish three basic questions usually asked by fungal population geneticists: (1) Is population structure consistent with random mating? (2) Is there evidence for recombination? (3) What is the degree of relatedness between mates? In practice, answers to these questions are not independent and investigators often take the inability to reject a random mating hypothesis as an evidence for recombination.

The identification of populations and species is an essential prerequisite to the study of the reproductive mode and mating system. Hidden population subdivision or cryptic species within the units defined to perform analyses can indeed lead to erroneous conclusions on the reproductive biology of a fungus. This causes deviations from random mating or from random association among alleles. A well-known example is the Wahlund effect, where the failure to detect population subdivision influences measures of inbreeding and association among alleles at different loci and leads to the same signal as inbreeding. Although they form a prerequisite to the study of reproduction, methods to analyze population subdivision generally do not provide a genuine assessment of the characteristics of the mating system (Gao et al., 2007), and therefore specific analyses are needed.

The most immediate consequence of asexual reproduction is the occurrence of repeated identical genotypes. The ratio of the number of multilocus genotypes found over the sample size can give an idea as to the rate of asexual reproduction, ranging from zero for a completely clonal population to one for a sexually reproducing population. Many populations of plant pathogens actually fall between the two extremes, having annual sexual cycles and asexual epidemic phases that amplify clones (Milgroom, 1996). One approach to analyzing the reproductive biology of these pathogens is to include a single representative of each multilocus genotype (an approach referred to as "clone correction"). However, one should ensure prior to clone correction that repeated genotypes do not simply result from insufficient discriminative power of the molecular markers assayed. This can be tested by calculating the likelihood that a multilocus genotype observed more than once in a sample is the result of sexual reproduction, given the observed allele frequencies and assuming random mating (Stenberg et al., 2003). The GENODIVE (Meirmans and Van Tienderen, 2004) and GENCLONE (Arnaud-Haond and Belkhir, 2007) programs offers user-friendly implementations of clone correction methods.

Under random mating, the frequency of multilocus genotypes is expected to be equal to the product of the allelic frequencies. Deviation from this expectation (or linkage disequilibrium) can hence serve as a test for random mating. A first approach is to analyze linkage disequilibrium between pairs of loci. The lack of association among pairs of loci in two isolated groups of the agent responsible for gray mold (*Botrytis cinerea*), for example, supported regular events recombination despite the absence of a sexual structure in field observations (Giraud et al., 1997). The existence of linkage equilibrium was also taken as evidence for sexual reproduction in populations of the wheat pathogen *Mycosphaerella graminicola* in regions where the teleomorph is rare or absent (Chen and McDonald, 1996; Zhan et al., 2003). Another, more powerful approach is to analyze linkage disequilibrium over multiple loci. This forms the basis of the test based on the index of association (I_A; Maynard-Smith et al., 1993). The I_A statistic relies on the variance of the number of differences among individual allelic profiles. This variance is higher than expected if mating is nonrandom due to an excess of very close and very large distances among individuals. The statistical significance of the I_A statistic can be established using the program MULTILOCUS, by comparing the observed value of the statistics to the distribution obtained from datasets for which alleles at each locus are resampled without replacement to simulate the effect of random mating (Agapow and Burt, 2001). This procedure has been applied to investigate the reproductive mode of *Penicillium marneffei*, the causal agent of biverticilliate mycosis in mammals. Analyses revealed very high and significant values of the I_A statistic (Fisher et al., 2005), providing one of the very rare cases of a fungus showing no evidence of recombination by population genetic criteria (Taylor et al., 2006). Another striking example of a fungus displaying a highly clonal population structure is provided by the European populations of the yellow rust of wheat (*Puccinia striiformis* f. sp. *tritici*). There are also several examples where the I_A suggested the existence of cryptic sexual reproduction in fungal pathogens in species where sex has not been observed in nature, such as the human pathogens *Coccidioides immitis* (Burt et al., 1996), *Aspergillus nidulans* (Pringle et al., 2005), and the alfatoxin producing *A. flavus* (Geiser et al., 1998).

Several tests for recombination or random mating were adapted from phylogenetic methods. A popular implementation of this approach is the parsimony tree length permutation test (PTLPT; Burt et al., 1996) that tests for random mating. The statistic used is the tree length, in number of steps, and the data are allelic profiles. The rationale for using PTLPT is that asexual populations produce few short, well-resolved genealogies, whereas the contrary is expected for recombining populations (Taylor et al., 1999b). The significance of the PTLPT can be assessed using the same method and program as the I_A. Significance is calculated based on the proportion of trees in simulated datasets that are at least as long as those built from data. Other phylogenetic approaches search for the presence of recombination, based on sequence data. These tests exploit the predictions that in the absence of recombination, alleles at different regions are associated and all gene trees should therefore be congruent (Maynard-Smith, 1999) and have mutation (reversals, parallelisms, or convergence) as the sole possible cause of homoplasy (Hudson and Kaplan, 1985; Maynard-Smith and Smith, 1998). For instance, Matute et al. (2005) analyzed gene genealogies from eight regions to search for recombination in the pathogen *Paracoccidioides brasiliensis*. Incongruence among gene genealogies was examined by comparing the sum of the lengths of the most parsimonious trees inferred for each region to the sum of the length of trees obtained by permuting characters among regions (the incongruence length difference test; Farris et al., 1994). The null hypothesis of congruence for all isolated and all regions could be rejected, consistent with a lack of association among alleles, and thus with recombination. Another elegant application of methods inspired by phylogenetic analysis to search for recombination is provided by Couch et al. (2005). On the basis of a worldwide collection of isolates of the rice pathogen *Magnaporthe oryzae*, they investigated the reproductive mode of the fungus using a pairwise compatibility matrix for polymorphic sites built from the combination of nine sequence loci. A compatibility matrix is a visual representation of Hudson and Kaplan's four gamete test (Hudson and Kaplan, 1985) for all possible pairs of sites (the program SITES, available from Jody Hey's website at Rutgers University, can be used to perform such analyses). Incompatible sites are sites that support conflicting genealogies (and therefore introduce homoplasy in tree reconstructions) due to recombination or recurrent mutations. The finding of large blocks of incompatibility among loci from the same chromosome and from different chromosomes was interpreted as a sign of recombination on some, but not all, hosts of this pathogen.

In diploids or dikaryotic fungi, insights into the reproductive mode can be provided by the use of Wright's F-statistics (Halkett et al., 2005). F-statistics are hierarchical measures of the correlations of alleles within individuals and populations. A very informative F-statistic in this context is F_{IS}, a measure of the deviation from random mating. F_{IS} corresponds to the identity of alleles within individuals relative to the identity of alleles randomly drawn from two different individuals within the same population (Balloux and Lugon-Moulin, 2002; De Meeus et al., 2006). The value of F_{IS} can vary between -1 (all individuals being heterozygous) and $+1$ (all individuals being homozygous). Large negative values are expected for asexuals (Goyeau et al., 2007), and large positive values for selfers (Giraud, 2004). Several programs implement the calculation and test of F_{IS} (e.g., GENODIVE, Meirmans and Van Tienderen, 2004; GENEPOP, Rousset, 2008b; FSTAT, Goudet, 1995). For instance, the finding of F_{IS} values nonsignificantly different from zero allowed Mboup et al. (2009) to conclude to the existence of sexuality in Chinese populations of *P. striiformis* f. sp. *tritici*, a fungus showing a highly clonal population structure in other regions of the world. In another application of this approach, James et al. (2009) revealed a significant excess of heterozygous genotypes for half of the loci surveyed (i.e., $F_{IS}<0$) in worldwide samples of the amphibian-killing fungus *Batrachochytrium dendrobatidis*, suggesting a predominantly asexual mode of reproduction. Another remarkable particularity of diploids is the "Meselson effect," where the absence of sex over long evolutionary times allows alleles at a single locus to become highly divergent within individuals as the two gene copies accumulate mutations independently in the absence of recombination (Birky, 1996; Welch and Meselson, 2000). Meselson effects have been evidenced in European populations of *P. striiformis* f. sp. *tritici* (Enjalbert et al., 2002; Mboup et al., 2009) and in *Scutellospora castanea*, an arbuscular mycorrhizal (nonpathogen) fungus (Kuhn et al., 2001).

A number of methods have also been developed to estimate the population recombination rate (ρ) from haplotype data representing multiple positions in the genome (i.e., typically, moderate to large genomic dataset) (Hudson, 1987) and Wakeley (1997) developed moment estimators of the population recombination rate from the variance in pairwise differences. By making use of only a summary of the data, these methods are not computationally demanding at the expense of some loss in accuracy (Wall, 2000). Other methods use coalescent models to relate genetic variation in random population samples to the underlying recombination rate. Some approaches use conditioning on the complete dataset to obtain a maximum likelihood of the recombination rate (Griffiths and Marjoram, 1996; Kuhner et al., 2000; Nielsen, 2000; Fearnhead and Donnelly, 2001). These full likelihood methods have the advantage of making use of all of the information available in the data, but they become impractical for genomic regions of moderate size

(Fearnhead and Donnelly, 2001). Hudson (2001), Fearnhead and Donnelly (2002), and McVean et al. (2002) proposed a "composite-likelihood" approach for estimating rates of recombination for large genomic regions. The principle of the composite-likelihood approach is to calculate likelihoods for subsets of data (pairs of sites or small genomic regions) and multiplying likelihoods obtained for each subset. A promising recent development is the approximate likelihood approach developed by Li and Stephens (2003). This approach relies on a "copying model" to represent haplotypes (Davison et al., 2009) and overcome several limitations of the approaches described earlier (see Li and Stephens [2003] for details). A population genomics study has recently been conducted to gain insights into the reproductive biology of the wild yeast *Saccharomyces paradoxus*. Tsai et al. (2008) analyzed DNA sequence variation on the third chromosome (containing the mating-type locus) among 20 isolates. By comparing estimates of population size obtained from the population mutation rate θ (Watterson, 1975) and from three distinct estimates of the population recombination rate ρ (using methods of Wakeley, 1997; McVean et al., 2002; Li and Stephens, 2003), in addition to using estimates of the rates of mutation and recombination per base pair per generation from *S. cerevisiae*, the author inferred from this discrepancy that sexual reproduction occurs once every 1000–3000 generations in this species. They also estimated the frequency of intra-tetrad mating to be approximately 94%. This was accomplished by comparing values of the population recombination rate for regions located near the mating-type locus and for the whole chromosome.

4.7.2 Dispersal, Migration, and Gene Flow

Dispersal is the movement of gametes or individuals. Parameters of dispersal can be estimated by: (1) direct methods, relying on direct observation of dispersing individuals at particular life-history stages, which provides a measure of actual dispersal; or (2) by indirect methods that use the changes in some characteristics of populations caused by movement of individuals and provide a measure of effective dispersal (Slatkin, 1985; Broquet and Petit, 2009). Because the movement of individuals obviously leads to movement of genes, the study of dispersal is tightly related to the study of gene flow (direct methods) and the monitoring of particular genotypes (indirect methods). The two types of methods are treated together here. We use the term "gene flow" synonymously with "migration," as is often the case in population genetics (though migrants that do not reproduce in a new environment do not contribute to gene flow).

Gene flow can be defined as the change (in gene frequency) due to movements of gametes, individuals, or groups of individuals (Slatkin, 1987). Implications of gene flow among populations and species are so manifold that it is difficult to provide a synthetic and concise overview. Generally speaking, gene flow can be regarded as either a constraining force that prevents adaptation to local conditions or a creative force that promotes evolution by spreading new genes and combinations within and between species (Slatkin, 1985, 1987). For fungal pathogens, in practical terms, some of the most unfortunate consequences of gene flow for human affairs include immigration of genotypes capable of defeating a resistance gene, exchanges of alleles allowing resistance to antifungal molecules (and more generally the spread of variants with increased pathogenicity), increase in population size which in turns increases the probability of accumulating mutations, and increase the efficacy of selection (and the possibility of selective sweeps). The degree of gene flow is also of central importance in the formation and maintenance of pathogen species. Humans have moved many pathogens far beyond their natural dispersal limits, and it is a safe bet that many pathogens are still transported among continents today (Yarwood, 1970; McDonald and Linde, 2002). These introductions likely have set the stage for the formation of reproductively isolated populations adapting to local hosts or environments (Giraud et al., 2010) or for secondary contacts followed by introgression or hybridization among species (Stukenbrock and McDonald, 2008). Gene flow is thus a critical target for disease management tactics. The objective is dual: first to set up or maintain barriers to gene flow among fungal pathogen populations, and second to prevent the emergence of new diseases. This requires a comprehensive understanding of transmission pathways and of the processes that govern gene flow in focal species, as well as the role of gene flow in species formation or maintenance. Three different aspects can be distinguished in analyses of gene flow and dispersal: rate and direction of gene flow, dispersal distance, and distribution of gene flow in time and in the genome.

Rate and Direction of Gene Flow

Pathogenic fungal species are often organized into discrete populations. Population genetics usually assumes a simple model of n populations, each of which is equally likely to receive and give migrants to and from each of the other populations (the n-island model; Wright, 1931; Latter, 1973). Under this model, providing additional simplifying assumptions, a relationship between $N_e m_e$ (N_e being the effective size of each population; m_e being the effective migration rate between populations) and F_{ST} (a F-statistic that measures genetic differentiation among populations by quantifying the differences in allele frequencies between populations) can be derived: $F_{ST} \approx 1/(1 + 4 N_e m_e)$. The same type of relationships can be established from other measures of differentiation, using the same assumptions (Slatkin, 1985; Excoffier, 2007). This approach has been severely criticized by some authors (Bossart and Prowell, 1998; Whitlock and McCauley, 1999) who raised concerns about the unrealistic assumptions under the n-island model (constant population sizes, symmetrical migration at constant rates, no selection, and persistence for periods of time long enough to achieve migration-drift equilibrium), leaving only little quantitative information to be gained about gene flow from measures of population differentiation. Hence, although $N_e m_e$ was taken as an effective number of migrants in the original model, it is safer to interpret its value as the per-generation number of migrants that would characterize an idealized island system having the same F_{ST} value as the study system (Broquet and Petit, 2009). For the fungal pathogens. a major pitfall of estimating migration rates from allele frequency data is the assumption that the whole population has reached equilibrium between migration and genetic drift. However, many fungal pathogens have been introduced recently into new areas, or have recently invaded continents. Therefore, populations may not have had sufficient time to reach equilibrium. The time to reach a new equilibrium can be extremely long if population sizes are large and migration rates are low (Whitlock and McCauley, 1999). Even though they do not provide reliable estimates of rates of gene flow, measures of population differentiation can nonetheless be used to gain information on the history of dispersal. Several studies reported very low differentiation among samples of fungal pathogens of agricultural crops or forestry trees from different localities across a continent, including *Fusarium verticillioides* (Reynoso et al., 2009), *M. graminicola* (Linde et al., 2002), *Venturia inaequalis* (Tenzer and Gessler, 1999; Gladieux et al., 2008, 2010c), *Gibberella zeae* (Zeller et al., 2004), *Phaeosphaeria nodorum* (Keller et al., 1997), and *Melampsora larici-populina* (Barres et al., 2008). These patterns of weak population structure within continent are likely due to the superimposition of moderate to high levels of contemporary gene flow and relatively recent colonization. However, other pathogens of cultivated crops show highly significant differentiation within continents. For example, North American populations of the chickpea blight pathogen *Ascochyta rabiei* are highly differentiated, possibly due to restricted dispersal and possible selection by host cultivar combined with some life cycle characteristics conducive to differentiation (Peever et al., 2004; Giraud, 2006). High population differentiation is also a footprint of the "founder effect" that occurs following introduction in a new region. Such patterns have been evidenced for newly virulent populations of *V. inaequalis* spreading in northern France (Guérin et al., 2007), for populations of *Mycosphaerella fijiensis* spreading on banana plantations in Africa (Rivas et al., 2004), and for populations of several species that spread across continents (Salamati et al., 2000; Engelbrecht et al., 2004; Ordonez and Kolmer, 2007; Zhou et al., 2007; Zaffarano et al., 2009).

The coalescent theory (Kingman, 1982) relates patterns of common ancestry within a set of genes to the structure of the populations from which they were sampled. In coalescent models, patterns of relationships among genes are represented by a genealogy, and the structure of the population is represented by parameters such as population size, rates of population growth, or rates and directions of gene flow (as is relevant in the present section). Both the genealogy and the parameters are generally unknown, and one usually wants to estimate the parameters of the model. It is generally impossible to jointly consider all possible ancestral relationships and parameter values and to search for the combinations that maximize the probability of the model. Instead, approaches have been developed that simultaneously explore many relatively probable genealogies (loosely speaking, irrelevant genealogies are disregarded) and parameter values (see Stephens, 2008; Kuhner, 2009 for reviews). These approaches are collectively referred to as "coalescent genealogy samplers"; there are two families of such samplers: Markov chain Monte Carlo (MCMC) algorithms and important sampling algorithms. Several methods relying on coalescent genealogy samplers were designed to estimate, among other parameters, rates of gene flow between species or populations (Griffiths and Tavare, 1994; Wang et al., 1997; Beerli and Felsenstein, 1999, 2001;

Nielsen and Wakeley, 2001; Hey and Nielsen, 2004, 2007). These methods offer the advantage of allowing less restrictive models than the more traditional methods presented earlier, and thus they accommodate complexities that are typical of real populations, such as nonsymmetrical gene flow. These methods have been successfully applied to infer the ancestral routes of colonization for several fungal globally distributed plant pathogens such as the wheat pathogens *M. graminicola* (Banke and McDonald, 2005) and *P. nodorum* (Stukenbrock et al., 2006), the barley scald pathogen *Rhynchosporium secalis* (Brunner et al., 2007; Zaffarano et al., 2009), and the apple scab pathogen *V. inaequalis* (Gladieux et al., 2008). Programs using coalescent genealogy samplers can also be applied to assess gene flow between well-defined species. The isolation-with-migration model implemented in the IMa programs (Hey and Nielsen, 2007) was used, for instance, to demonstrate unidirectional gene flow from *Microbotryum lychnidis-dioicae* into *M. silenes-dioicae*, respectively anther smut pathogens of the white and red campions (Gladieux et al., 2010b).

Additional information on historical migration routes can also be retrieved by examining the relationship between patterns of diversity and the putative history of introduction of the species in different regions from a source population. Under a model of serial founder effects, genetic variation is expected to decrease steadily from the earliest to the latest populations formed along the colonization route (Austerlitz et al., 1997; Ramachandran et al., 2005; Linz et al., 2007; Szpiech et al., 2008). The existence of such patterns has been uncovered for *V. inaequalis*, in which allelic richness tends to be lower in regions where apple has been introduced more recently, suggesting that the pathogen tracked its host during the spread of apple cultivation worldwide (Gladieux et al., 2008).

Methods based on coalescent genealogy samplers remain computationally demanding. For many datasets and models of population structure, they even remain computationally intractable. As a result, there is an increasing interest in developing alternative approaches that are faster and easier to implement, without loosing too much accuracy (Marjoram and Tavare, 2006; Stephens, 2008; Nielsen and Beaumont, 2009). The most promising approaches are rejection sampling and approximate Bayesian computation (Tavare et al., 1997; Li and Fu, 1999; Pritchard et al., 1999; Beaumont et al., 2002), composite-likelihood methods (Hudson, 2001), and product of approximate conditionals (Li and Stephens, 2003; Davison et al., 2009). These methods also potentially offer the advantage of being more easily tailored to the specificities of fungal pathogens, such as for life cycles with both sexual and asexual modes of reproduction, or histories of sequential introduction with exchanges of migrants among neighboring populations. Approximate Bayesian computation has been shown to be particularly powerful to determine the origin and routes of introduction of invading pest species (Miller et al., 2005; Cornuet et al., 2008; Guillemaud et al., 2010), and it is very likely that it will also provide important insights into the history of fungal pathogens.

A number of approaches have been developed for inferring recent gene flow by extracting information from the transient disequilibrium observed at individual multilocus genotypes of migrants or their recent descendants. These methods can serve as direct estimators of recent migration (Paetkau et al., 1995; Rannala and

Mountain, 1997; Cornuet et al., 1999; Pritchard et al., 2000; Dawson and Belkhir, 2001; Anderson and Thompson, 2002; Gaggiotti et al., 2002; Gao et al., 2007) or even rates of recent gene flow (Wilson and Rannala, 2003). More details on these methods are given in Section 4.7.3. Such an approach was used, for instance, to estimate the frequency of cross host-species disease transmission and hybridization between two species of anther smut fungi (Gladieux et al., 2010b), or to demonstrate intercontinental dispersal and admixture of populations of the destructive dry rot fungus *Serpula lacrymans*, possibly linked to the development of worldwide shipping activity with wooden sea vessels (Kauserud et al., 2007). For pathogens with very little genetic variation and also with life cycles violating assumptions of the above-cited methods, approaches based on measures of the percentage of identity between isolates from recently colonized areas and putative source populations can also provide an easy-to-use and rapid method for the identification of migration routes. Hovmoller et al. (2008) used this approach, in combination with pathotyping, to demonstrate the foreign incursion of particular strains of the wheat yellow rust fungus in regions of North America, Australia, and Europe where severe epidemics have been observed in recent years.

Dispersal Distance

There is considerable interest in estimating the distance fungal pathogens disperse at agriculturally relevant scales, such as fields or production areas. This information can be inferred from patterns of genetic variation by fitting a model of isolation by distance (Wright, 1943, 1946; Kimura, 1953). A general formulation of isolation by distance models is the infinite lattice model (Malécot, 1951), in which individuals or populations are distributed on a lattice with spatially homogenous demographic parameters (i.e., homogenous population sizes or density and dispersal; Broquet and Petit, 2009; Guillot et al., 2009). The slope of the regression of differentiation statistics (e.g., F_{ST}) onto the log-transformed geographic distance among individuals or populations allows estimation of the product of D, the population density, and σ^2, the second moment of dispersal distance (Rousset, 1997, 2000, 2008a; Vekemans and Hardy, 2004). For fungal pathogens that alternate asexual and sexual reproduction during their life cycle, these methods are not suitable due to the occurrence of repeated genotypes (Dutech et al., 2008). This specificity must be considered for correct inference of dispersal distance. Wagner et al. (2005) developed a weighting procedure that retains the spatial positions of all individuals but also applies a weighting to each genotype inversely proportional to its frequency. They also showed that variograms (i.e., plots of the semivariance in number of differences between genotypes against distance) are efficient tools to estimate the degree and extent of spatial genetic structure accounting for autocorrelation (which is the tendency that nearby observations to be more similar than distant ones). Variograms were used to study dispersal in the chestnut blight fungus (*Cryphonectria parasitica*), showing that asexual spores probably disperse over several hundred meters, which is a far larger spatial scale than previously thought (Dutech et al., 2008).

Distribution of Gene Flow in Time and Along the Genome

The coalescent-based implementation of the isolation-with-migration model in the IM and IMa program (Nielsen and Wakeley, 2001; Hey and Nielsen, 2004, 2007) offers the opportunity to gain valuable insights into the history of gene flow between species. An interesting feature of the program is that counts and dates of migration events in sampled genealogies can be recorded during the course of the MCMC at stationarity for each locus to obtain the migration time distribution. IM was used to demonstrate that the wheat pathogen *M. graminicola* emerged in the Fertile Crescent at the time of wheat domestication following a series of introgressions from populations infecting three different uncultivated grasses (Stukenbrock et al., 2007). Estimates of the time of gene flow events indicated that populations from wheat and uncultivated grasses diverged in the face of gene flow but are now genetically isolated. This approach was also used to show that the species of anther smut fungi *M. lychnidis-dioicae* and *M. silenes-dioicae* initially diverged in allopatry without gene flow and exchanged genes only recently following secondary contact (Gladieux et al., 2010b). Gene flow between the two species of *Microbotryum* appeared restricted to four loci, supporting the view that genomes are mosaics with respect to interspecific gene flow, with some regions more or less permeable to genetic exchanges (Wu and Ting, 2004).

4.7.3 Population Subdivision

Fungal pathogens, like all organisms, are not homogenously distributed across the environment, which can lead to genetic structure. There are two main sources of population subdivision in fungal pathogens: geography and hosts. While some species have very broad host ranges (e.g., the amphibian pathogen *B. dendrobatidis*, >350 host species, Fisher et al., 2009; or the gray mold *B. cinerea*, >235 host species, Fournier et al., 2005; Staats et al., 2005), others display clear subdivisions that correspond to the host of origin of populations (e.g., *Verticillium dahliae*, Atallah et al., 2010; *V. inaequalis*, Gladieux et al., 2010a, or *A. rabiei*, Frenkel et al., 2010). Such host-specific divergence may evolve as a consequence of limited dispersal or of trade-offs in adaptation (Timms and Read, 1999; Giraud et al., 2006). Among pathogen species found on a single host, some species display clear geographically distinct populations (e.g., the mammalian pathogen *Histoplasma capsulatum*, Kasuga et al., 2003); or the white campion smut *M. lychnidis-dioicae*, Vercken et al., 2010), while others appear to have global distributions such as the human pathogen *A. fumigates* (Pringle et al., 2005; Rydholm et al., 2006). These patterns of geographical subdivision result from a complex interplay between contemporary and historical gene flow processes.

Understanding the origin of population subdivision is fundamental to our knowledge of the mechanisms responsible both for disease emergence and for the biodiversity of fungi. Four main approaches are available to analyze population subdivision: measures of differentiation, evolutionary trees, multivariate methods, and model-based clustering algorithms.

Measures of Differentiation

Population subdivision can be assessed by calculating differentiation indices (e.g., F_{ST}) between pairs of populations (see Section 4.7.2 for a detailed discussion). A more general framework to study population subdivision has been developed by Cockerham (1969; 1973), who introduced the use of the analysis of variance (ANOVA) framework to decompose the total variance of gene frequencies into variance components associated to different subdivision levels (reviewed in Excoffier, 2007). Later, Cockerham's ANOVA of gene frequencies was extended to include information conveyed by the amount of differences (mutations) between alleles (the analysis of molecular variance, AMOVA; Excoffier et al., 1992). The AMOVA framework is widely used in population genetics studies of fungal pathogens. It is implemented in the ARLEQUIN (Excoffier et al., 2005) and GENETIC STUDIO packages (Dyer, 2009). The principle is to summarize population differentiation into F-statistics by partitioning molecular variance among the different hierarchically nested levels of sampling represented in a dataset (which can be localities, host species, regions, continents, etc.). The main drawback of this procedure is that the sampling units must be assigned into given hierarchical subdivisions by investigators, which may be a relevant issue. The main advantage is that it is a very fast way to get a summary representation of the differentiation existing among the different assumed levels of subdivision within a species. A useful application of the AMOVA procedure is for instance to determine whether the most important source of differentiation within a species is the host or the region of origin (e.g., see Morgan et al., 2007 or Gladieux et al., 2010c).

The issue of defining an a priori model of population arrangement can be sidestepped by using a multivariate graph-theoretic approach called population graphs (Dyer and Nason, 2004). The principle is to measure the genetic covariance relationships among all sampling units simultaneously and to represent these relationships graphically. Examples of application to fungal pathogens can be found in Guérin et al. (2007) and Fournier and Giraud (2008). Population graphs can be built using GENETIC STUDIO (Dyer, 2009).

Evolutionary Trees

The most traditional approach to track population subdivision from genetic data is to build an evolutionary tree. Such trees are often improperly called "phylogenetic trees," though a phylogeny describes the pattern of ancestry among species, rather the pattern of genetic ancestry among pieces of DNA sampled within a species or a set of closely related species (Hey and Machado, 2003). Two main classes of evolutionary tree construction methods are available: (1) clustering methods use an iterative method (e.g., neighbor-joining) to combine samples in a hierarchical fashion, (2) searching methods that consider a range of possible trees and choose the ones that best fit the data according to an optimality criterion (such as maximum parsimony, maximum likelihood, or maximum Bayesian probability; Holder and Lewis, 2003). In practice, clustering methods are often used for data from multiple loci (typically, microsatellite markers) summarized by a matrix of distances among samples (Kalinowski, 2002), while searching methods are used for data from individual loci (typically, sequence data). Note that a very useful application of evolutionary trees based on sequence data is for species identification, but this approach is developed in Section 4.8, and we therefore focus on intraspecific subdivision here.

Evolutionary trees are appealing because they provide a graphical representation of the relationships among samples (Hey and Machado, 2003). When constructed from multilocus data, evolutionary trees can be very useful for exploratory data analysis or for visualizing the main subdivisions within a dataset. When interpreting an evolutionary tree, there are two main reasons to be cautious: (1) the stochastic variance in evolutionary trees (the problem being greater for evolutionary trees based on a single locus), and (2) the inadequacy of a bifurcating model when applied at the intraspecific level. The stochastic variance in evolutionary trees is due to the fact that different loci that have passed through the same demographic history, leading to evolutionary trees that vary widely in topology and branch lengths (Hey and Machado, 2003). The cause of this variance is that the processes that produce treelike relationships among gene copies (i.e., birth, death, and Mendelian reproduction, in a neutral model) and mutations are stochastic (Felsenstein, 2007). The other potential issue is that bifurcating models may not be appropriate to represent relationships at the intraspecific level. Several phenomena lead to violation of the assumptions underlying reconstruction methods and lead to poor resolution or inadequately portray genealogic relationships (reviewed in Posada and Crandall, 2001). These phenomena include low divergence among alleles, the persistence of an ancestral allele together with its descendants within a population, the existence of multiple descendants for a single allele, recombination among alleles through crossing-over, and exchanges of alleles (gene flow) between lineages. An alternative to tree-based approaches for representing relationships among samples is to use a network. Several methods of network reconstruction have been developed. Networks offer the advantage over evolutionary trees of being able to incorporate persistent ancestral nodes, multifurcations, and reticulations (Posada and Crandall, 2001). Examples in fungal pathogens can be found in Morgan et al. (2007) or Couch et al. (2005).

Another caveat for the use of evolutionary trees concerns the so-called population trees. The distance among populations is often represented as a tree in studies analyzing the population structure of fungal pathogens. However, this approach is questionable, as there is no reason to think that a model in which populations split from a common ancestor and subsequently do not mix represents the reality of population evolution.

Model-Based Bayesian Clustering Algorithms

The aim of model-based Bayesian clustering algorithms (or assignment methods) is to infer groups of individuals (called clusters or populations) that fit some genetic criteria that define them as distinct groups (Guillot et al., 2009). The use of a clustering method is an almost unavoidable step in every population genetic study. This field has been flourishing for a decade, and we will not give an extensive description of all the methods currently available. The most popular program is STRUCTURE (Pritchard et al., 2000; Falush et al., 2003). The method assumes a model in which there are K clusters, each of which is characterized by a set of allele frequencies. Assuming Hardy–Weinberg and linkage equilibrium within clusters, the program simultaneously estimates allele frequencies in each cluster and then assigns every individual probabilistically to a single cluster ("no-admixture model"), or estimates the proportion of ancestry of every individual in every cluster ("admixture model"). It uses a variant of MCMC to approximate the probabilities of assigning individuals to clusters or membership proportions. The method has also been modified to allow for linkage of the loci (Falush et al., 2003). When using STRUCTURE, one usually wants to determine the number of clusters that is optimal for describing the population structure. The program does not estimate such an optimal number of clusters, but a heuristic method for selecting K can be used, based on the rate of change in the log-probability of data between successive K value (Evanno et al., 2005). Other Bayesian clustering programs can be used to obtain estimates of the optimal number of clusters based on various statistical methods (Dawson and Belkhir, 2001; Pella and Masuda, 2006; Corander and Tang, 2007; Huelsenbeck and Andolfatto, 2007), some of which are subject to debate (Durand et al., 2009; Guillot, 2009a,b). However, the biologic interpretation of any "best K" estimate may not be straightforward (Pritchard et al., 2007) and should not be taken at face value. This is all the more true as there may be several different relevant K numbers, in particular if the population structure is hierarchical. The best approach is therefore to provide a representation of several K values and not a single "optimal" one. Departures from the structure of the model due to isolation by distance or inbreeding can lead to spurious signals of population structure and artificially increase the number of inferred clusters (Gao et al., 2007; Guillot et al., 2009). This issue is partially alleviated when using the INSTRUCT program (Gao et al., 2007), an extension of the approach implemented in STRUCTURE that eliminated the assumption of Hardy–Weinberg equilibrium within clusters and instead estimates individual membership on the basis of inbreeding or selfing rates. Last, we note that in case of poorly informative datasets (too few loci or individuals, or not enough differentiation among populations), group information such as host or region of origin of samples or geographic coordinates can be incorporated to achieve better results in analyzes of population subdivision (briefly reviewed in Hubisz et al., 2009).

Multivariate Methods

The principle of multivariate analyses, when applied to genetic variation among individuals or populations, is to extract and summarize multivariate genetic information into a few synthetic variables (Jombart et al., 2009). Methods such as principal component analysis have long been applied to population genetics questions (Cavalli-Sforza, 1966). They benefit from a renewed interest thanks to recent results of theoretical statistics and the development of software packages specifically devoted to the multivariate analysis of genetic data (Patterson et al., 2006; Jombart, 2008). Multivariate methods offer three main advantages. A first advantage is that they perform much faster than methods that are based on evolutionary trees or Bayesian clustering algorithms. The speed of analytical tools will become an increasingly important criterion of choice with the development of population genomics datasets of hundreds of markers for hundreds of individuals. A second advantage is that these methods make no assumption of population structure, such as Hardy—Weinberg or linkage equilibrium. This can be particularly useful for fungal pathogens with asexual or partially asexual modes of reproduction, for which Bayesian clustering algorithms present a high risk of producing spurious assignments (Falush et al., 2003). A principal component analysis was applied to investigate the origin of French populations of the chestnut blight fungus, a species in which high rates of asexual reproduction and may be also of intrahaploid sexual reproduction (allowed by homothallism) result in high frequencies of repeated multilocus genotypes (Dutech et al., 2009). Analyses revealed three distinct genetic lineages with separate geographic distributions, suggesting independent introduction events with limited gene flow among lineages descending from the three original groups of founding strains.

4.7.4 Conclusion

Empirical population genetics studies have revolutionized our understanding of fungal pathogen evolutionary biology. The distribution range of pathogens (in space and on hosts), their reproductive system, and transmission pathways are crucial features of pathogen biology that would have remained inaccessible based solely on phenotypic data and without the powerful inferential framework of population genetics. How could have we showed that "everything is not everywhere" and that many broadly distributed fungal pathogens are actually subdivided into populations constrained to small geographical areas? How could have we known that only very few fungal pathogens are ancient strictly asexual species and that the deuteromycota do not constitute a formal phylum of fungi? The upcoming flood of genomic data should galvanize investigations on central topics such as the evolution of reproductive systems (Heitman et al., 2007; Billiard et al., 2010), the acquisition of virulence to new hosts, resistance to disease control strategies (Stukenbrock and McDonald, 2008; Hogenhout et al., 2009; Morris et al., 2009; Giraud et al., 2010), and the evolution of reproductive isolation (Kohn, 2005; Taylor et al., 2006; Giraud et al., 2008b). However, this technologic leap should be accompanied by the development of population genetic models tailored to the specificities of fungal pathogens, such as the possibility of complex life cycles or nonpanmictic mating systems. There is still much to discover using population genetics. Most existing work has been focused on fungal species causing disease of humans, agricultural crops, or domesticated animals in the developed world (Taylor and Fisher, 2003; Morris et al., 2009). Studies are needed of pathogens from wild species in natural settings, from developing areas, and of nonpathogenic species that might be the pathogens of the future.

4.8 Species and Speciation in Pathogenic Fungi

Understanding how the 1.5 million fungal species estimated to exist (Hawksworth, 1991) have arisen is of fundamental interest and has tremendous applied consequences for understanding emergent diseases on plants and animals (Giraud et al., 2010). We will briefly summarize here the main recent advances on fungal speciation, but more extensive reviews on speciation and species recognition in fungi have been written elsewhere (Taylor et al., 2000; Kohn, 2005; Giraud et al., 2008b).

4.8.1 Species Concept Versus Species Criteria

To study speciation, an obvious first step is to define species. The continual proposal of new species concepts may lead one to think that there is no general agreement about what species are. The apparently endless dispute about species concepts stems from the confusion between a species definition (describing the kind of entity that is a species) and species criteria (standard for judging or recognizing whether individuals should be considered members of the same species). Many so-called species concepts actually correspond to species criteria (i.e., practical means to recognize and delimit species) (Taylor et al., 2000; Hey, 2006; De Queiroz, 2007). The biological species concept (BSC) for instance emphasizes intersterility, the morphological species concept (MSC) emphasizes morphologic divergence, the ecological species concept (ESC) emphasizes adaptation to a particular ecologic niche, and the phylogenetic species concept (PSC) emphasizes nucleotide divergence. These species criteria correspond to the different events that occur during lineage separation and divergence, rather than to fundamental differences in what is considered to represent a species. To the contrary, it has been argued that all modern biologists agree on a common "species concept" or "species definition" that would be segments of evolutionary lineages that evolve independently from one another (de Queiroz, 1998).

One may wonder why there are conflicts over which species criterion we adopt. In fact, there are three main reasons why such criteria cannot be universal: (1) speciation is a temporally extended process, but one which varies tremendously in its pace among different types of organisms; (2) several modes of speciation can occur, during which the phenomena used for species recognition do not necessarily appear in the same chronologic order (Figure 4.1); and (3) characteristics of certain organisms render some criteria difficult to apply. The most useful criterion to apply to recognize species in nature thus necessarily depends on the type of organism, on its history of speciation, and on the degree of achieved divergence. Searching for a single species criterion that would be applicable to all cases thus appears fundamentally hopeless.

Until quite recently, the most commonly used species criterion for fungi has been the MSC. However, many cryptic species have been discovered within morphologic species, using the BSC (Anderson and Ullrich, 1978), or the genealogical concordance phylogenetic species recognition (GCPSR) (Taylor et al., 2000), an extension of the PSC. This latter species criterion uses the phylogenetic

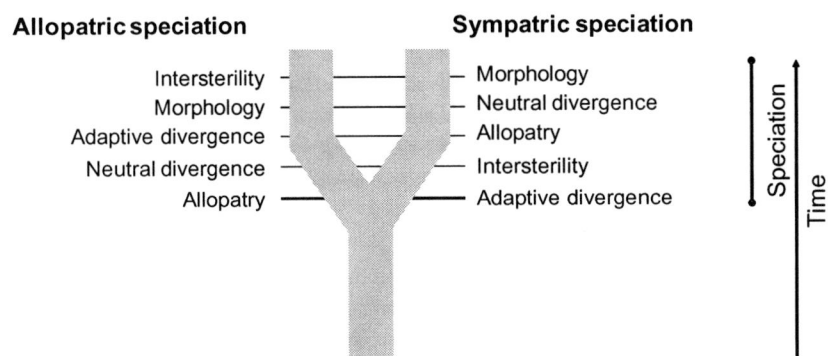

Figure 4.1 Schematic divergence of two species, in two hypothetical cases of respectively allopatric and sympatric speciation, with the progressive appearance of various criteria traditionally used to recognize species.

concordance of multiple unlinked genes to indicate a lack of genetic exchange, and thus evolutionary independence of lineages. Species can thus be identified that cannot be recognized using other species criteria due to the lack of morphological characters or incomplete intersterility. The GCPSR criterion has proved immensely useful in fungi, because it is more finely discriminating than the other criteria in many cases, or more convenient (e.g., for species that we are not able to cross), and is currently the most widely used within the fungal kingdom (Koufopanou et al., 2001; Dettman et al., 2003a; Fournier et al., 2005; Johnson et al., 2005; Pringle et al., 2005; Le Gac et al., 2007a).

4.8.2 Fungal Speciation

How new species arise in nature is still a highly active field of research. It has long been believed that species originate mostly through allopatric divergence (Mayr, 1963), because extrinsic geographic barriers seemed obvious impediments to gene flow. Fungi could appear as exceptions because eukaryotic microorganisms have long been considered to have global geographic ranges (ubiquitous dispersal hypothesis; Finlay, 2002), at least for those not dependent on a host having a restricted range. This was in particular true for airborne fungal pathogens because their spores can be dispersed over a very long distance (Brown and Hovmoller, 2002). Among the numerous complexes of sibling species recently uncovered using the GCPSR criterion, many however appear consistent with allopatric divergence because the cryptic species occupy nonoverlapping areas separated by geographic barriers (Taylor et al., 2006). This is the case for the species complexes of the model organism *Neurospora crassa* (Dettman et al., 2003a), the yeast *Saccharomyces paradoxus* (Kuehne et al., 2007), the plant pathogen *Fusarium graminearum* (O'Donnell et al., 2004), and the mushrooms *Schizophyllum commune* (James et al., 1999) and *Armillaria mellea* (Anderson et al., 1980, 1989).

In contrast to the wide acceptance of allopatric speciation, the possibility of sympatric speciation in sexual populations had long been dismissed. This is because recombination between different subsets of a population that are adapting to different resources or habitats counteracts natural selection for locally adapted gene combinations (Felsensein, 1981; Rice, 1984). Recombination indeed prevents both the building of linkage disequilibrium between adaptive alleles at different loci and divergence at loci not under disruptive selection.

Theoretical models have shown that the simplest way to eliminate the role of recombination in breaking down the effects of selection, and thereby allow sympatric speciation, is to have the same gene(s) controlling pleiotropically both enhanced fitness in a specialized habitat and assortative mating (mate choice, i.e., prezygotic isolation) or both fitness and habitat choice if mating takes place within habitats (Rice, 1984). Such "magic traits" (Gavrilets, 2004) have, however, proved difficult to find in nature. Another way to reduce recombination between two populations specialized on different niches is to build up an association (linkage disequilibrium) between habitat-based fitness genes and either assortative mating genes and/or habitat choice genes if mating is restricted within habitats (Dickinson and Antonovics, 1973; Johnson et al., 1996). Theoretical models have shown that this is plausible under certain conditions, although the limitations to the process are far from trivial.

Fungi are passively dispersed and cannot actively choose the habitat in which they will grow, but for many fungal species sex must occur in the habitat after mycelial development (e.g., on or within the host for fungal pathogens). A recent model has shown that, due to this important characteristic of the lifestyle (inability to disperse between development on the host or habitat and mating), mutations providing adaptation on a new habitat can affect pleiotropically both the fitness on the habitat and the ability to mate in this habitat. Adaptation to a new habitat can thus be sufficient to restrict gene flow in sympatry in fungi for which mating occurs within their specialized host or habitat, without requiring active assortative mating (i.e., prezygotic intersterility) (Giraud, 2006; Giraud et al., 2006). Specialization would act in these fungi as a "magic trait" (Gavrilets, 2004), pleiotropically allowing both adaptation to the new host or habitat and reproductive isolation, thus facilitating sympatric speciation (Giraud et al., 2010). This mechanism where a single gene controls both host adaptation and assortative mating is different from that seen in phytophagous insects mating onto their host plants, and where a linkage disequilibrium has to evolve between a gene controlling host choice and a gene controlling host adaptation (Johnson et al., 1996). The mechanism of speciation by host adaptation causing pleiotropically assortative mating is closer to the recently proposed "reduced viability of immigrants" barrier to gene flow (Nosil et al., 2005; Giraud, 2006; Giraud et al., 2006).

Sympatry is often said to be difficult to define for microorganisms and pathogens. For instance, pathogens specialized on different sympatric hosts are sometimes considered allopatric (Huyse et al., 2005). A simple, widely applicable definition of sympatry is available: "In sympatry, the probability of mating between two individuals depends only on their genotypes," (Kondrashov, 1986) and not on extrinsic barriers. Two populations of pathogens adapted to different hosts but able to disperse to both hosts are therefore truly in sympatry: only their genes responsible for host adaptation will prevent them meeting within their respective hosts to mate (Giraud et al., 2006).

Compelling evidence for the sympatric divergence is extremely difficult to provide because excluding a past period of allopatry is almost always impossible (Coyne and Orr, 2004, p. 142). Even the most famous candidate cases are still debated, such as the phytophagous insect *Rhagoletis pomonella* (Coyne and Orr, 2004; Feder et al., 2005, pp. 159–162) or the cichlid fishes in African lakes (Coyne and Orr, 2004, pp. 145–154). Evidence consistent with sympatric divergence of fungal populations driven by parasitic adaptation to different hosts has, however, been reported.

An example is provided by *Ascochyta* pathogens, where recent multilocus phylogenetic analyses of a worldwide sample of *Ascochyta* fungi causing blights of chickpea, faba bean, lentil, and pea have revealed that fungi causing disease on each of these hosts form distinct species (Peever, 2007). Experimental inoculations demonstrated that infection was highly host-specific, yet in vitro crosses showed that the species were completely interfertile. The host specificity of these fungi may therefore constitute a strong reproductive barrier, and the sole one (Peever, 2007), following a mechanism of sympatric divergence by host usage (Giraud, 2006; Giraud et al., 2006). More generally, there exist many close species of ascomycete pathogens that are sympatric but isolated by weak intersterility barriers (Le Gac and Giraud, 2008). The coexistence in sympatry of interfertile populations, specialized on different hosts, and that remain genetically differentiated cannot indeed be explained currently by models other than the reduced viability of immigrants (Nosil et al., 2005; Giraud, 2006; Giraud et al., 2006). This mechanism seems to be able to maintain the species differentiated in sympatry and could similarly have created the divergence in sympatry. It is difficult to exclude a period of allopatry in the past that would have facilitated specialization, such as the accumulation of different alleles beneficial on alternate hosts, as has been proposed for the well-studied case of the phytophagous insect *R. pomonella* (Coyne and Orr, 2004, pp. 159–162).

An elegant way to demonstrate the sympatric occurrence of speciation is to show that gene flow has occurred after initial divergence using approaches based on coalescence (Wu and Ting, 2004). Such approaches appear very promising and have been used in the fungal plant pathogen *Mycosphaerella graminicola*, showing that this wheat pathogen arose recently, most probably during wheat domestication in the fertile crescent, by differentiation from *Mycosphaerella* species pathogens of natural grasses in the face of gene flow (Stukenbrock et al., 2007).

Many examples exist in fungi of divergence of sibling pathogen species on different hosts. In disagreement with the long-prevailing view that coevolution between host and pathogens should lead to cospeciation, these radiations of different hosts most often involved hosts' shifts (Roy, 1998; Refrégier et al., 2008; Tellier et al., 2010). This also seems the case for some fungi causing diseases on crop plants, having arisen from shifts from natural plants (Couch et al., 2005; Brunner et al., 2007; Stukenbrock and McDonald, 2008; Zaffarano et al., 2008).

As seen earlier, a sine qua non of speciation in sexually reproducing organisms is the decrease of gene flow between incipient species due to the development of reproductive barriers. Two types of reproductive barriers are usually distinguished, premating and postmating. Premating isolation has been shown to prevent gene flow in fungi, with different types of barriers: (1) for organisms depending on biotic vectors, specialization of these vectors can prevent contact between two populations even if they lie close to one another, yielding ecological isolation. For example, the endophyte *Epichloë typhina* is preferentially chosen by its fly vectors *Botanophila* as opposed to *Epichloë clarkia* (Bultman and Leuchtmann, 2003), which may promote a certain degree of reproductive isolation. Another example is the complex *Microbotryum violaceum*, where the insect vectors are different to some extent between host species, leading to a reduction in mating opportunities among strains from different plants, although the barrier is not complete (van Putten et al., 2007); (2) specialization may also allow for ecological premating isolation if mating occurs within habitats (hosts for pathogens), as discussed earlier (Giraud, 2006; Giraud et al., 2006); (3) allochrony, or differences in the time of reproduction, may also be efficient to promote premating isolation. The sister species *Saccharomyces cerevisiae* and *S. paradoxus* exhibit different cell growth kinetics; this allows most individuals of one species to undergo homospecific crosses before or after reproduction of the individuals of the other species. Proportion of interspecific matings can therefore be significantly reduced without the need of incompatibility factors (Murphy et al., 2006); (4) as has been invoked in plants (Fishman and Wyatt, 1999), a high rate of selfing may be efficient in limiting interspecific matings. Selfing has been suggested to act as a reproductive barrier in the anther smut fungus *M. violaceum* (Giraud et al., 2008c); and (5) assortative mating due to mate recognition occurs if individuals or gametes are able to discriminate between conspecifics and heterospecifics. Assortative mating seems to be especially important in the reproductive isolation of *Homobasidiomycota*, where clamp connections between mycelia of opposite types are almost exclusively observed when the tested mycelia belong to the same species when in sympatry (Le Gac and Giraud, 2008).

Postmating isolation refers to barriers associated with hybrid inviability and sterility and is expected to arise as a result of the divergence of incipient species. In the case of postmating isolation, heterospecific crosses occur and lead to the production of unfit offspring. Hybrids may be inviable or sterile due to genetic incompatibilities if mutations fixed independently in the diverging lineages display negative epistatic interactions when brought together in the same individual, a phenomenon known as Dobzhansky—Müller incompatibilities (Orr and Turelli, 2001). This kind of intrinsic postmating reproductive isolation is responsible for the numerous reported cases in fungi of crosses that initiate and subsequently abort during in vitro experiments. For instance, heterospecific crosses among *Microbotryum* species produce in vitro fewer viable mycelia than conspecific ones (Le Gac et al., 2007b), and hybrids are sterile (Sloan et al., 2008; de Vienne et al., 2009); crosses among *Neurospora* species lead to few or abnormal perithecia or to few viable ascospores (Dettman et al., 2003b). Dobzhansky—Müller incompatibilities have been identified between a nuclear gene and a mitochondrial gene as causing sterility in hybrids between two yeast species (Lee et al., 2008). Chromosomal rearrangements have also been proposed as mechanisms generating sterility in fungi given their high genomic fluidity (Zolan, 1995; Poggeler et al., 2000; Delneri et al., 2003; de Vienne et al., 2009). However, the karyotypically differentiated strains may also have differed in their genic content, and showing that the rearrangements were a cause of the divergence and not only its consequence remains a challenging task.

Speciation in fungi can also occur by hybridization between species (see Brasier, 2001; Olson and Stenlid, 2002; Schardl and Craven, 2003; Giraud et al., 2008b; Aguileta et al., 2009 for extensive reviews). Allopolyploid hybrids (with higher ploidy level than the parental lines) have been identified in diverse genera. Examples include *Botrytis allii*, the agent of gray mold neck rot of onion and garlic (Staats et al., 2005), several *Neotyphodium* species, symbiotic endophytes of grasses (Moon et al., 2004), and several *Saccharomyces* species empirically selected for brewing (Masneuf et al., 1998). Homoploid speciation (with no change in chromosome number) has also been described in fungi. A well-known case is that of the rust *Melampsora* × *columbiana* that emerged from hybridization of *M. medusa*, pathogen of *Populus deltoides*, and *M. occidentalis*, pathogen of *P. trichocarpa* (Newcombe et al., 2000), and that is able to parasitize the hybrid trees.

In conclusion, important advances have been made recently on the speciation in fungi, and they have proved tractable biologic models for the general study of speciation. Fungi also exhibit some specific and interesting modes of speciation, and many open questions remain which will be fascinating to explore.

4.9 Mating and Pathogenesis

Within the fungal kingdom, the ability to undergo sexual reproduction is linked to pathogenesis from a number of different aspects. The four major links to pathogenesis include the ability to form infectious spores, the production of invasive hyphae, the generation of genetic diversity through meiotic recombination, and in some species an association of mating type with virulence.

Sexual reproduction in many fungal pathogens leads to the formation of infectious spores. Spores are one of the likely routes of infection for human pathogens, including *Cryptococcus*, and have recently been documented to be infectious like yeast cells (Giles et al., 2009; Velagapudi et al., 2009). In representative plant pathogens including the *Ustilago* and *Microbotryum* genera, sexual reproduction is stimulated in association with the host plant and the dikaryotic hyphae produced by mating are the infectious form, thus requiring sexual reproduction for successful infections (Bakkeren et al., 2008). In *Cryptococcus* (Xue, 2010), it has been shown that a full sexual cycle, leading to the production of infectious spores, can be completed in association with plants, in some cases triggering hyphal penetration and disease in leaves and triggering jasmonate-mediated plant defenses, further linking the animal and plant fungal pathogens (Xue et al., 2007). Following these studies, recent work has shown that the inositol permease family is expanded in *Cryptococcus* (Xue et al., 2010), and inositol stimulates mating and can be supplied by the plant. These findings in a plant-associated human pathogen complement analogous mechanisms defined in well-characterized plant pathogens. For example, a recent study described a novel sucrose transporter in *U. maydis* that is required for virulence, which allows the pathogen to scavenge sucrose from the host while avoiding immune recognition by enabling direct carbon source utilization without the production of extracellular monosaccharides (Wahl et al., 2010). In contrast to the plant pathogenic basidiomycetes, the infectious form of *Cryptococcus* in animals is the yeast cells, and hyphal forms are only rarely observed in tissue sections from infected patients and animals.

Sex can create diversity by enabling genetic exchange within a population of a given species, but can also impact the evolution of virulence via introgression of more limited genomic regions between closely related species (Kavanaugh et al., 2006) or hybridization between two different species (Cogliati et al., 2001; Lengeler et al., 2001). There are a series of reports on hybridization events, likely that occurred via sexual reproduction between distinct species, which have resulted in isolates with altered virulence properties in plant and insect pathogens, including both fungi and oomycetes (Viaud et al., 1998; Olson and Stenlid, 2002; Schardl and Craven, 2003). In these cases, hybrid vigor may be the cause of altered host range, environmental niche, or virulence properties. Studies in plant pathogens have also found that hybridizations between species can lead to mitochondrial transfer, impacting virulence (Olson and Stenlid, 2001). These mitochondrial findings have garnered interest from the *Cryptococcus* field, as recent studies of the *C. gattii* outbreak genotypes indicate a strong association between the mitochondrial genome and virulence, although studies in *C. neoformans* found no link associated with the exchange of serotype A and D mitochondrial genomes and virulence (Toffaletti et al., 2004; Ma et al., 2009).

While the mammalian fungal pathogens do not require sex for infections, sexual reproduction has been retained, and in a number of cases, shown to be critical for aspects of successful infections. A recent seminal discovery of sexual reproduction in *Aspergillus fumigatus* reported that sex occurred only after 6 months of incubation on oatmeal supplemented agar, and may lead to the production of aerosolized resistant infectious spores (O'Gorman et al., 2009). Through whole genome analysis, it was revealed that the dandruff causing basidiomycete *Malassezia* has retained many of the genes related to mating and meiosis, including the mating-type locus, leading to the current hypothesis that this organism may complete a sexual cycle in association with human skin (Xu et al., 2007). Similar bioinformatics studies in the ascomycete dermatophytes, the dimorphic pathogens, and an examination of *Pneumocystis* also found evidence for mating-type loci and meiotic genes, indicating likely roles of sexual reproduction in these human pathogens and possibly occurring with a commensal or infectious state in humans (Smulian et al., 2001; Bubnick and Smulian, 2007; Fraser et al., 2007; Mandel et al., 2007; Burgess et al., 2008, 2009; Li et al., 2010). In the case of *Pneumocystis*, a group of obligate pathogenic species that only proliferate in the lungs of their infected hosts and which are highly species-specific, pathologic studies have suggested that sexual reproduction may be occurring in the lung of the infected host. Genomic studies have revealed genes encoding meiotic gene homologs, and when heterologously expressed in the closely related fission yeast *Schizosaccharomyces pombe*, these genes can complement to restore meiotic function in mutants lacking the *S. pombe* ortholog (Burgess et al., 2008). This suggests the potential capacity for an extant sexual cycle that may remain to be defined in the *Pneumocystis* sp. Additionally, mating has been observed to occur in vivo during infections with *C. albicans*, with reports documenting mating on the skin, in the GI tract, and other regions, and might therefore influence the evolutionary trajectory in response to, for example, antifungal drug therapy (Hull et al., 2000; Lachke et al., 2003; Dumitru et al., 2007).

Phylogenetic studies reveal that the *Malassezia* species, which are specialized to survive as commensals on human skin and associated with a myriad of inflammatory skin disorders, are basidiomycetes closely related to the plant pathogen of corn, *Ustilago maydis*. Although no extant sexual cycle has been defined for the *Malassezia* species, as discussed earlier, the genome reveals a mating-type locus similar to that of the bipolar species *Ustilago hordei* that infects barley (Bakkeren et al., 2008), and the machinery for mating and meiosis. Thus, like *U. maydis*, this raises the possibility that sexual reproduction might occur in conjunction with and even be stimulated by the host, and could be associated with infection. It further raises the possibility that the infectious form in humans could be a filamentous dikaryon produced by mating, rather than the yeast form. Further studies to address this interesting hypothesis are clearly warranted, and might reveal that sex produces novel antigens associated with disease manifestations.

Another distinct way in which the sexual cycle of fungi is linked to virulence involves roles for the mating-type locus itself in promoting pathogenesis. This has been most clearly established in *C. neoformans*, in which the serotype D variety *neoformans* lineage strains of the α mating type are more pathogenic than congenic strains of the **a** mating type (Kwon-Chung et al., 1992; Nielsen et al., 2005a). There appears to be an influence of genetic background, and thus the α locus contributes to virulence to a greater extant compared to the **a** locus in some serotype D genetic backgrounds but not others (Nielsen et al., 2005b), consistent with models in which virulence is a quantitative trait and the mating-type locus represents one of several genomic loci that contribute to infection. In contrast to these findings in serotype D, in the serotype A variety *grubii* lineage congenic strains of the α and **a** mating type were found to be of equivalent virulence in several different animal models, including heterologous model hosts (Nielsen et al., 2003). Thus far this has only been addressed in congenic strains derived from the sequence reference strain H99, and whether there might be a virulence difference for strains of opposite mating type during solo infections of the host remains to be explored in other serotype A strain backgrounds. Thus, it may emerge that this is a serotype/variety-specific difference, or it may emerge that in both serotype D and A that the mating-type locus contributes to virulence but that the impact is dependent on the strain background. Construction of additional serotype A congenic strain pairs will be necessary to examine this question in further detail.

Interestingly, during co-infections of the serotype A congenic strains of opposite mating type, the α strain has a greater predilection to penetrate and infect the central nervous system (CNS) compared to a co-infecting strain of **a** mating type (Nielsen et al., 2005a). This involves crossing the blood–brain barrier, as there is no apparent difference in colonization of the lung, or dissemination to the spleen, but fewer **a** cells are observed in the CNS compared to α cells in most infected animals. When the inocula was directly delivered to the CNS rather than to the lung, there was no difference in survival of the α and **a** strains, providing evidence that their interaction occurs at the level of crossing the blood–brain barrier (Nielsen et al., 2005a). Recent findings provide evidence that the Ste3 pheromone receptor is involved in the interaction between α and **a** cells, in that **a** strains lacking the *ste3***a** receptor are enabled to compete with α cells for CNS penetration, suggesting that a form of pheromone-based quorum sensing may occur between the two mating types during co-infection (Okagaki et al., 2010).

Similar types of interactions may occur with other fungal pathogens during infection of the host, and which may involve a dialog between cells in the population involving mating machinery but which does not actually lead to sexual reproduction. For example, Soll and colleagues have presented evidence that long-distance communication occurs between **a/a** and α/α *MTL* homozygous strains of *C. albicans* in the context of a biofilm, and that this pheromone communication alters the formation and adhesiveness of the resulting biofilm (Daniels et al., 2006).

Sexual reproduction among eukaryotes is pervasive and a major driving force in evolution. Among the fungi, sexual reproduction often occurs at a lower frequency than asexual propagation, although all species examined thus far appear to retain genes encoding meiotic machinery. Thus, there may be few if any true asexual fungi but many that are cryptically or covertly sexual, enabling recombination and increasing genetic diversity. Among the fungal pathogens, sexual reproduction also leads to the production of infectious spores and invasive hyphae. While sex is not yet known to be directly necessary for successful animal infections, it is in several examples linked to aspects of pathogenicity. In the plant pathogens the links between pathogenicity and sex are often obligatory, illustrating a requirement of sex for infection. Overall, sexual reproduction remains a central aspect of fungal virulence in both the plant and animal kingdoms, and whether sex plays a more intimate role in fungal infection of animals remains a provocative hypothesis to be explored in future investigations, as in fungi infecting plants.

4.10 Genomics of Fungi: What makes a Fungus Pathogenic?

4.10.1 Comparative Genomics of Plant Pathogens

In this section, we are interested in exploring the genomic characteristics that allow some fungi to infect plants and, more rarely, animals (for a thorough review see Aguileta et al., 2009). The pathogenic fungi are most often opportunistic (Richardson, 1991; Pfaller and Diekema, 2004). Their capacity to derive nutrients from a large range of plant hosts appears to rely on a battery of genomic resources that are the result of different evolutionary processes. Perhaps the most important source of new genes and gene functions that are specific of fungal pathogens are derived via expansions of gene families that facilitate the infection of the host (Sidhu, 2002; Keller et al., 2005). Typically, these gene families include cell surface receptors such as the G-protein-coupled receptors (GPCRs), which bind exogenous ligands and participate in signaling cascades (Dean et al., 2005; Cuomo et al., 2007); secreted proteins, which constitute a diverse group of small peptides such as toxins, proteinaceous effectors, hydrolytic and degrading enzymes (Machida et al., 2005; Hane et al., 2007; Xu et al., 2007); protein effectors that suppress plant defenses and alter cellular metabolism (Kamper et al., 2006; Hane et al., 2007); and secondary metabolites such as nonspecific and host-specific toxins (Soanes et al., 2007). Key gene families involved in the biosynthesis of toxins include polyketide synthases (PKS) (Hopwood, 1997; Shen, 2000), nonribosomal peptide synthesis genes (NRPS) (Yuen et al., 2003), hybrid PKS-NRPSs (Kroken et al., 2003; Bohnert et al., 2004), and cytochrome P450 (Deng et al., 2007). Other genomic elements that have expanded include genes that trigger regulatory cascades (Martin et al., 2007; Martin et al., 2008). Gene families typically expand by gene duplication, which in fungal genomes range from whole-genome duplications (Dujon et al., 2004; Kellis et al., 2004; Scannell et al., 2006) to several instances of tandem duplications, such as events involving pathogenicity-related gene families including adhesins (Verstrepen and Fink, 2009), cellular motors called kinesins (Schoch et al., 2003), the ABC transporters and MFS drug efflux systems that help fungi detoxify products from the plants defenses (Howlett, 2006), the multidrug resistance transporter families (Gbelska et al., 2006), major surface glycoproteins, hexose uptake (Dulermo et al., 2009), TRK potassium transporters (Miranda et al., 2009), related proteins, and proteases (Keely et al., 2005). Gene duplications related to adaptations to the pathogenic lifestyle have also been documented, as in the case of the oxidative phosphorylation pathway, whose components have evolved by functional divergence with several instances of gene loss and duplication (Marcet-Houben et al., 2009). Following duplication, rapid rates of evolution and positive selection can give rise to novel gene functions that allow the fungus to coevolve with its host or to infect new hosts. In fungal genomes, positive selection has been found to act in the evolution of functionally important gene families, in particular those that confer an adaptation to a pathogenic lifestyle. These include genes coding for defense systems or for evading host resistance mechanisms, toxic protein genes, and other virulence-related genes (Staats et al., 2007). Particular examples of genes under positive selection that have been identified in fungal genomes include the mycotoxin gene cluster in *Fusarium* (Ward et al., 2002; Cuomo et al., 2007), various phytotoxin genes in *Botrytis* (Staats et al., 2007) and *Phytophthora infestans* (Liu et al., 2005), the aflatoxin gene cluster in *Aspergillus* (Carbone et al., 2007), host-specific toxin the wheat pathogen *Phaeosphaeria nodorum* (Stukenbrock and McDonald, 2007), antigens in *Coccidioides* human pathogens (Johannesson et al., 2004), and serine proteases in 10 fungal species (Hu and Leger, 2004). Positive selection in the plant defense R-genes is frequently followed by coevolution in the avirulence genes of the fungal pathogen (Jones and Jones, 1997; Parniske et al., 1997; Meyers et al., 1998). This gene-for-gene interaction with corresponding responses in both the host and the pathogen genomes is referred to as an "arms-race" process (Dawkins and Krebs, 1979).

In terms of the structure of fungal genomes, it has been shown that genes encoding biochemical products aiding in infection are often clustered together (Jargeat et al., 2003). Clustering of important gene families appears to offer several advantages for pathogenicity (Sidhu, 2002; Keller et al., 2005). Indeed, evidence shows that fungal genes interacting in the same metabolic pathway tend to be clustered together (Keller and Hohn, 1997). Another mechanism linking genes in the genome is the suppression of recombination that occurs surrounding the genes that determine mating compatibility, which are clustered at the mating-type loci (Herskowitz, 1989). Protein products of mating-type loci additionally may serve functions for mating and virulence through common G-protein-mediated environmental sensing and response pathways (Bolker, 1998). Interestingly, and in contrast with gene clustering, genetic variation created by chromosomal rearrangements has been reported to favor the adaptation to novel hosts or nutritional environments (Larriba, 2004), thus contributing to pathogenicity. Transposable elements are another class of genomic elements that have also been shown to play a significant role in enhancing the pathogenic capacities of fungi (Wostemeyer and Kreibich, 2002). In several pathogenic fungi, including *Leptosphaeria maculans* and *Magnaporthe grisea*, sequences coding for avirulence genes are found in genomic regions dense with transposable elements (Kang et al., 2001; Gout et al., 2006; Rehmeyer et al., 2006; Fudal et al., 2007), potentially contributing to the extreme variability of avirulence genes that are associated with host—pathogen coevolution. Telomeres are rapidly evolving genomic regions particularly prone to the accumulation of transposable elements, and they sometimes contain avirulence genes, thereby playing a role in host adaptation (Rehmeyer et al., 2006; Chen et al., 2007; Sánchez-Alonso and Guzman, 2008). Sometimes the genes that confer pathogenicity to fungi come from other species, either via HGT or hybridization. Although HGT is not as pervasive in fungal genomes as it is in bacteria, it appears to have occurred multiple independent times (Penalva et al., 1990, Kavanaugh et al., 2006). Occasionally, complete clusters are speculated to have been horizontally transferred (Walton, 2000). Finally, hybridization is another way to mix genes and produce new crosses with increased pathogenic capacities. There has been growing concern during the past decade over the number of reported hybridizations in fungi, particularly among pathogenic species (see Olson and Stenlid, 2002, for a review). Several hybridization events have thus been identified among pathogens and mutualists from all clades of fungi. Some have been related to an increase in virulence or host range, a shift in host spectrum or even a switch toward mutualism (Olson and Stenlid, 2002).

Fungal genomes are extremely plastic. This is highlighted by the different genomic processes that have generated a versatile repertoire of biochemical functions that allow fungi to colonize a diverse range of environments and to also establish relationships with other species, either by infection or symbiosis, with an extensive array of partners. New genomic data will continue to fascinate us with examples of amazing potentials for adaptation.

4.10.2 Comparing Animal and Plant Pathogens

Pathogenic fungi are mostly intracellular pathogens, indicating that at some point during the interaction between the host and the invading species the pathogen lives inside the host cell. Despite the variety of intracellular fungal pathogens infecting both plant and animal cells in seemingly unique ways, there are only a few general solutions to the challenge of penetrating and surviving inside host cells (Casadevall, 2008). Indeed, the problem represented by intracellular infection has been tackled by convergent solutions that have evolved in parallel in the different fungal lineages (Morris et al., 2009) of both plant and animal pathogens. It is interesting to note that among fungi there appear to be many more species that parasitize plants than animals (Desprez-Loustau et al., 2010). The reasons for this imbalance are not very clear and deserve further attention. Typically, fungi maintain closer relationships with plants and have evolved many biochemical functions to take advantage of their plant hosts to obtain nutrients (Pallen and Wren, 2007; Boller and He, 2009; Burdon and Thrall, 2009; Pringle et al., 2009). An intriguing hypothesis posits that fungi became pathogenic through the evolution of dual-use traits. It is hypothesized that this first appeared as a mechanism to defend from environmental aggressions, such as predation by amoebae, and later allowed invasion of plant and animal cells alike (Morris et al., 2009). *C. neoformans* provides a good example of dual-use traits that have helped this species to defend itself from amoebal predation and infect animal cells. Dual-use traits include, but are not limited to, capsule formation, and the production of melanin, laccase, phospholipase, proteases, and ureases (Casadevall et al., 2003). Also, according to this thesis, originally saprophytic interactions existed among plants, animals, and fungi that allowed the evolution of interkingdom biochemical exchanges by different strategies (Dodds et al., 2009; Oldroyd et al., 2009). Later on, these innovations were exploited to invade plant and animal cells and derive benefits (Brun et al., 2009; Grant and Jones, 2009). Examples of the new uses of previous adaptations to saprophytic lifestyles abound, and include toxin production, adhesins to adhere to host cells, injectors to penetrate the cell, interaction of fungal cells with host effector cells, and microbial efflux pumps for managing toxic environmental compounds that also help detoxify the fungal cell from defense plant products (Morris et al., 2009; Panstruga and Dodds, 2009).

The genomes of fungal animal pathogens have not been as extensively studied as phytopathogens. However, work has been published on a few of the best known infect animals. These include *Aphanomyces astaci* (Bangyeekhun et al., 2001; Oidtmann et al., 2004), which causes the crayfish plague; *Cordyceps tuberculata* (Sung et al., 2007) and *Beauveria bassiana* (Coates et al., 2002), both of which are entomopathogens; *Batrachochytrium dendrobatidis* (Kilpatrick et al., 2010), a chytridiomycete fungus that is suspected to have caused the demise of many frog species; and recently an emergent disease of bats resulting in a population decline in excess of 75% has been associated with the pathogenic fungus *Geomyces destructans* (Blehert et al., 2008; Puechmaille et al., 2010). Also, there are the documented cases of *Malassezia globosa* (Xu et al., 2007), the causative agent of dandruff; the infamous species *A. fumigatus* (Nierman et al., 2005; O'Gorman et al., 2009), which is responsible for aspergillosis in immunosuppressed human patients; and the well-known infections caused by *C. albicans* (http://www.candidagenome.org/), *C. neoformans* (Loftus et al., 2005), *Coccidioides immitis* (Johannesson et al., 2004), and *Histoplasma capsulatum* (Magrini et al., 2004). In all of these examples, lineage-specific gene family expansions have played a significant role in pathogenicity.

More research needs to be conducted and more animal pathogens need to be sequenced before we have a comprehensive view of the genetic basis, if any, of the differences between the fungal genomes of plant and animal pathogens. Most mechanisms and gene functions may be shared, as has been shown by a study of the NLP toxin whose fold is conserved and shows similarities with that of bacteria (Ottmann et al., 2009), so we can speculate about lineage- and host-specific genes and gene functions in each case.

4.11 Conclusion

Comparative genomic studies in plant pathogenic and symbiotic fungi, although still in the early stages and limited to a few pathogens, have already brought many insights into the evolution of the pathogenic lifestyle, in particular into the mechanisms of virulence and host adaptations. There is a marked bias in the sequencing efforts toward pathogenic fungi, but current projects are covering the fungal genomes of species with very diverse lifestyles, that will hopefully allow us to gain further insights into the genomics of pathogenicity.

Regarding epidemiology, molecular methods have much to offer to the study of fungal pathogens, allowing elucidation of ecological and microevolutionary processes. Population genetic approaches have provided important insights for some fungal pathogens on their mating systems, dispersal, and population structure. However, much wider employment of these methods is warranted to study fungal pathogens, where it is still too restricted, although much progress has been made recently. Microsatellite markers in particular are very powerful tools (Jarne and Lagoda, 1996) and should be more widely used for population studies in fungi, despite the technical challenges of their isolation in this kingdom (Dutech et al., 2007). Further, new methods to analyze data are being developed at a rapid pace, some using the Bayesian or coalescence frameworks, or coupling geography and genetics to unravel migration and speciation histories, which should allow even more powerful inferences on the evolutionary processes. However, further theoretical development is badly needed to apply the extant molecular methods to the variety and specificities of the fungal life cycles, such as pervasive clonality and alternation between haplo- and diploid phases (Balloux and Lugon-Moulin, 2002; Halkett et al., 2005).

Important advances have also been made recently on the speciation in fungi. Recently developed analytical methods for studying past gene flow and differentiation should be useful to determine in which cases fungal speciation by specialization onto novel hosts has occurred in sympatry (Hey and Nielsen, 2004; Hey et al., 2004). Deciphering the genetics of speciation should also prove to be fascinating; for instance, by finding markers segregating with inviability or sterility in interspecific progeny.

Acknowledgements

We thank Michel Tibayrenc for his cordial invitation to write this review. We acknowledge the grants ANR 06-BLAN-0201, ANR 07-BDIV-003, ANR-09-0064-01, and NIH/NIAID AI39115. Parts of this chapter are derivatives of articles previously published (Giraud et al., 2008a, 2008b, 2010; Aguileta et al., 2009). We thank the Emerfundis group. We apologize to all those colleagues whose work we might have missed to cite in this chapter.

References

Achtman, M., 1996. A surfeit of YATMs? J. Clin. Microbiol. 34, 1870.
Agapow, P.-M., Burt, A., 2001. Indices of multilocus linkage disequilibrium. Mol. Ecol. Notes 1, 101–102.
Aguileta, G., Hood, M., Refrégier, G., Giraud, T., 2009. Genome evolution in pathogenic and symbiotic fungi. Adv. Bot. Res. 49, 151–193.
Anagnostakis, S., 2001. The effect of multiple importations of pets and pathogens on a native tree. Biol. Invasions 3, 245–254.
Anderson, E.C., Thompson, E.A., 2002. A model-based method for identifying species hybrids using multilocus genetic data. Genetics 160, 1217–1229.
Anderson, J.B., Ullrich, R.C., 1978. Biological species of *Armillaria mellea* in North America. Mycologia 71, 402–414.
Anderson, J.B., Korhonen, K., Ullrich, R.C., 1980. Relationships between European and North American biological species of *Armillaria mellea*. Exp. Mycol. 4, 87–95.
Anderson, J.B., Bailey, S.S., Pukkila, P.J., 1989. Variation in ribosomal DNA among biological species of *Armillaria*, a genus of root-infecting fungi. Evolution 43, 1652–1662.
Anderson, P.K., Cunningham, A.A., Patel, N.G., Morales, F.J., Epstein, P.R., Daszak, P., 2004. Emerging infectious diseases of plants: pathogen pollution, climate change and agrotechnology drivers. Trends Ecol. Evol. 19, 535–544.
Arnaud-Haond, S., Belkhir, K., 2007. GENCLONE: a computer program to analyse genotypic data, test for clonality and describe spatial clonal organization. Mol. Ecol. Notes 7, 15–17.
Atallah, Z.K., Maruthachalam, K., Toit, L.d., Koike, S.T., Michael Davis, R., Klosterman, S.J., et al., 2010. Population analyses of the vascular plant pathogen *Verticillium dahliae* detect recombination and transcontinental gene flow. Fungal Genet. Biol. Vol. 47. pp. 416–422.
Austerlitz, F., Jung-Muller, B., Godelle, B., Gouyon, P.-H., 1997. Evolution of coalescence times, genetic diversity and structure during colonization. Theor. Popul. Biol. 51, 148–164.
Bain, J.M., Tavanti, A., Davidson, A.D., Jacobsen, M.D., Shaw, D., Gow, N.A.R., et al., 2007. Multilocus sequence typing of the pathogenic fungus *Aspergillus fumigatus*. J. Clin. Microbiol. 45, 1469–1477.
Bakkeren, G., Kamper, J., Schirawski, J., 2008. Sex in smut fungi: structure, function and evolution of mating-type complexes. Fungal Genet. Biol. 45 (Suppl. 1), S15–21.
Balloux, F., Lugon-Moulin, N., 2002. The estimation of population differentiation with microsatellite markers. Mol. Ecol. 11, 155–165.

Bangyeekhun, E., Cerenius, L., Soderhall, K., 2001. Molecular cloning and characterization of two serine proteinase genes from the crayfish plague fungus, *Aphanomyces astaci*. J. Invertebr. Pathol. 77, 206–216.

Banke, S., McDonald, B.A., 2005. Migration patterns among global populations of the pathogenic fungus *Mycosphaerella graminicola*. Mol. Ecol. 14, 1881–1896.

Barres, B., Halkett, F., Dutech, C., Andrieux, A., Pinon, J., Frey, P., 2008. Genetic structure of the poplar rust fungus *Melampsora larici-populina*: evidence for isolation by distance in Europe and recent founder effects overseas. Infect. Genet. Evol. 8, 577–587.

Bartlett, K.H., Kidd, S.E., Kronstad, J.W., 2008. The emergence of *Cryptococcus gattii* in British Columbia and the Pacific Northwest. Curr. Infect. Dis. Rep. 10, 58–65.

Beaumont, M.A., Zhang, W.Y., Balding, D.J., 2002. Approximate Bayesian computation in population genetics. Genetics 162, 2025–2035.

Beerli, P., Felsenstein, J., 1999. Maximum-likelihood estimation of migration rates and effective population numbers in two populations using a coalescent approach. Genetics 152, 763–773.

Beerli, P., Felsenstein, J., 2001. Maximum likelihood estimation of a migration matrix and effective population sizes in n subpopulations by using a coalescent approach. Proc. Natl. Acad. Sci. U.S.A. 98, 4563–4568.

Bell, G., 1982. The Masterpiece of Nature: The Evolution and Genetics of Sexuality. University of California Press, Berkeley, CA.

Berbee, M.L., 2001. The phylogeny of plant and animal pathogens in the Ascomycota. Physiol. Mol. Plant Pathol. 59, 165–187.

Berbee, M.L., Taylor, J.W., 2001. Fungal molecular evolution. Gene Trees and Geologic Time The Mycota: A Comprehensive Treatise on Fungi as Experimental Systems for Basic and Applied Research. New York, Springer-Verlag, Volume VII: Systematics and Evolution Part B. pp. 229–245.

Berger, L., Speare, R., Daszak, P., Green, D., Cunningham, A., Goggin, C.L., et al., 1998. Chytridiomycosis causes amphibian mortality associated with population declines in the rain forests of Australia and Central America. Proc. Natl. Acad. Sci. U.S.A. 95, 9031–9036.

Billiard, S., López-Villavicencio, M., Devier, B., Hood, M.E., Fairhead, C., Giraud, T., 2010. Having sex, yes, but with whom? Inferences from fungi on the evolution of anisogamy and mating types. Biol. Rev. In Press.

Birch, P.R.J., Whisson, S.C., 2001. *Phytophthora infestans* enters the genomics era. Mol. Plant Pathol. 2, 257–263.

Birky, C.W., 1996. Heterozygosity, heteromorphy, and phylogenetic trees in asexual eukaryotes. Genetics 144, 427–437.

Blankenship, J.R., Singh, N., Alexander, B.D., Heitman, J., 2005. *Cryptococcus neoformans* isolates from transplant recipients are not selected for resistance to calcineurin inhibitors by current immunosuppressive regimens. J. Clin. Microbiol. 43, 464–467.

Blehert, D.S., Hicks, A.C., Behr, M., Meteyer, C.U., Berlowski-Zier, B.M., Buckles, E.L., et al., 2008. Bat white-nose syndrome: an emerging fungal pathogen? Science 323, 227.

Bohnert, H.U., Fudal, I., Dioh, W., Tharreau, D., Notteghem, J.L., Lebrun, M.H., 2004. A putative polyketide synthase peptide synthetase from *Magnaporthe grisea* signals pathogen attack to resistant rice. Plant Cell 16, 2499–2513.

Bolker, M., 1998. Sex and crime: heterotrimeric G-proteins in fungal mating and pathogenesis. Fungal Genet. Biol 25, 143–156.

Boller, T., He, S.Y., 2009. Innate immunity in plants: an arms race between pattern recognition receptors in plants and effectors in microbial pathogens. Science 324, 742–744.

Bossart, J.L., Prowell, D.P., 1998. Genetic estimates of population structure and gene flow: limitations, lessons and new directions. Trends Ecol. Evol. 13, 202–206.
Bovers, M., Hagen, F., Kuramae, E.E., Boekhout, T., 2008. Six monophyletic lineages identified within *Cryptococcus neoformans* and *Cryptococcus gattii* by multi-locus sequence typing. Fungal Genet. Biol. 45, 400–421.
Brasier, C.M., 2001. Rapid evolution of introduced plant pathogens via interspecific hybridization. BioScience 51, 123–133.
Broquet, T., Petit, E.J., 2009. Molecular estimation of dispersal for ecology and population genetics. Annu. Rev. Ecol. Evol. Syst. 40, 193–216.
Brown, J.K.M., 1994. Chance and selection in the evolution of barley mildew. Trends Microbiol. 2, 461–501.
Brown, J.K.M., Hovmoller, M.S., 2002. Aerial dispersal of pathogens on the global and continental scales and its impact on plant disease. Science 297, 537–541.
Brun, S., Malagnac, F., Bidard, F., Lalucque, H., Silar, P., 2009. Functions and regulation of the Nox family in the filamentous fungus *Podospora anserina*: a new role in cellulose degradation. Mol. Microbiol. 74, 480–496.
Brunner, P.C., Schurch, S., McDonald, B.A., 2007. The origin and colonization history of the barley scald pathogen *Rhynchosporium secalis*. J. Evol. Biol. 20, 1311–1321.
Bubnick, M., Smulian, A.G., 2007. The MAT1 locus of *Histoplasma capsulatum* is responsive in a mating type-specific manner. Eukaryot. Cell 6, 616–621.
Bultman, T.L., Leuchtmann, A., 2003. A test of host specialization by insect vectors as a mechanism for reproductive isolation among entomophilous fungal species. Oikos 103, 681–687.
Burdon, J.J., Thrall, P.H., 2009. Coevolution of plants and their pathogens in natural habitats. Science 324, 755–756.
Burgess, J.W., Kottom, T.J., Limper, A.H., 2008. *Pneumocystis carinii* exhibits a conserved meiotic control pathway. Infect. Immun. 76, 417–425.
Burgess, J.W., Kottom, T.J., Villegas, L.R., Lamont, J.D., Baden, E.M., Ramirez-Alvarado, M., et al., 2009. The *Pneumocystis* meiotic PCRan1p kinase exhibits unique temperature-regulated activity. Am. J. Respir. Cell Mol. Biol. 41, 714–721.
Burt, A., Carter, D.A., Koenig, G.L., White, T.J., Taylor, J.W., 1996. Molecular markers reveal cryptic sex in the human pathogen *Coccidioides immitis*. Proc. Natl. Acad. Sci. U.S.A. 93, 770–773.
Byrnes 3rd, E.J., Heitman, J., 2009. *Cryptococcus gattii* outbreak expands into the Northwestern United States with fatal consequences. F1000 Biol. Rep. 1, 62.
Byrnes 3rd, E.J., Bildfell, R.J., Dearing, P.L., Valentine, B.A., Heitman, J., 2009a. *Cryptococcus gattii* with bimorphic colony types in a dog in western Oregon: additional evidence for expansion of the Vancouver Island outbreak. J. Vet. Diagn. Invest. 21, 133–136.
Byrnes 3rd, E.J., Bildfell, R.J., Frank, S.A., Mitchell, T.G., Marr, K.A., Heitman, J., 2009b. Molecular evidence that the range of the Vancouver Island outbreak of *Cryptococcus gattii* infection has expanded into the Pacific Northwest in the United States. J. Infect. Dis. 199, 1081–1086.
Byrnes 3rd, E.J., Li, W., Lewit, Y., Ma, H., Voelz, K., et al., 2010. Emergence and pathogenicity of highly virulent *Cryptococcus gattii* genotypes in the northwest United States. PLoS Pathog 6, e1000850.
Byrnes 3rd, E.J., Li, W., Lewit, Y., Perfect, J.R., Carter, D.A., Cox, G.M., et al., 2009c. First reported case of *Cryptococcus gattii* in the Southeastern USA: implications for travel-associated acquisition of an emerging pathogen. PLoS One 4, e5851.

Calderon, E.J., 2010. *Pneumocystis* infection: seeing beyond the tip of the iceberg. Clin. Infect. Dis. 50, 354–356.
Carbone, I., Ramirez-Prado, J.H., Jakobek, J.L., Horn, B.W., 2007. Gene duplication, modularity and adaptation in the evolution of the aflatoxin gene cluster. BMC Evol. Biol. 7, 111.
Carlile, M., Watkinson, S., Gooday, G., 2001. The Fungi. second ed. Academic Press, London.
Casadevall, A., 2008. Evolution of intracellular pathogens. Annu. Rev. Microbiol. 62, 19–33.
Casadevall, A., Perfect, J., 1998. Cryptococcus Neoformans. ASM Press, Washington DC.
Casadevall, A., Steenbergen, J.N., Nosanchuk, J.D., 2003. '"Ready made" virulence and "dual use" virulence factors in pathogenic environmental fungi—the *Cryptococcus neoformans* paradigm. Curr. Opin. Microbiol. 6, 332–337.
Cavalli-Sforza, L.L., 1966. Population structure and human evolution. Proc. R. Soc. Lond. B. Biol. Sci. 164, 362–379.
Chen, Q.H., Wang, Y.C., Li, A.N., Zhang, Z.G., Zheng, X.B., 2007. Molecular mapping of two cultivar-specific avirulence genes in the rice blast fungus *Magnaporthe grisea*. Mol. Genet. Genomics 277, 139–148.
Chen, R.S., McDonald, B.A., 1996. Sexual reproduction plays a major role in the genetic structure of populations of the fungus *Mycosphaerella graminicola*. Genetics 142, 1119–1127.
Chen, S., Sorrell, T., Nimmo, G., Speed, B., Currie, B., Ellis, D., et al., 2000. Epidemiology and host- and variety-dependent characteristics of infection due to *Cryptococcus neoformans* in Australia and New Zealand. Australasian Cryptococcal Study Group. Clin. Infect. Dis. 31, 499–508.
Cheng, P.Y., Sham, A., Kronstad, J.W., 2009. *Cryptococcus gattii* isolates from the British Columbia cryptococcosis outbreak induce less protective inflammation in a murine model of infection than *Cryptococcus neoformans*. Infect. Immun. 77, 4284–4294.
Cleaveland, S., Laurenson, M.K., Taylor, L.H., 2001. Diseases of humans and their domestic mammals: pathogen characteristics, host range and the risk of emergence. Philos. Trans. R. Soc. Lond. B. Biol. Sci. 356, 991–999.
Clement, M., Posada, D., Crandall, K.A., 2000. TCS: a computer program to estimate gene genealogies. Mol. Ecol. 9, 1657–1659.
Coates, B.S., Hellmich, R.L., Lewis, L.C., 2002. Allelic variation of a *Beauveria bassiana* (Ascomycota : Hypocreales) minisatellite is independent of host range and geographic origin. Genome 45, 125–132.
Cockerham, C.C., 1969. Variance of gene frequencies. Evolution 23, 72–83.
Cockerham, C.C., 1973. Analyses of gene frequencies. Genetics 74, 679–700.
Cogliati, M., Esposto, M.C., Clarke, D.L., Wickes, B.L., Viviani, M.A., 2001. Origin of *Cryptococcus neoformans* var. *neoformans* diploid strains. J. Clin. Microbiol. 39, 3889–3894.
Cooke, D.E.L., 2007. Tracking the sudden oak death pathogen. Mol. Ecol. 16, 3735–3736.
Corander, J., Tang, J., 2007. Bayesian analysis of population structure based on linked molecular information. Math. Biosci. 205, 19–31.
Cornuet, J.M., Piry, S., Luikart, G., Estoup, A., Solignac, M., 1999. New methods employing multilocus genotypes to select or exclude populations as origins of individuals. Genetics 153, 1989–2000.
Cornuet, J.M., Santos, F., Beaumont, M.A., Robert, C.P., Marin, J.M., Balding, D.J., et al., 2008. Inferring population history with DIY ABC: a user-friendly approach to approximate Bayesian computation. Bioinformatics 24, 2713–2719.

Couch, B.C., Fudal, I., Lebrun, M.-H., Tharreau, D., Valent, B., van Kim, P., et al., 2005. Origins of host-specific populations of the blast pathogen, *Magnaporthe oryzae*, in crop domestication with subsequent expansion of pandemic clones on rice and weeds of rice. Genetics 170, 613–630.

Coyne, J.A., Orr, H.A., 2004. Speciation. Sinauer Associates, Sunderland, MA.

Criscione, C.D., Poulin, R., Blouin, M.S., 2005. Molecular ecology of parasites: elucidating ecological and microevolutionary processes. Mol. Ecol. 14, 2247–2257.

Cuomo, C.A., Gueldener, U., Xu, J.R., Trail, F., Turgeon, B.G., Di Pietro, A., et al., 2007. The *Fusarium graminearum* genome reveals a link between localized polymorphism and pathogen specialization. Science 317, 1400–1402.

Cushion, M.T., 2004. *Pneumocystis*: unraveling the cloak of obscurity. Trends Microbiol. 12, 243–249.

Cushion, M.T., Walzer, P.D., 1984a. Cultivation of *Pneumocystis carinii* in lung-derived cell lines. J. Infect. Dis. 149, 644.

Cushion, M.T., Walzer, P.D., 1984b. Growth and serial passage of *Pneumocystis carinii* in the A549 cell line. Infect. Immun. 44, 245–251.

Daniels, K.J., Srikantha, T., Lockhart, S.R., Pujol, C., Soll, D.R., 2006. Opaque cells signal white cells to form biofilms in *Candida albicans*. Embo J. 25, 2240–2252.

Dasbach, E.J., Davies, G.M., Teutsch, S.M., 2000. Burden of aspergillosis-related hospitalizations in the United States. Clin. Infect. Dis. 31, 1524–1528.

Daszak, P., Cunningham, A.A., Hyatt, A.D., 2000. Emerging infectious diseases of wildlife—threats to biodiversity and human health. Science 287, 443–449 .

Datta, K., Bartlett, K.H., Baer, R., Byrnes, E., Galanis, E., Heitman, J., et al., 2009. Spread of *Cryptococcus gattii* into Pacific Northwest region of the United States. Emerg. Infect. Dis. 15, 1185–1191.

Davison, D., Pritchard, J.K., Coop, G., 2009. An approximate likelihood for genetic data under a model with recombination and population splitting. Theor. Popul. Biol. 75, 331–345.

Dawkins, R., Krebs, J.R., 1979. Arms race between and within species. Proc. R. Soc. Lond. B 205, 489–511.

Dawson, K.J., Belkhir, K., 2001. A Bayesian approach to the identification of panmictic populations and the assignment of individuals. Genet. Res. 78, 59–77.

De Meeus, T., Lehmann, L., Balloux, F., 2006. Molecular epidemiology of clonal diploids: a quick overview and a short DIY (do it yourself) notice. Infect. Genet. Evol. 6, 163–170.

de Queiroz, K., 1998. The general lineage concept of species, and the process of speciation. In: Howard, D.J., Belocher, S.H. (Eds.), Endless Forms: Species and Speciation. Oxford University Press, Oxford, pp. 57–75.

de Queiroz, K., 2007. Species Concepts and Species Delimitation. *Systematic Biology* 56, 879–886.

de Vienne, D.M., Refrégier, G., Hood, M., Guigue, A., Devier, B., Vercken, E., et al., 2009. Hybrid sterility and inviability in the parasitic fungal species complex *Microbotryum*. J. Evol. Biol. 22, 683–698.

Dean, R.A., et al., 2005. The genome sequence of the rice blast fungus *Magnaporthe grisea*. Nature 434, 980–986.

Delmotte, F., Giresse, X., Richard-Cervera, S., M'Baya, J., Vear, F., Tourvieille, J., et al., 2008. Single nucleotide polymorphisms reveal multiple introductions into France of *Plasmopara halstedii*, the plant pathogen causing sunflower downy mildew. Infect. Genet. Evol. 8, 534–540.

Delneri, D., Colson, I., Grammenoudi, S., Roberts, I.N., Louis, E.J., Oliver, S.G., 2003. Engineering evolution to study speciation in yeasts. Nature 422, 68–72.

Deng, J.X., Carbone, I., Dean, R.A., 2007. The evolutionary history of Cytochrome P450 genes in four filamentous Ascomycetes. BMC Evol. Biol. 7, 30.

Desjardins, A.E., Hohn, T.M., McCormick, S.P., 1993. Trichothecene biosynthesis in *Fusarium* species: chemistry, genetics and significance. Microbiol. Rev. 57, 595–604.

Desprez-Loustau, M., Robin, C., Buee, M., Courtecuisse, R., Garbaye2, J., Suffert, F., et al., 2007. The fungal dimension of biological invasions. Trends Ecol. Evol. 22, 472–480.

Desprez-Loustau, M.L., Courtecuisse, R., Robin, C., Husson, C., Moreau, P.A., Blancard, D., et al., 2010. Species diversity and drivers of spread of alien fungi (sensu lato) in Europe with a particular focus on France. Biol. Invasions 12, 157–172.

Dettman, J.R., Jacobson, D.J., Taylor, J.W., 2003a. A multilocus genealogical approach to phylogenetic species recognition in the model eukaryote *Neurospora*. Evolution 57, 2703–2720.

Dettman, J.R., Jacobson, D.J., Turner, E., Pringle, A., Taylor, J.W., 2003b. Reproductive isolation and phylogenetic divergence in *Neurospora*: comparing methods of species recognition in a model eukaryote. Evolution 57, 2721–2741.

Dickinson, H., Antonovics, J., 1973. Theoretical considerations of sympatric divergence. Am. Nat. 107, 256–274.

Dodds, P.N., Rafiqi, M., Gan, P.H.P., Hardham, A.R., Jones, D.A., Ellis, J.G., 2009. Effectors of biotrophic fungi and oomycetes: pathogenicity factors and triggers of host resistance. New Phytol. 183, 993–999.

Douhan, G.W., Peever, T.L., Murray, T.D., 2002. Multilocus population structure of *Tapesia yallundae* in Washington State. Mol. Ecol. 11, 2229–2239.

Dujon, B., Sherman, D., Fischer, G., Durrens, P., Casaregola, S., Lafontaine, I., et al., 2004. Genome evolution in yeasts. Nature 430, 35–44.

Dulermo, T., Rascle, C., Chinnici, G., Gout, E., Bligny, R., Cotton, P., 2009. Dynamic carbon transfer during pathogenesis of sunflower by the necrotrophic fungus *Botrytis cinerea*: from plant hexoses to mannitol. New Phytol. 183, 1149–1162.

Dumitru, R., Navarathna, D.H., Semighini, C.P., Elowsky, C.G., Dumitru, R.V., Dignard, D., et al., 2007. In vivo and in vitro anaerobic mating in *Candida albicans*. Eukaryot Cell 6, 465–472.

Durand, E., Chen, C., Francois, O., 2009. Comment on "On the inference of spatial structure from population genetics data". Bioinformatics 25, 1802–1804.

Dutech, C., Enjalbert, E., Fournier, E., Delmotte, F., Barrès, B., Carlier, J., et al., 2007. Challenges of microsatellite isolation in fungi. Fungal Genet. Biol. 44, 933–949.

Dutech, C., Rossi, J.P., Fabreguettes, O., Robin, C., 2008. Geostatistical genetic analysis for inferring the dispersal pattern of a partially clonal species: example of the chestnut blight fungus. Mol. Ecol. 17, 4597–4607.

Dutech, C., Fabreguettes, O., Capdevielle, X., Robin, C., 2009. Multiple introductions of divergent genetic lineages in an invasive fungal pathogen, *Cryphonectria parasitica*, in France. Heredity.

Dyer, R.J., 2009. GeneticStudio: a suite of programs for spatial analysis of genetic-marker data. Mol. Ecol. Resour. 9, 110–113.

Dyer, R.J., Nason, J.D., 2004. Population graphs: the graph theoretic shape of genetic structure. Mol. Ecol. 1713–1727.

Engelbrecht, C.J.B., Harrington, T.C., Steimel, J., Capretti, P., 2004. Genetic variation in eastern North American and putatively introduced populations of *Ceratocystis fimbriata* f. *platani*. Mol. Ecol. 13, 2995–3005.

Enjalbert, J., Duan, X., Giraud, T., Vautrin, D., de Vallavieille-Pope, C., Solignac, M., 2002. Isolation of twelve microsatellite loci, using an enrichment protocol, in the phytopathogenic fungus *Puccinia striiformis* f.sp. *tritici*. Mol. Ecol. Notes 2, 563–565.

Ergin, C., Ilkit, M., Kaftanoglu, O., 2004. Detection of *Cryptococcus neoformans* var. *grubii* in honeybee (*Apis mellifera*) colonies. Mycoses 47, 431–434.

Evanno, G., Regnaut, S., Goudet, J., 2005. Detecting the number of clusters of individuals using the software structure: a simulation study. Mol. Ecol. 14, 2611–2620.

Excoffier, L., 2007. Analysis of population subdivision. In: Balding, D., Bishop, M., Cannings, C. (Eds.), Handbook of Statistical Genetics, third ed. John Wiley & Sons, Ltd, Chichester, pp. 980–1020.

Excoffier, L., Smouse, P.E., Quattro, J.M., 1992. Analysis of molecular variance inferred from metric distances among DNA haplotypes: application to human mitochondrial DNA restriction data. Genetics 131, 479–491.

Excoffier, L., Laval, G., Schneider, S., 2005. Arlequin (version 3.0): an integrated software package for population genetics data analysis. Evol. Bioinform. Online 1, 47–50.

Falush, D., Stephens, M., Pritchard, J.K., 2003. Inference of population structure using multi-locus genotype data: linked loci and correlated allele frequencies. Genetics 164, 1567–1587.

Farris, J.S., Kallersjo, M., Kluge, A.G., Bult, C., 1994. Testing significance of incongruence. Cladistics 10, 315–319.

Fearnhead, P., Donnelly, P., 2001. Estimating recombination rates from population genetic data. Genetics 159, 1299–1318.

Fearnhead, P., Donnelly, P., 2002. Approximate likelihood methods for estimating local recombination rates. J. R. Stat. Soc. Series B Stat. Methodol. 64, 657–680.

Feder, J.L., Xie, X., Rull, J., Velez, S., Forbes, A., Leung, B., et al., 2005. Mayr, Dobzhansky, and Bush and the complexities of sympatric speciation in *Rhagoletis*. PNAS 102, 6573–6580.

Feil, E.J., Li, B.C., Aanensen, D.M., Hanage, W.P., Spratt, B.G., 2004. eBURST: inferring patterns of evolutionary descent among clusters of related bacterial genotypes from multilocus sequence typing data. J. Bacteriol. 186, 1518–1530.

Felsenstein, J., 1981. Skepticism towards Santa Rosalia, or why are there so few kinds of animals? Evolution 35, 124–138.

Felsenstein, J., 2007. Trees of genes in populations. In: Gascuel, O., Steel, M. (Eds.), Reconstructing Evolution. Oxford University Press, New York, NY, pp. 3–29.

Findley, K., Rodriguez-Carres, M., Metin, B., Kroiss, J., Fonseca, A., Vilgalys, R., et al., 2009. Phylogeny and phenotypic characterization of pathogenic *Cryptococcus* species and closely related saprobic taxa in the Tremellales. Eukaryot Cell 8, 353–361.

Finlay, B., 2002. Global dispersal of free-living microbial eukaryote species. Science 296, 1061–1063.

Fisher, D., Burrow, J., Lo, D., Currie, B., 1993. *Cryptococcus neoformans* in tropical northern Australia: predominantly variant *gattii* with good outcomes. Aust. N. Z. J. Med. 23, 678–682.

Fisher, M., Garner, T., 2007. The relationship between the introduction of *Batrachochytrium dendrobatidis*, the international trade in amphibians and introduced amphibian species. Fungal Biol. Rev. 21, 2–9.

Fisher, M., Bosch, J., Yin, Z., Stead, D., Walker, J., et al., 2010. Proteomic and phenotypic profiling of an emerging pathogen of amphibians, *Batrachochytrium dendrobatidis*, shows that genotype is linked to virulence. Mol. Ecol. (2009) 18: 415–429.

Fisher, M.C., Koenig, G.L., White, T.J., Taylor, J.W., 2000. Pathogenic clones versus environmentally driven population increase: analysis of an epidemic of the human fungal pathogen *Coccidioides immitis*. J. Clin. Microbiol. 38, 807–813.

Fisher, M.C., Koenig, G.L., White, T.J., San-Blas, G., Negroni, R., Alvarez, I.G., et al., 2001. Biogeographic range expansion into South America by *Coccidioides immitis* mirrors New World patterns of human migration. Proc. Natl. Acad. Sci. U.S.A. 98, 4558–4562.

Fisher, M.C., Koenig, G.L., White, T.J., Taylor, J.W., 2002. Molecular and phenotypic description of *Coccidioides posadasii* sp nov., previously recognized as the non-California population of *Coccidioides immitis*. Mycologia 94, 73–84.

Fisher, M.C., De Hoog, S., Vanittanakom, N., 2004. A highly discriminatory multilocus microsatellite typing (MLMT) system for *Penicillium marneffei*. Mol. Ecol. Notes 4, 515–518.

Fisher, M.C., Hanage, W.P., de Hoog, S., Johnson, E., Smith, M.D., White, N.J., et al., 2005. Low effective dispersal of asexual genotypes in heterogeneous landscapes by the endemic pathogen *Penicillium marneffei*. Plos Pathog. 1, 159–165.

Fisher, M.C., Garner, T.W.J., Walker, S.F., 2009. Global emergence of *Batrachochytrium dendrobatidis* and amphibian chytridiomycosis in space, time, and host. Annu. Rev. Microbiol. 63, 291–310.

Fishman, L., Wyatt, R., 1999. Pollinator-mediated competition, reproductive character displacement, and the evolution of selfing in *Arenaria uniflora* (Caryophyllaceae). Evolution 53, 1723–1733.

Fournier, E., Giraud, T., 2008. Sympatric genetic differentiation of a generalist pathogenic fungus, *Botrytis cinerea*, on two different host plants, grapevine and bramble. J. Evol. Biol. 21, 122–132.

Fournier, E., Giraud, T., Albertini, C., Brygoo, Y., 2005. Partition of the *Botrytis cinerea* complex in France using multiple gene genealogies. Mycologia 97, 1251–1267.

Fraser, J.A., Subaran, R.L., Nichols, C.B., Heitman, J., 2003. Recapitulation of the sexual cycle of the primary fungal pathogen *Cryptococcus neoformans* var. *gattii*: implications for an outbreak on Vancouver Island, Canada. Eukaryot. Cell 2, 1036–1045.

Fraser, J.A., Giles, S.S., Wenink, E.C., Geunes-Boyer, S.G., Wright, J.R., Diezmann, S., et al., 2005. Same-sex mating and the origin of the Vancouver Island *Cryptococcus gattii* outbreak. Nature 437, 1360–1364.

Fraser, J.A., Stajich, J.E., Tarcha, E.J., Cole, G.T., Inglis, D.O., Sil, A., et al., 2007. Evolution of the mating type locus: insights gained from the dimorphic primary fungal pathogens *Histoplasma capsulatum*, *Coccidioides immitis*, and *Coccidioides posadasii*. Eukaryot Cell 6, 622–629.

Frenkel, O., Peever, T.L., Chilvers, M.I., Ozkilinc, H., Can, C., Abbo, S., et al., 2010. Ecological genetic divergence of the fungal pathogen *Didymella rabiei* on sympatric wild and domesticated *Cicer* spp. (Chickpea). Appl. Environ. Microbiol. 76, 30–39.

Friesen, T.L., Stukenbrock, E.H., Liu, Z.H., Meinhardt, S., Ling, H., Faris, J.D., et al., 2006. Emergence of a new disease as a result of interspecific virulence gene transfer. Nature Genet. 38, 953–956.

Fuchs, B.B., Mylonakis, E., 2006. Using non-mammalian hosts to study fungal virulence and host defense. Curr. Opin. Microbiol. 9, 346–351.

Fudal, I., Ross, S., Gout, L., Blaise, F., Kuhn, M.L., Eckert, M.R., et al., 2007. Heterochromatin-like regions as ecological niches for avirulence genes in the *Leptosphaeria maculans* genome: map-based cloning of AvrLm6. Mol. Plant Microbe Interact. 20, 459–470.

Gaggiotti, O.E., Jones, F., Lee, W.M., Amos, W., Harwood, J., Nichols, R.A., 2002. Patterns of colonization in a metapopulation of grey seals. Nature 416, 424−427.
Galagan, J.E., Henn, M.R., Ma, L.J., Cuomo, C.A., Birren, B., 2005. Genomics of the fungal kingdom: Insights into eukaryotic biology. Genome Res. 15, 1620−1631.
Galanis, E., MacDougall, L., 2010. Epidemiology of *Cryptococcus gattii*, British Columbia, Canada, 1999−2007. Emerg. Infect Dis. 16, 251−257.
Gao, H., Williamson, S., Bustamante, C.D., 2007. A Markov chain Monte Carlo approach for joint inference of population structure and inbreeding rates from multilocus genotype data. Genetics 176, 1635−1651.
Garner, T.W.J., Perkins, M.W., Govindarajulu, P., Seglie, D., Walker, S., Cunningham, A.A., Fisher, M.C., 2006. The emerging amphibian pathogen Batrachochytrium dendrobatidis globally infects introduced populations of the North American bullfrog, Rana catesbeiana. BIOLOGY LETTERS. Genetics 2, 455−459.
Gavrilets, S., 2004. Fitness Landscapes and the Origin of Species. Princeton University Press, Princeton.
Gbelska, Y., Krijger, J.J., Breunig, K.D., 2006. Evolution of gene families: the multidrug resistance transporter genes in five related yeast species. FEMS Yeast Res. 6, 345−355.
Geiser, D.M., Pitt, J.I., Taylor, J.W., 1998. Cryptic speciation and recombination in the aflatoxin-producing fungus *Aspergillus flavus*. Proc. Natl. Acad. Sci. U.S.A. 95, 388−393.
Gezuele, E., Calegari, L., Sanabría, D., Davel, G., Civila, E., 1993. Isolation in uruguay of *Cryptococcus neoformans* var. *gattii* from nest of the wasp *Polybia occidentalis*. Rev. Iberoam. Micol. 10, 5−6.
Giles, S.S., Dagenais, T.R., Botts, M.R., Keller, N.P., Hull, C.M., 2009. Elucidating the pathogenesis of spores from the human fungal pathogen *Cryptococcus neoformans*. Infect. Immun. 77, 3491−3500.
Giraud, T., 2004. Patterns of within population dispersal and mating of the fungus *Microbotryum violaceum* parasitising the plant *Silene latifolia*. Heredity 93, 559−565.
Giraud, T., 2006. Speciation: selection against migrant pathogens: the immigrant inviability barrier in pathogens. Heredity 97, 316−318.
Giraud, T., Fortini, D., Levis, C., Leroux, P., Brygoo, Y., 1997. RFLP markers show genetic recombination in *Botryotinia fuckeliana* (*Botrytis cinerea*) and transposable elements reveal two sympatric species. Mol. Biol. Evol. 14, 1177−1185.
Giraud, T., Villareal, L.M.M.A., Austerlitz, F., Le Gac, M., Lavigne, C., 2006. Importance of the Life Cycle in Sympatric Host Race Formation and Speciation of Pathogens. Phytopathology 96, 280−287.
Giraud, T., Enjalbert, J., Fournier, E., Delmotte, F., Dutech, C., 2008a. Population genetics of fungal diseases of plants. Parasite 15, 449−454.
Giraud, T., Refregier, G., de Vienne, D.M., Le Gac, M.a., Hood, M.E., 2008b. Speciation in fungi. Fungal Genet. Biol. 45, 791−802.
Giraud, T., Yockteng, R., Lopez-Villavicencio, M., Refregier, G., Hood, M.E., 2008c. Mating system of the anther smut fungus *Microbotryum violaceum*: selfing under heterothallism. Eukaryot Cell 7, 765−775.
Giraud, T., Gladieux, P., Gavrilets, S., 2010. Linking emergence of fungal plant diseases and ecological speciation. Trends Ecol. Evol. 25, 387−395.
Gladieux, P., Zhang, X.-G., Afoufa-Bastien, D., Valdebenito Sanhueza, R.-M., Sbaghi, M., Le Cam, B., 2008. On the origin and spread of the scab disease of apple: out of central Asia. PLoS ONE 3, e1455.

Gladieux, P., Caffier, V., Devaux, M., Le Cam, B., 2010a. Host-specific differentiation among populations of *Venturia inaequalis* causing scab on apple, pyracantha and loquat. Fungal Genet. Biol. 47, 511–521.

Gladieux P., Vercken E., Fontaine M.C., Hood, M.E., Jonot O., Couloux A., Giraud T. 2010b. Maintenance of fungal pathogen species that are specialized to different hosts: allopatric divergence and introgression through secondary contact. Molecular Biology and Evolution in press. doi:10.1093/molbev/msq235.

Gladieux, P., Zhang, X.G., Roldan-Ruiz, I., Caffier, V., Leroy, T., Devaux, M., et al., 2010c. Evolution of the population structure of *Venturia inaequalis*, the apple scab fungus, associated with the domestication of its host. Mol. Ecol. 19, 658–674.

Gleason, F., Kagami, M., Lefevre, E., Sime-Ngando, T., 2008. The ecology of chytrids in aquatic ecosystems: roles in food web dynamics. Fungal Biol. Rev. 17–25.

Goka, K., Yokoyama, J., Une, Y., Kuroki, T., Suzuki, K., Nakahara, M., et al., 2009. Amphibian chytridiomycosis in Japan: distribution, haplotypes and possible route of entry into Japan. Mol. Ecol. 18, 4757–4774.

Goudet, J., 1995. FSTAT (Version 1.2): a computer program to calculate F-statistics. J. Hered. 86, 485–486.

Gout, L., Fudal, I., Kuhn, M.L., Blaise, F., Eckert, M., Cattolico, L., et al., 2006. Lost in the middle of nowhere: the AvrLm1 avirulence gene of the Dothideomycete *Leptosphaeria maculans*. Mol. Microbiol. 60, 67–80.

Goyeau, H., Halkett, F., Zapater, M.F., Carlier, J., Lannou, C., 2007. Clonality and host selection in the wheat pathogenic fungus *Puccinia triticina*. Fungal Genet. Biol. 44, 474–483.

Grant, M.R., Jones, J.D.G., 2009. Hormone (dis)harmony moulds plant health and disease. Science 324, 750–752.

Greene, D.R., Koenig, G., Fisher, M.C., Taylor, J.W., 2000. Soil isolation and molecular identification of *Coccidioides immitis*. Mycologia 92, 406–410.

Griffiths, R.C., Marjoram, P., 1996. Ancestral inference from samples of DNA sequences with recombination. J. Comput. Biol. 3, 479–502.

Griffiths, R.C., Tavare, S., 1994. Sampling theory for neutral alleles in a varying environment. Philos. Trans. R. Soc. Lond. B Biol. Sci. 344, 403–410.

Gudlaugsson, O., Gillespie, S., Lee, K., Vande Berg, J., Hu, J., Messer, S., et al., 2003. Attributable mortality of nosocomial candidemia, revisited. Clin. Infect. Dis. 37, 1172–1177.

Guérin, F., Gladieux, P., Le Cam, B., 2007. Origin and colonization history of newly virulent strains of the phytopathogenic fungus *Venturia inaequalis*. Fungal Genet. Biol. 44, 284–292.

Guillemaud, T., Beaumont, M.A., Ciosi, M., Cornuet, J.M., Estoup, A., 2010. Inferring introduction routes of invasive species using approximate Bayesian computation on microsatellite data. Heredity 104, 88–99.

Guillot, G., 2009a. On the inference of spatial structure from population genetics data. Bioinformatics 25, 1796–1801.

Guillot, G., 2009b. Response to comment on "On the inference of spatial structure from population genetics data". Bioinformatics 25, 1805–1806.

Guillot, G., Leblois, R., Coulon, A., Frantz, A.C., 2009. Statistical methods in spatial genetics. Mol. Ecol. 18, 4734–4756.

Halkett, F., Simon, J.-C., Balloux, F., 2005. Tackling the population genetics of clonal and partially clonal organisms. Trends Ecol. Evol. 20, 194–201.

Hane, J.K., Lowe, R.G.T., Solomon, P.S., Tan, K.-C., Schoch, C.L., Spatafora, J.W., et al., 2007. Dothideomycete plant interactions illuminated by genome sequencing and EST analysis of the wheat pathogen *Stagonospora nodorum*. Plant Cell 19, 3347–3368.
Hardham, A.R., 2005. Phytophthora cinnamomi. Mol. Plant Pathol. 6, 589–604.
Hawksworth, D.L., 1991. The fungal dimension of biodiversity: magnitude, significance, and conservation. Mycol. Res. 95, 641–655.
Heitman, J., Kronstad, J.W., Taylor, J.W., Casselton, L.A., 2007. Sex in Fungi: Molecular Determination and Evolutionary Implications. American Society for Microbiology Press, Washington DC.
Herskowitz, I., 1989. A regulatory hierarchy for cell specialization in yeast. Nature 342, 749–757.
Hey, J., 2006. On the failure of modern species concepts. Trends Ecol. Evol. 21, 447–450.
Hey, J., Machado, C.A., 2003. The study of structured populations—new hope for a difficult and divided science. Nat. Rev. Genet. 4, 535–543.
Hey, J., Nielsen, R., 2004. Multilocus methods for estimating population sizes, migration rates and divergence time, with applications to the divergence of *Drosophila pseudoobscura* and *D. persimilis*. Genetics 167, 747–760.
Hey, J., Nielsen, R., 2007. Integration within the Felsenstein equation for improved Markov chain Monte Carlo methods in population genetics. Proc. Natl. Acad. Sci. U.S.A. 104, 2785–2790.
Hey, J., Won, Y.-J., Sivasundar, A., Nielsen, R., Markert, J., 2004. Using nuclear haplotypes with microsatellites to study gene flow between recently separated *Cichlid* species. Mol. Ecol. 13, 909–919.
Hogenhout, S.A., Van der Hoorn, R.A.L., Terauchi, R., Kamoun, S., 2009. Emerging concepts in effector biology of plant-associated organisms. Mol. Plant Microbe Interact. 22, 115–122.
Holder, M., Lewis, P.O., 2003. Phylogeny estimation: traditional and Bayesian approaches. Nat. Rev. Genet. 4, 275–284.
Hopwood, D.A., 1997. Genetic contributions to understanding polyketide synthases. Chem. Rev. 97, 2465–2497.
Horsfall, J., 1983. Impact of introduced pests on man. In: Wilson, C., Graham, CL (Eds.), Exotic Plant Pests and North American Agriculture. Academic Press, New York, NY, pp. 1–13.
Hovmoller, M.S., Yahyaoui, A.H., Milus, E.A., Justesen, A.F., 2008. Rapid global spread of two aggressive strains of a wheat rust fungus. Mol. Ecol. 17, 3818–3826.
Howlett, B.J., 2006. Secondary metabolite toxins and nutrition of plant pathogenic fungi. Curr. Opin. Plant Biol. 9, 371–375.
Hsueh, Y.P., Fraser, J.A., Heitman, J., 2008. Transitions in sexuality: recapitulation of an ancestral tri- and tetrapolar mating system in *Cryptococcus neoformans*. Eukaryotic Cell 7, 1847–1855.
Hu, G., Leger, R.J.S., 2004. A phylogenomic approach to reconstructing the diversification of serine proteases in fungi. J. Evol. Biol. 17, 1204–1214.
Hubisz, M.J., Falush, D., Stephens, M., Pritchard, J.K., 2009. Inferring weak population structure with the assistance of sample group information. Mol. Ecol. Resour. 9, 1322–1332.
Hudson, R.R., 1987. Estimating the recombination parameter of a finite population-model without selection. Genet. Res. 50, 245–250.
Hudson, R.R., 2001. Two-locus sampling distributions and their application. Genetics 159, 1805–1817.

Hudson, R.R., Kaplan, N.L., 1985. Statistical properties of the number of recombination events in the history of a sample of DNA-sequences. Genetics 111, 147–164.
Huelsenbeck, J.P., Andolfatto, P., 2007. Inference of population structure under a Dirichlet process model. Genetics 175, 1787–1802.
Hull, C.M., Raisner, R.M., Johnson, A.D., 2000. Evidence for mating of the "asexual" yeast *Candida albicans* in a mammalian host. Science 289, 307–310.
Huson, D.H., Bryant, D., 2006. Application of phylogenetic networks in evolutionary studies. Mol. Biol. Evol. 23, 254–267.
Huyse, T., Poulin, R., Théron, A., 2005. Speciation in parasites: a population genetics approach. Trends Parasitol. 21, 469–475.
Idnurm, A., Walton, F., Floyd, A., Heitman, J., 2008. Identification of the sex genes in an early diverged fungus. Nature 451, 193–198.
Ioos, R., Andrieux, A., Marçais, B., Frey, P., 2006. Genetic characterization of the natural hybrid species *Phytophtora alni* as inferred from nuclear and mitochondrial analyses. Fungal Genet. Biol 43, 511–529.
Ivors, K., Garbelotto, M., Vries, I., Ruyter-Spira, C., Hekkert, B., Rosenzweig, N., et al., 2006. Microsatellite markers identify three lineages of *Phytophthora ramorum* in US nurseries, yet single lineages in US forest and European nursery populations. Mol. Ecol. 15, 1493–1505.
Ivors, K.L., Hayden, K.J., Bonants, P.J.M., Rizzo, D.M., Garbelotto, M., 2004. AFLP and phylogenetic analyses of North American and European populations of *Phytophthora ramorum*. Mycol. Res. 108, 378–392.
James, T., et al., 2006. Reconstructing the early evolution of fungi using a six-gene phylogeny. Nature 443, 818–822.
James, T.Y., Porter, D., Hamrick, J.L., Vilgalys, R., 1999. Evidence for limited intercontinental gene flow in the cosmopolitan mushroom, *Schizophyllum commune*. Evolution 53, 1665–1677.
James, T.Y., Litvintseva, A.P., Vilgalys, R., Morgan, J.A.T., Taylor, J.W., Fisher, M.C., et al., 2009. Rapid global expansion of the fungal disease chytridiomycosis into declining and healthy amphibian populations. Plos Pathog. 5, e1000458.
Jargeat, P., Rekangalt, D., Verner, M.C., Gay, G., Debaud, J.C., Marmeisse, R., et al., 2003. Characterisation and expression analysis of a nitrate transporter and nitrite reductase genes, two members of a gene cluster for nitrate assimilation from the symbiotic basidiomycete *Hebeloma cylindrosporum*. Curr. Genet. 43, 199–205.
Jarne, P., Lagoda, P.J.L., 1996. Microsatellites, from molecules to populations and back. Trends Ecol. Evol. 11, 424–429.
Ji, Z.-H., Jia, J.H., Bai, F.-Y., 2009. Four novel *Candida* species in the *Candida albicans/Lodderomyces elongisporus* clade isolated from the gut of flower beetles. Antonie Van Leeuwenhoek Int. J. Gen. Mol. Microbiol. 95, 23–32.
Johannesson, H., Vidal, P., Guarro, J., Herr, R.A., Cole, G.T., Taylor, J.W., 2004. Positive directional selection in the proline-rich antigen (PRA) gene among the human pathogenic fungi *Coccidioides immitis, C. posadasii* and their closest relatives. Mol. Biol. Evol. 21, 1134–1145.
Johnson, J.A., Harrington, T.C., Engelbrecht, C.J.B., 2005. Phylogeny and taxonomy of the North American clade of the *Ceratocystis fimbriata* complex. Mycologia 97, 1067–1092.
Johnson, P.A., Hoppensteadt, F.C., Smith, J.J., Bush, G.L., 1996. Conditions for sympatric speciation: a diploid model incorporating habitat fidelity and non-habitat assortative mating. Evol. Ecol. 10, 187–205.

Jombart, T., 2008. adegenet: a R package for the multivariate analysis of genetic markers. Bioinformatics 24, 1403−1405.
Jombart, T., Pontier, D., Dufour, A.B., 2009. Genetic markers in the playground of multivariate analysis. Heredity 102, 330−341.
Jones, D.A., Jones, J.D.G., 1997. The role of leucine-rich repeat proteins in plant defences. Advances in Botanical Research Incorporating Advances in Plant Pathology 24, 89−167.
Kalinowski, S.T., 2002. Evolutionary and statistical properties of three genetic distances. Mol. Ecol. 11, 1263−1273.
Kamper, J., Kahmann, R., Bolker, M., Ma, L.-J., Brefort, T., Saville, B.J., et al., 2006. Insights from the genome of the biotrophic fungal plant pathogen *Ustilago maydis*. Nature 444, 97−101.
Kang, S., Lebrun, M.H., Farrall, L., Valent, B., 2001. Gain of virulence caused by insertion of a Pot3 transposon in a *Magnaporthe grisea* avirulence gene. Mol. Plant Microbe Interact. 14, 671−674.
Kareiva, P., Watts, S., McDonald, R., Boucher, T., 2007. Domesticated nature: shaping landscapes and ecosystems for human welfare. Science 316, 1866−1869.
Kasuga, T., White, T.J., Koenig, G., McEwen, J., Restrepo, A., Castaneda, E., et al., 2003. Phylogeography of the fungal pathogen *Histoplasma capsulatum*. Mol. Ecol. 12, 3383−3401.
Kauserud, H., Svegarden, I.B., Saetre, G.P., Knudsen, H., Stensrud, O., Schmidt, O., et al., 2007. Asian origin and rapid global spread of the destructive dry rot fungus *Serpula lacrymans*. Mol. Ecol. 16, 3350−3360.
Kavanaugh, L.A., Fraser, J.A., Dietrich, F.S., 2006. Recent evolution of the human pathogen *Cryptococcus neoformans* by intervarietal transfer of a 14-gene fragment. Mol. Biol. Evol. 23, 1879−1890.
Keeling, P.J., Burger, G., Durnford, D.G., Lang, B.F., Lee, R.W., Pearlman, R.E., et al., 2005. The tree of eukaryotes. Trends Ecol. Evol. 20, 670−676.
Keely, S.P., Renauld, H., Wakefield, A.E., Cushion, M.T., Smulian, A.G., Fosker, N., et al., 2005. Gene arrays at *Pneumocystis carinii* telomeres. Genetics 170, 1589−1600.
Keller, N.P., Hohn, T.M., 1997. Metabolic pathway gene clusters in filamentous fungi. Fungal Genet. Biol. 21, 17−29.
Keller, N.P., Turner, G., Bennett, J.W., 2005. Fungal secondary metabolism—from biochemistry to genomics. Nat. Rev. Micro. 3, 937−947.
Keller, S.M., McDermott, J.M., Pettway, R.E., Wolfe, M.S., McDonald, B.A., 1997. Gene flow and sexual reproduction in the wheat glume blotch pathogen *Phaeosphaeria nodorum* (*Anamorph Stagonospora nodorum*). Phytopathology 87, 353−358.
Kellis, M., Birren, B.W., Lander, E.S., 2004. Proof and evolutionary analysis of ancient genome duplication in the yeast *Saccharomyces cerevisiae*. Nature 428, 617−624.
Kidd, S.E., Sorrell, T.C., Meyer, W., 2003. Isolation of two molecular types of *Cryptococcus neoformans* var. *gattii* from insect frass. Med. Mycol. 41, 171−176.
Kidd, S.E., Hagen, F., Tscharke, R.L., Huynh, M., Bartlett, K.H., Fyfe, M., et al., 2004. A rare genotype of *Cryptococcus gattii* caused the cryptococcosis outbreak on Vancouver Island (British Columbia, Canada). Proc. Natl. Acad. Sci. U.S.A. 101, 17258−17263.
Kidd, S.E., Guo, H., Bartlett, K.H., Xu, J., Kronstad, J.W., 2005. Comparative gene genealogies indicate that two clonal lineages of *Cryptococcus gattii* in British Columbia resemble strains from other geographical areas. Eukaryot. Cell 4, 1629−1638.

Kidd, S.E., Bach, P.J., Hingston, A.O., Mak, S., Chow, Y., MacDougall, L., et al., 2007a. *Cryptococcus gattii* dispersal mechanisms, British Columbia, Canada. Emerg. Infect. Dis. 13, 51−57.

Kidd, S.E., Chow, Y., Mak, S., Bach, P.J., Chen, H., Hingston, A.O., et al., 2007b. Characterization of environmental sources of the human and animal pathogen *Cryptococcus gattii* in British Columbia, Canada, and the Pacific Northwest of the United States. Appl. Environ. Microbiol. 73, 1433−1443.

Kilpatrick, A.M., Briggs, C.J., Daszak, P., 2010. The ecology and impact of chytridiomycosis: an emerging disease of amphibians. Trends Ecol. Evol. 25, 109−118.

Kimura, M., 1953. "Stepping stone" model of population structure and the decrease of genetic correlation with distance. Annu. Rep. Natl. Inst. Genet., 3, 62−63, Japan.

Kingman, J.F.C., 1982. The coalescent. Stochastic process. Appl 13, 235−248.

Kiss, L., 2005. Powdery mildew as invasive plant pathogens: new epidemics caused by two North American species in Europe. Mycol. Res. 109, 259−260.

Klee, J., Besana, A.M., Genersch, E., Gisder, S., Nanetti, A., Tam, D.Q., et al., 2007. Widespread dispersal of the microsporidium *Nosema ceranae*, an emergent pathogen of the western honey bee, *Apis mellifera*. J. Invertebr. Pathol. 96, 1−10.

Kohn, L.M., 2005. Mechanisms of fungal speciation. Annu. Rev. Phytopathol. 43, 279−308.

Kolar, C.S., Lodge, D.M., 2001. Progress in invasion biology: predicting invaders. Trends Ecol. Evol. 16, 199−204.

Kondrashov, A., 1986. Sympatric speciation: when is it possible? Biol. J. Linn. Soc. 27, 201−223.

Koufopanou, V., Burt, A., Szaro, T., Taylor, J.W., 2001. Gene genealogies, cryptic species, and molecular evolution in the human patho*gen Coccidioides immi*tis and relatives (Ascomycota, Onygenales). Mol. Biol. Evol. 18, 1246−1258.

Kroken, S., Glass, N.L., Taylor, J.W., Yoder, O.C., Turgeon, B.G., 2003. Phylogenomic analysis of type I polyketide synthase genes in pathogenic and saprobic ascomycetes. Proc. Natl. Acad. Sci. U.S.A. 100, 15670−15675.

Kuehne, H.A., Murphy, H.A., Francis, C.A., Sniegowski, P.D., 2007. Allopatric divergence, secondary contact and genetic isolation in wild yeast populations. Curr. Biol. 17, 407−411.

Kuhn, G., Hijri, M., Sanders, I.R., 2001. Evidence for the evolution of multiple genomes in arbuscular mycorrhizal fungi. Nature 414, 745−748.

Kuhner, M.K., 2009. Coalescent genealogy samplers: windows into population history. Trends Ecol. Evol. 24, 86−93.

Kuhner, M.K., Yamato, J., Felsenstein, J., 2000. Maximum likelihood estimation of recombination rates from population data. Genetics 156, 1393−1401.

Kwon-Chung, K.J., Bennett, J.E., 1984a. Epidemiologic differences between the two varieties of *Cryptococcus neoformans*. Am. J. Epidemiol. 120, 123−130.

Kwon-Chung, K.J., Bennett, J.E., 1984b. High prevalence of *Cryptococcus neoformans* var. *gattii* in tropical and subtropical regions. Zentralbl. Bakteriol. Mikrobiol. Hyg. A. 257, 213−218.

Kwon-Chung, K.J., Varma, A., 2006. Do major species concepts support one, two or more species within *Cryptococcus neoformans*? FEMS Yeast Res. 6, 574−587.

Kwon-Chung, K.J., Edman, J.C., Wickes, B.L., 1992. Genetic association of mating types and virulence in *Cryptococcus neoformans*. Infect. Immun. 60, 602−605.

Kwon-Chung, K.J., Boekhout, T., Fell, J.W., Diaz, M., 2002. Proposal to conserve the name *Cryptococcus gattii* against *C. hondurianus* and *C. bacillisporus* (Basidiomycota, Hymenomycetes, Tremellomycetidae). Taxon 51, 804−806.

Lachke, S.A., Lockhart, S.R., Daniels, K.J., Soll, D.R., 2003. Skin facilitates *Candida albicans* mating. Infect. Immun. 71, 4970–4976.

Larriba, G., 2004. Genome instability, recombination, and adaptation in *Candida albicans*. In: San-Blas, G., Calderone, R. (Eds.), Pathogenic Fungi: Host Interactions and Emerging Strategies for Control. Horizon Press, UK, pp. 285–334.

Latter, B.D.H., 1973. Island model of population differentiation—general solution. Genetics 73, 147–157.

Le Gac, M., Giraud, T., 2008. Existence of a pattern of reproductive character displacement in Basidiomycota but not in Ascomycota. J. Evol. Biol. 21, 761–772.

Le Gac, M., Hood, M.E., Fournier, E., Giraud, T., 2007a. Phylogenetic evidence of host-specific cryptic species in the anther smut fungus. Evolution 61, 15–26.

Le Gac, M., Hood, M.E., Giraud, T., 2007b. Evolution of reproductive isolation within a parasitic fungal complex. Evolution 61, 1781–1787.

Lee, H.-Y., Chou, J.-Y., Cheong, L., Chang, N.-H., Yang, S.-Y., Leu1, J.-Y., 2008. Incompatibility of nuclear and mitochondrial genomes causes hybrid sterility between two yeast species. Cell 135, 1065–1073.

Lengeler, K.B., Cox, G.M., Heitman, J., 2001. Serotype AD strains of *Cryptococcus neoformans* are diploid or aneuploid and are heterozygous at the mating-type locus. Infect. Immun. 69, 115–122.

Li, N., Stephens, M., 2003. Modeling linkage disequilibrium and identifying recombination hotspots using single-nucleotide polymorphism data. Genetics 165, 2213–2233.

Li, W., Metin, B., White, T.C., Heitman, J., 2010. Organization and evolutionary trajectory of the mating type (*MAT*) locus in dermatophyte and dimorphic fungal pathogens. Eukaryot Cell 9, 46–58.

Li, W.H., Fu, Y.X., 1999. Coalescent theory and its application in population genetics. In: Halloran, M.E., Geiser, S. (Eds.), Statistics in Genetics. Springer, Berlin, pp. 45–80.

Linde, C.C., Zhan, J., McDonald, B.A., 2002. Population structure of *Mycosphaerella graminicola*: from lesions to continents. Phytopathology 92, 946–955.

Linz, B., Balloux, F., Moodley, Y., Manica, A., Liu, H., Roumagnac, P., et al., 2007. An African origin for the intimate association between humans and *Helicobacter pylori*. Nature 445, 915–918.

Lips, K., Brem, F., Brenes, R., Reeve, J., Alford, R., Voyles, J., et al., 2006. Emerging infectious disease and the loss of biodiversity in a neotropical amphibian community. Proc. Natl. Acad. Sci. U.S.A. 103, 3165–3170.

Lips, K., Diffendorfer, J., Mendelson, I.J., Sears, M., 2008. Riding the wave: reconciling the roles of disease and climate change in amphibian declines. PLoS Biol. 6, 441–454.

Liu, Z., Bos, J.I.B., Armstrong, M., Whisson, S.C., da Cunha, L., Torto-Alalibo, T., et al., 2005. Patterns of diversifying selection in the phytotoxin-like scr74 gene family of *Phytophthora infestans*. Mol. Biol. Evol. 22, 659–672.

Lizarazo, J., Linares, M., de Bedout, C., Restrepo, A., Agudelo, C.I., Castaneda, E., 2007. [Results of nine years of the clinical and epidemiological survey on cryptococcosis in Colombia, 1997–2005]. Biomedica 27, 94–109.

Loftus, B.J., Fung, E., Roncaglia, P., Rowley, D., Amedeo, P., Bruno, D., et al., 2005. The genome of the basidiomycetous yeast and human pathogen *Cryptococcus neoformans*. Science 307, 1321–1324.

London, R., Orozco, B.S., Mylonakis, E., 2006. The pursuit of cryptococcal pathogenesis: heterologous hosts and the study of cryptococcal host–pathogen interactions. FEMS Yeast Res. 6, 567–573.

Longcore, J., Pessier, A., Nichols, D., 1999. *Batrachochytrium dendrobatidis* gen et sp nov, a chytrid pathogenic to amphibians. Mycologia 91, 219–227.

Lutzoni, F., Kauff, F., Cox, C.J., McLaughlin, D., Celio, G., Dentinger, B., et al., 2004. Where are we in assembling the fungal tree of life, classifying the fungi, and understanding the evolution of their subcellular traits? Am. J. Bot. 91, 1446–1480.

Ma, H., Hagen, F., Stekel, D.J., Johnston, S.A., Sionov, E., Falk, R., et al., 2009. The fatal fungal outbreak on Vancouver Island is characterized by enhanced intracellular parasitism driven by mitochondrial regulation. Proc. Natl. Acad. Sci. U.S.A. 106, 12980–12985.

MacDougall, L., Fyfe, M., 2006. Emergence of *Cryptococcus gattii* in a novel environment provides clues to its incubation period. J. Clin. Microbiol. 44, 1851–1852.

MacDougall, L., Kidd, S.E., Galanis, E., Mak, S., Leslie, M.J., Cieslak, P.R., et al., 2007. Spread of *Cryptococcus gattii* in British Columbia, Canada, and detection in the Pacific Northwest, USA. Emerg. Infect. Dis. 13, 42–50.

Machida, M., Asai, K., Sano, M., Tanaka, T., Kumagai, T., Terai, G., et al., 2005. Genome sequencing and analysis of *Aspergillus oryzae*. Nature 438, 1157–1161.

Magrini, V., Warren, W.C., Wallis, J., Goldman, W.E., Xu, J., Mardis, E.R., et al., 2004. Fosmid-based physical mapping of the *Histoplasma capsulatum* genomne. Genome Res. 14, 1603–1609.

Maiden, M.C.J., Bygraves, J.A., Feil, E., Morelli, G., Russell, J.E., Urwin, R., et al., 1998. Multilocus sequence typing: a portable approach to the identification of clones within populations of pathogenic microorganisms. Proc. Natl. Acad. Sci. U.S.A. 95, 3140–3145.

Malécot, G., 1951. Un traitement stochastique des problèmes linéaires (mutation, linkage, migration) en génétique des populations. Annales de l'Université de Lyon A 14, 79–117.

Mandel, M.A., Barker, B.M., Kroken, S., Rounsley, S.D., Orbach, M.J., 2007. Genomic and population analyses of the mating type loci in *Coccidioides* species reveal evidence for sexual reproduction and gene acquisition. Eukaryot. Cell 6, 1189–1199.

Marcet-Houben, M., Marceddu, G., Gabaldon, T., 2009. Phylogenomics of the oxidative phosphorylation in fungi reveals extensive gene duplication followed by functional divergence. BMC Evol. Biol. 9, 295.

Marjoram, P., Tavare, S., 2006. Modern computational approaches for analysing molecular genetic variation data. Nat. Rev. Genet. 7, 759–770.

Marr, K.A., Carter, R.A., Boeckh, M., Martin, P., Corey, L., 2002a. Invasive aspergillosis in allogeneic stem cell transplant recipients: changes in epidemiology and risk factors. Blood 100, 4358–4366.

Marr, K.A., Patterson, T., Denning, D., 2002b. Aspergillosis. Pathogenesis, clinical manifestations, and therapy. Infect. Dis. Clin. North Am. 16, 875–894, vi.

Martin, F., Kohler, A., Duplessis, S., 2007. Living in harmony in the wood underground: ectomycorrhizal genomics. Curr. Opin. Plant Biol. 10, 204–210.

Martin, F., Aerts, A., Ahren, D., Brun, A., Danchin, E.G.J., Duchaussoy, F., et al., 2008. The genome of *Laccaria bicolor* provides insights into mycorrhizal symbiosis. Nature 452, 88–92.

Mascheretti, S., Croucher, P.J.P., Vettraino, A., Prospero, S., Garbelotto, M., 2008. Reconstruction of the sudden oak death epidemic in California through microsatellite analysis of the pathogen *Phytophthora ramorum*. Mol. Ecol. 17, 2755–2768.

Masneuf, I., Hansen, J., Groth, C., Piskur, J., Dubourdieu, D., 1998. New hybrids between *Saccharomyces* sensu stricto yeast species found among wine and cider production strains. App. Environ. Microbiol. 64, 3887–3892.

Matute, D., McEwen, J., Puccia, R., Montes, B., San-Blas, G., Bagagli, E., et al., 2005. Cryptic speciation and recombination in the fungus *Paracoccidioides brasiliensis* as revealed by gene genealogies. Mol. Biol. Evol. 23, 65–73.

Matute, D.R., Sepulveda, V.E., Quesada, L.M., Goldman, G.H., Taylor, J.W., Restrepo, A., et al., 2006. Microsatellite analysis of three phylogenetic species of *Paracoccidioides brasiliensis*. J. Clin. Microbiol. 44, 2153–2157.

Maynard-Smith, J., 1999. The detection and measurement of recombination from sequence data. Genetics 153, 1021–1027.

Maynard-Smith, J., Smith, N.H., 1998. Detecting recombination from gene trees. Mol. Biol. Evol. 15, 590–599.

Maynard-Smith, J., Smith, N.H., O'Rourke, M., Spratt, B.G., 1993. How clonal are bacteria? Proc. Natl. Acad. Sci. U.S.A. 90, 4384–4388.

Mayr, E., 1963. Animal Species and Evolution. Harvard University Press, Cambridge, MA.

Mboup, M., Leconte, M., Gautier, A., Wan, A.M., Chen, W., de Vallavieille-Pope, C., et al., 2009. Evidence of genetic recombination in wheat yellow rust populations of a Chinese oversummering area. Fungal Genet. Biol. 46, 299–307.

McDonald, B.A., Linde, C., 2002. Pathogen population genetics, evolutionary potential, and durable resistance. Annu. Rev. Phytopathol. 40, 349–379.

McVean, G., Awadalla, P., Fearnhead, P., 2002. A coalescent-based method for detecting and estimating recombination from gene sequences. Genetics 160, 1231–1241.

Meirmans, P.G., Van Tienderen, P.H., 2004. GENOTYPE and GENODIVE: two programs for the analysis of genetic diversity of asexual organisms. Mol. Ecol. Notes 4, 792–794.

Meteyer, C., Buckles, E., Blehert, D., Hicks, A., Green, D., Shearn-Bochsler, V., et al., 2009. Histopathologic criteria to confirm white-nose syndrome in bats. J. Vet. Diagn. Invest. 21, 411–414.

Meyer, W., Aanensen, D.M., Boekhout, T., Cogliati, M., Diaz, M.R., Esposto, M.C., et al., 2009. Consensus multi-locus sequence typing scheme for *Cryptococcus neoformans* and *Cryptococcus gattii*. Med. Mycol. 47, 561–570.

Meyers, B.C., Chin, D.B., Shen, K.A., Sivaramakrishnan, S., Lavelle, D.O., Zhang, Z., et al., 1998. The major resistance gene cluster in lettuce is highly duplicated and spans several megabases. Plant Cell 10, 1817–1832.

Milgroom, M.G., 1996. Recombination and the multilocus structure of fungal populations. Annu. Rev. Phytopathol. 34, 457–477.

Milgroom, M.G., Peever, T.L., 2003. Population biology of plant pathogens—the synthesis of plant disease epidemiology and population genetics. Plant Dis. 87, 608–617.

Milgroom, M.G., Sotirovski, K., Risteski, M., Brewer, M.T., 2009. Heterokaryons and parasexual recombinants of *Cryphonectria parasitica* in two clonal populations in southeastern Europe. Fungal Genet. Biol. 46, 849–854.

Miller, N., Estoup, A., Toepfer, S., Bourguet, D., Lapchin, L., Derridj, S., et al., 2005. Multiple transatlantic introductions of the western corn rootworm. Science 310, 992.

Miranda, M., Bashi, E., Vylkova, S., Edgerton, M., Slayman, C., Rivetta, A., 2009. Conservation and dispersion of sequence and function in fungal TRK potassium transporters: focus on *Candida albicans*. FEMS Yeast Res. 9, 278–292.

Moon, C.D., Craven, K.D., Leuchtmann, A., Clement, S.L., Schardl, C.L., 2004. Prevalence of interspecific hybrids amongst asexual fungal endophytes of grasses. Mol. Ecol. 13, 1455–1467.

Morehouse, E.A., James, T.Y., Ganley, A.R.D., Vilgalys, R., Berger, L., Murphy, P.J., et al., 2003. Multilocus sequence typing suggests the chytrid pathogen of amphibians is a recently emerged clone. Mol. Ecol. 12, 395–403.

Morgan, J., Meltzer, M.I., Plikaytis, B.D., Sofair, A.N., Huie-White, S., Wilcox, S., et al., 2005. Excess mortality, hospital stay, and cost due to candidemia: a case–control study using data from population-based candidemia surveillance. Infect. Control Hosp. Epidemiol. 26, 540–547.

Morgan, J.A.T., Vredenburg, V.T., Rachowicz, L.J., Knapp, R.A., Stice, M.J., Tunstall, T., et al., 2007. Population genetics of the frog-killing fungus *Batrachochytrium dendrobatidis*. Proc. Natl. Acad. Sci. U.S.A. 104, 13845–13850.

Morris, C.E., Bardin, M., Kinkel, L.L., Moury, B., Nicot, P.C., Sands, D.C., 2009. Expanding the paradigms of plant pathogen life history and evolution of parasitic fitness beyond agricultural boundaries. PLoS Pathog 5, e1000693.

Murphy, H.A., Kuehne, H.A., Francis, C.A., Sniegowski, P.D., 2006. Mate choice assays and mating propensity differences in natural yeast populations. Biol. Lett 2, 553–556.

Mylonakis, E., Moreno, R., El Khoury, J.B., Idnurm, A., Heitman, J., Calderwood, S.B., et al., 2005. *Galleria mellonella* as a model system to study *Cryptococcus neoformans* pathogenesis. Infect. Immun. 73, 3842–3850.

Neal, P.R., Anderson, G.J., 2005. Are "mating systems" "breeding systems" of inconsistent and confusing terminology in plant reproductive biology? Or is it the other way around? Plant Syst. Evol. 250, 173–185.

Newcombe, G., Stirling, B., McDonald, S., Vradshaw Jr., H.D., 2000. *Melampsora xcolumbiana*, a natural hybrid of *M. medusae* and *M. occidentalis*. Mycol. Res. 104, 261–274.

Nguyen, N.H., Suh, S.O., Marshall, C.J., Blackwell, M., 2006. Morphological and ecological similarities: wood-boring beetles associated with novel xylose-fermenting yeasts, *Spathaspora passalidarum* gen. sp. nov. and *Candida jeffriesii* sp. nov. Mycol. Res. 110, 1232–1241.

Nguyen, N.H., Suh, S.O., Blackwell, M., 2007. Five novel *Candida* species in insect-associated yeast clades isolated from *Neuroptera* and other insects. Mycologia 99, 842–858.

Nielsen, K., Cox, G.M., Wang, P., Toffaletti, D.L., Perfect, J.R., Heitman, J., 2003. Sexual cycle of *Cryptococcus neoformans* var. *grubii* and virulence of congenic a and alpha isolates. Infect. Immun. 71, 4831–4841.

Nielsen, K., Cox, G.M., Litvintseva, A.P., Mylonakis, E., Malliaris, S.D., Benjamin Jr., D.K., et al., 2005a. *Cryptococcus neoformans* α strains preferentially disseminate to the central nervous system during coinfection. Infect. Immun. 73, 4922–4933.

Nielsen, K., Marra, R.E., Hagen, F., Boekhout, T., Mitchell, T.G., Cox, G.M., et al., 2005b. Interaction between genetic background and the mating-type locus in *Cryptococcus neoformans* virulence potential. Genetics 171, 975–983.

Nielsen, R., 2000. Estimation of population parameters and recombination rates from single nucleotide polymorphisms. Genetics 154, 931–942.

Nielsen, R., Beaumont, M.A., 2009. Statistical inferences in phylogeography. Mol. Ecol. 18, 1034–1047.

Nielsen, R., Wakeley, J., 2001. Distinguishing migration from isolation: a Markov chain Monte Carlo approach. Genetics 158, 885–896.

Nierman, W.C., Pain, A., Anderson, M.J., Wortman, J.R., Kim, H.S., Arroyo, J., et al., 2005. Genomic sequence of the pathogenic and allergenic filamentous fungus *Aspergillus fumigatus*. Nature 438, 1151–1156.

Nosil, P., Vines, T.H., Funk, D.J., 2005. Perspective: reproductive isolation caused by natural selection against immigrants from divergent habitats. Evolution 59, 705–719.

Nunn, M.A., Schaefer, S.M., Petrou, M.A., Brown, J.R., et al., 2007. Environmental source of *Candida dubliniensis*. Emerg. Infect. Dis. 13, 747–750.

O'Donnell, K., Ward, T.J., Geiser, D.M., Kistler, H.C., Aoki, T., 2004. Genealogical concordance between the mating type locus and seven other nuclear genes supports formal

recognition of nine phylogenetically distinct species within the *Fusarium graminearum* clade. Fungal Genet. Biol. 41, 600–623.
O'Gorman, C.M., Fuller, H.T., Dyer, P.S., 2009. Discovery of a sexual cycle in the opportunistic fungal pathogen *Aspergillus fumigatus*. Nature 457, 471–474.
Odds, F., Bougnoux, M., Shaw, D., Bain, J., Davidson, A., Diogo, D., et al., 2007. Molecular phylogenetics of *Candida albicans*. Eukaryot. Cell 6, 1041–1052.
Okagaki, L.H., Strain, A.K., Nielsen, J.N., Charlier, CK., Baltes, N.J., et al., 2010. Cryptococcal cell morphology affects host cell interactions and pathogenicity. PLoS Pathog 6, e1000953.
Oidtmann, B., Schaefers, N., Cerenius, L., Soderhall, K., Hoffmann, R.W., 2004. Detection of genomic DNA of the crayfish plague fungus *Aphanomyces astaci* (Oomycete) in clinical samples by PCR. Vet. Microbiol. 100, 269–282.
Oldroyd, G.E.D., Harrison, M.J., Paszkowski, U., 2009. Reprogramming plant cells for endosymbiosis. Science 324, 753–754.
Olson, A., Stenlid, J., 2001. Plant pathogens. Mitochondrial control of fungal hybrid virulence. Nature 411, 438.
Olson, A., Stenlid, J., 2002. Pathogenic fungal species hybrids infecting plants. Microbes Infect. 4, 1353–1359.
Ordonez, M.E., Kolmer, J.A., 2007. Simple sequence repeat diversity of a worldwide collection of *Puccinia triticina* from durum wheat. Phytopathology 97, 574–583.
Orr, H.A., Turelli, M., 2001. The evolution of postzygotic isolation: accumulating Dobzhansky-Muller incompatibilities. Evolution 55, 1085–1094.
Ottmann, C., Luberacki, B., Kufner, I., Koch, W., Brunner, F., Weyand, M., et al., 2009. A common toxin fold mediates microbial attack and plant defense. Proc. Natl. Acad. Sci. U.S.A. 106, 10359–10364.
Otto, S.P., 2009. The evolutionary enigma of sex. Am. Nat. 174, S1–S14.
Paetkau, D., Calvert, W., Stirling, I., Strobeck, C., 1995. Microsatellite analysis of population-structure in canadian polar bears. Mol. Ecol. 4, 347–354.
Pallen, M.J., Wren, B.W., 2007. Bacterial pathogenomics. Nature 449, 835–842.
Palm, M.E., 2001. Systematics and the impact of invasive fungi on agriculture in the United States. BioScience 51, 141–147.
Panstruga, R., Dodds, P.N., 2009. Terrific protein traffic: the mystery of effector protein delivery by filamentous plant pathogens. Science 324, 748–750.
Pappas, P.G., Rex, J.H., Lee, J., Hamill, R.J., Larsen, R.A., Powderly, W., et al., 2003. A prospective observational study of candidemia: epidemiology, therapy, and influences on mortality in hospitalized adult and pediatric patients. Clin. Infect. Dis. 37, 634–643.
Pappas, P.G., Kauffman, C.A., Andes, D., Benjamin Jr., D.K., Calandra, T.F., Edwards Jr., J.E., et al., 2009. Clinical practice guidelines for the management of candidiasis: 2009 update by the Infectious Diseases Society of America. Clin. Infect. Dis. 48, 503–535.
Park, B.J., Wannemuehler, K.A., Marston, B.J., Govender, N., Pappas, P.G., Chiller, T.M., 2009. Estimation of the current global burden of cryptococcal meningitis among persons living with HIV/AIDS. AIDS 23, 525–530.
Parniske, M., Hammond-Kosack, K.E., Goldstein, C., Thomas, C.M., Jones, D.A., Harrisson, K., et al., 1997. Novel disease resistance specificities result from sequence exchange between tandemly repeated genes at the Cf-4/9 locus of tomato. Cell 91, 821–932.
Paterson, D.L., Singh, N., 1999. Invasive aspergillosis in transplant recipients. Medicine (Baltimore) 78, 123–138.
Patterson, N., Price, A.L., Reich, D., 2006. Population structure and eigenanalysis. Plos Genet. 2, 2074–2093.

Peever, T., 2007. Role of host specificity in the speciation of *Ascochyta* pathogens of cool season food legumes. Eur. J. Plant Pathol 119, 119–126.
Peever, T.L., Zeigler, T.S., Dorrance, A.E., Correa-Victoria, F.J., Martin, S.S., 2000. Pathogen Population Genetics and Breeding for Disease Resistance (http://www.apsnet.org/online/feature/PathPopGenetics/). pp.
Peever, T.L., Salimath, S.S., Su, G., Kaiser, W.J., Muehlbauer, F.J., 2004. Historical and contemporary multilocus population structure of *Ascochyta rabiei* (teleomorph: *Didymella rabiei*) in the Pacific Northwest of the United States. Mol. Ecol. 13, 291–309.
Pella, J., Masuda, M., 2006. The Gibbs and split-merge sampler for population mixture analysis from genetic data with incomplete baselines. Can. J. Fish. Aquat. Sci. 63, 576–596.
Penalva, M.A., Moya, A., Dopazo, J., Ramon, D., 1990. Sequences of isopenicillin-*N* synthetase genes suggest horizontal gene-transfer from prokaryotes to eukaryotes. Proc. R. Soc. Lond. B Biol. Sci. 241, 164–169.
Perfect, J.R., 1989. Cryptococcosis. Infect. Dis. Clin. North Am. 3, 77–102.
Pessier, A., Nichols, D., Longcore, J., Fuller, M., 1999. Cutaneous chytridiomycosis in poison dart frogs (*Dendrobates* spp.) and white's tree frogs (*Litoria caerulea*). J. Vet. Diagn. Invest. 11, 194–199.
Pfaller, M.A., Diekema, D.J., 2004. Rare and emerging opportunistic fungal pathogens: concern for resistance beyond *Candida albicans* and *Aspergillus fumigatus*. J. Clin. Microbiol. 42, 4419–4431.
Pimentel, D., McNair, S., Janecka, J., Wightman, J., Simmonds, C., O'Connell, C., et al., 2001. Economic and environmental threats of alien plant, animal, and microbe invasions. Agric. Ecosyst. Environ. 84, 1–20.
Piotrowski, J., Annis, S., Longcore, J., 2004. Physiology of *Batrachochytrium dendrobatidis*, a chytrid pathogen of amphibians. Mycologia 96, 9–15.
Piry, S., Alapetite, A., Cornuet, J.M., Paetkau, D., Baudouin, L., Estoup, A., 2004. GENECLASS2: a software for genetic assignment and first-generation migrant detection. J. Hered. 95, 536–539.
Poggeler, D., Masloff, S., Jacobsen, S., Kuck, U., 2000. Karyotype polymorphism correlates with intraspecific infertility in the homothallic ascomycete *Sordaria macrospora*. J. Evol. Biol. 13, 281–289.
Posada, D., Crandall, K.A., 2001. Intraspecific gene genealogies: trees grafting into networks. Trends Ecol. Evol. 16, 37–45.
Power, A.G., Mitchell, C.E., 2004. Pathogen spillover in disease epidemics. Am. Nat. 164, S79–S89.
Pringle, A., Baker, D., Platt, J., Wares, J., Latgé, J., Taylor, J., 2005. Cryptic speciation in the cosmopolitan and clonal human pathogenic fungus *Aspergillus fumigatus*. Evolution 59, 1886–1899.
Pringle, A., Bever, J.D., Gardes, M., Parrent, J.L., Rillig, M.C., Klironomos, J.N., 2009. Mycorrhizal symbioses and plant invasions. Annu. Rev. Ecol. Evol. Syst. 40, 699–715.
Pritchard, J.K., Seielstad, M.T., Perez-Lezaun, A., Feldman, M.W., 1999. Population growth of human Y chromosomes: a study of Y chromosome microsatellites. Mol. Biol. Evol. 16, 1791–1798.
Pritchard, J.K., Stephens, M., Donnelly, P., 2000. Inference of population structure using multilocus genotype data. Genetics 155, 945–959.
Pritchard, J.K., Wen, X., Falush, D., 2007. Documentation for structure software: Version 2.2. pp.

Prospero, S., Black, J.A., Winton, L.M., 2004. Isolation and characterization of microsatellite markers in *Phytophthora ramorum*, the causal agent of sudden oak death. Mol. Ecol. Notes 4, 672–674.

Prospero, S., Hansen, E.M., Grunwald, N.J., Winton, L.M., 2007. Population dynamics of the sudden oak death pathogen *Phytophthora ramorum* in Oregon from 2001 to 2004. Mol. Ecol. 16, 2958–2973.

Puechmaille, S.J., Verdeyroux, P., Fuller, H., Gouilh, M., Beckaert, M., Teeling, E.C., 2010. White-nose syndrome fungus (*Geomyces destructans*) in Bat, France. Emerg. Infect. Dis. 16, 290–293, DOI: 10.3201/eid1602.091391.

Ramachandran, S., Deshpande, O., Roseman, C.C., Rosenberg, N.A., Feldman, M.W., Cavalli-Sforza, L.L., 2005. Support from the relationship of genetic and geographic distance in human populations for a serial founder effect originating in Africa. Proc. Natl. Acad. Sci. U.S.A. 102, 15942–15947.

Rannala, B., Mountain, J.L., 1997. Detecting immigration by using multilocus genotypes. Proc. Natl. Acad. Sci. U.S.A. 94, 9197–9201.

Rappleye, C.A., Goldman, W.E., 2006. Defining virulence genes in the dimorphic fungi. Annu. Rev. Microbiol. 60, 281–303.

Refrégier, G., Le Gac, M., Jabbour, F., Widmer, A., Hood, M., Yockteng, R., et al., 2008. Cophylogeny of the anther smut fungi and their Caryophyllaceous hosts: prevalence of host shifts and importance of delimiting parasite species. BMC Evol. Biol. 8, 100.

Rehmeyer, C., Li, W.X., Kusaba, M., Kim, Y.S., Brown, D., Staben, C., et al., 2006. Organization of chromosome ends in the rice blast fungus, *Magnaporthe oryzae*. Nucleic Acids Res. 34, 4685–4701.

Reis, R.S., Almeida-Paes, R., Muniz Mde, M., Tavares, P.M., Monteiro, P.C., Schubach, T.M., et al., 2009. Molecular characterisation of *Sporothrix schenckii* isolates from humans and cats involved in the sporotrichosis epidemic in Rio de Janeiro, Brazil. Mem. Inst. Oswaldo Cruz 104, 769–774.

Reynoso, M., Chulze, S., Zeller, K., Torres, A., Leslie, J., 2009. Genetic structure of *Fusarium verticillioides* populations isolated from maize in Argentina. Eur. J. Plant Pathol. 123, 207–215.

Rice, W.R., 1984. Disruptive selection on habitat preference and the evolution of reproductive isolation: a simulation study. Evolution 38, 1251–1260.

Richardson, M.D., 1991. Opportunistic and pathogenic fungi. J. Antimicrob. Chemother. 28, 1–11.

Rimek, D., Haase, G., Luck, A., Casper, J., Podbielski, A., 2004. First report of a case of meningitis caused by *Cryptococcus adeliensis* in a patient with acute myeloid leukemia. J. Clin. Microbiol. 42, 481–483.

Rippon, J.W., 1988. Medical Mycology. WB Saunders, Philadelphia, PA.

Rivas, G.-G., Zapater, M.-F., Abadie, C., Carlier, J., 2004. Founder effects and stochastic dispersal at the continental scale of the fungal pathogen of bananas *Mycosphaerella fijiensis*. Mol. Ecol. 13, 471–482.

Rizzo, D.M., Garbelotto, M., Hansen, E.A., 2005. *Phytophthora ramorum*: integrative research and management of an emerging pathogen in California and Oregon forests. Annu. Rev. Phytopathol. 43, 309–335.

Rousset, F., 1997. Genetic differentiation and estimation of gene flow from F-statistics under isolation by distance. Genetics 145, 1219–1228.

Rousset, F., 2000. Genetic differentiation between individuals. J. Evol. Biol. 13, 58–62.

Rousset, F., 2008a. Dispersal estimation: demystifying Moran's I. Heredity 100, 231–232.

Rousset, F., 2008b. Genepop'007: a complete re-implementation of the genepop software for Windows and Linux. Mol. Ecol. Res. 8, 103–106.
Roy, B.A., 1998. Cryptic species in the *Puccinia monoica* complex. Mycologia 90, 846–853.
Rydholm, C., Szakacs, G., Lutzoni, F., 1998. Low genetic variation and no detectable population structure in *Aspergillus fumigatus* compared to closely related Neosartorya species. Eukaryotic Cell 5, 650–657.
Salamati, S., Zhan, J., Burdon, J.J., McDonald, B.A., 2000. The genetic structure of field populations of *Rhynchosporium secalis* from three continents suggests moderate gene flow and regular recombination. Phytopathology 90, 901–908.
Sampaio, J.P., Inacio, J., Fonseca, A., Gadanho, M., Spencer-Martins, I., Scorzetti, G., et al., 2004. *Auriculibuller fuscus* gen. nov., sp. nov. and *Bullera japonica* sp. nov., novel taxa in the Tremellales. Int. J. Syst. Evol. Microbiol. 54, 987–993.
Sánchez-Alonso, P., Guzman, P., 2008. Predicted elements of telomere organization and function in *Ustilago maydis*. Fungal Genet. Biol. 45, S54–S62.
Scannell, D.R., Byrne, K.P., Gordon, J.L., Wong, S., Wolfe, K.H., 2006. Multiple rounds of speciation associated with reciprocal gene loss in polyploid yeasts. Nature 440, 341–345.
Schaffer, R., 1975. The major groups of Basidiomycetes. Mycologia 66, 1–18.
Schardl, C.L., Craven, K.D., 2003. Interspecific hybridization in plant-associated fungi and oomycetes: a review. Mol. Ecol. 12, 2861–2873.
Schoch, C.L., Aist, J.R., Yoder, O.C., Gillian Turgeon, B., 2003. A complete inventory of fungal kinesins in representative filamentous ascomycetes. Fungal Genet. Biol. 39, 1–15.
Schulze, J., Sonnenborn, U., 2009. Yeasts in the gut: from commensals to infectious agents. Dtsch. Arztebl. Int. 106, 837–842.
Schurko, A.M., Neiman, M., Logsdon, J.M., 2009. Signs of sex: what we know and how we know it. Trends Ecol. Evol. 24, 208–217.
Seaton, R.A., 1996. The management of cryptococcal meningitis in Papua New Guinea. P. N. G. Med. J. 39, 67–73.
Sharpton, T.J., Stajich, J.E., Rounsley, S.D., Gardner, M.J., Wortman, J.R., Jordar, V.S., et al., 2009. Comparative genomic analyses of the human fungal pathogens *Coccidioides* and their relatives. Genome Res. 19, 1722–1731.
Shen, B., 2000. Biosynthesis of aromatic polyketides, Biosynthesis: Aromatic Polyketides, Isoprenoids, Alkaloids, vol. 209. pp. 1–51 Topics in Current Chemistry.
Sidhu, G.S., 2002. Mycotoxin genetics and gene clusters. Eur. J. Plant Pathol. 108, 705–711.
Singh, N., Alexander, B.D., Lortholary, O., Dromer, F., Gupta, K.L., John, G.T., et al., 2008. Pulmonary cryptococcosis in solid organ transplant recipients: clinical relevance of serum cryptococcal antigen. Clin. Infect. Dis. 46, e12–8.
Slatkin, M., 1985. Gene-flow in natural populations. Annu. Rev. Ecol. Syst. 16, 393–430.
Slatkin, M., 1987. Gene flow and the geographic structure of natural populations. Science 236, 787–792.
Sloan, D., Giraud, T., Hood, M., 2008. Maximized virulence in a sterilizing pathogen: the anther-smut fungus and its co-evolved hosts. J. Evol. Biol. 21, 1544–1554.
Smulian, A.G., Sesterhenn, T., Tanaka, R., Cushion, M.T., 2001. The ste3 pheromone receptor gene of *Pneumocystis carinii* is surrounded by a cluster of signal transduction genes. Genetics 157, 991–1002.
Soanes, D.M., Richard, T., Talbot, N., 2007. Insights from sequencing fungal and oomycete genomes: what can we learn about plant disease and the evolution of pathogenicity? The Plant Cell 19, 3318–3326.

Sorrell, T.C., 2001. *Cryptococcus neoformans* variety *gattii*. Med. Mycol. 39, 155–168.
Spratt, B.G., Hanage, W.P., Li, B., Aanensen, D.M., Feil, E.J., 2004. Displaying the relatedness among isolates of bacterial species—the eBURST approach. FEMS Microbiol. Lett. 241, 129–134.
Staats, M., van Baarlen, P., van Kan, J.A.L., 2005. Molecular phylogeny of the plant pathogenic genus *Botrytis* and the evolution of host specificity. Mol. Biol. Evol. 22, 333–346.
Staats, M., van Baarlen, P., Schouten, A., van Kan, J.A.L., Bakker, F.T., 2007. Positive selection in phytotoxic protein-encoding genes of *Botrytis* species. Fungal Genet. Biol. 44, 52–63.
Staples, R.C., 2000. Research on the rust fungi during the twentieth century. Annu. Rev. Phytopathol. 38, 49–69.
Stenberg, P., Lundmark, M., Saura, A., 2003. MLGsim: a program for detecting clones using a simulation approach. Mol. Ecol. Notes 3, 329–331.
Stephens, M., 2008. Inference under the coalescent. In: Balding, D.J., Bishop, M., Cannings, C. (Eds.), Handbook of Statistical Genetics, third ed. John Wiley & Sons, Ltd, Chichester, pp. 878–908.
Stukenbrock, E., Banke, S., McDonald, B., 2006. Global migration patterns in the fungal wheat pathogen *Phaeosphaeria nodorum*. Mol. Ecol. 15, 2895–2904.
Stukenbrock, E.H., McDonald, B.A., 2007. Geographical variation and positive diversifying selection in the host-specific toxin SnToxA. Mol. Plant Pathol. 8, 321–332.
Stukenbrock, E.H., McDonald, B.A., 2008. The origins of plant pathogens in agro-ecosystems. Annu. Rev. Phytopathol. 46, 75–100.
Stukenbrock, E.H., Banke, S., Javan-Nikkhah, M., McDonald, B.A., 2007. Origin and domestication of the fungal wheat pathogen *Mycosphaerella graminicola* via sympatric speciation. Mol. Biol. Evol. 24, 398–411.
Suh, S.O., Blackwell, M., 2004. Three new beetle-associated yeast species in the *Pichia guilliermondii* clade. FEMS Yeast Res. 5, 87–95.
Suh, S.O., Blackwell, M., 2005. Four new yeasts in the *Candida mesenterica* clade associated with basidiocarp-feeding beetles. Mycologia 97, 167–177.
Suh, S.O., Marshall, C.J., McHugh, J.V., Blackwell, M., 2003. Wood ingestion by passalid beetles in the presence of xylose-fermenting gut yeasts. Mol. Ecol. 12, 3137–3145.
Suh, S.O., Gibson, C.M., Blackwell, M., 2004a. *Metschnikowia chrysoperlae* sp. nov., *Candida picachoensis* sp. nov. and *Candida pimensis* sp. nov., isolated from the green lacewings *Chrysoperla comanche* and *Chrysoperla carnea* (Neuroptera: Chrysopidae). Int. J. Syst. Evol. Microbiol. 54, 1883–1890.
Suh, S.O., McHugh, J.V., Blackwell, M., 2004b. Expansion of the *Candida tanzawaensis* yeast clade: 16 novel *Candida* species from basidiocarp-feeding beetles. Int. J. Syst. Evol. Microbiol. 54, 2409–2429.
Suh, S.O., Nguyen, N.H., Blackwell, M., 2005. Nine new *Candida* species near *C. membranifaciens* isolated from insects. Mycol. Res. 109, 1045–1056.
Suh, S.O., Nguyen, N.H., Blackwell, M., 2006. A yeast clade near *Candida kruisii* uncovered: nine novel *Candida* species associated with basidioma-feeding beetles. Mycol. Res. 110, 1379–1394.
Suh, S.O., Nguyen, N.H., Blackwell, M., 2008. Yeasts isolated from plant-associated beetles and other insects: seven novel *Candida* species near *Candida albicans*. FEMS Yeast Res. 8, 88–102.
Sukroongreung, S., Nilakul, C., Ruangsomboon, O., Chuakul, W., Eampokalap, B., 1996. Serotypes of *Cryptococcus neoformans* isolated from patients prior to and during the AIDS era in Thailand. Mycopathologia 135, 75–78.

Sung, G.H., Hywel-Jones, N.L., Sung, J.M., Luangsa-Ard, J.J., Shrestha, B., Spatafora, J.W., 2007. Phylogenetic classification of *Cordyceps* and the clavicipitaceous fungi. Stud. Mycol. 5–59.

Szpiech, Z.A., Jakobsson, M., Rosenberg, N.A., 2008. ADZE: a rarefaction approach for counting alleles private to combinations of populations. Bioinformatics 24, 2498–2504.

Tampakakis, E., Peleg, A.Y., Mylonakis, E., 2009. Interaction of *Candida albicans* with an intestinal pathogen, *Salmonella enterica* serovar *typhimurium*. Eukaryot Cell 8, 732–737.

Tatem, A.J., Rogers, D.J., Hay, S.I., 2006. Global transport networks and infectious disease spread. Adv. Parasitol. 62, 293–343.

Tavare, S., Balding, D.J., Griffiths, R.C., Donnelly, P., 1997. Inferring coalescence times from DNA sequence data. Genetics 145, 505–518.

Taylor, J.W., Fisher, M.C., 2003. Fungal multilocus sequence typing—it's not just for bacteria. Curr. Opin. Microbiol. 6, 351–356.

Taylor, J.W., Geiser, D.M., Burt, A., Koufopanou, V., 1999a. The evolutionary biology and population genetics underlying fungal strain typing. Clin. Microbiol. Rev. 12, 126–146.

Taylor, J.W., Jacobson, D., Fisher, M., 1999b. The evolution of asexual fungi: reproduction, speciation and classification. Annu. Rev. Phytopathol. 37, 197–246.

Taylor, J.W., Jacobson, D.J., Kroken, S., Kasuga, T., Geiser, D.M., Hibbett, D.S., et al., 2000. Phylogenetics species recognition and species concepts in fungi. Fungal Genet. Biol. 31, 21–32.

Taylor, J.W., Turner, E., Townsend, J.P., Dettman, J.R., Jacobson, D., 2006. Eukaryotic microbes, species recognition and the geographic limits of species: examples from the kingdom fungi. Philos. Trans. R. Soc. B Biol. Sci. 361, 1947–1963.

Tellier, A., de Vienne, D., Giraud, T., Shykoff, J., Hood, M., Refrégier, G., 2010. Theory and examples of host–parasite reciprocal influence: from short-term to long term interactions, from coevolution to cospeciation and host shifts. Host–Pathogen Interactions: Genetics, Immunology and Physiology. Nova Science Publishers, New York, NY, pp.

Tenzer, I., Gessler, C., 1999. Genetic diversity of *Venturia inaequalis* across Europe. Eur. J. Plant Pathol. 105, 545–552.

Thrall, P., Hochberg, M., Burdon, J., Bever, J., 2006. Coevolution of symbiotic mutualists and parasites in a community context. Trends in Ecology and Evolution 22, 120–126.

Tibayrenc, M., 1998. Genetic epidemiology of parasitic protozoa and other infectious agents: the need for an integrated approach. Int. J. Parasitol. 28, 85–104.

Tibayrenc, M., 2005. Bridging the gap between molecular epidemiologists and evolutionists. Trends Microbiol. 13, 575–580.

Timms, R., Read, A.F., 1999. What makes a specialist special? Trends Ecol. Evol. 14, 333–334.

Toffaletti, D.L., Nielsen, K., Dietrich, F., Heitman, J., Perfect, J.R., 2004. *Cryptococcus neoformans* mitochondrial genomes from serotype A and D strains do not influence virulence. Curr. Genet 46, 193–204.

Tsai, I.J., Bensasson, D., Burt, A., Koufopanou, V., 2008. Population genomics of the wild yeast *Saccharomyces paradoxus*: quantifying the life cycle. Proc. Natl. Acad. Sci. U.S. A. 105, 4957–4962.

Upton, A., Fraser, J.A., Kidd, S.E., Bretz, C., Bartlett, K.H., Heitman, J., et al., 2007. First contemporary case of human infection with *Cryptococcus gattii* in Puget Sound: evidence for spread of the Vancouver Island outbreak. J. Clin. Microbiol. 45, 3086–3088.

Urwin, R., Maiden, M.C.J., 2003. Multi-locus sequence typing: a tool for global epidemiology. Trends Microbiol. 11, 479–487.
van Putten, W.F., Elzinga, J.A., Biere, A., 2007. Host fidelity of the pollinator guilds of *Silene dioica* and *Silene latifolia*: possible consequences for sympatric host race differentiation of a vectored plant disease. Int. J. Plant Sci 168, 421–434.
Vanittanakom Jr., N., Cooper, C.R., Fisher, M.C., Sirisanthana, T., 2006. *Penicillium marneffei* infection and recent advances in the epidemiology and molecular biology aspects. Clin. Microbiol. Rev. 19, 95–110.
Vekemans, X., Hardy, O.J., 2004. New insights from fine-scale spatial genetic structure analyses in plant populations. Mol. Ecol. 13, 921–935.
Velagapudi, R., Hsueh, Y.P., Geunes-Boyer, S., Wright, J.R., Heitman, J., 2009. Spores as infectious propagules of *Cryptococcus neoformans*. Infect. Immun. 77, 4345–4355.
Vercken, E., Fontaine, M.C., Gladieux, P., Hood, M.E., Jonot, O., Giraud, T. Glacial refugia in pathogens: European genetic structure of anther smut pathogens on *Silene latifolia* and *S. dioica*. PloS Pathogens in press.
Verstrepen, K.J., Fink, G.R., 2009. Genetic and epigenetic mechanisms underlying cell-surface variability in protozoa and fungi. Annu. Rev. Genet. 43, 1–24.
Viaud, M., Couteaudier, Y., Riba, G., 1998. Molecular analysis of hypervirulent somatic hybrids of the entomopathogenic fungi *Beauveria bassiana* and *Beauveria sulfurescens*. Appl. Environ. Microbiol. 64, 88–93.
Voyles, J., Young, S., Berger, L., Campbell, C., Voyles, W., Dinudom, A., et al., 2009. Pathogenesis of chytridiomycosis, a cause of catastrophic amphibian declines. Science 326, 582–585.
Wagner, H.H., Holderegger, R., Werth, S., Gugerli, F., Hoebee, S.E., Scheidegger, C., 2005. Variogram analysis of the spatial genetic structure of continuous populations using multilocus microsatellite data. Genetics 169, 1739–1752.
Wahl, R., Wippel, K., Goos, S., Kamper, J., Sauer, N., 2010. A novel high-affinity sucrose transporter is required for virulence of the plant pathogen *Ustilago maydis*. PLoS Biol. 8, e1000303.
Wakeley, J., 1997. Using the variance of pairwise differences to estimate the recombination rate. Genet. Res. 69, 45–48.
Wall, J.D., 2000. A comparison of estimators of the population recombination rate. Mol. Biol. Evol. 17, 156–163.
Walsh, T.J., Anaissie, E.J., Denning, D.W., Herbrecht, R., Kontoyiannis, D.P., Marr, K.A., et al., 2008. Treatment of aspergillosis: clinical practice guidelines of the Infectious Diseases Society of America. Clin. Infect. Dis. 46, 327–360.
Walton, J.D., 2000. Horizontal gene transfer and the evolution of secondary metabolite gene clusters in fungi: an hypothesis. Fungal Genet. Biol. 30, 167–171.
Wang, R.L., Wakeley, J., Hey, J., 1997. Gene flow and natural selection in the origin of *Drosophila pseudoobscura* and close relatives. Genetics 147, 1091–1106.
Ward, T.J., Bielawski, J.P., Kistler, H.C., Sullivan, E., O'Donnell, K., 2002. Ancestral polymorphism and adaptative evolution in the trichothecene mycotoxin gene cluster of phytopathogenic *Fusarium*. Proc. Natl. Acad. Sci. U.S.A. 99, 9278–9283.
Watterson, G.A., 1975. Number of segregating sites in genetic models without recombination. Theor. Popul. Biol. 7, 256–276.
Welch, D.M., Meselson, M., 2000. Evidence for the evolution of *Bdelloid rotifers* without sexual reproduction or genetic exchange. Science 288, 1211–1215.
Weldon, C., du Preez, L., Hyatt, A., Muller, R., Speare, R., 2004. Origin of the amphibian chytrid fungus. Emerg. Infect. Dis. 10, 2100–2105.

Whitlock, M.C., McCauley, D.E., 1999. Indirect measures of gene flow and migration: F-ST not equal 1/(4Nm + 1). Heredity 82, 117–125.
Williams, P.D., 2009. Darwinian interventions: taming pathogens through evolutionary ecology. Trends Parasitol. 26, 83–92.
Wilson, G.A., Rannala, B., 2003. Bayesian inference of recent migration rates using multilocus genotypes. Genetics 163, 1177–1191.
Wisplinghoff, H., Bischoff, T., Tallent, S.M., Seifert, H., Wenzel, R.P., Edmond, M.B., 2004. Nosocomial bloodstream infections in US hospitals: analysis of 24,179 cases from a prospective nationwide surveillance study. Clin. Infect. Dis. 39, 309–317.
Wostemeyer, J., Kreibich, A., 2002. Repetitive DNA elements in fungi (Mycota): impact on genomic architecture and evolution. Curr. Genet. 41, 189–198.
Wright, S., 1931. Evolution in Mendelian populations. Genetics 16, 0097–0159.
Wright, S., 1943. Isolation by distance. Genetics 28, 114–138.
Wright, S., 1946. Isolation by distance under diverse systems of mating. Genetics 31, 39–59.
Wu, C., Ting, C., 2004. Genes and speciation. Nat. Rev. Genet. 5, 114–122.
Xu, J., Saunders, C.W., Hu, P., Grant, R.A., Boekhout, T., Kuramae, E.E., et al., 2007. Dandruff-associated *Malassezia* genomes reveal convergent and divergent virulence traits shared with plant and human fungal pathogens. Proc. Natl. Acad. Sci. U.S.A. 104, 18730–18735.
Xue, C., Tada, Y., Dong, X., Heitman, J., 2007. The human fungal pathogen *Cryptococcus* can complete its sexual cycle during a pathogenic association with plants. Cell Host Microbe 1, 263–273.
Xue, C., Liu, T., Chen, L., Li, W., Liu, I., Kronstad, J.W., Seyfang, A., Heitman, J., 2010. Role of an Expanded Inositol Transporter Repertoire in *Cryptococcus neoformans* Sexual Reproduction and Virulence. MBio. 1, e00084-10.
Yarwood, C.E., 1970. Man-made plant diseases. Science 168, 218–220.
Yuen, K., Pascal, G., Wong, S.S.Y., Glaser, P., Woo, P.C.Y., Kunst, F., et al., 2003. Exploring the *Penicillium marneffei* genome. Arch. Microbiol. 179, 339–353.
Zaffarano, P.L., McDonald, B.A., Linde, C.C., 2008. Rapid speciation following recent host shifts in the palnt pathogenic fungus *Rhynchosporium*. Evolution 62, 1418–1436.
Zaffarano, P.L., McDonald, B.A., Linde, C.C., 2009. Phylogeographical analyses reveal global migration patterns of the barley scald pathogen *Rhynchosporium secalis*. Mol. Ecol. 18, 279–293.
Zaoutis, T.E., Argon, J., Chu, J., Berlin, J.A., Walsh, T.J., Feudtner, C., 2005. The epidemiology and attributable outcomes of candidemia in adults and children hospitalized in the United States: a propensity analysis. Clin. Infect. Dis. 41, 1232–1239.
Zeller, K.A., Bowden, R.L., Leslie, J.F., 2004. Population differentiation and recombination in wheat scab populations of *Gibberella zeae* from the United States. Mol. Ecol. 13, 563–571.
Zhan, J., Pettway, R.E., McDonald, B.A., 2003. The global genetic structure of the wheat pathogen *Mycosphaerella graminicola* is characterized by high nuclear diversity, low mitochondrial diversity, regular recombination, and gene flow. Fungal Genet. Biol. 38, 286–297.
Zhou, X.D., Burgess, T.I., Beer, Z.W., Lieutier, F., Yart, A., Klepzig, K., et al., 2007. High intercontinental migration rates and population admixture in the sapstain fungus *Ophiostoma ips*. Mol. Ecol. 16, 89–99.
Zolan, M.E., 1995. Chromosome-length polymorphism in fungi. Microbiol. Rev. 59, 686–698.

5 Clonal Evolution

Thierry de Meeûs[1,2,*] *and Franck Prugnolle*[3]

[1]Interactions Hôtes-Vecteurs-Parasites dans les infections par des Trypanosomatidae (TRYPANOSOM), Centre International de Recherche-Développement sur l'Elevage en zone Subhumide (CIRDES), Bobo-Dioulasso, Burkina-Faso, [2]CNRS, Délégation Languedoc-Roussillon, Montpellier, France, [3]Génétique et Evolution des Maladies Infectieuses (GEMI), UMR CNRS/IRD 2724, Centre IRD de Montpellier, Montpellier, France

5.1 Introduction

Asexual reproduction is probably the most widespread means of biological propagation (De Meeûs et al., 2007b, 2009b) and is probably the oldest one, though recombination might be almost as old (Cavalier-Smith, 2002). But this of course depends on what is meant and what is understood (not always the same thing) by clonality and recombination.

Asexual reproduction has been the subject of numerous studies and reviews from diverse biological disciplines (Bell, 1982; Jackson et al., 1985; Hughes, 1989; Asker and Jerling, 1992; Savidan, 2000; Otto and Lenormand, 2002). The issue appears to be perceived differently for specialists working on Bacteria, Archaea, Eukaryota, unicellular, or pluricellular animals or plants. In this review, we will therefore first deal with specific definitions, as this subject area is littered with vocabulary that sometimes has ambiguous meanings. We will then try to go back in time to the origin of asexual reproduction and recombination and attempt to describe the diversity of ways in which prokaryotes and eukaryotes reproduce asexually and recombine. Following this, we will describe the various ways that asexual reproduction is incorporated in eukaryotic life cycles. After a brief attempt to quantify the importance of asexuality in living organisms, the genetic consequences of asexuality are reviewed, followed by a section on the evolution and the paradox of sex. What evolutionary advantages are brought by clonality? What disadvantages result from clonality? What is the so-called twofold cost of sex? The last section will deal with clonal microevolution. It will consist of two parts: the first one treating neutral gene variability in clonal populations (population genetics structure), and the second addressing selective issues like the evolution of resistance or virulence in clonal populations. Finally, we will conclude with economic and medical issues linked to asexual organisms.

*E-mail: thierry.demeeus@ird.fr

5.2 Definitions

Asexual reproduction is a process of genetic propagation of genomes, following which the genomes that descend from this process are strictly identical to the parental genome, in terms of quantity and quality, with the exception of uncorrected errors during the duplication process (i.e., mutations) (De Meeûs et al., 2007b). Besides cell division (e.g., mitosis in unicellular eukaryotes), many other processes correspond to clonal propagation as agametic (animals) or vegetative (plants) reproduction, ameiotic thelytokous parthenogenesis, endomitotic automictic parthenogenesis with pair formation of sister chromatids occurring before meiosis, automictic parthenogenesis with fusion of two polar bodies, deuterokous parthenogenesis, gynogenesis, apomixy, or agamospermy (reviewed in De Meeûs et al., 2007b).

Sexual reproduction is not initially a propagation mode even if it is now 100% correlated with the multiplication of many organisms (e.g., mammals). It is a recombinational repair tool (Cavalier-Smith, 2002; Ramesh et al., 2005; Glansdorff et al., 2009a), hence the use of sexual recombination (SR) in the rest of this paper as a synonymous for meiotic sex. Recombination in the wide sense is present in the three domains of life (Archaea, Bacteria, and Eukaryota), although through very different means (Cavalier-Smith, 2002), while SR is a eukaryotic hallmark (Cavalier-Smith, 2002; Solari, 2002; Glansdorff et al., 2009a). Recombination can take three forms in Bacteria and Archaea: conjugation, transformation, and transduction (Luo and Wasserfallen, 2001; Cavalier-Smith, 2002; Poole, 2009). Conjugation concerns plasmid exchange through a specialized structure called pilus. It is unidirectional in Bacteria (donor and recipient) and apparently bidirectional in Archaea (Luo and Wasserfallen, 2001). Transformation is the absorption of soluble naked DNA present in the microenvironment by a recipient cell and its further inclusion (recombination), if compatible, in the chromosome. Transduction is a horizontal gene transfer (HGT) mediated by viruses. Calling transduction, transformation and conjugation sex is unsound and true sex, with meiosis and syngamy, is only found in eukaryotes and never in prokaryotes (Cavalier-Smith, 2002).

Panmixia defines a population where zygotes (eggs) are produced by the random syngamy (union) of available sexual cells. It can thus only occur in eukaryotes, if any. Then, talking about panmictic bacteria is inappropriate as well. The genetic consequence of panmixia is the establishment of the famous Hardy–Weinberg (HW) genotypic proportions of the form p^2, $2pq$ and q^2 (for two alleles of frequencies p and q). These proportions are only expected to be approximately met in populations of highly mobile monoecious individuals with panmictic sex. Consequently, talking of panmixia for a microbe is also fairly unsound.

Linkage disequilibrium (LD) reflects the statistical association between different alleles at different loci in the genome. LD can be generated by virtually all evolutionary forces. Besides the obvious physical linkage, selection, population structure (small subpopulation sizes and migration), mutation, and reproductive system (except panmixia) all have a positive impact on LD. Estimation and testing of positive LD is a hard task and only very strong signals are expected to be detected, the variance of which is expected to be substantial (De Meeûs and Balloux, 2004; De Meeûs et al., 2009a). Furthermore, very strong interactions between sampling design, reproductive system, and population structure can considerably bias LD perception (Prugnolle and De Meeûs, 2010). Consequently, assessing reproductive systems through LD measures is at best risky, and measuring it through the proportion of significant LD tests found is definitely flawed.

5.3 The Origin of Life, the Origin of Propagation and Recombination

Whether a RNA phase came before the DNA world will not be discussed here. There is nevertheless a large consensus on the fact that all extant life is the descent of a single ancestor (Glansdorff et al., 2009b). The last universal common ancestor (LUCA), also known as the cenancestor (Cavalier-Smith, 2002), originated some 3–3.5 billion years ago (Vaneechoutte and Fani, 2009). The emergence of LUCA probably followed a phase of extensive HGT between the different arising entities (Glansdorff et al., 2009a,b). The order of branching of Bacteria, Eukaryota, and Archaea domains is controversial, one interesting hypothesis being that eukaryotes emerged as the result of a symbiotic fusion of some bacterial and archaeal lineages (Gargaud et al., 2009). Confusion finds its origin in the potential important disturbing HGT believed to occasionally or often occur between prokaryotic organisms (Gribaldo and Brochier, 2009). Evolution of meiosis is viewed by certain as a defense mechanism that evolved against HGT to promote the best coordination between coevolved functions. When chromosomes pair during meiosis, a number of mechanisms such as repair, conversion, and recombination are triggered, allowing the elimination of deleterious differences, which is viewed as a protection against HGT (Glansdorff et al., 2009a). Nevertheless, meiosis probably arose from mitosis, which is also specific to eukaryotes (Cavalier-Smith, 2002). According to this author, SR appeared about 850 million years ago as a cell cycle repair mechanism to correct accidental polyploidy. Many of the enzymes involved in meiosis have related enzymes in prokaryotic toolkits for controlling replication fidelity (rescue of broken or stalled replication forks, recombination or mismatch corrections) (Cavalier-Smith, 2002; Solari, 2002).

Consequently, clonality evolved first (whether prokaryotes appeared first or not), but recombination probably arose soon after or at the same time to control for intensive HGT and/or polyploidy, and this was then followed by SR in eukaryotes. It is noteworthy that SR emergence is not presented as a response to a changing environment (red queen hypothesis) or to prevent Muller's ratchet of deleterious allele accumulation (e.g., Otto and Lenormand, 2002; De Meeûs et al., 2007b for review) but as a mechanism for restoring genomic harmony after replication mistakes or any DNA damage. The fact SR did not evolve in prokaryotes probably comes from the constraints resulting from their particular peptidoglycan envelope said to act as a "chastity belt" (Cavalier-Smith, 2002). It is nevertheless a proof that SR is by no means a necessity to adapt to variable environments or fight against Muller's ratchet.

Microbes represent the major part of genetic diversity on earth, most of which is still represented by uncultivated organisms (Gribaldo and Brochier, 2009). Clonality is thus as old as life. It does not evolve in competition with recombination or SR but coevolves with it in most situations.

5.4 Clonal Modes

Prokaryotes have various ways to recombine and only one way to divide (Cavalier-Smith, 2002). On the contrary, eukaryotes, and in particular pluricellular ones, have barely a single way for recombination (if we exclude possible gene transfer through viruses or with endosymbionts) and many different ways to propagate clonally. Reviewing all these modes would be tedious and unnecessary as most was already presented in a recent review (De Meeûs et al., 2007b). It is interesting, though, to focus briefly on a particular family of clonal modes that diverted SR to, so to speak, reintegrate back clonal reproduction. The different forms of parthenogenesis that produce daughters identical to their mother (see earlier) correspond to that. It is obvious that these cases attracted the most attention of the evolutionary biologists working on the evolution of sex, in particular the famous asexual scandal of bdelloid rotifers (Judson and Normark, 1996; Mark Welch and Meselson, 2000). In fact, fixed clonality has rarely been demonstrated, but the coexistence of both systems is much more the rule as in aphids, other rotifers (except purely sexual acanthocephalans), cyclophorans, and many others (De Meeûs et al., 2007b). The fact that it must have been a real challenge to divert meiosis apparatus and that this nevertheless evolved many times in complex eukaryotes appears as a spectacular illustration of how costly SR must be, hence the impressive amount of works dedicated to this issue (see later).

In recent reviews, De Meeûs et al. (2007b, 2009b) found it convenient to classify organisms according to the kind of cycle they are involved in with regard to clonal propagation. We will stick to this classification in the following. This classification separates four kinds of cycles: (1) the purely sexual cycle (Sex) corresponds to organisms that can only reproduce through SR; (2) complex life cycles with an instantaneous clonal phase with only one (I) clonal generation per cycle; (3) complex life cycles with several generations of asexuality (S) where the clonal phase involves more than one clonal generation; and (4) life cycles where sexual reproduction is more or less frequent (or even absent) with an acyclic pattern (A). In cases (2) and (3), and for all surviving individuals, SR must intervene at one point in the cycle to form zygotes. In case (4) the life cycle is not defined by a regular pattern of sexual or asexual reproduction. Case (1) is typical of vertebrates, especially mammals and birds but also cestodes, lice, or nematodes. Cycle (2) applies to all species with polyembryony and many budding species. For example, this cycle is typical of trematodes (flukes). Case (3) is typical of aphids, monogonont rotifers, cladocerans, many fungi, and most Sporozoa (parasitic unicellular organisms, including the malaria agents *Plasmodium* spp.). Finally, case (4) is common in plants and unicellular organisms. In particular, it is found for strictly clonal organisms, or at least those organisms for which sex is unknown, such as bdelloid rotifers, imperfect fungi (e.g., *Candida albicans*), Parabasalia (*Trichomonas vaginalis*), Metamonadina (*Giardia lamblia*), parasitic amoebas, and kinetoplastid parasites (*Leishmania, Trypanosoma*).

5.5 Quantifying the Importance of Asexuality in the Biosphere

There are two ways to comprehend this issue. In terms of described (known) species, purely sexual species are the most represented (De Meeûs et al., 2009b). Nevertheless, there is an obvious bias in accounting biological diversity through described species (De Meeûs and Renaud, 2002; De Meeûs et al., 2003). As quoted earlier, microbes (cycles S or A) represent the major part of genetic diversity on earth, most of which is still represented by uncultivated organisms (Gribaldo and Brochier, 2009). It can thus be safely postulated that organisms with a clonal phase represent the major part of biodiversity. If this was accounted for in terms of energy devoted to clonality and SR on earth per second, SR would probably look like an epiphenomenon. This should be trivial as the real way to propagate for life is through cell (hence asexual) division while SR is in fact meant to DNA repair and/or control DNA replication fidelity.

The numeric importance of clonal parasitic eukaryotes was already reviewed by De Meeûs et al. (2009b). Whole described species again give a biased advantage to purely sexual species. Nevertheless, a glance at the most documented human parasitic fauna completely reverses the tendency, thus suggesting that: (1) parasites represent the most important part of eukaryotic biodiversity, and (2) that clonal species (i.e., using this mode at one stage of their life cycle) are the majority among them. If Archaea and Bacteria are included, known species number is useless. There are indeed more known bird species than the sum of known Archaea and Bacteria, which is nonsense. Prokaryotes are so numerous everywhere that estimating how much of their diversity specialized in parasitism looks impossible. We can, however, suspect this number to be tremendous regarding all bacterial diseases that can affect mankind (around 43 after a quick and dirty look in the web). For eukaryotic parasites, it was recently estimated that more than a billion people are affected by such diseases (De Meeûs et al., 2009b), some of which are the most severe ones (e.g., malaria). Clonality in infectious disease cannot thus be treated lightly.

5.6 Genetic Consequences of Asexuality

This issue was reviewed many times (e.g., in Suomalainen et al., 1976; Jackson et al., 1985; Tibayrenc et al., 1990, 1991; Maynard-Smith et al., 1993; Carvalho,

1994; Tibayrenc, 1995, 1998, 1999; Judson and Normark, 1996; Milgroom, 1996; Taylor et al., 1999; Savidan, 2000; Tibayrenc and Ayala, 2002; Halkett et al., 2005; De Meeûs et al., 2006, 2007a,b, 2009b), so we will be brief and stick to the essential. In haploid organisms, clonality tends to create and maintain statistical associations between the different loci of the genome irrespective of their location. In purely asexuals this should end with the presence of numerous repetitions of a certain clone, and hence of the same multilocus genotype (MLG). Depending on population structure, MLG diversity will vary from low (e.g., a single MLG) to high variability (several MLGs). As linkage is total, MLGs can be considered as the different alleles of a single locus. If no SR is involved it is expected that the different MLGs that can be maintained can potentially be highly divergent. This may represent a problem because at a given level of divergence it is probable that adaptive differences will arise. Moreover, especially in small subpopulations that are not expected to maintain much equivalent different MLGs, the stable maintenance of highly diverged MLGs of the same "species" might lead to interpret it as an ecological divergence. When some SR is involved, the combination between drift, reproduction, and sampling renders difficult the interpretation of the patterns of genetic variability in haploids. This is also true for diploids even if, when the amount of SR is large enough, populations display patterns of genetic variability close to that observed for a sexual population.

In diploids, haplotypic consequences are similar, but here in the absence of SR, the two alleles of a lineage will continuously diverge since the last SR event. Consequently, as illustrated in Figure 5.1, divergence between the two alleles of the same individual will be higher than mean divergence between lineages. This is the Meselson effect (Judson and Normark, 1996; Mark Welch and Meselson, 2000). Another way to see it is that in lineages that have stayed clonal for a sufficient amount of time, all loci will be heterozygous for all individuals. Genomic fixed heterozygosity can thus represent an unambiguous signature of full clonality.

5.7 Evolution and the Paradox of Sex

The paradox of sex essentially concerns parthenogenetic multicellular organisms and, as explained earlier, microbes are not concerned. This has been the subject of an impressive amount of literature and, except for plant parasitic arthropods (insects, mites) and nematodes, very few animal parasites are parthenogenetic (some nematodes, gyrodactilid monogens, rare cestodes, and trematodes) (De Meeûs et al., 2007b). It would be useless to do something more than a short reminder here. Parthenogenetic females produce twice as many offspring as sexually reproducing females that need to produce half "useless" males, which themselves cannot produce eggs. This has been called the twofold cost of sex (Hurst and Peck, 1996). Consequently, parthenogenetic females should quickly invade the whole planet. There are several reasons why this is not so, most of which are not exclusive and probably account together for the maintenance of sex in such situations.

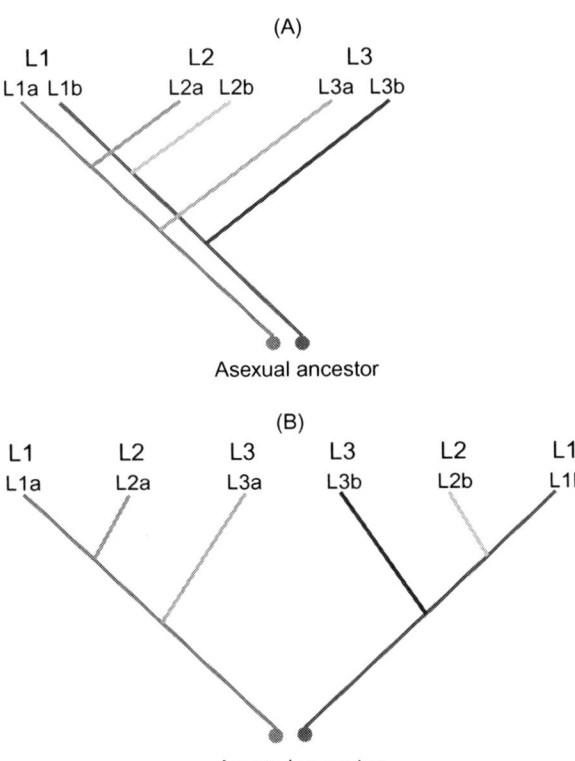

Figure 5.1 Illustration of the Meselson effect. In (A) the evolutionary relationships among three asexual diploid lineages are represented (L1—3). The genetic divergence is also represented with varying colors providing the two alleles present in each taxon (alleles a and b). If we develop the tree corresponding to all DNA sequences (all alleles) as in (B), it is easily seen that the maximum divergence is obtained between the two alleles of the same lineage. This is what is expected in ancient clones and can be used as a criterion for detecting a long absence of sex in a group of taxa (Meselson method). (For interpretation of the references to color in this figure legend, the reader is referred to the web version of this book.)

First of all, as mentioned earlier, the hijacking of SR for producing clonal descents is probably extremely difficult, and the diversity of tricks that evolved to achieve it, sometimes through extremely (at least in appearance) odd means, can be the sign of how difficult it is to reach that point. For instance, automictic parthenogenesis with fusion of two polar bodies illustrates this last point (see Figure 3b in De Meeûs et al., 2007b). The rarity of emergence of parthenogenesis, apparently restricted in few lineages (but this can be misleading because of biases in the intensity of work devoted to certain groups), can thus largely be explained by such constraints. For instance, it seems impossible to evolve in mammals or in birds.

Secondly, the problem only arises for populations that exclusively reproduce either sexually or parthenogenetically and for which these two morphs compete for the same resources. This might be rare. Some aphids might correspond to this, as for instance *Rhopalosiphum padi* (Delmotte et al., 2002), though it is not well established how similar the ecological niche of these two morphs is.

According to the red queen hypothesis (Judson and Normark, 1996), pure parthenogenetic females cannot efficiently fight against the continuously evolving aggressors (parasites and predators) or victims (preys or hosts) as compared to sexual females that produce many different combinations of offspring at each generation. This hypothesis alone has two important drawbacks. First, in pure sexuals, the best combination is lost in the next generation. Second, most populations are not that polymorphic and often are small and thus inbred. The possible combinations created by SR might not be that diverse or new.

Muller's ratchet (Kondrashov, 1993) imposes to parthenogenetic lineages an accumulation of deleterious mutations that could lead to an eventual collapse of such lineages as compared to sexual lineages where deleterious mutations are more efficiently removed. This model alone also has two drawbacks. First it requires several generations to work efficiently, and might even be almost silent in diploids. Second, as above, small sexually reproducing populations might also be affected by Muller's ratchet.

Finally, as mentioned earlier and elsewhere (Schaefer et al., 2006), SR may also be viewed as a resetting process that evolved to restore the best combinations, a purpose for which it indeed evolved for in the first eukaryotes. Such a view also has the advantage to explain why SR often concerns genetically related partners, hence the evolution of reproductive isolation often observed in pluricellular eukaryotes (De Meeûs et al., 2003).

5.8 Clonal Microevolution

This aspect can be tackled differently depending on what kind of genetic information we are dealing with: neutral variation, and its use as a signature of demographic events, and variation under selection.

5.8.1 Neutral Loci Variability in Clonal Populations (Population Genetics Structure)

Neutral variation and its distribution in time and space can be used to make useful inferences on the population biology of the targeted organisms. Under certain hypotheses, several inferences can be made as regard to population size, dispersal, and reproductive mode. Most tools were developed for sexual species but recent works have made available equivalent tools for clonal populations (see De Meeûs et al., 2006, 2007a, 2009b for reviews). In that case special care must be given to how to deal with MLGs. For A cycles complete datasets must be kept. For I cycles, it was shown that besides analyzing complete datasets, population subdivision is better assessed if only a single representative of each MLG is kept (Prugnolle et al., 2005; Caillaud et al., 2006). For S cycles, it all depends where in the cycle individuals are sampled. A strategy similar to the one used for I cycles is to be used if individuals are sampled early after the last SR event. If individuals are sampled after a substantial amount of clonal generations, then a strategy similar to the one used for A cycles is preferred.

For A cycles, if clonal reproduction is so prevalent that no perceptible signature of any SR can be noticed, then tools specific to that situation should be used for ecological inferences. This of course must take into account some basic knowledge of the population. When the population can be assumed to be strongly subdivided

Clonal Evolution

in numerous demes, it was shown that the number of migrants can be estimated through the formula (De Meeûs and Balloux, 2005; Nébavi et al., 2006):

$$N(m + u) = - \frac{1 + F_{IS}}{4F_{IS}} \quad (5.1)$$

where N is the clonal subpopulation size, m the proportion of migrants that each subpopulation contain, u the mutation rate, and F_{IS} the Wright's fixation index (Wright, 1965; De Meeûs et al., 2007a) measuring inbreeding within individuals relative to inbreeding between individuals. In that case, estimating independently N and m, even if we assume u negligible as compared to m, is not easy and will require further studies. When the population can be assumed to comprise only two subpopulations, then more precise estimates can be made (Koffi et al., 2009):

$$N = - \frac{1 + F_{IS}}{8uF_{IS}} \quad (5.2)$$

and

$$m = \frac{1}{2} \left[1 - \sqrt{\frac{F_{ST}}{F_{ST} - 4uF_{IS}}} \right] \quad (5.3)$$

where F_{ST} is Wright's fixation index measuring the between individuals inbreeding within subpopulations relative to inbreeding between subsamples. It also requires knowledge of u. Finally, when subpopulations are assumed completely isolated, their clonal size can be estimated as (Simo et al., 2010):

$$N = - \frac{1 + F_{IS}}{4uF_{IS}} \quad (5.4)$$

Now if some SR influences the distribution of genetic diversity, then it is usually wiser to use classical population genetics tools (De Meeûs et al., 2007a) except for cases of extremely rare SR events where the behavior of most parameters is odd and thus inferences can only be very general (De Meeûs et al., 2006). Similar advice can be given for I and S cycles if individuals studied are sampled just after SR.

5.8.2 Selection and Adaptation in Clonal Populations

The vast majority of mutations are neutral or deleterious (Loewe and Hill, 2010). Extensive study of such mutations has explained the genetic diversity in many populations and has been useful for inferring population parameters and histories from data as explained earlier. Yet beneficial mutations, despite their rarity, are what cause long-term adaptation and can also dramatically alter the genetic diversity at linked sites (see Nielsen, 2005 for a review). Unfortunately, our understanding of their dynamics remains poor, especially in asexual populations.

Adaptation by natural selection occurs through the spread and substitution of mutations that improve the performance of an organism and its reproductive success in a particular environment. For example, this happens in a pathogen when an allele increases in frequency in the population because it confers a certain degree of resistance against a particular drug. Most early works on the dynamics of adaptation in asexual populations considered that beneficial mutations only occurred very rarely (Atwood et al., 1951a,b). Under such circumstances, the rates of adaptation of asexual populations is the same (all else being equal) as that of sexual populations and depends only on the time separating the appearance of two beneficial mutations. This conventional model, known as the "periodic selection" model, remained a very influential theory until the 1990s despite the classic works of Muller (1932) that clearly showed that the dynamic of adaptation in sexual and asexual populations could be very different when beneficial mutations were common.

One particularity of the dynamic of adaptation of asexual populations when beneficial mutations are common is that beneficial mutations that have arisen independently in different individuals cannot recombine and therefore have to compete for fixation. This effect is called "clonal interference" (Gerrish and Lenski, 1998; Desai and Fisher, 2007; Desai et al., 2007). To date, two main models of clonal interference have been proposed: the one-by-one mutation model (Gerrish and Lenski, 1998) and the multiple mutations model (Desai and Fisher, 2007; Desai et al., 2007). These two models differ in how and where new beneficial mutations appear. We will not enter into the details of these models here and we advise readers to refer to recent reviews for more details. We simply want to stress that, under the two models, beneficial mutations enter into competition and some beneficial mutations are therefore "wasted" during the process of adaptation (Gerrish and Lenski, 1998; Gerrish, 2001; Rozen et al., 2002; Wilke, 2004). This leads to a slowdown in the rate of adaptation in purely asexual populations as compared to sexual populations. Note that a similar effect was described for sexual populations in the case of physically linked genes, which is known as the Hill–Robertson effect (Hill and Robertson, 1966).

Clearly, a complete picture of adaptation in asexual populations should also include the impact of deleterious mutations. They indeed play an important role in adaptation because their presence influences the fate of beneficial mutations, and consequently affects the strength of clonal interference (Felsenstein, 1974; Charlesworth, 1994; Bachtrog and Gordo, 2004). It is indeed well established that deleterious mutations can cause a severe reduction in the adaptation rate, as a consequence of reducing the effective population size. The simplest situation corresponds to the case in which only beneficial mutations that occur in individuals that are mutation-free contribute to the adaptive process.

Here, we have mainly focused on complete clonal organisms (life cycle A with 100% clonality). As shown here, clonal reproduction occurs under several forms and in several life cycles. Models analyzing the dynamic of adaptation under such life cycles have not been done yet, but we think that as soon as a bit of recombination occurs the dynamic of adaptation will be similar to the one described by models dealing with the problem of interference (or Hill–Robertson effect) in sexual organisms. However, since pure sexuals tend to lose the most beneficial combinations built in previous generations, clonal populations with rare sex probably display much more efficient adaptive dynamics. A rare sexual event can build an "optimal" combination that will be easily and faithfully propagated by clonal reproduction. This might help understanding the formidable adaptive speed of microbes and, in particular, pathogenic microbes.

5.9 Conclusions

Clonal reproduction is as old as life itself and is widespread in the living world. SR appeared in Eukaryota, after this group evolved mitosis, not as a propagation tool alternative to clonal reproduction but as a repairing tool to preserve the most harmonious combinations of the numerous genes necessary to build a eukaryotic cell and because of the mitosis apparatus that evolved only in this lineage, a necessary prerequisite for meiosis. Sex is totally linked to propagation only in two pluricellular lineages (Metazoa and Metabionta). Only in those complex lineages SR can be in competition with clonal reproduction under certain precise circumstances. Clonality is the most important propagation mode used by pathogenic agents and its genetic consequences must be understood precisely, though SR or recombination is also very important to take into account for those diseases that practice it. When SR is so rare that no signature can be found in the genetic architecture of populations, some specific patterns arise as presence of multilocus repeated genotypes and, for diploids, fixed heterozygosity. These patterns can be exploited for demographic inferences using specific tools. If SR has even a small influence, then classical tools of population genetics can be used to infer subpopulation sizes and dispersal. It is thus possible to infer population sizes and dispersal for clonal parasites with the study of variable molecular markers, which is good news as the populations of such organisms are difficult to study directly. Such information can reveal very important to understand the epidemiology of diseases.

Though purely sexual populations are at a theoretical advantage as compared to purely asexual lineages as regards the dynamics of adaptation, things become less clear if the most general case is taken into account. Clones with more or less rare sex (or recombination) may indeed represent an extremely efficient (and hence widespread) way to adapt to the environment. This helps explain the speed at which pathogenic agents respond to defense mechanisms, including pharmacologically mediated ones, of their victims.

Abbreviation list:

HGT Horizontal gene transfer
HW Hardy–Weinberg
LD Linkage disequilibrium
LUCA Last universal common ancestor
MLG Multilocus genotype
SR Sexual recombination

Acknowledgement

We thank Joseph Carl Robnett Licklider and Leonard Kleinrock, MIT initiators of Internet, without whom this article (and many others) could not have come to light. TDM and FP are financed by the CNRS and IRD. FP is also supported by ANR MGANE SEST 012 2007.

References

Asker, S.E., Jerling, L., 1992. Apomixis in Plants. CRC Press Inc, Boca Raton, FL.
Atwood, K.C., Schneider, L.K., Ryan, F.J., 1951a. Periodic selection in *Escherichia coli*. Proc. Natl. Acad. Sci. U.S.A. 37, 146–155.
Atwood, K.C., Schneider, L.K., Ryan, F.J., 1951b. Selective mechanisms in bacteria. Cold Spring Harb. Symp. Quant. Biol. 16, 345–355.
Bachtrog, D., Gordo, I., 2004. Adaptive evolution of asexual populations under Muller's ratchet. Evolution 58, 1403–1413.
Bell, G., 1982. The Masterpiece of Nature. University of California Press, Berkeley, CA.
Caillaud, D., Prugnolle, F., Durand, P., Théron, A., De Meeûs, T., 2006. Host sex and parasite genetic diversity. Microbes Infect. 8, 2477–2483.
Carvalho, G.C., 1994. Genetics of aquatic clonal organisms. In: Beaumont, A. (Ed.), Genetics and Evolution of Aquatic Organisms. Chapman and Hall, London, pp. 291–319.
Cavalier-Smith, T., 2002. Origins of the machinery of recombination and sex. Heredity 88, 125–141.
Charlesworth, B., 1994. The effect of background selection against deleterious mutations on weakly selected, linked variants. Genet. Res. 63, 213–227.
De Meeûs, T., Balloux, F., 2004. Clonal reproduction and linkage disequilibrium in diploids: a simulation study. Infect. Genet. Evol. 4, 345–351.
De Meeûs, T., Balloux, F., 2005. F-statistics of clonal diploids structured in numerous demes. Mol. Ecol. 14, 2695–2702.
De Meeûs, T., Renaud, F., 2002. Parasites within the new phylogeny of eukaryotes. Trends Parasitol. 18, 247–251.
De Meeûs, T., Durand, P., Renaud, F., 2003. Species concepts: what for? Trends Parasitol. 19, 425–427.
De Meeûs, T., Lehmann, L., Balloux, F., 2006. Molecular epidemiology of clonal diploids: a quick overview and a short DIY (do it yourself) notice. Infect. Genet. Evol. 6, 163–170.
De Meeûs, T., McCoy, K.D., Prugnolle, F., Chevillon, C., Durand, P., Hurtrez-Boussès, S., et al., 2007a. Population genetics and molecular epidemiology or how to "débusquer la bête". Infect. Genet. Evol. 7, 308–332.
De Meeûs, T., Prugnolle, F., Agnew, P., 2007b. Asexual reproduction: genetics and evolutionary aspects. Cell. Mol. Life Sci. 64, 1355–1372.
De Meeûs, T., Guégan, J.F., Teriokhin, A.T., 2009a. MultiTest V.1.2, a program to binomially combine independent tests and performance comparison with other related methods on proportional data. BMC Bioinformatics 10, 443.
De Meeûs, T., Prugnolle, F., Agnew, P., 2009b. Asexual reproduction in infectious diseases. In: Schön, I., Martens, K., van Dijk, P. (Eds.), Lost Sex: The Evolutionary Biology of Parthenogenesis. Springer, NY, pp. 517–533.

Delmotte, F., Leterme, N., Gauthier, J.P., Rispe, C., Simon, J.C., 2002. Genetic architecture of sexual and asexual populations of the aphid *Rhopalosiphum padi* based on allozyme and microsatellite markers. Mol. Ecol. 11, 711–723.
Desai, M.M., Fisher, D.S., 2007. Beneficial mutation selection balance and the effect of linkage on positive selection. Genetics 176, 1759–1798.
Desai, M.M., Fisher, D.S., Murray, A.W., 2007. The speed of evolution and maintenance of variation in asexual populations. Curr. Biol. 17, 385–394.
Felsenstein, J., 1974. The evolutionary advantage of recombination. Genetics 78, 737–756.
Gargaud, M., Martin, H., López-García, P., Montmerle, T., Pascal, R., 2009. Le Soleil, La Terre ... La Vie: La Quête des Origines. Belin-Pour la Science, Paris.
Gerrish, P.J., 2001. The rhythm of microbial adaptation. Nature 413, 299–302.
Gerrish, P.J., Lenski, R.E., 1998. The fate of competing beneficial mutations in an asexual population. Genetica 102–103, 127–144.
Glansdorff, N., Xu, Y., Labedan, B., 2009a. The conflict between horizontal gene transfer and the safeguard of identity: origin of meiotic sexuality. J. Mol. Evol. 69, 470–480.
Glansdorff, N., Xu, Y., Labedan, B., 2009b. The origin of life and the last universal common ancestor: do we need a change of perspective? Res. Microbiol. 160, 522–528.
Gribaldo, S., Brochier, C., 2009. Phylogeny of prokaryotes: does it exist and why should we care? Res. Microbiol. 160, 513–521.
Halkett, F., Simon, J.C., Balloux, F., 2005. Tackling the population genetics of clonal and partially clonal organisms. Trends. Ecol. Evol. 20, 194–201.
Hill, W.G., Robertson, A., 1966. The effect of linkage on limits to artificial selection. Genet. Res. 8, 269–294.
Hughes, R.N., 1989. A Functional Biology of Clonal Animals. Chapman and Hall, London and New York.
Hurst, L.D., Peck, J.R., 1996. Recent advances in understanding of the evolution and maintenance of sex. Trends. Ecol. Evol. 11, 46–52.
Jackson, J.B.C., Buss, L.W., Cook, R.E., 1985. Population Biology and Evolution of Clonal Organisms. Yale University Press, New Haven, CT.
Judson, O.P., Normark, B.B., 1996. Ancient asexual scandals. Trends Ecol. Evol. 11, 41–46.
Koffi, M., De Meeûs, T., Bucheton, B., Solano, P., Camara, M., Kaba, D., et al., 2009. Population genetics of *Trypanosoma brucei gambiense*, the agent of sleeping sickness in Western Africa. Proc. Natl. Acad. Sci. U.S.A. 106, 209–214.
Kondrashov, A.S., 1993. Classification of hypotheses on the advantage of amphimixis. J. Hered. 84, 372–387.
Loewe, L., Hill, W.G., 2010. The population genetics of mutations: good, bad and indifferent. Philos. Trans. R. Soc. Lond. B. Biol. Sci. 365, 1153–1167.
Luo, Y.N., Wasserfallen, A., 2001. Gene transfer systems and their applications in Archaea. Syst. Appl. Microbiol. 24, 15–25.
Mark Welch, D.B., Meselson, M., 2000. Evidence for the evolution of bdelloid rotifers without sexual reproduction or genetic exchange. Science 288, 1211–1215.
Maynard-Smith, J., Smith, N.H., Orourke, M., Spratt, B.G., 1993. How clonal are bacteria. Proc. Natl. Acad. Sci. U.S.A. 90, 4384–4388.
Milgroom, M.G., 1996. Recombination and the multilocus structure of fungal populations. Annu. Rev. Phytopathol. 34, 457–477.
Muller, H.J., 1932. Some genetic aspects of sex. Am. Nat. 66, 118–138.
Nébavi, F., Ayala, F.J., Renaud, F., Bertout, S., Eholié, S., Moussa, K., et al., 2006. Clonal population structure and genetic diversity of *Candida albicans* in AIDS patients from Abidjan (Cote d'Ivoire). Proc. Natl. Acad. Sci. U.S.A. 103, 3663–3668.

Nielsen, R., 2005. Molecular signatures of natural selection. Annu. Rev. Genet. 39, 197–218.
Otto, S.P., Lenormand, T., 2002. Resolving the paradox of sex and recombination. Nat. Rev. Genet. 3, 252–261.
Poole, A.M., 2009. Horizontal gene transfer and the earliest stages of the evolution of life. Res. Microbiol. 160, 473–480.
Prugnolle, F., De Meeûs, T., 2010. Apparent high recombination rates in clonal parasitic organisms due to inappropriate sampling design. Heredity 104, 135–140.
Prugnolle, F., Théron, A., Pointier, J.P., Jabbour-Zahab, R., Jarne, P., Durand, P., et al., 2005. Dispersal in a parasitic worm and its two hosts: consequence for local adaptation. Evolution 59, 296–303.
Ramesh, M.A., Malik, S.B., Logsdon, J.M., 2005. A phylogenomic inventory of meiotic genes: evidence for sex in *Giardia* and an early eukaryotic origin of meiosis. Curr. Biol. 15, 185–191.
Rozen, D.E., de Visser, J.A.G.M., Gerrish, P.J., 2002. Fitness effects of fixed beneficial mutations in microbial populations. Curr. Biol. 12, 1040–1045.
Savidan, Y., 2000. Apomixis: genetics and breeding. Plant Breed. Rev. 18, 13–86.
Schaefer, I., Domes, K., Heethoff, M., Schneider, K., Schon, I., Norton, R.A., et al., 2006. No evidence for the "Meselson effect" in parthenogenetic oribatid mites (Oribatida, Acari). J. Evol. Biol. 19, 184–193.
Simo, G., Njiokou, F., Tume, C., Lueong, S., De Meeûs, T., Cuny, G., et al., 2010. Population genetic structure of Central African *Trypanosoma brucei gambiense* isolates using microsatellite DNA markers. Infect. Genet. Evol. 10, 68–76.
Solari, A.J., 2002. Primitive forms of meiosis: the possible evolution of meiosis. Biocell 26, 1–13.
Suomalainen, E., Saura, A., Lokki, J., 1976. Evolution of parthenogenetic insects. Evol. Biol. 9, 209–257.
Taylor, J.W., Geiser, D.M., Burt, A., Koufopanou, V., 1999. The evolutionary biology and population genetics underlying fungal strain typing. Clin. Microbiol. Rev. 12, 126–146.
Tibayrenc, M., 1995. Population genetics of parasitic protozoa and other micro-organisms. Adv. Parasitol. 36, 47–115.
Tibayrenc, M., 1998. Genetic epidemiology of parasitic protozoa and other infectious agents: the need for an integrated approach. Int. J. Parasitol. 28, 85–104.
Tibayrenc, M., 1999. Toward an integrated genetic epidemiology of parasitic protozoa and other pathogens. Annu. Rev. Genet. 33, 449–477.
Tibayrenc, M., Ayala, F.J., 2002. The clonal theory of parasitic protozoa: 12 years on. Trends Parasitol. 18, 405–410.
Tibayrenc, M., Kjellberg, F., Ayala, F.J., 1990. A clonal theory of parasitic protozoa: the population structures of *Entamoeba*, *Giardia*, *Leishmania*, *Naegleria*, *Plasmodium*, *Trichomonas*, and *Trypanosoma* and their medical and taxonomical consequences. Proc. Natl. Acad. Sci. U.S.A. 87, 2414–2418.
Tibayrenc, M., Kjellberg, F., Arnaud, J., Oury, B., Brenière, S.F., Dardé, M.L., et al., 1991. Are eukaryotic microorganisms clonal or sexual? A population genetics vantage. Proc. Natl. Acad. Sci. U.S.A. 88, 5129–5133.
Vaneechoutte, M., Fani, R., 2009. From the primordial soup to the latest universal common ancestor. Res. Microbiol. 160, 437–440.
Wilke, C.O., 2004. The speed of adaptation in large asexual populations. Genetics 167, 2045–2053.
Wright, S., 1965. The interpretation of population structure by F-statistics with special regard to system of mating. Evolution 19, 395–420.

6 Coevolution of Host and Pathogen

Andrew D. Morgan[1,*] and Britt Koskella[2]

[1]Institute for Evolutionary Biology, University of Edinburgh, Edinburgh, UK
[2]Department of Zoology, University of Oxford, Oxford, UK

6.1 Coevolution of Host and Pathogen

6.1.1 Introduction to Coevolution of Host and Pathogen

No species is an island: every individual organism is in constant interaction with other species around it, whether it is with prey, predators, herbivores, competitors, mutualists, pollinators, or pathogens. These biotic interactions often have large effects on individual fitness and can significantly alter the evolutionary trajectory of a population. Importantly, selection imposed by species interactions can drive genetic divergence between populations and maintain diversity both locally (Haldane, 1949; Parker, 1989; Salvaudon et al., 2008) and globally (Buckling and Rainey, 2002b; Thompson, 2005; Laine, 2009; Paterson et al., 2010). This is because a given genotype might have a very different fitness in the context of one environment/community than another, as the species and genotypes with which it will interact in each environment/community are likely to differ. When biotic interactions drive reciprocal change in both populations, as one species imposes selection on the other and vice versa, the species are said to be coevolving (Janzen, 1980).

Coevolutionary dynamics between hosts and pathogens has been perhaps the most well-studied interspecific interaction. This is due to the tight coupling of the two players and the implications of these dynamics for understanding the structure of communities (Hudson et al., 2006), population dynamics (Lively, 1999), the maintenance of sexual recombination (Jaenike, 1978), and the trajectory of species invasions (Prenter et al., 2004). Recent research on host−parasite interactions has indicated that coevolution occurs in relatively short time periods (Buckling and Rainey, 2002a; Forde et al., 2004; Morgan et al., 2005; Koskella and Lively, 2007; Jokela et al., 2009) and that the trajectories of coevolution are strongly influenced by the spatial structure of populations (Gandon and Van Zandt, 1998; Nuismer et al., 2003; Thompson, 2005; Nuismer and Goodnight, 2009). For those host−pathogen interactions in which there is an underlying genetic basis to infection, both the size and genetic makeup of the pathogen population at any point in time will be a function of the frequency, and in many cases density, of susceptible host

*E-mail: andrew.morgan@ed.ac.uk

genotypes in previous generations. Similarly, the probability that a given host will become infected is a function of the frequency of pathogen genotypes in the population that can infect it, which is again determined by past genotype frequencies in both populations. Accordingly, each population acts as a moving target for the other and it is these dynamic changes of one population in response to another that can maintain polymorphism over time, as different alleles will be favored in one generation relative to the next (Haldane, 1949; Hamilton, 1982; Nee, 1989).

In this chapter, we first discuss the process of host–pathogen coevolution. We then outline common methods for examining pathogen adaptation to hosts, and host response to pathogens and highlight a few key examples to illustrate that this process is both common in nature and critically important in explaining the amount of genetic variation found on the planet. Finally we discuss the implications of coevolution and summarize the importance of studying coevolution.

6.1.2 Antagonistic Coevolution

Pathogens, by definition, have deleterious fitness effects on their hosts, and thus have the capacity to act as major selective forces on host populations. At the same time, pathogens are often reliant on their hosts for some stage of their life cycle, and so any change in the host population will have strong impacts on the pathogen population. This interaction between host and pathogen will have different outcomes depending on factors ranging from the degree of pathogen specialization to the abiotic environment in which the interaction occurs. The interaction is not always a coevolutionary one; in some cases selection only acts on one partner. For example, a generalist pathogen may sweep through a small population of a rare host species and significantly alter the host dynamics without being changed itself. However, due to the tight genetic interaction between many hosts and pathogens, an evolutionary change in one partner is likely to cause evolutionary change in the other, leading to ongoing coevolution. Therefore, a common definition of host–pathogen coevolution is "the reciprocal evolution of interacting hosts and pathogens."

Host–pathogen coevolution is usually antagonistic, since an increase in fitness of one player typically leads to a decrease in fitness of the other. For example, hosts may evolve resistance (incurring higher fitness in the face of harmful pathogens) and pathogens may evolve counter infectivity. Such antagonistic coevolution may be either directional or cyclical (see Box 6.1). If it is directional, hosts and parasites evolve ever-mounting resistance and infectivity in the form of an "arms-race": where future types are more resistant and infective than their ancestors (Thompson and Burdon, 1992; Parker, 1994; Buckling and Rainey, 2002a). This type of coevolution is typical of the interaction between bacteria and bacteriophage, and plants and their pathogens. In its simplest form, this type of directional, arms-race coevolution will lead to the extinction of one player or the other, as genetic variation is ultimately exhausted. However, in cases where there are significant costs to resistance and infectivity, these dynamics can be continuous and cyclical, as costs build up, and the "arms-race" crashes (Sasaki, 2000). An example of such a crash may occur with the modification of a host cell receptor to stop a pathogen binding. The modification of the receptor may have some negative effect on the function of the receptor, and therefore affect the fitness of the host organism. The pathogen would have increasing costs associated with a reduction in its ability to bind to the receptor, and so fewer successful infections. The modification of the host receptor may continue up to a point where the negative fitness effects would be so great that sensitive hosts with fully functioning receptors would be fitter than the resistant host. The cycle would then restart (Sasaki, 2000). Although it remains unclear how ubiquitous these costs to resistance and infectivity might be, there is strong evidence that host resistance is costly from at least a few studies (Buckling et al., 2006; Morgan et al., 2009; Boots and Begon, 1993; Ferdig et al., 1993; Fellowes et al., 1998; Langand et al., 1998); and that parasite virulence is costly from a few others (Bahri et al., 2009; Grim et al., 2009; Huang et al., 2010).

Cyclical coevolution, on the other hand, occurs when successful infection of a host requires specific genotypic matching of pathogens. For example, host A is susceptible to pathogen A but not pathogen B, and host B is susceptible to pathogen B but not pathogen A (see Box 6.1). Under this scenario, resistance and infectivity do not increase through time, as no parasite is universally virulent and no host is inherently more resistant than another. Instead, all fitnesses are determined by the frequency of "matching" genotypes in the population. Under this scenario, pathogens will evolve to infect the most common host genotype, giving rare hosts an advantage (Hamilton, 1980; Nee, 1989; Frank, 1994). These rare genotypes might increase in frequency until they become common, and eventually the target of local pathogens. These cyclical dynamics are often referred to as "Red Queen" dynamics (Bell, 1982) after the character in Lewis Carroll's *Through the Looking Glass* who explains to Alice that, in Wonderland, "it takes all the running you can do, to keep in the same place" (Carroll, 1871). Similarly, populations of hosts and pathogens are engaged in a constant coevolutionary battle but are, on average, maintaining the same fitness with respect to one other. The Red Queen metaphor is also used more generally to describe antagonistic coevolution whether dynamics are cyclical or directional (Woolhouse and Webster, 2000).

6.1.3 The Evolution of Pathogen Virulence

Although it is intuitively clear that pathogens might harm their hosts as a by-product of passing themselves on from one generation to the next (*e.g.*, by redirecting host resources away from host reproduction and into pathogen reproduction), it is less clear why there are more virulent pathogens that kill or sterilize their hosts. The dilemma arises because an increase in pathogen fitness via greater within-host reproduction might lead to a decrease in fitness via lower rates of transmission if the host becomes too sick to interact with other hosts or spread infectious propagules into the environment. This "trade-off hypothesis" is the most popular evolutionary explanation for why pathogens often do not reach their maximum reproductive potential (Anderson and May, 1982; de Roode et al., 2008b; Frank, 1996; Day, 2001). Increases in virulence can accompany shifts to new host populations or species (Bolker et al., 2010), drastic changes in host population size or

Box 6.1 Infection Genetics

One critical determinate of host—pathogen coevolutionary dynamics is the underlying genetic interaction between them. Theoretical work has shown that tight genetic specificity for infection can lead to oscillations in genotype frequencies (*i.e.*, Red Queen dynamics), and the long-term maintenance of genetic diversity (Seger, 1988; Morand et al., 1996). These oscillatory dynamics are key to many central theories regarding host—parasite coevolution, including both local adaptation and the maintenance of sexual reproduction (Hamilton, 1980; Bell, 1982; Price and Waser, 1982; Hamilton et al., 1990). Two models describing infection specificity in host—parasite interactions have been most commonly used, although numerous others exist.

The first model is the "matching alleles model" (MAM); based upon a system of self/nonself recognition molecules where hosts can successfully defend against any parasite genotype that does not match their own (Hamilton, 1980; Peters and Lively, 1999; Grosberg and Hart, 2000). A parasite must specifically match host alleles at infection loci in order for it to evade detection by the immune system and successfully infect the host. The MAM assumes that one parasite genotype will have a different subset of susceptible hosts than another parasite genotype such that infection success is determined by both host and parasite genotype. The tight specificity leads to cyclical "Red Queen" Dynamics.

The second model, referred to as the "gene-for-gene model," (GFGM) predicts that the interaction between parasite virulence loci and host resistance loci determines successful infection (Flor, 1956). The GFGM is based on resistance and virulence genes found in plants and their pathogens, respectively, and is characterized by directional "arms-race" dynamics (Thompson and Burdon, 1992; Parker, 1994; Sasaki, 2000). At an interacting locus, pathogens can have either an avirulence or virulence gene, and the host will have either a susceptible or resistance gene. A pathogen with an avirulence gene at an interacting locus can infect a host with a susceptible gene, but not a host with a resistance gene. A pathogen with a virulence gene can infect a host with either a susceptible or resistance gene. There may be several loci involved, so initially at the start of the coevolutionary interaction a parasite may have several avirulence genes and the host several susceptible genes. During the course of coevolution a host susceptible gene would evolve to be a resistance gene at one locus and the corresponding parasite locus would counter evolve from an avirulence gene to a virulence gene. This process would continue at other loci until, in the absence of costs associated with infectivity and resistance, parasites become super-generalists, infecting a wider and wider range of host genotypes, and hosts become generally resistant to wider and wider range of parasite genotypes (Thompson and Burdon, 1992; Parker, 1994; Sasaki, 2000; Buckling and Rainey, 2002a).

The MAM and the GFGM are probably two ends of a spectrum, and the interaction between most hosts and pathogens is likely to lie somewhere between the two extremes with some degree of specialization and some generalization. This may be due to costs in the GFGM. Where gaining several virulence and resistance genes may be costly to the parasite or host, which may prevent super-generalists fixing in the population with virulence or resistance genes at every locus (Sasaki, 2000). The cost would give a fitness advantage to a host with susceptibility genes in the presence of a pathogen with the corresponding virulence loci. Once hosts with susceptible genes increase in frequency, selection will favor pathogens with avirulence genes as these can also infect the common susceptible hosts, but do not carry any costs associated with virulence. This will lead to cyclical dynamics like those seen in the MAM (Sasaki, 2000). Recent theory have suggested that a combination of the two models might capture more biological realism and relax the assumptions required for the maintenance of genetic diversity by parasites (Agrawal and Lively, 2002, 2003; Salathé et al., 2005). For example, in one model the pathogens have full infectivity on matching genotypes, as assumed under the MAM, but there is a continuum where other genotypes can be infected as under the GFGM except the parasites have lower fitness and the host suffers less than they would if the genotypes fully matched (Agrawal and Lively, 2002). In this model any departure from a pure GFGM leads to cyclical dynamics, as under the MAM.

Understanding how successful infection is determined at the genotypic level is critical in understanding how disease spreads through a population. Specifically, if infection success is based solely on host resistance or parasite virulence, as is true under the strict GFGM, virulent parasites should quickly sweep through any susceptible host populations and infect most of the host population (Sasaki, 2000). Alternatively, if infection success is determined by an interaction between host and parasite genotype, as is true under the MAM, only a subset of host genotypes will be infected, and only a subset of parasite genotypes will be infective at any given time. Testing the underlying assumption of tight genetic specificity for infection has thus far produced mixed results. Although it is clear that there exists a great deal of natural variation in host resistance and parasite infectivity (Henter and Via, 1995; Kraaijeveld et al., 1998; Webster and Woolhouse, 1998; Little and Ebert, 1999; Salvaudon et al., 2005), it is less clear whether specific host-genotype by parasite-genotype interactions typically govern the outcome of infection (Blanford et al., 2003).

Evidence from natural populations of hosts and parasites has shown that invertebrate host resistance is often highly specific to parasite genotype (Carius et al., 2001; Schulenburg and Ewbank, 2004; Lambrechts et al., 2005; Rauch et al., 2006). However, results from experiments in which parasite specificity is selected upon via experimental passaging on single host genotypes have produced mixed results. For example, when the trypanosome parasite, *Crithidia bombi*, was passaged through individuals from a colony of worker bees, the parasite did not gain infectivity on its own colony but did lose infectivity to

> **Box 6.1 (cont'd)**
>
> other, allopatric colonies (Yourth and Schmid-Hempel, 2006). Similar results were found when an RNA bacteriophage was passaged through novel genotypes of bacterial hosts (Duffy et al., 2007). It is also becoming clear that increased specificity is not always indicative of genotype by genotype interactions. When a bacterial parasite, *Holospora undulate*, was passaged on host lines of the protozoan host *Paramecium caudatum*, for example, no host-line by parasite-line interactions were found despite evidence for increased infection success on sympatric host–parasite combinations (Nidelet and Kaltz, 2007).

structure (Boots et al., 2004), or competition with other pathogens (Bremermann and Pickering, 1983; Brown et al., 2002). However, ongoing coevolution between host and parasite populations is expected to lead to decreased virulence, as fitness of both populations is optimized.

Evidence for decreased virulence over time has been demonstrated in experimental populations of Red Flour beetles, *Tribolium castaneum*, and the microsporidian parasite, *Nosema whitei*. After only 11 generations of experimental coevolution, parasite lines became less virulent, as measured by host mortality, without losing their ability to infect hosts (Bérénos et al., 2009). Further evidence comes from experimental systems of bacteria and plasmids (circular strands of DNA often carried by bacteria that can carry beneficial genes, such as those conferring antibiotic resistance). Plasmids can be considered parasitic in that hosts harboring these elements suffer a reduction in growth rate, possibly due to the additional expression of plasmid products which compete for the host ribosomes with the expression of host genes (Zund and Lebek, 1980). In an experimental study, Bouma and Lenski (1988) show that the costs of carrying a plasmid were reduced during experimental evolution, albeit via changes in the host only. A different study further demonstrated that genetic changes in both the host cell and the plasmid lead to increases in reproductive fitness of the host cell (Modi and Adams, 1991).

Aside from the trade-off model there have been several other theories to explain the evolution of virulence. Some have suggested that mixed infections of different pathogen genotypes within a single host may have important effects upon virulence, in some cases decreasing virulence while in others increasing it (Brown et al., 2002; Nowak and May, 1994; Frank, 1996). Models where virulence increases have a similar assumption to the trade-off model: selection will favor the parasite with the fastest within-host growth rate, rather than a more prudent host exploiter. The parasite with the faster growth rate is predicted to outcompete the slower growing parasite and to have a higher probability of transmission, leading to the evolution of higher virulence than that expected for single infections. Alternatively, if parasites produce a "public good" that are utilized by all the parasites within a host, mixed infections may select for cheating behavior because of low relatedness (*i.e.*, they are different genotypes) between parasites (Turner and

Chao, 1999). Examples of public goods include siderophores (Griffin et al., 2004), which are iron-scavenging molecules in bacteria, and coat proteins in viruses (Turner and Chao, 1999). Such molecules may be costly for a parasite to produce. If a parasite "cheats" and does not produce them, but uses the molecules produced by a competing parasite, it does not pay the costs but gains the benefits, giving it a higher growth rate or competitive advantage. Such cheating behavior will therefore have a selective advantage, and the cheats will increase in frequency (Griffin et al., 2004). However if there are too many cheats there won't be enough parasites producing the "public goods" to support all the cheats, decreasing the growth rate of the parasite population, and ultimately its virulence (Harrison et al., 2006).

At an even greater extreme, initially parasitic organisms may evolve to benefit the host by increasing the host's fitness, changing the interaction to a mutualistic one (Frank, 1995). There is evidence for this type of transition between a grain weevil *Sitophilus zeamais* and a bacterial mutualist, *Sitophilus zeamais* primary endosymbiont (SZPE). The genome of SZPE encodes a type III secretion system, and expression of these genes coincides with the timing of bacteriome infection within a developing weevil (Dale et al., 2002). There is evidence that the ancestor of SZPE was originally pathogenic, as type III secretion systems are found in a diverse range of bacteria pathogenic to plants or animals, including *Salmonella* spp. and *Pseudomonas* spp. (Buttner and Bonas, 2002) and are used by these pathogens to invade the host cell (Galan and Collmer, 1999). It is likely that through the course of evolution, SZPE has evolved to become a mutualist, but still uses the same method to enter its host as its ancestral pathogenic bacterium did (Dale et al., 2002).

It is important to note that virulence is not necessarily a fixed characteristic of a pathogen. Rather, virulence is often context-dependent and can be influenced by host condition (Brown et al., 2000; Pulkkinen and Ebert, 2004), host density (Bedhomme et al., 2005; Lively, 2006), or interactions with species at other trophic levels (De Roode et al., 2008a). Understanding the evolution of virulence is critical to understanding the process of host−pathogen coevolution because the magnitude of parasite-mediated selection on host populations is a direct function of both pathogen prevalence, which determines the likelihood of becoming infected, and pathogen virulence, which determines the fitness cost of being infected.

6.2 The Process of Antagonistic Coevolution

6.2.1 Introduction to the Process of Antagonistic Coevolution

There are several factors that are thought to affect the dynamics of antagonistic coevolution, including both biotic and abiotic factors. The biotic factors include the genetic basis of host−pathogen interactions (Box 6.1) (Hamilton, 1980; Sasaki, 2000), mutation and recombination rates (Gandon and Michalakis, 2002), relative generation times of host and parasite (Lively, 1999; Gandon and Michalakis, 2002), and interactions with other parasites. Abiotic factors include environmental productivity and barriers to gene flow. Other factors, such as migration rate, may be a combination of biotic and abiotic effects (Lively, 1999; Gandon, 2002). Together, these factors may affect the rate of coevolution, or may give either the host or pathogen an advantage over the other. When either the host or parasite population has an evolutionary advantage over the other, it can rapidly adapt to changes in its local coevolving partner.

6.2.2 Migration, Mutation, and Recombination

The supply of new genetic diversity plays a crucial role in shaping coevolution. For hosts and pathogens to coevolve, there needs to be a constant input of new alleles upon which selection can act, as one population responds to changes in the other. Genetic diversity may be increased by mutation, recombination, or migration rates, all of which can be affected by population size. Mutation and recombination have the potential to generate novel genetic diversity within a population. Migration can also introduce novel alleles if there is spatial structuring. For example, populations are often thought to exist as metapopulations (populations divided into discrete subpopulations), resulting from environmental factors such as differences in productivity or geographic barriers. Coevolution may then drive divergence between subpopulations, as they follow different coevolutionary trajectories (Thompson, 1999; Gomulkiewicz et al., 2000; Buckling and Rainey, 2002b). Low rates of migration will introduce variation from one subpopulation to another but high rates of migration might decrease genetic diversity as the metapopulation becomes homogenized. Population size is also related to diversity: a large population will have a higher total number of mutants and migrants than a smaller population when the mutation and migration rates are equal; and it will also reduce the chances of beneficial mutations being lost by drift (Gandon and Michalakis, 2002).

If mutation, recombination, migration rates, and population sizes are equivalent between hosts and parasites, then they are predicted to coevolve together at similar rates. An increase in any of these factors for both coevolving organisms is predicted to increase the rate of coevolution, as they will increase the genetic supply rate, shortening the time for reciprocal adaptation to occur. It is more likely, however, that these factors will differ between host and parasite populations, giving one of the coevolving partners an evolutionary advantage. Since parasites typically have higher mutation, recombination, migration rates, and larger population sizes, they can rapidly respond to changes in local host populations and are predicted to be ahead in the coevolutionary race (Price, 1980; Ebert, 1994; Lively, 1999; Gandon and Michalakis, 2002; Greischar and Koskella, 2007).

6.2.3 Generation Time

Generation time is also thought to be an important determinant of rate and strength of coevolution. A shorter generation time allows favorable genotypes that have arrived in the population by mutation, recombination, or migration to rapidly increase in frequency (Gandon and Michalakis, 2002). In most cases parasites have shorter generation times than their hosts. Although conventional wisdom suggests that the coevolving partner with the fastest generation time gains an evolutionary advantage, theoretical predictions and empirical data suggest that this may not always be the case (Gandon and Michalakis, 2002; Morgan and Buckling, 2006; Greischar and Koskella, 2007). A faster generation may allow an organism to become rapidly adapted to host, but this may come at a cost of purging the genetic diversity of a population, if the supply of new genetic diversity is limited by low mutation, migration, or recombination rates. If the host subsequently adapts to the parasite, the parasite is less able to counteradapt due to its low genetic diversity.

6.2.4 Environmental and Community Context

In addition to the factors influencing the rate of population change outlined earlier, the trajectory and outcome of host−pathogen coevolution will be strongly influenced by both the community context and the abiotic environment in which it occurs. The Geographic Mosaic theory states that coevolution is shaped by three genetic and ecological attributes of species interactions: coevolutionary hot spots and cold spots, whereby the intensity of reciprocal selection among populations differs; selection mosaics, whereby the structure of the interaction differs among environments; and remixing of coevolved traits, whereby gene flow, mutation, genetic drift, and local extinction result in a continual reshuffling of coevolved genes among populations (Thompson, 1994, 2005). This geographic variation can result from genetic divergence among populations and/or by differing abiotic or biotic environments.

Among the more obvious examples of biotic factors that might influence coevolution are (1) the presence of alternate host species for more generalist pathogens, (2) the prevalence of other parasite species within a community, and (3) the presence or absence of final host species for parasites with complex life cycles or hyperparasites (*i.e.*, parasites that infect parasites). For example, coevolution between polyphagous insects and their parasites is likely to be influenced by the plant upon which the insect feeds. The plant environments may differ in regard to chemistry, architecture, or palatability; all of which could influence the fitness of hosts, fitness of parasites, and the interaction between them (reviewed in Cory and Myers, 2003). Host plant environment has also been shown to influence the infectivity, virulence, and transmission probability of nucleopolyhedrovirus among island populations of western tent caterpillars, *Malacosoma californicum pluviale* (Cory and Myers, 2004). A similar result was found for the interaction between protozoan parasites, *Ophryocystis elektroscirrha*, and monarch butterflies, *Danaus plexippus* L. across two milkweed species (De Roode et al., 2008a). Variation in host plants is also likely to influence coevolution between bacterial pathogens and hyperparasites, such as bacteriophage. For example, a study of phage adaptation to natural populations of *Pseudomonas syringae* on horse chestnut trees suggests that the microenvironment within the tree host (surface versus interior of leaves) determines the magnitude of phage adaptation to local bacteria (Koskella et al., 2010). These studies emphasize that the biotic environment, in addition to the abiotic environment, can create selection mosaics across space (Thompson, 2005).

6.3 Testing for Host—Pathogen Coevolution

6.3.1 Introduction to Testing for Host—Pathogen Coevolution

Several different methodologies have been used to test for coevolution between hosts and pathogens. Coevolution can be directly measured, but to successfully do this, a system must allow for measurement of changes that have occurred through time and testing of whether these changes can be attributed to coevolution. Furthermore, the coevolutionary change must be rapid enough to be detected by the chosen methodology within the timescale of the experiment. The direct measurement of coevolution has been achieved in several different ways, including the simultaneous measurement of host resistance and parasite infectivity over time and of population genetic changes. For systems in which direct testing is not feasible, either due to timescale or difficulty of experimental manipulation, evidence of coevolution can be gleaned from studies of adaptation across space by studying reciprocal adaptation of parasites and hosts from multiple populations.

6.3.2 Direct Comparisons Between Coevolving Organisms Across Time

Perhaps the most straightforward way to test for host—pathogen coevolution comes from experimental systems in which reciprocal changes over time can be explicitly compared. These "time-shift" experiments, as they are sometimes known (Gaba and Ebert, 2009) have been achieved in several different ways, but are most commonly utilized in microbial systems. Here, we highlight how coevolution between bacteria and bacteriophage can be measured in the laboratory.

Microbial systems are highly amenable models for the study of coevolutionary processes (Elena and Lenski, 2003; Buckling et al., 2009). They have large population sizes and short generation times, which allow rapid coevolution in a short period of time: over a matter of days and weeks. Multiple populations can be kept in a laboratory enabling easy replication of experiments, and variables of interest can be directly manipulated while controlling for all other effects. Perhaps the key advantage of using microbes to study coevolution is that they can be frozen and stored in "suspended animation" at regular intervals during coevolution experiments. These frozen lines give a "living fossil record" where samples from different time points can be directly compared to show how the populations have changed over time.

The majority of these bacteria-phage studies use lytic phage that infect a given host bacterium, hijacking its cellular machinery and turning it into a "factory" that produces more phage progeny inside the cell. In order for phage to "escape" the host cell and infect other host cells, they must burst the host cell open, beginning the cycle again. Because phages are obligate killers, there is strong selection for bacteria to evolve resistance, and equally strong selection for counteradaptation by the obligatory parasitic phage to infect. Lysogenic phage have also been used as model organisms for host—pathogen evolution, and are an interesting contrast in that they are not always obligate killers and are often vertically transmitted between bacterial generations. After infecting a host cell, a lysogenic phage may go down one of two paths: either producing more copies of the phage and lysing the host cell (as the lytic phage do) or integrating into the host genome and being transmitted vertically to the next generation of the bacteria. Therefore, lysogenic phage can be used as a model to investigate the processes that favor horizontal versus vertical transmission and the subsequent evolution of virulence (Bull et al., 1991).

An example of how coevolution can be measured in a bacteria-lytic phage system is illustrated by the bacterium *Pseudomonas fluorescens* SBW25 and the lytic DNA phage SBW25Φ2 (Buckling and Rainey, 2002a). This system has been shown to coevolve in the laboratory for more than 500 bacterial generations (Morgan et al., 2005; Morgan and Buckling, 2006). The bacteria and phage (at least in the early stages of coevolution) typically follow a gene-for-gene model (GFGM) of coevolutionary interaction, where the bacteria and phage evolve to become evermore resistant and infective respectively to a wider range of genotypes (Buckling and Rainey, 2002a).

To measure coevolution during the course of an experiment, samples of bacteria and phage are frozen and stored at regular intervals. After the specified period of coevolution, bacterial colonies are isolated from each of the frozen samples. These colonies are streaked on an agar plate across samples of a population of phage isolated from either (1) a time point before the focal bacteria was isolated, (2) the same time point as the focal bacteria, or (3) a time point after the focal bacteria. After incubation, the bacterial colonies are scored as either sensitive or resistant to phage depending on ability to grow over the phage zone. As coevolution in these experiments is typically escalatory, the majority of colonies in a population are resistant to phage from previous time points, an intermediate number of colonies are resistant to phage from the same time point, and colonies are mostly susceptible to phage from later time points. This gives a negative change for the proportion of bacteria resistant to phage through time as shown in Box 6.2. The steepness of the slopes indicates the rate of coevolution, with steeper slopes indicating that coevolutionary change is occurring more rapidly (Buckling and Rainey, 2002a; Brockhurst et al., 2003). This allows for a comparison between different factors that may affect the rate of coevolution, such as mutation rate (Morgan et al., in press).

Another way to compare the rates of coevolution between populations of *P. fluorescens* and SBW25Φ2 is to measure resistance and infectivity ranges. Because coevolution is directional in the early stages and the bacteria and phage become more resistant and infective to a wider range of genotypes, they follow a predictable trajectory. Therefore, bacteria and phage from faster coevolving populations will have wider resistance and infectivity ranges as they will be further along this trajectory than slower coevolving populations (Buckling and Rainey, 2002a). To determine resistance and infectivity ranges in the *P. fluorescens*–SBW25Φ2 system, bacterial colonies from each population are streaked across phage from the same time point but from all the different populations. This gives an average measure of resistance of the bacterial population to all the phage and the infectivity for all phage populations on the bacteria. Typically comparisons are between different treatments where a factor that is predicted to change the rate of coevolution, such as migration

Box 6.2 Rates of Coevolution Between the Bacterium *Pseudomonas fluorescens* SBW25 and Phage SBW25Φ2

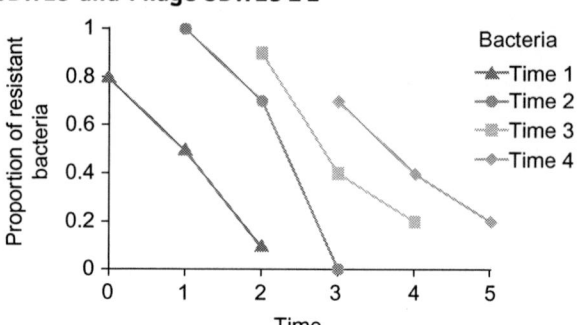

This figure shows a stylized example of the typical relationship between the resistance of bacteria to phage from different time points, thus indicating the rate of coevolution. All lines represent a single bacterial population, but are broken up into time points whereby each separate symbol represents the bacteria from a corresponding time point. The three points on each line (from left to right) represents the resistance of bacteria from one particular time point to phage from the same population but from the (1) previous time point, (2) same time point the bacteria under test were isolated from, and (3) next time point. Therefore, the lines show a negative slope as the infectivity of the phage has increased through time. The steepness of the slope indicates the rate of change through time and so consequently the rate of coevolution.

The graph also illustrates how bacterial resistance increases through time. Where there are two data points at the same point in time, the bacteria from two time points (two different data lines) are being compared on the same phage from one point in time (*i.e.*, the contemporory phage for bacteria from one point in time is the past phage for the bacteria from the subsequent point in time). In this graph the bacteria always have higher resistance to a phage population from a particular time point than bacteria from the previous time point, so we can conclude that the bacteria has evolved increased resistance through time.

rate (Morgan et al., 2005; Morgan et al., 2007) or generation time (Morgan and Buckling, 2006), is manipulated. In this case multiple replicates are used for each treatment, and the resistance of each replicate population of bacteria is measured against all phage replicates from all replicate populations and all treatments.

As an example of a time-shift experiment from the field, an eloquent study by Decaestecker et al. (2007) took sediment cores from a pond that contained dormant eggs of the waterflea Daphnia magna and dormant isolates of one of its parasites, the bacterium *Pasteuria ramosa*. These sediments contained about 39 years of coevolutionary history preserved in a sequential "living fossil record," which is effectively the same as bacteria and phage being stored in the freezer during coevolution experiments, yet over a much longer time scale. The authors examined the resistance of *Daphnia* to the parasites from one layer below (past), the same layer (contemporary), and the layer above (future). Contemporary parasites were found to be more infective than the past and future parasites, which was consistent with the matching alleles model (MAM) of Red Queen coevolution.

6.3.3 Measuring Genetic Change

In a coevolving system, genotypes of hosts and pathogen will change in frequency through time, as one responds to selection imposed by the other. Molecular methods can enable the measurement of genetic change over time, and give an indication of whether molecular evolution is influenced by the process of coevolution. In a recent study with the *P. fluorescens*–SBW25Φ2 bacteria-phage system, Paterson and colleagues (2010) allowed phage populations to experimentally evolve in the presence of either (1) coevolving bacterial host populations or (2) static, nonevolving bacterial populations, in which the bacteria was continually discarded and replaced with the ancestral strain. After, 24 days, the entire genome of each phage line was sequenced. The genome of SBW25Φ2 is approximately only 40 kbp long, which is 100 times smaller than the genome of *E. coli*, allowing for rapid sequencing and analysis. The authors then compared the number of nonsynonymous mutations in the coevolved and evolved phage, relative to the ancestor and found that coevolved phage had double the genetic divergence from the ancestor than the evolved phage; the coevolved phage had more mutated sites than the evolved phage; and there was more genetic divergence between populations of the coevolved lines than populations of the evolved phage. The results clearly show that coevolution, relative to directional evolution, leads to increased genetic divergence between populations and, ultimately, to the maintenance of genetic variation over space and time.

6.3.4 Pathogen-Mediated Rare Host Advantage

In 1949, Haldane suggested that parasites could be a significant evolutionary force, as they are under selection to infect the most common genotypes in a host population, thereby giving a fitness advantage to rare host genotypes. Specifically, when a host genotype is common in a population, any parasite able to infect that genotype will have a large fitness advantage, and will thus increase in frequency over time (Jaenike, 1978; Hamilton, 1980; Hutson and Law, 1981; Bell, 1982). This will, in turn, lead to a decrease in the frequency of the susceptible host genotype and a subsequent decrease in the corresponding parasite genotype, further driving populations apart via parasite-mediated selection. Although this hypothesis has fueled a good deal of theoretical investigation (Clarke, 1976; Hutson and Law, 1981; Bremermann and Fiedler, 1985; Hamilton, 1993), there have been relatively few empirical tests of host rare advantage (Dybdahl and Lively, 1998; Little and Ebert, 1999; Stahl et al., 1999; Koskella and Lively, 2009; Wolinska and Spaak, 2009). A key feature of Red Queen dynamics is the time lag between the rise in frequency of a recently rare and resistant host genotype and the subsequent chance introduction of a matching parasite genotype via migration, mutation, or recombination. Once introduced, this parasite would realize a significant fitness advantage and, after a time lag, drives down the frequency of its host in the population. This time lag (or phase difference) is essential for driving oscillatory dynamics and has therefore been the focus of much theoretical work (Hutson and Law, 1981).

Parasite-mediated negative frequency-dependent dynamics can be tested directly, using either an experimental evolution approach or a time-shift experiment. The process can also be examined indirectly by following infection dynamics over time in natural populations. One system that has proven ideal for these methods is the New Zealand mudsnail, *Potymopyrgus antipodarum*, and its trematode parasite, *Microphallus* sp. Upon successful infection, the trematode sterilizes, but does not kill, its snail host. Instead, the parasite reproduces within the snail and remains there, as metacercaria, until the snail is eaten by a duck, the trematode's final host. Given the parasite's high virulence (as a sterilized host has zero reproductive fitness), and thus strong potential as a selective agent, the lack of direct horizontal transmission and the relatively short generation times, this system is amenable to experimental coevolution methods in the laboratory. For example, a recent experiment, in which artificial populations of snails were evolved (1) with coevolving parasites, (2) with parasites that were lagged behind by one host generation, or (3) the absence of the trematode parasites, showed evidence for time lagged tracking of host populations by local parasites. After only six host generations, there was evidence for reciprocal change in both the host and the parasite populations, and hosts were found to be more resistant to parasites that were lagged behind than they were to coevolving parasites (Koskella and Lively, 2007).

This result was then followed up with a direct test of whether parasites were disproportionately infecting common host genotypes, thereby giving rare host genotypes a fitness advantage (Koskella and Lively, 2009). The genes involved in determining infection for this system are unknown, but the asexual reproductive mode of the snail means that any infection alleles will be necessarily linked to neutral allozyme markers. By comparing genotype frequencies of each experimental population across three time points, the start, midpoint, and end of the experiment, the authors were able to demonstrate that the initially common clone declined in frequency over time in the presence, but not in the absence, of parasites. These results are consistent with negative frequency-dependent dynamics, as predicted under the MAM of coevolution and support previous evidence, from the field, that the trematode can impose strong selection on host populations (Lively, 1989) and maintain host genetic diversity over time via rare advantage (Dybdahl and Lively, 1995; Jokela et al., 2009).

There have also been a number of key studies from *Daphnia* and their parasites that have directly measured the change in frequency of host and/or pathogen genotypes under both experimental (Capaul and Ebert, 2003) and field conditions (Duncan and Little, 2007; Duffy et al., 2008; Wolinska and Spaak, 2009). *Daphnia* has been used as a model host organism to examine coevolution with a number of different naturally occurring parasites (Ebert, 2008). In one study, genotypic compositions of natural *D. magna* populations were compared before and after epidemics of the bacterial pathogen *Pasteuria ramosa*. Resistant host genotypes were found to dominate the population after the parasite epidemic. Also, parasitism temporarily decreased genetic diversity within a population, as all susceptible genotypes were wiped out (Duncan and Little, 2007). Another elegant coevolution study used the *Daphnia galeata* x *hyalina* x *cucullata* species complex to investigate negative frequency-dependent selection imposed by four of its common parasites: a protozoan, a fungal-like oomycete, and two bacterial species (Wolinska and Spaak, 2009). The authors tracked changes in host genotypes through time across natural lake populations using allozyme analysis. By comparing these changes with dynamics in populations where no infections were found, the authors show that, on average, the most common host genotype was underinfected by parasites. This indicated that the host had an evolutionary advantage, which the authors suggested could be because hosts can migrate between lakes via birds transporting their eggs, while the parasites could not. Secondly, it was found that in most cases the common genotype declined through time in the presence of parasites, but not in the populations without parasites, clearly demonstrating parasite-mediated, negative frequency-dependent selection.

6.3.5 Pathogen Local Adaptation

"Red Queen" dynamics are considered to be one of the major driving forces of pathogen local adaptation (LA), defined as either (1) the better performance of a local parasite on its local host compared to other, allopatric parasites or (2) the better performance of a parasite on its local host compared to its performance on other, allopatric hosts (Parker, 1985, 1989; Roy, 1998; Kawecki and Ebert, 2004; Morgan et al., 2005). A host genotype that is common in one population, and thus being targeted by local parasites, is unlikely to be also common in another at a given point in time. However, since parasites are lagged in their tracking of host genotypes (*i.e.*, are always responding to changes in the host population), it is predicted that parasites will occasionally be locally maladapted and thus do better on a population of allopatric hosts (Morand et al., 1996). For many systems it is difficult, if not impossible, to measure coevolutionary change through time. Therefore, it is often easier to study the outcome, consequences, or signatures of coevolution rather than the process itself. Pathogen LA, as a signature of coevolution, is relatively easy to measure, and can be examined over space (*i.e.*, across multiple populations) as a way to understand what is likely happening over time.

The degree of parasite LA is essentially a measure of the strength of host−parasite coevolution. Parasites which are more closely able to track local host populations, and thus drive changes in local host dynamics, are expected to do better on their own hosts than they would on a randomly picked allopatric host source. The absence of LA, however, can be interpreted in one of three ways: first, that parasites are not currently successful in tracking the host population; second, that hosts are ahead in the coevolutionary game and responding to selection more effectively; or third, that coevolution is not occurring, or is weak, in that host−parasite population (*i.e.*, that the population is acting as a coevolutionary cold spot; Thompson, 1994). As predicted, parasites are often found to be locally adapted to host populations (reviewed in Kaltz and Shykoff, 1998; Greischar and Koskella, 2007; Hoeksema and Forde, 2008), suggesting that coevolutionary dynamics are driving population divergence in many natural systems. However, there are also many cases in which parasites are found to be maladapted, indicating that hosts can often be ahead in the coevolutionary battle.

6.4 Implications of Coevolution

6.4.1 Diversification and Speciation

Host—pathogen coevolution may cause rapid divergence between populations that are isolated or have minimal levels of migration between them (Thompson, 1999). Hosts and pathogens impose strong selection pressures upon each other, so host—pathogen coevolution happens relatively quickly in evolutionary time and represents an interesting case in which ecological and evolutionary timescales might overlap. By chance, different populations will follow divergent trajectories (Thompson, 1999), so different hosts and pathogens may dominate separate populations. In one study, coevolving populations of bacteria and bacteriophage were found to have higher allopatric diversity relative to control populations (Buckling and Rainey, 2002b). Such rapid between-population divergence is a prerequisite of local adaptation (LA), as populations must be different for differential performance of hosts and pathogens (Gandon et al., 1996; Lively, 1999; Gandon, 2002; Gandon and Michalakis, 2002; Kawecki and Ebert, 2004; Morgan and Buckling, 2006).

Ultimately, populations of hosts and pathogens may diverge so much that they become separate species. Although there is good evidence that host—pathogen coevolution can lead to sympatric diversification and speciation of parasites (*e.g.*, following a host shift; Weiblen and Bush, 2002; Zietara and Lumme, 2002; Sorenson et al., 2003), no direct evidence that host—parasite coevolution has caused host speciation exists. Several studies have shown that the phylogenies of hosts and parasites are congruent, suggesting cospeciation over time (Hafner and Nadler, 1988; Hafner and Page, 1995; Storfer et al., 2007; Shafer et al., 2009).

6.4.2 The Maintenance of Genetic Diversity

The evolution and maintenance of sex is a central theoretical problem in biology. This is because there is a "cost of males," who do not produce offspring themselves. Therefore, an asexual female would be able to produce twice as many reproducing offspring as a sexual female, and sexual reproduction should be severely disadvantageous (Maynard Smith, 1978; Hurst and Peck, 1996). Despite the high theoretical cost of sex, which is high, most eukaryotes are still sexual. Several theories have been suggested to explain why sexual reproduction is retained. Some simply suggest that it is physically impossible to revert back to asexuality, as sex may be an integral part of the organism's development, as for example, meiosis in ciliates allows an escape from senescence (Bell, 1988); or the benefits of male care outweigh the costs (Maynard Smith, 1997). One of the two major explanations is that the recombination associated with sexual reproduction purges deleterious mutations and reverses "Muller's Ratchet" (Kondrashov, 1988; De Visser et al., 1996). The other, the Red Queen hypothesis, is that parasites play a role in the maintenance of sex by selecting for rare or novel genotypes (Hamilton et al., 1990; Lively et al., 1990; Peters and Lively, 1999; Salathe et al., 2008). Sexual recombination brings together genes from two genomes and can create or recreate rare/novel genotypes. This allows the host to constantly change every generation and "keep up" with a rapidly evolving parasite. In addition to the evidence for parasite-mediated rare advantage discussed earlier in the chapter, there is direct evidence for increased meiotic recombination within experimental populations of the red flour beetle *Tribolium castaneum* in the presence of a parasitic microsporidian (Fischer and Schmid-Hempel, 2005).

Like sex, a high mutation rate may also introduce the required genotypic diversity to allow a host to keep up with a rapidly evolving parasite. Although an elevated mutation rate is typically disadvantageous when an organism is adapted to its environment (as deleterious mutations will outweigh beneficial mutations), it may be advantageous when an organism is in a new or changing environment (Giraud et al., 2001). For hosts, a parasite may act as a constantly changing environment, and thus a host with a higher mutation rate might benefit. This has been supported by an experimental evolution study that showed laboratory populations of bacteria were more likely to evolve a higher mutation rate, in the order of 50–100 times higher, when coevolving with phage than populations that were not exposed to phage. The mutations that conferred a higher mutation rate in genes were involved in the DNA repair pathway (Pal et al., 2007).

6.5 Summary/Future Outlook

Host–pathogen coevolution is a critical and rapidly paced evolutionary force, shaping both the diversity and population structure of hosts and their pathogens. Coevolution has been demonstrated in a diverse set of host–parasite systems and, due to the ubiquity of parasites, it is likely to be very widespread across ecosystems. Although there is a large body of literature on host–pathogen coevolution, there are still several open questions in need of empirical investigation. For example, the question of what makes a pathogen more virulent, instead of mutualistic, is far from being resolved. The commonly known trade-off model of the evolution of virulence is contested by some researchers (Ebert and Bull, 2003), but there is little evidence supporting alternative hypotheses. Another open question is why some pathogens evolve to be specialists and others to be generalists. Again, there is thought to be a trade-off underlying the polymorphism in pathogen strategies whereby pathogen specialization allows for increased infectivity on a given host but decreased infectivity at the community level (Wilson and Yoshimura, 1994; Regoes et al., 2000).

Perhaps the largest and most important open questions are regarding how human activity impacts host–pathogen coevolution. It remains to be determined whether knowledge of host—pathogen interactions can be beneficially applied to manipulate the outcome of coevolution. For example, theory and empirical evidence has shown that migration, particularly asymmetric migration of host and pathogen, can radically alter host resistance and pathogen virulence (Lively, 1999; Gandon, 2002; Morgan et al., 2005; Morgan et al., 2007). Recent centuries have seen increased movement of humans over greater distances, especially with the advent of air travel. With them have traveled their pathogens, pathogens of animals and plants, and animal and plant hosts. How this movement has impacted on the evolution of disease of humans, and diseases in natural ecosystems and economically important animals and crops, has received little investigation so far. For example, we need to understand how parasites influence species invasions (Tompkins et al., 2003; Prenter et al., 2004; Torchin and Mitchell, 2004) and how host shifts might change parasite virulence and transmission (Antonovics et al., 2002; López-Villavicencio et al., 2005; Duffy et al., 2007).

However, human intervention could deliberately alter host—pathogen coevolution in favor of one of the coevolving organisms. For example, when using pathogens as biocontrol agents to kill pests in agriculture, it may be advantageous to manipulate coevolution, such as increasing migration, to tilt the balance in favor of the pathogen, and away from the host pest. The medical field could also use knowledge of coevolution to reduce the effects of disease. For example, bacteriophage have been suggested as an alternative to antibiotics in treating bacterial infections (Levin and Bull, 2004; Wright et al., 2009). The advantage that bacteriophage have over antibiotics is that they can evolve to overcome resistance, whereas once a bacterium becomes resistant to an antibiotic, it is no longer of use. In this case, knowledge of the evolutionary theory behind host—pathogen coevolution could tip the evolutionary advantage toward the bacteriophage. Similarly, coevolutionary theory could be used to alter the outcome of coevolution in favor of the host in halting the spread of disease. Thus, the area where knowledge and manipulation of coevolution could have its most dramatic application is within the medical field. The approach has been dubbed "Darwinian Medicine" and has received a lot of attention since the late 1990s (Williams, 2010). Indeed it has been suggested that all medics should be obliged to study evolution (Nesse et al., 2010). To parasitize the famous phrase from Dobzhansky (1973), we suggest that "no disease makes sense except in the light of coevolution."

References

Agrawal, A.F., Lively, C.M., 2002. Infection genetics: gene-for-gene versus matching-allele models, and all points in between. Evol. Ecol. Res. 4, 1—12.
Agrawal, A.F., Lively, C.M., 2003. Modelling infection as a two-step process combining gene-for-gene and matching-allele genetics. Proc. R. Soc. Lon. B Biol. Sci. 270, 323—334.
Anderson, R.M., May, R.M., 1982. Coevolution of hosts and parasites. Parasitology 85, 411—426.
Antonovics, J., Hood, M., Partain, J., 2002. The ecology and genetics of a host shift: *Microbotryum* as a model system. Am. Nat. 160, S40—S53.

Bahri, B., Kaltz, O., Leconte, M., de Vallavieille-Pope, C., Enjalbert, J, 2009. Tracking costs of virulence in natural populations of the wheat pathogen, *Puccinia striiformis* f.sp. *tritici*. BMC Evol. Biol. 9 article 26.
Bedhomme, S., Agnew, P., Vital, Y., Sidobre, C., Michalakis, Y., 2005. Prevalence-dependent costs of parasite virulence. PLoS Biol. 3, 1403–1408.
Bell, G., 1982. The Masterpiece of Nature: The Evolution and Genetics of Sexuality. University of California Press, Berkeley, CA.
Bell, G., 1988. Recombination and the immortality of the germ line. J. Evol. Biol. 1, 67–82.
Bérénos, C., Schmid-Hempel, P., Wegner, K.M., 2009. Evolution of host resistance and trade-offs between virulence and transmission potential in an obligately killing parasite. J. Evol. Biol. 22, 2049–2056.
Blanford, S., Thomas, M.B., Pugh, C., Pell, J.K., 2003. Temperature checks the Red Queen? Resistance and virulence in a fluctuating environment. Ecol. Lett. 6, 2–5.
Bolker, B.M., Nanda, A., Shah, D. 2010. Transient virulence of emerging pathogens. J. R. Soc. Interface 7, 811–822.
Boots, M., Begon, M., 1993. Trade-offs with resistance to a granulosis-virus in the Indian Meal Moth, examined by a laboratory evolution experiment. Funct. Ecol. 7, 528–534.
Boots, M., Hudson, P.J., Sasaki, A., 2004. Large shifts in pathogen virulence relate to host population structure. Science 303, 842–844.
Bouma, J.E., Lenski, R.E., 1988. Evolution of a bacteria/plasmid association. Nature 335, 351–352.
Bremermann, H.J., Fiedler, B., 1985. On the stability of polymorphic host-pathogen populations. J. Theor. Biol. 117, 621–631.
Bremermann, H.J., Pickering, J., 1983. A game-theoretical model of parasite virulence. J. Theor. Biol. 100, 411–426.
Brockhurst, M.A., Morgan, A.D., Rainey, P.B., Buckling, A., 2003. Population mixing accelerates coevolution. Ecol. Lett. 6, 975–979.
Brown, M.J.F., Loosli, R., Schmid-Hempel, P., 2000. Condition-dependent expression of virulence in a trypanosome infecting bumblebees. Oikos 91, 421–427.
Brown, S.P., Hochberg, M.E., Grenfell, B.T., 2002. Does multiple infection select for raised virulence? Trends in Microbiology 10: 401–405.
Buckling, A., Rainey, P.B., 2002a. Antagonistic coevolution between a bacterium and a bacteriophage. Proc. R. Soc. Lond. B Biol. Sci. 269, 931–936.
Buckling, A., Rainey, P.B., 2002b. The role of parasites in sympatric and allopatric host diversification. Nature 420, 496–499.
Buckling, A., Maclean, R.C., Brockhurst, M.A., Colegrave, N., 2009. The Beagle in a bottle. Nature 457, 824–829.
Buckling, A., Wei, Y., Massey, R.C., Brockhurst, M.A., Hochberg, M.E., 2006. Antagonistic coevolution with parasites increases the cost of host deleterious mutations. Proc. R. Soc. Lond. B Biol. Sci. 273, 45–49.
Bull, J.J., Mollineux, I.J., Rice, W.R., 1991. Selection of benevolence in a host–parasite system. Evolution 45, 875–882.
Buttner, D., Bonas, U., 2002. Port of entry—the type III secretion translocon. Trends Microbiol. 10, 186–192.
Capaul, M., Ebert, D., 2003. Parasite-mediated selection in experimental *Daphnia magna* populations. Evolution 57, 249–260.
Carius, H.-J., Little, T.J., Ebert, D., 2001. Genetic variation in a host–parasite association: potential for coevolution and frequency dependent selection. Evolution 55, 1136–1145.
Carroll, L., 1871. Through the Looking Glass. Macmillan, London.

Clarke, B., 1976. The ecological relationships of host–parasite relationships. In: Taylor, A. E.R., Muller, R. (Eds.), Genetic Aspects of Host–Parasite Relationships. Blackwell, Oxford, pp. 87–103.

Cory, J.S., Myers, J.H., 2003. The ecology and evolution of insect baculoviruses. Annu. Rev. Ecol. Evol. Syst. 34, 239–272.

Cory, J.S., Myers, J.H., 2004. Adaptation in an insect host–plant pathogen interaction. Ecol. Lett. 7, 632–639.

Dale, C., Plague, G.R., Wang, B., Ochman, H., Moran, N.A., 2002. Type III secretion systems and the evolution of mutualistic endosymbiosis. Proc. Natl. Acad. Sci. U.S.A. 99, 12397–12402.

Day, T., 2001. Parasite transmission modes and the evolution of virulence. Evolution 55, 2389–2400.

De Roode, J.C., Pedersen, A.B., Hunter, M.D., Altizer, S., 2008a. Host Plant Species Affects Virulence in Monarch Butterfly Parasites. Blackwell, Oxford, UK.

De Roode, J.C., Yates, A., Altizer, S., 2008b. Virulence-transmission trade-offs and population divergence in virulence in a naturally occurring butterfly parasite. Proc. Natl. Acad. Sci. U.S.A. 105, 7489–7494.

De Visser, J.A.G.M., Hoekstra, R.F., van den Ende, H., 1996. The effect of sex and deleterious mutations on fitness in *Chlamydomonas*. Proc. R. Soc. Lond. B Biol. Sci. 263, 193–200.

Decaestecker, E., Gaba, S., Raeymaekers, J.A.M., Stoks, R., Van Kerckhoven, L., Ebert, D., et al., 2007. Host-parasite 'Red Queen' dynamics archived in pond sediment. Nature 450, 870–873.

Dobzhansky, T., 1973. Nothing in biology makes sense except in the light of evolution. Am. Biol. Teach. 35, 125–129.

Duffy, S., Burch, C.L., Turner, P.E., 2007. Evolution of host specificity drives reproductive isolation among RNA viruses. Evolution 61, 2614–2622.

Duffy, M.A., Brassil, C.E., Hall, S.R., Tessier, A.J., Caceres, C.E., Conner, J.K., 2008. Parasite-mediated disruptive selection in a natural *Daphnia* population. BMC Evol. Biol. 8, 80.

Duncan, A.B., Little, T.J., 2007. Parasite-driven genetic change in a natural population of *Daphnia*. Evolution 61, 796–803.

Dybdahl, M.F., Lively, C.M., 1995. Host–parasite interactions: infection of common clones in natural populations of a freshwater snail (*Potamopyrgus antipodarum*). Proc. R. Soc. Lond. B Biol. Sci. 260, 99–103.

Dybdahl, M.F., Lively, C.M., 1998. Host–parasite coevolution: evidence for rare advantage and time-lagged selection in a natural population. Evolution 52, 1057–1066.

Ebert, D., 1994. Virulence and local adaptation of a horizontally transmitted parasite. Science 265, 1084–1086.

Ebert, D., 2008. Host–parasite coevolution: insights from the Daphnia-parasite model system. Curr. Opin. Microbiol. 11, 290–301.

Ebert, D., Bull, J.J., 2003. Challenging the trade-off model for the evolution of virulence: is virulence management feasible? Trends Microbiol. 11, 15–20.

Elena, S.F., Lenski, R.E., 2003. Evolution experiments with microorganisms: the dynamics and genetic bases of adaptation. Nat. Rev. Genet. 4, 457–469.

Fellowes, M.D.E., Kraaijeveld, A.R., Godfray, H.C.J., 1998. Trade-off associated with selection for increased ability to resist parasitoid attack in *Drosophila melanogaster*. Proc. R. Soc. Lon. B Biol. Sci. 265, 1553–1558.

Ferdig, M.T., Beerntsen, B.T., Spray, F.J., Li, J.Y., Christensen, B.M., 1993. Reproductive costs associated with resistance in a mosquito-filarial worm system. Am. J. Trop. Med. Hyg. 49, 756–762.

Fischer, O., Schmid-Hempel, P., 2005. Selection by parasites may increase host recombination frequency. Biol. Lett. 1, 193–195.

Flor, H.H., 1956. The complementary genetic systems in flax and flax rust. Adv. Genet. 8, 29–54.

Forde, S.E., Thompson, J.N., Bohannan, B.J.M., 2004. Adaptation varies through space and time in a coevolving host–parasitoid interaction. Nature 431, 841–844.

Frank, S.A., 1994. Recognition and polymorphism in host–parasite genetics. Philos. Trans. R. Soc. Lond., Ser. B: Biol. Sci. 346, 283–293.

Frank, S.A., 1995. The origin of synergistic symbiosis. J. Theor. Biol. 176, 403–410.

Frank, S.A., 1996. Models of parasite virulence. Q. Rev. Biol. 71, 37–78.

Gaba, S., Ebert, D., 2009. Time-shift experiments as a tool to study antagonistic coevolution. Trends Ecol. Evol. 24, 226–232.

Galan, J.E., Collmer, A., 1999. Type III secretion machines: bacterial devices for protein delivery into host cells. Science 284, 1322–1328.

Gandon, S., 2002. Local adaptation and the geometry of host–parasite coevolution. Ecol. Lett. 5, 246–256.

Gandon, S., Capowiez, Y., DuBois, Y., Michalakis, Y., Olivieri, I., 1996. Local adaptation and gene-for-gene coevolution in a metapopulation model. Proc. R. Soc. Lond. B Biol. Sci. 263, 1003–1009.

Gandon, S., Michalakis, Y., 2002. Local adaptation, evolutionary potential and host–parasite coevolution: interactions between migration, mutation, population size and generation time. J. Evol. Biol. 15, 451–462.

Gandon, S., Van Zandt, P.A., 1998. Local adaptation and host–parasite interactions. Trends Ecol. Evol. 13, 214–216.

Giraud, A., Matic, I., Tenaillon, O., Clara, A., Radman, M., Fons, M., et al., 2001. Costs and benefits of high mutatation rates: adaptive evolution of bacteria in the mouse gut. Science 291, 2606–2608.

Gomulkiewicz, R., Thompson, J.N., Holt, R.D., Nuismer, S.L., Hochberg, M.E., 2000. Hot spots, cold spots, and the geographic mosaic theory of coevolution. Am. Nat. 156, 156–174.

Greischar, M.A., Koskella, B., 2007. A synthesis of experimental work on parasite local adaptation. Ecol. Lett. 10, 418–434.

Griffin, A.S., West, S.A., Buckling, A., 2004. Cooperation and competition in pathogenic bacteria. Nature 430, 1024–1027.

Grim, T., Rutila, J., Cassey, P., Hauber, M.E., 2009. The cost of virulence: an experimental study of egg eviction by brood parasitic chicks. Behav. Ecol. 20, 1138–1146.

Grosberg, R.K., Hart, M.W., 2000. Mate selection and the evolution of highly polymorphic self/nonself recognition genes. Science 289, 2111–2114.

Hafner, M.S., Nadler, S.A., 1988. Phylogenetic trees support the coevolution of parasites and their hosts. Nature 332, 258–259.

Hafner, M.S., Page, R.D.M., 1995. Molecular phylogenies and host–parasite cospeciation— gophers and lice as a model. Philos. Trans. R. Soc. Lond., Ser. B: Biol. Sci. 349, 77–83.

Haldane, J.B.S., 1949. Disease and evolution. Ric. Sci. 19 (Suppl.), 68–76.

Hamilton, W.D., 1980. Sex versus non-sex versus parasite. Oikos 35, 282–290.

Hamilton, W.D., 1982. Pathogens as causes of genetic diversity in their host populations. In: Anderson, R.M., May, R.M. (Eds.), Population Biology of Infectious Diseases. Springer-Verlag, New York, pp. 269–296.

Hamilton, W.D., 1993. Haploid dynamic polymorphism in a host with matching parasites: effects of mutation/subdivision, linkage, and patterns of selection. J. Hered. 84, 328–338.

Hamilton, W.D., Axelrod, R., Tanese, R., 1990. Sexual reproduction as an adaptation to resist parasites (a review). Proc. Natl. Acad. Sci. U.S.A. 87, 3566–3573.

Harrison, F., Browning, L.E., Vos, M., Buckling, A., 2006. Cooperation and virulence in acute *Pseudomonas aeruginosa* infections. BMC Biol. 4, article 21.

Henter, H.J., Via, S., 1995. The potential for coevolution in a host-parasitoid system. I. Genetic variation within an aphid population in susceptibility to a parasitic wasp. Evolution 49, 257–265.

Hoeksema, J.D., Forde, S.E., 2008. A meta-analysis of factors affecting local adaptation between interacting species. Am. Nat. 171, 275–290.

Huang, Y.J., Balesdent, M.H., Li, Z.Q., Evans, N., Rouxel, T., Fitt, B.D.L., 2010. Fitness cost of virulence differs between the AvrLm1 and AvrLm4 loci in *Leptosphaeria maculans* (phoma stem canker of oilseed rape). Eur. J. Plant Pathol. 126, 279–291.

Hudson, P., Dobson, A., Lafferty, K., 2006. Is a healthy ecosystem one that is rich in parasites? Trends Ecol. Evol. 21, 381–385.

Hurst, L.D., Peck, J.R., 1996. Recent advances in understanding of the evolution and maintenance of sex. Trends Ecol. Evol. 11, 46–53.

Hutson, V., Law, R., 1981. Evolution of recombination in populations experiencing frequency-dependent selection with time delay. Proc. R. Soc. Lond. B Biol. Sci. 213, 345–359.

Jaenike, J., 1978. An hypothesis to account for the maintenance of sex within populations. Evol. Theory 3, 191–194.

Janzen, D.H., 1980. When is it coevolution? Evolution 34, 611–612.

Jokela, J., Dybdahl Mark, F., Lively Curtis, M., 2009. The maintenance of sex, clonal dynamics, and host–parasite coevolution in a mixed population of sexual and asexual snails. Am. Nat. 174, S43–S53.

Kaltz, O., Shykoff, J.A., 1998. Local adaptation in host–parasite systems. Heredity 81, 361–370.

Kawecki, T.J., Ebert, D., 2004. Conceptual issues in local adaptation. Ecol. Lett. 7, 1225–1241.

Kondrashov, A., 1988. Deleterious mutations and the evolution of sexual reproduction. Nature 336, 435–440.

Koskella, B., Lively, C.M., 2007. Advice of the rose: experimental coevolution of a trematode parasite and its snail host. Evolution 61, 152–159.

Koskella, B., Lively, C.M., 2009. Evidence for negative frequency-dependent selection during experimental coevolution of a freshwater snail and a sterilizing trematode. Evolution 63, 2213–2221.

Koskella, B., Thompson, J.N., Preston, G.M., Buckling, A. 2010. Parasite host environment determines the scale of hyperparasite adaptation.

Kraaijeveld, A.R., Van Alphen, J.J., Godfray, H.C., 1998. The coevolution of host resistance and parasitoid virulence. Parasitology 116, S29–S45.

Laine, A.-L., 2009. Role of coevolution in generating biological diversity: spatially divergent selection trajectories. J. Exp. Bot. 60, 2957–2970.

Langand, J., Jourdane, J., Coustau, C., Delay, B., Morand, S., 1998. Cost of resistance, expressed as a delayed maturity, detected in the host–parasite system *Biomphalaria glabrata Echinostoma caproni*. Heredity 80, 320–325.

Lambrechts, L., Halbert, J., Durand, P., Gouagna, L., Koella, J., 2005. Host genotype by parasite genotype interactions underlying the resistance of anopheline mosquitoes to *Plasmodium falciparum*. Malar. J. 4, 3.

Levin, B.R., Bull, J.J., 2004. Population and evolutionary dynamics of phage therapy. Nat. Rev. Microbiol. 2, 166–173.

Little, T.J., Ebert, D., 1999. Associations between parasitism and host genotype in natural populations of *Daphnia* (Crustacea: Cladocera). J. Anim. Ecol. 68, 134–149.

Lively, C.M., 1989. Adaptation by a parasitic trematode to local populations of its snail host. Evolution 43, 1663–1671.

Lively, C.M., 1999. Migration, virulence, and the geographic mosaic of adaptation by parasites. Am. Nat. 153, S34–S47.

Lively, C.M., 2006. The ecology of virulence. Ecol. Lett. 9, 1089–1095.

Lively, C.M., Craddock, C., Vrijenhoek, R.C., 1990. Red Queen hypothesis supported by parasitism in sexual and clonal fish. Nature 344, 864–866.

López-Villavicencio, M., Enjalbert, J., Hood, M.E., Shykoff, J.A., Raquin, C., Giraud, T., 2005. The anther smut disease on *Gypsophila repens*: a case of parasite sub-optimal performance following a recent host shift? J. Evol. Biol. 18, 1293–1303.

Maynard Smith, J., 1978. The Evolution of Sex. Cambridge University Press, Cambridge.

Maynard Smith, J., 1997. The maintenance of sex. In: Ridley, M. (Ed.), Evolution. Oxford University Press, Oxford.

Modi, R.I., Adams, J., 1991. Coevolution in bacterial-plasmid populations. Evolution 45, 656–667.

Morand, S., Manning, S.D., Woolhouse, M.E.J., 1996. Parasite–host coevolution and geographic patterns of parasite infectivity and host susceptibility. Proc. R. Soc. Lond. B Biol. Sci. 263, 119–128.

Morgan, A.D., Gandon, S., Buckling, A., 2005. The effect of migration on local adaptation in a coevolving host-parasite system. Nature 437, 253–256.

Morgan, A.D., Buckling, A., 2006. Relative number of generations of hosts and parasites does not influence parasite local adaptation in coevolving populations of bacteria and phages. J. Evol. Biol. 19, 1956–1963.

Morgan, A.D., Brockhurst, M.A., Lopez-Pascua, L.D.C., Pal, C., Buckling, A., 2007. Differential impact of simultaneous migration on coevolving hosts and parasites. BMC Evol. Biol. 7, 1.

Morgan, A.D., Maclean, R.C., Buckling, A., 2009. Effects of antagonistic coevolution on parasite-mediated host coexistence. J. Evol. Biol. 22, 287–292.

Morgan, A.D., Bonsall, M.B., Buckling, A., in press. Impact of bacterial mutation rates in coevolutionary dynamics between bacteria and phages. Evolution.

Nee, S., 1989. Antagonistic coevolution and the evolution of genotypic randomization. J. Theor. Biol. 140, 499–518.

Nesse, R.M., Bergstrom, C.T., Ellison, P.T., Flier, J.S., Gluckman, P., Govindaraju, D.R., et al., 2010. Making evolutionary biology a basic science for medicine. Proc. Natl. Acad. Sci. U.S.A. 107, 1800–1807.

Nidelet, T., Kaltz, O., 2007. Direct and correlated responses to selection in a host-parasite system: testing for the emergence of genotype specificity. Evolution 61, 1803–1811.

Nowak, M.A., May, R.M., 1994. Superinfection and the evolution of parasite virulence. Proc. R. Soc. Lond. B Biol. Sci. 255, 81–89.

Nuismer, S.L., Goodnight, C., 2009. Parasite local adaptation in a geographic mosaic. Evolution 60, 24–30.

Nuismer, S.L., Thompson, J.N., Gomulkiewicz, R., 2003. Coevolution between hosts and parasites with partially overlapping geographic ranges. J. Evol. Biol. 16, 1337–1345.

Pal, C., Macia, M.D., Oliver, A., Schachar, I., Buckling, A., 2007. Coevolution with viruses drives the evolution of bacterial mutation rates. Nature 450, 1079–1081.

Parker, M.A., 1985. Local population differentiation for compatibility in an annual legume ant its host-specific fungal pathogen. Evolution 39, 713–723.

Parker, M.A., 1989. Disease impact and local genetic diversity in the clonal plant *Podophyllum peltatum*. Evolution 43, 540–547.

Parker, M.A., 1994. Pathogens and sex in plants. Evol. Ecol. 8, 560–584.

Paterson, S., Vogwill, T., Buckling, A., Benmayor, R., Spiers, A.J., Thomson, N.R., et al., 2010. Antagonistic coevolution accelerates molecular evolution. Nature 464, 275–278.

Peters, A.D., Lively, C.M., 1999. The Red Queen and fluctuating epistasis: a population genetic analysis of antagonistic coevolution. Am. Nat. 154, 393–405.

Prenter, J., MacNeil, C., Dick, J.T.A., Dunn, A.M., 2004. Roles of parasites in animal invasions. Trends Ecol. Evol. 19, 385–390.

Price, P.W., 1980. Evolutionary Biology of Parasites. Princeton University Press, Princeton, NJ.

Price, M.V., Waser, N.M., 1982. Population structure, frequency-dependent selection, and the maintenance of sexual reproduction. Evolution 36, 35–43.

Pulkkinen, K., Ebert, D., 2004. Host starvation decreases parasite load and mean host size in experimental populations. Ecology 85, 823–833.

Rauch, G., Kalbe, M., Reusch, T.B.H., 2006. One day is enough: rapid and specific host–parasite interactions in a stickleback-trematode system. Biol. Lett. 2, 382–384.

Regoes, R.R., Nowak, M.A., Bonhoeffer, S., 2000. Evolution of virulence in a heterogeneous host population. Evolution 54, 64–71.

Roy, B.A., 1998. Differentiating the effects of origin and frequency in reciprocal transplant experiments used to test negative frequency-dependent selection hypothesis. Oecologia 115, 73–83.

Salathe, M., Scherer, A., Bonhoeffer, S., 2005. Neutral drift and polymorphism in gene-for-gene systems. Ecol. Lett. 8, 925–932.

Salathe, M., Kouyos, R.D., Regoes, R.R., Bonhoeffer, S., 2008. Rapid parasite adaptation drives selection for high recombination rates. Evolution 62, 295–300.

Salvaudon, L., Heraudet, V., Shykoff, J.A., 2005. Parasite-host fitness trade-offs change with parasite identity: genotype-specific interactions in a plant-pathogen system. Evolution 59, 2518–2524.

Salvaudon, L., Giraud, T., Shykoff, J.A., 2008. Genetic diversity in natural populations: a fundamental component of plant–microbe interactions. Curr. Opin. Plant Biol. 11, 135–143.

Sasaki, A., 2000. Host–parasite coevolution in a multilocus gene-for-gene system. Proc. R. Soc. Lond. B Biol. Sci. 267, 2183–2188.

Schulenburg, H., Ewbank, J., 2004. Diversity and specificity in the interaction between *Caenorhabditis elegans* and the pathogen *Serratia marcescens*. BMC Evol. Biol. 4, 49.

Seger, J., 1988. Dynamics of some simple host–parasite models with more than two genotypes in each species. Philos. Trans. R. Soc. Lond., Ser. B: Biol. Sci. 319, 541–555.

Shafer, A.B.A., Williams, G.R., Shutler, D., Rogers, R.E.L., Stewart, D.T., 2009. Cophylogeny of Nosema (Microsporidia: Nosematidae) and bees (Hymenoptera: Apidae) suggests both cospeciation and a host-switch. J. Parasitol. 95, 198–203.

Sorenson, M.D., Sefc, K.M., Payne, R.B., 2003. Speciation by host switch in brood parasitic indigobirds. Nature 424, 928–931.
Stahl, E.A., Dwyer, G., Mauricio, R., Kreitman, M., Bergelson, J., 1999. Dynamics of disease resistance polymorphism at the Rpm1 locus of *Arabidopsis*. Nature 400, 667–671.
Storfer, A., Alfaro, M.E., Ridenhour, B.J., Jancovich, J.K., Mech, S.G., Parris, M.J., et al., 2007. Phylogenetic concordance analysis shows an emerging pathogen is novel and endemic. Ecol. Lett. 10, 1075–1083.
Thompson, JN., 1994. The coevolutionary process. University of Chicago Press, Chicago.
Thompson, J.N., 1999. Specific hypotheses on the geographic mosaic of coevolution. Am. Nat. 153, S1–S14.
Thompson, J.N., 2005. The Geographic Mosaic of Coevolution. University of Chicago Press, Chicago.
Thompson, J.N., Burdon, J.J., 1992. Gene-for-gene coevolution between plants and parasites. Nature 360, 121–125.
Tompkins, D.M., White, A.R., Boots, M., 2003. Ecological replacement of native red squirrels by invasive greys driven by disease. Ecol. Lett. 6, 189–196.
Torchin, M.E., Mitchell, C.E., 2004. Parasites, pathogens, and invasions by plants and animals. Front. Ecol. Environ. 2, 183–190.
Turner, P.E., Chao, L., 1999. Prisoner's dilemma in an RNA virus. Nature 398, 441–443.
Webster, J.P., Woolhouse, M.E.J., 1998. Cost of resistance to *Schistosoma mansoni* in the snail intermediate host *Biomphalaria glabrata*. Trans. R. Soc. Trop. Med. Hyg. 92, 367.
Weiblen, G.D., Bush, G.L., 2002. Speciation in fig pollinators and parasites. Mol. Ecol. 11, 1573–1578.
Williams, P.D., 2010. Darwinian interventions: taming pathogens through evolutionary ecology. Trends Parasitol. 26, 83–92.
Wilson, D.S., Yoshimura, J., 1994. On the coexistence of specialists and generalists. Am. Nat. 144, 692.
Wolinska, J., Spaak, P., 2009. The cost of being common: evidence from natural *Daphnia* populations. Evolution 63, 1893–1901.
Woolhouse, M.E.J., Webster, J.P., 2000. In search of the red queen. Parasitol. Today 16, 506–508.
Wright, A., Hawkins, C.H., Anggard, E.E., Harper, D.R., 2009. A controlled clinical trial of a therapeutic bacteriophage preparation in chronic otitis due to antibiotic-resistant *Pseudomonas aeruginosa*; a preliminary report of efficacy. Clin. Otolaryngol. 34, 349–357.
Yourth, C.P., Schmid-Hempel, P., 2006. Serial passage of the parasite *Crithidia bombi* within a colony of its host, *Bombus terrestris*, reduces success in unrelated hosts. Proc. R. Soc. Lond. B Biol. Sci. 273, 655–659.
Zietara, M.S., Lumme, J., 2002. Speciation by host switch and adaptive radiation in a fish parasite genus *Gyrodactylus* (*Monogenea, Gyrodactylidae*). Evolution 56, 2445–2458.
Zund, P., Lebek, G., 1980. Generation time-prolonging R plasmids: correlation between increases in the generation time of *Escherichia coli* caused by R plasmids and their molecular size. Plasmid 3, 65–69.

7 Elucidating Human Migrations by Means of their Pathogens

Aude Gilabert[1,2] and Thierry Wirth[1,2],*

[1]Muséum National d'Histoire Naturelle, Laboratoire Origine Structure Evolution de la Biodiversité, Paris, France, [2]Muséum National d'Histoire Naturelle, Laboratoire de Biologie intégrative des populations, Ecole Pratique des Hautes Etudes, Paris, France

7.1 Introduction

A major aspect of human evolutionary biology consists of disentangling origins and migrations. To this aim, human population genetics has been directly investigated using polymorphic markers such as proteins, mitochondrial DNA (mtDNA), Y-chromosome, microsatellites, or single nucleotide polymorphisms (SNPs) (see Cavalli-Sforza, 1998; Cavalli-Sforza and Feldman, 2003, for reviews). These studies, combined to morphologic, anthropologic, and linguistic ones, have led to the formulation of a standard model of modern human evolution. This theory advocates that humans originated in East Africa and dispersed first throughout much of Africa ~100,000–150,000 years ago and subsequently—between 60,000 and 40,000 years—into Asia and then Europe (Cavalli-Sforza and Feldman, 2003). However, recent genetic analyses suggested that the out of Africa was more recent than expected, i.e., 60,000 years ago. The settlements of the Americas and Oceania seemed to occur later through several migrations out of Asia (Eshleman et al., 2003; Hurles et al., 2003). These successive waves of migrations resulted in a relatively low genetic diversity within modern human populations and in a decrease of genetic diversity from the horn of Africa. The genetic differentiation increases with geographic distances following an isolation by distance (IBD) model, but remains low ($F_{st} < 2\%$). Therefore, human genetic studies are often weakly resolved and moderately informative (Prugnolle et al., 2005; Ramachandran et al., 2005; Liu et al., 2006).

As stated earlier, the use of human genetic markers has contributed to the understanding of human evolution but has also failed to elucidate some recent features. Several issues still remain controversial such as the timing, source, and number of migrations to America (O'Rourke and Raff, 2010) and to Oceania (Moodley et al., 2009). In addition, relationships between closely related populations are difficult to decipher because of their too-recent divergence. Indeed, direct inference of human

*E-mail: wirth@mnhn.fr

evolutionary history is limited because of the low genetic variability due to the strong genetic bottlenecks that humans were subjected to during migrations. Other techniques such as microsatellites or SNPs, supposedly more variable than other markers, also present technical limitations to resolve human migrations (Wirth et al., 2004).

To overcome these hindrances, an alternative is to focus on human pathogens since they have coevolved with humans and reflect their evolutionary histories (Wirth et al., 2005). Pathogens present generally higher mutation rates and shorter generation times than humans (Whiteman and Parker, 2005). Thus, their populations are more diversified genetically, making the study of their population structure more informative than that of human ones. However, not all microbes are good candidates to infer host evolution and their efficacy depends on several factors (Whiteman and Parker, 2005). In addition, the choice of the pathogen will also be influenced by the timescale of the study (Nieberding and Olivieri, 2007).

Several pathogens have proven their usefulness in deciphering humans migrations and origins (Wirth et al., 2005). In particular, their study allows the distinction between closely related groups of humans, which previously was not directly possible, due to a lack of resolution of markers and/or a sampling failure for instance (Wirth et al., 2004; Moodley et al., 2009). In this review, after a brief section on some advantages and disadvantages of pathogens in the context of human host history inference, we shall illustrate pathogen utility with some relevant examples that pointed out congruence or discrepancies with human migratory history.

7.2 Using Pathogens as Genetic Tracers for Host History

Parasites have often been used to infer their host evolutionary history (Whiteman and Parker, 2005), usually using phylogeographical analyses, due to their narrow relationships with their hosts as well as their generally higher levels of genetic diversity. However, even if both protagonists share a common history, their genealogies are not necessarily similar (Holmes, 2004; Nieberding and Olivieri, 2007) and several evolutionary mechanisms can lead to identical gene trees (Rannala and Michalakis, 2003). Therefore, microbes have to be carefully chosen to be relevant in this context, some pathogen traits being of particular importance to correctly infer host history. Crucial features and parameters that will determine their usefulness are degree of intimacy with host species and mode of transmission, as well as mutation and recombination rates (Rannala and Michalakis, 2003; Whiteman and Parker, 2005; Nieberding and Olivieri, 2007).

First of all, parasites without any secondary hosts or free-living stages are preferable for those with complex life cycles that may evolve independently from the host of interest (Nieberding and Olivieri, 2007). Also, pathogens are more relevant to infer host history if they are persistent and transmitted vertically (from parents to children). When pathogens are transmitted through an epidemic, their population structure tends to reflect their own demographic history (frequent bottlenecks followed by rapid expansions) than those of their host (Holmes, 2004). However, low rates of horizontal transmission, if occurring within and not among divergent populations, in parallel with vertical transmission, will not lead to incongruence between parasite and host trees (Wirth et al., 2005; Nieberding and Olivieri, 2007). Therefore, in the selection of a pathogen species as an inferential tool, one has to know its life cycle and mode of transmission, which can be achieved by means of ecologic surveys, experimentations, and within-family studies (Schwarz et al., 2008).

The rate of molecular evolution also greatly determines the efficacy of microbes as tracers. If the mutation rate is too low, the resolution of phylogenies can be crude and recent events may not be detected due to a lack of signal. Thus, in such cases, the use of parasites to infer host genealogies is not obvious. The opposite is also questionable: with a mutation rate that is too high, one can overlook information due to saturation at informative sites and homoplasy. In addition, depending on the mutation rate of their DNA, studies of pathogens give insight on past or recent events in their host histories. A wrong estimation of the mutation rate may lead to misinterpretations.

Another important parameter is the recombination rate. Indeed, recombination, even at a low rate, leads to intermediate genotypes, larger terminal branches, and an underestimation of the time to the most recent common ancestor (MRCA) (Schierup and Hein, 2000). However, recombination generally occurs between related populations which permits the maintenance of the genetic structure even if the signal is weaker. Several methods are now available to estimate the recombination rate (see, e.g., Martin et al., 2005) including coalescent and Monte Carlo-based simulation methods (Gibbs et al., 2000; Nielsen, 2000; Worobey, 2001; McVean et al., 2002). Otherwise, the homoplasy test (Maynard Smith and Smith, 1998) or the compatibility matrices test (Jakobsen and Easteal, 1996) detect and estimate the frequency of recombination events (see, e.g., Achtman et al., 1999). When evidence of recombination is found, other approaches such as a Bayesian clustering method, which can deal with recombination, are preferable.

Finally, selection can also dramatically reduce the reliability of molecular phylogenies since populations can be clustered together because they are under identical selection regimes despite the distinct histories they present (see Wirth et al., 2005, for a comparison of phylogenies based on neutral sequences and on sequences under selection). Another illustration can be found in a study by Devi et al. (2006) in which *H. pylori* population structure has been investigated from housekeeping genes and the *cag* pathogenecity island (*cag*PAI), which is under selective pressure. In this study, all Peruvian strains harbored a "western" type *cag*PAI, suggesting a European origin, while the analysis based on the housekeeping genes revealed that some clustered with the hpAmerind population (see later), suggesting an Asian origin for these strains. Hence, an analysis based on *cag*PAI only would not include all information. Selection can be detected in protein-coding genes by comparing the number of nonsynonymous amino acid changes (d_N) with the number of synonymous amino acid changes (d_S) with standard (DNASP) or more sophisticated (phylogenetic analysis by maximum likelihood [PAML]) tools. If the ratio d_N/d_S is equal to 1, the gene is under neutral evolution while if it is different from 1, the gene is either under purifying (<1) or directional (>1) selection.

To overcome some of these discrepancies (such as mutation and recombination rates or selection), reconstructing phylogenies from several genes is preferable. Incongruence among individual gene signals can be tested with the partition homogeneity test (Farris et al., 1994). When this test reveals homogeneity among datasets, the different genes can be concatenated. In addition, to confirm pathogen efficiency to infer host evolutionary history, it would be advisable to directly compare host and pathogen phylogenies by collecting data from the same material.

7.3 Candidates

7.3.1 Bacteria

Helicobacter pylori

The ubiquitous bacterium *H. pylori* has been shown to be a powerful tracer of human population structure (Falush et al., 2003; Wirth et al., 2004) and it is one of the most studied pathogens in human history inferences. *H. pylori* is a Gram-negative bacteria that infects human stomachs and is associated with gastrointestinal diseases such as gastritis, ulcers, or cancers although infections are mostly benign (Dunn et al., 1997). Prevalence of the infection exceeds 50% of the human population but decreases in industrialized countries (Dunn et al., 1997; Schwarz et al., 2008). Until recently, *H. pylori* was thought to be mainly transmitted vertically during childhood (Suerbaum et al., 1998; Suerbaum and Michetti, 2002; Kivi et al., 2003). However, recent investigations on the transmission of this bacterium in both developed and developing countries revealed that horizontal transmission might not be negligible and might depend on socioeconomic status (Delport et al., 2006; Schwarz et al., 2008). Mixed infections of *H. pylori* are not rare but generally involve one dominant strain (Raymond et al., 2004; Delport et al., 2006; Schwarz et al., 2008). This bacterium species shows an unusually high level of genetic diversity which may result from a combination of high mutation rates, frequent recombination events, and a continuous acquisition of new strains (Achtman et al., 1999; Kraft and Suerbaum, 2005; Delport et al., 2006; Kraft et al., 2006). Falush et al. (2001) implemented a Bayesian model to estimate the mutation and the recombination rates from 10 gene fragments. They estimated the rates to be less than 4×10^{-5} and around 7×10^{-5} respectively. *H. pylori* appeared to be nearly panmictic (Suerbaum et al., 1998) although several studies revealed phylogeographical differences. This apparent contradiction could be explained by frequent recombinations between geographically related strains so that the population structure can be maintained (Delport et al., 2006).

The suggestion that *H. pylori* has coevolved with humans and that its population structure reflects human migrations was first reported in studies that investigated the genetic differences between bacterial populations from distinct areas (Achtman et al., 1999; Covacci et al., 1999). Since then, several studies allowed a fine timing of the relationships between *H. pylori* and humans and the elucidation of ancient human migrations. Linz et al. (2007) documented a linear relationship between geographic distance from Africa and the microbial genetic diversity. IBD patterns was also observed both at a global and a local (European) scale. Similar correlations were obtained in humans, highlighting an intimate association between *H. pylori* and humans over a long period of time. Ramachandran et al. (2005) observed, in addition to such an IBD pattern, a decrease in the expected heterozygosity (estimated from 783 microsatellite loci) with distance from Addis Ababa. Using simulations, the authors evidenced that this pattern can be explained by serial founder effects starting at a single origin, thus confirming sequential waves of migration during modern human history. Linz et al. (2007) also investigated the origin and demography of *H. pylori* by means of demographically explicit genetic simulations. Three alternative scenarios were tested for the origin of *H. pylori*: an East African, a South African, and an East Asian origin. The best model (based on the proportion of total variance explained by the model and the Akaike information criterion) argued in favor of an East African origin. The simulations also indicated that *H. pylori* spread from Africa about 58,000 years ago, which is consistent with the dating of human migrations out of Africa (Liu et al., 2006) and suggests that humans were already infected before their initial migrations (Linz et al., 2007). Linz et al. (2007) concluded that all these parallels between bacteria and humans, observed at both global and local scales, reflect an expansion of *H. pylori* via ancient human migrations and genetic admixture after horizontal transmission or through recent migrations.

Using Bayesian clustering analyses performed on concatenated sequences from seven housekeeping gene fragments and one virulence gene, *H. pylori* from a global sample split into seven populations and subpopulations characterized by clear geographical distributions reflected in their name: hpAfrica1 subdivided into two subpopulations, hspWAfrica and hspSAfrica; hpAfrica2; hpEastAsia containing hspAmerind, hspEAsia, and hspMaori subpopulations; and hpEurope (Falush et al., 2003). Later on, three additional populations, hpAsia2 and hpNEAfrica (Linz et al., 2007) and hpSahul (Moodley et al., 2009) were described using extended datasets. All these populations and subpopulations are derived from six ancestral populations (Ancestral Africa1, Ancestral Africa2, Ancestral East Asia, Ancestral Europe1 [AE1], Ancestral Europe2 [AE2], and Ancestral Sahul; Falush et al., 2003; Linz et al., 2007; Moodley et al., 2009). The geographical distribution and genetic relationships between these populations are consistent with the classical model of human migrations (i.e., two subsequent waves of migration from Africa into Asia and Europe and a colonization of America from Asia through the Bering Strait and more recently from Europe) (Falush et al., 2003; Figure 7.1).

The division of the hpEurope population into subpopulations failed. This is probably due to its complex history, namely colonization by several independent waves of migration of genetically distinct populations. This is supported by the observation of two opposite clines, namely AE1 and AE2 within European populations that correlated with the first two principle components of European human diversity (Falush et al., 2003; Linz et al., 2007).

Hence, the geographical and genetic structures of *H. pylori* populations at a global scale are consistent with ancient human migrations. These studies are

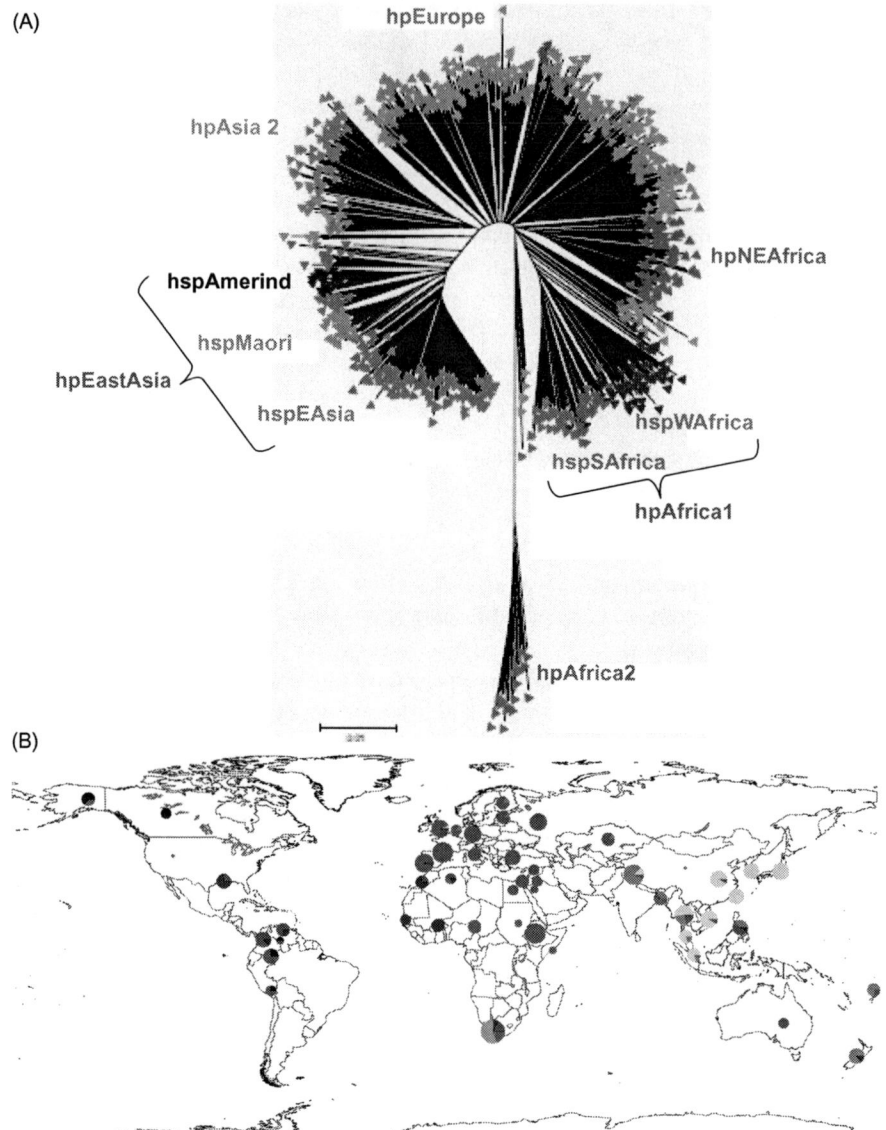

Figure 7.1 (A) Neighbor-joining tree of 769 concatenated sequences from *H. pylori* strains. Colors indicate the population assignment of each strain into one of the nine subpopulations defined by the Bayesian clustering analysis in Linz et al. (2007): hpAfrica, hspSAfrica, hspWAfrica, hpNEAfrica, hpEurope, hpAsia2, hspAmerind, hspMaori, and hspEasia.
(B) Geographical distribution of the nine *H. pylori* subpopulations. At each sampling location, the proportion of strains assigned to different bacterial subpopulations are represented by pie charts. (For interpretation of the references to color in this figure legend, the reader is referred to the web version of this book.)
Source: From Linz et al. (2007).

congruent with the accepted human history scenario but do not necessarily supply more information than former human genetic studies. However, the seminal study of Falush et al. (2003) on *H. pylori* populations clearly showed that its spread could be attributed to human migrations. They initiated a series of studies that provided evidence of the usefulness of this bacterium species to infer global and small scale evolutionary history of their host. We describe hereunder the main results gathered from some of these studies that improved our understanding of human population structure. Most of these studies are based on the same seven housekeeping gene fragments and implemented their own data in addition of those available on the *H. pylori* multilocus sequence typing (MLST) database. Unless we precise it, the results we relate below are based on these gene fragments and Bayesian clustering methods and phylogenetic analyses.

One of the most devious human settling is that of the Pacific. Several scenarios have been suggested depending on the evidence (archaeological, linguistic, or genetic; see Gray et al., 2009 for a description of the main models) but, until recently, the details have remained unclear. Recently, *H. pylori* isolates from native inhabitants in Taiwan, Papua New Guinea, New Caledonia, and Australia allowed the clarification of Pacific settlement and supplied proof of the utility of this bacterium species to infer host history (Moodley et al., 2009). This study advocates, for the Pacific peopling via two major waves of migration: the first one, from Asia to New Guinea and Australia, which was accompanied by hpSahul strains and occurred 31,000—37,000 years ago and the second one, from Taiwan through the Pacific 5,000 years ago, which led to the Austronesian expansion and hspMaori dispersal (Figure 7.2). Interestingly, these results are consistent with another study that aimed at testing Austronesian expansion using language phylogenies (Gray et al., 2009; Renfrew, 2009). With a large dataset based on language similarities (400 Austronesian languages) and Bayesian phylogenetic methods, Gray et al. (2009) resolved the peopling of the Pacific by Austronesian speakers. Like Moodley et al. (2009) in the genetic study, Gray et al. (2009) observed that Austronesian people originated from Taiwan ~5,200 years ago. The linguistic study also described several expansion pulses and pauses after the first migration from Taiwan that led to the actual distribution of Pacific people (Gray et al., 2009; Figure 7.2).

Another attractive aspect of *H. pylori* is that its population structure reflects human history at a local or fine scale (Wirth et al., 2004; Linz et al., 2007; Latifi-Navid et al., 2010). For instance, Wirth et al. (2004) were able to detect differences in population genetic structure of *H. pylori* from Ladakhi Buddhists and Muslims, two major ethnic groups in Ladakh socially separated in this province since 500 to 1,000 years ago due to cultural and religious differences. The *H. pylori* isolates from the two ethnic groups presented different ancestries: while isolates from Buddhists derived from AE1 and Ancestral East Asia populations, isolates from Muslims mostly derived from AE1 and a few strains from AE2. Therefore, Buddhism was introduced in Ladakh by Tibetans (carrying hpEastAsia bacteria) into an ancestral Ladakhi population (carrying AE1 bacteria) and Islamism by a few people carrying AE2 bacteria (Wirth et al., 2004). Altogether, these results

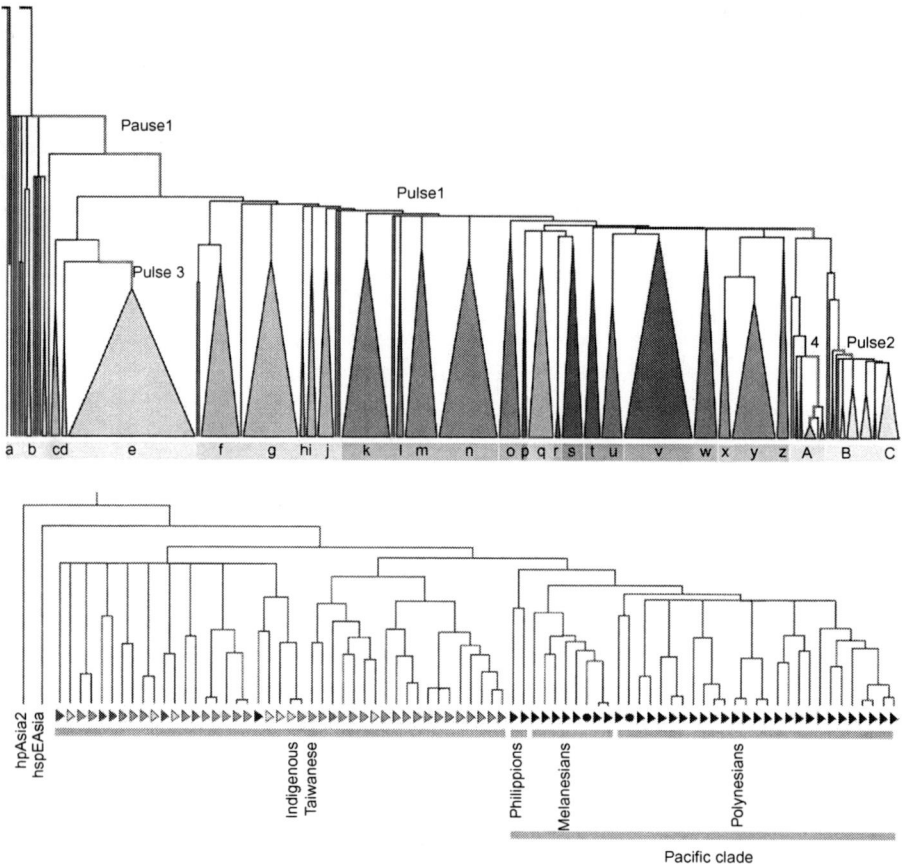

Figure 7.2 The peopling of the Pacific Islands from human bacterial (bottom) and language phylogenies (top).
Source: From Renfrew (2009).

reflect cultural differences and the recent history of population movements in Ladakh, which were not detectable with classical human markers (microsatellites and the mitochondrial D-loop). In the same vein, Latifi-Navid et al. (2010) observed ethnic and geographic differences in *H. pylori* isolates from Iranian patients from seven ethnic groups. Using the Bayesian clustering method, Iranian *H. pylori* isolates fall into the hpEurope population and are derived from AE1 and AE2. However, at the population level, most of Iranian and European populations were genetically differentiated. A neighbor-joining tree based on pairwise F_{st} revealed that *H. pylori* isolated from the different Iranian populations were not clustered together but clustered with strains isolated (ethnically or geographically) from nearby countries reflecting historical contacts during the past 2,000 years (Latifi-Navid et al., 2010). Thus, the Iranian-Arab population, which reached Iran during the Islamic conquest of Persia in the seventh century, was close to the

Palestinian and Israeli isolates; two populations from West Central Iran were close to Turkish strains, probably reflecting contacts with Turks during the Ottoman Empire and, later, two populations from the northeastern part of Iran were close to isolates from Uzbekistan, which could be the consequence of contacts during the fight for the control of the region between Uzbeks and Iranians in the sixteenth century (Latifi-Navid et al., 2010).

Other studies focused on even finer scales. Either within areas that have complicated history because of their localization, at the boundaries of several continents, or where multiple ethnic groups were present. For instance, investigations on *H. pylori* isolated from the three major ethnic groups in Malaysia (Chinese, Indians, and Malays), revealed a common origin of Malaysian Indian and Malay strains, which was different from that of Chinese isolates (Tay et al., 2009). The authors observed that most Indian isolates clustered within the hpAsia2 population and most Chinese within the hspEAsia population. In contrast, Malay strains were not assigned to one particular population but belonged to five different (sub)populations, the majority of them belonging to the hspAsia2 one. By completing the hpAsia2 population with Malaysian isolates, these authors identified a subdivision of this population into two subpopulations, hspLadakh, which contained Ladakhi strains only, and hspIndia, comprising the majority of both Malaysian Indian and Malay isolates. Tay et al. (2009) advocated for a common origin of Malaysian Indian and Malay strains, which was different from that of Chinese isolates and a recent acquisition of *H. pylori* by Malay populations from other populations. Devi et al. (2006) detected an hpEuropean genetic signature for the majority of strains from native Peruvians but also detected an hpAmerind signature for some of them ($\sim 20\%$). At the same time, the analysis of the *cag*PAI revealed that all native Peruvian strains, even the hpAmerind ones, presented a "western" type. The authors concluded that European strains that have spread into South America during the last 500 years might have outcompeted the hpAmerind ones, probably as a result of a selective advantage. In addition, they suggested that lateral gene transfer occurred between hpEuropean and hpAmerind strains during the colonization of Peru, since some isolates presented two different genetic signatures depending on the markers (housekeeping genes or *cag*PAI). Finally, Devi et al. (2007) observed that 63 Indian isolates showed European genetic signatures, suggesting a common ancestral origin between the two populations. The authors suggested that *H. pylori* strains might have arrived with the Indo-Europeans $\sim 4,000-10,000$ years ago.

To conclude, *H. pylori*, with its unexpected high diversity, has been proven to be a good human migration tracer, both in short and long timescales. Despite frequent recombination events, its populations were geographically structured, suggesting that recent migrations did not completely obscure the signatures left by geographic isolations and therefore rather reflect ancient human history (Suerbaum and Achtman, 2004). Moreover, the presence of recombination allows the detection of admixture between several ancestral populations, revealing multiple independent waves of migration and sometimes multiple ethnicity signatures within a single genome. Today, the frequency of infections has decreased, in industrialized countries in particular (in the USA, less than 10% of the children are infected; Blaser,

2006), highlighting the need for other candidates to infer human evolutionary history and to urgently collect specimens from endangered ethnic groups.

Mycobacterium

Mycobacterium tuberculosis, one of the most important human pathogens, belongs to the *M. tuberculosis* complex (MTB complex), which includes seven other closely related species and subspecies infecting both humans and animals. Each subspecies of the MTB complex shows a distinct host preference without being dependent on this sole host (Smith et al., 2009). *M. tuberculosis* is a Gram-positive bacterium and is the etiologic agent of the human tuberculosis that killed ~1.3 million persons in 2008 (World Health Organization, 2009). Bacilli, which disseminate through air and dust, are inhaled and penetrate into the lung via the respiratory tract. *M. tuberculosis* infects one-third of the world's population, although prevalence and mortality are higher in developing countries. Most often, infections are latent but 5—10% of the infections expand into disease or even cause death (Smith et al., 2009). The MTB complex presents a strictly clonal population structure with none or few recombinations (Smith et al., 2003; Supply et al., 2003; Hirsh et al., 2004; Wirth et al., 2008). Initially, studies on the MTB complex showed a low genetic diversity and weak or starlike phylogenies (Gagneux and Small, 2007) but this was based on studies with important limitations due to the choice of the markers and/or to problems linked with the ascertainment bias (see Hershberg et al., 2008 for more details). Recent advances in mycobacterial genomics have revealed higher levels of genetic diversity than previously thought and documented the detection of geographical and/or ethnical structures within *M. tuberculosis* populations (Gagneux and Small, 2007; Achtman, 2008). These include several recent studies based on either gene sequences (Dos Vultos et al., 2008; Hershberg et al., 2008; Pepperell et al., 2010) or mycobacterial interspersed repetitive units (MIRUs that contained variable number of tandem repeats [VNTR]; Dou et al., 2008; Mokrousov, 2008; Wirth et al., 2008).

Using VNTR markers, Wirth et al. (2008) tackled the origin, timing, and spread of the MTB complex and estimated that *M. tuberculosis* and humans probably coevolved for at least 40,000 years. They drew a *M. tuberculosis* phylogeny from 24 MIRU loci and from 355 isolates representative of the MTB complex distribution and estimated the divergence time between main clades using two approaches: probabilistic and distance-based. Both individual and population-based phylogenies evidenced two major lineages, which were confirmed by a Bayesian clustering analysis: clade1 encompassed all *M. tuberculosis* sensu stricto (all from humans) except for the East African Indian (EAI) population. Clade2 grouped all strains from animals plus the West African populations. The EAI population was basal and suggested a human origin for this mainly animal-associated clade. Clade1 presented a geographical substructure with an African, an Asian, a Latin American—Mediterranean, and an African-European cluster. Wirth et al. (2008) estimated the mutation rate of VNTR loci to be $\sim 10^{-4}$ and the emergence of the two clades was dated about ~40,000 years ago, which is consistent with the first human migrations from Africa. Clade1 emerged from the MTB complex ~30,000 years ago and dispersed with humans through the other continents through several waves of migration. The second clade is dated at ~20,000 years and descended from an EAI-like population that has been transmitted from humans to animals (and not the opposite), probably in Mesopotamia some 13,000 years ago, when domestication began. Using Bayesian statistics, Wirth et al. (2008) showed that all human *M. tuberculosis* populations exhibited consistent expansion rates—the highest expansion was detected for the Beijing population, which had a 500-fold increase—and that the expansion began 180 years ago, at the same time as the modern demographic explosion of humans, the industrial area, and modern intercontinental movements. Hence, this study highlights the noteworthy parallel demographic evolution between humans and *M. tuberculosis*.

Whichever genetic markers are used (large sequence polymorphisms, SNPs, indel analyses), global phylogenies are concordant and lead to a biogeographic consensus, namely the existence of six lineages that appeared to be associated to particular areas and may reflect human history (Gagneux and Small, 2007; Achtman, 2008). For instance, Hershberg et al. (2008) constructed maximum parsimony and neighbor-joining phylogenies of MTB complex using 89 concatenated gene sequences from 108 strains comprising a representative sample of the MTB complex. Both phylogenies were congruent and showed similar topologies to the phylogeny from Wirth et al. (2008), that is to say, the existence of two primary branches splitting "ancient" and "modern" lineages according to Brosch et al. (2002); Figure 7.3(A). The presence of the six main lineages in Africa and the deeply rooted West African branches (the only deeply rooted ones) argued in favor of an African origin of *M. tuberculosis* and its sequential spread into Europe and Asia. According to them, land in the Indian Ocean would have been colonized by modern humans from Africa quite early on. Therefore, *M. tuberculosis* seemed to reflect as well past human history, but interestingly, it also appears to mirror recent colonization movements and demographic changes in human populations (Figure 7.3B–E). For instance, Euro-American strains were found almost everywhere, possibly reflecting the numerous migrations after the European population increase from Europe to America during the nineteenth century and to Africa, Asia, and Middle East during the post-Columbian era. Similarly, the presence of East Asian strains in South Africa might be explained by the import of slaves from Southeast Asia by Dutch colonialists during the seventeenth and eighteenth centuries, or by Chinese migrants who came into South Africa in the early 1900s to work on gold mines.

MTB therefore appears to be of particular interest in the inference of recent host history and this might also be true for the livestock that are affected by this disease. Mokrousov (2008) collected the data of 11 VNTR loci from 1302 Beijing strains (a particularly successful *M. tuberculosis* lineage), mainly from Eurasia, and performed phylogenetic network and multidimensional scaling (MDS) analyses. He observed that the geographic distribution of this particular *M. Tuberculosis* lineage in Eurasia mirrors geographical, political, and sociocultural differences (Mokrousov et al., 2005; Mokrousov, 2008). Another study that focused on Taiwan documented associations between *M. tuberculosis* genotypes and ethnic and

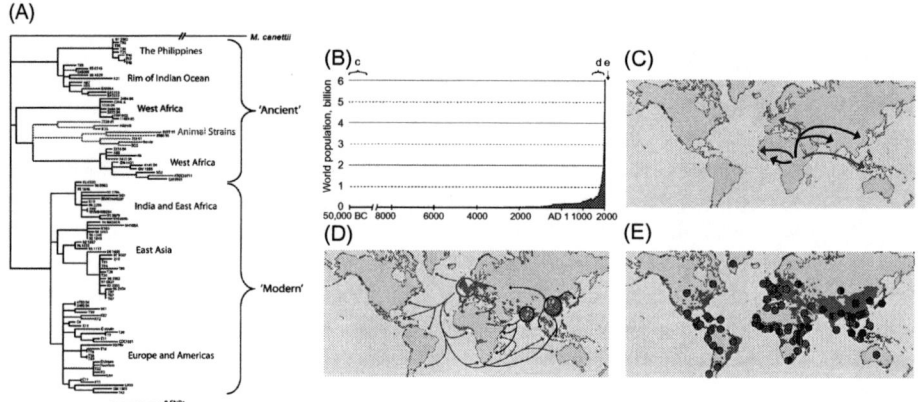

Figure 7.3 Evolutionary history of human MTB complex. (A) Maximum parsimony tree of 89 concatenated sequences from *M. tuberculosis* isolates. (B) Global human population size during the last 50,000 years. The time periods correspond to graphs (C), (D), and (E) are indicated above the graph. (C) Putative ancient migrations of MTB complex lineages. (D) Increase of global human population size during the last century (one gray dot corresponds to 1 million people) and global spread of modern lineages of the MTB complex with recent human migrations, trades, and conquests. (E) Actual human population size and distribution of the main MTB lineages.
Source: From Hershberg et al. (2008).

migratory populations (Dou et al., 2008). The authors studied strains from three Taiwanese populations differing in their ethnicity and in the time of arrival in Taiwan: the first population was composed of aborigines who descended from Austronesian people and inhabited Taiwan at least since the sixteenth century, while the others were two Han Chinese populations. The two latter populations were composed of veterans who were born in Mainland China and migrated into Taiwan during the Chinese civil war from 1946 to 1950, and the general Taiwanese population composed of individuals whose ancestors migrated from China 200–400 years ago. Hence, the three populations differed in their ethnicity and/or in their time of arrival in Taiwan. The authors discriminated the different genotypes by combining several genetic tools (VNTR, spoligotyping, and indel analyses), which provided a high discriminatory power. They performed an unweighted pair group method with arithmetic mean (UPGMA) tree and looked at the separation of the genotypes. Their analyses showed that the genotypes differed between populations and that isolates from aborigines were comparatively more ancient than those from veterans and from the general Taiwanese population, thus arguing in favor of an association between genotypes, ethnicity, and migratory movements (i.e., the length of migratory time in Taiwan; Dou et al., 2008).

Hence, as in the case of *H. pylori*, *M. tuberculosis* seems to harbor a population genetic structure, even at a small scale, that correlates with human history. Past and recent human migrations and expansions influenced *M. tuberculosis* population structure probably because of the strong and stable association between the two protagonists (Hirsh et al., 2004; Gagneux et al., 2006). Some discrepancies have, however, been highlighted. This is the case in the study of Mokrousov (2008), where proximities between Japanese and Russian populations were found. Moreover, the pattern observed in Chinese strains was unclear, suggesting that *M. tuberculosis* structure might reflect unknown human migrations, epidemiologic links between these populations, or that *M. tuberculosis* population structure is affected by high and rapid horizontal transmission events during epidemics in these regions.

Likewise for *M. tuberculosis*, it has been shown that *M. leprae* reflects human migrations (Monot et al., 2005, 2009). This pathogen causes chronic dermatologic and neurologic diseases and has humans as a unique known reservoir. The *M. leprae* genome contains an amazing number of pseudogenes and is exceptionally well-conserved (Cole et al., 2001; Monot et al., 2005, 2009). Like its close relatives, *M. leprae* is highly clonal (Monot et al., 2005). Monot et al. (2005) first described four subtypes that were geographically structured from three informative SNPs (SNP type1 were found in Asia, Pacific, and East Africa, SNP type4 in West Africa and the Caribbean, SNP type3 in Europe, North Africa, and the Americas, and SNP type2, the rarest, in Ethiopia, Malawi, North India, and New Caledonia). They proposed a scenario for the origin of leprosy and its spread though ancient human migrations, colonization, and the slave trade. However, the origin of leprosy was unclear. In a second study with an extended dataset, they confirmed their initial results and clarified the origin of leprosy. According to these two studies, SNP type2 from South Africa was the ancestral type and gave rise to SNP type1 and SNP type3, which dispersed into Asia and Europe, respectively. However, two independent introductions of leprosy seemed to occur in Asia: the first one occurred when humans left Africa and a second one came from Europe and was associated to the Silk Route. SNP type4 appeared to originate from SNP type3 and was found in West Africa and countries linked to the slave trade. Finally, the authors observed that leprosy in America most likely originates from Europe rather than from Asia through the Bering Strait reflecting the relatively recent massive migrations from Europe to America. Complete genomes' approaches are certainly advisable to clarify these preliminary results.

Streptococcus mutans

Streptococcus mutans belongs to the mutans streptococci group and is associated with human caries. This bacterium has a ubiquitous distribution and seems to be indigenous to humans. The transmission of *S. mutans* is mainly vertical (Lapirattanakul et al., 2008), the colonization is stable and probably lifelong, which makes the bacterium a good candidate to infer human evolutionary history although its population structure and phylogeny has been poorly investigated (Caufield et al., 2007). A recent study presented results in agreement with this idea albeit it emphasizes the importance of the choice of markers (Caufield et al.,

2007; Caufield, 2009). Indeed, the authors investigated the population structure of *S. mutans* from five continents using several genetic markers to reconstruct molecular phylogenies. They observed a certain number of incongruences between the different trees, possibly due to horizontal gene transfers. One phylogeny was based on a 600 bp hypervariable region (HVR) of a plasmid because this noncoding region presented high polymorphism. However, the HVR phylogenies showed evolutionary incongruence with ethnic or geographic human hosts. The second phylogeny was based on the intergenic spacer region (IGSR) between the 16S and 23S rRNA genes, which showed only few polymorphisms (nine informative sites over 388 bp). Using a polyphasic approach, which combines genetic traits (here the serotype and mutacin types) and DNA sequences, Caufield et al. (2007) reconstructed a phylogeny in which geographical substructures were visible but apart from the Asian group, the other groups (African1, African2, and Caucasian) were supported by low bootstrap values (Caufield, 2009). In addition, the phylogenetic position of the Caucasians was intriguing. An important shortcoming in Caufield's studies (Caufield et al., 2007; Caufield, 2009) is that they were restricted to *S. mutans* strains that harbored a plasmid, dismissing 95% of the natural occurring strains that do not harbor this genetic element (Do et al., 2010). Nakano et al. (2007) developed a MLST scheme to address *S. mutans* population structure and observed no genetic structure in strains lacking the plasmid. However, this study only contained strains from Japan and Finland and does not reflect the true diversity. Therefore, before using *S. mutans* as an inferential tool, complementary studies with extended datasets based on MLST are required and both plasmid-positive and plasmid-negative strains should be investigated.

7.3.2 Viruses

Many viruses have been proposed as providing valuable insights on human population history. Among them we can cite human T-cell lymphotrophic viruses 1 and 2 (HTLV-1 and -2), the human polyomavirus JC (JC polyomavirus [JCV]) and its closely related BK virus (BKV), the human papillomavirus (HPV), and the hepatitis G virus (GBV-C/HGV). However, most often, the hypothesis of viral and human codivergences is not well supported or evidenced and/or is founded only on geographical associations that may be coincidental or may result from other factors such as natural selection (see Wirth et al., 2005). Indeed, most of these viruses seemed to suffer from drawbacks, making them poor candidates to elucidate past human migrations (Holmes, 2004, 2008). For instance, a majority of viruses are often transmitted horizontally, which leads to fast genetic admixture. Here, we shall detail the cases of JCV, GBV-C/HGV, and HPV to illustrate some of the kinds of problems we can encounter when studying viruses with regard to human history.

Human Polyomavirus

One of the most investigated viruses in the context of human migrations is the JCV. JCV is a double-stranded DNA virus responsible for harmless kidney infections, except for immunocompromised patients where leukoencephalopathy can develop (Weber and Major, 1997). The virus is acquired during childhood and persists in renal tissues for life (Chesters et al., 1983). JCV is human-specific and ubiquitous with an estimated seroprevalence of at least 70% in the human population (Padgett and Walker, 1973). Evidence for both vertical and horizontal transmission have been found, although horizontal transmission seems to occur preferentially between closely related populations (Kunitake et al., 1995; Kato et al., 1997; Suzuki et al., 2002). In addition, Kitamura et al. (1997) detected identical strains from sequential samples from the same patients taken ~ 6 years apart, suggesting that multiple infections might rarely occur, thus limiting recombination events between divergent JCV strains. All these features have led to the hypothesis that JCV might be useful to reconstruct human migrations.

Several major viral strains and subtypes showing geographical associations have been described from partial gene (Agostini et al., 1997b; Sugimoto et al., 1997) and whole-genome sequences (Jobes et al., 1998). Since then, numerous studies at global and local scales have documented the genetic relationships and the geographical distributions of these genotypes. Briefly, type 1 was found mainly in Europe (Agostini et al., 2001; Sugimoto et al., 2002), types 2 and 7 in Asia (Agostini et al., 1997b, 1998; Sugimoto et al., 1997, 2002), types 3 and 6 in Africa (Guo et al., 1996; Agostini et al., 1997a; Sugimoto et al., 2002), type 4 in the USA and Europe (Stoner et al., 2000; Agostini et al., 2001), and type 8 has been detected in Papua New Guinea and the Pacific Islands (Jobes et al., 2001; Yanagihara et al., 2002). Type 2 was subdivided into several subtypes presenting variations in abundance according to area: subtype 2A was preponderant in the Japanese and Native American populations, subtype 2B in Eurasians, subtype 2D in Indians, and 2E in Australians and populations from the West Pacific (Yanagihara et al., 2002). In the same way, type 7 included subtype 7A preponderant in Southern China and Southeast Asia, subtype 7B which was found in higher proportion in Northern China, Mongolia, and Japan (Cui et al., 2004). Cui et al. (2004) detected a third subtype called 7C in northern and southern China. Finally, type 5 was shown to combine type 6 and type 2B sequences and is the unique example of recombinant JCV strain (Hatwell and Sharp, 2000).

Interestingly, the multiple origins of American people was detectable by analyzing JCV genotypes (Stoner et al., 2000). Native Americans represented by two ethnic groups (Flathead People and Navaho) were mostly infected by subtype 2A, a genotype found in East Asia and Japan, which may reflect an Asian origin through the Bering Strait (Agostini et al., 1997b). In contrast, European Americans carried type 1 (European genotype) for a majority and in lower proportion type 4 (14%) and type 2 (less than 10%) (Stoner et al., 2000). Surprisingly, no type 6 was found in the African-American population but type 1 (32%), type 4 (44%), and type 3 (18%), suggesting a genetic admixture between African and European types and reflecting both past and recent migratory movements (Chima et al., 2000). Stoner et al. (2000) suggested that the high frequency of European strains in African-European populations could be due to a selective advantage of these strains compared to African ones.

JCV populations also appears to be geographically structured in the Pacific Islands, probably due to multiple human migration waves (Jobes et al., 2001; Yanagihara et al., 2002). Four subtypes were identified within JCV populations from western Pacific Islands: subtype 8A restricted to Papua New Guinea, subtypes 8B from non-Austronesians and 2E from Austronesians widely distributed through Pacific Islands, and subtype 7A rarely found. Yanagihara et al. (2002) proposed that subtype 8A first arrived in Papua New Guinea or Sahul followed by subtype 8B. Later ($\sim$5,000 years ago), Austronesian expansion might have led to the spread of subtype 2E. Recent migrations from South China or Taiwan might have brought subtype 7A into Guam. Surprisingly, Australian JCV strains belonged to subtype 2E, which is genetically close to the subtype found in East Asia (subtype 2A). This is in sharp contradiction with the known history of Pacific peopling which was confirmed by *H. pylori* population studies and language phylogeny (see Section *Helicobacter pylori*). Indeed, the first wave of migrations from Asia into the Pacific Islands led to the peopling of both Australia and New Guinea. Therefore, we expected to find in native Australian strains the same subtypes as those found in New Guinea.

In accordance with these results, Pavesi (2003, 2004, 2005) tackled the evolution of JCV genotypes by means of principal component analyses based on JCV sequences from the five continents. These analyses evidenced that type 6 (the African one) might be the ancestral genotype that gave rise to two major independent lineages, one clustering types 1 and 4 while the other contained types 2, 3, 7, and 8. This analysis has led the author to propose an alternative to the classical model of human migrations namely "the two-migration model." This model hypothesizes two early routes of expansion out of Africa: one route into Asia and the second one into Europe. However, phylogenies based on whole JCV genome sequences showed some discrepancies (Wooding, 2001). For instance, the basal European clade position was paradoxal (Sugimoto et al., 1997; Hatwell and Sharp, 2000; Agostini et al., 2001; Jobes et al., 2001). This is inconsistent with the hypothesis of an infection of humans by JCV before their expansion from Africa. Pavesi (2003) handled this question by reconstructing two phylogenies based on slow- and fast-evolving sites defined from the Shannon entropy index. Phylogenies based on invariants plus slow-evolving sites and on invariants plus fast-evolving sites were different. When invariants and slow-evolving sites were used to reconstruct phylogeny, the topology was similar to topologies obtained from the whole-genome sequences with the European clade at the basal position and type 6 as the ancestral type of all other types. In contrast, the phylogeny based on invariants and slow-evolving sites placed the type 6 on the deepest branch. This is consistent with an African ancestry. However, other questionable findings remain to be clarified, such as the higher genetic diversity observed in European and Asian than in African JCV (Wooding, 2001). Coincidences between geographical association between JCV and human populations may result from other factors such as natural selection or specific viral life-history traits. More studies on JVC are therefore needed before concluding with regards to human migrations. In addition, the molecular clock needs to be carefully reevaluated (Wooding, 2001).

One debatable point of all these studies is that they have relied on the hypothesis of JVC and human codivergence and on a slow mutation rate, which has not been tested independently from the coevolution hypothesis. Mutation rates were first estimated to range between 1 and 4×10^{-7} per site per year (Hatwell and Sharp, 2000; Fernandez-Cobo et al., 2002). These estimations were founded on the assumption of a longtime coevolution between JCV and humans (at least since the expansion from Africa ~150,000 years ago) and estimations were calibrated from host divergence times. Hence, this approach is somewhat tautologic. In contradiction with the above, two recent studies found much faster mutation rates using a Bayesian Markov chain Monte Carlo (MCMC) approach which is free from the assumption of codivergence and is based on coalescent analysis of sequences sampled at different times (Shackelton et al., 2006; Kitchen et al., 2008). Shackelton et al. (2006) tested congruence between JCV and human phylogenetic trees by mapping consensus JCV trees onto three possible human trees, thus creating "tanglegrams." From each of these tanglegrams, the potentially optimal solutions were determined by evaluating the noncoevolutionary events (i.e., duplication, horizontal transfer, loss of a virus by a host population) required to reconcile JCV and human trees. Randomizations of the branches of the viral tree were used to test the hypothesis that the viral tree was more congruent with the human tree than a random tree would be. In both studies, no evidence for codivergence between human and virus phylogenies was found (Shackelton et al., 2006; Kitchen et al., 2008). Shackelton et al. (2006) estimated for humans the age of the MRCA to be in accordance with the accepted estimates (i.e., between 100,000 and 150,000 years) and provided evidence for an expansion starting 50,000 years ago when major cultural changes occurred. In contrast, the MRCA for JCV was estimated not to exceed 3,100 years ago. Both studies found a significantly higher mean substitution rate for JCV than previous estimations (more than 100-fold faster: 1.7×10^{-5} [Shackelton et al., 2006] and 3.6×10^{-5} [Kitchen et al., 2008]). Considering this faster mutation rate, skyline plots, a coalescent method for estimating past population dynamics (Drummond et al., 2005), revealed that the global viral population increased during the last 350 years (Shackelton et al., 2006) and that posterior population estimates for viruses and humans differed totally (Figure 7.4; Shackelton et al., 2006; Kitchen et al., 2008).

These last two studies demonstrated that JCV populations should not be used to infer past human history because their population dynamics occurred at timescales that are too recent. It seems more likely that JCV population phylogenies and dynamics reflect recent societal and epidemiologic shifts in human behavior or technological innovations (Kitchen et al., 2008). In agreement with this, skyline plots indicated expansions of JCV in Africa, Europe, and Japan that began ~50 years ago (Kitchen et al., 2008). Expansions in Europe and Japan 50 years ago may be due to societal changes that occurred after the World War II.

Hepatitis G Virus

GBV-C/HGV, a positive-strand RNA virus, was proposed to be a tracer for prehistoric human migrations because it showed an African origin and its genotypes

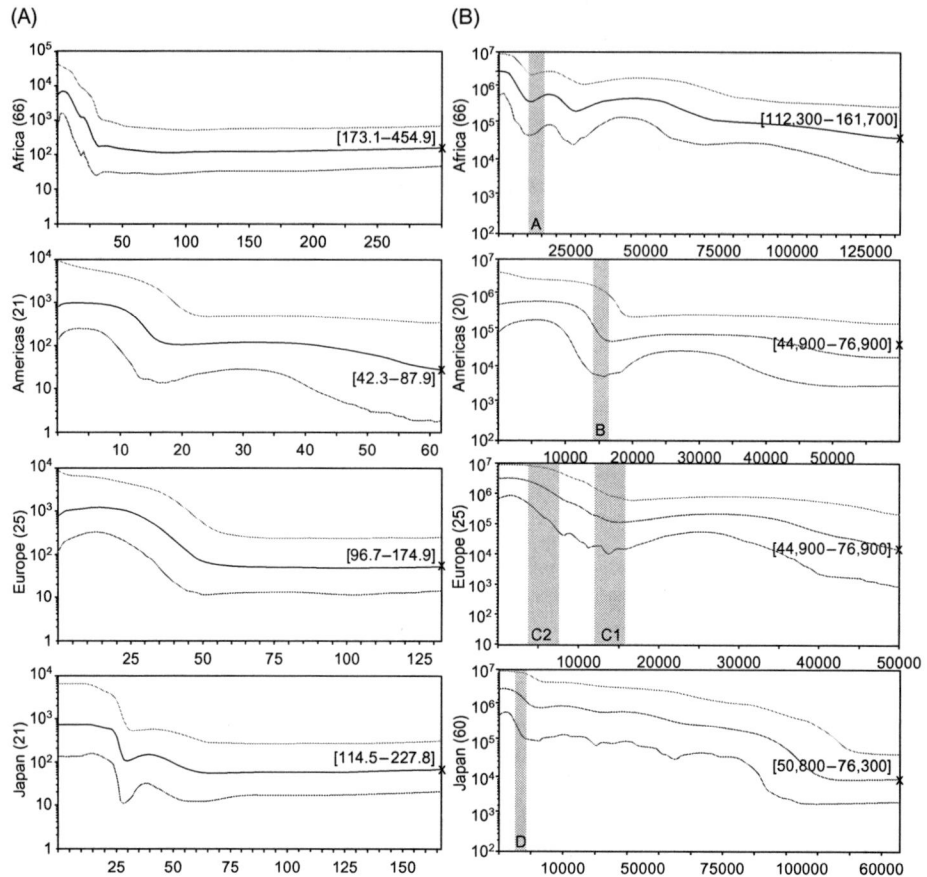

Figure 7.4 Bayesian skyline plots performed on JCV genomes (A) and on human mtDNA sequences (B) with nucleotide substitution rates of 3.642×10^{-5} and 1.7×10^{-8} for virus and human sequences respectively. The x-axis is the number of years before present and the y-axis is the scaled population size ($= N_e \times g$). The median estimate of the population size ($N_e \times g$; black line) and the 95% confident intervals (light gray line) are given.
Source: From Kitchen et al. (2008).

presented geographical or ethnic associations (Pavesi, 2001; Muerhoff et al., 2005). However, trees differed depending on the regions used to reconstruct phylogenies and supports for distinct groups of genotypes were generally low (Smith et al., 2000; Muerhoff et al., 2005). In addition, estimates of the mutation rate for GBV-C/HGV gave very different results. According to the selected approach, estimations ranged between ~4.10^{-4} and 3.10^{-3} when analyzing viral sequences sampled at different periods (Nakao et al., 1997; Sarrazin et al., 2000) and ~10^{-7} when assuming the hypothesis of coevolution between humans and GBV-C/HGV (Suzuki et al., 1999). Using the same Bayesian approach as Shackelton et al. (2006),

Romano et al. (2008) estimated the mean mutation rate for GBV-C/HGV to range between 10^{-2} and 10^{-3}, much faster than previous estimations but more similar to other RNA viruses. This high substitution rate is opposed to the weak degree of divergence between GBV-C/HGV sequences and is incompatible with GBV-C/HGV epidemiologic distributions (Simmonds, 2001). Following the observation of an excess of invariant synonymous sites in the coding sequences, it was suggested that genomes might be under selective pressures, probably because of functional constraints on RNA secondary structures, resulting in a lower variability at sites involved in secondary structure (Simmonds and Smith, 1999). Therefore, if we also consider the large viral population sizes, we may conclude that GBV-C/HGV evolve apart from humans (Simmonds, 2001).

Human Papillomavirus

Papillomaviruses (PVs) are a family of viruses that comprise more than 180 types clustered into at least 16 genera (types and genera being defined according to their nucleotide sequence similarities and biologic features) and present a wide host range (de Villiers et al., 2004; Bernard et al., 2010). Infections cause mucosal and cutaneous lesions. They are most often benign but they can cause cancers. PVs are mainly sexually transmitted (Vaccarella et al., 2006), although a direct contact can be sufficient. There are growing evidences of mixed infections (Liaw et al., 2001; Clifford et al., 2005; Angulo and Carvajal-Rodriguez, 2007; Antonishyn et al., 2008) and inter- and intra-types recombination (Angulo and Carvajal-Rodriguez, 2007; Carvajal-Rodríguez, 2008; Jiang et al., 2009). The PV genomes are double-stranded circular DNA of ~ 8 kb with 8 protein-coding genes (genes E1, E2, E4—E7 involved in viral replication, transcription, and transformation, and genes L1 and L2 encoding capsid proteins) and a noncoding regulatory long control region (LCR). Using a Bayesian approach to analyze 108 nucleotide sequences of PV isolated from several hosts, Shah et al. (2010) estimated independently from the coevolution hypothesis the mean substitution rates of the E1 and L1 genes to be respectively 7.1×10^{-9} ($\pm 1.49 \times 10^{-9}$) and 9.57×10^{-9} ($\pm 2.08 \times 10^{-9}$) per site per year.

Until recently, PVs were assumed to be host specific and to have coevolved with their hosts although no clear evidences were found. In addition, host and virus phylogenies were rarely congruent and several recent studies argued that incongruences between host and PV gene trees might reveal a complex evolutionary history of PV involving several mechanisms (ancient recombination, viral divergence before host speciation, host transfer, and/or selection on particular genes or gene fragments) in addition to cospeciation (Narechania et al., 2005; Varsani et al., 2006; Gottschling et al., 2007b; Shah et al., 2010). Shah et al. (2010) applied important sampling techniques to Bayesian methods to test whether virus divergence times were consistent with host divergence times (cospeciation hypothesis) or occurred before (viral radiations before host speciation) or after (host transfer) host divergence times. The authors observed that most often the cospeciation hypothesis was not rejected, but they also evidenced radiations of some PV lineages prior to host speciation and a single example of host transfer. Using this method, prior viral divergence and host transfers can be underestimated if they occurred too close from host divergence time. Analyses of L1 genes of nonhuman and human β-PV revealed that nonhuman PV were polyphyletic and nested within human β-PV suggesting host transfer events (Gottschling et al., 2007a). These studies suggest that horizontal transfers can arise between closely related species and much rarely between phylogenetic distant species (Gottschling et al., 2007b). However, these studies highlight that cospeciation only cannot explain PV evolutionary history at a deep taxonomic level but can be assumed at a shallower level such as the species one (Gottschling et al., 2007b).

HPV, represented by more than 100 types, showed some evidences of coevolution with modern humans since their origin (Bernard et al., 2006). Two particular HPV types, HPV-16 and HPV-18, have been extensively studied because of their high worldwide prevalence and their importance in human health. The distribution of their variants presented some geographic and ethnic specificities arguing in favor of their usefulness to elucidate human migrations. The analysis of a 364 pb segment of the LCR of 301 HPV-16 variants isolated from individuals that belonged to 25 different geographic locations and ethnic groups revealed five main HPV-16 lineages. Two lineages, Af1 and Af2, were predominantly found in Africa, one in Asia and Americas (AA), one in East Asia (As), and another one mainly collected in Europe (E) but also found elsewhere except in Africa (Ho et al., 1993). The authors suggested an African origin of these HPV-16 variants although they could not root the tree because of difficulties to align HPV-16 variants with their closest related types (HPV-31 and HPV-35). These five distinct lineages were also detected when analyzing the E6 gene, a region of L1 and the LCR of a worldwide sample by PCR-based hybridization methods and a phylogenetic analysis (Yamada et al., 1997). Analyses of the LCR of HPV-18 variants gave consistent results. Three main phylogenetic branches of HPV-18 variants were detected: an African, an Asian-American, and a European one (Ong et al., 1993). These main lineages were confirmed when HPV-18 variants were analyzed using other regions (LCR-E6,E2 and L1, Arias-Pulido et al., 2005; LCR,E6 and E1, Cerqueira et al., 2008). HPV-18 tree was rooted against a closely related type HPV-45, which revealed an African origin for HPV-18 variants. These results suggest that HPV-16 and HPV-18 have coevolved with the three main human lineages (Africans, Caucasians, and Asians). Some discrepancies with the geographic/ethnic repartition of HPV-16 and HPV-18 variants were observed but they may illustrate some recent migration events or founder effects (Wirth et al., 2005; Bernard et al., 2006). The correlation between ethnicity and HPV-16 and HPV-18 variants is unclear and seems to depend on the degree of admixture of the investigated populations (Sichero et al., 2007; Junes-Gill et al., 2008). For instance, HPV-16 and HPV-18 variants from women from the USA were found to be race-associated and to persist longer when host ethnicity and geographical origin of the variants matched (Schlecht et al., 2005; Xi et al., 2006). Similarly, Chimeddorj et al. (2008) investigated HPV-16 diversity among Mongolian women and detected a correlation between virus variants and the small ethnic groups. On the contrary, no significant association between host ethnicity and HPV-16 and HPV-18 variants were detected in populations from Brazil and Costa Rica (Villa et al., 2000; Hildesheim et al., 2001; Sichero et al., 2007; Junes-Gill et al., 2008). However, the lack of significant correlation between ethnicity and HPV variants may be due in some studies to a small sample size or to ethnic categorization (Junes-Gill et al., 2008). In addition, *P*-values of statistical tests are often close to the significant threshold (0.05). The question of an association between HPV variants and host ethnicity therefore remains open. Worldwide distribution of other less predominant HPV types have also been investigated and showed specific patterns according to the studied type, in accordance or not with human migratory history (Heinzel et al., 1995; Stewart et al., 1996; Calleja-Macias et al., 2004, 2005a,b; Prado et al., 2005; Li et al., 2009). The differences between HPV patterns of evolution may result from distinct rates of molecular evolution (Prado et al., 2005; Bernard et al., 2006). Moreover, the lack of correlation between geography and HPV variants observed within these types may also result from their lower prevalence and/or differences in susceptibility between ethnicities.

7.4 Conclusion

In this chapter, we made a brief overview in the emerging field of the use of microbes or parasites as proxies for human migrations. Now, nearly two decades after the first attempts using viruses for this purpose, we have reached the age of maturity. Though many candidates have been tested, the most convincing and innovative results were obtained with *H. pylori*. This was mainly driven by the intrinsic qualities of this bug, but also by a decent sampling scheme, the accumulation of large datasets and a good geographic coverage. Indeed, *H. pylori* MLST brought a lot of information in terms of genetic diversity and structure of this pathogen and indirectly of its human host. However, with the advent of the new sequencing technologies (pyrosequencing, Solexa, and Solid) and the increasing facility to generate complete microbial genome at reasonable costs the paradigm of *H. pylori* might shift toward other bacterial species that displayed too few mutations in an MLST scheme but that might unravel precious information at the genomic scale. Therefore, the shift from a multiple gene approach to a genomic approach will potentially give raise to a new golden age in this field and will rely on new population genomic tools and algorithms. The next targets will certainly belong to the cluster of the human pathogens where financial support are easier to obtain, however, commensals should not be dismissed, and the strategies presented here might be applied for all kinds of nonhuman hosts. The evolutionary history of our little "companions" can be seen like a multilayer information box; different species might certainly not provide us with a unique congruent scenario but will unravel the complexity of host–parasite interaction from neutral coinciding genetic patterns to extreme selection biases. Extending the use of bugs to infer host migrations of other mammals or organisms is certainly one of the future challenges we might face. The origin and dispersal of our animal stocks could be revisited; the demography and "ghost" genetic structure of endangered species (big cats) could be evaluated. The limits are our imagination and the difficulty to handle the new generation metadata.

List of Abbreviations

AE1 Ancestral Europe1
AE2 Ancestral Europe2
BKV BK polyomavirus
***cag*PAI** *cag* pathogenecity island
EAI East African Indian
GBV-C/HGV Hepatitis G virus
HTLV Human T-cell lymphotrophic virus
HPV Human papillomavirus
HVR Hypervariable region
IBD Isolation by distance
IGSR Intergenic spacer region
JCV JC polyomavirus
LCR Long control region
MCMC Markov chain Monte Carlo
MIRUs Mycobacterial interspersed repetitive units
MLST Multilocus sequence typing
MRCA Most recent common ancestor
MTB *Mycobacterium tuberculosis*
mtDNA mitochondrial DNA
SNP Single nucleotide polymorphism
UPGMA Unweighted pair group method with arithmetic mean
VNTR Variable number of tandem repeats

References

Achtman, M., 2008. Evolution, population structure, and phylogeography of genetically monomorphic bacterial pathogens. Annu. Rev. Microbiol. 62, 53–70.
Achtman, M., Azuma, T., Berg, D.E., Ito, Y., Morelli, G., Pan, Z.-J., et al., 1999. Recombination and clonal groupings within *Helicobacter pylori* from different geographical regions. Mol. Microbiol. 32, 459–470.
Agostini, H.T., Ryschkewitsch, C.F., Brubaker, G.R., Shao, J., Stoner, G.L., 1997a. Five complete genomes of JC virus Type 3 from Africans and African Americans. Arch. Virol. 142, 637–655.
Agostini, H.T., Yanagihara, R., Davis, V., Ryschkewitsch, C.F., Stoner, G.L., 1997b. Asian genotypes of JC virus in Native Americans and in a Pacific Island population: markers of viral evolution and human migration. Proc. Natl. Acad. Sci. U.S.A. 94, 14542–14546.
Agostini, H.T., Shishido-Hara, Y., Baumhefner, R.W., Singer, E.J., Ryschkewitsch, C.F., Stoner, G.L., 1998. JC virus Type 2: definition of subtypes based on DNA sequence analysis of ten complete genomes. J. Gen. Virol. 79, 1143–1151.

Agostini, H.T., Deckhut, A., Jobes, D.V., Girones, R., Schlunck, G., Prost, M.G., et al., 2001. Genotypes of JC virus in East, Central and Southwest Europe. J. Gen. Virol. 82, 1221–1331.
Angulo, M., Carvajal-Rodriguez, A., 2007. Evidence of recombination within human alphapapillomavirus. Virol. J. 4, 33.
Antonishyn, N.A., Horsman, G.B., Kelln, R.A., Saggar, J., Severini, A., 2008. The impact of the distribution of human papillomavirus types and associated high-risk lesions in a colposcopy population for monitoring vaccine efficacy. Arch. Pathol. Lab. Med. 132, 54–60.
Arias-Pulido, H., Peyton, C.L., Torrez-Martínez, N., Anderson, D.N., Wheeler, C.M., 2005. Human papillomavirus type 18 variant lineages in United States populations characterized by sequence analysis of LCR-E6, E2, and L1 regions. Virology 338, 22–34.
Bernard, H.-U., Calleja-Macias, I., Dunn, S.T., E., 2006. Genome variation of human papillomavirus types: phylogenetic and medical implications. Int. J. Cancer 118, 1071–1076.
Bernard, H.-U., Burk, R.D., Chen, Z., van Doorslaer, K., Hausen, H.Z., de Villiers, E.-M., 2010. Classification of papillomaviruses (PVs) based on 189 PV types and proposal of taxonomic amendments. Virology 401, 70–79.
Blaser, M.J., 2006. Who are we? Indigenous microbes and the ecology of human diseases. EMBO Rep. 7, 956–960.
Brosch, R., Gordon, S.V., Marmiesse, M., Brodin, P., Buchrieser, C., Eiglmeier, K., et al., 2002. A new evolutionary scenario for the *Mycobacterium tuberculosis* complex. Proc. Natl. Acad. Sci. U.S.A. 99, 3684–3689.
Calleja-Macias, I.E., Kalantari, M., Huh, J., Ortiz-Lopez, R., Rojas-Martinez, A., Gonzalez-Guerrero, J.F., et al., 2004. Genomic diversity of human papillomavirus-16, 18, 31, and 35 isolates in a Mexican population and relationship to European, African, and Native American variants. Virology 319, 315–323.
Calleja-Macias, I.E., Kalantari, M., Allan, B., Williamson, A.-L., Chung, L.-P., Collins, R.J., et al., 2005a. Papillomavirus subtypes are natural and old taxa: phylogeny of human papillomavirus types 44 and 55 and 68a and -b. J. Virol. 79, 6565–6569.
Calleja-Macias, I.E., Villa, L.L., Prado, J.C., Kalantari, M., Allan, B., Williamson, A.-L., et al., 2005b. Worldwide genomic diversity of the high-risk human papillomavirus types 31, 35, 52, and 58, four close relatives of human papillomavirus type 16. J. Virol. 79, 13630–13640.
Carvajal-Rodríguez, A., 2008. Detecting recombination and diversifying selection in human alpha-papillomavirus. Infect. Genet. Evol. 8, 689–692.
Caufield, P.W., 2009. Tracking human migration patterns through the oral bacterial flora. Clin. Microbiol. Infect. 15, 37–39.
Caufield, P.W., Saxena, D., Fitch, D., Li, Y., 2007. Population structure of plasmid-containing strains of *Streptococcus mutans*, a member of the human indigenous biota. J. Bacteriol. 189, 1238–1243.
Cavalli-Sforza, L.L., 1998. The DNA revolution in population genetics. Trends Genet. 14, 60–65.
Cavalli-Sforza, L.L., Feldman, M.W., 2003. The application of molecular genetic approaches to the study of human evolution. Nat. Genet. 33, 266–275.
Cerqueira, D., Raiol, T., Véras, N., von Gal Milanezi, N., Amaral, F., de Macedo Brígido, M., et al., 2008. New variants of human papillomavirus type 18 identified in central Brazil. Virus Genes 37, 282–287.

Chesters, P.M., Heritage, J., McCance, D.J., 1983. Persistence of DNA sequences of BK virus and JC virus in normal human tissues and in diseased tissues. J. Infect. Dis. 147, 676–684.

Chima, S.C., Ryschkewitsch, C.F., Fan, K.J., Stoner, G.L., 2000. Polyomavirus JC genotypes in an urban United States population reflect the history of African origin and genetic admixture in modern African Americans. Hum. Biol 72, 837–850.

Chimeddorj, B., Pak, C.Y., Damdin, A., Okamoto, N., Miyagi, Y., 2008. Distribution of HPV-16 intratypic variants among women with cervical intraepithelial neoplasia and invasive cervical cancer in Mongolia. Asian Pac. J. Cancer Prev. 9, 563–568.

Clifford, G.M., Gallus, S., Herrero, R., Muñoz, N., Snijders, P.J.F., Vaccarella, S., et al., 2005. Worldwide distribution of human papillomavirus types in cytologically normal women in the International Agency for Research on Cancer HPV prevalence surveys: a pooled analysis. Lancet 366, 991–998.

Cole, S.T., Eiglmeier, K., Parkhill, J., James, K.D., Thomson, N.R., Wheeler, P.R., et al., 2001. Massive gene decay in the leprosy bacillus. Nature 409, 1007–1011.

Covacci, A., Telford, J.L., Giudice, G.D., Parsonnet, J., Rappuoli, R., 1999. *Helicobacter pylori* virulence and genetic geography. Science 284, 1328–1333.

Cui, X., Wang, J., Deckhut, A., Joseph, B., Eberwein, P., Cubitt, C.L., et al., 2004. Chinese strains (Type 7) of JC virus are Afro-Asiatic in origin but are phylogenetically distinct from the Mongolian and Indian strains (Type 2D) and the Korean and Japanese strains (Type 2A). J. Mol. Evol. 58, 568–583.

de Villiers, E.-M., Fauquet, C., Broker, T.R., Bernard, H.-U., zur Hausen, H., 2004. Classification of papillomaviruses. Virology 324, 17–27.

Delport, W., Cunningham, M., Olivier, B., Preisig, O., van der Merwe, S.W., 2006. A population genetics pedigree perspective on the transmission of *Helicobacter pylori*. Genetics 174, 2107–2118.

Devi, S.M., Ahmed, I., Khan, A., Rahman, S., Alvi, A., Sechi, L., et al., 2006. Genomes of *Helicobacter pylori* from native Peruvians suggest admixture of ancestral and modern lineages and reveal a western type *cag*-pathogenicity island. BMC Genomics 7, 191.

Devi, S.M., Ahmed, I., Francalacci, P., Hussain, M.A., Akhter, Y., Alvi, A., et al., 2007. Ancestral European roots of *Helicobacter pylori* in India. BMC Genomics 8, 184.

Do, T., Gilbert, S.C., Clark, D., Ali, F., Fatturi Parolo, C.C., Maltz, M., et al., 2010. Generation of diversity in *Streptococcus mutans* genes demonstrated by MLST. PLoS One 5, e9073.

Dos Vultos, T., Mestre, O., Rauzier, J., Golec, M., Rastogi, N., Rasolofo, V., et al., 2008. Evolution and diversity of clonal bacteria: the paradigm of *Mycobacterium tuberculosis*. PLoS One 3, e1538.

Dou, H.-Y., Tseng, F.-C., Lu, J.-J., Jou, R., Tsai, S.-F., Chang, J.-R., et al., 2008. Associations of *Mycobacterium tuberculosis* genotypes with different ethnic and migratory populations in Taiwan. Infect. Genet. Evol. 8, 323–330.

Drummond, A.J., Rambaut, A., Shapiro, B., Pybus, O.G., 2005. Bayesian coalescent inference of past population dynamics from molecular sequences. Mol. Biol. Evol. 22, 1185–1192.

Dunn, B.E., Cohen, H., Blaser, M.J., 1997. Helicobacter pylori. Clin. Microbiol. Rev. 10, 720–741.

Eshleman, J.A., Malhi, R.S., Smith, D.G., 2003. Mitochondrial DNA studies of Native Americans: conceptions and misconceptions of the population prehistory of the Americas. Evol. Anthropol. 12, 7–18.

Falush, D., Kraft, C., Taylor, N.S., Correa, P., Fox, J.G., Achtman, M., et al., 2001. Recombination and mutation during long-term gastric colonization by *Helicobacter pylori*: estimates of clock rates, recombination size, and minimal age. Proc. Natl. Acad. Sci. U.S.A. 98, 15056−15061.

Falush, D., Wirth, T., Linz, B., Pritchard, J.K., Stephens, M., Kidd, M., et al., 2003. Traces of human migrations in *Helicobacter pylori* populations. Science 299, 1582−1585.

Farris, J.S., Kallersjo, M., Kluge, A.G., Bult, C., 1994. Testing significance of incongruence. Cladistics 10, 315−319.

Fernandez-Cobo, M., Agostini, H.T., Britez, G., Ryschkewitsch, C.F., Stoner, G.L., 2002. Strains of JC virus in Amerind-speakers of North America (Salish) and South America (Guaraní), Na-Dene-speakers of New Mexico (Navajo), and modern Japanese suggest links through an ancestral Asian population. Am. J. Phys. Anthropol 118, 154−168.

Gagneux, S., Small, P.M., 2007. Global phylogeography of *Mycobacterium tuberculosis* and implications for tuberculosis product development. Lancet Infect. Dis. 7, 328−337.

Gagneux, S., DeRiemer, K., Van, T., Kato-Maeda, M., de Jong, B.C., Narayanan, S., et al., 2006. Variable host−pathogen compatibility in *Mycobacterium tuberculosis*. Proc. Natl. Acad. Sci. U.S.A. 103, 2869−2873.

Gibbs, M.J., Armstrong, J.S., Gibbs, A.J., 2000. Sister-Scanning: a Monte Carlo procedure for assessing signals in recombinant sequences. Bioinformatics 16, 573−582.

Gottschling, M., Köhler, A., Stockfleth, E., Nindl, I., 2007a. Phylogenetic analysis of beta-papillomaviruses as inferred from nucleotide and amino acid sequence data. Mol. Phylogenet. Evol 42, 213−222.

Gottschling, M., Stamatakis, A., Nindl, I., Stockfleth, E., Alonso, A., Bravo, I.G., 2007b. Multiple evolutionary mechanisms drive papillomavirus diversification. Mol. Biol. Evol. 24, 1242−1258.

Gray, R.D., Drummond, A.J., Greenhill, S.J., 2009. Language phylogenies reveal expansion pulses and pauses in Pacific settlement. Science 323, 479−483.

Guo, J., Kitamura, T., Ebihara, H., Sugimoto, C., Kunitake, T., Takehisa, J., et al., 1996. Geographical distribution of the human polyomavirus JC virus types A and B and isolation of a new type from Ghana. J. Gen. Virol. 77, 919−927.

Hatwell, J.N., Sharp, P.M., 2000. Evolution of human polyomavirus JC. J. Gen. Virol. 81, 1191−1200.

Heinzel, P.A., Chan, S.Y., Ho, L., O'Connor, M., Balaram, P., Campo, M.S., et al., 1995. Variation of human papillomavirus type 6 (HPV-6) and HPV-11 genomes sampled throughout the world. J. Clin. Microbiol. 33, 1746−1754.

Hershberg, R., Lipatov, M., Small, P.M., Sheffer, H., Niemann, S., Homolka, S., et al., 2008. High functional diversity in *Mycobacterium tuberculosis* driven by genetic drift and human demography. PLoS Biol. 6, e311.

Hildesheim, A., Schiffman, M., Bromley, C., Wacholder, S., Herrero, R., Rodriguez, A.C., et al., 2001. Human papillomavirus type 16 variants and risk of cervical cancer. J. Natl. Cancer Inst. 93, 315−318.

Hirsh, A.E., Tsolaki, A.G., DeRiemer, K., Feldman, M.W., Small, P.M., 2004. Stable association between strains of *Mycobacterium tuberculosis* and their human host populations. Proc. Natl. Acad. Sci. U.S.A. 101, 4871−4876.

Ho, L., Chan, S.Y., Burk, R.D., Das, B.C., Fujinaga, K., Icenogle, J.P., et al., 1993. The genetic drift of human papillomavirus type 16 is a means of reconstructing prehistoric viral spread and the movement of ancient human populations. J. Virol. 67, 6413−6423.

Holmes, E.C., 2004. The phylogeography of human viruses. Mol. Ecol. 13, 745−756.

Holmes, E.C., 2008. Evolutionary history and phylogeography of human viruses. Annu. Rev. Microbiol. 62, 307–328.
Hurles, M.E., Matisoo-Smith, E., Gray, R.D., Penny, D., 2003. Untangling Oceanic settlement: the edge of the knowable. Trends Ecol. Evol. 18, 531–540.
Jakobsen, I.B., Easteal, S., 1996. A program for calculating and displaying compatibility matrices as an aid in determining reticulate evolution in molecular sequences. Comput. Appl. Biosci 12, 291–295.
Jiang, M., Xi, L.F., Edelstein, Z.R., Galloway, D.A., Olsem, G.J., Lin, W.C.-C., et al., 2009. Identification of recombinant human papillomavirus type 16 variants. Virology 394, 8–11.
Jobes, D.V., Chima, S.C., Ryschkewitsch, C.F., Stoner, G.L., 1998. Phylogenetic analysis of 22 complete genomes of the human polyomavirus JC virus. J. Gen. Virol. 79, 2491–2498.
Jobes, D.V., Friedlaender, J.S., Mgone, C.S., Agostini, H.T., Koki, G., Yanagihara, R., et al., 2001. New JC virus (JCV) genotypes from Papua New Guinea and Micronesia (Type 8 and Type 2E) and evolutionary analysis of 32 complete JCV genomes. Arch. Virol. 146, 2097–2113.
Junes-Gill, K., Sichero, L., Maciag, P.C., Mello, W., Noronha, V., Villa, L.L., 2008. Human papillomavirus type 16 variants in cervical cancer from an admixtured population in Brazil. J. Med. Virol. 80, 1639–1645.
Kato, A., Kitamura, T., Sugimoto, C., Ogawa, Y., Nakazato, K., Nagashima, K., et al., 1997. Lack of evidence for the transmission of JC polyomavirus between human populations. Arch. Virol. 142, 875–882.
Kitamura, T., Sugimoto, C., Kato, A., Ebihara, H., Suzuki, M., Taguchi, F., et al., 1997. Persistent JC virus (JCV) infection is demonstrated by continuous shedding of the same JCV strains. J. Clin. Microbiol. 35, 1255–1257.
Kitchen, A., Miyamoto, M.M., Mulligan, C.J., 2008. Utility of DNA viruses for studying human host history: case study of JC virus. Mol. Phylogenet. Evol. 46, 673–682.
Kivi, M., Tindberg, Y., Sorberg, M., Casswall, T.H., Befrits, R., Hellstrom, P.M., et al., 2003. Concordance of *Helicobacter pylori* strains within families. J. Clin. Microbiol. 41, 5604–5608.
Kraft, C., Suerbaum, S., 2005. Mutation and recombination in *Helicobacter pylori*: mechanisms and role in generating strain diversity. Int. J. Med. Microbiol. 295, 299–305.
Kraft, C., Stack, A., Josenhans, C., Niehus, E., Dietrich, G., Correa, P., et al., 2006. Genomic changes during chronic *Helicobacter pylori* infection. J. Bacteriol. 188, 249–254.
Kunitake, T., Kitamura, T., Guo, J., Taguchi, F., Kawabe, K., Yogo, Y., 1995. Parent-to-child transmission is relatively common in the spread of the human polyomavirus JC virus. J. Clin. Microbiol. 33, 1448–1451.
Lapirattanakul, J., Nakano, K., Nomura, R., Hamada, S., Nakagawa, I., Ooshima, T., 2008. Demonstration of mother-to-child transmission of *Streptococcus mutans* using multilocus sequence typing. Caries Res. 42, 466–474.
Latifi-Navid, S., Ghorashi, S.A., Siavoshi, F., Linz, B., Massarrat, S., Khegay, T., et al., 2010. Ethnic and geographic differentiation of *Helicobacter pylori* within Iran. PLoS One 5, e9645.
Li, Y., Li, Z., He, Y., Kang, Y., Zhang, X., Cheng, M., et al., 2009. Phylogeographic analysis of human papillomavirus 58. Sci. China C Life Sci. 52, 1164–1172.
Liaw, K.-L., Hildesheim, A., Burk, R.D., Gravitt, P., Wacholder, S., Manos, M.M., et al., 2001. A prospective study of human papillomavirus (HPV) type 16 DNA detection by

polymerase chain reaction and its association with acquisition and persistence of other HPV types. J. Infect. Dis. 183, 8−15.
Linz, B., Balloux, F., Moodley, Y., Manica, A., Liu, H., Roumagnac, P., et al., 2007. An African origin for the intimate association between humans and *Helicobacter pylori*. Nature 445, 915−918.
Liu, H., Prugnolle, F., Manica, A., Balloux, F., 2006. A geographically explicit genetic model of worldwide human-settlement history. Am. J. Hum. Genet. 79, 230−237.
Martin, D.P., Williamson, C., Posada, D., 2005. RDP2: recombination detection and analysis from sequence alignments. Bioinformatics 21, 260−262.
Maynard Smith, J., Smith, N.H., 1998. Detecting recombination from gene trees. Mol. Biol. Evol. 15, 590−599.
McVean, G., Awadalla, P., Fearnhead, P., 2002. A coalescent-based method for detecting and estimating recombination from gene sequences. Genetics 160, 1231−1241.
Mokrousov, I., 2008. Genetic geography of *Mycobacterium tuberculosis* Beijing genotype: a multifacet mirror of human history? Infect. Genet. Evol. 8, 777−785.
Mokrousov, I., Ly, H.M., Otten, T., Lan, N.N., Vyshnevskyi, B., Hoffner, S., et al., 2005. Origin and primary dispersal of the *Mycobacterium tuberculosis* Beijing genotype: clues from human phylogeography. Genome Res. 15, 1357−1364.
Monot, M., Honore, N., Garnier, T., Araoz, R., Coppee, J.-Y., Lacroix, C., et al., 2005. On the origin of leprosy. Science 308, 1040−1042.
Monot, M., Honore, N., Garnier, T., Zidane, N., Sherafi, D., Paniz-Mondolfi, A., et al., 2009. Comparative genomic and phylogeographic analysis of *Mycobacterium leprae*. Nat. Genet. 41, 1282−1289.
Moodley, Y., Linz, B., Yamaoka, Y., Windsor, H.M., Breurec, S., Wu, J.-Y., et al., 2009. The peopling of the Pacific from a bacterial perspective. Science 323, 527−530.
Muerhoff, A.S., Leary, T.P., Sathar, M.A., Dawson, G.J., Desai, S.M., 2005. African origin of GB virus C determined by phylogenetic analysis of a complete genotype 5 genome from South Africa. J. Gen. Virol. 86, 1729−1735.
Nakano, K., Lapirattanakul, J., Nomura, R., Nemoto, H., Alaluusua, S., Gronroos, L., et al., 2007. *Streptococcus mutans* clonal variation revealed by multilocus sequence typing. J. Clin. Microbiol. 45, 2616−2625.
Nakao, H., Okamoto, H., Fukuda, M., Tsuda, F., Mitsui, T., Masuko, K., et al., 1997. Mutation rate of GB virus C/Hepatitis G virus over the entire genome and in subgenomic regions. Virology 233, 43−50.
Narechania, A., Chen, Z., DeSalle, R., Burk, R.D., 2005. Phylogenetic incongruence among oncogenic genital alpha human papillomaviruses. J. Virol. 79, 15503−15510.
Nieberding, C.M., Olivieri, I., 2007. Parasites: proxies for host genealogy and ecology? Trends Ecol. Evol. 22, 156−165.
Nielsen, R., 2000. Estimation of population parameters and recombination rates from single nucleotide polymorphisms. Genetics 154, 931−942.
Ong, C.K., Chan, S.Y., Campo, M.S., Fujinaga, K., Mavromara-Nazos, P., Labropoulou, V., et al., 1993. Evolution of human papillomavirus type 18: an ancient phylogenetic root in Africa and intratype diversity reflect coevolution with human ethnic groups. J. Virol. 67, 6424−6431.
O'Rourke, D.H., Raff, J.A., 2010. The human genetic history of the Americas: the final frontier. Curr. Biol. 20, R202−R207.
Padgett, B.L., Walker, D.L., 1973. Prevalence of antibodies in human sera against JC virus, an isolate from a case of progressive multifocal leukoencephalopathy. J. Infect. Dis. 127, 467−470.

Pavesi, A., 2001. Origin and evolution of GBV-C/hepatitis G virus and relationships with ancient human migrations. J. Mol. Evol 53, 104−113.
Pavesi, A., 2003. African origin of polyomavirus JC and implications for prehistoric human migrations. J. Mol. Evol 56, 564−572.
Pavesi, A., 2004. Detecting traces of prehistoric human migrations by geographic synthetic maps of polyomavirus JC. J. Mol. Evol. 58, 304−313.
Pavesi, A., 2005. Utility of JC polyomavirus in tracing the pattern of human migrations dating to prehistoric times. J. Gen. Virol. 86, 1315−1326.
Pepperell, C., Hoeppner, V.H., Lipatov, M., Wobeser, W., Schoolnik, G.K., Feldman, M.W., 2010. Bacterial genetic signatures of human social phenomena among *M. tuberculosis* from an aboriginal Canadian population. Mol. Biol. Evol. 27, 427−440.
Prado, J.C., Calleja-Macias, I.E., Bernard, H.-U., Kalantari, M., Macay, S.A., Allan, B., et al., 2005. Worldwide genomic diversity of the human papillomaviruses-53, 56, and 66, a group of high-risk HPVs unrelated to HPV-16 and HPV-18. Virology 340, 95−104.
Prugnolle, F., Manica, A., Balloux, F., 2005. Geography predicts neutral genetic diversity of human populations. Curr. Biol. 15, R159−R160.
Ramachandran, S., Deshpande, O., Roseman, C.C., Rosenberg, N.A., Feldman, M.W., Cavalli-Sforza, L.L., 2005. Support from the relationship of genetic and geographic distance in human populations for a serial founder effect originating in Africa. Proc. Natl. Acad. Sci. U.S.A. 102, 15942−15947.
Rannala, B., Michalakis, Y., 2003. Population genetics and cospeciation: from process to pattern. In: Page, R.D.M. (Ed.), Tangled Trees: Phylogeny, Cospeciation, and Coevolution. University of Chicago Press, Chicago, IL, pp. 120−143.
Raymond, J., Thiberge, J.-M., Chevalier, C., Kalach, N., Bergeret, M., Labigne, A., et al., 2004. Genetic and transmission analysis of *Helicobacter pylori* strains within a family. Emerging Infect. Dis. 10, 1816−1821.
Renfrew, C., 2009. Where bacteria and languages concur. Science 323, 467−468.
Romano, C., Zanotto, P.M.D.A., Holmes, E., 2008. Bayesian coalescent analysis reveals a high rate of molecular evolution in GB virus C. J. Mol. Evol. 66, 292−297.
Sarrazin, C., Rüster, B., Lee, J.-H., Kronenberger, B., Roth, W.K., Zeuzem, S., 2000. Prospective follow-up of patients with GBV-C/HGV infection: specific mutational patterns, clinical outcome, and genetic diversity. J. Med. Virol. 62, 191−198.
Schierup, M.H., Hein, J., 2000. Consequences of recombination on traditional phylogenetic analysis. Genetics 156, 879−891.
Schlecht, N.F., Burk, R.D., Palefsky, J.M., Minkoff, H., Xue, X., Massad, L.S., et al., 2005. Variants of human papillomaviruses 16 and 18 and their natural history in human immunodeficiency virus-positive women. J. Gen. Virol. 86, 2709−2720.
Schwarz, S., Morelli, G., Kusecek, B., Manica, A., Balloux, F., Owen, R.J., et al., 2008. Horizontal versus familial transmission of *Helicobacter pylori*. PLoS Path. 4, e1000180.
Shackelton, L.A., Rambaut, A., Pybus, O.G., Holmes, E.C., 2006. JC virus evolution and its association with human populations. J. Virol. 80, 9928−9933.
Shah, S.D., Doorbar, J., Goldstein, R.A., 2010. Analysis of host-parasite incongruence in papillomavirus evolution using importance sampling. Mol. Biol. Evol. 27, 1301−1314.
Sichero, L., Trottier, H., Ferreira, S., Duarte-Franco, E., Franco, E.L., Villa, L.L., 2007. Re: Human papillomavirus type 16 and 18 variants: race-related distribution and persistence. J. Natl. Cancer Inst. 99, 653−654.
Simmonds, P., 2001. Reconstructing the origins of human hepatitis viruses. Philos. Trans. R. Soc. Lond. Ser. B Biol. Sci. 356, 1013−1026.

Simmonds, P., Smith, D.B., 1999. Structural constraints on RNA virus evolution. J. Virol. 73, 5787−5794.
Smith, D.B., Basaras, M., Frost, S., Haydon, D., Cuceanu, N., Prescott, L., et al., 2000. Phylogenetic analysis of GBV-C/hepatitis G virus. J. Gen. Virol. 81, 769−780.
Smith, N.H., Dale, J., Inwald, J., Palmer, S., Gordon, S.V., Hewinson, R.G., et al., 2003. The population structure of *Mycobacterium bovis* in Great Britain: clonal expansion. Proc. Natl. Acad. Sci. U.S.A. 100, 15271−15275.
Smith, N.H., Hewinson, R.G., Kremer, K., Brosch, R., Gordon, S.V., 2009. Myths and misconceptions: the origin and evolution of *Mycobacterium tuberculosis*. Nat. Rev. Microbiol. 7, 537−544.
Stewart, A.C., Eriksson, A.M., Manos, M.M., Munoz, N., Bosch, F.X., Peto, J., et al., 1996. Intratype variation in 12 human papillomavirus types: a worldwide perspective. J. Virol. 70, 3127−3136.
Stoner, G.L., Jobes, D.V., Fernandez Cobo, M., Agostini, H.T., Chima, S.C., Ryschkewitsch, C.F., 2000. JC virus as a marker of human migration to the Americas. Microb. Infect. 2, 1905−1911.
Suerbaum, S., Achtman, M., 2004. *Helicobacter pylori*: recombination, population structure and human migrations. Int. J. Med. Microbiol. 294, 133−139.
Suerbaum, S., Michetti, P., 2002. *Helicobacter pylori* infection. New Engl. J. Med. 347, 1175−1186.
Suerbaum, S., Smith, J.M., Bapumia, K., Morelli, G., Smith, N.H., Kunstmann, E., et al., 1998. Free recombination within *Helicobacter pylori*. Proc. Natl. Acad. Sci. U.S.A. 95, 12619−12624.
Sugimoto, C., Kitamura, T., Guo, J., Al-Ahdal, M.N., Shchelkunov, S.N., Otova, B., et al., 1997. Typing of urinary JC virus DNA offers a novel means of tracing human migrations. Proc. Natl. Acad. Sci. U.S.A. 94, 9191−9196.
Sugimoto, C., Hasegawa, M., Kato, A., Zheng, H.-Y., Ebihara, H., Taguchi, F., et al., 2002. Evolution of human polyomavirus JC: implications for the population history of humans. J. Mol. Evol. 54, 285−297.
Supply, P., Warren, R.M., Bañuls, A.-L., Lesjean, S., van der Spuy, G.D., Lewis, L.-A., et al., 2003. Linkage disequilibrium between minisatellite loci supports clonal evolution of *Mycobacterium tuberculosis* in a high tuberculosis incidence area. Mol. Microbiol. 47, 529−538.
Suzuki, M., Zheng, H.-Y., Takasaka, T., Sugimoto, C., Kitamura, T., Beutler, E., et al., 2002. Asian genotypes of JC virus in Japanese-Americans suggest familial transmission. J. Virol. 76, 10074−10078.
Suzuki, Y., Katayama, K., Fukushi, S., Kageyama, T., Oya, A., Okamura, H., et al., 1999. Slow evolutionary rate of GB virus C/hepatitis G virus. J. Mol. Evol. 48, 383−389.
Tay, C., Mitchell, H., Dong, Q., Goh, K.-L., Dawes, I., Lan, R., 2009. Population structure of *Helicobacter pylori* among ethnic groups in Malaysia: recent acquisition of the bacterium by the Malay population. BMC Microbiol. 9, 126.
Vaccarella, S., Franceschi, S., Herrero, R., Muñoz, N., Snijders, P.J.F., Clifford, G.M., et al., 2006. Sexual behavior, condom use, and human papillomavirus: pooled analysis of the IARC human papillomavirus prevalence surveys. Cancer Epidemiol. Biomarkers Prev. 15, 326−333.
Varsani, A., van der Walt, E., Heath, L., Rybicki, E.P., Williamson, A.L., Martin, D.P., 2006. Evidence of ancient papillomavirus recombination. J. Gen. Virol. 87, 2527−2531.

Villa, L.L., Sichero, L., Rahal, P., Caballero, O., Ferenczy, A., Rohan, T., et al., 2000. Molecular variants of human papillomavirus types 16 and 18 preferentially associated with cervical neoplasia. J. Gen. Virol. 81, 2959–2968.

Weber, T., Major, E.O., 1997. Progressive multifocal leukoencephalopathy: molecular biology, pathogenesis and clinical impact. Intervirology 40, 98–111.

Whiteman, N.K., Parker, P.G., 2005. Using parasites to infer host population history: a new rationale for parasite conservation. Anim. Conserv. 8, 175–181.

Wirth, T., Wang, X., Linz, B., Novick, R.P., Lum, J.K., Blaser, M., et al., 2004. Distinguishing human ethnic groups by means of sequences from *Helicobacter pylori*: lessons from Ladakh. Proc. Natl. Acad. Sci. U.S.A. 101, 4746–4751.

Wirth, T., Meyer, A., Achtman, M., 2005. Deciphering host migrations and origins by means of their microbes. Mol. Ecol. 14, 3289–3306.

Wirth, T., Hildebrand, F., Allix-Béguec, C., Wölbeling, F., Kubica, T., Kremer, K., et al., 2008. Origin, spread and demography of the *Mycobacterium tuberculosis* complex. PLoS Path. 4, e1000160.

Wooding, S., 2001. Do human and JC virus genes show evidence of host–parasite codemography? Infect. Genet. Evol. 1, 3–12.

World Health Organization, 2009. Global tuberculosis control: a short update to the 2009 report.

Worobey, M., 2001. A novel approach to detecting and measuring recombination: new insights into evolution in viruses, bacteria, and mitochondria. Mol. Biol. Evol. 18, 1425–1434.

Xi, L.F., Kiviat, N.B., Hildesheim, A., Galloway, D.A., Wheeler, C.M., Ho, J., et al., 2006. Human papillomavirus type 16 and 18 variants: race-related distribution and persistence. J. Natl. Cancer Inst. 98, 1045–1052.

Yamada, T., Manos, M.M., Peto, J., Greer, C.E., Munoz, N., Bosch, F.X., et al., 1997. Human papillomavirus type 16 sequence variation in cervical cancers: a worldwide perspective. J. Virol. 71, 2463–2472.

Yanagihara, R., Nerurkar, V.R., Scheirich, I., Agostini, H.T., Mgone, C.S., Cui, X., et al., 2002. JC virus genotypes in the western Pacific suggest Asian mainland relationships and virus association with early population movements. Hum. Biol. 74, 473–488.

8 Phylogenetic Analysis of Pathogens

David A. Morrison*

Section for Parasitology (SWEPAR), Department of Biomedical Sciences and Veterinary Public Health, Swedish University of Agricultural Sciences, Uppsala, Sweden

8.1 Introduction

One of the most famous statements in the biological sciences is that "nothing in biology makes sense except in the light of evolution" (Dobzhansky, 1973). This was a deliberate overstatement by the author, intended to emphasize an important point. That point is as relevant today as it was 40 years ago: if biological evolution has occurred, then we need to interpret biological phenomena in the light of the history of how those phenomena arose. Which phenomena are possible and which are impossible is determined by their historical development; and our ability to understand, explain, and (most importantly) predict the phenomena thus depends heavily on our ability to correctly determine the evolutionary history.

This being so, it is clearly important to be able to reconstruct the phylogeny (the branching sequence of the lineages during their evolutionary history), both for a group of species and also for the individuals within those species. Unfortunately, this is one of the hardest forms of data analysis known. The events under study are unobservable historical accidents, and so we can neither make direct observations of them nor perform experiments to investigate them. Indeed, some researchers see phylogenetics as being outside of "normal" empirical science. Nevertheless, phylogenetics is based on the use of observable characteristics of contemporary organisms to try to deduce the sequence of events that occurred during the descent of those organisms. That is, we use what we can see now to infer the events that led to what we can see.

Charles Darwin's main contribution to biology was to recognize that there are two distinct types of biological evolution: (i) transformational evolution, in which individual objects each change through time, and (ii) variational evolution, in which groups of variable objects change their relative proportions through time. Transformational evolution is common in the physical sciences as well as in biology, but variational evolution has a special place in the biological sciences because isolated changes in variation will ultimately lead to new species. Both types of evolution can best be represented as a tree-like diagram (Figure 8.1), because this can

---

*E-mail: David.Morrison@slu.se

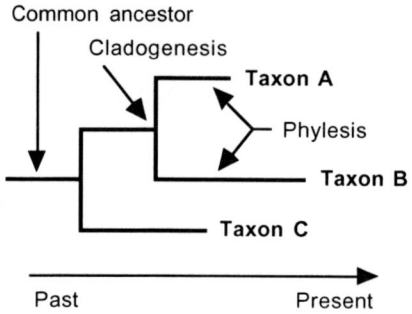

Figure 8.1 Phylogenetic tree for three contemporary taxa (A−C), showing the various relevant characteristics used for interpretation of the diagram.

show the phylesis (changes through time) as relative branch lengths and the cladogenesis (speciation) as the relative branching order. The base of the tree represents the common ancestor, while the bifurcating branches represent a successive series of descendants, arriving finally at the contemporary organisms at the twigs.

This chapter introduces the important facets of the topic of producing phylogenetic trees, discussing those points that are of most direct relevance to the study of infectious organisms. This chapter provides an overview and an introduction to the recent literature. For more general issues, there are a number of essays (e.g., Baldauf, 2003; Stevens and Schofield, 2003; Delsuc et al., 2005; Morrison, 2006a) as well as excellent books, varying from the introductory (Bromham, 2008; Hall, 2008; Lemey et al., 2009) to the detailed (Nei and Kumar, 2000; Felsenstein, 2004; Yang, 2006).

8.2 The Uses of Phylogenies

The importance of knowing the phylogenetic history of all living organisms has become increasingly widely recognized. There are now many biological studies based on phylogenies in all sorts of fields, including such diverse subjects as immunology, epidemiology, and conservation. Also, many new areas of study have recently arisen that make extensive use of phylogenies, such as evolutionary development (known as evodevo) and evolutionary ecology. Elucidating phylogenies is therefore now an important component of nearly all of the biological sciences.

Phylogenies form the framework within which we can best arrange our knowledge of all aspects of biology (Lecointre and Le Guyader, 2006). It has taken biologists a long time to explicate the simple idea that it is the study of biodiversity that makes biology different from other sciences—the nature and scale of the interrelationships among organisms is something that has never been conceived of within physics and chemistry. Evolutionary history is our explanation for the origin of that biodiversity, and so the best way to present biodiversity is in the context of phylogenetic trees. For pathogens, we are interested in the evolution of the diseases at the genetic level and what this can tell us about their past and present diversity.

Thus, we care about the phylogenetic study of diversity because the phylogenetic tree organizes our knowledge of biodiversity. It tells us how closely organisms are related to each other (e.g., all of the animals are more closely related to each other than they are to plants), and we can recognize groups of organisms (e.g., animals form a group). These groups are most conveniently arranged into a hierarchical taxonomic scheme.

The scientific study of biology requires a stable taxonomy based on a robust phylogeny. The groups named in the taxonomy should represent real evolutionary groups (called clades), so that these groups can be used for what is now termed comparative biology. In comparative biology, the evolutionary groups identify the pertinent experimental comparisons that are needed for the quantitative study of biodiversity patterns and evolutionary processes.

If taxonomy is to reflect phylogeny (as it should), then phylogenetic evidence is the crucial evidence. That is, the key to a robust and well-accepted classification is a robust and well-accepted phylogeny, in which the evolutionary relationships are not subject to change as new evidence accumulates, and which can therefore be accepted by nonexperts. Too often, traditional taxonomic schemes have been utilitarian rather than phylogenetic, which often means that they are very anthropocentric. This applies particularly to the study of pathogens.

The term "pathogen" was devised about 1880, which was a time of great activity in the attempt to depict organismal relationships as trees, notably with the work of Ernst Haeckel. If we consider pathogenic organisms such as viruses, bacteria, microfungi, protists, and helminths, then it is clear that the members of some of these groups are not closely related to each other in the evolutionary sense, notably the organisms grouped as protists and helminths. Recognizing and understanding that these are utilitarian groupings based on nonevolutionary criteria has been one of the major contributions of phylogenetic analyses to modern biology.

Pathogens are often grouped on the basis of phenotypic similarity (e.g., hosts, predilection sites, infection route, microscopy) or similarity of disease (e.g., symptoms, diagnostic procedures). However, these criteria do not automatically imply similarity of evolutionary history. For example, the traditionally recognized group helminths ("worms") consists of the Platyhelminthes (flatworms) and the Nematoda (roundworms). However, the latter have a body cavity enclosed by mesoderm (called the coelom) whereas the former do not, and so we infer that the roundworms are more closely related to (for example) insects than they are to the flatworms.

Perhaps the most valuable uses of a phylogeny are for explaining and predicting organismal features. The strongest argument for a phylogenetic classification scheme is that it organizes our knowledge in a way that maximizes information content by being both explanatory and predictive (Farris, 1979). For example, *Cryptosporidium* (Apicomplexa) causes cryptosporidiosis in mammals, and its life cycle and ultrastructure are similar to those of the agents causing coccidiosis, toxoplasmosis, and neosporosis in vertebrates; thus it has traditionally been placed within the Coccidia. Recent molecular-based phylogenies contradict this placement (Barta and Thompson, 2006), however, indicating instead that it is related to the Gregarina, which infect invertebrates. This revised placement helps explain why anticoccidial treatments are ineffective on members of this genus: if susceptibility to anticoccidial agents is a trait inherited from the common ancestor of the Coccidia, then any unrelated organisms will lack this trait.

Similarly, *Sarcocystis* is also part of the Coccidia, causing sarcocystosis in vertebrates. It has a two-host (indirect) life cycle, with the definitive host a carnivore and the intermediate host usually a herbivore. Sometimes species have been collected only from the intermediate host, and we thus need to predict the definitive host. Our best prediction will be that the host is the same as for the closely related species. For example, *Sarcocystis alces* was originally collected only from European elk, the intermediate host, but Dahlgren and Gjerde (2008) suggested that it is part of an evolutionary group that has canids as their definitive host; and so this would be our best prediction. This hypothesis was tested successfully by Dahlgren and Gjerde (2010), who demonstrated that both red foxes and arctic foxes (canids) can act as definitive hosts.

There are many other important uses of phylogenies (Harvey et al., 1996), including the study of co-phylogeny of hosts and pathogens (e.g., understanding the role of hosts in pathogen evolution) and pathogen biogeography (e.g., understanding the spread of pathogens). Different pathogens have different distributions, different patterns of spread, and different rates of evolution. This results in very different characteristics at the genetic and geographic levels. For example, the phylogenetic tree can be compared to the geographic locations of the samples in order to investigate the spread of disease, or so-called molecular clocks can be applied to estimate the age of important events in the origin and spread of new pathogens. These relationships are discussed in more detail in other chapters of this book.

8.3 The Logic of Phylogeny Reconstruction

Reconstructing a phylogenetic history is conceptually straightforward, although it took a long time for someone (Hennig, 1966) to explicate the most appropriate approach. Interestingly, the study of historical linguistics has developed the same methodology (Platnick and Cameron, 1977; Atkinson and Gray, 2005), thus independently arriving at exactly the same solution to what is, in effect, exactly the same problem. From this point of view, the methodology itself is uncontroversial, and its generic nature means that it can be used for any objects with characteristics that can be identified and measured, and that follow a history of descent with modification. It has thus been applied to several other types of historical studies (see Barbrook et al., 1998; Mace et al., 2005; Lipo et al., 2006).

The objective is to infer the ancestors of the contemporary organisms, and the ancestors of those ancestors, etc., all the way back to the most recent common ancestor of the group of organisms being studied. Ancestors can be inferred because the organisms share unique characteristics. That is, they have features that they hold in common and that are not possessed by any other organisms. The simplest explanation for this observation is that the features are shared because they were inherited from an ancestor. The ancestor acquired a set of inheritable (i.e., genetically controlled) characteristics, and passed those characteristics on to its offspring. We observe the offspring, and from the resulting observations we infer the existence of the unobserved ancestor(s).

For example, we might note that a subset of our organisms all have an internal (bony) skeleton. No other organisms are known to possess this complex structure. There are only two realistic explanations for this observation: the organisms developed this structure independently, or they inherited it from their common ancestor. The second explanation is the simplest one, and so it constitutes our working hypothesis of the evolutionary history of the organisms.

If we collect a number of such observations, what we usually find is that they form a set of nested groupings of the organisms. For example, one subset of the organisms with an internal skeleton also possesses feathers, thus leading us to infer that this subgroup has a more recent common ancestor than does the skeleton group.

These nested sets and subsets of organisms can be represented in a tree diagram (Figure 8.1), which has been the conventional way to denote hypotheses of phylogenetic history since the work of Charles Darwin. Each internal branch on such a tree indicates an inferred ancestor, and each terminal branch (or leaf) represents an observed organism. The branching order of the tree indicates the order of the historical events leading to divergence of the organisms, often called the "sister-group" relationships of the organisms. The length of the branches is commonly (but not always) used as a convention to represent the amount of evolutionary change that occurred in each ancestor, so that the length of a particular branch is proportional to the number of unique characteristics inferred to have been acquired by that ancestor (and passed on to its offspring).

These hypotheses of ancestry (both branching order and relative branch lengths) are open to testing by acquiring observations of other features of the organisms. These may support the previous observations or they may conflict with them. The practical process of reconstructing the phylogenetic history of a group of organisms consists of evaluating the (often) contradictory nature of the evidence. We collect as many observations as is practicable (given time, money, and other resources), and we compare the various pieces of evidence in order to arrive at the most plausible scenario for the historical events.

As a specific example that this logic can work in practice, Lemey et al. (2005) studied the transmission history of the HIV-1 virus among a particular group of people. In this case there was independent evidence concerning the transmission history, based on interviews with the nine people concerned, so that we have a pretty good idea who passed the virus to whom and when. This known transmission history constitutes the true evolutionary history (Figure 8.2). Some of the genes of the virus were also sequenced in these same people at varying time intervals. This means that we can independently attempt to reconstruct the evolutionary history (phylogeny) using these sequence data. In this case, the known history and the reconstructed phylogeny turn out to be identical, for at least some of the known types of phylogenetic analysis, and so we can justifiably conclude that our phylogenetic methods are valid.

As an example of an experimentally produced evolutionary history, we can consider the work of Sanson et al. (2002). These workers used known errors in gene copying within *Trypanosoma cruzi* (Kinetoplastida), to mutate a single rRNA gene sequence into a set of eight descendant sequence clones, where all of the intermediate ancestral clones were sequenced as well. In this case, we have molecular data for all of the ancestors and all of the descendants, and we know the true historical relationship among them all. Here, *all* of the known methods for reconstructing phylogenetic trees from molecular data produce exactly the same solution, which perfectly matches the known history.

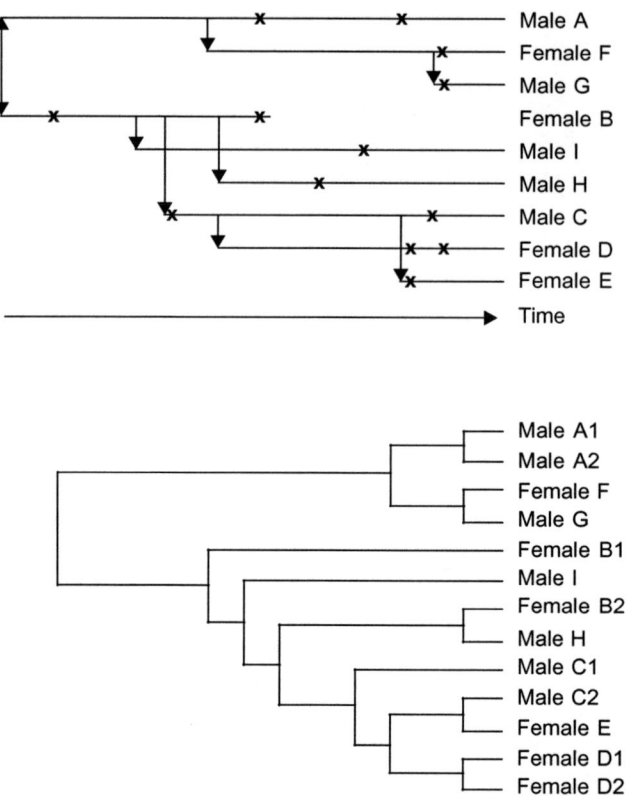

Figure 8.2 The known transmission history (upper panel) along with the phylogeny (lower panel) reconstructed from the *env* gene of the HIV-1 virus, based on the data of Lemey et al. (2005). In the transmission history the arrows indicate the direction and time of transmission. So, for example, male A infected female F, who subsequently infected male G. The times at which samples were taken for DNA analysis are indicated by **x**. In the reconstructed evolutionary history, different samples from the same individual are indicated by labels 1 and 2. The inferred phylogeny is identical to the transmission history. For example, individuals F and G have a very closely related form of the virus, which is also closely related to the isolates A1 and A2.
Source: Modified with permission from Morrison (2011).

Phylogeneticists therefore feel confident that these phylogenetic methods also apply to situations where direct knowledge of the past is absent. If this were not so, then there would be no purpose to the analyses.

8.4 Characters and Samples

Phylogenetic analysis can be used for any objects with characteristics that can be identified and measured, and that follow a history of descent with modification.

The objects being sampled are usually referred to as "taxa" and the characteristics being measured are "characters."

The taxa can be any part of the standard taxonomic hierarchy (species, genera, families, etc.) or they can be individuals or populations. They can also be cultures or pathology samples, or even fossils. It is expected that the characters will be measurable on most of these taxa, although some of the taxa may lack some of the characters.

The sample of taxa used to construct the phylogenetic tree needs to be adequate in order to provide a convincing case for particular phylogenetic relationships. Showing that a problem exists is easy with a small sample size, but revealing the solution takes much more effort. Furthermore, a biased sample usually leads to biased estimates of phylogenetic relationship. Relationships cannot be detected if the related groups have not been sampled, for example.

Unfortunately, pathogens are difficult to study because they are usually hard to find. Access to hosts can be difficult, and endoparasites often can be found only in symptomatic individuals. Therefore, sampling to date for many taxonomic groups has been almost entirely opportunistic (Barta, 2001). Opportunities for sampling arise principally from case studies of medical and veterinary diseases, rather than from purposive experimental designs. Phylogenetic relevance has often not been the criterion for sample choice, which leads to a small and biased sample.

As a result of the large biodiversity of many pathogen groups, we need to choose exemplar taxa for a worthwhile phylogenetic analysis (e.g., at least one species from each genus). This is difficult if the biodiversity has not been well cataloged. In particular, the Apicomplexa, bacteria, and viruses are the three worst-known groups in terms of their named biodiversity, each with <1% of their estimated number of species having been described to date (Morrison, 2009a). This compares very unfavorably with all other taxonomic groups. Even the Insecta, which is usually considered to be the prime example of a poorly known group, has ~1 million species known out of an estimated total of 4.5–30 million. This situation creates several possible impediments for the phylogenetic study of pathogens, which are discussed by Morrison (2008) for the Apicomplexa as an example.

Obviously, the characters measured must be heritable, which means that they must be genotypic characteristics rather than merely phenotypic ones (i.e., those greatly influenced by environmental variables). Most pathogens are unicellular or multicellular without specialized tissues, which severely limits the number and range of available characters. Traditionally, the characters used for phylogenetic and taxonomic analyses have been based mainly on life cycle features, disease characteristics, and ultrastructure. It may be rather difficult to determine homologies among such characters (i.e., their evolutionary comparability), so that related character states are being compared, and the data are also regrettably incomplete for most species. Consequently, phylogenies based solely on these characters have been rare, and they have not been particularly robust.

For this reason, molecular data have now become the predominant character data for phylogenetic studies of pathogens. DNA mutates, the sequences change, and as pathogens spread they bring these changes with them. Molecular characters include allozymes, DNA-DNA hybridization, randomly amplified polymorphic DNA (RAPD), restriction fragment length polymorphism (RFLP), amplified fragment length polymorphism (AFLP), protein sequences, and nucleotide sequences. Of these, nucleotide sequences are now by far the most common, for all taxa not just pathogens. Indeed, many microbiology journals have guidelines stating that a phylogenetic context is required for the publication of new taxa, so that their nucleotide sequences and organismal phylogenies are part of the "publication pipeline."

Many generalist and specialist databases exist to store these sequences. Most of these sequence databases are uncurated, so that the sequences submitted are entirely the responsibility of the submitter. Unfortunately, there is cause for concern about the quality of these data. There are four main potential problems with uncurated public-access sequence databases: wrongly annotated genes (location, orientation, boundaries); poor sequence quality; inconsistencies between the database data and the published data; and misidentified and mislabeled taxa. Experience with the use of these databases shows that all four of these problems will be encountered repeatedly in practice. The unknown (and perhaps unknowable) quality of the data raises the serious question as to whether these databases are of any use for high-quality phylogenetic analyses. Indeed, it is obvious that increasing numbers of authors are becoming very selective about which, if any, database sequences will be included in their own phylogenetic analyses.

This does not mean that we should despair, of course. Misannotation, for example, can often be dealt with by careful comparative analysis. Poor sequencing can be corrected if the DNA is still available, although there are no generally accessible depositories for preserving DNA. Inconsistencies can be dealt with by contacting the authors, although this is sometimes a frustrating experience. Taxon misidentification and mislabeling can be straightforward to deal with by having voucher specimens, cultures, or microscope slides for every sequence. Sadly, deposition of vouchers in any of the many available depositories has been infrequent for many groups of pathogens (particularly protists). It is therefore important to emphasize that without an extant voucher of the sequence source the identification can never be verified, and the sequence might as well be labeled as "unidentified."

We need to reconstruct the phylogenetic tree of the organisms involved. For nucleotide sequences, only concordance between the phylogenies derived from several molecular sequences will be accepted as evidence for the organism phylogeny. A tree from a single molecular sequence represents only the phylogeny of that one gene, which does not necessarily reflect the phylogeny of the organism (Doyle, 1992). Just how many genes might be required to reconstruct the organismal phylogeny is an open question (Gatesy et al., 2007), however. Many other molecular data types (e.g., AFLPs) sample widely from the genome, and so they should naturally represent the organismal phylogeny.

To date, most pathogen phylogenies have been based on the sequence of a single gene, usually the nucleotide sequence of the small-subunit (16S or 18S) ribosomal RNA gene. Indeed, most of the reclassification of the bacteria since the late 1970s has been based principally on this gene (Sapp, 2009). Many of the other genes sequenced are those for host recognition or for dealing with the host immune system (perhaps sequenced as part of projects producing new drugs or vaccines), which are often unique to each taxonomic group or are subject to heavy selection and are thus not necessarily useful for phylogeny. In particular, bacterial genomes often have clusters of functionally related genes such as those for antibiotic resistance (Hedges, 1972), which can affect phylogenetic analysis. Consequently, the character data are rather fragmentary for many taxonomic groups. A multigene phylogeny is therefore unlikely to be produced from these current data.

Missing characters occur either because some taxa could not be completely sequenced for a particular gene or they could not be sequenced at all for that gene. Characters with missing data should be excluded from the analysis if they result from technical problems that can cause ambiguity, such as sequence repeats or inadequate priming. However, including incomplete taxa in a tree-building analysis is more likely to increase the accuracy of the tree building than to decrease it (Wiens et al., 2005), although the degree depends on the proportion of missing data. Taxa with lots of missing data are more problematic than are characters with lots of missing data, as these can be positively misleading (de la Torre-Bárcena et al., 2009). Supertree methods (Cotton and Wilkinson, 2008) provide one possible approach to the phylogenetic analysis of disparate datasets.

An obvious source of multigene sequence data is complete genomes, the necessary techniques being routinely feasible nowadays for prokaryotic organisms, at least. Thus, there are now available hundreds of complete genomes for bacterial taxa, although less than a dozen exist for the eukaryotic Apicomplexa, for example.

Sequences of complete genomes have contributed much to comparative genomics, which assumes that the phylogeny is known and can be used as the basis for comparisons among species. However, these genomes might never prove to be useful for phylogeny reconstruction itself. The only situation where they are likely to be useful is where the original gene samples were biased, because the genomes might then correct the sampling error. However, if the genes previously examined were a representative sample of the genome, then the complete genomes will only confirm what was already known in terms of both confident and problematic relationships (for an example, see Kuo et al., 2008). Of particular importance here is the possibility of horizontal gene exchange (as opposed to the vertical inheritance assumed by phylogenetics), which will be discussed below.

We therefore need to be realistic about what we can expect from the phylogenetic analysis of sequence data, especially genomic sequences. Of particular importance will be our ability to locate representative genes that are appropriate to the evolutionary timescale being examined, rather than merely the quantity of the data per se. There needs to be a widespread base of people actively collecting a purposive sample of phylogenetically relevant multigene data (Tenter et al., 2002). Without this base, both the taxon and character sampling will be inadequate, in the sense that data will not be available for the critical exemplar taxa. This leads to uncertainties about organismal relationships, and concordance of multiple gene sequences cannot be demonstrated.

8.5 The Practice of Phylogeny Reconstruction

Even though we cannot examine evolutionary history directly in any experimental way, scientists have developed sophisticated methods that allow us to use contemporary data, such as phenotypic or genotypic characteristics, to reconstruct phylogenetic trees. While the logic of phylogeny reconstruction is straightforward, applying this logic in practice, in the face of conflicting evidence, is far from straightforward.

Phylogenetic analysis of molecular sequences (the predominant type of data these days) usually consists of three distinct procedures: (i) sequence alignment, (ii) character coding, and (iii) tree building. These steps are performed in this order, and all three of them need to be fully described for a phylogenetic analysis to be repeatable (in the scientific sense). Also, there are many known artifacts potentially associated with each of these procedures, and they need to be seriously considered in all analyses. Some of these issues and ways of dealing with them are illustrated by Morrison (2006a, 2009a), using pathogens as the examples.

Alignment is the process of establishing the possible homology relationships among the sequence residues (Morrison, 2006b, 2009b). Homology refers to the relationships of features that are shared among taxa due to common ancestry (i.e., they all inherited the feature from their most recent common ancestor). That is, we hypothesize that each of the aligned residues has descended from a common ancestral residue. Unfortunately, the term "homology" has been used historically to refer to a wide variety of concepts, and it is important to understand its strict evolutionary definition (Reeck et al., 1987).

Sequence similarity is often used as strong evidence for potential homology, and this is the basis of all automated alignment procedures (i.e., computer programs). However, sequence similarity decreases rapidly as taxa become evolutionarily more distant, so that processes causing sequence length variation become more probable (such as duplication, translocation, deletion, and insertion). Under these circumstances, similarity cannot be treated as homology (see Figure 8.3). In evolutionary terms similarity = homology + analogy, and analogy (chance similarity due to parallelism, convergence, or reversal) increases with increasing evolutionary distance. This exacerbates the problems of poor taxon sampling. It also exacerbates the problems caused by distant outgroups, which can be very difficult to align with the ingroup (see later).

For the degree of sequence similarity that commonly occurs in phylogenetic analyses, automated alignment methods have proved to be inadequate. For this reason, more than three-quarters of phylogeneticists manually intervene in the alignment process (Morrison, 2009c), either by manually adjusting the alignment output by the computer program or by producing a completely manual alignment. This reflects the simple fact that there is not yet any automated procedure capable of producing a multiple sequence alignment that reflects homology. Personal judgment may not be perfect, but at least it can consciously be based on homology as a concept.

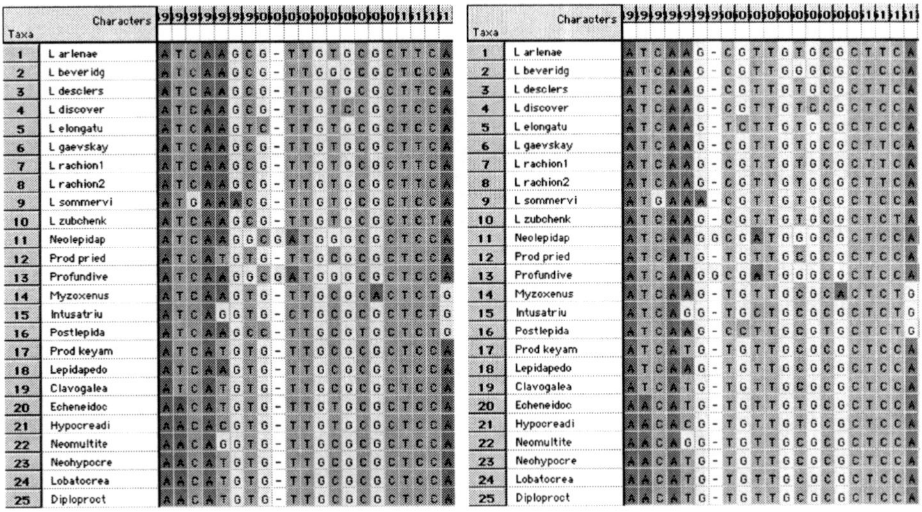

Figure 8.3 Two alternative alignments of part of the nuclear large-subunit ribosomal RNA gene for the Lepocreadioidea (Platyhelminthes). Two of the sequences have an extra nucleotide in length compared to the others, and this length variation needs to be addressed. The alignment on the left is the similarity-based alignment produced by the ClustalX computer program, which simply puts the extra nucleotides as far to the right as possible based on the relative "gap weights." The alignment on the right is based directly on putative homologies, implying in this case that the extra nucleotides have been formed as tandem repeats (or microsatellites). Note that there are considerable differences in the phylogenetic information, concerning the relationships of the taxa "Neolepidap" and "Profundive" to the other taxa, in two of the three columns that differ between the two alignments.
Source: The data are from Bray et al. (2009).

For molecular data types other than sequences, homology often refers to homology of the bands appearing on the gels, and thus to the primers used. For example, AFLP data are based on a set of randomly chosen primers, and we must hypothesize that bands in the same position on the gels are homologous (i.e., they represent amplification by the primer pair of the same genetic sequence). The inability to assess these hypotheses is sometimes listed as a major limitation of nonsequence character data.

Character coding (Müller, 2006; Ochoterena, 2009) is often overlooked as an important step in sequence analyses. Those parts of the sequence alignment involving length variation (where there are so-called gaps) are sometimes considered to be uncertainly aligned, and most computer programs treat gaps as missing data. Furthermore, some regions in the sequence alignment might be considered to be ambiguously aligned across the dataset, even if some subsets of the sequences have been aligned with certainty. These regions are often excluded from the tree-building analysis (Lee, 2001). In both cases, phylogenetic information is lost (see Figure 8.4). This issue can be dealt with by coding the length-variable regions as a

(A) Alignment
Caecincola   CAGCAATGAGTACGGTTATATTGACTTGGC
Siphodera   CAGCAATGAGTACGGTAATATTGACTTGGC
Paracrypto   CAGCAATGAGTACGGTAATGCTGACATGGC
Mitotrema   CAGCAATGAGTACGGTAATGCTGACATGGC
Schistorch   CAGCACTGAGTACGGTAATCTGGAAATGGC
Callohelmi   CAGCATTGAGTACGGT-TTATGGACATGGC
Homal_arma   CAGCATTGAGTACGGT---ATGGACATGGC
Homal_syna   CAGCATTGAGTACGGT---ATGGACATGGC
N_splende1   CAGCATTGAGTACGGT---ATGGACATGGC
N_splende2   CAGCATTGAGTACGGT---ATGGACATGGC

(B) Coding 1— standard
Caecincola   CAGCAATGAGTACGGTTATATTGACTTGGC
Siphodera   CAGCAATGAGTACGGTAATATTGACTTGGC
Paracrypto   CAGCAATGAGTACGGTAATGCTGACATGGC
Mitotrema   CAGCAATGAGTACGGTAATGCTGACATGGC
Schistorch   CAGCACTGAGTACGGTAATCTGGAAATGGC
Callohelmi   CAGCATTGAGTACGGT?TTATGGACATGGC
Homal_arma   CAGCATTGAGTACGGT???ATGGACATGGC
Homal_syna   CAGCATTGAGTACGGT???ATGGACATGGC
N_splende1   CAGCATTGAGTACGGT???ATGGACATGGC
N_splende2   CAGCATTGAGTACGGT???ATGGACATGGC

(C) Coding 2— gaps deleted
Caecincola   CAGCAATGAGTACGGT   ATTGACTTGGC
Siphodera   CAGCAATGAGTACGGT   ATTGACTTGGC
Paracrypto   CAGCAATGAGTACGGT   GCTGACATGGC
Mitotrema   CAGCAATGAGTACGGT   GCTGACATGGC
Schistorch   CAGCACTGAGTACGGT   CTGGAAATGGC
Callohelmi   CAGCATTGAGTACGGT   ATGGACATGGC
Homal_arma   CAGCATTGAGTACGGT   ATGGACATGGC
Homal_syna   CAGCATTGAGTACGGT   ATGGACATGGC
N_splende1   CAGCATTGAGTACGGT   ATGGACATGGC
N_splende2   CAGCATTGAGTACGGT   ATGGACATGGC

(D) Coding 3—indels informative
Caecincola   CAGCAATGAGTACGGTTATATTGACTTGGC 00
Siphodera   CAGCAATGAGTACGGTAATATTGACTTGGC 00
Paracrypto   CAGCAATGAGTACGGTAATGCTGACATGGC 00
Mitotrema   CAGCAATGAGTACGGTAATGCTGACATGGC 00
Schistorch   CAGCACTGAGTACGGTAATCTGGAAATGGC 00
Callohelmi   CAGCATTGAGTACGGT?TTATGGACATGGC 10
Homal_arma   CAGCATTGAGTACGGT???ATGGACATGGC 11
Homal_syna   CAGCATTGAGTACGGT???ATGGACATGGC 11
N_splende1   CAGCATTGAGTACGGT???ATGGACATGGC 11
N_splende2   CAGCATTGAGTACGGT???ATGGACATGGC 11

Figure 8.4 Alignment of part of the nuclear large-subunit ribosomal RNA gene for the Lepocreadioidea (Platyhelminthes). The alignment (A) has several taxa with a gap that might be phylogenetically informative, and which can be coded in any of several ways that do not represent the same phylogenetic information (B—D). Most phylogeny programs treat the gaps as missing data (B), so that each alignment column independently contributes information only for those taxa with nucleotides in that column. Here, the gaps are not

set of independent characters, which are then included in the tree-building analysis (Müller, 2006).

A cautionary note is warranted here. When dealing with nonmolecular character data, it is common to decide a priori which characters will be sampled and which ones will not. However, when collecting molecular data, this only applies to the choice of genes to be sequenced or to the primers to be used. It does not apply to the actual data collected. This leaves the experimenter open to choose to include or exclude the observations at will *after* the data have been collected. This applies when we decide to exclude characters that cannot be aligned unambiguously, alignment positions that appear to be overly variable or saturated (such as third-codon positions), or even simply positions where gaps have been introduced into the alignment (Morrison, 2006a). Although there are objective criteria for deleting regions of variable or ambiguous alignment in phylogenetic analyses (Castresana, 2000), a posteriori data exclusion should be treated with caution, as it has the obvious potential to introduce bias as well as to alleviate it.

Tree building is the third step of a phylogenetic analysis, and it simply displays the information obtained from the sequence alignment and coding steps as a branching diagram (Morrison, 2006a). That is, conceptually all it should do is change the tabular data (the alignment) into a picture of the data (the tree), all of the hard work having been done in the previous two steps. In practice, it is rarely this simple.

A number of different types of analysis have been developed, based on different mathematical optimality criteria. Some of these are based on estimated genetic distances while others are based directly on the characters, such as parsimony, likelihood, and Bayesian analysis. The latter try to maximize the amount of inferred homology on the tree (or minimize the amount of inferred homoplasy) as part of their optimality criterion, which gives them a theoretical advantage (and one that also appears in practice). Choosing among such methods is discussed later.

Unfortunately, different tree-building methods often add to the tree artifactual information that does not reflect evolutionary history. For example, substitutional saturation is an almost universal problem (due to superimposed substitutions; Xia et al., 2003) and compositional heterogeneity is a recurring problem (e.g., $A + T$ bias or codon bias; Jermiin et al., 2004), as are juxtaposed long and short branches (resulting in what is known as long-branch attraction; Bergsten, 2005). It is worth noting that many of the currently recognized practical problems (e.g., long-branch attraction, compositional bias, and saturation) are merely specific examples of how analogy appears in molecular biology. Analogy exacerbates the problems caused by poor taxon sampling and distant outgroups.

Figure 8.4, (Caption Continued) treated as indels, but as missing information.
Alternatively, many researchers simply delete alignment columns that contain gaps (C), thus losing all of the potential phylogenetic information. Here, the indels do not exist at all. Other researchers code the gapped columns as separate indels (D). Here, extra characters are added that represent the sharing of the indel patterns among the taxa, which are then phylogenetically informative when analyzed.
Source: The data are from Bray et al. (2009).

While it is impossible to make generalizations about the phylogenetic problems of pathogens, because the different groups are not closely related, there are recurring themes. For example, the main cause of substitutional saturation and long-branch attraction is large evolutionary distances among the taxa, which is a common situation for unicellular organisms such as most pathogens. Similarly, nucleotide composition biases reflect mutational as well as selective forces, so that AT-richness often characterizes mutation-prone genomes such as those of intracellular bacteria, although there are also bacteria (such as the Actinobacteria) that are GC rich instead. Nucleotide bias is also associated with the parasitic lifestyle, such as in the AT-richness of *Plasmodium falciparum* (Apicomplexa), where it is presumably advantageous as it permits rapid genetic selection in response to survival threats.

Computationally, artifacts arise because one or more of the assumptions of the analysis have been violated. All data analyses are based on some form of underlying model, whether explicit or implicit, which specifies the assumptions that need to be met by the data in order for the results of the analyses to be reliable (Penny et al., 1994). The choice among phylogenetic models should be quantitatively assessed rather than arbitrarily chosen (Johnson and Omland, 2004), as this is the only proactive way of dealing with artifacts. These issues often can be dealt with by deleting length-variable regions and autapomorphies from the alignment, or by choosing appropriate evolutionary models for the analysis (Morrison, 2006a).

The most basic assumption of the models is that the model does not change through time along the evolutionary lineages (i.e., in different subtrees). If this is so, then mathematically the model is said to be stationary. Biologically, stationarity is an unlikely assumption, because the physical constraints on the macromolecule coded for by the gene are likely to have varied through time, and so the DNA sequence is expected to have been subjected to temporal variation as well. Suggestions have recently been made that allow for temporal variation in parameters of likelihood models (see Morrison, 2006a; Gascuel and Guindon, 2007). Unfortunately, few of the current suggestions have yet been incorporated into the most commonly used computer programs, mainly because they do not fit easily as extensions of the current simple models.

Phylogenetic analysis of all organisms is usually treated as being rather similar, except for viruses and perhaps bacteria. Otherwise, the differences between different pathogen groups are quantitative rather than qualitative. Some groups have certain genotypic characteristics more strongly than do others, and these will thus affect the analyses to varying degrees. Bacteria often are subject to horizontal gene flow of some sort, as well as hierarchical inheritance, and this can confound phylogenetic inferences. This is discussed in more detail in a later section. For viruses, it is often possible to study the genotypic changes occurring during the course of infection from serial samples.

The phylogenetic techniques that are presented here were developed to investigate what are sometimes called "well-behaved" evolutionary problems, where historical relationships are typically represented by a bifurcating tree with a small number of taxa appearing as the terminal branches. In viral epidemiology the picture can be more complex than this. Even if there is a single underlying phylogenetic tree, it may have thousands of branches, and many of these may have multifurcations. Furthermore, when the evolution of a virus is studied by serial sampling in patients, data will be available for taxa distributed throughout the tree, not just for the taxa at the terminal branches. This is particularly true for RNA viruses, which have a very high substitution rate, and whose molecular evolution may be up to two orders of magnitude more rapid than that of eukaryotic or prokaryotic genes (Holmes, 2009). Suitable methods for the phylogenetic analysis of serial samples are currently under development (e.g., Rodrigo et al., 2007; Hasegawa et al., 2009).

8.6 Choosing a Method

It is possible to perform all three procedures of a phylogenetic analysis (sequence alignment, character coding, and tree building) simply by choosing some popular computer programs and then using the default parameter values of those programs. For example, one strategy popular in the literature is to choose Clustal for alignment, to ignore any explicit coding, and then to choose PAUP* for tree building. Unfortunately, this is a very naïve approach, because it does not consider the possible unsuitability of the analyses for the specific dataset at hand, which may lead to results that are artifacts (e.g., Roger and Hug, 2006).

This notwithstanding, several "analysis pipelines" have appeared recently, mostly aimed at microbiologists, which do indeed combine several computer programs together to perform a single phylogenetic analysis at the press of a button. These include BIBI (Devulder et al., 2003), PhyloGena (Hanekamp et al., 2007), WASABI (Kauff et al., 2007), AMPHORA (Wu and Eisen, 2008), and ASAP (Sarkar et al., 2008). Probably a better approach is provided by services that allow you to mix and match various programs (e.g., Dereeper et al., 2008; Gouy et al., 2010). Attempts have even been made to provide descriptions of "standard procedures" for phylogenetic analysis (Peplies et al., 2008).

Which phylogenetic method you choose depends both on what you want to do and on the data you have at hand. Phylogenetic analysis is based on assessing unobservable historical patterns. Direct empirical evidence is usually lacking for independently verifying the performance of phylogenetic methods (except where sequences have been sampled through time, as in some experimental and epidemiological studies). Consequently, the only protection that we can have against false conclusions is the quality of the data and the quality of the data analysis. A phylogenetic analysis is only as good as the steps taken to ensure the highest quality of data and to evaluate and use the most appropriate mathematical model for the data analysis.

Unfortunately, in some areas of biology overly simplistic analyses still seem to be the order of the day for many practitioners. In the modern world, however, with the advent of more realistic models of character evolution, phylogenetic analyses need no longer be treated as "black boxes" into which data are fed and from which a tree spontaneously emerges. We need to be aware of what assumptions are made by different analyses and how to interpret the information that comes out. This knowledge will help to choose an appropriate phylogenetic analysis for the data.

This chapter is not the place to review the pros and cons of each method, and this can be found in several books (e.g., Felsenstein, 2004). You will find that there are several important concepts to bear in mind when considering different methodologies: efficiency; the objective function used; the search strategy used (exhaustive, branch-and-bound, heuristic); robustness; power; consistency; reconstruction probability; and falsifiability. The method chosen will probably be a compromise from among these criteria, as no method has yet shown itself to be superior on more than a few of them.

There are two distinct types of error that will affect a phylogenetic analysis: (i) random or stochastic error, and (ii) systematic error. Stochastic error is error that results from sampling. That is, we cannot make a complete inventory of all of the data that could be collected, and so we collect a sample instead. That sample may not be representative of the complete collection of data, and this results in random error. Systematic error, on the other hand, results from mismatches between our goal and our sampling and analytical procedures. That is, we may (unintentionally) collect data from taxa that are inappropriate (e.g., diseased), or choose to analyze the data with an inappropriate evolutionary model. Systematic error is thus associated with the accuracy of the answer (i.e., how close to the truth we get), while random error is associated with the precision with which we can present that answer (i.e., how repeatable it is).

In a phylogenetic study, random error is always expected to occur but we can attempt to reduce its impact, while systematic error is something that we actively strive to avoid if we can. Random error can usually be dealt with by increasing the sample size, either of characters or taxa as appropriate. Systematic error, however, cannot be fixed by increases in sample size because the same bias will exist throughout the genome (Jeffroy et al., 2006). For example, several of the gene trees of the Microsporidia have been shown to suffer problems with long-branch attraction due to fast-evolving lineages (Thomarat et al., 2004), and this is not alleviated by studying whole genomes because these fast-evolving genes occur genome-wide. If systematic bias affects many or most of the genes then the reconstructed organismal tree will be wrong, and adding new genes will not resolve the issue. Similar problems have been reported for whole genomes of the Apicomplexa, where incongruent phylogenetic relationships based on a small number of genes were simply confirmed as incongruent by whole-genome phylogenies (Kuo et al., 2008).

As the number of multigene datasets increases, an important methodological decision is how best to derive the organismal phylogeny from the collection of gene phylogenies (i.e., how to get the species tree from the gene trees). Note that there are actually two separate issues here. First, a tree produced from any one dataset may or may not represent the true history of the taxa in that dataset (e.g., the reconstructed gene tree might not be the true gene tree). Second, even if we have the true tree for the dataset, it still may or may not represent the true history of the taxa (e.g., the gene tree might not be the same as the species tree). Indeed, there are compelling reasons to expect that most gene trees will not match the species tree (Avise and Robinson, 2008). Dealing with both of these issues simultaneously is no mean task.

There are two basic strategies for analyzing combined data from multiple datasets (Morrison, 2006a): (i) combine the data into one set and then produce a single tree from it; and (ii) produce a tree from each of the datasets and then combine these into a single tree. That is, we can do the combining either before or after we do the tree building. The first strategy can be called concatenation (since we concatenate the data) while the second can be called consensus (since we produce a consensus of the trees), although these strategies have been called many different things in the literature (e.g., supermatrix and supertree, respectively). These two strategies may produce mutually contradictory answers, although they often do not, and there is a long history of unresolved debate concerning their relative merits (Rannala and Yang, 2008). Indeed, methods are under constant development to improve upon these approaches by estimating the organismal tree directly rather than indirectly (e.g., Degnan and Rosenberg, 2009; Knowles, 2009; Ren et al., 2009).

8.7 Phylogenetic Trees

The idea of a tree as the appropriate representation of phylogenetic relationships has been with us for 150 years now, and yet is quite clear from the literature that many biologists have still not fully grasped this idea and its consequences (Gregory, 2008). That is, misinterpretation of trees, and the taxon groupings (clades) represented by those trees, is endemic in comparative biology (Baum et al., 2005). Indeed, this failure of "tree thinking" seems to be deep-seated in the general public as well (O'Hara, 1997).

In particular, the distinction between an unrooted tree and a rooted tree is often not made, or is wrongly made, leading to blatantly incorrect interpretations of the trees. For example, authors sometimes write about "branching order" on a tree when the tree has no root, and yet it is the root that determines the order; or they write about "groups" or "clusters" of taxa on an unrooted tree, when it is the root that determines the groupings. These logical contradictions seem to pass unnoticed by the authors, the referees, and the editors of the papers, indicating that we have here a very serious problem.

As far as relationships among organisms are concerned, evolutionary trees are intended to replace the traditional Aristotelian ladder (also known as the Great Chain of Being), which arranges organisms in a linear sequence ending with human beings as the ultimate biological form. Instead, we now use a branching diagram in which humans are merely one twig among many, thus supplanting the traditional anthropocentric viewpoint. Unfortunately, many people seem to be imagining a pine tree (O'Hara, 1997), with a single central axis leading to the "most derived" species and many side branches leading to "lesser" organisms, rather than picturing a continuously branching bush-like structure (Figure 8.5).

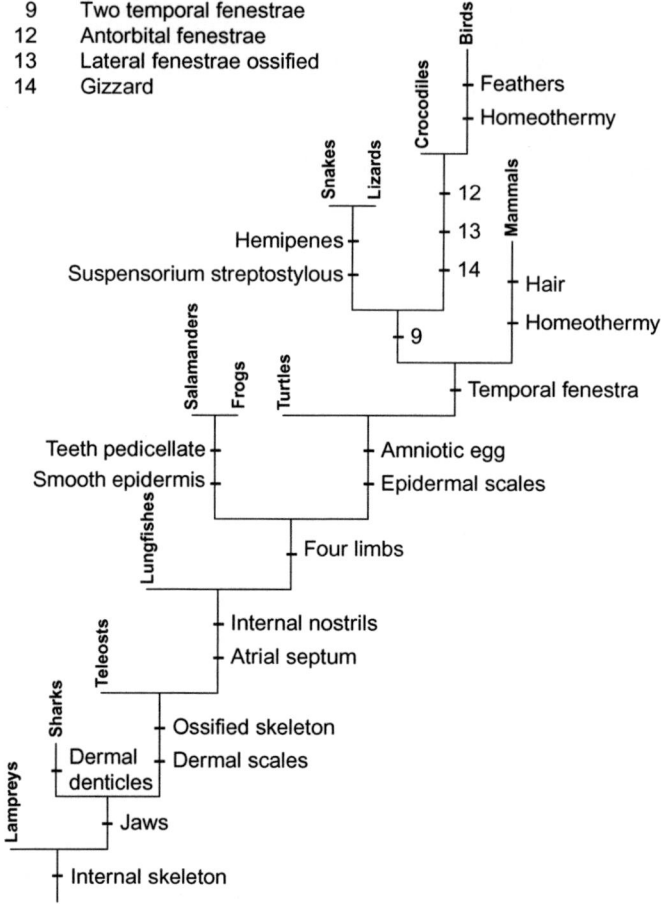

Figure 8.5 Phylogenetic tree showing some of the known characteristics of a selection of extant vertebrate groups. Each of the morphological changes indicates a derived character state that appeared in the ancestor represented by the relevant branch. All of the characters shown are compatible except for homeothermy, which is indicated as having arisen twice (i.e., in two unrelated groups). The branches show only some of the known vertebrates, as no fossil groups have been included (e.g., dinosaurs). A tree such as this shows sister-group relationships; for example, birds are the sister group to crocodiles rather than crocodiles being the ancestor to birds.
Source: Modified with permission from Morrison (2011).

An evolutionary tree obviously must have a time direction (from ancestors to descendants), which is provided by the root. That is, the internal nodes of the tree represent ancestors and the external nodes represent the final descendants. If the taxa were species, each node would then represent a speciation event and the branch lengths would represent the amount of change in the sequences.

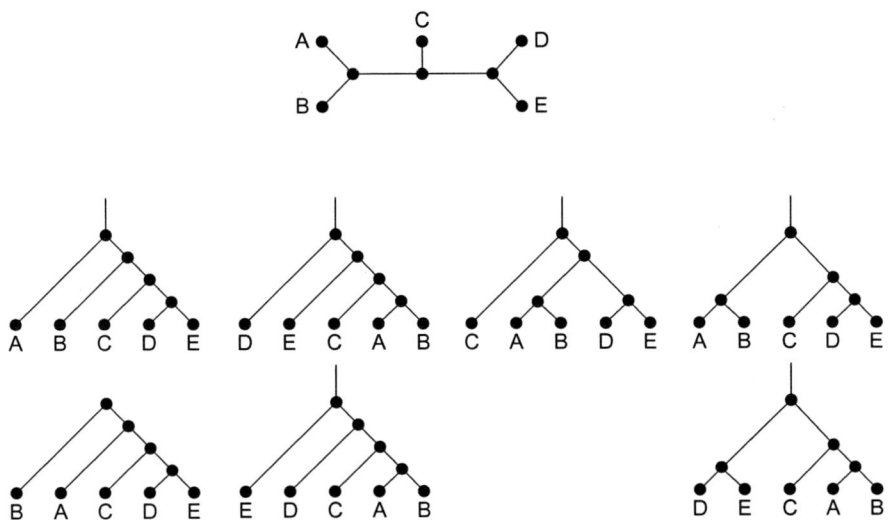

Figure 8.6 An unrooted tree of five taxa (top), which can be rooted on any of its seven branches, yielding seven rooted trees (below). It should be clear which branch of the unrooted tree has been used as the root for each of the seven rooted trees. Thus, there are more rooted trees than unrooted trees because every unrooted tree can potentially be rooted on any of its branches.

An unrooted tree cannot be a picture of evolutionary history because there is no indication of the direction of evolutionary change across the tree (which would be away from the root). However, an unrooted tree can be an important step toward obtaining a picture of evolutionary history. For example, for nine taxa there are 135,135 unrooted binary trees, each of which can be rooted in any one of 15 different places (Figure 8.6), yielding 2,027,025 possible rooted trees. Finding the unrooted tree thus eliminates 2,027,010 of these trees, leaving us with only 15 possible trees. This is clearly a major step, even if we never work out the precise location of the root.

Nevertheless, almost all of the questions being asked by biologists, which they are trying to answer by performing a phylogenetic analysis, can only be answered using a rooted diagram. It is inappropriate to identify evolutionary "groups" of taxa on an unrooted tree (Morrison, 2006a; Wilkinson et al., 2007), because only monophyletic groups (called clades) make any sense in an evolutionary context. A clade includes the most recent common ancestor of the group plus all of its descendants; and so, by definition, a clade cannot be determined from an unrooted tree. An unrooted tree only indicates partitions (or splits) in the collection of taxa. For example, there are three possible ways to split four taxa into partitions of two taxa each, and an unrooted tree will show only one of them. Thus, an unrooted tree contains information that allows us to eliminate possible groups from consideration. However, it does not contain positive information about groups because both of the partitions do not necessarily form evolutionary groups.

Furthermore, relationships among clades are equal, in the sense that each clade is the sister to some other clade and vice versa. Thus, clades cannot be "basal" or "crown'" (Krell and Cranston, 2004), because each single clade branches from some other single clade, rather than each clade being a side-branch from a main stem. Logically, at each speciation event two new species arise, rather than one species producing an extra offshoot species. There is no main stem in an evolutionary tree, but instead there is a series of branches leading to a series of twigs, even if some of the branches do have more twigs than others. Furthermore, neither of the sisters represents the ancestor; instead they share a common ancestor, which may not look like either of them.

Finally, characters change through time (Figure 8.5), and so character states can be either ancestral (the original form) or derived (modified in some descendant). However, clades themselves cannot be either ancestral/primitive ("lower") or derived/advanced ("higher"), as each clade will have a combination of ancestral and derived character states. There is no chain leading from ancestral species to derived species. Instead, each species (or group) is the sister to some other species (or group), with which it shares some characters inherited from their ancestors and from which it differs by some unique characters. Any group that is interpreted to be ancestral is paraphyletic (since it does not contain all of the descendants from the common ancestor) rather than monophyletic, and thus has no phylogenetic relevance.

All of this leads us inevitably to the question of how best to root a phylogenetic tree. For molecular data, there are basically six ways that have been proposed (Huelsenbeck et al., 2002; Morrison, 2006a): (i) a priori polarizing of the character states; (ii) via reversible substitution models; (iii) midpoint rooting; (iv) using the molecular clock, or minimizing tip–root variance; (v) coalescence theory for population samples; and (vi) using an outgroup. Some of these methods have been more popular than the others, and not all of them are equally effective.

Use of an outgroup (vi) is far and away the most widely used method of rooting, and rightly so. The outgroup consists of one or (preferably) more taxa that are not part of the study group (i.e., the ingroup). The root of the tree is then simply the branch that connects the outgroup taxa to the ingroup taxa. The main limitation of this method is the choice of the taxa to be included in the outgroup. For robust phylogenetic analysis (Smith, 1994) the outgroup needs to consist of several members of the sister taxon to the ingroup (i.e., the most closely related group to the ingroup), preferably ones with relatively short branch lengths to the ingroup. Evolutionarily more distant species can end up rooting the ingroup at what is effectively a random location, due to the lack of relevant phylogenetic signal involved in the long branch lengths leading to the outgroup. Alternatively, evolutionarily close species may not be reciprocally monophyletic with the ingroup, due to incomplete separation of their gene flows; this means that there will be multiple "true" root locations on the tree. So, choosing an appropriate outgroup is a balancing act between too close and too distant, even for genomic datasets (de la Torre-Bárcena et al., 2009).

The only way to root the tree of life, which is of some interest when dealing with pathogens, since many of these groups were intimately involved in the origin of life, is to use method (i). This has been a topic of long-standing interest in evolutionary biology (Valas and Bourne, 2009).

It is also worth mentioning that it is possible to bias the presentation of a phylogenetic tree (O'Hara, 1992; Sandvik, 2009), either inadvertently or deliberately. Such biases include: (i) presenting a sequence of contemporary taxa so that an axis passes through the diagram; (ii) the left−right ordering of the taxa at the tips; (iii) the selective pruning of side branches; (iv) the use of paraphyletic groups; and (v) the differential resolution of branches.

8.8 Phylogenetic Networks

The view of phylogenetics described above assumes a hierarchy of bifurcating (or sometimes multifurcating) groups. Indeed, the assumption of a universal Tree of Life hinges upon the process of evolution being tree-like throughout history (Lecointre and Le Guyader, 2006). In eukaryotes, the molecular mechanisms and species-level population genetics of variation mainly do cause a tree-like structure over time, but in prokaryotes they often do not, as there are known to be many mechanisms for genetic exchange that disrupt a genealogical tree.

This has lead to an ongoing discussion about whether bacterial phylogenetics, in particular, should be based on the concept of a tree (Galtier and Daubin, 2008) or not (Bapteste et al., 2009). We have previously used a series to represent biodiversity (the Great Chain of Being) and we have used a tree (the Tree of Life)—does our increased understanding of molecular evolution mean that it is time to find a new representation?

To this end there has been interest in the use of networks rather than trees as the basis for phylogenetic analysis. The intention here is to replace the Darwinian model of a bifurcating tree by a "reticulating tree" (Ragan, 2009), with the reticulations representing evolutionary processes other than lineal descent with modification. Such processes involve gene flow of some sort, including: hybridization, introgression, recombination, horizontal (or lateral) gene transfer, genome fusion, ancestral polymorphism (also called deep coalescence or incomplete lineage sorting), and gene duplication−loss (or hidden paralogy). It is now more than 30 years since the difficulties of fitting bacteria (Sneath, 1975) and hybrids (Bremer and Wanntorp, 1979) into a phylogenetic tree were first aired, but the issues have only recently received widespread attention.

Unfortunately, this field is rather poorly developed at the moment (Morrison, 2010a). Networks that try explicitly to represent evolutionary history (called "evolutionary networks") all have serious restrictions on the types of patterns they can analyze, and on the allowed complexity of those patterns. As noted by Huson et al. (2009), there are many promising directions to follow and rudimentary software implementations, but there is no tool currently available that biologists can routinely use on real data. Of course, an evolutionary network must be rooted in order to form an hypothesized evolutionary history. All of the internal nodes should be (inferred) ancestors and all of the branches should represent inferred evolutionary events (with a direction of transformation). Nodes where two or more lineages converge (a reticulation node) indicate pooling of genetic material; and nodes with one branch coming in and two or more going out (a tree node) represent genetic divergence (see Figure 8.7).

What we have, instead, is a wide array of methods for displaying data conflict in phylogenetic datasets (called "data-display networks"). That is, compatible or congruent data patterns are displayed as a tree, while incompatibilities in the data are displayed as reticulations in the tree (Figure 8.8). Incompatibilities can also arise, in addition to the gene-flow processes listed earlier, from: (i) analogous rather than homologous characters (e.g., parallelism, convergence, reversal); or (ii) methodological issues in data collection (e.g., taxon sampling, character sampling, outgroups) or data analysis (e.g., model mis-specification, choice of optimality criterion). We cannot distinguish, from the network alone, the cause of the character incompatibility, and so the nodes do not necessarily represent ancestors (as they would in a rooted tree), and the branches do not necessarily represent biological character transformations (from ancestor to descendant). Data-display networks are very useful for exploratory data analysis (Morrison, 2010b) or estimating genetic diversity (Minh et al., 2009), but they should not be confused with (or treated as) evolutionary networks.

It is becoming increasingly obvious that the basic biological model for most evolutionary studies is a (relatively well-supported) tree on which is superimposed a (smaller) collection of nontree (reticulation) events; metaphorically, this is a tree obscured by vines. The choice of a tree or a network to display evolutionary history then depends on the extent to which the tree has been obscured by the vines. Since most gene trees are not expected to match the species tree, even when one exists, when is it worthwhile to reconstruct a species tree? Resolving this issue, and devising methods for constructing evolutionary networks, may be the biggest current challenges for bacterial phylogenetics (Galtier and Daubin, 2008; Bapteste et al., 2009; Doolittle, 2009; Koonin, 2009), in particular.

Much of the problem arises from the lack of sexual reproduction and lack of available macrocharacters in prokaryotes, so that molecular mechanisms loom large in their phylogenetic datasets, particularly horizontal gene transfer. Furthermore, sequences of the small-subunit rRNA gene have played the dominant role in microbiology, and one gene phylogeny cannot be used reliably to reconstruct the organismal evolutionary history. The sequences of the small-subunit rRNA gene may well have a tree-like history but that does not automatically entail that the genomes have a similar structure.

The rest of the problem comes from whether we see the Tree of Life primarily as a metaphor (i.e., a model) for the structure of the evolutionary past, or whether it is a specific hypothesis about that structure (i.e., the evolutionary process really does generate a tree). Obviously, there is a tree-like history generated by cell divisions of

Phylogenetic Analysis of Pathogens 225

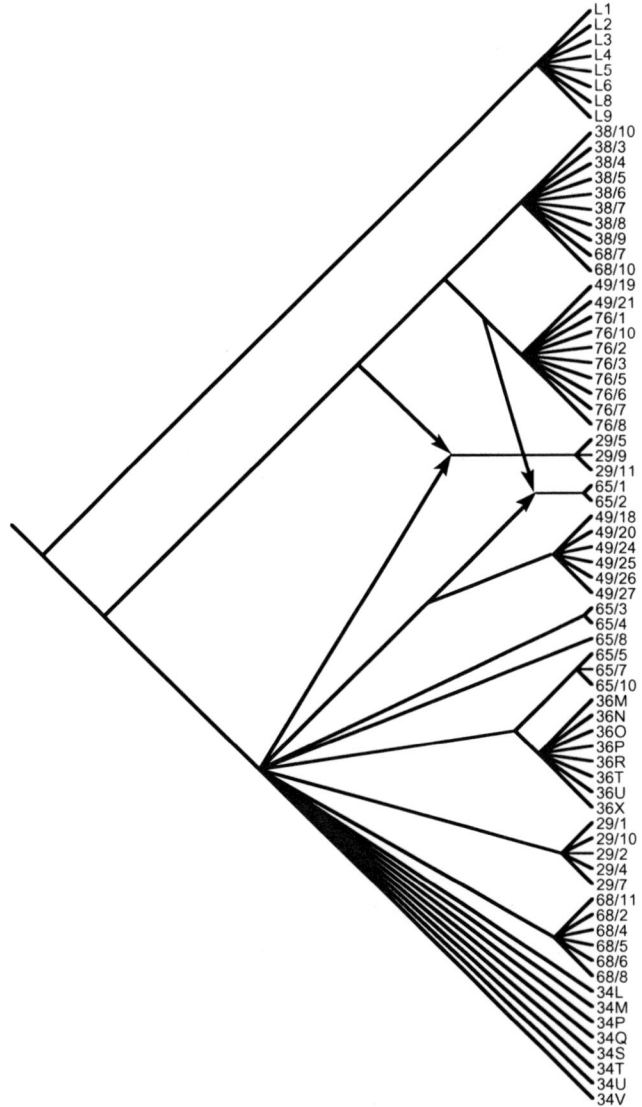

Figure 8.7 Recombination network for 1542 aligned nucleotides from 72 samples of *Dictyocaulus viviparus* (Nematoda). This evolutionary network shows the inferred historical relationships among 64 farm samples and 8 samples from a laboratory isolate (used to root the network, at the left). Most of the samples from each farm are closely related in a simple divergent fashion through time. However, two groups of samples descend from reticulation nodes (indicated by arrows), thus indicating the pooling of two distinct sources of genetic material. The farms involved (29 and 65) may thus have multiple sources of infection.
Source: The data are mitochondrial protein, rRNA and tRNA gene sequences, from Höglund et al. (2006).

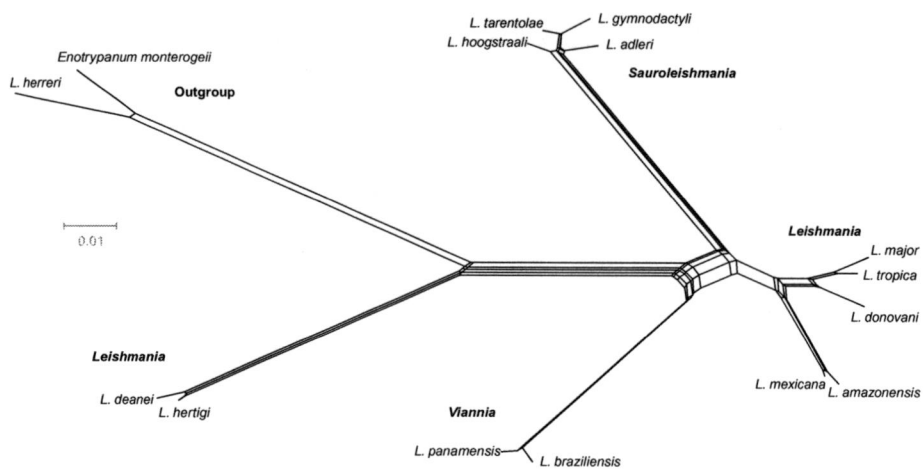

Figure 8.8 Split decomposition analysis of 2207 aligned nucleotides from 15 *Leishmania* species (Kinetoplastida). The subgenera are labeled; and the scale bar represents the split support for the branches. In this data-display network, the length of the branches represents the amount of supporting information and the "thickness" represents the amount of contradiction in the dataset. The network thus displays the conflicting data patterns, particularly the uncertain relationship of subgenus *Viannia* to *Leishmania* and *Sauroleishmania*.
Source: The data are nuclear DNA and RNA polymerase gene sequences, from Croan et al. (1997).

prokaryotes, but is this Tree of Cells the most useful way of organizing our knowledge of biodiversity? Microbiologists seem to have been at times wary of phylogenetic analysis, and much of the history of bacterial classification has unfolded by deliberately ignoring the basic principles that are summarized here (Sapp, 2009). Indeed, it may be that microbiology and phylogeny are incompossible. If so, then microbiologists need another paradigm; but those who object to trees do not yet seem to have one (i.e., they are anti-tree rather than pro-something-else).

References

Atkinson, Q.D., Gray, R.D., 2005. Curious parallels and curious connections—phylogenetic thinking in biology and historical linguistics. Syst. Biol. 54, 513–526.

Avise, J., Robinson, T., 2008. Hemiplasy: a new term in the lexicon of phylogenetics. Syst. Biol. 57, 503–507.

Baldauf, S.L., 2003. Phylogeny for the faint of heart: a tutorial. Trends Genet. 19, 345–351.

Bapteste, E., O'Malley, M.A., Beiko, R.G., Ereshefsky, M., Gogarten, J.P., Franklin-Hall, L., et al., 2009. Prokaryotic evolution and the tree of life are two different things. Biol. Direct 4, 34.

Barbrook, A.C., Howe, C.J., Blake, N., Robinson, P., 1998. The phylogeny of *The Canterbury Tales*. Nature 394, 839.

Barta, J.R., 2001. Molecular approaches for inferring evolutionary relationships among protistan parasites. Vet. Parasitol. 101, 175–186.

Barta, J.R., Thompson, R.C.A., 2006. What is *Cryptosporidium*? Reappraising its biology and phylogenetic affinities. Trends Parasitol. 22, 463–468.

Baum, D.A., Smith, S.D., Donovan, S.S., 2005. The tree-thinking challenge. Science 310, 979–980.

Bergsten, J., 2005. A review of long-branch attraction. Cladistics 21, 163–193.

Bray, R.A., Waeschenbach, A., Cribb, T.H., Weedall, G.D., Dyal, P., Littlewood, D.T.J., 2009. The phylogeny of the Lepocreadioidea (Platyhelminthes, Digenea) inferred from nuclear and mitochondrial genes: implications for their systematics and evolution. Acta Parasitol. 54, 310–329.

Bremer, K., Wanntorp, H.-E., 1979. Hierarchy and reticulation in systematics. Syst. Zool. 28, 624–627.

Bromham, L., 2008. Reading the Story in DNA: A Beginner's Guide to Molecular Evolution. Oxford University Press, Oxford.

Castresana, J., 2000. Selection of conserved blocks from multiple alignments for their use in phylogenetic analysis. Mol. Biol. Evol. 17, 540–552.

Cotton, J.A., Wilkinson, M., 2008. Supertrees join the mainstream of phylogenetics. Trends Ecol. Evol. 24, 1–3.

Croan, D.G., Morrison, D.A., Ellis, J.T., 1997. Evolution of the genus *Leishmania* revealed by comparison of DNA and RNA polymerase gene sequences. Mol. Biochem. Parasitol. 89, 149–159.

Dahlgren, S.S., Gjerde, B., 2008. *Sarcocystis* in moose (*Alces alces*): molecular identification and phylogeny of six *Sarcocystis* species in moose, and a morphological description of three new species. Parasitol. Res. 103, 93–110.

Dahlgren, S.S., Gjerde, B., 2010. The red fox (*Vulpes vulpes*) and the arctic fox (*Vulpes lagopus*) are definitive hosts of *Sarcocystis alces* and *Sarcocystis hjorti* from moose (*Alces alces*). Parasitology 137, 1547–1557.

Degnan, J.H., Rosenberg, N.A., 2009. Gene tree discordance, phylogenetic inference, and the multispecies coalescent. Trends Ecol. Evol. 24, 332–340.

de la Torre-Bárcena, J.E., Kolokotronis, S.O., Lee, E.K., Stevenson, D.W., Brenner, E.D., Katari, M.S., et al., 2009. The impact of outgroup choice and missing data on major seed plant phylogenetics using genome-wide EST data. PLoS One 4, e5764.

Delsuc, F., Brinkmann, H., Philippe, H., 2005. Phylogenomics and the reconstruction of the tree of life. Nat. Rev. Genet. 6, 361–375.

Dereeper, A., Guignon, V., Blanc, G., Audic, S., Buffet, S., Chevenet, F., et al., 2008. Phylogeny.fr: robust phylogenetic analysis for the non-specialist. Nucleic Acids Res. 36, W465–W469.

Devulder, G., Perrière, G., Baty, F., Flandrois, J.P., 2003. BIBI, a bioinformatics bacterial identification tool. J. Clin. Microbiol. 41, 1785–1787.

Dobzhansky, T., 1973. Nothing in biology makes sense except in the light of evolution. Am. Biol. Teach. 35, 125–129.

Doolittle, W.F., 2009. The practice of classification and the theory of evolution, and what the demise of Charles Darwin's tree of life hypothesis means for both of them. Philos. Trans. R. Soc. Lond. B Biol. Sci. 364, 2221–2228.

Doyle, J.J., 1992. Gene trees and species trees: molecular systematics as one-character taxonomy. Syst. Bot. 17, 144–163.

Farris, J.S., 1979. The information content of the phylogenetic system. Syst. Zool. 28, 483–519.

Felsenstein, J., 2004. Inferring Phylogenies. Sinauer Associates, Sunderland, MA.
Galtier, N., Daubin, V., 2008. Dealing with incongruence in phylogenomic analyses. Philos. Trans. R. Soc. Lond. B Biol. Sci. 363, 4023–4029.
Gascuel, O., Guindon, S., 2007. Modelling the variability of evolutionary processes. In: Gascuel, O., Steel, M. (Eds.), Reconstructing Evolution: New Mathematical Concepts and Computational Advances. Oxford University Press, Oxford, pp. 65–107.
Gatesy, J., Desalle, R., Whalberg, N., 2007. How many genes should a systematist sample? Conflicting insights from a phylogenomic matrix characterized by replicated incongruence. Syst. Biol. 56, 355–363.
Gouy, M., Guindon, S., Gascuel, O., 2010. SeaView version 4: a multiplatform graphical user interface for sequence alignment and phylogenetic tree building. Mol. Biol. Evol. 27, 221–224.
Gregory, T.R., 2008. Understanding evolutionary trees. Evol. Educ. Outreach 1, 121–137.
Hall, B.G., 2008. Phylogenetic Trees Made Easy: A How-To Manual, third ed. Sinauer Associates, Sunderland, MA.
Hanekamp, K., Bohnebeck, U., Beszteri, B., Valentin, K., 2007. PhyloGena—a user-friendly system for automated phylogenetic annotation of unknown sequences. Bioinformatics 23, 793–801.
Harvey, P.H., Brown, A.J.L., Maynard Smith, J., Nee, S. (Eds.), 1996. New Uses for New Phylogenies. Oxford University Press, New York.
Hasegawa, N., Sugiura, W., Shibata, J., Matsuda, M., Ren, F., Tanaka, H., 2009. Inferring within-patient HIV-1 evolutionary dynamics under anti-HIV therapy using serial virus samples with vSPA. BMC Bioinformatics 10, 360.
Hedges, R.W., 1972. The pattern of evolutionary change in bacteria. Heredity 28, 39–48.
Hennig, W., 1966. Phylogenetic Systematics. University of Illinois Press, Urbana, IL [Translated by Davis, D.D. and Zangerl, R. from Hennig, W., 1950. Grundzüge einer Theorie der Phylogenetischen Systematik. Deutscher Zentralverlag, Berlin.]
Höglund, J., Morrison, D.A., Mattsson, J.G., Engström, A., 2006. Population genetics of the bovine/cattle lungworm (*Dictyocaulus viviparus*) based on mtDNA and AFLP marker techniques. Parasitology 133, 89–99.
Holmes, E.C., 2009. The Evolution and Emergence of RNA Viruses. Oxford University Press, New York.
Huelsenbeck, J.P., Bollback, J.P., Levine, A.M., 2002. Inferring the root of a phylogenetic tree. Syst. Biol. 51, 32–43.
Huson, D.H., Rupp, R., Berry, V., Gambette, P., Paul, C., 2009. Computing galled networks from real data. Bioinformatics 25, i85–i93.
Jeffroy, O., Brinkmann, H., Delsuc, F., Philippe, H., 2006. Phylogenomics: the beginning of incongruence? Trends Genet. 22, 225–231.
Jermiin, L.S., Ho, S.Y.W., Ababneh, F., Robinson, J., Larkum, A.W.D., 2004. The biasing effect of compositional heterogeneity on phylogenetic estimates may be underestimated. Syst. Biol. 53, 638–643.
Johnson, J.B., Omland, K.S., 2004. Model selection in ecology and evolution. Trends Ecol. Evol. 19, 101–108.
Kauff, F., Cox, C.J., Lutzoni, F., 2007. WASABI: an automated sequence processing system for multigene phylogenies. Syst. Biol. 56, 523–531.
Knowles, L.L., 2009. Estimating species trees: methods of phylogenetic analysis when there is incongruence across genes. Syst. Biol. 58, 463–467.
Koonin, E.V., 2009. Darwinian evolution in the light of genomics. Nucleic Acids Res. 37, 1011–1034.

Krell, F.-T., Cranston, P.S., 2004. Which side of a tree is more basal? Syst. Entomol. 29, 279–281.
Kuo, C.-H., Wares, J.P., Kissinger, J.C., 2008. The apicomplexan whole genome phylogeny: an analysis of incongruence among gene trees. Mol. Biol. Evol. 25, 2689–2698.
Lecointre, G., Le Guyader, H., 2006. The Tree of Life: A Phylogenetic Classification. Belknap Press, Cambridge, MA.
Lee, M.S.Y., 2001. Unalignable sequences and molecular evolution. Trends Ecol. Evol. 16, 681–685.
Lemey, P., Derdelinckx, I., Rambaut, A., Van Laethem, K., Dumont, S., Vermeulen, S., et al., 2005. Molecular footprint of drug-selective pressure in a Human Immunodeficiency Virus transmission chain. J. Virol. 79, 11981–11989.
Lemey, P., Salemi, M., Vandamme, A.-M. (Eds.), 2009. The Phylogenetic Handbook: A Practical Approach to Phylogenetic Analysis and Hypothesis Testing, second ed. Cambridge University Press, Cambridge, UK.
Lipo, C.P., O'Brien, M.J., Collard, M., Shennan, S.J. (Eds.), 2006. Mapping Our Ancestors: Phylogenetic Approaches in Anthropology and Prehistory. AldineTransaction, New Brunswick, NJ.
Mace, R., Holden, C.J., Shennan, S.J. (Eds.), 2005. The Evolution of Cultural Diversity: A Phylogenetic Approach. UCL Press, London.
Minh, B.Q., Klaere, S., von Haeseler, A., 2009. Taxon selection and split diversity. Syst. Biol. 58, 586–594.
Morrison, D.A., 2006a. Phylogenetic analyses of parasites in the new millennium. Adv. Parasitol. 63, 1–124.
Morrison, D.A., 2006b. Multiple sequence alignment for phylogenetic purposes. Aust. Syst. Bot. 19, 479–539.
Morrison, D.A., 2008. Prospects for elucidating the phylogeny of the Apicomplexa. Parasite 15, 191–196.
Morrison, D.A., 2009a. Evolution of the Apicomplexa: where are we now? Trends Parasitol. 25, 375–382.
Morrison, D.A., 2009b. A framework for phylogenetic sequence alignment. Plant Syst. Evol. 282, 127–149.
Morrison, D.A., 2009c. Why would phylogeneticists ignore computerized sequence alignment? Syst. Biol. 58, 150–158.
Morrison, D.A., 2010a. Phylogenetic networks in systematic biology (and elsewhere). In: Mohan, R.M. (Ed.), Research Advances in Systematic Biology. Global Research Network, Trivandrum, India, pp. 1–48.
Morrison, D.A., 2010b. Using data-display networks for exploratory data analysis in phylogenetic studies. Mol. Biol. Evol. 27, 1044–1057.
Morrison, D.A., 2011. Introduction to Phylogenetic Networks. Dystenium LLC, New York.
Müller, K.F., 2006. Incorporating information from length-mutational events into phylogenetic analysis. Mol. Phylogenet. Evol. 38, 667–676.
Nei, M., Kumar, S., 2000. Molecular Evolution and Phylogenetics.. Oxford University Press, New York.
Ochoterena, H., 2009. Homology in coding and non-coding DNA sequences: a parsimony perspective. Plant Syst. Evol. 282, 151–168.
O'Hara, R.J., 1992. Telling the tree: narrative representation and the study of evolutionary history. Biol. Philos. 7, 135–160.
O'Hara, R.J., 1997. Population thinking and tree thinking in systematics. Zool. Scr. 26, 323–329.

Penny, D., Lockhart, P.J., Steel, M.A., Hendy, M.D., 1994. The role of models in reconstructing evolutionary trees. In: Scotland, R.W., Siebert, D.J., Williams, D.M. (Eds.), Models in Phylogeny Reconstruction. Clarendon Press, Oxford, pp. 211–230.

Peplies, J., Kottmann, R., Ludwig, W., Glöckner, F.O., 2008. A standard operating procedure for phylogenetic inference (SOPPI) using (rRNA) marker genes. Syst. Appl. Microbiol. 31, 251–257.

Platnick, N.I., Cameron, H.D., 1977. Cladistic methods in textual, linguistic, and phylogenetic analysis. Syst. Biol. 26, 380–385.

Ragan, M.A., 2009. Trees and networks before and after Darwin. Biol. Direct 4, 43.

Rannala, B., Yang, Z., 2008. Phylogenetic inference using whole genomes. Annu. Rev. Genomics Hum. Genet. 9, 217–231.

Reeck, G.R., de Haën, C., Teller, D.C., Doolittle, R.F., Fitch, W.M., Dickerson, R.E., et al., 1987. "Homology" in proteins and nucleic acids: a terminology muddle and a way out of it. Cell 50, 667.

Ren, F., Tanaka, H., Yang, Z., 2009. A likelihood look at the supermatrix–supertree controversy. Gene 441, 119–125.

Rodrigo, A., Ewing, G., Drummond, A., 2007. The evolutionary analysis of measurably evolving populations using serially sampled gene sequences. In: Gascuel, O., Steel, M. (Eds.), Reconstructing Evolution: New Mathematical Concepts and Computational Advances. Oxford University Press, Oxford, pp. 30–61.

Roger, A.J., Hug, L.A., 2006. The origin and diversification of eukaryotes: problems with molecular phylogenetics and molecular clock estimation. Philos. Trans. R. Soc. Lond. B Biol. Sci. 361, 1039–1054.

Sandvik, H., 2009. Anthropocentrisms in cladograms. Biol. Phil. 24, 425–440.

Sanson, G.F.O., Kawashita, S.Y., Brunstein, A., Briones, M.R.S., 2002. Experimental phylogeny of neutrally evolving DNA sequences generated by a bifurcate series of nested polymerase chain reactions. Mol. Biol. Evol. 19, 170–178.

Sapp, J., 2009. The New Foundations of Evolution: On the Tree of Life. Oxford University Press, New York.

Sarkar, I.N., Egan, M.G., Coruzzi, G., Lee, E.K., DeSalle, R., 2008. Automated simultaneous analysis phylogenetics (ASAP): an enabling tool for phylogenomics. BMC Bioinformatics 9, 103.

Smith, A.B., 1994. Rooting molecular trees: problems and strategies. Biol. J. Linn. Soc. 51, 279–292.

Sneath, P.H.A., 1975. Cladistic representation of reticulate evolution. Syst. Zool. 24, 360–368.

Stevens, J.R., Schofield, C.J., 2003. Phylogenetics and sequence analysis: some problems for the unwary. Trends Parasitol. 19, 582–588.

Tenter, A.M., Barta, J.R., Beveridge, I., Duszynski, D.W., Mehlhorn, H., Morrison, D.A., et al., 2002. The conceptual basis for a new classification of the *Coccidia*. Int. J. Parasitol. 32, 595–616.

Thomarat, F., Vivares, C.P., Gouy, M., 2004. Phylogenetic analysis of the complete genome sequence of *Encephalitozoon cuniculi* supports the fungal origin of microsporidia and reveals a high frequency of fast-evolving genes. J. Mol. Evol. 59, 780–791.

Valas, R.E., Bourne, P.E., 2009. Structural analysis of polarizing indels: an emerging consensus on the root of the tree of life. Biol. Direct 4, 30.

Wiens, J.J., Fetzner, J.W., Parkinson, C.L., Reeder, T.W., 2005. Hylid frog phylogeny and sampling strategies for speciose clades. Syst. Biol. 54, 778–807.

Wilkinson, M., McInerney, J.O., Hirt, R.P., Foster, P.G., Embley, T.M., 2007. Of clades and clans: terms for phylogenetic relationships in unrooted trees. Trends Ecol. Evol. 22, 114–115.
Wu, M., Eisen, J.A., 2008. A simple, fast, and accurate method of phylogenomic inference. Genome Biol. 9, R151.
Xia, X., Xie, Z., Salemi, M., Chen, L., Wang, Y., 2003. An index of substitution saturation and its application. Mol. Phylogenet. Evol. 26, 1–7.
Yang, Z., 2006. Computational Molecular Evolution. Oxford University Press, Oxford.

9 Evolutionary Responses to Infectious Disease

*Gregory Cochran and Henry Harpending**

Department of Anthropology University of Utah, Salt Lake City, UT, USA

9.1 Introduction

Humans have a typical mammalian immune system, with three components: external barriers, the innate immune system, and the adaptive immune system. External barriers include physical barriers such as skin and mucosal surfaces as well as antibacterial secretions like lysozyme and defensins.

The second component, the innate immune system, responds quickly to attack but is not tailored to the specific attacking pathogen. It has built-in features that recognize and attack pathogens using pattern recognition receptors, which are triggered by characteristic molecular signatures associated with certain classes of disease organisms, such as the lipopolysaccharides found in the cell walls of Gram-negative bacteria. It also includes some very specific defenses against particular pathogens. The innate immune system knows that certain molecules are danger signs, but it does not learn. Its knowledge was generated by natural selection rather than individual experience, rather like the fear of snakes that is especially easily invoked in humans.

The third component, the adaptive immune system, can tailor responses to a specific pathogen and retains the ability to rapidly respond to future visitations by that pathogen. It acts as an individual immunological memory.

The human immune system has defenses which work against practically any conceivable pathogen (the adaptive system), defenses that are pre-tuned against traditional classes of pathogens such as bacteria, RNA and DNA viruses, protozoa, parasitic worms, and fungi, and a number of other defenses that are specifically aimed at particular pathogens. For example, there are genes that defend against Herpes simplex (Sancho-Shimizu et al., 2007), Epstein–Barr virus (Rigaud et al., 2006), and certain dangerous strains of human papilloma virus (Ramoz et al., 2002). People with two broken copies of such a gene almost inevitably have serious or lethal infections of the associated pathogen.

Some human defenses protect against regional pathogens. Mainly that means malaria, which we will discuss at length later, but we also know of a built-in defense that is effective against *Trypanosoma brucei*, the cause of a common

*E-mail: harpend@xmission.com

African trypanosomal infection known as nagana in livestock. The molecule is APOL-1 (Pays et al., 2006), a lipoprotein: normally such molecules transport lipids, but this one also plays a role in killing trypanosomes. APOL-1 has, we think, played a role in human evolution. First, it is a pretty clear signal that human evolved in sub-Saharan Africa, since that is where nagana is found. It probably says something about our preferred habitats in the early days of human evolution, since the tsetse flies carrying the disease are mainly found near rivers and lakes, in forests along watercourses, and in wooded savanna. Tsetse flies are not common in arboreal environments, and interestingly, chimpanzees do not have this defense (Puente et al., 2005), although gorillas do.

Innate human nagana defenses are not just another signpost pointing to Africa. They also suggest possible ecological explanations for some patterns in the fossil record, for example, for the fact that Neanderthal remains have never been found in Africa. Given several hundred thousand years living in areas without tsetse flies, Neanderthals probably had nonfunctional versions of this defense, due to relaxed selection and mutation accumulation. The loss of this defense (and most likely of other defenses against specific African pathogens) would have made it almost impossible to expand back into Africa. Since Africa was pathogen-rich compared to Europe and southwest Asia, the main Neanderthal homeland, the situation was not symmetrical. Hominids could leave Africa, but they could not go home again.

Many known or suspected genetic responses to infectious disease in human are loss of function mutations, damaged or broken genes. This is the case for nearly all the known malaria adaptations: we will discuss falciparum malaria in some detail later. Other protective variants that are known or suspected are also damaged versions of the wild type. Most tropical Africans have the Duffy-negative chemokine receptor, which confers protection against vivax malaria. For many years it has been thought that the deletion had no negative consequences, but recently Reich et al. (2009) have shown that the Duffy-negative allele itself causes a significant reduction in neutrophil count in people of African descent. This reduced white cell count is almost certainly not a good thing for its bearers.

9.2 Parasites as Our Friends

Of course parasites can also be helpful. Invading modern humans may have carried diseases that hit archaic humans harder during the original modern human diaspora out of Africa. This effect helped the Europeans expand into the New World. Venereal diseases would be good candidates since they can propagate successfully at the low human densities typical of hunter-gatherers. Directly transmitted crowd diseases, on the other hand, would not have been particularly devastating to the low-density archaic populations encountered in the Levant and Europe during that early expansion (as opposed to the high-density agricultural populations in the New World encountered by the Spanish).

Parasites may also have contributed to human success in hunting. We picked up three species of taeniid tapeworms from African predators (another sign of our origin) several hundred thousand years ago (Hoberg et al., 2001), one from hyenas and two from lions. These tapeworms, like many other parasites, have a complex life cycle, forming cysts in herbivores (the intermediate host) and reaching maturity in the carnivorous definitive host. Obviously their interests conflict with those of their intermediate hosts, since they benefit when their host is eaten.

In such a situation, there is an evolutionary incentive for parasites to manipulate the behavior of their intermediate host, for example, by making them easier to catch (Lagrue and Poulin, 2010). This may be the case for toxoplasma, a protozoan that uses many herbivores as intermediate hosts and cats as the definitive host. Toxoplasma has been shown to cause fearlessness in rats and mice (Berdoy et al., 2000): who benefits? There is evidence that *Echinococcus*, another taeniid tape worm with canid definitive hosts, increases predation on its intermediate hosts (e.g., moose). Those human tapeworms may have played an important role in human hunting success, particularly in the olden days when human weapons and hunting skills were far less sophisticated than those used by contemporary hunter-gatherers. Before agriculture, those tapeworms used wild pigs and ungulates as intermediate hosts. Now they cycle through domesticated pigs and cows, suggesting another way in which those parasites could have aided humans by fostering domestication (Ivy Smith, personal communication).

Wild boars are quite formidable, but the auroch (the wild ancestor of domesticated cattle) was simply terrifying, being 2 m high at the shoulder and weighing over a ton. Domestication sounds difficult and dangerous—but it might have been easier if a parasite was, for its own reasons, reducing the aurochs' fear of humans.

9.3 Demography and Parasites

Pathogen dynamics can have a major influence on long-term demographics—and the other way around, of course. Pathogens typically require a minimum number of hosts in fairly close proximity (called the critical community size) in order to survive. Consider measles: it is infectious for no more than 10 days, and survivors have lifelong immunity. Clearly measles can only flourish in a situation where there is a steady supply of fresh, never-infected hosts; that is to say, children. Because of these facts, measles has a critical community size of roughly a quarter of a million people: it could not have existed in its present form back in hunter-gatherer times, since there were no such large population concentrations (Black, 1966).

At the opposite extreme, chickenpox, after infecting children, lingers in nerve ganglia. It often recurs much later in life as shingles, which cause excruciating pain. Children can catch chickenpox from their grandfather's case of shingles. So, due to its persistence and ability to wait, chickenpox has a critical community size around 1000 (Black et al., 1974).

These facts about infectious disease imply certain things about our ancestral demographics. For one thing, a population crash would have usually been followed by a boom, partly because resources become more abundant in such situations, but also because infectious diseases become less important at low population densities. A mega-crash, one in which humans had a brush with extinction, could thus have had a silver lining: some human-specific parasite could have gone extinct. If that parasite had imposed a heavy fitness burden, humans would have flourished after the crash. Something similar (a bottleneck in space rather than time) happens sometimes when a species colonizes a new continent—the settler population is too small to carry along key parasites and thrives to a surprising degree in its new home.

Africa is rich in human pathogens. Since we originated there, African pathogens have had a long time to adapt to humans and other primates. We mentioned that populations such as Neanderthals that spent a long time outside of Africa probably lost defenses that would have been necessary in Africa, and thus could not go back. The other side of this coin is that those vigorous defenses against African pathogens had costs, costs that were no longer necessary in cooler climates. Leaving Africa may have had substantial payoffs, first for archaic humans in Eurasia and later for anatomically modern humans.

9.4 Agriculture

The biggest demographic change ever experienced by humans was the population explosion made possible by the development of agriculture. Our numbers increased by factors of 50–100, which had a fundamental (and highly unpleasant) impact on human infectious disease. Pathogens that already infected humans became more common and had greater impacts on fitness, while new pathogens arose that could only spread in high-density populations—crowd diseases. We acquired most of these crowd diseases from other animals. Some originated in the animals we domesticated, while a number of others came from African primates. Some probably evolved from older human pathogens moving into newly available ecological niches.

The human genome responded to the new disease pressures, and we have observed the resulting changes in many components of the immune system. The 35delG mutation of connexin-26 causes deafness in homozygotes, but also changes characteristics of the skin (thicker) and sweat (saltier): it may protect against infections of the skin such as erysipelas (Meyer et al., 2002). It is also a common cause of deafness in homozygotes. There is evidence of selection on a number of genes in the innate immune system such as CR-1 (a malaria defense) and some of the Toll-like receptors (TLRs) that recognize characteristic pathogen molecules. Some changes, such as the mutations causing familiar Mediterranean fever (Chae et al., 2006) and alpha-1-antitrypsin deficiency (Lomas, 2006), loosen protective restrictions on some of the more aggressive components of the immune system—you might compare these to unleashing the police, always a dangerous thing to do. There have been changes in the adaptive immune system as well, particularly in the major histocompatibility complex (MHC). We have recognized many of these changes because they cause serious Mendelian diseases that would hardly have reached such high frequencies unless there was some form of heterozygote advantage. Genomic scans have discovered other adaptive changes that do not have such high costs.

It is an odd fact that we seem to see fewer of these expensive disease defenses in East Asia, particularly considering only those that defend against something other than falciparum malaria. We know of no obvious reason why this should be so: conceivably it might be a result of looking harder at European genetics (ascertainment bias) but right now it is something of a puzzle.

9.5 Some Lessons from Malaria

The case of malaria illustrates a number of general principles about the relationship between infectious disease, biological evolution, and social evolution in humans. We will discuss aspects of malaria biology in some detail, but much the same story could be told for other infectious diseases of humans, for example, yellow fever. Falciparum malaria is the most serious human infectious disease and has been the strongest and best understood selective force acting on humans over the past few thousand years. This selection pressure operated in the peoples of the Old World tropics and subtropics—but not elsewhere—and so caused those populations to diverge from the rest of humanity in some ways (Pennington et al., 2009). The most dramatic impact has been the rise to high frequency of many protective alleles. A number of those alleles (the best-studied ones) are overdominant and cause major health problems in homozygotes.

The sickle-cell mutation is the most famous protective allele. Heterozygotes gain substantial protection against falciparum malaria while homozygotes suffer from a severe anemia that is usually lethal in childhood without modern medical treatment (even with treatment, it continues to cause substantial morbidity and mortality). It is the most common lethal mutation in humans, with a gene frequency of around 10% or more in many populations of tropical Africa.

There are a number of similar protective polymorphisms, which are also disease alleles. Some change the hemoglobin molecule, either by amino acid changes (like hemoglobin C and hemoglobin E) or by changing the relative numbers of hemoglobin subunits, as in the thalassemias. Others change the red cell in different ways, interfering with its metabolism (G6PD deficiency) or altering membrane proteins (Melanesian ovalocytosis). It seems likely that falciparum malaria has existed in its present form for 5000 years or less. The approximate age of some of the protective polymorphisms has been determined, and they all seem to be younger than that (Ohashi et al., 2004; Saunders et al., 2005).

Increasingly, researchers are discovering alleles favored by malaria selection that apparently do not cause disease, not even in homozygotes. Some affect familiar targets like the red cell membrane, as we see with glycophorin C (Maier et al., 2003) and type O blood (Rowe et al., 2009). We also see variants of immune system molecules such as CD36 (Pain et al., 2001) and CR-1 (Cockburn et al., 2004).

This trend of discoveries is likely to continue, and we should eventually observe malaria-induced changes in the frequencies of many alleles, even those that have only weak effects on resistance. That is the typical pattern seen in artificial selection experiments. Strong selection for any trait other than fitness itself causes negative changes in other traits—so resistance to malaria has most likely had significant costs. Obviously we know of the costs of many that take the form of Mendelian diseases, but there are likely others as well.

Falciparum malaria's unusual virulence can be explained in part by its means of transmission. Natural selection favors low virulence in many infectious diseases that are spread directly from person to person, since immobilizing the host interferes with transmission. Since malaria is a vector-borne disease (spread by mosquitoes), a severe infection can still spread, even if the host is bedridden (Ewald, 1994). If high parasite blood counts increase the probability of transmission, severe infection may be a favored strategy. The other major kind of human malaria, *Plasmodium vivax*, is also mosquito-borne. It is a fairly serious disease, although much less so than falciparum. It is often found in temperate climates, where it must survive winters without active mosquitoes. In order to do so, it has the ability to hide in liver cells for long periods, in some cases for decades. Of course, this strategy would not work if the host died, which explains why vivax malaria has relatively low virulence. Falciparum mainly exists in warmer climates where mosquito transmission occurs throughout most or all of the year, so that it can keep moving to new hosts.

Malaria has another characteristic that increases its severity. Unlike most other pathogens, malaria repeatedly switches its surface proteins. A single parasite clone has about 60 antigenic variants and thus can stay ahead of the immune system for a year or more, while greater variety in the parasite population as a whole means that a single infection does not result in lasting immunity (Scherf et al., 2008). This defensive tactic of malaria has made the development of an effective vaccine very difficult: no such vaccine is clinically available at this time.

Selection for malaria resistance in humans illustrates several key evolutionary principles: some of these are very well known, while others are not so obvious. First, it shows that adaptive evolution is a continuing process in humans, one that can cause significant changes over historical time and whose direction is not the same in every population.

This may have been especially the case over the Holocene, during which humans experienced substantial climate change and were exposed to the selective pressures associated with agriculture, and greatly increased in number.

Malaria selection is also a clear example of convergent evolution. The protective alleles in Southeast Asia are entirely different from those in Africa: some are different mutations of genes that have produced defensive alleles in Africa (e.g., G6PD deficiency) while others involve different genes. One sees the same thing in artificial selection experiments: the phenotypic changes are similar in different lines experiencing the same selective pressures (people in both Africa and Southeast Asia are resistant to malaria) but the genetic details are in general different. Another point is that strong selection evokes changes in many genes, changes that are concentrated in a few metabolic paths. In this case we know of many polymorphisms that affect the red cell and hemoglobin, as well as a number that result in immunological changes. We have seen arguments that this pattern is somehow unparsimonious: one sweep might be caused by strong selective pressures, but surely not many! But in fact strong selection is likely to cause a number of sweeps—basically, every gene that significantly affects the trait under selection is a candidate for an adaptive mutation.

These convergent adaptations also show us something about the way in which advantageous alleles have spread through populations. Particular protective alleles have spread through much of sub-Saharan Africa, across New Guinea, or throughout the coastal regions of the Mediterranean, but few have managed to cross the Sahara Desert or move between India and Southeast Asia. Strong geographical barriers (and limited time) have prevented high-fitness alleles from spreading to all the places they would have worked, and thus local protective variants took their place. Evolution was faster than gene flow.

Many of these protective alleles are overdominant, since homozygotes suffer from serious disease. Overdominance means that the heterozygote has higher fitness than the homozygote: such alleles never go to fixation. A recessive lethal-like sickle cell is clearly overdominant, but some of the other defensive alleles that do not cause obvious disease may also have lower fitness in homozygotes. A number of domesticated animals also have overdominant alleles that are products of recent strong selection, such as myostatin mutations in whippets and cattle. This may be a general feature of strong selection: many of the sweeping alleles generated by such selection may therefore reach maximum frequencies well under 100%.

Another interesting point comes from a simple thought experiment: there must have been a time when falciparum malaria had not existed for long and protective alleles were as yet rare. In those days, sickle cell heterozygotes (for example) should have had a larger fitness advantage, relative to the population average, than they do today, since in those days the average person had no other protective alleles. Today, on the other hand, someone in Africa who does not carry the sickle cell allele is likely to have a number of other protective alleles: alpha thalassemia, G6PD deficiency, etc. Africans who do not carry sickle cell are still far more resistant to falciparum malaria than northern Europeans or Amerindians. So the fitness advantage of being a sickle cell carrier (which was as high as 20% in recent centuries) must have been even larger thousands of years ago. This means that the rate of growth, and the equilibrium frequency, if overdominant, of every allele that protected against malaria slowed down as time passed, as the population acquired more and more resistance to malaria from other alleles. This effect can also stop a selective sweep short of fixation.

We think that falciparum malaria has had another interesting effect on human evolution, in that it often kept populations below the Malthusian limit—that is, kept population density below the level at which resource limitations would have stopped further growth. In a Malthusian situation, resources are short and individuals compete for them. Selection in that situation favors efficient use of available resources, which would involve improvements in metabolic and work efficiency—basically, farmers who can plow more acres per calorie. It also favors paternal investment. At the limit, you end up with hardworking peasant couples (both father and mother) who can just barely manage to raise enough food for themselves and the two children who replace them in the next generation.

In a sub-Malthusian ecology, where factors like disease and/or violence keep the population well below the subsistence limit, selection pushes in a different direction. Here the limiting factor might be health rather than wealth. Disease resistance in a mate could be more valuable than land, so a father's genetic quality might be important than his provisioning ability. In much of Africa today, women do most of the farm work: this low level of paternal investment is only possible when resources are plentiful. Female self-sufficiency combined with a high value placed on genetic quality favors polygyny (multiple wives), since man's genes are more important than his wages. Polygyny is more common in West Africa than anywhere else.

9.6 Disease and Standard of Living in Preindustrial Societies: A Simple Model

We can elaborate the role of disease in shaping human cultural diversity with a simple model. Disease in a population that would otherwise be Malthusian, that is to say resource limited, can have the effect of reducing the population size, leading to an increase in the standard of living of those who remain. A familiar example is the prosperity and high wages in Europe following the massive human die off with the great plague epidemics.

We start with a small group of 1000 colonizers in an empty environment. Initially the population is at such a low density that there is no competition for resources among people. Births and deaths occur at a constant per person rate. There is no age structure, no youth nor old people, so everyone is subject to the same rates—these assumptions make algebraic models easy and they reflect well what happens in more realistic (but more complicated) models. Plausible generic values for low-technology human populations are 50 births per thousand people per year and 30 deaths per thousand people per year. The difference, 20 per thousand per year, is the intrinsic growth rate, 20 per thousand or 2 per hundred, 2% per year. In the absence of any limitation the population grows according to this rate exactly like money at compound interest. After a generation of 25 years the population size would be $1000 \times (1.02)^{25}$ or 1640 people. This population would double in about 35 years and would double slightly more than 14 times in 500 years to an implausibly large size of nearly 20 million people.

The customary models in population geneticists focus on gene frequency change, and mean fitness, population growth rate, is normalized away in the equations for genetic change. Here we need to acknowledge the demographic consequences of gene change and retain the mean fitness parameter in order to study the interaction of demography and genetics. The mean fitness of 1.02 per year (or 1.64

per generation) occurs at low population densities but gradually declines as the population grows to the carrying capacity. At this limit the mean fitness is just 1.0; the population remains stationary.

What are the long-term implications of this modest rate of growth? The rate of 2% per year is commonplace among human populations yet a growth of 2,000,000% over 500 years seems and is outlandish. Early in the process resources would become scarce and the rate of growth would slow. Assuming the initial colony occupied 100 square miles, the expanded population after five centuries would need to occupy nearly 2,000,000 (two million) square miles, about the area of Argentina or Kazakhstan. This is explosive growth in historical time but it corresponds to everyday population growth today in many low-technology societies. We know that over the long period from the modern human diaspora out of Africa about 45,000 years ago to the industrial revolution about 200 years ago human numbers grew but at long-term rates far below our modest 2% per year. On this long timescale, they hardly grew at all. It is likely that most of the time populations were growing at rates like our 2%, perhaps slower, but that there were frequent catastrophic events like wars, famines, and plagues that cut population sizes back.

9.7 Population Limitation

There is a convenient and standard way to make a model of population limited by resources called the logistic model. This may not be very accurate but it is simple and, given our poor understanding of detailed dynamics, more than good enough. The idea is that there is some carrying capacity K of the environment. Populations below the carrying capacity in size can grow while populations above the carrying capacity decline until they reach K. If we write P_t for population in some generation t and P_{t+1} for population the following generation then simple population growth like compound interest, called geometric, follows this formula:

$$P_{t+1} = P_t \times (1 + R)$$

where R is just the intrinsic growth rate and $(1 + R)$ is the mean fitness. We write $R = 0.641$ in the expression for the intrinsic growth rate since a rate of 2% per year corresponds to growth of 64.1% per generation. The logistic model specifies that the growth rate R is damped by the current ratio of population to carrying capacity:

$$P_{t+1} = P_t \times (1 + R \times (1 - P_t/K))$$

In an empty environment without intraspecific competition, population P is much less than carrying capacity K and population growth is almost the same as the simple geometric case. But as population increases the ratio P/K becomes significant, growth slows down, and eventually stops when population reaches carrying capacity, that is when $P = K$. If the carrying capacity of the environment into which our population moved were 10,000 people, then the population would grow at a decreasing rate to reach 10,000.

What if the carrying capacity is not static but increases with the number of people? For example, we might imagine that more people bring more farmland under cultivation so that K itself changes. It turns out (Cohen, 1995) that nothing much changes if the increase in carrying capacity K is proportionally less than the increase in population P as would happen if the best land were cleared first while lower and lower quality land were subsequently brought under cultivation. The population still approaches a (new, larger) carrying capacity so that as equilibrium is approached population P is equal to carrying capacity. The end result is that the standard of living, by which we mean the ratio of resources to people K/P, is still unity. There are more people but they are not living any better than they did before the new land was cleared. Such a population, limited by resources, is referred to as a Malthusian population.

An interesting variant of this model is to introduce a new source of mortality, perhaps disease or warfare (Armstrong and Gilpin, 1977; Keeley, 1997). In areas of central Africa with high levels of falciparum malaria the cost to fitness of an individual may be around 25%: that means that with malaria an average individual will leave 25% fewer living descendants one generation later. With a growth rate of 20 per thousand per year, an average individual has 1.64 daughters one generation later. If malaria now decreases fitness by 25%, the average individual will only have 75% of 1.64, or 1.23 daughters one generation later. In terms of annual rates, the malaria cuts population growth from 2% to 0.8% per year. (Notice that we count only daughters since our model is of a simple population that does not take into account sexual reproduction.)

Now we can consider the fixed carrying capacity K and examine the consequences for the population and for individual well-being. The algebraic model now becomes (writing M for the extra density-independent death rate, from malaria in our example but also likely to be from violence and local warfare):

$$P_{t+1} = P_t \times (1 + R - RP_t/K - M)$$

We can find the equilibrium population; that is, the population that would remain unchanging in this environment with the extra mortality. We simply set P_{t+1} equal to P_t, rearrange some terms, and find that the new equilibrium is:

$$P = K \times (R - M)/R$$

We substitute our assumed values, an intrinsic growth rate R of 0.64 and an extra mortality rate M of 0.25, to obtain:

$$P/K = 0.39/0.64 \sim 0.61$$

The population now equilibrates at 61% of the old carrying capacity. A more interesting way to summarize what we have found by manipulating the logistic model is in terms of the standard of living, where a value of 1 means the bare subsistence minimum compatible with life and the maintenance and population size and a value of, say, 5, means that there is five times the subsistence minimum amount of resources available to the average person. In our model population, the standard of living is the reciprocal of 0.61 or 1.6. There is more than half again as many resources per person as there were before malaria appeared. What this means on the ground is that people do not have to work very hard to get enough to eat, that there is fruit on the trees for plucking, and that there are not great labor demands on anyone. Those who survive the malaria enjoy a much higher standard of living. Figure 9.1 is a simulation of this process using plausible numbers for a low-technology human population. The population grows 10-fold from 1000 to 10,000 people in 10 generations, then quickly shrinks to the new equilibrium size of 6100 people after the introduction of falciparum malaria.

Clark (2007) points out that the medieval English had a higher standard than the medieval Japanese because there was much more sewage and filth in England and so a heavier burden of disease. This extra disease translated, as in our malaria example, to a lower population density and higher standard of living.

What are the social consequences of this new disease for low-technology human populations? The most important immediate consequence is that there are plentiful resources for everyone and so, following the nature of the creature, males withdraw from subsistence work as they find that they can simply parasitize women for food. In much of central Africa, Oceania, and the Americas, the result is or has been

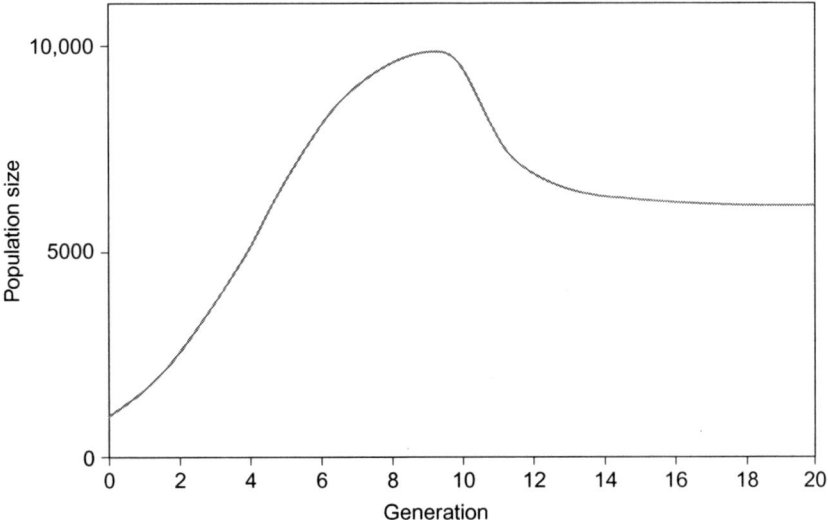

Figure 9.1 Model population size over time of a population of 1000 introduced into an empty area with a carrying capacity of 10,000. After 250 years (10 generations) falciparum malaria appears, and population size quickly drops to about 6500. Generation is on the x-axis, population size on the y-axis.

societies in which men do not do anything very useful and women provision themselves, their children, and the men. The euphemism in economics for this kind of society is "female farming system." Left free of the demands of subsistence the men start hanging out together, perhaps even all moving into a village men's house (not so common in Africa). This leads to local and regional raiding and warfare and an entrenched culture of local violence.

9.8 Disease, Mating, and Reproductive Strategy

Several decades ago Hamilton and Zuk (1982) showed a correlation in North American passerines between parasite burden and gaudiness. Their model was that a slowly changing parasite load leads to parent–offspring correlations in ducks in parasite resistance leading to mating preferences for bright colors as signals of that resistance. Subsequent literature suggests that a similar phenomenon occurs in human societies (Gangestad and Buss, 1993; Low, 1990).

The underlying logic is clear enough, much of it similar to that of the discussion of endemic malaria. Human females, like all mammals, may obtain provisioning from a male for herself and her offspring. This is the pattern in many settled agricultural societies where male subsistence labor is necessary for successful reproduction: these are so-called dad societies. On the other hand, in societies that are far below the Malthusian limit, female mate preferences are more likely to favor males other than good providers. In these societies, often characterized by chronic local raiding and warfare, males protect females from other males. In places with high endemic disease loads then, as in ducks, disease resistance can be heritable so that females prefer to mate with males with "good genes" rather than with males who are "good providers." Of course if females are selecting males with good genes rather than males who are good providers then this is an open door for polygyny. A peasant farmer would have great difficulty provisioning several families but no difficulty at all simply mating with several females. Traditional societies of tropical Africa are indeed mostly polygynous.

There seem to be several important ecological routes to quasi-stable high standard of living non-Malthusian societies. One is warfare: in many well-documented cases, deaths by violence are one quarter to one half of all deaths with the result that human densities remain below any purely subsistence population limit. Much of highland New Guinea appears to be a classic example of this cultural ecosystem. Nothing much seems to have changed there in many millenia. The rough broken terrain contributes to the persistence of this system, since the terrain makes it nearly impossible for an effective constabulary to suppress the chronic violence.

Another route to such a society is through a high burden of endemic disease, as in the malaria example. Malaria is, of course, the classic case, but yellow fever, hookworm, and many others should have similar effects. Much of sub-Saharan Africa is described by economists as "female farming systems," a euphemism for societies where men essentially parasitize women for subsistence while they commit more effort than males do elsewhere to subtle and not-so-subtle male–male competition. As females prefer males who appear healthy (for their "good genes") there may be selection in males to accommodate this preference, that is, there should be sexual selection for appearance. Several recently described myostatin mutants in Africa (Saunders et al., 2006) are probably recently evolved signals of male quality.

The worldwide fall of fertility rates following the Industrial Revolution in northern nations suggests that a stable non-Malthusian world is attainable without the unpleasantness and misery of violence and infectious disease. Meanwhile, it is important to understand that a disease like falciparum malaria not only causes much human misery directly, it also leaves in its wake damaging genetic traces that may take hundreds of generations to dissipate. It is also a social order likely not so well suited for modern industrial society. While fierce, physically attractive males may be favored in a social system where there are adequate resources for females to do all the provisioning, they will not do so well in a subsistence ecology that demands hard agricultural labor and actively sanctions violent behavior.

9.9 Prosperity and the Postindustrial Era Mortality Decline

It has become apparent in the last decade that evolution in humans is an ongoing process that is even speeding up in the face of drastic cultural changes and the large number of humans on earth, each of whom is a potential target for mutations, including favorable mutations (Hawks et al., 2007). Clark (2007) has proposed that genetic change during the millennium or so before the industrial revolution led to essentially a new version of humans that made the revolution possible. The greatest change in human economic history since the origins of agriculture, the industrial revolution of around 1800 released our species from Malthusian constraints as income growth suddenly outstripped population growth. This revolution in human society was accompanied by many changes, and no one has a very clear idea of how they are related to each other. Clark, whose focus is on Great Britain, emphasizes these changes:

- A decline in propensity to violence, especially male violence. In the nations of Europe, homicide rates fell by one to two orders of magnitude in the millennium before 1800 (Eisner, 2001, 2003). In many preindustrial societies, violent males enjoyed a reproductive advantage through greater access to mates, but that advantage turned into a severe disadvantage in settled agricultural societies with effective constabularies.
- Declining interest rates reflecting declining time preference. People were more and more inclined to delay gratification.
- An increasing affinity for work.
- A strong correlation between wealth and reproductive success of males.

At the same time, there were other equally profound changes:

- A striking mortality decline, the cause of which is not well understood. Civil engineering and vaccination are often suggested as causes of this decline but the evidence is not very

clear. The decline may also in part reflect genetic adaptation to new kinds of infectious disease.
- Birth rates fell drastically. The fall of birth rates lagged the fall in death rates by several decades. This is the so-called demographic transition from high mortality and high fertility to low mortality and low fertility. This led to the relationship between wealth and fertility to reverse, as it is today in industrial societies. Today, wealthier people have fewer surviving offspring, and this reversal was the immediate precursor of the popular eugenics movement of the late nineteenth and early twentieth centuries.

What is the role, if any, of the mortality decline in this seismic shift in the nature of society? A possibility is that the decline in pressure from infectious disease freed up much of the genome to evolve in new directions determined by the new social environment. Most of the genetic adaptations to infectious disease that we (think that we) understand involve major or minor damage to genes and to individuals. We discussed the sickle cell adaptation to malaria above with its high purely genetic death toll on homozygotes, but there are many parallel adaptations to falciparum malaria that are known and almost certainly many, many more that we do not yet understand. In aggregate, they impose a large genetic burden on populations with a history of living with malaria.

As the prevalence of infectious disease declined in pre- and post-industrial societies, there may have been widespread relaxation of the selection maintaining these damaging genetic polymorphisms, with the effect of releasing these constraints on the genome and facilitating selection to move phenotypes in different, more favorable directions in the phenotype space.

References

Armstrong, R.A., Gilpin, M.E., 1977. Evolution in a time-varying environment. Science 195, 591.
Berdoy, M., Webster, J.P., Macdonald, D.W., 2000. Fatal attraction in rats infected with *Toxoplasma gondii*. Proc. R. Soc. B. Biol. Sci. 267, 1591–1594.
Black, F.L., 1966. Measles endemicity in insular populations: critical community size and its evolutionary implication. J. Theor. Biol. 11, 207–211.
Black, F.L., Hierholzer, W.J., Pinheiro, F.D., Evans, A.S., Woodall, J.P., Opton, E.M., et al., 1974. Evidence for persistence of infectious agents in isolated human populations. Am. J. Epidemiol. 100, 230–250.
Chae, J.J., Wood, G., Masters, S.L., Richard, K., Park, G., Smith, B.J., et al., 2006. The B30.2 domain of pyrin, the familial Mediterranean fever protein, interacts directly with caspase-1 to modulate IL-1β production. Proc. Natl. Acad. Sci. U.S.A. 103, 9982–9987.
Clark, G., 2007. A Farewell to Alms. Princeton University Press, Princeton.
Cockburn, I.A., Mackinnon, M.J., O'Donnell, A., Allen, S.J., Moulds, J.M., Baisor, M., et al., 2004. A human complement receptor 1 polymorphism that reduces *Plasmodium falciparum* rosetting confers protection against severe malaria. Proc. Natl. Acad. Sci. U.S.A. 101, 272–277.
Cohen, J.E., 1995. How Many People Can the Earth Support? W W. Norton & Company, USA.

Eisner, M., 2001. Modernization, self-control and lethal violence. The long-term dynamics of European homicide rates in theoretical perspective. Br. J. Criminol. 41, 618.
Eisner, M., 2003. Long-term historical trends in violent crime. Crime Justice 30, 83.
Ewald, P.W., 1994. Evolution of Infectious Disease. Oxford University Press, USA.
Gangestad, S.W., Buss, D.M., 1993. Pathogen prevalence and human mate preferences. Ethol. Sociobiol. 14, 89–96.
Hamilton, WD, Zuk, M, 1982. Heritable true fitness and bright birds: a role for parasites? Science 218, 384.
Hawks, J., Wang, E.T., Cochran, G.M., Harpending, H.C., Moyzis, R.K., 2007. Recent acceleration of human adaptive evolution. Proc. Natl. Acad. Sci. U.S.A. 104, 20753.
Hoberg, E.P., Alkire, N.L., Queiroz, A.D., Jones, A, 2001. Out of Africa: origins of the taenia tapeworms in humans. Proc. R. Soc. Lond. B. Biol. Sci. 268, 781–787.
Keeley, L.H., 1997. War Before Civilization. Oxford University Press, USA.
Lagrue, C., Poulin, R., 2010. Manipulative parasites in the world of veterinary science: implications for epidemiology and pathology. Vet. J. 184, 9–13.
Lomas, D.A., 2006. The selective advantage of {alpha}1-antitrypsin deficiency. Am. J. Respir. Crit. Care Med. 173, 1072–1077.
Low, B.S., 1990. Marriage systems and pathogen stress in human societies. Integr. Comp. Biol. 30, 325.
Maier, A.G., Duraisingh, M.T., Reeder, J.C., Patel, S.S., Kazura, J.W., Zimmerman, P.A., et al., 2003. *Plasmodium falciparum* erythrocyte invasion through glycophorin c and selection for Gerbich negativity in human populations. Nat. Med. 9, 87–92.
Meyer, C.G., Amedofu, G.K., Brandner, J.M., Pohland, D., Timmann, C., Horstmann, R.D., 2002. Selection for deafness? Nat. Med. 8, 1332–1333.
Ohashi, J., et al., 2004. Extended linkage disequilibrium surrounding the hemoglobin E variant due to malarial selection. Am. J. Hum. Genet. 74, 1198–1208.
Pain, A., et al., 2001. A non-sense mutation in Cd36 gene is associated with protection from severe malaria. Lancet 357, 1502–1503.
Pays, E., Vanhollebeke, B., Vanhamme, L., Paturiaux-Hanocq, F., Nolan, D.P., Perez-Morga, D., 2006. The trypanolytic factor of human serum. Nat. Rev. Microbiol. 4, 477–486.
Pennington, R., Gatenbee, C., Kennedy, B., Harpending, H., Cochran, G., 2009. Group differences in proneness to inflammation. Infect. Genet. Evol. 9, 1371–1380.
Puente, X.S., et al., 2005. Comparative genomic analysis of human and chimpanzee proteases. Genomics 86, 638–647.
Ramoz, N., Rueda, L.-A., Bouadjar, B., Montoya, L.-S., Orth, G., Favre, M., 2002. Mutations in two adjacent novel genes are associated with epidermodysplasia verruciformis. Nat. Genet. 32, 579–581.
Reich, D., Nalls, M.A., Kao, W.H.L., Akylbekova, E.L., Tandon, A., Patterson, N., et al., 2009. Reduced neutrophil count in peopl of African dscent is due to a regulatory vaiant in th Duffy antigen receptor for chemokines gene. PLoS Genet 5, e1000360.
Rigaud, S., et al., 2006. XIAP deficiency in humans causes an X-linked lymphoproliferative syndrome. Nature 444, 110–114.
Rowe, J.A., Opi, D.H., Williams, T.N., 2009. Blood groups and malaria: fresh insights into pathogenesis and identification of targets for intervention. Curr. Opin. Hematol. 16, 480–487.
Sancho-Shimizu, V., et al., 2007. Genetic susceptibility to herpes simplex virus 1 encephalitis in mice and humans. Curr. Opin. Allergy Clin. Immunol. 7, 495–505.

Saunders, M.A., et al., 2005. The extent of linkage disequilibrium caused by selection on G6PD in humans. Genetics 171, 1219.
Saunders, M.A., Good, J.M., Lawrence, E.C., Ferrell, R.E., Li, W.H., Nachman, M.W., 2006. Human adaptive evolution at myostatin (GDF8), a regulator of muscle growth. The American Journal of Human Genatics 79, 1089–1097.
Scherf, A., Lopez-Rubio, J.J., Riviere, L., 2008. Antigenic variation in *Plasmodium falciparum*. Annu. Rev. Microbiol. 62, 445–470.

10 Infectious Disease Genomics

Yu-Tsueng Liu*

Division of Infectious Diseases, Department of Medicine, Moores Cancer Center, University of California San Diego, La Jolla, CA, USA

10.1 Introduction

The history and development of infectious disease genomics are tightly associated with the Human Genome Project (HGP) (Watson, 1990). A series of important discussions about the HGP were made in 1985 and 1986 (Dulbecco, 1986; Watson, 1990), which led to the appointment of a special National Research Council (NRC) committee by the National Academy of Sciences to address the needs and concerns, such as its impact, leadership, and funding sources. The committee recommended that the United States begin the HGP in 1988 (NRC, 1988). They emphasized the need for technological improvements in the efficiency of gene mapping, sequencing, and data analysis capabilities. In order to understand potential functions of human genes through comparative sequence analyses, they also advised that the HGP must not be restricted to the human genome and should include model organisms including mouse, bacteria, yeast, fruit fly, and worm. In the meantime, the Office of Technology Assessment (OTA) of the US Congress also issued a similar report to support the HGP (OTA, 1988). In 1990, the Department of Energy (DOE) and the National Institutes of Health (NIH) jointly presented an initial 5-year plan for the HGP (DHHS and DOE, 1990). In October 1993, the Sanger Center/Institute (Hinxton, UK) was officially open to join the HGP. The cost of DNA sequencing was about $2–5 per base in 1990, and the initial aim was to reduce the costs to less than $0.50 per base before large-scale sequencing (DHHS and DOE, 1990). The sequencing cost gradually declined during the subsequent years. In 2004, the National Human Genome Research Institute (NHGRI) challenged scientists to achieve a $100,000 human genome (3 Gb/haploid genome) by 2009 and a $1000 genome by 2014 to meet the need of genomic medicine.

    The first complete genome to be sequenced was the phiX174 bacteriophage (5.4 kb) by Sanger's group in 1977 (Sanger et al., 1977). The complete genome sequence of SV40 polyomavirus (5.2 kb) was published in 1978 (Fiers et al., 1978; Reddy et al., 1978). The human Epstein–Barr virus (170 kb) genome was determined in 1984 (Baer et al., 1984). The first completed free-living organism genome was

*E-mail: ytliu@ucsd.edu

Haemophilus influenza (1.8 Mb), sequenced through a whole-genome shotgun approach in 1995 (Fleischmann et al., 1995). The second sequenced bacterial genome, *Mycoplasma genitalium* (600 kb), was completed in less than a month in the same year using the same approach (Smith, 2004). The DOE was the first to start a microbial genome program (MGP) as a companion to its HGP in 1994 (DOE, 2009). The initial focus was on nonpathogenic microbes. Along with the development of the HGP, there was exponential growth of the number of completely sequenced free-living organism genomes. The Fungal Genome Initiative (FGI) (FGI, 2010) was established in 2000 to accelerate the slow pace of fungal genome sequencing since the report of the genome of *Saccharomyces cerevisiae* in 1996 (Goffeau et al., 1996). One of the major interests was to sequence organisms that are important in human health and commercial activities. As of September 2009, 1100 completed genome projects, a 1.7-fold increase from 2 years ago, were documented (Liolios et al., 2010). These include 914 bacterial, 68 archaeal, and 118 eukaryotic genomes. In addition, more than 4000 other ongoing sequencing projects were reported.

The genomes of human malaria parasite *Plasmodium falciparum* and its major mosquito vector *Anopheles gambiae* were published in 2002 (Gardner et al., 2002; Holt et al., 2002). The effort to sequence the malaria genome began in 1996 by taking advantage of a clone derived from laboratory-adapted strain (Hoffman et al., 1997). Many parasites have complex life cycles that involve both vertebrate and invertebrate hosts and are difficult to maintain in the laboratory. Currently, a few other important human pathogenic parasites, such as Trypansomes (Berriman et al., 2005; El-Sayed et al., 2005), Leishmania (Ivens et al., 2005), and Schistosomas (Berriman et al., 2009; Consortium, 2009), have been either completely or partially sequenced (Brindley et al., 2009; Aurrecoechea et al., 2010). In the meantime, the genome sequence of *Aedes aegypti*, the primary vector for yellow fever and dengue fever, was published in 2007 (Nene et al., 2007). The genome size (1376 Mb) of this mosquito vector is about 5 times larger than the previously sequenced genome of the malaria vector *Anopheles gambiae*. Approximately 50% of the genome consists of transposable elements. In 2010, the genome sequence of the body louse (*Pediculus humanus humanus*), an obligatory parasite of humans and the main vector of epidemic typhus (*Rickettsia prowazekii*), relapsing fever (*Borrelia recurrentis*), and trench fever (*Bartonella quintana*), was reported (Kirkness et al., 2010). Its 108 Mb genome is the smallest among the known insect genomes. Genome sequencing projects for other important human disease vectors are in progress (Lawson et al., 2009; Megy et al., 2009). These include *Culex pipiens* (mosquito vector of West Nile virus), *Ixodes scapularis* (tick vector of Lyme disease, babesia, and anaplasma), and *Glossina morsitans* (tsetse fly vector of African trypanosomiasis). The challenge to sequence the genome of an insect vector is much greater than a microbe. For example, the genomes of ticks were estimated to be between 1 and 7 Gb and may have a significant proportion of repetitive DNA sequences, which may be a problem for genome assembly (Pagel Van Zee et al., 2007). Furthermore, the evolutionary distances among insect species may also affect homology-based gene predictions.

It is as important to understand the sequence diversity within a species as to perform a de novo sequencing of a reference genome from the perspective of human health. This is true for both hosts and pathogens (Feero et al., 2008; Alcais et al., 2009). The goal of the 1000 Genomes Project is to find most genetic variants that have frequencies of at least 1% in the human populations studied (Kaiser, 2008). One of the similar efforts for human pathogens is the NIH Influenza Genome Sequencing Project. When this project began in November 2004, only seven human influenza H3N2 isolates had been completely sequenced and deposited in the GenBank database (Fauci, 2005; Ghedin et al., 2005). As of May 2010, more than 5000 human and avian isolates have been completely sequenced, including the 1918 "Spanish" influenza virus (Taubenberger et al., 2005). Databases for human immunodeficiency virus (HIV) and hepatitis C virus have also been established.

While most human studies of microbes have focused on the disease-causing organisms, interest in resident microorganisms has also been growing. In fact, it has been estimated that the human body is colonized by at least 10 times more prokaryotic and eukaryotic microorganisms than the number of human cells (Savage, 1977). It was suggested to have "the second human genome project" to sequence human microbiome (Relman and Falkow, 2001). Highly variable intestinal microbial flora among normal individuals has been well documented (Eckburg et al., 2005; Costello et al., 2009; Turnbaugh et al., 2009). Therefore, the Human Microbiome Project (HMP) was initiated by the NIH to study samples from multiple body sites from each of at least 250 "normal" volunteers to determine whether there are associations between changes in the microbiome and several different medical conditions, and to provide both standardized data resources and new technological approaches (Peterson et al., 2009).

The completed or ongoing genome projects (Table 10.1) will provide enormous opportunities for the discovery of novel vaccines and drug targets against human pathogens as well as the improvement of diagnosis and discovery of infectious agents and the development of new strategies for invertebrate vector control. Specific examples will be provided to illustrate how the information provided by various genome projects may help achieve the goal of promoting human health.

10.2 Vaccine Target

Meningococcal isolates produce 1 of 13 antigenically distinct capsular polysaccharides, but only 5 (A, B, C, W135, and Y) are commonly associated with disease (Lo et al., 2009). The polysaccharide capsule is important for meningococci to escape from complement-mediated killing. While conventional vaccines consisting of the conjugation of capsular polysaccharides to carrier proteins for meningococcus serogroups A, C, Y, and W-135 have been clinically successful, the same approach failed to produce clinically useful vaccine for serogroup B (MenB). The capsule polysaccharide (α2-8 N-acetylneuraminic acid) of MenB is identical to human polysialic acid and therefore is poorly immunogenic (Finne et al., 1987). Alternatively, vaccines consisting of outer membrane vesicles (OMV) have been successfully developed to control MenB outbreaks in areas where epidemics are dominated by one particular strain (Bjune et al., 1991; Sierra et al., 1991; Boslego

Table 10.1 The Completed or Ongoing Genome Projects

General
NCBI (Sayers et al., 2010) (http://www.ncbi.nlm.nih.gov/sites/genome)
ENSEMBL (Kersey et al., 2010) (http://www.ensemblgenomes.org/)
JCVI (Davidsen et al., 2010) (http://cmr.jcvi.org/)
GOLD (Liolios et al., 2010) (http://www.genomesonline.org)
Sanger Pathogen Genomics (http://www.sanger.ac.uk/Projects/Pathogens/)
GeMInA (Genomic Metadata for Infectious Agents) (Ecker et al., 2005; Schriml et al., 2010) (http://gemina.igs.umaryland.edu)
Bacteria
HMP (Nelson et al., 2010) (http://www.hmpdacc.org/)
Fungi
FGI (http://www.broadinstitute.org/science/projects/fungal-genome-initiative)
Parasites
Eukaryotic pathogens (Aurrecoechea et al., 2010) (http://EuPathDB.org)
Parasite genome projects (http://www.pasteur.fr/recherche/unites/tcruzi/minoprio/genomics/parasites.htm)
Invertebrate vectors
VectorBase (Lawson et al., 2009; Megy et al., 2009) (http://www.vectorbase.org)
Viruses
Influenza virus (Bao et al., 2008) (http://www.ncbi.nlm.nih.gov/genomes/FLU/)
HIV (http://www.hiv.lanl.gov/)
HCV (http://hcv.lanl.gov/)

et al., 1995; Jackson et al., 2009). The most significant limitation of this type of vaccine is that the immune response is strain-specific, mostly directed against the porin protein, PorA, which varies substantially in both expression level and sequence across strains (Martin et al., 2000; Pizza et al., 2000).

With the completion of the genome sequence of a virulent MenB strain, a "reverse vaccinology" approach was applied for the development of a universal MenB vaccine by Novartis (Pizza et al., 2000; Tettelin et al., 2000; Giuliani et al., 2006). Through bioinformatic searching for surface-exposed antigens, which may be the most suitable vaccine candidates due to their potential to be readily recognized by the immune system, 570 open reading frames (ORFs) were selected from a total of 2158 ORFs of the MC58 genome. Eventually, five antigens were chosen as the vaccine components based on a series of criteria including the ability of candidates to be expressed in *Escherichia coli* as recombinant proteins (350 candidates), the confirmation of surface exposure by immunological analyses, the ability of induced protective antibodies in experimental animals (28 candidates), and the conservation of antigens within a panel of diverse meningococcal strains, primarily the disease-associated MenB strains (Pizza et al., 2000; Giuliani et al., 2006; Rinaudo et al., 2009). The vaccine formulation consists of an fHBP-GNA2091 fusion protein, a GNA2132-GNA1030 fusion protein, NadA, and OMV from the New Zealand MeNZB vaccine strain, which contains the immunogenic PorA. Initial phase II clinical results in adults and infants showed that this vaccine could induce a protective immune response against three diverse MenB strains in

89–96% of subjects following three vaccinations and 93–100% after four vaccinations (Rinaudo et al., 2009). In 2010, a phase III trial for this vaccine (4CMenB) has met primary endpoint.

10.3 Drug Target

Targeting an essential pathway is a necessary but not sufficient requirement for an effective antimicrobial agent (Brinster et al., 2009). Identification of essential genes in a completely sequenced genome has been actively pursued with various approaches (Hutchison et al., 1999; Ji et al., 2001). The indispensable fatty acid synthase (FAS) pathway in bacteria has been regarded as a promising target for the development of antimicrobial agents (Wright and Reynolds, 2007). The subcellular organization of the fatty acid biosynthesis components is different between mammals (type I FAS) and bacteria (dissociated type II FAS), which raises the likelihood of host specificity of the targeting drugs. Comparison of the available genome sequences of various species of prokaryotes reveals highly conserved FAS II systems suggesting that the antimicrobial agent can be broad spectrum (Zhang et al., 2003). In addition, through computational analyses, new members of the FAS II system have been discovered in different bacterial species (Heath and Rock, 2000; Marrakchi et al., 2002). One of the protein components in this system, FabI, is the target of an anti-tuberculosis drug isoniazid and a general antibacterial and antifungal agent, triclosan (Banerjee et al., 1994; Levy et al., 1999; Zhang et al., 2006).

Through a systematic screening of 250,000 natural product extracts, a Merck team identified a potent and broad-spectrum antibiotic, platensimycin, which is derived from *Streptomyces platensis* and a selective FabF/B inhibitor in FAS II system (Wang et al., 2006). Treatment with platensimycin eradicated *Staphylococcus aureus* infection in mice. Platensimycin did not have cross-resistance to other antibiotic-resistant strains in vitro, including methicillin-resistant *S. aureus*, vancomycin-intermediate *S. aureus*, and vancomycin-resistant enterococci. No toxicity was observed using a cultured human cell line. The activity of platensimycin was not affected by the presence of human serum in this study. However, the FAS II system appears to be dispensable for another Gram-positive bacterium, *Streptococcus agalactiae*, when exogenous fatty acids are available, such as in human serum (Brinster et al., 2009; Balemans et al., 2010). The susceptibility to inhibitors targeting the FAS II system indicates heterogeneity in fatty acid synthesis or in acquiring exogenous fatty acids among Gram-positive pathogens (Balemans et al., 2010). Comparative genomic approaches may be useful to identify and develop a strategy to target the salvage pathway for *Streptococcus agalactiae*. Alternatively, similar approaches as described earlier for MenB vaccine may also be applied for *Streptococcus agalactiae* (Group B streptococcus) (Maione et al., 2005).

10.4 Vector Control

An early mathematical model for malaria control suggested that the most vulnerable element in the malaria cycle was survivorship of adult female mosquitoes

(Macdonald, 1957; Enayati and Hemingway, 2010). Therefore, insect control is an important part of reducing transmission. The use of DDT as an indoor residual spray in the global malaria eradication program from 1957 to 1969 reduced the population at risk of malaria to ~50% by 1975 compared with 77% in 1900 (Hay et al., 2004; Enayati and Hemingway, 2010). Engineering genetically modified mosquitoes refractory to malaria infection appeared to be an alternative approach (Curtis, 1968) given the environmental impact of DDT and the emergence of insecticide-resistant insects. The Vector Biology Network (VBN) was formed in 1989 and proposed a 20-year plan with the World Health Organization (WHO) in 2001 to achieve three major goals: (1) to develop basic tools for the stable transformation of anopheline mosquitoes by the year 2000; (2) to engineer a mosquito incapable of carrying the malaria parasite by 2005; and (3) to run controlled experiments to test how to drive the engineered genotype into wild mosquito populations by 2010 (Alphey et al., 2002; Morel et al., 2002; Beaty et al., 2009). While some proof-of-concept experiments were achieved for the first two aims in 2002 when the *Anopheles gambiae* genome was completely sequenced (Catteruccia et al., 2000; Ito et al., 2002), the progress has been relatively slow (Marshall and Taylor, 2009).

Genomic loci of the *Anopheles gambiae* responsible for *Plasmodium falciparum* resistance have been identified through surveying a mosquito population in a West African malaria transmission zone (Riehle et al., 2006). A candidate gene, Anopheles Plasmodium-responsive leucine-rich repeat 1 (*APL1*), was discovered. Subsequently, other resistant genes have also been identified (Blandin et al., 2009; Povelones et al., 2009). Studying the genetic basis of resistance to malaria parasites and immunity of the mosquito vector will be important to control malaria transmission.

10.5 Diagnostic Target and Pathogen Discovery

Perhaps the most immediate impact of a completely sequenced pathogen genome is for infectious disease diagnosis. The information may be of great importance to the public health when a newly emerged or re-emerged pathogen is discovered. The 2009 swine-origin influenza A virus (S-OIV) (Dawood et al., 2009) and 2003 SARS (severe acute respiratory syndrome) coronavirus (Ksiazek et al., 2003; Rota et al., 2003) are the two most recent examples.

S-OIV emerged in the spring of 2009 in Mexico and was also discovered in specimens from two unrelated children in the San Diego area in April 2009 (CDC, 2009; Dawood et al., 2009). Those samples were positive for influenza A but negative for both human H1 and H3 subtypes. The complete genome sequence and a real-time PCR-based diagnostic assay were released to the public in late April. The outbreak evolved rapidly and the WHO declared the highest Phase 6 worldwide pandemic alert on June 11, 2009. S-OIV has three genome segments (HA, NP, NS) from the classic North American swine (H1N1) lineage, two segments (PB2, PA) from the North American avian lineage, one segment (PB1) from the seasonal H3N2, and most notably, two segments (NA, M) from the Eurasian swine (H1N1) lineage (Dawood et al., 2009). With the available influenza genome database, diagnostic assays to distinguish previous seasonal H1N1, H3N2, and S-OIV can be easily accomplished (Lu et al., 2009).

A comprehensive pathogen genome database is not only useful for infectious disease diagnosis but also for novel pathogen discovery (Liu, 2008). Homologous sequences within the same family or among different family members are important for new pathogen identification even with the advent of third-generation sequencing technology (Munroe and Harris, 2010). De novo pathogen discovery may be also complicated by coexisting microorganisms, such as commensal bacteria in the human body. Without prior knowledge of these microorganisms, one may be misled.

In 2003, a microarray-based assay, designated Virochip, was used to help discover the SARS coronavirus (Wang et al., 2003). The Virochip contained the most highly conserved 70mer sequences from every fully sequenced reference viral genome in GenBank. The computational search for conservation was performed across all known viral families. A microarray hybridized with a reaction derived from a viral isolate cultivated from a SARS patient revealed that the strongest hybridizing array elements belong to families Astroviridae and Coronaviridae. Alignment of the oligonucleotide probes having the highest signals showed that all four hybridizing oligonucleotides from the Astroviridae and one oligonucleotide from avian infectious bronchitis virus, an avian coronavirus, shared a core consensus motif spanning 33 nucleotides. Interestingly, it had been known previously through bioinformatic analyses that this sequence is present in the 3' UTR of all astroviruses, avian infectious bronchitis virus, and an equine rhinovirus (Jonassen et al., 1998). Therefore, a new member of the coronavirus was identified through the unique hybridizing pattern and subsequent confirmations.

The finding of the seventh human oncogenic virus, Merkel cell polyomavirus (MCV) (Feng et al., 2008) in 2008 is another example of why conserved sequences are important for novel pathogen discovery. MCV is the etiological agent of Merkel cell carcinoma (MCC), which is a rare but aggressive skin cancer of neuroendocrine origin. Two cDNA libraries derived from MCC tumors were subjected to high-throughput sequencing by a next-generation Roche/454 sequencer. Nearly 400,000 sequence reads were generated. The majority (99.4%) of the sequences derived from human origin were removed from further analyses. Only one of the remaining 2395 cDNA was homologous to the T antigen of two known polyomaviruses. One additional cDNA was subsequently identified to be part of the MCV sequence when the complete viral sequence was known. Later analyses showed that 80% (8/10) of the MCC had integrated MCV in the human genome. Monoclonal viral integration was revealed by the patterns of Southern blot analysis. Only 8–16% of control tissues had low copy number of MCV infection.

10.6 Conclusion

While we can expect that the efforts of a variety of genome projects may improve human health, the socioeconomic issues that are not discussed in this chapter may be substantial. In addition, the tremendous amount of information derived from these projects will also be a challenge for scientists as well nonscientists to follow and understand.

References

Alcais, A., Abel, L., Casanova, J.L., 2009. Human genetics of infectious diseases: between proof of principle and paradigm. J. Clin. Invest. 119, 2506–2514.

Alphey, L., Beard, C.B., Billingsley, P., Coetzee, M., Crisanti, A., Curtis, C., et al., 2002. Malaria control with genetically manipulated insect vectors. Science 298, 119–121.

Aurrecoechea, C., Brestelli, J., Brunk, B.P., Fischer, S., Gajria, B., Gao, X., et al., 2010. EuPathDB: a portal to eukaryotic pathogen databases. Nucleic Acids Res. 38, D415–419.

Baer, R., Bankier, A.T., Biggin, M.D., Deininger, P.L., Farrell, P.J., Gibson, T.J., et al., 1984. DNA sequence and expression of the B95-8 Epstein–Barr virus genome. Nature 310, 207–211.

Balemans, W., Lounis, N., Gilissen, R., Guillemont, J., Simmen, K., Andries, K., et al., 2010. Essentiality of FASII pathway for *Staphylococcus aureus*. Nature 463, E3, discussion E4.

Banerjee, A., Dubnau, E., Quemard, A., Balasubramanian, V., Um, K.S., Wilson, T., et al., 1994. inhA, a gene encoding a target for isoniazid and ethionamide in *Mycobacterium tuberculosis*. Science 263, 227–230.

Bao, Y., Bolotov, P., Dernovoy, D., Kiryutin, B., Zaslavsky, L., Tatusova, T., et al., 2008. The influenza virus resource at the National Center for Biotechnology Information. J. Virol. 82, 596–601.

Beaty, B.J., Prager, D.J., James, A.A., Jacobs-Lorena, M., Miller, L.H., Law, J.H., et al., 2009. From Tucson to genomics and transgenics: the vector biology network and the emergence of modern vector biology. PLoS Negl. Trop. Dis. 3, e343.

Berriman, M., Ghedin, E., Hertz-Fowler, C., Blandin, G., Renauld, H., Bartholomeu, D.C., et al., 2005. The genome of the African trypanosome *Trypanosoma brucei*. Science 309, 416–422.

Berriman, M., Haas, B.J., LoVerde, P.T., Wilson, R.A., Dillon, G.P., Cerqueira, G.C., et al., 2009. The genome of the blood fluke *Schistosoma mansoni*. Nature 460, 352–358.

Bjune, G., Hoiby, E.A., Gronnesby, J.K., Arnesen, O., Fredriksen, J.H., Halstensen, A., et al., 1991. Effect of outer membrane vesicle vaccine against group B meningococcal disease in Norway. Lancet 338, 1093–1096.

Blandin, S.A., Wang-Sattler, R., Lamacchia, M., Gagneur, J., Lycett, G., Ning, Y., et al., 2009. Dissecting the genetic basis of resistance to malaria parasites in *Anopheles gambiae*. Science 326, 147–150.

Boslego, J., Garcia, J., Cruz, C., Zollinger, W., Brandt, B., Ruiz, S., et al., 1995. Efficacy, safety, and immunogenicity of a meningococcal group B (15:P1.3) outer membrane protein vaccine in Iquique, Chile. Chilean National Committee for Meningococcal Disease. Vaccine 13, 821–829.

Brindley, P.J., Mitreva, M., Ghedin, E., Lustigman, S., 2009. Helminth genomics: the implications for human health. PLoS Negl. Trop. Dis. 3, e538.

Brinster, S., Lamberet, G., Staels, B., Trieu-Cuot, P., Gruss, A., Poyart, C., 2009. Type II fatty acid synthesis is not a suitable antibiotic target for Gram-positive pathogens. Nature 458, 83–86.

Catteruccia, F., Nolan, T., Loukeris, T.G., Blass, C., Savakis, C., Kafatos, F.C., et al., 2000. Stable germline transformation of the malaria mosquito *Anopheles stephensi*. Nature 405, 959–962.
CDC, 2009. Swine influenza A (H1N1) infection in two children—Southern California, March–April 2009. MMWR Morb. Mortal. Wkly Rep. 58, 400–402.
Consortium, 2009. The *Schistosoma japonicum* genome reveals features of host–parasite interplay. Nature 460, 345–351.
Costello, E.K., Lauber, C.L., Hamady, M., Fierer, N., Gordon, J.I., Knight, R., 2009. Bacterial community variation in human body habitats across space and time. Science 326, 1694–1697.
Curtis, C.F., 1968. Possible use of translocations to fix desirable genes in insect pest populations. Nature 218, 368–369.
Davidsen, T., Beck, E., Ganapathy, A., Montgomery, R., Zafar, N., Yang, Q., et al., 2010. The comprehensive microbial resource. Nucleic Acids Res. 38, D340–D345.
Dawood, F.S., Jain, S., Finelli, L., Shaw, M.W., Lindstrom, S., Garten, R.J., et al., 2009. Emergence of a novel swine-origin influenza A (H1N1) virus in humans. N. Engl. J. Med. 360, 2605–2615.
DHHS and DOE, 1990. Understanding our genetic inheritance, the U.S. Human Genome Project: the first five years: fiscal years 1991–1995. <http://www.ornl.gov/sci/techresources/Human_Genome/project/5yrplan/summary.shtml>.
DOE, 2009. Microbial Genome Program. <http://microbialgenomics.energy.gov/mgp.shtml>.
Dulbecco, R., 1986. A turning point in cancer research: sequencing the human genome. Science 231, 1055–1056.
Eckburg, P.B., Bik, E.M., Bernstein, C.N., Purdom, E., Dethlefsen, L., Sargent, M., et al., 2005. Diversity of the human intestinal microbial flora. Science 308, 1635–1638.
Ecker, D.J., Sampath, R., Willett, P., Wyatt, J.R., Samant, V., Massire, C., et al., 2005. The microbial rosetta stone database: a compilation of global and emerging infectious microorganisms and bioterrorist threat agents. BMC Microbiol. 5, 19.
El-Sayed, N.M., Myler, P.J., Bartholomeu, D.C., Nilsson, D., Aggarwal, G., Tran, A.N., et al., 2005. The genome sequence of *Trypanosoma cruzi*, etiologic agent of Chagas disease. Science 309, 409–415.
Enayati, A., Hemingway, J., 2010. Malaria management: past, present, and future. Annu. Rev. Entomol. 55, 569–591.
Fauci, A.S., 2005. Race against time. Nature 435, 423–424.
Feero, W.G., Guttmacher, A.E., Collins, F.S., 2008. The genome gets personal—almost. JAMA 299, 1351–1352.
Feng, H., Shuda, M., Chang, Y., Moore, P.S., 2008. Clonal integration of a polyomavirus in human Merkel cell carcinoma. Science 319, 1096–1100.
FGI (2010). Fungal Genome Initiative. <http://www.broadinstitute.org/science/projects/fungal-genome-initiative>.
Fiers, W., Contreras, R., Haegemann, G., Rogiers, R., Van de Voorde, A., Van Heuverswyn, H., et al., 1978. Complete nucleotide sequence of SV40 DNA. Nature 273, 113–120.
Finne, J., Bitter-Suermann, D., Goridis, C., Finne, U., 1987. An IgG monoclonal antibody to group B meningococci cross-reacts with developmentally regulated polysialic acid units of glycoproteins in neural and extraneural tissues. J. Immunol. 138, 4402–4407.
Fleischmann, R.D., Adams, M.D., White, O., Clayton, R.A., Kirkness, E.F., Kerlavage, A.R., et al., 1995. Whole-genome random sequencing and assembly of *Haemophilus influenzae* Rd. Science 269, 496–512.

Gardner, M.J., Hall, N., Fung, E., White, O., Berriman, M., Hyman, R.W., et al., 2002. Genome sequence of the human malaria parasite *Plasmodium falciparum*. Nature 419, 498−511.

Ghedin, E., Sengamalay, N.A., Shumway, M., Zaborsky, J., Feldblyum, T., Subbu, V., et al., 2005. Large-scale sequencing of human influenza reveals the dynamic nature of viral genome evolution. Nature 437, 1162−1166.

Giuliani, M.M., Adu-Bobie, J., Comanducci, M., Arico, B., Savino, S., Santini, L., et al., 2006. A universal vaccine for serogroup B meningococcus. Proc. Natl. Acad. Sci. U.S.A. 103, 10834−10839.

Goffeau, A., Barrell, B.G., Bussey, H., Davis, R.W., Dujon, B., Feldmann, H., et al., 1996. Life with 6000 genes. Science 274, 546, 563−567.

Hay, S.I., Guerra, C.A., Tatem, A.J., Noor, A.M., Snow, R.W., 2004. The global distribution and population at risk of malaria: past, present, and future. Lancet Infect. Dis. 4, 327−336.

Heath, R.J., Rock, C.O., 2000. A triclosan-resistant bacterial enzyme. Nature 406, 145−146.

Hoffman, S.L., Bancroft, W.H., Gottlieb, M., James, S.L., Burroughs, E.C., Stephenson, J.R., et al., 1997. Funding for malaria genome sequencing. Nature 387, 647.

Holt, R.A., Subramanian, G.M., Halpern, A., Sutton, G.G., Charlab, R., Nusskern, D.R., et al., 2002. The genome sequence of the malaria mosquito *Anopheles gambiae*. Science 298, 129−149.

Hutchison, C.A., Peterson, S.N., Gill, S.R., Cline, R.T., White, O., Fraser, C.M., et al., 1999. Global transposon mutagenesis and a minimal *Mycoplasma* genome. Science 286, 2165−2169.

Ito, J., Ghosh, A., Moreira, L.A., Wimmer, E.A., Jacobs-Lorena, M., 2002. Transgenic anopheline mosquitoes impaired in transmission of a malaria parasite. Nature 417, 452−455.

Ivens, A.C., Peacock, C.S., Worthey, E.A., Murphy, L., Aggarwal, G., Berriman, M., et al., 2005. The genome of the kinetoplastid parasite, *Leishmania major*. Science 309, 436−442.

Jackson, C., Lennon, D.R., Sotutu, V.T., Yan, J., Stewart, J.M., Reid, S., et al., 2009. Phase II meningococcal B vesicle vaccine trial in New Zealand infants. Arch. Dis. Child. 94, 745−751.

Ji, Y., Zhang, B., Van, S.F., Horn, W.P., Woodnutt, G., Burnham, M.K., et al., 2001. Identification of critical staphylococcal genes using conditional phenotypes generated by antisense RNA. Science 293, 2266−2269.

Jonassen, C.M., Jonassen, T.O., Grinde, B., 1998. A common RNA motif in the 3′ end of the genomes of astroviruses, avian infectious bronchitis virus and an equine rhinovirus. J. Gen. Virol. 79 (Pt 4), 715−718.

Kaiser, J., 2008. DNA sequencing. A plan to capture human diversity in 1000 genomes. Science 319, 395.

Kersey, P.J., Lawson, D., Birney, E., Derwent, P.S., Haimel, M., Herrero, J., et al., 2010. Ensembl genomes: extending Ensembl across the taxonomic space. Nucleic Acids Res. 38, D563−D569.

Kirkness, E.F., Haas, B.J., Sun, W., Braig, H.R., Perotti, M.A., Clark, J.M., et al., 2010. Genome sequences of the human body louse and its primary endosymbiont provide insights into the permanent parasitic lifestyle. Proc. Natl. Acad. Sci. U.S.A. 107, 12168−12173.

Ksiazek, T.G., Erdman, D., Goldsmith, C.S., Zaki, S.R., Peret, T., Emery, S., et al., 2003. A novel coronavirus associated with severe acute respiratory syndrome. N. Engl. J. Med. 348, 1953−1966.

Lawson, D., Arensburger, P., Atkinson, P., Besansky, N.J., Bruggner, R.V., Butler, R., et al., 2009. VectorBase: a data resource for invertebrate vector genomics. Nucleic Acids Res. 37, D583−587.
Levy, C.W., Roujeinikova, A., Sedelnikova, S., Baker, P.J., Stuitje, A.R., Slabas, A.R., et al., 1999. Molecular basis of triclosan activity. Nature 398, 383−384.
Liolios, K., Chen, I.M., Mavromatis, K., Tavernarakis, N., Hugenholtz, P., Markowitz, V.M., et al., 2010. The Genomes OnLine Database (GOLD) in 2009: status of genomic and metagenomic projects and their associated metadata. Nucleic Acids Res. 38, D346−354.
Liu, Y.T., 2008. A technological update of molecular diagnostics for infectious diseases. Infect. Disord. Drug Targets 8, 183−188.
Lo, H., Tang, C.M., Exley, R.M., 2009. Mechanisms of avoidance of host immunity by *Neisseria meningitidis* and its effect on vaccine development. Lancet Infect. Dis. 9, 418−427.
Lu, Q., Zhang, X.Q., Pond, S.L., Reed, S., Schooley, R.T., Liu, Y.T., 2009. Detection in 2009 of the swine origin influenza A (H1N1) virus by a subtyping microarray. J. Clin. Microbiol. 47, 3060−3061.
Macdonald, G., 1957. The Epidemiology and Control of Malaria. Oxford University Press, Oxford.
Maione, D., Margarit, I., Rinaudo, C.D., Masignani, V., Mora, M., Scarselli, M., et al., 2005. Identification of a universal Group B streptococcus vaccine by multiple genome screen. Science 309, 148−150.
Marrakchi, H., Choi, K.H., Rock, C.O., 2002. A new mechanism for anaerobic unsaturated fatty acid formation in *Streptococcus pneumoniae*. J. Biol. Chem. 277, 44809−44816.
Marshall, J.M., Taylor, C.E., 2009. Malaria control with transgenic mosquitoes. PLoS Med. 6, e20.
Martin, S.L., Borrow, R., van der Ley, P., Dawson, M., Fox, A.J., Cartwright, K.A., 2000. Effect of sequence variation in meningococcal PorA outer membrane protein on the effectiveness of a hexavalent PorA outer membrane vesicle vaccine. Vaccine 18, 2476−2481.
Megy, K., Hammond, M., Lawson, D., Bruggner, R.V., Birney, E., Collins, F.H., 2009. Genomic resources for invertebrate vectors of human pathogens, and the role of VectorBase. Infect. Genet. Evol. 9, 308−313.
Morel, C.M., Toure, Y.T., Dobrokhotov, B., Oduola, A.M., 2002. The mosquito genome—a breakthrough for public health. Science 298, 79.
Munroe, D.J., Harris, T.J., 2010. Third-generation sequencing fireworks at Marco Island. Nat. Biotechnol. 28, 426−428.
Nelson, K.E., Weinstock, G.M., Highlander, S.K., Worley, K.C., Creasy, H.H., Wortman, J.R., et al., 2010. A catalog of reference genomes from the human microbiome. Science 328, 994−999.
Nene, V., Wortman, J.R., Lawson, D., Haas, B., Kodira, C., Tu, Z.J., et al., 2007. Genome sequence of *Aedes aegypti*, a major arbovirus vector. Science 316, 1718−1723.
NRC, 1988. Mapping and sequencing the human genome. <http://www.nap.edu/catalog. php?record_id=1097>.
OTA, 1988. Mapping our genes—genome projects: how big? how fast? <http://www.ornl.gov/sci/techresources/Human_Genome/publicat/OTAreport.pdf>.
Pagel Van Zee, J., Geraci, N.S., Guerrero, F.D., Wikel, S.K., Stuart, J.J., Nene, V.M., et al., 2007. Tick genomics: the *Ixodes* genome project and beyond. Int. J. Parasitol. 37, 1297−1305.

Peterson, J., Garges, S., Giovanni, M., McInnes, P., Wang, L., Schloss, J.A., et al., 2009. The NIH Human Microbiome Project. Genome Res. 19, 2317–2323.

Pizza, M., Scarlato, V., Masignani, V., Giuliani, M.M., Arico, B., Comanducci, M., et al., 2000. Identification of vaccine candidates against serogroup B meningococcus by whole-genome sequencing. Science 287, 1816–1820.

Povelones, M., Waterhouse, R.M., Kafatos, F.C., Christophides, G.K., 2009. Leucine-rich repeat protein complex activates mosquito complement in defense against *Plasmodium* parasites. Science 324, 258–261.

Reddy, V.B., Thimmappaya, B., Dhar, R., Subramanian, K.N., Zain, B.S., Pan, J., et al., 1978. The genome of simian virus 40. Science 200, 494–502.

Relman, D.A., Falkow, S., 2001. The meaning and impact of the human genome sequence for microbiology. Trends Microbiol. 9, 206–208.

Riehle, M.M., Markianos, K., Niare, O., Xu, J., Li, J., Toure, A.M., et al., 2006. Natural malaria infection in *Anopheles gambiae* is regulated by a single genomic control region. Science 312, 577–579.

Rinaudo, C.D., Telford, J.L., Rappuoli, R., Seib, K.L., 2009. Vaccinology in the genome era. J. Clin. Invest. 119, 2515–2525.

Rota, P.A., Oberste, M.S., Monroe, S.S., Nix, W.A., Campagnoli, R., Icenogle, J.P., et al., 2003. Characterization of a novel coronavirus associated with severe acute respiratory syndrome. Science 300, 1394–1399.

Sanger, F., Air, G.M., Barrell, B.G., Brown, N.L., Coulson, A.R., Fiddes, C.A., et al., 1977. Nucleotide sequence of bacteriophage phi X174 DNA. Nature 265, 687–695.

Savage, D.C., 1977. Microbial ecology of the gastrointestinal tract. Annu. Rev. Microbiol. 31, 107–133.

Sayers, E.W., Barrett, T., Benson, D.A., Bolton, E., Bryant, S.H., Canese, K., et al., 2010. Database resources of the National Center for Biotechnology Information. Nucleic Acids Res. 38, D5–D16.

Schriml, L.M., Arze, C., Nadendla, S., Ganapathy, A., Felix, V., Mahurkar, A., et al., 2010. GeMInA, Genomic Metadata for Infectious Agents, a geospatial surveillance pathogen database. Nucleic Acids Res. 38, D754–D764.

Sierra, G.V., Campa, H.C., Varcacel, N.M., Garcia, I.L., Izquierdo, P.L., Sotolongo, P.F., et al., 1991. Vaccine against group B *Neisseria meningitidis*: protection trial and mass vaccination results in Cuba. NIPH Ann. 14, 195–207, discussion 208–210.

Smith, H.O., 2004. History of microbial genomics. In: Fraser, C.M., Read, T.D., Nelson, K.E. (Eds.), Microbial Genomes. Humana, Totowa, NJ, pp. 3–16.

Taubenberger, J.K., Reid, A.H., Lourens, R.M., Wang, R., Jin, G., Fanning, T.G., 2005. Characterization of the 1918 influenza virus polymerase genes. Nature 437, 889–893.

Tettelin, H., Saunders, N.J., Heidelberg, J., Jeffries, A.C., Nelson, K.E., Eisen, J.A., et al., 2000. Complete genome sequence of *Neisseria meningitidis* serogroup B strain MC58. Science 287, 1809–1815.

Turnbaugh, P.J., Hamady, M., Yatsunenko, T., Cantarel, B.L., Duncan, A., Ley, R.E., et al., 2009. A core gut microbiome in obese and lean twins. Nature 457, 480–484.

Wang, D., Urisman, A., Liu, Y.T., Springer, M., Ksiazek, T.G., Erdman, D.D., et al., 2003. Viral discovery and sequence recovery using DNA microarrays. PLoS Biol. 1, E2.

Wang, J., Soisson, S.M., Young, K., Shoop, W., Kodali, S., Galgoci, A., et al., 2006. Platensimycin is a selective FabF inhibitor with potent antibiotic properties. Nature 441, 358–361.

Watson, J.D., 1990. The Human Genome Project: past, present, and future. Science 248, 44–49.

Wright, H.T., Reynolds, K.A., 2007. Antibacterial targets in fatty acid biosynthesis. Curr. Opin. Microbiol. 10, 447–453.

Zhang, Y.M., Marrakchi, H., White, S.W., Rock, C.O., 2003. The application of computational methods to explore the diversity and structure of bacterial fatty acid synthase. J. Lipid Res. 44, 1–10.

Zhang, Y.M., White, S.W., Rock, C.O., 2006. Inhibiting bacterial fatty acid synthesis. J. Biol. Chem. 281, 17541–17544.

11 Proteomics and Host—Pathogen Interactions: A Bright Future?

David G. Biron[1,], Dobrin Nedelkov[2], Dorothée Missé[3] and Philippe Holzmuller[4]*

[1]HPMCT Proteome Analysis, St-Elphège, QC, Canada, [2]Institute for Population Proteomics and Intrinsic Bioprobes, Inc., Tempe, AZ, USA, [3]GEMI, Centre IRD de Montpellier, Montpellier, France, [4]CIRAD UMR 17 Trypanosomes, Campus International de Baillarguet, Montpellier, France

11.1 Introduction

Living organisms are constantly exposed to pathogens. In any **environment**, a molecular war begins when a host encounters a pathogen. In many host—pathogen associations, the molecular war was in progress a long time ago. Nevertheless, a disease as an outcome of a pathogen attack remains an exception rather a rule. Most host species have acquired strategies by selective pressure to mislead the pathogen and to win fight during their crosstalk (i.e., molecular dialogue). However many pathogen species have acquired strategies by selective pressure to bypass the host defenses to win the molecular war and to ensure the completion of its life cycle. Pathogens remain a significant threat to any host species. Critical to the mitigation of this threat is the ability to rapidly detect, respond to, treat, and contain the pathogen transmission. For many centuries, some scientific fields (i.e., agroecology, evolutionary ecology, evolutionary medicine, biochemistry, microbiology, medicine, veterinary medicine, immunology, and molecular biology) have surveyed host—parasite interactions to improve our understanding of pathogenic diseases and to prevent pathogen transmission in host populations.

During the course of human history, pathogenic diseases have seriously affected many societies worldwide. In Europe, one of the most dramatic disease events was the great plague pandemic of the mid-fourteenth century (Watts, 1997; Achtman et al., 2004). Notably, pathogenic diseases are a leading cause of premature death in the world. Pathogenic diseases result from an intimate relationship between a host and a pathogen which involves molecular "crosstalk." Clearly, elucidation of this complex molecular dialogue between host and pathogen is desirable in order to improve our understanding of pathogen virulence, to develop pathogen-specific

*E-mail: dbiron@clermont.inra.fr

host biomarkers, and to define novel therapeutic and vaccine targets. **Proteomics** applications to decipher host–parasite interactions are in their infancy and should lead to new insights on host **specificity** and on the evolution of pathogen virulence. In this chapter, we present the interest of proteomics to survey host–pathogen interactions, a synthetic review of previous proteomics studies, the pitfalls of the current approach in surveys, new conceptual approaches to decipher host–parasite interactions, a new avenue to decipher the crosstalk diversity involved in trophic interactions in an habitat (i.e., the **population proteomics**), and 5-year view for future prospects on proteomics and host–pathogen interactions.

11.2 Interest of Proteomics to Study Host–Pathogen Interactions

Since the start of the **genomic** era in the early 1990s, many parasitologists and molecular biologists are confident that complete sequencing of the **genome** of the partners in host–pathogen associations for pathogens with simple life cycles (i.e., one host) and in host–vector–pathogen associations for pathogens with complex life cycles (i.e., at least two hosts) will enable total understanding of the molecular mechanisms involved in most pathogenic diseases and will contribute to finding new drugs for treating them (Hochstrasser, 1998; Degrave et al., 2001). Unfortunately, little progress has been achieved in the control of such diseases as malaria and sleeping sickness, despite decades of intensive genomic projects on host–pathogen interactions, vaccines, and chemotherapeutics. Pathogens continue to be a major cause of morbidity and mortality in humans and domestic livestock, especially in developing countries (Ouma et al., 2001; Ryan, 2001; Guzman and Kouri, 2002; Gelfand and Callahan, 2003).

Until now, many parasitologists and molecular biologists have focused their studies on DNA analyses based on the central dogma of molecular biology—that is to say, the general pathway for the expression of genetic information stored in DNA. Although the basic blueprint of life is encoded in DNA, the execution of the genetic plan is carried out by the activities of proteins. The fabric of biological diversity is therefore protein-based, and natural selection acts at the protein level (Karr, 2008). At the end of the twentieth century it had become clear to many parasitologists and molecular biologists that knowing genome sequences, whilst technically mandatory, was not in itself enough to fully understand complex biological events like the immune response of a host to a pathogen infection or the molecular strategies used by pathogens to thwart the host defenses during their interaction (Barret et al., 2000; Ashton et al., 2001; Fell, 2001; Fields, 2001; Schmid-Hempel, 2008).

The evolution of any given species has tremendously increased complexity at the level of pre- (gene splicing, mRNA editing) and posttranslational (phosphorylation, glycosylation, acetylation, etc.) gene–protein interaction. The genomics era has revealed that: (1) DNA sequences may be "fundamental," but can provide little information on the dynamic processes within and between host and parasite during their physical and molecular interaction (Barret et al., 2000; Ashton et al., 2001); (2) the correlation between the expressed "transcriptome" (i.e., total mRNA transcription pattern) and the levels of translated proteins is poor (Anderson and Seilhaver, 1997; Gygi et al., 1999; Maniatis and Tasic, 2002); and (3) a single gene can produce different protein products (Fell, 2001; Fields, 2001; Maniatis and Tasic, 2002). Moreover, the structure, function, abundance, and even the number of proteins in an organism cannot yet be predicted from the DNA sequence alone (Anderson and Anderson, 1996; Gygi et al., 1999; Barret et al., 2000). Also, posttranslational modifications such as phosphorylation and glycosylation are often extremely important for the function of many proteins, although most of these modifications cannot yet be predicted form genomic or mRNA sequences (Gygi et al., 1999). Thus, the biological phenotype of an organism is not directly related to its genotype (i.e., DNA sequences).

Epigenetic systems control and modify gene expression. Almost all the elements of epigenetic control systems are proteins (Anderson and Anderson, 1996). The cells of an organism are reactive systems in which information flows not only from genes to proteins but in the reverse direction as well (Hochstrasser, 1998). The **proteome** is the genome-operating system by which the cells of an organism react to environmental signals (Anderson and Anderson, 1996). It comprises an afferent arm, the cytosensorium (i.e., many cellular proteins are sensors, receptors, and information transfer units from environmental signals) and an efferent arm, the cytoeffectorium (i.e., in cells, reaction of the genome via regulation of either individual proteins or a group of proteins in response to environmental changes).

Proteomics is the study of the proteome. In a broad sense, the proteome (i.e., the genome-operating system) means all the proteins produced by a cell or tissue. Proteomics will contribute to bridge the gap between our understanding of genome sequence and cellular behavior. Proteomics offers an excellent way to study the reaction of the host and pathogen proteomes (i.e., genome-operating systems) during their complex biochemical crosstalk (Biron et al., 2005a,b). Using the first generation proteomics approach, two-dimensional electrophoresis (2-DE) and **mass spectrometry** (MS), posttranslational modifications of host and pathogen proteins (such as phosphorylation, glycosylation, acetylation, and methylation) in reaction to their interaction can be detected. Such modifications are vital for the correct activity of numerous proteins and are being increasingly recognized as a major mechanism in cellular regulation. Although 2-DE offers a high-quality approach for the study of host and/or pathogen proteomes, several proteomics tools have been developed that complement this approach (Gygi et al., 1999; Fung et al., 2001; Lopez and Pluksal, 2003; Wu et al., 2003; Bischoff and Luider, 2004). Table 11.1 shows a comparison of the most popular proteomics tools.

11.3 Retrospective Analysis of Previous Proteomics Studies

The host—pathogen crosstalks reflect the balance of host defenses and pathogen virulence mechanisms. Postgenomic technology promises to revolutionize many

Table 11.1 A Comparison of Proteomics Tools

Name of Technique	Separation	Quantification	Identification of Candidate Protein Spots	Hydrophic Proteins	Low Expressed Proteins	Requirement for Protein Identification	Potential for Discovering New Proteins	Detection of Specific Isoforms	Relative Assay Time	Cost to Acquire and to Use
2-DE	Electrophoresis: IEF PAGE	Densitometry of stains	Mass spectrometry (PMF;MS/MS)	Dependent on detergents used	Marginal	No	Yes	Yes	Moderate	Cheap
2-DIGE	Electrophoresis: IEF PAGE	Densitometry of Cy3- and Cy5-labeled proteins normalize to Cy2	Mass spectrometry (PMF;MS/MS)	Dependent on detergents used	Moderate (especially with scanning gels)	No	Yes	Yes	Moderate	Expensive
MuDPIT	LC–LC of peptides	None	Mass spectrometry (MS/MS)	Theoretically better than electrophoresis but not systematically examined	Moderate, often used with large sample amounts	No	Yes	Yes	Rapid	Moderate
ICAT™	LC of peptides	Through use of heavy and light tags	Mass spectrometry (MS/MS)	No better than 2-DE	Moderate	No	Yes	No	Rapid	Moderate
SELDI-TOF-MS	Binding of proteins based on their chemical and physical characteristics	Comparison of MS peaks	Requires series of samples or coupling to second MS instrument	Moderate	Marginal to moderate	No	Yes	No	Rapid	Expensive
Protein arrays	Antibody-based chips (binding to affinity reagent)	Densitometry of binding	Binding to particular affinity reagent	Unknown	Unknown	Yes	No	Yes	Rapid	Cheap

2-DE: two-dimensional electrophoresis; 2-DIGE: two-dimensional difference in gel electrophoresis; MuDPIT: multidimensional protein identification technology; LC: liquid chromatography; LC–LC: tandem liquid chromatography; PAGE: polyacrylamide gel electrophoresis; ICAT: isotopecoded affinity tags; SELDI-TOF-MS: spectrum-enhanced laser desorption ionization—time of flight—mass spectrometry; PMF: peptide mass fingerprint; MS/MS: tandem mass spectrometry.

fields in biology by providing enormous amounts of genetic data from model and nonmodel organisms. Proteomics promises to bridge the gap between our understanding of genome sequences and cellular behavior involved in host–pathogen interactions. Proteomics offers the possibility to characterize host–pathogen interactions from a global proteomic view. To date, most proteomics surveys on host–parasite interactions have focused on cataloguing protein content of pathogens and identifying virulence-associated proteins or proteomic alterations in host response to a pathogen. Also, many parasitologists and molecular biologists have used proteomics to find pathogen-specific host biomarkers for rapid pathogen detection and characterization of host–pathogen crosstalks during the infection process. In this section, a synthetic retrospective of previous proteomics studies on host–pathogen interactions and some pitfalls of these surveys are presented.

11.3.1 Deciphering of the Molecular Strategies Involved in Parasite Immune Evasion

To elude the vigilance of the immune system of a host, particularly a mammal, a causative microorganism must actually act as a double agent. Indeed, the broad immunity has a natural or innate and adaptive component. Innate immunity constitutes the first antimicrobial defense and rapidly induces soluble mediators such as complement, inflammatory cytokines, and chemokines together with effector cells such as macrophages and natural killers, in order to control or delay the spreading of the infectious agent. Then a specific response of adaptive immunity will take place to eliminate pathogens that would have survived innate immune response (Roitt and Delves, 2001). These immune selective pressures have conducted pathogens to develop mechanisms to modulate and alter host responses or to evade phagocytosis. As a result of these host pathogen interactions, protein expression profiles of the host immune system (susceptibility/tolerance factors: antibodies, cell receptors, biochemical pathway, ...) and of the pathogen (virulence/pathogenicity factors: **antigens**, immunomodulators) are mutually modified (Zhang et al., 2005; Coiras et al., 2008; Holzmuller et al., 2008).

Depending on the pathogen type (virus, bacteria, fungi, unicellular or multicellular parasites), strategies of interactions will be different and the subversion of the host immune responses will exhibit specificities at the protein level (for reviews, see Walduck et al., 2004; Biron et al., 2005a; Viswanathan and Früh, 2007). In fact, these molecular dialogues and conflicts can be seen as a chess game between the host immune cell populations and the pathogen populations, in which the pathogen plays with the whites (i.e., it starts the game). Because of differences in host–pathogen organisms' size and ratio, leading to size differences of respective proteomes, the pathogen proteome could be considered as overwhelmed by the host proteome during the interactions. But in terms of immune evasion, this is not limiting because the immune system works on a qualitative basis, which constitutes a second advantage for the pathogen that can induce large-scale damages with low amounts of molecules. By contrast, this represents one major limitation to characterize host–pathogen interactions, but also a challenging perspective for proteomics technology. This is why retrospectively proteomics studies were mainly conducted to evidence pathogenic virulence and pathogenicity factors (Ouellette et al., 2003; Texier et al., 2005; Sims and Hyde, 2006; Van Hellemond et al., 2007; Liu et al., 2008a,b; Bird and Opperman, 2009; Jagusztyn-Krynicka et al., 2009; Premsler et al., 2009; Weiss et al., 2009; Steuart, 2010; Bhavsar et al., 2010; Holzmuller et al., 2010).

Independently of the proteomics workflow used for analysis, parasite immune evasion could be illustrated by at least three strategies that are commonly widespread among pathogens: (1) immune evasion based on antigenic variation, (2) inhibition of adaptive immunity activation systems, and (3) host mimicry. In African trypanosomes, the antigenic variation of the variant surface glycoprotein (VSG) constituting the surface coat of the parasite is well described (Morrison et al., 2009). But as in proteomics study, the parasite population, which has switched the VSG, is so poorly represented it goes undetected, and therefore always keeps one step ahead of host immune responses. Also, in trypanosomatids, *Leishmania amastigotes*, which establish within macrophage (a major immune effector cell), developed the ability to degrade class II **major histocompatibility complex** molecules to prevent Th1-type immunity to be induced (Antoine et al., 2004). This strategy can be likened to the concept of histocompatibility testing in the case of transplant to avoid rejection of non-self by the recipient. Another protozoan parasite, *Toxoplasma gondii*, generates its parasitophorous vacuole with elements of the plasma membrane from the targeted host cells, thus using the host "self" to evade immune recognition (Plattner and Soldati-Favre, 2008). These few examples perfectly illustrate how difficult it is to decipher, at the protein level during interactions, the pathogen molecular components involved in immune evasion. Nevertheless, advances in proteomics offer challenging perspectives to decipher the molecular war in host—pathogen interactions.

11.3.2 Host Proteome Responses to Parasite Infection

While it seems obvious to say that when a pathogen will infect a host, it will react by expressing molecules that can be characterized by clinical proteomics, it is surprising how few studies are devoted to this research. Yet the discovery for biomarkers differentiating an infected state from a healthy state is the heart of the Infectious Disease Research (te Pas and Claes, 2004; Azad et al., 2006), and expression proteomics has quickly developed to characterize the differential expression of proteins encoded by a particular gene and their posttranslational modifications in biological fluids and tissues (Fournier et al., 2008; Hood et al., 2009; Wilm, 2009). In characterizing the host proteome responses to a pathogen infection, different levels of analysis have to be considered: soluble biomarkers expressed in biological fluids (e.g., serum, saliva, urine, and cerebrospinal fluid), tissue biomarkers indicative of an organ response and cellular biomarkers indicative of a cell-type response (e.g., immune cells).

Interestingly, the majority of the proteomics studies on host response to infection were performed on viral deregulation of host cells proteome ex-vivo (Liu et al.,

2008a,b; Sun et al., 2008; Zheng et al., 2008; Antrobus et al., 2009; Lee et al., 2009; Pastorino et al., 2009; Vester et al., 2009; Zhang et al., 2009, 2010a,b). These works allowed characterization at the molecular level, the overall modifications in protein profiles of the target cells, and were of high interest to the better understanding of the pathogen influence on its host. In bacteria, studies have evaluated the mode of action of known toxins or bacterial components on host cells (Kuhn, et al., 2006; Shui et al., 2009). Concerning parasites, ex-vivo experiments on host cell–parasite interactions have highlighted molecular details of manipulation strategies suffered by target cells during toxoplasmosis Chagas disease or malaria (Teixeira et al., 2006; Nelson et al., 2008; Wu et al., 2009b). Curiously, few works directly focused on the subversion of the immune system, mainly through monocyte/macrophage deregulation (Oura et al., 2006; Fischer et al., 2007).

As a paradox, the most striking studies on host proteome response to parasite infection were performed on arthropod (infectious diseases vectors)–parasite interactions, probably because the parasite induced a strong phenotype modification (Biron et al., 2005c; Rachinsky et al., 2007), particularly in the case of insect behavior manipulation (Lefèvre et al., 2007a,b). Although few in number, taken together these pioneering analyses of the response of the proteome of the host to a pathogen paved the way for the dynamic analysis of host–pathogen interactions. These approaches deserve to be strengthened and extended to all infectious diseases to increase and improve knowledge of the molecular dialogue and conflict that govern host–pathogen interactions.

On the other hand, the clinical aspect is important in infectious diseases, and a number of studies have sought to characterize more comprehensively the proteome response of the host to infection in biological fluids, with a purpose diagnosis. One interesting pioneering study was performed in rabbits and allowed detection of intra-amniotic infection by proteomic-based amniotic fluid analysis (Klein et al., 2005). For human diseases or those of livestock, the biological fluid, which should enable the detection of infection in the host serum linked to host proteome response. Several studies performed on this biological sample have allowed discriminating host–commensal from host–pathogen interactions in *Candida albicans* (Pitarch et al., 2009) and determining the immunome of pathogens (Sakolvaree et al., 2007; Ju et al., 2009). Moreover, in African trypanosomiasis, proteomics analysis of the serum not only was indicative of the host response to infection, but also was promising for characterizing disease progression toward neurological disorder (Papadopoulos et al., 2004; Agranoff et al., 2005). This illustrates how proteomics will help in considering at different analytical levels the host proteome response to a pathogen infection, with the prospect of benefits in improving diagnostic and therapeutic.

11.3.3 Biomarkers Linked to Infection Process by a Pathogen Using SELDI-TOF-MS Technology

High-throughput proteomic technology offers promise for the discovery of disease biomarkers and has extended our ability to unravel proteomes. In this section, we will focus on the surface-enhanced laser desorption time-of-flight mass spectrometry (SELDI-TOF-MS) technology. This mass spectrometric-based method requires a minimal amount of sample for analysis and allows the rapid high-throughput analysis of complex protein samples (De Bock et al., 2010). SELDI-TOF-MS differs from conventional matrix-assisted laser desorption ionization (MALDI)-TOF-MS because the target surfaces to which the proteins and matrices are applied are coated with various chemically active ProteinChip® surfaces (ion exchange, immobilized metal affinity capture, and reverse phase arrays). Therefore, it is possible to fractionate proteins within a mixture, or particular classes of proteins, on the array surface prior to analysis. As with MALDI, different matrices can be used to facilitate the ionization and desorption of proteins from the SELDI array surface (Merchant and Weinberger, 2000).

This technology was initially applied to the discovery of early diagnostic or prognostic biomarkers of cancer (Petricoin and Liotta, 2004; Xiao et al., 2005; Yang et al., 2005). Recently, this technology has been used to discover fluid or tissue protein biomarkers for infectious diseases such as HIV-1 (Luo et al., 2003; Sun et al., 2004; Missé et al., 2007; Luciano-Montalvo et al., 2008; Toro-Nieves et al., 2009; Wiederin et al., 2008), hepatitis B and C viruses (Poon et al., 2005; Kanmura et al., 2007; Fujita et al., 2008; Molina et al., 2008; Wu et al., 2009a,b), severe acute respiratory syndrome (Pang et al., 2006) and BK virus (Jahnukainen et al., 2006), African Trypanosomiasis (Stiles et al., 2004; Agranoff et al., 2005), infection of Artemia by cestodes (Sánchez et al., 2009), tuberculosis (Liu et al., 2010), bacterial endocarditis (Fenollar et al., 2006), and *Helicobacter pylori* infection (Das et al., 2005).

Certain individuals are resistant to HIV-1 infection, despite repeated exposure to the virus. The analysis of resistance to HIV infection is one of a research avenue which has the hope of resulting in the development of a more effective treatment or a successful preventive vaccine against HIV infection. However, the molecular mechanism underlying resistance in repeatedly HIV-1-exposed, uninfected individuals is unclear. A complementary transcriptome and SELDI-TOF-MS analyses on peripheral blood T cells, and plasma or serum from EU, their HIV-1- infected sexual partners, and healthy controls have been performed (Missé et al., 2007). This study detected a specific biomarker associated with innate host resistance to HIV infection, as an 8.6-kDa A-SAA cleavage product.

In the same vein, understanding the virus−host interactions that lead to patients with acute Hepatitis C virus (HCV) infection to viral clearance is a key toward the development of more effective treatment and prevention strategies. SELDI-TOF-MS technology have been used to compare, at a proteomic level, plasma samples respectively from donors who had resolved their HCV infection after seroconversion, from donors with chronic HCV infection and from unexposed healthy donors (Molina et al., 2008). A candidate marker of about 9.4 kDa was found to be higher in donors with HCV clearance than in donors with chronic infection. This biomarker was identified by nanoLC-Q-TOF-MS/MS as Apolipoprotein C-III and validated by Western Blot analysis.

11.4 Pitfalls of the Current Conceptual Approach in many "Parasito-Proteomics" Studies

During the postgenomics era, the "**parasito-proteomics**" have been suggested to study host–pathogen interactions. The "parasito-proteomics" is the study of the reaction of the host and parasite genomes through the expression of the host and parasite proteomes during their biochemical crosstalk. Studies in "parasito-proteomics" are performed either by following the expression of the parasite proteome during infection by a given parasite (Langley et al., 1987; Moura and Visvesvara, 2001; Cohen et al., 2002; Boonmee et al., 2003), by the reaction of the host proteome following an invasion by a parasite species (Moskalyk et al., 1996; Thiel and Bruchhaus, 2001; Cohen et al., 2002), or by the injection of immune **elicitors** (Han et al., 1999; Vierstraete et al., 2004).

Some elegant studies on the proteome responses of insect hosts during their **molecular crosstalk** with pathogens concluded that insects could rapidly react to infection by a given pathogen (i.e., bacteria or fungi) by producing immune-induced proteins including peptide or polypeptides (Hoffman, 1995). However, a key point is to define whether the host genomic responses elicited through activation of immune constitutive proteins, induction, and/or suppression of proteins during the infection by a parasite represent a nonspecific response that might be induced by any pathogen. Many "parasite-proteomics" studies dealt with a limited framework by deciphering the host proteome response during the infection process by a specific pathogen (Huang et al., 2002).

The classical approach in parasito-proteomics makes it possible to identify proteins of interest for a given host–parasite association. For example, Wattam and Christensen (1992) associated some polypeptides with the genome response of the host mosquito, *Aedes aegypti* (Diptera: Culicidae), with the invasion of the filarial worm, *Brugia malayi*. This pioneering study provided important new information on the response of the host insect to invasion by a specific parasite species. Nevertheless, it was *not* possible to determine whether that the response detected in *Ae. aegypti* is specific to *B. malayi* (Spiruria: Filariidae), or whether it can be observed for any parasitic worm species invading a dipterous host.

Other studies have revealed the limitations of the current approach in parasito-proteomics by showing that in the host–parasite interaction, many immune mechanisms are involved (constitutive, induced, specific, or otherwise) (Haab et al., 2001; Levy et al., 2004; Vierstraete et al., 2004). By using two treatments, the injection of lipopolysaccharides (LPS) and a sterile injury, Vierstraete et al. (2004) were able to disentangle proteome modifications induced by immunity from those induced by a physical stress. Levy et al. (2004) studied the immune response of the fruitfly *Drosophila* to bacterial (*Micrococcus luteus* and *Escherichia coli*) and fungal (*Beauveria bassiana*) infections. The data revealed that 70 of the 160 protein spots detected were differentially expressed at least fivefold after a bacterial or fungal challenge. In addition, the majority of these spots were specifically regulated by one pathogen, whereas only a few spots corresponded to proteins altered in all cases of infection. In summary, the classical approach in parasito-proteomics has many benefits in terms of understanding fundamental aspects of gene–protein functional interactions. Unfortunately, it is not applicable to different pathogen species (and as such, does not encourage the growth of knowledge of general host proteome responses), nor does it necessarily help in the creation of a proteomic database with a holistic relevance to the understanding of host–parasite interactions.

Moreover one classical pitfall in "parasite-proteomics" surveys is the use of a single proteomic technique. Recently, Bridges et al. (2008) have demonstrated by using a battery of proteomic techniques to characterize the plasma membrane subproteome of bloodstream form of *T. brucei* that these techniques are complementary since each one has identified a unique subset of proteins of the plasma membrane. Although 2-DE offers a high-quality approach for studying the host and/or parasite proteomes, several proteomic tools have been developed that will complement this traditional proteomic tool (see Table 11.1). In the same way, MS analysis could take advantages of combined techniques. For example, the widely used analysis of peptides via collisionally activated dissociation (CAD) is rapid and results in reproducible predictable fragmentation behavior for a given peptide sequence, a substantial proportion of peptide product ion mass spectra does not result in successful sequence identification (Steen and Mann, 2004). This is well illustrated by Hart et al. (2009), who showed that a substantial number of proteins from trypanosome flagellum were identified in their three independent flagellar proteome investigations, but also that combining electron transfer dissociation (ETD) with CAD allowed the identification of 168 proteins that were not recognized in their first analysis. This strengthens the idea of integrating both approaches and technologies to reach exhaustive protein datasets from a given biological compartment.

Finally, although 2-DE and MS have been very successfully employed to identify proteins involved in host–parasite crosstalks, many recent papers have emphasized the pitfalls of 2-DE experiments, especially in relation to experimental design, poor statistical treatment, and the high rate of "false positive" results with regard to protein identification (Barret et al., 2005; Biron et al., 2006a; Holzmuller et al., 2008). It is necessary to be careful in the interpretation of results for the previous "parasito-proteomics" surveys on host–parasite interactions (see Biron et al., 2006a; Holzmuller et al., 2008).

11.5 Toward New Conceptual Approaches to Decipher the Host–Parasite Interactions for Parasites with Short or Complex Life Cycle

One main goal of "parasite-proteomics" surveys is to find proteins for use as pathogen-specific host biomarkers and to decipher host–pathogen crosstalks. Some recent papers emphasize that a significant number of surveys were done with a nonrigorous experimental design and without a conceptual approach to disentangle a general host proteome response from a specific host proteome response during the interaction with a pathogen (Tastet et al., 1999; Ashton et al., 2001; Huang et al., 2002; Biron et al., 2005a; Holzmuller et al., 2008, 2010). A new attitude is essential to improve the reliability of proteomics data on host–pathogen interactions. Lately, some conceptual approaches have been proposed to researchers working on host–pathogen interactions to improve the reliability of "parasite-proteomics" results and to stimulate the creation of proteomic database with a holistic view of host–pathogen interactions. Thus, in this section, three new avenues to decipher host–pathogen interactions for any pathogen species (i.e., with simple or complex life cycle) are presented.

11.5.1 A Holistic Approach to Disentangle the Host and Parasite Genome Responses During Their Interactions

Some proteomics studies have shown common features in the innate response of plants, insects, and mammals (Broekaert et al., 1995; Rock et al., 1998; Cao et al., 2001; Taylor et al., 2003). The plant defense response is mediated by disease resistance genes (R genes), which are abundant throughout the genome and confer resistance to many microorganisms, nematodes, and/or insects (Dixon et al., 2000). R genes of several families of plants studied to date show homology with the Drosophila receptor Toll and the mammalian interleukin-1 receptor (Rock et al., 1998). In addition, plants, invertebrates, and vertebrates produce the "**defensins**" class of peptides, which are pathogen-inducible (Broekaert et al., 1995; Hoffman, 1995). Some peptides and/or proteins used by phytophagous or animal parasites to modify the genome expression of their host, share many structural and functional homologies. Thus, for example, phytoparasitic root-knot nematodes of the genus *Meloidogyne* secrete substances into their plant hosts in order to make a giant cell used as a feeding site (Abad et al., 2003; Doyle and Lambert, 2003). A similar system is observed for the zooparasite *Trichinella spiralis* (Stichosomida: Trichinellidae) (Jasmer, 1993). Furthermore, the injection of a peptide isolated from nematode secretions to either plant protoplasts or human cells enhances cell division (Goverse et al., 2000). The mechanism is not yet well-known but protein induction is considered as a strong possibility.

Currently, many data are obtained by genomic and proteomics projects concerned with host–parasite interactions. Nevertheless, as mentioned earlier, generally little effort is made to elaborate such projects with respect to a holistic view of the goal to increase knowledge concerning immune responses of a host along with the biochemical crosstalk between host and pathogen/parasite. Thus far, parasito-proteomics studies are in their infancy but have already led to new insights concerning molecular pathogenesis and microorganism identification (Moura et al., 2003; Vierstraete et al., 2004; Levy et al., 2004; Biron et al., 2005d). However, many parasito-proteomics studies have been done with powerful tools but without a conceptual approach to disentangle the host and parasite genome responses during their interactions.

A new holistic approach proposed to parasitologists and molecular biologists based on evolutionary concepts of the immune response of a host to an invading parasite (for more details, see Biron et al., 2005a). For instance, this new conceptual approach enables the classification of the host genomic response to infection by a parasite according to the immune mechanisms used (constitutive versus induced) and the degree of specificity. From an evolutionary-ecological point of view, host immune responses to a particular parasite can be plotted on a chart according to the immune mechanisms used (constitutive versus induced) and degree of specificity. The first axis of the defense chart refers to the immune mechanisms employed by the host with the two extreme cases: (1) a constitutive immune mechanism used by the host to rapidly impair the invasion by a parasite; and (2) an induced immune mechanism which has the advantage of avoiding a costly defense system, yet meanwhile has the disadvantage that the parasite might escape host control (Schmid-Hempel and Ebert, 2003). The second axis of the defense chart refers to the degree of specificity of the host immune response.

Whatever the tactics used and the degree of specificity, the host genome ensures the adequate operation of the immune response via the proteome (genome-operating system). For each immune tactic, many proteins are implicated. Consequently, any researcher in parasito-proteomics working with the immune defense chart will be able to categorize the host genome reaction for any given parasite at any given time. Also, for the pathogen, from an evolutionary-ecological point of view, parasite molecular strategies used to counteract host immune system can be plotted on a chart according to the infection mechanisms used (constitutive versus induced) and degree of specificity. This type of approach should be as much hypothesis generating for parasito-proteomics as for evolutionary ecology itself.

Lately, pioneer proteomics studies on parasite-induced alteration of host behavior (widespread transmission strategy among pathogens) have been carried out on six arthropod host—parasite associations: two orthoptera—hairworm associations, two insect vector—pathogen associations, and two gammarid—parasite associations (Lefèvre et al. 2009). These parasito-proteomics studies were based on the conceptual approach suggested by Biron et al. (2005a,b). Thus, in each study, many biological treatments have been effected to control the potential confusion resulting from proteins that are nonspecific to the manipulative process and to find the protein potentially linked with host behavioral changes. Also, for each study, to limit the possible effects of multiple infection and/or host sex-specific factors on the host proteome response, only mono-infected host males were used for the protoemics analysis. These parasito-proteomics surveys on the parasitic manipulation hypothesis showed that proteomic tools and the conceptual approach suggested by Biron et al. (2005a,b) are sensitive enough to disentangle host proteome alterations and also the parasite proteome alterations linked to many factors such as the circadian cycle, parasitic status, parasitic emergence, quality of a habitat, and manipulative process.

11.5.2 Pathogeno-proteomics: A "New" Avenue to Decipher Host—Vector—Pathogen Interactions

Relationships between pathogens and their hosts and vectors depend on a molecular dialogue being tightly regulated. The reciprocal influence of a pathogen with its host or vector will affect the level of their genomes and their expression, respectively (Holzmuller et al., 2008). Variability and cross-regulation increase from genomic DNA (mutations, rearrangement, methylation) through RNA transcripts (initiation, splicing, maturation, editing, stability) to functional proteins (initiation, folding, posttranslational modifications, localization, function). Pathogeno-proteomics is a new approach to decipher host—vector—pathogen interactions, which integrate modifications at all analytical levels (genome, transcriptome, proteome: whole cell content, and secretome: naturally excreted—secreted molecules) through the analysis of their end-products' profile (Figure 11.1).

The concept is based on management with drawers of the analytic workflow, from the determination of number of experimental treatments and design of the biological material preparation to the dedicated proteomics and **bioinformatics** tools needed to answer a research question in cell immunobiology (directly involved in host—pathogen interactions) but also in ecology and evolution, population's biology and adaptive processes (Biron et al., 2006b; Holzmuller et al., 2008; Karr, 2008). Moreover, it has been proved that the results of this type of integrated approach has a concrete impact on the discovery of the causes of infectious diseases, as well as on improving the diagnosis, vaccine development, and rational drug design (Doytchinova et al., 2003; Bansal, 2005; Chautard et al., 2009). Despite a theoretical aspect (Kint et al., 2010), the pathogeno-proteomics concept brought new insights into important aspects of cell signaling (Kleppe et al., 2006) and molecular medicine (Ahram and Petricoin, 2008; Ostrowski and Wyrwicz, 2009). As an example, proteomics and bioinformatics tools enable the formulation of relevant biological

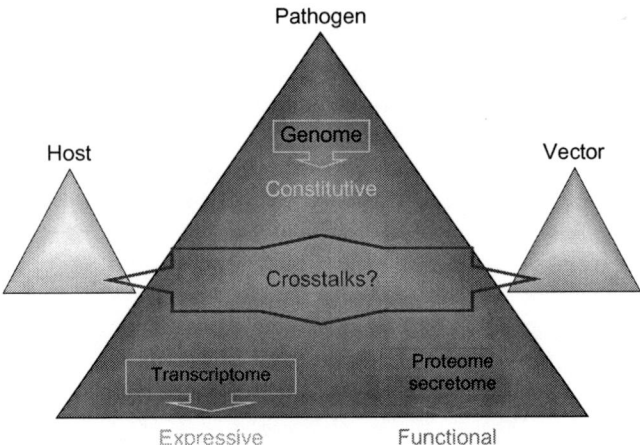

Figure 11.1 Pathogeno-proteomics: integrating analytical levels in host—vector—pathogen interactions.

hypothesis on why part of the fungal population is killed while a significantly high percentage survives in *C. albicans*–macrophage interactions (Diez-Orejas and Fernández-Arenas, 2008), leading to edition of a specific database for studying *C. albicans*–host interactions (Vialás et al., 2009). Direct applications in terms of discovery of antifungal drug targets or design of new effective antibacterial vaccines become reality (Tournu et al., 2005; Jagusztyn-Krynicka et al., 2009). Other studies have also highlighted the pathogenic changes in the brain of SIV-infected monkeys (Pendyala et al., 2009), adaptive metabolic changes in *Trypanosoma cruzi* and *Trypanosoma congolense* (Grébaut et al., 2009; Roberts et al., 2009) or molecular biomarkers of intestinal disorder induced by *H. pylori* or *Tritrichomonas muris* (Wu et al., 2008; Kashiwagi, et al., 2009). More recently, the use of model organisms interacting with infectious agent of medical importance emphasized the complexity and pathogen-specificity of the worm's immune response (Bogaerts et al., 2010). Taken together these examples demonstrate the potential of the concept of pathogeno-proteomics and promote this new research avenue.

11.5.3 Host–Pathogen Interactomes

The past few years have witnessed the birth of new biological entities named interactomes. They correspond in an "ideal world" to the complete set of protein–protein interactions existing between all the proteins of an organism. In reality they are far from complete since an unknown number of interactions are yet to be discovered. Current interactomes are only a part of the whole set of possible interactions occurring within an organism or between organisms. They are generally assembled from the results of large-scale two hybrid screens (LS-Y2H) around 6000, 4000, 23,000, and 5500 interactions for yeast (Uetz et al., 2000; Ito et al., 2001), the nematode, *Caenorhabditis elegans* (Maupas) (Li et al., 2004), *Drosophila* spp. (Giot et al., 2003; Formstecher et al., 2005), and humans (Rual et al., 2005, Stelzl et al., 2005), respectively; and the interactions identified by low-scale experiments described in the literature that may be eventually compiled in specialized databases (e.g., INTACT [Hermjakob et al., 2004], MINT [Zanzoni et al., 2002], HPRD [Peri et al., 2003], BIND [Alfarano et al., 2005]). Consequently, they do not reflect temporal influences because interactions are gathered from different cell types, tissues, development stages, and types of experiment.

Interactomes form large intricate networks leading to a renewed vision of cell biology as an integrated system. However, extracting and revealing the functional information they contain depends on our ability to analyze them in detail. For this, bioinformatics methods that partition the interaction network into functional modules have been proposed. These modules usually correspond to group of proteins involved in the same pathway, the same protein complex, or the same cellular process. Since interaction networks are represented by complex graphs in which nodes correspond to proteins and edges to the interactions, a number of these network analysis methods have been grounded on principles that derive from graph theory.

Noticeably, a functional module or a class of protein that is functionally related and based on network analysis can be deduced from: (i) a search for graph regions particularly densely populated by interactions (Bader and Hogue, 2003; Spirin and

Mirny, 2003); (ii) the similarity between the shortest paths in the graph (Rives and Galitski, 2003); (iii) the progressive disconnection of the graph using a calculation of edge "betweenness" (Girvan and Newman, 2002; Dunn et al., 2005); and (iv) the sharing of interactors (Brun et al., 2003; Samanta and Liang, 2003) or a combination thereof (Brun et al., 2004). Some of these methods use the functional annotations of the protein (such as Gene Ontology annotations) to annotate the functional modules they predict. Based on the characteristics of a protein of unknown function to some of these annotated modules or classes, a putative function for such proteins can be proposed (Brun et al., 2003). Currently, specialized methods are being developed to investigate the interactomes. But it is clear that the field starts to move forward. Thus some softwares, plug-ins, or servers are available for free use by the proteomics research community (for instance, MCODE in Cytoscape [Shannon et al., 2003]; Prodistin [Baudot et al., 2006]).

Although the deciphering of the interactomes of the main model organisms is not yet complete, studies of the interactomes of pathogens are increasing. The first pathogens to be investigated in terms of their interactomes were the hepatitis C virus (Flajolet et al., 2000) and the bacterium, *H. pylori* (Rain et al., 2001). More recently, the interactomes of the herpes viruses (Uetz et al., 2006) and the malaria parasite, *Plasmodium falciparum* (LaCount et al., 2005), have been determined. This makes one believe that in the near future, as initiated by Uetz et al. (2006), the docking of the interactomes of pathogens onto those of their hosts will be possible. The analysis of the host–pathogen interactome (i.e., "docked interactomes") during the host–pathogen crosstalk is certainly a very promising and exciting aspect of interactomics because of its obvious potential impacts on human and animal health (Figure 11.2). The host–pathogen interactome will permit to identify host and pathogen protein networks linked to specific functions during their interaction.

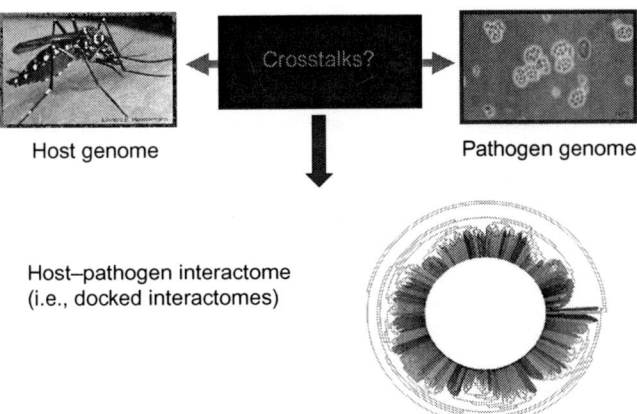

Figure 11.2 A new biological entities named host–pathogen interactome corresponding to complete set of protein–protein interactions existing between all the proteins of a host and a pathogen during their interaction.

11.6 Population Proteomics, an Emerging Discipline, to Study Host−Parasite Interactions

The host susceptibility to a pathogen and/or the pathogen virulence are often fluctuating within a host population even when infected hosts are collected in the same habitat and at the same time. This host phenotypic variability can be caused by three factors: (i) host genotype and/or pathogen genotype, (ii) different environmental experiences (habitat fragmented in microclimates), and (iii) host coinfection by pathogens (competition or mutualism among co-infecting pathogens within hosts). What are the host−pathogen crosstalks at individual and population scales in a habitat? Is it possible to detect and to decipher the host proteome variability within a habitat for the molecular mechanisms and for the protein networks involved in the host−pathogen interactions? In this section, a new emerging discipline in proteomics, the population proteomics, and its prospects are presented with results of some pioneer studies on this topic, especially in human population proteomics.

11.6.1 Prospects with Population Proteomics for Any Living Organisms

One limiting factor for the first generation of proteomics tools (such as 2-DE) is the amount of proteins required to study the host and/or pathogen proteome expression(s) during their interactions. Most surveys in "parasite-proteomics" were done by pooling many individuals for any treatment (such as infected and noninfected hosts) required to answer to a query. Thus, with this kind of experimental protocol, no data can be acquired on the interindividual variation in expression of host and pathogen proteomes during their crosstalk. New proteomics tools and methods have been developed as 2D-LC/MS that can permit to study the interindividual variation of molecular crosstalk in host−pathogen associations (Nedelkov et al., 2004; Predel et al., 2004; Brand et al., 2005).

At beginning of the century, Dobrin Nedelkov proposed a new scientific field in proteomics: the population proteomics (Nedelkov et al., 2004). Population proteomics was defined as the study of protein diversity in human populations, or more specifically, targeted investigation of human proteins across and within populations to define and understand protein diversity with the main aim to discover disease-specific protein modulations (Nedelkov, 2008). Biron et al. (2006b) have proposed to broaden the "population proteomics" concept to all living organisms with the aim to complement **population genetics** and to offer a new avenue to decipher the crosstalk diversity involved in trophic interactions in a habitat, since the execution of the genetic plan is carried out by the activities of proteins and natural selection acts at the protein level (Karr, 2008; Cieslak and Ribera, 2009).

The apparent separation between genomics and proteomics that leads to different perspectives on the same ecological reality is a fundamental limitation that needs to be overcome if complex processes, like adaptation, pathogen virulence, and host susceptibility are to be understood. Population proteomics coupled with population genetics has a great potential to resolve issues specific to the ecology, the evolution of natural populations, the dynamic of host susceptibility to pathogens, the evolution of pathogen virulence, and the range of host genotypes that can be infected with a given pathogen genotype in host–parasite interactions. Some perspectives for the population proteomics are resumed in Figure 11.3. Even if we are yet far from this "promised land", a better understanding of the information contained in proteomics markers should permit an impressive amount of information to be gathered on the past as well as current environmental conditions experienced by a given population of a species, something that could be summarized as "show me your proteome and I will tell you who you are, where you are from, and where you should go from here."

Lately, pioneer surveys on population proteomics have been carried out with classical proteomic tools (i.e., 2-DE and MS) to determine the genetic variability between species and between populations of a given species (Chevalier et al., 2004; Diz and Skibinski, 2007; Valcu et al., 2008), to identify biochemical signatures linked to particular habitat and/or environmental conditions (Thiellement et al., 1999; Pedersen et al., 2010) and phylogenetic studies (Navas and Albar 2004, Dorus et al., 2006). Nedelkov et al. (2005, 2007) have investigated the human plasma proteins' diversity using approaches similar to enzyme-linked immunosorbent assay but utilizing MS as method of detection (Nedelkov, 2008). These pioneer results should help to discover disease-specific protein modulations but also to find pathogen-specific protein biomarkers.

The next subsection will present more in details the Nedelkov results on protein diversity in human populations.

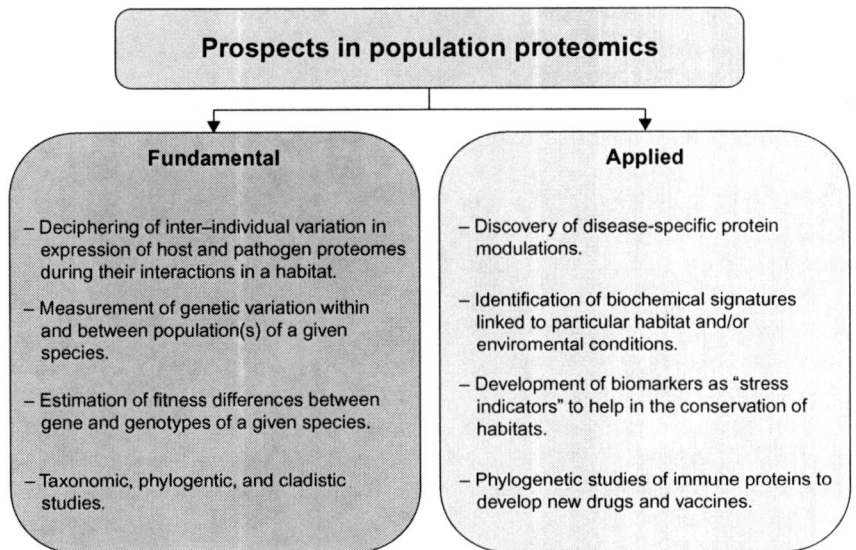

Figure 11.3 Potential of population proteomics as an emerging discipline in proteomics.

11.6.2 Human Population Proteomics

Human population proteomics deciphers protein diversity in human populations. In broader terms, human population proteomics can be compared to human population genomics, where individuals are interrogated with the aim of cataloguing common genetic variants and determining how they are distributed among people within populations and among populations in different parts of the world (Nedelkov et al., 2004, 2006; Nedelkov, 2005). Although human population proteomics cannot (yet) claim such outreach and goals, it has the potential to become an important proteomics subdiscipline as the tools and approaches that enable it become more embraced and practiced.

Human population proteomics does not engage the study of entire proteomes because it is very likely that, for a specific cell or tissue proteome, there is no definitive set and number of proteins that is common to all within a group or a larger population. Instead, human population proteomics focuses on interrogation of a selected number of proteins but from a large number of individuals, to delineate the distribution of specific protein modifications within these subpopulations. Hence, targeted protein analysis approaches utilizing MS as detection method are employed. MS measures a unique feature of each fully expressed protein—its molecular mass. Changes in the protein structure resulting from structural modifications are reflected in its molecular mass and can be detected via MS, without a priori knowledge of the modification. The MS methods utilized in human population proteomics must be capable of analyzing hundreds if not thousands of samples per day, with high reproducibility and sensitivity. Hence, top-down MS approaches utilizing affinity ligands are the most likely methods of choice for population proteomics (Nedelkov, 2006). Surface-immobilized ligands can be utilized to affinity to retrieve a protein of interest from a biological sample, after which the protein (with or without the affinity ligand) is introduced in a mass spectrometer. One of the first affinity MS methods developed was mass spectrometric immunoassay (MSIA) (Nelson et al., 1995). The approach combines targeted protein affinity extraction with rigorous characterization using MALDI-TOF MS (Figure 11.4). Protein(s) are extracted from a biological sample with the help of affinity pipettes derivatized with polyclonal antibodies. The proteins are eluted from the affinity pipettes with a MALDI matrix, and are MS-analyzed. Enzymatic digestion, if needed, is performed on the MALDI target itself. Specificity and sensitivity, as in traditional immunoassays, are dictated by the affinity-capture reagents—the antibodies.

However, a second measure of specificity is incorporated in the resulting mass spectra, wherein each protein registers at specific m/z value. During data analysis, the major signal in the mass spectrum that corresponds to the targeted protein is initially evaluated; it should be within a reasonable range (e.g., error of measurement of <0.05%) from the value of the empirically calculated mass obtained from the sequence of the protein deposited in the Swiss-Prot databank. Once this mass value is confirmed (or observed to be shifted), the presence of protein modifications is noted by the appearance of other signals in the mass spectra (usually in the vicinity of the native protein peaks), or by mass shifts of the major protein signal.

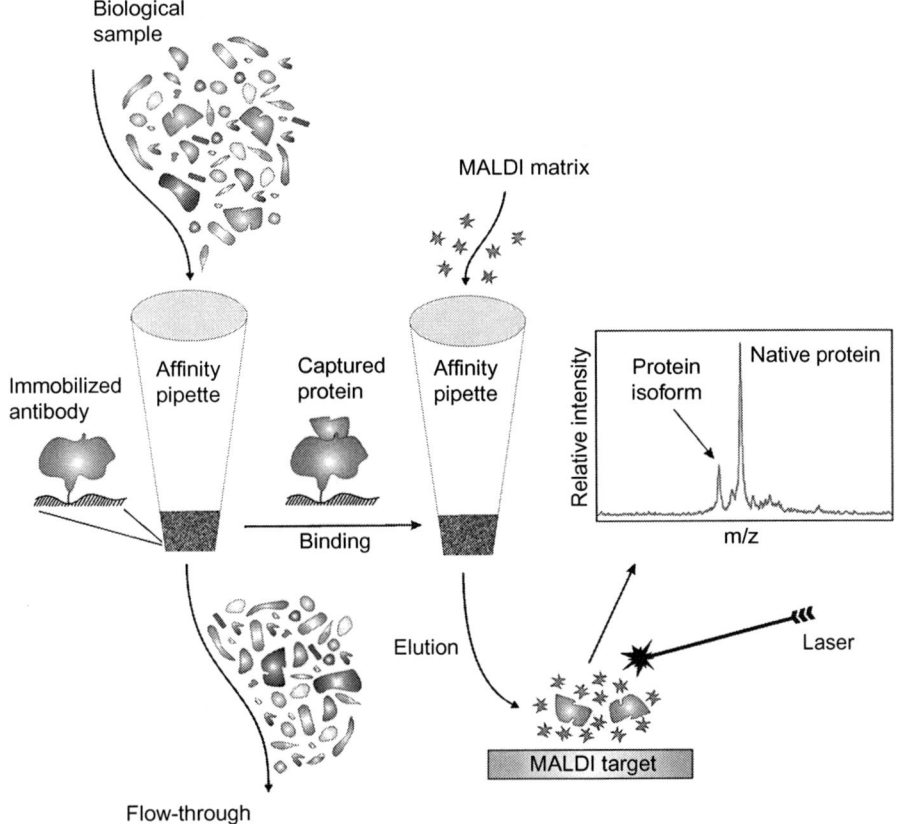

Figure 11.4 Schematics of the MSIA approach.

Modifications can be tentatively assigned by accurate measurement of the observed mass shifts (from the wild-type protein signals and/or in-silico calculated mass) and knowledge of the protein sequence and possible modifications. The identity of the modifications is then verified using proteolytic digestion and mass mapping approaches in combination with high-performance MS.

In an initial study of human protein diversity using mass spectrometric methods of detection, 25 plasma proteins from a cohort of 96 healthy individuals were investigated via MSIA (Nedelkov et al., 2005). The protocol and an example of the data generated for one of the protein, transthyretin (TTR), are outlined in Figure 11.5.

The TTR MSIA were performed in parallel on the 96 human plasma samples using affinity pipettes derivatized with anti-TTR antibody. Following mass spectrometric analysis, data matrix containing all tentatively assigned modifications was assembled. Then, peptide-mapping experiments were performed on selected number of samples to identify the specific modifications and finalize the modifications database. The data for all 25 proteins is presented in Figure 11.6, which lists the

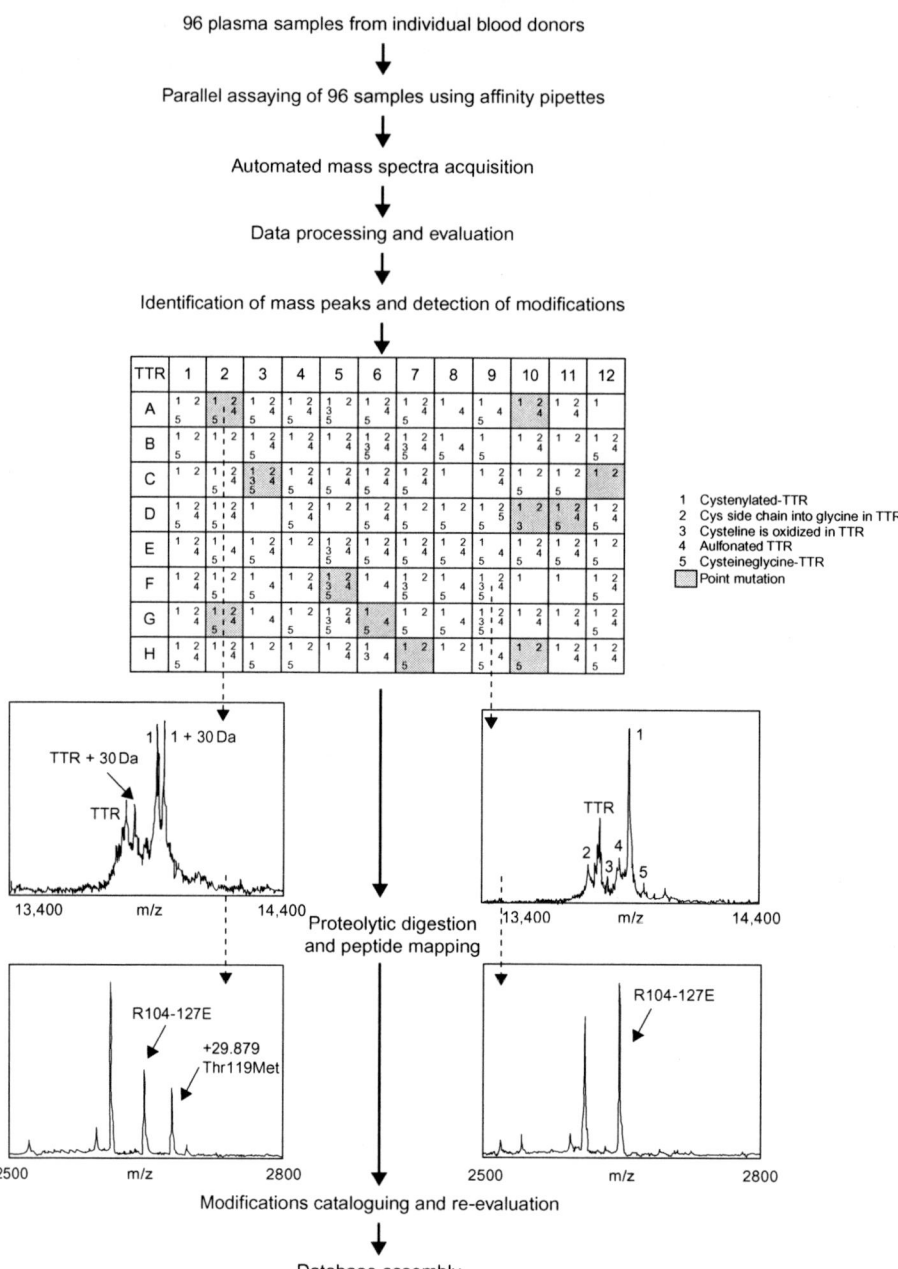

Figure 11.5 An outline of a population proteomics approach using TTR as an example. m/z: mass-to-charge ratio; TTR: transthyretin.

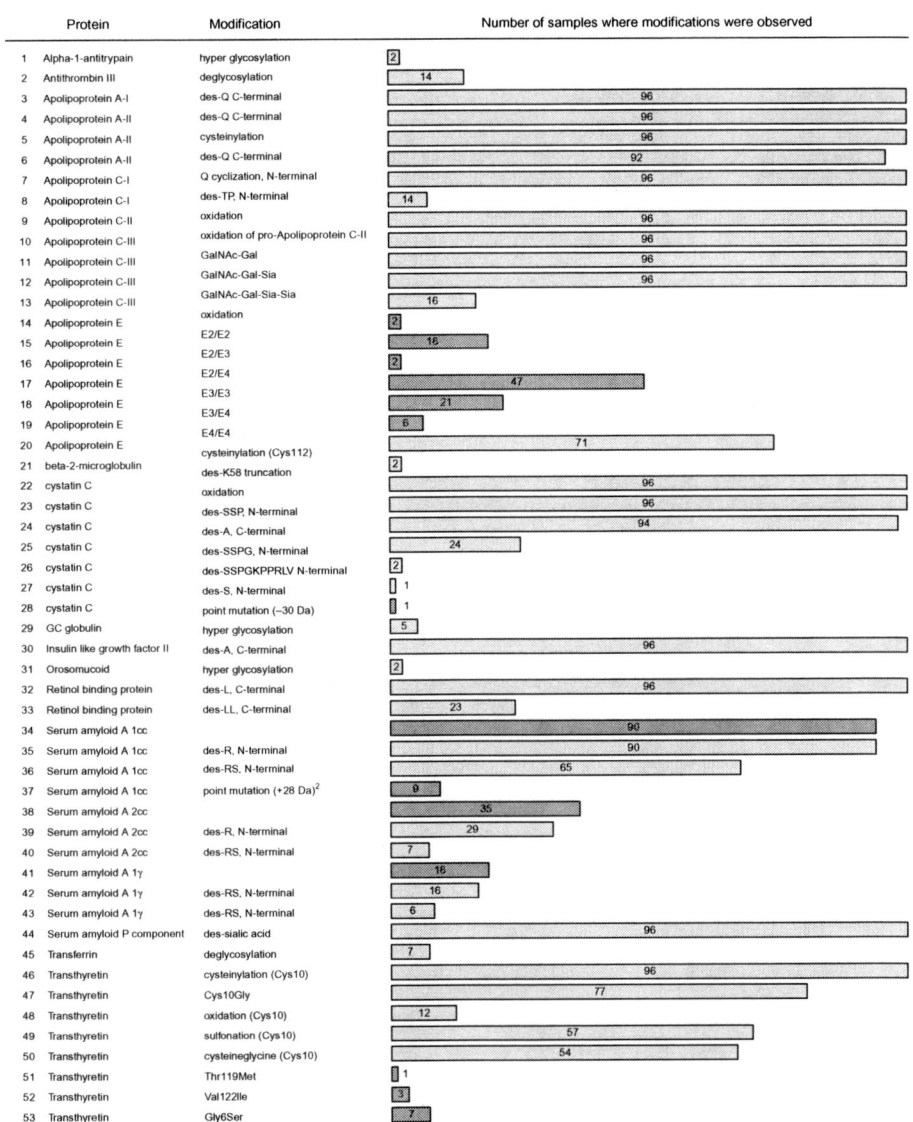

Figure 11.6 Modifications observed in 18 of the 25 proteins analyzed from 96 human plasma samples.

modifications observed for 18 of the 25 proteins studied (modifications were not observed for 7 proteins), and shows the frequency of each modification in the 96-samples cohort. A total of 53 protein variants were observed for these 18 proteins, stemming from posttranslational modifications and point mutations.

The largest number of posttranslationally modified protein variants was found to be C- or N-terminal truncated protein isoforms. Deglycosylation, oxidation, and cysteinylation were also observed among several of the proteins. Among the point mutations detected for four of the proteins, notable was the high incidence of point mutations for apolipoprotein E and TTR, which is consistent with genomic studies that have found these proteins to be highly polymorphic. The overall frequency of the modifications in the 96-sample cohort was wide ranged. Fourteen modifications were observed in all 96 samples, suggesting that they must be regarded as wild-type protein forms. Others, such as most of the point mutations, were present in only few of the samples. Overall, 23 of the modifications were observed in more than 65% of the samples, and 20 in less than 15% of the 96 samples analyzed. Upon further data analysis, and taking into the consideration the gender, age, and ethnicity of the individuals who provided the samples, it was determined that the Gly6Ser mutation in TTR was detected only in individuals of Caucasian origin, which is consistent with existing knowledge about the occurrence of this common nonamyloidogenic population polymorphism in Caucasians (Connors et al., 2003). Another correlation was observed in regard to interprotein variations in specific individuals: all seven individuals for which carbohydrate deficient transferrin was detected were also characterized with deglycosylated antithrombin III.

Following this small scale protein diversity study, a second study of human protein diversity was recently carried out wherein the number of samples was greatly expanded in order to get an accurate view of the distribution of some of the protein modifications in the general population (Nedelkov et al., 2007). One thousand individuals from four geographical regions in the USA (California, Florida, Tennessee, and Texas) were selected and the protein modifications for beta-2-microglobulin (b2m), cystatin C (cysC), retinol binding protein (RBP), transferrin (TRFE), and TTR were delineated (in the 96-sample study, these five proteins accounted for 19 of the 53 protein variants observed). The results of the study are summarized in Figure 11.7, which lists the protein modifications observed and the frequency of each in the 1000-samples cohort.

A total of 27 protein modifications (20 posttranslational modifications and 7 point mutations) were detected, with various frequencies in the cohort of samples. Variants resulting from oxidation were observed most frequently, along with single amino acid truncations. Least frequent were variants arising from point mutations and extensive sequence truncations. In total, six modifications were observed with high frequency (present in >80% of the samples), five were of medium frequency (20–50% of the samples), and sixteen were low frequency modifications observed in <7% of the samples. Nine of the low frequency modifications were not observed in the 96 individuals studied. Thus, by increasing the size of the population, it became possible to detect these low-occurrence protein modifications. When the frequencies of the modifications in the two studies were compared, an excellent correlation was obtained. For example, in both cohorts ~7% of the individuals were characterized with carbohydrate deficient transferrin. Upon further data analysis based on the gender, age, and geographical origin of the individuals who provided the samples, it was determined that the samples obtained from California contained significantly less protein modifications than the samples obtained from Florida, Tennessee, and Texas, even though the samples from all four states were

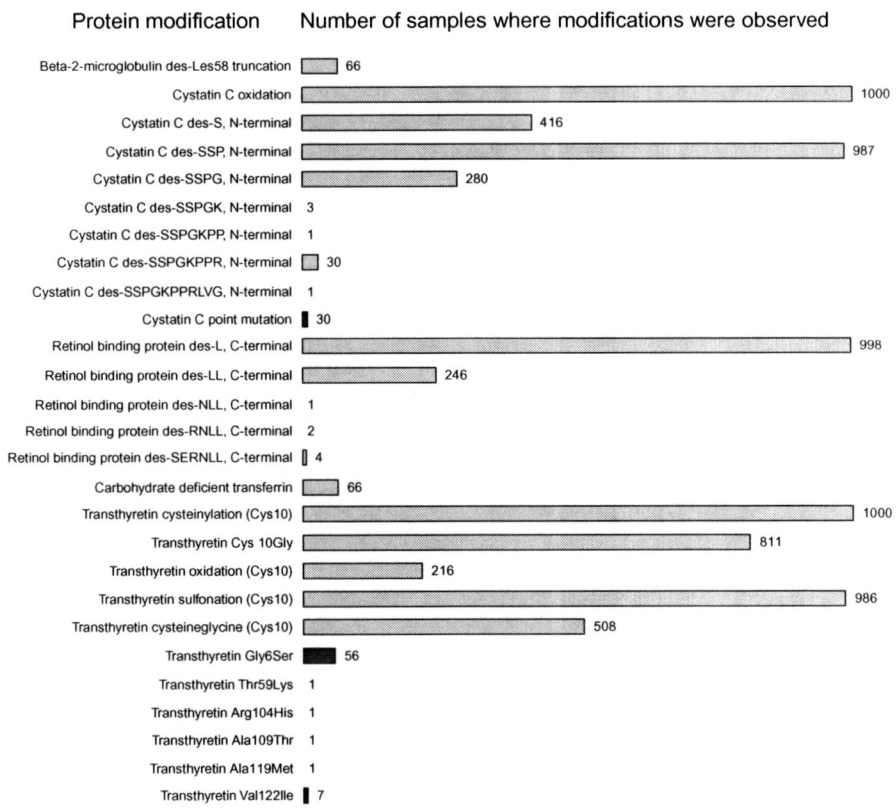

Figure 11.7 Modifications observed for five proteins studied from 1000 human plasma samples.

collected in the same way within a 3-month window in the spring of 2005, and stored under identical conditions until analysis. Correlations were also made in regard to the gender distribution of two protein modifications. Carbohydrate deficient transferrin was observed in ~1% of the females and ~10% of the males in the 1000-samples cohort. Carbohydrate deficient transferrin is an FDA-approved clinical biomarker for alcoholism, and this gender correlation can partially be explained by the higher prevalence of alcohol dependence in males than in females. The second gender correlation was related to cystatin C: all 10 of the cystatin C point mutations were found in males.

Two conclusions can be made from these two systematic studies of protein modifications and variants. First, MS is capable of detecting structural protein modifications, and, when coupled to immunoaffinity separations, it can be employed in a high-throughput systematic study of human protein diversity. Second, the human protein diversity is far more complex than the variation observed at the genetic level. While it might be premature to declare the human proteins variation "the next big thing," it is reasonable to predict that assessing human proteome variations among and within populations will be a paramount effort that can facilitate biomarker discovery. Such an endeavor would represent a paradigm shift in proteomics with significant clinical and diagnostic implications, as protein variations, quantitative and qualitative, begin to be associated with specific diseases.

11.7 5-Year View

11.7.1 Metabolomics, a Key "Omics" Tools to Decipher Host–Parasite Crosstalks

Metabolomics (i.e., metabolic profiling) is concerned with the measurement of global sets of low-molecular-weight metabolites to detect changes in cell behavior and organ function. The term "metabolome" refers to the complete set of metabolites found in an organism (Peltonen and McKusick, 2001). Metabolomics approaches use high-throughput analytical techniques such as chromatography, NMR spectrometry, and MS to measure populations of low-molecular-weight metabolites in biological samples. These large-scale efforts involve the identification and quantification of known and unknown metabolites in host tissues and fluids. Metabolite profiles can be important indicators of pathological states of a host and raise the possibility of identifying novel markers linked to the infection process of a specific pathogen.

Pathogens, especially extracellular pathogens with a complex life cycle, such as malaria and sleeping sickness, must constantly monitor and respond to environmental changes in their intermediate (i.e., insect vector) and final (human and/or animal) hosts. How pathogens detect these changes is a black box, but they must have the ability to sample changes in nutrients and other small molecules. Afterward pathogens reprogram their gene expression profile in response to host environmental changes. The metabolomics is likely to bridge data from other "omics" tools. Correlation between the pathogen metabolites expressed during its life cycle in its hosts with the global view of genome and proteome expression profiles may lead to new insight into how a pathogen interacts with host cells during a host–vector–pathogen crosstalk.

11.7.2 New Diagnostic Tools and Identification of New Therapeutic Targets

Despite the efforts of recent years, we still lack reliable biomarkers for diagnosis, prediction of clinical outcomes for many infectious diseases, and therapeutic follow-up of human diseases. Specific proteomic fingerprints might be present in biological fluids or tissues in response to the infection and could be useful for early detection of the disease, by noninvasive (saliva, urine, serum) or invasive (cerebrospinal fluid, tissue biopsies) sampling, or could constitute therapeutic targets. Before identifying of a biomarker having great potential to become an important diagnostic tool, it is important to verify the clinical applicability of a technique.

In fact, different criteria are necessary for a clinically applicable technique: reproducibility, sensitivity, specificity, and rapidity.

Over the past few years, several studies have demonstrated that comparative protein profiling using a newly developed, high-throughput technique, without a priori knowledge of the proteins present, namely SELDI-TOF-MS, breaks new ground in diagnostics (Fenollar et al., 2006; Buhimschi et al., 2007; Kanmura et al., 2007; Wu et al., 2009a,b). This technique is a potentially powerful investigative tool to improve the understanding as well as the diagnostic/prognostic capabilities for many human diseases. Nevertheless, if using such an approach, it is important that potentially new biomarkers for early diagnosis are validated in a larger number of samples to avoid the risk of false-significant results.

Inspired by genome exploration, a metaproteomic strategy, namely proteomics shotgun, has also emerged to facilitate and accelerate the discovery of novel protein biomarkers with potential diagnostic and therapeutic potential (Swanson and Washburn, 2005; Spivey, 2009). Based on different MS workflows (e.g., capillary isotachophoresis[CITP]-based multidimensional or multidimensional LC separation systems coupled tandem MS), it promotes integration of proteomics and metabolomics datasets from direct analysis of a tissue or a biological fluid (Dowell et al., 2008; Pan et al., 2008; Fang et al., 2009; Kim and Moon, 2009; Mangé et al., 2009; Maccarrone et al., 2010). Although strategic, these technologies offer a "naïve" global approach of potential biomarkers by taking into accounts the level of expression, the posttranslational modifications, and the maturation state (Ahrné et al., 2010; Park and Yates, 2010; Zhang et al., 2010a,b). These technologies are in their infancy in the analysis of host–pathogen interactions, but some pioneering studies have already highlighted some new biomarkers with diagnostic and therapeutic potential (Florens et al., 2004; Athanasiadou et al., 2008; Lal et al., 2009; Walters and Mobley, 2009; Vaezzadeh et al., 2010).

Understanding the three-dimensional structure of proteins—their posttranslational modifications, their biological functions, and the interaction between host and pathogen molecular constituents—is necessary to validate a potential biomarker in its natural molecular environment. Advances in proteomics can consider a more functional understanding of molecular dialogue and conflict that governs host–pathogen interactions, and thus develop more efficient tools to improve diagnosis, and drug design to improve therapy of infectious diseases.

11.7.3 Bioterrorism and Proteomics

Following the events of September 11, 2001, and the subsequent postal anthrax attacks in the USA, the possibility of further bioterrorism attacks became all too real. As a direct consequence of this, the US government expanded its biodefense program, with studies ranging from basic research to applications in detection, prevention, and treatment of diseases caused by such microbiological agents. The net result of this has been great progress in understanding their genomics (Fauci et al., 2005). Efforts were focused on the three major categories of critical biological agents classified by the Centers for Disease Control and Prevention (http://www.bt.

cdc.gov/agent/agentlist-category.asp#catdef). In the postgenomic era, the benefit of having the full sequence of the genomes of these agents is obvious. There are now genome sequences for a few isolates of each species, which have made studies of comparative genomics a reality and led to important discoveries, including the diversity of closely related isolates and the identification of new putative virulence genes (Fraser, 2004). In this way, breakthrough transcriptomics and proteomics studies promise further exciting results and surprises over the next few years, which hopefully will have highly beneficial applications in terms of combating the scourge of global bioterrorism (see Morse 2007).

11.7.4 Environment and Host–Parasite Interactions

For host–pathogen interactions, the main assumption is that, over ecological time-scales, host susceptibility and pathogen virulence are fixed at the onset of the crosstalk (Bull, 1994; Dieckmann et al., 2002). Also, environmental factors are traditionally viewed as "setting the scene" for the crosstalk rather than having any explicit role once it is underway. As a result, the effect of extrinsic factors on host susceptibility and pathogen virulence during a crosstalk has received little attention. However, it is common to find in populations of a pathogen species a substantial variation in the virulence, even when pathogens are collected in the same environment and at the same time. When a biological characteristic such as the virulence is variable for both genetic and environmental reasons, two individuals may differ because they differ in genotype, because they have had different environmental experiences, or both (Elliot et al., 2002). Unfortunately, the extent to which different individual pathogens and pathogen ecotypes display different virulence abilities is poorly documented and deciphered.

Life-history traits of hosts and pathogens are shaped by coevolution processes (Wolinska and King, 2009). Infections measured under laboratory conditions have shown that the environment in which hosts and pathogens interact may affect the range of host genotypes that can be infected with a given pathogen genotype in host pathogen associations (i.e. the **specificity of selection**). Despite this important fact, environmental fluctuations are often excluded in surveys on host–pathogen interactions. Since most host–pathogen interactions are in heterogeneous environments, there is a crucial need to take into account environmental conditions in proteomics surveys. The population proteomics would be a promising prospect to resolve interesting issues specific to host–pathogen crosstalks in a varying environment (Figure 11.8). This kind of survey would bring pioneer molecular data to decipher the **reaction norm** of a genotype and to understand why pathogens sometimes evolve in a given environment toward high virulence and hosts toward high resistance. Also, these surveys would permit to assess the fixity or not of host–parasite interactomes involved in a host–pathogen association in a varying environment.

11.7.5 Human Population Proteomics

While healthy population protein diversity cataloguing is a pretty straightforward proposition, the ultimate question is how human population proteomics can enable

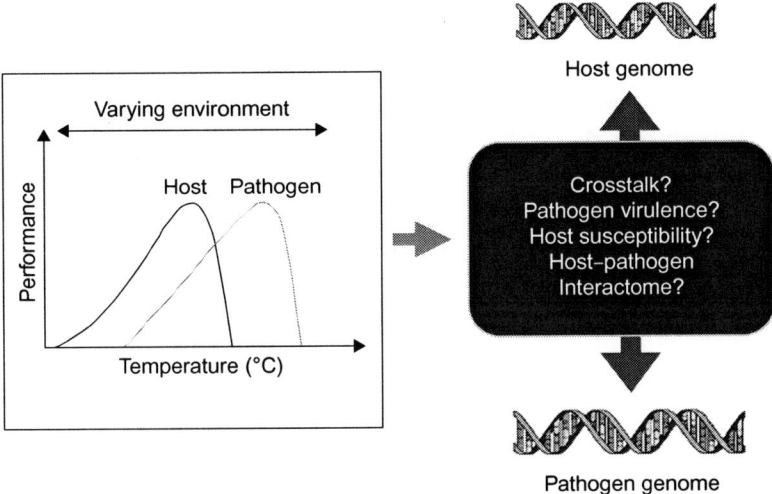

Figure 11.8 Host–pathogen interactions in a varying environment.

better disease diagnosis and management. Because of the one-protein-at-a-time approach, most studies involving cohorts of sick versus healthy control samples must first make an "educated guess" what proteins to analyze via population proteomics. To start with, there is plenty of existing data on specific plasma proteins and their quantitative modulations with specific diseases. Hence, these proteins would be the first to benefit from qualitative reassessment via human population proteomics; next in line are proteins that are within the biomarker proteins' interaction network and pathways. To illustrate this point, structural isoforms have recently been discovered for some well-established biomarkers such as cardiac troponin I (Jaffe and Van Eyk, 2006; Peronnet et al., 2006) and B-type natriuretic peptide (Brandt et al., 2006; Lam et al., 2007). These structural variants might prove to be better sensitivity and specificity biomarkers than the native proteins themselves. To assess their modulation with disease, quantitative assessments of the isoforms abundance can be made by comparing the ratios of the isoforms and the native protein signals, or via standard curve approaches. The MSIA present straightforward means of looking into the protein microheterogeneity using the well-established methods of immunoaffinity separation and mass spectrometric detection. As such, it is expected that MS-based immunoassays will readily be accepted into the clinical and diagnostic laboratories to study the effects of protein modifications in pathological processes and evaluate their potential as new biomarkers of disease.

11.8 Conclusion

From the dawn of human evolution to the influenza and **HIV/AIDS** pandemics of the twentieth and early twenty-first centuries, infectious diseases have continued to emerge and re-emerge with great ferocity and by so doing, seriously affect populations as well as challenge our abilities to fight the responsible agents. Over the past decade, strains of many common pathogens have continued to develop resistance to the drugs that once were effective against them. In the battle against pathogens, humankind has created new megatechnologies such as massive sequencing, proteomics, and bioinformatics, but without conceptual approaches based on the evolutionary concepts. Parasite genome sequences do not themselves provide a full explanation of the biology of an organism or on the molecular war involved in host−pathogen associations. Since the 1990s, proteomic tools have been successfully employed in a large number of studies to find and identify proteins involved in biological phenomena, such as host−parasite interactions. Even so, many studies have, as outlined earlier, revealed pitfalls in the approaches used. Thus, whatever the new technological advancements, it is apparent that parasitologists and molecular biologists should attempt to improve their experimental design. This new attitude will surely improve the reliability of the data deriving from proteomics studies and will open the way for an enhanced comprehension of many biological mechanisms. In this chapter, new ways based on evolutionary concepts are suggested to enable further elucidation of the molecular complexities of host−pathogen genome interactions. These new ways could help to increase the knowledge about the molecular war involved in host−pathogen associations, taking into account the environmental factors.

Glossary

AIDS acronym for acquired immune deficiency syndrome, the gravest of the sexually transmitted diseases, or STDs. It is caused by the human immunodeficiency virus (HIV), now known to be a retrovirus, an agent first identified in 1983. HIV is transmitted in body fluids, mainly blood and genital secretions.

Antigen substances that are foreign to the body and cause the production of antibodies.

Bioinformatics the use of mathematical and informational techniques, including statistics, to solve biological problems, usually by creating or using computer programs, mathematical models, or both. One of the main areas of bioinformatics is the data mining and analysis of the data gathered by the various genome projects. Other areas are sequence alignment, protein structure prediction, systems biology, protein−protein interactions, and virtual evolution.

Defensin a substance with natural antibiotic effects found in human blood cells. There are three types of defensins. Other animal species have similar substances.

Elicitors molecules produced by the host (or pathogen) that induce a response from the pathogen (or host).

Environment elements of a habitat. In this text, "environment" refers to a broad range of biotic and abiotic conditions, including interactions with other species.

Genomics the study of an organism's genome and the use of the genes. It deals with the systematic use of genome information, associated with other data, to provide answers in biology, medicine, and industry.

Genome the full complement of genes carried by a single (haploid) set of chromosomes. The term may be applied to the genetic information carried by an individual or to the range of genes found in a given species.

Major histocompatibility complex two classes of molecules on cell surfaces. MHC class I molecules exist on all cells and hold and present foreign antigens to CD8 cytotoxic T lymphocytes if the cell is infected by a virus or other microbe. MHC class II molecules are the billboards of the immune system. Peptides derived from foreign proteins are inserted into the MHC's binding groove and displayed on the surface of antigen-presenting cells. These peptides are then recognized by T lymphocytes so that the immune system is alerted to the presence of foreign material. See Histocompatibility Testing.

Mass spectrometry a technique for separating ions by their mass-to-charge (m/z) ratios.

Molecular crosstalk molecular communications in a host−parasite system during their interaction.

Parasito-proteomics the study of the reaction of the host and parasite genomes through the expression of the host and parasite proteomes (genome-operating systems) during their biochemical crosstalk.

Pathogenicity the capability of a pathogen to cause disease.

Population genetics the study of the distribution of genes (the units of genetic inheritance) and genotypes (the genetic complement at one or more loci), and the mechanisms determining genetic variability within a population.

Population proteomics the study of protein diversity in populations of any species; or more specifically, targeted investigation of proteins across and within populations of a species to define and understand protein diversity and to facilitate the discovery of disease- or pathogen-specific protein modulation.

Proteome the term proteome was first used in 1995 and has been applied to several different types of biological systems. A cellular proteome is the collection of proteins found in a particular cell type under a particular set of environmental conditions, such as exposure to hormone stimulation. It can also be useful to consider an organism's complete proteome. The complete proteome for an organism can be conceptualized as the complete set of proteins from all of the various cellular proteomes. This is very roughly the protein equivalent of the genome. The term "proteome" has also been used to refer to the collection of proteins in certain subcellular biological systems. For example, all of the proteins in a virus can be called a viral proteome.

Proteomics the large-scale study of proteins, particularly their structures and functions. This term was coined to make an analogy with genomics, and is often viewed as the "next step," but proteomics is much more complicated than genomics.

Reaction norm the set of phenotypes that can be produced by a genotype under various environmental settings.

Specificity an alternative concept to explain why hosts vary in their susceptibility to parasites is that host−parasite interactions have some degree of specificity.

Specificity of selection the range of host genotypes that can be infected with a given parasite genotype in host−parasite interactions.

Transcriptome the whole set of mRNA species in one or a population of cells.

Transcriptomics Techniques to identify mRNA from actively transcribed genes.

Two-dimensional gel electrophoresis proteomics, the study of the proteome, has largely been practiced through the separation of proteins by two-dimensional gel electrophoresis. In the first dimension, proteins are separated by isoelectric focusing (separation of proteins according to their isoelectric point in a pH gradient gel), resolved on the basis of charge. In the second dimension, they are separated by molecular weight using

SDS-PAGE. To visualize the proteins, the gel is dyed with Coomassic Blue, silver, or other reagents. Spots on the gel are proteins that have migrated to specific locations.

References

Abad, P., Favery, B., Ross, M.N., Castagnone-Sereno, P., 2003. Root-knot nematode parasitism an host response: molecular basis of a sophisticated interaction. Mol. Plant. Pathol. 4, 217–224.

Achtman, M., Morelli, G., Zhu, P., Wirth, T., Diehl, I., Kusecek, B., et al., 2004. Microevolution and history of the plague bacillus, *Yersinia pestis*. Proc. Natl. Acad. Sci. U.S.A. 101, 17837–17842.

Agranoff, D., Stich, A., Abel, P., Krishna, S., 2005. Proteomic fingerprinting for the diagnosis of human African trypanosomiasis. Trends Parasitol. 21, 154–157.

Ahram, M., Petricoin, E.F., 2008. Proteomics discovery of disease biomarkers. Biomark Insights 23, 325–333.

Ahrné, E., Müller, M., Lisacek, F., 2010. Unrestricted identification of modified proteins using MS/MS. Proteomics 10, 671–686.

Alfarano, C., Andrade, C.E., Anthony, K., Bahroos, N., Bajec, M., Bantoft, K., et al., 2005. The biomolecular interaction network database and related tools 2005 update. Nucleic Acids Res. 33, D418–D424.

Anderson, N.G., Anderson, N.L., 1996. Twenty years of two-dimensional electrophoresis: past, present and future. Electrophoresis 17, 443–453.

Anderson, L., Seilhaver, J., 1997. A comparison of selected mRNA and protein abundance in human liver. Electrophoresis 18, 533–537.

Antoine, J.C., Prina, E., Courret, N., Lang, T., 2004. *Leishmania* spp.: on the interactions they establish with antigen-presenting cells of their mammalian hosts. Adv. Parasitol. 58, 1–68.

Antrobus, R., Grant, K., Gangadharan, B., Chittenden, D., Everett, R.D., Zitzmann, N., et al., 2009. Proteomic analysis of cells in the early stages of herpes simplex virus type-1 infection reveals widespread changes in the host cell proteome. Proteomics 9, 3913–3927.

Ashton, P.D., Curwen, R.S., Wilson, R.A., 2001. Linking proteome and genome: how to identify parasite proteins. Trends Parasitol. 17, 198–202.

Athanasiadou, S., Pemberton, A., Jackson, F., Inglis, N., Miller, H.R., Thévenod, F., et al., 2008. Proteomic approach to identify candidate effector molecules during the in vitro immune exclusion of infective *Teladorsagia circumcincta* in the abomasum of sheep. Vet. Res. 39, 58.

Azad, N.S., Rasool, N., Annunziata, C.M., Minasian, L., Whiteley, G., Kohn, E.C., 2006. Proteomics in clinical trials and practice: present uses and future promise. Mol. Cell. Proteomics 5, 1819–1829.

Bader, G.D., Hogue, C.W., 2003. An automated method for finding molecular complexes in large protein interaction networks. BMC Bioinformatics 4, 2.

Bansal, A.K., 2005. Bioinformatics in microbial biotechnology—a mini review. Microb. Cell. Fact. 4, 19.

Barret, J., Jefferies, J.R., Brophy, P.M., 2000. Parasite proteomics. Parasitol. Today 16, 400–403.

Barret, J., Brophy, P.M., Hamilton, J.V., 2005. Analysing proteomic data. Int. J. Parasitol. 35, 543–553.

Baudot, A., Martin, D., Mouren, P., Chevenet, F., Guénoche, A., Jacq, B., et al., 2006. PRODISTIN web site: a tool for the functional classification of proteins from interaction networks. Bioinformatics 22, 248–250.
Bhavsar, A.P., Auweter, S.D., Finlay, B.B., 2010. Proteomics as a probe of microbial pathogenesis and its molecular boundaries. Future Microbiol. 5, 253–265.
Bird, D.M., Opperman, C.H., 2009. The secret(ion) life of worms. Genome Biol. 10, 205.
Biron, D.G., Moura, H., Marché, L., Hughes, A.L., Thomas, F., 2005a. Towards a new conceptual approach to "Parasitoproteomics". Trends Parasitol. 21, 162–168.
Biron, D.G., Joly, C., Galéotti, N., Ponton, F., Marché, L., 2005b. The proteomics: a new prospect for studying parasitic manipulation. Behav. Process. 68, 249–253.
Biron, D.G., Marché, L., Ponton, F., Loxdale, H.D., Galéotti, N., Renault, L., et al., 2005c. Behavioural manipulation in a grasshopper harbouring hairworm: a proteomics approach. Proc. R. Soc. Lond. B 272, 2117–2126.
Biron, D.G., Agnew, P., Marché, L., Renault, L., Sidobre, C., Michalakis, Y., 2005d. Proteome of *Aedes aegypti* larvae in response to infection by the intracellular parasite *Vavraia culicis*. Int. J. Parasitol. 35, 1385–1397.
Biron, D.G., Brun, C., Lefèvre, T., Lebarbenchon, C., Loxdale, H.D., Chevenet, F., et al., 2006a. The pitfalls of proteomics experiments without the correct use of bioinformatics tools. Proteomics 6, 5577–5596.
Biron, D.G., Loxdale, H.D., Ponton, F., Moura, H., Marché, L., Brugidou, C., et al., 2006b. Population proteomics: an emerging discipline to study metapopulation ecology. Proteomics 6, 1712–1715.
Bischoff, R., Luider, T.M., 2004. Methodological advances in the discovery of protein and peptide disease markers. J. Chromatogr. B 803, 27–40.
Bogaerts, A., Beets, I., Temmerman, L., Schoofs, L, Verleyen, P., 2010. Proteome changes of *Caenorhabditis elegans* upon a *Staphylococcus aureus* infection. Biol. Direct. 5, 11.
Boonmee, S., Imtawil, K., Wongkham, C., Wongkham, S., 2003. Comparative proteomic analysis of juvenile and adult liver fluke, *Opisthorchis viverrini*. Act. Trop. 88, 233–238.
Brand, S., Hahner, D., Ketterlinus, R., 2005. Protein profiling and identification in complex biological samples using LC-MALDI. Drug Plus International September, 6–8.
Brandt, I., Lambeir, A.M., Ketelslegers, J.M., Vanderheyden, M., Scharpe, S., De Meester, I., 2006. Dipeptidyl-peptidase IV converts intact B-type natriuretic peptide into its des-SerPro form. Clin. Chem. 52, 82–87.
Bridges, D.J., Pitt, A.R., Hanrahan, O., Brennan, K., Voorheis, H.P, Herzyk, P., et al., 2008. Characterisation of the plasma membrane subproteome of bloodstream form *Trypanosoma brucei*. Proteomics 8, 83–99.
Broekaert, W.F., Terras, F.R.G., Cammue, B.P.A., Osborn, R.W., 1995. Plants defensins: novel antimicrobial peptides as components of the host defense system. Plant Physiol. 108, 1353–1358.
Brun, C., Martin, D., Chevenet, F., Wojcik, J., Guénoche, A., Jacq, B., 2003. Functional classification of proteins for the prediction of cellular function from a protein–protein interaction network. Genome Biol. 5, R6.
Brun, C., Herrmann, C., Guenoche, A., 2004. Clustering proteins from interaction networks for the prediction of cellular functions. BMC Bioinformatics 5, 95.
Buhimschi, C.S., Bhandari, V., Hamar, B.D., Bahtiyar, M.O., Zhao, G., Sfakianaki, A.K., et al., 2007. Proteomic profiling of the amniotic fluid to detect inflammation, infection, and neonatal sepsis. PLoS Med. 4, e18.
Bull, J.J., 1994. Perspective: virulence. Evolution 48, 1423–1437.

Cao, H., Baldini, R.L., Rahme, L.G., 2001. Common mechanism for pathogens of plant and animals. Annu. Rev. Phytopath. 39, 259−284.
Chautard, E., Thierry-Mieg, N., Ricard-Blum, S., 2009. Interaction networks: from protein functions to drug discovery. A review. Pathol. Biol. 57, 324−333.
Chevalier, F., Martin, O., Rofidal, V., Devauchelle, A.D., Barteau, S., Sommerer, N., et al., 2004. Proteomic investigation of natural variation between *Arabidopsis* ecotypes. Proteomics 4, 1372−1381.
Cieslak, A., Ribera, I., 2009. Aplicaciones de proteómica en ecología y evolución. Ecosistemas 18, 34−43.
Cohen, A.M., Rumpel, K., Coombs, G.H., Wastling, J.M., 2002. Characterisation of global protein expression by two-dimensional electrophoresis and mass spectrometry: proteomics of *Toxoplasma gondii*. Int. J. Parasitol. 32, 39−51.
Coiras, M., Camafeita, E., López-Huertas, M.R., Calvo, E., López, J.A., Alcamí, J., 2008. Application of proteomics technology for analyzing the interactions between host cells and intracellular infectious agents. Proteomics 8, 852−873.
Connors, L.H., Lim, A., Prokaeva, T., Roskens, V.A., Costello, C.E., 2003. Tabulation of human transthyretin (TTR) variants, 2003. Amyloid 10, 160−184.
Das, S., Sierra, J.C., Soman, K.V., Suarez, G., Mohammad, A.A., Dang, T.A., Luxon, B.A., Reyes, V.E., 2005. Differential protein expression profiles of gastric epithelial cells following Helicobacter pylori infection using ProteinChips. J. Proteome Res. 4, 920−930.
De Bock, M., de Seny, D., Meuwis, M.A., Chapelle, J.P., Louis, E., Malaise, M., et al., 2010. Challenges for biomarker discovery in body fluids using SELDI-TOF-MS. J. Biomed. Biotechnol. 2010, 906082.
Degrave, W.M., Melville, S., Ivens, A., Aslett, M., 2001. Parasite genome initiatives. Int. J. Parasitol. 31, 532−536.
Dieckmann, U., Metz, J.A.J., Sabelis, M.W., Sigmund, K. (Eds.), 2002. Adaptive Dynamics of Infectious Diseases: In Pursuit of Virulence Management. Cambridge University Press, Cambridge, England.
Diez-Orejas, R, Fernández-Arenas, E., 2008. *Candida albicans*−macrophage interactions: genomic and proteomic insights. Future Microbiol. 3, 661−681.
Dixon, M.S., Golstein, C., Thomas, C.M., Van Der Biezen, E.A., Jones, J.D., 2000. Genetic complexity of pathogen perception by plants: the example of Rcr3, a tomato gene required specifically by Cf-2. Proc. Nat. Acad. Sci. U.S.A. 97, 8807−8814.
Diz, A.P., Skibinski, D.O.F., 2007. Evolution of 2-DE protein patterns in a mussel hybrid zone. Proteomics 7, 2111−2120.
Dorus, S., Busby, S.A., Gerike, U., Shabanowitz, J., Hunt, D.F., Karr, T.L., 2006. Genomic and functional evolution of the *Drosophila melanogaster* sperm proteome. Nat. Genet. 38, 1440−1445.
Dowell, J.A., Frost, D.C., Zhang, J., Li, L., 2008. Comparison of two-dimensional fractionation techniques for shotgun proteomics. Anal. Chem. 80, 6715−6723.
Doyle, E.A., Lambert, K.N., 2003. *Meloidogyne javanica* chorismate mutase 1 alters plant cell development. Mol. Plant Microbe Interact. 16, 123−131.
Doytchinova, I.A., Taylor, P., Flower, D.R., 2003. Proteomics in vaccinology and immunobiology: an informatics perspective of the immunone. J. Biomed. Biotechnol. 2003, 267−290.
Dunn, R., Dudbridge, F., Sanderson, C.M., 2005. The use of edge-betweenness clustering to investigate biological function in protein interaction networks. BMC Bioinformatics. 6, 39.

Elliot, S.L., Blanford, S., Thomas, M.B., 2002. Host−pathogen interactions in a varying environment: temperature, behavioural fever and fitness. Proc. R. Soc. Lond. B 269, 1599−1607.

Fang, X., Balgley, B.M., Wang, W., Park, D.M., Lee, C.S., 2009. Comparison of multidimensional shotgun technologies targeting tissue proteomics. Electrophoresis 30, 4063−4070.

Fauci, A.S., Touchette, N.A., Folkers, G.K., 2005. Emerging infectious diseases: a 10-Year perspective from the National Institute of Allergy and Infectious Diseases. Emerging Infect. Dis. (www.cdc.gov/eid). 11, 519−525.

Fell, D.A., 2001. Beyond genomics. Trends Genet. 17, 680−682.

Fenollar, F., Goncalves, A., Esterni, B., Azza, S., Habib, G., Borg, J.P., et al., 2006. A serum protein signature with high diagnostic value in bacterial endocarditis: results from a study based on surface-enhanced laser desorption/ionization time-of-flight mass spectrometry. J. Infect. Dis. 194, 1356−1366.

Fields, S., 2001. Proteomics in genomeland. Science 291, 1221−1224.

Fischer, J., West, J., Agochukwu, N., Suire, C., Hale-Donze, H., 2007. Induction of host chemotactic response by *Encephalitozoon* spp. Infect. Immun. 75, 1619−1625.

Flajolet, M., Rotondo, G., Daviet, L., Bergametti, F., Inchauspé, G., Tiollais, P., et al., 2000. A genomic approach of the hepatitis C virus generates a protein interaction map. Gene. 242, 369−379.

Florens, L., Liu, X., Wang, Y., Yang, S., Schwartz, O., Peglar, M., et al., 2004. Proteomics approach reveals novel proteins on the surface of malaria-infected erythrocytes. Mol. Biochem. Parasitol. 135, 1−11.

Formstecher, E., Aresta, S., Collura, V., Hamburger, A., Meil, A., Trehin, A., et al., 2005. Protein interaction mapping: a *Drosophila* case study. Genome Res. 15, 376−384.

Fournier, I., Wisztorski, M., Salzet, M., 2008. Tissue imaging using MALDI-MS: a new frontier of histopathology proteomics. Expert Rev. Proteomics 5, 413−424.

Fraser, C.M., 2004. A genomics-based approach to biodefence preparedness. Nat. Rev. Genet. 5, 23−33.

Fujita, N., Sugimoto, R., Motonishi, S., Tomosugi, N., Tanaka, H., Takeo, M., et al., 2008. Patients with chronic hepatitis C achieving a sustained virological response to peginterferon and ribavirin therapy recover from impaired hepcidin secretion. J. Hepatol. 49, 702−710.

Fung, E.T., Thulasiraman, V., Weinberger, S.R., Dalmasso, E.A., 2001. Protein biochips for differential profiling. Curr. Opin. Biotechnol. 12, 65−69.

Gelfand, J.A., Callahan, M.V., 2003. Babesiosis: an update on epidemiology and treatment. Curr. Infect. Dis. Rep. 5, 53−58.

Giot, L., Bader, J.S., Brouwer, C., Chaudhuri, A., Kuang, B., Li, Y., et al., 2003. A protein interaction map of *Drosophila melanogaster*. Science 302, 1727−1736.

Girvan, M., Newman, M.E., 2002. Community structure in social and biological networks. Proc. Natl. Acad. Sci. U.S.A. 99, 7821−7826.

Goverse, A., De Engler, J.A., Verhees, J., Van der Krol, S., Helder, J.H., Gheysen, G., 2000. Cell cycle activation by plant parasitic nematodes. Plant. Mol. Biol. 43, 747−776.

Grébaut, P., Chuchana, P., Brizard, J,P., Demettre, E, Seveno, M, Bossard, G., et al., 2009. Identification of total and differentially expressed excreted-secreted proteins from *Trypanosoma congolense* strains exhibiting different virulence and pathogenicity. Int. J. Parasitol. 10, 1137−1150.

Guzman, M.G., Kouri, G., 2002. Dengue: an update. The Lancet Inf. Dis. 2, 33−42.

Gygi, S.P., Rochon, Y., Franza, B.R., Aebersold, R., 1999. Correlation between protein and mRNA abundance in yeast. Mol. Cell. Biol. 19, 1720−1730.

Haab, B.B., Dunham, M.J., Brown, P.O., 2001. Protein microarrays for highly parallel detection and quantification of specific proteins and antibodies in complex solutions. Genome Biol. 2, Research004, 1–13.

Han, Y.S., Chun, J., Schwartz, A., Nelson, S., Paskewitz, S.M., 1999. Induction of mosquito hemolymph proteins in response to immune challenge and wounding. Dev. Comp. Immunol. 23, 553–562.

Hart, S.R., Lau, K.W., Hao, Z., Broadhead, R., Portman, N., Hühmer, A., et al., 2009. Analysis of the trypanosome flagellar proteome using a combined electron transfer/collisionally activated dissociation strategy. J. Am. Soc. Mass. Spectrom. 20, 167–175.

Hermjakob, H., Montecchi-Palazzi, L., Bader, G., Wojcik, J., Salwinski, L., Ceol, A., et al., 2004. The HUPO PSI's molecular interaction format—a community standard for the representation of protein interaction data. Nat. Biotechnol. 22, 177–183.

Hochstrasser, D.F., 1998. Proteome in perspective. Clin. Chem. Lab. Med. 36, 825–836.

Hoffman, J.A., 1995. Innate immunity of insects. Curr. Opin. Immunol. 7, 4–10.

Holzmuller, P., Grébaut, P., Brizard, J.P., Berthier, D., Chantal, I., Bossard, G., et al., 2008. "Pathogeno-proteomics": towards a new approach of host–vector–pathogen interactions. Ann. N.Y. Acad. Sci. 1149, 66–70.

Holzmuller, P., Grébaut, P., Cuny, G., Biron, D.G., 2010. Tsetse flies, trypanosomes, humans and animals: what is proteomics revealing about their crosstalks? Expert Rev. Proteomics. 7, 113–126.

Hood, B.L., Malehorn, D.E., Conrads, T.P., Bigbee, W.L., 2009. Serum proteomics using mass spectrometry. Methods Mol. Biol. 520, 107–128.

Huang, S.H., Triche, T., Jong, A.Y., 2002. Infectomics: genomics and proteomics of microbial infections. Funct. Integr. Genomics 1, 331–344.

Ito, T., Chiba, T., Ozawa, R., Yoshida, M., Hattori, M., Sakaki, Y., 2001. A comprehensive two-hybrid analysis to explore the yeast protein interactome. Proc. Natl. Acad. Sci. U.S.A. 98, 4569–4574.

Jaffe, A.S., Van Eyk, J.E., 2006. Degradation of cardiac troponins. Implications for clinical practise. In: Morrow, D.A. (Ed.), Cardiovascular Biomarkers—Pathophysiology and Disease Management. Humana Press, Totowa, pp. 161–174.

Jagusztyn-Krynicka, E.K., Dadlez, M., Grabowska, A, Roszczenko, P., 2009. Proteomic technology in the design of new effective antibacterial vaccines. Expert Rev. Proteomics 6, 315–330.

Jahnukainen, T., Malehorn, D., Sun, M., Lyons-Weiler, J., Bigbee, W., Gupta, G., et al., 2006. Proteomic analysis of urine in kidney transplant patients with BK virus nephropathy. J. Am. Soc. Nephrol. 17, 3248–3256.

Jasmer, D.P., 1993. *Trichinella spiralis* infected skeletal muscle cells arrest in G2/M and cease muscle gene expression. J. Cell. Biol. 121, 785–793.

Ju, J.W., Joo, H.N., Lee, M.R., Cho, S.H., Cheun, H.I., Kim, J.Y., et al., 2009. Identification of a serodiagnostic antigen, legumain, by immunoproteomic analysis of excretory–secretory products of *Clonorchis sinensis* adult worms. Proteomics 9, 3066–3078.

Kanmura, S., Uto, H., Kusumoto, K., Ishida, Y., Hasuike, S., Nagata, K., et al., 2007. Early diagnostic potential for hepatocellular carcinoma using the SELDI ProteinChip system. Hepatology 45, 948–956.

Karr, T.L., 2008. Application of proteomics to ecology and population biology. Heredity 100, 200–206.

Kashiwagi, A., Kurosaki, H., Luo, H., Yamamoto, H., Oshimura, M., Shibahara, T., 2009. Effects of *Tritrichomonas muris* on the mouse intestine: a proteomic analysis. Exp. Anim. 58, 537–542.

Kim, K.H., Moon, M.H., 2009. High speed two-dimensional protein separation without gel by isoelectric focusing-asymmetrical flow field flow fractionation: application to urinary proteome. J. Proteome Res. 8, 4272−4278.
Kint, G., Fierro, C., Marchal, K., Vanderleyden, J., De Keersmaecker, S.C., 2010. Integration of "omics" data: does it lead to new insights into host−microbe interactions? Future Microbiol. 5, 313−328.
Klein, L.L., Freitag, B.C., Gibbs, R.S., Reddy, A.P., Nagalla, S.R., Gravett, M.G., 2005. Detection of intra-amniotic infection in a rabbit model by proteomics-based amniotic fluid analysis. Am. J. Obstet. Gynecol. 193, 1302−1306.
Kleppe, R., Kjarland, E., Selheim, F., 2006. Proteomic and computational methods in systems modeling of cellular signaling. Curr. Pharm. Bizotechnol. 7, 135−145.
Kuhn, J.F., Hoerth, P., Hoehn, S.T., Preckel, T., Tomer, K.B., 2006. Proteomics study of anthrax lethal toxin-treated murine macrophages. Electrophoresis 27, 1584−1597.
LaCount, D.J., Vignali, M., Chettier, R., Phansalkar, A., Bell, R., Hesselberth, J.R., et al., 2005. A protein interaction network of the malaria parasite *Plasmodium falciparum*. Nature 438, 103−107.
Lal, K., Prieto, J.H., Bromley, E., Sanderson, S.J., Yates, J.R. 3rd, Wastling, J.M., et al., 2009. Characterisation of Plasmodium invasive organelles; an ookinete microneme proteome. Proteomics 9, 1142−1151.
Lam, C.S., Burnett Jr., J.C., Costello-Boerrigter, L., Rodeheffer, R.J., Redfield, M.M., 2007. Alternate circulating pro-B-type natriuretic peptide and B-type natriuretic peptide forms in the general population. J. Am. Coll. Cardiol. 49, 1193−1202.
Langley, R.C., Cali, A., Somberg, E.W., 1987. Two-dimensional electrophoretic analysis of spore proteins of the microsporida. J. Parasit. 73, 910−918.
Lee, S.R., Nanduri, B., Pharr, G.T., Stokes, J.V., Pinchuk, L.M., 2009. Bovine viral diarrhea virus infection affects the expression of proteins related to professional antigen presentation in bovine monocytes. Biochim. Biophys. Acta 1794, 14−22.
Lefèvre, T., Thomas, F., Ravel, S., Patrel, D., Renault, L., Le Bourligu, L., et al., 2007a. *Trypanosoma brucei brucei* induces alteration in the head proteome of the tsetse fly vector *Glossina palpalis gambiensis*. Insect Mol. Biol. 16, 651−660.
Lefèvre, T., Thomas, F., Schwartz, A., Levashina, E., Blandin, S., Brizard, J.P., et al., 2007b. Malaria *Plasmodium* agent induces alteration in the head proteome of their *Anopheles* mosquito host. Proteomics 7, 1908−1915.
Lefèvre, T., Adamo, S.A., Biron, D.G., Missé, D., Hughes, D., Thomas, F., 2009. Invasion of the body snatchers: the diversity and evolution of manipulative strategies in host−parasite interactions. In: Webster, J.P., Rollinson, D., Hay, S.I. (Eds.), Advances in Parasitology, 68. Academic Press, London, pp. 46−84.
Levy, F., Bulet, P., Ehret-Sabatier, L., 2004. Proteomic analysis of the systemic immune response of Drosophila. Mol. Cell. Proteomics 3, 156−166.
Li, S., Armstrong, C.M., Bertin, N., Ge, H., Milstein, S., Boxem, M., et al., 2004. A map of the interactome network of the metazoan *C. elegans*. Science 303, 540−543.
Liu, H.C., Hicks, J., Yoo, D., 2008a. Proteomic dissection of viral pathogenesis. Dev. Biol. 132, 43−53.
Liu, N., Song, W., Wang, P., Lee, K., Chan, W., Chen, H., et al., 2008b. Proteomics analysis of differential expression of cellular proteins in response to avian H9N2 virus infection in human cells. Proteomics 8, 1851−1858.
Liu, Q., Chen, X., Hu, C., Zhang, R., Yue, J., Wu, G., et al., 2010. Serum protein profiling of smear-positive and smear-negative pulmonary tuberculosis using SELDI-TOF mass spectrometry. Lung. 188, 15−23.

Lopez, M.F., Pluskal, M.G., 2003. Protein micro- and macroarrays: digitizing the proteome. J. Chromatogr. B 787, 19−27.

Luciano-Montalvo, C., Ciborowski, P., Duan, F., Gendelman, H.E., Meléndez, L.M., 2008. Proteomic analyses associate cystatin B with restricted HIV-1 replication in placental macrophages. Placenta 29, 1016−1023.

Luo, X., Carlson, K.A., Wojna, V., Mayo, R., Biskup, T.M., Stoner, J., et al., 2003. Macrophage proteomic fingerprinting predicts HIV-1-associated cognitive impairment. Neurology 60, 1931−1937.

Maccarrone, G., Turck, C.W., Martins-de-Souza, D., 2010. Shotgun mass spectrometry workflow combining IEF and LC-MALDI-TOF/TOF. Protein J. 29, 99−102.

Mangé, A., Chaurand, P., Perrochia, H., Roger, P., Caprioli, R.M., Solassol, J., 2009. Liquid chromatography-tandem and MALDI imaging mass spectrometry analyses of RCL2/CS100-fixed, paraffin-embedded tissues: proteomics evaluation of an alternate fixative for biomarker discovery. J. Proteome Res. 8, 5619−5628.

Maniatis, T., Tasic, B., 2002. Alternative pre-mRNA splicing and proteome expression in metazoans. Nature 418, 236−243.

Merchant, M., Weinberger, S.R., 2000. Recent advancements in surface-enhanced laser desorption/ionization-time of flight-mass spectrometry. Electrophoresis 21, 1164−1177.

Missé, D., Yssel, H., Trabattoni, D., Oblet, C., Lo Caputo, S., Mazzotta, F., et al., 2007. IL-22 participates in an innate anti-HIV-1 host-resistance network through acute-phase protein induction. J. Immunol. 178, 407−415.

Molina, S., Misse, D., Roche, S., Badiou, S., Cristol, J.P., Bonfils, C., et al., 2008. Identification of apolipoprotein C-III as a potential plasmatic biomarker associated with the resolution of hepatitis C virus infection. Proteomics Clin. Appl. 2, 751−761.

Morrison, L.J., Marcello, L., McCulloch, R., 2009. Antigenic variation in the African trypanosome: molecular mechanisms and phenotypic complexity. Cell Microbiol. 11, 1724−1734.

Morse, S.A., 2007. The challenge of bioterrorism. In: Tibayrenc, M. (Ed.), Encyclopedia of Infectious Diseases: Modern Methodologies. John Wiley & Sons, New Jersey, pp. 619−638.

Moskalyk, L.A., Oo, M.M., Jacobs-Lorena, M., 1996. Peritrophic matrix proteins of *Anopheles gambiae* and *Aedes aegypti*. Ins. Mol. Biol. 5, 261−268.

Moura, H., Visvesvara, G.S., 2001. A proteome approach to host−parasite interaction of the microsporidian *Encephalitozoom intestinalis*. J. Eukaryot. Microbiol. (Suppl.), 56S−59S.

Moura, H., Ospina, M., Woolfitt, A.R., Barr, J.R., Visvesvara, G.S., 2003. Analysis of four human microsporidian isolates by MALDI-TOF mass spectrometry. J. Eukaryot. Microbiol. 50, 156−163.

Navas, A., Albar, J.P., 2004. Application of proteomics in phylogenetic and evolutionary studies. Proteomics 4, 299−302.

Nedelkov, D., 2005. Population proteomics: addressing protein diversity in humans. Expert Rev. Proteomics. 2, 315−324.

Nedelkov, D., 2006. Mass spectrometry-based immunoassays for the next phase of clinical applications. Expert Rev. Proteomics 3, 631−640.

Nedelkov, D., 2008. Population proteomics: investigation of protein diversity in human populations. Proteomics 8, 779−786.

Nedelkov, D., Tubbs, K.A., Niederkofler, E.E., Kiernan, U.A., Nelson, R.W., 2004. High-throughput comprehensive analysis of human plasma proteins: a step toward population proteomics. Anal. Chem. 76, 1733−1737.

Nedelkov, D., Kiernan, U.A., Niederkofler, E.E., Tubbs, K.A., Nelson, R.W., 2005. Investigating human plasma proteins diversity. Proc. Natl. Acad. Sci. U.S.A. 102, 10852–10857.

Nedelkov, D., Kiernan, U.A., Niederkofler, E.E., Tubbs, K.A., Nelson, R.W., 2006. Population Proteomics: The Concept, Attributes, and Potential for Cancer Biomarker Research. Mol. Cell. Proteomics 5, 1811–1818.

Nedelkov, D., Phillips, D.A., Tubbs, K.A., Nelson, R.W., 2007. Investigation of human protein variants and their frequency in the general population. Mol. Cell. Proteomics 6, 1183–1187.

Nelson, R.W., Krone, J.R., Bieber, A.L., Williams, P, 1995. Mass-spectrometric immunoassay. Anal. Chem. 67, 1153–1158.

Nelson, M.M., Jones, A.R., Carmen, J.C., Sinai, A.P., Burchmore, R., Wastling, J.M., 2008. Modulation of the host cell proteome by the intracellular apicomplexan parasite *Toxoplasma gondii*. Infect. Immun. 76, 828–844.

Ostrowski, J., Wyrwicz, L.S., 2009. Integrating genomics, proteomics and bioinformatics in translational studies of molecular medicine. Expert Rev. Mol. Diagn. 6, 623–630.

Ouellette, M., Olivier, M., Sato, S., Papadopoulou, B., 2003. Studies on the parasite *Leishmania* in the post-genomic era. Med. Sci. 19, 900–909.

Ouma, J.H., Vennervald, B.J., Butterworth, A.E., 2001. Morbidity in schistosomiasis: an update. Trends Parasitol. 17, 117–118.

Oura, C.A., McKellar, S., Swan, D.G., Okan, E., Shiels, B.R., 2006. Infection of bovine cells by the protozoan parasite *Theileria annulata* modulates expression of the ISGylation system. Cell. Microbiol. 8, 276–288.

Pan, J., Chen, H.Q., Sun, Y.H., Zhang, J.H., Luo, X.Y., 2008. Comparative proteomic analysis of non-small-cell lung cancer and normal controls using serum label-free quantitative shotgun technology. Lung. 186, 255–261.

Pang, R.T., Poon, T.C., Chan, K.C., Lee, N.L., Chiu, R.W., Tong, Y.K., et al., 2006. Serum proteomic fingerprints of adult patients with severe acute respiratory syndrome. Clin. Chem. 52, 421–429.

Papadopoulos, M.C., Abel, P.M., Agranoff, D., Stich, A., Tarelli, E., Bell, B.A., et al., 2004. A novel and accurate diagnostic test for human African trypanosomiasis. Lancet 363, 1358–1363.

Park, S.K., Yates, 3rd. J.R., 2010. Census for proteome quantification. Curr. Protoc. Bioinformatics (Chapter 13), Unit 13.12, 1–11.

Pastorino, B., Boucomont-Chapeaublanc, E., Peyrefitte, C.N., Belghazi, M., Fusaï, T., Rogier, C., et al., 2009. Identification of cellular proteome modifications in response to West Nile virus infection. Mol. Cell. Proteomics 8, 1623–1637.

Pedersen, K.S., Codrea, M.C., Vermeulen, C.J., Loeschcke, V., Bendixen, E., 2010. Proteomic characterization of a temperature-sensitive conditional lethal in *Drosophila melanogaster*. Heredity 104, 125–134.

Peltonen, L., McKusick, V.A., 2001. Genomics and medicine. Dissecting human disease in the postgenomic era. Science 291, 1224–1229.

Pendyala, G., Trauger, SA, Kalisiak, E, Ellis, RJ, Siuzdak, G, Fox, HS., 2009. Cerebrospinal fluid proteomics reveals potential pathogenic changes in the brains of SIV-infected monkeys. J. Proteome Res. 5, 2253–2260.

Peri, S., Navarro, J.D., Amanchy, R., Kristiansen, T.Z., Jonnalagadda, C.K., Surendranath, V., et al., 2003. Development of human protein reference database as an initial platform for approaching systems biology in humans. Genome Res. 13, 2363–2371.

Peronnet, E., Becquart, L., Poirier, F., Cubizolles, M., Choquet-Kastylevsky, G., Jolivet-Reynaud, C., 2006. SELDI-TOF MS analysis of the Cardiac Troponin I forms present in plasma from patients with myocardial infarction. Proteomics 6, 6288−6299.

Petricoin, E.F., Liotta, L.A., 2004. SELDI-TOF-based serum proteomic pattern diagnostics for early detection of cancer. Curr. Opin. Biotechnol. 15, 24−30.

Pitarch, A., Nombela, C., Gil, C., 2009. Proteomic profiling of serologic response to Candida albicans during host−commensal and host−pathogen interactions. Methods Mol. Biol. 470, 369−411.

Plattner, F., Soldati-Favre, D., 2008. Hijacking of host cellular functions by the Apicomplexa. Annu. Rev. Microbiol. 62, 471−487.

Poon, T.C., Hui, A.Y., Chan, H.L., Ang, I.L., Chow, S.M., Wong, N., et al., 2005. Prediction of liver fibrosis and cirrhosis in chronic hepatitis B infection by serum proteomic fingerprinting: a pilot study. Clin. Chem. 51, 328−335.

Predel, R., Wegener, C., Russell, W.K., Tichy, S.E., Russell, D.H., Nachman, R.J., 2004. Peptidomics of CNS-associated neurohemal systems of adult Drosophila melanogaster: a mass spectrometric survey of peptides from individuals' flies. J. Comp. Neurol. 474, 379−392.

Premsler, T., Zahedi, R.P., Lewandrowski, U., Sickmann, A., 2009. Recent advances in yeast organelle and membrane proteomics. Proteomics 9, 4731−4743.

Rachinsky, A., Guerrero, F.D., Scoles, G.A., 2007. Differential protein expression in ovaries of uninfected and Babesia-infected southern cattle ticks, Rhipicephalus (Boophilus) microplus. Insect Biochem. Mol. Biol. 37, 1291−1308.

Rain, J.C., Selig, L., De Reuse, H., Battaglia, V., Reverdy, C., Simon, S., et al., 2001. The protein−protein interaction map of Helicobacter pylori. Nature 409, 553.

Rives, A.W., Galitski, T., 2003. Modular organization of cellular networks. Proc. Natl. Acad. Sci. U.S.A. 100, 1128−1133.

Roberts, S.B., Robichaux, J.L., Chavali, A.K., Manque, P.A., Lee, V., Lara, AM, et al., 2009. Proteomic and network analysis characterize stage-specific metabolism in Trypanosoma cruzi. BMC Syst. Biol. 3, 52.

Rock, F.L., Hardiman, G., Timans, J.C., Kastelein, R.A., Bazan, J.F., 1998. A family of human receptors structurally related to Drosophila Toll. Proc. Natl. Acad. Sci. U.S.A. 95, 588−593.

Roitt, I.M, Delves, P.J., 2001. Roitt's Essential Immunology. Blackwell Publishing, Inc., Malden, Massachusetts, U.S.A., 496 pages.

Rual, J.F., Venkatesan, K., Hao, T., Hirozane-Kishikawa, T., Dricot, A., Li, N., et al., 2005. Towards a proteome-scale map of the human protein−protein interaction network. Nature 437, 1173−1178.

Ryan, E.T., 2001. Malaria: epidemiology, pathogenesis, diagnosis, prevention, and treatment— an update. Curr. Clin. Top. Infect. Dis. 21, 83−113.

Sakolvaree, Y., Maneewatch, S., Jiemsup, S., Klaysing, B., Tongtawe, P., Srimanote, P., et al., 2007. Proteome and immunome of pathogenic Leptospira spp. revealed by 2DE and 2DE-immunoblotting with immune serum. Asian Pac. J. Allergy Immunol. 25, 53−73.

Samanta, M.P., Liang, S., 2003. Predicting protein functions from redundancies in large-scale protein interaction networks. Proc. Natl. Acad. Sci. U.S.A. 100, 12579−12583.

Sánchez, M.I., Thomas, F., Perrot-Minnot, M.J., Biron, D.G., Bertrand-Michel, J., Missé, D., 2009. Neurological and physiological disorders in Artemia harboring manipulative cestodes. J. Parasitol. 95, 20−24.

Schmid-Hempel, P., 2008. Parasite immune evasion: a momentous molecular war. Trends Ecol. Evol. 23, 318−326.

Schmid-Hempel, P., Ebert, D., 2003. On the evolutionary ecology of specific immune defence. Trends Ecol. Evol. 18, 27−32.

Shannon, P., Markiel, A., Ozier, O., Baliga, N.S., Wang, J.T., Ramage, D., et al., 2003. Cytoscape: a software environment for integrated models of biomolecular interaction networks. Genome Res. 13, 2498−2504.

Shui, W., Gilmore, S.A., Sheu, L., Liu, J., Keasling, J.D., Bertozzi, C.R., 2009. Quantitative proteomic profiling of host−pathogen interactions: the macrophage response to *Mycobacterium tuberculosis* lipids. J. Proteome Res. 8, 282−289.

Sims, P.F., Hyde, J.E., 2006. Proteomics of the human malaria parasite *Plasmodium falciparum*. Expert Rev. Proteomics. 3, 87−95.

Spirin, V., Mirny, L.A., 2003. Protein complexes and functional modules in molecular networks. Proc. Natl. Acad. Sci. U.S.A. 100, 12123−12128.

Spivey, A., 2009. Amplify, amplify: shotgun proteomics boosts the signal for biomarker discovery. Environ. Health Perspect. 117, A206−A209.

Steen, H., Mann, M., 2004. The ABC's (and XYZ's) of peptide sequencing. Nat. Rev. Mol. Cell. Biol. 5, 699−711.

Stelzl, U., Worm, U., Lalowski, M., Haenig, C., Brembeck, F.H., Goehler, H., et al., 2005. A human protein−protein interaction network: a resource for annotating the proteome. Cell 122, 957−968.

Steuart, R.F., 2010. Proteomic analysis of *Giardia*: studies from the pre- and post-genomic era. Exp. Parasitol. 124, 26−30.

Stiles, J.K., Whittaker, J., Sarfo, B.Y., Thompson, W.E., Powell, M.D., Bond, V.C., 2004. Trypanosome apoptotic factor mediates apoptosis in human brain vascular endothelial cells. Mol. Biochem. Parasitol. 133, 229−240.

Sun, B., Rempel, H.C., Pulliam, L., 2004. Loss of macrophage-secreted lysozyme in HIV-1-associated dementia detected by SELDI-TOF mass spectrometry. AIDS 18, 1009−1012.

Sun, J., Jiang, Y., Shi, Z., Yan, Y., Guo, H., He, F., et al., 2008. Proteomic alteration of PK-15 cells after infection by classical swine fever virus. J. Proteome Res. 7, 5263−5269.

Swanson, S.K., Washburn, M.P., 2005. The continuing evolution of shotgun proteomics. Drug Discov. Today 10, 719−725.

Tastet, C., Bossis, M., Gauthier, J.P., Renault, L., Mugniéry, D., 1999. *Meloidogyne chitwoodi* and *M. fallax* protein variation assessed by two-dimensional electrophoregram computed analysis. Nematology 1, 301−314.

Taylor, J.E., Hatcher, P.E., Paul, N.D., 2003. Crosstalk between plant responses to pathogens and herbivores: a view from the outside in. J. Exp. Bot. 55, 159−168.

te Pas, M.F., Claes, F., 2004. Functional genomics and proteomics for infectious diseases in the post-genomics era. Lancet 363, 1337.

Teixeira, P.C., Iwai, L.K., Kuramoto, A.C., Honorato, R., Fiorelli, A., Stolf, N., et al., 2006. Proteomic inventory of myocardial proteins from patients with chronic Chagas' cardiomyopathy. Braz. J. Med. Biol. Res. 39, 1549−1562.

Texier, C., Brosson, D., El Alaoui, H., Méténier, G., Vivarès, C.P., 2005. Post-genomics of microsporidia, with emphasis on a model of minimal eukaryotic proteome: a review. Folia. Parasitol. 52, 15−22.

Thiel, M., Bruchhaus, I., 2001. Comparative proteome analysis of *Leishmania donovani* at different stages of transformation from promastigotes to amastigotes. Med. Microbiol. Immunol. 190, 33−36.

Thiellement, H., Bahrman, N., Damerval, C., Plomion, C., Rossignol, M., Santoni, V., et al., 1999. Proteomics for genetic and physiological studies in plants. Electrophoresis 20, 2013−2026.

Toro-Nieves, D.M., Rodriguez, Y., Plaud, M., Ciborowski, P., Duan, F., Pérez Laspiur, J., et al., 2009. Proteomic analyses of monocyte-derived macrophages infected with human immunodeficiency virus type 1 primary isolates from Hispanic women with and without cognitive impairment. J. Neurovirol. 15, 36–50.

Tournu, H., Serneels, J., Van Dijck, P., 2005. Fungal pathogens research: novel and improved molecular approaches for the discovery of antifungal drug targets. Curr. Drug Targets 6, 909–922.

Uetz, P., Giot, L., Cagney, G., Mansfield, T.A., Judson, R.S., Knight, J.R., et al., 2000. A comprehensive analysis of protein–protein interactions in *Saccharomyces cerevisiae*. Nature 403, 623–627.

Uetz, P., Dong, Y.A., Zeretzke, C., Atzler, C., Baiker, A., Berger, B., et al., 2006. Herpesviral protein networks and their interaction with the human proteome. Science 311, 239–242.

Vaezzadeh, A.R., Briscoe, A.C., Steen, H., Lee, R.S., 2010. One-Step Sample Concentration, Purification, and Albumin Depletion Method for Urinary Proteomics. J. Proteome Res. [Epub ahead of print].

Valcu, C.M., Lalanne, C., Muller-Starck, G., Plomion, C., Schlink, K., 2008. Protein polymorphism between 2 *Picea abies* populations revealed by 2-dimensional gel electrophoresis and tandem mass spectrometry. Heredity 99, 364–375.

Van Hellemond, J.J., van Balkom, B.W., Tielens, A.G., 2007. Schistosome biology and proteomics: progress and challenges. Exp. Parasitol. 117, 267–274.

Vester, D., Rapp, E., Gade, D., Genzel, Y., Reichl, U., 2009. Quantitative analysis of cellular proteome alterations in human influenza A virus-infected mammalian cell lines. Proteomics 9, 3316–3327.

Vialás, V., Nogales-Cadenas, R., Nombela, C., Pascual-Montano, A., Gil, C., 2009. Proteopathogen a protein database for studying *Candida albicans*–host interaction. Proteomics 9, 4664–4668.

Vierstraete, E., Verleyen, P., Baggerman, G., D'Hertog, W., Van den Bergh, G., Arckens, L., et al., 2004. A proteomic approach for the analysis of instantly released wound and immune proteins in *Drosophila melanogaster* hemolymph. Proc. Natl. Acad. Sci. U.S.A. 101, 470–475.

Viswanathan, K., Früh, K., 2007. Viral proteomics: global evaluation of viruses and their interaction with the host. Expert Rev. Proteomics 4, 815–829.

Walduck, A., Rudel, T., Meyer, T.F., 2004. Proteomic and gene profiling approaches to study host responses to bacterial infection. Curr. Opin. Microbiol. 7, 33–38.

Walters, M.S., Mobley, H.L., 2009. Identification of uropathogenic Escherichia coli surface proteins by shotgun proteomics. J. Microbiol. Methods 78, 131–135.

Wattam, A.R., Christensen, B.M., 1992. Induced polypetides associated with filarial worm refractoriness in *Aedes aegypti*. Proc. Natl. Acad. Sci. U.S.A. 89, 6502–6505.

Watts, S., 1997. Epidemics and History: Disease, Power and Imperialism. Yale University Press, New Haven, CT.

Weiss, L.M., Fiser, A., Angeletti, R.H, Kim, K., 2009. *Toxoplasma gondii* proteomics. Expert Rev. Proteomics 6, 303–313.

Wiederin, J., Rozek, W., Duan, F., Ciborowski, P., 2008. Biomarkers of HIV-1 associated dementia: proteomic investigation of sera. Proteome Sci. 17, 7–8.

Wilm, M., 2009. Quantitative proteomics in biological research. Proteomics 9, 4590–4605.

Wolinska, J., King, K.C., 2009. Environment can alter selection in host–parasite interactions. Trends Parasitol. 25, 236–244.

Wu, C.C., MacCoss, M.J., Howell, K.E., Yates, III. J.R., 2003. A method for the comprehensive proteomics analysis of membrane proteins. Nat. Biotechnol. 21, 532–538.

Wu, M.S., Chow, L.P., Lin, J.T., Chiou., S.H., 2008. Proteomic identification of biomarkers related to *Helicobacter pylori*-associated gastroduodenal disease: challenges and opportunities. J. Gastroenterol. Hepatol. 23, 1657–1661.

Wu, C., Wang, Z., Liu, L., Zhao, P., Wang, W., Yao, D., et al., 2009a. Surface enhanced laser desorption/ionization profiling: new diagnostic method of HBV-related hepatocellular carcinoma. J. Gastroenterol. Hepatol. 24, 55–62.

Wu, Y., Nelson, M.M., Quaile, A., Xia, D., Wastling, J.M., Craig, A., 2009b. Identification of phosphorylated proteins in erythrocytes infected by the human malaria parasite *Plasmodium falciparum*. Malar. J. 8, 105.

Xiao, Z., Prieto, D., Conrads, T.P., Veenstra, T.D., Issaq, H.J., 2005. Proteomic patterns: their potential for disease diagnosis. Mol. Cell. Endocrinol. 230, 95–106.

Yang, S.Y., Xiao, X.Y., Zhang, W.G., Zhang, L.J., Zhang, W., Zhou, B., et al., 2005. Application of serum SELDI proteomic patterns in diagnosis of lung cancer. BMC Cancer 5, 83.

Zanzoni, A., Montecchi-Palazzi, L., Quondam, M., Ausiello, G., Helmer-Citterich, M., Cesareni, G., 2002. MINT: a molecular INTeraction database. FEBS Lett. 513, 135–140.

Zhang, C.G., Chromy, B.A., McCutchen-Maloney, S.L., 2005. Host–pathogen interactions: a proteomic view. Expert Rev. Proteomics 2, 187–202.

Zhang, X., Zhou, J., Wu, Y., Zheng, X., Ma, G., Wang, Z., et al., 2009. Differential proteome analysis of host cells infected with porcine circovirus type 2. J. Proteome Res. 8, 5111–5119.

Zhang, L., Jia, X., Zhang, X., Sun, J., Peng, X., Qi, T., et al., 2010a. Proteomic analysis of PBMCs: characterization of potential HIV-associated proteins. Proteome Sci. 8, 12.

Zhang, Y., Wen, Z., Washburn, M.P., Florens, L., 2010b. Refinements to label free proteome quantitation: how to deal with peptides shared by multiple proteins. Anal. Chem. 82, 2272–2281.

Zheng, X., Hong, L., Shi, L., Guo, J., Sun, Z., Zhou, J., 2008. Proteomics analysis of host cells infected with infectious bursal disease virus. Mol. Cell. Proteomics 7, 612–625.

12 The Evolution of Antibiotic Resistance

Fernando González-Candelas[1,2,*], Iñaki Comas[3], José Luis Martínez[2,4], Juan Carlos Galán[2,5] and Fernando Baquero[2,5]

[1]Unidad Mixta de Investigación "Genómica y Salud" CSISP-UV/Instituto Cavanilles, Valencia, Spain, [2]CIBER en Epidemiología y Salud Pública, Spain, [3]National Institute for Medical Research, London, UK, [4]Centro Nacional de Biotecnología (CNB), CSIC, Madrid, Spain, [5]IRYCIS, Servicio de Microbiología, Hospital Ramón y Cajal, Madrid, Spain

12.1 Introduction

What is antibiotic resistance? This expression was obviously coined first in relation to medical microbiology and the therapy of infections. Antibiotic resistance refers to the property of bacteria which prevents the inhibition of their growth by antimicrobial agents used in the clinical setting. During the past decade many research, editorial, and review articles have focused on antibiotic resistance (Levy and Marshall, 2004; Pitout and Laupland, 2008; Livermore, 2009). The problem is dramatic in some countries (Vatopoulos, 2008) and especially worrying in highly pathogenic species such as *Mycobacterium tuberculosis* (Wright et al., 2009), methicillin-resistant *Staphylococcus aureus* (MRSA) (De Lencastre and Tomasz, 2008), *Acinetobacter baumanii* (Karageorgopoulos and Falagas, 2008), enterococci (Arias and Murray, 2008), or *Klebsiella pneumoniae* (Souli et al., 2008). Antibiotic resistance represents one of the best examples of natural selection, the basic process of evolutionary change, in action; and also one of the major hurdles in humankind's fight against infectious diseases. By taking a look at antibiotic resistance from a dual perspective, evolutionary and clinical, we hope to contribute to a better understanding of the principles and processes that result in the emergence of this undesirable character and to suggest strategies to, ideally, prevent or at least delay its extension. Human actions may not prevent evolution, but we can try to drive it through less damaging pathways.

*E-mail: gonzalef@uv.es

Antibiotic resistance can be "natural," when all the strains of the same bacterial species are resistant to a particular drug (also known as intrinsic resistance), or "acquired," when there are susceptible and resistant strains in the same species, with resistant strains having evolved from susceptible ones by selection after mutation or lateral genetic transfer events. It is of note that susceptible bacteria may have some type of "intrinsic resistance" along their life; for instance, high-density, slow-growing, stationary-phase bacteria are often refractory to inhibition by antimicrobial agents. This type of resistance is often qualified as "phenotypic" or "non-inheritable," as the organisms that are able to resist the drug under these circumstances give rise to a susceptible progeny under different conditions. Quite naturally, this view is controversial because when the original conditions are restored the resistant phenotype will occur again. The issue is even more complex, as "resistance" or "susceptibility" is not an absolute property.

A bacterial organism may be resistant to a particular concentration of an antibiotic and susceptible to a higher one. This is why we need breakpoints for defining susceptibility or resistance, based on minimal inhibitory concentration (MIC, the lowest amount of antibiotic able to inhibit bacterial visible growth) values. Clinical breakpoints consider a bacterial organism as resistant if its MIC is higher than the available concentration of the antibiotic at the site of infection, and/or higher than the concentration associated with favorable clinical outcomes. "Natural" (in contrast to clinical) breakpoints consider a bacterial organism as resistant if its MIC is significantly higher than the modal MIC of a collection of strains of the same bacterial species, thus considering resistance as an "abnormally higher" MIC, which indicates some kind of acquired genetic change.

Antibiotic resistance and MIC values have been consistently rising for many bacterial species and populations, even nonpathogenic ones, since the start of the industrial production of antimicrobial agents. The total annual production of antibiotics can be estimated between 100,000 and 200,000 tons and about half of them are not used in humans (mostly in farming) (Gootz, 2010). Considering that most antibiotics act at concentrations close to 1 μg/mL, such amount of antimicrobials is enough to cover the entire surface of the Earth with inhibitory concentrations; in other words, they are capable of altering the population genetic structure of microbes. Moreover, antibiotics are not easily removed from the environment, and some of them can remain active for extended periods of time. The release of antibiotics is probably one of the major anthropogenic effects on the microbiosphere, altering microbial systems. Part of these alterations are predictable, such as extended antibiotic resistance, but unpredictable effects are most likely to occur, as changes in the interactions between microbes or with animals, plants, or influencing basic cycles of life in the biosphere (Baquero, 2009a,b; Martínez, 2009).

The use of industrial antibiotics in human and veterinary medicine, agriculture, farming, and other areas converges to a single, cooperative effect, changing bacterial ecology not only in different environments but also in the *common* environment. This effect can be observed as selection for antibiotic-resistant organisms and a faster evolution of antibiotic resistance. The main problem is the existing connectivity between all environments, human, farming, and agricultural, so that the antibiotic-derived effects in one of them has consequences in all the others. The connection between different environments with regard to the undesirable consequences of antibiotic resistance occurs essentially in two ways, which will be considered in the next paragraphs.

First, dispersal and migration of biological units, such as bacterial communities, bacterial clones, mobile genetic elements (MGEs), and, in general, genes, play a major role in antibiotic resistance. Second, there is also dispersal of antimicrobial ecotoxic agents, which results in the production of selective mixed gradients and stressor effects, and in an acceleration of microbial evolutionary rates. The combination of migration of antibiotics and antibiotic-resistance biological units results in evolutionary activating interactions that occur in four main genetic reactors: (i) the intestinal microbiota of humans and animals; (ii) the highly antibiotic-exposed areas with high rates of bacterial transmission, such as hospitals (particularly newborn wards and intensive-care units), (iii) wastewater, effluents, and sewage treatment plants, and (iv) soil, sediments, surface and ground waters (Baquero et al., 2008), all of which contribute to the escalation of the emergence and spread of antimicrobial resistance.

The most evident (visible) threat of antimicrobial resistance for humankind is the failure of therapy against infectious diseases. The decrease in the incidence of infectious diseases in the Western world started in the beginning of the nineteenth century, by reasons related to social progress, better nutrition, and housing and hygienic procedures, but in the absence of antibiotics. Nevertheless, the discovery and subsequent industrial production of antimicrobials between 1935 and 1960 was followed by a significant reduction in the morbidity and mortality by infections, particularly the more severe ones, and has probably contributed to the increase in the expected duration of lifetime of human populations. At the same time, antibiotics facilitated the progress in Medicine at large, allowing interventions (complex surgery, intensive-care units, immunosuppressive and anti-cancer chemotherapy, or transplantation) that expose the impaired host to both pathogenic and opportunistic bacterial infections.

If antibiotic resistance were surpassing a threshold limit, the consequences on the current standards of hospital-based medicine (including long-term-care facilities for the elderly) could become severely compromised. With the emergence of multi-resistant Gram-positive organisms, such as MRSA or *Enterococcus faecium*, or Gram-negatives, such as pan-resistant *Escherichia coli*, *Klebsiella pneumoniae*, or *Pseudomonas aeruginosa* producing extended-spectrum β-lactamases (ESBL) and carbapenemases are currently very close to such a threshold. A transient equilibrium was reached from 1950 to 1980 because resistance was counteracted by the continued discovery of novel antimicrobial agents active on resistant strains. Unfortunately, during the last quarter of the twentieth century no significant advances occurred in this field as a result of the interest of a number of pharmaceutical companies in investing more in chronic, non-curable diseases. In our days, resistance to the newest antibiotics continues evolving mostly on the bases of the old genetic mobile structures (plasmids, transposons, integrons) that became prevalent by the selective effect of the old antibiotics. The effect that the anthropogenic release of antibiotics has already caused on the genetic structure of bacterial populations is probably irreversible and will influence the evolutionary future of microbes on Earth.

Cleaning nature of this resistance gene pool is impossible. The best we can do is trying to control the emergence, selection, and spread of antibiotic-resistance genes in bacterial organisms interacting with humans, animals, or plants. The classical strategies for controlling the emergence of resistance are based on the reduction of chronic antibiotic-promoted bacterial mutagenic-stress associated with low dosages, the use of combinations of drugs, early-intensive therapy, maintaining low the bacterial density, and, more recently, the surveillance of hypermutable organisms and the suppression of phenotypic resistance. A number of these strategies have been explored by population and mathematical modeling (Levin et al., 2000; Levin, 2001; Bergstrom et al., 2004). Controlling the selection of antibiotic resistance is a major practical goal, which can be addressed again by the development of novel anti-infective drugs and the appropriate use of antibiotics, avoiding low dosages able to select low-level mutations serving as stepping-stones for high-level resistance. A classic approach to prevent the spread of antibiotic resistance is based on general hygiene and containment (infection control) measures, for instance by decreasing contacts between patients contaminated (infected or carriers) with resistant bacteria and non-contaminated ones.

Unfortunately, these measures are becoming increasingly insufficient in the current global landscape of antibiotic resistance. Avoiding the emergence of resistance in the individual patient is obviously important for the individual, but has minimal effects at the community level. The efficacy of classical ways of controlling selection and spread is inversely proportional to the density and penetration of resistant organisms and their MGEs in particular environments. Measures that might be successful in the early stages of resistance development, or in hospitals or countries with low rates of antibiotic resistance, are worthless in areas where resistance is already well established. Even in low antibiotic-resistance polluted areas, such as Sweden, recent studies have shown that a 2-year discontinuation in the use of trimethoprim did not reduce significantly the *E. coli* resistance rates to this compound (Sundqvist et al., 2010). This was probably due to the dispersion of trimethoprim-resistance genes in a multiplicity of bacterial organisms and MGEs frequently harboring other resistance determinants, thus assuring co-selection of *dfr* genes with other resistance genes.

Some regions of the world are densely polluted with antibiotic resistance. In a global world, sooner or later, resistance originated in these "source of resistance" areas will invade still clean environments. Resistant organisms are constantly diluted and potentially extinguish in competition with constant immigration of susceptible bacteria in local environments, but such a trend might collapse by the increase of resistant populations. Moreover, the success of resistant organisms will contribute to the constant accumulation in the bacterial world of genetic platforms and vehicles able to efficiently recruit and spread novel resistance genes. Antibiotic resistance increases bacterial evolvability; resistance calls for more resistance, in a phenomenon described as "genetic capitalism" (Baquero, 2004). In other words, resistance might be reversible when rare; if frequent, reversibility is not to be expected.

12.2 Mechanisms and Sources of Antibiotic Resistance

To produce an effect in a bacterial cell, an antibiotic has to cross different cell envelopes, in some occasions has to be activated by bacterial enzymes, and reach its target at a concentration high enough to allow a successful interaction and the inhibition of bacterial growth or killing (Figure 12.1A). Resistance can thus be achieved either if the antibiotic concentration reaching the target is too low or if the interaction between the antibiotic and the target is not efficient enough to produce the inhibition of bacterial growth. This includes intrinsic and acquired resistance (Figure 12.1).

The most classical mechanisms of intrinsic resistance are the absence of the target and a reduced permeability to a given antibiotic. These two mechanisms are passive systems of resistance. However, bacterial populations also present active mechanisms of resistance based on the detoxification of the antibiotic. These include chromosomally encoded antibiotic-inactivating enzymes and multidrug (MDR) efflux pumps. More recently, the analysis of comprehensive libraries of mutants from *E. coli* (Tamae et al., 2008) and *P. aeruginosa* (Breidenstein et al.,

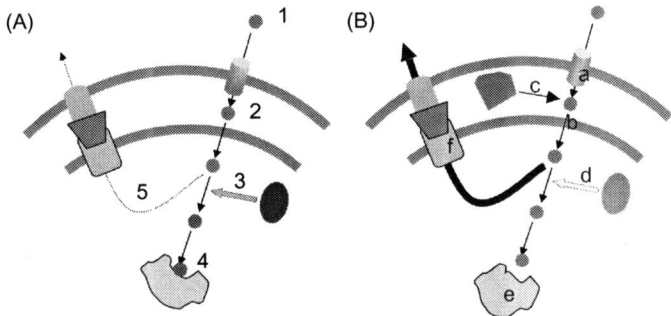

Figure 12.1 Basic mechanisms of antibiotic action and resistance. In order to inhibit bacterial growth an antibiotic requires to successfully interact with its target at concentrations high enough for inhibiting its activity. For this (A), the antibiotic (1) requires to traverse cellular envelopes (2), in some occasions to be activated by an intracellular enzyme (3), and reach its target (4). The activity of constitutively expressed MDR efflux pumps (5) can decrease the intracellular concentration of the antibiotic. As shown in (B), resistance is achieved by interfering with this pathway, either by changes that impede the entrance of the antibiotic (a, b), by the activity of antibiotic-inactivating enzymes (c) or overexpression of MDR efflux pumps (f) that reduce the effective intracellular concentration, or by mutations in the enzyme that activates the pre-antibiotic (d) or in its target (e), which preclude target/antibiotic interactions.

2008; Fajardo et al., 2008; Dotsch et al., 2009) has demonstrated that several genes participate in the intrinsic resistance phenotype of a bacterial species. This suggests that intrinsic resistance is not just a consequence of the adaptation of a bacterial species to the presence of a given antibiotic, but rather a phenotypic consequence of the general physiological characteristics of each species.

Contrary to intrinsic resistance, which is an ancient phenotype of bacterial populations, acquired resistance is the consequence of adaptive evolution to the recent selective pressure exerted by the utilization of antibiotics by humankind (Martinez, 2009). Resistance can be achieved by avoiding the activity of the antibiotic at different levels (Figure 12.1B). These include changes in bacterial targets, which prevent the efficient action of the antibiotic, and the reduction of the effective intracellular concentration of the antibiotic, achieved by changes in the permeability of the cell envelopes (including a reduced expression of specific membrane transporters), by the expression of antibiotic-inactivating enzymes or MDR efflux pumps capable of efficiently expelling the antibiotics out or, in the case of intracellularly activated pre-antibiotics, because the enzyme involved in their activation is not expressed or is not active as a consequence of mutations.

A formerly susceptible organism can acquire resistance by two main genetic mechanisms: mutation (Martinez and Baquero, 2000) or acquisition of foreign DNA (Davies, 1994). Mutation is the major cause in originating resistant bacterial pathogens during infections in the absence of a donor of antibiotic-resistance genes (Macia et al., 2005). Mutations involved in the development of resistance produce structural changes in the targets (for instance, quinolone resistance due to mutations in genes coding for bacterial topoisomerases; Piddock, 1999), in the enzymes that activate the pre-antibiotic (for instance, in the resistance to isoniazid due to mutations in the *Mycobacterium* catalase gene; Zhang et al., 1992), or in antibiotic transporters. Mutations in regulatory elements are also relevant for acquiring resistance. They act by changing the level of expression of antibiotic transporters (for instance, the porin OprD2 that transports imipenem inside *P. aeruginosa* and is not expressed in imipenem-resistant mutants; Yoneyama and Nakae, 1993) or by increasing the expression of antibiotic-detoxifying systems such as chromosomally encoded antibiotic-inactivating enzymes (Nelson and Elisha, 1999) and MDR efflux pumps (Martinez et al., 2009b). As we will see later, mutation is also important for the evolution of antibiotic-resistance genes acquired by bacterial pathogens through horizontal gene transfer (HGT).

Resistance can be also achieved as a consequence of the incorporation of DNA from other bacteria. In some occasions, this DNA recombines with homologous genes present in the chromosome of the new host, rendering novel mosaic genes that make the host resistant to antibiotics (for instance, the formation of recombinant penicillin-binding proteins; Spratt et al., 1992; Sibold et al., 1994). In other occasions, resistance is attained by the acquisition of an element that confers resistance on its own (antibiotic-resistance genes) (Davies, 1997). Given that bacterial pathogens were susceptible to antibiotics prior to their use as therapeutic agents for treating infections (Datta and Hughes, 1983), one intriguing question concerns the origin of antibiotic-resistance determinants.

Since natural (i.e., non-synthetic) antibiotics are produced by environmental microorganisms (Waksman and Woodruff, 1940), it was proposed early that antibiotic-producing microorganisms would be the most likely source of resistance determinants (Benveniste and Davies, 1973). The rationality of this proposal comes from the need of producers to protect themselves from the activity of their own antimicrobials. It was proposed that the actual source of resistance could be the DNA from antibiotic producers that might contaminate antibiotic preparations, which could be incorporated by bacterial pathogens simultaneously to treatment (Webb and Davies, 1993). This origin from antibiotic producers might be true on occasion, and indeed a recent search of resistance elements in *Actinomycetes*, the most important group of antibiotic producers, demonstrates that they harbor in their genomes a very large number of genes whose expression can trigger resistance in other hosts (D'Acosta et al., 2006). Nevertheless, in spite of the efforts at tracking the origin of resistance genes currently present in human pathogens, yet little is known about the source of these determinants, and in most occasions the comparison of resistance genes present in human pathogens to those located in the chromosomes of the microorganisms selected by the pharmaceutical industry for producing antibiotics show that they are not exactly the same but rather belong to the same structural family, thus indicating a phylogenetic relationship but not a direct origin.

Notably, a direct origin of some specific resistance genes has been demonstrated for bacterial species that do not produce antibiotics. This is the case for Qnr determinants, which contribute to plasmid-acquired low-level resistance to quinolones in Enterobacteriaceae (Robicsek et al., 2006b). It is worth mentioning that quinolones are synthetic antibiotics, so that it was proposed that resistance could be achieved only by mutations and that the existence of quinolone-resistance genes that could be acquired by human pathogens would be unlikely. Contrary to this hypothesis, the existence of plasmids carrying quinolone-resistance genes was described in 1998 (Martinez-Martinez et al., 1998) and, as stated above, these plasmids are currently disseminated among Enterobacteriaceae. The search for *qnr* elements in the chromosomes of fully sequenced bacteria has shown that these determinants are mainly present in aquatic bacteria, which do not produce antibiotics (Sanchez et al., 2008). As these genes are conserved among members of the same species and the level of synteny of the flanking regions in the chromosome is high, this indicates that the analyzed aquatic bacteria have not acquired those elements from other bacteria but rather those microorganisms are the source of Qnr determinants. Indeed, the origin of the *qnrA1* gene, the most abundant in Enterobacteriaceae plasmids, is *Shewanella algae* (Poirel et al., 2005). Similarly, *Kluyvera ascorbata*, a non-producer organism, is the most likely origin of the β-lactamases belonging to the CTX-M family (Humeniuk et al., 2002).

The functional role of resistance genes in producers seems clear: they might be detoxification elements needed to avoid the activity of the antibiotic. However, their function in non-producers is less apparent. It can be postulated that these elements might serve to resist the activity of inhibitory compounds produced by competitors in complex microbial ecosystems. As we will describe later, this might be the function of some of these resistance elements. However, on other occasions, these elements have not evolved to specifically counteract the activity of antibiotic producers. For instance, Enterobacteriaceae harbor chromosomal β-lactamases (Lindberg and Normark, 1986), which have been present in their genomes for several hundred million years. However, the natural habitat of these bacterial species is the gut, an ecosystem that does not contain β-lactam producers and, thus, did not contain β-lactam antibiotics until the recent use of these compounds for the treatment of infections. Similarly, quinolones are among the most frequent substrates of MDR efflux pumps (Alonso et al., 1999) despite the synthetic origin of these antibiotics, which implies that bacterial populations have not been exposed to their action until recently (in evolutionary terms). A suitable hypothesis for explaining the origin of these elements might be that they have been selected during bacterial evolution to play other roles than resistance (Fajardo et al., 2008), but their natural substrates present structural similarities to antibiotics currently used for therapy, in such a way that they can detoxify bacteria from these drugs even though this is not their original physiological function.

For instance, all bacterial species harbor in their genomes genes coding for MDR efflux pumps. These genes are highly conserved (the same MDR elements are present in all the strains of a given bacterial species) and redundant (bacterial species usually harbor several different MDR efflux pumps), indicating that they are ancient elements relevant for bacterial physiology (Martinez et al., 2009b). Given their genomic redundancy and their overlap in substrate usage, it is unlikely that the original function of these elements would be resistance to the drugs currently used in the clinical setting. This is the case of AcrAB-TolC, the major MDR efflux pump in *E. coli* and *Salmonella*. This element can extrude, besides antibiotics, bile salts (Thanassi et al., 1997), which are toxic compounds present in the natural ecosystem of these bacterial species, the gut. Similarly, it has been described that other MDR efflux pumps (that contribute to antibiotic resistance) can have a primary role in the trafficking of bacterial signals (Kohler et al., 2001), in the bacterial response to plant-produced signals (Burse et al., 2004; Maggiorani Valecillos et al., 2006), in the detoxification of intracellular toxic intermediary metabolites (Aendekerk et al., 2002, 2005) or in the response to non-antibiotic bacterial inhibitors such as heavy metals (Nies, 2003) or solvents (Ramos et al., 2002).

All this indicates that the universe of elements that can confer resistance to a heterologous host upon their transfer by HGT is larger than previously thought. As an example, the study of the **resistome** of the human gut microbiota (Sommer et al., 2009) indicates the existence of a large number of elements that are not disseminated among human pathogens and that can confer resistance despite the fact that this ecosystem does not contain the regular producers of antibiotics, Actinomyces. The potential universe of resistance genes will thus include those that first evolved to confer resistance (for instance, in producers) but also those that can recognize the antibiotic even though their original substrate (and thus their functional role) was a different one.

Given the large number of resistance genes present in natural ecosystems, it is paradoxical that, in comparison, the number and variability of HGT-acquired resistance determinants currently present in human bacterial pathogens are relatively low. This might indicate that there are some restrictions for the transfer of a potential resistance element to a human pathogen. The first impediment would be ecological connectivity. The presence of resistance elements has been demonstrated in bacteria from the deep terrestrial subsurface (Brown and Balkwill, 2009) and in the deep Greenland ice core (Miteva et al., 2004), where the presence of human pathogens is not expected. The probability of transfer of these elements to human pathogens will thus be very low, but chances will increase for bacteria whose natural habitats are closer to those of human pathogens. One example of this type or "reactors" for resistance are wastewater treatment plants, where human-linked microbiota (recipients of resistance genes) can get in contact with environmental microorganisms (potential donors) in the presence of residues of antibiotics which act as selecting agents (Baquero et al., 2008).

A second obstacle for the transfer of a resistance gene will be its integration into an efficient dissemination vector or into a bacterial epidemic clone, which will allow a fast spread for the resistance determinant (Baquero, 2004; Martinez et al., 2007).

The third obstacle consists of the fitness costs associated with the acquisition of resistance (Andersson, 2006). It is generally accepted that the development of resistance by formerly susceptible bacteria might confer a metabolic burden such that resistant populations might be outcompeted by susceptible ones in the absence of antibiotic selective pressure. It is worth mentioning that, unless the fitness costs are unaffordable, their relevance during antibiotic treatment will be negligible because in these conditions being resistant is a prerequisite for producing an efficient infection (Martinez and Baquero, 2002). Nevertheless, fitness costs may be highly relevant for the persistence and spread of resistance in the absence of selection (for instance, in nonclinical, natural ecosystems). It has been described that the fitness costs associated to the development of resistance to a specific antibiotic might be different depending on the involved mechanisms (Balsalobre and de la Campa, 2008). Furthermore, fitness costs can be compensated by secondary mutations in the bacterial genome (Bjorkman et al., 2000; Paulander et al., 2007; Lofmark et al., 2008). The acquisition of a resistance-determinant by a gene-recruitment element such as an integron, which might harbor other resistance elements (for instance, heavy metals), might allow the maintenance of resistance by co-selection mechanisms. Finally, the incorporation of a resistance-determinant into a plasmid encoding toxin-antitoxin systems allows the persistence of resistance even in the absence of selective pressure (Moritz and Hergenrother, 2007).

Overcoming all these biological and ecological obstacles is almost a necessary condition for the establishment of a resistance-determinant in a bacterial population. Nevertheless, several resistance-determinants besides those already disseminated among human pathogens might fulfill the requirements to overcome those obstacles. However, they are not currently disseminated. To explain this restriction further, **founder effects** must be considered (Martinez et al., 2009a). If one resistance element enters a formerly susceptible population through an efficient vector (or clone) and does not confer a high-fitness cost, it will soon spread in environments, like clinical settings, with a high antibiotic load. Once this element is established, the antibiotic no longer exerts a selective pressure because the bacteria are already resistant to it, so that the acquisition of a new resistance-determinant will not represent an adaptive advantage. As a result, human antibiotic usage has rendered a strong increase in the prevalence of a few resistance elements that were previously present in the chromosomes of environmental microorganisms and are now located in gene-transfer units spreading not just in bacteria in clinical settings but also in environmental ecosystems.

The release of human pathogens harboring gene-transfer units containing resistance elements, eventually simultaneously with antibiotic-containing wastes, might have a deep impact on the evolution of the microbiota from natural ecosystems, and this can influence also the evolution of clinically relevant mechanisms of antibiotic resistance (Baquero et al., 2008). As an example, the same antibiotic-resistance elements currently present in human pathogens can be found in wild animals (Livermore et al., 2001; Pallecchi et al., 2008) or in environmental locations without a history of antibiotic pollution (Pei et al., 2006). Furthermore, the study of historical soils has demonstrated that the introduction of antibiotics has produced an increase in the prevalence of specific resistance determinants in environmental, nonclinical ecosystems (Knapp et al., 2010).

The recent use (in evolutionary terms) of antibiotics by humankind has produced a strong enrichment in the distribution of a few specific antibiotic-resistance elements in clinical and nonclinical ecosystems. The impact of this enrichment in specific genes, and eventually bacterial clones, on the composition and activity of the microbiosphere remains to be fully understood. Given that natural ecosystems are the source of resistance genes (Martinez, 2008) and the reservoirs for their maintenance (Simoes et al., 2010), more studies on the ecological behavior of resistance in nonclinical habitats are required to unveil how these changes might impact on the acquisition of antibiotic resistance by human pathogens.

12.3 Evolution of Antibiotic-Resistance Genes

12.3.1 Antibiotic-Resistance Genes as Targets of Evolution

Antibiotic resistance is not only a clinical problem. It also represents a unique opportunity for observing evolution almost in real time, and therefore a meeting point between clinical and evolutionary microbiologists. The evolution of antibiotic resistance is a consequence of the selection of resistant organisms. Thanks to the acquisition of antibiotic-resistance genes, these bacteria are able to express a set of functions (resistance phenotypes) in a particular physiological and ecological context. On the other hand, the evolution of antibiotic resistance does not depend only on genes encoding mechanisms of resistance. The acquisition, expression, variability, and persistence of these genes depend on the genetic environment in which they are hosted, as integrons, transposons, plasmids, phages, or bacterial clones.

The evolution of resistance genes is therefore dependent on cell physiology and ecology. In consequence, antibiotic selection of antibiotic-resistant organisms implies the selection of organisms with particular physiological or ecological abilities.

A classic theoretical problem in evolutionary biology is whether or not genes are the units of selection. According to the preceding paragraph, and the conventional wisdom in evolution, selection acts on organisms that exhibit particular (selectable) phenotypes, and genes encoding those phenotypes will be selected by second-order processes of selection on the organisms that harbor them. A different (more reductionist) view was proposed by Dawkins (1976), who defended the preeminence of the gene as a selfish evolutionary element, an idea derived from the differentiation between replicators and vehicles. This view is weakened in the microbial world, as the "vehicles" (the organisms) are not really as perishable as higher sexual organisms are. Indeed, only HGT restores the importance of gene evolution in the microbial world. Genes may spread among bacterial populations and therefore are able to "replicate" independently from their vehicles. Indeed, genes involved in antibiotic resistance frequently migrate and spread among commensal and pathogenic organisms, and therefore have an evolutionary history which is independent of that of their bacterial hosts. In the following paragraphs we will illustrate a number of features related to the evolution of antibiotic-resistance genes, focusing only on those genes able to provide resistance phenotypes, leaving carrier genetic structures (such as integrons, transposons, phages, or plasmids) out of the scope of this review.

After the initial views that antibiotic-resistance genes had their origin in the environment (Benveniste and Davies, 1973) and that producers spread antibiotic-resistance genes to bacterial pathogens of humans (Marshall et al., 1998), it was accepted that genes encoding mechanisms of resistance arise in potentially any bacteria, because the ancestral precursors are present in all of them, in most cases as house-keeping genes involved in the physiological functions required for daily bacterial life. There are well-documented examples, such as β-lactamases, proteins capable of inactivating penicillin and cephalosporins, derived from protein-binding proteins (PBPs), which are essential in the construction of the peptidoglycan layer (Massova and Mobashery, 1998) or the essential Ser/Thr/Tyr protein kinases, the ancestral proteins of aminoglycoside or macrolide inactivating enzymes (Shakya and Wrigh, 2010). More recently, the sequencing of many complete bacterial genomes has provided a more complex scenario, showing that bacteria can have multiple alternative pathways to develop low levels of antibiotic resistance (Fajardo et al., 2008), thus leading to the concept of resistome (Wright, 2007). Examples such as GadA and GadB proteins (glutamate decarboxylase) as well as AmpC and HdeB, proteins that increase ampicillin resistance in *E. coli* (Adam et al., 2008), show the possibilities of different evolutionary pathways for developing antibiotic resistance in bacteria.

Therefore, resistant phenotypes can occur and even evolve in the absence of antibiotic selection; conversely, antibiotics may influence the evolution of bacterial functions associated to the adaptation to particular environments. In any case, it is essential to understand that there is a wealth of potential mechanisms of resistance contained in bacterial chromosomes and in MGEs. In this section we will illustrate a number of issues related to the evolution of genes directly involved in antibiotic-resistance phenotypes.

The main mechanisms of gene variation leading to variation and diversification of antibiotic-resistance genes are mutation, recombination, and amplification. The frequency of these mechanisms is variable in normal populations, being typically from 10^{-9} to 10^{-6} in the case of mutation, from 10^{-7} to 10^{-13} for recombination, and from 10^{-5} to 10^{-2} for tandem gene amplification. Note that considering the large sizes of bacterial populations (normally exceeding 10^8 cells/mL in their niches) these mechanisms of variation should be sufficient to provide enough variants of potential adaptive value in the presence of antibiotics. This is not necessarily the case if these populations are reduced during the infective process by innate or acquired immunity, by antimicrobial drugs, or during tissue-to-tissue or host-to-host transmission bottlenecks. In these cases population sizes might be insufficient to provide resistant mutants.

Bacteria with increased mutation rates are usually known as mutators. They present 10- to 1000-fold more chances of introducing changes in their DNA sequence during each replication cycle and therefore have a high probability of generating selectively advantageous mutations giving rise to the acquisition of antimicrobial resistance (Blázquez, 2003). Hypermutation is generally a consequence of a defective mismatch-repair system; as expected, mutators are easily selected when bacterial populations are exposed to recurrent selection pressures (such as during antibiotic treatments) (Mao et al., 1997). Mutators have been described, for example, in clinical strains of *S. aureus*, *S. pneumoniae*, *H. influenzae*, *E. coli*, *P. aeruginosa*, or *S. maltophilia* (Oliver et al., 2000; Morosini et al., 2003; Baquero et al., 2004; Turrientes et al., 2010). Mutators have been used in order to experimentally predict the emergence and selection of resistant variants (Galán et al., 2003; Novais et al., 2008; Stepanova et al., 2008), even though other methods are even more powerful to generate diversity (Orencia et al., 2001; Rasila et al., 2009). If predictions based on the use of **mutator strains** can provide insights about the type of variants conferring resistance to a particular antibiotic, they do not give us certainty about which of these variants will be more successful in the real world. This suggests that other factors apart from single antibiotic selective pressure also affect the evolution of the resistance phenotype. For instance, it has been recently demonstrated that the huge diversity of CTX-M β-lactamases can only be explained by the simultaneous exposure to two antibiotics, cefotaxime and ceftazidime, **genetic drift** also playing a relevant role in some stages of the process (Novais et al., 2010).

Antibiotics might act as accelerators of mutation-based evolution of resistance (Couce and Bláquez, 2009) not only by selecting mutator strains (Oliver et al., 2000), but also inducing intracellular redox imbalance, thus increasing intracellular superoxides. Accumulation of hydroxyl radicals (ROS formation) is highly toxic for the cell, and has a potent mutagenic power following the induction of the SOS response resulting from DNA damage. Elicited SOS-repair processes also increase the expression of error-prone polymerases such as polymerases IV (DinB) and V

(UmuCD), which present 100-fold less fidelity than the canonical polymerase II (PolB). In consequence, bactericidal antibiotics (bacteriostatic antibiotics are not inducers of hydroxyl radicals) have a double effect, inducing a process leading to cellular death by blocking their target (PBPs, DNA gyrase, etc.) while promoting the generation of genetic diversity and, therefore, of antibiotic resistance. An elegant model is the fluoroquinolone-resistance mediated by *qnr* genes: treatment with ciprofloxacin induces the SOS response, increasing the expression of error-prone polymerases (in a similar way to β-lactams or aminoglycosides), and promotes the cleavage of the LexA protein, a negative repressor of the *qnrB2* gene, thus leading to QnrB overexpression. QnrB binds to DNA gyrase, protecting it from quinolone inhibition (Da Re et al., 2009).

Recombination is a powerful mechanism for the evolution of antibiotic-resistance genes, particularly relevant for organisms able to exchange DNA by transformation, such as *S. pneumoniae*, *Neisseria meningitidis*, or *H. influenzae*, the typical example being recombination between sequences of *pbp* genes (penicillin-binding proteins) leading to different β-lactam resistance phenotypes in *S. pneumoniae* (Brückner et al. 2004). In this same species, hyper-recombinant strains have been associated to the generation of antibiotic resistance (Hanage et al., 2009). We should admit that the extent by which transformation and subsequent gene recombination occurs in many bacterial organisms is widely unknown; for instance, population biology studies on *Enterococcus faecalis* indicate high frequency of recombination among clones, despite the fact that transformation has not been documented in this species (Ruiz-Garbajosa et al., 2006). Indeed intraorganismal gene recombination is also a powerful mechanism for antibiotic resistance gene evolution, particularly relevant to rapidly spread adaptive mutations within a genome when they occur in a copy of otherwise repeated homologous genes. This phenomenon, known as gene conversion (Prammananan et al., 1999), increases for instance the efficiency of antibiotic-resistance mutations within *rrn* genes.

Finally, tandem gene amplification has been suggested as a possibility for evolving novel antibiotic-resistance genes. Amplification might immediately provide an adaptive advantage (increased gene dosage might increase the level of resistance protein), and later, eventually, one of the copies might serve as basis for further gene evolution without the risk of losing the ancient function (Pettersson et al., 2009; Kugelberg et al., 2010). Interestingly, all these mechanisms of variation (mutation, recombination, amplification) can increase under antibiotic stress at sub-inhibitory concentrations (López and Blázquez, 2009).

Resistance genes are frequently able to further evolve under variable antibiotic selection, allowing bacteria to grow when exposed to increased antibiotic concentrations, or enlarging the number of antibiotics within a family to which they provide protection. A good example is illustrated by a mutation in position 83 (Ser-83) of the A subunit of the DNA gyrase which provides a ciprofloxacin MIC of about 1 μg/mL in clinical isolates; a double change in Ser-83 and Asp-87 was found to increase MICs reaching values equal or higher than 8 μg/mL (Vila et al., 1994). Another example is provided by the cumulative mutational events driving the diversification of CTX-M β-lactamases (Coque et al., 2008; Novais et al., 2010).

12.3.2 Potential Evolution of the Enzymes Involved in Antibiotic Resistance

From a structural point of view, proteins involved in antibiotic resistance frequently show a high molecular plasticity, easily generating variants with an enlarged spectrum of action for new drugs within a chemical family, particularly under intense exposure. One of the first evidences on the functional versatility of antibiotic-resistance genes was described using TEM-1 β-lactamase. It was found that only 43 out of the 263 amino acid residues did not tolerate substitutions without loss of function (Huang et al., 1996). Large numbers of different variants were selected when bacteria carrying TEM-1 were exposed to cefotaxime (a synthetic β-lactam antibiotic active against TEM-1 producing organisms) (Zaccolo and Gherardi, 1999). This plasticity is also common in other determinants conferring resistance to other families of antibiotics, such as tetracyclines (Thaker et al., 2010) or aminoglycosides (Smith and Baker, 2002). However, β-lactamases are still the best model to understand the evolutionary potential of antibiotic-resistance elements (Bush and Jacoby, 2010), as β-lactams are the most extensively used antibiotics in the clinical setting (Goossens, 2009) and the family for which largest number of chemical molecules have been developed (due to their low level of toxicity and relatively low costs of production). Simultaneous strong selective pressure and changing selector (different β-lactam) has thus allowed the evolutionary radiation of β-lactamases during decades.

From 1940, when penicillin was commercialized, to 2007, when the last carbapenem (doripenem) was approved, β-lactamases have been in continuous evolution. In 1944 the *Staphylococcus aureus* PC1 penicillinase was the first enzyme known to confer resistance to the natural compound, penicillin. Subsequently, plasmid-mediated broad-spectrum β-lactamases (TEM-1 β-lactamase) were found in 1963 and ESBL in 1983 (Sirot et al., 1987), 5 years after cefotaxime was commercialized. The inhibitor-resistant TEM-type or SHV-type variant β-lactamases (IRT or IRS) were detected in 1992 (Blázquez et al., 1993), 8 years after the introduction of amoxicillin-clavulanate in the clinical setting. Plasmid-mediated carbapenemases (both metallo-β-lactamases and serine-β-lactamases) were detected in 1999 and 2003 (Lauretti et al., 1999; Smith Moland et al., 2003), when the consumption of carbapenems increased. Now almost 900 different β-lactamases are known. They are divided in four groups (A−D) according to their enzymatic properties and evolutionary relationships; class A β-lactamases being the most widely distributed. In fact, the number of variant class A β-lactamases with clinical importance is enormous and they include about 190 TEM and OXA-variants, 135 SHV variants, 100 CTX-M variants, or 35 carbapenemases. It is important to remark that, although the genes at the evolutionary root of these groups of β-lactamases had an origin in environmental bacteria, their subsequent mutation-driven evolution has likely taken place in clinical environments as the consequence of strong selection pressure by changing β-lactams. Among TEM variants, 66 amino acid positions (∼25% of all) have changed at least once. Nevertheless, this overwhelming diversity in antibiotic-resistance protein evolution follows only a few mutational pathways (Weinreich et al., 2006), as either some mutations are antagonistic when they are present in the same enzyme, or some mutations provide a phenotypic advantage only in specific backgrounds (sign epistasis) (Poelwijk et al., 2007).

An example of the first case is the simultaneous presence of P167S and D240G changes, which are involved in extended-spectrum activity of the CTX-M β-lactamases toward hydrolyzing ceftazidime efficiently, and both confer a lower hydrolytic activity against ceftazidime than each change alone (Novais et al., 2008). When the mutational pathway toward the highest cefotaxime hydrolytic activity was reconstructed by site-directed mutagenesis, some mutations yielded either an increase or decrease in resistance depending on the previous mutations, indicating that there is an order in the selection of mutations (Novais et al., 2010). Also, in the case of TEM β-lactamase, the M182T change conferred positive effects on the hydrolytic activity of TEM β-lactamase on eight occasions, with five neutral and three negative (Weinreich et al., 2006).

The plasticity of the antibiotic-resistance gene plays an important role in the selection of antibiotic resistance, but such plasticity may have a limit related to the possible loss of function. In any case, in a landscape dominated by HGT, clonal dispersal into different hosts, and under a highly selective exposure, a rapid diversification of antibiotic-resistance genes is expected, as it has recently occurred with plasmid-carried fosfomycin-resistance or quinolone-resistance genes (Martinez-Martinez et al., 1998; Wachino et al., 2010). When the local prevalence of determinants of resistance is high, two or more determinants are likely to be simultaneously present in the same strain, thus favoring recombination events originating hybrid proteins (Barlow et al., 2009). These hybrid enzymes would be selected if they confer selective advantages, such as increasing MIC values to antibiotics, reducing the fitness cost of resistance, or increasing the spectrum of activity.

Mosaic genes have been described among tetracycline-resistant determinants, *tet* genes, especially between *tet*(O) and *tet*(W) in Firmicutes (Patterson et al., 2007) showing higher protection against tetracycline than non-recombinant *tet* genes (Stanton and Humphrey, 2003). Recently, a hybrid β-lactamase (609 amino acids) with two active centers, resulting from the fusion between class C (346 amino acids) and class D (253 amino acids) β-lactamases (Allen et al., 2009), has been described. Probably the best examples known are the hybrid bifunctional enzymes of aminoglycoside resistance AAC(6′)/APH(2′), an enzyme able to perform acetylation and phosphorylation, and ANT(3″)-Ii/AAC(6′)-IId, an enzyme that catalyzes acetylation and adenylation reactions (Zhang et al., 2009). However, the most worrying discovery was that the bifunctional enzyme AAC(6′)-Ib-cr was able to confer resistance to aminoglycosides and fluoroquinolones (Robicsek et al., 2006a). Recently, Maurice et al. (2008) have shown that substitutions W92R and D169Y in the AAC(6′)-Ib sequence are responsible for increasing its spectrum of action to quinolones. This enzyme is a good example of the surprising capacity for adaptive evolution of antibiotic-resistance genes.

We should have learned the lesson about the capacities of bacteria to develop antibiotic-resistance mechanisms when β-lactam plus β-lactamase-inhibitor antibiotic combinations and broad-spectrum cephalosporins were discovered in the

1980s. In those days, the original idea that antibiotic-resistant Gram-negative bacteria would be unable to develop resistance to β-lactam plus β-lactamase-inhibitor combinations was accepted by the scientific community (Medeiros, 1997). This idea arose because mutant β-lactamases able to confer resistance to broad-spectrum cephalosporins (ESBL) proved to be hypersusceptible to β-lactamase inhibitors, and vice versa. This mutational **pleiotropic antagonism** was confirmed in several in vitro experiments and, indeed, the combinations of mutations leading to resistance to both β-lactam plus β-lactamase inhibitor were not observed in nature. However, in 2003 a laboratory mutant of ROB-1 β-lactamase harboring changes S130G, involved in inhibitor resistance, and R164W and A237T, involved in ESBL resistance, was obtained, the host bacteria being resistant to both amoxicillin-clavulanate and ceftazidime (Galán et al., 2003). Later, in 2004, a new β-lactamase CMT (a complex mutant TEM), which confers a high level of resistance to ceftazidime combined with a reduced susceptibility to amoxicillin-clavulanate, was reported in a clinical isolate (Poirel et al., 2004).

12.4 Limitations to Adaptation and the Cost of Resistance

12.4.1 The Genetics of Adaptation

Genetic variability in a population does not increase inevitably along time, since it is the result of factors acting in opposite directions: some processes introduce new variation in the populations while others remove it. Two main processes deplete variation from bacterial populations: selection and drift. By increasing the proportion of cells that carry particular high-fitness variants, selection may transitorily reduce genetic variability in populations, while the effect of drift is continuous and affects equally all variants in the population, regardless of their effect on fitness.

Fitness is a relative measure of the contribution of an organism to the next generation. In its simplest formulation, fitness is an individual property but it can be, and usually is, extended to groups of organisms that share some common features such as belonging to the same lineage, sharing a specific mutation, or displaying a given phenotype. Despite its simplicity, fitness can be a very elusive concept. In the context of antibiotic resistance, fitness can be defined as the relative capacity of bacteria to survive and reproduce within an infected individual and to spread to infect others. The epidemiological component of this definition emphasizes the need for considering all the meaningful biological levels at which fitness can be analyzed. A very successful variant which can resist an antibiotic will be of very little relevance if it fails to be transmitted to other individuals. Both fitness components, intrahost and interhost, are usually correlated but this is not necessarily so. Evaluation of intrahost components can be approached using in vitro systems but the transmission fitness is only possible from epidemiological observations.

Fitness is not a fixed property of individuals or groups: it is contextual and it can change dramatically when the environment or the genetic background is altered. This is readily exemplified when we talk about the cost of resistance. This is the reduction of the fitness of antibiotic-resistant bacteria in the absence of the drug. In the presence of that particular antibiotic, and occasionally in that of others as well, the increase in fitness associated to survival, reproduction, and transmission reveals the environmental dependence of the concept. Similarly, compensatory mutations can alter the fitness value of a certain resistance mutation by modifying the genetic context in which they are expressed. Naturally, both genetic and environmental changes can interact in synergistic or antagonistic ways, making more difficult the prediction of the phenotypic value under particular combinations of the two components.

Fitness is a relative measure of reproductive success and is usually evaluated as the difference between the population growth rate of a reference and the target strain (Elena and Lenski, 2003). Nevertheless, other measures of fitness are more useful for the study of antibiotic-resistant bacteria at the within-population level. For instance, survival under a given concentration of a drug is, at the individual level, a binary variable, but the proportion of individuals from a given population surviving under such conditions is a quantitative value. This allows us to rank strains or genotypes according to their fitness. Similarly, a given population may be resistant to higher concentrations of an antibiotic or the necessary concentration of this to inhibit completely bacterial growth (MIC value) can be higher. The values of these variables are often taken as indirect measures of fitness and they usually correlate with increased risk or potential harm in the clinical practice. A higher dose of an antibiotic may have serious side effects or may not be easily tolerated by some patients, thus posing at higher risk their survival from an infection.

At the population level, bacterial fitness is measured through R_0, the basic reproductive number. This is defined as the number of secondarily infected hosts from any infected individual. For any infection to persist in a population its R_0 must necessarily be larger than 1, although temporary persistence is possible even if $R_0 < 1$. The value of R_0 is computed in a population with a large proportion of susceptible hosts and, in consequence, R_0 cannot be constant in a real population, where it will decrease through time.

Genetic drift is the result of the sampling process that occurs in every population in which the total number of individuals is limited. This limit (which is equivalent to the infection dose in this context) can be very high (millions or billions, in the case of bacterial populations) or very low, as when an individual is infected by a single bacterial cell, as in some tuberculosis (TB) infections. In the former case, the reduction in genetic variability is almost imperceptible and it is easily compensated by the continuous generation of new genetic variation. On the contrary, extreme reductions in population size, especially during the transmission from one infected host to a new one result in a drastic elimination of genetic variability after which only a few of the initially present variants are represented in the newly established population. In this case, the variants that originate the new population are drawn at random from those initially present and the particular variants transmitted are not necessarily associated with increased fitness: they can be more, equal, or less fit than those of the average population they derive from.

Although usually overlooked, if not ignored, in the study of genetic variation in microorganisms, the neutral theory of molecular evolution (Kimura, 1968; Kimura, 1983) sustains that most variation found at the molecular level does not have an impact on fitness and, as a consequence, is neutral in terms of natural selection. The original proposal was subsequently expanded (Ohta and Kimura, 1971; Ohta, 1992) by incorporating the evolutionary consequences of slightly deleterious mutations whose fate does not depend exclusively on the relative reduction in fitness they produce, but on the size of the population where they arise. Stochastic processes, usually associated with genetic drift, will dominate the fate of these mutations if effective population size is lower than the reciprocal of the corresponding selection coefficients. When population sizes or selection coefficients are larger and the above inequality no longer holds, then deterministic processes will dominate and selection will be the main evolutionary force in the population. Given the large population sizes associated with bacteria, it is usually considered that genetic drift is not as important as selection in determining evolutionary change in bacterial populations. But this is not the case during transmission or during chronic infection. Effective sizes for parasitic or pathogenic bacterial populations have been estimated to be much lower than free-living bacteria (Hughes, 2005). This implies that stochastic factors may have an important role in the evolution of bacterial pathogens at this level. One additional, often overlooked, aspect of the quasi-neutral theory is that it also applies to slightly favorable mutations. While some mutations may confer increased fitness, their dynamics (stochastic or deterministic) will be determined by the relationship between the effective population size and the selection coefficient: a slightly advantageous mutation may easily disappear from a small population while it will likely increase in frequency in a large one.

The interplay between selection and drift can have consequences and leave imprints at different levels. The study of the evolution of CTX-M β-lactamases toward higher MIC values for cefotaxime and ceftazidime (Novais et al., 2010) demonstrates that some critical steps in some of the evolutionary trajectories revealed in the analysis were only possible if drift had played an important role, since the fitness of a necessary new genotype in a pathway was lower than that of the preceding variant. Apart from invoking evolution in alternative environments (with different fitness landscapes than those considered), this is only possible by the action of stochastic factors, among which drift is the major player. At a different level, reduced population sizes in *M. tuberculosis* may explain the higher relative rates of non-synonymous substitutions in their genes when compared with other free-living bacteria (Comas et al., 2010). In consequence, although selection may be the dominant factor in the evolution of bacterial populations, explaining almost perfectly the observed dynamics of antibiotic resistance in the presence of the selective drug, other evolutionary processes cannot be dismissed completely as irrelevant. Since these dynamics do not depend on fitness advantages, it is not necessary to invoke a cost of adaptation in every case and, most especially, in the absence of antibiotic.

12.4.2 From Genotype to Phenotype: The Many Ways Toward Fitness Compensation

While it is true that there are examples of drug-resistance mutations with no associated fitness cost, it is clear from their usually low frequencies that a fitness cost in the absence of the drug tends to be associated with resistance. This observation has led some researchers to argue that removal of antibiotics will leave room for drug-susceptible strains and that these will outcompete those harboring drug-resistance mutations (Austin et al., 1999). Although the strategy of drug removal seems to have some success in particular settings (Seppala et al., 1997; Guillemot et al., 2005) it is clear that other factors apart from the total levels of drug use influence the frequency of drug resistance. One of these factors, as shown by several clinical, experimental, and epidemiological data, indicates that compensation of fitness cost occurs, so that the advantage of drug-susceptible strains in an antibiotic-free environment is reduced or even disappears (Andersson and Hughes, 2010). As discussed previously, the fitness cost can be ameliorated through reversion, which is a very unlikely process, particularly for chromosomal mutations where exactly the same codon has to change to the wild-type allele (Levin et al., 2000). It is more likely that, in the absence of the antibiotic, the low-fit drug-resistant strains either become extinct or find ways of recovering fitness while keeping a drug-resistant phenotype. This process is usually known as compensation. Compensation is much more likely than reversion because there are usually many more loci that potentially can restore, at least partially, fitness costs. These loci can be in the same gene harboring the drug-resistance (intragenic) mutation, in other genes that somewhat interact with the drug-resistance mutated gene (intergenic), in plasmids or in another chromosome, depending on the mechanisms giving more chances to compensate than to revert a drug-susceptible phenotype (Maisnier-Patin and Andersson, 2004).

Mechanisms leading to compensation of drug resistance can be grouped in three categories: (i) those based on chromosomal compensatory mutations, (ii) those based on some kind of regulation alteration of the expression, and (iii) those based on the so-called bypass mechanisms. Chromosomal mutations leading to compensation represent the case in which fitness loss is compensated by a second, or more, mutations. These mutations can occur either in the same protein affected by the drug-resistance gene (intragenic mutation) or in other proteins that interact with it (intergenic mutations). But the complexity of compensation mutational pathways can go far beyond the accumulation of one or two mutations. Marcusson et al. (2009) showed how in isogenic, lab-constructed strains of *E. coli* resistant to fluoroquinolones, sometimes higher fitness effects are only attainable when four or five mutations are combined in the same strain and always depend on the loci mutated. They even showed how in the absence of the drug lower susceptibility can be achieved as a by-product of mutations increasing fitness. Interaction between mutations can occur also among drug-resistance mutations for different antibiotics, sometimes leading to a higher (positive) or lower (negative) fitness than the mere sum of their individual effects, a phenomenon usually known as epistasis.

Epistasis, when positive, can explain why the frequency of high-cost drug-resistant strains in clinical settings is higher than expected. A clear example of these epistatic interactions is shown by Trindade et al. (2009). They introduced mutations to different drugs in isogenic strains, thus creating multidrug-resistant strains and focused on combinations of drug-resistant mutations to rifampicin (*rpoB* gene), nalidixic acid (*gyrA*), and streptomycin (*rpsL*). They found that several combinations of these mutations led to fitter than expected mutants. Furthermore, these mutations were not gene- but allele-specific, and therefore epistasis and compensation depended on combinations of particular codon changes. In some cases, the double mutants not only were fitter than expected from their individual mutant's fitness but also were fitter than at least one of the two individual mutants. This phenomenon is called sign epistasis and means that there is not only amelioration of the fitness cost between drug-resistant mutations (positive epistasis) but also compensation leading to partial restoration of fitness. It is interesting that combinations of *gyrA*, *rpsL*, and *rpoB* drug-resistance mutations have been shown to be present in different bacterial backgrounds, which suggests that epistasis among drug-resistance mutations can be present in many pathogens. In fact, multidrug clinical resistant strains of *M. tuberculosis* have been reported to have higher fitness than their rifampicin-susceptible counterparts, thus indicating that compensation during treatment and/or epistatic effects between different drug-resistance mutations ameliorate, or even revert, the fitness cost of individual changes (Gagneux et al., 2006a).

Another way to compensate for the fitness cost of drug-resistance mutations is at the level of the expression of a protein (Maisnier-Patin and Andersson, 2004). There are examples of upregulation, through mutation, of a gene to counteract the negative effects on the expression of drug-resistance mutations, processes usually known as bypass mechanisms. The most typical example is that of KatG and the upregulation of *ahpC* in *M. tuberculosis* (Sherman et al., 1996). Isoniazid is a prodrug and it needs the catalase-peroxidase activity of KatG to become active. Mutations in KatG confer isoniazid drug resistance. A whole spectrum of mutations altering KatG function have been described (Ando et al., 2010) and because the gene has an important role in the bacterial response to oxidative stress, it has been assumed that all of them have an associated fitness cost. It has been reported that upregulation of the *ahpC* gene due to a mutation in its promoter can partially compensate for the loss of activity of KatG. Although *ahpC* and *katG* remain as the best example of upregulation to compensate fitness loss, it is worth mentioning that there are conflicting reports about the occurrence of the promoter mutation of *ahpC* in nonresistant strains (Borrell and Gagneux, 2009).

Another common way to increase the expression of a particular product is by gene duplication and amplification (GDA) (Sandegren and Andersson, 2009). GDA is a common mechanism to confer resistance to many antibiotics. However, it has been shown to be also a way of compensating for the fitness loss associated to resistance. GDA as a compensatory mechanism has been demonstrated more clearly in experimental evolution tests with *Salmonella enterica* (Nilsson et al., 2006). Tandem duplications of the *metZ* and *metW* genes compensate for the loss of methionyl-tRNA formyl-transferase by increasing levels of the non-formylated tRNA inhibitor, the one used by eukaryotes for translation initiation.

12.4.3 Beyond Model Organisms: Epidemiological and Experimental Fitness Cost in M. tuberculosis

Experimental evolution with model microorganisms has been a successful approach to test evolutionary hypotheses. These experiments allow studying evolution in real time, producing accurate measures of key parameters like fitness, generation times, population sizes, or mutation rates (Elena and Lenski, 2003). As we have seen above, drug resistance can be approached within an evolutionary framework given that antibiotics are the main evolutionary pressure a microorganism can face jointly with the host's immune system. However, many studies on antibiotic resistance have been done with organisms with limited public health impact like nonpathogenic laboratory strains of *E. coli* or *Pseudomonas* spp. These works have given insights on how microorganisms cope with antibiotics mainly because model organisms are easier to work with than clinical strains of pathogens, which usually have longer generation times and are more difficult to culture.

A paradigmatic, or even extreme, case in this respect is *M. tuberculosis*, the causative agent of TB, with a colony-forming time of 3–4 weeks and which requires working in BSL3 facilities. This is why alternative model organisms, like *M. smegmatis*, are frequently used to test hypotheses in TB research. However, *M. smegmatis*, which is a fast-growing nonpathogenic mycobacteria, has a genome size much larger than *M. tuberculosis* and many phenotypic differences, therefore caution must be taken when trying to establish parallelisms with *M. tuberculosis* (Barry, 2001; Reyrat and Kahn, 2001). Experimental work on drug resistance with *M. tuberculosis* has been successfully completed, corroborating many conclusions drawn from model organisms and justifying a constant feedback between model organism and real pathogens. A clear example is the case of the evolution of drug resistance against rifampicin. Rifampicin targets the β unit of the DNA-dependent RNA polymerase of microorganisms (encoded by the *rpoB* gene) by competing for the union to DNA and inhibiting RNA synthesis (Campbell et al., 2001). Therefore, it is a wide-spectrum antibiotic as it affects, with different efficiencies, many bacteria. Early work with *E. coli* (Ezekiel and Hutchins, 1968) and other bacteria identified homologous positions mutated in drug-resistant strains, both in experimental and clinical settings, something expected given the high conservation of the *rpoB* gene among bacteria. A screening of gene mutations in *rpoB* from clinical strains of *M. tuberculosis* was also able to identify many of them, as well as other mutations (O'Sullivan et al., 2005). However, these mutations varied in frequency, suggesting a possible difference in the degree of antibiotic resistance conferred and/or the fitness associated to them. Experimental evolution of two different lineages of *M. tuberculosis* revealed the existence of two main factors affecting drug-resistance fitness cost in this species: the genetic background of the strain and specific codon mutations (Gagneux et al., 2006b). Different codon mutations were found to have different fitness costs. Furthermore, these fitness costs varied among two lineages of *M. tuberculosis*, although in both cases the change, S531L, was the one associated with less fitness reduction. Reinforcing these results, mutations with lower fitness costs were found to be the most frequent among clinical strains, suggesting a correlation between in vitro and epidemiological fitness cost. Finally, the fitness of pair isolates of rifampicin-susceptible (RIFs) and rifampicin-resistant (RIFr) strains coming from 10 different patients who converted to drug resistance during treatment were screened, showing not only comparable results to the experimental findings but also cases in which the fitness of the drug-resistant strain was higher than that of the drug-susceptible counterpart, a result that was explained by the occurrence of compensatory mutations and/or by epistatic effects of drug-resistance mutations, as we have seen earlier.

12.5 Can the Evolution of Antibiotic Resistance be Predicted?

Conventional scientific wisdom dictates that evolution is a process that is sensitive to many unforeseeable events and influences and, therefore, is essentially unpredictable. On the other hand, considering the tremendous amounts of recent knowledge about bacterial genetics and genomics, population genetics and ecology of bacterial organisms, and their subcellular elements involved in HGT, we should eventually face the possibility of predicting the evolution of bacterial populations and traits (Martínez et al., 2007). The importance of such type of approach is self-evident in the case of the evolution of antibiotic resistance and bacterial–host interactions, including infections. The prediction of bacterial evolution could provide similar clues as weather prediction, with higher probabilities of success in the closer and more local frames. Indeed there is a *local* evolutionary biology based on local selective constraints that shape the possible local trajectories, even though in our global world, some of these locally originated trends might result in global influences. In the case of adaptive functions (as antibiotic-resistance genes in pathogenic bacteria) some of the elements whose knowledge is critical for predicting evolutionary trajectories are: (i) the origin and function of these genes in the source environmental bacterial organisms; (ii) their ability to be captured (mobilized) by different genetic platforms and to integrate in particular MGEs; (iii) the ability of these MGEs to be selected, transferred, and spread among bacterial populations; (iv) the probability of intrahost mutational variation and recombination; (v) the probability of recombinatorial events among these and other mobile elements, with consequences in selectable properties and bacterial host ranges, (vi) the original and resulting fitness of the bacterial clones in which the new functions are hosted, including their colonization power and capacity to spread in an epidemic form; (vii) the results of the interactions between these bacterial hosts and the microbial environments in which they are inserted; and (viii) the selective events, such as the patterns of local antibiotic consumption or industrial pollution and, in general, the structure of the environment that might influence the success of particular genetic configurations in which the adaptive genes are hosted. Dealing simultaneously with all these sources of evolutionary variation is certainly a challenge at present.

Such a type of complex structure has evolved along all hierarchical levels of biology, creating specific "Chinese-boxes" or "Russian-dolls" patterns of stable (preferential) combinations, for instance, encompassing bacterial species, phylogenetic subspecific groups, clones, plasmids, transposons, insertion sequences, and genes encoding adaptive traits. Assuming a relatively high frequency of combinatorial events, the existing transhierarchical combinations are probably the result on the local availability of the different elements (pieces) in particular locations (local biology), the local advantage provided by particular combinations, and also the biological cost in fitness of some of them. More research is needed to draw the interactive pattern of biological pieces in particular environments (grammar of affinities). Such a complex framework required for predicting evolutionary trajectories will be analyzed (and integrated) by considering heuristic techniques for the understanding of multilevel selection. The application of new methods, based on covariance, and contextual analysis, for instance using Price's equation (Price, 1970), should open an entirely new synthetic way of approaching the complexity of the living world.

12.6 Conclusions and Perspectives

In the absence of new antibiotics, most efforts have focused in protecting the few current ones that maintain activity, trying to reduce their strong selective effects by reducing antibiotic consumption in animals and humans while maintaining their efficiency. In a number of countries this collective policy has proven insufficient. It has been proposed that the control of antibiotic exposure should be considered by society as an individual-based attitude to reduce individual risks, using similar approaches to those for controlling tobacco-associated diseases, hypercholesterolemia, or hypertension (Baquero, 2007). Reductions in the host-to-host transmission of resistant organisms through innovative approaches trying to influence the ecology and evolution of resistant organisms might represent alternative ways to limit the spread of antibiotic resistance in the microbiosphere. In this respect, the possibility of applying in the future eco-evo drugs—drugs acting *not* to cure the individual patient but to "cure" specific environments from antibiotic resistance, and to prevent or weaken the evolutionary possibilities (the evolvability) of the biological elements involved in it—should be considered. In other words, this strategy proposes to combat (decontaminate, de-evolve) resistance not in infected patients, but rather in the whole population, including infected and noninfected people alike, as it occurs in hospitals, nurseries, elderly facilities, etc. By extension, other environments that can be successfully treated are farms, fish factories, or sewage facilities. Indeed, the notion of "ill environment" should be increasingly encouraged, and medical-like approaches might be increasingly applied to prevent and cure biologically altered environments (Baquero, 2009a).

The targets of these future drugs, some of them in early development, are not only resistant, "high-risk" clones but also the interbacterial transmissibility, the maintenance of bacterial plasmids and integrative-conjugative elements carrying resistance, the ability of transposons and integrons to move between genomes, or the mechanisms of bacterial adaptation to antibiotic stress, including control of mutation and recombination rates.

Acknowledgments

We would like to thank M. Tibayrenc for providing the opportunity to contribute to this volume and to the following financial supporters of our research: Instituto de Salud Carlos III FIS PI080624 (JCG), MICINN BFU2008-03000 and GVACOMP 2010-0148 (FGC), and BIO2008-00090 (JLM) and EU KBBE-227258 and HEALTH-F3-2010-241476 (JLM).

Glossary

Founder effect the random change in genetic composition of a population due to a extreme reduction in its size during a colonization or infection episode.
Genetic drift the random change in the genetic composition of a population due to its finite size. Every population experiences genetic drift but its effects, a reduction in genetic variation eventually leading to fixation of a variant, are more intense, both in magnitude and speed, the smaller its population size.
Mutator strains bacterial strains with an increased mutation rate usually due to a defective mismatch-repair system.
Pleiotropic antagonism the effect of a gene on two different traits with opposite consequences on fitness.
Resistome the set of antibiotic-resistance genes or proteins found in a given environment.

List of Abbreviations

ESBL extended-spectrum β-lactamases
GDA gene duplication and amplification
HGT horizontal gene transfer
MDR multidrug resistance
MGE mobile genetic element
MIC minimal inhibitory concentration
MRSA methicillin-resistant *Staphylococcus aureus*
R_0 basic reproductive number
RIFr rifampicin-resistant
RIFs rifampicin-susceptible
TB tuberculosis

References

Adam, M., Murali, B., Glenn, N.O., 2008. Epigenetic inheritance based evolution of antibiotic resistance in bacteria. BMC Evol. Biol. 8, 52.

Aendekerk, S., Ghysels, B., Cornelis, P., Baysse, C., 2002. Characterization of a new efflux pump, MexGHI-OpmD, from *Pseudomonas aeruginosa* that confers resistance to vanadium. Microbiology 148, 2371–2381.

Aendekerk, S., Diggle, S.P., Song, Z., Hoiby, N., Cornelis, P., Williams, P., et al., 2005. The MexGHI-OpmD multidrug efflux pump controls growth, antibiotic susceptibility and virulence in *Pseudomonas aeruginosa* via 4-quinolone-dependent cell-to-cell communication. Microbiology 151, 1113–1125.

Allen, H.K., Moe, L.A., Rodbumner, J., Gaarder, A., Handelsman, J., 2009. Functional metagenomics reveals diverse β-lactamases in a remote Alaskan soil. ISME J. 3, 243–251.

Alonso, A., Rojo, F., Martinez, J.L., 1999. Environmental and clinical isolates of *Pseudomonas aeruginosa* show pathogenic and biodegradative properties irrespective of their origin. Environ. Microbiol. 1, 421–430.

Andersson, D.I., 2006. The biological cost of mutational antibiotic resistance: any practical conclusions? Curr. Opin. Microbiol. 9, 461–465.

Andersson, D.I., Hughes, D., 2010. Antibiotic resistance and its cost: is it possible to reverse resistance? Nat. Rev. Microbiol. 8, 260–271.

Ando, H., Kondo, Y., Suetake, T., Toyota, E., Kato, S., Mori, T., et al., 2010. Identification of katG mutations associated with high-level isoniazid resistance in *Mycobacterium tuberculosis*. Antimicrob. Agents Chemother. 54, 1793–1799.

Arias, C.A., Murray, B.E., 2008. Emergence and management of drug-resistant enterococcal infections. Expert. Rev. Anti. Infect. Ther. 6, 637–655.

Austin, D.J., Kristinsson, K.G., Anderson, R.M., 1999. The relationship between the volume of antimicrobial consumption in human communities and the frequency of resistance. Proc. Natl. Acad. Sci. U.S.A. 96, 1152–1156.

Balsalobre, L., de la Campa, A.G., 2008. Fitness of *Streptococcus pneumoniae* fluoroquinolone-resistant strains with topoisomerase IV recombinant genes. Antimicrob. Agents Chemother. 52, 822–830.

Baquero, F., 2004. From pieces to patterns: evolutionary engineering in bacterial pathogens. Nat. Rev. Microbiol. 2, 510–518.

Baquero, F., 2007. Evaluation of risks and benefits of consumption of antibiotics: from individual to public health. In: Tibayrenc, M. (Ed.), Encyclopedia of Infectious Diseases. Wiley and Sons, Inc., Hoboken, NJ, pp. 509–520.

Baquero, F., 2009a. Prediction: evolutionary trajectories and planet medicine. Microb. Biotech. 2, 130–132.

Baquero, F., 2009b. Environmental stress and evolvability in microbial systems. Clin. Microbiol. Infect. 15 (Suppl. 1), 5–10.

Baquero, M.R., Nilsson, A.I., Turrientes, M.C., Sandvang, D., Galán, J.C., Martínez, J.L., et al., 2004. Polymorphic mutation frequencies in *Escherichia coli*: emergence of weak mutator in clinical isolates. J. Bacteriol. 186, 5538–5542.

Baquero, F., Martínez, J.L., Cantón, R., 2008. Antibiotics and antibiotic resistance in water environments. Curr. Opin. Biotechnol. 19, 260–265.

Barlow, M., Fatollahi, J., Salverda, M., 2009. Evidence for recombination among the alleles encoding TEM and SHV β-lactamases. J. Antimicrob. Chemother. 63, 256–259.

Barry III, C.E., 2001. *Mycobacterium smegmatis*: an absurd model for tuberculosis? Trends Microbiol. 9, 473–474.
Benveniste, R., Davies, J., 1973. Aminoglycoside antibiotic-inactivating enzymes in actinomycetes similar to those present in clinical isolates of antibiotic-resistant bacteria. Proc. Natl. Acad. Sci. U.S.A. 70, 2276–2280.
Bergstrom, C.T., Lo, M., Lipsitch, M., 2004. Ecological theory suggests that antimicrobial cycling will not reduce antimicrobial resistance in hospitals. Proc. Natl. Acad. Sci. U.S.A. 101, 13285–13290.
Bjorkman, J., Nagaev, I., Berg, O.G., Hughes, D., Andersson, D.I., 2000. Effects of environment on compensatory mutations to ameliorate costs of antibiotic resistance. Science 287, 1479–1482.
Blázquez, J., 2003. Hypermutation as a factor contributing to the acquisition of antimicrobial resistance. Clin. Infect. Dis. 37, 1201–1209.
Blázquez, J., Baquero, M.R., Canton, R., Alos, I., Baquero, F., 1993. Characterization of a new TEM-type β-lactamase resistant to clavulanate, sulbactam, and tazobactam in a clinical isolate of *Escherichia coli*. Antimicrob. Agents Chemother. 37, 2059–2063.
Borrell, S., Gagneux, S., 2009. Infectiousness, reproductive fitness and evolution of drug-resistant *Mycobacterium tuberculosis*. Int. J. Tuberc. Lung Dis. 13, 1456–1466.
Breidenstein, E.B., Khaira, B.K., Wiegand, I., Overhage, J., Hancock, R.E., 2008. Complex ciprofloxacin resistome revealed by screening a *Pseudomonas aeruginosa* mutant library for altered susceptibility. Antimicrob. Agents Chemother. 52, 4486–4491.
Brown, M.G., Balkwill, D.L., 2009. Antibiotic resistance in bacteria isolated from the deep terrestrial subsurface. Microb. Ecol. 57, 484–493.
Brückner, R., Nuhn, M., Reichmann, P., Weber, B., Hakenbeck, R., 2004. Mosaic genes and mosaic chromosomes-genomic variation in *Streptococcus pneumoniae*. Int. J. Med. Microbiol. 294, 157–168.
Burse, A., Weingart, H., Ullrich, M.S., 2004. The phytoalexin-inducible multidrug efflux pump AcrAB contributes to virulence in the fire blight pathogen, *Erwinia amylovora*. Mol. Plant Microbe Interact. 17, 43–54.
Bush, K., Jacoby, G.A., 2010. Update functional classification of β-lactamase. Antimicrob. Agents Chemother. 54, 969–976.
Campbell, E.A., Korzheva, N., Mustaev, A., Murakami, K., Nair, S., Goldfarb, A., Darst, S.A., 2001. Structural mechanism for rifampicin inhibition of bacterial RNA polymerase. Cell 104, 901–912.
Comas, I., Chakravarti, J., Small, P.M., Galagan, J., Niemann, S., Kremer, K., et al., 2010. Human T cell epitopes of *Mycobacterium tuberculosis* are evolutionarily hyperconserved. Nat. Genet. 42, 498–503.
Coque, T.M., Novais, A., Carattoli, A., Poirel, L., Pitout, J., Peixe, L., et al., 2008. Dissemination of clonally related *Escherichia coli* strains expressing extended-spectrum β-lactamase CTX-M-15. Emerg. Infect. Dis. 14, 195–200.
Couce, A., Blázquez, J., 2009. Side effects of antibiotics on genetic variability. FEMS Microbiol. Rev. 33, 531–538.
Da Re, S., Garnier, F., Guérin, E., Campoy, S., Denis, F., Ploy, M.C., 2009. The SOS response promotes qnrB quinolone-resistance determinant expression. EMBO Rep. 10, 929–933.
D'Acosta, V.M., McGrann, K.M., Hughes, D.W., Wright, G.D., 2006. Sampling the antibiotic resistome. Science 311, 374–377.

Datta, N., Hughes, V.M., 1983. Plasmids of the same Inc groups in Enterobacteria before and after the medical use of antibiotics. Nature 306, 616−617.

Davies, J., 1994. Inactivation of antibiotics and the dissemination of resistance genes. Science 264, 375−382.

Davies, J.E., 1997. Origins, acquisition and dissemination of antibiotic resistance determinants. Ciba Found. Symp. 207, 15−27.

Dawkins, R., 1976. The Selfish Gene. Oxford University Press, New York.

De Lencastre, H., Tomasz, A., 2008. Multiple stages in the evolution of the methicillin resistant *Staphylococcus aureus*. In: Baquero, F., Nombela, C., Cassell, G.H., Gutierrez-Fuentes., J.A. (Eds.), Evolutionary Biology of Bacterial and Fungal Pathogens. ASM Press, Washington, DC.

Dotsch, A., Becker, T., Pommerenke, C., Magnowska, Z., Jansch, L., Haussler, S., 2009. Genome-wide identification of genetic determinants of antimicrobial drug resistance in *Pseudomonas aeruginosa*. Antimicrob. Agents Chemother. 53, 2522−2531.

Elena, S.F., Lenski, R.E., 2003. Evolution experiments with microorganisms: the dynamics and genetic bases of adaptation. Nat. Rev. Genet. 4, 457−469.

Ezekiel, D.H., Hutchins, J.E., 1968. Mutations affecting RNA polymerase associated with rifampicin resistance in *Escherichia coli*. Nature 220, 276−277.

Fajardo, A., Martinez-Martin, N., Mercadillo, M., Galan, J.C., Ghysels, B., Matthijs, S., et al., 2008. The neglected intrinsic resistome of bacterial pathogens. PLoS ONE 3, e1619.

Gagneux, S., Burgos, M.V., DeRiemer, K., Encisco, A., Munoz, S., Hopewell, P.C., et al., 2006a. Impact of bacterial genetics on the transmission of isoniazid-resistant *Mycobacterium tuberculosis*. PLoS Pathog. 2, e61.

Gagneux, S., Long, C.D., Small, P.M., Van, T., Schoolnik, G.K., Bohannan, B.J.M., 2006b. The competitive cost of antibiotic resistance in *Mycobacterium tuberculosis*. Science 312, 1944−1946.

Galán, J.C., Morosini, M.I., Baquero, M.R., Reig, M., Baquero, F., 2003. *Haemophilus influenzae bla*(ROB-1) mutations in hypermutagenic ΔampC *Escherichia coli* conferring resistance to cefotaxime and β-lactamase inhibitors and increased susceptibility to cefaclor. Antimicrob. Agents Chemother. 47, 2551−2557.

Goossens, H., 2009. Antibiotic consumption and link to resistance. Clin. Microbiol. Infect. (Suppl. 3), 12−15.

Gootz, T.D., 2010. The global problem of antibiotic resistance. Crit. Rev. Immunol. 30, 79−103.

Guillemot, D., Varon, E., Bernède, C., Weber, P., Henriet, L., Simon, S., et al., 2005. Reduction of antibiotic use in the community reduces the rate of colonization with penicillin G-nonsusceptible *Streptococcus pneumoniae*. Clin. Infect. Dis. 41, 930−938.

Hanage, W.P., Fraser, C., Tang, J., Connor, T.R., Corander, J., 2009. Hyper-recombination, diversity, and antibiotic resistance in *Pneumococcus*. Science 324, 1454−1457.

Huang, W., Petrosino, J., Hirsch, M., Shenkin, P.S., Palzkill, T., 1996. Amino acid sequence determinants of β-lactamase structure and activity. J. Mol. Biol. 258, 688−703.

Hughes, A.L., 2005. Evidence for abundant slightly deleterious polymorphisms in bacterial populations. Genetics 169, 533−538.

Humeniuk, C., Arlet, G., Gautier, V., Grimont, P., Labia, R., Philippon, A., 2002. β-lactamases of *Kluyvera ascorbata*, probable progenitors of some plasmid-encoded CTX-M types. Antimicrob. Agents Chemother. 46, 3045−3049.

Karageorgopoulos, D.E., Falagas, M.E., 2008. Current control and treatment of multidrug-resistant *Acinetobacter baumanii* infections. Lancet Infect. Dis. 8, 751–762.
Kimura, M., 1968. Evolutionary rate at the molecular level. Nature 217, 624–626.
Kimura, M., 1983. The Neutral Theory of Molecular Evolution. Cambridge University Press, Cambridge.
Knapp, C.W., Dolfing, J., Ehlert, P.A., Graham, D.W., 2010. Evidence of increasing antibiotic resistance gene abundances in archived soils since 1940. Environ. Sci. Technol. 44, 580–587.
Kohler, T., van Delden, C., Curty, L.K., Hamzehpour, M.M., Pechere, J.C., 2001. Overexpression of the MexEF-OprN multidrug efflux system affects cell-to-cell signalling in *Pseudomonas aeruginosa*. J. Bacteriol. 183, 5213–5222.
Kugelberg, E., Kofoid, E., Andersson, D.I., Lu, Y., Mellor, J., Roth, F.P., et al., 2010. The tandem inversion duplication in *Salmonella enterica*: selection drives unstable precursors to final mutation types. Genetics 185, 65–80.
Lauretti, L., Riccio, M.L., Mazzariol, A., Cornaglia, G., Amicosante, G., Fontana, R., et al., 1999. Cloning and characterization of *bla*VIM, a new integron-borne metallo-β-lactamase gene from a *Pseudomonas aeruginosa* clinical isolate. Antimicrob. Agents Chemother. 43, 1584–1590.
Levin, B.R., 2001. Minimizing potential resistance: a population dynamics view. Clin. Infect. Dis. 33 (Suppl. 3), S161–S169.
Levin, B.R., Perrot, V., Walker, N., 2000. Compensatory mutations, antibiotic resistance and the population genetics of adaptive evolution in bacteria. Genetics 154, 985–997.
Levy, S.B., Marshall, B., 2004. Antibacterial resistance worldwide: cause, challenges and responses. Nat. Med. 10 (Suppl. 12), S122–S129.
Lindberg, F., Normark, S., 1986. Contribution of chromosomal beta-lactamases to β-lactam resistance in enterobacteria. Rev. Infect. Dis. 8 (Suppl. 3), S292–S304.
Livermore, D.M., 2009. Has the era of untreatable infections arrived? J. Antimicrob. Chemother. (Suppl. 1), i29–i36.
Livermore, D.M., Warner, M., Hall, L.M., Enne, V.I., Projan, S.J., Dunman, P.M., et al., 2001. Antibiotic resistance in bacteria from magpies (*Pica pica*) and rabbits (*Oryctolagus cuniculus*) from west Wales. Environ. Microbiol. 3, 658–661.
Lofmark, S., Jernberg, C., Billstrom, H., Andersson, D.I., Edlund, C., 2008. Restored fitness leads to long-term persistence of resistant Bacteroides strains in the human intestine. Anaerobe 14, 157–160.
López, E., Blázquez, J., 2009. Effect of subinhibitory concentrations of antibiotics on intrachromosomal homologous recombination in *Escherichia coli*. Antimicrob. Agents Chemother. 53, 3411–3415.
Macia, M.D., Blanquer, D., Togores, B., Sauleda, J., Perez, J.L., Oliver, A., 2005. Hypermutation is a key factor in development of multiple-antimicrobial resistance in *Pseudomonas aeruginosa* strains causing chronic lung infections. Antimicrob. Agents Chemother. 49, 3382–3386.
Maggiorani Valecillos, A., Rodriguez Palenzuela, P., Lopez-Solanilla, E., 2006. The role of several multidrug resistance systems in *Erwinia chrysanthemi* pathogenesis. Mol. Plant Microbe Interact. 19, 607–613.
Maisnier-Patin, S., Andersson, D.I., 2004. Adaptation to the deleterious effects of antimicrobial drug resistance mutations by compensatory evolution. Res. Microbiol. 155, 360–369.
Mao, E., Lane, L., Lee, J., Miller, J.H., 1997. Proliferation of mutators in a cell population. J. Bacteriol. 179, 417–422.

Marcusson, L.L., Frimodt-Möller, N., Hughes, D., 2009. Interplay in the selection of fluoroquinolone resistance and bacterial fitness. PLoS Pathog. 5, e1000541.
Marshall, C.G., Lessard, I.A., Park, I., Wright, G.D., 1998. Glycopeptide antibiotic resistance genes in glycopeptide-producing organisms. Antimicrob. Agents Chemother. 42, 2215–2220.
Martinez, J.L., 2008. Antibiotics and antibiotic resistance genes in natural environments. Science 321, 365–367.
Martinez, J.L., 2009. The role of natural environments in the evolution of resistance traits in pathogenic bacteria. Proc. Biol. Sci. 276, 2521–2530.
Martinez, J.L., Baquero, F., 2000. Mutation frequencies and antibiotic resistance. Antimicrob. Agents Chemother. 44, 1771–1777.
Martinez, J.L., Baquero, F., 2002. Interactions among strategies associated with bacterial infection: pathogenicity, epidemicity, and antibiotic resistance. Clin. Microbiol. Rev. 15, 647–679.
Martinez, J.L., Baquero, F., Andersson, D.I., 2007. Predicting antibiotic resistance. Nat. Rev. Microbiol. 5, 958–965.
Martinez, J.L., Fajardo, A., Garmendia, L., Hernandez, A., Linares, J.F., Martinez-Solano, L., et al., 2009a. A global view of antibiotic resistance. FEMS Microbiol. Rev. 33, 44–65.
Martinez, J.L., Sanchez, M., Martinez-Solano, L., Hernandez, A., Garmendia, L., Fajardo, A., et al., 2009b. Functional role of bacterial multidrug efflux pumps in microbial natural ecosystems. FEMS Microbiol. Rev. 33, 430–449.
Martinez-Martinez, L., Pascual, A., Jacoby, G.A., 1998. Quinolone resistance from a transferable plasmid. Lancet 351, 797–799.
Massova, I, Mobashery, S., 1998. Kinship and diversification of bacterial penicillin-binding proteins and β-lactamases. Antimicrob. Agents Chemother. 42, 1–17.
Maurice, F., Broutin, I., Podglajen, I., Benas, P., Collatz, E., Dardel, F., 2008. Enzyme structural plasticity and the emergence of broad-spectrum antibiotic resistance. EMBO Rep. 9, 344–349.
Medeiros, A.A., 1997. Evolution and dissemination of β-lactamases accelerated by generations of β-lactam antibiotics. Clin. Infect. Dis. 24 (Suppl. 1), S19–S45.
Miteva, V.I., Sheridan, P.P., Brenchley, J.E., 2004. Phylogenetic and physiological diversity of microorganisms isolated from a deep Greenland glacier ice core. Appl. Environ. Microbiol. 70, 202–213.
Moritz, E.M., Hergenrother, P.J., 2007. Toxin-antitoxin systems are ubiquitous and plasmid-encoded in vancomycin-resistant enterococci. Proc. Natl. Acad. Sci. U.S.A. 104, 311–316.
Morosini, M.I., Baquero, M.R., Sánchez-Romero, J.M., Negri, M.C., Galán, J.C., del Campo, R., et al., 2003. Frequency of mutation to rifampicin resistance in *Streptococcus pneumoniae* clinical strains: *hex*A and *hex*B polymorphisms do not account for hypermutation. Antimicrob. Agents Chemother. 47, 1464–1467.
Nelson, E.C., Elisha, B.G., 1999. Molecular basis of AmpC hyperproduction in clinical isolates of *Escherichia coli*. Antimicrob. Agents Chemother. 43, 957–959.
Nies, D.H., 2003. Efflux-mediated heavy metal resistance in prokaryotes. FEMS Microbiol. Rev. 27, 313–339.
Nilsson, A.I., Zorzet, A., Kanth, A., Dahlstrom, S., Berg, O.G., Andersson, D.I., 2006. Reducing the fitness cost of antibiotic resistance by amplification of initiator tRNA genes. Proc. Natl. Acad. Sci. U.S.A. 103, 6976–6981.
Novais, Â., Cantón, R., Coque, T.M., Moya, A., Baquero, F., Galán, J.G., 2008. Mutational events in cefotaximase extended-spectrum β-lactamases of the CTX-M-1 cluster involved in ceftazidime resistance. Antimicrob. Agents Chemother. 52, 2377–2382.

Novais, Â., Comas, I., Baquero, F., Cantón, R., Coque, T.M., Moya, A., et al., 2010. Evolutionary trajectories of β-lactamase CTX-M-1 cluster enzymes: predicting antibiotic resistance. PLoS Pathog. 6, e1000735.

Ohta, T., 1992. The nearly neutral theory of molecular evolution. Ann. Rev. Ecol. Syst. 21, 263–286.

Ohta, T., Kimura, M., 1971. On the constancy of the evolutionary rate of cistrons. J. Mol. Evol. 1, 18–25.

Oliver, A., Cantón, R., Campo, P., Baquero, F., Blázquez, J., 2000. High frequency of hypermutable *Pseudomonas aeruginosa* in cystic fibrosis lung infection. Science 288, 1251–1254.

Orencia, M.C., Yoon, J.S., Ness, J.E., Stemmer, W.P., Stevens, R.C., 2001. Predicting the emergence of antibiotic resistance by directed evolution and structural analysis. Nat. Struct. Biol. 8, 238–242.

O'Sullivan, D.M., McHugh, T.D., Gillespie, S.H., 2005. Analysis of *rpoB* and *pncA* mutations in the published literature: an insight into the role of oxidative stress in *Mycobacterium tuberculosis* evolution? J. Antimicrob. Chemother. 55, 674–679.

Pallecchi, L., Bartoloni, A., Paradisi, F., Rossolini, G.M., 2008. Antibiotic resistance in the absence of antimicrobial use: mechanisms and implications. Expert. Rev. Anti. Infect. Ther. 6, 725–732.

Patterson, A.J., Rincon, M.T., Flint, H.J., Scott, K.P., 2007. Mosaic tetracycline resistance genes are widespread in human and animal fecal samples. Antimicrob. Agents Chemother. 51, 1115–1118.

Paulander, W., Maisnier-Patin, S., Andersson, D.I., 2007. Multiple mechanisms to ameliorate the fitness burden of mupirocin resistance in *Salmonella typhimurium*. Mol. Microbiol. 64, 1038–1048.

Pei, R., Kim, S.C., Carlson, K.H., Pruden, A., 2006. Effect of river landscape on the sediment concentrations of antibiotics and corresponding antibiotic resistance genes (ARG). Water Res. 40, 2427–2435.

Pettersson, M.E., Sun, S., Andersson, D.I., Berg, O.G., 2009. Evolution of new gene functions: simulation and analysis of the amplification model. Genetica 135, 309–324.

Piddock, L.J., 1999. Mechanisms of fluoroquinolone resistance: an update 1994–1998. Drugs 58, 11–18.

Pitout, J.D., Laupland, K.B., 2008. Extended-spectrum β-lactamase-producing Enterobacteriaceae: an emerging public-health concern. Lancet Infect. Dis. 8, 159–166.

Poelwijk, F.J., Kiviet, D.J., Weinreich, D.M., Tans, S.J., 2007. Empirical fitness landscapes reveal accessible evolutionary paths. Nature 445, 383–386.

Poirel, L., Mammeri, H., Nordmann, P., 2004. TEM-121, a novel complex mutant of TEM-type β-lactamase from *Enterobacter aerogenes*. Antimicrob. Agents Chemother. 48, 4528–4531.

Poirel, L., Rodriguez-Martinez, J.M., Mammeri, H., Liard, A., Nordmann, P., 2005. Origin of plasmid-mediated quinolone resistance determinant QnrA. Antimicrob. Agents Chemother. 49, 3523–3525.

Prammananan, T., Sander, P., Springer, B., Böttger, E.C., 1999. RecA-mediated gene conversion and aminoglycoside resistance in strains heterozygous for rRNA. Antimicrob. Agents Chemother. 43, 447–453.

Price, G., 1970. Selection and covariance. Nature 227, 520–521.

Ramos, J.L., Duque, E., Gallegos, M.T., Godoy, P., Ramos-Gonzalez, M.I., Rojas, A., et al., 2002. Mechanisms of solvent tolerance in Gram-negative bacteria. Ann. Rev. Microbiol. 56, 743−768.

Rasila, T.S., Pajunen, M.I., Savilahti, H., 2009. Critical evaluation of random mutagenesis by error-prone polymerase chain reaction protocols, *Escherichia coli* mutator strain, and hydroxylamine treatment. Anal. Biochem. 388, 71−80.

Reyrat, J.M., Kahn, D., 2001. *Mycobacterium smegmatis*: an absurd model for tuberculosis? Trends Microbiol. 9, 472−473.

Robicsek, A., Strahilevitz, J., Jacoby, G.A., Macielag, M., Abbanat, D., Park, C.H., et al., 2006a. Fluoroquinolone-modifying enzyme: a new adaptation of a common aminoglycoside acetyltransferase. Nat. Med. 12, 83−88.

Robicsek, A., Jacoby, G.A., Hooper, D.C., 2006b. The worldwide emergence of plasmid-mediated quinolone resistance. Lancet Infect. Dis. 6, 629−640.

Ruiz-Garbajosa, P., Bonten, M.J., Robinson, D.A., Top, J., Nallapareddy, S.R., Torres, C., et al., 2006. Multilocus sequence typing scheme for *Enterococcus faecalis* reveals hospital-adapted genetic complexes in a background of high rates of recombination. J. Clin. Microbiol. 44, 2220−2228.

Sanchez, M.B., Hernandez, A., Rodriguez-Martinez, J.M., Martinez-Martinez, L., Martinez, J.L., 2008. Predictive analysis of transmissible quinolone resistance indicates *Stenotrophomonas maltophilia* as a potential source of a novel family of Qnr determinants. BMC Microbiol. 8, 148.

Sandegren, L., Andersson, D.I., 2009. Bacterial gene amplification: implications for the evolution of antibiotic resistance. Nat. Rev. Microbiol. 7, 578−588.

Seppala, H., Klaukka, T., Vuopio-Varkila, J., Muotiala, A., Helenius, H., Lager, K., et al., 1997. The effect of changes in the consumption of macrolide antibiotics on erythromycin resistance in group A streptococci in Finland. N. Engl. J. Med. 337, 441−446.

Shakya, T., Wrigh, G.D., 2010. Nucleotide selectivity of antibiotic kinase. Antimicrob. Agents Chemother. 54, 1909−1913.

Sherman, D.R., Mdluli, K., Hickey, M.J., Arain, T.M., Morris, S.L., Barry III, C.E., et al., 1996. Compensatory *ahpC* gene expression in isoniazid-resistant *Mycobacterium tuberculosis*. Science 272, 1641−1643.

Sibold, C., Henrichsen, J., Konig, A., Martin, C., Chalkley, L., Hakenbeck, R., 1994. Mosaic *pbpX* genes of major clones of penicillin-resistant *Streptococcus pneumoniae* have evolved from *pbpX* genes of a penicillin-sensitive *Streptococcus oralis*. Mol. Microbiol. 12, 1013−1023.

Simoes, R.R., Poirel, L., Da Costa, P.M., Nordmann, P., 2010. Seagulls and beaches as reservoirs for multidrug-resistant *Escherichia coli*. Emerg. Infect. Dis. 16, 110−112.

Sirot, D., Sirot, J., Labia, R., Morand, A., Courvalin, P., Darfeuille-Michaud, A., et al., 1987. Transferable resistance to third-generation cephalosporins in clinical isolates of *Klebsiella pneumoniae*: identification of CTX-1, a novel β-lactamase. J. Antimicrob. Chemother. 20, 323−334.

Smith, C.A., Baker, E.N., 2002. Aminoglycoside antibiotic resistance by enzymatic deactivation. Curr. Drug Targets Infect. Disord. 2, 143−160.

Smith Moland, E., Hanson, N.D., Herrera, V.L., Black, J.A., Lockhart, T.J., Hossain, A., et al., 2003. Plasmid-mediated, carbapenem-hydrolysing β-lactamase, KPC-2, in *Klebsiella pneumoniae* isolates. J. Antimicrob. Chemother. 51, 711−714.

Sommer, M.O., Dantas, G., Church, G.M., 2009. Functional characterization of the antibiotic resistance reservoir in the human microflora. Science 325, 1128−1131.

Souli, M., Galani, I., Giamarellou, H., 2008. Emergence of extensively drug-resistant and pandrug-resistant Gram-negative bacilli in Europe. Euro Surveill. 13, 19045.

Spratt, B.G., Bowler, L.D., Zhang, Q.Y., Zhou, J., Smith, J.M., 1992. Role of interspecies transfer of chromosomal genes in the evolution of penicillin resistance in pathogenic and commensal *Neisseria* species. J. Mol. Evol. 34, 115–125.

Stanton, T.B., Humphrey, S.B., 2003. Isolation of tetracycline-resistant *Megasphaera elsdenii* strains with novel mosaic gene combinations of *tet*(O) and *tet*(W) from swine. Appl. Environ. Microbiol. 69, 3874–3882.

Stepanova, M.N., Pimkin, M., Nikulin, A.A., Koryreva, V.K., Agapova, E.D., Edelstein, M.V., 2008. Convergent in vivo and in vitro selection of ceftazidime resistance mutations at position 167 of CTX-M-3 β-lactamase in hypermutable *Escherichia coli* strains. Antimicrob. Agents Chemother. 52, 1297–1301.

Sundqvist, M., Geli, P., Andersson, D.I., Sjölund-Karlsson, M., Runehagen, A., Cars, H., et al., 2010. Little evidence for reversibility of trimethoprim resistance after a drastic reduction in trimethoprim use. J. Antimicrob. Chemother. 65, 350–360.

Tamae, C., Liu, A., Kim, K., Sitz, D., Hong, J., Becket, E., et al., 2008. Determination of antibiotic hypersensitivity among 4,000 single-gene-knockout mutants of *Escherichia coli*. J. Bacteriol. 190, 5981–5988.

Thaker, M., Spanogiannopoulos, P., Wright, G.D., 2010. The tetracycline resistome. Cell Mol. Life Sci. 67, 419–431.

Thanassi, D.G., Cheng, L.W., Nikaido, H., 1997. Active efflux of bile salts by *Escherichia coli*. J. Bacteriol. 179, 2512–2518.

Trindade, S., Sousa, A., Xavier, K.B., Dionisio, F., Ferreira, M.G., Gordo, I., 2009. Positive epistasis drives the acquisition of multidrug resistance. PLoS Genet. 5, e1000578.

Turrientes, M.C., Baquero, M.R., Sánchez, M.B., Valdezate, S., Escudero, E., Berg, G., et al., 2010. Polymorphic mutation frequencies of clinical and environmental *Stenotrophomonas maltophilia* populations. Appl. Environ. Microbiol. 76, 1746–1758.

Vatopoulos, A., 2008. High rates of metallo-β-lactamase-producing *Klebsiella pneumoniae* in Greece—a review of the current evidence. Euro Surveill. 13, 8023.

Vila, J., Ruiz, J., Marco, F., Barcelo, A., Goni, P., Giralt, E., et al., 1994. Association between double mutation in *gyr*A gene of ciprofloxacin-resistant clinical isolates of *Escherichia coli* and MICs. Antimicrob. Agents Chemother. 38, 2477–2479.

Wachino, J., Yamane, K., Suzuki, S., Kimura, K., Arakawa, Y., 2010. Prevalence of fosfomycin resistance among CTX-M-producing *Escherichia coli* clinical isolates in Japan and identification of novel plasmid-mediated fosfomycin-modifying enzymes. Antimicrob. Agents Chemother. 54, 3061–3064.

Waksman, S.A., Woodruff, H.B., 1940. The soil as a source of microorganisms antagonistic to disease-producing bacteria. J. Bacteriol. 40, 581–600.

Webb, V., Davies, J., 1993. Antibiotic preparations contain DNA: a source of drug resistance genes? Antimicrob. Agents Chemother. 37, 2379–2384.

Weinreich, D.M., Delaney, N.F., Depristo, M.A., Hartl, D.L., 2006. Darwinian evolution can follow only very few mutational paths to fitter proteins. Science 312, 111–114.

Wright, G.D., 2007. The antibiotic resistome: the nexus of chemical and genetic diversity. Nat. Rev. Microbiol. 5, 175–186.

Wright, A., Zignol, M., Van Deun, A., Falzon, D., Gerdes, S.R., Feldman, K., et al., 2006. Epidemiology of antituberculosis drug resistance 2002–07: an updated analysis of the Global Project on Anti-Tuberculosis Drug Resistance Surveillance. Lancet 373, 1861–1873.

Yoneyama, H., Nakae, T., 1993. Mechanism of efficient elimination of protein D2 in outer membrane of imipenem-resistant *Pseudomonas aeruginosa*. Antimicrob. Agents Chemother. 37, 2385−2390.

Zaccolo, M., Gherardi, E., 1999. The effect of high-frequency random mutagenesis on in vitro protein evolution: a study on TEM-1 β-lactamase. J. Mol. Biol. 285, 775−783.

Zhang, Y., Heym, B., Allen, B., Young, D., Cole, S., 1992. The catalase-peroxidase gene and isoniazid resistance of *Mycobacterium tuberculosis*. Nature 358, 591−593.

Zhang, W., Fisher, J.F., Mobashery, S., 2009. The bifunctional enzymes of antibiotic resistance. Curr. Opin. Microbiol. 12, 505−511.

13 General Mechanisms of Antiviral Resistance

Anthony Vere Hodge[1], and Hugh J. Field[2]*

[1]Vere Hodge Antivirals Ltd, Old Denshott, Reigate, Surrey, RH2 8RD, UK,
[2]Department of Veterinary Medicine, Cambridge University, Madingley Road, Cambridge, CB3 0ES, UK

13.1 Introduction

Mammalian viruses represent a diverse group of infectious agents. The viruses that cause the common diseases of man and domestic animals comprise approximately 25 known families, which fall into groups according to their genome and replication strategies. Some important examples of these viruses are summarized in Table 13.1.

Over recent years, knowledge of the complete nucleotide sequences has enhanced our understanding of the interrelationships between virus nucleic acids and relevant host genes. Such molecular studies (see Chapter 3) indicate that there are homologies between viral and host proteins; of particular interest to the antiviral field are those involved with genome replication and other virus enzymes. Although no host enzymes exist in eukaryotes to replicate RNA (a prerequisite for all RNA viruses) and reverse transcriptase (RT) has no corresponding host function, many viral and host genes appear to have common origins. These findings support the view that, during the coevolution of virus and host cell, there have been exchanges of functional modules, mediated by several forms of genetic recombination. Further evolution of modern viruses is continuing with mutations (substitutions, additions, deletions), recombinations, or reassortments (Holland and Domingo, 1998).

Modern viruses have developed, through their evolutionary history, an extraordinary diversity of strategies for their efficient replication and survival, counteracting both innate and adaptive immune responses. For example, herpes simplex virus (HSV) and varicella-zoster virus (VZV) have a latent state enabling the virus to survive lifelong in the host. Reactivation at intervals allows the herpesvirus to spread to new individuals, enabling transmission through space and time. In contrast, influenza virus is usually cleared from the host within days or weeks, but has the ability to spread rapidly from person to person, potentially worldwide. Only in the relatively recent past, since Jenner's first use of cowpox as a vaccine for human smallpox, have viruses faced a new threat to their replication—human intervention. Faced with vaccines and specific antiviral compounds, some viruses appear to be

*E-mail: averehodge@aol.com

Table 13.1 Examples of Human Viruses and their Genome Structures

DNA	Double strand	**Adenoviruses**
		Herpesviruses
		HSV
		Cytomegalovirus
		VZV
		Papillomaviruses
		Poxviruses
	Single strand	**Parvoviruses**
	Partial double strand, replicating via RNA	**Hepadnavirus**
		HBV
RNA	Double strand, segmented	**Reoviruses**
		Rotavirus
	Single strand, positive strand	**Flaviviruses**
		HCV
		West Nile virus
		Yellow fever virus
		Dengue virus
		Picornaviruses
		hepatitis A or HAV
		Rhinovirus
		Togaviruses
		Rubella virus
	Single strand, negative strand, segmented	**Orthomyxoviruses**
		Influenza virus
	Single strand, negative strand, nonsegmented	**Rabies virus**
		Paramyxoviruses
		Mumps, measles, RSV
	Replicating via DNA	**Retroviruses**
		HIV

poorly adapted to survive. A well-known example is smallpox, which was eliminated from the human population by vaccination. Similarly, poliovirus has been eradicated from most countries of the world. The use of vaccines has greatly reduced the burden of human disease caused by several other human viruses (e.g., rubella, mumps, measles, hepatitis A [HAV], and hepatitis B [HBV]). Nevertheless, there seems to be no prospect of eradicating these viruses in the near future and patients with chronic HBV cannot be cured by vaccination. In complete contrast, several human viruses have, through their evolutionary history, developed survival strategies which happen to enable them to resist vaccines (e.g., human immunodeficiency virus [HIV] and hepatitis C [HCV]). Specific antiviral compounds have been developed for several of those viral infections that have not been adequately controlled by vaccines. Examples of widely licensed compounds are given in Table 13.2.

This review aims to explore general mechanisms of virus resistance. We start by summarizing those evolutionary outcomes that have enabled human viruses to resist mankind's best efforts at control. The review then focuses on how viruses acquire resistance to compounds that specifically target a virus protein (e.g., polymerase,

Table 13.2 Illustrative Examples of Commonly Used Antiviral Compounds

A: Primarily active against Herpes viruses

Generic Name (Abbreviation) Trade Name (Company)	Structure[a]	Mechanism Viral Target	Target Viruses
Valaciclovir (VACV) Valtrex® (GSK)	Prodrug of acyclovir (ACV) Zovirax®	Activated by viral TK, inhibits viral polymerase	HSV-1 and 2 VZV
Famciclovir (FCV) Famvir® (Novartis)	Prodrug of penciclovir (PCV) Denavir®/Vectavir®	Activated by viral TK, inhibits viral polymerase	HSV-1 and 2 VZV
Foscarnet (PFA) Foscavir® (Astra Zeneca)	Pyrophosphate analog	Polymerase inhibitor	HSV-1 and 2 VZV
Valganciclovir (VGCV) Valcyte® (Roche)	Prodrug of ganciclovir (GCV), Cymmevene®	Activated by kinase encoded by UL 97, polymerase inhibitor	CMV

[a] For prodrugs, parent antiviral drug name, abbreviated name and trade name are given.

B: Primarily active against RNA viruses

Generic Name Trade Name (Company)	Structure	Mechanism Viral Target	Target Viruses
Zanamivir Relenza® (GSK)	Sialic acid analog	Neuraminidase inhibitor	Influenza A and B
Oseltamivir Tamiflu® (Roche)	Sialic acid analog	Neuraminidase inhibitor	Influenza A and B
Ribavirin Virazole® (Schering-Plough)	Nucleoside analog	Possibly no direct viral target	HCV RSV

C: Primarily active against Hepadnaviruses (HBV)

Generic Name (Abbreviation) Trade Name	Structure	Mechanism Viral Target	Company
Lamivudine (LMV or 3TC) Zeffix®, Heptovir®	Nucleoside analog	Polymerase	GSK
Adefovir dipivoxil (ADV) Hepsera®	Prodrug of adefovir nucleotide analog	Polymerase	Gilead
Entecavir Baraclude®	Nucleoside analog	Polymerase	BMS

D: Primarily active against Retroviruses (HIV)

Generic Name (Abbreviation) Trade Name	Structure	Mechanism Viral Target	Company
Zidovudine/Lamivudine (AZT/3TC) Combivir®	Two NRTIs	Polymerase	GSK
AZT/3TC/abacavir Trizivir®	Three NRTIs	Polymerase	GSK
Emtricitabine/tenofovir/ efavirenz Atripla®	Two NRTIs and one NNRTI	Polymerase	Gilead and BMS (Jointly)
Nevirapine Viramune®	NNRTI	Polymerase	Boehringer
Fosamprenavir Lexiva®	gag cleavage site mimic	Protease	GSK
Saquinavir mesylate Invirase® Fortovase®	gag cleavage site mimic	Protease	Roche
Lopinavir/ritonavir Kaletra®	gag cleavage site mimic/PK enhancer	Protease	Abbott Lab
Indinavir Crixivan®	gag cleavage site mimic	Protease	Merck
Darunavir Prezista®	gag cleavage site mimic	Protease	Tibotec
Raltegravir Isentress®	Dihydropyrimidine derivative	Integration of HIV DNA into chromosome	Merck

D: Primarily active against Retroviruses (HIV) (Continued)

Generic Name (Abbreviation) Trade Name	Structure	Mechanism Viral Target	Company
Efuviritide (T-20) Fuzeon®	Polypeptide	Envelope protein gp41	Roche
Maraviroc Selzentry®	CCR5 ligand mimic	Blocks receptor on host cell membrane	Pfizer

Source: Table adapted from Field and Vere Hodge (2008).

protease, integrase, sialidase). For the therapy of chronic viral infections, such as HIV, the concept of the genetic barrier has emerged as key factor for delaying antiviral resistance. In some cases, the price to the virus of gaining resistance may be reduced "fitness." There may, however, be other less obvious effects.

Conceptually, one way to avoid virus resistance is to use a compound to target a host protein rather than a viral protein. Such an approach seems to risk causing unacceptable toxicity, although recently it has been shown that there can be specificity for the virus-infected cell. We end by asking the question, how will viruses respond to such an indirect challenge?

13.2 Evolutionary Outcomes that have Enabled Viruses to Resist Control

Viral resistance is usually discussed in the context of antiviral therapy. However, through the long process of evolution, viruses have acquired various attributes that happen to limit our ability to control the burden of disease and resist mankind's best efforts to control the viral infections. Examples of such attributes are the following.

13.2.1 The Virus Has Developed the Ability to Enter Latency

Herpesviruses establish a latent state that enables the virus to remain in the host for a lifetime despite normal adaptive immune responses. The latent virus can reactivate at intervals with or without clinical signs. Antivirals are effective at reducing virus replication during an acute episode but, currently, there are no therapies that remove latent herpesvirus infections. HSV, VZV, and human cytomegalovirus (HCMV) have been major and successful antiviral targets for three decades, acyclovir (ACV) being the first antiviral drug to be both potent and selective. ACV was followed by famciclovir (FCV), prodrug of penciclovir (PCV), and valaciclovir (VACV), prodrug of ACV, which are used to treat or suppress HSV-1, HSV-2, and VZV and ganciclovir (GCV) for HCMV. Although these have been used clinically worldwide and have helped patients manage their herpes infections, the latent virus remains as a potential source of reactivation.

13.2.2 Integration

Another way in which a virus can establish a form of latency is by means of integration of a DNA copy of the genome. Soon after infecting a cell, HIV RNA is the template for the viral RT to synthesize HIV DNA, which is then integrated into cellular DNA. Those current therapies, which inhibit the viral RT (RTIs) or which target the viral protease (PIs), have no direct effect on integrated viral DNA. The integrase inhibitors prevent the integration process but have no effect on viral DNA already integrated into host DNA. Because some cells contain integrated HIV DNA which remains quiescent (latent), it has been impossible to "cure" HIV-infected patients by clearing the HIV completely, notwithstanding the fact that combination therapy has given good clinical control of the symptomatic disease.

13.2.3 The Virus Has Over 100 Serotypes/Genotypes

There are two well-studied examples: rhinovirus (common cold) and papillomavirus (wart virus).

Even at the research stage with rhinoviruses, no compound showed activity against all serotypes (e.g., pleconaril is active against about 70% of serotypes). Although this compound was selected for development, one of the potential problems was that it could not be clinically effective against all rhinovirus serotypes, let alone against all viruses causing similar symptoms. (The development of pleconaril was terminated due to toxicological considerations.)

For the second example, papillomavirus vaccines (designed to prevent carcinoma of the cervix) seem very effective against the targeted virus strains (papilloma types 16 and 18) but give little or no protection against those remaining strains which are associated with a minority of carcinomas. Types 16 and 18 are associated with 70–75% of cervical cancers, 70% of vaginal cancers, and 50% of vulvar cancers. To protect against essentially all these cancers, it would be necessary to have vaccines for about 13 types of papilloma, with types 45, 31, 33, 52, 58, and 35 being the most important after 16 and 18. The current vaccines, Cervarix (GSK) and Gardasil (MSD), contain antigenic proteins from types 16 and 18. In addition, Gardisil includes types 6 and 11, which cause 90% of genital warts not associated with cancers. Will other strains, not countered by the vaccines, now become more prominent?

13.2.4 Rapid Mutation Rates and Quasi-Species

The mutation rate of a virus has been described as the probability that during a single replication of the virus genome a particular nucleotide position is altered (Smith and Inglis, 1987). While mutation frequencies are directly measurable, in practice, it is extremely difficult to convert this to a "rate" (Drake and Holland,

1999). The "rate" may, however, be reduced if many potential mutations lead to nonviable virions. For example, HBV has overlapping reading frames for the surface antigen and the polymerase. As a consequence, some mutations in the surface antigen may cause the polymerase to be nonfunctional.

There is a consensus that RNA viruses have relatively high mutation rates compared with DNA viruses by two orders of magnitude or more. Average misincorporations per nucleotide base in RNA viruses have been reported to be of the order 10^{-4} to 10^{-5} (Holland and Domingo, 1998). This is thought, at least in part, to be a consequence of the lack of proofreading and mismatch repair.

Conventionally, DNA viruses are considered to have low mutation rates relative to RNA viruses; even so, this may be perhaps a 100-fold higher than that of host DNA. As a consequence, there will be low proportions of mutant viruses, sometimes referred to as polymorphisms, within an infected individual. It is becoming recognized that pre-existing polymorphisms may include resistant mutants that greatly increase the rate (in tissue culture) at which DNA viruses develop antiviral resistance compared with the appearance of resistance due to spontaneous mutations. However, with rapidly mutating RNA viruses (e.g., HIV or HCV), there may be no practical distinction between pre-existing and *de novo* resistance mutations.

Mutation rate alone does not determine how soon resistant virus will appear in clinical practice. There are other important factors including the number of virions formed per day in the patient and the proportion of progeny that are "fit." Furthermore, in some cases "fitness" may require compensating mutations (section 13.5). This combination of factors we shall call "resistance rate."

HIV and HCV are two viruses that produce huge numbers of virions each day (ca. 10^9 and 10^{12} virions/day, respectively) in untreated patients (Field and Vere Hodge, 2008). The large population of new virions, coupled with high rates of mutation (ca. 10^{-4}), can quickly lead to enormous genetic diversity within a single infected host. For example, HIV has a single-strand RNA genome of approximately 9,000 nucleotides. The replication rate in an infected individual has been estimated to be approximately 10^9 daily, thus $10^{-4} \times 9,000 \times 10^9 = 9 \times 10^8$ mutants occur each day. This means that, in theory, every point mutation occurs 10^5 times per day in an HIV-infected individual and every double mutant 10 times per day! As a result, HIV actually exists as a quasi-species or "swarm" around a particular consensus sequence. Similarly, HCV exists as quasi-species; it has the fastest known daily replication rate of 10^{12} virions daily.

Among different viruses, there is a huge range of "resistance rates" and this has clinical consequences (Table 13.3).

With the production rate of new virions being a key factor in resistance rate; this emphasizes the importance of reducing viral replication (e.g., HIV and HCV) as quickly as possible after commencing therapy, since a large (e.g., 8 $\log_{10}$) reduction in replication will give a corresponding reduction in the formation and selection of resistant mutants. It is crucially important to keep viral replication at the lowest possible level both throughout a single day and during a long course of therapy. Missed doses and "drug holidays" can give the virus a better chance to mutate and so become resistant.

Table 13.3 Clinical Consequences due to Varying "Resistance" Rates

Virus (DNA/RNA)	"Resistance Rate"[a]	Clinical Outcome
Vaccinia (DNA)	Very slow	Vaccine has eliminated virus from human population. Selective antiviral agents (e.g., ST 246) being developed as anti-bioterrorism agent. Resistance can be obtained in the laboratory but no clinical data available.
Polio (RNA)	Very slow	Vaccine has eliminated virus in most countries.
Varicella zoster (DNA)	Moderately slow	Vaccine expected to be effective for decades; antiviral therapy has not led to an increase (<1%) of resistant isolates among the immunocompetent patients (no increase in three decades) but some increase in immunocompromised patients.
Herpes simplex types 1 and 2 (DNA)	Moderately slow	No efficacious vaccine yet available but resistance to antiviral therapy similar to that with VZV.
Rubella, mumps, measles, HAV (RNA viruses), and HBV[b]	Slow	Vaccines have remained clinically effective for years; antiviral resistance to therapy of HBV with single antiviral agents may occur (within one or a few years).
Influenza (RNA)	Fast	Vaccine needs to be updated at least annually. Resistance to antiviral compounds occurs in the population at various rates for different compounds (days to years).
HIV (RNA)	Very fast	No vaccine successful. Monotherapy leads quickly to resistance in individual patients. Combination therapy (3 or 4) gives low "resistance rate" (several years).
HCV (RNA)	Very fast	No vaccine successful. As for HIV, monotherapies lead to quick appearance of resistance. Antiviral combinations being evaluated.

[a] See text for definition of "Resistance rate"
[b] Hepatitis B is a DNA virus but replicates via an RNA intermediate.

13.3 One Principle Mechanism for Development of Resistance

In spite of the many strategies for viral replication and transmission, as summarized briefly above, all viruses have one main mechanism for development of resistance to antiviral compounds and vaccines—the selection of random mutations. Darwin's theory of evolution—random changes followed by survival of the fittest—is well illustrated in the virus field. At least for some viruses, the evolution of resistance can be followed in days or weeks as the genome replication and mutation rates, leading to random changes, are so high. Sequence analysis of the DNA or RNA shows that a particular resistant variant may have one or more base changes that account for resistance (usually confirmed by the introduction of the same mutation(s) by means of site-directed mutagenesis into a wild type (wt) background). Other mutations may be random changes with no particular consequences for the viability of the virus. For example, there may be base changes that neither alter the encoded amino acid nor cause significant change to the RNA secondary structure. In DNA viruses, such variants are commonly referred to as polymorphisms. In those RNA viruses which mutate rapidly, the huge number of variants are called quasi-species. In a natural infection, under pressure from antiviral therapy, the proportion of wt virus decreases markedly whereas the resistant variant, either as a pre-existing minor variant or a newly formed mutant, becomes dominant.

13.4 Viruses with Segmented Genomes: Additional Resistance Mechanism

Several families of RNA virus have segmented genomes (Table 13.1). The clinically most important is influenza, which has eight segments. These viruses have an additional mechanism of acquiring resistance. When two strains co-infect a cell, in theory, the gene segments may re-assort in every possible combination. This gives the possibility for a drug-resistant virus, which has a poor ability to transmit, to re-assort with a highly infectious but drug-sensitive virus, so that some of the progeny viruses will be highly infectious and drug-resistant.

In the 2009 pandemic H_1N_1 influenza, the eight RNA segments or genes were recently derived from avian (two segments), swine from two continents (five segments) and human (one segment) viruses, presumably in a series of re-assortments. In this case, the resulting virus was not a drug-resistant strain but one to which the general human population, at least those under about 60 years old, did not have effective immunity. It was fortunate that the initial widespread transmission of this virus did not cause devastating burden of illness and deaths. There were only a few reports of oseltamivir-resistant influenza (section 13.6.4) during the first year of the pandemic.

13.5 Evolution of Resistant Mutants

In some instances, a single mutation leads to high-level resistance to the antiviral compound and the virus appears to remain fully "fit." An example is the M2 channel-blocking inhibitors, amantadine and rimantadine, which had activity against influenza viruses. Resistant variants are selected so quickly that a treated person can pass on resistant virus to contacts. Being fully fit, the resistant virus is easily spread. During the 2005–2006 season in the United States, 109/120 (91%) of H3N2 clinical isolates were resistant to amantadine and rimantadine. This has severely limited the clinical usefulness of these drugs.

In contrast to amantadine and rimantadine, there is often a slower evolution of resistance to antiviral compounds that act as substrate mimics for a viral enzyme and so bind to the catalytic site. Most potential mutations will give a nonfunctional enzyme and a nonviable or "less fit" virus. There may be very few (even a single) specific mutation(s) that interfere with the binding of the antiviral compound to the enzyme yet does not reduce by too much the catalytic activity. Such mutations, which usually appear first, are therefore termed "primary mutations." Initially, the degree of resistance may be modest and so there is pressure to create additional "secondary mutations," which enhance the level of resistance to the drug. These structural changes often result in reduced catalytic activity and probably will affect the fitness of the virus (the term "fitness" embraces not only the viability or replication rate of the virus but may also include effectiveness of immune evasion genes, transmission, etc.). So yet further mutations may appear which are apparently unrelated to the protein sites that interact with the drug. Such "tertiary mutations" may have no direct effect on biochemical drug–protein binding but may increase enzyme efficiency so as to compensate for the deleterious effects of the primary and secondary mutations. Many compensating mutations are suspected but often their precise role has yet to be elucidated. Of course, virus mutations do not always fit tidily into these human concepts. There may be a "step" mutation which then allows further mutations. In the case of HIV protease, both the protease and the corresponding cleavage sites can co-mutate to give cross-resistance to PIs (section 13.6.3).

13.6 Illustrative Examples of Resistance to Specific Antiviral Drugs

13.6.1 Poxvirus

Cidofovir (HPMPC) is used clinically to treat AIDS-associated cytomegalovirus retinitis but has also been shown in cell culture and animal tests to be an effective therapy against poxviruses. It has been suggested that cidofovir (or a less toxic prodrug with improved bioavailability) could be stockpiled for use in the event of malicious introduction of smallpox. A study by Andrei et al. (2006) investigated the mutations giving resistance to HPMPC and if the drug resistance was inextricably linked to reduced virulence. If this were the case, then there would be no reason for malicious introduction of mutations conferring resistance.

Drug-resistant vaccinia virus (VV) was obtained by serial passage of the virus in cell cultures with increasing concentrations of HPMPC. In parallel, wt virus was passaged in drug-free cultures. From the final passage, seven plaque-purified HPMPC-resistant (HPMPC^R) isolates and five plaque-purified wt isolates were obtained. As it was thought that resistance to HPMPC would be due to mutations in the viral DNA polymerase gene (*E9L*), this gene was sequenced for each of these isolates. The results are summarized in Table 13.4.

It was known that the original stock of VV contained polymorphisms, in particular at amino acid residue 420. A second polymorphic locus was found at positions 936 to 938. Two clones suffered a small in-frame deletion. However, all wt clones were equally sensitive to HPMPC and so these polymorphisms were unlikely to be related to drug resistance. Therefore it seemed likely that only two point mutations, A314T and A684V, were potentially associated with resistance. These are within the 3′–5′exonuclease proofreading domain and the polymerase catalytic domain,

Table 13.4 Mutations in the E9L Gene of HPMPC^R Vaccinia Virus (strain Lederle)

Virus	Amino Acid Present at Position(s) (Vaccinia Virus)^a						
	246	314	420	684	845	857	936-937-938
HPMPC^R 7 clones^b	R	T	S	V	M	R	A-N-V
Lederle wt clones							
Clone 1	R	A	S	A	M	R	N-Δ^c-G
Clone 2	R	A	L	A	M	R	N-Δ-G
Clone 7	R	A	S	A	M	R	A-N-V
Clone 8	Q	A	L	A	M	R	A-N-V
Clone 11	R	A	S	A	M	R	A-N-V
VV strains							
Ankara	R	A	S	A	T	G	A-N-V
Copenhagen	R	A	L	A	T	G	A-N-V
WR	R	A	L	A	T	G	A-N-V

^{a}The amino acid numbering refers to the numbering system for vaccinia E9L gene. The residue numbering differs slightly for homologous residues in other orthopoxvirus genes.
^{b}All seven clones were identical.
$^c\Delta$ symbol for deletion.
Source: Table adapted from Andrei et al. (2006).

respectively. Marker rescue methods were used to investigate the role of each of these in drug resistance. The cloned DNA encoded A314T, A684V, or both A314T and A684V mutants. As a control, wt DNA was included to test for any of the polymorphisms having an effect on drug resistance. The results showed that A314T and A684V contributed to resistance but both together gave the greater resistance (about 10-fold) to HPMPC. During the serial passaging, it seems that the A314T mutation appeared first followed by the A684V mutation, an example of primary and secondary mutations.

All three HPMPC^R recombinant viruses exhibited reduced virulence in mice (i.e., the mutants were "less fit"). With both mutations together, the reduction in virulence was about 100-fold. To test for efficacy of HPMPC against this resistant virus, mice were challenged with 4,000 pfu, which caused considerable loss in body weight but nearly all mice survived. Mice treated with HPMPC at 10 mg/kg daily had a small transient weight loss whereas those treated with 50 mg/kg daily were similar to uninfected controls. Although one must be cautious that this result has been shown only for one animal species, it is encouraging that HPMPC would be expected to give useful cover against malicious vaccinia release even if the resistant mutations were introduced.

13.6.2 Herpesvirus

The nucleoside analogs, VACV, FCV, and GCV, owe their high selectivity to the fact that their activity requires phosphorylation by a viral enzyme. For ACV and PCV, it is the viral thymidine kinase (TK) enzyme, for GCV the UL97 kinase. The corresponding monophosphate is then further phosphorylated to the triphosphate by cellular enzymes. It is the triphosphate that interacts with the viral DNA polymerase and terminates viral DNA replication. The pyrophosphate analog, foscarnet, and the cyclic phosphonates (e.g., adefovir) do not require the initial phosphorylation step by a viral enzyme and so their selectivity depends solely on their preferential inhibition of the viral DNA polymerase. The recently described VZV inhibitor, FV100, requires phosphorylation by the VZV TK, but too little of the triphosphate is formed to account for its activity; its mechanism remains a puzzle.

Mutant viruses with acquired resistance to all these compounds can be selected in tissue culture. Resistance-conferring mutations can be detected in the target proteins involved in the mechanism of action of each of the drugs. Mutations in the TK gene may lead to an ablation of this enzyme, thus conferring cross-resistance between ACV, PCV, and other compounds that require this phosphorylation step. A wide variety of different mutations can give rise to a truncated or nonfunctional TK polypeptide and loss of enzyme activity. Clinical and laboratory isolates of HSV typically contain of $\geq 10^{-4}$ TK-defective variants. Since a single plaque produced in a tissue culture contains $\geq 10^5$ infectious virions, it may be seen that TK-mediated resistance develops rapidly in tissue culture. Early work on ACV demonstrated that clinical isolates also contain TK-defective variants at high frequency (Paris and Harrington, 1982).

Another mechanism for the development of resistance are mutations leading to single amino acid residue substitutions in TK or DNA polymerase, which reduces the affinity of the drug to the enzyme but maintains, at least in part, the enzymic activity. Such changes occur at about two orders of magnitude less frequently (ca. 10^{-6}) in tissue culture-grown virus but have been shown to account for clinical drug resistance in HSV, VZV, and HCMV (Andrei et al., 2007).

The helicase-primase inhibitors (HPIs) represent a new generation of antiviral compounds active against HSV and VZV. It was shown that resistance mutations to BAY 57-1293 occur in the helicase gene, most being located to a group of residues just downstream from the fourth functional motif. For example, the substitution K356N accounts for >5,000-fold resistance. Such mutations are apparent at a frequency of $\leq 10^{-6}$ in tissue culture for many virus isolates. Surprisingly, it was observed that both laboratory isolates and some recent clinical isolates contain HPI resistance mutations at 100-fold higher frequency (Sukla et al., 2010). PCR amplification experiments and other evidence shows beyond doubt that such mutations exist at high frequency as polymorphisms in the virus population prior to drug exposure (Biswas et al., 2007).

While herpesvirus drug-resistance occurs at relatively high frequency in tissue culture, the widespread clinical use of herpesvirus antivirals is rarely limited by the emergence of resistance in immunocompetent patients. Indeed, large-scale screening of isolates of both HSV-1 and HSV-2 obtained from typical lesions of labial or genital herpes show no obvious trends to resistance (<1%) after extensive use over the period from the early 1980s to date (Bacon et al., 2003). However, resistance to nucleoside analogs does appear to be more common in patients with neonatal HSV and herpes keratitis. In the former, resistance may be observed in 5% of patients and for ocular isolates from herpes keratitis, Duan et al. (2008) reported that 11/173 (6.4%) patients yielded resistant isolates and 10/11 of these isolates mapped to TK, with these authors arguing that the cornea represents an immunologically privileged site. In immunocompromised patients, resistance to ACV and similar drugs has commonly been reported in about 5% of patients and in some cases up to 20%. Often, such viruses comprise mixtures of wild-type together with one or more different resistant mutants. There may be two reasons underlying the apparent divergence between results in immunocompetent and immunocompromised patients. Most important, many resistance mutations clearly result in loss of virus "fitness." This is most easily demonstrated for TK-defective strains in a variety of animal models. Such strains are much less neuropathogenic (Field and Wildy, 1978) and, while they can establish a latent infection, these seem unable to reactivate efficiently to produce infectious virus (see 13.8.3). Also, other resistance mutations undoubtedly result in subtle defects that may diminish the ability of the virus to reactivate efficiently from latency and/or replicate successfully. Secondly, the establishment of neuronal latency with wt virions during primary infection means that subsequent reactivations originate from the pool of sensitive virus.

The discovery of high frequency of resistance mutations (or polymorphisms) in clinical isolates of HSV challenges the dogma that large DNA viruses display high genetic stability. While the genome is generally highly conserved, it is still not clear why frequency of particular resistance mutations leading to single amino acid substitution at defined loci may be as high as 10^{-4} in some strains. Perhaps herpesviruses are able to generate some constrained genetic flexibility during DNA replication to overcome host heterogeneity, provide tropism for biological sites, and/or enable immune avoidance? However, the mechanism for this intriguing ability has yet to be determined.

Finally, where herpesvirus resistance has become a recognized problem, such as in herpes keratitis, neonatal herpes, and herpes in the immunocompromised patient, the lessons learned from HIV and hepatitis viruses will be applied in the form of drug combinations. These will most likely involve nucleoside analogs in combination with the ether-lipid analog of cidofovir, CMX001, HPIs, and other novel compounds.

13.6.3 HIV: Protease Inhibitors

Resistance to HIV protease inhibitors provides good examples of stepwise mutations (Molla et al., 1996): primary, secondary, and tertiary as defined above (section 13.3). At the time of a review by Schafer (2002), there were six protease inhibitors (PIs) approved in the United States—amprenavir, indinavir, lopinavir, nelfinavir, ritonavir, and saquinavir. For all these, the primary mutation occurs in the substrate cleft of the protease, thereby reducing the affinity of the inhibitor. These primary mutations generally reduce the activity of the compounds by only 2- to 5-fold, not enough to be clinically resistant but sufficient to confer a selective advantage to the virus. These variants will overtake the wt HIV and allow other mutations to develop. For indinavir and ritonavir, the first mutation is V82A/T/F/S, for nelfinavir, D30N. For other PIs, especially saquinavir, the first mutation is often L90M (Clavel and Hance, 2004). The second mutation for indinavir, ritonavir, saquinavir, and amprenavir is I84V and for nelfinavir N88D/S. Further mutations can occur in the protease flap, generally I54V but also I54T/L/M. By the time there are about six mutations, the new strain is likely to be highly resistant and have cross-resistance against several PIs.

There are many other reported mutations which may be considered to be tertiary mutations helping to restore viral "fitness" but their role has not been defined. Remarkably, viral "fitness" can also be increased by mutations in HIV gag, the main viral substrate of the protease. Such gag mutations, A431V and L449F, can improve the ability of the protease of resistant strains to interact with the substrate.

13.6.4 Influenza Virus

Rational design programs led to development of the neuraminidase (NA) (or sialidase) inhibitors, zanamivir, and oseltamivir. Both compounds block the action of the essential virus function, NA, which is required by influenza for efficient release of infectious progeny. These compounds are generally held to be efficacious (Dutowski, 2010).

Oseltamivir has been prescribed far more often in Japan than elsewhere, thus oseltamivir-resistance has been investigated in this population. There have been several reports of NA-inhibitor resistance among clinical isolates. For example, a study in Japan found that 9 of 50 children with influenza A (H3N2) virus infection who had been treated with oseltamivir had a virus with drug resistance, although it was suggested that these mutations were less fit than the wt viruses from which they were derived (Kiso et al., 2004). One study followed resistance from 1996 through 2007 (Tashiro et al., 2009). During the period 1996−2002, influenza A N2 viruses were circulating but no resistant viruses were detected (0/175). During the season 2003−2004, 0.3% (3/1180) of N2 samples were resistant. During the following three seasons, no N2 resistant viruses were detected but N1 virus started circulating. In 2004−2005 and 2006−2007, no resistant viruses were detected but in 2005−2006, 3% (4/132) were found. This survey confirms that resistance to oseltamivir occurs in the normal population far less readily than does resistance to amantadine. Unfortunately, the situation changed dramatically and globally within 3 years (Okomo-Adhiambo et al., 2010). Among seasonal H_1N_1 influenza, the proportions of oseltamivir-resistant viruses were low (ca. 1%) in 2006−2007 but then resistance emerged rapidly worldwide; in the United States, high-level resistance (100- to 3,000-fold) was found in about 20% of samples tested in 2007−2008 and about 90% in 2008−2009. Sequencing confirmed the H275Y mutation in resistant strains. It appears that a natural, spontaneously arising variant had spread globally, without drug selection pressure, during 2007−2008. Early work in cell culture and animals suggested that the H275Y mutant virus was somewhat disabled, but there seems to have been co-selection of other compensating mutations, perhaps to the hemagglutin gene (HA), to give a "fit" virus enabling this variant to spread globally.

Among the circulating human influenza viruses, there are three subtypes of neuraminidase (NA), influenza A types N1 and N2, and influenza B NA. These NAs differ in the structure of a pocket adjacent to the active site of the enzyme. Oseltamivir makes use of this pocket in binding to the NA and so resistance can occur with mutations. With N1, just a single mutation, H275Y, gives high-level resistance but the corresponding mutation in N2, H274Y, does not give resistance. Instead two mutations, E119V and R292K, give high-level resistance with N2. With Influenza B NA, R152K and D198N give resistance. Generally, these mutations affecting the pocket do not give cross-resistance to zanamivir but there are some NA mutations which do so, for example, R371K. With influenza B, R152K and D198N give cross-resistance. However, it seems that such strains may be disabled as zanamivir resistance has been isolated only rarely in the clinic.

In clinical studies on oseltamivir resistance, it has been noted that resistance occurs at a higher rate in influenza with N1 than N2, presumably because it takes just a single mutation, H275Y, to give high-level resistance with N1. For example, in a study of oseltamivir-treated children during 2005−2007 (Stephenson et al., 2009) resistance was detected in 3/11 (27%) with influenza A H_1N_1, 1/34 (3%) H3N2, and 0/19 with influenza B.

During the H_1N_1 pandemic of 2009−2010, resistance to oseltamivir has been reported in case studies of seriously ill patients. Fortunately, human-to-human spread has occurred sporadically in geographically dispersed regions. Virtually all the resistant viruses have had the H275Y mutation. Other than these few cases, oseltamivir has been used widely during the pandemic and seems to have not been associated with resistance in the general population. This situation could easily change. With such high proportion of resistance among the seasonal N1 influenza, and with influenza viruses having the ability to re-assort, it seems likely that the threat of resistant pandemic H_1N_1 is ever-present.

13.7 Optimizing Drug Combinations to Avoid Resistance

13.7.1 Genetic Barrier

When a virus is being inhibited by an antiviral compound, resistance mutations are selected, but the ease with which this is done depends upon how many potential mutations can give resistance without compromising virus fitness. This has become known as the genetic barrier.

The anti-influenza M2 channel blockers, amantadine and rimantadine, and antipicornarvirus capsid-binding compounds, such as pleconaril, are examples of agents which present too low a genetic barrier to become useful clinical monotherapies. Such compounds may give added benefit if always used in drug combinations without ever being used alone.

The importance of the genetic barrier concept has been emphasized by experience from testing combinations of drugs active against HIV. There are three major classes of anti-HIV compounds: nucleoside/nucleotide reverse transcriptase inhibitors (NRTIs), non-nucleoside reverse transcriptase inhibitors (NNRTIs), and PIs.

More recently, a fusion inhibitor, HIV integrase inhibitors, and a receptor-binding blocker have also become available. With monotherapies, resistance appears quickly with NNRTIs but more slowly with NRTIs, which also target RT but at its catalytic site. It appears that mutations at an allosteric site are more readily accommodated than mutations at the catalytic site.

Similarly, for PIs, which target the protease catalytic site, the rate of appearance of resistance is about comparable to that for NRTIs. It was thought that combining one NRTI and one PI would delay the appearance of resistance greatly but clinical practice showed that the delay was modest. The gain from such combinations of drugs is probably due to the faster reduction in the rate of virus replication (virions/day), thus reducing the opportunities for creating resistant mutants. However, aside from this factor, it is as easy to form resistant mutants, one in the RT and one in the protease, as for the monotherapies. It is more effective to combine two or three compounds that target the same HIV enzyme. An ideal situation is when the resistance mutations to one drug confer enhanced sensitivity to the other. Even with drugs which have differing mutation patterns, the aim is to have no possible mutations without causing a large reduction in enzyme efficiency. In summary, one high genetic barrier is more effective in delaying resistance than two low genetic barriers.

An example of a commonly used combination therapy for HIV is Atripla® (Gilead and BMS jointly). This single pill contains three compounds: emtricitabine, tenofovir, efavirenz, or two NRTIs and a NNRTI, respectively. Another example is Trizivir® (GSK), combining the three NRTIs, zidovudine, lamivudine, and abacavir. These combination pills, targeting the HIV RT, are often used with an HIV protease inhibitor. When used correctly, these multi-drug therapies provide good control of HIV replication and symptoms and have prevented resistance development for at least several years.

The experience with HIV should guide rational choice of compounds for combination therapies for influenza. Of the two neuraminidase inhibitors (NIs), oseltamivir has been much the more widely prescribed than zanamivir (inhaled) but, fortunately, zanamivir retains activity against the H274Y (H275Y in N1 numbering) mutant resistant to oseltamivir. Zanamivir is now being developed as an IV drug for use in seriously ill patients. Peramivir, another NI, is also being developed as an IV drug. There is cross-resistance between oseltamivir and peramivir, so this combination would not increase the genetic barrier. However, zanamivir and peramivir have differing resistance mutations although Q136K alone gives reduced sensitivity to both zanamivir (36-fold) and to peramivir (80-fold). As these two compounds are being developed as IV therapies, it may be beneficial to combine these into one IV product.

Favipiravir (T-705) is the first influenza RNA polymerase inhibitor to reach phase III clinical trials. In combination with oseltamivir and zanamivir, T-705 has given additive to synergistic activity in a mouse model. The combination of oseltamivir and T-705 has been evaluated in a Phase I study. Clearly, it is hoped that this combination would give highly effective control of influenza in seriously ill patients. But experience from HIV indicates that this would not be an optimum combination to delay resistance, especially in immunocompromised patients. The greater reduction in virus replication with the combination therapy will help to delay resistance but the genetic barrier is the same as for the two individual therapies. For an optimum combination, we need a second viral RNA polymerase inhibitor, with resistance mutations differing from those for T-705. However, in otherwise healthy individuals, one may hope that the immune system would clear a small population of resistant influenza virus. Combination therapy is a strategy that was proposed by Hayden (1986). However, while there is a paucity of candidate compounds, obtaining combinations with an appropriate virological and pharmacological match remains a major challenge to this day.

Generally, vaccines could be considered as "drug combinations" as they may induce many antibodies, each specific for a different part (epitope) of the viral protein surface. To become resistant, the virus would have to change many parts of the protein surface (i.e., it presents a high genetic barrier). In contrast, therapy with a monoclonal antibody (e.g., palivizumab for respiratory syncytial virus [RSV]) which targets a single epitope, would be more susceptible to virus resistance. In a study in immunosuppressed cotton rats (Zhao and Sullender, 2005), palivizumab resistance was detected in 3/5 animals. Within the F gene, one mutation, A816T was sufficient to give resistance. This is similar to the situation following drug monotherapy.

13.8 Unexpected Consequences of Resistance Mutations

13.8.1 Multiple Changes Arising from One Mutation

Some viruses make very efficient use of their small genome size by using not just one of the three possible reading frames but two or all three reading frames. An example is HBV, in which the polymerase and capsid protein reading frames overlap. For treating HBV infections, the commonly used antiviral compounds are lamivudine (LMV or 3TC), adefovir dipivoxil (ADV), and telbuvidine. All these three inhibit the viral polymerase, and so resistant mutations arise in the gene coding for the polymerase. The same mutation can, however, also change the viral surface protein due to the overlap of the reading frames. Conversely, the immune system would exert pressure on the virus to generate mutations in the surface protein but such mutations may give rise to nonfunctional viral polymerase.

13.8.2 Carbohydrate-Binding Agent Leads to Greater Immunogenicity

Although many viral infections are short-lived and the host is able to clear the virus, HIV infections continue despite a vigorous antibody response. Could the antibody response be made more effective in clearing the virus and perhaps limit the progression of HIV infection? A novel approach to therapy uses compounds that bind to carbohydrate moiety of the HIV gp120. The concept is that this would lead to a high genetic barrier to resistance whilst making resistant mutants more susceptible to neutralizing antibody (Balzarini, 2005).

HIV gp120 is highly glycosylated (~50%), many of the carbohydrate chains being high-mannose type, which are rare on human cells. The virus envelope glycosylation is required for the proper folding of the gp120, for efficient entry of the virus into target cells, and for hiding the potentially highly immunogenic protein surface of gp 120. Dendritic cells have a receptor, DC-SIGN, which captures HIV via the (high-mannose) glycans and then directs transmission of HIV to T-lymphocytes. The expression of gp120 in the cell membrane of virus-infected cells allows fusion with uninfected cells, resulting in giant multinucleated cells. Carbohydrate-binding agent (CBA) have the potential to inhibit all these steps. As proof of concept, several plant lectins, with binding preference to mannose-containing glycans, have been shown to inhibit all the above steps. More encouraging for drug potential, Pradimicin A (PRMA) is a non-peptidic, small molecular weight CBA. Although less active, on a μM basis, than the plant lectins, it has similar broad-spectrum activity against a variety of HIV-1, HIV-2, and SIV strains.

Selection of CBA-resistant HIV can be achieved in cell culture but only after many passages. The mutations are predominantly in gp120, notably not in gp41, and result in loss of glycosylation sites, mainly the high-mannose glycan sites. There is a high genetic barrier due to the possibility that many PRMA molecules can bind to each single gp120 molecule and there has to be multiple glycan deletions for significant phenotypic resistance. When in the presence of an immune system, it is hoped that these CBA-resistant strains of HIV will be rendered susceptible to neutralizing antibody due to the exposure of the gp120 protein surface which is normally hidden under a protective glycan cover. There is now some evidence for this. Hu et al. (2007) used cyanovirin-N (CV-N), a CBA, to generate strains of HIV resistant to CV-N and other CBAs so that they could investigate the impact of the immune system on these CBA-resistant strains. One of the isolated resistant clones, GCV4, had five mutations resulting in the loss of glycosylation at amino acid residues 289, 332, 339, 392, and 448, all these being in the constant regions C2, C3, and C4 of gp 120. When used to infect cells with control serum, wt and GCV4 infectivities were not changed by the concentration of the serum. In contrast, when serum from HIV^{+ve} patient was used, the serum had a greater potency against GCV4 than wt HIV. Furthermore, GCV4 was more sensitive to monoclonal antibodies (MAbs) directed to the V3 loop of gp120, a major determinant of viral entry. There was about an 8-fold higher sensitivity to MAb 1101 and over 200-fold for MAb 447-52D. As controls, there were no changes in sensitivities to MAbs directed at other parts of gp120. Furthermore, when wt and mutant SIV (lacking several gp120 glycans) were compared in monkeys; the wt gave long-lasting viremia (about 7 $\log_{10}$) whereas the mutant virus gave a short period of high virus levels which dropped as antibody levels rose.

CBA may be effective therapy, not just for HIV but also several other families of enveloped viruses. For example, CBA have shown marked activity against HCV, influenza and coronaviruses (but not HSV, VZV, RSV, or parainfluenza). The lack of activity against the latter viruses may be due to less mannose-rich glycans being present. The challenge now is to discover a low-molecular-weight CBA with the right properties for a good and specific antiviral agent.

13.8.3 Herpesvirus Latency Potential

As mentioned in section 13.6.2, a characteristic of herpesviruses is that they establish a latent form which remains viable for the rest of the host's life. The commonly used antiherpesvirus compounds, ACV, VACV, and FCV, are activated only in herpesvirus-infected cells; the crucial first step in that activation requires the viral TK. The TK function is not required for efficient HSV replication in cell culture; wt and TK^{-ve} strains replicate with similar rates to similar titers (Field and Wildy, 1978). This allows the virus in vitro an easy option to become resistant to all the TK-mediated compounds. Furthermore, clinical resistant strains in immunocompromised patients and herpes keratitis patients are most commonly TK^{-ve} strains.

However, TK^{-ve} strains exhibit a marked reduction in viral "fitness." In mice, this is manifested by a large ($>1,000$-fold) reduction in lethality and altered latency. It appears that TK^{-ve} strains are able to establish latent infection but their ability to reactivate greatly impaired. It seems that in humans reactivation is similarly impaired. TK^{-ve} viruses do not readily spread among the population. In contrast, wt virus often transmits while the subject is unaware of a sub-clinical reactivation. For an individual immunocompetent subject, if treatment of an episode leads to resistant TK^{-ve} herpesvirus appearing, then the next episode of recurrent herpes will not be from the new TK^{-ve} strain but from the original wt virus. This seems to account for the continuing low rate ($<1\%$) of resistant herpesvirus in the general population even after several decades of antiviral therapy. As may be expected, resistant herpesviruses are a concern in immunocompromised patients.

13.8.4 Reduced Replication Fitness

A common consequence of resistant mutations in various viruses, such as influenza virus, HBV, or HIV, is that the resistant virus has reduced ability to replicate in the patient. For example, when only lamivudine was available for the treatment of HBV infections, highly resistant virus was sometimes present within a year of starting therapy but it was better to continue therapy as the resistant virus was partially disabled. Now that other drugs are available, switching to another drug, such as adefovir, is usually the best option. Were it available, however, a combination would be preferable, since in principle, sequential switching from drug to drug is undesirable as it may more readily lead to multiple resistance.

13.9 A Role for Compounds Targeting Host Proteins for Antiviral Therapy

For some years there have been attempts to target a host function essential for virus replication. This seemed to provide an attractive way of circumventing virus resistance mutations, although this approach risks unacceptable toxicity. Recently this problem seems to have been addressed for influenza and other viruses by targeting only virus-infected cells.

For example, TSG101 is a host protein which is part of the system regulating transport within the cell. Importantly, TSG101 is normally found only in the cytoplasm of uninfected cells, but an influenza viral protein binds to TSG101, and this results in TSG101 being localized to the cell membrane. Using TSG101-specific antibodies, it was shown that at time of infection with influenza virus, there was no TSG101 on the surface of the cell but by 24 hours after infection, TSG101 was on the surface. This was confirmed with different cells and various strains of influenza. Furthermore, TSG101 monoclonal antibodies reduced the release of influenza virus from infected cells, indicating that TSG101 plays a vital role in the replication cycle of influenza virus. In cell culture, it has been possible to add TSG101 antibody and natural killer cells to target specifically influenza infected cells.

This approach could provide a broad-spectrum therapy against many different strains of influenza. As TSG101 is normally resident within the cytoplasm of the cell, it is envisaged that TSG101 antibody would be safe to use. The same approach could be useful with other enveloped viruses (HIV, RSV, HSV-1, and 2, Ebola and parainfluenza) which, like influenza, "hijack" TSG101 to help transport the virus from the cell interior to the outer membrane. It seems remarkable that viruses from different families have evolved to use this single mechanism.

In a search for potential compounds, Kinch et al. (2009) used computer modeling to select a panel of low-molecular-weight compounds that may disrupt the binding of TSG101 to viral proteins. These compounds were screened for activity in a range of viruses. One compound, FGI-104, was active against all the tested viruses (including HBV, HCV, HIV, Ebola, and cowpox) in cell culture assays. FGI-104 was then evaluated in a mouse model of Ebola virus; dosing at 10 mg/kg daily gave 100% survival of the treated mice whereas there were 90% deaths in the control group. Although it remains to be demonstrated that FGI-104 is acting via TSG101, it seems that this is an encouraging result.

The budding of HIV has been shown to be dependent on the binding of HIV gag to TSG101. The binding site on TSG101 is highly conserved, Pro-Thr-Ala-Pro (Chen et al., 2010). Using similar strategy as for research leading to PIs, compounds which mimic the protein structure at the gag-TSG101 binding site are being evaluated for inhibition of HIV budding. A disadvantage of this approach is that the selectivity for the infected cell is lost when the anti-HIV compound binds to TSG101 inside the cell rather than on the cell surface.

How would viruses counter such an attack on their replication? To think that the blocking of TSG101 would permanently inhibit virus replication is both overly optimistic and unwise. The virus may mutate to increase its binding to the host protein so that it outcompetes the inhibitor. Alternatively, it is possible that there is a secondary mechanism for the release of virions from the cell. The efficiency of such a secondary mechanism could be enhanced by mutations, the new virus variant then becoming dominant.

13.10 Conclusion

The origins and evolution of viruses may be shrouded in mystery but one current aspect is certain. Only in the relatively recent past, since Jenner's first use of cowpox as a vaccine for human smallpox, have viruses faced a new threat to their replication—mankind's intervention. Through the course of evolution, viruses have developed many hugely varying strategies for their highly successful survival. Now faced with this new threat posed by vaccines and specific antiviral compounds, some viruses are poorly adapted to survive. The human smallpox virus has been eliminated from the human population. The global polio eradication initiative has been highly successful. Its aim, to eradicate polio worldwide, seems achievable but remains elusive. The use of vaccines has been successful in preventing many viral infections, including rubella, mumps, measles, HAV and HBV. However, there seems to be no prospect of eliminating these viruses in the near future. More recently, specific antiviral compounds have been developed to control those human viruses for which, generally, no effective vaccines are available. ACV, active against herpesviruses (HSV-1 and -2, VZV), was the first truly active and selective antiviral agent. Some three decades later, it is still being used but has been joined by just two other drugs, the prodrug of ACV, VACV, and FCV. Although these drugs can limit the symptoms of acute infection, the incidence of latent infection has not been reduced. The mainstay for therapy and prevention of HCMV is just one drug, valganciclovir (VGCV). The spread of HIV has prompted a huge search for effective drugs and combinations of three or four drugs are providing at least several years of clinical control. From this research, drugs against HBV were developed. As for HIV, a similar combination approach is being developed for HCV. The very high mutation rates for HIV and HCV, combined with their high replication rates (10^9 and 10^{12} virions/day, respectively, in an infected patient without therapy), means that the threat of breakthrough remains ever-present especially if drug doses are missed. The concept of genetic barrier has been helpful in guiding combination therapies to give effective control of patient symptoms for at least several years. Although HIV has spread across the world and caused so much human disease in just a few decades, perhaps the virus with most potential to cause a rapid pandemic is influenza virus. The 2009–2010 pandemic, caused by H_1N_1 strain of influenza is the first influenza pandemic that may have been constrained by the use of antiviral drugs and the rapid development of a vaccine.

The last three decades have seen many advances but also highlighted the limitations of mankind's attempts to control viruses. Truly active and safe antiviral compounds seemed rather a remote possibility until ACV was discovered. Even then, when HIV was identified as the cause of AIDS, a vaccine approach was seen as the preferable way forward with an effective vaccine expected in 2 years. Instead, it has been remarkable how combination pills have given HIV patients an easy-to-use, once-daily dosing regimen, which is well tolerated. In too many publications, the *Introductions* state that virus resistance is limiting the use current antiviral compounds, and therefore new compounds with a different virus target and new mode of action are required. Even better to delay resistance, look for compounds which raise the genetic barrier. So far, the best combinations have been with compounds that target the virus polymerase. Similar combinations with protease inhibitors have not been so successful because the protease and the virus polypeptide cleavage sites can co-mutate, an option not available to the virus polymerase. Although the genetic barrier needs to be increased for long-term delay in resistance in chronic infections, with any drug combination used in naturally self-limiting infections, the extra effect in reducing viral load quickly may well be a useful benefit.

Our current antiviral therapies have been successful in reducing the burden of human diseases but many viruses have evolved strategies for countering new threats to their replication. These strategies pose an ever-present threat to our modern human therapies. We need to use our antivirals wisely.

Abbreviations

Viruses/virus enzymes
HAV hepatitis A
HBV hepatitis B
HCV hepatitis C
HSV herpes simplex virus
HCMV human cytomegalovirus
HIV human immunodeficiency virus
HK herpes keratitis
RSV respiratory syncytial virus
VV vaccinia virus
VZV varicella-zoster virus
TK thymidine kinase
RT reverse transcriptase
HA hemagglutin
NA neuraminidase

Antiviral compounds/inhibitor type
ACV acyclovir
VACV valaciclovir
ADV adefovir dipivoxil
HPMPC cidofovir
CV-N cyanovirin-N
FCV famciclovir
PCV penciclovir
T-705 favipiravir
VGCV valganciclovir
GCV ganciclovir
HPIs helicase-primase inhibitors
LMV or 3TC lamivudine
PRMA pradimicin A
CBA carbohydrate-binding agent
NI neuraminidase inhibitor
NRTIs nucleoside/nucleotide reverse transcriptase inhibitors
NNRTIs non-nucleoside reverse transcriptase inhibitors
PIs protease inhibitors

Others
PK pharmacokinetics
wt wild type
iv intravenous

Cross References to Other Chapters

(1) Molecular epidemiology and species definition of pathogens
Michel Tibayrenc

(2) Virus species
Marc Van Regenmortel (University of Strasbourg, France)

(3) Viral evolution
Hiroshi Haeno and Yoh Iwasa (Department of Biology, Faculty of Sciences, Kyushu University, Fukuoka, Japan)

(23) Genomics of infectious diseases and private industry
Guy Vernet (Biomerieux company, Marcy-L'Etoile, France)

(30) The origins of human immunodeficiency virus and implications for global epidemics
Eric Delaporte, Martine Peeters (IRD, Montpellier, France)

(31) Evolution of SARS coronavirus and the relevance of modern molecular epidemiology
Zhengli Shi (Key Laboratory of Virology, Wuhan Institute of Virology, Chinese Academy of Sciences, Wuhan 430071, China) and Lin-fa Wang (CSIRO Livestock Industries, Australian Animal Health Laboratory, Geelong, Victoria, Australia)

(32) Ecology and evolution of avian influenza: the risk of a major pandemics
Ron. A.M. Fouchier (Department of Virology, Erasmus Medical Center, Rotterdam, The Netherlands)

References

Andrei, G., Gammon, D.B., Fiten, P., De Clercq, E., Opdenakker, G, Snoeck, R, et al., 2006. Cidofovir resistance in vaccinia virus is linked to diminished virulence in mice. J. Virol. 80, 9391–9401.

Andrei, G., Fiten, P, Froeyen, M., De Clercq, E., Opdenakker, G., Snoeck, R., 2007. DNA polymerase mutations in drug-resistant herpes simplex virus mutants determine in vivo neurovirulence and drug-enzyme interactions. Antivir. Ther. 12, 719–732.

Bacon, T.H., Levin, M.J., Leary, J.J., Sarisky, R.T., Sutton, D., 2003. Herpes simplex virus resistance to acyclovir and penciclovir after two decades of antiviral therapy. Clin. Microbiol. Rev. 16, 114–128.

Balzarini, J., 2005. Targeting the glycans of gp120: a novel approach aimed at the Achille heel of HIV. Lancet Infect. Dis. 5, 726–731.

Biswas, S., Smith, C., Field, H.J., 2007. Detection of HSV-1 variants highly resistant to the helicase-primase inhibitor BAY 57-1293 at high frequency in two of ten recent clinical isolates of HSV-1. J. Antimicrob. Chemother. 60, 274–279.

Chen, H., Liu, X., Li, Z., Zhan, P., De Clercq, E., 2010. TSG101: a novel anti-HIV-1 drug target. Curr. Med. Chem. 17, 750–758.

Clavel, F., Hance, A.J., 2004. HIV drug resistance. N. Engl. J. Med. 350, 1023–1035.

Drake, J.W., Holland, J.J., 1999. Mutation rates among RNA viruses. Proc. Natl. Acad. Sci. U.S.A. 96, 13910−13913.

Duan, R., de Vries, R.D., Osterhaus, A.D.M.E., Remeijer, L., Verjans, G.M.G.M., 2008. Acyclovir-resistant corneal HSV-1 isolates from patients with herpetic keratitis. J. Infect. Dis. 198, 659−663.

Dutkowski, R., 2010. Oseltamivir in seasonal influenza: cumulative experience in low- and high-risk patients. J. Antimicrob. Chemother. 65 (Suppl. 2), ii11−ii24.

Field, H.J., Vere Hodge, R.A., 2008. Antiviral agents. In: Mahy, B.W., Van Regenmortel., M.H.V. (Eds.), Encyclopedia of Virology, third ed. I. Elsevier, Oxford, pp. 142−154.

Field, H.J., Wildy, P., 1978. The pathogenicity of thymidine kinase-deficient mutants of herpes simplex virus in mice. J. Hyg. 81, 267−277.

Hayden, F.G., 1986. Combinations of antiviral agents for treatment of influenza virus infections. J. Antimicrob. Chemother. Suppl. B 18, 177−183.

Holland, J., Domingo, E., 1998. Origin and evolution of viruses. Virus Genes 16, 13−21.

Hu, Q., Mahmood, N., Shattock, R.J., 2007. High-mannose-specific deglycosylation of HIV-1 gp120 induced by resistance to cyanovirin-N and the impact on antibody neutralization. Virology 368, 145−154.

Kinch, M.S., Yunus, A.S., Lear, C., Mao, H., Chen, H., Fesseha, Z., et al., 2009. FGI-104: a broad-spectrum small molecule inhibitor of viral infection. Am. J. Transl. Res. 1, 87−98.

Kiso, M., Mitamura, K., Sakai-Tagawa, Y., Shiraishi, K., Kawakami, C., Kimura, K., et al., 2004. Resistant influenza A viruses in children treated with oseltamivir: descriptive study. Lancet 364, 759−765.

Molla, A., Korneyeva, M., Gao, Q., Vasavanonda, S., Schipper, P.J., Mo, H-M., et al., 1996. Ordered accumulation of mutations in HIV protease confers resistance to ritonavir. Nat. Med. 2, 760−766.

Okomo-Adhiambo, M., Nguyen, H.T., Sleeman, K., Sheu, T.G., Deyde, V.M., Garten, R.J., et al., 2010. Host cell selection of influenza neuraminidase variants: implications for drug resistance monitoring in A(H1N1) viruses. Antiviral Res. 85, 381−388.

Paris, D.S., Harrington, J.E., 1982. Herpes simplex virus variants resistant to high concentrations of acyclovir exist in clinical isolates. Antimicrob. Agents Chemother. 22, 71−77.

Shafer, R.W., 2002. Genotypic testing for human immunodeficiency virus type 1 drug resistance. Clin. Microbiol. Rev. 15, 247−277.

Smith, D.B., Inglis, S.C., 1987. The mutation rate and variability of eukaryotic viruses: an analytical review. J. Gen. Virol. 68, 2729−2740.

Stephenson, I., Democratis, J., Lackenby, A., McNally, T., Smith, J., Pareek, M., et al., 2009. Neuraminidase inhibitor resistance after oseltamivir treatment of acute influenza A and B in children. Clin. Infect. Dis. 48, 389−396.

Sukla, S., Biswas, S., Birkmann, A., Lischka, P., Zimmermann, H., Field, H.J., 2010. Mismatch primer-based PCR reveals that helicase-primase inhibitor resistance mutations pre-exist in herpes simplex virus type 1 clinical isolates and are not induced during incubation with the inhibitor. J. Antimicrob. Chemother. 65, 1347−1352.

Tashiro., M., McKimm-Breschkin, JL, Saito, T., Klimov, A., Macken, C., Zambon, M., et al., 2009. Surveillance for neuraminidase-inhibitor-resistant influenza viruses in Japan, 1996−2007. Antivir. Ther. 14, 751−761.

Zhao, X., Sullender, W.M., 2005. In vivo selection of respiratory syncytial viruses resistant to palivizumab. J. Virol. 79, 3962−3968.

14 Evolution of Resistance to Insecticide in Disease Vectors

Pierrick Labbé[1], Haoues Alout[1], Luc Djogbénou[2], Nicole Pasteur[1] and Mylène Weill[1,*]

[1]Institut des Sciences de l'Evolution (CNRS UMR 5554) Université Montpellier II, 34095 Montpellier, France, [2]Institut de Recherche en Santé Publique (IRSP), Université Abomey-Calavi, Ouidah, Bénin

14.1 Introduction

The control of vector-borne diseases represents one of the greatest global public health challenges of the twenty-first century. They contribute substantially to the global burden of infectious diseases ($\sim 17\%$) and their prevalence tends to increase. Human population growth in many areas has led to extensive deforestation, irrigation, and urbanization, and these environmental modifications have created conditions that favor the proliferation of many arthropod vectors, such as mosquitoes, ticks, flies, and so on. More than a billion people, primarily in developing countries, are now at risk for contracting many new or re-emerging diseases (WHO, 2006).

Mosquitoes are probably the most common vectors of infectious diseases (review in Tolle, 2009). About 3500 species are found throughout the world and, in almost all species, the female finds the proteins she needs for developing eggs through blood-feeding on vertebrates. This makes mosquitoes particularly prone to transfer viruses and other parasites between humans and animal hosts.

Malaria is a human plague documented since Greek Antiquity (it is mentioned in the Iliad) and may have afflicted humanity even earlier, as indicated by Neolithic bones' pathologic modifications (see Reiter, 2001). It is responsible for 350–500 million annual cases with more than 1 million deaths, mostly in children in Africa. It is caused by the parasitic protists (*Plasmodium* sp.) vectored by mosquitoes of the *Anopheles* genus.

Dengue, yellow fever, and chikungunya are viral diseases vectored by mosquitoes of the genus *Aedes* (*Ae. aegypti* and *Ae. albopictus*) that recently expanded their range due to increased human population growth and travel, together with poor sanitary conditions. The importance of yellow fever, with 200,000 cases and 30,000 deaths reported annually, is probably underestimated. Dengue (50 million

*E-mail: mylene.weill@univ-montp2.fr

infections, >12,000 deaths per year, mostly children) is the most widespread arthropod-borne virus infection: 9 countries were affected in 1970 and 60 in 1999. It is caused by four distinct viruses and is mainly vectored by *Ae. aegypti* (Lambrechts et al., 2009). Finally, chikungunya, a formerly obscure arbovirus endemic to East Africa, recently caught attention after several outbreaks in countries of the Indian Ocean and Southeast Asia, where millions of cases were documented. A mutation of the original virus (making it better adapted to *Ae. albopictus* than *Ae. aegypti*, the original vector) is probably responsible for recent chikungunya outbreaks in La Réunion Island and in Italy (Rezza et al., 2007; Tsetsarkin et al., 2007; Vazeille et al., 2007). Over the past 20–30 years, the geographic distribution of *Ae. albopictus* has considerably increased through worldwide commerce of used tires and because of its capacity of diapausing and the resistance of its eggs to dessication (Enserink, 2008). It is presently replacing *Ae. aegypti* in tropical regions and causes the diffusion of the chikungunya in temperate regions.

Other mosquito-borne diseases include the West Nile virus, now endemic in the USA (Campbell et al., 2002), the Japanese encephalitis virus, which is expanding in the Indian subcontinent and Australasia (both transmitted by *Culex* genus), and filarial nematodes, causing elephantiasis vectored by the *Culex* sp. and *Ae. polynesiensis*.

Other major vector-borne diseases are transported by nonmosquito arthropods. This is the case of the sleeping sickness, vectored by the tsetse flies (*Glossina* sp.), a neglected disease that imposes a burden close to that of malaria on humans (>300,000 cases per year, but severe morbidity), mainly in African isolated and underserved rural areas, where it also affects the cattle, the main local resource (see Welburn et al., 2006). Other dipteran as sand flies (Phlebotominae) and black flies (Simulidae) are vectors of leishmaniasis and onchocercosis, respectively, as well as of several viruses (review in Alexander and Maroli, 2003; Surendran et al., 2005); houseflies transmit diarrheal diseases (WHO, 2006). Other public health pests include fleas (plague, Bartonella, rickettsioses), ticks (Lyme disease, ehrlichiosis, babesiosis, and anaplasmosis), lices (Bartonella, rickettsioses), cockroaches, bedbugs, and triatomine bugs (trypanosomiasis, Chagas disease).

In addition to an increase of "airport malaria" (Europe and USA, Guillet et al., 1998; Tatem et al., 2006), some tropical vector-borne diseases have recently been observed in developed countries: chikungunya (Italy, La Réunion) or West Nile virus (Europe, USA). It is often assumed that this expansion of tropical vector-diseases could reflect the influence of climate change on vector range. However, if climate (temperature, rainfall, and humidity) does influence disease transmission, expansion of disease range is mostly due to human factors such as forest clearing, increase travelling, and transport activities. Overall, it seems that the main determinants of vector-borne disease prevalence are socioeconomic (see Reiter, 2001; Kay and Vu, 2005; Ooi et al., 2006; Morrison et al., 2008). They disproportionately affect poor and underserved populations living in tropical and subtropical regions. For example, dengue vector is present in both Mexico and Texas, which have a similar climate but because of distinct human factors (air conditioning, layout of cities, building structures), dengue is frequent in Mexico but almost absent in Texas (Reiter, 2001). Unfortunately, the burden that vector-borne diseases impose directly impairs the public health and socioeconomic development of many of the poor areas. Controlling these diseases is thus a necessity. This ideally entails active case detection and treatment of human infections (vaccines, antiparasitic drugs). However, few vaccines are currently available (e.g., yellow fever, Japanese encephalitis) and many pathogens are now resistant to antiparasitic drugs. Moreover, populations from endemic countries struggle to get access to them due to economic impediments. Thus, in many instances, the control of vectors is the only affordable measure.

The first documented attempts to control malaria by limiting the densities of vectors go back to the Roman times: in an attempt to control the "Roman fever" (the name of malaria at that time), Julius Caesar himself had the Codetan swamp around Rome drained and planted with trees (Varro approx. 40 BC, in Cheesman, 1964; Kelly, 2009). Environmental modifications aimed at reducing the number of breeding sites have shown great success: for example, the construction of the Panama Canal was possible only after US Army Surgeon General William Gorgas stopped yellow fever transmission among workers by eliminating *Ae. aegypti* breeding sites (in Morrison et al., 2008). However, today the most common and affordable way of fighting the major disease vectors is the use of insecticides (Roberts and Andre, 1994; Hemingway and Ranson, 2000; Beier et al., 2008). Many scientific investigations and reports show that the use of synthetic insecticides can dramatically reduce the risk of insect-borne diseases. These approaches, combined with extensive use of drugs, have rapidly led to the eradication of many diseases (e.g., malaria) from most nontropical areas of the world, but in spite of initial successes, eradication has proven more elusive in the tropics (Dialynas et al., 2009). However, mechanisms allowing survival to insecticide exposures have been selected in many species of arthropod vectors. Resistance to all classes of synthetic insecticides is now widespread among pests of public health importance, and it is considered to be the most important impediment in the successful control of vector-borne diseases.

The general aim of this chapter is to provide a global overview of insecticide resistance mechanisms, their evolution in disease vectors, and to explore some antivector strategies. The first part will describe various aspects of insecticide resistance from an historical point of view. The second part will describe the different mechanisms and genetic modifications leading to resistance, and their evolution for different insecticides in various species. In the third part, the different strategies implemented to prevent vector-borne diseases through insecticide treatment will be described, to show how resistance is taken into account.

14.2 Part 1: Insecticide Resistance: Definition and History

Insecticide resistance in pest populations affects both economy and public health on a worlwide scale: it decreases crop yields (and thus profitability), induces the need to increase the quantity of insecticide and to develop new insecticides (thereby having a strong impact on costs and on the environment), and is responsible for higher incidence of human or animal diseases (Georghiou and Lagunes-Tejeda, 1991; Whalon et al., 2008). This general society problem, however, provides evolutionary biologists with a unique contemporary model that is ideal for studying how new adaptations evolve by natural selection. The selecting agent is known (insecticides), evolution is recent and rapid (few years after insecticide selection), and the biological and genetic mechanisms are often known (see Part 2). This explains why it has been the subject of such a large body of work over the years.

Resistance is defined as a heritable decrease of the susceptibility to an insecticide (Nauen, 2007). Three categories of resistance can be distinguished: behavioral (avoidance of contact with insecticide), physiological (e.g., increased cuticule thickness), and biochemical (enhanced insecticide detoxification and/or decreased insecticide target sensitivity). Only a few examples of behavioral and physiological resistances have been reported because they provide weak protection (e.g., hornflies and Anophelines, Roberts and Andre, 1994; *Triatoma infestans*, Pedrini et al., 2009), while biochemical resistances typically result in relatively high levels of protection. Resistance arises from the selection of individuals able to survive and reproduce in presence of insecticide. Resistant individuals carry one or several genetic mutations that prevent insecticide disruption of the target functionning. As a result, the frequency of resistance gene(s) increases in the population over time. Insecticide resistance is confirmed by toxicological tests (bioassays) establishing resistance ratio (or RR, corresponding to the number by which an insecticide dose must be multiplied in order to obtain the same mortality in resistant than in susceptible insects). It can be investigated at many levels, from the molecular characterization of genes conferring resistance and their biochemical products, to the effect of these genes on the fitness (i.e., reproductive capacity) of the individuals carrying resistance alleles, to the dynamics and evolution of these resistance alleles in natural vector populations and their effect on pests and disease control.

The first recorded attempt of insect pest control is found in the litterature of the eighteenth century with the application of tobacco juice against sheep scabs (Lisle, 1757, in Wood, 1981). The first case of resistance was reported in 1908, in a population of San Jose scale (*Aspidiotus perniciosus*) resistant to lime sulfur (Melander, 1914). A century later (2007), 553 arthropod species are reported as resistant to at least one insecticide, among which are many disease vectors such as flies, mosquitoes, lice, bedbugs, triatomines, fleas, cockroaches, and ticks. Some species can be resistant to a large array of compounds: *Tetranychus urticae* (a nonvector Acari that is a pest for crops) has been found resistant to 80 different compounds. More than 100 mosquito species are resistant to at least one insecticide (56 Anopheline species, 39 Culicine species); *Cx. pipiens pipiens* and *An. albimanus* are resistant to more than 30 different compounds (Whalon et al., 2008).

14.2.1 Synthetic Insecticides

Originally only inorganic insecticides (such as lime sulfur) and natural products were available, for example, flower-extracted pyrethrum for malaria control in the

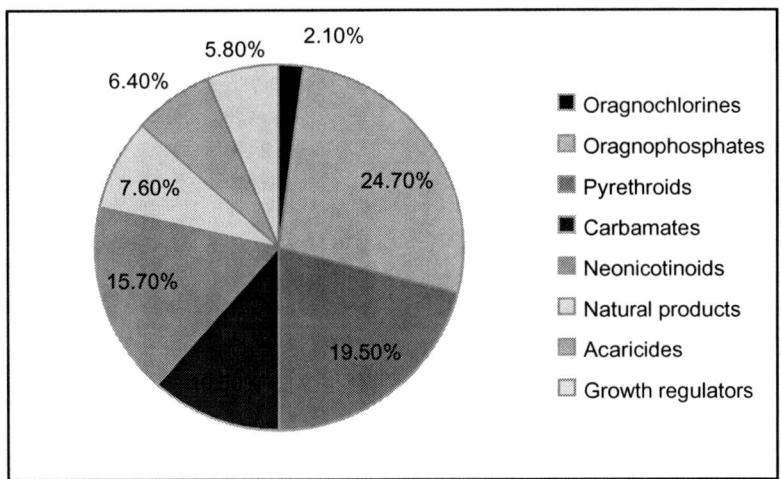

Figure 14.1 Main classes of insecticides and their respective world scale market share. *Source*: Modified from Nauen (2006).

1930s (in Brooke et al., 2001). Today, four classes of organic (synthetic) insecticides are essentially used: the organochlorines (OC), the organophosphates (OP), the carbamates (CX), and the pyrethroids (PYR) (Figure 14.1).

The first synthetic insecticides, introduced during World War II for malaria control, belonged to the OC class. The first one was the dichlorodiphenyltrichloroethane or DDT (introduced in 1943), which target the voltage-gated sodium-channels (Na-channels). The other one was cyclodiene dieldrin, which targets the γ-aminobutyric acid (GABA) receptor, both targets being essential in the insect nervous system (see Part 2). In addition to their public health applications, enormous tonnages of DDT and dieldrin were used worldwide in agriculture. It was at first a great success with large WHO-led campaigns leading to reduction of morbidity and mortality from malaria in many endemic regions after World War II. Widely acclaimed, DDT and dieldrin rapidly selected resistance in insect vectors. In *An. gambiae*, resistance to DDT was first noted 11 years after its introduction (WHO, 1957), while a population from northern Nigeria was reported resistant to dieldrin soon after (Davidson, 1956). DDT resistance has now been reported in mosquitoes (*Aedes* sp., *Anopheles* sp., and *Culex* sp.), houseflies, sand flies, body lice, and head lice, while resistance to dieldrin (60% of reported cases of resistance before 1990) has been detected in more than 277 arthropods, including mosquitoes (*Aedes* sp., *Anopheles* sp., and *Culex* sp.), fleas, ticks, biting flies, bedbugs, coackroaches, and human lice (ffrench-Constant et al., 1993, 2000; Roberts and Andre, 1994; Hemingway and Ranson, 2000; WHO, 2006).

An important issue with these insecticides was their environmental impact. Rachel Carson's book *Silent Spring* (1962) was a seminal work publicizing and politicizing the toxic effects of the accumulation of DDT and its metabolites in the food chain. For vertebrates, DDT can interfer with reproduction, and in humans it can have neurologic, carcinogenic, and reproductive effects, although the evidences remain debated. These insecticides are also extremely stable in the environment, contaminating groundwaters and remaining in soil samples long after their use. In the 1970s, the Persistent Organic Pollution Treaty led to total banning of dieldrin and to the banning of DDT for all uses except malaria control when this disease is very frequent and there is no alternative. DDT use rapidly declined in the 1970s (it is no longer used in Latin America; van den Berg, 2009), but it recently gained new advocates due to the development of resistance to the alternative insecticides, and to its low cost (Brooke et al., 2001; Rogan and Chen, 2005; WHO, 2006; Coleman et al., 2008; van den Berg, 2009).

Since the late 1970s, OCs were replaced by the PYR class for vector control, and these became widely used in agriculture and public health, and more particularly against malaria vector (in Hemingway et al., 2004). As DDT, these insecticides target the Na-channels (i.e., neurotoxic effect). Their rapid popularity comes from their very low toxicity to humans and their rapid knockdown effect (KD) (i.e., mosquitoes are rapidly incapacitated, which diminishes their potential of blood-feeding and thus of transmitting pathogens) associated with an excito-repellancy effect (i.e., most of the mosquitoes fly away from the treated material, which diminishes their transmission potential). Finally, PYRs have also cyto- and geno-toxic effects (i.e., they disturb the transmission of nervous influx, as well as the normal functioning of the cells). PYR-based indoor residual spraying (IRS) and insecticide-treated nets and curtains (ITNs) are currently advocated as standard malaria vector control strategies (WHO, 2006).

Unfortunately, PYR resistance was reported in 1993, in *An. gambiae* populations from Côte d'Ivoire (Elissa et al., 1993) and later in *Cx. p. quinquefasciatus* also in West Africa (Chandre et al., 1998). Resistance is now widespread in mosquitoes (*Aedes* sp., *Anopheles* sp., and *Culex* sp.,) (see Liu et al., 2006 for a review), body and head lice, ticks (e.g., *Boophilus microplus*), and fleas (Roberts and Andre, 1994; Hemingway and Ranson, 2000; Hernandez et al., 2002; Thomas et al., 2006; WHO, 2006). For example, resistance to PYRs in the *Cx. pipiens* complex is now found in Martinique, Cuba, USA, China, Saudi Arabia, Tanzania, West Africa, and Tunisia (Hardstone et al., 2007). As PYR resistance developed, many control programs attempted to revert to DDT for disease control. However, because these insecticide classes share a common target site, there is cross-resistance to both insecticide classes in many locations (Brooke et al., 2001; Coleman et al., 2008).

Finally, two other classes of synthetic insecticides are used on a large scale worldwide: the OP and the CX, which were first used in the 1940s and the 1950s, respectively (Nauen, 2006; WHO, 2006). OPs and CXs target the synaptic acetylcholinesterase (AChE), an essential enzyme in the nervous system (Massoulié and Bon, 1993). They are usually used as larvicids (although some are now considered for ITN impregnation, as an alternative to PYR, Kolaczinski et al., 2000; Sharp et al., 2007a; Oxborough et al., 2008), and are particularly well-suited for species with delimited breeding sites (most *Culex* and *Aedes* sp.). However, they have a short half-life and 2–3 rounds of IRS are needed per year. This, combined in some instances with their high price, can make these insecticides too costly for most malaria control programs, despite fewer reports of resistance (Coleman et al., 2008). Early resistance to these insecticides has been detected shortly after their first application: for example, first OP treatments in the Montpellier area (Southern France) started in 1969, the first resistance being detected only 3 years later (Pasteur and Sinègre, 1975). Resistance has now been recorded in mosquitoes (*Aedes* sp., *Anopheles* sp., and *Culex* sp.), biting flies (e.g., *Simulium damnosum*, vector of onchocerciasis), sand flies, houseflies, and fleas (reviews in Roberts and Andre, 1994; Hemingway and Ranson, 2000; Oakeshott et al., 2005; WHO, 2006).

Despite intense research to find insecticides with different modes of action, for agricultural use which is more profit-making, the molecules available today essentially belong to the four classes described earlier (OP, CX, PYR, and OC), for which resistance is now widespread. Some species are even resistant to most or all classes: for example, *An. arabiensis* is resistant to OPs, dieldrin, PYRs, and DDT in Sudan (Matambo et al., 2007; Abdalla et al., 2008), *Cx. p. quinquefaciatus* is resistant to DDT, dieldrin, OPs, and PYRs in Côte d'Ivoire and Burkina Faso (Chandre et al., 1998) and *Cx. pipiens* is resistant to DDT, OPs, CXs, and PYRs in China (Cui et al., 2006). A fifth class was discovered in the 1990s, the neonicotinoids that target the nicotinic acetylcholine (ACh) receptors in the central nervous system; however, they are used mostly for agricultural pests, not for disease vectors (Nauen, 2006; Whalon et al., 2008). Recently, phtaleic acid diamides or anthranilic acid diamides that target the ryanodine-sensitive intracellular Ca^{2+} release channels (that mediate many cellular and physiological activities) were described, but thus far they are also used only for agriculture (Nauen, 2006). Finally, another type of synthetic insecticides is growth regulators (GR). It regroups synthetic products called juvenoids that mimic the juvenile hormone (JH) (review in Hollingworth and Dong, 2008) and chitine inhibitors (see Hirose et al., 2010). By simulating a continuous high level of JH or limiting the chitine synthesis, they disrupt the insect development, particularly affecting the transitions between larval stages, the larva-to-pupa molt and the adult's emergence. So far, only few cases of resistance have been reported in houseflies and mosquitoes (e.g., resistance to methoprene, a JH analogue in the mosquito *Ochlerotatus nigromaculis*, Cornel et al., 2002).

In summary, most often only PYRs are available, essentially for economic cost reasons (Kelly-Hope et al., 2008): the most recent PYR has been introduced 30 years ago and no new synthetic insecticide has been found in the past 20 years (Kelly-Hope et al., 2008). The shrinking availability of insecticides as a result of resistance is exacerbated by the removal from the market of insecticides that are no longer registered for public health use: some compounds are too costly, and insecticide use is restricted by regulatory agencies due to environmental concerns. Consequently, new environment-proof products (highly specific, no effects on nontargets) are now required for sustainable vector control (Farenhorst et al., 2009). Some alternative means of control are emerging through GR and biological insecticides, but they represent only a small fraction of the insecticide market (Figure 14.1).

14.2.2 Alternative Insecticides

Environmental pollution concerns and unresolved issues pertaining to the toxicity of synthetic insecticides to humans and nontarget species have led to investigation of alternative "biological" insecticides (Mittal, 2003; Scholte et al., 2003). Three main types of these alternative insecticides are bacterial toxins, essential oils, and fungi.

There are two main sources of bacterial toxins: *Bacillus sphaericus* (Bs) and *Bacillus thuriengiensis* (Bt). They kill insect larvae by binding to various receptors on midgut epithelial cells (review in Hollingworth and Dong, 2008). Bs toxicity is due to a binary toxin, whereas Bt toxicity is due to the interaction of four different toxins. Both Bs and Bt induce a cytolysis, although the involved mechanisms are different. Despite being called "biological," the usual form of these insecticides is a purified extract of the toxins. These larvicides are presented as highly specific and effective at low doses, and are thus expected to be safe for the environment. Toxins extracted from Bs and a variety of Bt (*Bacillus thuringiensis* var *israelensis* or Bti) are used for mosquito control. In these species, bacterial toxins show some differences in specificity: Bti is more effective against *Aedes* and *Culex* species than against *Anopheles*, whereas Bs is more effective against *Culex* than *Anopheles* species, and has no effect on *Aedes* sp. that lack receptors. While the presence of several toxins was expected to delay resistance apparition, Bs and Bti resistances have been detected in various mosquitoes (Nielsen-Leroux et al., 1997; Mittal, 2003; Paul et al., 2005) and resistance to Bt has been detected in several agricultural pests (Tabashnik et al., 1997a; Gahan et al., 2001; Griffitts et al., 2001).

Although less documented, essential oils are investigated as potential biological larvicides. They are advocated to be more specific than synthetic insecticides, and biodegradable, thus with reduced impact on the environment. Three essential oils (*Satureja hortensis*, *Thymus satureoides*, *Thymus vulgaris*) have showed relatively good efficacy at low dose against *Cx. p. quinquefasciatus* larvae: they appear to increase larval mortality and decrease adults' emergence and oviposition; they also show a repellent effect (review in Pavela, 2009).

Finally, fungi can be used as biological insecticides: they target the adult stage of mosquitoes and are used essentially for malaria control. The fungus *Metarhizium anisopliae* has been shown to reduce *An. gambiae* adult lifespan in the laboratory and in the field in Tanzania (Scholte et al., 2003, 2005), while *Beauveria bassiana* decreases the survival of another malaria vector, *An. stephensi* (Blanford et al., 2005). These agents have several advantages: they are cheap, easily stored for the long term, and specific to insects (thus innocuous to humans). They kill their host later than other insecticides, but before the time required by the malaria agents, *Plasmodium* sp., to reach the infectious stage. Moreover, experiments have shown that they can decrease the female mosquito blood meal size, its feeding propensity, and its fecundity (Scholte et al., 2006). Consequently these fungal insecticides have a direct effect on *Plasmodium* transmission and are expected to decrease malaria prevalence. Finally, their acting late in life is considered by several authors to be an important advantage, as it will decrease selective pressure and reduce the risk of resistance development (potentially "evolution-proof" insecticides; Farenhorst et al., 2009; Michalakis and Renaud, 2009; Read et al., 2009).

It should be noted that all treated species do not develop insecticide resistance. This can be linked to the particular life cycle of the species or to molecular constraints preventing the evolution of resistance mechanism. For example, after decades of treatment, the tsetse flies (*Glossina* sp.) have not yet developed resistance to DDT or PYRs, probably due to their very small number of young, which limits their evolutionary reactivity (Hemingway and Ranson, 2000; Welburn et al., 2006; WHO, 2006). Similarly, *Ae. aegypti* did not develop the most efficient resistance mechanism to OPs and CXs because its particular codon usage prevents the apparition of the required mutation (Weill et al., 2004, 2005). This last example shows that understanding why resistance occurs or not also requires elucidating the mechanisms of insecticide resistance at the molecular and biochemical levels.

14.3 Part 2: Mechanisms of Resistance

The targets of most insecticides are critical molecules of the insect nervous system. Insecticides bind to specific sites on their target and disrupt its function. Any mechanism that decreases the insecticide effect will lead to resistance. This encompasses reduced penetration of the insecticide, increased excretion or sequestration of the insecticide, increased metabolism of the insecticide, and target modification that limits the binding of the insecticide.

The three first mechanisms are poorly documented and do not seem to play a prominent role in resistance. Reduced uptake of PYRs through the cuticle has been observed in the cockroach, inducing 2−3 times resistance (i.e., the PYR dose needed to be multiply 2−3 times to achieve the same mortality as in susceptibles). Similarly, reduced uptake through cuticle (9−10 times) has also been identified in the housefly (Scott, 1999). Transporters (efflux pumps) that excrete the insecticide have been identified in *D. melanogaster*, but they have only a limited effect on resistance (Hollingworth and Dong, 2008). Finally, sequestration of the insecticide has been described as playing a role in resistance (see Section Carboxyl-esterases), although the extent of this role is poorly known.

Most studies aiming at understanding the mechanisms and the genetic bases of insecticide resistance focus on metabolic resistance and target site modification. Increased rates of insecticide detoxification and reduced sensitivity of insecticide targets may be due either to point mutations in structural genes, to gene amplification (i.e., increased number of gene copies leading to an increased number of produced proteins) or to transcriptional regulation (i.e., increased production of transcripts and then of proteins). Usually these resistances are explained by a limited number of mechanisms, which usually are monogenic.

In this chapter, we will present the various documented mechanisms of resistance. We will specifically focus on disease vector species, although many mechanisms are common to agricultural pests. We will insist on the evolutionary aspects of resistance, while the detailed mechanisms will be treated more succintly. More comprehensive reviews can be found (e.g., Hemingway et al., 2004; Liu et al., 2006; Hollingworth and Dong, 2008).

14.3.1 Metabolic Resistance

Metabolic resistance regroups the various defense mechanisms that degrade the insecticide in less or nontoxic products, thus decreasing the quantity of toxic molecules that reach the target. Three major families of enzymes are involved in this type of resistance, although recent genomic studies have suggested that other types of enzymes may be implicated (Oakeshott et al., 2003; Vontas et al., 2005; Waterhouse et al., 2008; Awolola et al., 2009).

Glutathione S-transferases

Glutathione S-transferases (GSTs) catalyze the reaction of the sulfydryl group of the tripeptide glutathione of various xenobiotics. This sulfydryl group reacts with electrophilic sites on xenobiotics, leading to formation of conjugates that are more readily excreted and typically less toxic than the parent insecticide. In addition to this direct detoxication, GSTs catalyze the secondary metabolism of compounds oxidized by other enzymes (see later).

GST enzymes are present in most insects. They represent a large family of generalists detoxifying enzymes (six classes of GSTs have been identified in the genome of *An. gambiae*) and have thus broad substrate specificities. The GST family expands either by alternative splicing or by local gene duplication, the last leading to clusters of GST genes. Quantitative genetics analyses identified a quantitative trait locus (QTL) for resistance to DDT in *An. gambiae*, within which there is a cluster of eight GSTs (Ortelli et al., 2003).

However, GSTs usually provide limited levels of resistance ($\sim 10\times$) and are primarily associated with resistance to OCs, particularly DDT, and OPs (Table 14.1). They are also suspected to play a role in the resistance to PYRs, although this issue is still debated.

GST-based resistance seems to be associated with an increased amount of enzyme resulting from gene duplication or, more often, increased transcription rates. For example, a constitutive GST over-expression has been found in resistant strains of *An. gambiae* (Hemingway and Ranson, 2000). GST enzymes display specific activity: (e.g., the enzyme GST2-2 shows a specific DDT dechlorinase activity in *An. gambiae*) (Ortelli et al., 2003; Enayati et al., 2005). Finally, GST may have an indirect role in resistance to PYRs by detoxifying the lipid peroxidation products induced by this class of insecticides (i.e., an antioxidant effect such as in the brown planthopper *Nilaparvata lugens*) (Vontas et al., 2001).

Multifunctional Monooxygenases

The multifunctional monooxygenases (MFOs) catalyzed by the cytochrome P450 enzymes (the terminal oxidases of the system) are found in all aerobic organisms.

Table 14.1 Metabolic Resistance Mechanisms

Gene Family	Insecticides	Species
MFO	OP	M. domestica (Scott, 1999), D. melanogaster (Daborn et al., 2007), An. gambiae (Muller et al., 2008), Cx. pipiens (Hardstone et al., 2007), Ae. aegypti (Marcombe et al., 2009)
	CX	C. pipiens (Shrivast et al., 1970)
	OC	D. melanogaster, D. simulans (Daborn et al., 2007)
	PYR	An. stephensi, An. albopictus, An. gambiae (Chandre et al., 1998; Hemingway and Ranson, 2000), Cx. p. quinquefasciatus (Kasai et al., 1998), Cx. p. pallens (Shen et al., 2003), M. domestica, B. germanica (Scott, 1999)
	Neonicotinoid	D. melanogaster (Daborn et al., 2002)
COE	OP	Anopheles sp., M. domestica, L. cuprina (Claudianos et al., 1999; Hemingway and Ranson, 2000; Oakeshott et al., 2005; Perera et al., 2008), Cx. pipiens (Pasteur and Raymond, 1996), Cx. tarsalis (Prabhaker et al., 1987), Ae. aegypti (Brengues et al., 2003)
	Malathion (OP)	Anopheles sp., M. domestica, L. cuprina (Claudianos et al., 1999; Hemingway and Ranson, 2000; Oakeshott et al., 2005), Cx. tarsalis (Hemingway et al., 2004)
	PYR	Bo. microplus (Hernandez et al., 2002), Ae. aegypti (Brengues et al., 2003)
GST	OC	An. gambiae, An. arabiensis (Hemingway and Ranson, 2000; Hemingway et al., 2004), An. dirus (Enayati et al., 2005), An. subpictus (Perera et al., 2008), An. albimanus (Penilla et al., 2006), Ae. aegypti (Enayati et al., 2005)
	OP	An. subpictus (Perera et al., 2008), M. domestica (Scott, 1999)
	PYR	P. h. capitis (Thomas et al., 2006), Bl. germanica (Scott, 1999)

Note: Cx. pipiens is a species complex grouping Cx. p. pipiens, Cx. p. quinquefasciatus, and Cx. p. pallens. Based on COE (carboxyl-esterases), MFO (multifunctionnal oxidases), and GST (glutathione S-transferases) for resistance to organophosphates (OP), organochlorides (OC), pyrethroids (PYR), and carbamates (CX) in various species.

Cytochrome P450 activates an oxygen atom, which is then inserted in a large variety of substrates. The oxidation products of MFOs are often unstable and breakdown further. These P450 MFOs can oxidize a large variety of substrates, and this defense mechanism is implicated in many reactions against a large array of xenobiotics, notably insecticides and toxins.

MFOs are the members of a very large family of enzymes in insects, with an average of about 100 genes in the different insect genomes analyzed so far. They display high constitutive levels of expression in strategic tissues (midgut, fat body, Malpighian tubules), and are also inducible by the presence of xenobiotics. They are probably the most frequent metabolic resistance mechanism, although the level of resistance conferred is often low: the resistance ratio is usually around 5.

Cytochrome P450 associated MFOs have been reported as responsible for resistance to PYRs, OPs, CXs, OCs (DDT and CDs), and neonicotinoid insecticides (Table 14.1), mostly in *Drosophila melanogaster*, with indications that different genes of the cytochrome P450 (CYP) MFO family induce resistance to different classes of insecticides (Scott, 1999; Brooke et al., 2001; Daborn et al., 2002, 2007; Hemingway et al., 2004; Feyereisen, 2005). The implication of MFOs in insecticide resistance is usually detected by bioassays, including a nonspecific inhibitor of some cytochrome P450s, the synergist piperonyl butoxide (PBO). The observation that toxicity is significantly more reduced in a resistant than in a susceptible strain exposed to the insecticide associated with PBO is indicating a role of MFOs. Nevertheless the large diversity of MFOs and the less than perfect specificity of PBO impose further evidences to verify that MFOs have a role in the observed resistance, for example gene silencing.

Resistance is generally associated with the over-expression of one or several P450 associated MFOs. This over-expression usually results of enhanced transcription rather than gene amplification, although the large diversity of MFOs makes it difficult to pinpoint the exact mechanisms at the origin of the observed resistance. The responsible genes are consequently rarely identified directly, although a certain specificity is reported: recent transcriptome analyses have shown that particular over-expressed CYP genes may have a direct role for resistance to some insecticides (e.g., *Cyp6g1* for resistance to DDT and *Cyp6g2* for resistance to diazon [OP] in *D. melanogaster*) but the results are often conflicting (Daborn et al., 2007). Similarly, using microarray and expression in *Escherichia coli*, the *Cyp6p3* gene has been shown to be involved in the resistance metabolism of permethrin and deltamethrin (PYRs) in *An. gambiae* (Muller et al., 2008). Interestingly, when a link between insectide resistance and particular CYP genes is suggested, most of the time these genes belong to the *Cyp6* family (Hemingway et al., 2004). No common mutation has been identified, but increased transcription might be due to the loss of function in some transacting suppressors, like in *D. melanogaster*, the insertion of a retrotransposon in the regulatory area sequence (absent in susceptible but present in all resistant individuals) appears to be responsible for over-transcription of the *Cyp6g1* gene (Daborn et al., 2002). However, despite some knowledge of the genes that are over-expressed, the molecular basis for this over-expression and the final proof of their implication in resistance is most of the time lacking (Hollingworth and Dong, 2008; but see Muller et al., 2008).

Finally, MFOs can play a role in activating certain insecticides. This is notably the case with OPs like diazon or malathion; in this case, resistance may be achieved by down-regulating the expression of a particular MFO.

Carboxyl-esterases

More than 30 genes coding carboxyl-esterases (COE) are found in insects (see detailed review in Oakeshott et al., 2005; Hollingworth and Dong, 2008). Most COEs are serine esterases (i.e., they have a serine residue within a catalytic triad necessary for hydrolysis). COEs bind to an ester group and then break the ester bond by a process of acylation-deacylation. Multiple forms of COEs are found in insects, with broad and overlapping substrate specificities.

The majority of insecticides, including almost all CXs and OPs, most PYRs, and some GRs are esters: in most cases, hydrolysis of the ester group leads to reduced toxicity of the insecticide. Consequently, COEs are often implicated in metabolic mechanisms leading to resistance, although the level of resistance conferred is often relatively low ($\sim 10 \times$). As for MFOs, the role of COEs in resistance is usually diagnosed by the addition of a synergist, the S,S,S-tributyl phosphorotrithioate or DEF to bioassays. DEF is an inhibitor of the COEs (but also inhibits GSTs): if COEs are responsible of resistance, toxicity is expected to undergo a significantly higher decrease in resistant than in susceptible insects in presence of DEF (Pasteur et al., 1984). COEs have been detected in various species (Table 14.1) as a mechanism of resistance to OPs and to a lesser extent to PYRs (Hemingway and Ranson, 2000; Nauen, 2007).

Resistance in mosquitoes of the *Culex* genus is generally caused by an elevated COE protein quantity, up to 80 times the level found in susceptible individuals (Mouchès et al., 1987). This over-expression is usually caused by upregulation or by an increased gene number (amplification) coding for one or two different esterases, named A and B (Mouchès et al., 1986; Rooker et al., 1996; Guillemaud et al., 1997; Vaughan et al., 1997; Callaghan et al., 1998; Paton et al., 2000). The loci coding for the esterases A and B (*Est-3* and *Est-2*, respectively) are very close (e.g., in the mosquito *Cx. pipiens*, less than 1% recombination, 2–6 kb; Pasteur et al., 1981a) and behave as a single locus named *Ester* (Wirth et al., 1990; Tomita et al., 1996; Guillemaud et al., 1997; Labbé et al., 2005). The number of genes within an amplification of the *Ester* locus can vary greatly, potentially in relation with the intensity of insecticide treatments (Pasteur et al., 1984; Guillemaud et al., 1999; Weill et al., 2000a).

However, qualitative changes may also be responsible for COE-mediated resistance to particular insecticides. For example, resistances specific to malathion (OP) in Anophelinae, *M. domestica*, and *Lucilia cuprina* are due to a particular point mutation that induces a faster hydrolysis, with no elevation of the esterase quantity (McKenzie et al., 1992; Claudianos et al., 1999; Hemingway and Ranson, 2000; Hemingway et al., 2004; Oakeshott et al., 2005).

Because over-expressed COEs can represent a large percentage of the total protein of the insect (up to 12% of the soluble proteins in some resistant mosquitoes, Fournier et al., 1987), it is difficult to disentangle their sequestration effect (i.e., when the esterase is bound to the insecticide, the insecticide cannot reach its target) from the direct hydrolysis of the insecticide. It depends on the species and the esterase allele: hydrolysis is major in aphid E4 esterase, while in mosquitoes the $Ester^{B1}$ and $Ester^2$ allele rapidly sequesters the insecticide and then degrades it slowly (Cuany et al., 1993; Feyereisen, 1995; Karunaratne et al., 1995). This large over-expression can also affect pathogen transmission: in the mosquito *Cx. p. quinquefasciatus*, over-expressed COE have been shown to have a negative effect on the parasite burden of the filarial worm *Wucheria bancrofti* (McCarroll et al., 2000; McCarroll and Hemingway, 2002).

COE resistance to OPs in *Cx. pipiens* is probably one of the best-studied cases. In this mosquito, several *Ester* alleles confer OP resistance: the *Ester^1* allele results in overproduction by transcription upregulation of the *Est-3* gene (esterase A), whereas *Ester^{B1}* results from a *Est-2* (esterase B) gene amplification, and *Ester^2* and *Ester^4* from the coamplification of both A and B genes (Mouchès et al., 1986; Rooker et al., 1996; Guillemaud et al., 1997; Weill et al., 2000a). *Cx. pipiens'* resistance to OPs is monitored since their first application (1969) in Montpellier area, Southern France (Pasteur and Sinègre, 1975; Pasteur et al., 1981b; Chevillon et al., 1995; Guillemaud et al., 1998; Raymond et al., 2001; Labbé et al., 2009). More than 40 years of monitoring showed that several *Ester* resistance alleles have been replacing each other: *Ester^1* was the first detected allele in 1972, then *Ester^4* in 1986, and finally *Ester^2* arrived by migration in 1991. These alleles are selected in insecticide-treated areas (i.e., a selective advantage) as they survive better in this environment, but they are costly as they confer a fitness disadvantage (lower mating success, lower survival, etc., Berticat et al., 2002a,b, 2004; Bourguet et al., 2004; Duron et al., 2006b) in absence of treatment, and are thus selected against in nontreated areas (Lenormand et al., 1998b, 1999). Their frequencies follow a clinal shape, which has been recently used to quantify the fitness cost and advantage of the three alleles (Labbé et al., 2009). It has been shown that *Ester^4* was favored over *Ester^1* because of a lower cost (selection for a generalist allele), while *Ester^2* is replacing the two first alleles thanks to its higher resistance in the current treatment practices, despite its relatively high cost (selection for a specialist allele).

The *Cx. pipiens* example shows that resistance is an evolutive dynamic process, as new mutations can appear that improve the previous adaptation. Several metabolic mechanisms can be present in the same species for resistance to the same class of insecticides, as for example in *Ae. aegypti* (Brengues et al., 2003; Strode et al., 2008) or in *An. gambiae* from Cameroon (Etang et al., 2007) where MFOs, GSTs, and COEs contribute to the observed resistance to DDT and PYRs. However, these metabolic resistance mechanisms most often confer relatively low resistance levels, particularly when compared to target site modifications.

14.3.2 Target Site Modification

Resistance by target site modification is due to point mutations in the insecticide target that limits insecticide binding, rather than to a change in expression level. Because most insecticide targets are vital molecules, there is generally only a limited number of mutations in the target able to decrease insecticide affinity without impeding its original function to an unsustainable degree (see detailed review in Hollingworth and Dong, 2008). A mutation conferring resistance while partly impairing the target's normal function leads to a fitness cost.

GABA Receptors

GABA is a major neurotransmitter in the insect central and peripheral nervous system and in neuromuscular junctions. The GABA receptors are linked to Cl^--gated channels, causing hyperpolarization that blocks the nervous influx. GABA receptors are the target of cyclodienes (CD). CDs are noncompetitive inhibitors that bind to a site on the receptor close to the Cl^--gated channel, stabilizing it in an inactive closed state. This induces an overexcitation by removal of the inhibition, and leads to convulsions and death of the insect. GABA receptors have also secondary binding sites for some PYRs or insecticides of the avermectin family (Hemingway and Ranson, 2000).

Resistance to CDs is due to a decreased sensibility to insecticide of the GABA receptor A, through the mutation of a single amino acid in the receptor-coding gene. This gene, called *Rdl* (resistance to dieldrin, the most used CD), has been first cloned in *D. melanogaster* (ffrench-Constant et al., 1993). It is composed of a 2-kb open reading frame encoding one of the GABA receptor subunits. In all *D. melanogaster* resistant individuals, the *Rdl* locus displays a similar mutation at postion 302 in the channel-lining domain sequence, changing an alanine into a serine (A302S). The role of this mutation in CD resistance was confirmed by directed mutagenesis. The serine residue occupies the insecticide-binding site of the GABA receptor and destabilizes its conformation, which in turn increases the opening time of the Cl^--gated channel, and thus limits the nervous influx transmission (review in ffrench-Constant et al., 2000). The resistance allele (Rdl^R) is semidominant and can confer some cross-resistance (e.g., between dieldrin and fipronil, ffrench-Constant et al., 2000; Hemingway et al., 2004). Interestingly, the *Rdl* gene is duplicated in the greenbug *Myzus persicae* (Anthony et al., 1998).

Due to an extensive use of CDs before their banning in the 1980s, resistance has been selected in several insect species (Table 14.2), which all display a mutation at the same position (A302S or A302G) (ffrench-Constant et al., 1993; Hemingway et al., 2004). For example, *An. gambiae* (A302G) became resistant to dieldrin everywhere in Africa in the 1950s and 1960s (Chandre et al., 1999a). Whether these mutations are costly depends on species: a fitness cost associated with resistance has been shown both in the laboratory and in the field for *L. cuprina* (McKenzie, 1996) and has been recently suggested for *Cx. pipiens* and *Ae. albopictus* (Tantely et al., 2010), but no cost has been found for *D. melanogaster* (ffrench-Constant et al., 2000).

Voltage-Dependant Sodium Channels

Nerve action potentials are transmitted by a wave of depolarization along the neural axone. It is due to the movement of sodium ions (Na^+) across the axonal membrane through the opening of voltage-dependent sodium channels (Na-channels), and stops when they are inactivated. Na-channels are glycoproteins with a pore for ion transport and can adopt three different states: resting, open, or inactivated; the Na^+ ions pass only when they are open (Lund, 1984).

Na-channels are the targets of DDT and PYRs. When these insecticides bind to the Na-channels, they slow their closing speed, prolonging the depolarization (Lund 1984; Vais et al., 2001; Soderlund and Knipple, 2003). The intensity of the effect is dose-dependent, proportional to the number of Na-channels inactivated

Table 14.2 Target Site Modification Resistance Mechanisms. Mutations of Insecticide Targets Are Presented in Various Species

Target	Mutation		Species
GABA receptor	Rdl^R	A302S	*D. melanogaster* (ffrench-Constant et al., 1993), *D. simulans, M. domestica, Bl. germanica, L. cuprina, Ae. aegypti* (ffrench-Constant et al., 2000), *An. gambiae* (Davidson, 1956) *An. stephensi, An. arabiensis* (Du et al., 2005), *Cx. p. quinquefasciatus* and *Ae. albopictus* (Tantely et al., 2010)
		A302G	*D. simulans* (ffrench-Constant et al., 2000), *An. gambiae* (Du et al., 2005)
Na-channels	*kdr*	L1014F	*M. domestica* (Williamson et al., 1996), *Bl. germanica* (Hollingworth and Dong, 2008), *An. gambiae* (Martinez-Torres et al., 1998), *An. stephensi* (Enayati et al., 2003), *An. arabiensis* (Diabate et al., 2004), *An. sacharovi* (Luleyap et al., 2002), *An. subpictus* (Perera et al., 2008), *An. culicifacies* (Singh et al., 2009), *An. arabiensis* (Stump et al., 2004), *Cx. pipiens* (Martinez-Torres et al., 1999; Xu et al., 2006), *H. irritans* (Soderlund and Knipple, 2003)
		L1014S	*Cx. pipiens* (Martinez-Torres et al., 1999), *An. arabiensis* (Ranson et al., 2000), *An. sacharovi* (Luleyap et al., 2002)
		Others	*P. h. capititis* (Thomas et al., 2006), *D. melanogaster, Bo. microplus* (Soderlund and Knipple, 2003), *Ae. aegypti* (Brengues et al., 2003; Saavedra-Rodriguez et al., 2007)
	super-*kdr*	M918T	*M. domestica* (Williamson et al., 1996), *H. irritans* (Soderlund and Knipple, 2003)
		Others	*Bl. germanica* (Soderlund and Knipple, 2003)
	?		*Bo. microplus* (Hernandez et al., 2002)
AchE	*ace-1^R*	G119S	*An. albimanus, An. nigerimus, An. atroparvus* (Hemingway et al., 2004), *An. subpictus* (Perera et al., 2008), *An. gambiae, Cx. pipiens* (Weill et al., 2003), *Cx. vishnui* (Alout et al., 2007)
		F331W	*Cx. tritaeniorhynchus* (Nabeshima et al., 2004; Alout et al., 2007)
		F290V	*Cx. pipiens* (Alout et al., 2007, 2009)
	ace-1^D	G119S	*Cx. pipiens* (Lenormand et al., 1998a; Labbé et al., 2007a), *An. gambiae* (Djogbénou et al., 2009)
		F290V	*Cx. pipiens* (Alout et al., 2009)
	ace-2		*D. melanogater, M. domestica* (Fournier and Mutéro, 1994)

Note: Cx. pipiens is a species complex grouping *Cx. p. pipiens, Cx. p. quinquefasciatus,* and *Cx. p. pallens.*

(Lund, 1984). For PYRs, the magnitude of the effect depends on the type of insecticide molecules, type I (e.g., permethrin) or type II (e.g., lambda-cyhalothrin and deltamethrin), which respectively lack or not a cyano group. During action potential, type II PYRs prolong the sodium flux more than type I, and thus usually display a more intense effect (Vais et al., 2001). At the phenotypic level, Na-channels inactivation results in a rapid KD, the insect being incapacited for some time, with an eventual recovery or death depending on species and development stages (in mosquitoes, the adults tend to recover, while larvae will drown).

One major mechanism, named knockdown resistance (kdr), is responsible for PYR and DDT resistance, by reducing the receptors' sensitivity (binding capacity) to these insecticides and modifying the action potential of the channel (Soderlund and Knipple, 2003; Hollingworth and Dong, 2008). First discovered in *M. domestica*, this mechanism has been described in many agricultural pests and vectors (see Table 14.2). This resistance mechanism has several consequences: it decreases the irritant and the repellent effect, and either cancels or reduces the KD effect (Chandre et al., 2000).

By extension, the gene encoding the Na-channels has been called *kdr*. By sequencing the Na-channel protein (>2000 amino acids), the two first mutations conferring the kdr phenotype were identified in *M. domestica*, both in the protein second domain. The first one (L1014F) is associated with moderate (10−30×) PYR resistance; the second (M918T, also called *super-kdr*) is always associated with the L1014F to confer a higher resistance (up to 500×) (Williamson et al., 1996). Substitution of the L1014 is found in a large variety of species (L1014F or L1014S, Table 14.2, and also L1014H in *H. virescens*) and corresponds to the kdr^R alleles (Martinez-Torres et al., 1999; Vais et al., 2001; Soderlund and Knipple, 2003; Etang et al., 2009). The phenotype conferred by kdr^R is recessive or semirecessive (Chandre et al., 2000; Hemingway and Ranson, 2000; Corbel et al., 2004), with higher resistance to type I than type II PYRs (Chandre et al., 1999a). However, the various mutations show some specificity, as L1014F confers a high resistance to both DDT and permethrin (PYR), while L1014S confers a lower resistance to permethrin than to DDT (Martinez-Torres et al., 1999; Ranson et al., 2000). Some other mutations (about 30 in total) have also been described in various species, including super-kdr mutations (Vais et al., 2001; Soderlund and Knipple, 2003). Some of these mutations are conserved over a large array of organisms, while others are more specific and unique. In *Ae. aegypti*, the kdr phenotype has been observed, but it appears that a codon bias does not allow the appearance of any L1014 mutation (Brengues et al., 2003). Several mutations were observed to be associated with resistance, the V1016I and V1016G mutations being particularly interesting candidates in Latin America and the Carribbean (Saavedra-Rodriguez et al., 2007; Garcia et al., 2009; Kawada et al., 2009; Marcombe et al., 2009). However, the causal effect of these mutations remains to be confirmed.

The role of the L1014F/S mutations (kdr^R) as the sole cause of the kdr phenotype is still discussed (Brooke, 2008). The kdr^R is clearly associated with PYR and DDT resistance in *Bl. germanica*, *Cx. pipiens*, houseflies, hornflies, and some moths (review in Xu et al., 2006). In *An. gambiae*, although metabolic resistance is often present, high resistance to PYR and DDT is often associated with a high kdr^R frequency and resistant insects carry at least one kdr^R copy (Ranson et al., 2000; Awolola et al., 2007; Reimer et al., 2008; Dabiré et al., 2009; Ramphul et al., 2009). Moreover, kdr^R frequency usually increases when PYRs are used (Stump, 2004, but see Corbel et al., 2004). Similarly, in West African *Cx. p. quinquefasciatus*, resistance frequency follows a gradient of treatment intensity (Chandre et al., 1998). A recent study showed that a *Cx. p. quinquefasciatus* strain resistant to PYR contained 87% kdr^R heterozygotes, a surprising proportion that remained stable despite selection. It appears that alternative splicing and RNA editing could explain the discrepancies between kdr^R frequencies and PYR/DDT resistance level (Xu et al., 2006).

In the field, *An. gambiae* resistance to PYRs through *kdr* can lead to reduced repellent effect and decreased mortality. For example, kdr^R frequency is high in Benin and Côte d'Ivoire while no other PYR resistance mechanism is found: studies have shown strong diminution of vector control with PYR-treated bednets in these countries (Kolaczinski et al., 2000; N'Guessan et al., 2007). In contrast, other studies have found that despite the high correlation between *kdr* mutations and PYR resistance, PYR-treated bednets remained somewhat efficient against resistant *An. gambiae* (Casimiro et al., 2007; Brooke, 2008). This could be due to the ability of resistant mosquitoes to stay on a treated bednet longer than susceptibles, and thus absorb a high enough quantity of insecticide to be killed (Chandre et al., 2000). For example, in Kenya, the use of PYR-treated bednets increased kdr^R frequency, but had no impact on malaria and mosquito population densities, as both decreased in treated and untreated villages (Stump et al., 2004). Similarly, two studies found that *kdr* alone (i.e., in absence of metabolic resistance) did not reduce bednet efficiency against resistant *An. stephensi*, despite a reduced KD effect (Hodjati and Curtis, 1997; Enayati and Hemingway, 2006). The issue of the impact of *kdr* resistance on PYR-treated bednet efficiency to control malaria remains thus hotly debated.

Evolution of the kdr phenotype is best described in *An. gambiae* in Africa. The L1014F mutation was first detected in West Africa (Côte d'Ivoire, Burkina Faso) (Martinez-Torres et al., 1998), while only the L1014S mutation was observed soon after in East Africa (Kenya) (Ranson et al., 2000). These mutations appear to have spread from their center of origin: L1014F is now present everywhere in West and Central Africa, from Senegal to Angola (it is almost fixed in Côte d'Ivoire and Burkina Faso; Dabiré et al., 2009), while L1014S is absent from West Africa but present in Central and East Africa (Pinto et al., 2007; Santolamazza et al., 2008; Etang et al., 2009). However, analyses of the noncoding regions of the *kdr* gene suggest that the two alleles occurred several times independently (at least 3 times for L1014F and 2 for L1014S) (Weill et al., 2000b; Pinto et al., 2007; Etang et al., 2009).

Acetylcholinesterase

In the cholinergic synapses of invertebrate and vertebrate central nervous system, AChE terminates the synaptic transmission by rapidly hydrolyzing the neurotransmitter ACh. AChE is the target of OP and CX insecticides, which are competitive inhibitors of ACh with a low turnover: when they bind to AChE, their very low release prevents hydrolysis of the natural substrate. Consequently, ACh remains active in the synapse and the nervous influx is continued, leading to the insect death by tetany (see Massoulié and Bon, 1993).

In most insects there are two genes, *ace-1* and *ace-2*, coding for AChE1 and AChE2, respectively. In these species, AChE1 is the main synaptique enzyme, and the physiological role of AChE2 is unknown. Diptera of the Cyclorrapha group or "true" flies (such as *D. melanogaster* and *M. domestica*) possess a single AChE, which is encoded by the *ace-2* gene and is the synaptic enzyme in that case. Phylogenetic analyses have shown that the presence of two *ace* genes is probably the ancestral insect state (Weill et al., 2002; Huchard et al., 2006).

The first molecular studies on an insensitive AChE conferring resistance to OPs and CXs were carried out on *D. melanogaster*. Several mutations were identified, each giving a low resistance when alone, and a higher resistance when in combination (Fournier et al., 1989; Fournier and Mutéro, 1994). Similar results were later found with other Diptera that have only the *ace-2* gene (e.g., *M. domestica*, Oakeshott et al., 2005).

In mosquitoes where AChE1 is the synaptique enzyme, the most common resistance mutation (G119S) in the *ace-1* gene is situated near the catalytic site. In *Cx. pipiens*, G119S occurred at least 3 times independently, once in *Cx. p. pipiens* and twice in *Cx. p. quinquefasciatus* (Weill et al., 2003, 2004; Labbé et al., 2007a). However, two other mutations in *ace-1* have been identified, both close to the active site: (i) F331W has been observed only in *Cx. tritaeniorhynchus* (Nabeshima et al., 2004; Alout et al., 2007), (ii) F290V has been observed only *in Cx. p. pipiens* (Alout et al., 2009). The type of mutation is highly constrained by the codon use: the G119S mutation has never been found in *Ae. aegypti*, *Ae. albopictus*, or *Cx. tritaeniorhynchus*, probably because it requires two mutation steps (Weill et al., 2004).

The *ace* mutations are responsible for a decreased inhibition of the AChE by the insecticides (Alout et al., 2008). There are only few resistance mutations observed in various species (see Table 14.2), suggesting high constraints: those observed in the field are within the active gorge of the enzyme and cause steric problems with bulkier side-chains, while other substitutions (lab-engineered) often result in the inability of enzyme to degrade ACh (Oakeshott et al., 2005). In mosquitoes, these mutations confer a high resistance to OPs and CXs, respectively up to 100 times (e.g., chlorpyrifos) and $>9000\times$ (e.g., propoxur); OP resistance conferred by *ace* alleles is usually higher than COE metabolic resistance (Raymond et al., 1986; Poirié et al., 1992; Severini et al., 1993).

The evolution of insensitive AChE1 has been studied in depth in the mosquitoes *Cx. pipiens* and *An. gambiae*. In *Cx. pipiens*, it was first detected in Southern France in 1978, 9 years after the beginning of OP treatments (Raymond et al., 1986). The gene coding for this G119S mutated AChE1 ($ace-1^R$) rapidly spread in treated natural populations. However, its frequency remained low in adjacent untreated areas connected by migration, indicating a fitness cost associated with $ace-1^R$ (Lenormand et al., 1999). The $>60\%$ reduction of AChE1 activity in G119S resistant mosquitoes (Bourguet et al., 1997) may explain, at least partially, this cost, which is expressed phenotypically through various developmental and behavioral problems in individuals carrying $ace\text{-}1^R$ (Berticat et al., 2002a, 2004; Bourguet et al., 2004; Duron et al., 2006b). Similarly, the F290V mutation is probably associated with a fitness cost, although it does not appear to be due to activity reduction (Alout et al., 2009). Recently, several independant duplications of the $ace\text{-}1$ gene, putting a susceptible and a resistant copy in tandem ($ace\text{-}1^D$), have been identified in *Cx. p. pipiens* and *Cx. p. quinquefasciatus* (Table 14.2; Lenormand et al., 1998a; Labbé et al., 2007a). These alleles are thought to be selected because they reduce the cost of the $ace\text{-}1^R$ allele, although not always successfully (Labbé et al., 2007b). Several other duplications have been observed recently in the Mediterranean area, with a F290V copy instead of a G119S copy (Alout et al., 2009). In *An. gambiae*, the recent occurrence of $ace\text{-}1^R$ has been detected in West Africa, probably spreading from Côte d'Ivoire to Benin (Weill et al., 2003; Djogbénou et al., 2008). A duplication carrying a G119S copy has also been found, and appears to follow the same trajectory as in *Cx. pipiens* (Djogbénou et al., 2009).

14.3.3 Other Resistances

Growth Regulators

Juvenoids mimic JH and disrupt the target insect development. Few resistances have been detected but they have been reported in various species (review in Hollingworth and Dong, 2008). High resistance to methoprene has been described in the mosquito *Ochlerotatus nigromaculis* in California, potentially through target site mutation (Cornel et al., 2002), while a 7.7× resistance to the same insecticide has been reported in *Cx. p. pipiens* from New York (Paul et al., 2005).

Toxin Receptors

Bt toxins have a complex mode of action not clearly understood. Moreover, Bt resistance is rare in the field (Griffitts et al., 2001; Mittal, 2003). It has been mainly studied in the agricultural pest moth *Plutella xylostella*: in a laboratory strain from Hawaii, a single mutation confers resistance to at least four toxins by decreased binding on a common receptor, but it is not the only responsible, as strains from various places show complementation, which indicates the epistasis between several genes (Tabashnik et al., 1997a,b). In another pest moth, *Heliothis virescens*, QTL mapping showed that Bt resistance in a laboratory strain is probably due to a modification of the cadherin gene BtR-4 (Gahan et al., 2001). Presently, the only report of field resistance for a vector is a 33× resistance to Bti (Bt var. *israelensis*, the only Bt variety active on mosquitoes) detected in a natural population of *Cx. p. pipiens* from New York. However, the mechanism of this resistance is still unknown (Paul et al., 2005).

For Bs toxins, resistance has been described essentially in mosquitoes of the *Cx. pipiens* complex. It developed very rapidly within the first year of treatment in

India (10 to 155 × resistance; Mittal, 2003) and in Tunisia (*Sp-T* gene, >5000× resistance, Nielsen-Leroux et al., 2002). Similarly, control using Bs toxins started in the early 1990s in Southern France and first failure was reported in 1994 in Port Louis (near Marseille). This resistance (>10,000×) was due to a recessive sex-linked gene, named *sp-1*. In 1996, Bs resistance was reported close to the Spain border (Perpignan, 200 km away from Port St. Louis); it was due to a second gene, *sp-2*, which was recessive and sex-linked (Chevillon et al., 2001). Now Bs resistance has been observed worldwide in the *Cx. pipiens* complex (Mittal, 2003). Two of the genes identified (*sp-2^R* and a gene selected in a laboratory strain from California, Nielsen-Leroux et al., 1995) change the toxin receptor binding properties and were found to be due to "stop" mutations or mobile element insertion in the toxin receptor (Darboux et al., 2002, 2007), while the effect of the others is unknown (Nielsen-Leroux et al., 2002). Bs resistance has also been selected in the laboratory in *An. stephensi* (Mittal, 2003).

Other

Most studies focus on small sets of genes. Recently developed genomic and transcriptomic techniques allowed access to mechanisms that had previously proven intractable. They allowed the deeper description of known resistance gene families and helped to find new candidate genes (Daborn et al., 2002; Oakeshott et al., 2003; Vontas et al., 2005). For example in *An. funestus,* genomic positional cloning identified a major QTL including a cluster of 11 P450 MFOs for PYR resistance, two of them being over-expressed in a resistant strain (Wondji et al., 2009). Detox chips were designed (including P450 MFOs, GSTs, COEs, redox genes, and partners of P450 in oxidative metabolic complex) for *An. gambiae* and *Ae. aegypti* (David et al., 2005; Strode et al., 2008). In *An. gambiae*, several GST and MFO genes showed over-expression in mosquitoes resistant to PYR and DDT, but also other candidate genes (COEs, transferases, aldehyde dehydrogenase, NADH-cytochrome *b* reductase, NADH dehydrogenase, NADH-ubiquinone oxidoreductase, nitrilase thioredoxin peroxidase, and cuticular genes, Vontas et al., 2005; Awolola et al., 2009). Genomic analyses in *Ae. aegypti* show an expansion of certain gene families (e.g., MFO and COE) or increased alternative splicing (e.g., GST) (Strode et al., 2008), and identified new candidates for resistance (aldehyde oxidase and xanthine deshydrogenase, Waterhouse et al., 2008). However, in most cases the causal role of the candidates remains to be formally validated.

14.3.4 Resistance Generalities

Some general patterns can be identified from the variety of mechanisms observed for insecticide resistance.

A first characteristic is that resistance evolves rapidly, with fast selective sweeps in field populations. Although clear evidence is scarce, resistance often seems absent before insecticide treatments (Andreev et al., 1999), but there are some contradictory examples like malathion resistance in *L. cuprina* (Hartley et al., 2006) or PYR resistant *kdr* L1014S in Kenya in *An. gambiae* (Stump et al., 2004).

Second, resistance appears locally but can spread very rapidly (Brogdon and McAllister, 1998). A single resistance gene may have a large distribution (ffrench-Constant et al., 2000, 2004; Etang et al., 2009) (e.g., the worldwide migration of *Ester²* in *Cx. pipiens*) (Raymond et al., 2001). Alternatively, other resistance genes have multiple origins: *ace-1^R* mutations in *Cx. pipiens* (G119S, Weill et al., 2003; F290V, Alout et al., 2009) or *kdr* mutations in *Ae. aegypti* (Saavedra-Rodriguez et al., 2007; Garcia et al., 2009; Kawada et al., 2009; Marcombe et al., 2009).

It also seems that resistance evolution is quite constrained. For target site resistance, most mutations are costly and compromise the performance of the native protein function, so that codon use may prevent resistance apparition (Brengues et al., 2003; Weill et al., 2004).

Another issue is the cross-resistance between different insecticides (Figure 14.2). Cross-resistances between families of insecticides are associated with the sharing

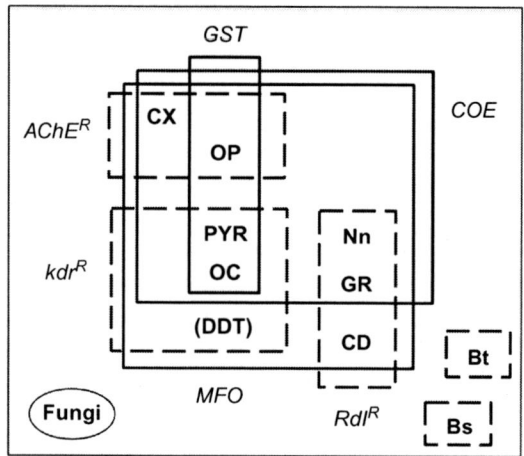

Figure 14.2 Cross-resistance between main insecticide classes. Metabolic resistances are indicated by continuous lines, while resistances by target site modification are represented by interrupted lines (*Note*: no resistance has been yet identified for fungi). Resistance mechanisms (in italic): *COE*: carboxyl-esterases, *MFO*: multifunctionnal oxidases, *GST*: glutathione *S*-transferases, *AChE^R*: resistant acetylcholinesterase, *kdr^R*: knockdown resistance allele, *Rdl^R*: resistance-to-dieldrin allele. Insecticide classes (in bold): CX: carbamates, OP: organophosphates, PYR: pyrethroids, OC: organochlorins, Nn: neonicotinoids, GR: growth regulators, CD: cyclodienes, Bti and Bs: bacterial toxins. Note that while target mutations confer cross-resistance to insecticides of different classes (interrupted lines), detoxifying enzymes give only "insecticide family cross-resistance" (i.e., a single gene does not confer cross-resistance between classes of insecticides, but different genes of the same family can provide resistance to different insecticide classes; continuous lines).
Source: Modified from Brogdon and McAllister (1998).

of target sites. For example, kdr^R causes cross-resistance between DDT and PYRs in *An. gambiae* (Chandre et al., 1999b), and $ace\text{-}1^R$ between OPs and CXs (Alout et al., 2008). In contrast, a particular metabolic resistance gene usually does not confer cross-resistance between insecticide families. However, different genes belonging to a same metabolic family can cause resistance to several insecticides of this family (gene family cross-resistance): for example, different COE and MFO genes cause resistance to DDT, others to PYRs, OPs, and CXs in Anophelines (Brogdon and McAllister, 1998; Brooke et al., 2001; Etang et al., 2007), and different MFO genes are responsible for resistance to DDT, neonicotinoids, and GRs in *D. melanogaster* (Daborn et al., 2002). The consequences of these cross-resistances are to reduce the alternative insecticides available, thereby gravely endangering vector control.

Finally despite advances, a full analysis of resistance remains challenging due to interactions and pleiotropy when several resistance mechanisms and/or resistance genes are present in the same insect (Hollingworth and Dong, 2008). This is unfortunately a common case: almost all known resistance mechanisms are present in the Anopheline species from Sri Lanka (Perera et al., 2008), Latin America (Penilla et al., 1998), and from various parts of Africa (Vulule et al., 1999; Diabate et al., 2002; Hougard et al., 2003; Stump et al., 2004; East Africa Casimiro et al., 2006b; Enayati and Hemingway, 2006; Brooke et al., 2001; Central Africa Etang et al., 2006; Matambo et al. 2007; Corbel et al., 2007; Dabiré et al., 2008; Okoye et al., 2008; Awolola et al., 2009; Ndjemai et al., 2009; Ramphul et al., 2009; West Africa Yadouleton et al., 2009). Multiple resistances with multiple mechanisms are also observed in *Cx. p. quinquefasciatus* (Chandre et al., 1998; Kasai et al., 1998; Corbel et al., 2007; Hardstone et al., 2007), *Ae. aegypti* (Brengues et al., 2003), sand flies (Alexander and Maroli, 2003; Surendran et al., 2005), and head lice (Thomas et al., 2006). Interactions between resistance loci have been studied in houseflies or mosquitoes, and most of them appear to be synergic. Such synergies have been observed between COE and *ace-1* for OP resistance and between *kdr* and P450 MFO for PYR resistance in *Cx. pipiens* (Raymond et al., 1989; Hardstone and Scott, 2009), between repellants (DEET) and CXs in *Ae. aegypti* (Pennetier et al., 2005; Bonnet et al., 2009) or between PYR resistance and susceptibility to fungus applications in three *Anopheles* species (Farenhorst et al., 2009). Moreover, these interactions may vary with environmental conditions (positive synergism for resistance in treated areas but negative synergism for cost in nontreated areas) or with the genetic background of the insect (Hardstone and Scott, 2009). For example, the presence of kdr^R decreases the cost of $ace\text{-}1^R$ in *Cx. pipiens* (Berticat et al., 2008).

In conclusion, any disease control strategy should take into account these various aspects of resistance as they can greatly impact its success (vector control failures) and may have a direct effect on pathogens transmission (McCarroll et al., 2000; McCarroll and Hemingway, 2002; Vontas et al., 2004). The various strategies for vector control and how they deal with resistance will be the subject of the following section.

14.4 Part 3: Treatment Practices and Resistance Management Strategies

As mentioned earlier, disease control relies greatly on vector control. Vector control strategies are designed to reduce adult vector populations (and thus pathogen transmission), but are faced with many constraints. They must be environmentally, socially, and politically acceptable (i.e., they must prevent any adverse effect on the environment, nontarget species, and humans) and must be economically realistic both in the short and long terms. To be successful, knowledge of the life history and ecology of vector species is critical (breeding sites, life cycle, preferred blood source, etc.; Alexander and Maroli, 2003; WHO, 2006; Walker and Lynch, 2007). Another key to success for these strategies is to prevent the development of insecticide resistance by monitoring resistance genes and adapting the practices. Ideally, vector control strategies should be integrated into national health and community systems to be sustainable (Kay and Vu, 2005; Ooi et al., 2006; Beier et al., 2008; Morrison et al., 2008).

In this part, we will review the current treatment practicies, their limits and constraints and how vector resistance is accounted for.

14.4.1 Treatment Practices

Adult Spray

Large scale adult spraying, where the insecticide is pulverized in the air, presents a low efficacy for a high operational cost, and is consequently recommended only in cases of severe epidemic. Preferred insecticides for this type of treatments are OPs and PYRs (WHO, 2006). They are used locally to control tsetse flies (Welburn et al., 2006) and *Ae. aegypti* for dengue control (Morrison et al., 2008; Luz et al., 2009). They are also used against *Anopheles* sp., domestic flies, sand flies, midges (*Simulium* sp.), and fleas (Walker and Lynch, 2007).

Larvicidal Treatment

Another common control procedure is larvicidal treatments. Their efficacy is maximal when larvae are restricted to breeding sites accessible and limited in size and numbers. Various insecticides are used (oils, Bti, GRs, CX, or OPs), mostly against *Aedes* and *Culex* mosquito species, but also against Anophelines, *Simulium* flies (midges), Ceratopogonidae, and Phlebotominae (see details in WHO, 2006; Walker and Lynch, 2007).

Indoor Residual Spraying

IRS (on walls) has long been the preferred vector control, because of its simplicity. The first insecticide used (and still used in case of high malaria prevalence) was DDT. It has been presently replaced by PYRs and to a lesser extent by CXs and

OPs (WHO, 2006). PYRs are preferred because of their excellent contact action, their rapid KD effect and their safe use (low toxicity on mammalian). IRS is used against mosquitoes, houseflies, sand flies, and tsetse flies (*Glossina* sp.) (Walker and Lynch, 2007). It has contributed to eradication of malaria in several countries and is still an important tool for vector control. It is very efficient if the target vector populations retain their susceptibility to insecticide exposure and if the insecticides are used reliably, efficiently, and sufficiently in terms of coverage (Okoye et al., 2008; Read et al., 2009). New insecticides as fungi have a high potential for IRS, because they infect through the insects' legs when they are resting (Scholte et al., 2003, 2005, 2006; Blanford et al., 2005).

Insecticide-Treated Materials

The use of insecticide-impregnated fabrics (bednets, curtains, sheeting, hammocks) may also produce significant levels of protection against disease vectors, if appropriately used with sufficient coverage on susceptible vector populations (Okoye et al., 2008). The use of ITNs impregnated with PYRs is the most common protective measure against malaria vectors (WHO, 2006): they act as physical barriers, which combine with the repellent and killing effects of PYRs. ITN are always preferred to IRS by local users (Curtis et al., 1998). They can be highly cost-effective (Enayati and Hemingway, 2006), and have a large effect if used appropriately by reducing the vectors and pathogen prevalence in populations (in these cases people who do not use a net are also protected) (Stump et al., 2004). The problem is that ITN are often torn or poorly fitted (Roberts and Andre, 1994; Pages et al., 2007). The necessity to regularly re-impregnate the nets with insecticide is a major constraint: washing decreases their efficacy by removing the insecticides, so that their effective usable life is 2 or 3 years in field conditions, after which they are dirty and holed (Curtis et al., 1998; Erlanger et al., 2004). In consequence, proper use of ITNs is often highly correlated to the socioeconomic status (Fillinger et al., 2008; Goesch et al., 2008). In response to these constraints, long-lasting impregnated nets (LLIN) have been conceived, for which there is no need for retreatment for 5 years. They showed good efficacy in rural Côte d'Ivoire where they were first tested, although some appear better than others. In particular, washing induces no significant decrease of their killing efficacy by contact (Kilian et al., 2008), although their repellent effect disappears quickly (N'Guessan et al., 2001; Dabiré et al., 2006; Kilian et al., 2008). Finally, new impregnated fabrics are being developed, such as plastic sheetings to cover ceilings or walls, or as tents to control malaria in refugee camps (Graham et al., 2002; Diabate et al., 2006; Djenontin et al., 2009).

If ITNs have shown good efficacy against malaria vectors (Anophelines) and various flies (Walker and Lynch, 2007), they have little impact on some species as the dengue vector *Ae. aegypti* which bites during the day, or sand flies which are active at sunset, before people are inside a net (Reiter, 2001; Alexander and Maroli, 2003). Moreover, ITNs are often perceived as inefficient by users, because despite the protection they confer against *Anopheles* mosquitoes, they often fail at preventing bitting by other mosquito species as *Cx. p. quinquefasciatus*, which is more resistant and bites during the day (Fillinger et al., 2008).

Other Treatments

IRS and ITNs are the most common means of vector control. They usually work well in countries with a developed health system (N'Guessan et al., 2007). In Tanzania both ITNs and IRS were efficient at decreasing vector densities and prevalence of *Plasmodium* in mosquitoes; however, no decrease in the cases of malaria was observed, which may show the limits of such strategies (Curtis et al., 1998).

Other methods are available to decrease vector populations (WHO, 2006; Pages et al., 2007), such as environment management. This is probably the most efficient of all and has been used for centuries. It consists in modifying the habitat by eliminating potential breeding sites of vectors. These modifications can be permanent (drainage, land levelling, and filling) or temporary (vegetation modification, intermittent flushing or irrigation, changing water salinity, using polystyrene beads). Modifying the habitat may also have secondary positive effects, such as sanitation, increase of agriculture lands, and reduction of water-borne diseases. However, two problems limit these applications: first, environment modification can potentially be problematic for the ecology of nontarget organisms, and second, these modifications, particularly the permanent ones, are often extremely costly. Nevertheless, such a cheap procedure as vegetation clearing is very efficient against *Glossina* flies, and hygiene can greatly reduce domestic flies-borne diseases (Walker and Lynch, 2007).

Another feature is biological control, such as with baits like ovitraps, which are cheap and efficient means of control for *Glossina* and domestic flies (Walker and Lynch, 2007) and are frequently used to control *Ae. aegypti* and dengue (Morrison et al., 2008). Another method consists in rendering the potential breeding sites unsuitable for vectors with an aquatic larval phase by adding some predators (copepods or fishes). It has shown great success for controlling dengue and *Ae. aegypti* populations in Vietnam. Local copepods (*Mesocyclops* sp.) were identified in the field, reproduced and then shared at the community level. They were put into water storage which led to dengue disappearance in treated towns (while it was still present in close untreated areas) (Kay and Vu, 2005). However, although relatively efficient, the use of introduced fishes (*Gambusia* sp.) on a world scale is now criticized for its impact on local ecology (Walker and Lynch, 2007).

Another option is sterile males and genetically modified organisms (GMOs) releases: these methods may be used to carry out two objectives: (1) decreasing vector populations by massive introduction of sterile males or of GMOs with a dominant gene killing females (Alphey et al., 2007); when crossed with field females, they will produce no offspring or only male offsprings; (2) introducing genes conferring resistance to the pathogens in the vector natural populations (review in Sinkins and Gould, 2006). These methods require a fitness advantage for the released insects, which could be provided by various gene-drives: transposable elements, homing endonucleases genes, meiotic drive, or endosymbiotic bacteria such as *Wolbachia* (Bourtzis and Robinson, 2003; Sinkins and Gould, 2006; Engelstadter and Telschow, 2009). Release of *Culex pipiens* males incompatible with field females due to *Wolbachia* have been attempted in the early 1970s in France and India (Laven, 1967; Curtis and Adak, 1974; Curtis, 1976; Curtis et al., 1982) but was short lasting, a fact that is not surprising since we know today that these endosymbionts have an extremely high genetic variability (Duron et al., 2006a, 2007). The only satisfactory results concern the American screwworm *Cochliomyia hominivorax*, a cattle parasite, which is controlled by releasing every year several hundred thousand irradiated sterile males and has been eradicated from the USA, and the control of tsetse flies and trypanosomiasis in Zanzibar, again using release of sterile males (review in Bowman, 2006).

There is an array of methods available to control vector populations and, through these, the diseases they carry. Some are presently available, and others (sterile male and GMO releases) remain to be worked out, tested, and validated. For maximizing the chances of success of available methods, all authors agree that they should be implemented in conjunction. For example, tsetse flies are controlled by modifying the habitat (vegetation clearing), spraying insecticides (PYRs) locally, and trapping adults (review in Welburn et al., 2006). They also emphasize the necessity of cohesion between the various actors of vector control (scientists, operationals, managers, politics, and local populations) and of long-term plannings (Beier et al., 2008; Morrison et al., 2008). For example, apart from three examples (Singapore, Cuba, and Vietnam), lack of cohesion between these different actors resulted in a global disaster in broad-scale dengue vector control, with an increase in dengue cases since the 1960s (Kay and Vu, 2005).

14.4.2 Impact of Agriculture

Another pitfall awaiting control programs is the impact of agriculture. Urban agriculture is developing in many large cities across Africa, where it is necessary to prevent food shortage and to provide a balanced diet (Akogbéto et al., 2006; Yadouleton et al., 2009). However, it creates new breeding sites for vectors as mosquitoes or livestock parasites: such as in Accra (Ghana), where urban agriculture development led to an increase in both quantities of vectors and malaria cases (Klinkenberg et al., 2008). Moreover, most agricultural pesticides are toxic for disease vectors: more than 90% of all insecticides produced have indeed been devised and used for agriculture, vector control being of very low rentability for chemical firms. When vectors are exposed to these pesticides, they can potentially develop resistance, threatening the success of current, and later control attemps (Georghiou, 1990; Roberts and Andre, 1994; Brogdon and McAllister, 1998; Boyer et al., 2006; Poupardin et al., 2008). Several examples have been suggested in the litterature:

1. In Sri Lanka, CX resistance has been detected in *An. subpictus*, *An. nigerrimus*, and *An. peditaniatus*, while these insecticides are only used for agriculture (Perera et al., 2008), so resistance is present before insecticide use for public health. Similarly, a high frequency of the Rdl^R allele (resistance to OCs) has been recently found in La Réunion

Island in *Cx. p. quinquefasciatus* and *Ae. albopictus*, while this insecticide has never been used for public health (Tantely et al., 2010).
2. Higher vector resistance in agricultural areas, as in Benin where *An. gambiae* resistance is present in urban vegetable areas where PYRs, OPs, and OCs are used as agropesticides (Corbel et al., 2007; Yadouleton et al., 2009). This could result in malaria outbreaks in cities (Pages et al., 2007).
3. A correlation between the quantities of insecticide for agriculture and the level of resistance. In Mexico, a reduction of *An. albimanus* insecticide resistance was observed after decrease in insecticide use for agriculture (Penilla et al., 1998). In Burkina Faso, a change from pure PYRs to use of a mixture of OPs and PYRs to control *H. armigera* in cotton fields in the late 1990s is probably the cause of OP resistance increase in *An. gambiae* (Dabiré et al., 2008). Similarly, intensification of cotton cultivation and consecutive insecticide use was associated with an increased insecticide resistance in both *An. gambiae* and *An. arabiensis* (Dabiré et al., 2009). Finally in Spain, the end of OP use in public health did not decrease OP resistance frequency in *Cx. p. pipiens* (despite its fitness cost) due to the large use of OPs in agriculture close to breeding sites (Eritja and Chevillon, 1999).
4. The impact of agropesticides on vector resistance can also be revealed by fluctuations of resistance frequency with period of crops spraying, as in Burkina Faso where PYR treatment intensification during cotton growth is correlated to an increase of *An. gambiae* resistance to these insecticides (Diabate et al., 2002).

Effects of agriculture and forestry on resistance spread have been reported also in Cameroon for *An. gambiae* (Ndjemai et al., 2009), and in China for *Cx. p. pipiens*, *An. siniensis*, *Ae. aegypti*, and *Ae. albopictus* (Cui et al., 2006). However, a specific public health use of insecticides can also select for resistance in nontargeted but potential desease vector organisms: in Sri Lanka, CX and OP resistance in sand flies is believed to be due to their use in agriculture and for mosquito control, respectively (Surendran et al., 2005), and in Benin, *Cx. p. quinquefasciatus* may have become resistant to PYRs due to *An. gambiae* treatment for public health (Corbel et al., 2007).

14.4.3 Sustainable Resistance Management

All treatment practices tend to select for insecticide resistance, a selection that is reinforced by other pesticide uses (agriculture, forestry, etc.). In malaria control for example, as the use of IRS and ITN is scaling up, so will the selection pressure for insecticide resistance (Farenhorst et al., 2009). Moreover, due to these resistance mechanisms and to environmental concerns, the number of insecticide molecules available for vector control is decreasing and no new molecule is expected soon (Nauen, 2007; Kelly-Hope et al., 2008).

Consequently, a large international effort is required for devising new and sustainable strategies to manage resistance and prolonging the lifespan of the few available insecticides. Resistance management requires deep practice changes (Coleman and Hemingway, 2007; Hollingworth and Dong, 2008; Whalon et al., 2008), away from the common practice of using an insecticide until resistance becomes a limiting factor, which rapidly erodes the number of suitable insecticides

(Penilla et al., 1998). A striking example is *An. gambiae* control: in the 1950s and 1960s it became resistant to dieldrin (OC) in all Africa, leading to a shift to the mass use of PYRs in the late 1970s, in turn leading to wildspread PYR and DDT resistance (Chandre et al., 1999a).

Once insecticide resistance is established in a population, there is indeed a real danger of re-emergence of vector-borne diseases that had been presumed under control, even if in some instances resistance does not seem to compromise disease control on the short term (Brogdon and McAllister, 1998; Asidi et al., 2005). Therefore, focusing on surveillance wherever possible is essential in order to react proactively once a regional population manifests shift in its susceptibility toward a class of insecticides used in public health (Nauen, 2007). Resistance surveillance means (i) providing baseline data for program planning and pesticide selection before the start of control operations; (ii) detecting resistance at an early stage, and (iii) continuously monitoring the effect of these resistance mechanisms on vector control strategies, so that timely management can be implemented (Brogdon and McAllister, 1998).

Resistance Management Strategies

Resistance management strategies' goal is to prevent and delay the spread of resistance while maintaining a good control of vector populations (WHO, 2006; Nauen, 2007). The susceptibility of vectors and pests should be considered a valuable resource that must be preserved as long as possible. It is critical that the strategies of resistance management must be implemented preventively to preserve the efficacy of the few insecticides available for public health purposes (WHO, 2006).

Different strategies can be implemented (reviews in Roberts and Andre, 1994; Coleman and Hemingway, 2007; Hollingworth and Dong, 2008; Whalon et al., 2008; Hardstone and Scott, 2009):

1. It is possible to vary the dose or frequency of pesticide applications, but increasing the dose of insecticides should be done with great caution, as it can increase the speed of resistance spread. However, in early stages when resistance genes are mainly at the heterozygote state, it can possibly delay resistance spread if it is recessive. For example, high doses have been recommended for PYR-based *An. gambiae* control when *kdr* is at low frequency, because this gene is partially recessive (Corbel et al., 2004).
2. Insecticide application may also be narrowly focused on a small area or a short time period (e.g., only when endemic vector-borne diseases are present). For example, the onchocerciasis control program in West Africa began in 1975, with OP-based larviciding against the black fly (*Simulium damnosum* complex). To manage OP resistance, they are only used at the beginning of rivers' high waters, to prevent the pic of black fly; other insecticides (PYRs and Bti) are used at other periods (Roberts and Andre, 1994).
3. Such alternative use of various pesticides is probably the most efficient way of managing resistance (Nauen, 2007). Several pesticides can be used in mixtures (all together), in mosaics (different insecticides in different sites), in sequences (different insecticides at different times), or in rotations (different insecticides both at different times and sites) (Roberts and Andre, 1994; Hemingway and Ranson, 2000). However, due to cross-resistance (Figure 14.2), it requires using insecticides with different modes of action,

rather than merely alternating members of one chemical class or different chemical classes that address the same target site (Nauen, 2007). For example, for *An. gambiae* and *Cx. p. quinquefasciatus* control in highly PYR resistant populations of West and Equatorial Africa, successful trials have been implemented to use OPs or CXs as alternative to PYRs for ITN impregnation, in mixtures or mosaics (Kolaczinski et al., 2000; Hougard et al., 2003; Asidi et al., 2005; Sharp et al., 2007a; Oxborough et al., 2008).

All these resistance management strategies mainly rely on the fitness cost often associated with resistance alleles. Each resistance allele indeed corresponds to a recent adaptation, and is thus often associated with a diminution of the fitness (i.e., reduced reproductive ability in absence of insecticide [cost]). This has been shown for *Cx. pipiens Ester* and *ace-1* OP resistance alleles (Lenormand et al., 1999; Berticat et al., 2002a, 2004; Bourguet et al., 2004; Duron et al., 2006b), and for PYR resistance in *Ae. aegypti* (Kumar et al., 2009). Theoretical expectations are that costly resistance should rapidly disappear when insecticides cease to be used, natural selection favoring susceptible insects (Lenormand and Raymond, 1998; Eritja and Chevillon, 1999). However, as this adaptation evolves rapidly, new mutations that reduce or suppress the fitness cost have also been described (McKenzie, 1996; Labbé et al., 2007a; e.g., duplications of the *ace-1* gene in the *Cx. pipiens* and *An. gambiae* complexes, Alout et al., 2009; Djogbénou et al., 2009; or allele replacement and epistasy, Labbé et al., 2009), and may dramatically endanger resistance management and disease-vector control.

Examples

Resistance management is often planned using mathematical models, which can be very useful tools (e.g., Lenormand and Raymond, 1998; Luz et al., 2009). In the field, however, proper detection and monitoring of resistance in vector populations are essential components of any resistance management program (Roberts and Andre, 1994) and these field studies are often missing (Penilla et al., 1998; Hemingway and Ranson, 2000). Some examples are available, which show the benefits of evidence-based vector control with insecticide resistance management (Coleman and Hemingway, 2007).

In Singapore, a sustainable control of dengue through *Ae. aegypti* control has been implemented successfully for 35 years (review in Ooi et al., 2006). It is based on entomological surveillance and reduction of *Aedes* larval habitat availability (no risk of resistance), mainly through environmental measures that require the implication of the population. This experience shows that public education and law, continuous entomological surveillance (rather than emergency reactions), and a regional level collaboration between several countries are required for success.

In Vietnam, successful elimination of dengue and *Ae. aegypti* was achieved (review in Kay and Vu, 2005). It relied on four key elements for its success: the continuous evaluation and quantification of *Ae. aegypti* larvae in the breeding sites, the use of local predatory copepods (cheap, no risk of resistance), the implication of the local community, and the education of the population.

In Dar-es-salaam (Tanzania), a community level larviciding was established to control *An. gambiae*-vectored malaria, and *Aedes* and *Culex* biting nuisance. They used Bti and Bs, as alternative to PYRs for which resistance was present. It resulted in a 96% reduction of *An. gambiae* populations, 40% reduction of *P. falciparum* prevalence and 31% reduction of malaria cases, but was less successful with the other Culicidae (*Culex* and *Aedes*) nuisance. This strategy thus showed good potential, but was costly and evidenced the need for community implication and experienced actors (Fillinger et al., 2008).

Constant monitoring allows early detection of insecticide resistance and changes in policies. It prevented program failures in at least two initiatives: Bioko Island malaria control project and Lubombo spatial development initiative. In Bioko Island (Equatorial Guinea), PYRs were used for *An. gambiae* control. The continuous monitoring kdr resistance showed an increase from 50% in 2003 to 85% in 2005. Consequently, the withdrawal of PYRs, replaced by CXs prevented *An. gambiae* and malaria control failure (Sharp et al., 2007a). In Mozambic, DDT was used to control Anopheline vectors from 1946 until 1988, when it changed for PYRs. A systematic survey of *An. funestus* and *An. arabiensis* was engaged in 1999 by the Lubombo spatial development initiative (Mozambique, Swaziland, and South Africa). It showed a high level of PYR resistance in these mosquitoes and led to a policy change: CXs replaced PYRs (no CX resistance in 2000). In 2003, high $ace\text{-}1^R$ frequency was detected, so that DDT was reintroduced in 2006 to control CX resistance and for cost problems; in 2006, PYR resistance in *An. arabiensis* was no longer detectable (Casimiro et al., 2006a,b). Overall, resistance baseline data and monitoring allowed effective adaptation of policies keeping malaria under control (it decreased from 88% in 2000 to 33% in 2005; Sharp et al., 2007b).

14.5 Conclusion

The natural history of mosquito-borne diseases is complex, and the interplay of climate, ecology, vector biology, and many other factors defies simplistic analyses. The recent resurgence of many of these diseases is a major cause for concern. Its principal determinants are politics, economics, and human activities (rather than climate change). In order to control these diseases and ameliorate the socioeconomic burden they cause in developing countries, vector control remains a powerful and accessible tool. However, it is urgently required to change the treatment strategies to manage the development of insecticide resistance. This includes using alternative tools to insecticides for vector control, preserving the remaining insecticides by carefully planning their use to minimize resistance selection, and finally establishing continuous survey of resistance at a local scale by implicating the local population, a difficult but essential task to set goals and evaluate success. Several survey sites in different conditions are required for sentinel purposes, together with some baseline information, to rapidly detect resistance, identify the mechanisms, and change the policies adequately (Kelly-Hope et al., 2008). In order to achieve this survey, basic tools like bioassays remains most powerful, and should always be a preliminary step before more complex and more costly analyses. Clearly, the greatest challenge for successful vector and disease control is the coordination of the different actors (chemical industries, researchers, politics, control agencies, and local populations), which do not have the same agendas, motivations or economical interests.

Besides its implications in public health and development, insecticide resistance remains a powerful evolutionary biology model to study the contemporary adaptation of organisms to a changing environment. It indeed allows a complete and integrative study, from the molecular mechanisms to the fitness consequences at the individual level and their impacts on insect population dynamics and interactions with pathogens. Moreover, it is for once pleasant to constat that these rather fundamental approaches of evolutionary biology may have a direct impact in the society and help design new strategies for the successful control of some of the most threatening human diseases (Michalakis and Renaud, 2009).

References

Abdalla, H., Matambo, T.S., Koekemoer, L.L., Mnzava, A.P., Hunt, R.H., Coetzee, M., 2008. Insecticide susceptibility and vector status of natural populations of *Anopheles arabiensis* from Sudan. Trans. R Soc. Trop. Med. Hyg. 102, 263–271.

Akogbéto, M., Djouaka, R., Kindé-Gazard, D., 2006. Screening of pesticide residues in soil and water samples from agricultural settings. Malar. J. 5, 22.

Alexander, B., Maroli, M., 2003. Control of phlebotomine sandflies. Med. Vet. Entomol. 17, 1–18.

Alout, H., Berthomieu, A., Cui, F., Tan, Y., Berticat, C., Qiao., C., et al., 2007. Different amino-acid substitutions confer insecticide resistance through acetylcholinesterase 1 insensitivity in *Culex vishnui* and *Culex tritaeniorhynchus* (Diptera: Culicidae) mosquitoes from China. J. Med. Entomol. 44, 463–469.

Alout, H., Djogbénou, L., Berticat, C., Chandre, F., Weill, M., 2008. Comparison of *Anopheles gambiae* and *Culex pipiens* acetycholinesterase 1 biochemical properties. Comp. Biochem. Physiol. B, Biochem. Mol. Biol. 150, 271–277.

Alout, H., Labbé, P., Berthomieu, A., Pasteur, N., Weill, M., 2009. Multiple duplications of the rare ace-1 mutation F290V in *Culex pipiens* natural populations. Insect Biochem. Mol. Biol. 39, 884–891.

Alphey, N., Coleman, P.G., Donnelly, C.A., Alphey, L., 2007. Managing insecticide resistance by mass release of engineered insects. J. Econ. Entomol. 100, 1642–1649.

Andreev, D., Kreitman, M., Phillips, T.W., Beeman, R.W., ffrench-Constant, R.H., 1999. Multiple origins of cyclodiene insecticide resistance in *Tribolium castaneum* (Coleoptera: Tenebrionidae). J. Mol. Evol. 48, 615–624.

Anthony, N., Unruh, T., Ganser, D., ffrench-Constant, R., 1998. Duplication of the Rdl GABA receptor subunit gene in an insecticide-resistant aphid, *Myzus persicae*. Mol. Gen. Genet. 260, 165–175.

Asidi, A., N'Guessan, R., Koffi, A., Curtis, C., Hougard, J.-M., Chandre, F., et al., 2005. Experimental hut evaluation of bednets treated with an organophosphate (chlorpyrifos-methyl) or a pyrethroid (lambdacyhalothrin) alone and in combination against insecticide-resistant *Anopheles gambiae* and *Culex quinquefasciatus* mosquitoes. Malar. J. 4, 25.

Awolola, T.S., Oduola, A.O., Oyewole, I.O., Obansa, J.B., Amajoh, C.N., Koekemoer, L.L., et al., 2007. Dynamics of knockdown pyrethroid insecticide resistance alleles in a field population of *Anopheles gambiae s.s.* in southwestern Nigeria. J. Vector Borne Dis. 44, 181–188.

Awolola, T.S., Oduola, O.A., Strode, C., Koekemoer, L.L., Brooke, B., Ranson, H., 2009. Evidence of multiple pyrethroid resistance mechanisms in the malaria vector *Anopheles gambiae sensu stricto* from Nigeria. Trans. R. Soc. Trop. Med. Hyg. 103, 1139–1145.

Beier, J., Keating, J., Githure, J., Macdonald, M., Impoinvil, D., Novak, R., 2008. Integrated vector management for malaria control. Malar. J. 7, S4.

Berticat, C., Boquien, G., Raymond, M., Chevillon, C., 2002a. Insecticide resistance genes induce a mating competition cost in *Culex pipiens* mosquitoes. Genet. Res. Cambrige 79, 41–47.

Berticat, C., Rousset, F., Raymond, M., Berthomieu, A., Weill, M., 2002b. High *Wolbachia* density in insecticide resistant mosquitoes. Proc. R. Soc. Lond. B 269, 1413–1416.

Berticat, C., Duron, O., Heyse, D., Raymond, M., 2004. Insecticide resistance genes confer a predation cost on mosquitoes, *Culex pipiens*. Genet. Res. 83, 189–196.

Berticat, C., Bonnet, J., Duchon, S., Agnew, P., Weill, M., Corbel, V., 2008. Costs and benefits of multiple resistance to insecticides for *Culex quinquefasciatus* mosquitoes. BMC Evol. Biol. 8, 104–112.

Blanford, S., Chan, B.H.K., Jenkins, N., Sim, D., Turner, R.J., Read, A.F., et al., 2005. Fungal pathogen reduces potential for malaria transmission. Science 308, 1638–1641.

Bonnet, J., Pennetier, C., Duchon, S., Lapied, B., Corbel, V., 2009. Multi-function oxidases are responsible for the synergistic interactions occurring between repellents and insecticides in mosquitoes. Parasit. Vectors 2, 17.

Bourguet, D., Roig, A., Toutant, J.P., Arpagaus, M., 1997. Analysis of molecular forms and pharmacological properties of acetylcholinesterase in several mosquito species. Neurochem. Int. 31, 65–72.

Bourguet, D., Guillemaud, T., Chevillon, C., Raymond, M., 2004. Fitness costs of insecticide resistance in natural breeding sites of the mosquito *Culex pipiens*. Evolution 58, 128–135.

Bourtzis, K., Robinson, A.S., 2003. Insect pest control using *Wolbachia* and/radiation. In: Bourtzis, K., Miller, T.A. (Eds.), Insect Symbiosis. Taylor and Francis Group, Boca Raton, USA.

Bowman, D.D., 2006. Successful and currently ongoing parasite eradication programs. Vet. Parasitol. 139, 293–307.

Boyer, S., Sérandour, J., Lempérière, G., Raveton, M., Ravanel, P., 2006. Do herbicide treatments reduce the sensitivity of mosquito larvae to insecticides? Chemosphere 65, 721–724.

Brengues, C., Hawkes, N.J., Chandre, F., Mccarroll, L., Duchon, S., Guillet, P., et al., 2003. Pyrethroid and DDT cross-resistance in *Aedes aegypti* is correlated with novel mutations in the voltage-gated sodium channel gene. Med. Vet. Entomol. 17, 87–94.

Brogdon, W.G., McAllister, J.C., 1998. Insecticide resistance and vector control. Emerg. Infect. Dis. 4, 605–613.

Brooke, B.D., 2008. kdr: can a single mutation produce an entire insecticide resistance phenotype? Trans. R. Soc. Trop. Med. Hyg. 102, 524–525.

Brooke, B.D., Kloke, G., Hunt, R.H., Koekemoer, L.L., Tem, E.A., Taylor, M.E., et al., 2001. Bioassay and biochemical analyses of insecticide resistance in southern African *Anopheles funestus* (Diptera: Culicidae). Bull. Entomol. Res. 91, 265–272.

Callaghan, A., Guillemaud, T., Makate, N., Raymond, M., 1998. Polymorphisms and fluctuations in copy number of amplified esterase genes in *Culex pipiens* mosquitoes. Insect Mol. Biol. 7, 295–300.

Campbell, G.L., Marfin, A.A., Lanciotti, R.S., Gubler, D.J., 2002. West Nile virus. Lancet Infect. Dis. 2, 519–529.

Carson, R., 1962. Silent Spring. Houghton Mifflin, Boston, MA.

Casimiro, S., Coleman, M., Hemingway, J., Sharp, B., 2006a. Insecticide resistance in *Anopheles arabiensis* and *Anopheles gambiae* from Mozambique. J. Med. Entomol. 43, 276–282.

Casimiro, S., Coleman, M., Mohloai, P., Hemingway, J., Sharp, B., 2006b. Insecticide resistance in *Anopheles funestus* (Diptera: Culicidae) from Mozambique. J. Med. Entomol. 43, 267–275.

Casimiro, S.L.R., Hemingway, J., Sharp, B.L., Coleman, M., 2007. Monitoring the operational impact of insecticide usage for malaria control on *Anopheles funestus* from Mozambique. Malar. J. 6, 142–148.

Chandre, F., Darriet, F., Darder, M., Cuany, A., Doannio, J.M., Pasteur, N., et al., 1998. Pyrethroid resistance in *Culex quinquefasciatus* from West Africa. Med. Vet. Entomol. 12, 359–366.

Chandre, F., Darrier, F., Manga, L., Akogbeto, M., Faye, O., Mouchet, J., et al., 1999a. Status of pyrethroid resistance in *Anopheles gambiae sensu lato*. Bull. World Health Organ. 77, 230–234.

Chandre, F., Darriet, F., Manguin, S., Brengues, C., Carnevale, P., Guillet, P., 1999b. Pyrethroid cross resistance spectrum among populations of *Anopheles gambiae s.s.* from Cote d'Ivoire. J. Am. Mosq. Control Assoc. 15, 53–59.

Chandre, F., Darriet, F., Duchon, S., Finot, L., Manguin, S., Carnevale, P., et al., 2000. Modifications of pyrethroid effects associated with *kdr* mutation in *Anopheles gambiae*. Med. Vet. Entomol. 14, 81–88.

Cheesman, D.F., 1964. Varro and the small beasts: a bimillennium for microbiologists. Nature 203, 911–912.

Chevillon, C., Pasteur, N., Marquine, M., Heyse, D., Raymond, M., 1995. Population structure and dynamics of selected genes in the mosquito *Culex pipiens*. Evolution 49, 997–1007.

Chevillon, C., Bernard, C., Marquine, M., Pasteur, N., 2001. Resistance to Bacillus sphaericus in Culex pipiens (Diptera: Culicidae): Interaction Between Recessive Mutants and Evolution in Southern France. J. Med. Entomol. 38, 657–664.

Claudianos, C., Russell, R.J., Oakeshott, J.G., 1999. The same amino acid substitution in orthologous esterases confers organophosphate resistance on the house fly and a blowfly. Insect Biochem. Mol. Biol. 29, 675–686.

Coleman, M., Hemingway, J., 2007. Insecticide resistance monitoring and evaluation in disease transmitting mosquitoes. J. Pest. Sci. 32, 69–76.

Coleman, M., Casimiro, S., Hemingway, J., Sharp, B., 2008. Operational impact of DDT reintroduction for malaria control on *Anopheles arabiensis* in Mozambique. J. Med. Entomol. 45, 885–890.

Corbel, V., Chandre, F., Brengues, C., Akogbéto, M., Lardeux, F., Hougard, J., et al., 2004. Dosage-dependent effects of permethrin-treated nets on the behaviour of *Anopheles gambiae* and the selection of pyrethroid resistance. Malar. J. 3, 22.

Corbel, V., N'Guessan, R., Brengues, C., Chandre, F., Djogbenou, L., Martin, T., et al., 2007. Multiple insecticide resistance mechanisms in *Anopheles gambiae* and *Culex quinquefasciatus* from Benin, West Africa. Acta Trop. 101, 207–216.

Cornel, A.J., Stanich, M.A., McAbee, R.D., Mulligan, F.S., 2002. High level methoprene resistance in the mosquito *Ochlerotatus nigromaculis* (Ludlow) in Central California. Pest Manag. Sci. 58, 791–798.

Cuany, A., Handani, J., Berge, J., Fournier, D., Raymond, M., Georghiou, G.P., et al., 1993. Action of Esterase B1 on chlorpyrifos in organophosphate-resistant *Culex* mosquitoes. Pestic. Biochem. Physiol. 45, 1–6.

Cui, F., Raymond, M., Qiao, C.-L., 2006. Insecticide resistance in vector mosquitoes in China. Pest Manag. Sci. 62, 1013–1022.

Curtis, C.F., 1976. Population replacement in *Culex fatigans* by means of cytoplasmic incompatibility. Bull. World Health Organ. 53, 107–119.

Curtis, C.F., Adak, T., 1974. Population replacement in *Culex fatigans* by means of cytoplasmic incompatibility. Bull. World Health Organ. 51, 249–255.

Curtis, C.F., Brooks, G.D., Ansari, M.A., Grover, K.K., Krishnamurthy, B.S., Rajagopolan, P.K., et al., 1982. A field trial on control of *Culex quinquefasciatus* by release of males of a strain integrating cytoplasmic incompatibility and a translocation. Entomol. Exp. Appl. 31, 181–190.

Curtis, C.F., Maxwell, C.A., Maxwell, C.A., Finch, R.J., Finch, R.J., Njunwa, K.J., 1998. A comparison of use of a pyrethroid either for house spraying or for bednet treatment against malaria vectors. Trop. Med. Int. Health 3, 619–631.

Dabiré, R., Diabaté, A., Baldet, T., Paré-Toé, L., Guiguemdé, R., Ouédraogo, J.-B., et al., 2006. Personal protection of long lasting insecticide-treated nets in areas of *Anopheles gambiae s.s.* resistance to pyrethroids. Malar. J. 5, 12.

Dabiré, K., Diabaté, A., Djogbénou, L., Ouari, A., N'Guessan, R., Ouédraogo, J.-B., et al., 2008. Dynamics of multiple insecticide resistance in the malaria vector *Anopheles gambiae* in a rice growing area in South-Western Burkina Faso. Malar. J. 7, 188.

Dabiré, K.R., Diabaté, A., Namountougou, M., Toe, K.H., Ouari, A., Kengne, P., et al., 2009. Distribution of pyrethroid and DDT resistance and the L1014F kdr mutation in *Anopheles gambiae s.l.* from Burkina Faso (West Africa). Trans. R. Soc. Trop. Med. Hyg. 103, 1113–1120.

Daborn, P.J., Yen, J.L., Bogwitz, M.R., Le Goff, G., Feil, E., Jeffers, S., et al., 2002. A single P450 allele associated with insecticide resistance in *Drosophila*. Science 297, 2253–2256.

Daborn, P.J., Lumb, C., Boey, A., Wong, W., ffrench-Constant, R.H., Batterham, P., 2007. Evaluating the insecticide resistance potential of eight *Drosophila melanogaster* cytochrome P450 genes by transgenic over-expression. Insect Biochem. Mol. Biol. 37, 512–519.

Darboux, I., Pauchet, Y., Castella, C., Silva-Filha, M.H., Nielsen-Leroux, C., Charles, J.-F., et al., 2002. Loss of the membrane anchor of the target receptor is a mechanism of bioinsecticide resistance. Proc. Natl. Acad. Sci. U.S.A. 99, 5830–5835.

Darboux, I., Charles, J.F., Pauchet, Y., Warot, S., Pauron, D., 2007. Transposon-mediated resistance to *Bacillus sphaericus* in a field-evolved population of *Culex pipiens* (Diptera: Culicidae). Cell. Microbiol. 9, 2022–2029.

David, J.-P., Strode, C., Vontas, J., Nikou, D., Vaughan, A., Pignatelli, P.M., et al., 2005. The *Anopheles gambiae* detoxification chip: a highly specific microarray to study

metabolic-based insecticide resistance in malaria vectors. Proc. Natl. Acad. Sci. U.S.A. 102, 4080–4084.
Davidson, G., 1956. Insecticide resistance in *Anopheles gambiae* Giles: a case of simple Mendelian inheritance. Nature 178, 863–864.
Diabate, A., Baldet, T., Chandre, F., Akoobeto, M., Guiguemde, T.R., Darriet, F., et al., 2002. The role of agricultural use of insecticides in resistance to pyrethroids in *Anopheles gambiae s.l.* in Burkina Faso. Am. J. Trop. Med. Hyg. 67, 617–622.
Diabate, A., Baldet, T., Chandre, E., Dabiré, K.R., Simard, F., Ouedraogo, J.B., et al., 2004. First report of a kdr mutation in *Anopheles arabiensis* from Burkina Faso, West Africa. J. Am. Mosq. Control Assoc. 20, 195–196.
Diabate, A., Chandre, F., Rowland, M., N'Guessan, R., Duchon, S., Dabiré, K.R., et al., 2006. The indoor use of plastic sheeting pre-impregnated with insecticide for control of malaria vectors. Trop. Med. Int. Health 11, 597–603.
Dialynas, E., Topalis, P., Vontas, J., Louis, C., 2009. MIRO and IRbase: IT tools for the epidemiological monitoring of insecticide resistance in mosquito disease vectors. PLoS Negl. Trop. Dis. 3, e465.
Djenontin, A., Chabi, J., Baldet, T., Irish, S., Pennetier, C., Hougard, J.M., et al., 2009. Managing insecticide resistance in malaria vectors by combining carbamate-treated plastic wall sheeting and pyrethroid-treated bed nets. Malar. J. 8, 233–241.
Djogbénou, L., Chandre, F., Berthomieu, A., Dabiré, R., Koffi, A., Alout, H., et al., 2008. Evidence of introgression of the ace-1R mutation and of the ace-1 duplication in West African *Anopheles gambiae s. s.* PLoS ONE 3, e2172, 2171–2177.
Djogbénou, L., Labbe, P., Chandre, F., Pasteur, N., Weill, M., 2009. Ace-I duplication in *Anopheles gambiae*: a challenge for malaria control. Malar. J. 8.
Du, W., Awolola, T.S., Howell, P., Koekemoer, L.L., Brooke, B.D., Benedict, M.Q., et al., 2005. Independent mutations in the *Rdl* locus confer dieldrin resistance to *Anopheles gambiae* and *An. arabiensis*. Insect Mol. Biol. 14, 179–183.
Duron, O., Fort, P., Weill, M., 2006a. Hypervariable prophage WO sequences describe an unexpected high number of *Wolbachia* variants in the mosquito *Culex pipiens*. Proc. R. Soc. B Biol. Sci. 273, 495–502.
Duron, O., Labbe, P., Berticat, C., Rousset, F., Guillot, S., Raymond, M., et al., 2006b. High *Wolbachia* density correlates with cost of infection for insecticide resistant *Culex pipiens* mosquitoes. Evolution 60, 303–314.
Duron, O., Boureux, A., Echaubard, P., Berthomieu, A., Berticat, C., Fort, P., et al., 2007. Variability and expression of ankyrin domain genes in *Wolbachia* variants infecting the mosquito *Culex pipiens*. J. Bacteriol. 189, 4442–4448.
Elissa, N., Mouchet, J., Riviere, F., Meunier, J.Y., Yao, K., 1993. Resistance of *Anopheles gambiae s.s.* to pyrethroids in Cote d'Ivoire. Ann. Soc. Belg. Med. Trop. 73, 291–294.
Enayati, A.A., Hemingway, J., 2006. Pyrethroid insecticide resistance and treated bednets efficacy in malaria control. Pestic. Biochem. Physiol. 84, 116–126.
Enayati, A.A., Vatandoost, H., Ladonni, H., Townson, H., Hemingway, J., 2003. Molecular evidence for a kdr-like pyrethroid resistance mechanism in the malaria vector mosquito *Anopheles stephensi*. Med. Vet. Entomol. 17, 138–144.
Enayati, A.A., Ranson, H., Hemingway, J., 2005. Insect glutathione transferases and insecticide resistance. Insect Mol. Biol. 14, 3–8.
Engelstadter, J., Telschow, A., 2009. Cytoplasmic incompatibility and host population structure. Heredity 103, 196–207.
Enserink, M., 2008. Entomology: a mosquito goes global. Science 320, 864–866.

Eritja, R., Chevillon, C., 1999. Interruption of chemical mosquito control and evolution of insecticide resistance genes in *Culex pipiens* (Diptera: Culicidae). J. Med. Entomol. 36, 41–49.

Erlanger, T.E., Enayati, A.A., Hemingway, J., Mshinda, H., Tami, A., Lengeler, C., 2004. Field issues related to effectiveness of insecticide-treated nets in Tanzania. Med. Vet. Entomol. 18, 153–160.

Etang, J., Fondjo, E., Chandre, F., Morlais, I., Brengues, C., Nwane, P., et al., 2006. First report of knochdown mutations in the malaria vector *Anopheles gambiae* from Cameroon. Am. J. Trop. Med. Hyg. 74, 795–797.

Etang, J., Manga, L., Toto, J.C., Guillet, P., Fondjo, E., Chandre, F., 2007. Spectrum of metabolic-based resistance to DDT and pyrethroids in *Anopheles gambiae s.l.* populations from Cameroon. J. Vector Ecol. 32, 123–133.

Etang, J., Vicente, J.L., Nwane, P., Chouaibou, M., Morlais, I., Rosario, V.E.D., et al., 2009. Polymorphism of intron-1 in the voltage-gated sodium channel gene of *Anopheles gambiae s.s.* populations from Cameroon with emphasis on insecticide knockdown resistance mutations. Mol. Ecol. 18, 3076–3086.

Farenhorst, M., Mouatcho, J.C., Kikankie, C.K., Brooke, B.D., Hunt, R.H., et al., 2009. Fungal infection counters insecticide resistance in African malaria mosquitoes. Proc. Natl. Acad. Sci. 106, 17443–17447.

Feyereisen, R., 1995. Molecular biology of insecticide resistance. Toxicol. Lett. 82-83, 83–90.

Feyereisen, R., 2005. Insect cytochrome P450. In: Gilbert, L.I., Iatrou, K., Gill, S.S. (Eds.), Comprehensive Insect Molecular Science. Elsevier, Oxford, UK, pp. 1–77.

ffrench-Constant, R.H., Steichen, J.C., Rocheleau, T.A., Aronstein, K., Roush, R.T., 1993. A single-amino acid substitution in a gamma-aminobutyric acid subtype A receptor locus is associated with cyclodiene insecticide resistance in *Drosophila* populations. Proc. Natl. Acad. Sci. U.S.A. 90, 1957–1961.

ffrench-Constant, R.H., Anthony, N., Aronstein, K., Rocheleau, T., Stilwell, G., 2000. Cyclodiene insecticide resistance: from molecular to population genetics. Annu. Rev. Entomol. 45, 449–466.

ffrench-Constant, R.H., Daborn, P.J., Goff, G.L., 2004. The genetics and genomics of insecticide resistance. Trends Genet. 20, 163–170.

Fillinger, U., Kannady, K., William, G., Vanek, M., Dongus, S., Nyika, D., et al., 2008. A tool box for operational mosquito larval control: preliminary results and early lessons from the Urban Malaria Control Programme in Dar es Salaam, Tanzania. Malar. J. 7, 20.

Fournier, D., Mutéro, A., 1994. Modification of acetylcholinesterase as a mechanism of resistance to insecticides. Comp. Biochem. Physiol. 108C, 19–31.

Fournier, D., Bride, J.M., Mouchès, C., Raymond, M., Magnin, M., Bergé, J.B., et al., 1987. Biochemical characterization of the esterases A1 and B1 associated with organophosphate resistance in the *Culex pipiens* complex. Pestic. Biochem. Physiol. 27, 211–217.

Fournier, D., Karch, F., Bride, J.M., Hall, L.M.C., Bergé, J.-B., Spierer, P., 1989. *Drosophila melanogaster* acetylcholinesterase gene, structure evolution and mutations. J. Mol. Evol. 210, 15–22.

Gahan, L.J., Gould, F., Heckel, D.G., 2001. Identification of a gene associated with Bt resistance in *Heliothis virescens*. Science 293, 857–860.

Georghiou, G.P., 1990. The effect of agrochemicals on vector populations. In: Roush, R.T., Tabashnik, B.E. (Eds.), Pesticide Resistance in Arthropods. Chapman & Hall, New York, NY.

Georghiou, G.P., Lagunes-Tejeda, A., 1991. The Occurrence of Resistance to Pesticides in Arthropods. Food and Agriculture Organization, Rome.

Garcia, G.P., Flores, A.E., Fernandez-Salas, I., Saavedra-Rodriguez, K., Reyes-Solis, G., Lozano-Fuentes, S., et al., 2009. Recent rapid rise of a permethrin knock down resistance allele in *Aedes aegypti* in Mexico. Plos Negl. Trop. Dis. 3, e531.

Goesch, J., Schwarz, N., Decker, M.-L., Oyakhirome, S., Borchert, L., Kombila, U., et al., 2008. Socio-economic status is inversely related to bed net use in Gabon. Malar. J. 7, 60.

Graham, K., Mohammad, N., Rehman, H., Nazari, A., Ahmad, M., Kamal, M., et al., 2002. Insecticide-treated plastic tarpaulins for control of malaria vectors in refugee camps. Med. Vet. Entomol. 16, 404–408.

Griffitts, J.S., Whitacre, J.L., Stevens, D.E., Aroian, R.V., 2001. Bt toxin resistance from loss of a putative carbohydrate-modifying enzyme. Science 293, 860–864.

Guillemaud, T., Makate, N., Raymond, M., Hirst, B., Callaghan, A., 1997. Esterase gene amplification in *Culex pipiens*. Insect Mol. Biol. 6, 319–327.

Guillemaud, T., Lenormand, T., Bourguet, D., Chevillon, C., Pasteur, N., Raymond, M., 1998. Evolution of resistance in *Culex pipiens*: allele replacement and changing environment. Evolution 52, 443–453.

Guillemaud, T., Raymond, M., Tsagkarakou, A., Bernard, C., Rochard, P., Pasteur, N., 1999. Quantitative variation and selection of esterase gene amplification in *Culex pipiens*. Heredity 83, 87–99.

Guillet, P., Germain, M.C., Giacomini, T., Chandre, F., Akogbeto, M., Faye, O., et al., 1998. Origin and prevention of airport malaria in France. Trop. Med. Int. Health 3, 700–705.

Hardstone, M.C., Scott, J.G., 2009. A review of the interactions between multiple insecticide resistance loci. Pestic. Biochem. Physiol. 97, 123–128.

Hardstone, M.C., Leichter, C., Harrington, L.C., Kasai, S., Tomita, T., Scott, J.G., 2007. Cytochrome P450 monooxygenase-mediated permethrin resistance confers limited and larval specific cross-resistance in the southern house mosquito, *Culex pipiens quinquefasciatus*. Pestic. Biochem. Physiol. 89, 175–184.

Hartley, C.J., Newcomb, R.D., Russell, R.J., Yong, C.G., Stevens, J.R., Yeates, D.K., et al., 2006. Amplification of DNA from preserved specimens shows blowflies were preadapted for the rapid evolution of insecticide resistance. Proc. Natl. Acad. Sci. U.S.A. 103, 8757–8762.

Hemingway, J., Ranson, H., 2000. Insecticide resistance in insect vectors of human disease. Annu. Rev. Entomol. 45, 371–391.

Hemingway, J., Hawkes, N.J., McCarroll, L., Ranson, H., 2004. The molecular basis of insecticide resistance in mosquitoes. Insect Biochem. Mol. Biol. 34, 653–665.

Hernandez, R., Guerrero, F.D., George, J.E., Wagner, G.G., 2002. Allele frequency and gene expression of a putative carboxylesterase-encoding gene in a pyrethroid resistant strain of the tick *Boophilus microplus*. Insect Biochem. Mol. Biol. 32, 1009–1016.

Hirose, T., Sunazuka, T., Omura, S., 2010. Recent development of two chitinase inhibitors, Argifin and Argadin, produced by soil microorganisms. Proc. Jpn. Acad. Ser. B-Phys. Biol. Sci 86, 85–102.

Hodjati, M.H., Curtis, C.F., 1997. Dosage differential effects of permethrin impregnated into bednets on pyrethroid resistant and susceptible genotypes of the mosquito *Anopheles stephensi*. Med. Vet. Entomol. 11, 368–372.

Hollingworth, R.M., Dong, K., 2008. The biochemical and molecular genetic basis of resistance in arthropods. In: Whalon, M.E., Mota-Sanchez, D., Hollingworth, R.M. (Eds.), Global Pesticide Resistance in Arthropods. CAB International, Cambridge, MA, p. 192.

Hougard, J.M., Corbel, V., N'Guessan, R., Darriet, F., Chandre, F., Akogbéto, M., et al., 2003. Efficacy of mosquito nets treated with insecticide mixtures or mosaics against insecticide resistant *Anopheles gambiae* and *Culex quinquefasciatus* (Diptera: Culicidae) in Côte d'Ivoire. Bull. Entomol. Res. 93, 491–498.

Huchard, E., Martinez, M., Alout, H., Douzery, E.J., Lutfalla, G., Berthomieu, A., et al., 2006. Acetylcholinesterase genes within the Diptera: takeover and loss in true flies. Proc. Biol. Sci. 273, 2595–2604.

Karunaratne, S.H.P.P., Hemingway, J., Jayawardena, K.G.I., Dassanayaka, V., Vaughan, A., 1995. Kinetic and molecular differences in the amplified and non-amplified esterases from insecticide-resistant and susceptible *Culex quinquefasciatus* mosquitoes. J. Biol. Chem. 270, 31124–31128.

Kasai, S., Weerashinghe, I.S., Shono, T., 1998. P450 monooxygenases are an important mechanism of permethrin resistance in *Culex quinquefasciatus* Say larvae. Arch. Insect Biochem. Physiol. 37, 47–56.

Kawada, H., Higa, Y., Komagata, O., Kasai, S., Tomita, T., Thi Yen, N., et al., 2009. Widespread distribution of a newly found point mutation in voltage-gated sodium channel in pyrethroid-resistant *Aedes aegypti* populations in Vietnam. PLoS Negl. Trop. Dis. 3, e527.

Kay, B., Vu, S.N., 2005. New strategy against *Aedes aegypti* in Vietnam. Lancet 365, 613–617.

Kelly, K., 2009. The History of Medecine. Early Civilizations: Prehistoric Times to 500 C.E. Facts On File, USA.

Kelly-Hope, L., Ranson, H., Hemingway, J., 2008. Lessons from the past: managing insecticide resistance in malaria control and eradication programmes. Lancet Infect. Dis. 8, 387–389.

Kilian, A., Byamukama, W., Pigeon, O., Atieli, F., Duchon, S., Phan, C., 2008. Long-term field performance of a polyester-based long-lasting insecticidal mosquito net in rural Uganda. Malar. J. 7, 49.

Klinkenberg, E., McCall, P.J., Wilson, M., Amerasinghe, F., Donnelly, M., 2008. Impact of urban agriculture on malaria vectors in Accra, Ghana. Malar. J. 7, 151.

Kolaczinski, J.H., Fanello, C., Hervé, J.P., Conway, D.J., Carnevale, P., Curtis, C.F., 2000. Experimental and molecular genetic analysis of the impact of pyrethroid and non-pyrethroid insecticide impregnated bednets for mosquito control in an area of pyrethroid resistance. Bull. Entomol. Res. 90, 125–132.

Kumar, S., Thomas, A., Samuel, T., Sahgal, A., Verma, A., Pillai, M.K.K., 2009. Diminished reproductive fitness associated with the deltamethrin resistance in an Indian strain of dengue vector mosquito, *Aedes aegypti* L. Trop. Biomed. 26, 155–164.

Labbé, P., Lenormand, T., Raymond, M., 2005. On the worldwide spread of an insecticide resistance gene: a role for local selection. J. Evol. Biol. 18, 1471–1484.

Labbé, P., Berthomieu, A., Berticat, C., Alout, H., Raymond, M., Lenormand, T., et al., 2007a. Independent duplications of the acetylcholinesterase gene conferring insecticide resistance in the mosquito *Culex pipiens*. Mol. Biol. Evol. 24, 1056–1067.

Labbé, P., Berticat, C., Berthomieu, A., Unal, S., Bernard, C., Weill, M., et al., 2007b. Forty years of erratic insecticide resistance evolution in the mosquito *Culex pipiens*. PLoS Genet. 3, e205.

Labbé, P., Sidos, N., Raymond, M., Lenormand, T., 2009. Resistance gene replacement in the mosquito *Culex pipiens*: fitness estimation from long term cline series. Genetics 182, 303–312.

Lambrechts, L., Chevillon, C., Albright, R., Thaisomboonsuk, B., Richardson, J., Jarman, R., et al., 2009. Genetic specificity and potential for local adaptation between dengue viruses and mosquito vectors. BMC Evol. Biol. 9, 160.

Laven, H., 1967. Eradication of *Culex pipiens fatigans* through cytoplasmic incompatibility. Nature 216, 383–384.

Lenormand, T., Raymond, M., 1998. Resistance management: the stable zone strategie. Proc. R. Soc. Lond. B 265, 1985–1990.

Lenormand, T., Guillemaud, T., Bourguet, D., Raymond, M., 1998a. Appearance and sweep of a gene duplication: adaptive response and potential for new functions in the mosquito *Culex pipiens*. Evolution 52, 1705–1712.

Lenormand, T., Guillemaud, T., Bourguet, D., Raymond, M., 1998b. Evaluating gene flow using selected markers: a case study. Genetics 149, 1383–1392.

Lenormand, T., Bourguet, D., Guillemaud, T., Raymond, M., 1999. Tracking the evolution of insecticide resistance in the mosquito *Culex pipiens*. Nature 400, 861–864.

Lisle, E., 1757. Observations on Husbandry. J. Hughs, London.

Liu, N., Xu, Q., Zhu, F., Zhang, L., 2006. Pyrethroid resistance in mosquitoes. Insect Sci. 13, 159–166.

Luleyap, H.U., Alptekin, D., Kasap, H., Kasap, M., 2002. Detection of knockdown resistance mutations in *Anopheles sacharovi* (Diptera: Culicidae) and genetic distance with *Anopheles gambiae* (Diptera: Culicidae) using cDNA sequencing of the voltage-gated sodium channel gene. J. Med. Entomol. 39, 870–874.

Lund, A.E., 1984. Pyrethroid modification of sodium channel: current concepts. Pestic. Biochem. Physiol. 22, 161–168.

Luz, P.M., Codeco, C.T., Medlock, J., Struchiner, C.J., Valle, D., Galvani, A.P., 2009. Impact of insecticide interventions on the abundance and resistance profile of *Aedes aegypti*. Epidemiol. Infect. 137, 1203–1215.

Marcombe, S., Poupardin, R., Darriet, F., Reynaud, S., Bonnet, J., Strode, C., et al., 2009. Exploring the molecular basis of insecticide resistance in the dengue vector *Aedes aegypti*: a case study in Martinique Island (French West Indies). BMC Genomics 10, 494.

Martinez-Torres, D., Chandre, F., Williamson, M.S., Darriet, F., Bergé, J.B., Devonshire, A.L., et al., 1998. Molecular characterization of pyrethroid knockdown resistance (*kdr*) in the major malaria vector *Anopheles gambiae s.s.* Insect Mol. Biol. 7, 179–184.

Martinez-Torres, D., Chevillon, C., Brun-Barale, A., Bergé, J.B., Pasteur, N., Pauron, D., 1999. Voltage-dependent Na+ channels in pyrethroid-resistant *Culex pipiens* L. mosquitoes. Pestic. Sci. 55, 1012–1020.

Massoulié, J., Bon, S., 1993. L'acétylcholinestérase: une structure originale pour une fonction vitale. Ann. Inst. Pasteur (actualités) 4, 35–49.

Matambo, T.S., Abdalla, H., Brooke, B.D., Koekemoer, L.L., Mnzava, A., Hunt, R.H., et al., 2007. Insecticide resistance in the malarial mosquito *Anopheles arabiensis* and association with the kdr mutation. Med. Vet. Entomol. 21, 97–102.

McCarroll, L., Hemingway, J., 2002. Can insecticide resistance status affect parasite transmission in mosquitoes? Insect Biochem. Mol. Biol. 32, 1345–1351.

McCarroll, L., Paton, M.G., Karunaratne, S.H.P.P., Jayasuryia, H.T.R., Kalpage, K.S.P., Hemingway, J., 2000. Insecticides and mosquito-borne disease. Nature 407, 961–962.

McKenzie, J.A., 1996. Ecological and Evolutionnary Aspects of Insecticide Resistance. Academic Press, Austin, Texas, USA.

McKenzie, J.A., Parker, A.G., Yen, J.L., 1992. Polygenic and single gene responses to selection for resistance to diazinon in *Lucilia cuprina*. Genetics 130, 613–620.

Melander, A., 1914. Can insects become resistants to sprays? J. Econ. Entomol. 7, 167−172.

Michalakis, Y., Renaud, F., 2009. Malaria: evolution in vector control. Nature 462, 298−300.

Mittal, P.K., 2003. Biolarvicides in vector control: challenges and prospects. J. Vector Borne Dis. 40, 20−32.

Morrison, A.C., Zielinski-Gutierrez, E., Scott, T.W., Rosenberg, R., 2008. Defining challenges and proposing solutions for control of the virus vector *Aedes aegypti*. PLoS Med. 5, e68.

Mouchès, C., Pasteur, N., Bergé, J.B., Hyrien, O., Raymond, M., Robert de Saint Vincent, B., et al., 1986. Amplification of an esterase gene is responsible for insecticide resistance in a California *Culex* mosquito. Science 233, 778−780.

Mouchès, C., Magnin, M., Bergé, J.-B., Silvestri, M.D., Beyssat, V., Pasteur, N., et al., 1987. Overproduction of detoxifying esterases in organophosphate-resistant *Culex* mosquitoes and their presence in other insects. Proc. Natl. Acad. Sci. U.S.A. 84, 2113−2116.

Muller, P., Warr, E., Stevenson, B.J., Pignatelli, P.M., Morgan, J.C., Steven, A., et al., 2008. Field-caught permethrin-resistant *Anopheles gambiae* overexpress CYP6P3, a P450 that metabolises pyrethroids. PLoS Genet. 4, 10.

N'Guessan, R., Darriet, F., Doannio, J.M.C., Chandre, F., Carnevale, P., 2001. Olyset Net[(R)] efficacy against pyrethroid-resistant *Anopheles gambiae* and *Culex quinquefasciatus* after 3 years field use in Côte d'Ivoire. Med. Vet. Entomol. 15, 97−104.

N'Guessan, R., Corbel, V., Akogbéto, M., Rowland, M., 2007. Reduced efficacy of insecticide-treated nets and indoor residual spraying for malaria control in pyrethroid resistance area, Benin. Emerg. Infect. Dis. 13, 199−206.

Nabeshima, T., Mori, A., Kozaki, T., Iwata, Y., Hidoh, O., Harada, S., et al., 2004. An amino acid substitution attributable to insecticide-insensitivity of acetylcholinesterase in a Japanese encephalitis vector mosquito, *Culex tritaeniorhynchus*. Biochem. Biophys. Res. Commun. 313, 794−801.

Nauen, R., 2006. Insecticide mode of action: return of the ryanodine receptor. Pest Manag. Sci. 62, 690−692.

Nauen, R., 2007. Insecticide resistance in disease vectors of public health importance. Pest Manag. Sci. 63, 628−633.

Ndjemai, H.N.M., Patchoke, S., Atangana, J., Etang, J., Simard, F., Bilong, C.F.B., et al., 2009. The distribution of insecticide resistance in *Anopheles gambiae* s.l. populations from Cameroon: an update. Trans. R. Soc. Trop. Med. Hyg. 103, 1127−1138.

Nielsen-Leroux, C, Charles, J.F., Thiéry, I., Georghiou, G.P., 1995. Resistance in a laboratory population of *Culex quinquefasciatus* (Diptera: Culicidae) to *Bacillus sphaericus* binary toxin is due to a change in the receptor on midgut brush-border membranes. Eur. J. Biochem. 228, 206−210.

Nielsen-Leroux, C., Pasquier, F., Charles, J.F., Sinègre, G., Gaven, B., Pasteur, N., 1997. Resistance to *Bacillus sphaericus* involves different mechanisms in *Culex pipiens* (Diptera: Culicidae) larvae. J. Econ. Entomol. 34, 321−327.

Nielsen-Leroux, C., Pasteur, N., Prètre, J., Charles, J.-f., Sheikh, H.B., Chevillon, C., 2002. High resistance to *Bacillus sphaericus* binary toxin in *Culex pipiens* (Diptera: Culicidae): the complex situation of West Mediterranean countries. J. Med. Entomol. 39, 729−735.

Oakeshott, J., Home, I., Sutherland, T., Russell, R., 2003. The genomics of insecticide resistance. Genome Biol. 4, 202.

Oakeshott, J.G., Devonshire, A.L., Claudianos, C., Sutherland, T.D., Horne, I., Campbell, P.M., et al., 2005. Comparing the organophosphorus and carbamate insecticide resistance mutations in cholin- and carboxyl-esterases. Chem. Biol. Interact. 157−158, 269−275.

Okoye, P.N., Brooke, B.D., Koekemoer, L.L., Hunt, R.H., Coetzee, M., 2008. Characterisation of DDT, pyrethroid and carbamate resistance in *Anopheles funestus* from Obuasi, Ghana. Trans. R. Soc. Trop. Med. Hyg. 102, 591–598.

Ooi, E.-E., Goh, K.-T., Gubler, D.J., 2006. Dengue prevention and 35 years of vector control in Singapore. Emerg. Infect. Dis. 12, 887–893.

Ortelli, F., Rossiter, L.C., Vontas, J., Ranson, H., Hemingway, J., 2003. Heterologous expression of four glutathione transferase genes genetically linked to a major insecticide-resistance locus from the malaria vector *Anopheles gambiae*. Biochem. J. 373, 957–963.

Oxborough, R.M., Mosha, F.W., Matowo, J., Mndeme, R., Feston, E., Hemingway, J., et al., 2008. Mosquitoes and bednets: testing the spatial positioning of insecticide on nets and the rationale behind combination insecticide treatments. Ann. Trop. Med. Parasitol. 102, 717–727.

Pages, F., Orlandi-Pradines, E., Corbel, V., 2007. Vecteurs du paludisme : biologie, diversite, controle et protection individuelle. Med. Mal. Infect. 37, 153–161.

Pasteur, N., Raymond, M., 1996. Insecticide resistance genes in mosquitoes. Their mutations, migration and selection in field populations. J. Hered. 87, 444–449.

Pasteur, N., Sinègre, G., 1975. Esterase polymorphism and sensitivity to Dursban organophosphorous insecticide in *Culex pipiens pipiens* populations. Biochem. Genet. 13, 789–803.

Pasteur, N., Sinègre, G., Gabinaud, A., 1981a. Est-2 and Est-3 polymorphism in *Culex pipiens* L. from southern France in relation to organophosphate resistance. Biochem. Genet. 19, 499–508.

Pasteur, N., Iseki, A., Georghiou, G.P., 1981b. Genetic and biochemical studies of the highly active esterases A' and B associated with organophosphate resistance in mosquitoes of the *Culex pipiens* complex. Biochem. Genet. 19, 909–919.

Pasteur, N., Georghiou, G.P., Iseki, A., 1984. Variation in organophosphate resistance and esterase activity in *Culex quinquefasciatus* Say from California. Génét. Sél. Evol. 16, 271–284.

Paton, M.G., Karunaratne, S.H., Giakoumaki, E., Roberts, N., Hemingway, J., 2000. Quantitative analysis of gene amplification in insecticide-resistant *Culex* mosquitoes. Biochem. J. 346, 17–24.

Paul, A., Harrington, L.C., Zhang, L., Scott, J.G., 2005. Insecticide resistance in *Culex pipiens* from New York. J. Am. Mosq. Control Assoc. 21, 305–309.

Pavela, R., 2009. Larvicidal property of essential oils against *Culex quinquefasciatus* Say (Diptera: Culicidae). Ind. Crops Prod. 30, 311–315.

Pedrini, N., Mijailovsky, S.J., Girotti, J.R., Stariolo, R., Cardozo, R.M., Gentile, A., et al., 2009. Control of pyrethroid-resistant Chagas disease vectors with entomopathogenic fungi. Plos Negl. Trop. Dis. 3, 11.

Penilla, P.R., Rodriguez, A.D., Hemingway, J., Torres, J.L., Arredondo-Jimenez, J.I., Rodriguez, M.H., 1998. Resistance management strategies in malaria vector mosquito control: baseline data for a large-scale field trial against *Anopheles albimanus* in Mexico. Med. Vet. Entomol. 12, 217–233.

Penilla, R.P., Rodriguez, A.D., Hemingway, J., Torres, J.L., Solis, F., Rodriguez, M.H., 2006. Changes in glutathione S-transferase activity in DDT resistant natural Mexican populations of *Anopheles albimanus* under different insecticide resistance management strategies. Pestic. Biochem. Physiol. 86, 63–71.

Pennetier, C., Corbel, V., Hougard, J.-M., 2005. Combination of a non-pyrethroid insecticide and a repellent. A new approach for controlling knockdown resistant mosquitoes. Am. J. Trop. Med. Hyg. 72, 739–744.

Perera, M.D.B., Hemingway, J., Karunaratne, S., 2008. Multiple insecticide resistance mechanisms involving metabolic changes and insensitive target sites selected in anopheline vectors of malaria in Sri Lanka. Malar. J. 7.

Pinto, J., Lynd, A., Vicente, J.L., Santolamazza, F., Randle, N.P., Gentile, G., et al., 2007. Multiple origins of knockdown resistance mutations in the Afrotropical mosquito vector *Anopheles gambiae*. PLoS ONE 2, e1243.

Poirié, M., Raymond, M., Pasteur, N., 1992. Identification of two distinct amplifications of the esterase B locus in *Culex pipiens* (L.) mosquitoes from Mediterranean countries. Biochem. Genet. 30, 13–26.

Poupardin, R., Reynaud, S., Strode, C., Ranson, H., Vontas, J., David, J.P., 2008. Cross-induction of detoxification genes by environmental xenobiotics and insecticides in the mosquito *Aedes aegypti*: impact on larval tolerance to chemical insecticides. Insect Biochem. Mol. Biol. 38, 540–551.

Prabhaker, N., Georghiou, G.P., Pasteur, N., 1987. Genetic association between highly active esterases and organophosphate resistance in *Culex tarsalis*. J. Am. Mosq. Control Assoc. 3, 473–475.

Ramphul, U., Boase, T., Bass, C., Okedi, L.M., Donnelly, M.J., Muller, P., 2009. Insecticide resistance and its association with target-site mutations in natural populations of *Anopheles gambiae* from eastern Uganda. Trans. R. Soc. Trop. Med. Hyg. 103, 1121–1126.

Ranson, H., Jensen, B., Vulule, J.M., Wang, X., Hemingway, J., Collins, F.H., 2000. Identification of a point mutation in the voltage-gated sodium channel gene of Kenyan *Anopheles gambiae* associated with resistance to DDT and pyrethroids. Insect Mol. Biol. 9, 491–497.

Raymond, M., Fournier, D., Bride, J.-M., Cuany, A., Bergé, J., Magnin, M., et al., 1986. Identification of resistance mechanisms in *Culex pipiens* (Diptera: Culicidae) from southern France: insensitive acetylcholinesterase and detoxifying oxidases. J. Econ. Entomol. 79, 1452–1458.

Raymond, M., Heckel, D.G., Scott, J.G., 1989. Interactions between pesticide genes. Model and experiment. Genetics 123, 543–551.

Raymond, M., Callaghan, A., Fort, P., Pasteur, N., 1991. Worldwide migration of amplified insecticide resistance genes in mosquitoes. Nature 350, 151–153.

Raymond, M., Berticat, C., Weill, M., Pasteur, N., Chevillon, C., 2001. Insecticide resistance in the mosquito *Culex pipiens*: what have we learned about adaptation? Genetica 112-113, 1–10.

Read, A.F., Lynch, P.A., Thomas, M.B., 2009. How to make evolution-proof insecticides for malaria control. PLoS Biol. 7, e58.

Reimer, L., Fondjo, E., Patchok, S., Diallo, B., Lee, Y., Ng, A., et al., 2008. Relationship between kdr mutation and resistance to pyrethroid and DDT insecticides in natural populations of *Anopheles gambiae*. J. Med. Entomol. 45, 260–266.

Reiter, P., 2001. Climate change and mosquito-borne disease. Environ. Health Perspect. 109 (Suppl. 1), 141–161.

Rezza, G., Nicoletti, L., Angelini, R., Romi, R., Finarelli, A.C., Panning, M., et al., 2007. Infection with chikungunya virus in Italy: an outbreak in a temperate region. Lancet 370, 1840–1846.

Roberts, D.R., Andre, R.G., 1994. Insecticide resistance issues in vector-borne disease control. Am. J. Trop. Med. Hyg. 50, 21−34.
Rogan, W.J., Chen, A., 2005. Health risks and benefits of bis(4-chlorophenyl)-1,1,1-trichloroethane (DDT). Lancet 366, 763−773.
Rooker, S., Guillemaud, T., Bergé, J., Pasteur, N., Raymond, M., 1996. Coamplification of esterase A and B genes as a single unit in Culex pipiens mosquitoes. Heredity 77, 555−561.
Saavedra-Rodriguez, K., Urdaneta-Marquez, L., Rajatileka, S., Moulton, M., Flores, A.E., Fernandez-Salas, I., et al., 2007. A mutation in the voltage-gated sodium channel gene associated with pyrethroid resistance in Latin American Aedes aegypti. Insect Mol. Biol. 16, 785−798.
Santolamazza, F., Calzetta, M., Etang, J., Barrese, E., Dia, I., Caccone, A., et al., 2008. Distribution of knock-down resistance mutations in Anopheles gambiae molecular forms in west and west-central Africa. Malar. J. 7, 74.
Scholte, E.-J., Njiru, B., Smallegange, R., Takken, W., Knols, B., 2003. Infection of malaria Anopheles gambiae s.s. and filariasis Culex quinquefasciatus vectors with the entomopathogenic fungus Metarhizium anisopliae. Malar. J. 2, 29.
Scholte, E.-J., Ng'habi, K., Kihonda, J., Takken, W., Paaijmans, K., Abdulla, S., et al., 2005. An entomopathogenic fungus for control of adult African malaria mosquitoes. Science 308, 1641−1642.
Scholte, E.-J., Knols, B.G.J., Takken, W., 2006. Infection of the malaria mosquito Anopheles gambiae with the entomopathogenic fungus Metarhizium anisopliae reduces blood feeding and fecundity. J. Invertebr. Pathol. 91, 43−49.
Scott, J.G., 1999. Cytochromes P450 and insecticide resistance. Insect Biochem. Mol. Biol. 29, 757−777.
Severini, C., Romi, R., Marinucci, M., Raymond, M., 1993. Mechanisms of insecticide resistance in field populations of Culex pipiens from Italy. J. Am. Mosq. Control Assoc. 9, 164−168.
Sharp, B., Ridl, F., Govender, D., Kuklinski, J., Kleinschmidt, I., 2007a. Malaria vector control by indoor residual insecticide spraying on the tropical island of Bioko, Equatorial Guinea. Malar. J. 6, 52.
Sharp, B.L., Kleinschmidt, I., Streat, E., Maharaj, R., Barnes, K.I., Durrheim, D.N., et al., 2007b. Seven years of regional malaria control collaboration—Mozambique, South Africa and Swaziland. Am. J. Trop. Med. Hyg. 76, 42−47.
Shen, B., Dong, H.-Q., Tian, H.-S., Ma, L., Li, X.-L., Wu, G.-L., et al., 2003. Cytochrome P450 genes expressed in the deltamethrin-susceptible and -resistant strains of Culex pipiens pallens. Pestic. Biochem. Physiol. 75, 19−26.
Shrivast, S.P., Georghiou, G.P., Metcalf, R.L., Fukuto, T.R., 1970. Carbamate resistance in mosquitoes—metabolism of propoxur by susceptible and resistant larvae of Culex pipiens fatiguans. Bull. World Health Organ. 42, 931−942.
Singh, O., Bali, P., Hemingway, J., Subbarao, S., Dash, A., Adak, T., 2009. PCR-based methods for the detection of L1014 kdr mutation in Anopheles culicifacies sensu lato. Malar. J. 8, 154.
Sinkins, S.P., Gould, F., 2006. Gene drive systems for insect disease vectors. Nat. Rev. Genet. 7, 427−435.
Soderlund, D.M., Knipple, D.C., 2003. The molecular biology of knockdown resistance to pyrethroid insecticides. Insect Biochem. Mol. Biol. 33, 563−577.

Strode, C., Wondji, C.S., David, J.-P., Hawkes, N.J., Lumjuan, N., Nelson, D.R., et al., 2008. Genomic analysis of detoxification genes in the mosquito *Aedes aegypti*. Insect Biochem. Mol. Biol. 38, 113−123.

Stump, A.D., Atieli, F.K., Vulule, J.M., Besansky, N.J., 2004. Dynamics of the pyrethroid knockdown resistance in Western Kenya populations of *Anophles gambiae* in response to insecticide-treated bed net trials. Am. J. Trop. Med. Hyg. 70, 591−596.

Surendran, S.N., Karunaratne, S.H.P.P., Adams, Z., Hemingway, J., Hawkes, N.J., 2005. Molecular and biochemical characterization of a sand fly population from Sri Lanka: evidence for insecticide resistance due to altered esterases and insensitive acetylcholinesterase. Bull. Entomol. Res. 95, 371−380.

Tabashnik, B.E., Liu, Y.-B., Finson, N., Masson, L., Heckel, D.G., 1997a. One gene in diamondback moth confers resistance to four *Bacillus thuringiensis* toxins. Proc. Natl. Acad. Sci. U.S.A. 94, 1640−1644.

Tabashnik, B.E., Liu, Y.-B., Malvar, T., Heckel, D.G., Masson, L., Ballester, V., et al., 1997b. Global variation in the genetic and biochemical basis of diamondback moth resistance to *Bacillus thuringiensis*. Proc. Natl. Acad. Sci. U.S.A. 94, 12780−12785.

Tantely, M.L., Tortosa, P., Alout, H., Berticat, C., Berthomieu, A., Rutee, A., et al., 2010. Insecticide resistance in *Culex pipiens quinquefasciatus* and *Aedes albopictus* mosquitoes from La Réunion Island. Insect Biochem. Mol. Biol. 40, 317−324.

Tatem, A., Rogers, D., Hay, S., 2006. Estimating the malaria risk of African mosquito movement by air travel. Malar. J. 5, 57.

Thomas, D.R., McCarroll, L., Roberts, R., Karunaratne, P., Roberts, C., Casey, D., et al., 2006. Surveillance of insecticide resistance in head lice using biochemical and molecular methods. Arch. Dis. Child. 91, 777−778.

Tolle, M.A., 2009. Mosquito-borne diseases. Curr. Probl. Pediatr. Adolesc. Health Care 39, 97−140.

Tomita, T., Kono, Y., Shimada, T., 1996. Chromosomal localization of amplified esterase genes in insecticide resistant *Culex* mosquitoes. Insect Biochem. Mol. Biol. 26, 853−857.

Tsetsarkin, K.A., Vanlandingham, D.L., McGee, C.E., Higgs, S., 2007. A single mutation in Chikungunya virus affects vector specificity and epidemic potential. PLoS Pathog 3, e201.

Vais, H., Williamson, M.S., Devonshire, A.L., Usherwood, P.N.R., 2001. The molecular interactions of pyrethroid insecticides with insect and mammalian sodium channels. Pest Manag. Sci. 57, 877−888.

van den Berg, H., 2009. Global status of DDT and its alternatives for use in vector control to prevent disease. Environ. Health Perspect. 117, 1656−1663.

Varro. approx. 40 BC. *De re rustica libri*.

Vaughan, A., Hawkes, N., Hemingway, J., 1997. Co-amplification explains linkage disequilibrium of two mosquito esterase genes in insecticide-resistant *Culex quinquefasciatus*. Biochem. J. 325, 359−365.

Vazeille, M., Moutailler, S., Coudrier, D., Rousseaux, C., Khun, H., Huerre, M., et al., 2007. Two Chikungunya isolates from the outbreak of La Reunion (Indian Ocean) exhibit different patterns of infection in the mosquito, *Aedes albopictus*. PLoS ONE 2, e1168.

Vontas, J.G., Small, G.J., Hemingway, J., 2001. Glutathione *S*-transferases as antioxidant defence agents confer pyrethroid resistance in *Nilaparvata lugens*. Biochem. J. 357, 65−72.

Vontas, J.G., McCarroll, L., Karunaratne, S.H.P.P., Louis, C., Hurd, H., Hemingway, J., 2004. Does environmental stress affect insect-vectored parasite transmission? Physiol. Entomol. 29, 210–213.

Vontas, J., Blass, C., Koutsos, A.C., David, J.P., Kafatos, F.C., Louis, C., et al., 2005. Gene expression in insecticide resistant and susceptible *Anopheles gambiae* strains constitutively or after insecticide exposure. Insect Mol. Biol. 14, 509–521.

Vulule, J.M., Beach, R.F., Atieli, F.K., Mcallister, J.C., Brogdon, W.G., Roberts, J.M., et al., 1999. Elevated oxidase and esterase levels associated with permethrin tolerance in *Anopheles gambiae* from Kenyan villages using permethrin-impregnated nets. Med. Vet. Entomol. 13, 239–244.

Walker, K., Lynch, M., 2007. Contributions of *Anopheles* larval control to malaria suppression in tropical Africa: review of achievements and potential. Med. Vet. Entomol. 21, 2–21.

Waterhouse, R.M., Wyder, S., Zdobnov, E.M., 2008. The *Aedes aegypti* genome: a comparative perspective. Insect Mol. Biol. 17, 1–8.

Weill, M., Berticat, C., Raymond, M., Chevillon, C., 2000a. Quantitative polymerase chain reaction to estimate the number of amplified esterase genes in insecticide-resistant mosquitoes. Anal. Biochem. 285, 267–270.

Weill, M., Chandre, F., Brengues, C., Manguin, S., Akogbeto, M., Pasteur, N., et al., 2000b. The *kdr* mutation occurs in the Mopti form of *Anopheles gambiae s.s.* through introgression. Insect Mol. Biol. 9, 451–455.

Weill, M., Fort, P., Berthomieu, A., Dubois, M.P., Pasteur, N., Raymond, M., 2002. A novel acetylcholinesterase gene in mosquitoes codes for the insecticide target and is nonhomologous to the ace gene in *Drosophila*. Proc. R. Soc. Lond. B 269, 2007–2016.

Weill, M., Lutfalla, G., Mogensen, K., Chandre, F., Berthomieu, A., Berticat, C., et al., 2003. Insecticide resistance in mosquito vectors. Nature 423, 136–137.

Weill, M., Berthomieu, A., Berticat, C., Lutfalla, G., Negre, V., Pasteur, N., et al., 2004. Insecticide resistance: a silent base prediction. Curr. Biol. 14, R552–R553.

Weill, M., Labbé, P., Duron, O., Pasteur, N., Fort, P., Raymond, M., 2005. Insecticide resistance in the mosquito *Culex pipiens*: towards an understanding of the evolution of *ace* genes. In: Fellowes, M.D.E., Holloway, G.J., Rolff, J. (Eds.), Insect Evolutionary Ecology. CABI publishing, Oxon, UK, pp. 393–404.

Welburn, S.C., Coleman, P.G., Maudlin, I., Fevre, E.M., Odiit, M., Eisler, M.C., 2006. Crisis, what crisis? Control of Rhodesian sleeping sickness. Trends Parasitol. 22, 123–128.

Whalon, M.E., Mota-Sanchez, D., Hollingworth, R.M., 2008. Analysis of global pesticide resistance in arthropods. In: Whalon, M.E., Mota-Sanchez, D., Hollingworth, R.M. (Eds.), Global Pesticide Resistance in Arthropods. CAB International, Cambridge, MA, p. 192.

Wirth, M., Marquine, M., Georghiou, G.P., Pasteur, N., 1990. Esterase A2 and B2 in *Culex quinquefasciatus* (Diptera: Culicidae): role in organophosphate resistance and linkage. J. Econ. Entomol. 27, 202–206.

WHO, 1957. Malaria section. Bull. World Health Organ. 16, 874.

WHO. 2006. Pesticides and their application for the control of vectors and pests of public health importance. World Health Organization, Geneva, Switzerland.

Williamson, M.S., Martinez-Torres, D., Hick, C.A., Devonshire, A.L., 1996. Identification of mutations in the housefly para-type sodium channel gene associated with knockdown resistance (kdr) to pyrethroid insecticides. Mol. Gen. Genet. 252, 51–60.

Wondji, C.S., Irving, H., Morgan, J., Lobo, N.F., Collins, F.H., Hunt, R.H., et al., 2009. Two duplicated P450 genes are associated with pyrethroid resistance in *Anopheles funestus*, a major malaria vector. Genome Res. 19, 452–459.

Wood, R.J., 1981. Insecticide resistance: genes and mechanisms. In: Bishop, J.A., Cook, L.M. (Eds.), Genetic Consequence of Man Made Change. Academic Press, London, pp. 53–96.

Xu, Q., Wang, H., Zhang, L., Liu, N., 2006. *Kdr* allelic variation in pyrethroid resistant mosquitoes, *Culex quinquefasciatus* (S.). Biochem. Biophys. Res. Commun. 345, 774–780.

Yadouleton, A., Asidi, A., Djouaka, R., Braima, J., Agossou, C., Akogbeto, M., 2009. Development of vegetable farming: a cause of the emergence of insecticide resistance in populations of *Anopheles gambiae* in urban areas of Benin. Malar. J. 8, 103.

15 Genetics of Major Insect Vectors

Patricia L. Dorn[1,], François Noireau[2], Elliot S. Krafsur[3], Gregory C. Lanzaro[4] and Anthony J. Cornel[5]*

[1]Loyola University New Orleans, New Orleans, LA, USA, [2]IRD, Montpellier, France, [3]Iowa State University, Ames, IA, USA, [4]University of California at Davis, Davis, CA, USA, [5]University of California at Davis, Davis, CA, USA and Mosquito Control Research Lab, Parlier, CA, USA

15.1 Introduction

15.1.1 Significance and Control of Vector-Borne Disease

Vector-borne diseases are responsible for a substantial portion of the global disease burden causing ~1.4 million deaths annually (Campbell-Lendrum et al., 2005; Figure 15.1) and 17% of the entire disease burden caused by parasitic and infectious diseases (Townson et al., 2005). Control of insect vectors is often the best, and sometimes the only, way to protect the population from these destructive diseases. Vector control is a moving target with globalization and demographic changes causing changes in infection patterns (e.g., rapid spread, urbanization, appearance in nonendemic countries); and the current unprecedented degradation of the global environment is affecting rates and patterns of vector-borne disease in still largely unknown ways.

15.1.2 Contributions of Genetic Studies of Vectors to Understanding Disease Epidemiology and Effective Disease Control Methods

Studies of vector genetics have much to contribute to understanding vector-borne disease epidemiology and to designing successful control methods. Geneticists have performed phylogenetic analyses of major species; have identified new species, subspecies, **cryptic species**, and introduced vectors; and have determined which taxa are epidemiologically important. Cytogeneticists have shown that the evolution of chromosome structure is important in insect vector speciation. Population geneticists have uncovered the complex population structures of insect vector populations. Monitoring gene flow among populations has revealed the geographic coverage needed for control and the source of insects appearing following pesticide applications. Genetic control methods such as the sterile insect technique (SIT) (Krafsur, 2002), or introduction of refractory traits or transgenic symbionts

*E-mail: dorn@loyno.edu

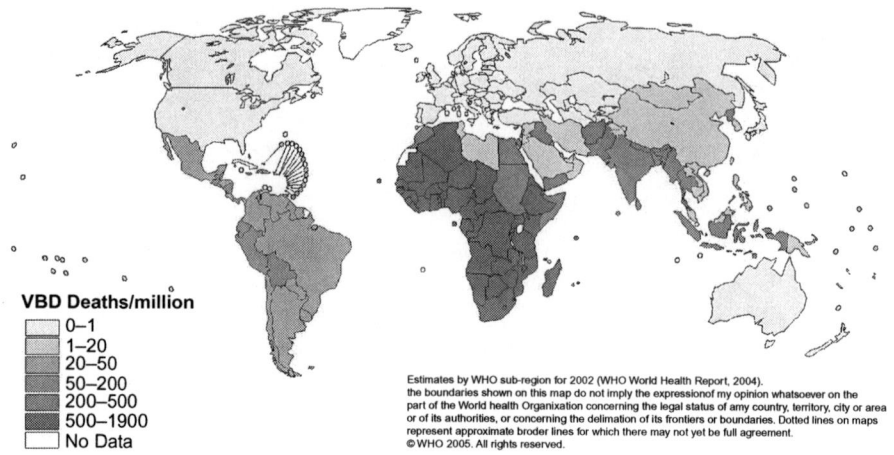

Figure 15.1 Estimates of vector-borne disease deaths per million inhabitants, with permission, copyright WHO.
Source: Available from http://www.who.int/heli/risks/vectors/vector/en/index.html

carrying molecules toxic to pathogens, although still largely theoretical, can add to the arsenal. Molecular genetics and new genome and proteome tools promise advances in understanding the genetic basis of vector capacity including habitat and host preference, innate immunity, drought tolerance, and insecticide resistance, among other phenomena, and the development of new attractants and repellents.

15.1.3 Chapter Overview

In this chapter we review what is known about the genetics of the vectors of three of the most important vector-borne diseases: African sleeping sickness, Chagas disease, and malaria.

15.2 Genetics of Tsetse Flies, *Glossina* spp. (Diptera: Glossinidae), and African Trypanosomiasis

15.2.1 Introduction

Tsetse flies (Diptera: Glossinidae) are among the most important insects in sub-Sahara Africa because they are obligate blood feeders and the vectors of African trypanosomiasis, caused by trypanosomes (Mastigophora: Kinetoplastida: Trypanosomatidae) that kill humans and domestic mammals. It is hard to overestimate the importance of African trypanosomiasis. More than 50 million people are said to be at risk for human African trypanosomiasis (HAT) in 37 countries (WHO, 2006). Trypanosomiasis in domestic animals is termed Nagana, and was estimated to cost African agriculture US $4.5 billion per year (Reinhardt, 2002) via loss of food, dung, and drafting power. Conservative estimates claim approximately

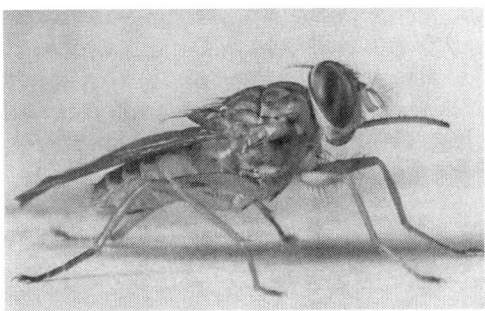

Figure 15.2 Resting tsetse.
Source: Photo courtesy of the DFID Animal Health Program.

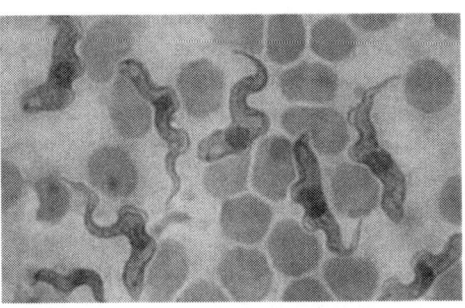

Figure 15.3 *Trypanosoma brucei* in blood.
Source: Photo courtesy of the DFID Animal Health Program.

450,000 HAT cases in 1997 declined to 10,880 by 2007 (Barrett, 2006; WHO, 2006). Simarro et al. (2008) offer a comprehensive epidemiological review.

Tsetse (pronounced "tsee-tsee") flies (Figure 15.2) and trypanosomes (Figure 15.3) have many interesting and unusual biological features. Here I briefly review the principle biological and ecological features of tsetse flies with an emphasis on their population structures and relationships with trypanosomes. Citations are kept to a minimum, with emphasis on recent literature. Much fuller treatments of tsetse flies, trypanosomiasis, and African trypanosomes are available in texts by Buxton (1955), Ford (1971), Jordan (1993), Leak (1998), and Maudlin et al. (2004). Specialist reviews include Gooding and Krafsur (2004, 2005) for genetics, Krafsur (2003, 2009) for population genetics, and Walshe et al. (2009) for tsetse–trypanosome–symbiont interactions.

15.2.2 Systematics and Biology of Tsetse Flies in a Nutshell

The Family Glossinidae

Systematics and unresolved taxonomic problems were reviewed by Jordan (1993) and Gooding and Krafsur (2005). Tsetse flies are assigned to the family Glossinidae (McAlpine, 1989). All extant tsetse flies are classified into a single genus *Glossina* Wiedemann 1830. Four subgenera have been described. *Austenina*, *Nemorhina*, and *Glossina* correspond to the Fusca, Palpalis, and Morsitans species groups, respectively. In general, Fusca group flies are forest inhabitants, Palpalis are riverine and lacustrine (living near lakes), and Morsitans group flies are savanna inhabitants. Subgenus *Machadomyia* consists only of two *G. austeni* subspecies. There is an unnamed, extinct sister group known from the Florissant shale of Colorado, dating some 38 mya and similar tsetse-like fossils were uncovered in Oligocene strata in Germany indicating formerly a much wider geographic distribution. Thirty-four taxa have been described, consisting of 23 species and 7 species complexes of 17 named subspecies that differ slightly morphologically, if at all. Most subspecies are allopatric. Morphological species identification is not easy.

Species Complexes of Medical Importance: G. morsitans s.l., G. palpalis s.l., and G. fuscipes s.l.

Three species complexes are geographically widespread and of much medical and economic importance. Some constituent taxa have been examined genetically in *G. morsitans* s.l., *G. palpalis* s.l., and *G. fuscipes* s.l. The most thoroughly examined is *G. morsitans* s.l. and its close relative, *G. swynnertoni*. *G. morsitans* s.l. comprises *G. morsitans morsitans*, *G. m. centralis*, and *G. submorsitans*. Genetic data suggest longstanding isolation. *G. palpalis* s.l. comprises *G. p. palpalis* and *G. p. gambiense*; recent studies suggest incipient speciation in *G. p. palpalis* (Ravel et al., 2007; Dyer et al., 2009). The foregoing taxa are allopatric and hybrid males are sterile, the females typically sterile or semisterile (Gooding, 1993). *G. fuscipes* s.l. consists of *G. f. fuscipes*, *G. f. martini*, and *G. f. quanzensis*. There is no known cross-breeding work on this species complex.

Glossina Life History

All *Glossina* are exclusively hematophagous. Reproduction is viviparous by adenotrophic viviparity. In this unusual birthing process, found in several insect taxa, an egg develops into a larva within the mother, the mother deposits the mature larva in an appropriate microhabitat, and the animal pupariates in the soil (Figures 15.4 and 15.5). In tsetse flies, one egg develops at a time within the "uterus" and the subsequent larva develops to maturity within its mother over a 9- or 10-day interval. Adult development from the deposited larva requires about 28 days at tropical temperatures. The youngest larvapositing female is at least 15 days old, so the minimum time necessary to produce two offspring is 25 days. Compensating for slow reproduction are their survival rates, which for adult females typically exceed a mean 97% per day in stable and growing populations. Generation time for savanna species is 45–50 days and population-doubling time is at least 35 days. Hargrove

Figure 15.4 Tsetse larviposition.
Source: Photo courtesy of the DFID Animal Health Program.

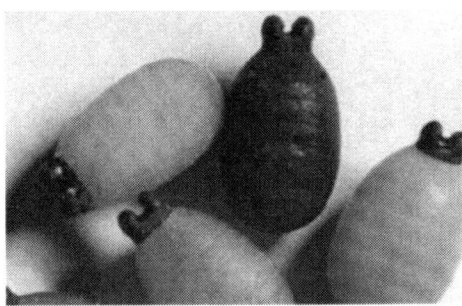

Figure 15.5 Tsetse pupae: light are young and dark are older.
Source: Photo courtesy of the DFID Animal Health Program.

(2003, 2004) authoritatively reviewed the essentials of tsetse demography and its consequences for fly population management.

15.2.3 Biogeography

Distribution

Most tsetse fly populations occur within latitudes 12°N to 25°S, about one-third of the African continent (Figures 15.6–15.8). To oversimplify, moisture availability is

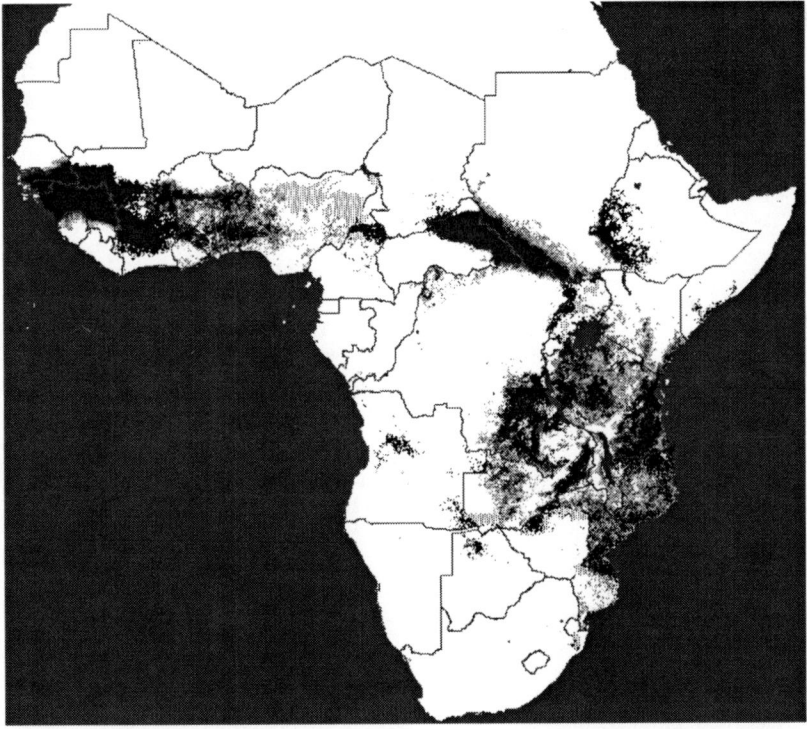

Figure 15.6 Distribution of Morsitans group tsetse flies predicted from satellite imagery courtesy of ERGO Ltd and TALA, Oxford, with permission.
Source: Available from http://ergodd.zoo.ox.ac.uk/.

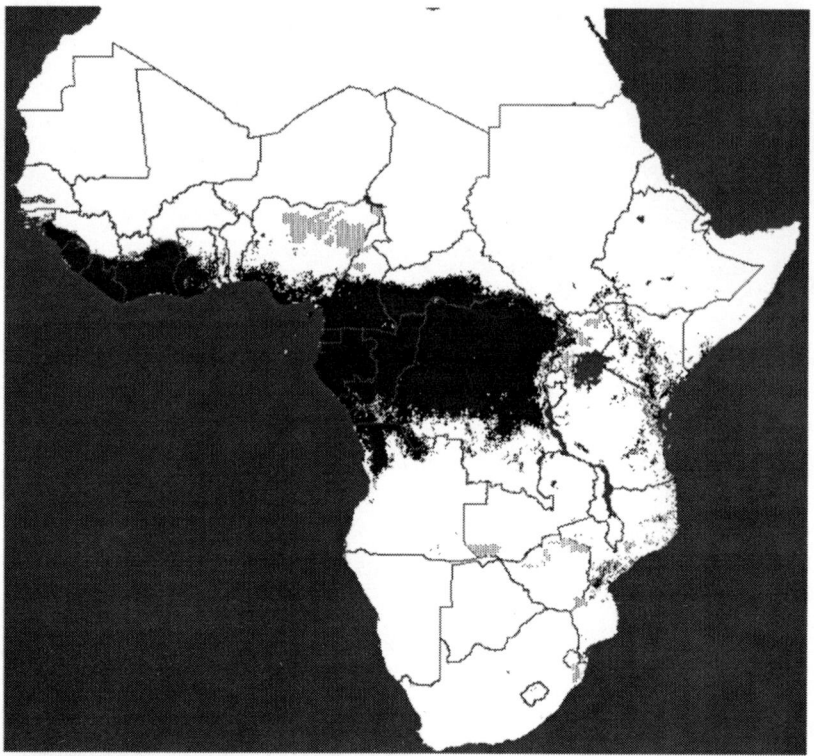

Figure 15.7 Distribution of Fusca group tsetse flies predicted from satellite imagery courtesy of ERGO Ltd and TALA, Oxford, with permission.
Source: Available from http://ergodd.zoo.ox.ac.uk/.

limiting to the north and low temperatures limit southern distribution. Thus tsetse flies are largely confined to sub-Saharan Africa, although relict populations of *G. pallidipes* and *G. palpalis* have been recorded in the southwestern corner of the Arabian Peninsula. They are likely survivors from earlier times when more equable climates prevailed. It would be interesting to examine their genetic affinities. Distribution maps of 30 tsetse taxa are available and correlations of distribution with satellite imagery indices confirm that each has rather specific temperature and precipitation requirements (Rogers and Robinson, 2004).

15.2.4 Genetics and Population Genetics

Cytogenetics

All tsetse flies examined cytologically have two pairs of metacentric autosomes and a sex bivalent: $2N = 4 + XY$. Many also have heterochromatic supernumerary chromosomes that vary in number within and among taxa. Sex chromosome

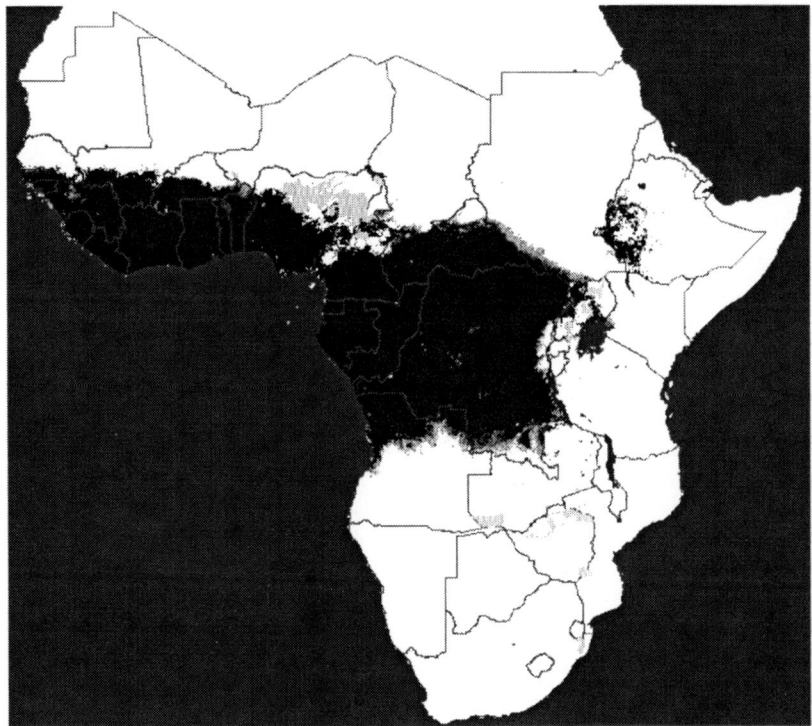

Figure 15.8 Distribution of Palpalis group tsetse flies predicted from satellite imagery courtesy of ERGO Ltd and TALA, Oxford, with permission.
Source: Available from http://ergodd.zoo.ox.ac.uk/.

polymorphisms have been recorded in wild *G. p. palpalis*. Well-banded polytene chromosomes can be obtained from pupal trichogen cells. Examination of polytenes in Morsitans and Palpalis flies has demonstrated that taxa can be separated by pericentric and paracentric inversions (Figure 15.9). Surveys of polytene chromosome **diversity** in natural populations remain to be undertaken and should prove highly informative.

Genetic Variability Based on Allozymes

Allozyme, microsatellite, and mitochondrial variation has been assessed in all Morsitans and most medically important Palpalis group taxa (Table 15.1). Allozyme diversities have also been assessed in some Fusca taxa. Allozyme diversities are fairly homogeneous among the taxa examined. Diversity, defined as the mean heterozygosity over loci, is about 6% in tsetse and compares with 10—20% in other Diptera such as *Musca*, *Stomoxys*, *Haematobia*, and *Drosophila*. The best explanation for the difference is that genetic diversity is inversely proportional to **effective population sizes**. Effective population size, N_e, is the harmonic mean number of reproducing individuals in an ideal population that have the same allele

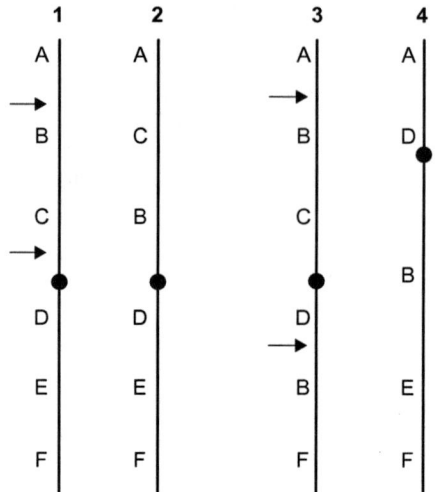

Figure 15.9 Chromosome inversions arise by two simultaneous breaks, indicated by arrows, followed by inversion of the interior segments and rejoins with the "wrong" partners. A paracentric inversion (chromosomes 1→2) occurs in one arm of a chromosome and a pericentric inversion (3→4) includes the spindle fiber attachment region (dark circles) and both arms of a chromosome. Note that pericentric inversions can result in a change in chromosome morphology.

Table 15.1 Allozyme Variation in Tsetse Flies Compared with Other Diptera

Taxa	Number of Loci	Percent Polymorphic	Mean Alleles per Locus	Mean Diversity, H_e	Diversity at Polymorphic Loci
G. m. morsitans	45	20.0	1.4	6.6	29.9
G. m. centralis	31	32.3	1.4	6.0	18.7
G. swynnertoni	34	17.6	1.2	7.2	40.5
G. pallidipes	38	26.3	1.5	6.8	25.4
Musca domestica	68	53.4	2.5	18.3	36.7
Musca autumnalis	50	62.0	2.4	18.6	30.5
Stomoxys calcitrans	38	52.6	1.8	9.6	18.2

frequencies as an actual population under study. *Glossina* population numbers averaged over time and space are much less than the common Diptera with which they have been compared. Diversities at polymorphic allozyme loci, on the other hand, are of the same magnitude as those in other Diptera. There is good empirical evidence in *G. pallidipes* for **balancing selection** that favors allozyme heterozygotes.

Genetic Variability Based on Microsatellite Loci

Microsatellite loci in genomic DNA have been identified in Morsitans and Palpalis group flies (Table 15.2). Many loci are shared across subgenera. Microsatellite diversities, averaged over loci, varied from a low 0.43 in *G. f. fuscipes* up to 0.81 in *G. m. submorsitans*. The mean over 6 taxa and 58 population samples is 0.70

Table 15.2 Microsatellite Diversities and Tests for Random Matings F in *Glossina* spp. and the Housefly

	Number of Populations	Number of Loci	Alleles per Locus	Diversity H_e	Within Demes F_{IS}	Among Demes F_{ST}
G. m. morsitans	6	6	11.0 ± 5.6	0.73 ± 0.06	0.03 ± 0.03	0.19 ± 0.05
G. m. morsitans	9	7	28.6 ± 4.5	0.74 ± 0.05	0.17 ± 0.07	0.13 ± 0.01
G. m. centralis	7	7	8.8 ± 3.7	0.70 ± 0.09	−0.12 ± 0.04	0.19 ± 0.04
G. m. submorsitans	7	7	12.7 ± 6.2	0.81 ± 0.04	0.03 ± 0.03	0.17 ± 0.07
G. pallidipes	21	8	26.8 ± 8.7	0.80 ± 0.03	0.07 ± 0.03	0.18 ± 0.02
G. f. fuscipes [a]	8	5	8.2 ± 3.6	0.43 ± 0.07	0.11 ± 0.05	0.22 ± 0.07
Musca domestica	14	7	7.9 ± 1.1	0.86 ± 0.02	0.08 ± 0.02	0.13 ± 0.02

[a](Abila et al., 2008)

Table 15.3 Mitochondrial Diversities and Genetic Differentiation in Wild *Glossina* Species

	Method	Number of Populations	Number of Flies	Number of Haplotypes	Haplotype Diversity, H_S	F_{ST}
G. m. morsitans	SSCP	5	111	25	0.81 ± 0.04	0.09 ± 0.02
G. m. morsitans	ABI 3730	7	96	33	0.81	0.40 ± 0.08
G. m. centralis	SSCP	6	265	7	0.54 ± 0.16	0.81 ± 0.07
G. m. submorsitans	SSCP	7	282	26	0.51 ± 0.12	0.35
G. pallidipes	SSCP	21	624	39	0.42 ± 0.02	0.52 ± 0.001
G. pallidipes	ABI 3730	23	873	181	0.73 ± 0.09	0.47 ± 0.07
G. p. gambiensis	SSCP	13	372	9	0.18	0.68
G. swynnertoni	ABI 3730	8	149	18	0.59 ± 0.10	0.04 ± 0.003
G. f. fuscipes [a]	ABI 3730	22	284	36	0.91 ± 0.008	0.60 ± 0.07

[a](Beadell et al., 2010)

and compares with a mean 0.86 over 14 samples of the cosmopolitan housefly, *Musca domestica*. Here, again, we see that a larger effective population size results in greater diversity.

Genetic Variability Based on mtDNA

Mitochondrial variation has also been assessed in *Glossina*. Examination of cytochrome oxidase subunit I and ribosomal 16S2 loci in large samples of Morsitans and Palpalis group flies disclosed many variants (Table 15.3). Mitochondrial diversity (only 18 haplotypes) was least in *G. swynnertoni* although allozyme variation was about the same as in other taxa. This tsetse fly occupies a relatively small region of north-central Tanzania. Further, it has been subject to extensive control procedures and its range has contracted. Mitochondrial variation is single copy and inherited matrilineally and more sensitive to demographic flux than diploid, nuclear variation. As already discussed with regard to *G. pallidipes*, natural selection favors allozyme heterozygotes.

Rationale for Diversity Levels Observed

Low diversities in *G. m. centralis* and southern African *G. pallidipes* reflect earlier demographic events. The rinderpest epizootic that began in the late nineteenth century is said to have killed 90% or more of the mammalian population of central and southern Africa with the virtual elimination of Morsitans group flies, that only slowly recovered from undetectably small numbers (Ford, 1971). In *G. pallidipes*, microsatellite and mitochondrial diversities were less in southern Africa than in East Africa and both were strongly correlated with each other, but neither was correlated with allozyme diversity. In contrast to the allozyme diversities, mitochondrial and microsatellite variation was consistent with a severe and prolonged reduction in population sizes in southern Africa. Allozyme variation was conserved, however, and differentiation at allozyme loci among populations was much less than at microsatellite and mitochondrial loci providing a good example of balanced polymorphism, discussed further later.

15.2.5 Population Structure

Breeding structure and gene flow have been examined in *G. morsitans* s.l., *G. pallidipes*, *G. swynnertoni*, *G. f. fuscipes*, *G. p. palpalis*, and *G. p. gambiense*. These are among the chief vectors of African trypanosomes. Some of the foregoing taxa were sampled over much of their geographic ranges and lab cultures of some were compared with genetic data from field samples. Two generalities emerged from the investigations: random **genetic drift** was pronounced in all taxa, leading to highly significant levels of genetic differentiation among conspecific populations. Most population samples were differentiated even when within 25–50 km of each other, and genetic diversity in lab cultures was only mildly attenuated compared with their field cousins, with the possible exception mitochondrial diversity (8 haplotypes, $H_s = 0.36$) in a longstanding *G. austeni* culture. Estimates of genetic differentiation among subpopulations are provided by F_{ST} and its analogs. F_{ST} may be defined as the among-subpopulation variance in gene frequencies as a fraction of the variance among all individuals and is interpreted as departures from random mating among subpopulations. The accompanying tables provide mean estimates of F_{ST} taken over all samples and indicate low levels of gene flow (Tables 15.2 and 15.3). The theoretical, mathematical relationship between genetic differentiation and gene flow is a reciprocal one such that the number of reproducing migrants exchanged among demes is proportional to the inverse of F_{ST}. For most tsetse taxa examined, the mean numbers of reproductive flies exchanged among populations is generally less than one to two per generation, indicative of strong genetic drift. No work has been reported on the breeding structures of vectors *G. brevipalpis*, *G. tachinoides*, *G. longipalpis*, and *G. f. quanzensis*.

Explanations for Population Structure Observed

How may we best interpret the significant levels of genetic differentiation among conspecific tsetse populations? Mark-recapture studies have shown strong propensities to disperse (reviewed by Rogers, 1977; Leak, 1998; Hargrove, 2003). Thus, mean displacement rates of savanna-inhabiting tsetse flies were a mean 252 m d^{-1}. The foregoing rate predicts rapid dispersion and a frontal advance of approximately 5 km y^{-1}, thereby counteracting genetic drift. Theory indicates that numerically little gene flow is necessary to do so (Wright, 1978). Assuming numerically significant rates of dispersion among demes, how may we account for the high levels of genetic drift typically recorded? Spatial variation in natural selection offers an explanation.

Agents of natural selection include prevailing temperature and moisture regimes that govern the distribution of tsetse flies. It has been shown that among regional, conspecific *G. pallidipes* and *G. morsitans* populations, different discriminant models and predictor variables best describe their distributions (Rogers and Robinson, 2004). This provides empirical evidence that spatially separated demes have adapted to their different environments, perhaps accounting for a measure of the genetic differentiation typically observed. Tests of hypotheses were performed on *G. pallidipes* in five ecologically diverse sites in Kenya and Zambia where the prevailing climates substantially differ. Hypotheses tested included homogeneity in among-population thermal tolerances, desiccation tolerances, body water amounts, lipid amounts, and spontaneous locomotory activities. Population responses were nearly homogeneous and little potential for evolutionary change was detected (Terblanche et al., 2006).

Cuticular hydrocarbons and lipids are important in water conservation and sex recognition and could, in principle, provide evidence of local adaptation. The chief sex pheromone in *G. pallidipes* did not differ among flies from Ethiopia, Zimbabwe, and five independently derived lab cultures (Carlson et al., 2000), nor in four diverse Kenya populations (Jurenka et al., 2007). Hydrocarbon quantities differed among populations, but did not correlate with environmental variables.

The foregoing genetic studies indicate that genetic drift in tsetse is typically stronger than gene flow. We have no direct evidence to support selection as an isolating mechanism. Drift is inversely proportional to effective population size, and tsetse maximum reproductive rates are low. For now, it seems we cannot reject the hypothesis that random genetic drift operates principally as a consequence of small effective population sizes and spatial isolation.

15.2.6 Commensals and Symbionts in Tsetse Flies

Three maternally transmitted microorganisms can be found in tsetse (Aksoy, 2000; Walshe et al., 2009, for reviews). These are *Wolbachia pipientis*, an intracellular parasitic Rickettsia that causes **cytoplasmic incompatibility** in some insect species, but cytoplasmic incompatibility has not been demonstrated experimentally in tsetse flies; *Wigglesworthia glossinidia* is an intracellular, obligatory symbiont that produces essential vitamins and other substances absent in the host fly's blood meals. This bacterium has been sequenced and has a small genome a bit less than

700 kb. Studies indicate an association of *W. glossinidia* with vector competence in host flies in the laboratory. *Sodalis glossinidius* occurs in some species and strains of tsetse flies and it demonstrates much genotypic diversity. Its genome has also been sequenced. *Wolbachia* and *Sodalis* incidence vary among natural tsetse species and populations. In natural conspecific tsetse populations, *Sodalis glossinidius* prevalence varies spatially and its presence in fly colonies can favor susceptibility to trypanosome infection. There is empirical evidence from Cameroon of a statistical association of *Sodalis* with trypanosome (*T. brucei* s.l., *T. congolense*) infections (Farikou et al., 2010).

15.2.7 Population Genetics of African Trypanosomes

Trypanosome Systematics of T. b. brucei, T. b. gambiense, *and* T. b. rhodesiense

Trypanosome systematics was reviewed by Stevens and Brisse (2004). Only the medically and economically most significant taxa are dealt with here because they are the best known. *Trypanosoma brucei* s.l. consists of *T. b. brucei*, *T. b. gambiense*, and *T. b. rhodesiense*. They cannot be distinguished morphologically. *T. b. brucei* is a widely distributed parasite of ungulates and a chief agent of nagana. It cannot infect people. HAT is caused in east and southern Africa by *T. b. rhodesiense*, and in west and central Africa by *T. b. gambiense*. These cause clinically distinct diseases but both are invariably lethal unless carefully treated by expert medical help. *T. b. rhodesiense* infection is clinically acute; its reservoirs include humans and cattle. Infection with *T. b. gambiense* is chronic and much slower to kill, however, this form accounts for most human cases. In the decade between 1997 and 2006, 97% of reported HAT cases were caused by *T. b. gambiense* (Simarro et al., 2008). Its reservoirs are principally humans but there is tangible evidence for other mammalian hosts (Njiokou et al., 2006). In mammals, trypanosomes undergo continuous antigenic variation, thereby defying host immune responses. Most HAT foci, about 200 in number, are localized and longstanding, but epidemics occur, some of great magnitude (Buxton, 1955; Ford, 1971).

Identification of Different Species of Trypanosome

In addition to *T. b. brucei*, economically significant agents of animal trypanosomiasis include *T. congolense* and *T. vivax*. Identification of trypanosome infection in the field depends on clinical criteria and microscopic diagnosis of trypanosomes, however, lacks adequate sensitivities. There is a useful card agglutination test for *T. b. gambiense* good for mass screening but no such test for *T. b. rhodesiense*. PCR-based specific diagnostic tools are available for experimental uses but sensitive, inexpensive, diagnostic tests for field use are urgently required. Chappuis et al. (2005) reviewed diagnostic options.

Trypanosome Genetic Diversity

Much genetic diversity has been demonstrated within *T. brucei* subspecies; Koffi et al. (2009) used microsatellite loci to demonstrate high levels of genetic differentiation among geographic isolates of *T. b. gambiense* from patients and extraordinary levels of genetic homogeneity in isolates from single patients (i.e., strong linkage disequilibrium and negative F_{IS}), indicative of strong clonal structure. Simo et al. (2010) also demonstrated strong population structure in *T. brucei* among 72 isolates (67 were *T. b. gambiense*) from nine foci in west and central Africa with the same results. Indeed, HAT epidemics usually demonstrate clonal etiologies, including those caused by *T. b. rhodesiense*.

Trypanosome Population Structure

Although trypanosome population structure is typically clonal, superinfections of two or more strains have been found in ~9% of 137 cryopreserved isolates of the three *T. brucei* subspecies from mammalian hosts across Africa (e.g., Balmer and Caccone, 2008). Meiosis and genetic recombination of *T. b. brucei* in *G. morsitans* have been conclusively demonstrated experimentally and take place in the fly's salivary glands (Gibson et al., 2008). In principle, genetic recombination is important epidemiologically because it would allow the evolution of new strains; its frequency and extent among natural populations is unknown but likely to be quite low. For now, genetically and spatially representative sampling of *Trypanosoma brucei* s.l. in the field is technically and operationally problematic because the technology of trypanosome recovery is inadequate and survey costs great.

Short vector dispersal distances could explain, in principle, spatially differentiated trypanosome populations. The question of trypanosome adaptation to local vector populations is relevant. Low trypanosome infection rates in tsetse (<1%) are typical because only a small fraction of ingested trypanosomes establish infective forms in the insect's salivary glands and there are multiple mechanisms of tsetse immunity to trypanosome infection (Aksoy and Rio, 2005, and Walshe et al., 2009, for reviews).

15.2.8 Genomics

Glossina *Genomics*

The genome of *G. m. morsitans* is being sequenced via the shotgun approach and its transcriptome made available. Annotated expressed sequence tag libraries in *G. m. morsitans* are available and continue to grow. There is also a bacterial artificial chromosome library (http://bacpac.chori.org). A *G. m. morsitans* gene database is available from http://old.genedb.org/genedb/glossina/, ftp://ftp.sanger.ac.uk/pub/pathogens/Glossina/morsitans/ and http://www.vectorbase.org/index.php. Sequencing the *G. p. palpalis* genome has been proposed (Askoy, 2010). Its genome is c. 7000 mb, compared with c. 600 mb in *G. m. morsitans* and c. 278 mb in *Anopheles gambiae*.

Trypanosome Genomics

Trypanosoma b. brucei has been sequenced (Berriman et al., 2005). Its genome is c. 26 mb. Partial shotgun sequencing of *T. b. gambiense* is underway and whole genome shotgun sequencing of *T. congolense* and *T. vivax* has begun. Details and sequences are available at the Wellcome Trust Sanger Institute: http://www.sanger.ac.uk/Projects/G_morsitans. The genomes are now available of tsetse symbiotic enteric bacteria *Sodalis glossinidia* (Toh et al., 2006) and *Wigglesworthia glossinidia* (Akman et al., 2002). The *Wolbachia pipientis* genome is also available (Wu et al., 2004).

The foregoing developments are promising because they greatly facilitate research by providing a framework for functional studies and may lead ultimately to more effective drug chemotherapies for trypanosomiasis and provide important insights into trypanosome morphological differentiation and cellular adaptations to its vector, much of which are poorly understood (Sharma et al., 2009). Additional practical applications for genomic research have been identified (Askoy et al., 2010) and discussed (Walshe et al., 2009).

15.2.9 Tsetse Population Management and Relevance of Genetic Studies

A Brief History of Attempts at Population Management

There are no vaccines for HAT and pharmaceutical treatment is expensive, dangerous, and unavailable to millions of people most at risk. Thus, African trypanosomiasis is best controlled by eliminating its insect vectors. Older methods of tsetse population management included destruction of wildlife that serve as trypanosome reservoirs, removal of vegetation that can harbor tsetse flies, and insecticidal applications. None of these methods taken separately or together have afforded long-term control because of invasion from nearby, untreated populations. Genetic methods have been advocated and tested (see Gooding and Krafsur, 2005, for review and critique). The SIT has been applied experimentally to several tsetse taxa (reviews in Hargrove, 2003; Klassen and Curtis, 2005). In theory, cytoplasmic incompatibility can be used to introduce high genetic loads (leading to reduced reproductive capacity) into a natural population and replace it with innocuous conspecific forms. Current research, elegant in concept and design, involves developing transgenic *Sodalis glossinidius* to express trypanolytic substances and building a *Wolbachia*-infected line of tsetse containing the transformed symbionts (Aksoy et al., 2001; Aksoy, 2003). The *Wolbachia* is then used to introduce cytoplasmic incompatibility when released into an uninfected natural population. Uninfected, wild-type females are at a reproductive disadvantage because they are sterile when mated with *Wolbachia*-infected males. In theory, the engineered, trypanosome-refractory line eventually replaces the wild-type, trypanosome- susceptible tsetse population.

The Promise and Limitations of Genetic Population Management

How applicable are genetic methods of tsetse population management and trypanosomiasis control likely to become? An eradication program utilizing population reduction by insecticide-laced targets and SIT was carried out against *G. austeni* on Unguja Island (Zanzibar). Sterile male releases, begun in August 1994, were ended in December 1997. *G. austeni* has not reappeared since. Hargrove (2003) authoritatively evaluated the Zanzibar results and concluded that application of SIT on the continent is contraindicated because of its very high dollar costs, poor sterile fly competitiveness, and the availability of other proven cost-effective methods. Examination of the SIT for tsetse flies, via a useful simulation model, has shown the method is economically and ecologically inappropriate even under the best-case scenarios modeled (Vale and Torr, 2005). Replacement of natural vector populations with conspecific nonvectors, while elegant in theory, seems impractical because it depends on: (1) the successful development, production, and release of adequate numbers of reproductively competitive flies carrying *Wolbachia* and genetically transformed commensals, (2) a thorough knowledge of vector-trypanosome ecology throughout the region to be treated, (3) absence of trypanosome strains resistant to the trypanocidal substance(s), and (4) the age structure of target populations will cause lengthy population response times such that long generation times and high survival rates require continuous releases of an engineered tsetse strain over long intervals to expose target population virgins to the engineered released forms. Thus, the financial costs of field application and follow-up are likely to be enormous and difficult to sustain.

Conventional Tsetse Population Management

Applicable conventional methods include reduction of vector survival probabilities and densities via traps and targets, sequential aerial applications of low-volume insecticides, and targeted applications of pour-on insecticides to cattle at risk (Torr et al., 2005). Odor-baited, deltamethrin-impregnated targets and carefully plotted and timed aerial applications of deltamethrin eliminated *G. m. centralis* from the Okavango delta in Botswana during 2001−2002 (Kgori et al., 2006). The tsetse has not been detected since in the treated 15,900 km^2 (6139 mi^2) area even though untreated populations continue in nearby Zambia and Angola.

15.2.10 Further Work Needed

Barriers to developing further scientific knowledge of tsetse fly and trypanosome biology include a severe paucity of laboratory cultures representative of natural populations. Much work on trypanosomes is without reference to their vectors in nature and some may be irrelevant, therefore, to field conditions. Culture of tsetse flies is extraordinarily labor intensive, often difficult, and expensive. Few colonies now exist and most have been extant for many years, subject to inadvertent selection and inbreeding. Of 34 taxa, only nine are in culture: *G. austeni, G. pallidipes, G. m. morsitans, G. m. centralis, G. m. submorsitans, G. tachinoides, G. p. palpalis, G. p. gambiense*, and *G. f. fuscipes*. The pronounced genetic variation among natural tsetse populations argues for geographically more extensive sampling across the geographic range to assess genetic variation and reciprocal crossing of different lines to assay fertilities and uncover sibling species. Because of the importance of commensal organisms in tsetse physiology and vector capacity, a better understanding of their prevalence, genetic diversity, and geographic distribution is also required. Vector−parasite coadaptations have important epidemiological and economic consequences but how they vary spatially is unknown.

Regarding tsetse fly population management, it is noteworthy that their historical distribution and abundance is unchanged except for their elimination in relatively small areas on the northern and southern margins—southwestern Zambia, northeastern Zimbabwe, northern Nigeria, and the Okavango in Botswana. Proposed genetic approaches to trypanosomiasis control are interesting and related research is yielding important scientific insights, but laboratory elegance alone would seem unlikely to overcome the dynamic nature of tsetse fly populations in their natural habitats. Nevertheless, exploitation of the rapidly accumulating genomic databases may lead eventually to highly efficacious and affordable trypanosomiasis control.

15.2.11 Concluding Remarks

Current knowledge of pathogenic trypanosome genetic variation in the field and corresponding variation in vector tsetse fly species and populations are insufficient to support a thorough understanding of trypanosomiasis epidemiology and epizootiology. Better methods are required for assessing trypanosome prevalence and genetic variation in the field. Comprehensive geographical sampling is required of trypanosome variation vis-à-vis that of their tsetse vectors. The breeding structures of *G. brevipalpis, G. f. quanzensis*, and *G. longipalpis* are unknown and these are the vectors in Mozambique and much of central Africa. The present view that most, if not all, Morsitans and many Palpalis group populations are local may serve to define areas in which systematic vector management schemes may be applied without massive immigration from untreated, conspecific populations. An effective and affordable genetically based area-wide tsetse fly population management is unlikely to be developed in the foreseeable future.

15.3 Genetics of the Triatominae (Hemiptera, Reduviidae) and Chagas Disease

15.3.1 Introduction

Distribution and Importance

American trypanosomiasis is a zoonosis caused by the flagellate protozoa *Trypanosoma cruzi* and is restricted to the New World. The distribution of animal trypanosomiasis is widespread from the southern United States to Patagonia (roughly 40°N to 45°S). Foci of the human disease (also called Chagas disease) range from Mexico to the northern half of Argentina, mostly in poor rural areas where houses are infested with insect vectors belonging to the sub-family Triatominae (Hemiptera, Reduviidae). Chagas disease is amongst the most serious of the so-called neglected tropical diseases. It can be fatal in its early acute stage, but more usually progresses to a debilitating chronic disease that affects up to 30% of infected people and involves severe cardiac and intestinal lesions (Schofield, 1994). The World Bank (1993) ranked Chagas disease as by far the most serious of the parasitic diseases affecting Latin America, with a considerable social and economic impact far in excess of the combined impact of malaria, leishmaniasis, and schistosomiasis.

After a peak estimate of 24 million people infected and 100 million at risk in the second part of the twentieth century (Walsh, 1984), regional Chagas control initiatives (Southern Cone, Andean, Central American, and Amazonian) have resulted in a dramatic reduction of disease prevalence due to decreased vector transmission and an increase in blood donation screening. A 1990 estimate of 16–18 million people infected has been reduced to a current estimate of ~9 million infected (Schofield et al., 2006).

Vector Transmission

The Triatominae, also known as kissing or conenose bugs, are the sole vectors of American trypanosomiasis (Figure 15.10). A triatomine vector becomes infected with *T. cruzi* by feeding on the blood of an infected person or animal. Following the blood meal, it generally deposits urine and feces on the host skin and the parasite is transmitted through the feces of infected vectors. *T. cruzi* is a stercorian trypanosome, meaning the infective forms develop in the digestive tract of the vector.

Genetic Structure of Trypanosoma cruzi

The genetic structure of *T. cruzi* is predominantly clonal, with restricted recombination, as various cellular strains persist as stable genotypes that can spread through large geographic regions (Tibayrenc et al., 1986; Tibayrenc and Breniere, 1988). Rare genetic exchange, over evolutionary time, has had a profound impact on the adaptation of *T. cruzi* to new environments, vectors, and hosts, including humans (Vallejo et al., 2009). The biological, biochemical, and genetic diversities of

Figure 15.10 Sylvan *T. infestans* caught on a sticky trap (Noireau trap). Note the smaller nymphal stages as well as the larger adult on the right. Photographer, François Noireau.

T. cruzi strains has long been recognized along with their eco-epidemiological complexity, and numerous approaches have been used to characterize the population structure of *T. cruzi*. Recently, a consensus was reached that *T. cruzi* strains should be referred to by six discrete typing units (*T. cruzi* I–VI) (Zingales et al., 2009).

15.3.2 Systematics and Biogeography

Phylogeny of the Triatominae

There is consensus that the Triatominae are derived from predatory Reduviids. On the other hand, their monophyletic or polyphyletic origin is still in dispute (Hypsa et al., 2002; de Paula et al., 2005). According to Lent and Wygodzinsky (1979), all triatomines derived from a single ancestral form, with the defining synapomorphic trait being the bloodsucking habit and associated morphological and physiological adaptations; and this monophyletic hypothesis was supported by results of a comprehensive study of 16S rDNA sequence over a large number of taxa (Hypsa et al., 2002) and recently, by cladistic analyses of Reduviidae based on genes sequences and morphological characters (Weirauch, 2008; Weirauch and Munro, 2009). This hypothesis is difficult to reconcile with the wide geographical distribution and "nest-dwelling" habit of Triatominae, because it implies that a specialized ancestral nest-dweller would subsequently adapt to range of different habitats associated with a range of different hosts (Schofield, 1988). By contrast, several arguments suggest that the Triatominae represent a polyphyletic group derived from different lineages within the predatory Reduviidae. Evidence is adduced from comparative differences between tribes in form (different tribes in Triatominae have a body form that matches that of distinct subfamilies in Reduviidae), and also at the level of salivary gland physiology and parasite susceptibility (Schofield and Galvão, 2009). More recent studies based on DNA sequence showed disruption of the Rhodniini + Triatomini clades by predatory taxa, supporting the idea of polyphyletic origins (de Paula et al., 2005).

The Five Triatominae Tribes

Triatominae are generally classified into 5 tribes and 15 genera and include, at present, some 140 species. One species (*Triatoma rubrofasciata*) is widespread both in the New World and some tropical regions of Asia and Africa. A few others are found only in Asia and India (Galvão et al., 2003). However, most species (∼125) occur exclusively in the New World, between latitude 42°N (northeast United States) and 46°S (Argentine Patagonia) (Lent and Wygodzinsky, 1979; Schofield and Galvão, 2009). The species of epidemiological importance belong to the two tribes Rhodniini and Triatomini that seem to be relatively phylogenetically distant (de Paula et al., 2005). Cryptic species, which are thought to arise through morphological convergence of two different species in competition for similar habitats, seem uncommon in Triatominae (Dujardin et al., 2009). They were apparently detected within *Triatoma sordida* (Noireau et al., 1998; Panzera et al., 2006), *R. robustus* (Monteiro et al., 2003), and *Triatoma dimidiata* (Marcilla et al., 2001;

Panzera et al., 2006), although the latter species showed phenotypic differences as well as genetic (Bustamante et al., 2004; Dorn et al., 2007).

The Rhodniini Tribe

The tribe Rhodniini contains two genera, *Rhodnius* and *Psammolestes*. Mitochondrial cytochrome *b* gene (*cyt b*) genealogies tend to support the existence of two lineages within the tribe Rhodniini: the *pictipes* and the *robustus* lineages (Abad-Franch and Monteiro, 2007). The *pictipes* lineage includes species from both the eastern (*pictipes, stali, brethesi,* and some sylvan species) and western sides of the Andes (*pallescens, colombiensis,* and *ecuadoriensis*). The members of the *robustus* lineage, also including the genus *Psammolestes*, are found east of the Andes (Abad-Franch et al., 2009). Within the *robustus* lineage, the species forming the *prolixus* group (*R. prolixus, R. robustus, R. neglectus,* and *R. nasutus*) are particularly difficult to distinguish. One of these (*R. prolixus*) is a primary domestic vector of Chagas disease (Figure 15.11) while the three others are synanthropic species that can invade or even sporadically colonize man-made ecotopes. The specific status of *R. robustus*, which is virtually indistinguishable from *R. prolixus*, has been clarified by the sequence analysis of a fragment of the mitochondrial *cyt b* gene. *R. robustus* currently includes four cryptic species. *R. robustus* I occurs in Venezuela (Orinoco

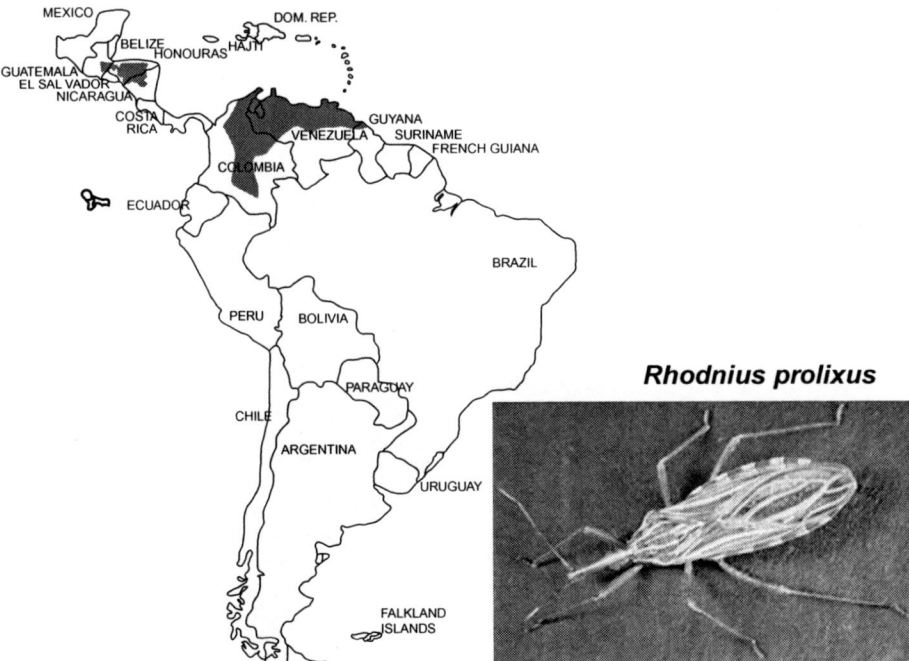

Figure 15.11 Approximate current distribution of *Rhodnius prolixus*.
Source: Data from WHO (1989). Linda Waller, artist; James Gathany, photographer (Centers for Disease Control and Prevention).

region) and is more closely related to *R. prolixus* than to the other three cryptic species found in the Amazon region (Monteiro et al., 2003).

The Triatomini Tribe

Two genera belonging to the Triatomini tribe (*Panstrongylus* and *Triatoma*) are of epidemiological importance. The phylogeny of *Panstrongylus* is currently under discussion, although there is evidence supported by cladistic, molecular, morphometric, and cytogenetic studies, suggesting the existence of northern and southern clades (Patterson et al., 2009). Only the placement of *P. megistus* is discordant. Indeed, it is largely southern in its distribution and is unusual cytogenetically, having only 18 autosomal chromosomes compared to the 20 autosomes of other *Panstrongylus* (Crossa et al., 2002). *Triatoma* is the most numerous genus of Triatominae, with 80 formally recognized species (Schofield and Galvão, 2009). North American, Central American, and Caribbean species as a whole seem genetically separate from the South American species, that is, in parallel to the northern and southern clades of *Panstrongylus* species.

The "Infestans" Subcomplex

The subcomplex *infestans* sensu (Schofield and Galvão, 2009) includes the species *infestans*, *delpontei*, and *platensis*. *T. infestans* was and still remains the most important and widespread vector of Chagas disease (Figure 15.12) in the Southern

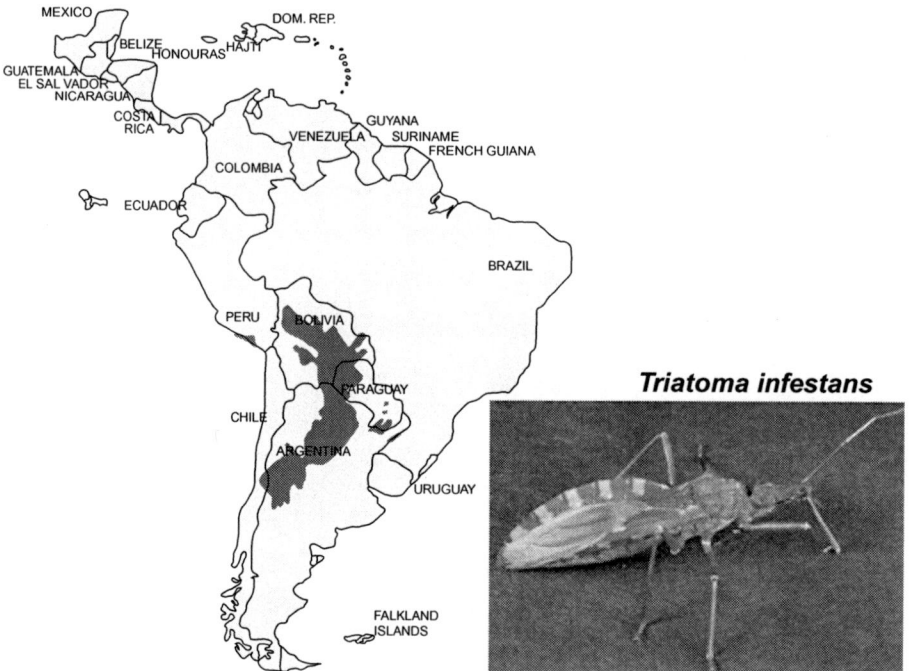

Figure 15.12 Approximate current distribution of *Triatoma infestans*.
Source: Data from Schofield (2006). Linda Waller, artist; James Gathany, photographer (Centers for Disease Control and Prevention).

Cone countries, whereas the latter two species are closely associated with bird nests. These three species are very closely related and have the same diploid chromosome number 2N = 22 (20 autosomes + XX/XY). They also have several cytogenetic traits that differ from all other triatomines, including large autosomes, C-heterochromatic blocks, and meiotic heteropycnotic chromocentres formed by autosomes and sex chromosomes (Panzera et al., 1995). All evidence indicates that, within this subcomplex, *T. platensis* is the closest relative to *T. infestans* (Bargues et al., 2006). The status of *T. infestans* and *T. platensis* as two distinct species is almost entirely based upon their ecological niche separation.

T. dimidiata, the Major Vector in Mesoamerica

T. dimidiata has become the most important vector of Chagas disease in Mexico, Central America, and northern Andean countries of South America, since control activities to eliminate *R. prolixus* have made substantial progress (Figure 15.13). *T. dimidiata* has extensive phenotypic, genotypic, and behavioral diversity in sylvan, peridomestic, and domestic habitats across its geographic range. Recent studies strongly suggest that this taxon, which has been historically regarded as a single species, includes a cryptic species and perhaps as many as three subspecies distributed in specific geographic areas with different epidemiological importance (Marcilla

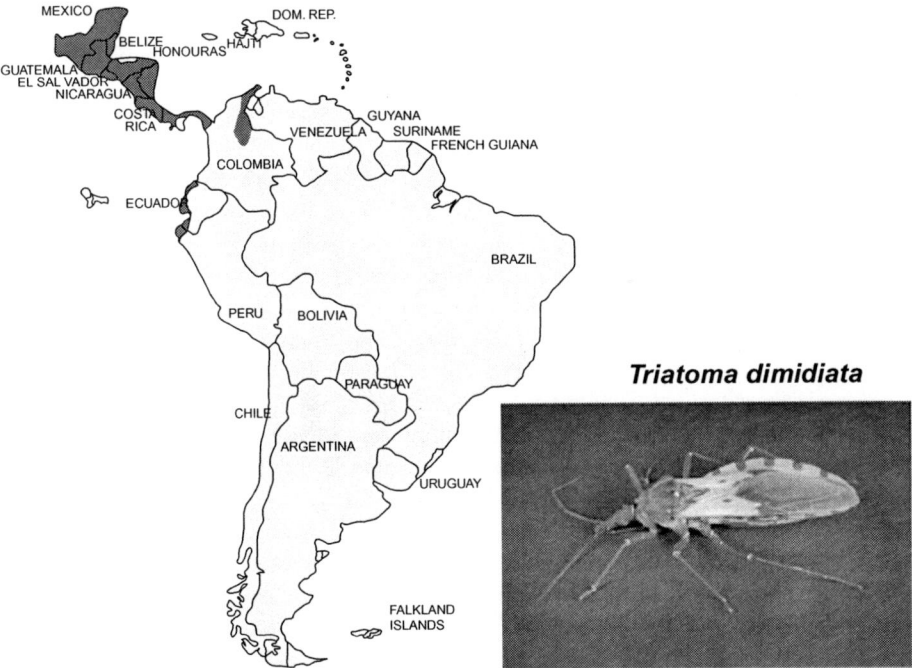

Figure 15.13 Approximate current distribution of *Triatoma dimidiata*.
Source: Data from WHO (1989). Linda Waller, artist; James Gathany, photographer (Centers for Disease Control and Prevention).

et al., 2001; Panzera et al., 2006; Dorn et al., 2007, 2009; Bargues et al., 2008). These current distinctions between taxa have major implications for transmission capacity and vector control.

15.3.3 Basic Biological and Ecological Features

Life Cycle of the Triatominae

The life cycle of the Triatominae is composed of egg and five nymph stages, which will reach sexual maturity from a few months (*R. prolixus, T. infestans*) to more than 1 year (*T. dimidiata, P. megistus*). All the stages feed on vertebrate blood but deprived of access to blood, some species can feed on other arthropods, a relic ancestral trait (Schofield, 1988). Bugs may survive for more than a month without access to food, depending on other environmental conditions, such as temperature. In most Triatominae, including strictly sylvan ones, it is worth noting a relative lack of host specificity in the feeding habits. Exceptions are observed in some ornithophylic species (*Psammolestes* spp. and *T. delpontei*) or the tropicopolitan *T. rubrofasciata* that feed preferentially on rodents.

Life History of the Triatominae

During the day, in sylvan as well as in the domestic environment, the Triatominae remain motionless, hidden inside their refuges. At nightfall, they may walk or fly (adults) to look for blood (Figueiras and Lazzari, 2000). Beyond the locomotor capacity, it is important to distinguish two dispersion modes in the Triatominae: active and passive dispersion. Both walking and flying constitute active dispersion. Flight orientation is apparently random, but it seems that during its flight the bug can be attracted to light. Passive dispersion is transportation of generally immature stages by an animal host or with the familiar objects carried or worn by the human host. This latter mode of dispersion is the most important to explain the territorial expansion of the main vectors (*T. infestans, R. prolixus, T. dimidiata*) (Schofield, 1988).

As a general but not absolute rule, the habitat of the Triatominae offers shelter and easy access to a blood source with a stable availability of hosts. According to Schofield (1988) and Gaunt and Miles (2000), each of the three most epidemiological important genera of Triatominae is virtually always associated with a particular habitat. So, species of the genus *Rhodnius* are primarily associated with palms, the genus *Panstrongylus* has predominantly evolved in burrows and tree cavities, and the genus *Triatoma* in terrestrial rocky habitats or rodent burrows. If this assumption is generally true for the genera *Rhodnius* and *Panstrongylus*, it is more questionable for the genus *Triatoma*, in which species may be found exclusively in trees or in both habitats (arboreal and terrestrial) in different localities (e.g., for *T. infestans, T. sordida,* and *T. guasayana,* Andean populations live in rockpiles and lowland populations in hollow trees).

Effects of Anthropogenic Change and Domicilation

Environmental disturbance caused by humans and resulting in damage to triatomine habitat may lead to changes in insect habitats, breeding behavior, and host preference. These disturbances and subsequent starvation are certainly the main causes of flight dispersal, which can lead to settlement in artificial structures. Domesticity is the key determinant of vector capacity in the Triatominae. Only the species adapted to human dwellings, which represent less than 5% of the total number of species, are actively contributing to transmission. The distinction between intrusion, domiciliation, and domestication may help in defining the epidemiological importance of populations or species of Triatominae (Dujardin et al., 2000). Intrusion occurs when adult specimens of sylvan species are present inside human dwellings, probably attracted by light or introduced by passive carriage. In this situation, there is no evidence of colonization (eggs, nymphs, exuviae are not found). Domiciliation represents a tentative adaptation to the house and corresponds to the finding of eggs and exuviae, and small colonies of adults and nymphs in the house (which means the complete cycle of the insect is occurring inside the house). For example, this situation has been described for *Triatoma sordida* in Bolivia (Noireau et al., 1997). The definition of domestication includes the above criteria for domiciliation, with an additional one related to the geographic scope. It is no longer a local, geographically restricted observation, but concerns a more widely extended territory with obvious arguments supporting migration by passive carriage. It is important to recall that the existence of "domesticated" species does not exclude the existence of sylvan foci. Wild populations of *T. infestans* and *R. prolixus* were recorded in Bolivia and Argentina, and Venezuela, respectively (Noireau et al., 2005; Fitzpatrick et al., 2008). More research is needed to understand the factors allowing a species to reach a high level of adaptation to the domestic habitat. We know that the adaptation systematically reduces the size of the insect (Schofield et al., 1999; Caro-Riaño et al., 2009). Another sort of plasticity, that is behavioral and related to habitat selection, might also factor into the adaptation. In the sylvan environment, species can exclusively inhabit a particular microhabitat (e.g., rockpiles or trees). However, when they invade the peridomestic/domestic environment (Figure 15.14), they can adapt to different habitats and occupy substrates that they do not colonize in the sylvan environment (de la Fuente et al., 2008). In order to provide an initial operational evaluation of each species, it may be useful to adopt a classification system based on a scale describing the degree of adaptation: (1) restricted to the sylvan environment, (2) infestation of peridomestic structures, (3) intrusion into houses, (4) domiciliation, and finally, (5) domestication; and the epidemiological importance (proved or putative vector vs. species without a documented role in the transmission of *T. cruzi* to humans).

MacArthur and Wilson (1967) distinguished "r" and "K" demographic strategies defining populations occupying unstable or stable habitats, respectively. If the "r" strategy is typically the one of mosquitoes, the "K" strategy is congruent with domestic Triatominae that are relatively large insects, produce small quantity of descendants (about 100 eggs during the life in a female of *T. infestans*), have an

Figure 15.14 House and peridomestic area (rock walls, granery) in Bolivian Andes. Photographer, François Noireau.

extended developmental cycle, poor active dispersal capacity, and are timid feeders. Triatominae do not exhaust the resources of their environment; rather seem to opt for an optimal use. Finally, they would probably be unable to recover or to escape to other richer environments after a catastrophic mortality (Rabinovich, 1974). Thus, extinction would be the fate of "K" strategist when confronted with an adverse environment, as observed for the domestic species of Triatominae, which have been targeted by international vector control programs (Schofield and Dias, 1999).

15.3.4 Triatominae Genetics

Cytogenetics

Triatomine chromosomes are holocentric; that is, there is not a single identifiable centromere, but rather the centromere function is spread out along the entire chromosome (see review of triatomine cytogenetics in Panzera et al., 2010). Therefore, even if a chromosome is fragmented, the fragments will be appropriately aligned and segregated during mitosis and meiosis. All but three species of triatomines examined cytologically contain 20 autosomes. The majority of the species also contain a single sex bivalent, XY for males and XX for females ($2N = 20 + XY$). However, in some species the X chromosome is fragmented (in both males and females) into two or (rarely) three fragments. For example, a *T. dimidiata* male contains 23 chromosomes, consisting of 20 autosomes, plus an X fragmented into two pieces (males, $2N = 20 + X_1X_2Y$); and in *T. vitticeps* the X is in three pieces (e.g., females) ($2N = 20 + X_1X_1X_2X_2X_3X_3$). The fragmented X is found in nearly all North American *Triatoma* species, whereas most South American species have the XY system. Variation in the amount and location of heterochromatin has been observed, most notably with a significantly greater amount of C-banded heterochromatin in Andean *T. infestans* as compared to non-Andean (Panzera et al., 2004). In the Andean, thought to represent the ancestral form, C-heterochromatic blocks are present on the majority of the 22 chromosomes, whereas in non-Andean specimens, heterochromatic regions are restricted to only 4–7 autosomes. This additional heterochromatin correlates with ~30% more DNA in the specimens from the Andes. *T. infestans* is thought to have lost DNA during the dispersal process from an origin in the Bolivian Andes. Chromosomal differences were also important in uncovering a cryptic species in *T. dimidiata* (Panzera et al., 2006). The power of chromosome diversity analysis in resolving taxonomic and phylogenetic questions is just beginning to be realized. Recent advances such as fluorescent in situ hybridization and comparative genome hybridization will add to this power.

Population Genetics

Genetic Diversity in Triatomine Populations

Variability of triatomine populations has been assessed by allozymes (reviewed in Dujardin et al., 2002), nuclear and mitochondrial DNA sequence, and by microsatellite markers. In general, two main vectors in South America, *T. infestans* and *R. prolixus*, show less genetic diversity than Central and North American species by isozyme analyses (Table 15.4). This lower genetic diversity is also seen with very low levels of variability in *R. prolixus* mtDNA (*cyt b*, Table 15.5); with only three haplotypes observed in 41 specimens (Monteiro et al., 2003) or 14 haplotypes/~550 specimens (Fitzpatrick et al., 2008). *T. infestans* also shows a similar *cyt b* diversity (7 haplotypes/62 specimens, Giordano et al., 2005), whereas Central American *T. dimidiata* shows high (15 *cyt b* haplotypes/58 specimens (Blandon-Naranjo et al., 2010)) or extremely high levels (21 haplotypes/23 specimens, Dorn et al., 2009); however, the latter is from a region containing the cryptic species and so likely reflects interspecific variability. A single North American population of *T. sanguisuga* showed the astonishing number of 37 haplotypes/54 specimens; a possible cryptic species in this population is being investigated (de la Rua, et al., unpublished data). Nuclear (ITS2) sequence data also supports higher diversity in *T. dimidiata* populations as compared to *T. infestans* and *R. prolixus* (Table 15.5). This may be due to the fact that over most of their range *T. infestans* and *R. prolixus* are exclusively domestic; it has been suggested that domestication results in a diminished genetic repertoire perhaps due to founder effects and genetic drift in isolated populations (Schofield et al., 1999). However, some studies report no difference in allele frequencies (allozymes, *T. infestans*, Dujardin et al., 1987) or haplotype diversity (*cyt b*, *R. prolixus*, Fitzpatrick et al., 2008) between domestic and sylvatic populations, and in other studies, a higher haplotype diversity was noted in domestic as compared to sylvatic populations (*cytochrome oxidase* sequence, *T. infestans*, Piccinali et al.,

Table 15.4 Isozyme Diversities of Triatomine Populations

North and Central American Species	n	Number of Loci	% Polymorphic	Mean Diversity, H_e	Reference
T. barberi Mexico	35	17	18	0.038	Flores et al. (2001)
T. dimidiata Mexico	18	16	50	0.187	Flores et al. (2001)
T. longipennis Mexico	39	17	53	0.109	Flores et al. (2001)
T. pallidipennis Mexico	31	17	29	0.072	Flores et al. (2001)
T. picturata Mexico	28	17	41	0.124	Flores et al. (2001)
South American Species					
R. prolixus Colony, Venezuela	90	22	22	0.06	Harry et al. (1992)
R. prolixus Colony, Venezuela	181	19	10.6	0.017	Harry et al. (1993)
R. prolixus Colombia	41	10	9.1	0.011	Lopez et al. (1995)
R. prolixus Colombia	8	17	0	0.0	Dujardin et al. (1998b)
R. prolixus Colony, Venezuela	12	12	8.0	0.01	Monteiro et al. (2002)
T. infestans Bolivia	37	19	15.7	0.04	Dujardin and Tibayrenc (1985a)
T. infestans Peru	NS	19	5	$H_o = 0.015$	Dujardin and Tibayrenc (1985b)
T. infestans Chile	36	18	15.3	$H = 0.178$	Frias and Kattan (1989)
T. infestans Bolivia	177	19	16	$H = 0.049$	Dujardin et al. (2002)
T. infestans Colony, Argentina	50	17	52.9	0.074	Garcia et al. (1995)
T. infestans Bolivia and Uruguay	1044	24	4–13	0.050	Pereira et al. (1996)
T. infestans Bolivia, Peru, Uruguay, and Brazil	1154	19	5	0.049	Dujardin et al. (1998a)

Table 15.5 Mitochondrial and Nuclear Diversities in Triatomine Populations

North and Central American Species	Method	Number of Populations	Number of Insects	Number of Haplotypes or [a]Mean Alleles per Locus	Mean Diversity, H_S or H_e	Reference
T. dimidiata Costa Rica	cyt b	7	58	15	NS	Blandon-Naranjo et al. (2010)
T. dimidiata Mexico and Guatemala	cyt b	NS	23	21[b]	NS	Dorn et al. (2009)
T. dimidiata Costa Rica	16S rDNA	7	58	6	NS	Blandon-Naranjo et al. (2010)
T. dimidiata 7 countries	ITS2	47	113	24[c]	$H_S = 0.822$[f]	Bargues et al. (2008)
T. dimidiata Mexico and Guatemala	ITS2	26	47	11[c]	NS	Dorn et al. (2009)
T. dimidiata Mexico	4 ML	12	295	10.3[ab]	$H_e = 0.805$[‡]	Dumonteil et al. (2007)
T. sanguisuga United States	cyt b	1	54	37	$H_S = 0.978$	de la Rua, et al., unpublished data
South American Species						
R. prolixus 4 countries	cyt b	12	41	3	NS	Monteiro et al. (2003)
R. prolixus Venezuela	cyt b	33	551	14	$H_S = 0.432$	Fitzpatrick et al. (2008)
T. infestans Brazil, Bolivia, and Argentina	cyt b	9	33	4	NS	Monteiro et al. (1999)
T. infestans Bolivia	cyt b	30	62	7	$H_S = 0.448$	Giordano et al. (2005)
T. infestans Peru, Bolivia, Uruguay, Argentina	co I	54	244	37	$H_S = 0.766$	Piccinali et al. (2009)

(Continued)

Table 15.5 (Continued)

North and Central American Species	Method	Number of Populations	Number of Insects	Number of Haplotypes or [a]Mean Alleles per Locus	Mean Diversity, H_S or H_e	Reference
T. infestans Argentina	16S rDNA	16	130	18	$H_S = 0.030$	Segura et al. (2009)
T. infestans Argentina	12S + 16S rDNA	5	40	13	$H_S = 0.538$	Garcia et al. (2003)
T. infestans Paraguay	ITS2	7	7	2	NS	Marcilla et al. (2000)
T. infestans 7 countries	ITS2	31	35	5	NS	Bargues et al. (2006)
T. infestans 7 countries	ITS1, 5.8S, ITS2	31	35	7	$H_S = 0.298$	Bargues et al. (2006)
R. prolixus Venezuela	9–10 ML	33	555	3.1[a]	$H_e = 0.476$	Fitzpatrick et al. (2008)
T. infestans Argentina	10 ML	1	34	10.7[a]	$H_e = 0.753$	García et al. (2004)
T. infestans Argentina	10 ML	19	598	6.8[a]	$H_e = 0.588$	Pérez de Rosas et al. (2007)
T. infestans Argentina	10 ML	21	352	17[a]	$H_e = 0.72$	Marcet et al. (2008)
T. infestans Bolivia	9 ML	6	111	7.3[a]	$H_e = 0.656$	Richer et al. (2007)
T. infestans Bolivia	10 ML	6	238	5.6[a]	$H_e = 0.432$	Pérez de Rosas et al. (2008)
T. infestans Bolivia	10 ML	23	253	14.5[a]	NS	Pizarro et al. (2008)

ML = microsatellite loci; H_S = haplotype diversity, H_e = expected heterozygosity, H_o = observed heterozygosity, H = heterozygosity. NS = not stated.
[a]Mean number of alleles per locus.
[b]From area with cryptic species present.
[c]Cryptic species removed.

2009). An alternative hypothesis is that low genetic variability reflects rapid and recent expansion due to passive dispersal (Dujardin et al., 1998b). *T. dimidiata* and other Central and North American species are widespread in sylvan, peridomestic, and domestic habitats. Survival in these very diverse habitats may have resulted in distinct selective pressures, thus maintaining genetic diversity. Interestingly, the strong selective pressure of pesticide application can result in diminished genetic variability (Garcia et al., 2003) or no effect or even an increase in the genetic variability in a population (Pérez de Rosas et al., 2007, 2008). The authors speculate that the reduction in population may not be as severe as expected (i.e., less genetic bottleneck effect) or that the high variability may be due to genetic drift in subpopulations, where each subpopulation retains a different combination of alleles. If that is followed by rapid growth of the population, it may be preserving the genetic diversity.

Importance of Population Genetic Studies for Taxonomy and Control

Effective control methods require correct identification of the target population, and genetic studies have been important in taxonomic clarification in the triatomines, including an understanding of the epidemiological importance of sylvan populations. For example, genetic studies demonstrated that domestic *R. prolixus* is not a derivative of the sylvan *R. robustus* but, in fact, is a separate species, diminishing the concern that *R. robustus* might be epidemiologically important (Lyman et al., 1999). To date genetic studies have not identified cryptic species in *T. infestans*. However, it is a **polytypic species**, meaning that genetically and morphologically distinguishable forms have been described, which when mated result in fertile offspring. In addition to the cytogenetic differences in the Andean compared to non-Andean *T. infestans*, these taxa can be distinguished by nuclear (with one exception, Bargues et al., 2006) and mtDNA (Giordano et al., 2005) with a significant F_{ST} between Andean and non-Andean populations. Interestingly, despite all these differences, offspring resulting from laboratory crosses between the two taxa are fertile (Panzera et al., 2004). Distinct chromatic *T. infestans* variants have also been identified. A "dark morph" was indistinguishable from other *T. infestans* by allozyme analysis, however, differs in nuclear (Bargues et al., 2006) and mtDNA sequence (Monteiro et al., 1999). Another dark variant found in Argentina, first identified as *T. melanosoma*, is now considered a subspecies of *T. infestans, T. i. melanosoma*, based on several characteristics (Monteiro et al., 1999; Bargues et al., 2006). Recent mtDNA analyses suggest a divergent taxon in *T. infestans* in Argentina (Piccinali et al., 2009). *T. dimidiata* also appears to be a polytypic species, with the number of divergent taxa still under investigation (reviewed in Dorn et al., 2007). Genetic studies also revealed a cryptic species in *T. dimidiata* (Marcilla et al., 2001) as well as in *T. sordida, T. brasiliensis, R. robustus*, and *R. ecuadoriensis* (reviewed in Abad-Franch and Monteiro, 2005). Studies are underway to understand the epidemiological importance of the distinct taxa. The finding of these morphologically similar but genetically divergent cryptic species has led to the suggestion that triatomines show "morphological plasticity," that is, rapid morphological change in response to environmental selection that may reflect convergent evolution rather than shared ancestry (Dujardin et al., 1999).

Hybrids, Cryptic Species, and Introgression in the Triatominae

Natural hybrids between species also occur in triatomines such as *T. infestans* and *T. platensis* (Abalos, 1948) and *T. dimidiata* and the cryptic species from the Yucatan Peninsula, Mexico (Herrera-Aguilar et al., 2009). In the former case the two species occupy distinct ecological niches and the rare hybrids are fertile over many generations. In contrast, for *T. dimidiata*, populations are sympatric, however, decreased fitness of hybrids may be what is maintaining >5 million year separation of the species. (The time of separation is based on a molecular clock using ITS2 divergence rates; Bargues et al., 2008.) **Introgression** of mtDNA was suggested because of similar mtDNA sequence in *T. infestans* and *T. platensis* (Garcia and Powell, 1998). Discordant phylogenies based on nuclear and mtDNA provide stronger evidence of introgression between *R. prolixus* and *R. robustus* (Fitzpatrick et al., 2008), *T. platensis* and *T. delpontei* (Mas-Coma and Bargues, 2009), *T. dimidiata* and the cryptic species (Herrera-Aguilar et al., 2009), and *Mepraia spinolai* and *M. gajardoi* (Calleros et al., 2010).

Population Structure

Studies investigating gene flow among populations have been important in understanding the geographic coverage needed for successful control, the source of triatomines appearing following pesticide treatment, the role of peridomestic and sylvan populations in human transmission; and would be important to understand the potential spread of an introduced genetically modified symbiont. Triatomines are generally poor flyers; however, some species can fly up to 1 km and even wingless nymphs can walk tens of meters (Núñez, 1987). Passive transport by humans and perhaps even migratory birds (via eggs or nymphs) has been important in spreading and mixing populations. The overall results of population genetic studies in *T. dimidiata* and *T. infestans* show that at larger geographic scales (populations >50 km apart) there is generally a gradient of allele frequency differences among populations consistent with an "**isolation by distance**" model (Wright, 1943). At smaller geographic scales, the picture is more complicated, varies geographically, and sometimes is affected by the pesticide application history.

Where *R. prolixus* is present in both domestic and sylvan ecotopes, shared mtDNA haplotypes and a nonsignificant F_{ST} using 10 microsatellite loci demonstrated gene flow between these populations and indicated that sylvan *R. prolixus* poses a risk for reinfestation of treated houses (Fitzpatrick et al., 2008).

Population Structure of *T. dimidiata* Across its extensive range, *T. dimidiata* shows quite different behaviors ranging from living exclusively in domestic and **peridomestic habitats**, to seasonally entering homes, to living exclusively in sylvan areas. Therefore, substantially different degrees of migration and gene flow might be expected among different populations. For domestic *T. dimidiata* populations, there appears to be high gene flow among houses within a village and among nearby villages (up to 27 km distance) by RAPD-PCR ($F_{ST} = 0.025$ and 0.019, respectively, Table 15.6) giving an estimated 9.7−12 mating migrants per generation (Nm) (Dorn et al., 2003). With the same technique Ramirez et al. (2005)

Table 15.6 Subdivision of *T. dimidiata*, *R. prolixus*, and *T. infestans* Populations by Microsatellite Analyses

Species/ Country	Number of Loci	Number of Insects per Population	Geographic Distance	F_{ST}	Reference
Among villages					
T. dimidiata Mexico	4	11–34	<280 km	0.06[a]	Dumonteil et al. (2007)
T. dimidiata Guatemala	8	28–30	<13 km	0.07	Stevens, et al., unpublished data
T. infestans Argentina	10	20–37	<30 km	θ = 0.135	Pérez de Rosas et al. (2007)
T. infestans Argentina	10	28–70	<220 km	θ = 0.169	Pérez de Rosas et al. (2008)
T. infestans Argentina	10	12–99	<31 km	0.02–0.2	Marcet et al. (2008)
T. infestans Bolivia	10	9–78	<100 km	0.06	Pizarro et al. (2008)
Among houses					
T. dimidiata Guatemala	8	2–7	NS	0.07–0.27	Stevens, et al., unpublished data
R. prolixus Venezuela	9–10	14	NS	−0.02	Fitzpatrick et al. (2008)
T. infestans Argentina	10	11–24	NS	θ = 0.05–0.07	Pérez de Rosas et al. (2007)
T. infestans Argentina	10	9–30	<2.5 km	0.02–0.16	Marcet et al. (2008)
T. infestans Argentina	10	15–31	<1.3 km	0.10	Pérez de Rosas et al. (2008)
T. infestans Bolivia	10	3–14	<1.5 km	0.07	Pizarro et al. (2008)
Among ecotopes					
R. prolixus Venezuela	9–10	12–39	Among ecotopes	0.002–0.2	Fitzpatrick et al. (2008)
T. infestans Argentina	9	6–31	<1.1 km, among sylvan	0.002–0.02	Richer et al. (2007)
T. infestans Bolivia	9	6–32	<1.1 km sylvan vs. domestic or peridomestic	0.03, 0.11	Richer et al. (2007)
T. infestans Bolivia	10	36–42	<750 m domestic vs. peridomestic	0.03	Pizarro et al. (2008)

[a] From area with cryptic species present.

showed movement of at least two or three mating migrants per generation ($F_{ST} = 0.07$) among nearby (within 10–200 m) domestic, peridomestic, and sylvan (cave) populations. Also by RAPD-PCR, nearly half the individuals within a single house were unrelated, again supporting substantial migration by *T. dimidiata* (Melgar et al., 2007). In localities where *T. dimidiata* seasonally enters houses, results of microsatellite analyses showed a high gene flow among nearby houses ($F_{ST} = 0.037$), villages ($F_{ST} = 0.055$, within 250 km), and between the forest and houses ($F_{ST} = 0.01-0.03$, 1–280 km), for an Nm = 5–25 (Dumonteil et al., 2007). However, the cryptic species was later discovered in this area, so these results need to be confirmed with the identity of the samples known. More recently, in an area of domestic *T. dimidiata*, microsatellite analyses showed a weak but statistically significant genetic structure among populations within 12 km ($F_{Sg} = 0.07$ or about three mating migrants per generation between towns) and among houses ($F_{ST} = 0.07-0.27$) (Stevens et al., unpublished data). Twenty-four genetic clusters were spread across seven villages, and most insects were only distantly related to others in the same house, again supporting substantial movement among houses and villages. This high migration among nearby houses, villages, and even sylvan locations by *T. dimidiata* means that simply treating individual houses or even villages with pesticides is not likely to be effective (Figure 15.15).

Population Structure of *T. infestans* Results using allozyme analyses show gene flow between nearby (~20–50 km) *T. infestans* domestic, and domestic and sylvan populations with a smaller panmictic unit observed in some localities (Dujardin et al., 1987, 1998b; Brenière et al., 1998). With higher resolution microsatellite markers, geographic variation in the degree of subdivision of populations was again observed, and in certain areas pesticide application appeared to affect population subdivision. In most localities movement was quite limited as there was significant genetic differentiation among villages and houses within a village (Pérez de Rosas et al., 2007, 2008; Marcet et al., 2008) (Table 15.6). The number of genetically related clusters identified is close to the number of villages studied in several studies (8 clusters in 10 villages, Marcet et al., 2008; 5 in 5 villages, Pizarro et al., 2008; 7 clusters in 6 villages, Pérez de Rosas et al., 2008) and generally most clusters were localized to just one or a few villages. Taken together results indicate that *T. infestans* shows limited gene flow, likely due to limited migration.

Results of most, but not all, studies indicate gene flow among nearby ecotopes for *T. infestans* and *R. prolixus* (Dujardin et al., 1987; Monteiro et al., 1999; Giordano et al., 2005; Fitzpatrick et al., 2008; Table 15.6).

So overall, *T. dimidiata* populations present more diversity than do those of *T. infestans* or *R. prolixus*. *T. dimidiata* also shows more migration than does *T. infestans* based on the differentiation among populations (by F_{ST}) and the number and distribution of genetically related clusters in villages.

Implications for Control

These results have important implications for control. With a long life span and low genetic diversity, one might hope to avoid development of insecticide

Figure 15.15 Fumigation using residual pesticides in Guatemala. Photographer, Patricia Dorn.

resistance in *T. infestans*. Unfortunately, insecticide resistance has been noted in several populations, perhaps partly due to inadequate pesticide application (Gonzalez Audino et al., 2004; Picollo et al., 2005; Toloza et al., 2008). Although not presently known, it is likely just a matter of time before resistance will also appear in the more diverse *T. dimidiata*.

Understanding the amount of gene flow among populations can inform the geographic coverage needed for control. The low gene flow detected among most domestic *T. infestans* populations made it an excellent target and is probably the reason for the success of the Southern Cone Initiative, which has achieved the remarkable average reduction of 94% in Chagas disease incidence in the Southern Cone countries (WHO, 2002). Application of residual pesticides in houses has also been extremely effective in the elimination of most populations of the exclusively domestic *R. prolixus* in Central America (Yamagata and Nakagawa, 2006). However, where sylvan and peridomestic *T. infestans* populations occur in addition to domestic, gene flow among the three ecotopes can at least partially explain the control failures in the Gran Chaco region (northern Argentina, Bolivia, and Paraguay), along with emergence of resistance to pyrethroids (Noireau et al., 2005). Understanding the source of insects appearing after pesticide treatment, if they are not simply resistant, is enormously important to designing effective control strategies. Population genetic studies have identified nearly all "reinfestants" as survivors or migrants from nearby peridomestic sites (Dujardin et al., 1996; Garcia et al., 2003; Pérez de Rosas et al., 2007; Pizarro et al., 2008). Peridomestic sites, with their extremely heterogeneous microhabitats, are particularly challenging for control. Gene flow among sylvan and domestic *R. prolixus* in Venezuela and the substantial migration seen among *T. dimidiata* populations means that simply spraying houses is unlikely to be effective for these species. Indeed, even in areas where sylvan *T. dimidiata* populations are unknown, reinfestation confounds control efforts (Yamagata and Nakagawa, 2006). Where peridomestic and sylvan populations are present, methods such as the use of screens and house improvements will be necessary to control transmission (Ferral et al., 2010; Monroy et al., 1998). And certainly if the genetically modified gut symbiont, developed to kill *T. cruzi* in the triatomine gut (CRUZIGUARD), is to be applied in the field (Beard et al., 2002), the migration pattern of the target vector population needs to be understood to predict the potential spread of the symbiont.

15.3.5 Further Work Needed

Genetic studies have been important in understanding the Triatominae at all levels: phylogenetic, taxonomic, and population. Results have contributed to an understanding of the epidemiology of Chagas disease and provided important information toward the design of effective control strategies. However, much remains to be done.

There is a need to reevaluate phylogenetic relationships and taxonomy of the Triatominae that were previously based on morphology and isozymes using new phenotypic and genetic markers and sampling methods that avoid biases. Phylogenetic and taxonomic questions remain at the tribe (Alberproseniini, Bolboderini, Cavernicolini, Linshcosteini), genus (*Meccus, Mepraia, Eratyrus*) and species level (*T. dimidiata* [and the phyllosoma complex], *T. sordida, T. brasiliensis*). Species classification originally based on phenotypic characters, which is now in some cases contradicted by genetic data, needs to be revised. Species of Triatominae that occupy substantial geographical ranges may display clinal variation (revealed by genetic differences) along a major latitudinal axis. If populations of such species are sampled over a wide area, clinal variation can be assessed. On the other hand, if populations are sampled much more focally, the sampled subsets may be considered in isolation—and receive specific designation—when they may in fact represent components of an unknown continuous population (Schofield and Galvão, 2009). Correct identification of organisms is critical to effective control and to implying shared behavioral traits. As divergent taxa are uncovered, it is important to understand their epidemiological importance.

Of particular interest is to understand what forces (geological, ecological, anthropogenic) have resulted in the subdivision and maintenance of genera, species, and divergent taxa. Combining new tools in genetics, global information systems, and mathematical modeling will make it possible to address these questions and potentially predict future distribution of vector populations, including secondary vectors showing synanthropic tendencies. Human activities such as control efforts, global travel, and those resulting in deforestation and climate change are likely to have a major impact on future vector and Chagas disease distribution.

The diversity of triatomine chromosome structure, even within a species, makes it a model system for understanding chromosome structure and its evolution, and the role of heterochromatin. Comparative genomics and proteomics are in their infancy in triatomine biology. Completion of the *R. prolixus* genome: http://genome.wustl.edu/genomes/view/rhodnius_prolixus/#sequences_maps will provide many new tools for identifying important genes and markers. A *Triatoma* species genome should be added to this toolbox. Comparative studies promise advances in understanding the genetic basis of vector competence/capacity, reproductive isolation among sympatric species, as well as genes and proteins involved in hematophagy, habitat preference especially the important process of domestication, pesticide resistance, and identification of genes/haplotypes/proteins that are common to triatomines or perhaps important for species-specific behavior. These tools can also help to unravel vector and parasite interactions and coevolution. Perhaps as in other vectors, parasite infection could affect vector behavior in ways that increase transmission, or passage through particular triatomine species could affect parasite virulence. Where molecular markers provided geneticists with dramatically increased resolution over isozyme markers, development of new tools such as single nucleotide polymorphisms (SNP) assays and whole genome sequencing will provide a new leap in resolution over current molecular methods used in triatomines.

15.3.6 Concluding Remarks

So, 100 years after its discovery, Chagas disease remains the most serious of the parasitic diseases affecting Latin America. It is now evident that we will miss the World Health Organization's goal of elimination of Chagas disease by 2010. Intergovernmental control initiatives have been enormously successful at interrupting transmission caused by domestic triatomines. However, substantial challenges remain such as vectors with more varied habitats, emergence of insecticide resistance, and demographic and habitat changes that have resulted in the spread of Chagas to urban and nonendemic regions. New hope for eventual elimination comes from an integrative approach combining fields such as genetics with Global Information Systems and mathematical modeling, moving the field from descriptive to predictive, and prevention rather than reaction.

15.4 The *Anopheles gambiae* Complex

15.4.1 Introduction

Background and Brief History of Anopheles gambiae *Complex Classification*

The *Anopheles gambiae* species complex was initially described as containing six cryptic species: *A. gambiae* s.s. Giles, *A. arabiensis* Patton, *A. bwambae* White, *A. melas* Theobald, *A. merus* Dönitz, and *A. quadriannulatus* Theobald. The status of these species was established via the demonstration of F_1 hybrid sterility among crosses between different *A. gambiae* s.l. populations (Davidson, 1956, 1964a,b; Davidson and Hunt, 1973). Subsequent studies revealed that these six species could be distinguished on the basis of fixed differences in chromosomal inversions (Davidson et al., 1967; Davidson and Hunt, 1973; Coluzzi et al., 1979). Keys to distinguish species on the basis of allozyme differences were later developed (Mahon et al., 1976; Miles, 1979). Species in the complex are distributed throughout sub-Saharan Africa (Figure 15.16). Two additional species were later described: *A. comorensis* Brunhes, le Goff and Geoffroy based on subtle morphological features (Brunhes et al., 1997) and *A. quadriannulatus* species B Hunt, Coetzee and Fettene based on hybrid male sterility in crosses with *A. quadriannulatus* (Hunt et al., 1998). Of the eight species, two, *A. gambiae* s.s. and *A. arabiensis*, have the broadest geographic distribution and are the most important vectors of human malaria (Gillies and De Meillon, 1968; Coetzee et al., 2000). *A. gambiae* has been the most studied with respect to molecular and population genetics, and its whole genome sequence was published in 2002 (Holt et al., 2002). Natural populations of *A. gambiae* s.s. have an extremely complex genetic structure that has been the subject of a great deal of research. Despite these efforts the population genetics of this species remains poorly understood.

15.4.2 Levels of Population Genetic Structure in A. gambiae

The ideal gene pool in population genetics is a panmictic population, a homogeneous, randomly mating group of individuals that remains the same through time. To the extent that real populations depart from this they are said to be structured. *A. gambiae* s.l. is structured in at least 3 ways: (1) temporal: there are seasonal variations in population size and composition; (2) geographical: they mate locally, with little migration among villages; and (3) nondimensional: even within the same location and time, mating is nonrandom.

Temporal Structure

There are seasonal differences in abundance and composition of *A. gambiae* s.l. For example, in Banambani, Mali, *A. arabiensis* and *A. gambiae* s.s. are present in large numbers during the rainy season, with a progressive increase of *A. gambiae* s.s. during the rainy season and *A. arabiensis* in the drier months. The bulk of evidence suggests that they are present, but simply in low numbers (Holstein, 1954). One

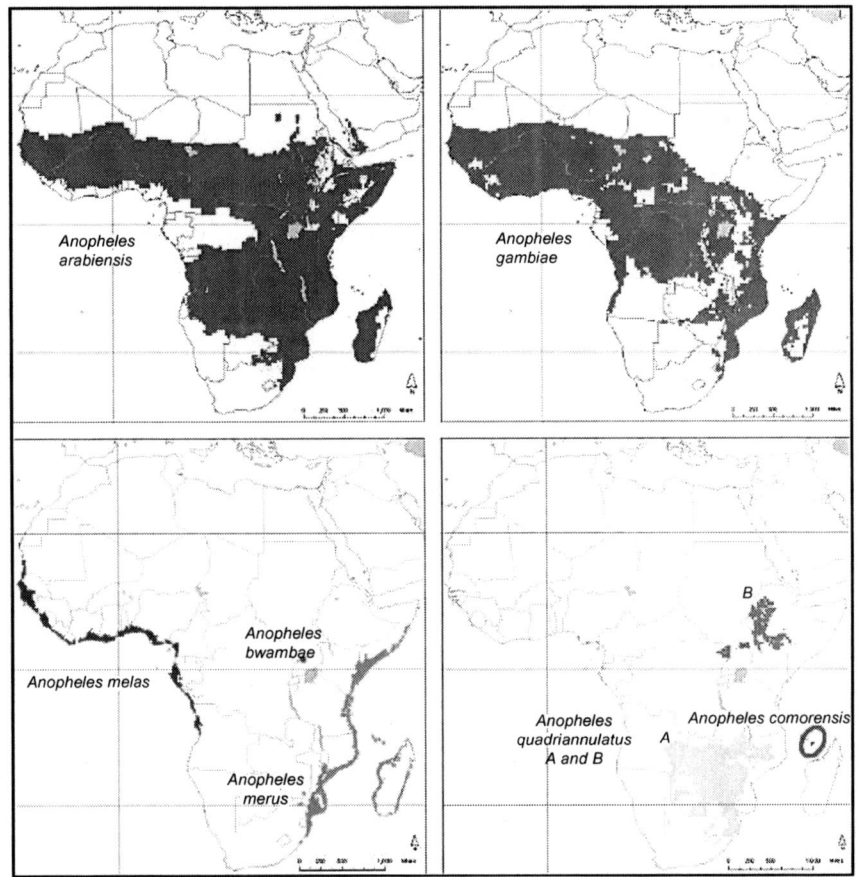

Figure 15.16 Geographic distribution of species in the *Anopheles gambiae* complex. *Source*: Adapted from Ayala and Coluzzi (2005).

cannot, however, absolutely exclude the possibility that they go locally extinct and are re-colonized from neighboring areas where permanent water is available (Taylor et al., 1993). The pattern varies somewhat from place to place, and is especially different in irrigated areas (Diuk-Wasser et al., 2005, 2006).

Geographic Structure

The geographic structure seems to be complex and is poorly understood through much of the species range. Populations of *A. gambiae* s.l., as with many species in the *Anopheles*, carry high levels of polymorphism in the form of **paracentric chromosome inversions**. Because inversion frequencies are certainly subject to selection, most recent attention has been focused on using microsatellite DNA variation, which is assumed to be neutral with respect to selection, to describe the genetic structure of populations. Utilizing gene frequencies at 9 microsatellite loci, Lehmann et al. (1999) showed that the Rift Valley in eastern Africa imposes a huge barrier to gene flow among populations of *A. gambiae* s.s. there. Lehmann et al. (2003) subsequently conducted a more extensive study that included 16 sites in 10 countries spanning continental Africa. A cluster analysis based on F_{ST} values based on gene frequencies for 11 microsatellite loci revealed a major subdivision among *A. gambiae* populations in Africa. They identified a northwestern (NW) population group, containing populations in Senegal, Ghana, Nigeria, Cameroon, Gabon, Democratic Republic of Congo, and western Kenya, and a southeastern (SE) group including populations in eastern Kenya, Tanzania, Malawi, and Zambia (Figure 15.17). Differentiation between these two population groups was relatively high (F_{ST} >0.1). Genetic differentiation among populations within the two groups was substantially lower and a significant relationship between genetic distance and geographic distance was observed, consistent with an isolation by distance model of population structure.

The classic studies of Coluzzi et al. (1979) and the comprehensive survey of Mali by Touré et al. (1998) have shown widespread geographic variation for chromosome inversion frequencies in populations of *A. gambiae* s.s. and in *A. arabiensis*. Lehmann et al. (1996) and Lanzaro et al. (1998) demonstrated geographic structure for microsatellite DNA for populations in Mali, and Besansky et al. (1994) (with Lehmann et al., 2000) showed the same to be true of mtDNA.

Nondimensional Structure

There is extensive nonrandom mating among genetically distinct subpopulations of *A. gambiae* s.s., known as chromosomal and **molecular forms** (described later). Taylor et al. (2001) measured the amount of gene flow among these populations and between species in Mali in a variety of ways. Their measures of gene flow between forms in Mali seem internally consistent. However, the amount of hybridization between forms varies considerably from location to location. Because some forms are more persistently present than others, and even absent at some locations, the amount of crossing will vary from place to place. There are apparently intrinsic factors that also play a role in the degree of between-form hybridization.

15.4.3 Polytene Chromosomes and Chromosomal Forms of A. gambiae s.s.

Structure and Methods

Discernable banding patterns in polytene chromosomes of late fourth instar larval salivary gland cells (Coluzzi, 1966; Coluzzi, 1968; Coluzzi and Sabatini, 1968, 1969) and in nurse cells of developing ovules in half gravid females reviewed in Coluzzi et al. (2002) served a primary role as a diagnostic technique to separate members of the *A. gambiae* complex. The full chromosome compliment that polytenizes consists of one sex chromosome (X) and 4 autosomal arms (2R + 2L and

Genetics of Major Insect Vectors 449

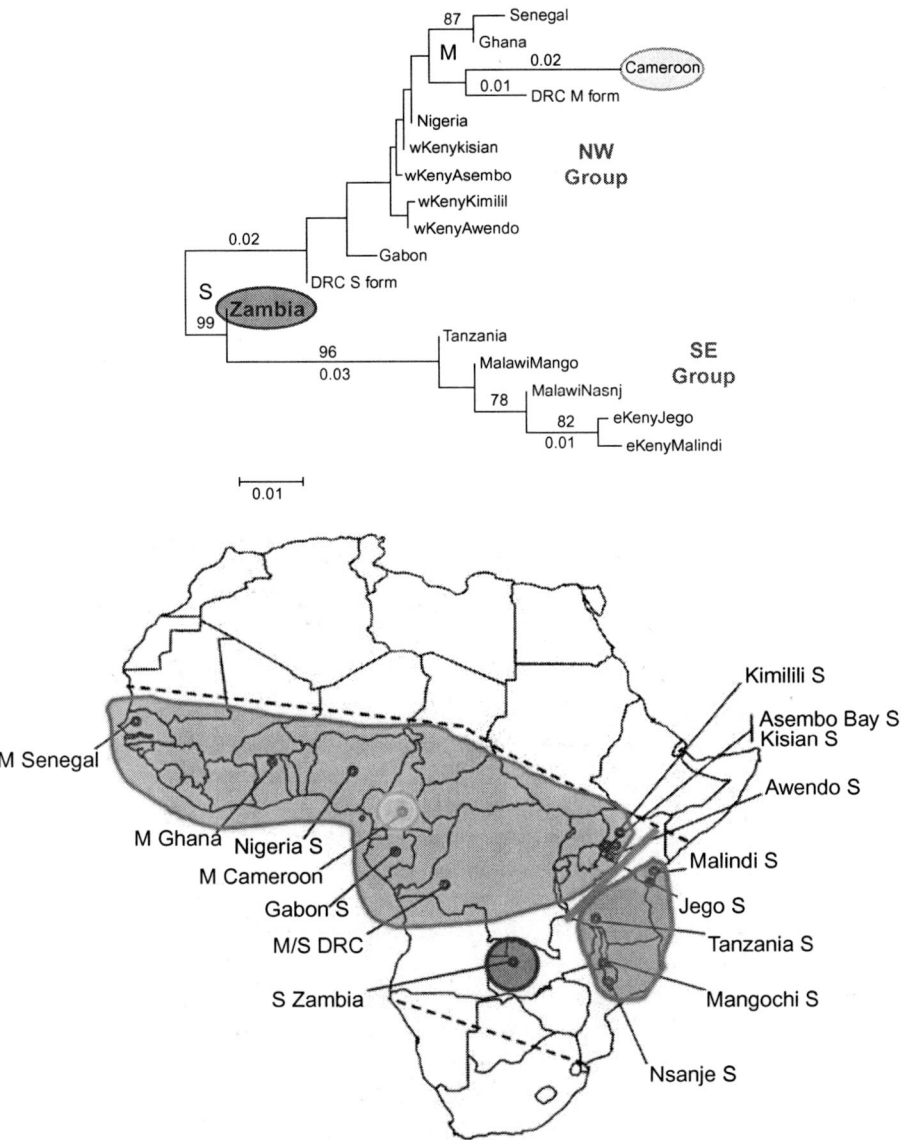

Figure 15.17 Top. Unrooted neighbor-joining population tree based on mean F_{ST} across nine microsatellite loci. M and S populations are denoted at the bases of the clades. The Northwest and Southeast population groups are indicated. Fractions denote branch length (over 0.01) and integers denote biologically significant bootstrap support values. Bottom: Map roughly indicating the boundaries of the different population groups. The orange line separating the Southeast group (in red) from the Northwest group (in green) represents the location of the Great Rift Valley. (For interpretation of the references to color in this figure legend, the reader is referred to the web version of this book.)
Source: Figures adapted from Lehmann et al. (2003).

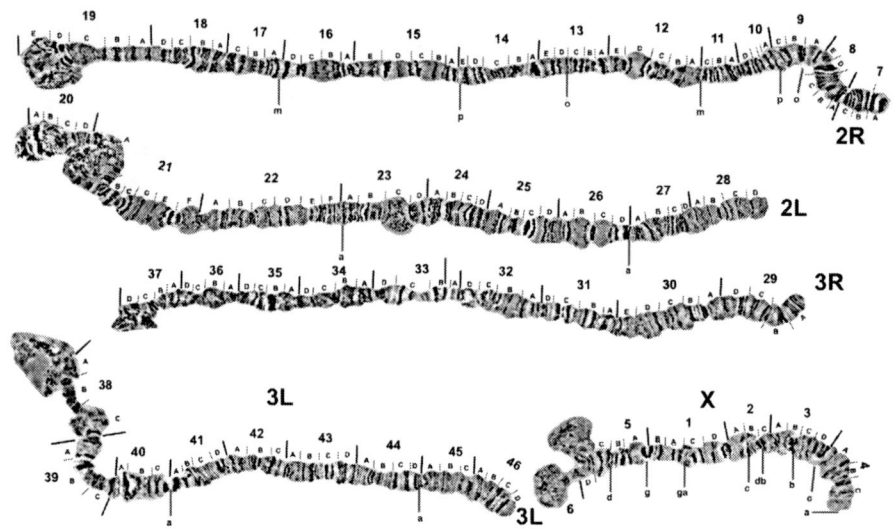

Figure 15.18 Photomap of full polytene chromosome complement of *A. gambiae* s.s. Forest-M form (collected from Tiko, Cameroon) depicting positions of breakpoints of major inversions.

3R + 3L). All members of the complex except for *A. quadriannulatus A* and *B* can be distinguished by fixed paracentric chromosome inversion differences on the X, 2R, 2L, and 3L chromosome arms (bold in Figure 15.18; see Figure 15.9 for an explanation of pericentric and paracentric inversions).

Details of methods used to prepare mosquito polytene chromosomes, originally derived from those used in *Drosophila,* appear in a number of publications (Coluzzi and Sabatini, 1967; Hunt, 1973). Comprehensive methods for salivary gland and nurse cell polytene chromosome extractions, spreading and staining are also posted online at the MR4 website, in *Anopheles* protocols chapters 5.6 and 5.7 (http://www.mr4.org/Portals/3/Methods_in_Anopheles_Research.pdf).

These days, salivary gland chromosomes are seldom used because nurse cell chromosome spreads are of higher quality. Drawings depicting banding patterns of partial and full compliments of salivary gland and ovarian nurse cell polytene chromosome with locations of divisions, subdivisions, and positions of the primary paracentric inversion breakpoints of *A. gambiae* s.l. are available in numerous publications (Coluzzi and Sabatini, 1967; Green, 1972; Coluzzi et al., 2002; della Torre et al., 2002). A photomap of the full chromosome compliment with positions of divisions, subdivisions, and major inversion breakpoints is provided in Figure 15.18. To circumvent the need for actual examination of chromosome spreads to identify inversions, PCR-based (molecular karyotyping) assays have been developed that can identify individuals in any developmental stage and in both sexes that carry inversions 2La (White et al., 2007) and 2Rj (Coulibaly et al., 2007). Complications in interpreting the results using the 2La PCR, due to polymorphism within the region being amplified, have been reported (Ng'habi et al., 2008).

Chromosome Inversion Polymorphism

Multiple polymorphic inversions within all species except for *A. quadriannulatus* B, *A. merus*, and *A. comorensis* have been recorded (Touré, 1985; Coluzzi et al., 2002). Inversion polymorphisms are particularly high in *A. gambiae* s.s. (Pombi et al., 2008) and a number of subpopulations termed "**chromosomal forms**" based on the inversions that characterize them have been described (Coluzzi et al., 1985; Touré et al., 1998). These include the Bamako, Bissau, Forest, Mopti, and Savanna forms (Table 15.7). The Savanna form has the broadest distribution occurring throughout sub-Saharan Africa, the Mopti form predominates in drier habitats in West Africa, the Forest form occurs in wetter habitats in both East and West Africa, the Bamako form occurs in habitats along the Niger River in West Africa and the Bissau form is restricted to West Africa (della Torre et al., 2002). There is general agreement that inversions represent coadapted gene complexes that allow individuals carrying them to occupy different ecological niches. The nonrandom distribution of inversion breakpoints along the chromosomes (Figure 15.18; Pombi et al., 2008) and the distribution of inversion frequencies throughout the geographical ranges of the species strongly suggest that at least some of the inversions are the product of selection that allow different species and, in the case of *A. gambiae*, s.s., populations to survive and exploit a wide variety of habitats (Touré et al., 1998; Coluzzi et al., 2002; Lee et al., 2009). The best example is the strong association of inversions 2La and 2Rb with aridity with the frequency of these inversions being high in dry areas and even increasing in frequency during the dry season at places that experience distinct wet and dry seasons (Touré et al., 1998; Lee et al., 2009). This has led to the term "**ecophenotype**" being frequently applied to describe chromosomal forms of *A. gambiae* s.s. (Coluzzi et al., 1977). It has furthermore been suggested that the chromosomal forms are to some extent reproductively isolated and represent distinct species or **incipient species** that have evolved or are evolving via a process described as "**ecotypic speciation**" (Coluzzi et al., 1977; Manoukis et al., 2008). Studies of the distribution of the knockdown insecticide resistance gene (*kdr*) in sympatric Bamako-Savanna populations in Mali revealed the gene is present in the Savanna form, but absent in sympatric Bamako populations, which was taken as strong support that the two are reproductively isolated and the authors concluded that the two represent "incipient species" (Fanello et al., 2003). However, in a later study, also based on populations in Mali, the *kdr* gene was in fact found in both Bamako and Savanna forms, even from collections as far back as 1996, suggesting some level of gene flow between the two (Tripet et al., 2007).

Using the chromosomal form concept to define discreet, reproductively isolated populations is problematic because there is substantial overlap in inversions that define them, probably due to some level of contemporary gene flow. This creates ambiguities in assigning individuals to form, diminishing the utility of the

Table 15.7 Fixed and Most Common Polymorphic (floating) Paracentric Inversions Observed in the Seven Members of the *Anopheles gambiae* Complex

Chromosome	Arabiensis	Bwambae	Melas	Quadriannulatus B	Quadriannulatus A	Gambiae	Merus
X	bcd/bcd e/+	+/+	+/+	+/+	+/+ f/+	ag/ag	ag/ag
2R	a/+ b/+ bc/+ be/+ bf/+ d¹/+ s/+ br/+ q/+	l/+	m/m n/+ n¹/+ m¹/+	+/+	i/+	+/+ (Sav, Mop)[a] b/+ (For, Sav) cu/+ (Sav) bcu/+ (Sav) jcu/jbcu (Bam) bc/u (Mop) j/+ (Sav) d/+ (Bis) jbd/+ (Sav) bd/+ (Sav) jb/+ (Sav) bcd/+ (Mop) bk/+ (Sav)	op/op
2L	**a/a**	+/+	a²/+	+/+	+/+	a/+	a¹/a¹
3R	a/+	b/+	c/+ e/+ e¹/+	+/+	+/+	+/+	+/+
3L	+/+	a/a	a/a	+/+	+/+	+/+	+/+

[a]Chromosomal forms: Bamako (Bam), Bissau (Bis), Forest (For), Mopti (Mop), and Savanna (Sav).

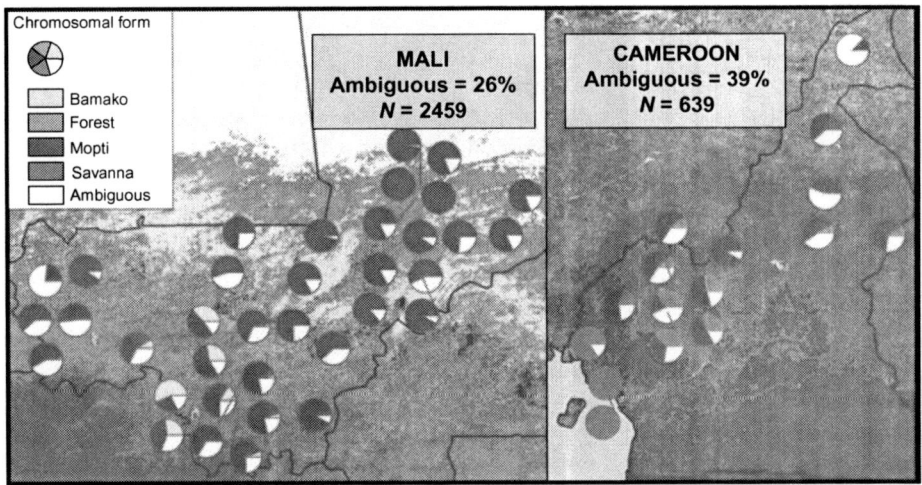

Figure 15.19 The distribution of chromosomal forms of *A. gambiae* s.s. at 36 sites in Mali (left) and 15 sites in Cameroon (right). The chromosomal form concept originated with observations made in West Africa and is based largely on populations in Mali. Even in Mali it is difficult to apply the concept to all individuals and at all locations. We found that 26% of individuals sampled in Mali and 38% from Cameroon could not be classified into chromosomal form.
Source: VectorBase, http://www.vectorbase.org/PopulationData/

chromosomal form concept in defining the reproductive boundaries among populations. For example, in a recent survey of populations in Mali we found that 26% of 2,459 individuals could not be assigned to a chromosomal form and in Cameroon 39% of 632 individuals could likewise not be assigned (Figure 15.19, data available at VectorBase, http://www.vectorbase.org/PopulationData/).

15.4.4 Molecular Diagnostics and the M and S Molecular Forms of A. gambiae s.s.

Identification of five of the eight species in the *A. gambiae* complex (the exception are *A. bwambae*, *A. quadriannulatus B*, and *A. comorensis*) was greatly simplified with the development of a PCR diagnostic based on fixed differences in the intergenic spacer (IGS) sequence of the ribosomal gene (rDNA) family (Scott et al., 1993). This breakthrough allowed rapid and accurate identification of species in the complex without the severe restrictions inherent in cytological examination of polytene chromosomes, which is labor intensive, requires a significant level of training, and is restricted to adult females at a specific stage of ovarian development. A similar molecular approach was utilized in an attempt to develop a diagnostic for the chromosomal forms of *A. gambiae* s.s. Favia et al. (1997) first found diagnostic RFLPs also within the rapidly evolving noncoding regions of the rDNA.

They identified 10 nucleotide residues that differ between the Mopti and the Savanna or Bamako chromosomal forms in a 2.3 kb fragment of the 5' end of the rDNA IGS region, which is located on the X chromosome (Favia et al., 2001). These findings were notable because they were the first fixed molecular genetic differences found between chromosomal forms of *A. gambiae* s.s. and they led to the development of a PCR-based diagnostic to differentiate Mopti individuals carrying the M-form of rDNA from Bamako and Savanna individuals carrying the S-form of rDNA (Favia et al., 1997). The diagnostic was developed using samples from Mali and among those early samples there were a few equivocal cases where karyotyping did not match the molecular diagnostic (Favia et al., 1997; della Torre et al., 2001). The diagnostic was also used to identify between-form hybrid-like karyotypes. M/S hybrids produced in the laboratory did yield clearly distinguishable hybrid patterns (of course only in females since the rDNA is located on the X chromosome). Surprisingly, however, field collected individuals carrying "hybrid" karyotypes did not produce results consistent with their being hybrid, but rather produced either M or S patterns (Favia et al., 1997). This observation supports the notion that certain karyotypes, thought to be fixed in one chromosomal form or another, are in fact shared, occurring commonly in one form and rarely in another, the result of ancestry and/or ongoing gene flow. The M/S diagnostic now forms the basis of recognizing two distinct subpopulations of *A. gambiae* s.s., known as "molecular forms" (M and S). Understanding the relationship between these two forms has been the focus of an intense and ongoing research effort. The S form has the broadest distribution occurring throughout sub-Saharan Africa, whereas the M form occurs throughout West and parts of Central Africa but, with the exception of a single site in northern Zimbabwe, is absent from eastern Africa (Figure 15.20; della Torre et al., 2005).

There is good correspondence between the M molecular form and the Mopti chromosomal form in Burkina Faso and Mali, however, the Bamako and Savanna chromosomal forms cannot be distinguished (both are of the S molecular form). The association of M and S molecular forms and chromosomal forms breaks down at other locations in West Africa. For example, in western Senegal and Gambia the association between the Savanna chromosomal form and S molecular form does not hold (della Torre et al., 2005) and the Forest form contains both M and S individuals.

The Relationship Between Chromosomal and Molecular Forms of A. gambiae *s.s.*

In summary, the M and S molecular forms are associated with chromosomal forms only in some locations, and so therefore they largely fail as a diagnostic for chromosomal form. However, the significance of the M and S forms of *A. gambiae* goes well beyond their utility as proxies for identifying chromosomal forms. M and S forms occur in sympatry at many sites in West and Central Africa, and typically there is a high degree of reproductive isolation between the two forms. Hybridization between the forms occurs rarely (<1%) in Mali (Tripet et al., 2001; Edillo et al., 2002) and reproductive isolation between M and S appears to be

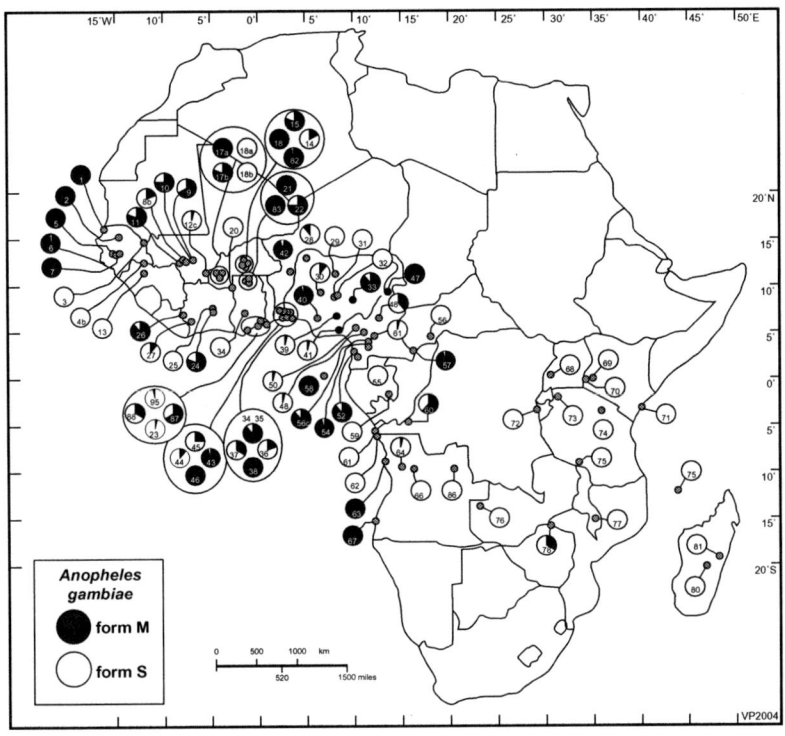

Figure 15.20 Relative frequencies of M-form and S-form of *A. gambiae* s.s. at 87 sites in 24 African countries, with permission, copyright Elsevier.
Source: From della Torre et al. (2005).

complete in Cameroon (Wondji et al., 2005). The M and S "alleles" are based on two base pair substitutions in the IGS sequence of the rDNA gene family on the X chromosome. Studies aimed at describing genetic differentiation between the M and S form populations revealed that microsatellite DNA differentiation was exceptionally high in a region of the genome proximal to the centromere on the X chromosome, near the M/S locus (Wang et al., 2001; Lehmann et al., 2003). High levels of M/S form divergence in this portion of the X chromosome was substantiated through detailed examination of the region using microsatellites and DNA sequencing (Stump et al., 2005a,b; Slotman et al., 2006). Studies aimed at describing genome-wide divergence between M and S using the Affymetrix *Plasmodium/Anopheles* Genome Microarray (Turner et al., 2005; White et al., 2009; White et al., 2010) likewise revealed the same X chromosome region, but also revealed divergence on chromosomes 2 and 3. These regions of divergence have been considered to represent "**islands of speciation**" because it is thought that they contain genes that are directly involved in reproductive isolation. It appears that the M and S "alleles" are linked to genes located within these "islands of speciation" and that two largely reproductively isolated populations of *A. gambiae* (M form and S form) exist in nature. The M and S forms are commonly referred to in the literature as "incipient species" (della Torre et al., 2001, 2002, 2005; Fanello et al., 2003; Stump et al., 2005b; Manoukis et al., 2008; White et al., 2010, and many others). But there are problems that suggest that the M and S form concept does not represent the full level of complexity in the genetic structure of *A. gambiae* s.s.

Geographic Variation in the Association of Chromosomal Forms and Molecular Forms of *A. gambiae* s.s.

Although the M and S forms are largely reproductively isolated in many places where they occur together, this is not true everywhere. In the Gambia M/S hybrids were identified from a number of sites at frequencies as high as 16.7% of the *A. gambiae* s.s. individuals sampled (Caputo et al., 2008) and in Guinea-Bissau hybrids were recovered in 24% of the individuals assayed (Oliveira et al., 2008). These results suggest that in localities covering a relatively large geographic area the linkage between the M and S alleles and those genes that directly affect reproductive isolation has broken down. Therefore the notion of an M form and an S form that are largely reproductively isolated (incipient species) is an oversimplification. As described above in Cameroon, the M and S forms appear to be completely isolated reproductively, whereas in Mali the reproductive barrier between them appears to be "leaky" (e.g., some hybridization occurs). A comparison of the M form in Mali and the M form in Cameroon has revealed that the two are very different genetically, in fact, divergence between these two is higher than the level of divergence between the M and S forms (Figure 15.21). This

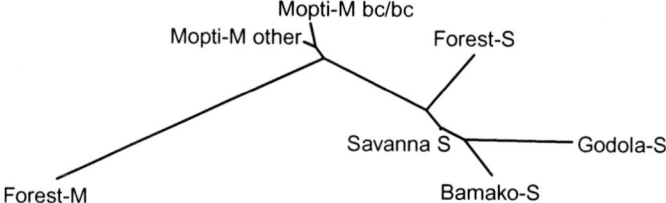

Figure 15.21 Unrooted phylogenetic tree (neighbor-joining) of seven groups of *An. gambiae* identified by a Bayesian analysis (using *Structure* software), these include Forest-M = M form individuals with the standard karyotype (no inversions) collected from the villages of Tiko and Mutengene, Cameroon; Mopti-M bc/bc = M form individuals homozygous for the 2R b/c inversions, collected from the villages of Nara, Banikane and Dire in Mali; Mopti-M other = M form individuals with inversion polymorphism excluding the bc/bc karyotype, collected from the villages of Banikane, Nara, Banambani and Kela in Mali; Forest-S = S form individuals with the standard karyotype, collected from Tiko, Mutengene, Foumbot and Ndop in Cameroon; Savanna-S = S form individuals with typical Savanna form inversion polymorphism including 2R b, c and u, collected from Mutengene and Foumbot in Cameroon and Pemperena, Kela and Banambani in Mali; Godola-S = S form individuals with the 2Rbcd karyotype. The tree is based on pair-wise F_{ST} values derived from allele frequencies at 20 microsatellite loci on chromosomes 2 and 3. Distances between all branches are significant. *Source*: From Lee et al. (2009).

observation has led to a recognition of two, distinct M form groups, the Mopti-M form, which is polymorphic with respect to the 2R b, c, and u chromosome inversions and the Forest-M form which lacks inversions on chromosome 2R (Slotman et al., 2007; Lee et al., 2009). It is likely that additional subdivision of both the M and S forms will be resolved, for example, in places like the Gambia and Guinea-Bissau where the M/S alleles are apparently less tightly linked to speciation genes.

15.4.5 Significance of A. gambiae Population Genetics to Malaria Transmission and Control

Ultimately malaria control efforts in Africa will have to be conducted on a large geographical scale. Although the development of a strategic plan for such an undertaking will require coordinating efforts along political boundaries, the success of such an effort will depend largely on identifying regions of operation based on biologically meaningful boundaries. Sub-Saharan Africa includes a wide variety of ecological zones. It is not surprising that the mosquito, *Anopheles gambiae* s.s., with a distribution across the continent, is highly diverse. The success of malaria control strategies aimed at controlling or manipulating its vectors will have to include knowledge of diversity in vector populations, how this diversity is distributed in time and space and the forces limiting gene flow and maintaining diversity among populations.

Studies of the population genetics of *A. gambiae* and of vector species in general have resulted in numerous and significant contributions to our understanding of the biology of this group of organisms. The earliest contributions focused on clarifying the taxonomic status of populations. Species in the genus *Anopheles* have frequently evolved without acquiring obvious morphological differences. As described above, cryptic or sibling species within the *A. gambiae* complex were initially described on the basis of male sterility in hybrid individuals resulting from between-species mating experiments. Conducting crosses to detect hybrid sterility as a method for routine identification has obvious limitations. Efficient and readily applicable methods were soon developed including the development of effective cytotaxonomic and later molecular tools that are used to distinguish species in the complex. However, many questions remain regarding the taxonomic status of populations, especially within *A. gambiae* s.s., and the resolution of these problems will be achieved by applying population genetics approaches to the problem.

Analyses of the genetic structure of *A. gambiae* s.l. populations have contributed to understanding the distribution of phenotypic variation in the complex and at the subspecific level. The description of *A. quadriannulatus* as a distinct species and recognition that it is primarily zoophyllic provided an explanation for variation in host preference (Gillies and De Meillon, 1968). The association of specific chromosome inversions in *A. gambiae* s.s. with endo- and exophily (Coluzzi et al., 1977) and with habitat preferences with respect to aridity (Coluzzi et al., 1979; Touré et al., 1998; Lee et al., 2009) have been established. The evolution of insecticide resistance in populations of *A. gambiae* s.s. poses a serious challenge to current malaria control programs that rely on impregnated bednets and indoor spraying for vector control (N'Guessan et al., 2010). The differential distribution of resistance genes, such as *kdr*, among M and S form populations establishes the importance of recognizing population structure to insecticide resistance monitoring (Chandre et al., 1999; Weill et al., 2000; Tripet et al., 2007).

The availability of the *A. gambiae* s.s. whole genome sequence (Holt et al., 2002) has ushered in the advent of **population genomics** in vector biology. The promise of establishing the relationship between phenotype and genotype is attainable through the powerful new approach of **association mapping**. Early work aimed at identifying genes directly responsible for phenotypes of interest involved the use of laboratory strains selected for those phenotypes (Sinden, 2004; Aguilar et al., 2005). Recently it has been pointed out that this approach has serious limitations and that studies based on natural populations provides far more useful information (Tripet et al., 2008; Boëte, 2009). It is well known that the presence of population structure can result in "spurious associations" between a phenotype and markers that are not linked to any causative loci (e.g., Lander and Schork, 1994; Ewens and Spielman, 1995; Pritchard and Rosenberg, 1999; Kang et al., 2008). This becomes a problem when these subpopulations are not recognized so that a sample being used in an association mapping study consists of a mixture of individuals originating from two or more diverged subpopulations.

The movement of genes, including the use of gene drive vehicles (e.g., transposable elements), from one lineage or population to another depends on mating between an individual carrying the gene and one which does not. Although designs for novel approaches to target vector populations of mosquitoes are interesting and potentially useful, the population genetics component is very poorly understood. Critically, most conceptual models for genetic control assume that the mosquito population into which a refractory gene system is to be released represents a single, randomly mating unit. We have summarized the evidence that natural populations of *A. gambiae* are subdivided by barriers to reproduction and that gene flow via migration among geographic populations is limited. Field studies designed to estimate levels and patterns of gene flow within and among natural vector populations are needed to provide a foundation for predicting the potential utility of new molecular-level approaches, and for designing field trials to evaluate their efficacy under natural conditions in Africa.

15.4.6 Conclusions

With respect to current concepts toward describing the genetics of populations of *A. gambiae* s.s.:

- The chromosomal form concept is only valid in a very restricted geographical region.
- The distribution of at least some chromosome inversions supports the idea that they represent "coadapted gene complexes."
- The chromosome inversions themselves do not appear to mediate reproductive isolation.
- Reproductive isolation among molecular forms appears to be associated with relatively small genomic "islands of divergence" located near the centromere on the X chromosome and possibly other "islands" located on chromosomes 2 and 3.

- The M/S molecular form concept is valid only where these "alleles" are linked to "speciation genes," probably on the X chromosome. This linkage appears to be absent in some places (e.g., Guinea-Bissau, the Gambia) and in some populations (e.g., Savanna and Bamako populations).
- Natural populations of *A. gambiae* s.s. are structured into subpopulations that are to varying degrees reproductively isolated. The boundaries among these populations are poorly defined by the current concepts of chromosomal and molecular form. It is possible, perhaps likely, that some of these populations represent species or incipient species, however, current concepts of molecular and chromosomal forms fail to define them.

Glossary

Allopatric a description of taxa occupying distinctly different geographic ranges.

Association mapping it is a method of gene mapping that utilizes historic linkage disequilibrium (linkage) to associate phenotypes to genotypes.

Balancing selection maintenance of a polymorphism or polymorphisms above the frequency established by its mutation rate. Heterozygote advantage is one manifestation of balancing selection, meaning that a heterozygote has a fitness advantage over the corresponding homozygotes.

Chromosomal forms reproductively isolated populations of *A. gambiae* s.s. characterized by different paracentric inversion chromosome arrangements on the right arm of chromosome 2.

Cryptic species those species that are identical or nearly identical in appearance but can be discovered by genetic divergence as indicated by mating incompatibilities or sterile matings in interspecific crosses.

Cytoplasmic incompatibility sterile matings between *Wolbachia*-infected males and uninfected females. Females carrying reproductive parasite *Wolbachia* are fertile when mated with uninfected or *Wolbachia*-infected flies. In this way the maternal host lineages with *Wolbachia* can displace uninfected, maternal lineages if their Darwinian fitnesses are adequate.

Diversity at diploid genetic loci, it is heterozygosity averaged over loci. For mitochondrial genomes, which are single copy and therefore haploid, it is the probability that two randomly chosen individuals of a species have different haplotypes. Mathematically, diversity h of a nuclear or mitochondrial polymorphism is, $h = 1 - \Sigma x_i^2$, where x is the frequency of allele (or haplotype) i. Diversity over r loci is $H = \Sigma h/r$.

Ecophenotype a phenotype, in the case of *A. gambiae* s.s. chromosomal of molecular form, that shows a strong association with regional and seasonal environments.

Ecotypic speciation ecological and adaptive divergence among populations leading to reproductive isolation and speciation.

Effective population size and gene flow effective population size N_e is the hypothetical number of reproducing organisms in an ideal population (i.e., a population obeying Hardy–Weinberg assumptions) that corresponds with the population under investigation. It is unrelated to census numbers but always much less. Its relationship to gene flow is the number of effective migrants, $N_e m$, exchanged per generation among demes-$N_e m = (1 - F_{ST})/4F_{ST}$ for diploid loci.

F_{ST} and F_{IS} inbreeding coefficients estimated by variance statistics. Biologically, F_{ST} measures departures from random mating among demes (subpopulations). It estimates genetic differentiation and is the among-deme variance in allelic frequencies as a fraction of the variance among all individuals in a population. F_{ST} is inversely related to gene flow. F_{IS} measures departures from random mating within demes. F statistics can also be defined as correlations between uniting gametes: F_{IS} is the correlation between gametes in an individual relative to its subpopulation; F_{ST} is the correlation between random gametes in a deme relative to the correlation of pooled gametes in the whole population.

Genetic bottleneck a great reduction in effective population size. Genetic diversity becomes reduced in proportion to the magnitude of population reduction and the number of generations over which the bottleneck occurs. Recovery from a bottleneck, in terms of increased genetic diversity, requires tens of thousands of generations in the absence of inward gene flow.

Genetic drift change in gene frequencies from one generation to the next caused by random sampling. Its magnitude is inversely proportional to effective population size.

Incipient species populations evolving toward complete reproductive isolation, and therefore distinct biological species status,

Introgression introduction of genetic material from another species or variant into a population by hybridization followed by repeated backcrossing.

Islands of speciation discrete segments of a genome that are highly diverged and thought to contain genes affecting reproductive isolation between individuals within different populations.

Isolation by distance the probability of mating between two individuals decreases with increasing distance between them, resulting in a direct relationship between geographic and genetic distance.

Molecular forms populations of *A. gambiae* s.s. that differ with respect to specific sequence in a region of the intergenic spacer segment of the ribosomal gene, as visualized using diagnostic PCR methods.

Monophyletic describes a group of organisms that includes the most recent common ancestor of all those organisms and all the descendents of that common ancestor.

Paracentric chromosome inversion an inversion of a segment of a chromosome that does not include the centromere.

Peridomestic habitats the area surrounding houses (e.g., wood piles, animal corrals, fences, walls).

Polyphyletic describes a group of organisms derived from two or more parental lineages.

Polytypic species a species that contains several variant forms, especially geographically or temporally differentiated subspecies or varieties, which would normally interbreed if present in the same time and place.

Population genomics sampling of numerous, or all, variable gene loci within a genome to infer those evolutionary forces responsible for observed patterns of variation.

Sylvan refers to wild areas, as in "sylvan habitats."

Synapomorphic a description of a shared, derived trait found among two or more taxa and their most recent common ancestor, whose ancestor in turn does not possess the trait.

Transgenic an organism that contains foreign genes that were introduced into its genome.

References

Abad-Franch, F., Monteiro, F.A., 2005. Molecular research and the control of Chagas disease vectors. An. Acad. Bras. Cienc. 77, 437–454.

Abad-Franch, F., Monteiro, F.A., 2007. Biogeography and evolution of Amazonian triatomines (Heteroptera: Reduviidae): implications for Chagas disease surveillance in humid forest ecoregions. Mem. Inst. Oswaldo Cruz 102 (Suppl. 1), 57–70.

Abad-Franch, F., Monteiro, F.A., Jaramillo, O.N., Gurgel-Goncalves, R., Dias, F.B., Diotaiuti, L., 2009. Ecology, evolution, and the long-term surveillance of vector-borne Chagas disease: a multi-scale appraisal of the tribe Rhodniini (Triatominae). Acta Trop. 110, 159–177.

Abalos, J.W., 1948. Sobre híbridos naturales y experimentales de *Triatoma*. Anales Inst. Med. Reg. 2, 209–223.

Abila, P.P., Slotman, M.A., Parmakelis, A., et al., 2008. High levels of genetic differentiation between Ugandan *Glossina fuscipes fuscipes* populations separated by Lake Kyoga. PLoS Negl. Trop. Dis. 2, e242.

Aguilar, R., Dong, Y., Warr, E., Dimopoulos, G., 2005. *Anopheles* infection responses; laboratory models versus field malaria transmission systems. Acta Trop. 95, 285–291.

Akman, L., Yamashita, A., Watanabe, H., Oshima, K., Shiba, T., Hattori, M., et al., 2002. Genome sequence of the endocellular obligate symbiont of tsetse flies, *Wigglesworthia glossinidia*. Nat. Genet. 32, 402–407.

Aksoy, S., 2000. Tsetse—a haven for microorganisms. Parasitol. Today 16, 114–118.

Aksoy, S., 2003. Control of tsetse flies and trypanosomes using molecular genetics. Vet. Parasitol. 115, 125–145.

Aksoy, S., Rio, R.V., 2005. Interactions among multiple genomes: tsetse, its symbionts and trypanosomes. Insect Biochem. Mol. Biol. 35, 691–698.

Aksoy, S., Maudlin, I., Dale, C., Robinson, A.S., O'Neill, S.L., 2001. Prospects for control of African trypanosomiasis by tsetse vector manipulation. Trends Parasitol. 17, 29–35.

Askoy, S., 2010. White paper. A proposal for tsetse fly (*Glossina*) genome projects. http://www.genome.gov/26525388#al-3 (accessed 29.10.10).

Ayala, F.J., Coluzzi, M., 2005. Chromosome speciation: humans, *Drosophila*, and mosquitoes. Proc. Natl. Acad. Sci. U.S.A. 102 (Suppl. 1), 6535–6542.

Balmer, O., Caccone, A., 2008. Multiple-strain infections of *Trypanosoma brucei* across Africa. Acta Trop. 107, 275–279.

Bank, W., 1993. World development report. Investing in Health. Oxford University Press, New York, NY.

Bargues, M.D., Klisiowicz, D.R., Panzera, F., et al., 2006. Origin and phylogeography of the Chagas disease main vector *Triatoma infestans* based on nuclear rDNA sequences and genome size. Infect. Genet. Evol. 6, 46–62.

Bargues, M.D., Klisiowicz, D.R., Gonzalez-Candelas, F., et al., 2008. Phylogeography and genetic variation of *Triatoma dimidiata*, the main Chagas disease vector in central America, and its position within the Genus *Triatoma*. PLoS Negl. Trop. Dis. 2, e233.

Barrett, M.P., 2006. The rise and fall of sleeping sickness. Lancet 367, 1377–1378.

Beadell, J.S., Hyseni, C., Abila, P.P., et al., 2010. Phylogeography and population structure of *Glossina fuscipes fuscipes* in Uganda: implications for control of tsetse. PLoS Negl. Trop. Dis. 4, e636.

Beard, C.B., Cordon-Rosales, C., Durvasula, R.V., 2002. Bacterial symbionts of the triatominae and their potential use in control of Chagas disease transmission. Annu. Rev. Entomol. 47, 123–141.

Berriman, M., Ghedin, E., Hertz-Fowler, C., et al., 2005. The genome of the African trypanosome *Trypanosoma brucei*. Science 309, 416–422.

Besansky, N.J., Powell, J.R., Caccone, A., Hamm, D.M., Scott, J.A., Collins, F.H., 1994. Molecular phylogeny of the *Anopheles gambiae* complex suggests genetic introgression between principal malaria vectors. Proc. Natl. Acad. Sci. U.S.A. 91, 6885–6888.

Blandon-Naranjo, M., Zuriaga, M.A., Azofeifa, G., Zeledon, R., Bargues, M.D., 2010. Molecular evidence of intraspecific variability in different habitat-related populations of *Triatoma dimidiata* (Hemiptera: Reduviidae) from Costa Rica. Parasitol. Res. 106, 895–905.

Boëte, C., 2009. Anopheles mosquitoes: not just flying malaria vectors. Especially in the field. Trends Parasitol. 25, 53–55.

Brenière, S.F., Bosseno, M.F., Vargas, F., et al., 1998. Smallness of the panmictic unit of *Triatoma infestans* (Hemiptera: Reduviidae). J. Med. Entomol. 35, 911–917.

Brunhes, J., Le Goff, G., Geoffroy, B., 1997. *Anophèles* Afro-Tropicaux. I.—Desciptions D'Espèces nouvelles et changements de statuts taxonomiques (Diptera: Culicidae). Ann. Soc. Entomol. Fr. (N. S.) 33, 173–183.

Bustamante, D.M., Monroy, C., Menes, M., et al., 2004. Metric variation among geographic populations of the Chagas vector *Triatoma dimidiata* (Hemiptera: Reduviidae: Triatominae) and related species. J. Med. Entomol. 41, 296–301.

Buxton, P.A., 1955. The Natural History of Tsetse Flies. H.K. Lewis, London.

Calleros, L., Panzera, F., Bargues, M.D., et al., 2010. Systematics of *Mepraia* (Hemiptera-Reduviidae): cytogenetic and molecular variation. Infect. Genet. Evol. 10, 221–228.

Campbell-Lendrum, D., Molyneux, D., Amerasinghe, F., et al., 2005. Ecosystems and vector-borne disease control. In: Epstein, P., Githeko, A., Rabinovich, J., Weinstein, P. (Eds.), Ecosystems and Human Well-being, Vol 3: Policy responses. Findings of the responses working group of the Millenium Ecosystem Assessment. Island Press, Washington, DC.

Caputo, B., Nwakanma, D., Jawara, M., et al., 2008. *Anopheles gambiae* complex along the Gambia river, with particular reference to the molecular forms of *An. gambiae* s.s. Malar J. 7, 182.

Carlson, D., Sutton, B., Bernier, U., 2000. Cuticular hydrocarbons in *Glossina austeni* and *G. pallidipes*: similarities between populations. Insect Sci. Applics 20, 281–294.

Caro-Riaño, H., Jaramillo, N., Dujardin, J.P., 2009. Growth changes in *Rhodnius pallescens* under simulated domestic and sylvatic conditions. Infect. Genet. Evol. 9, 162–168.

Chandre, F., Manguin, S., Brengues, C., et al., 1999. Current distribution of a pyrethroid resistance gene (*kdr*) in *Anopheles gambiae* complex from west Africa and further evidence for reproductive isolation of the Mopti form. Parassitologia 41, 319–322.

Chappuis, F., Loutan, L., Simarro, P., Lejon, V., Buscher, P., 2005. Options for field diagnosis of human African trypanosomiasis. Clin. Microbiol. Rev. 18, 133–146.

Coetzee, M., Craig, M., le Sueur, D., 2000. Distribution of African malaria mosquitoes belonging to the *Anopheles gambiae* complex. Parasitol. Today 16, 74–77.

Coluzzi, M., 1966. Oservazioni comparative sul cromosoma X nelle specie A e B del complesso *Anophles ganbiae*. Rendiconti Accademia Nazionale dei Lincei 40, 671–678.

Coluzzi, M., 1968. Cromosomi poltenici delle cellule nutrice ovariche nel complesso Gambiae del genere *Anopheles*. Parassitologia X, 179–184.

Coluzzi, M., Sabatini, A., 1967. Cytogenetic observations on species A and B of the *Anopheles gambiae* complex. Parassitologia IX, 73–88.

Coluzzi, M., Sabatini, A., 1968. Cytogenetic observations on species C of the *Anopheles gambiae* complex. Parassitologia X, 155–165.

Coluzzi, M., Sabatini, A., 1969. Cytogenetic observations on the salt water species, *Anopheles merus* and *Anopheles melas*, of the Gambiae complex. Parassitologia XI, 177–187.

Coluzzi, M., Sabatini, A., Petrarca, V., Di Deco, M.A., 1977. Behavioural divergences between mosquitoes with different inversion karyotypes in polymorphic populations of the *Anopheles gambiae* complex. Nature 266, 832–833.

Coluzzi, M., Sabatini, A., Petrarca, V., Di Deco, M.A., 1979. Chromosomal differentiation and adaptation to human environments in the *Anopheles gambiae* complex. Trans. R. Soc. Trop. Med. Hyg. 73, 483–497.

Coluzzi, M., Petrarca, V., Di Deco, M.A., 1985. Chromosomal inversion intergradation and incipient speciation in *Anopheles gambiae*. Bollettino di Zoologia 52, 45–63.

Coluzzi, M., Sabatini, A., della Torre, A., Di Deco, M.A., Petrarca, V., 2002. A polytene chromosome analysis of the *Anopheles gambiae* species complex. Science 298, 1415–1418.

Coulibaly, M.B., Pombi, M., Caputo, B., et al., 2007. PCR-based karyotyping of *Anopheles gambiae* inversion 2Rj identifies the BAMAKO chromosomal form. Malar J. 6, 133.

Crossa, R.P., Hernandez, M., Caraccio, M.N., et al., 2002. Chromosomal evolution trends of the genus *Panstrongylus* (Hemiptera, Reduviidae), vectors of Chagas disease. Infect. Genet. Evol. 2, 47–56.

Davidson, G., 1956. Insecticide resistance in *Anopheles gambiae* Giles: a case of simple mendelian inheritance. Nature 178, 863–864.

Davidson, G., 1964a. *Anopheles gambiae*, a complex of species. Bull. World Health Org. 31, 625–634.

Davidson, G., 1964b. The five mating-types in the *Anopheles gambiae* complex. Riv. Malariol. 43, 167–183.

Davidson, G., Hunt, R.H., 1973. The crossing and chromosome characteristics of a new, sixth species in the *Anopheles gambiae* complex. Parassitologia 15, 121–128.

Davidson, G., Paterson, H.E., Coluzzi, M., Mason, G.F., Micks, D.W., 1967. The *Anopheles gambiae* complex. In: Wright, J.W., Pal, R. (Eds.), Genetics of Insect Vectors of Disease. Elsevier, Amsterdam.

de Paula, A.S., Diotaiuti, L., Schofield, C.J., 2005. Testing the sister-group relationship of the Rhodniini and Triatomini (Insecta: Hemiptera: Reduviidae: Triatominae). Mol. Phylogenet. Evol. 35, 712–718.

de la Fuente, A.L., Dias-Lima, A., Lopes, C.M., et al., 2008. Behavioral plasticity of Triatominae related to habitat selection in northeast Brazil. J. Med. Entomol. 45, 14–19.

della Torre, A., Fanello, C., Akogbeto, M., et al., 2001. Molecular evidence of incipient speciation within *Anopheles gambiae* s.s. in West Africa. Insect Mol. Biol. 10, 9–18.

della Torre, A., Costantini, C., Besansky, N.J., Caccone, A., Petrarca, V., Powell, J.R., et al., 2002. Speciation within *Anopheles gambiae*-the glass is half full. Science 298, 115–117.

della Torre, A., Tu, Z., Petrarca, V., 2005. On the distribution and genetic differentiation of *Anopheles gambiae* s.s. molecular forms. Insect. Biochem. Mol. Biol. 35, 755–769.

Diuk-Wasser, M.A., Toure, M.B., Dolo, G., et al., 2005. Vector abundance and malaria transmission in rice-growing villages in Mali. Am. J. Trop. Med. Hyg. 72, 725–731.

Diuk-Wasser, M.A., Dolo, G., Bagayoko, M., et al., 2006. Patterns of irrigated rice growth and malaria vector breeding in Mali using multi-temporal ERS-2 synthetic aperture radar. Int. J. Remote Sens. 27, 535–548.

Dorn, P.L., Melgar, S., Rouzier, V., et al., 2003. The Chagas vector, *Triatoma dimidiata* (Hemiptera: Reduviidae), is panmictic within and among adjacent villages in Guatemala. J. Med. Entomol. 40, 436–440.

Dorn, P.L., Monroy, C., Curtis, A., 2007. *Triatoma dimidiata* (Latreille, 1811): a review of its diversity across its geographic range and the relationship among populations. Infect. Genet. Evol. 7, 343–352.

Dorn, P.L., Calderon, C., Melgar, S., et al., 2009. Two distinct *Triatoma dimidiata* (Latreille, 1811) taxa are found in sympatry in Guatemala and Mexico. PLoS Negl. Trop. Dis. 3, e393.

Dujardin, J.P., Tibayrenc, M., 1985a. [Isoenzymatic studies of the principal vector of Chagas disease: *Triatoma infestans* (Hemiptera: Reduviidae)]. Ann. Soc. Belg. Med. Trop. 65 (Suppl. 1), 165–169.

Dujardin, J.P., Tibayrenc, M., 1985b. [Study of 11 enzymes and formal genetic findings for 19 enzymatic loci in Triatoma infestans (Hemiptera: Reduviidae)]. Ann. Soc. Belg. Med. Trop. 65, 271–280.

Dujardin, J.P., Tibayrenc, M., Venegas, E., Maldonado, L., Desjeux, P., Ayala, F.J., 1987. Isozyme evidence of lack of speciation between wild and domestic *Triatoma infestans* (Heteroptera: Reduviidae) in Bolivia. J. Med. Entomol. 24, 40–45.

Dujardin, J.P., Cardozo, L., Schofield, C., 1996. Genetic analysis of *Triatoma infestans* following insecticidal control interventions in central Bolivia. Acta Trop. 61, 263–266.

Dujardin, J.P., Munoz, M., Chavez, T., Ponce, C., Moreno, J., Schofield, C.J., 1998a. The origin of *Rhodnius prolixus* in Central America. Med. Vet. Entomol. 12, 113–115.

Dujardin, J.P., Schofield, C.J., Tibayrenc, M., 1998b. Population structure of Andean *Triatoma infestans*: allozyme frequencies and their epidemiological relevance. Med. Vet. Entomol. 12, 20–29.

Dujardin, J.P., Panzera, P., Schofield, C.J., 1999. Triatominae as a model of morphological plasticity under ecological pressure. Mem. Inst. Oswaldo Cruz 94 (Suppl. 1), 223–228.

Dujardin, J.P., Schofield, C.J., Panzera, F., 2000. Los Vectores de la Enfermedad de Chagas. Academie Royale des Sciences D'Outre-Mar, Brussels.

Dujardin, J.P., Schofield, C.J., Panzera, F., 2002. Los vectores de la enfermedad de Chagas. Academie Royal des Sciences D'outre-Mer, Brussels.

Dujardin, J.P., Costa, J., Bustamante, D., Jaramillo, N., Catala, S., 2009. Deciphering morphology in Triatominae: the evolutionary signals. Acta Trop. 110, 101–111.

Dumonteil, E., Tripet, F., Ramirez-Sierra, M.J., Payet, V., Lanzaro, G., Menu, F., 2007. Assessment of *Triatoma dimidiata* dispersal in the Yucatan Peninsula of Mexico by morphometry and microsatellite markers. Am. J. Trop. Med. Hyg. 76, 930–937.

Dyer, N.A., Furtado, A., Cano, J., et al., 2009. Evidence for a discrete evolutionary lineage within Equatorial Guinea suggests that the tsetse fly *Glossina palpalis palpalis* exists as a species complex. Mol. Ecol. 18, 3268–3282.

Edillo, F.E., Toure, Y.T., Lanzaro, G.C., Dolo, G., Taylor, C.E., 2002. Spatial and habitat distribution of *Anopheles gambiae* and *Anopheles arabiensis* (Diptera: Culicidae) in Banambani village, Mali. J. Med. Entomol. 39, 70–77.

Ewens, W.J., Spielman, R.S., 1995. The transmission/disequilibrium test: history, subdivision, and admixture. Am. J. Hum. Genet. 57, 455–464.

Fanello, C., Petrarca, V., della Torre, A., et al., 2003. The pyrethroid knock-down resistance gene in the *Anopheles gambiae* complex in Mali and further indication of incipient speciation within *An. gambiae* s.s. Insect Mol. Biol. 12, 241–245.

Farikou, O., Njiokou, F., Mbida Mbida, J.A., et al., 2010. Tripartite interactions between tsetse flies, *Sodalis glossinidius* and trypanosomes—an epidemiological approach in two historical human African trypanosomiasis foci in Cameroon. Infect. Genet. Evol. 10, 115–121.

Favia, G., della Torre, A., Bagayoko, M., Lanfrancotti, A., Sagnon, N., Toure, Y.T., et al., 1997. Molecular identification of sympatric chromosomal forms of *Anopheles gambiae* and further evidence of their reproductive isolation. Insect Mol. Biol. 6, 377–383.

Favia, G., Lanfrancotti, A., Spanos, L., Siden-Kiamos, I., Louis, C., 2001. Molecular characterization of ribosomal DNA polymorphisms discriminating among chromosomal forms of *Anopheles gambiae* s.s. Insect Mol. Biol. 10, 19–23.

Ferral, J., Chavez-Nunez, L., Euan-Garcia, M., Ramirez-Sierra, M.J., Najera-Vazquez, M.R., Dumonteil, E., 2010. Comparative field trial of alternative vector control strategies for non-domiciliated *Triatoma dimidiata*. Am. J. Trop. Med. Hyg. 82, 60–66.

Figueiras, A.N., Lazzari, C.R., 2000. Temporal change of the aggregation response in *Triatoma infestans*. Mem. Inst. Oswaldo Cruz 95, 889–892.

Fitzpatrick, S., Feliciangeli, M.D., Sanchez-Martin, M.J., Monteiro, F.A., Miles, M.A., 2008. Molecular genetics reveal that silvatic *Rhodnius prolixus* do colonise rural houses. PLoS Negl. Trop. Dis. 2, e210.

Flores, A., Gastelum, E.M., Bosseno, M.F., et al., 2001. Isoenzyme variability of five principal triatomine vector species of Chagas disease in Mexico. Infect. Genet. Evol. 1, 21–28.

Ford, J., 1971. The Role of Trypanosomiases in African Ecology. Clarendon Press, Oxford, UK.

Frias, D., Kattan, F., 1989. Molecular taxonomic studies in *Triatoma infestans* (Klug, 1934) and *Triatoma spinolai* Porter, 1933 populations (Hemiptera: Triatominae). Acta Entomol. Chilena 15, 205–210.

Galvão, C., Carcavallo, R., Da Silva Rocha, D., Jurberg, J., 2003. A checklist of the current valid species of the subfamily Triatominae Jeannel, 1919 (Hemiptera, Reduviidae) and their geographical distribution, with nomenclatural and taxonomic notes. Zootaxa 202, 1–36.

Garcia, B.A., Powell, J.R., 1998. Phylogeny of species of *Triatoma* (Hemiptera: Reduviidae) based on mitochondrial DNA sequences. J. Med. Entomol. 35, 232–238.

Garcia, B.A., Canale, D.M., Blanco, A., 1995. Genetic structure of four species of *Triatoma* (Hemiptera: Reduviidae) from Argentina. J. Med. Entomol. 32, 134–137.

Garcia, B.A., Manfredi, C., Fichera, L., Segura, E.L., 2003. Short report: variation in mitochondrial 12S and 16S ribosomal DNA sequences in natural populations of *Triatoma infestans* (Hemiptera: Reduviidae). Am. J. Trop. Med. Hyg. 68, 692–694.

García, B.A., Zheng, L.O., Pérez De Rosas, A.R., Segura, E.L., 2004. Primer note. Isolation and characterization of polymorphic microsatellite loci in the Chagas' disease vector *Triatoma infestans* (Hemiptera: Reduviidae). Mol. Ecol. Notes 4, 568–571.

Gaunt, M., Miles, M., 2000. The ecotopes and evolution of triatomine bugs (triatominae) and their associated trypanosomes. Mem. Inst. Oswaldo Cruz 95, 557–565.

Gibson, W., Peacock, L., Ferris, V., Williams, K., Bailey, M., 2008. The use of yellow fluorescent hybrids to indicate mating in *Trypanosoma brucei*. Parasit Vectors 1, 4.

Gillies, M.T., De Meillon, B., 1968. The Anophelinae of Africa south of the Sahara (Ethiopian Zoogeographical Region). Publ. S. Afr. Inst. Med. Res. 54, 1–343.

Giordano, R., Cortez, J.C., Paulk, S., Stevens, L., 2005. Genetic diversity of *Triatoma infestans* (Hemiptera: Reduviidae) in Chuquisaca, Bolivia based on the mitochondrial cytochrome b gene. Mem. Inst. Oswaldo Cruz 100, 753–760.

Gonzalez Audino, P., Vassena, C., Barrios, S., Zerba, E., Picollo, M.I., 2004. Role of enhanced detoxication in a deltamethrin-resistant population of Triatoma infestans (Hemiptera, Reduviidae) from Argentina. Mem. Inst. Oswaldo Cruz 99, 335–339.

Gooding, R., 1993. Genetic analysis of sterility in hybrids from crosses of *Glossina morsitans submorsitans* and *Glossina morsitans centralis* (Diptera: Glossinidae). Can. J. Zool. 71, 963–972.

Gooding, R.H., Krafsur, E.S., 2004. Tsetse genetics: applications to biology and systematics. In: Maudlin, I., Homes, P.H., Miles, M.A. (Eds.), The Trypanosomiases. CABI, Oxfordshire, UK, pp. 95–111.

Gooding, R.H., Krafsur, E.S., 2005. Tsetse genetics: contributions to biology, systematics, and control of tsetse flies. Annu. Rev. Entomol. 50, 101–123.

Green, C.A., 1972. Cytological maps for the practical identification of females of the three freshwater species of the *Anopheles gambiae* complex. Ann. Trop. Med. Parasitol. 66, 143–147.

Hargrove, J.W., 2003. Tsetse Eradication: Sufficiency, Necessity and Desirability. DFID Animal Health Programme, Centre for Tropical Veterinary Medicine. University of Edinburgh, Edinburgh.

Hargrove, J., 2004. Tsetse population dynamics. In: Maudlin, I., Holmes, P., Miles, M. (Eds.), The Trypanosomiases. CABI, Oxford, UK, pp. 113–137.

Harry, M., Galindez, I., Cariou, M.L., 1992. Isozyme variability and differentiation between *Rhodnius prolixus*, *R. robustus* and *R. pictipes*, vectors of Chagas disease in Venezuela. Med. Vet. Entomol. 6, 37–43.

Harry, M., Moreno, G., Goyffon, M., 1993. Isozyme variability in *Rhodnius prolixus* populations, vectors of Chagas disease in Venezuela. Evol. Biol. 6, 175–194.

Herrera-Aguilar, M., Be-Barragan, L.A., Ramirez-Sierra, M.J., Tripet, F., Dorn, P., Dumonteil, E., 2009. Identification of a large hybrid zone between sympatric sibling species of *Triatoma dimidiata* in the Yucatan peninsula, Mexico, and its epidemiological importance. Infect. Genet. Evol. 9, 1345–1351.

Holstein, M.H., 1954. Biology of *Anopheles gambiae*: research in French West Africa. Monograph Series No. 9. World Health Organization, Geneva, p. 172.

Holt, R.A., Subramanian, G.M., Halpern, A., et al., 2002. The genome sequence of the malaria mosquito *Anopheles gambiae*. Science 298, 129–149.

Hunt, R.H., 1973. A cytological technique for the study of *Anopheles gambiae* complex. Parassitologia 15, 137–139.

Hunt, R.H., Coetzee, M., Fettene, M., 1998. The *Anopheles gambiae* complex: a new species from Ethiopia. Trans. R. Soc. Trop. Med. Hyg. 92, 231–235.

Hypsa, V., Tietz, D.F., Zrzavy, J., Rego, R.O., Galvao, C., Jurberg, J., 2002. Phylogeny and biogeography of Triatominae (Hemiptera: Reduviidae): molecular evidence of a New World origin of the Asiatic clade. Mol. Phylogenet. Evol. 23, 447–457.

Jordan, A.M., 1993. Tsetse flies (Glossinidae). In: Lane, R.P., Crosskey, R.W. (Eds.), Medical Insects and Arachnids. Chapman & Hall, London, pp. 333–388.

Jurenka, R., Terblanche, J.S., Klok, C.J., Chown, S.L., Krafsur, E.S., 2007. Cuticular lipid mass and desiccation rates in *Glossina pallidipes*: interpopulation variation. Physiol. Entomol. 32, 287–293.

Kang, H.M., Zaitlen, N.A., Wade, C.M., Kirby, A., Heckerman, D., Daly, M.J., et al., 2008. Efficient control of population structure in model organism association mapping. Genetics 178, 1709–1723.

Kgori, P.M., Modo, S., Torr, S.J., 2006. The use of aerial spraying to eliminate tsetse from the Okavango Delta of Botswana. Acta Trop. 99, 184–199.

Klassen, W., Curtis, C.F., 2005. History of the sterile insect technique. In: Dyck, V.A., Hendrichs, J., Robinson, A.S. (Eds.), Sterile Insect Technique. Springer, Dordrecht, Netherlands, pp. 3–36.

Koffi, M., De Meeus, T., Bucheton, B., et al., 2009. Population genetics of *Trypanosoma brucei gambiense*, the agent of sleeping sickness in Western Africa. Proc. Natl. Acad. Sci. U.S.A. 106, 209–214.

Krafsur, E.S., 2002. The sterile insect technique. In: Pimentel, D. (Ed.), Encyclopedia of Pest Management. Taylor & Francis, London, pp. 788–791.

Krafsur, E.S., 2003. Tsetse fly population genetics: an indirect approach to dispersal. Trends Parasitol. 19, 162–166.

Krafsur, E.S., 2009. Tsetse flies: genetics, evolution, and role as vectors. Infect. Genet. Evol. 9, 124–141.
Lander, E.S., Schork, N.J., 1994. Genetic dissection of complex traits. Science 265, 2037–2048.
Lanzaro, G.C., Toure, Y.T., Carnahan, J., et al., 1998. Complexities in the genetic structure of *Anopheles gambiae* populations in west Africa as revealed by microsatellite DNA analysis. Proc. Natl. Acad. Sci. U.S.A. 95, 14260–14265.
Leak, S.G.A., 1998. Tsetse Biology and Ecology: Their Role in the Epidemiology and Control of Trypanosomiasis. CABI Publicatons, New York, NY.
Lee, Y., Cornel, A.J., Meneses, C.R., et al., 2009. Ecological and genetic relationships of the Forest-M form among chromosomal and molecular forms of the malaria vector *Anopheles gambiae* sensu stricto. Malar J. 8, 75.
Lehmann, T., Hawley, W.A., Kamau, L., Fontenille, D., Simard, F., Collins, F.H., 1996. Genetic differentiation of *Anopheles gambiae* populations from East and west Africa: comparison of microsatellite and allozyme loci. Heredity 77 (Pt 2), 192–200.
Lehmann, T., Hawley, W.A., Grebert, H., Danga, M., Atieli, F., Collins, F.H., 1999. The Rift Valley complex as a barrier to gene flow for *Anopheles gambiae* in Kenya. J. Hered. 90, 613–621.
Lehmann, T., Blackston, C.R., Besansky, N.J., Escalante, A.A., Collins, F.H., Hawley, W.A., 2000. The Rift Valley complex as a barrier to gene flow for *Anopheles gambiae* in Kenya: the mtDNA perspective. J. Hered. 91, 165–168.
Lehmann, T., Licht, M., Elissa, N., et al., 2003. Population Structure of *Anopheles gambiae* in Africa. J. Hered. 94, 133–147.
Lent, H., Wygodzinsky, P., 1979. Revision of the Triatominae (Hemiptera, Reduviidae) and their significance as vectors of Chagas disease. Bull. Am. Mus. Nat. Hist. 163, 123–520.
Lopez, G., Moreno, J., 1995. Genetic variability and differentiation between populations of *Rhodnius prolixus* and *R. pallescens*, vectors of Chagas' disease in Colombia. Mem. Inst. Oswaldo Cruz 90, 353–357.
Lyman, D.E., Monteiro, F.A., Escalante, A.A., Cordon-Rosales, C., Wesson, D.M., Dujardin, J.P., et al., 1999. Mitochondrial DNA sequence variation among triatomine vectors of Chagas' disease. Am. J. Trop. Med. Hyg. 60, 377–386.
MacArthur, R.H., Wilson, E.O., 1967. The Theory of Island Biogeography. Princeton University Press, Princeton, USA.
Mahon, R.J., Green, C.A., Hunt, R.H., 1976. Diagnostic allozymes for routine identification of adults of the *Anopheles gambiae* complex. Bull. Entomol. Res. 68, 25–31.
Manoukis, N.C., Powell, J.R., Toure, M.B., et al., 2008. A test of the chromosomal theory of ecotypic speciation in *Anopheles gambiae*. Proc. Natl. Acad. Sci. U.S.A. 105, 2940–2945.
Marcet, P.L., Mora, M.S., Cutrera, A.P., Jones, L., Gürtler, R.E., Kitron, U., Dotson, E.M., 2008. Genetic structure of *Triatoma infestans* populations in rural communities of Santiago Del Estero, northern Argentina. Infect. Genet. Evol. 8, 835–846.
Marcilla, A., Canese, A., Acosta, N., Lopez, E., Rojas de Arias, A., Bargues, M.D., et al., 2000. Populations of *Triatoma infestans* (Hemiptera:Reduviidae) from Paraguay: a molecular analysis based on the second internal transcribed spacer of the rDNA. Res. Rev. Parasitol. 60, 99–105.
Marcilla, A., Bargues, M.D., Ramsey, J.M., et al., 2001. The ITS-2 of the nuclear rDNA as a molecular marker for populations, species, and phylogenetic relationships in

Triatominae (Hemiptera: Reduviidae), vectors of Chagas disease. Mol. Phylogenet. Evol. 18, 136–142.
Mas-Coma, S., Bargues, M.D., 2009. Populations, hybrids and the systematic concepts of species and subspecies in Chagas disease triatomine vectors inferred from nuclear ribosomal and mitochondrial DNA. Acta Trop. 110, 112–136.
Maudlin, I., Homes, P.H., Miles, M.A., 2004. The Trypanosomes. CABI Publications, Wallingford, UK.
McAlpine, J.F., 1989. Phylogeny and classification of the Muscomorpha. In: McAlpine, J.F. (Ed.), Manual of Nearctic Diptera, Vol. 3. Res. Branch Agric., Can, Ottawa, pp. 1397–1518.
Melgar, S., Chavez, J.J., Landaverde, P., et al., 2007. The number of families of *Triatoma dimidiata* in a Guatemalan house. Mem. Inst. Oswaldo Cruz 102, 221–223.
Miles, S.J., 1979. A biochemical key to adult members of the *Anopheles gambiae* group of species (Diptera: Culicidae). J. Med. Entomol. 15, 297–299.
Monroy, C., Rodas, A., Mejia, M., Tabaru, Y., 1998. Wall plastering and paints as methods to control vectors of Chagas disease in Guatemala. Med. Entomol. Zool. 49, 187–193.
Monteiro, F.A., Perez, R., Panzera, F., et al., 1999. Mitochondrial DNA variation of *Triatoma infestans* populations and its implication on the specific status of *T. melanosoma*. Mem. Inst. Oswaldo Cruz 94, 229–238.
Monteiro, F.A., Lazoski, C., Noireau, F., Sole-Cava, A.M., 2002. Allozyme relationships among ten species of Rhodniini, showing paraphyly of *Rhodnius* including *Psammolestes*. Med. Vet. Entomol. 16, 83–90.
Monteiro, F.A., Barrett, T.V., Fitzpatrick, S., Cordon-Rosales, C., Feliciangeli, D., Beard, C.B., 2003. Molecular phylogeography of the Amazonian Chagas disease vectors *Rhodnius prolixus* and *R. robustus*. Mol. Ecol. 12, 997–1006.
N'Guessan, R., Boko, P., Odjo, A., Chabi, J., Akogbeto, M., Rowland, M., 2010. Control of pyrethroid and DDT-resistant *Anopheles gambiae* by application of indoor residual spraying or mosquito nets treated with a long-lasting organophosphate insecticide, chlorpyrifos-methyl. Malar J. 9, 44.
Ng'habi, K.R., Meneses, C.R., Cornel, A.J., Slotman, M.A., Knols, B.G., Ferguson, H.M., et al., 2008. Clarification of anomalies in the application of a 2La molecular karyotyping method for the malaria vector *Anopheles gambiae*. Parasit Vectors 1, 45.
Njiokou, F., Laveissiere, C., Simo, G., Nkinin, S., Grebaut, P., Cuny, G., et al., 2006. Wild fauna as a probable animal reservoir for *Trypanosoma brucei gambiense* in Cameroon. Infect. Genet. Evol. 6, 147–153.
Noireau, F., Brenière, F., Ordoñez, J., et al., 1997. Low probability of transmission of *Trypanosoma cruzi* to humans by domiciliary *Triatoma sordida* in Bolivia. Trans. R. Soc. Trop. Med. Hyg. 91, 653–656.
Noireau, F., Gutierrez, T., Zegarra, M., Flores, R., Breniere, F., Cardozo, L., et al., 1998. Cryptic speciation in *Triatoma sordida* (Hemiptera:Reduviidae) from the Bolivian Chaco. Trop. Med. Int. Health 3, 364–372.
Noireau, F., Cortez, M.G., Monteiro, F.A., Jansen, A.M., Torrico, F., 2005. Can wild *Triatoma infestans* foci in Bolivia jeopardize Chagas disease control efforts? Trends Parasitol. 21, 7–10.
Núñez, J.A., 1987. Behavior of Triatominae bugs. Chagas' Disease Vectors. CRC Press, Boca Raton, FL, pp. 1-27.
Oliveira, E., Salgueiro, P., Palsson, K., et al., 2008. High levels of hybridization between molecular forms of *Anopheles gambiae* from Guinea Bissau. J. Med. Entomol. 45, 1057–1063.

Panzera, F., Perez, R., Panzera, Y., Alvarez, F., Scvortzoff, E., Salvatella, R., 1995. Karyotype evolution in holocentric chromosomes of three related species of triatomines (Hemiptera-Reduviidae). Chromosome Res. 3, 143–150.

Panzera, F., Dujardin, J.P., Nicolini, P., et al., 2004. Genomic changes of Chagas disease vector, South America. Emerg. Infect. Dis. 10, 438–446.

Panzera, F., Ferrandis, I., Ramsey, J., et al., 2006. Chromosomal variation and genome size support existence of cryptic species of *Triatoma dimidiata* with different epidemiological importance as Chagas disease vectors. Trop. Med. Int. Health 11, 1092–1103.

Panzera, F., Perez, R., Panzera, Y., Ferrandis, I., Ferreiro, M.J., Calleros, L., 2010. Cytogenetics and genome evolution in the subfamily Triatominae (Hemiptera, Reduviidae). Cytogenet. Genome Res. 128, 77–87.

Patterson, J.S., Barbosa, S.E., Feliciangeli, M.D., 2009. On the genus *Panstrongylus* Berg 1879: evolution, ecology and epidemiological significance. Acta Trop. 110, 187–199.

Pereira, J., Dujardin, J.P., Salvatella, R., Tibayrenc, M., 1996. Enzymatic variability and phylogenetic relatedness among *Triatoma infestans, T. platensis, T. delpontei* and *T. rubrovaria*. Heredity 77, 47–54.

Pérez de Rosas, A.R., Segura, E.L., García, B.A., 2007. Microsatellite analysis of genetic structure in natural *Triatoma infestans* (Hemiptera: Reduviidae) populations from Argentina: its implication in assessing the effectiveness of Chagas' disease vector control programmes. Mol. Ecol. 16, 1401–1412.

Pérez de Rosas, A.R., Segura, E.L., Fichera, L., Garcia, B.A., 2008. Macrogeographic and microgeographic genetic structure of the Chagas' disease vector *Triatoma infestans* (Hemiptera: Reduviidae) from Catamarca, Argentina. Genetica 133, 247–260.

Piccinali, R.V., Marcet, P.L., Noireau, F., Kitron, U., Gürtler, R.E., Dotson, E.M., 2009. Molecular population genetics and phylogeography of the Chagas disease vector *Triatoma infestans* in South America. J. Med. Entomol. 46, 796–809.

Picollo, M.I., Vassena, C., Santo Orihuela, P., Barrios, S., Zaidemberg, M., Zerba, E., 2005. High resistance to pyrethroid insecticides associated with ineffective field treatments in *Triatoma infestans* (Hemiptera: Reduviidae) from Northern Argentina. J. Med. Entomol. 42, 637–642.

Pizarro, J.C., Gilligan, L.M., Stevens, L., 2008. Microsatellites reveal a high population structure in *Triatoma infestans* from Chuquisaca, Bolivia. PLoS Negl. Trop. Dis. 2, e202.

Pombi, M., Caputo, B., Simard, F., et al., 2008. Chromosomal plasticity and evolutionary potential in the malaria vector *Anopheles gambiae* sensu stricto: insights from three decades of rare paracentric inversions. BMC Evol. Biol. 8, 309.

Pritchard, J.K., Rosenberg, N.A., 1999. Use of unlinked genetic markers to detect population stratification in association studies. Am. J. Hum. Genet. 65, 220–228.

Rabinovich, J.E., 1974. Demographic strategies in animal populations: a regression analysis. In: Golloy, F.B., Medina, E. (Eds.), Tropical Ecological Systems. Springer Verlag, New York, NY, pp. 19–40.

Ramírez, C.J., Jaramillo, C.A., del Pilar Delgado, M., Pinto, N.A., Aguilera, G., Guhl, F., 2005. Genetic structure of sylvatic, peridomestic and domestic populations of *Triatoma dimidiata* (Hemiptera: Reduviidae) from an endemic zone of Boyaca, Colombia. Acta Trop. 93, 23–29.

Ravel, S., de Meeus, T., Dujardin, J.P., et al., 2007. The tsetse fly *Glossina palpalis palpalis* is composed of several genetically differentiated small populations in the sleeping sickness focus of Bonon, Cote d'Ivoire. Infect. Genet. Evol. 7, 116–125.

Reinhardt, E., 2002. Travailler ensemble: la mouche tsé-tsé et la pauvreté rurale, in [*Internet*]. Available from http://www.un.org/french/pubs/chronique/2002/numero2/0202p17_la_mouche_tsetse.html. Chronique ONU, ONU Editor-September 02.

Richer, W., Kengne, P., Cortez, M.R., Perrineau, M.M., Cohuet, A., Fontenille, D., et al., 2007. Active dispersal by wild *Triatoma infestans* in the Bolivian Andes. Trop. Med. Int. Health 12, 759−764.

Rogers, D.J., 1977. Study of a natural population of tsetse flies and a model for fly movement. J. Anim. Ecol. 46, 309−330.

Rogers, D.J., Robinson, T.P., 2004. Tsetse distribution. In: Maudlin, I., Holmes, P., Miles, M. (Eds.), The Trypanosomiases. CABI, Oxford, UK, pp. 139−179.

Schofield, C.J., 1994. Triatominae: Biology and Control. Eurocommunica Publications, West Sussex, UK.

Schofield, C.J., 1988. Biosystematics of the Triatominae. In: Sevice, M.W. (Ed.), Biosystematics of Haematophagous Insects. Clarendon Press, Oxford, UK, pp. 287−312.

Schofield, C.J., Dias, J.C., 1999. The Southern Cone Initiative against Chagas disease. Adv. Parasitol. 42, 1−27.

Schofield, C.J., Galvão, C., 2009. Classification, evolution, and species groups within the Triatominae. Acta Trop. 110, 88−100.

Schofield, C.J., Diotaiuti, L., Dujardin, J.P., 1999. The process of domestication in Triatominae. Mem. Inst. Oswaldo Cruz 94 (Suppl. 1), 375−378.

Schofield, C.J., Jannin, J., Salvatella, R., 2006. The future of Chagas disease control. Trends Parasitol. 22, 583−588.

Scott, J.A., Brogdon, W.G., Collins, F.H., 1993. Identification of single specimens of the *Anopheles gambiae* complex by the polymerase chain reaction. Am. J. Trop. Med. Hyg. 49, 520−529.

Segura, E.L., Torres, A.G., Fusco, O., Garcia, B.A., 2009. Mitochondrial 16S DNA variation in populations of *Triatoma infestans* from Argentina. Med. Vet. Entomol. 23, 34−40.

Sharma, R., Gluenz, E., Peacock, L., Gibson, W., Gull, K., Carrington, M., 2009. The heart of darkness: growth and form of *Trypanosoma brucei* in the tsetse fly. Trends Parasitol. 25, 517−524.

Simarro, P.P., Jannin, J., Cattand, P., 2008. Eliminating human African trypanosomiasis: where do we stand and what comes next? PLoS Med. 5, e55.

Simo, G., Njiokou, F., Tume, C., Lueong, S., De Meeus, T., Cuny, G., et al., 2010. Population genetic structure of Central African *Trypanosoma brucei gambiense* isolates using microsatellite DNA markers. Infect. Genet. Evol. 10, 68−76.

Sinden, E.E., 2004. Mosquito−malaria interactions: a reappraisal of the concepts of susceptibility and refractoriness. Insect Biochem. Mol. Biol. 34, 625−629.

Slotman, M.A., Mendez, M.M., Torre, A.D., Dolo, G., Toure, Y.T., Caccone, A., 2006. Genetic differentiation between the BAMAKO and SAVANNA chromosomal forms of *Anopheles gambiae* as indicated by amplified fragment length polymorphism analysis. Am. J. Trop. Med. Hyg. 74, 641−648.

Slotman, M.A., Tripet, F., Cornel, A.J., et al., 2007. Evidence for subdivision within the M molecular form of *Anopheles gambiae*. Mol. Ecol. 16, 639−649.

Stevens, J.R., Brisse, S., 2004. Systematics of trypanosomes of medical and veterinary importance. In: Maudlin, I., Holmes, P., Miles, M. (Eds.), The Trypanosomiases. CAB International, Oxford, UK, pp. 1−24.

Stump, A.D., Fitzpatrick, M.C., Lobo, N.F., et al., 2005a. Centromere-proximal differentiation and speciation in *Anopheles gambiae*. Proc. Natl. Acad. Sci. U.S.A. 102, 15930−15935.

Stump, A.D., Shoener, J.A., Costantini, C., Sagnon, N., Besansky, N.J., 2005b. Sex-linked differentiation between incipient species of *Anopheles gambiae*. Genetics 169, 1509–1519.

Taylor, C.E., Touré, Y.T., Coluzzi, M., Petrarca, V., 1993. Effective population size and persistence of *Anopheles arabiensis* during the dry season in West Africa. Med. Vet. Entomol. 7, 351–357.

Taylor, C., Toure, Y.T., Carnahan, J., et al., 2001. Gene flow among populations of the malaria vector, *Anopheles gambiae*, in Mali, West Africa. Genetics 157, 743–750.

Terblanche, J.S., Klok, C.J., Krafsur, E.S., Chown, S.L., 2006. Phenotypic plasticity and geographic variation in thermal tolerance and water loss of the tsetse *Glossina pallidipes* (Diptera: Glossinidae): implications for distribution modelling. Am. J. Trop. Med. Hyg. 74, 786–794.

Tibayrenc, M., Breniere, S.F., 1988. *Trypanosoma cruzi*: major clones rather than principal zymodemes. Mem. Inst. Oswaldo Cruz 83 (Suppl. 1), 249–255.

Tibayrenc, M., Ward, P., Moya, A., Ayala, F.J., 1986. Natural populations of *Trypanosoma cruzi*, the agent of Chagas disease, have a complex multiclonal structure. Proc. Natl. Acad. Sci. U.S.A. 83, 115–119.

Toh, H., Weiss, B.L., Perkin, S.A., Yamashita, A., Oshima, K., Hattori, M., et al., 2006. Massive genome erosion and functional adaptations provide insights into the symbiotic lifestyle of *Sodalis glossinidius* in the tsetse host. Genome Res. 16, 149–156.

Toloza, A.C., Germano, M., Cueto, G.M., Vassena, C., Zerba, E., Picollo, M.I., 2008. Differential patterns of insecticide resistance in eggs and first instars of *Triatoma infestans* (Hemiptera: Reduviidae) from Argentina and Bolivia. J. Med. Entomol. 45, 421–426.

Torr, S.J., Hargrove, J.W., Vale, G.A., 2005. Towards a rational policy for dealing with tsetse. Trends Parasitol. 21, 537–541.

Touré, Y.T., 1985. Université de Droit, d'Economie et des Sciences Aix-Marseille III, Marseille, France.

Touré, Y.T., Petrarca, V., Traoré, S.F., et al., 1998. The distribution and inversion polymorphism of chromosomally recognized taxa of the *Anopheles gambiae* complex in Mali, West Africa. Parassitologia 40, 477–511.

Townson, H., Nathan, M.B., Zaim, M., Guillet, P., Manga, L., Bos, R., et al., 2005. Exploiting the potential of vector control for disease prevention. Bull. World Health Org. 83, 942–947.

Tripet, F., Toure, Y.T., Taylor, C.E., Norris, D.E., Dolo, G., et al., 2001. DNA analysis of transferred sperm reveals significant levels of gene flow between molecular forms of *Anopheles gambiae*. Mol. Ecol. 10, 1725–1732.

Tripet, F., Wright, J., Cornel, A., et al., 2007. Longitudinal survey of knockdown resistance to pyrethroid (*kdr*) in Mali, West Africa, and evidence of its emergence in the Bamako form of *Anopheles gambiae* s.s. Am. J. Trop. Med. Hyg. 76, 81–87.

Tripet, F., Aboagye'antwi, F., Hurd, H., 2008. Ecological immunology of mosquito–malaria interactions. Trends Parasitol. 24, 219–227.

Turner, T.L., Hahn, M.W., Nuzhdin, S.V., 2005. Genomic islands of speciation in *Anopheles gambiae*. PLoS Biol. 3, e285.

Vale, G.A., Torr, S.J., 2005. User-friendly models of the costs and efficacy of tsetse control: application to sterilizing and insecticidal techniques. Med. Vet. Entomol. 19, 293–305.

Vallejo, G.A., Guhl, F., Schaub, G.A., 2009. Triatominae—*Trypanosoma cruzi*/*T. rangeli*: vector–parasite interactions. Acta Trop. 110, 137–147.

Walsh, J.A., 1984. Estimating the burden of illness in the tropics. In: Warren, K., Mahmoud, A.A.F. (Eds.), Tropical and Geographical Medicine. McGraw-Hill, New York, NY, pp. 1073–1085.

Walshe, D.P., Ooi, C.P., Lehane, M.J., Haines, L.R., 2009. The enemy within: interactions between tsetse, trypanosomes and symbionts. Adv. Insect Physiol. 37, 120–175.

Wang, R., Zheng, L., Toure, Y.T., Dandekar, T., Kafatos, F.C., 2001. When genetic distance matters: measuring genetic differentiation at microsatellite loci in whole-genome scans of recent and incipient mosquito species. Proc. Natl. Acad. Sci. U.S.A. 98, 10769–10774.

Weill, M., Chandre, F., Brengues, C., et al., 2000. The *kdr* mutation occurs in the Mopti form of *Anopheles gambiae* s.s. through introgression. Insect Mol. Biol. 9, 451–455.

Weirauch, C., 2008. Cladistic analysis of Reduviidae (Heteroptera:Cimicomorpha) based on morphological characters. Syst. Entomol. 33, 229–274.

Weirauch, C., Munro, J.B., 2009. Molecular phylogeny of the assassin bugs (Hemiptera: Reduviidae), based on mitochondrial and nuclear ribosomal genes. Mol. Phylogenet. Evol. 53, 287–299.

White, B.J., Santolamazza, F., Kamau, L., et al., 2007. Molecular karyotyping of the 2La inversion in *Anopheles gambiae*. Am. J. Trop. Med. Hyg. 76, 334–339.

White, B.J., Cheng, C., Sangare, D., Lobo, N.F., Collins, F.H., Besansky, N.J., 2009. The population genomics of trans-specific inversion polymorphisms in *Anopheles gambiae*. Genetics 183, 275–288.

White, B.J., Cheng, C., Simard, F., Costantini, C., Besansky, N.J., 2010. Genetic association of physically unlinked islands of genomic divergence in incipient species of *Anopheles gambiae*. Mol. Ecol. 19, 925–939.

WHO, 1989. Geographical Distribution of Arthropod-Borne Diseases and their Principal Vectors. WHO/VBC/89.967, Geneva.

WHO, 2002. Control of Chagas disease (Second Report). In: WHO (Ed.), WHO Technical Report Series. World Health Organization, Geneva, p. 905.

WHO, 2006. Human African Trypanosomiasis (sleeping sickness): epidemiological update, Weekly Epidemiology Record 81. WHO, Geneva, pp. 71–80.

Wondji, C., Frederic, S., Petrarca, V., Etang, J., Santolamazza, F., Della Torre, A., et al., 2005. Species and populations of the *Anopheles gambiae* complex in Cameroon with special emphasis on chromosomal and molecular forms of *Anopheles gambiae* s.s. J. Med. Entomol. 42, 998–1005.

Wright, S., 1943. Isolation by distance. Genetics 28, 114–138.

Wright, S., 1978. Variability within and among natural populations. University of Chicago Press, Chicago.

Wu, M., Sun, L.V., Vamathevan, J., et al., 2004. Phylogenomics of the reproductive parasite *Wolbachia pipientis* wMel: a streamlined genome overrun by mobile genetic elements. PLoS Biol. 2, E69.

Yamagata, Y., Nakagawa, J., 2006. Control of Chagas disease. In: Molyneux, D.H. (Ed.), Advances in parasitology: control of human parasitic diseases. Elsevier & Academic Press, New York, NY, p. 662.

Zingales, B., Andrade, S.G., Briones, M.R., et al., 2009. A new consensus for *Trypanosoma cruzi* intraspecific nomenclature: second revision meeting recommends TcI to TcVI. Mem. Inst. Oswaldo Cruz 104, 1051–1054.

16 Modern Morphometrics of Medically Important Insects

*Jean-Pierre Dujardin**

GEMI, IRD, Montpellier, France

16.1 Introduction

The phenotype is the product of the interaction between genes and environment. Phenotypic variation is then an expected outcome of more than one factor. It can be scored by measurable changes in anatomy, morphology, physiology, life history, behavior, etc. (West-Eberhard, 1989; Gadagkar and Chandrashekara, 2005). This chapter focuses on the phenotype as a set of metric properties and their variations. Morphometric changes are generally recorded as variation in size and shape, although these two metric traits are not independent ones. Their interdependence (allometry) is worth considering in intraspecific studies, but is never complete. Therefore, the two metric properties are often considered separately for their genetic determinism, their heritability, their sensitivity to the environment, and their capacity to provide indirect information about the genetic differentiation of natural populations.

16.1.1 Modern and Traditional Morphometrics

Morphometric techniques aim at measuring size, shape, and the relation between size and shape (allometry). Before the so-called "revolution" (Rohlf and Marcus, 1993), shape was an abstraction, a residue after scaling for size, and it was not possible to visualize the "residue." The replacement of initial variables describing a distance between two anatomical points by the coordinates of these points, and the subsequent visualizing techniques, represented a giant step in the direct study of forms. The shift from traditional morphometrics to more complex geometric functions was facilitated by the development of image processing tools. Not only landmark methods but also "outline methods" (Rohlf and Marcus, 1993) and other techniques exploring textures and surface patterning (Lestrel, 2000) are used today. This chapter deals mainly with landmark-based geometric morphometry.

*E-mail: dujardinbe@gmail.com

16.2 Landmark-Based Geometric Morphometry

The coordinates of anatomical landmarks contain not only size, such as distances between landmarks, but also shape, such as their relative position.

In common practice, size and shape are derived from a configuration of landmarks collected on a non-articulated part, often a single organ (but see Adams, 1999). A few anatomical landmarks available on a wing (or any measurable part of the body) do not completely describe the wing, nor do they describe the complete body. However, provided there is anatomical correspondence among individual landmarks, only a partial capture of shape is needed to allow valid comparisons among populations and species.

The choice of suitable landmarks relies on their operational homology. In the morphometrics practice, homology is "correspondence of parts" with no specification about whether the parts correspond with respect to structure, development, or phylogeny (Smith, 1990). If individuals belong to a single species, homologous landmarks are probably similar due to common descent because all members of the species come from a common ancestor. If they belong to different species, there is no guarantee that homologous landmarks are similar due to common descent, except if they are known to be descending from a common ancestor (Lele and Richtsmeyer, 2001). This homology is one of the criteria making landmark-based morphometrics a suitable tool for systematics (see Section 16.6.1). Bookstein (1991) described various categories of landmarks with decreasing levels of precision.

16.2.1 Size

To avoid the problem of multidimensionality, traditional systematists often selected one single dimension to represent body size. For an insect, the length of the wing along its largest axis is frequently used as an estimator of body size (Nasci, 1990; Siegel et al., 1992; Lehmann et al., 2006). Such a relationship is often assumed rather than demonstrated (Siegel et al., 1992; Morales et al., 2010).

Size Variable: The Centroid Size

The centroid size (CS) is the square root of the sum of the squared distances from the centroid to each landmark (see Gower, 1971 in Rohlf, 1990). It thus can detect change in various directions. In the case of small, circular variation at each landmark, this estimator of isometric change of size is not correlated to shape variation (Bookstein, 1991). It is expressed in pixels, or units relative to the resolution of the viewing device (most often a computer display). As a scalar it is less sensible to small digitization errors and can be shared among systematists provided the pixels have been converted into absolute length units (inches, centimeters, millimeters, etc.). Thus, to allow for exchangeability of CS, an image on which landmarks may be collected should contain a scale for size (for instance the picture of a reticule) allowing the conversion of pixels to absolute units.

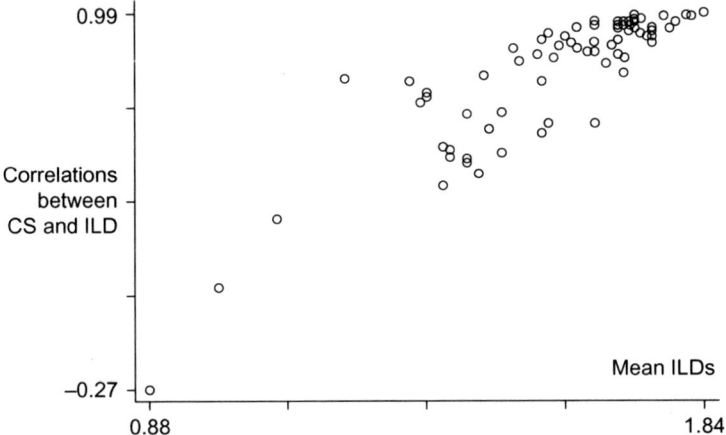

Figure 16.1 Each circle represents on the vertical axis the correlation found between an ILD and the CS of the wing computed from the total set of landmarks. The value of each ILD on the horizontal axis is an average (mean ILD) obtained from the total number of wings (78) examined in this sample. The wings belong to *Ae. aegypti*.

The relationship of CS values and the traditional wing length in the mosquito *Aedes aegypti* showed good correlation (Morales et al., 2010). Actually, the correlation of CS values with traditional interlandmark distances (ILD) is itself correlated to the relative dimensions of ILD: the largest the ILD, the highest its correlation with CS (see Figure 16.1).

16.2.2 Shape

Not only in medical entomology but also in many fields where morphometrics is applied, shape has been traditionally described as the ratio of one dimension to another. Although intuitively the ratio may appear as capable of scaling for size, it often does not (Burnaby, 1966; Albrecht et al., 1993; Klingenberg, 1996; Dujardin and Slice, 2007). Moreover, the ratios introduce some well-known statistical drawbacks (Albrecht et al., 1993). Angles do not improve the situation since they are another kind of ratio (Burnaby, 1966).

Shape Variables: The Procrustes Residuals, the Partial Warps, the Relative Warps

In geometric morphometrics, the shape of a configuration of landmarks is represented by their relative positions as contained in their coordinates. However, these coordinates also contain artifactual variation due to position, size, and orientation. Thus, shape must be described by new variables having removed these artifacts. This is obtained through the Procrustes[1] superimposition on a consensus configuration. If using the least squares fit as an optimality criterion, the statistical procedure of superposition is called Generalized Procrustes Analysis (GPA). It is currently the most common procedure, but other techniques also exist (Zelditch et al., 2004). The residual coordinates after a GPA provide a shape description relative to the consensus configuration of landmarks, they thus depend on the composition of the group under study. If other specimens (i.e., coordinates) are added to the analysis, shape variables must be recomputed accordingly (Rohlf and Marcus, 1993; Adams et al., 2004).

Furthermore, the residual coordinates lie in a weird mathematical shape space. When working on a two-dimensional space, the residual coordinates have lost 4 degrees of freedom (Rohlf, 1996). They lie in the Kendall space or not depending on the kind of Procrustes distance used, full or partial one (Slice, 2001), but they lie in a curved, non-Euclidean space unsuitable for standard statistical tests (except resampling methods). Since Procrustes residuals[2] lie in a non-Euclidean space, they must be further modified by a rigid rotation so that they can be studied using classical statistical tools (Rohlf and Bookstein, 2003).

Using for rigid rotation the eigenvectors of the variance—covariance matrix of the Procrustes residuals, the resulting projection is described by the principal components scores ("Procrustes components") and can be used for standard statistical analyzes and comparison tests. Using the eigenvectors of the bending-energy matrix (Bookstein, 1991), the resulting shape variables are called "partial warps" scores (PW). The PW, or their principal components, namely the "relative warps[3]" (RW), may be used in classical statistical analyzes and in visualization of shape changes through the deformation grids (i.e., D'Arcy Thompson-like plots showing the geometry of shape changes between objects; Bookstein, 1991).

The obvious interest of using principal components (either Procrustes components or RW) is that the number of input variables can be reduced: the few first RW generally represent a significant fraction of shape. The subset of first principal components to use is rather subjective, but it can follow some rules (Baylac and Frieß, 2005).

16.2.3 Allometry

Since each shape can be explained by the change in linear dimensions, it is obvious that size and shape are not independent attributes. The relationship between size and shape is called allometry. Geometric shape variables (see previous paragraph) are not allometry-free variables: they remove the isometric component of size change.

---

[1] Procrustes, whose name means "he who stretches", was a thief in Greek mythology (the myth of Theseus). He preyed on travelers along the road to Athens. He offered his victims hospitality on a magical bed that would fit any guest. As soon as the guest lay down Procrustes went to work upon him, either stretching the guest or cutting off his limbs to make him fit perfectly onto the bed (Grose Educational Media, 1997–1998, http://www.groseducationalmedia.ca/greekm/mythproc.html).

[2] Procrustes residuals are the differences between the residual coordinates of each object and the residual coordinates of the consensus configuration.

[3] A complete glossary of the many technical terms related to GM can be found at http://life.bio.sunysb.edu/morph/.

Assuming a common model of growth is not rejected, one can use the growth model to predict allometry-free shape variation among groups after fixing size to one value (MANCOVA). The tentative removal of the allometric effect on shape can be justified for intraspecific studies (Klingenberg, 1996; Caro-Riaño et al., 2009; Morales et al., 2010). It is less justified for interspecific comparisons, where allometric variation is likely to be part of the evolutionary differences relevant to systematics.

16.2.4 Measurement Error

As explained above, the extraction of shape information from raw coordinates is computed relative to the consensus configuration derived from a specific group of samples; this thwarts mixing the final variables with other such variables computed from other individuals. Only raw coordinates could be shared, to the condition there was no error introducing artifactual differences between two sets of homologous landmarks. The measurement error exists at various steps of morphometric analysis (Arnqvist and Mårtensson, 1998). The mounting technique of specimens or organs, the photographing conditions, and the user's skill in collecting landmark coordinates may produce artifactual variation. Generally, similar techniques are used to process similar organisms, and digital techniques of modern photography provide adequate resolution for correct recognition of landmarks under different conditions. Whatever the quality and reproducibility of landmark digitization, the recommended way to perform morphometric comparisons is to allow one single user to produce the data.

Even when performed by a single user, digitization should be repeated at least once, allowing one to measure the precision and to reduce the error by averaging the two digitizations. The precision is estimated by the "repeatability" (R) index as described by Arnqvist and Mårtensson (1998), which is a Model II one-way ANOVA on repeated measures, where "R" is provided by the ratio of the between individual variance and the total variance.

16.3 Nonenvironmental Sources of Metric Change

16.3.1 Shape as a Polygenic Character

Shape appears as a classical polygenic character (Klingenberg and Leamy, 2001). Evidence for strong genetic determinism of shape was suggested by significant association with chromosome polymorphism (Bitner-Mathé and Klaczko, 1999; Orengo and Prevosti, 2002; Hatadani and Klaczko, 2008), and confirmed by quantitative genetic studies (Breuker et al., 2006; Patterson and Klingenberg, 2007). When studies on quantitative trait loci (QTL) were applied to the shape and size of mouse mandible, many QTL were identified for shape (Klingenberg et al., 2004), many more than for size (Klingenberg et al., 2001; Workman et al., 2002). Few studies are found in insects, also fitting the idea of genetic determinism (Iriarte et al., 2003) and polygenic inheritance (Shrimpton and Robertson, 1988; Long et al., 1995).

16.3.2 Genetic Drift

Since shape seems the output of a cascade of genes, it is expected that in natural conditions genetic drift will be a common factor of shape variation. Field observation has frequently observed significant shape differences between geographic areas (De la Riva et al., 2001; Dujardin et al., 2003; Gumiel et al., 2003; Dujardin and Le Pont, 2004a; Camara et al., 2006.; Aytekin et al., 2007; Henry et al., 2010). Laboratory experiments reproducing conditions favoring genetic drift between lines sharing the same environment were performed in *Ae. aegypti*. Using a set of three isofemale lines of *Ae. aegypti* monitored during 10 generations, a significant shift of shape appeared in one line, with nonsignificant changes in corresponding size (Jirakanjanakit et al., 2008). In this experiment, the change apparently produced by genetic drift did not affect the same landmarks as those affected by larval food or density variation (Jirakanjanakit et al., 2007).

16.3.3 Heritability

Heritability is depending on the genetic variability related to the trait under study, it is then depending on the population under study. Its measurement is not indispensable to the interpretation of natural metric variation, but it can provide valuable information about the adaptiveness of metric traits. In insects, morphological traits commonly have the highest heritability values compared to other trait categories such as life history, probably because the former are less concerned with fitness.

Geometric techniques allow separate estimations of size and shape heritabilities. Size in insects may show consistent heritability values (Daly, 1992; Lehmann et al., 2006), so that they can be experimentally selected to constitute subpopulations genetically distinct for size (Anderson, 1973; Partridge et al., 1994). Various studies examining cross-environment heritability of wing shape in Diptera produced high and stable heritability, reaching 60% or more (Roff and Mousseau, 1987; Bitner-Mathé and Klaczko, 1999; Gilchrist and Partridge, 2001; Hoffman and Shirriffs, 2002). The consistent values of shape heritability suggest that a large fraction of morphometric divergence seen between natural populations of insects (Camara et al., 2006.; Henry et al., 2010; Morales et al., 2010) may be due to additive effects of genes.

In *Ae. aegypti*, shape appears to be more heritable than size. When comparing size and shape cross-environment heritability on the same populations in *Ae. aegypti*, much higher values for shape (Figure 16.2) than for size were found, providing indirect evidence for different genetic sources of variation (Morales et al., unpublished data).

16.3.4 Hybridism

CS was increased in hybrids obtained from two close species, initially considered as two subspecies (Costa and Felix, 2007), *T. brasiliensis* and *T. juazeirensis*. It was larger than the mid-parent size, and larger than the largest parent's size, suggesting heterosis pointing to a consistent genetic divergence of the parents.

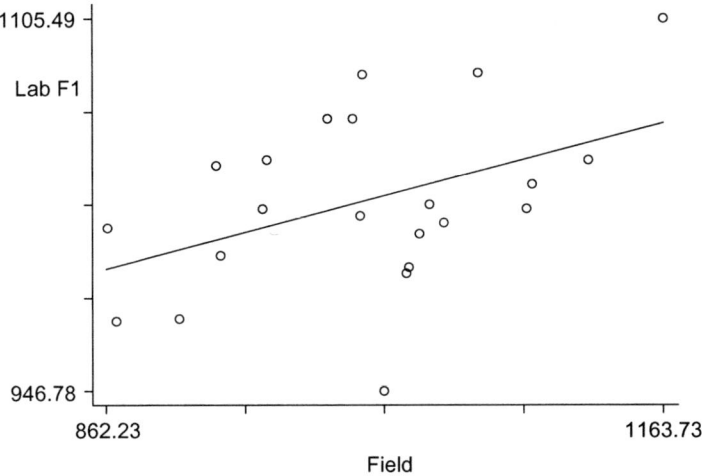

Figure 16.2 *Ae. aegypti*: regression of the first relative warps (RW1) of laboratory daughters on the RW1 of corresponding field-collected mothers in a cross-environment study of the heritability of the wing shape at 18 landmarks (Morales et al., unpublished data). Lab F1, female specimens obtained after crossing field-collected specimens.

Extending this study to experimental hybrids among the four members of the Brasiliensis complex, it was possible to show a linear relationship between the genetic divergence of the parents (Costa et al., 2001; Costa and Felix, 2007) and the increase in size of their offspring. Contrary to size, shape of the hybrids remained intermediate between parents (Costa et al., 2008).

Between cryptic species of *Diachasmimorpha longicaudata*, a hymenopteran parasitoid of fruit flies, hybrids showed intermediate shape on the morphological space obtained from the two first RW. The size of the hybrids was larger than that of mid-parents, although not significantly larger (Kitthawee and Dujardin, 2009).

These two studies indicate different genetic mechanisms affecting size and shape, suggesting size as a character prone to show heterosis in case of genetically differentiated parents. This behavior of size could help exploring the degree of genetic differentiation between populations, especially when they are suspected to undergo speciation. Thus, when shape and size remain at intermediate values between parents, the latter are probably not genetically distinct taxa. Between seven laboratory colonies of *T. protracta* (Dujardin et al., 2007), or between each of the five subspecies of *T. protracta*, each hybrid had an intermediate CS between parents. Accordingly, geometric shape was also intermediate between parents (Dujardin, unpublished data). These observations were in agreement with allopatric conspecific *T. protracta* populations instead of with distinct taxa. Indeed, as observed in this group of insects, like in *T. platensis*–*T. infestans* hybrids or in *T. brasiliensis* (Costa et al., 2008), genetically differentiated parents, or distinct but phylogenetically close species, would have produced an exaltation of some body dimensions and other phenotypic traits in their progeny.

16.4 Environmental Sources of Metric Changes

Environment often affects metric properties. However, it is important to specify which metric property is affected, size, and/or shape, and how it is affected. No simple rule can be formulated. Moreover, since there is no detectable polarity in metric changes, the present aspect of the phenotype does not contain any signal of its own history. As stated by Losos (2000), "... we cannot go back in time and expressly determine why a trait evolved ... the best we can do is enumerate tests suggested by a particular adaptive hypothesis."

There are different hypotheses about the way the environment affects shape. They describe situations which can be partially reproduced in laboratory experiments and/or manipulative field experiments. We will briefly consider only the following ones: phenotypic plasticity (PP), character displacement (CD), genetic assimilation, and accommodation.

16.4.1 Phenotypic Plasticity

The genotype does not give rise to the phenotype, but to a range of phenotypes. The "reaction norm" is the whole repertoire of possible phenotypes that may occur for a given genotype in all environments (Dobzhansky, 1971; Schlichting and Pigliucci, 1998). The reaction norm can easily be explored in laboratory experiments (Hillesheim and Stearns, 1991; David et al., 1994; Debat et al., 2003; Jirakanjanakit et al., 2007; Caro-Riaño et al., 2009). By definition, PP is the occurrence of phenotypic variation of a single genotype interacting with different environments (Schlichting and Pigliucci, 1998).

A new phenotype expressed in a new environment may be adaptive. To this condition, PP can aid speciation by making available a different phenotype upon which natural selection can act. In such scenario, speciation would start with PP, not reproductive isolation (Görür, 2005). In Triatominae, such a scenario is apparent (Dujardin et al., 1999b) and many examples exist of morphologically and ecologically recognized species which can still interbreed (Dujardin et al., 2009). Understanding the causes and consequences of phenotypic variation is important for understanding the mechanisms of evolution. However, the genetic mechanisms underlying the evolutionary importance of PP (Waddington, 1953; Thompson, 1971; West-Eberhard, 1989; Schlichting and Pigliucci, 1998) have to date received few experimental confirmations (Rutherford and Lindquist, 1998; Suzuki and Nijhout, 2006).

Contrary to its evolutionary importance, the ecological importance of PP is easy to understand: populations or species having wider adaptive plastic responses can enlarge their ecological niches. For instance, among the more than 140 species of Triatominae, the vectors of Chagas disease, a few species have been able to colonize human structures. Within some of these species, the comparison of "domestic" and sylvatic subpopulations highlighted significant size differences, sylvatic insects being generally larger. Were these species more plastic than the others so that they could reduce their size as apparently required by survival in artificial ecotopes? Was the size a secondary event selected by the domestic environment? Using emerging adults of an inbred line of *R. pallescens*, Caro-Riaño et al. (2009) evaluated size and shape variation of the heads and the wings under simulated conditions of sylvatic (low population density, low feeding frequency) and domestic (higher density, higher feeding frequency) habitats. Results demonstrated that selection was not needed to account for observed changes between sylvatic and domestic ecotopes. The significant size reduction was shown to be a plastic response to combined population density and feeding frequency.

In addition to explain diversity and adaptation, PP also impacts our understanding of taxonomy, because it suggests that species characteristics are not immutable, but are influenced by the environment and can be highly variable (Ananthakrishnan, 2005).

16.4.2 Character Displacement

The initial definition of the "character displacement" concept (Wilson and Brown, 1955; Brown and Wilson, 1956) did not predicate the real complexity of its demonstration: "the situation in which, when two species overlap geographically, the differences between them are accentuated in the zone of sympatry." Specifically, in sympatry selection was supposed to minimize attempts at hybridization (by mistaken identity) as well as competition between the two species.

The difficulties of obtaining unambiguous evidence from natural observations have been discussed by Grant (1972), and, more recently, by Losos (2000). Typically, CD was suspected when more difference was observed between species developing in sympatry than in allopatry; "soon after the theory was promulgated, ecologists and evolutionary biologists were seeing evidence for character displacement everywhere" (Losos, 2000). Various other conditions must be satisfied to assess CD, among which are the level of differences in sympatry (greater than expected by chance), the evolutionary history of sympatry (original or derived situation), the genetic nature of phenotypic differences and, importantly, the connection between characters and competition for resources (Grant and Grant, 2006). CD was demonstrated for behavioral and ecological characters more often than for morphological characters (Grant, 1972; Losos, 2000). Morphologically, the displaced character is expected to be part of the feeding apparatus (Adams and Rohlf, 2000). If the mouthparts have a species recognition function, then displacement may have consequences on speciation as well.

Thus, if PP could be analyzed raising the same genotypes of one species in different environments, CD would then be studied in one single environment, raising a genotype of one species alone and in combination to a genotype of another species. Nevertheless, if one considers for a given species the surrounding ones as making part of the environment, then CD could be considered as a particular case of PP induced by species competition. Medically important insects did not receive much attention.

16.4.3 Genetic Assimilation

Unexpectedly, PP can produce heritable changes. The mechanisms of such phenomenon, which is reminiscent of Lamarkian "inheritance of acquired characteristics," do not depart from orthodox genetics. They have been named "genetic assimilation" (Waddington, 1953) or "autonomization" according to Schmalhausen (Levit et al., 2006), and more recently "genetic accommodation" (West-Eberhard, 1989).

Waddington defined genetic assimilation as "a process by which a phenotypic character, which initially is produced only in response to some environmental influence, becomes, through a process of selection, taken over by the genotype, so that it is found even in the absence of the environmental influence which had at first been necessary" (Waddington, 1953).

Laboratory experiments, old ones (Waddington, 1953, 1956; Anderson, 1973) and more recent ones (Gibson and Hogness, 1996; Rutherford and Lindquist, 1998; Sollars et al., 2003), unambiguously demonstrated genetic assimilation. Indirect evidence was provided from natural populations of the medically important triatomine bug *Rhodnius pallescens*. Five lines reared at the same temperature, some of them reared over more than 40 generations, have been shown to harbor distinct sizes in accordance with the temperature of their region of origin in Colombia (Jaramillo, Ph.D. thesis). The absence of size convergence at the same laboratory temperature, and the correlation with temperature of initial field conditions, suggested a genetic determinism for size variation (Dujardin et al., 2009). This latter example illustrates that Bergmann size clines in natural populations may have more complex causes (Davidowitz et al., 2004) than a merely developmental process (David et al., 1994; Vanvoorhies, 1996).

16.4.4 Genetic Accommodation

Genetic accommodation (West-Eberhard, 1989, 2003) is a concept very close to genetic assimilation, but wider for two reasons: the nature of the trigger and the outcome of the process (Görür, 2005; Braendle and Flatt, 2006).

First, the hypothesis of genetic accommodation assumes that the trigger at the onset of the phenotypic change is either genetic or environmental, whereas the concept of genetic assimilation typically assumes only an environmental trigger. Second, the expected outcome of genetic assimilation is a new, heritable phenotype insensitive to environmental change (see hereunder the concept of "canalization"), while genetic accommodation can produce both insensitive and sensitive new, heritable phenotypes.

Thus, genetic accommodation is a generalization of genetic assimilation. In this general hypothesis, it is argued that environmentally triggered novelties may have greater evolutionary potential than mutationally induced ones, mainly because of two features (Görür, 2005): (i) they concern populations rather than individuals (while mutations are individual events) and (ii) they represent optimal, or close to optimal, adaptations (while mutations are often counter-selected). With time (the concept of "recurrence"), these two features increase the likelihood of genetic assimilation (selection of genotypes accidentally producing the same phenotype as the adapted one), ending up in local genetic changes.

Genetic accommodation has been experimentally demonstrated by Suzuki and Nijhout (2006) studying a color polyphenism in *Manduca sexta* (Braendle and Flatt, 2006).

Hidden Genetic Variability

The mechanisms by which an environmentally induced phenotype may become heritable are entirely compatible with concepts of classical neo-Darwinian evolutionary biology. Indeed, the environmental trigger (since this is the disputable one) just uncovers previously cryptic genetic variation (Gibson and Dworkin, 2004).

Thus, there are genetic mutations that can remain masked until the environment (or another mutation) reveals them (Bergman and Siegal, 2003). In case of an environmental trigger, the external stimulus has to be recurrent and consistent in time so that selection can lead to genetic accommodation.

The trigger, either an environmental or mutational one, acts through its effects on "capacitors," which are proteins able to buffer genotypic variation under normal conditions, thereby promoting the accumulation of hidden polymorphism. Published examples of capacitors for morphological evolution are the heat shock protein Hsp90 (Rutherford and Lindquist, 1998; Debat et al., 2006) or the genes regulating hormonal titers (Pennisi, 2006; Suzuki and Nijhout, 2006).

Hidden genetic variation can also be revealed through epigenetic mechanisms (Sollars et al., 2003), which are heritable changes in gene function that occur without a change in the sequence of nuclear DNA (Jablonka et al., 1992; Jablonka and Lamb, 2002). Epigenetic mechanisms such as DNA methylation, histone acetylation (producing changes to the chromatin packaging of DNA), and RNA interference (regulation of gene-expression control by non-coding RNA), and their effects in gene activation and silencing are increasingly understood to play a role in phenotype transmission and development (Bird, 2007).

16.5 The Regulation of Phenotype

To the many sources of phenotypic changes the organism opposes homeostatic processes. Two components of this homeostasis are canalization and developmental stability. The two components seem to be independent processes (Debat et al., 2000; Réale and Roff, 2003), and are easy to distinguish: canalization is the stability of development in different environments, while developmental stability refers to stability in the same environment. Canalization is thus a buffering process against external and/or mutational perturbation from one environment to another, while developmental stability allows the organism to withstand random accidents during development in the same environment (Graham et al., 1993).

16.5.1 Canalization

The term "canalization" also is due to Waddington, corresponding to the "stabilizing selection" of Schmalhausen (Levit et al., 2006). Here are the terms used by Schmalhausen himself: "Every adaptive modification is an expression of a norm of reaction, which went the long way of historical development under changing conditions. It is connected with the establishment of 'canals' through which a certain modification develops (Waddington talks about the 'canalization' of development). An external factor operates only to switch the development into one of the existing canals" (Schmalhausen in Levit et al., 2006).

Thus, as for PP, canalization is not a property of a species or of a population, but of a genotype (Dworkin, 2005). However, different traits of a single organism can be examined for their relative canalization by studying their natural variation in different lines, populations, or species. For instance, contrary to size changes, shape changes of the wings induced by striking altitudinal variation as found between the Andes and the Amazon basin could not interfere with species differences in sandflies (Dujardin et al., 2003). A similar study comparing the wing shape of transcontinental populations of two close mosquito species, *Ae. aegypti* and *Ae. albopictus*, showed that species differentiation based on wing shape, but not on its size, was not altered by transcontinental migration during the last decades. Thus, in spite of the many possible situations supposed to affect shape, like environmental changes, possible environmental stress, likely founder effect, possible genetic drift, and species competition, both species were still distinguishable at the same landmark locations (Henry et al., 2010). This relative constancy of shape patterns within each species contrasted with the lability of size. For the same species (*Ae. aegypti*), size was significantly affected by a simple change in the food concentration or in the larval density (Jirakanjanakit et al., 2007). Another example comparing size and shape responses is found in highly inbred lines of *R. pallescens* (Triatominae): the plastic response scored for the CS of the wing to the laboratory conditions of "domesticity" was not observed for the shape, except as an allometric change (Caro-Riaño et al., 2009).

Incidently, the apparently higher canalization of shape makes this trait a suitable character for populations and species distinction (Dujardin and Le Pont, 2004b).

16.5.2 Developmental Stability

Although development is an individual attribute, its stability is estimated at the population level and can be compared with other populations. The use of morphometrics as an indicator of environmental stress is generally performed by estimating the frequency of abnormal phenotypes (phenodeviants) or the amount of fluctuating asymmetry (FA) (Palmer and Strobeck, 1986). Bilateral symmetry is not supposed to change during development; it is a developmental invariant. Other measures of developmental stability could be used that also are developmental invariants, like fractal dimensions, although they were described for plants and vertebrates only (Graham et al., 1993).

Stress may have many different interpretations, and is probably not the only explanation for increased FA. Stress can be the infection by a virus or parasite, or a difficult conquest of a new habitat. In the few domestic populations of *T. sordida*, a potential vector of Chagas disease in Bolivia and Argentina, significant FA was found, whereas no FA at all could be disclosed in their sylvatic counterparts (Dujardin et al., 1999b).

Because of its relatively strong canalization, geometric shape of the insect wings is not prone to show significant changes under the normal range of developmental conditions encountered by organisms (Birdsall et al., 2000). Nevertheless, the use of shape variation in response to environmental stress has been advocated for insects, although these changes were considered as very subtle ones (Hoffmann et al., 2005). An advantage it could have on asymmetry analyzes would be possible signature changes in landmarks characteristic of a specific environmental stress. A disadvantage is that, contrary to symmetry, which is expected to be perfect, and contrary to the frequency of phenodeviants, which is expected to be zero, there is no "expected shape" and thus no way to use shape changes to measure the degree of stress.

16.6 Applications in Medical Entomology

16.6.1 *Species Identification and Detection*

The most important objection to the morphological concept of species is the existence of sibling (or isomorphic) species (Mayr, 2000). Sibling (or also cryptic) species are morphologically identical or nearly identical entities recognized as different species according to other, modern concept(s) of species. However, this objection to the typological concept (i.e., to "morphospecies") is weakened by the possibilities of modern quantitative shape comparisons (Baylac et al., 2003; Becerra and Valdecasas, 2004; Dujardin, 2008). Shape comparisons detect minimal morphological variations, which often are undetectable by traditional morphological studies and even by classical morphometric approaches. Cryptic species of insects showed distinct shapes in kissing bugs (Matias et al., 2001; Villegas et al., 2002; Dujardin et al., 2009), sandflies (De la Riva et al., 2001), scythridids (Roggero and Passerin d'Entrèves, 2005), parasitoid hymenoptera (Baylac et al., 2003; Villemant et al., 2007; Kitthawee and Dujardin, 2009), syrphids (Francuski et al., 2009), fruit flies (Kitthawee and Dujardin, 2010), and screwworm flies (Lyra et al., 2009). Although morphometric discrimination does not necessarily mean species determination, it has also been used to question species boundaries (Aytekin et al., 2007), or to synonymize controversial taxa (Gumiel et al., 2003).

Geometric morphometry is becoming a fast and low-cost alternative to identify cryptic species that often need the molecular machinery to be distinguished. However, the diagnostic metric features cannot be shared. Because geometric shape is defined relative to the consensus of the specimens under study, shape variables derived from one set of coordinates cannot be compared with shape variables derived from another set. Coordinates themselves could be used for such comparisons, but the measurement error may represent a significant obstacle, especially when the objective is to distinguish very similar species.

The "User Effect"

Among the sources of measurement error (see Section 16.2.4), user intervention is often the most important. The error is generally due to small but systematic differences in pointing to the exact localization of some landmarks. These subtle discrepancies are amplified by the power of multivariate analysis like the discriminant analysis. Their impact can be reduced averaging repeated collections of the data (Arnqvist and Mårtensson, 1998). However, such correction might not be satisfactory when comparing very close specimens or groups, and measurement error may become a significant obstacle for different users (Jordaens et al., 2002; Rasmussen et al., 2001). As a consequence, user A should not enter his own measurements in a database of coordinates collected by user B, and vice versa.

The Need for a Bank of Reference Images

To circumvent the lack of exchangeability of the morphometric variables, an alternative geometric descriptive system should be developed that separates data gathering and analyzes. It goes through the creation of a bank of reference images from which one can extract raw data and compare it to external, unknown specimens. The chances of successful identification would then depend on the relevance of reference images, on their level of shape divergence and on the classification techniques. Such an initiative is ongoing at http://www.mpl.ird.fr/morphometrics/clic/index.html under the name CLIC (Collection of Landmarks for Identification and Characterization). The need for such a database is underestimated, because the power of morphometrics to identify taxa is itself probably underestimated.

16.6.2 Characterization Tool at the Individual Level

In humans, some metric traits allow highly reliable individual identification (fingerprint, iris pattern, etc.). We can expect similar situation in animals. In medical entomology, it might be useful to assign a single individual to its origins. Two applications can be considered, one in systematics, the other one in population structure.

Species are generally well distinguished thanks to qualitative morphological characters, but close species might be difficult to confidently identify based on one single individual. Using geometric shape comparisons, one single individual can generally be accurately classified using a database of images of the candidate species (Matias et al., 2001). As an example, we show here unpublished data about mosquito identification. Each single individual has been allocated to its closest group (according to Mahalanobis distance) without using that individual to help determine a group center ("validated reclassification"). The wing venation patterns appear to be roughly the same among the genera of Culicidae, they allowed however an almost perfect reclassification of them all (Table 16.1). Within some genera like *Aedes* or *Anopheles*, the species discrimination was also very satisfactory; it was less convincing in the genus *Culex* (Table 16.2). The possibility to perform satisfactory identifications without being an expert in taxonomy is very attractive, but

Table 16.1 Morphometric Identification of Culicidae Based on 13 Landmarks of the Wing

Genera	Ur, Ma	An	Mi	Cu	Ae, Ar, Co
Scores	100%	97%	96%	95%	100%
N	508 (8)	446 (6)	348 (5)	317 (4)	127 (3)

The first column indicates that 100% of the genus *Uranotaenia* (Ur) and 100% of the genus *Mansonia* (Ma) could be recognized when mixed with the six other genera: *Anopheles* (An), *Mimomyia* (Mi), *Culex* (Cu), *Aedes* (Ae), *Armigeres* (Ar), and *Coquilliettidia* (Co). The second column indicates that 97% of the genus *Anopheles* could be recognized when mixed with the genera *Mimomyia*, *Culex*, *Aedes*, and *Armigeres*. The third column indicates that 96% of the genus *Mimomyia* could be recognized when mixed with the genera *Culex*, *Aedes*, and *Armigeres*. The fourth column indicates 95% of the *Culex* could be distinguished from the genera *Aedes*, *Armigeres*, and *Coquilliettidia*. The last column indicates that these three genera were perfectly discriminated by their wing geometry. N, total number of individuals in each analysis; number of genera in the analysis is given in parenthesis. Mosquito collection by A. Henry and P. Thongsripong (University of Hawaii). Morphological identification of the genera by Dr. R. Rattanarithikul (AFRIMS, Thailand). Digitization of wings by J.-F. Lasnes (University of Montpellier).

Table 16.2 Correct Species Attribution Scores Based on the Geometry of the Wings

Species	Scores (%)	n/N
Aedes		
(Stegomyia) aegypti	100	12/12
(Neomelaniconion) lineatopennis	66	10/15
(Aedimorphus) mediolineatus	100	12/12
(Aedimorphus) vexans	83	20/24
Anopheles		
(Anopheles) barbirostris	100	14/14
(Cellia) tessellatus	88	8/9
(Cellia) vagus	91	34/37
Culex		
(Culex) vishnui	55	29/52
(Culex) gelidus	61	11/18
(Culex) quinquefasciatus	91	11/12
(Oculeomyia) bitaeniorynchus	78	18/23
(Oculeomyia) sinensis	62	18/29
(Culiciomyia) nigropunctatus	91	11/12

Thirteen species belonging to three genera, *Anopheles* (An.), *Culex* (Cx.), and *Aedes* (Ae.), were analyzed for species identification, namely: *Ae. aegypti*, *Ae. lineatopennis*, *Ae. mediolineatus*, *Ae. vexans*, *An. barbirostris*, *An. tessellatus*, *An. vagus*, *Cx. bitaeniorhynchus*, *Cx. gelidus*, *Cx. nigropunctatus*, *Cx. quinquefasciatus*, *Cx. sinensis*, and *Cx. vishnui*. Scores, correct attributions in percentages by species after validated reclassification; n, number of individuals correctly assigned to the species; N, total number of individuals in the species. Mosquito collection by A. Henry and P. Thongsripong (University of Hawaii). Species morphological identification by Dr. R. Rattanarithikul (AFRIMS, Thailand). Digitization of wings by J.-F. Lasnes (University of Montpellier).

more studies are needed to evaluate the full interest of this identification approach in many groups of medically important insects.

More difficult is the identification when comparing few conspecific individuals. Reinfestant specimens after vector control measure may be few, and classical morphology could be unable to suggest their origin (see Section 16.6.4). Provided a database exists of specimens collected before control measures, shape can be used for quantitative comparisons of local and external individuals (Dujardin et al., 2007).

16.6.3 Biodiversity

The transmission of vector-borne diseases has obvious links with the environment. Studies exploring these links suggested that the reduction in global biodiversity is likely to contribute to vector-borne disease transmission through the "dilution effect"[4] (Chivian and Bernstein, 2004; Keesing et al., 2006). It is therefore highly desirable to quantify the environment. In this kind of study, geometric morphometrics has two advantages: the ability to help identify taxa and its own addition to knowledge about biodiversity.

Biodiversity is expressed as the combination of both species richness (SR), the number of species in a specific environment, and species evenness, the proportion of each of them. Different indexes have been suggested to take into account both richness and evenness, from which the most commonly used are the Shannon–Wiener's (Shannon and Weaver, 1949) and the Simpson's (Simpson, 1949) indexes.

In addition to these estimates of biodiversity, complementary information has been searched for in the morphological disparity of organisms. The morphological disparity has been expressed in two ways, one considering the range of shape variation (the difference between extreme forms), the other one the amount of shape variation (the variance of shape). Modern morphometrics is giving these estimations a powerful quantitative tool for accurate measurements and comparisons (Roy and Balch, 2001; Neige, 2003).

The relationship between morphological disparity and biodiversity differs according to the way biodiversity is measured (i.e., taking into account or not the evenness).

Metric Disparity and SR

Does metric disparity increase with the number of species? One could expect greater richness to be the cause of higher morphometric variation, but no such relationship could be confirmed. Trends in SR generally did not match trends in metric disparity (MD). However, one could argue that if selection targets forms rather than species, some relationship is predictable. For a given clade's history,

---

[4] In the "dilution effect" hypothesis, locales with few species capable of sustaining vectors will have higher disease risk because vectors feed more frequently on the species that serve as hosts of the pathogen. In contrast, "dilution" occurs in areas with high biodiversity because more species (not all of which harbor parasites) are available to sustain vectors.

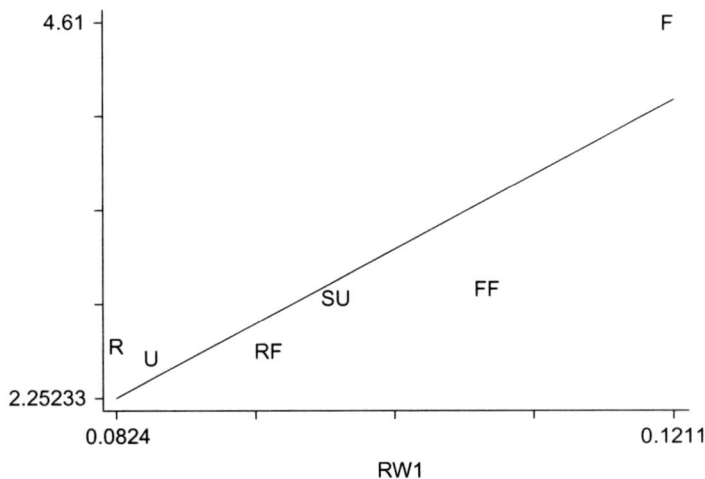

Figure 16.3 Relationship between the Shannon index of biodiversity (vertical axis) and the metric disparity (horizontal axis) computed as the range of the first RW in a region of Thailand (Henry et al., unpublished data). There were 584 mosquitoes defined by their wings at 13 landmarks in different environments: F, forest; FF, fragmented forest; RF, rice field; R, rural; SU, suburban; U, urban. Statistical significance: $P = 0.02$ (down to $P = 0.06$ if forest is removed from the data).

Foote (1993) predicts a high or low ratio MD/SR as depending mainly on the kind of selection during evolution: a selection precluding either intermediate or extreme forms, respectively.

When high, the ratio MD/SR is probably the result of combining a relatively low number of species with a relatively high morphometric variation. What could be the origin of such pattern? In addition to the possible effect of selection promoting extreme forms (Foote, 1993), the answer is probably to be found into what generates morphological heterogeneity: PP, species competition (Ricklefs and Miles, 1994), and of course phylogenetic diversity (Richman and Price, 1992; Shepard, 1998). Based on some idealized scenarios, the ratio MD/SR can help discussion about the geographic origin of some group of species in relatively isolated regions (Neige, 2003).

Taking into Account the Evenness

In addition to being poorly related to metric disparity, trends in SR do not necessarily match trends disclosed by other biodiversity metrics (Roy and Balch, 2001), either Shannon's index (H) or Simpson's index (D). These estimates take into account the proportion of each species (evenness), not only their number. In some occasions, they can show some relationship with morphological disparity. An ongoing study in Thailand about mosquito diversity according to different environments allowed the capture of most of the tribes and many genera of Culicidae (Henry et al., unpublished data). In this study, neither the richness nor the biodiversity indexes could show any significant correlation with the metric disparity estimates, except the Shannon–Wiener's index and the range of metric variation as estimated by the first RW (see Figure 16.3).

Heterozygosity

Interestingly, some correlation is observed more frequently when relating the metric disparity of conspecific populations and their genetic diversity (heterozygosity)[5]. The relationship is often, but not always, a negative one. The negative relationship has generally been interpreted as evidence for a higher developmental homeostasis in heterozygotes, but simple inbreeding has been also suggested (David, 1999).

16.6.4 Reinfestation

As long as geometric shape is able to identify the parental generation and to distinguish it satisfactorily from other subpopulations (Falconer, 1981; Dujardin et al., 2007), it might be able to provide relevant information in studies of reinfestation after treatment (Dujardin et al., 1997, 1999a).

Provided that samples were available from the population before insecticide application, relative similarities could suggest the origin of reinfesting specimens: they would be either the descendants of previously killed bugs, or immigrants from an external focus. Shape as extracted from traditional morphometrics (head measurements) of the Chagas disease vector *Triatoma infestans* provided information that could identify the source of reinfesting specimens (Dujardin et al., 1997), and such information has been shown to be in agreement with genetic markers (Dujardin et al., 1999a). The geometry of the wing of the North American *T. protracta* was tested on laboratory populations and was shown to be an interesting candidate to assess the origin of a given individual (Dujardin et al., 2007).

Since a residual population is assumed to be the same generation as or the next generation to the individuals subjected to insecticide spraying, the reinfestation analysis is based on the assumption that an insect is more similar to its parents than to other insects. But the reasons for successful results are not only the high heritability of shape (see Section 16.3.3). In the reinfestation studies, the objective is not to measure gene flow or levels of migration, but rather to distinguish local "inhabitants" from "immigrants." Thus, the environmental effect on metric traits (the environmental covariance) is a welcome effect. Insects reared in the same microenvironmental conditions (a few houses, a village) would share a significantly larger amount of metric similarity, making a residual population easier to recognize.

---

[5] Actually, the Simpson's index of biodiversity and the heterozygosity or genetic diversity index are obtained from the same probabilistic approach: both indexes consider the probability to sample at random two identical species or genotypes. This probability is actually estimating the homogeneity of the individuals (homozygosity), and to get the opposite, it is subtracted from 1, the total frequency.

Of course, the level of population structuring is an important condition for morphometric characters to be applicable in reinfestation studies. They would be less applicable to highly dispersive insects breaking the population structure at each new generation.

16.6.5 Population Structure

A recurrent need in medical entomology is to quantify the current exchanges of individuals among subpopulations. This quantification would inform on "population" structure, to be distinguished from "genetic" structure, which is defined by the level of gene flow among subpopulations. Although mark-release-recapture studies might be a valid option to evaluate the frequency of active migrants among subpopulations (Tapis and Hausermann, 1975; Harrington et al., 2005), it cannot account for passive migration of nonflying stages of the insect, so that this frequency is currently evaluated by indirect methods; the measurement of gene flow is the technique of choice (Slatkin, 1981, 1985).

Gene Flow and the Flow of Migrants

Gene flow measurement provides indirect information on the level of migration among subpopulations. However, this information is of unequal value depending on its output, either "lack of gene flow" or "complete gene flow." Lack of gene flow is valid information since in that circumstance (genetic divergence) migrants are highly unlikely. Less valid information is the case of complete gene flow, since no one can affirm that such (lack of) genetic structure is a reflection of the current level of migration. How contemporaneous or recent it depends on the effective size of the populations under study and the evolutionary rate of the genetic marker (McKay and Latta, 2002). Additional problems with genetic markers are that they are relatively costly and they need appropriate infrastructures. As an unfortunate consequence, genetic markers often remain inside research laboratories and have not yet found their way into routine medical entomology.

Environmental Variance of Size Versus Shape

Modern morphometrics is tempting as a candidate population marker because it is a fast, low-cost, easily spread tool; it is informative about current or very recent population events (Falconer, 1981); and it contains information on genetic variation. However, as long as morphometric traits have much higher environmental variance than genetic markers, they are not appropriate for gene flow estimation.

How then to interpret geographic variation of metric properties? After what we learned from natural and experimental studies on PP, an entirely environmental origin of phenotypic variation is not likely among natural populations. Metric variation can be decomposed into size and shape variation, and even if the two properties are not completely independent, their environmental variance can be examined separately. The importance of diversifying selection inflating size or shape variation among natural populations can be quantified by comparing on the same material (i) the *Fst* index as derived from neutral molecular markers and (ii) the *Qst* index as computed from metric characters. *Qst* separates quantitative genetic variation in a manner analogous to *Fst* for single gene markers (Spitze, 1993): if the quantitative characters and the molecular characters are neutral, *Qst* and *Fst* should converge to the same value (Hiernaux, 1977; Rogers and Harpending, 1983; Whitlock, 1999). Data comparing molecular *Fst* and quantitative *Qst* are few. They tend to show the following trends: (i) *Qst* is generally higher, or much higher, than *Fst*, and (ii) the value of *Qst* depends on character fitness (McKay and Latta, 2002). Within species, traits experiencing the strongest local selection pressures (diverging, or diversifying selection) are expected to be the most divergent from molecular *Fst* (McKay and Latta, 2002).

The small set of comparisons reported by Dujardin (2008) in medically important insects between *Qst* and *Fst* confirmed the importance of selection modifying the geometric variation among subpopulations. These comparisons allowed two more observations: (i) in agreement with the idea of shape having less environmental variance than size, they confirmed the lower sensitivity of shape (relative to size) in response to diversifying selection, and (ii) in agreement with the infrequent report of a *Qst* lower than *Fst*, which would suggest homogenizing selection acting on the quantitative trait, no such situation was observed in medically important insects.

Biogeographical Islands

Local elimination of an insect vector of disease is generally held to be feasible only for geographically constrained situations such as islands. The task of population genetics studies is, in a sense, to find and define those biogeographical "islands" (Patterson and Schofield, 2005) of the vector distribution on the mainland. Can modern morphometrics help defining these target areas?

To discuss this application, it is important again to insist on which metric property is considered, either geometric shape or size. Here, modern morphometrics means "geometric shape." Moreover, since the populations compared are conspecific ones, and especially if size variation is important, allometry-free shape should be preferred to just shape. As seen in the previous sections, geometric shape is made of homologous characters, it is more heritable than size and more canalized than size against environmental disturbances. The following two propositions are related to the geometric shape of the wings, but probably also would apply for the shape of other organs:

1. Because of genetic drift and the polygenic nature of shape, isolation in natural conditions will tend to quickly generate shape changes.
2. When frequent exchanges occur between populations (i.e., when genetic drift is not possible), shape is hardly different even if habitats are different.

The first proposition helps to decide a likely isolation between populations using knowledge about their respective environments. Between truly isolated populations, differences in shape should develop because of two main reasons: (i) genetic drift is likely to be a major force affecting shape, and (ii) homogenizing selection seems infrequent (or unable to counteract the effects of genetic drift). For the shape of the wing in isofemale lines of *Ae. aegypti*, laboratory observations suggested that genetic drift could occur after a few generations (Jirakanjanakit et al., 2008). However, because of the possible importance of diversifying selection, a decision is not possible when shape differences coexist with habitat heterogeneity.

The second proposition refers to lack of shape changes. The interpretation of shape homogeneity in nature does not require information about the environments of the populations under study. As long as homogenizing selection on geometric shape can be discarded, similarity is suggestive of exchanges between compared populations. This proposition is supported by various studies. Even between populations of houses and of palm trees, the shape of the wing of *R. prolixus* was similar (Feliciangeli et al., 2007). Such similarity strongly suggested exchange of individuals, which was confirmed later by genetic markers (Fitzpatrick et al., 2008). Lack of isolation was also described by both genetic and metric markers for tsetse flies along the Mouhoun River in Burkina Faso (Bouyer et al., 2007). Recently, tsetse flies were collected in the city of Abidjan (Ivory Coast) from a primary forest relict, from the zoological garden, and from the nearby university campus of Abobo Adjame. The three sites are a few miles away from each other. The question was: could the flies from the forest reinvade the university campus or the zoological garden after vector control there? The wings of the flies, either males or females, in spite of size differences in one sex, could not show allometry-free shape differences. The likely connection among three sites was confirmed by microsatellite markers applied on the same specimens (Kaba Dramane, Ph.D. thesis).

The Need for a Heuristic

Considering the cost represented by the molecular machinery in developing countries, these examples suggest that a faster and less expensive morphometric approach could be helpful, even as an orientation technique only. Thus, geometric shape variation could be our guide to quickly identify at low-cost areas where isolation is possible and where it is unlikely. Two directives helping interpretation could be the following:

- If shape does not show differences between populations, the most likely explanation is that populations are not isolated ones.
- If shape shows strong differences, one must consider also the habitats which are compared: isolation is a valid interpretation only in case of similar environments.

These guidelines are based on the hypothesis that genetic drift is the main force in nature producing fast differences in shape among conspecific populations. They refer to contemporaneous time, not to an undefined evolutionary past. They are easy to falsify, so that they invite more natural observations related to population structure and shape variation.

16.7 Conclusion

Because coordinates of anatomical landmarks contain information about both size and shape, modern morphometrics could convert the abstract quantification of shape into a direct visual representation. In the same process the geometric approach provided a global estimate of size independent from shape in the absence of allometry. These two metric properties were examined for their insight in evolutionary biology studies, as well as in relation to medical entomology. Their usefulness in quantifying PP was shown to help in the understanding of the evolutionary and ecological importance of the phenotype. In relation to medical entomology, the following needs were considered: species identification, biodiversity estimation, reinfestation analyzes, and "biogeographical islands" detection. It is suggested that geometric shape, as opposed to size, appears as the property of choice to meet these needs. Concrete propositions are made, one is a bank of reference images, the other one is a heuristic to population structure interpretation.

Acknowledgment

I thank M. Tibayrenc (IRD, France), S. Kitthawee, C. Apiwhatnasorn, and P. Kittayapong (Mahidol University, Bangkok, Thailand), who encouraged this work. Unpublished data were produced thanks to the collaboration of R. Rattanarithikul (AFRIMS, Bangkok, Thailand), R. Morales and N. Phumala-Morales (Mahidol University, Bangkok, Thailand), A. Henry and P. Thongsripong (University of Hawaii, USA), and J.-F. Lasnes (University of Montpellier, France). This study has been supported by IRD grants number HC3165-3R165-GABI-ENT2 and HC3165-3R165-NV00-THA1.

References

Adams, D.C., 1999. Methods for shape analysis of landmark data from articulated structures. Evol. Ecol. Res. 1 (8), 959–970.

Adams, D.C., Rohlf, F.J., 2000. Ecological character displacement in *Plethodon*: biomechanical differences found from a geometric morphometric study. Proc. Natl. Acad. Sci. U.S.A. 97, 4106–4111.

Adams, D.C., Rohlf, F.J., Slice, D.E., 2004. Geometric morphometrics: ten years of progress following the "revolution". Ital. J. Zool. 71, 5–16.

Albrecht, G.H., Gelvin, B.R., Hartman, S.E., 1993. Ratios as a size adjustment in morphometrics. Am. J. Phys. Anthropol. 4 (91), 441–468.

Ananthakrishnan, T.N., 2005. Perspectives and dimensions of phenotypic plasticity in insects. pp. 1–23. In: Ananthakrishnan, T.N, Whitman, D. (Eds.), Insect Phenotypic Plasticity. Diversity of responses. Science Publishers Inc., Enfield, NH, p. 213.

Anderson, W., 1973. Genetic divergence in body size among experimental populations of *Drosophila pseudoobscura* kept at different temperatures. Evolution 2 (27), 278–284.

Arnqvist, G., Mårtensson, T., 1998. Measurement error in geometric morphometrics: empirical strategies to assess and reduce its impact on measure of shape. Acta Zool. Academ. Sci. Hung. 44 (1–2), 73–96.

Aytekin, A.M., Alten, B., Caglar, S., Ozbel, Y., Kaynas, S., Simsek, F.M., et al., 2007. Phenotypic variation among local populations of phlebotomine sand flies (Diptera: Psychodidae) in southern Turkey. J. Vector Ecol. 32 (2), 226–234.

Baylac, M., Frieß, M., 2005. Fourier descriptors, Procrustes superimposition, and data dimensionality: an example of cranial shape analysis in modern human populations. Modern Morphometrics in Physical Anthropology, Chapter 6, pp. 145–165.

Baylac, M., Villemant, C., Simbolotti, G., 2003. Combining geometric morphometrics with pattern recognition for the investigation of species complexes. Biol. J. Linn. Soc. 80 (1), 89–98.

Becerra, J.M., Valdecasas, A.G., 2004. Landmark superimposition for taxonomic identification. Biol. J. Linn. Soc. 81, 267–274.

Bergman, A., Siegal, M.L., 2003. Evolutionary capacitance as a general feature of complex gene networks. Nature 424 (6948), 501–504.

Bird, A., 2007. Perceptions of epigenetics. Nature 447, 396–398.

Birdsall, K., Zimmerman, E., Teeter, K., Gibson, G., 2000. Genetic variation for the positioning of wing veins in *Drosophila melanogaster*. Evol. Dev. 2, 16–24.

Bitner-Mathé, B.C., Klaczko, L.B., 1999. Size and shape heritability in natural populations of *Drosophila mediopunctata*: temporal and microgeographical variation. Genetica 105, 35–42.

Bookstein, F.L., 1991. Morphometric Tools for Landmark Data: Geometry and Biology. Cambridge University Press, Cambridge, 435pp.

Bouyer, J., Ravel, S., Dujardin, J.P., de Meeus, T., Vial, L., Thevenon, S., et al., 2007. Population structuring of *Glossina palpalis gambiensis* (Diptera: Glossinidae) according to landscape fragmentation in the Mouhoun river, Burkina Faso. J. Med. Entomol. 44 (5), 788–795.

Braendle, C., Flatt, T., 2006. A role for genetic accommodation in evolution? BioEssays 28, 868–873.

Breuker, C., Patterson, J., Klingenberg, C., 2006. A single basis for developmental buffering of *Drosophila* wing shape. PLoS ONE 1 (1), e7.

Brown, W.L., Wilson, E.O., 1956. Character displacement. Syst. Zool. 5, 49–64.

Burnaby, T.P., 1966. Growth-invariant discriminant functions and generalized distances. Biometrics 22, 96–110.

Camara, M., Caro-Riaño, H., Ravel, S., Dujardin, J.P., Hervouet, J.P., de Meeus, T., et al., 2006. Genetic and morphometric evidence for isolation of a tsetse (Diptera: Glossinidae) population (Loos Islands, Guinea). J. Med. Entomol. 43 (5), 853–860.

Caro-Riaño, H., Jaramillo, N., Dujardin, J.P., 2009. Growth changes in *Rhodnius pallescens* under simulated domestic and sylvatic conditions. Infect. Genet. Evol. 9 (2), 162–168.

Chivian, E., Bernstein, A.S., 2004. Embedded in nature: human health and biodiversity. Environ. Health Perspect. 112, A12–A13.

Costa, J., Felix, M., 2007. *Triatoma juazeirensis* sp. nov. from Bahia State, northeastern Brazil (Hemiptera: Reduviidae: Triatominae). Mem. Inst. Oswaldo Cruz 102, 87–90.

Costa, J., Monteiro, F., Beard, C.B., 2001. *Triatoma brasiliensis* Neiva, 1911 the most important Chagas disease vector in Brazil—phylogenetic and population analyzes correlated to epidemiologic importance. Am. J. Trop. Med. Hyg. 65, 280.

Costa, J., Peterson, A., Dujardin, J.P., 2008. Indirect evidences suggest homoploid hybridization as a possible mode of speciation in Triatominae (Hemiptera, Heteroptera, Reduviidae). Infect. Genet. Evol. 9 (2), 263–270.

Daly, H.V., 1992. A statistical and empirical evaluation of some morphometric variables of honey bee classification. pp. 127–156. In: Foottit, R.G., Sorensen, J.T. (Eds.), Ordination

in the Study of Morphology, Evolution and Systematics of Insects: Applications and Quantitative Genetic Rationales. Elsevier, New York, p. 418.
David, J.R., Moreteau, B., Gauthier, J.P., Petavy, G., Stockel, A., Imasheva, A.G., 1994. Reaction norms of size characters in relation to growth temperature in *Drosophila melanogaster*: an isofemale lines analysis. Gen. Sel. Evol. 26, 229–251.
David, P., 1999. A quantitative model of the relationship between phenotypic variance and heterozygosity at marker loci under partial selfing. Genetics 153, 1463–1474.
Davidowitz, G., D'Amico, L.J., Nijhout, H.F., 2004. The effects of environmental variation on a mechanism that controls insect body size. Evol. Ecol. Res. 6, 49–62.
De la Riva, J., Le Pont, F., Ali, V., Matias, R., Mollinedo, S., Dujardin, J.P., 2001. Wing geometry as a tool for studying the *Lutzomyia longipalpis* (Diptera: Psychodidae) complex. Mem. Inst. Oswaldo Cruz 96 (8), 1089–1094.
Debat, V., Alibert, P., David, P., Paradis, E., Auffray, J.C., 2000. Independence between developmental stability and canalization in the skull of the house mouse. Proc. R. Soc. Lond. B Biol. Sci. 267 (1442), 423–430.
Debat, V., Begin, M., Legout, H., David, J.R., 2003. Allometric and nonallometric components of *Drosophila* wing shape respond differently to developmental temperature. Evolution 57 (12), 2773–2784.
Debat, V., Milton, C.C., Rutherford, S., Klingenberg, C.P., Hoffmann, A.A., 2006. Hsp90 and the quantitative variation of wing shape in *Drosophila melanogaster*. Evolution 60, 2529–2538.
Dobzhansky, T., 1971. Genetics of Evolutionary Process. Colombia University Press, New York, 505pp.
Dujardin, J.P., 2008. Morphometrics applied to medical entomology. Infect. Genet. Evol. 8, 875–890.
Dujardin, J.P., Le Pont, F., 2004a. Geographic variation of metric properties within neotropical sandflies. Infect. Genet. Evol. 4 (4), 353–359.
Dujardin, J.P., Le Pont, F., 2004b. Geographic variation of metric properties within the neotropical sandflies. Infect. Genet. Evol. 4 (4), 353–359.
Dujardin, J.P., Slice, D., 2007. Geometric morphometrics. Contributions to medical entomology. Chapter 25. In: Tibayrenc, M. (Ed.), Encyclopedia of Infectious Diseases. Modern Methodologies. Wiley & Sons, Hoboken, NJ, pp. 435–447.
Dujardin, J.P., Bermúdez, H., Schofield, C.J., 1997. The use of morphometrics in entomological surveillance of silvatic foci of *Triatoma infestans* in Bolivia. Acta Trop. 66, 145–153.
Dujardin, J.P., Bermúdez, H., Gianella, A., Cardozo, L., Ramos, E., Saravia, R., et al., 1999a. Uso de marcadores genéticos en la vigilancia entomológica de la enfermedad de Chagas. In: Alfred Cassab, J., Noireau, F., Guillen, G. (Eds.), La Enfermedad de Chagas en Bolivia—Conocimientos científicos al inicio del Programa de Control (1998–2002). Ministerio de Salud y Prevision social, OMS/OPS, IRD and IBBA, La Paz, pp. 157–169.
Dujardin, J.P., Panzera, P., Schofield, C.J., 1999b. Triatominae as a model of morphological plasticity under ecological pressure. Mem. Inst. Oswaldo Cruz 94, 223–228.
Dujardin, J.P., Le Pont, F., Baylac, M., 2003. Geographic versus interspecific differentiation of sandflies: a landmark data analysis. Bull. Entomol. Res. 93, 87–90.
Dujardin, J.P., Beard, C.B., Ryckman, R., 2007. The relevance of wing geometry in entomological surveillance of Triatominae, vectors of Chagas disease. Infect. Genet. Evol. 7 (2), 161–167.
Dujardin, J.P., Costa, J., Bustamante, D., Jaramillo, N., Catalá, S., 2009. Deciphering morphology in Triatominae: the evolutionary signals. Acta Trop. 110, 101–111.

Dworkin, I., 2005. Canalization, cryptic variation, and developmental buffering: a critical examination and analytical perspective. In: Hallgrimsson, B., Hall, B.K. (Eds.), Variation: A Central Concept in Biology. Academic Press, Oxford, pp. 131–158.

Falconer, D.S., 1981. Introduction to Quantitative Genetics. Longman, London, 300pp.

Feliciangeli, M.D., Sanchez-Martin, M., Marrero, R., Davies, C., Dujardin, J.P., 2007. Morphometric evidence for a possible role of *Rhodnius prolixus* from palm trees in house re-infestation in the State of Barinas (Venezuela). Acta Trop. 101, 169–177.

Fitzpatrick, S., Feliciangeli, M.D., Sanchez-Martin, M.J., Monteiro, F.A., Miles, M.A., 2008. Molecular genetics reveal that silvatic *Rhodnius prolixus* do colonise rural houses. PLoS Negl. Trop. Dis. 2, e210. 10.1371/journal.pntd.0000210.

Foote, M., 1993. Contributions of individual taxa to overall morphological disparity. Paleobiology 19, 403–419.

Francuski, L., Jasmina Ludoki, J., Vujić, A., Milankov, V., 2009. Wing geometric morphometric inferences on species delimitation and intraspecific divergent units in the *Merodon ruficornis* group (Diptera, Syrphidae) from the Balkan peninsula. Zool. Sci. 26(4), 301–308.

Gadagkar, R., Chandrashekara, K., 2005. Behavioral diversity and its apportionment in a primitively eusocial wasp. pp. 108–124. In: Ananthakrishnan, T.N., Whitman, D. (Eds.), Insect Phenotypic Plasticity. Diversity of Responses. Science Publishers, Inc., Enfield, NH, p. 213.

Gibson, G., Hogness, D.S., 1996. Effect of polymorphism in the *Drosophila* regulatory gene *Ultrabithorax* on homeotic stability. Science 271, 200–203.

Gibson, G., Dworkin, I., 2004. Uncovering cryptic genetic variation. Nat. Rev. Genet. 5, 681–690.

Gilchrist, A.S., Partridge, L., 2001. The contrasting genetic architecture of wing size and shape in *Drosophila melanogaster*. Heredity 86, 144–152.

Görür, G., 2005. The importance of phenotypic plasticity in herbivorous insect speciation. In: Anantakrishnan, T.N., Whitman, D. (Eds.), Insect Phenotypic Plasticity. Diversity of Responses. Science Publishers, Inc., Enfield, NH, pp. 145–172.

Graham, J., Freeman, D., Emlen, J., 1993. Developmental stability: a sensitive indicator of populations under 'stress'. In: Landis, W.G., Hugues, J.S., Lewis, M.A. (Eds.), Environmental Toxicology and Risk Assessment, ASTM STP 1179. American Society for Testing and Materials, Philadelphia, PA, pp. 136–158.

Grant, P.R., 1972. Convergent and divergent character displacement. Biol. J. Linn. Soc. 4, 39–68.

Grant, P.R., Grant, B.R., 2006. Evolution of character displacement in Darwin's finches. Science 313, 224.

Gumiel, M., Catalá, S., Noireau, F., de Arias, A.R., Garcia, A., Dujardin, J.P., 2003. Wing geometry in *Triatoma infestans* (Klug) and *T. melanosoma* Martinez, Olmedo and Carcavallo (Hemiptera: Reduviidae). Syst. Entomol. 28 (2), 173–179.

Harrington, L., Scott, T.W., Lerdthusnee, K., Coleman, R.C., Costero, A., Clark, G.G., et al., 2005. Dispersal of the dengue vector *Aedes aegypti* within and between rural communities. Am. J. Trop. Med. Hyg. 72 (2), 209–220.

Hatadani, L.M., Klaczko, L.B., 2008. Shape and size variation on the wing of *Drosophila mediopunctata*: influence of chromosome inversions and genotype–environment interaction. Genetica 133, 335–342.

Henry, A., Thongsripong, P., Fonseca-Gonzalez, I., Jaramillo-Ocampo, N., Dujardin, J.P., 2010. Wing shape of dengue vectors from around the world. Infect. Genet. Evol. 10, 207–214. 10.1016/j.meegid.2009.12.001.

Hiernaux, J., 1977. Long-term biological effects of human migration from the African savanna to the equatorial forest: a case study of human adaptation to a hot and wet climate. In: Harrison, G.A. (Ed.), Population Structure and Human Variation. Cambridge University Press, Cambridge, pp. 187–218.

Hillesheim, E., Stearns, S.C., 1991. The responses of *Drosophila melanogaster* to artificial selection on body weight and its phenotypic plasticity in two larval food environments. Evolution 45 (8), 1909–1923.

Hoffman, A.A., Shirriffs, J., 2002. Geographic variation for wing shape in *Drosophila serrata*. Evolution 56, 1068–1073.

Hoffmann, A.A., Woods, R.E., Collins, E., Wallin, K., White, A., McKenzie, J.A., 2005. Wing shape versus asymmetry as an indicator of changing environmental conditions in insects. Aust. J. Entomol. 44, 233–243.

Iriarte, P.F., Norry, F.M., Hasson, E.R., 2003. Chromosomal inversions effect body size and shape in different breeding resources in *Drosophila buzzatii*. Heredity 91, 51–59.

Jablonka, E., Lamb, M., 2002. The changing concept of epigenetics. Ann. N. Y. Acad. Sci. 981, 82–96.

Jablonka, E., Lamb, M.J., Lachmann, M., 1992. Evidence, mechanisms and models for the inheritance of acquired characteristics. J. Theor. Biol. 158 (2), 245–268.

Jirakanjanakit, N., Leemingsawat, S., Thongrungkiat, S., Apiwathnasorn, C., Singhaniyom, S., Bellec, C., et al., 2007. Influence of larval density or food variation on the geometry of the wing of *Aedes (Stegomyia) aegypti*. Trop. Med. Int. Health 12(11), 1354–1360.

Jirakanjanakit, N., Leemingsawat, S., Dujardin, J.P., 2008. The geometry of the wing of *Aedes (Stegomyia) aegypti* in isofemale lines through successive generations. Infect. Genet. Evol. 8, 414–421.

Jordaens, K., Van Dongen, S., Van Riel, P., Geenen, S., Verhagen, R., Backeljau, T., 2002. Multivariate morphometrics of soft body parts in terrestrial slugs: comparison between two datasets, error assessment and taxonomic implications. Biol. J. Linn. Soc. 75(4), 533–542.

Keesing, F., Holt, R.D., Ostfeld, R.S., 2006. Effects of species diversity on disease risk. Ecol. Lett. 9, 485–498.

Kitthawee, S., Dujardin, J.P., 2009. *Diachasmimorpha longicaudata*: reproductive isolation and geometric morphometrics of the wings. Biol. Control 51 (1), 191–197.

Kitthawee, S., Dujardin, J.P., 2010. The geometric approach to explore the *Bactrocera tau* complex (Diptera: Tephritidae) in Thailand. Zoology 113, 243–249.

Klingenberg, C.P., 1996. Multivariate allometry. In: Marcus, L.F., Corti, M., Loy, A., Naylor, G.J.P., Slice., D. (Eds.), Advances in Morphometrics Proceedings of the 1993 NATO-ASI on Morphometrics. Plenum Publ NATO ASI, Ser. A, Life Sciences, New York, pp. 23–49.

Klingenberg, C.P., Leamy, L.J., 2001. Quantitative genetics of geometric shape in the mouse mandible. Evolution 55, 2342–2352.

Klingenberg, C.P., Leamy, L.J., Routman, E.J., Cheverud, J.M., 2001. Genetic architecture of mandible shape in mice: effects of quantitative trait loci analyzed by geometric morphometrics. Genetics 157 (2), 785–802.

Klingenberg, C., Leamy, L.J., Cheverud, J.M., 2004. Integration and modularity of quantitative trait locus effects on geometric shape in the mouse mandible. Genetics 166, 1909–1921.

Lehmann, T., Dalton, R., Kim, E., Dahl, E., Diabate, A., Dabire, R., et al., 2006. Genetic contribution to variation in larval development time, adult size, and longevity of starved adults of *Anopheles gambiae*. Infect. Genet. Evol. 6 (5), 410–416.

Lele, S.R., Richtsmeyer, J., 2001. An Invariant Approach to Statistical Analysis of Shape. Chapman and Hall/CRC, Boca Raton, FL, pp. VIII + 308.

Lestrel, P.E., 2000. Morphometrics for Life Sciences. World Scientific Publishing, Singapore.

Levit, G.S., Hossfeld, U., Olssonzã, L., 2006. From the "Modern Synthesis" to cybernetics: Ivan Ivanovich Schmalhausen (1884–1963) and his research program for a synthesis of evolutionary and developmental biology. J. Exp. Zool. (Mol. Dev. Evol.) 306B, 89–106.

Long, A., Mullaney, S., Reid, L., Fry, J., Langley, C., Mackay, T.F.C., 1995. High resolution mapping of genetic factors affecting abdominal bristle number in *Drosophila melanogaster*. Genetics 139, 1273–1291.

Losos, J.B., 2000. Ecological character displacement and the study of adaptation. PNAS 97 (11), 5693–5695.

Lyra, M.L., Hatadani, L.M., de Azeredo-Espin, A.M., Klaczko, L.B., 2009. Wing morphometry as a tool for correct identification of primary and secondary new world screwworm fly. Bull. Entomol. Res. 23, 1–8.

Matias, A., De la Riva, J.X., Torrez, M., Dujardin, J.P., 2001. *Rhodnius robustus* in Bolivia identified by its wings. Mem. Inst. Oswaldo Cruz 96 (7), 947–950.

Mayr, E., 2000. The biological species concept. In: Wheeler, Q.D., Meier, R. (Eds.), Species Concepts and Phylogenetic Theory. A Debate. Columbia University Press, New York, pp. 17–29.

McKay, J.K., Latta, R.G., 2002. Adaptive population divergence: markers, QTL and traits. Trends Ecol. E. 17, 285–291.

Morales, V.E.R., Ya-umphan, P., Phumala-Morales, N., Komalamisra, N., Dujardin, J.P., 2010. Climate associated size and shape changes in *Aedes aegypti* (Diptera: Culicidae) populations from Thailand. Infect. Genet. Evol. 10 (4), 580–585.

Nasci, R.S., 1990. Relationship of wing length to adult dry weight in several mosquito species (Diptera: Culicidae). J. Med. Entomol. 27, 716–719.

Neige, P., 2003. Spatial patterns of disparity and diversity of the recent cuttlefishes (cephalopoda) across the Old World. J. Biogeogr. 30, 1125–1137.

Orengo, D.J., Prevosti, A., 2002. Relationship between chromosomal polymorphism and wing size in a natural population of *Drosophila subobscura*. Genetica 115, 311–318.

Palmer, A.R., Strobeck, C., 1986. Fluctuating asymmetry: measurement, analysis, patterns. Annu. Rev. Ecol. Syst. 17, 391–421.

Partridge, L., Barrie, B., Fowler, K., French, V., 1994. Evolution and development of body size and cell size in *Drosophila melanogaster* in response to temperature. Evolution 48, 1269–1276.

Patterson, J., Klingenberg, C., 2007. Developmental buffering: how many genes? Evol. Dev. 9 (6), 525–526.

Patterson, J.S., Schofield, C.J., 2005. Preliminary study of wing morphometry in relation to tsetse population genetics. S. Afr. J. Sci. 101, 132–134.

Pennisi, E., 2006. Evolution: hidden genetic variation yields caterpillar of a different color. Science 311, 591.

Rasmussen, P., Wheeler, W., Moser, T., Vine, L., Sullivan, B., Rusch, D., 2001. Measurements of Canada goose morphology—Sources of error and effects on classification of subspecies. J. Wildl. Manag. 65 (4), 716–725.

Réale, D., Roff, D.A., 2003. Inbreeding, developmental stability, and canalization in the sand cricket *Gryllus firmus*. Evolution 57 (3), 597–605.

Richman, A.D., Price, T., 1992. Evolution of ecological differences in the old world leaf warblers. Nature 355, 817–821.
Ricklefs, R.E., Miles, D.B., 1994. Ecological and evolutionay inferences from morphology: an ecological perspective. In: Wainwright, P.C., Reilly, S.M. (Eds.), Ecological Morphology. University of Chicago Press, Chicago, IL, pp. 13–41.
Roff, D.A., Mousseau, T.A., 1987. Quantitative genetics and fitness: lessons from *Drosophila*. Heredity 58, 103–118.
Rogers, A., Harpending, H.C., 1983. Population structure and quantitative characters. Genetics 105, 985–1002.
Roggero, A., Passerin d'Entrèves, P., 2005. Geometric morphometric analysis of wings variation between two populations of the *Scythris obscurella* species-group: geographic or interspecific differences? (Lepidoptera: Scythrididae). SHILAP Revista Lepid. 33 (130), 101–112.
Rohlf, F.J., 1990. Rotational fit (Procrustes) methods. In: Rohlf, F., Bookstein, F. (Eds.), Proceedings of the Michigan Morphometrics Workshop. Special Publication Number 2. The University of Michigan Museum of Zoology. University of Michigan Museums, Ann Arbor, MI, pp. 227–236. pp. 380.
Rohlf, F.J., 1996. Morphometric spaces, shape components and the effects of linear transformations. In: Marcus, L.F., Corti, M., Loy, A., Naylor, G., Slice, D. (Eds.), Advances in Morphometrics. Proceedings of the 1993 NATO-ASI on Morphometrics.. Plenum Publ. NATO ASI, Ser. A, Life Sciences, New York, pp. 117–129.
Rohlf, F.J., Marcus, L.F., 1993. A revolution in morphometrics. TREE 8 (4), 129–132.
Rohlf, F.J., Bookstein, F.L., 2003. Computing the uniform component of shape variation. Syst. Biol. 52 (1), 66–69.
Roy, K., Balch, D.P., Hellberg, M.E., 2001. Spatial patterns of morphological diversity across the Indo-Pacific: analyses using strombid gastropods. Proc. R. Soc. Lond. 268, 2503–2508.
Rutherford, S.L., Lindquist, S., 1998. Hsp90 as a capacitor for morphological evolution. Nature 396, 336–342.
Schlichting, C.D., Pigliucci, M., 1998. Phenotypic Evolution: A Reaction Norm Perspective. Sinauer Associates, Inc., Sunderland, MA, 387pp.
Shannon, C.E., Weaver, W., 1949. The Mathematical Theory of Communication. University of Illinois Press, Urbana, IL.
Shepard, U.L., 1998. A comparison of species diversity and morphological diversity across the North American latitudinal gradient. J. Biogeogr. 25, 19–29.
Shrimpton, A., Robertson, A., 1988. The isolation of polygenic factors controlling bristle score in *Drosophila melanogaster*: I. Allocation of third chromosome bristle effects to chromosome sections. Genetics 118, 437–443.
Siegel, J.P., Novak, R.J., Lampman, R.L., Steinly, B.A., 1992. Statistical appraisal of the weight-wing length relationship of mosquitoes. J. Med. Entomol. 29 (4), 711–714.
Simpson, E.H., 1949. Measurement of diversity. Nature 163, 688.
Slatkin, M., 1981. Estimating levels of gene flow in natural populations. Genetics 99, 323–335.
Slatkin, M., 1985. Rare alleles as indicators of gene flow. Evolution 39 (1), 53–65.
Slice, D.E., 2001. Landmark coordinates aligned by Procrustes analysis do not lie in Kendall's shape space. Syst. Biol. 50 (1), 141–149.
Smith, G.R., 1990. Homology in morphometrics and phylogenetics. In: Rohlf, F.J., Bookstein, FL (Eds.), Proceedings of the Michigan Morphometrics Workshop. The

University of Michigan Museum of Zoology, Ann Arbor, MI, pp. 325–338. , Special Publication Number 2.

Sollars, V., Lu, X., Xiao, L., Wang, X., Garfinkel, M.D., Ruden, D.M., 2003. Evidence for an epigenetic mechanism by which Hsp90 acts as a capacitor for morphological evolution. Nat. Genet. 33, 70–74.

Spitze, K., 1993. Population structure in *Daphnia obtusa*: quantitative genetic and allozymic variation. Genetics 135, 367–374.

Suzuki, Y., Nijhout, H.F., 2006. Evolution of a polyphenism by genetic accommodation. Science 5761 (311), 650–652.

Tapis, M., Hausermann, W., 1975. Demonstration of differential domesticity of *Aedes aegypti* (L.) (Diptera: Culicidae) in Africa by mark-release-recapture. Bull. Entomol. Res. 65, 199–208.

Thompson, J.D., 1971. Phenotypic plasticity as a component of evolutionary change. Trends Ecol. Evol. 6, 246–249.

Vanvoorhies, W., 1996. Bergmann size clines—a simple explanation for their occurrence in ectotherms. Evolution 50 (3), 1259–1264.

Villegas, J., Feliciangeli, M.D., Dujardin, J.P., 2002. Wing shape divergence between *Rhodnius prolixus* from Cojedes (Venezuela) and *R. robustus* from Mérida (Venezuela). Infect. Genet. Evol. 2, 121–128.

Villemant, C., Simbolotti, G., Kenis, M., 2007. Discrimination of *Eubazus* (Hymenoptera, Braconidae) sibling species using geometric morphometrics analysis of wing venation. Syst. Entomol. 32 (4), 625–634.

Waddington, C.H., 1953. Genetic assimilation for an acquired character. Evolution 7, 118–126.

Waddington, C.H., 1956. Genetic assimilation of the bithorax phenotype. Evolution 10, 1–13.

West-Eberhard, M.J., 1989. Phenotypic plasticity and the origins of diversity. Annu. Rev. Ecol. Syst. 20, 249–278.

West-Eberhard, M.J., 2003. Developmental Plasticity and Evolution. Oxford University Press, Oxford, United Kingdom, 794 pp.

Whitlock, M.C., 1999. Neutral additive variance in a metapopulation. Genet. Res. 74, 215–221.

Wilson, E.O., Brown, W.L., 1955. Revisionary notes on the *sanguinea* and *neogagates* groups of the ant genus *Formica*. Psyche 62, 108–129.

Workman, M.S., Leamy, L.J., Routman, E.J., Cheverud, J.M., 2002. Analysis of quantitative trait locus effects on the size and shape of mandibular molars in mice. Genetics 160 (4), 1573–1586.

Zelditch, M.L., Swiderski, D.L., Sheets, H.D., Fink, W.L., 2004. Geometric morphometrics for biologists: a primer. Elsevier, Academic Press, New York.

17 Multilocus Sequence Typing of Pathogens

Marcos Pérez-Losada[1,]*, Megan L. Porter[2], Raphael P. Viscidi[3] and Keith A. Crandall[4]

[1]CIBIO, Centro de Investigação em Biodiversidade e Recursos Genéticos, Universidade do Porto, Campus Agrário de Vairão, Vairão, Portugal, [2]Department of Biological Sciences, University of Maryland, Baltimore County, Baltimore, MD, USA, [3]Department of Pediatrics, Johns Hopkins University School of Medicine, Baltimore, MD, USA, [4]Department of Biology, Brigham Young University, Provo, UT, USA

17.1 Introduction

The ability to accurately distinguish between strains of infectious pathogens is crucial for efficient epidemiological and surveillance analysis, studying microbial population structure and dynamics and, ultimately, developing improved public health control strategies (2004). To further such general goals, several molecular typing methods have been proposed that can identify isolates worldwide (global epidemiology) and/or in localized disease outbreaks (local epidemiology) (see Foley et al., 2009 for a review). Nonetheless, since 1998, the proposed "gold standard" for molecular typing is multilocus sequence typing (MLST) (Maiden et al., 1998). MLST was built on the well-established population genetic concepts and methods of the multilocus enzyme electrophoresis (MLEE) technique, but provides significant advantages over this and other typing approaches (see Section 17.4 for advantages and caveats). MLST examines nucleotide variation in sequences of internal fragments of usually seven housekeeping genes, (i.e., those encoding fundamental metabolic functions) (see Section 17.2 for molecular design and development of MLST). For each gene, the different sequences present within a species are assigned as distinct alleles and, for each isolate, the alleles at each of the seven loci define the allelic profile or sequence type (ST). Each isolate is therefore unambiguously characterized by a series of seven integers, which correspond to the alleles at the seven housekeeping loci. Most bacterial species have sufficient variation within housekeeping genes to provide many alleles per locus, allowing billions of distinct allelic profiles to be distinguished using just seven loci. Alternatively, isolate

*E-mail: mlosada323@gmail.com

identification and tracking can be performed using the nucleotide data directly, although this approach is more frequently used for population studies (see Section 17.5 for methods of analyses).

MLST has become the universal approach for molecular typing (Maiden, 2006). Numerous examples exist of their use for describing the population structure of pathogens, vaccine studies, tracking transmission of epidemic strains, and identifying species and virulent strains associated with disease (see Section 17.6 for applications). This was made possible by three improvements in molecular microbiology (Maiden, 2006) involving: (1) bacterial evolution and population biology knowledge (below); (2) high-throughput nucleotide sequencing (see Section 17.2 for molecular basis); and (3) Internet databases (see Section 17.3). The bacterial population studies undertaken from the 1980s onwards were central to the development of the MLST. Those studies showed that genetic exchange among bacteria was more common than previously supposed, leading to a reassessment of the role of sexual processes in the structuring of bacterial populations. Using sequence data they showed that recombination (mosaic genes) was not only frequent in genes under diversifying selection (e.g., antigen-encoding and antibiotic resistant determinant genes), but also in genes under purifying selection (housekeeping genes), which suggested the clonal model (variation can only arise by mutation) was not universal and led to the proposal of new nonclonal or panmictic (variation is mainly generated by recombination) and partially clonal models of bacterial population structure. Consequently, typing methods needed to accommodate a broader spectrum of population structures and be able to distinguish among them, hence providing not only discriminatory power but also information about the clonal structure of the organism under study. Therefore, only molecular techniques that can contrast results across independent markers (such as MLST) would be adequate for bacterial typing and population genetic inferences.

In the following sections, we will describe in more detail all the epigraphs mentioned in this introduction. We also refer the reader to other reviews on MLST for complementary information (Urwin and Maiden, 2003; Cooper and Feil, 2004; Sullivan et al., 2005; Maiden, 2006).

17.2 Molecular Design and Development of MLST

The principal element in the design of an MLST scheme is the choice of genetic loci. The selection and number of loci are based on principle, precedent, and practice. Since MLST was developed as an updated version of MLEE, which indexes variation of multiple core metabolic or housekeeping genes at the protein level, the selected loci should be within housekeeping genes encoding proteins for core metabolic functions. Furthermore, housekeeping genes are expected to be subject to the rules of neutral evolution and thus should reveal genetic relationships among strains without concern for the influence of host or environmental factors, such as might occur with a gene encoding a hypervariable surface protein subject to host immune response driven by diversifying selection or a gene under antibiotic selection. The genes should be physically spaced around the genome in order to minimize genetic linkage of loci.

As a matter of principle and practicality, multiple loci of sufficient length need to be surveyed in order to provide a high level of discrimination. The first MLST scheme was designed by Maiden and colleagues (Maiden et al., 1998) and included six, later expanded to seven, loci. Most investigators have followed this precedent and developed schemes of seven loci. The length of nucleotide sequence amplified for each loci is generally in the range of 400−600 bp and is determined largely by the parameters of automated sequencing instruments available at the time the first MLST scheme was developed in 1998. Although MLST nucleotide sequence data are currently generated by the Sanger sequencing method, pyrosequencing (Margulies et al., 2005) will likely be the method of choice in the future. The new GS FLX titanium reagents for use on the Genome Sequencer FLX instrument from the 454 Life Sciences Company are capable of generating accurate read lengths of 250−350 bp, which is sufficient for most MLST size fragments if reads are generated from both ends of the DNA strand. Improvements in the technology should extend read length (up to 1000 bp) and increase capacity (over 1 million reads per run) in the near future. Pyrosequencing of PCR amplicons has already found numerous applications in microbiology and human genetics (Bordoni et al., 2008; Rozera et al., 2009). The design of bar codes that are incorporated at the 5′ end of amplicons allows simultaneous sequencing of homologous products from multiple samples in the same run (Binladen et al., 2007; Meyer et al., 2008). Since the complexity of PCR products for MLST is very low, pyrosequencing is only cost efficient if many PCR products can be processed simultaneously. The number of bar codes that can be designed using even a few bases is very large. For example, a bar coding method has been described that used a class of error-correcting codes called Hamming codes to design 1544 unique 8-base bar codes (Hamady et al., 2008).

The development of a new MLST scheme from scratch involves four initial steps (Table 17.1): (1) identification of loci, (2) PCR primer design, (3) survey of a small number of representative strains, and (4) analysis of nucleotide sequence data to establish neutral evolution of loci and level of strain discrimination. For many bacterial species, the selection of loci is greatly aided by the availability of annotated whole genomes, which allow ready identification of housekeeping genes and their physical location in the genome. An absolute requirement for loci included in an MLST is that there is only a single copy of the gene in the genome. It is advisable to choose more than seven loci because not all loci will pass subsequent tests of utility, and typically 12−18 loci are selected for subsequent tests. As near as possible, the loci should be evenly spaced across the genome and certainly separated by several tens of thousands of base pairs, although no rules allow a precise estimate of the maximum size of bacterial genomic fragments that can undergo recombination. The physical location of loci within genomes may differ among strains so use of a single reference strain, which is often all that is available, is at best an approximation. The design of primers is greatly assisted by the availability of open access and commercial software for primer design but ultimately depends on trial and error. Most MLST schemes use a nested PCR design both to increase

Table 17.1 Stages in the Design of an MLST Scheme

Actions	Criteria
• Analyze reference genome to identify 12–18 candidate loci	• Single copy gene • Putative core housekeeping gene • Genes evenly spaced in genome
• Design nested PCRs using primer select software	• Outer PCR product ~1000–1500 bp • Inner PCR product ~400–600 bp
• Select 15–20 representative strains	• Isolated in different years and different geographic sites • No known epidemiological linkage by transmission or shared phenotypic characteristics
• Perform nested PCRs of the 15–20 strains and redesign primers as needed	
• Analyze nucleotide sequence data	• Rank loci by level of nucleotide polymorphisms and select 7–9 loci with high levels of polymorphism
• Select 75–100 strains using above criteria, type using the 7–9 loci and perform analysis	• Confirm loci are under purifying selection • Assign each unique sequence an allele number • Assign each isolate an ST • The greater the number of STs, the greater the discriminatory power of the MLST

sensitivity for samples with a low bacterial DNA copy number and, more importantly, to provide a high-quantity and quality PCR product for sequencing. The initial evaluation of candidate loci is most easily accomplished with a small number of strains, perhaps 20, and the strains should not be epidemiologically linked or share defining characteristics, such as antibiotic resistance, that might lead to oversampling of a clonal population. Temporally and geographically separated strains provide one likely basis for accomplishing this goal. The data from this small set of strains should allow the stratification of loci on the basis of efficiency of detection by nested PCR and level of genetic variation. They also provide the opportunity to optimize primer design Table 17.1.

At least seven to nine loci that could be amplified from all 20 strains and showed a reasonably high level of genetic diversity (see Pérez-Losada et al., 2007b, for a review of methods for calculating genetic diversity and other population genetic parameters from MLST data) should then be evaluated with a larger dataset of 70–100 strains to accomplish the initial data analysis for evolutionary neutrality and level of strain discrimination. The same rules for selection of strains apply here as above. A representative collection of strains should be used, but in practice it is only possible to avoid obvious pitfalls such as selecting strains from a known outbreak. The purpose of the initial analysis of MLST data is to confirm that the chosen loci are under purifying selection, to assess the level of polymorphism at each locus, and to determine whether a sufficient level of discrimination is achieved for epidemiological studies. The number of unique nucleotide sequences among the 70–100 strains tested establishes the level of polymorphism, and alleles that are the most polymorphic will provide the greatest degree of discrimination among strains. While low levels of polymorphism are a reason to reject an allele for inclusion in an MLST scheme because they will provide little discriminatory power, the seven most polymorphic alleles are not necessarily the best choice. Ideally, all seven loci will contribute equally to the discriminatory power of the method and a very high level of variation may be indicative of diversifying selection pressure. On the other hand, evolutionary neutrality is a desirable but not absolutely necessary characteristic of loci used in an MLST typing scheme. In fact, most other methods for strain typing use highly polymorphic loci, which are often known to be subject to selection pressure. If one or more of the initially selected loci fail the test of neutrality or no combination of 6–7 loci provides sufficient strain discrimination, other loci surveyed in the smaller dataset of 20 isolates can be evaluated with the larger dataset and a new 6–7 loci MLST scheme can be designed. Finding the right balance in terms of efficiency of PCR amplification, locus neutrality, strain discrimination, and comparability of polymorphisms across loci is ultimately a matter of judgment rather than the application of precise rules.

Once candidate loci have been chosen and the MLST scheme defined, application of the method in the context of epidemiological studies will establish its reliability in typing large numbers of diverse strains and its ability to provide sufficient strain discrimination to address epidemiological questions of interest. For strains that cannot be typed using the initial PCR primers, it is generally easy to design new primers. Although the choice of loci used in the MLST scheme could be changed as more strains are typed, for example to increase discrimination, one of the strengths of MLST as a typing method would be sacrificed, namely, the comparability of data generated over time and by multiple investigators. Because the sequence type or ST is defined by the set of distinctly numbered alleles at the seven loci, changing loci would result in new STs that could not be directly compared to STs defined using the older scheme. If an epidemiological study requires discrimination of closely related strains, as may be necessary to examine short-term transmission of antibiotic resistant isolates, rather than add to or change the loci in an MLST, a better strategy is to supplement MLST with additional highly polymorphic markers, such as genes encoding antigens or cell surface proteins. Examples of such strategies are the combination of MLST and staphylococcal chromosomal cassette mec (SCCmec) typing for *S. aureus* (Cookson et al., 2007), MLST and *emm* sequencing for Group A *Streptococcus* (Metzgar et al., 2009), and MLST and porB sequencing for *Neisseria gonorrhoeae* (Pérez-Losada et al., 2007a).

17.3 MLST Databases

One of the aims of the MLST approach was the development of a website containing MLST databases to which public health and research communities could both have access and contribute and from which clinical, epidemiological, and population studies could benefit (Maiden et al., 1998; Urwin and Maiden, 2003; Maiden, 2006). The first MLST websites were based on single databases implemented in the MLSTdB software (Chan et al., 2001); but as MLST schemes began to expand several limitations became apparent: redundant information (each record contained the ST designation and the allelic profile), isolate bias (single databases were dominated by specific studies), and access (all databases were stored at a single location). To overcome these limitations, a new network-based database software, MLSTdBNet (Jolley et al., 2004) was developed and implemented on the PubMLST site (http://pubmlst.org/). This site is served by two databases: (i) a profiles database that contains the sequences of each MLST allele for each locus linked to an allele number, and (ii) an allelic profiles database with their ST designations. The profiles database can then serve other isolate databases. For each scheme on the PubMLST site there is a PubMLST isolate database that aims to include at least one isolate for each ST. MLST databases are hence different from other depository databases such as GenBank not only in organization but also in that they are actively curated for accuracy. It is important to highlight that MLST databases do not embody the global diversity of an organism but the extent of its diversity at the time they are accessed. Moreover, the stored data is unstructured and does not necessarily represent natural populations either.

Several other websites are accessible through the PubMLST site. The PubMed (NCBI) is linked to PubMLST databases so original publications describing MLST schemes can be retrieved. The AgdbNet—antigen sequence database software for bacterial typing (Jolley and Maiden, 2006) is also integrated into the system. Other websites are available for the storage and access of MLST data. At the time of writing, 54 MLST schemes (48 for bacteria and 6 for eukaryotes) with available websites were hosted at http://pubmlst.org/, http://www.mlst.net/ (Aanensen and Spratt, 2005) and http://mlst.ucc.ie/ and could be accessed via the PubMLST site. The PubMLST primary site is also mirrored in four locations, three in the United Kingdom and one in the United States. This provides access to MLST data globally and assures that databases are stored in multiple locations. A detailed description of the MLST databases, their structure, and most of the published MLST schemes can be found in Maiden (2006).

17.4 Advantages and Disadvantages of MLST

As the number of schemes available has increased, MLST has become the most commonly used method of pathogen typing. Previous methods were based on phenotypic (e.g., serotyping) or electrophoretic (e.g., MLEE) variation. In comparison, the use of sequence variation gives MLST the advantage of producing data that contain more variation and that are universally comparable, easily validated, and readily shared across laboratories. The use of generic sequencing technology makes MLST a broadly applicable methodology that can be fully automated and scalable from single isolates to thousands of samples. Because the material needed for MLST analysis—DNA or dead cells—are easily transported among labs without the problems associated with infective materials, both the necessary biological material and the resulting data are highly portable. Furthermore, the use of online electronic databases (see Section 17.3) to store and curate MLST schemes makes them a globally accessible resource.

MLST targets variation at multiple housekeeping loci. The number of loci that need to be evaluated to be able to confidently assign a ST has been minimized to reduce the expense and time required for characterization, with most studies using 6–10 loci. If performed manually, evaluating even this many loci can be time-consuming. However, fully automated systems (e.g., robotics; Jefferies et al., 2003) provide a high-throughput system for data collection that can run large volumes of samples with increased reliability. As sequencing technology progresses, the cost of this automation will decrease and data interpretation, rather than data generation, will be the limiting factor of our understanding of pathogen population genetics.

By focusing on sequence variation, MLST provides a highly replicable and reproducible typing method. Additionally, the focus on housekeeping genes provides significant amounts of genetic data that can be used to calculate pathogen population genetic parameters (see Section 17.5) on a global scale. These analyses of population parameters can be used to construct more sophisticated models of pathogen evolution and epidemiology that will improve our understanding of how to control the spread of these diseases. However, there is no single set of universal housekeeping genes that can be used with all pathogens as the recombination rates, substitution rates, and levels of selection vary across loci and across species for the same locus (Pérez-Losada et al., 2006). Therefore, a unique set of loci must be identified for each novel, untyped pathogen under study. The rapid expansion of available whole genomes will make data mining for useful housekeeping genes more feasible, reducing the time and cost required for constructing new schemes.

Unfortunately, not all pathogens are suitable for MLST methods. Some pathogens (e.g., *M. tuberculosis, Y. pestis*) contain very little diversity throughout their entire genome, most likely representing evolutionarily young pathogens that have not yet accumulated sufficient genetic variation to differentiate strains. For these pathogens, more rapidly evolving loci (e.g., insertion sequences or antibiotic-resistance determinants) are needed. Conversely, some bacterial genomes have accumulated enough variation that MLST using housekeeping genes does not provide any information useful for typing. As we enter the era of genomics, the comparison of genomes to identify variable regions/genes to use for typing will become easier. As sequencing technology continues to improve in the future, comparison of whole genomes will even become possible.

17.5 Analytical Approaches

There are two basic strategies to the analysis of MLST data (Figure 17.1), one relies on allele and ST designations to estimate relatedness among isolates (allele-based methods) and so ignores the number of nucleotide differences between alleles, and the other relies on nucleotide sequences directly to estimate relatedness and population parameters (nucleotide-based methods). The allele-based approach has been adopted from the analysis of MLEE data and so methods based on this strategy were the first applied to the analysis of MLST data (Maiden et al., 1998; Enright and Spratt, 1999). The allele-based approach is thought to work well in nonclonal organisms (e.g., *Helicobacter pylori*), while nucleotide-based approaches are preferable for clonal organisms (e.g., *Escherichia coli*) since the former are likely misleading (Maiden, 2006) Figure 17.1.

In practice, most microbes show some degree of clonality (clonal complex) in their populations, hence, in our opinion, both types of analyses can be conducted in population and epidemiological studies (e.g., Tazi et al., 2010). In this section, we present a brief description of some of the most commonly used approaches for

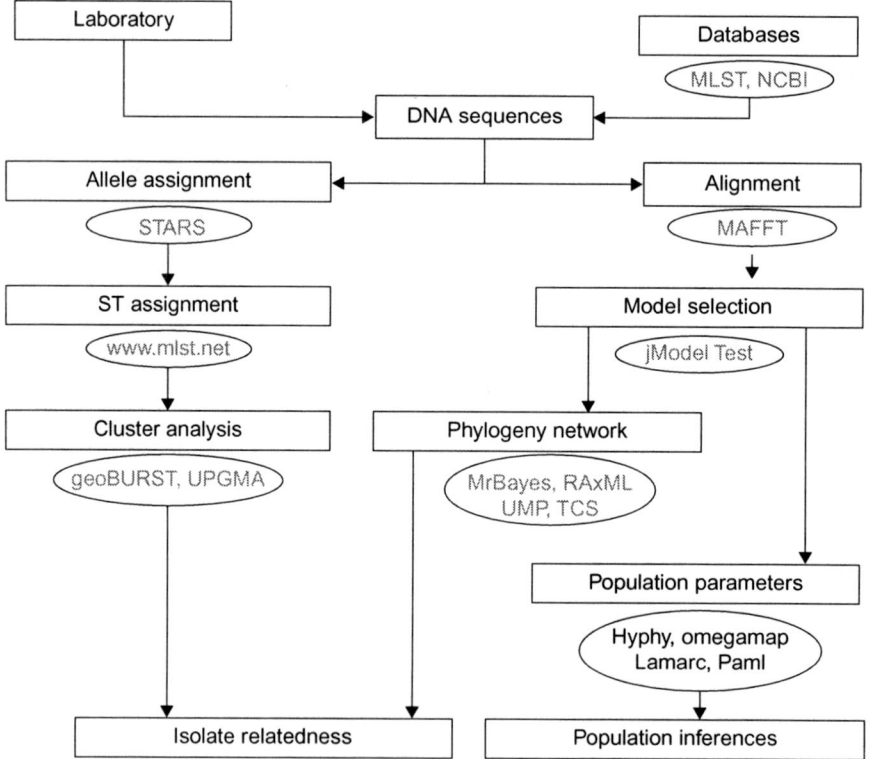

Figure 17.1 Pipeline analysis of MLST data.

analyzing MLST data. We refer the reader to previous reviews (Sullivan et al., 2005; Pérez-Losada et al., 2007b) for more detailed descriptions.

17.5.1 Allele-Based Methods

Since alleles are the unit of analysis, all these methods first require assigning an allele number to each DNA sequence from each locus. This is done by matching our sequences against those stored in public MLST databases (see Section 17.3). If no match is found, a new number is assigned in order of discovery. Several computational programs have been developed for this task, although sequence typing analysis and retrieval system (STARS) seems to be the best (Sullivan et al., 2005). The STARS interface was specifically designed for typing and allows the assembly of large number of sequences at once.

Once alleles have been assigned, data are entered in the MLST websites to acquire an ST profile. At this point exploratory analysis (e.g., allele and profile frequencies, polymorphism estimates, codon usage, etc.) of the data can be performed. The software package Sequence Type Analysis and Recombinational Tests (START2) can perform all these tasks (Jolley et al., 2001). Relatedness among STs can then be displayed using methods of cluster reconstruction such as the based upon related sequences types (eBURST) approach and the simple unweighted pair group method with arithmetic mean (UPGMA). eBURST (Feil et al., 2004) is based on a simple model of clonal expansion and diversification. It first identifies mutually exclusive groups of related STs and attempts to identify the founding ST of each group. Bootstrap estimates are also calculated to assess confidence in the groupings. The algorithm then predicts the descent from the predicted founding ST to the other STs in the group, displaying the output as a radial diagram, centered on the predicted founding ST. Recently, a new globally optimized version (goeBURST) has been released that identifies alternative patterns of descent using a graphic matroid approach (Francisco et al., 2009).

The traditional UPGMA method relies on a matrix of distances to estimate isolate relatedness. Distances are calculated for each pair of STs based of the number of allele differences and groups are then sequentially clustered in order of similarity (i.e., allelic matches).

Allele-based methods have the advantage of simplicity and speed, which are crucial for efficient epidemiological surveillance and public health management, but disregard much of the evolutionary information contained at the nucleotide level. A larger and more sophisticated plethora of nucleotide-based methods exist to estimate isolate relationships and key population parameters.

17.5.2 Nucleotide-Based Methods

Any analysis of nucleotide data usually begins with an alignment (i.e., estimation of the homologous nucleotide sites). Since the loci used for MLST usually evolve very slowly and code for proteins, this step becomes trivial, particularly at the amino acid level. If needed, several fast and accurate iterative aligning strategies are implemented in MAFFT (Katoh et al., 2005).

Once an alignment has been generated, we have to determine the model of evolution that fits the data the best. Model choice is a critical issue and the implemented model (or lack thereof) will affect all subsequent phylogenetic and population analyses (next two sections below). This issue is usually assessed within a phylogenetic framework (see Posada and Buckley, 2004). Over the past two decades, substitution models have increased in complexity, as parameters reflecting new information on nucleotide substitution processes are added to candidate models. Furthermore, model selection is moving towards using confidence sets of models (model averaging) (Posada and Buckley, 2004). Several criteria have been proposed for choosing models, although the Akaike Information Criterion is one of the best for evaluating model fit (Posada and Buckley, 2004). This and many other model choice strategies are implemented in JModeltest (Posada, 2009).

Phylogenetic Relatedness

Phylogenetic reconstruction methods can be divided into two types, those that proceed algorithmically (e.g., UPGMA, Neighbor-joining) and those based on optimality criteria. Here we will focus on those that implement maximum likelihood and Bayesian optimality criteria and allow for the implementation of multiple data partitions each under its best-fit model. We find this feature particularly important for analyzing MLST data.

Maximum likelihood (ML) inference attempts to identify the topology that explains the evolution of a set of aligned sequences under a given model of evolution with the greatest likelihood (Felsenstein, 1981). RAxML (Stamatakis, 2006) implements the ML criterion efficiently and accurately and can handle large datasets of >1000 sequences. Confidence in the estimated relationships (i.e., clade support) is usually assessed using a nonparametric bootstrap procedure (Felsenstein, 1985), which must be repeated >1000 times to achieve reasonable precision. Bootstrap proportions can be also rapidly estimated in RAxML.

Although similar to ML inference, Bayesian inference (BI) combines the prior probability of a phylogeny with the likelihood to produce a posterior probability distribution of trees, which can be interpreted as the probability that the tree(s) is (are) correct (Huelsenbeck et al., 2001). Clade support is estimated by summarizing this distribution of trees through a consensus analysis clade. Bayesian phylogenies are estimated using Metropolis-coupled Markov chain Monte Carlo (MCMC) methods and both are implemented in programs like MrBayes (Ronquist and Huelsenbeck, 2003). The output of the BI analysis must be evaluated to assure the MCMC chains have mixed well and converged; such tasks can be performed in Tracer (Rambaut and Drummond, 2009).

Often gene trees differ even when sampled from the same population. This can be the result of molecular processes (e.g., recombination) or stochastic variation (e.g., lineage sorting). Whatever the case, one may want to check if individual gene topologies are significantly different. Multiple ML topological tests have been developed for such purposes and several are implemented in CONSEL (Shimodaira and Hasegawa, 2001).

New coalescent approaches have been developed to deal with stochastic variation in gene trees. Among such, BEST (Liu, 2008) implements a Bayesian hierarchical model in MrBayes that estimates the joint posterior distribution of gene trees and the organism tree using multilocus molecular data.

When estimating evolutionary relationships among microbes using DNA sequences, the reticulating impact of recombination becomes a significant issue. If recombination is substantial, the evolutionary history of those sequences no longer fits a bifurcating model as those described before, and therefore a tree representation may fail to accurately portray a reasonable genealogy. Under such circumstances, network approaches (Posada and Crandall, 2001) can be used instead. Recently Woolley et al. (2008) have revised the most common algorithms for building networks and concluded that the union of maximum parsimonious (UMP) trees (Cassens et al., 2005) performed the best. TCS (Templeton et al., 1992) also performed well at estimating network gene genealogies.

Population Dynamics

The evolution of DNA sequences in natural populations can be described with parameters like recombination, mutation, growth, and selection rates. Indeed, the accurate estimation of these parameters is key for understanding the dynamics and evolutionary history of those populations, their epidemiology, the potential for and mode of evolution of antibiotic resistance, and ultimately for applying efficient public health control strategies. Population parameters are more efficiently estimated using explicit statistical models of evolution such as the coalescent approach, hence here we describe some population parameter estimators based on such models.

Recombination is generally defined as the exchange of genetic information between two nucleotide sequences. It influences biological evolution at many different levels as well as affects the estimation of other parameters. A comprehensive assessment of statistical methods for detecting and estimating recombination rates is presented in Posada et al. (2002). These studies indicate that one should not rely on a single method to detect or estimate recombination. With this idea in mind, software packages such as RDP (Martin et al., 2005) have been developed to implement a variety of methods for the same dataset. RDP is a package that includes eight recombination estimators and allows the user to draw conclusions based on the outcome of multiple tests.

Genetic diversity can be interpreted as two times the neutral mutation rate times the number of heritable gene copies in the population. Since this is a composite parameter, you need outside information for one parameter so the other can be estimated. However, even lacking such information, one can estimate the rate of recombination to mutation, a key parameter to understand how genetic diversity is generated in microbes (Feil et al., 1999). A review of classical and recent statistical methods for estimating genetic diversity is presented in Pearse and Crandall (2004).

Another key parameter for characterizing microbial population dynamics is the growth rate, which reflects the variation of genetic diversity along time. Growth rates can be estimated under a certain demographic model (e.g., exponential) or without dependence on a prespecified model (e.g., Bayesian skyline plot; Drummond et al., 2005). The latter approach is implemented in BEAST (Drummond and Rambaut, 2007). Interestingly, BEAST also allows for the analysis of temporally spaced sequence data. Recombination, genetic diversity, and exponential growth rates can all be estimated in LAMARC (Kuhner, 2006).

The standard method for estimating selection in protein-coding DNA sequences is through the nonsynonymous (d_N) to synonymous (d_S) amino acid substitution ratio d_N/d_S (ω). $\omega > 1$ indicates adaptive or diversifying selection, $\omega < 1$ purifying selection and $\omega \approx 0$ lack of selection. ω is usually estimated within a ML phylogenetic framework and assuming an explicit model of codon substitution. Such models can be very complex, allowing, for example, ω to vary across amino acid sites and/or tree branches (e.g., Yang, 2007). If significant evidence (usually obtained through likelihood ratio tests) of adaptive selection is obtained, then Bayesian tests can be applied to detect amino acid sites under selection (e.g., Yang et al., 2005). Such methods are implemented and described in more detail in the software package PAML (Yang, 2007).

If recombination is suspected in the data, other methods exist that can estimate recombination and selection rates simultaneously (OmegaMap; Wilson and McVean, 2006), or account for the former while estimating the latter (HYPHY; Kosakovsky Pond et al., 2005).

Most of the nucleotide-based methods described above and others have been compiled and can be executed online at the CBSU (http://cbsuapps.tc.cornell.edu), Cipres (http://www.phylo.org) and Datamonkey (http://www.datamonkey.org) websites.

17.6 Applications of MLST

The popularity of MLST is driven by its ease of use and discriminating power, but also by the broad array of applications for the MLST data. Although principally developed for high resolution typing, the genealogical information inherent in the DNA sequences themselves has made MLST data useful for studies of species boundaries and speciation, population dynamics, evolution of drug resistance, and molecular epidemiology.

17.6.1 Population Dynamics and Molecular Epidemiology

MLST has gained widespread popularity as a typing method and its use has advanced understanding of bacterial evolution and provided insights into the epidemiology of bacterial diseases. One major contribution of MLST to bacterial population genetics has been the elucidation of the relative role of recombination and point mutation in the evolutionary history of different bacterial species. Feil et al. (1999) developed a method to estimate the relative impact of recombination compared with point mutation in the diversification of clonal complexes using MLST data. As a quantitative measure, they estimated a ratio of recombination to mutation (r/m) per allele and r/m per site. Estimates of these parameters for *N. meningitides*, *S. pneumoniae*, and *S. aureus* gave values of 3.6:1, 8.9:1, and 6.5:1 for r/m per allele, respectively, and 100:1, 61:1, and 24:1 for the r/m per site, respectively (Feil et al., 1999, 2000; Day et al., 2001). Another example of a bacterial population structured largely by point mutation is the *Bacillus cereus* group, which includes the etiological agent of anthrax. An analysis of congruence performed on ML trees generated from concatenated housekeeping gene loci and from each of the loci individually showed that the *B. cereus* group is largely clonal, with evidence for some recombination (Priest et al., 2004). Analysis of MLST data for *S. epidermidis*, a cause of nosocomial infection in immunocompromised hosts and patients with indwelling medical devices, generated a per-allele r/m parameter of 2.5:1 and a per site r/m of 10:1 (Miragaia et al., 2007). Genetic diversity within populations of human pathogens may be shaped by the ecology of host-parasite interactions. Certain meningococci are overrepresented among disease-associated isolates, but the genetic basis for persistence of these hyperinvasive lineages is unclear. Jolley and colleagues analyzed MLST data and concluded that selective forces are responsible for structuring meningococcal populations (Jolley et al., 2005).

MLST are also useful tools for inferring population structure and the evolutionary dynamics of drug resistance. For example, MLST have been used to diagnose human-associated population structure in the opportunistic pathogen *Ochrobactrum anthropi*. Romano et al. (2009) developed a MLST scheme for this pathogen and used the evolutionary information inherent in the DNA sequences to identify a human-associated subpopulation from their collection of clinical and environmental isolates. Likewise, MLST have been used to track drug resistance variants through patients. Oteo et al. (2009) collected 162 isolates of *Klebsiella pneumoniae* from five hospitals in Spain and used the MLST data to demonstrate the spreading of *K. pneumoniae* as pathogens and colonizers among newborns and adult patients with multilocus resistance acquired through recombination. Similarly, Lee et al. (2010) used MLST to identify epidemic and virulent ciprofloxacin-resistant *E. coli* clones and their population structure in Korea causing urinary tract infections.

In a number of studies, MLST data have been used to reveal the epidemiological history of infectious diseases. For example, MLST has been successful in identifying clinically important strains of *N. meningitides* (i.e., hyperinvasive lineages) (Yazdankhah et al., 2004). MLST has been applied to a number of clinically important bacterial populations, including hospital-acquired strains of *E. faecalis* and *E. faecium* (Ruiz-Garbajosa et al., 2006; Leavis et al., 2006), and *S. pneumoniae* strains associated with invasive disease (Enright and Spratt, 1998). In some cases MLST has failed to distinguish clinically relevant populations. For example,

S. aureus isolates from persons with nasal carriage, community-acquired pneumonia, and hospital-acquired invasive disease are evenly distributed among clonal complexes (Feil et al., 2003). Similarly, there is a poor correlation between MLST data and tissue tropisms (throat or skin) of *S. pyogenes* isolates (Kalia et al., 2002). For phenotypes that are based on one or a few genes, such as antibiotic resistance, correlations with MLST data have been good. The evolutionary history of methicillin-resistant *S. aureus* (MRSA) has been clarified by MLST data, including the typing of the methicillin resistance genetic element, SCCmec (Robinson and Enright, 2003). Genotyping was used to identify outbreaks of gonorrhea in a recent study by Choudhury and colleagues (Choudhury et al., 2006). They typed consecutive gonococcal strains from London STI clinics over a 9-month period. Clusters of patients with the same strain showed similarities in behavioral and demographic features, suggesting that different strain clusters represent localized transmission chains.

17.6.2 Species Diagnosis and Phylogenetics

MLST data have been used to distinguish similar species, to inform the division of a genus into species, and to ask whether bacterial species exist. The MLST data are especially useful for species diagnoses as they provide both genealogical information as well as information on recombination, which is critical to consider with bacterial species diagnosis (Fraser et al., 2007). Indeed, even when the MLST are not as discriminating as other approaches, the phylogenetic information available through the MLST provides novel insights into species and strain relatedness that impacts public health decisions. In a study of *Clostridium difficile*, for example, Marsh et al. (2010) found the MLST less discriminatory compared to multilocus variable-number tandem-repeat analysis (MLVA) or restriction endonuclease analysis (REA) although concordant, but the combination of MLST with MLVA provided novel insights into the origins and evolutionary relationships bearing clinical and public health importance. Similarly, a phylogenetic analysis of concatenated sequences of seven MLST loci for *B. psuedomallei* and *B. thailandensis*, both soil saprophytes, and *B. mallei*, the cause of glanders, showed that all *B. pseudomallei* strains were tightly clustered and well resolved from all *B. thailandensis* strains (Godoy et al., 2003). However, *B. mallei* clustered with *B. pseudomallei* and, although designated as "species", it can be considered to be a strain (or clone) of *B. pseudomallei*. Other examples where MLST analysis has revealed that bacteria that are associated with human or animal disease and have been given species names are in reality clones with distinctive biology and ecology within a "mother" species include *B. anthracis* (Priest et al., 2004) and *S. typhi* (Kidgell et al., 2002). *N. gonorrhoeae* strains form a tight cluster at the end of a long branch arising from the meningococcal cluster (Hanage et al., 2006), supporting the hypothesis that gonococci arose relatively recently as a strain of human pharyngeal *Neisseria* species that acquired the ability to colonize the genital tract and be transmitted by the sexual route (Vazquez et al., 1993). This brief summary highlights the multiple applications of MLST.

17.7 Conclusions and Prospects

MLST has become the standard approach for characterizing bacteria and some eukaryotes because it is more variable, portable, and reproducible than other typing methods. As currently used, MLST has already achieved high levels of discrimination, but as new, more extensive, sequencing technologies (e.g., pyrosequencing) continue to develop, even higher levels of resolution can be expected. More importantly, sequencing additional loci and larger genomic regions will also provide more accurate and robust estimates of population genetic parameters under more complex and sophisticated statistical models, such as the coalescent. Within this framework, epidemiological data can be also integrated, hence more comprehensive and faster assessments of pathogen dynamics can be achieved. Additionally, full-genome sequencing data can be used to identify common phylogenetically informative loci across species. Such markers will greatly help to understand bacterial systematics and taxonomy and better define species boundaries.

Acknowledgments

This work was supported by a grant from NIH R01 AI50217 (to KAC and RPV). We thank Adam Bracken for his assistance in gathering references for this review.

References

Aanensen, D.M., Spratt, B.G., 2005. The multilocus sequence typing network: mlst.net. Nucleic Acids Res. 33, W728–733.

Binladen, J., Gilbert, M.T., Bollback, J.P., Panitz, F., Bendixen, C., Nielsen, R., et al., 2007. The use of coded PCR primers enables high-throughput sequencing of multiple homolog amplification products by 454 parallel sequencing. PLoS One 2, e197.

Bordoni, R., Bonnal, R., Rizzi, E., Carrera, P., Benedetti, S., Cremonesi, L., et al., 2008. Evaluation of human gene variant detection in amplicon pools by the GS-FLX parallel Pyrosequencer. BMC Genomics 9, 464.

Cassens, I., Mardulyn, P., Milinkovitch, M.C., 2005. Evaluating intraspecific "network" construction methods using simulated sequence data: do existing algorithms outperform the global maximum parsimony approach? Syst. Biol. 54, 363–372.

Chan, M.S., Maiden, M.C., Spratt, B.G., 2001. Database-driven multi locus sequence typing (MLST) of bacterial pathogens. Bioinformatics 17, 1077–1083.

Choudhury, B., Risley, C.L., Ghani, A.C., Bishop, C.J., Ward, H., Fenton, K.A., Ison, C.A., Spratt, B.G., 2006. Identification of individuals with gonorrhoea within sexual networks: a population-based study. Lancet 368, 139–146.

Cookson, B.D., Robinson, D.A., Monk, A.B., Murchan, S., Deplano, A., de Ryck, R., et al., 2007. Evaluation of molecular typing methods in characterizing a European collection of epidemic methicillin-resistant *Staphylococcus aureus* strains: the HARMONY collection. J. Clin. Microbiol. 45, 1830–1837.

Cooper, J.E., Feil, E.J., 2004. Multilocus sequence typing—what is resolved? Trends Microbiol. 12, 373–377.

Day, N.P., Moore, C.E., Enright, M.C., Berendt, A.R., Smith, J.M., Murphy, M.F., et al., 2001. A link between virulence and ecological abundance in natural populations of *Staphylococcus aureus*. Science 292, 114–116.

Drummond, A.J., Rambaut, A., 2007. BEAST: Bayesian evolutionary analysis by sampling trees. BMC Evol. Biol. 7, 214.

Drummond, A.J., Rambaut, A., Shapiro, B., Pybus, O.G., 2005. Bayesian coalescent inference of past population dynamics from molecular sequences. Mol. Biol. Evol. 22, 1185–1192.

Enright, M.C., Spratt, B.G., 1998. A multilocus sequence typing scheme for *Streptococcus pneumoniae*: identification of clones associated with serious invasive disease. Microbiology 144 (Pt 11), 3049–3060.

Enright, M.C., Spratt, B.G., 1999. Multilocus sequence typing. Trends Microbiol. 7, 482–487.

Feil, E.J., Cooper, J.E., Grundmann, H., Robinson, D.A., Enright, M.C., Berendt, T., et al., 2003. How clonal is *Staphylococcus aureus*? J. Bacteriol. 185, 3307–3316.

Feil, E.J., Maiden, M.C.J., Achtman, M., Spratt, B.G., 1999. The relative contributions of recombination and mutation to the divergence of clones of *Neisseria meningitides*. Mol. Biol. Evol. 16, 1496–1502.

Feil, E.J., Li, B.C., Aanensen, D.M., Hanage, W.P., Spratt, B.G., 2004. eBURST: inferring patterns of evolutionary descent among clusters of related bacterial genotypes from multilocus sequence typing data. J. Bacteriol. 186, 1518–1530.

Feil, E.J., Smith, J.M., Enright, M.C., Spratt, B.G., 2000. Estimating recombinational parameters in *Streptococcus pneumoniae* from multilocus sequence typing data. Genetics 154, 1439–1450.

Felsenstein, J., 1981. Evolutionary trees from DNA sequences: a maximum likelihood approach. J. Mol. Evol. 17, 368–376.

Felsenstein, J., 1985. Confidence limits on phylogenies: an approach using the bootstrap. Evolution 39, 783–791.

Foley, S.L., Lynne, A.M., Nayak, R., 2009. Molecular typing methodologies for microbial source tracking and epidemiological investigations of Gram-negative bacterial foodborne pathogens. Infect. Genet. Evol. 9, 430–440.

Francisco, A.P., Bugalho, M., Ramirez, M., Carrico, J.A., 2009. Global optimal eBURST analysis of multilocus typing data using a graphic matroid approach. BMC Bioinformatics 10, 152.

Fraser, C., Hanage, W.P., Spratt, B.G., 2007. Recombination and the nature of bacterial speciation. Science 315, 476–480.

Godoy, D., Randle, G., Simpson, A.J., Aanensen, D.M., Pitt, T.L., Kinoshita, R., et al., 2003. Multilocus sequence typing and evolutionary relationships among the causative agents of melioidosis and glanders, *Burkholderia pseudomallei* and *Burkholderia mallei*. J. Clin. Microbiol. 41, 2068–2079.

Hamady, M., Walker, J.J., Harris, J.K., Gold, N.J., Knight, R., 2008. Error-correcting barcoded primers for pyrosequencing hundreds of samples in multiplex. Nat. Methods 5, 235–237.

Hanage, W.P., Fraser, C., Spratt, B.G., 2006. Sequences, sequence clusters and bacterial species. Philos. Trans. R. Soc. Lond. B Biol. Sci. 361, 1917–1927.

Huelsenbeck, J.P., Ronquist, F., Nielsen, R., Bollback, J.P., 2001. Bayesian inference of phylogeny and its impact on evolutionary biology. Science 294, 2310–2314.

Jefferies, J., Clarke, S.C., Diggle, M.A., Smith, A., Dowson, C., Mitchell, T., 2003. Automated pneumococcal MLST using liquid-handling robotics and a capillary DNA sequencer. Mol. Biotechnol. 24, 303–307.

Jolley, K.A., Feil, E.J., Chan, M.S., Maiden, M.C., 2001. Sequence type analysis and recombinational tests (START). Bioinformatics 17, 1230–1231.

Jolley, K.A., Chan, M.S., Maiden, M.C., 2004. mlstdbNet—distributed multi-locus sequence typing (MLST) databases. BMC Bioinformatics 5, 86.

Jolley, K.A., Maiden, M.C., 2006. AgdbNet—antigen sequence database software for bacterial typing. BMC Bioinformatics 7, 314.

Jolley, K.A., Wilson, D.J., Kriz, P., McVean, G., Maiden, M.C., 2005. The influence of mutation, recombination, population history, and selection on patterns of genetic diversity in *Neisseria meningitidis*. Mol. Biol. Evol. 22, 562–569.

Kalia, A., Spratt, B.G., Enright, M.C., Bessen, D.E., 2002. Influence of recombination and niche separation on the population genetic structure of the pathogen *Streptococcus pyogenes*. Infect. Immun. 70, 1971–1983.

Katoh, K., Kuma, K., Toh, H., Miyata, T., 2005. MAFFT version 5: improvement in accuracy of multiple sequence alignment. Nucleic Acids Res. 33, 511–518.

Kidgell, C., Reichard, U., Wain, J., Linz, B., Torpdahl, M., Dougan, G., et al., 2002. *Salmonella typhi*, the causative agent of typhoid fever, is approximately 50,000 years old. Infect. Genet. Evol. 2, 39–45.

Kosakovsky Pond, S.L., Frost, S.D.W., Muse, S.V., 2005. HyPhy: hypothesis testing using phylogenies. Bioinformatics 21, 676–679.

Kuhner, M.K., 2006. LAMARC 2.0: maximum likelihood and Bayesian estimation of population parameters. Bioinformatics 22, 768–770.

Leavis, H.L., Bonten, M.J., Willems, R.J., 2006. Identification of high-risk enterococcal clonal complexes: global dispersion and antibiotic resistance. Curr. Opin. Microbiol. 9, 454–460.

Lee, M.Y., Choi, H.J., Choi, J.Y., Song, M., Song, Y., Kim, S.W., et al., 2010. Dissemination of ST131 and ST393 community-onset, ciprofloxacin-resistant *Escherichia coli* clones causing urinary tract infections in Korea. J. Infect. 60, 146–153.

Liu, L., 2008. BEST: Bayesian estimation of species trees under the coalescent model. Bioinformatics 24, 2542–2543.

Maiden, M.C., Bygraves, J.A., Feil, E., Morelli, G., Russell, J.E., Urwin, R., et al., 1998. Multilocus sequence typing: a portable approach to the identification of clones within populations of pathogenic microorganisms. Proc. Natl. Acad. Sci. U.S.A. 95, 3140–3145.

Maiden, M.C., 2006. Multilocus sequence typing of bacteria. Annu. Rev. Microbiol. 60, 561–588.

Margulies, M., Egholm, M., Altman, W.E., Attiya, S., Bader, J.S., Bemben, L.A., et al., 2005. Genome sequencing in microfabricated high-density picolitre reactors. Nature 437, 376–380.

Marsh, J.W., O'Leary, M.M., Shutt, K.A., Sambol, S.P., Johnson, S., Gerding, D.N., et al., 2010. Multilocus variable-number tandem-repeat analysis and multilocus sequence typing reveal genetic relationships among *Clostridium* difficile isolates genotyped by restriction endonuclease analysis. J. Clin. Microbiol. 48, 412–418.

Martin, D.P., Williamson, C., Posada, D., 2005. RDP2: recombination detection and analysis from sequence alignments. Bioinformatics 21, 260–262.

Metzgar, D., Baynes, D., Hansen, C.J., McDonough, E.A., Cabrera, D.R., Ellorin, M.M., et al., 2009. Inference of antibiotic resistance and virulence among diverse group A

Streptococcus strains using emm sequencing and multilocus genotyping methods. PLoS One 4, e6897.
Meyer, M., Stenzel, U., Hofreiter, M., 2008. Parallel tagged sequencing on the 454 platform. Nat. Protoc. 3, 267–278.
Miragaia, M., Thomas, J.C., Couto, I., Enright, M.C., de Lencastre, H., 2007. Inferring a population structure for Staphylococcus epidermidis from multilocus sequence typing data. J. Bacteriol. 189, 2540–2552.
Oteo, J., Cuevas, O., Lopez-Rodriguez, I., Banderas-Florido, A., Vindel, A., Perez-Vazquez, M., et al., 2009. Emergence of CTX-M-15-producing *Klebsiella pneumoniae* of multilocus sequence types 1, 11, 14, 17, 20, 35 and 36 as pathogens and colonizers in newborns and adults. J. Antimicrob. Chemother. 64, 524–528.
Pearse, D.E., Crandall, K., 2004. Beyond Fst: analysis of population genetic data for conservation. Conserv. Genet. 5, 585–602.
Pérez-Losada, M., Browne, E.B., Madsen, A., Wirth, T., Viscidi, R.P., Crandall, K.A., 2006. Population genetics of microbial pathogens estimated from Multilocus Sequence Typing (MLST) data. Infect. Genet. Evol. 6, 97–112.
Pérez-Losada, M., Crandall, K.A., Zenilman, J., Viscidi, R.P., 2007a. Temporal trends in gonococcal population genetics in a high prevalence urban community. Infect. Genet. Evol. 7, 271–278.
Pérez-Losada, M., Porter, M.L., Tazi, L., Crandall, K.A., 2007b. New methods for inferring population dynamics from microbial sequences. Infect. Genet. Evol. 7, 24–43.
Posada, D., 2009. Selection of models of DNA evolution with JModelTest. Methods Mol. Biol. 537, 93–112.
Posada, D., Buckley, T.R., 2004. Model selection and model averaging in phylogenetics: advantages of akaike information criterion and Bayesian approaches over likelihood ratio tests. Syst. Biol. 53, 793–808.
Posada, D., Crandall, K.A., 2001. Evaluation of methods for detecting recombination from DNA sequences: computer simulations. Proc. Natl. Acad. Sci. U.S.A. 98, 13757–13762.
Posada, D., Crandall, K.A., Holmes, E.C., 2002. Recombination in evolutionary genomics. Annu. Rev. Genet. 36, 75–97.
Priest, F.G., Barker, M., Baillie, L.W., Holmes, E.C., Maiden, M.C., 2004. Population structure and evolution of the *Bacillus cereus* group. J. Bacteriol. 186, 7959–7970.
Rambaut, A., Drummond, A.J., 2009. Tracer: MCMC trace analysis tool (Edinburgh, Institute of Evolutionary Biology). http://tree.bio.ed.ac.uk/software/tracer/.
Robinson, D.A., Enright, M.C., 2003. Evolutionary models of the emergence of methicillin-resistant *Staphylococcus aureus*. Antimicrob. Agents Chemother. 47, 3926–3934.
Romano, S., Aujoulat, F., Jumas-Bilak, E., Masnou, A., Jeannot, J.L., Falsen, E., et al., 2009. Multilocus sequence typing supports the hypothesis that *Ochrobactrum anthropi* displays a human-associated subpopulation. BMC Microbiol. 9, 267.
Ronquist, F., Huelsenbeck, J.P., 2003. MrBayes 3: Bayesian phylogenetic inference under mixed models. Bioinformatics 19, 1572–1574.
Rozera, G., Abbate, I., Bruselles, A., Vlassi, C., D'Offizi, G., Narciso, P., et al., 2009. Massively parallel pyrosequencing highlights minority variants in the HIV-1 env quasispecies deriving from lymphomonocyte sub-populations. Retrovirology 6, 15.
Ruiz-Garbajosa, P., Bonten, M.J., Robinson, D.A., Top, J., Nallapareddy, S.R., Torres, C., et al., 2006. Multilocus sequence typing scheme for *Enterococcus faecalis* reveals hospital-adapted genetic complexes in a background of high rates of recombination. J. Clin. Microbiol. 44, 2220–2228.

Shimodaira, H., Hasegawa, M., 2001. CONSEL: for assessing the confidence of phylogenetic tree selection. Bioinformatics 17, 1246−1247.
Stamatakis, A., 2006. RAxML-VI-HPC: maximum likelihood-based phylogenetic analyses with thousands of taxa and mixed models. Bioinformatics 22, 2688−2690.
Sullivan, C.B., Diggle, M.A., Clarke, S.C., 2005. Multilocus sequence typing: data analysis in clinical microbiology and public health. Mol. Biotechnol. 29, 245−254.
Tazi, L., Pérez-Losada, M., Gu, W., Yang, Y., Xue, L., Crandall, K.A., Viscidi, R.P., 2010. Population dynamics of *Neisseria gonorrhoeae* in Shanghai, China: a comparative study. BMC Infect. Dis. 10, 13.
Templeton, A.R., Crandall, K.A., Sing, C.F., 1992. A cladistic analysis of phenotypic associations with haplotypes inferred from restriction endonuclease mapping and DNA sequence data. III. Cladogram estimation. Genetics 132, 619−633.
Urwin, R., Maiden, M.C., 2003. Multi-locus sequence typing: a tool for global epidemiology. Trends Microbiol. 11, 479−487.
Vazquez, J.A., de la Fuente, L., Berron, S., O'Rourke, M., Smith, N.H., Zhou, J., et al., 1993. Ecological separation and genetic isolation of *Neisseria gonorrhoeae* and *Neisseria meningitidis*. Curr. Biol. 3, 567−572.
Wilson, D.J., McVean, G., 2006. Estimating diversifying selection and functional constraint in the presence of recombination. Genetics 172, 1411−1425.
Woolley, S.W., Posada, D., Crandall, K.A., 2008. A comparison of phylogenetic network methods using computer simulation. PLoS Comput. Biol. 3, e1913.
Yang, Z., 2007. PAML 4: phylogenetic analysis by maximum likelihood. Mol. Biol. Evol. 24, 1586−1591.
Yang, Z., Wong, W.S., Nielsen, R., 2005. Bayes empirical bayes inference of amino acid sites under positive selection. Mol. Biol. Evol. 22, 1107−1118.
Yazdankhah, S.P., Kriz, P., Tzanakaki, G., Kremastinou, J., Kalmusova, J., Musilek, M., et al., 2004. Distribution of serogroups and genotypes among disease-associated and carried isolates of *Neisseria meningitidis* from the Czech Republic, Greece, and Norway. J. Clin. Microbiol. 42, 5146−5153.

18 Omics, Bioinformatics, and Infectious Disease Research

Konrad H. Paszkiewicz and Mark van der Giezen*

Biosciences, College of Life and Environmental Sciences, University of Exeter, Exeter, UK

18.1 The Need for Bioinformatics

Although bioinformatics is generally perceived to be a modern science, the term had been put forward over thirty years ago by Paulien Hogeweg and Ben Hesper for "the study of informatic processes in biotic systems" (Hogeweg, 1978; Hogeweg and Hesper, 1978). It is necessarily nebulous—bioinformatics spans many disciplines and can have many shades of meaning. Indeed it can be argued that it is the collation and analysis of data from different disciplines that has provided some of the greatest insights. In the field of genomics and transcriptomics, bioinformatics is an incredibly diverse field. Evolution, epidemiology, ecology, and the response of an organism to its environment are all fields that require bioinformatics to accurately process and place into context various sources of data. At the heart of genomics and transcriptomics is the generation and analysis of vast quantities of sequence data. DNA sequencing took off in the late 1980s when Applied Biosystems developed the first automated sequencing machine. The subsequent development of more efficient ways to sequence resulted in the phenomenal growth of the number of sequences deposited in GenBank (Figure 18.1). Obviously, with over 100 million sequences deposited in GenBank, it is not feasible to do any serious manual work with such a large dataset. Data obtained from modern second-generation sequencers is on the order of 1000 times greater than capillary-based sequencers. It is now possible to routinely generate many gigabases of sequence data. Bioinformatics is tasked with making sense of it, mining it, storing it, disseminating it, and ensuring valid biological conclusions can be drawn from it. Many of the recent high-throughput functional genomics technologies rely on a bioinformatics component, though bioinformatics is just one part of the process. For example, identification of proteins by mass spectroscopy, quantitative analysis of expression data, phylogenetics, and so on all make use of bioinformatics tools, methods, and databases. Bioinformatics plays a key role at several steps in genomics,

---

*Email: m.vandergiezen@exeter.ac.uk

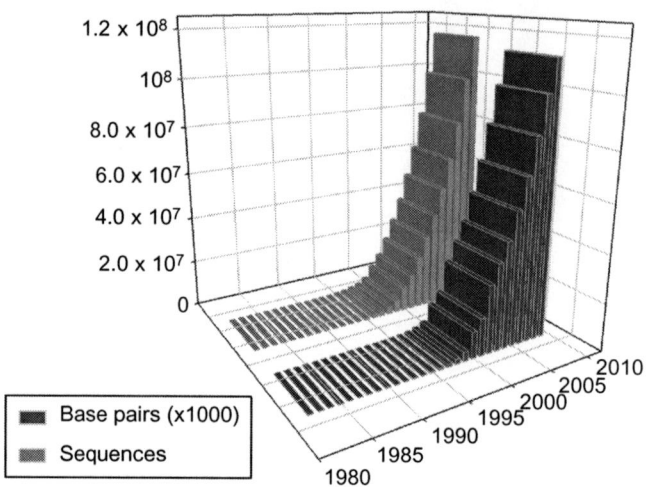

Figure 18.1 The growth of sequences submitted to GenBank. For further info, see http://www.ncbi.nlm.nih.gov/genbank.

comparative genomics, and functional genomics: sequence alignment, assembly, identification of single nucleotide polymorphisms (SNP), gene prediction, quantitative analysis of transcription data, etc. In this chapter, we will discuss the current state of play of bioinformatics related to genomics and transcriptomics and use relevant examples from the field of infectious diseases.

18.2 Metagenomics

The term "metagenomics" was originally used to describe the sequencing of genomes of uncultured microorganisms in order to explore their abilities to produce natural products (Handelsman et al., 1998, Rondon et al., 2000) and subsequently resulted in novel insights into the ecology and evolution of microorganisms on a scale not imagined possible before (see Cardenas and Tiedje, 2008; Hugenholtz and Tyson, 2008 for an overview). However, metagenomics now finds use in infectious disease research as well as the random sequencing of genomes from a variety of organisms from, for example, patient material that could lead to the identification of the cause of disease.

In a quite straightforward metagenomics approach to identify pathogens in sputa from cystic fibrosis patients, standard microbiological culture techniques were compared to molecular methods using 16S rDNA PCR (Bittar et al., 2008). The well-known disadvantage of the microbiological methods is that they normally employ "selective" media that are designed to pick up those bacterial pathogens that are thought to be present. Emerging pathogens will be missed using traditional culture techniques. Indeed, Bittar et al. identified 33 bacteria using cultivation while 53 bacterial species were detected using molecular methods (based on BLAST comparisons; Altschul et al., 1990), interestingly, 30% of the latter were anaerobes, organisms missed in the routine cultivation methods. Many bacteria identified using the molecular methods are traditionally not thought to be associated with cystic fibrosis. Whether these novel species are associated with the physiopathology of disease remains to be studied. Bittar et al. (2008) also noted that the number of bacteria detected increased with increased numbers of clones sequenced, a well-known phenomenon in environmental sequencing that relates to sample depth (Huber et al., 2007; Huse et al., 2010). However, with the increased use of next-generation sequencing methods in infectious disease research, the lessons learned from environmental studies relating to diversity and relative abundance of different microbes can be put to effective use.

An example of the use of second-generation sequencing in a metagenomics approach of patient material is the study by Nakamura et al. (2009) to identify viruses in nasal and fecal material. In this study, RNA was isolated from patient material obtained during seasonal influenza infections and norovirus outbreaks. This RNA was reverse transcribed into cDNA, which was subsequently subjected to large-scale parallel pyrosequencing resulting in 25,000 reads on average per sample. Although the influenza samples were mainly (>90%) human in origin, it was nonetheless possible to identify the influenza subtypes in each sample (Nakamura et al., 2009). As the fecal samples were cleared of human and bacterial cells, yields were much better and the complete norovirus GII.4 subtype genome was sequenced with an average cover depth of up to 258×. In addition to being able to identify the influenza and noroviruses, two recently identified human viruses were also identified: WU polyomavirus and human coronavirus HKU1 (Nakamura et al., 2009). Major bacterial species normally found in the respiratory tract were also identified. Although Nakamura et al. suggest that the high-throughput sequencing is more sensitive than standard PCR-based analysis and might result in the detection of additional possible pathogens, they also warn that the increased sensitivity might necessitate follow-up work to decide which of the detected pathogens is the actual cause of the disease.

Important results are expected from the Human Microbiome Project (http://www.hmpdacc.org/), which will obtain metagenomic information from various human microenvironments such as the gastrointestinal, nasooral, and urogenital cavities as well as the skin. Understanding the human microbiome is thought to answer questions such as whether changes in the human microbiome are related to human health. However, large-scale metagenomics projects that include eukaryotic genomes have thus far been quite costly and laborious due to the generally large genomes of eukaryotes. The lowering of sequencing costs may alleviate part of the problem, but sequence data are still accumulating at a faster rate than developments in computational analysis (Hugenholtz and Tyson, 2008).

18.3 Comparative Genomics

Organisms that have attracted the attention of genome centers are those that cause disease followed by those from model organisms such as *Saccharomyces cerevisiae* (Goffeau et al., 1996) and *Caenorhabditis elegans* (the C. elegans Sequencing

Consortium, 1998), for example. Indeed, the first bacterial genomes sequenced were those from pathogens (Fleischmann et al., 1995; Fraser et al., 1995; Tomb et al., 1997), and these were preceded by many bacteriophage genomes such as bacteriophage MS2 (Fiers et al., 1976) and φX174 (Sanger et al., 1977) and viral genomes (Fiers et al., 1978). Currently, pathogen genomes represent at least one third of all sequenced genomes.

Obviously, for comparative genomics two genomes are required, and indeed, when the second bacterial pathogen was sequenced (*Mycoplasma genitalium* by Fraser et al., 1995), it was immediately compared with the first one (*Haemophilus influenzae* by Fleischmann et al., 1995). Interestingly, the *H. influenzae* genome was completed using a "bioinformatics" approach. Unlike previous sequencing projects, the used shotgun approach relied on a computational justification that sufficient random sequencing of small fragments would result in a complete coverage of the whole genome. Comparing the *M. genitalium* genome with the *Haemophilus* genome suggested that the percentage of the total genome dedicated to genes is similar albeit that *M. genitalium* has far fewer genes (Fraser et al., 1995). Although the genome of *M. genitalium* is about three times smaller than that of *H. influenzae*, its smaller genome has not resulted in an increase in gene density or decrease in gene size. Detection of several repeats of components of the *Mycoplasma* adhesin, which elicits a strong immune response in humans, suggests that recombination might underlie its ability to evade the human immune response. That this initial genome study was only the tip of the comparative genomics iceberg was already clear from Fleischmann et al. (1995) last sentence: "Knowledge of the complete genomes of pathogenic organisms could lead to new vaccines." A whole-genome effort at identifying vaccine candidates appeared some 5 years later when Pizza et al. (2000) employed bioinformatics to extract putative surface-exposed antigens by genome analysis. Although effective vaccines against *Neisseria meningitidis*, the causative agent of meningococcal meningitis and sepsis, did exist, these vaccines did not cover all pathogenic serogroups. Serogroup B had evaded the development of a good vaccine as its capsular polysaccharide (against which the vaccines of the other serogroups were developed) is identical to a human carbohydrate. In order to identify putative candidates for vaccine development, Pizza et al. decided to sequence the whole genome of a serogroup B strain. All potential open reading frames (ORFs) were analyzed for putative cellular locations using BLASTX. Those ORFs likely to be cytosolic were excluded from further analysis. The remaining ORFs were analyzed to determine whether they encoded proteins that contained transmembrane domains, leader peptides, and outer membrane anchoring motives using a variety of databases such as Pfam (Finn et al., 2010) and ProDom (Servant et al., 2002). This resulted in 570 ORFs encoding putative exposed antigens. These 570 putative genes were cloned in *Escherichia coli* and Pizza et al. successfully expressed 350 ORFs. These 350 recombinant proteins were used to generate antisera that were tested in enzyme-linked immunosorbent assay (ELISA) and fluorescence-activated cell sorter (FACS) analyses to test whether they detected proteins on the outer surface of serogroup B meningococcus strains. In addition, the sera were tested for bactericidal activity. Of the 350 proteins, 85 reacted positively in at least one assay but only 7 were positive in all three assays. These 7 were subsequently tested on a large variety of strains to analyze their efficacy. A total of 5 seemed able to provide protection against 31 *N. meningitidis* strains and in addition, those 5 proteins are 95–99% similar to the homologous *N. gonorrhoeae* proteins, suggesting they might provide successful protection against that pathogen as well (Pizza et al., 2000). Arguably the most striking aspect of this study is that in 18 months the authors identified more vaccine candidates than in the preceding 40 years using a novel genomics/bioinformatics approach (Seib et al., 2009). This study resulted in a vaccine that is currently in Phase III clinical trials (Giuliani et al., 2006).

Protozoan infections are a major burden on developing nations; they take 8 of the 13 diseases targeted by the World Health Organization's Special Program for Research and Training in Tropical Diseases (http://www.who.int/tdr). Over the last 5 years or so, more than 10 parasitic genomes have been sequenced in the hope that their sequences would reveal weak spots to target these pathogens. The trypanosomatids cause serious disease in Africa and South America. *Trypanosoma brucei* causes sleeping sickness in humans and wasting disease in cattle. *Trypanosoma cruzi* is the causative agent of Chagas disease and *Leishmania major* leads to skin lesions. The completion of their genomes (Berriman et al., 2005, El-Sayed et al., 2005a, Ivens et al., 2005) and the comparative analysis of all three genomes (El-Sayed et al., 2005b) may be able to focus efforts toward obtaining vaccines, as current drugs have serious toxicity issues. Although their genomes encode a different number of protein-encoding genes (around 8100 in *T. brucei*; 8300 in *L. major*; 12,000 in *T. cruzi*), comparative analysis resulted in the identification of about 6200 genes that entail the trypanosomatid core proteome. All protein coding genes were compared in a three-way manner using BLASTP (El-Sayed et al., 2005b) and the mutual best hits were grouped as clusters of orthologous genes or COGs (Figure 18.2).

Trypanosomatid specific proteins from these 6200 might be used in a broadscale vaccine. The remainder of the protein-encoding genes from each parasite (26% of the genes in *T. brucei*; 12% in *L. major*; 32% in *T. cruzi*) consists of species-specific genes. Interestingly, a large proportion of these genes encode surface antigens and this might relate to the different mechanisms these parasites employ to evade the host immune system. In addition, it was noted that many genes

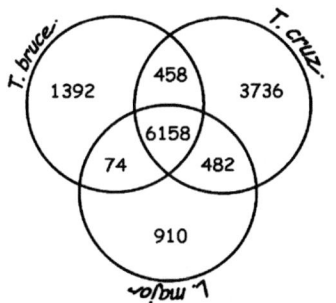

Figure 18.2 Kinetoplastid comparative genomics. A three-way comparison of all protein coding genes from *Trypanosoma cruzi*, *Trypanosoma brucei*, and *Leishmania major* resulted in the discovery of 6200 core proteins that all three kinetoplastids share and various dually shared and unique proteins.
Source: Adapted from El-Sayed et al. (2005b).

encoding surface antigens are found at or near telomeres and that many retroelements seem to be present in these regions as well. This might be related to the enormous antigenic variation observed in both *Trypanosoma* species. The presence of novel genes in these areas might suggest that their products play an unknown role in antigenic variation as well which warrants further studies into these uncharacterized genes (El-Sayed et al., 2005b).

Detailed knowledge of well-studied pathogens might be successfully used to understand the biology of closely related emerging pathogens. This was the driving force for the sequencing of six *Candida* species (Butler et al., 2009). *Candida* species are the most common opportunistic fungal infections in the world and *C. albicans* is the most common of all *Candida* species causing infection. However, *C. albicans* incidence is declining while other species are emerging. Comparison of eight *Candida* species indicated that although genome size was variable, gene content was nearly identical across all species. As the analysis included pathogenic and nonpathogenic species, Butler et al. (2009) specifically studied differences between these two groups. Of the over 9000 gene families analyzed, 21 were significantly enriched in pathogenic species. Many gene families known to be involved in pathogenesis were present in these 21 families (e.g., lipases, oligopeptide transporters, and adhesins). More interestingly, several poorly characterized gene families were also identified, suggesting these might play an unexpected role in pathogenesis as well. This comparative study revealed a wealth of new avenues to explore, which, combined with the large body of work performed on *C. albicans*, will aid understanding the newly emerging pathogenic *Candida* species (Butler et al., 2009).

18.4 Pan-Genomics

Although comparative studies using multiple species can reveal hitherto unknown features as evidenced from the mentioned trypanosomatid and *Candida* studies, they can also reveal something unexpected. Because the definition of a bacterial species has been debated for a long time, Tettelin et al. (2005) set out to address this question by sequencing multiple strains from *Streptococcus agalactiae*, the most common cause of illness or death among newborns. Unexpectedly, despite the presence of a "core-genome" shared between all 8 genomes, mathematical modeling suggested that each additional sequenced genome would add 33 new genes to the "dispensable genome." An additional analysis using *S. pyogenes* also suggested that sequencing additional genomes would continue to add new genes to the pool resulting in a pan-genome that can be defined as the global gene repertoire of a species (Medini et al., 2005). This cannot be extrapolated *ad infinitum*, as a similar analysis of *Bacillus anthracis* indicated that after the fourth genome, no additional genes were identified (Tettelin et al., 2005) in agreement with its known limited genetic diversity (Keim and Smith, 2002). Subsequent analyses have confirmed the presence of pan-genomes for many bacterial species (Hiller et al., 2007; Lefébure and Stanhope, 2007; Rasko et al., 2008; Schoen et al., 2008; Lefébure and Stanhope, 2009) and the ultimate gene repertoire of a bacterial species is much larger than generally perceived. Whether this would be the case for eukaryotes remains to be shown.

Despite the apparently ever-expanding possibilities of the pan-genome, it has also resulted in a universal vaccine candidate for group B *Streptococcus* (GBS). Because various GBS serotypes exist, current vaccines only offer protection against a limited set of serotypes. Eight genomes from six serotypes were compared resulting in the identification of a core-genome of 1811 genes and a dispensable genome of 765 genes, which were not present in each strain (Maione et al., 2005). Both genomes were analyzed for the presence of putative surface-associated and secreted proteins. Of the 598 identified genes, one third were part of the dispensable genome (193 genes). The authors subsequently produced recombinant tagged proteins in *E. coli* that were used to immunize mice. Ultimately, a combination of four antigens turned out to be highly effective against all major GBS serotypes. Three of these antigens were part of the dispensable genome. In addition, this bioinformatics approach highlights the importance of not dismissing unidentified ORFs on genomes (generally up to 50% of sequenced genomes) as all four antigens had no assigned function. Because of their identification using this method, it became obvious they were part of a pilus-like structure that had never seen before in Group B *Streptococcus* (Lauer et al., 2005). The presence of antigens that provide protection on these pilus-like structures suggest that these might play a role in pathogenicity.

18.5 Transcriptomics

Genomic information is useful as a scaffold. However, in a given environment pathogens and hosts only express a subset of their genes at any one time. The presence of pan-genomes only complicates matters even more. To investigate the response of an organism to an environmental or other stress it is necessary to examine the expression pattern of proteins. At present, this is not possible to accomplish directly on a large scale, but a good approximation can be made by sequencing and counting mRNA molecules. At present the process involves converting the RNA to cDNA, which can introduce biases but nonetheless sequencing has a great many advantages over traditional microarrays (Ledford, 2008). These include high specificity with little or no background noise and one also gains nucleotide level resolution of expression. Despite such drawbacks, microarrays are still extremely powerful tools to understand levels of gene expression, and this is obvious from the study by Toledo-Arana et al., who discovered novel regulatory mechanisms in *Listeria* (Toledo-Arana et al., 2009). *L. monocytogenes* is normally harmless but can lead to serious food-borne infections. Environmental change, from the soil through the stomach to the intestinal lumen and ultimately into the bloodstream, is thought to be responsible for the up- and downregulation of a plethora of genes. Comparative genomics of the nonpathogenic *L. innocua* has resulted in the identification of a virulence locus (Glaser et al., 2001). Using microarrays, transcripts of one strain grown at 37°C in rich medium were compared to three different conditions: stationary phase, hypoxia, and low temperature (30°C). In addition, knockout mutants in three known regulators of *Listeria* virulence gene expression (PrfA, SigB, and Hfq) were compared to the control strain as well. RNA was also extracted from the intestine of inoculated mice and from blood from healthy human donors that were both infected with three different strains (control and PrfA and sigB knockouts). This analysis resulted in the discovery of massive transcriptional reshaping under the control of SigB when *Listeria* enters the intestines. However, in the bloodstream, gene expression is under control of PrfA. Various noncoding RNAs were uncovered, which show the same expression patters as virulence genes suggesting a potential role in virulence (Toledo-Arana et al., 2009).

Because microarray data are based on a comparative difference in hybridization, high-throughput next-generation sequencing is seen as more quantitative as it based on number of hits for each sequenced transcript (van Vliet, 2010). However, when making cDNA for next-generation sequencing transcriptomics in prokaryotes, there are several difficulties not found in eukaryotes, such as high levels of rRNA and tRNA molecules as well as a lack of poly-A tails, making extraction difficult. Nontheless, it is possible to overcome these by either reducing the amount of rRNA and tRNA using commercially available kits or by bioinformatic removal of such sequences postsequencing (van Vliet, 2010). To date, some 20 RNA-seq style experiments have been performed on prokaryotes. To give an example of the sort of novel insights that can be gleaned using such technology, Passalacqua et al. (2009) sequenced the *Bacillus anthracis* transcriptome using SOLiD and Illumina sequencing and clearly showed the polycistronic nature of many transcripts on a whole genome scale. Although known for individual operons, this had never been shown on a genome-wide scale. They were also able to test the current genome annotations and discovered that 36 loci that were removed as nongenes showed significant transcriptional activity. In addition, 21 nonannotated regions had clear levels of transcription and should therefore be considered as genes (Passalacqua et al., 2009). As internal methionines could have incidentally been identified as start codons, they also checked whether upstream regions were included in the transcribed region. In 11 cases this proved to be the case suggesting the original start codons were incorrectly annotated. Reassuringly, when comparing their data with microarray data, a strong correlation was observed. Interestingly, because of the very high resolution of sequence-based transcriptomics studies, it is possible to identify novel regulatory elements. For example, when comparing expression levels under O_2- and CO_2-rich conditions, the first gene of an eight-gene operon did not show a marked difference in expression level while all the others were significantly upregulated under CO_2 (Passalacqua et al., 2009). Indeed, a bioinformatics approach had suggested the presence of a T-box riboswitch between genes 1 and 2 of this operon (Griffiths-Jones et al., 2005).

A similar approach to study how *Burkholderia cenocepacia*, an opportunistic cystic fibrosis pathogen, responds to environmental changes revealed several new potential virulence factors (Yoder-Himes et al., 2009). As *B. cenocepacia* is routinely isolated from soil, two strains (one isolated from a cystic fibrosis patient and one from soil) were analyzed in their response to changes from growth at synthetic human sputum medium and soil medium. Although their overall nucleotide identity is 99.8%, 179 and 120 homologous genes showed a significant difference in expression between the two strains when grown in synthetic sputum medium and soil medium, respectively. This suggests that despite the high level of relatedness, differential gene expression plays a large role in adaptation to their ecological niche (Yoder-Himes et al., 2009). Interestingly, similar to Passalacqua et al. (2009), several expressed noncoding RNAs were uncovered with different expression levels depending on environmental condition. The significance of this needs to be investigated but highlights the ability of second-generation sequencing to unearth novel findings.

18.6 Proteomics

Despite the fact that a species' genome could well be larger than the actual genome content of one member of that species due to the pan-genome concept, an organism's proteome is by far much more complex. As discussed earlier, transcriptomics will reveal which subset of the genome is expressed under a given condition. However, posttranslational modifications of proteins make the actual proteome far more complex than the transcriptome. This is also the strength of proteomics, as can be seen in a study of the obligate intracellular parasite *Chlamydia pneumonia*. *C. pneumonia* is the third-most-common cause of respiratory infections in the world, which, in part, is made possible due to the unique bi-phasic life cycle of this bacterial pathogen. *Chlamydia* spread via a metabolically inert infectious particle called the elementary body. These elementary bodies enter the host cell where they differentiate into reticulate bodies. As the elementary body is the infectious phase, proteins presented on the outer membrane would be ideal candidates for vaccine development, especially as effective vaccines are lacking and treatment is via antibiotic therapy. A large-scale genomics-proteomics study by Montigiani et al. (2002) systematically assessed putative exposed antigens for possible use in vaccine development. Of the 1073 *C. pneumonia* genes, 636 have assigned functions, 72 of the latter are predicted to be peripherally located and were therefore selected for follow-up studies. In addition, the remaining 437 ORFs were subjected to a series of search algorithms aimed at identifying putative surface-exposed antigens. In total, 141 ORFs were identified as being possibly located on the cell surface. These 141 were subsequently used to produce recombinant proteins in *E. coli*. Because both His-tagged as well as GST-tagged versions were made, a total of 173 recombinant proteins were produced and used for immunizations of mice. All antisera were used in FACS analysis to test if they could bind to the *C. pneumonia* cell surface. This resulted in the identification of 53 putative surface-exposed antigens. Interestingly, apart from well-known antigens, 14 antigens from unidentified ORFs were part of this group of potential vaccine candidates. All 53 candidates were tested on Western blots whether they generated a clean band of the expected size or whether they cross-reacted with other proteins; 33 of the 53 were specific. Finally, Montigiani et al. conducted a proteomic analysis of total protein from the elementary body phase identifying spots using mass spectrometry. Protein sequencing using MALDI-TOF identified 28 putative surface-exposed antigens on the

C. pneumonia 2D gels (Montigiani et al., 2002). A follow-up study by Thorpe et al. (2007) clearly showed that one of the identified candidates, LcrE, induced, amongst others, $CD4^+$ and $CD8^+$ T cell activation and completely cleared infection in a murine model. Interestingly, LcrE is homologous to a protein that is thought be part of the Type III secretion system of *Yersinia*. The exposed nature of LcrE on the *C. pneumonia* cell surface suggests that a Type III secretion system plays a role in *Chlamydia* infection (Montigiani et al., 2002).

The importance of exposed outer membrane proteins as potential vaccine candidates has prompted Berlanda Scorza et al. to assess the complement of outer membrane proteins from an extraintestinal pathogenic *E. coli* strain (Berlanda Scorza et al., 2008). Extraintestinal pathogenic *E. coli* is the leading cause of severe sepsis and current increases in drug resistance warrant the search for novel vaccine targets. In addition, current whole-cell vaccines suffer from undesired cross-reactions to commensal *E. coli* as well. The novel approach by Berland Scorza et al. is based on the observation that some Gram-negative bacteria release outer membrane vesicles (OMV) in the culture media, albeit in minute quantities. A TolR mutant appeared to release much more OMVs than wild-type cells and subsequent large-scale mass spectroscopic analysis of its protein content resulted in the identification of 100 proteins. The majority of these were outer membrane and periplasmic proteins. Intriguingly, three subunits from the cytolethal distending toxin (CDT) were included. This toxin is unusual in that one of its subunits is targeted to the eukaryotic host cell, where it breaks double-stranded DNA resulting in cell death (De Rycke and Oswald, 2001). To check whether the presence of CDT in the OMV was due to the TolR knockout, wild-type extraintestinal pathogenic *E. coli* was tested using Western blotting. Indeed, CDT was detected in wild-type OMV as well (Berlanda Scorza et al., 2008). This suggests that toxin delivery via vesicles might well be the key event in pathogenesis. Interestingly, 18 of the 100 identified proteins were not predicted to be targeted to the periplasm or outer membrane by PSORTb (Gardy et al., 2005). We see here excellent opportunities to train protein targeting algorithms with new wetbench data as these algorithms generally have been trained on a limited set of model organisms that do not reflect the diversity encountered in real life.

18.7 Structural Genomics/Proteomics

Despite the enormous progress in genomics of infectious diseases, the discovery of new drugs has not kept equal pace. For example, no candidate drugs have been identified after 70 high-throughput screens using validated bacterial drug targets (Payne et al., 2007). Although broad-spectrum drugs might be more desirable, there has been a recent trend in targeting specific proteins from specific pathogens using structural biology. Several structural genomics initiatives have been set up to target specific groups of pathogens. For example, the Seattle Structural Genomics Center for Infectious Diseases (http://ssgcid.org) and the Center for Structural Genomics of Infectious Diseases (http://www.csgid.org) work on category A to C agents listed by the National Institute for Allergy and Infectious Diseases (NIAID). Other centers focus on specific organisms such as *Mycobacterium tuberculosis*. Examples are the Mycobacterium Tuberculosis Structural Proteomics Project (http://xmtb.org) and the Mycobacterium Tuberculosis Structural Proteomics Consortium (http://www.doe-mbi.ucla.edu/TB). The field of structural genomics aims to solve as many protein structures as possible from human pathogens with the aim to come up with new drug targets or vaccines (Van Voorhis et al., 2009). Obviously, correct selection of candidates for structural genomics projects is paramount and various criteria have been put forward (Anderson, 2009; Van Voorhis et al., 2009). If a protein is already a validated drug target obviously aids in selection. The proteins need to be essential for the pathogen and ideally, absent in humans. Proteins involved in the uptake of essential nutrients are another target. Classically, drug design has been focusing on substrate binding sites. More recently, small molecules interfering with subunit binding have started to attract attention. As eukaryotic and prokaryotic inorganic pyrophosphatases differ in composition (the former are homodimers, while the latter are homohexamers), efforts are aimed at compounds that interfere with the oligomeric state of the enzyme. In contrast, the highly conserved active site of inorganic pyrophosphatase would not have been a good target (Van Voorhis et al., 2009). The 2003 SARS outbreak that caught the infectious diseases community (if not the whole world) by surprise is one example where structural genomics has made enormous progress. Despite knowing that coronaviruses caused serious diseases in animals, the fact that they only caused mild disease in humans meant that there was very little knowledge about coronavirus biology. The subsequent effort to understand viral assembly and replication/transcription, for example, has resulted in the elucidation of 12 SARS-CoV solved protein structures. Interestingly, the novel fold-discovery rate was nearly 50%, while it would normally be more close to 6% (Bartlam et al., 2007). In addition, one key protein, the SARS-CoV main protease, has since been at the center of structure-based drug discovery. Because of the nature of the discipline, structural genomics is dependent on various other disciplines such as biochemistry, microbiology, structural biology, computational biology, and bioinformatics and can only foster in a truly interdisciplinary environment (Anderson, 2009).

18.8 A "How-To" of Second-Generation Sequencing

It is now possible to sequence the entire genome of a bacterial pathogen, assemble the raw sequence reads, perform automated annotation, and visualize the results within 3 weeks. At the same time (indeed even on the same sequencer) it is also possible to selectively sequence the transcriptome (RNA-seq) regions of DNA bound to protein (ChIP-Seq) or for relevant species methylated DNA to study epigenetic effects as well as small RNA molecules. It is also possible to perform the very same sequencing on the host organism at the same time.

Bioinformatic algorithms and tools are a crucial tool in analyzing such unprecedented volumes of data. These data volumes have emerged as a result of second-generation sequencers such as the Roche/454, Illumina, and ABI/SoLID systems.

Although useful information can be extracted by single researchers by targeted analysis of the sequencer output, to gain the most information out of such data, it is becoming increasingly common for multiple researchers or research groups with widely differing areas of expertise to collaborate. This collaboration is absolutely crucial if relevant insights are to be gained from large-scale datasets. As a result a vast array of data is generated, which is required to be annotated and curated as well as analyzed for information relevant to any particular experiment. In addition this information needs to be stored, shared, and distributed in a manner that enables reanalysis if and when new hypotheses are generated.

Platforms as produced by the GMOD consortium (http://gmod.org), such as Gbrowse, and underlying databases are excellent web-based tools for visualizing and comparing datasets. However, they currently offer limited scope for collaborative annotation or curation of datasets where relevant expertise can be brought to bear from a variety of different research groups. This problem is magnified with the advent of second-generation sequencers since much smaller groups of researchers tend to be involved, meaning that the expertise that large collaborations can muster (such as the Influenza Research Database [FluDB], http://www.fludb.org/) is much smaller. Thus there is a need for integrated annotation and visualization pipelines to enable individual researchers to perform comparative genomics and transcriptomics.

The Broad Institute offers a number of useful visualization tools to the individual researcher such as ARGO (http://www.broadinstitute.org/annotation/argo/) and the Integrated Genome Viewer (IGV) (http://www.broadinstitute.org/igv/). ARGO offers the ability to manually annotate and visualize a genome as well as provide a good graphical overview for comparative genomics and transcriptomics.

Currently, there is no one standard for bioinformatics pipeline development for next-generation sequencing. Several efforts are underway or can be adapted from Sanger sequencing pipelines. These include the prokaryote annotation pipeline XBase and the ISGA server (Hemmerich et al., 2010). These enable de novo sequenced prokaryote genomes to be annotated automatically and corrected manually at a later date. Alternative Sanger adaptations such as Maker can also be used once an assembly has been generated.

18.9 Alignment or Assembly of Second-generation Sequences

A large array of programs is now available to either align reads to a reference genome or to assemble them de novo (Miller et al., 2010; Paszkiewicz and Studholme, 2010). They will not be listed in detail here as there are many considerations, including sequencing platform used, the read length in use, the expected genome size, length of longest repetitive elements, GC content, and whether paired-end reads are in use.

The proprietary Newbler software from Roche is the most popular method of de novo assembly of 454 reads (typically 400–500bp). Popular assemblers for short reads (i.e., mostly from Illumina or SoLID platforms) are Velvet (http://www.ebi.ac.uk/~zerbino/velvet) for the assembly of genomic DNA or Oases from the same group dealing with assembly of reads from transcriptomic cDNA (http://www.ebi.ac.uk/~zerbino/oases) (Zerbino and Birney, 2008). Other assemblers such as AbYSS (Simpson et al., 2009), ALLPATHS (Butler et al., 2008) or SOAPdenovo (http://soap.genomics.org.cn/soapdenovo.html) are also popular. AbYSS enables assembly to be parallelized, thus speeding up assembly. ALLPATHS has been shown to offer superior performance when multiple paired-end libraries are used. Independent of read length, it is crucial that paired-end libraries are used when constructing de novo assemblies of any genome. Note that the use of short-read sequences only can lead to significant gaps being left in the final assembly due to repetitive elements. However, for many analyses (especially for prokaryotic organisms) these gaps are generally not considered to be significant. In cases where closure of these gaps is more desirable than the addition of 454, Sanger or long-range PCR data can often help.

Where significant quantities of long- and short-read data are available, then a joint assembly can be attempted. A recommended protocol is to assemble the short and long reads separately using their respective packages and to then merge the two assemblers using programs such as Minimus (Sommer et al., 2007). Another option is to use a template sequence from a related organism to help guide the assembly (note—this is distinct from remapping as described). The amosCMP package is useful for this purpose (Pop et al., 2004). Finally, whatever assembly method is used, it is important to remember that a longer assembly is not necessarily a better one. Examining the reads making up a contig (e.g., using the AMOS package (http://amos.sourceforge.net) or the Tablet viewer (http://bioinf.scri.ac.uk/tablet) and alignment to a core-conserved group of genes should be standard practice to ensure that blatant errors are corrected. Remapping of short reads to a reference genome is also a valid method of comparison. Although software such as BLAT (Kent, 2002) can be used with longer 454 reads, it is not an ideal tool for shorter read technologies where data volumes are much greater. Where such a genome is available, software such as MAQ, its successor, BWA, Bowtie, SOAP, and others offer a wealth of tools to identify indels, SNPs, and other variants which may be of interest. Crucially in these cases it is important to have sufficient depth of coverage to ensure SNP calls are valid. Paired-end data is also valuable to have to highlight the presence of indels. After remapping it is also common practice to assemble unmapped reads using the de novo assembly software to reveal any novel sequence variants, which may be absent in the reference. In the case where pathogens and hosts are sequenced together, if the sequence of at least one is known, then it is relatively straightforward to separate the two using bioinformatic techniques. To deal with transcriptomic data where a reference sequence is available, softwares, such as ERANGE (http://woldlab.caltech.edu/rnaseq/), Tophat (Trapnell et al., 2009), and Cufflinks (http://cufflinks.cbcb.umd.edu/), are extremely useful. The Cufflinks module in particular offers the ability to predict the most likely exon isoform expression pattern using a combination of Bayesian statistics and graph-based algorithms.

18.10 Concluding Remarks

We are aware that our treatment of the use of "omics" and bioinformatics in infectious disease research is not exhaustive. As mentioned in the introduction, what constitutes bioinformatics is not entirely clear and arguably varies depending on who tries to define it. However, we have attempted to show the considerable progress in infectious diseases research that has been made in recent years using various "omics" case studies. In addition, the last section is an attempt to provide a brief overview of the problems and (bioinformatics) solutions that current-day scientists face who embark on second-generation sequencing strategies. This is a fast-moving field, but the provided references and websites should be a good first approach for those who wish to make further strides toward eradicating infectious diseases from our planet.

Acknowledgments

We would like to acknowledge our colleague Dr. David J. Studholme for his suggestions and feedback.

References

Altschul, S.F., Gish, W., Miller, W., Myers, E.W., Lipman, D.J., 1990. Basic local alignment search tool. J. Mol. Biol. 215, 403–410.

Anderson, W.F., 2009. Structural genomics and drug discovery for infectious diseases. Infect. Disord. Drug. Targets 9, 507–517.

Bartlam, M., Xu, Y., Rao, Z., 2007. Structural proteomics of the SARS coronavirus: a model response to emerging infectious diseases. J. Struct. Funct. Genomics 8, 85–97.

Berlanda Scorza, F., Doro, F., Rodríguez-Ortega, M.J., Stella, M., Liberatori, S., Taddei, A.R., et al., 2008. Proteomics characterization of outer membrane vesicles from the extraintestinal pathogenic *Escherichia coli* ΔtolR IHE3034 mutant. Mol. Cell. Proteomics 7, 473–485.

Berriman, M., Ghedin, E., Hertz-Fowler, C., Blandin, G., Renauld, H., Bartholomeu, D.C., et al., 2005. The genome of the African trypanosome *Trypanosoma brucei*. Science 309, 416–422.

Bittar, F., Richet, H., Dubus, J.-C., Reynaud-Gaubert, M., Stremler, N., Sarles, J., et al., 2008. Molecular detection of multiple emerging pathogens in sputa from cystic fibrosis patients. PLoS ONE 3, e2908.

Butler, J., MacCallum, I., Kleber, M., Shlyakhter, I.A., Belmonte, M.K., Lander, E.S., et al., 2008. ALLPATHS: de novo assembly of whole-genome shotgun microreads. Genome Res. 18, 810–820.

Butler, G., Rasmussen, M.D., Lin, M.F., Santos, M.A.S., Sakthikumar, S., Munro, C.A., et al., 2009. Evolution of pathogenicity and sexual reproduction in eight *Candida* genomes. Nature 459, 657–662.

Cardenas, E., Tiedje, J.M., 2008. New tools for discovering and characterizing microbial diversity. Curr. Opin. Biotechnol. 19, 544–549.

De Rycke, J., Oswald, E., 2001. Cytolethal distending toxin (CDT): a bacterial weapon to control host cell proliferation? FEMS Microbiol. Lett. 203, 141–148.

El-Sayed, N.M., Myler, P.J., Bartholomeu, D.C., Nilsson, D., Aggarwal, G., Tran, A.-N., et al., 2005a. The genome sequence of *Trypanosoma cruzi*, etiologic agent of Chagas disease. Science 309, 409–415.

El-Sayed, N.M., Myler, P.J., Blandin, G., Berriman, M., Crabtree, J., Aggarwal, G., et al., 2005b. Comparative genomics of trypanosomatid parasitic protozoa. Science 309, 404–409.

Fiers, W., Contreras, R., Duerinck, F., Haegeman, G., Iserentant, D., Merregaert, J., et al., 1976. Complete nucleotide sequence of bacteriophage MS2 RNA: primary and secondary structure of the replicase gene. Nature 260, 500–507.

Fiers, W., Contreras, R., Haegeman, G., Rogiers, R., Van de Voorde, A., Van Heuverswyn, H., et al., 1978. Complete nucleotide sequence of SV40 DNA. Nature 273, 113–120.

Finn, R.D., Mistry, J., Tate, J., Coggill, P., Heger, A., Pollington, J.E., et al., 2010. The Pfam protein families database. Nucl. Acids Res. 38, D211–222.

Fleischmann, R.D., Adams, M.D., White, O., Clayton, R.A., Kirkness, E.F., Kerlavage, A.R., et al., 1995. Whole-genome random sequencing and assembly of *Haemophilus influenzae* Rd. Science 269, 496–512.

Fraser, C.M., Gocayne, J.D., White, O., Adams, M.D., Clayton, R.A., Fleischmann, R.D., et al., 1995. The minimal gene complement of *Mycoplasma genitalium*. Science 270, 397–403.

Gardy, J.L., Laird, M.R., Chen, F., Rey, S., Walsh, C.J., Ester, M., et al., 2005. PSORTb v.2.0: expanded prediction of bacterial protein subcellular localization and insights gained from comparative proteome analysis. Bioinformatics 21, 617–623.

Giuliani, M.M., Adu-Bobie, J., Comanducci, M., Aricò, B., Savino, S., Santini, L., et al., 2006. A universal vaccine for serogroup B meningococcus. Proc. Natl. Acad. Sci. 103, 10834–10839.

Glaser, P., Frangeul, L., Buchrieser, C., Rusniok, C., Amend, A., Baquero, F., et al., 2001. Comparative genomics of *Listeria* species. Science 294, 849–852.

Goffeau, A., Barrell, B.G., Bussey, H., Davis, R.W., Dujon, B., Feldmann, H., et al., 1996. Life with 6000 genes. Science 274, 546–567.

Griffiths-Jones, S., Moxon, S., Marshall, M., Khanna, A., Eddy, S.R., Bateman, A., 2005. Rfam: annotating non-coding RNAs in complete genomes. Nucl. Acids Res. 33, D121–124.

Handelsman, J., Rondon, M.R., Brady, S.F., Clardy, J., Goodman, R.M., 1998. Molecular biological access to the chemistry of unknown soil microbes: a new frontier for natural products. Chem. Biol. 5, R245–R249.

Hemmerich, C., Buechlein, A., Podicheti, R., Revanna, K.V., Dong, Q., 2010. An Ergatis-based prokaryotic genome annotation web server. Bioinformatics 26, 1122–1124.

Hiller, N.L., Janto, B., Hogg, J.S., Boissy, R., Yu, S., Powell, E., et al., 2007. Comparative genomic analyses of seventeen *Streptococcus pneumoniae* strains: insights into the pneumococcal supragenome. J. Bacteriol. 189, 8186–8195.

Hogeweg, P., 1978. Simulating the growth of cellular forms. Simulation 31, 90–96.

Hogeweg, P., Hesper, B., 1978. Interactive instruction on population interactions. Comput. Biol. Med. 8, 319–327.

Huber, J.A., Mark Welch, D.B., Morrison, H.G., Huse, S.M., Neal, P.R., Butterfield, D.A., et al., 2007. Microbial population structures in the deep marine biosphere. Science 318, 97–100.

Hugenholtz, P., Tyson, G.W., 2008. Microbiology: metagenomics. Nature 455, 481–483.

Huse, S.M., Welch, D.M., Morrison, H.G., Sogin, M.L., 2010. Ironing out the wrinkles in the rare biosphere through improved OTU clustering. Environ. Microbiol. 12, 1889−1898.
Ivens, A.C., Peacock, C.S., Worthey, E.A., Murphy, L., Aggarwal, G., Berriman, M., et al., 2005. The genome of the kinetoplastid parasite, *Leishmania major*. Science 309, 436−442.
Keim, P., Smith, K.L., 2002. *Bacillus anthracis* evolution and epidemiology. Curr. Top. Microbiol. Immunol. 271, 21−32.
Kent, W.J., 2002. BLAT—the BLAST-like alignment tool. Genome Res. 12, 656−664.
Lauer, P., Rinaudo, C.D., Soriani, M., Margarit, I., Maione, D., Rosini, R., et al., 2005. Genome analysis reveals pili in group B *Streptococcus*. Science 309, 105.
Ledford, H., 2008. The death of microarrays? Nature 455, 847.
Lefébure, T., Stanhope, M., 2007. Evolution of the core and pan-genome of *Streptococcus*: positive selection, recombination, and genome composition. Genome Biol. 8, R71.
Lefébure, T., Stanhope, M.J., 2009. Pervasive, genome-wide positive selection leading to functional divergence in the bacterial genus *Campylobacter*. Genome Res. 19, 1224−1232.
Maione, D., Margarit, I., Rinaudo, C.D., Masignani, V., Mora, M., Scarselli, M., et al., 2005. Identification of a universal group B *Streptococcus* vaccine by multiple genome screen. Science 309, 148−150.
Medini, D., Donati, C., Tettelin, H., Masignani, V., Rappuoli, R., 2005. The microbial pan-genome. Curr. Opin. Genet. Dev. 15, 589−594.
Miller, J.R., Koren, S., Sutton, G., 2010. Assembly algorithms for next-generation sequencing data. Genomics. 95, 315−327.
Montigiani, S., Falugi, F., Scarselli, M., Finco, O., Petracca, R., Galli, G., et al., 2002. Genomic approach for analysis of surface proteins in *Chlamydia pneumoniae*. Infect. Immun. 70, 368−379.
Nakamura, S., Yang, C.-S., Sakon, N., Ueda, M., Tougan, T., Yamashita, A., et al., 2009. Direct metagenomic detection of viral pathogens in nasal and fecal specimens using an unbiased high-throughput sequencing approach. PLoS ONE 4, e4219.
Passalacqua, K.D., Varadarajan, A., Ondov, B.D., Okou, D.T., Zwick, M.E., Bergman, N.H., 2009. Structure and complexity of a bacterial transcriptome. J. Bacteriol. 191, 3203−3211.
Paszkiewicz, K.H., Studholme, D.J., 2010. De novo assembly of short sequence reads. Brief. Bioinformatics. 11, 457−472.
Payne, D.J., Gwynn, M.N., Holmes, D.J., Pompliano, D.L., 2007. Drugs for bad bugs: confronting the challenges of antibacterial discovery. Nat. Rev. Drug Discov. 6, 29−40.
Pizza, M., Scarlato, V., Masignani, V., Giuliani, M.M., Arico, B., Comanducci, M., et al., 2000. Identification of vaccine candidates against serogroup B Meningococcus by whole-genome sequencing. Science 287, 1816−1820.
Pop, M., Phillippy, A., Delcher, A.L., Salzberg, S.L., 2004. Comparative genome assembly. Brief. Bioinform. 5, 237−248.
Rasko, D.A., Rosovitz, M.J., Myers, G.S.A., Mongodin, E.F., Fricke, W.F., Gajer, P., et al., 2008. The pangenome structure of *Escherichia coli*: comparative genomic analysis of *E. coli* commensal and pathogenic isolates. J. Bacteriol. 190, 6881−6893.
Rondon, M.R., August, P.R., Bettermann, A.D., Brady, S.F., Grossman, T.H., Liles, M.R., et al., 2000. Cloning the soil metagenome: a strategy for accessing the genetic and functional diversity of uncultured microorganisms. Appl. Environ. Microbiol. 66, 2541−2547.

Sanger, F., Air, G.M., Barrell, B.G., Brown, N.L., Coulson, A.R., Fiddes, J.C., et al., 1977. Nucleotide sequence of bacteriophage phiX174 DNA. Nature 265, 687–695.
Schoen, C., Blom, J., Claus, H., Schramm-Glück, A., Brandt, P., Müller, T., et al., 2008. Whole-genome comparison of disease and carriage strains provides insights into virulence evolution in *Neisseria meningitidis*. Proc. Natl. Acad. Sci. **105**, 3473–3478.
Seib, K.L., Dougan, G., Rappuoli, R., 2009. The key role of genomics in modern vaccine and drug design for emerging infectious diseases. PLoS Genet. 5, e1000612.
Servant, F., Bru, C., Carrère, S., Courcelle, E., Gouzy, J., Peyruc, D., et al., 2002. ProDom: automated clustering of homologous domains. Brief. Bioinform. 3, 246–251.
Simpson, J.T., Wong, K., Jackman, S.D., Schein, J.E., Jones, S.J.M., Birol, Ä. n., 2009. ABySS: a parallel assembler for short read sequence data. Genome Res. 19, 1117–1123.
Sommer, D., Delcher, A., Salzberg, S., Pop, M., 2007. Minimus: a fast, lightweight genome assembler. BMC Bioinformatics 8, 64.
Tettelin, H., Masignani, V., Cieslewicz, M.J., Donati, C., Medini, D., Ward, N.L., et al., 2005. Genome analysis of multiple pathogenic isolates of *Streptococcus agalactiae*: implications for the microbial "pan-genome". Proc. Natl. Acad. Sci. U.S.A. 102, 13950–13955.
The C. elegans Sequencing Consortium, 1998. Genome sequence of the nematode *C. elegans*: a platform for investigating biology. Science 282, 2012–2018.
Thorpe, C., Edwards, L., Snelgrove, R., Finco, O., Rae, A., Grandi, G., et al., 2007. Discovery of a vaccine antigen that protects mice from *Chlamydia pneumoniae* infection. Vaccine 25, 2252–2260.
Toledo-Arana, A., Dussurget, O., Nikitas, G., Sesto, N., Guet-Revillet, H., Balestrino, D., et al., 2009. The *Listeria* transcriptional landscape from saprophytism to virulence. Nature 459, 950–956.
Tomb, J.-F., White, O., Kerlavage, A.R., Clayton, R.A., Sutton, G.G., Fleischmann, R.D., et al., 1997. The complete genome sequence of the gastric pathogen *Helicobacter pylori*. Nature 388, 539–547.
Trapnell, C., Pachter, L., Salzberg, S.L., 2009. TopHat: discovering splice junctions with RNA-Seq. Bioinformatics 25, 1105–1111.
van Vliet, A.H.M., 2010. Next generation sequencing of microbial transcriptomes: challenges and opportunities. FEMS Microbiol. Lett. 302, 1–7.
Van Voorhis, W.C., Hol, W.G.J., Myler, P.J., Stewart, L.J., 2009. The role of medical structural genomics in discovering new drugs for infectious diseases. PLoS Comput. Biol. 5, e1000530.
Yoder-Himes, D.R., Chain, P.S.G., Zhu, Y., Wurtzel, O., Rubin, E.M., Tiedje, J.M., et al., 2009. Mapping the *Burkholderia cenocepacia* niche response via high-throughput sequencing. Proc. Natl. Acad. Sci. 106, 3976–3981.
Zerbino, D.R., Birney, E., 2008. Velvet: algorithms for de novo short read assembly using de Bruijn graphs. Genome Res. 18, 821–829.

19 Genomics of Infectious Diseases and Private Industry

*Guy Vernet**

Fondation Mérieux, Lyon, France

19.1 Introduction

Genomics is the science of studying genomes of living organisms. The main driver of companies that specialize in genomics is human genome sequencing and most of the technology breakthroughs have been achieved to improve the speed and quality of human genome sequencing and to reduce its cost. Sanger sequencing was used for the 13-year-long Human Genome Project, which resulted in the first whole-genome sequence in 2003 for a budget of $2.7 billion. Five years later, the same result was obtained in 5 months for just $1.5 million (Voelkerding et al., 2009). The National Human Genome Research Institute (NHGRI) part of the NIH awarded in 2008 more than $20 million in grants (the $1,000 Genome grants) to develop innovative sequencing technologies inexpensive and efficient enough to sequence a person's DNA as a routine part of biomedical research and health care for less than $1000. The 1000 Genomes Project, a China, Germany, UK, and US collaboration, is currently sequencing the whole genome from 2000 individuals worldwide to identify genetic variations (Via et al., 2010). In the 1990s, microarray technologies have been developed to address mostly needs of human genomics. As of May 2010, 21,520 scientific papers using the Affymetrix (Santa Clara, CA, USA) proprietary microarray technology have been listed on the web site of the company (www.affymetrix.com). Microarrays have also been used for resequencing or gene expression monitoring of infectious agents. New sequencing technologies and instruments, referred to as Next-Generation Sequencing (NGS), have appeared during the last 5 years and interestingly were first used to sequence the whole genome of human pathogens *Mycoplasma genitalia* (Margulies et al., 2005) and *Escherichia coli* (Shendure et al., 2005).

Understanding hosts, vectors, and pathogens' genomes, as well as their transcriptomic and epigenomic modifications following infection, during the course of the disease and under treatment will ultimately lead to personalized medicine for which genome characteristics will be important to tailor treatments (Snyder et al., 2010).

Several biotech companies have been created in the last 15 years to develop and market systems for the analysis of genomes, many of them by scientists issued

*Email: guy.vernet@fondation-merieux.org

from universities. Since then, they have often been acquired by larger companies that are major players in the field of in vitro diagnostics.

19.2 Customers and Their Needs

19.2.1 Customers

Customers of companies involved in genomics are both public and private organizations. Research laboratories from universities or institutes have different needs, which were, up to now, impossible to address with a single technology or instrument. Due to high prices of equipment and maintenance and frequent needs for instrument upgrade, platforms serving the needs of several research laboratories and providing services to the scientific community have emerged, which provide a full set of equipments and dedicated human resources.

The Wellcome Trust Sanger Institute (Cambridge, UK) is an example where such a platform, linked to bioinformatics resources serves different research projects, including a pathogen genetic group exploring parasites especially malaria—and viruses genomics. The Broad Institute (Cambridge, MA, USA) is another example with a genome-sequencing platform that also considers fungi, bacteria, and virus genomes. The J. Craig Venter Institute, a not-for-profit private organization based in Rockville, Maryland (USA), has more than 500 scientists and staff, more than 250,000 square feet of laboratory and operates several resource centers (sequencing, genotyping, functional genomics, bioinformatics) for infectious disease genomics. In China, the Beijing Genomics Institute (BGI) from the Chinese Academy. of Sciences, based in Shenzhen and Hong Kong will have 3500 staff by the end of 2010 and has acquired 128 HiSeq 2000 NGS machines from Illumina (San Diego, CA, USA) for sequence service and its own research projects, some of them on infectious diseases like the severe acute respiratory syndrome (SARS)-Coronavirus (Cyranoski, 2010).

Such large institutes are not very numerous worldwide. However, they attract large international budgets that generate important sells for genomics companies in terms of instruments, maintenance, and reagents.

Genomics platforms often provide services to remain competitive through return on investment. Thus, they represent potential customers for genomics companies. A list of companies and public facilities that provide DNA-sequencing services in different parts of the world can be found at http://www.nucleics.com/DNA_sequencing_support/sequencing-service-reviews.html. It is quite complete, although it does not mention organizations like Illumina, Beckman Coulter, or the Wellcome Trust Sanger Institute. It is a very useful tool for scientists who need to choose a service company and includes, when available, prices, DNA-sequencing instruments used by each facility, specific DNA template and primer requirements, sample shipping and sequencing facility contact details, whether the DNA-sequencing facility is GLP/FDA certified, and the other DNA-sequencing-related services offered by the facility. There are more than 230 such facilities worldwide, including many universities in the USA and Canada. Many of them are compliant with good clinical practices (GCP), good laboratory practices (GLP), and good manufacturing practices (GMP) or, for clinical diagnostic services to the Clinical Laboratory Improvement Amendments (CLIA) regulations.

Pharmaceutical companies also have needs for complex genomic data for R&D purposes—new molecules or vaccine developments, toxicity and pharmacokinetic studies—although there is a trend for outsourcing upstream research to public laboratories or biotechnology companies.

At the other extremity of data complexity, high-level public or private laboratories, such as those in hospitals, use technologies based on genomic information for diagnostic, forensic, or surveillance needs. Although currently very centralized due to the requirement for sophisticated equipments and relatively high skills, these techniques will probably reach more and more customers in a near future. Finally, pharmaceutical and agro-food companies also use low-complexity data for quality control purposes ensuring product biosafety.

19.2.2 Research Needs

Whole-Genome Sequencing, Comparative-Genome Sequencing, and Targeted Resequencing

The first need of scientists is to sequence microbe genomes. The whole-genome sequence of 1987 viruses, 916 bacteria, and 67 archeal species were deposited to GenBank release 2.2.6 (Wooley et al., 2010), and these numbers grow rapidly, the emphasis being on species that are pathogenic for animals or plants. Sequencing will identify single nucleotide polymorphisms (SNPs), insertions, and deletions (indels) and large chromosomal rearrangements (structure or copy number changes).

Once a reference sequence as been established for a given species, the need is for resequencing (i.e., comparison of the sequence of new isolates to this reference sequence to identify differences) (Herring and Palsson, 2007). This comparison concerns the whole genome or targets genes of interest where modifications are expected to have consequences on the phenotype. Scientists will use these data for fundamental research: to annotate all genes, understand genome organization, classify species, and study their evolution. The knowledge of genome organization, combined with functional genomics is the basis to understand pathobiological mechanisms of infectious agents. Among downstream applications are the development of new drugs, vaccines, and diagnostics including the establishment of resistance and escape profiles. Surveillance networks will also use genomic information to monitor pathogens evolution: resistance to treatment, crossing of host speciation barriers, increased transmissibility, and virulence.

Metagenomics

A new field of investigation is the study of microbiomes or metagenomics (Wooley et al., 2010), which consists in the systematic sequencing of all nucleic acids in a given ecological "niche"—gut, upper respiratory tracts, and skin—to identify all microbes. This allows us to characterize the "normal" flora, for example at different ages in life, and interactions between this flora in pathogenic agents during infection, in particular the exchange of genetic material conferring resistance and virulence. Applications by pharmaceutical, diagnostic, and agro-food companies are very important. The study of microbiomes also have a potential application for forensics: sequencing the "bacteriome" in traces left on can be used to identify people as the flora present on the skin is a signature depending on food habits, environment, and diseases.

Functional Genomics

Transcriptome analysis is the characterization of all coding and noncoding transcriptional activity in any organism without a priori assumptions through annotation of SNPs and mapping to reference genomes, characterization of transcript isoforms, regulatory RNAs, or splice junctions and determination of the relative abundance of transcripts (gene expression analysis). Analysis of differential gene expression is important in hosts and pathogens as well as in vectors of transmissible diseases (mostly insects). Human, plant, or animal cells can be studied when they are confronted to infection to identify the mechanisms targeted by the pathogen and those by which they resist infections. Pathogens' gene expression during infection of the cell is an important area of investigation as it may reveal pathobiological mechanisms and targets for new drugs. Epigenome analysis is the study of chromatin structure and gene regulation by CpG methylation, histone modifications, or DNA−protein interactions. Besides fundamental research, functional genomics can find applications in pharmaceutical companies for new anti-infection drugs and diagnostic companies to identify new biomarkers for diagnostic or disease or treatment monitoring.

19.2.3 Diagnostic Needs

Genomics research generates tremendous amount of information. Among this are sequences that can be used to detect infection by a pathogen species by simple molecular techniques like polymerase chain reaction (PCR), transcription-mediated amplification (TMA), or nucleic acid sequence-based amplification (NASBA), which, if quantitative, can also help monitor treatment efficacy against, for example, human immunodeficiency virus (HIV) or hepatitis viruses. A syndrome-based approach (i.e., the detection of multiple pathogens responsible for a disease, such as pneumonia, fever, neurological diseases, diarrhea) may also be useful. Signature sequences can identify infection by a variant with a specific phenotype with given virulence, host specificity, or resistance to treatment and help patient care or epidemiological surveillance. The latter signatures can be entire genes (acquired by horizontal transmission, either plasmid or recombination), individual SNPs, groups of SNPs carried by a unique or different genes, indels, or even modified expression of a gene. A comprehensive assay associating pathogen identification, pathogen typing, identification of virulence, or resistance markers may prove valuable in chronic infections like HIV, hepatitis virus infections, or tuberculosis.

19.3 Technologies and Companies

Microarrays, Sanger sequencing, and NGS generate large quantities of data necessary for whole-genome resequencing, targeted resequencing, gene copy number variations, gene expression analysis, chromosome structural changes, or protein–DNA interactions. However, microarray analysis, by definition, requires a priori knowledge of sequences and only sequencing allows determination of unknown sequences. Only NGS allows rapid and low-cost whole-genome sequencing and metagenomics, as well as massive parallel processing of multiple specimens. Less complex technologies like pyrosequencing or low- to medium-density microarrays can be used for targeted sequencing or resequencing of one or a few genome regions which is of interest for patients management or for molecular epidemiology. Molecular diagnostics using PCR, TMA, NASBA, loop-mediated isothermal amplification (LAMP) have many more applications for infectious diseases research, epidemiological surveillance and treatment. However, we will not address this domain of activity which exploits data issued from genomics research.

19.3.1 Microarray Companies

At least 36 companies providing microarrays have been identified in 2009 in the USA and Europe (North Shore LIJ Research Institute; http://www.nslij-genetics.org/microarray/). Most of them propose low- to medium-density custom arrays, which are glass plates or beads with DNA probes either spotted or synthesized in situ using a variety of technologies, including printing with fine-pointed pins onto glass slides, photolithography using premade masks, photolithography using dynamic micromirror devices, ink-jet printing, or electrochemistry on microelectrode arrays. For a recent review on the use of microarrays for clinical microbiology, see Miller and Tang (2009). We will focus on companies providing high-density microarrays.

The Affymetrix GeneChip® technology, based on photolithography to synthesize probes in situ on the array, was invented in the late 1980s. More than 1500 publications can now be retrieved from the Affymetrix database with the key words "virus," "bacteria," "parasite," and "fungi." The current company platform GCS 3000Dx v.2 is 510(k) cleared and CE marked for in vitro diagnostic use and consists of a scanner, a fluidics station, and the Affymetrix Molecular Diagnostics Software (AMDS) for data interpretation. The Human Genome U133 Plus 2.0 array analyzes over 47,000 transcripts from human genome and allows gene expression profiling of cells infected by various pathogens. Several commercial arrays address human pathogens. The *E. coli* array contains probe sets to detect transcripts from the K12 strain of *E. coli* and three pathogenic strains of *E. coli*. It includes approximately 10,000 probe sets for all 20,366 genes present in four strains of *E. coli* over the entire open reading frame (ORF) of *E. coli*, over 700 intergenic regions as well as probe sets for various antibiotic resistance markers. The *Staphylococcus aureus* Genome Array allows the analysis of the expression of sequences in four strains of *S. aureus*. It contains probe sets to over 3300 *S. aureus* ORFs and to study both forward and reverse orientation of over 4800 intergenic regions. The *Pseudomonas aeruginosa* Genome Array represents the annotated genome of *P. aeruginosa* strain PA01 and includes 5549 protein-coding sequences, 18 tRNA genes, a representative of the ribosomal RNA cluster, and 117 genes present in strains other than PA01. In addition, 199 probe sets corresponding to all intergenic regions exceeding 600 base pairs have been included. The Plasmodium/Anopheles Genome Array includes probe sets to over 4300 *Plasmodium falciparum* transcripts and approximately 14,900 *Anopheles gambiae* transcripts.

The SARS Resequencing Array provides a standard assay for complete sequence analysis of the SARS coronavirus.

BioMérieux (Marcy-l'Etoile, France) has developed several resequencing microarrays covering pathogens genomes. A *Mycobacterium tuberculosis* array is based on two sequence databases: one for the species identification of mycobacteria (82 unique 16S rRNA sequences corresponding to 54 phenotypical species) and the other for detecting *M. tuberculosis* rifampin resistance in rpoB (Troesch et al., 1999; Sougakoff et al., 2004). An *S. aureus* array tiles 16S rDNA sequences to identify staphylococcus species (Couzinet, 2005a), grlA, gyrA, grlB, and gyrB genes for the presence of mutations involved in fluoroquinolone resistance (Couzinet, 2005b), and multilocus sequence typing (MLST) of *S. aureus* strains (van Leeuwen et al., 2003). An HIV microarray was designed to detect 204 antiretroviral resistance mutations simultaneously in Gag cleavage sites, protease, reverse transcriptase, integrase, and gp41 of HIV1 (Gonzalez et al., 2004). Similarly, a hepatitis B virus (HBV) microarray was designed to detect 245 mutations, 20 deletions, and 2 insertions at 151 positions and to determine the genotype of the HBV (Tran et al., 2006; Pas et al., 2008).

Roche NimbleGen (Madison, WI, USA) manufactures custom, high-density DNA arrays based on its proprietary Maskless Array Synthesizer (MAS) technology using a Digital Micromirror Device (DMD) combined with DNA synthesis chemistry allowing 385,000 to 2.1 million unique probe features in a single array. Arrays are synthesized on standard-sized glass microscope slides and are compatible with a range of hybridization, washing, and scanning instrumentation. With the new generation of HD2 arrays, oligos between 50 and 75 bases in length can be synthesized, increasing sensitivity, specificity, and reproducibility.

Recently, Roche NimbleGen introduced Sequence Capture arrays to produce targeted, sequencing-ready samples for use with NGS instruments. High density NimbleGen arrays with long oligonucleotides are used to hybridize either the whole human exome or human genomic regions of interest are hybridized. The purified human sequences are then eluted, amplified and sequenced. Although currently restricted to human the genome, and more recently, to wheat and rapeseed, this technology may prove valuable for infectious diseases genomics. Agilent Technologies (Santa Clara, CA, USA) and Febit (Heidelberg, Germany) also market capture microarrays.

19.3.2 NGS Companies

NGS is parallel sequencing of clonally amplified or single DNA molecules by iterative cycles of polymerase-based extension or oligonucleotide ligation that takes place in flow cells. These technologies have revolutionized all aspects of genomics: whole-genome sequencing, targeted resequencing, metagenomics, gene expression profiling, epigenomics, and DNA−protein interactions study (ChIP-Seq). For a recent review on NGS, see Holt and Jones (2008) and Voelkerding et al. (2009). Three companies have launched NGS platforms requiring clonal amplification: 454 Life Sciences, a Roche company (Branford, CT, USA), Illumina Inc. (San Diego, CA, USA), and Applied Biosystems, a division of Life Technologies (Carlsbad, CA, USA); one company has launched a system sequencing single DNA molecules: Helicos Biosciences Corporation (Cambridge, MA, USA). The main characteristics of these platforms are listed in Table 19.1.

454 Life Sciences was created in 2000 and released the first NGS platform, Genome Sequencer 20, in 2005. It was used to sequence the first human genome for less than US$1 million in 2006. 454 Life Sciences has been acquired by Roche Diagnostics in January 2007. The current 454 Life Sciences platform, the Genome Sequencer FLX system, is using the 454's sequencing-by-synthesis (SBS) technology for de novo sequencing, resequencing of whole genomes, and target DNA regions, metagenomics, and RNA analysis. The chemistry used for sequencing is described in Figure 19.1A. Automation using a magnetic beads technology simplifies emulsion-PCR and allows library preparation of genomics samples in hours in a single tube, eliminating cloning, and colony picking. The recognized advantage

Table 19.1 Characteristics of NGS Platforms

Platform	GS FLX	HiSeqTM2000	SOLiD™ 4	Genetic Analysis System
Company	Roche	Illumina	Applied Biosystems	Helicos
Throughput/run	1 million reads	1 billion reads	1.4 billion reads	1 billion reads
	0.4−0.6 Gb	Up to 200 Gb	Up to 180 Gb	Up to 35 Gb
Run duration	10 h	2.5 days	6 days	8 days
Multiplexing (samples/run)	8	200 (gene expression profiling)	48 RNA, 96 DNA	Up to 4800
Base call quality (manufacturer data)	99% of bases at QV20*	>90% at QV30*	80% at QV30*	0.2% error rate (SNPs)
Human genome coverage	>20	30	30	28
Read length	400 bp	100 bp	50 bp	35 bp
Paired end read	Yes	Yes	Yes	Yes

*Phred score quality value (QV): miscall probabilities of 10%, 1%, and 0.1% yield QV of 10, 20, and 30, respectively.

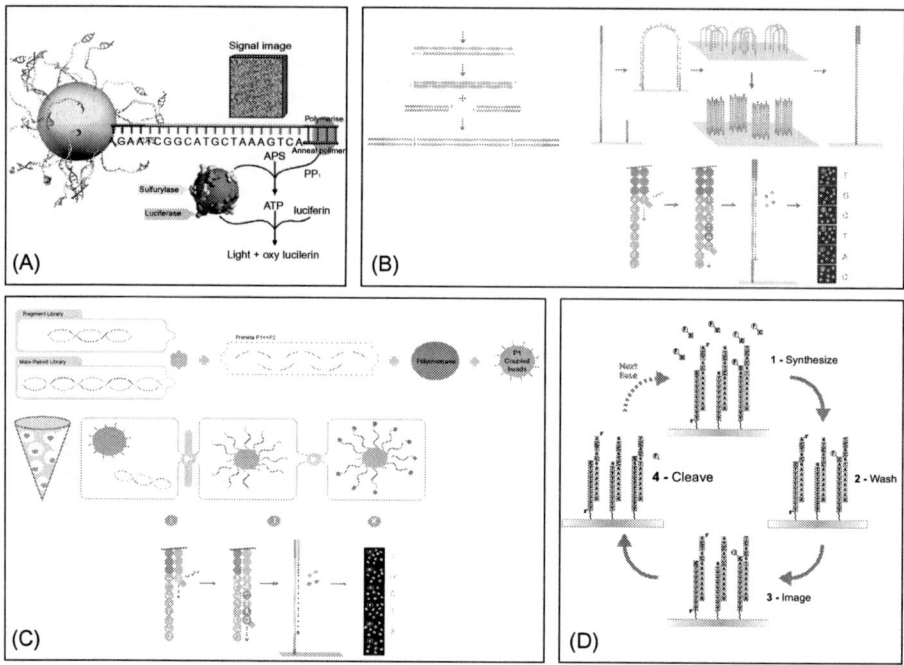

Figure 19.1 (A) 454 technology. Template DNA is fragmented with adapters added at both ends, and clonal amplification is done by emulsion-PCR using magnetic beads. Each single bead is added to a well of a picotiter plate and iterative pyrosequencing is used for sequencing. *Source:* 454 Sequencing. Copyright 2010. Roche Diagnostics. (B) Illumina HiSeq technology. Template DNA is fragmented, adapters are added at both ends, and DNA is attached to the flow cell. Bridge amplification generates clonal clusters and iterative SBS is performed. *Source:* Copyright 2010. Illumina Inc. (C) Applied Biosystems SOLiD technology. Template DNA is fragmented, adapters are added at both ends and clonal amplification is done by emulsion-PCR using magnetic beads. Sequencing is done by iterative ligation using a set of four fluorescently labeled di-base probes. *Source:* Copyright 2010. Life Technology Inc. (D) Helicos tSMS technology. Original DNA samples are first fragmented, the DNA double-helix is melted into single strands and a polyA tail is added to these DNA molecules. Billions of these single DNA molecules are captured on a proprietary surface within a flow cell and serve as templates for the SBS process. Genomic DNA is fragmented, polyA tail are added, and DNA molecules captured by oligo dT primers inside the flow cell. The tSMS process is a cyclical process involving multiple rounds of (1) synthesis using labeled nucleotides, (2) washing, (3) imaging, and (4) cleaving the fluorescent label until the desired read length is achieved. *Source:* Copyright 2010. Helicos Biosciences corp. All rights reserved.

of the FLX system compared to other NGS platforms is that it generates the longest single reads (400 bp) and long paired end reads of 20 kb, 8 kb, or 3 kb. In May 2010, the company introduced the GS Junior System which provides an integrated sequencing and bioinformatics solution, all in the size of a typical desktop laser printer. The company announces improvements of its sequencing chemistry to increase reads length to 1000 bp. There have been 748 published studies using the 454 technology, among which 219 concern infectious diseases. The technology has found applications in metagenomics, such as identification of a new arenavirus in transplantation patients (Palacios et al., 2008) or the characterization of microflora in oral cavity or guts (Keijser et al., 2008; Turnbaugh et al., 2009).

Illumina was created in 1998 to exploit rights on the BeadArray technology developed at Tufts University. The HiSeq TM 2000 platform is based on the SBS chemistry, which generates reads of 100 bp and paired end reads allowing the assembly of long scaffolds (Figure 19.1B). After library preparation, cluster generation is done on the cBot automated cluster generation system, which significantly reduces hands-on time compared to emulsion-PCR. Sequencing a genome can be done on one flow cell while, simultaneously, the other flow cell can analyze its epigenome and transcriptome. More than 500 publications illustrate the versatility of Illumina NGS technology. Recent publications on infectious diseases include *Trichinella spiralis* (Webb and Rosenthal, 2010), *P. falciparum* (Jiang et al., 2010), virus discovery in Drosophila cells and adult mosquitoes (Wu et al., 2010), methicillin-resistant *Staphylococcus aureus* (MRSA) molecular epidemiology (Harris et al., 2010), *Burkholderia cenocepacia* therapeutic targets (Yoder-Himes et al., 2010), and *Pasteurella multocida* virulence factors (Steen et al., 2010). The Illumina platform is the most versatile of the three NGS platforms requiring clonal amplification because it associates gigabases outputs and read lengths of 100 bp.

Life Technologies was created by the combination of Invitrogen Corporation and Applied Biosystems Inc. in 2008. Applied Biosystems commercializes a NGS platform called SOLiD 4. The chemistry used in this platform is described in Figure 19.1C. It is based on emulsion-PCR and oligonucleotides ligation. The SOLiD 4 platform has two flow cells allowing two independent experiments at the same time and can multiplex up to 96 samples. Applied biosystems proposes automated solutions for reproducible templated bead preparation with less than 1 h of hands-on time (EZ Bead™ System). The SOLiD platform generates the shortest read lengths among the three platforms, which makes it less versatile for the various applications. However, future evolutions of the SOLiD platforms, SOLiD 4hq, and SOLiD PI will generate longer reads (75 bp).

Helicos Biosciences Corporation (Cambridge, MA, USA) is commercializing the first platform that allows sequencing from single DNA molecules, thereby avoiding biases due to amplification. As the three companies described earlier, Helicos is a recipient of the "$1000 Genome" grant. The chemistry used on the Helicos Genetic Analysis System is based on tSMS technology (Figure 19.1D) in which single-stranded DNA molecules generated from a library and tagged with a polyA tail are attached to a proprietary surface at a density of up to 100×10^6 molecules per square centimeter and sequenced by synthesis. This system is now installed in several research centers in the USA and Europe. The Helicos platform was used for the first single-molecule whole-genome sequencing in 2009 (Pushkarev et al., 2009). This study achieved $28\times$ average coverage of the human genome and detected over 2.8 million SNPs, of which over 370,000 were novel. Validation with a genotyping array demonstrated 99.8% concordance. The unbiased nature of the single-molecule sequencing approach also allowed the detection of 752 copy number variations in this genome.

The NGS technologies generate a much higher amount of data than the well-established ABI Sanger platform. The analysis of gigabases or terabases requires complex software solutions. A list of software used with NGS platforms can be found in Voelkerding et al. (2009).

However, performances claimed by manufacturers may be overestimated. Harismendy et al. (2009) have compared the platforms of 454 Life Sciences, Illumina, and Applied Biosystems on a 260 kb human genome sample. Although the Illumina and Applied Biosystems produce the largest amounts of data, only 43% and 34% of them, respectively, are usable after quality filtration. In contrast, 95% of the data generated by the 454 platform are usable. All three technologies have biases that induce heterogeneous coverage of bases along the sequence: the 454 platform shows the lowest variability among unique and repetitive sequences, whereas the ABI and Illumina platforms tend to be affected by high Adenine/Thymine contents. NGS platforms tend to better detect indels than ABI Sanger platform as they sequence single strands. As expected, ABI Sanger sequencing has an error rate of approximately 7% and careful comparison with NGS reveals false positive and false negative rates of 0.9% and 3.1%. The overall sequencing accuracy of NGS platforms was very high (>99.99%), but the ability to detect variant was 95% for the 454 platform (which has the lowest sensitivity), 100% for the Illumina platform, and 96% for the ABI platform, the last two technologies being less specific. Overall, NGS platforms need to improve their uniformity of per-base sequence coverage as accuracy is lower in poorly covered regions.

New technologies based on nanopores are currently being explored to develop platforms for single-DNA molecule sequencing (Branton et al., 2008). A nanopore-based device provides single-molecule detection and analytical capabilities that are achieved by electrophoretically driving molecules in solution through a nano-scale pore. The nanopore provides a highly confined space within which single nucleic acid polymers can be analyzed at high throughput by one of a variety of means, and the perfect processivity that can be enforced in a narrow pore ensures that the native order of the nucleobases in a polynucleotide is reflected in the sequence of signals that is detected. Kilobase length polymers (single-stranded genomic DNA or RNA) or small molecules (e.g., nucleosides) can be identified and characterized without amplification or labeling, a unique analytical capability that makes inexpensive, rapid DNA sequencing a possibility. Further research and development to overcome current challenges to nanopore identification of each successive nucleotide in a DNA strand offers the prospect of third-generation instruments that will sequence a diploid mammalian genome for approximately $1000 in approximately 24 h. Oxford Nanopore Technologies (http://www.nanoporetech.com; Oxford, UK), our first generation of DNA-sequencing technology, uses a protein nanopore combined with a processive enzyme, multiplexed on a silicon chip. This elegant and scalable system has unique potential to transform the speed and cost of DNA sequencing. Future generations may interrogate single strands of DNA and may use "solid-state" nanopores for further improvements in speed and cost. Lingvitae (http://www.lingvitae.com; Oslo, Sweden) and Eid et al. (2009) from Pacific Biosciences (Menlo Park, CA, USA) have designed a prototype instrument able to sequence a single DNA molecule using DNA polymerase based on the observation of the temporal incorporation of labeled nucleotides, which takes place in a nanophotonic structure. This can drastically reduce the volume of observation. Currently, this prototype is able to multiplex 3000 such structures, allowing sequencing of small viral or bacterial genomes with high accuracy.

19.4 Conclusion and Perspectives

Microarrays and NGS have revolutionized genomics because they drastically increased scientists' access to tremendous amounts of information on genomes and gene expression and decreased time-to-result, hands-on time, and costs. Research needs for these technologies are well understood but still represent small (although rapidly growing) markets. The total market of NGS was evaluated to be $484 million in 2008 (http://www.researchandmarkets.com/reportinfo.asp?report_id=614823) and the market of functional genomics was about $2.2 billion in 2007 with an annual growth rate of 18% (http://www.researchandmarkets.com/reportinfo.asp?report_id=5545). The real development of sales of companies proposing technologies for genomics will mostly come from diagnostic applications in human or animal health. Genomics research can be translated into molecular diagnostics in the fields of human genetic, oncology, and infectious diseases. Several in vitro diagnostic companies propose instruments and reagents to detect sequence signatures for diagnosis or treatment monitoring applications. However, the current clinical needs mostly require low-complexity genetic information that can be covered by multiplex real-time amplification, reverse hybridization-based line probe assay, low-density microarrays or simple sequencing platforms like the PyroMark Q24 platform. The place of technologies generating more complex datasets in clinical applications is still to be defined. Developing such applications will require clinical validation of the value of complex sequence information and strong efforts for clinician education on its benefit. It will also require that costs of instruments and reagents further decrease and that the technologies become easier to use and more robust. All the major companies cited in this chapter make efforts to simplify equipments and reduce costs per analysis. Affymetrix recently launched a new, more affordable and smaller platform, GeneAtlas, along with a microarray strip format, which enables users to process up to two strips per day or eight strips per week. 454 Life Sciences announced the release of the GS Junior System scaled to suit the needs of individual laboratories for rapid sequencing of amplicons, targeted human resequencing studies, de novo sequencing of microbial and other small genomes, and for pathogen detection. Illumina recently launched the Genome Analyzer$_{IIe}$ with a lower cost, making the technology more accessible to laboratories of various sizes. Similarly, Applied Biosystems announces a less expensive, low-throughput benchtop platform (50 Gb per run) will allow shorter times-to-results. Qiagen (Hilden, Germany) is marketing a relatively simple, low footprint platform (PyroMark Q24) for real-time, sequence-based detection and quantification of sequence variants and epigenetic methylation that uses pyrosequencing technology. The instrument can process 1–24 samples in 15 min. This platform, which is affordable to mid-sized laboratories, allows analysis of CpG methylation, SNPs, insertion/deletions, short tandem repeats (STRs), and variable gene copy number.

References

Branton, D., Deamer, D.W., Marziali, A., Bayley, H., Benner, S.A., Butler, T., et al., 2008. The potential and challenges of nanopore sequencing. Nat. Biotechnol. 26, 1146–1153.

Couzinet., S., Yugueros, J., Barras, C., Visomblin, N., Francois, P., Lacroix, B., et al., 2005a. Evaluation of a high-density oligonucleotide array for characterization of grlA, grlB, gyrA and gyrB mutations in fluoroquinolone resistant *Staphylococcus aureus* isolates. J. Microbiol. Methods 60, 275–279.

Couzinet, S., Jay, C., Barras, C., Vachon, R., Vernet, G., Ninet, B., et al., 2005b. High-density DNA probe arrays for identification of staphylococci to the species level. J. Microbiol. Methods 61, 201–208.

Cyranoski, D., 2010. A primer on metagenomics. Nature 464, 22–24.

Eid, J., Fehr, A., Gray, J., Luong, K., Lyle, J., Otto, G., et al., 2009. Real-time DNA sequencing from single polymerase molecules. Science 323, 133–138.

Gonzalez, R., Masquelier, B., Fleury, H., Lacroix, B., Troesch, A., Vernet, G., et al., 2004. Detection of human immunodeficiency virus type 1 antiretroviral resistance mutations by high-density DNA probe arrays. J. Clin. Microbiol. 42, 2907–2912.

Harismendy, O., Ng, P.C., Strausberg, R.L., Wang, X., Stockwell, T.B., Beeson, K.Y., et al., 2009. Evaluation of next generation sequencing platforms for population targeted sequencing studies. Genome Biol. 10, R32.

Harris, S.R., Feil, E.J., Holden, M.T., Quail, M.A., Nickerson, E.K., Chantratita, N., et al., 2010. Evolution of MRSA during hospital transmission and intercontinental spread. Science 327, 469–474.

Herring, C.D., Palsson., B.Ø., 2007. An evaluation of Comparative Genome Sequencing (CGS) by comparing two previously-sequenced bacterial genomes. BMC Genomics. 14, 274.

Holt, R.A., Jones., S.J., 2008. The new paradigm of flow cell sequencing. Genome Res. 18, 839–846.

Jiang, L., López-Barragán, M.J., Jiang, H., Mu, J., Gaur, D., Zhao, K., et al., 2010. Epigenetic control of the variable expression of a *Plasmodium falciparum* receptor protein for erythrocyte invasion. Proc. Natl. Acad. Sci. U.S.A. 107, 2224–2229.

Keijser, B.J., Zaura, E., Huse, S.M., van der Vossen, J.M., Schuren, F.H., Montijn, R.C., et al., 2008. Pyrosequencing analysis of the oral microflora of healthy adults. J. Dent. Res. 87, 1016–1020.

Margulies, M., Egholm, M., Altman, W.E., Attiya, S., Bader, J.S., Bemben, L.A., et al., 2005. Genome sequencing in microfabricated high-density picolitre reactors. Nature 437, 376–380.

Miller, M.B, Tang, Y.W., 2009. Basic concepts of microarrays and potential applications in clinical microbiology. Clin Microbiol. Rev. 4, 611–613.

Palacios, G., Druce, J., Du, L., Tran, T., Birch, C., Briese, T., et al., 2008. A new arenavirus in a cluster of fatal transplant-associated diseases. N. Engl. J. Med. 358, 991–998.

Pas, S.D., Tran, N., de Man, R.A., Burghoorn-Maas, C., Vernet, G., Niesters, H.G., 2008. Comparison of reverse hybridization, microarray, and sequence analysis for genotyping hepatitis B virus. J. Clin. Microbiol. 46, 1268−1273.

Pushkarev, D., Neff, N.F., Quake., S.R., 2009. Single-molecule sequencing of an individual human genome. Nat. Biotechnol. 27, 847−852.

Shendure, J., Porreca, GJ, Reppas, NB, Lin, X, McCutcheon, JP, Rosenbaum, AM, et al., 2005. Accurate multiplex polony sequencing of an evolved bacterial genome. Science 309, 1728−1732.

Snyder, M., Du, J., Gerstein, M., 2010. Personal genome sequencing: current approaches and challenges. Genes Dev. 24, 423−431.

Sougakoff, W., Rodrigue, M., Truffot-Pernot, C., Renard, M., Durin, N., Szpytma, M., et al., 2004. Use of a high-density DNA probe array for detecting mutations involved in rifampicin resistance in *Mycobacterium tuberculosis*. Clin. Microbiol. Infect. 10, 289−294.

Steen, J.A., Harrison, P., Seemann, T., Wilkie, I., Harper, M., Adler, B., et al., 2010. Fis is essential for capsule production in *Pasteurella multocida* and regulates expression of other important virulence factors. PLoS Pathog. 6, e1000750.

Tran, N., Berne, R., Chann, R., Gauthier, M., Martin, D., Armand, M.A., et al., 2006. European multicenter evaluation of high-density DNA probe arrays for detection of hepatitis B virus resistance mutations and identification of genotypes. J. Clin. Microbiol. 44, 2792−2800.

Troesch, A., Nguyen, H., Miyada, C.G., Desvarenne, S., Gingeras, T.R., Kaplan, P.M., et al., 1999. *Mycobacterium* species identification and rifampin resistance testing with high-density DNA probe arrays. J. Clin. Microbiol. 37, 49−55.

Turnbaugh, P.J., Hamady, M., Yatsunenko, T., Cantarel, B.L., Duncan, A., Ley, R.E., et al., 2009. A core gut microbiome in obese and lean twins. Nat. Biotechnol. 27, 344−346.

van Leeuwen, W.B., Jay, C., Snijders, S., Durin, N., Lacroix, B., Verbrugh, H.A., et al., 2003. Multilocus sequence typing of *Staphylococcus aureus* with DNA array technology. J. Clin. Microbiol. 41, 3323−3326.

Via, M., Gignoux, C., Burchard, E.G., 2010. The 1000 Genomes Project: new opportunities for research and social challenges. Genome Med. 2, 3.

Voelkerding, K.V., Dames, S.A., Durtschi., J.D., 2009. Next-generation sequencing: from basic research to diagnostics. Clin. Chem. 55, 641−658.

Webb, K.M., Rosenthal., B.M., 2010. Deep resequencing of *Trichinella spiralis* reveals previously un-described single nucleotide polymorphisms and intra-isolate variation within the mitochondrial genome. Infect. Genet. Evol. 10, 304−310.

Wooley, J.C., Godzik, A., Friedberg, I., Miller, M.B., Tang, Y.W., 2010. Basic concepts of microarrays and potential applications in clinical microbiology. PLoS Comput. Biol. 26, e1000667.

Wu, Q., Luo, Y., Lu, R., Lau, N., Lai, E.C., Li, W.X., et al., 2010. Virus discovery by deep sequencing and assembly of virus-derived small silencing RNAs. Proc. Natl. Acad. Sci. U.S.A. 107, 1606−1611.

Yoder-Himes, D.R., Konstantinidis, K.T., Tiedje, J.M., 2010. Identification of potential therapeutic targets for *Burkholderia cenocepacia* by comparative transcriptomics. PLoS One. 5, 8724.

20 Current Progress in the Pharmacogenetics of Infectious Disease Therapy

Tabitha Mahungu[1] and Andrew Owen[2],*

[1]Hospital for Tropical Diseases, London, UK, [2]Department of Molecular and Clinical Pharmacology, Institute of Translational Medicine, University of Liverpool, Liverpool, UK

20.1 Introduction

Following ingestion of standard doses of medication, interindividual variation in both desired and toxic effects is often observed. Factors contributing to this variability include age, gender, ethnicity, body mass index, physiologic status, co-morbidity, dietary factors, and co-prescribed medication. The contribution of genetic variation to interindividual variability has been reported to range between 20% and 95% (Kalow et al., 1998).

The terms "pharmacogenetics" and "pharmacogenomics" are often used interchangeably, although strictly speaking they differ in meaning. The United States Food and Drug Administration (FDA) defines pharmacogenomics as the study of variations of DNA and RNA characteristics as related to drug response, while pharmacogenetics is defined as the study of variations in DNA sequence as related to drug response (U.S. Food and Drug Administration, 2006). In pharmacogenetic studies, the single nucleotide polymorphism (SNP) remains the measure of variability in most studies and is defined as a single nucleotide change occurring at an allele frequency of greater than 1% (Hoehe et al., 2003). In infectious diseases, the most commonly studied variants are SNPs in genes implicated in drug absorption, distribution, metabolism, and excretion (ADME) pathways. There is increasing interest in the impact of polymorphisms of nuclear receptors that regulate the expression of ADME genes, human leukocyte antigen (HLA) subtypes in hypersensitivity reactions, and in genes implicated in the development of metabolic toxicity.

The financial impact of standardized prescribing mainly through hospitalizations resulting from therapeutic failure or adverse drug events is increasingly being recognized. Hospitalizations secondary to adverse events are expensive and are estimated

*E-mail: aowen@liv.ac.uk

to cost $8000 per hospital bed per day in the United States of America (Bates et al., 1997), £4700 per hospital bed per year in mainland Europe (Moore et al., 1998), and £5000 per hospital bed per year in the United Kingdom (Davies et al., 2009). Pharmacogenetics aims to individualize therapies so that therapeutic benefits are maximized and toxic side effects are minimized. Although technological advancements have led to renewed interest in pharmacogenetics within the last decade, the concept of personalized medicine dates back to the 1950s (Meyer, 2004).

20.2 The Role of Pharmacogenetics in Communicable Diseases

Infectious diseases continue to account for a significant amount of morbidity and mortality worldwide, disproportionally affecting marginalized and resource-poor populations. In these populations, human immunodeficiency virus (HIV), malaria and tuberculosis account for a significant proportion of the communicable disease burden.

To date, most pharmacogenetic studies have been performed in patients receiving antiretroviral therapy. Antiretroviral therapy naturally lends itself to pharmacogenetic studies for a number of reasons. Most antiretroviral compounds exhibit wide interindividual variability in disposition and toxicity (Calmy et al., 2007). They are administered long-term in complex combinations to individuals from ethnically diverse backgrounds. Even when patients tolerate therapy, there is still concern of long-term metabolic side effects, especially now that HIV is considered a chronic disease (Mahungu et al., 2009a) and exposure to antiretroviral therapy has been shown to be associated with an increased cardiovascular risk (Friis-Moller et al., 2003; Friis-Moller et al., 2007; Sabin et al., 2008).

Unfortunately, there is limited data on the impact of genetic variants on ADME pathways of antimalarials and antituberculosis therapy. This is despite the fact that observations of interindividual variation in the disposition of both primaquine and isoniazid heralded the concept of pharmacogenetics over 50 years ago (Meyer, 2004). Compared to HIV infection, both malaria and tuberculosis have a limited armamentarium of effective therapy and significantly higher rates of drug resistance. Due to these limitations, there is increased interest in the mechanisms within the pathogen that result in resistant strains of plasmodium spp and mycobacteria spp.

20.3 Strategies for Investigating New Pharmacogenetic Associations

There are two over-arching strategies that have been used for investigating pharmacogenetic associations (Figure 20.1). First, in vitro studies have been employed to investigate the mechanisms involved for a particular phenotype. For example, if the phenotype is the variability in exposure to a particular drug, then the interaction of that drug with proteins that may influence its permeation across the gut or uptake

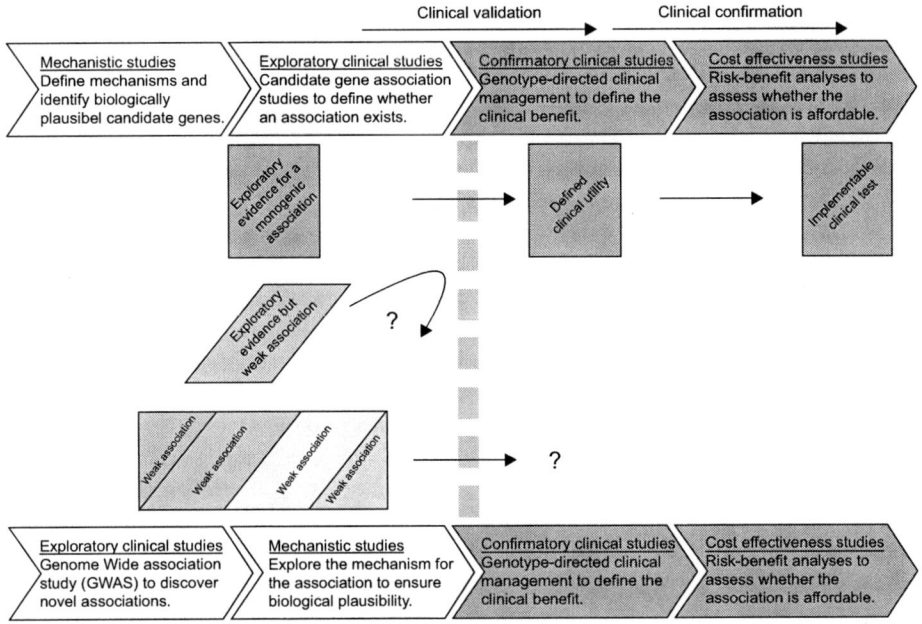

Figure 20.1 Schematic representation of the different strategies used to investigate novel pharmacogenetic associations. The top strategy involves analysis of mechanisms followed by candidate gene clinical studies. The bottom strategy employed more recently involves analysis of large numbers of polymorphisms in the clinic first, followed by mechanistic studies to confirm biological plausibility. The dotted line illustrates a theoretical barrier to implementation of pharmacogenetic testing due to the paucity of confirmatory clinical analyses and cost effectiveness studies. To date, only very strong, monogenic associations have translated into clinical practice and statistical analyses for assessing the contribution of multiple polymorphisms in unison continue to evolve and should provide a better basis for translating multigenic associations.

into the liver can be investigated in model systems. When novel interacting proteins (e.g., transporters) are elucidated then genetic variability in the genes encoding them can be investigated in candidate gene studies. Second, the expansion in technologies available for genotyping has allowed more thorough analysis of genetic variants across the genome. Genome-wide association studies (GWAS) are becoming more common and have yielded some important data. However, the mechanisms for associations discovered in this way still require confirmation in order to confer biological plausibility.

Candidate gene studies are already supported by a biologically plausible mechanism before the genetic analysis has been conducted. In many cases, the mechanisms are investigated independently of the pharmacogenetics as part of drug development or studies to understand variability or drug–drug interactions. However, only a very small number of the proteins coded for in the human genome have been functionally characterized in any detail. Developing the tools necessary to investigate them involves cloning and establishment of expression systems followed by functional characterization—this approach is therefore low throughput. GWAS, on the other hand, allows a lot of ground to be covered quickly. However, due to the number of polymorphisms being investigated (500,000 or more SNPs) they require very strict statistical corrections. The statistical methods are still evolving and it seems likely that in some cases the problem of false positive associations will be substituted for the problem of false negative associations.

A very clearly defined phenotype is required for any association study and analyzing continuous data such as pharmacokinetics is much more difficult than analyzing a categorical phenotype such as a particular toxicity or outcome. Irrespective of the strategy used, all pharmacogenetic associations need to have biologically plausible mechanisms and need to be replicated in multiple cohorts before their true value can be assessed. Only then is it justifiable to conduct clinical validation such as a prospective study followed by cost-effectiveness analyses.

Most associations that have been described to date have failed to progress through exploratory studies into clinical validation, and therefore into practice. This is likely to be influenced at least in part by the fact that most phenotypes are influenced by multiple gene products and are as such not monogenic. Therefore, single associations do not explain a sufficient degree of the variability in order to warrant expensive validation studies and thus be clinically implementable. An additional challenge therefore is the development of strategies to assess the contribution of multiple genetic contributors to a single clinical phenotype. Because many associations are not sufficiently predictive on their own and because of the expense associated with validation studies, there exists a virtual barrier between exploratory studies and studies necessary for clinical implementation. It therefore seems likely that characterizing multiple genetic influences for individual phenotypes and developing strategies for assessing the combined influence of multiple genotypes will facilitate the passage across this virtual barrier and into clinical practice (Figure 20.1).

20.4 Implementation of Pharmacogenetics

Before validation studies can be justified, it is important to consider carefully the optimum strategy for applying a pharmacogenetic test in the clinic (Figure 20.2). Probably the most commonly considered way to implement a test is if the association is with an adverse drug reaction. For example, if individuals susceptible to a particular toxicity can be identified, the physician can prescribe accordingly. Conversely, it may be that the association is with suboptimal drug exposure or a loss of efficacy. Since genetic variants may influence the absorption, clearance, and distribution into specific compartments (e.g., the CNS); this is certainly an area for study. However, there are also other ways in which a test might be applied. For example, some pharmacogenetic tests may be useful for predicting individuals more susceptible to drug–drug interactions or individuals better able to tolerate

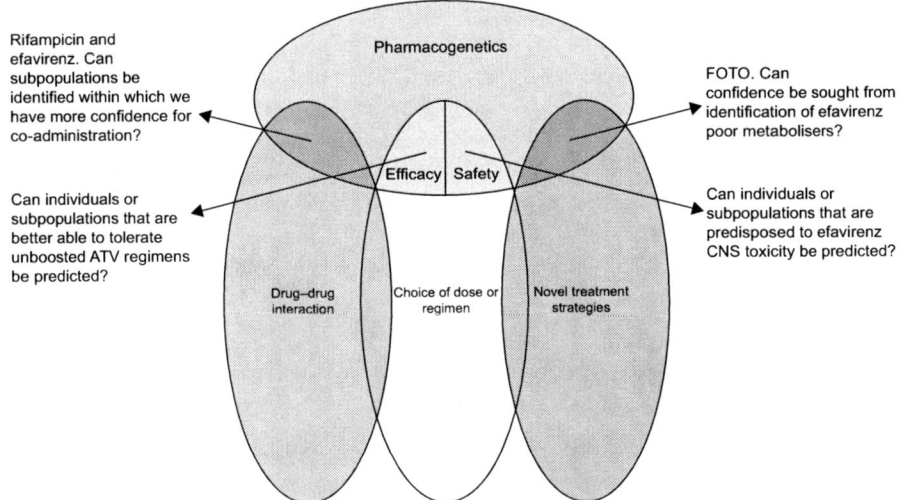

Figure 20.2 Clinical validation studies are expensive and time consuming and it is important to establish the optimum strategy for deploying a pharmacogenetic test prior to conducting them. A pharmacogenetic test could be assessed for its ability to direct the choice of dose or regimen and this could be driven by the need to optimise efficacy or reduce side effects (middle). However, it is also possible that some genetic associations may be better employed as a basis to reduce the clinical impact of a particular drug interaction (left) or to select novel treatment strategies for genetically characterised sub-populations (right).

novel treatment strategies. It's important to consider what is required from a test, and this is likely to be different for different genes and associations.

Not all pharmacogenetic associations are likely to be translatable into clinically worthwhile tests. For example, if the magnitude of the association is not of sufficient predictive power, the genetic variant is too rare to be of clinical utility or is not cost effective. However, in these cases, the observed associations are still of great mechanistic importance. In vitro functional assays can determine whether a particular drug is a substrate for a given transporter in an artificial system (e.g., at super-physiological expression) but they don't accurately determine that in an intact organ in an intact system on the backdrop of all other mechanisms, the protein is actually important in vivo. However, if there is a validated pharmacogenetic association in patients, then by definition the protein is important in vivo. This expands the knowledge base but could also potentially lead to novel pharmacological strategies.

There are also a number of clear ways in which pharmacogenetics may be useful in drug development. One example is when a genetic association validates a novel drug target. For example, the delta 32 polymorphism results in a truncated CCR5 protein and therefore partial resistance to HIV infection (Huang et al., 1996). This polymorphism is sufficiently frequent within Caucasian populations and it was known that there were unlikely to be adverse events related to the primary pharmacology of any agent. Consequently, maraviroc was developed as a CCR5 antagonist for treatment of HIV (Wood and Armour, 2005). Recently, IL28B variants were associated with natural clearance of genotype 1 hepatitis C virus (Ge et al., 2009) so other examples of this may also emerge. Finally, as functional variants are cataloged, pharmacogenetic tests can be useful to inform pre-clinical studies during drug development. For example, CYP2D6 metabolism is now routinely investigated by pharmaceutical companies at an early stage in discovery.

20.5 Pharmacogenetics of HIV Therapy

20.5.1 Nucleoside/Nucleotide Reverse Transcriptase Inhibitors

Nucleoside/nucleotide reverse transcriptase inhibitors are prescribed as pro-drugs that require intracellular phosphorylation to form their active metabolites (Anderson et al., 2004). Worldwide there is a move away from first-generation nucleosides such as stavudine (d4T), zalcitabine (ddC), didanosine (ddI), and zidovudine (AZT) toward better tolerated, less toxic NRTIs such as tenofovir (TDF) and abacavir. First-generation nucleoside analogs are associated with the potential development of peripheral lipoatrophy, nonalcoholic steato-hepatitis, lactic acidosis, pancreatitis, and peripheral neuropathy, through mitochondrial toxicity (Brinkman et al., 1999; Setzer et al., 2005). The newer-generation NRTIs are not without their toxicities—abacavir is associated with hypersensitivity reactions (HSR) and tenofovir is associated with renal tubular toxicity (Calmy et al., 2007). They often form the backbone of highly active antiretroviral therapy and are prescribed as coformulated compounds—Kivexa (Abacavir/Lamivudine) and Truvada (Tenofovir/Emtricitabine).

Abacavir and zidovudine (Veal and Back, 1995) are predominantly metabolized via hepatic glucuronidation while TDF and 3TC undergo renal excretion without hepatic metabolism. NRTIs are substrates for efflux transporters such as multidrug resistance protein 4 (MRP4/ABCC4), multidrug resistant protein 5 (MRP5/ABCC5), and breast cancer−related protein (BCRP/ABCG2) (Schuetz et al., 1999; Wijnholds et al., 2000; Takenaka et al., 2007). *MRP4* variants have been reported to be associated with higher intracellular levels of zidovudine triphosphate (*MRP4 3724G>A*) and lamivudine triphosphate (*MRP4 4131T>G*) (Anderson et al., 2006). The functional significance of these variants on MRP4 expression or function or treatment efficacy remains uncharacterized.

One of the key milestones in HIV pharmacogenetics was the discovery of HLA B*5701 as a strong predictor of the abacavir HSR (Mallal et al., 2002). Caucasian patients receiving abacavir have an 8% chance of developing a potentially fatal HSR within 6 weeks of initiating treatment (Calmy et al., 2007). The prospective screening for HLA B*5701 in a predominantly Caucasian population before commencing abacavir significantly reduces the incidence of abacavir HSR (Mallal et al., 2008). Although the carriage of HLA B*5701 and the subsequent rate of abacavir HSR is less frequent in Black populations (Cutrell et al., 2004; Hughes et al., 2004), HLA B*5701 carriage has been reported to be 100% sensitive as a marker of immunologically confirmed abacavir HSR in Black patients in the United States (Saag et al., 2008).

The excretion of tenofovir is facilitated by both influx and efflux transporters, such as human renal organic anion transporters (hOAT;SLC22A) (Uwai et al., 2007) located on the basolateral border of the proximal tubule and ABCC (MRP) transporters located on the brush border of the proximal renal tubule (Mallants et al., 2005; Ray et al., 2006). Toxicity is thought to result from increased tubular influx of tenofovir via SLC22As coupled with decreased efflux via ABCC2 transporters. An *ABCC2* haplotype (CATC) has been reported to be associated with an increased risk of proximal renal tubulopathy in a small, predominantly Caucasian cohort (Izzedine et al., 2006). In a more recent study, homozygosity for the C allele at position -24 of the ABCC2 gene was strongly associated with proximal tubular disease in a predominantly Caucasian population (Rodriguez-Novoa et al., 2009).

There is very limited data on the role of mitochondrial SNPs in the development of NRTI toxicity. However, a European haplogroup (mitochondrial haplogroup T) has been reported to be associated with significantly higher rates of peripheral neuropathy in Caucasian patients receiving both stavudine and didanosine (Hulgan et al., 2008). As both stavudine and didanosine remain mainstays of first-line regimens in resource-poor settings, further work was indicated in non-Caucasian populations. Recently, African mitochondrial subhaplogroup L1c was reported to be independently associated with peripheral neuropathy in Black patients receiving NRTI-based therapy (Canter et al., 2010).

20.5.2 Non-nucleoside Reverse Transcriptase Inhibitors

Non-nucleoside reverse transcriptase inhibitors (NNRTIs) are frequently prescribed components of highly active antiretroviral therapy (HAART). Efavirenz is the preferred third agent in first-line regimens across resource-rich settings (Clumeck et al., 2008; Gazzard et al., 2008) while nevirapine is the alternative third agent in first-line regimens in resource-poor settings where it is prescribed for the treatment of HIV-1 infected children and adults and in the prevention of mother to child transmission (PMTCT). Unfortunately, these drugs have a fragile genetic barrier to the development of resistance, and a single drug resistance mutation confers high-level resistance to all first-generation NNRTIs (Wainberg, 2003). The main difference between the two drugs lies in their toxicity profiles. The use of nevirapine is characterized by an idiosyncratic, potentially fatal, immune-mediated hypersensitivity (HSR) reaction which occurs in approximately five percent of treated individuals during the first six weeks of therapy. This HSR can manifest as hepatotoxicity, fever and/or Stevens-Johnson rash and occasionally death and is more likely in females with CD4 counts greater than 250 cells/mm^3 and in males with CD4 counts greater than 400 cells/mm^3 (Stern et al., 2003; Dieterich et al., 2004). The use of efavirenz is characterized by the development of central nervous system (CNS) side effects in approximately 40% of treated patients (Gazzard, 1999). Most of these symptoms (insomnia, dizziness, headache, and vivid dreams) are self-limiting and often disappear during the first 12 weeks of therapy (Lochet et al., 2003) but they can lead to treatment discontinuation and may also negatively impact upon compliance to therapy.

Nevirapine is primarily metabolized by the cytochrome P450 3A4 (CYP3A4) and 2B6 (CYP2B6) enzymes into its major metabolites 2-hydroxynevirapine and 3-hydroxynevirapine, respectively, with a minor contribution from CYP3A5 (Erickson et al., 1999). CYP3A4 (in concert with CYP2D6 and CYP3A5) is also involved in the biotransformation of nevirapine to 8-hydroxynevirapine and 12-hydroxynevirapine, both minor metabolites (Erickson et al., 1999). Efavirenz on the other hand is predominantly metabolized by CYP2B6 into 8-hydroxyefavirenz with a minor contribution from CYP3A4 (Ward et al., 2003). The biotransformation of EFV to 7-hydroxyefavirenz, its minor metabolite is via CYP2A6 with a minor contribution from CYP2B6 (Ward et al., 2003). The role of influx and efflux transporters in the disposition of NNRTIs remains largely uncharacterized. Isolated studies have shown that both drugs inhibit p-glycoprotein (MDR1; ABCB1) (Storch et al., 2007), ABCG2 (Weiss et al., 2007a), and ATP-binding cassette, sub-family C, member 1 (MRP1; ABCC1) (Weiss et al., 2007b). However, there are conflicting reports on whether either of these two NNRTIs are ABCB1 substrates (Stormer et al., 2002; Almond et al., 2005). The nuclear receptor, the constitutive androstane receptor (CAR) correlates with CYP2B6 (Chang et al., 2003) and CYP2A6 (Wortham et al., 2007) expression in liver, even in the absence of enzyme inducers and xenobiotics. Activators of CAR have also been shown to induce *UGT2B* genes in vivo (Shelby and Klaassen, 2006) and so CAR appears to play a role in basal and inducible regulation of all the enzymes involved in efavirenz metabolism. Recently, an SNP in *CAR* was shown to be associated with efavirenz plasma concentrations (Cortes et al., in press) and early treatment discontinuation of efavirenz-containing regimens (Wyen et al., in press).

The impact of the *CYP2B6* 516G>T SNP on the pharmacokinetics and pharmacodynamics of efavirenz is well described (Haas et al., 2004; Rodriguez-Novoa et al., 2005). A number of studies have also shown that the 516G>T SNP is also associated with nevirapine plasma concentrations (Rotger et al., 2005a; Penzak et al., 2007; Saitoh et al., 2007; Mahungu et al., 2009b). Heterozygous and homozygous carriers of the variant allele have been shown to express up to fourfold less CYP2B6 protein in comparison to the individuals homozygous for the common allele (Desta et al., 2007). The *CYP2B6* 983 T>C SNP, predominantly found in West African populations has also been shown to result in the reduced expression of CYP2B6 (Klein et al., 2005; Wang et al., 2006). This SNP is associated with up to threefold higher plasma concentrations of efavirenz which rises to up to fivefold if the CYP2B6 516 G>T SNP is also present (Wang et al., 2006). Heterozygosity for the 983T>C SNP has also been reported to be associated with significantly higher nevirapine plasma levels in Black patients (Wyen et al., 2008). Recently, a novel haplotype of *CYP2B6* consisting of three polymorphisms (rs10403955, rs2279345, and rs8192719) was shown to be more strongly associated with efavirenz concentrations above the reported minimum toxic concentration (4000 ng/mL) than the 516G>T polymorphism (Carr et al., 2010).

UDP-glucuronosyltransferase (UGT) 2B7, recently identified as the main enzyme involved in efavirenz N-glucuronidation, has been shown to predict efavirenz plasma concentrations (Kwara et al., 2009). There is a reported weak association between CYP3A4 (−392A>G) and CYP3A5 (6986A>G) variants and efavirenz exposure which is diminished when the study population is stratified by ethnicity (Haas et al., 2004). The Swiss HIV Cohort has also demonstrated that in the presence of defective CYP2B6 metabolism, there is a significant association between efavirenz exposure and *CYP3A4* and *CYP2A6* variants (di Iulio et al., 2009).

Higher plasma concentrations of efavirenz are also associated with central nervous system side effects (Marzolini et al., 2001; Nunez et al., 2001; Gallego et al., 2004; Kappelhoff et al., 2005). *CYP2B6* 516G>T is associated with efavirenz-induced CNS toxicity during the first week of therapy (Haas et al., 2004). More recently, a composite *CYP2B6* 516/983 "slow metabolizer" genotype has been shown to be associated with a higher rate of CNS side effects in Caucasians (Ribaudo et al., 2010). A number of studies have attempted to individualize efavirenz dosing using prospective *CYP2B6* 516G>T genotyping (Gatanaga et al., 2007a; Rotger and Telenti, 2008; Torno et al., 2008). Unfortunately, the clinical utility of this test is limited by the fact that homozygosity for the T allele of *CYP2B6* 516G>T does not always result in elevated efavirenz plasma concentrations and these studies were underpowered. It is important to recognize that high plasma concentrations of both efavirenz and nevirapine have also been associated with improved virological outcomes (Marzolini et al., 2001; Veldkamp et al., 2001; Csajka et al., 2003; Gonzalez de Requena et al., 2005). A recent study has shown that Black patients receiving efavirenz with a composite *CYP2B6* 516/983 slow metabolizer genotype had lower rates of virologic failure (Ribaudo et al., 2010). A retrospective study on a pediatric cohort on nevirapine-based regimens demonstrated a percentage increase in CD4 cell count was three times higher in patients with the *CYP2B6* 516TT genotype compared to those with the 516GG genotype (Saitoh et al., 2007). Therefore, future studies to define the utility of *CYP2B6* genetics for avoiding CNS side effects need to be conducted with caution and incorporate analysis of virological response.

HLA subtypes have been implicated in nevirapine hypersensitivity reactions. In one study HLA-DRB1*0101 predicted the development of nevirapine hepatotoxicity in a cohort consisting of predominantly Caucasian patients with CD4 cell percentages of greater than 25% (Martin et al., 2005). In this study the occurrence of an isolated rash was not associated with CD4 cell percentage or HLA-DRB1*0101. In another study, the occurrence of isolated rash in Caucasian patients on efavirenz and nevirapine-based regimens was associated with the presence of HLA-DRB1*0101 but was not associated with CD4 cell percentages (Vitezica et al., 2008). In this study, 83% of the participants presenting with isolated rashes were HLA-DRB1*0101 positive as compared to the 7% in the tolerant group. HLA-cw8 has been reported to be a significant predictor of nevirapine HSR in Sardinian and Japanese populations (Gatanaga et al., 2007b; Littera et al., 2006). Despite the lack of a defined role for ABCB1 in the disposition of NNRTIs, *ABCB1* 3435C>T has been associated with a decreased risk of NNRTI-associated hepatotoxicity (Haas et al., 2006; Ritchie et al., 2006).

20.5.3 Protease Inhibitors

Protease inhibitors (PIs) form the bulk of antiretoviral compounds licensed for treatment. Before the development of newer drug classes, they represented the last option and were often used in various combinations to offer patients salvage therapy. Most of them cause considerable gastrointestinal symptoms and are associated with a metabolic sydrome including dyslipidemia, impaired insulin resistance, and lypodystrophy (Carr et al., 1998). Patients receiving either atazanavir or indinavir have an increased risk of developing unconjugated bilirubinemia, a reversible phenomenon whereby most patients have elevated laboratory values without any clinical evidence of scleral icterus. Indinavir therapy is associated with the development of renal calculi while tipranavir and darunavir are associated with significant hepatotoxicity (Hughes et al., 2009).

PIs are principally metabolized by CYP3A enzymes (Ernest et al., 2005) and are normally administered with a low dose of ritonavir, a potent CYP3A4 inhibitor to improve their bioavailability (Cooper et al., 2003). They are highly protein bound to both albumin and α1-acid glycoprotein (AAG; orosomucoid; ORM1) (Boffito et al., 2003) and are ABCB1 substrates (Marzolini et al., 2004). They have also recently been shown to be substrates for influx transporters OATP1A2, OATP1B1, and OATP1B3 (Hartkoorn et al., 2010).

CYP3A5 expressors (defined as individuals with the A allele for the *CYP3A5* 6986A>G polymorphism) have been reported to show faster oral clearance of indinavir (Anderson et al., 2006) and saquinavir (Frohlich et al., 2004; Mouly et al., 2005; Josephson et al., 2007). Nelfinavir, is predominantly metabolized by CYP2C19 into its major metabolite M8 with a minor contribution from CYP3A4 (Zhang et al., 2001). There is a confirmed association between *CYP2C19* 681G>A and higher nelfinavir exposure (Haas et al., 2005; Burger et al., 2006). However, studies exploring the impact of *ABCB1* variants on the disposition of PIs have produced conflicting results. The most studied variant, *ABCB1* 3435C>T, is a synonymous SNP, which is believed to change substrate specificity (Kimchi-Sarfaty et al., 2007). The impact of other MDR1 variants on the expression and function of ABCB1 remains somewhat controversial. It has been speculated that the high protein binding associated with PIs may also contribute to the intra-individual variability of protease inhibitor disposition. AAG variants have been reported to increase the apparent clearance of both lopinavir and indinavir to varying degrees without impacting on the cellular exposure of either drug (Colombo et al., 2006). The significance of these findings remains unclear. Recently, *SLCO1B1* (coding for OATP1B1) polymorphisms have been associated with lopinavir plasma concentrations in two studies (Hartkoorn et al., 2010; Lubomirov et al., 2010). Also a pregnane X receptor (PXR) polymorphism was associated with plasma concentrations of unboosted atazanavir (Siccardi et al., 2008). PXR correlates with expression of ABCB1 and CYP3A4 in liver (even in the absence of enzyme inducers) and importantly therefore, there is a biologically plausible mechanism for this association (Owen et al., 2004; Albermann et al., 2005).

Apolipoprotein E (APOE), a lipid transport protein with three major isoforms (apo ε2, apo ε3, and apo ε4) and apolipoprotein C3 (APOC3) another polymorphic (-482C>T, 455T>C, 3238C>G) lipid transport protein have both been studied extensively in HIV-negative individuals (Li et al., 1995; Mahley et al., 2000). Furthermore, the Swiss HIV Cohort used scoring algorithms to correlate the degree of hyperlipidemia with the number of unfavorable polymorphisms. In one study individuals with unfavorable APOE isoforms (ε2 or ε4) as well as more than two of the APOC3 variants were observed to have significant hypertriglyceridemia (>6 mmol/L) if they received ritonavir containing antiretroviral regimens (Tarr et al., 2005). In a subsequent study, the same group added apolipoprotein A5 (*APOA5*; non*1/*1 haplotypes), ABC transporter A1 (*ABCA1*; 2962A>G) and cholesteryl ester transfer protein (*CETP*; 279G>A) variants to create a scoring algorithm where patients received a score based on their composite *ABCA1/APOA5/APOC3/APOE/CETP* genotype and the type of antiretroviral therapy they were on (Arnedo et al., 2007). Both longitudinal studies were performed in predominantly Caucasian cohorts and therefore validation studies should also explore the associations in other ethnicities.

Most patients who develop unconjugated bilirubinemia while taking atazanavir or indinavir only do so to a moderate degree through a mechanism that mimics that of Gilbert's syndrome. There are a few patients, however, who develop overt, stigmatizing hyperbilirubinemia sufficient to consider discontinuation of therapy. A promoter polymorphism in UDP-glucuronosyltransferase 1A1 (UGT1A1*28) has been associated with the occurrence of unconjugated hyperbilirubinemia in patients on atazanavir and indinavir (Zucker et al., 2001; Rotger et al., 2005b).

20.5.4 Entry and Integrase Inhibitors

The last decade has been characterized by the co-formulation of existing agents to ease adherence, the development of new agents from existing classes, and the development of new agents from new classes with novel mechanisms of action. Relatively few studies have been conducted on newer drugs, partly because cohorts do not yet contain sufficient numbers of patients for robust analyses. However, there is knowledge of the mechanisms involved in disposition of these drugs allowing hypotheses about which genes may be of importance. For example, maraviroc is a CCR5 antagonist which is metabolized by CYP3A4 and CYP3A5 (MacArthur and Novak, 2008). Therefore many of the associations for protease inhibitors may be relevant to this drug. Similarly, raltegravir is an integrase inhibitor that is predominantly metabolized by UDP-glucuronosyltransferase (Kassahun et al., 2007). However, initial work to investigate the influence of *UGT1A1* polymorphisms on the disposition of raltegravir has not revealed any significant associations (Wenning et al., 2009).

20.6 Pharmacogenetics of Antimalarial Therapy

Worldwide, the treatment of malaria is hindered by the significant development of parasite resistance against existing compounds as well as the limited range of effective therapies. Currently the most effective compounds for Plasmodium falciparum malaria are quinine and artemisinin preparations. Chloroquine remains effective against non-falciparum malarias. This situation, however, is ever changing, especially on the Thai-Cambodia border, a region which has been at the center of most reports of antimalaria resistance.

20.6.1 Quinine

Quinine is the drug of choice worldwide for severe Plasmodium falciparum malaria. Although it is potentially associated with the development of a prolonged QT interval, hypoglycemia and cinchonism, it is generally well tolerated (Taylor and White, 2004). It is primarily metabolized by CYP3A enzymes (Zhang et al., 1997). A study comparing the impact of *CYP3A5* genotypes on the hydroxylation of quinine between Tanzanians and Swedes found lower hydroxylation in Tanzanians that were homozygous for *CYP3A5*3* (Mirghani et al., 2006). This finding is yet to be confirmed in other populations.

20.6.2 Artemisinin Compounds

Artemisinins are the newest class of antimalarial agents (Woodrow et al., 2005). They are generally well tolerated. The four WHO-recommended oral artemisinin–based combination therapies (dihydroartemisinin–piperaquine, artemether–lumefantrine, artesunate–mefloquine, and artesunate–amodiaquine) have rapidly become first-line agents in the treatment of uncomplicated P. falciparum malaria in endemic countries. Intravenous artesunate, on the other hand, is fast becoming an alternative agent in the treatment of severe P. falciparum malaria, mainly because of its rapid action against the erythrocytic stages of the parasite (Rosenthal, 2008). The metabolism of artemether-based compounds is complex. Artesunate, artemether, and arteether are primarily metabolized by CYP3A4, CYP3A5, and CYP2A6 with a minor contribution from CYP2B6 to form dihydroartemisinin (Navaratnam et al., 2000). Dihydroartemisinin is subsequently inactivated via UGT1A9 and UGT2B7 (Ilett et al., 2002). Artemisinin on the other hand is primarily metabolized by CYP2B6 with a minor contribution from CYP3A4 and CYP2A6 (Svensson and Ashton, 1999). Artemether, artemisinin, arteether, and dihydroartemisinin have all been shown to induce CYP3A4, CYP2B6, and ABCB1 through activation of pregnane X receptor (PXR) and constitutive androstane receptor (CAR) (Burk et al., 2005; Simonsson et al., 2006). There is no pharmacogenetic data on artemesinin compounds but associations described with antiretoviral compounds may also be relevant given the similarity in drug disposition pathways.

20.6.3 Primaquine

Primaquine is used in patients with *Plasmodium ovale* and *Plasmodium vivax* infections to clear the latent hepatic hypnozoite stage of the parasite. It is primarily metabolized by CYP1A2 and CYP3A4 (Li et al., 2003). Although plasma levels of primaquine have been found to be significantly associated with ethnicity, the genetic polymorphisms underpinning this variability or their impact on efficacy or toxicity are yet to be characterised (Fletcher et al., 1981; Kim et al., 2004). The first records of variability in response to antimalarials dates back to World War II when African-American soldiers were found to experience higher rates of acute hemolysis when they received primaquine compared to their Caucasian counterparts (Clayman et al., 1952). The basis of these observed differences was later attributed to glucose 6-phosphate dehydrogenase (G6PD) deficiency, an x-linked recessive disorder (Alving et al., 1956). The most common variant in African populations is A− while the most common variant in Mediterranean populations is B−. The *G6PD* locus is highly polymorphic, so in clinical practice prospective qualitative and quantitative tests are performed in patients requiring primaquine (Hill et al., 2006).

20.6.4 Amodiaquine

Amodiaquine is commonly used in the treatment of uncomplicated malaria in combination with artesunate in Africa. It is primarily metabolized by CYP2C8 (Li et al., 2002) and exhibits huge interindividual variability in plasma drug concentrations (White et al., 1987; Winstanley et al., 1987). *CYP2C8*2*, the variant most common in African populations, has been reported to be associated with reduced clearance, as has *CYP2C8*3* the variant most common in Caucasian populations (Gil and Gil, 2007; Parikh et al., 2007). Despite these findings, no associations were found with treatment efficacy or toxicity.

20.6.5 Mefloquine

Mefloquine is primarily metabolized by CYP3A4 (Fontaine et al., 2000) and is suspected to be a ABCB1 substrate (Pham et al., 2000). It is used in chemoprophylaxis and the most common side effects are neuropsychiatric. The ABCB1 1236TT/2677TT/3455TT haplotype has been reported to be associated with increased neuropsychiatric events in a homogenous Caucasian cohort despite having no impact on the plasma levels of the compound (Aarnoudse et al., 2006).

20.6.6 Proguanil

Proguanil is a component of the widely prescribed malarone (atovaquone-proguanil) which is used in both chemoprophylaxis and in treatment. The bioactivation of proguanil is primarily through CYP2C19 with a minor role from CYP3A4. *CYP2C19* variants have been shown to predict proguanil plasma concentrations (Kaneko et al., 1997; Herrlin et al., 2000) but not treatment outcomes (Edstein et al., 1996; Kaneko et al., 1999).

20.7 Pharmacogenetics of Antituberculosis Therapy

Like malaria, the effective treatment of mycobacteria infections is also limited by the development of antimicrobial resistance and the paucity of effective treatment options. Resistance is a particular issue in tuberculosis (TB) especially with the emergence of multidrug resistant TB (MDR TB), and more recently, extensively drug-resistant TB (XDR TB). The unprecedented re-emergence of TB with the HIV pandemic complicates matters further. The drugs used in first-line therapy, namely isoniazid, rifampicin, pyrazinamide, and ethambutol, have remained unchanged for over half a century.

In the early 1950s, individuals receiving isoniazid for the treatment of tuberculosis were noted to have marked differences in the amount of isoniazid excreted in their urine (Hughes, 1953). The basis of these differences was later attributed to differences in an individual's ability to acetylate isoniazid (Hughes et al., 1954). Thus, slow acetylators are prone to isoniazid toxicity, which manifests itself as peripheral neuropathy. The variant alleles that account for most of the variability observed in slow acetylators are *NAT2*5* and *NAT2*6* (Blum et al., 1991). In routine clinical practice, *NAT2* testing is not done. Instead pyridoxine is prescribed along with isoniazid in all patients to prevent the development of peripheral neuropathy.

Hepatotoxicity is the most frequent side effect of first-line antituberculosis compounds. A meta-analysis looking at reported associations between antituberculosis drug-induced liver injury and drug-metabolizing variants identified homozygous variants of *NAT2 mt, CYP2E1*1A,* and *GSTM1 null* as significant predictors of hepatotoxicity (Sun et al., 2008). However, it is worth noting that most of these studies were performed in Asian populations on varying anti-TB medication with unstandardized definitions of hepatotoxicity and uncharacterized environmental factors.

Most recently, the pharmacokinetics of rifampicin were shown to be associated with a polymorphism within the *SLCO1B1* gene (Weiner et al., 2010). In this study, the rifampicin AUC was approximately 36% lower in patients with the *SLCO1B1 463C>A* polymorphism compared to patients homozygous for the C allele. Importantly, *SLCO1B1* polymorphisms associated with lower rifampicin exposure were also more frequent in Black patients.

20.8 Summary and Perspective

Ideally, pharmacogenetic testing should be implemented in conjunction with pathogen resistance testing and the characterization of a compound's pharmacokinetic and pharmacodynamic characteristics. Apart from the idiosyncratic HSRs seen with abacavir and nevirapine, most studied clinical phenotypes are much less subtle, develop over time, and are multifactorial in etiology. Current challenges within infectious diseases include antimicrobial resistance and limited effective therapies for emerging, re-emerging, and neglected infections. Pharmacogenetic studies of both human host and disease pathogens may help tackle the resistance issue and genome-wide studies to identify new drug targets for emerging, re-emerging, and neglected infections may also be of use.

It is clear that currently the technology for collecting information on genetic variation across the entire genome has surpassed current mechanistic knowledge and the technology to quickly generate mechanistic data. There are clear gaps in knowledge and the systems for studying all the necessary proteins *in vitro* have not yet been developed. With respect to associations that have been reliably re-produced in exploratory analyses, there is a lack of clinical validation. These are time consuming and very expensive, and funding streams for prospective studies are not well defined. There is currently debate over whether prospective studies are always necessary for implementation and it seems likely that this will ultimately depend on the specific association and how it can be applied.

References

Aarnoudse, A.L., van Schaik, R.H., Dieleman, J., Molokhia, M., van Riemsdijk, M.M., Ligthelm, R.J., et al., 2006. MDR1 gene polymorphisms are associated with neuropsychiatric adverse effects of mefloquine. Clin. Pharmacol. Ther. 80 (4), 367–374.

Administration USDoHaHSFaD, 2006. Guidance for Industry. Drug Interaction Studies— Study Design, Data Analysis, and Implications for Dosing and Labeling. [updated September 2006, 19/8/10].

Albermann, N., Schmitz-Winnenthal, F.H., Z'Graggen, K., Volk, C., Hoffmann, M.M., Haefeli, W.E., et al., 2005. Expression of the drug transporters MDR1/ABCB1, MRP1/ABCC1, MRP2/ABCC2, BCRP/ABCG2, and PXR in peripheral blood mononuclear cells and their relationship with the expression in intestine and liver. Biochem. Pharmacol. 70 (6), 949–958.

Almond, L.M., Hoggard, P.G., Edirisinghe, D., Khoo, S.H., Back, D.J., 2005. Intracellular and plasma pharmacokinetics of efavirenz in HIV-infected individuals. J. Antimicrob. Chemother. 56 (4), 738–744.

Alving, A.S., Carson, P.E., Flanagan, C.L., Ickes, C.E., 1956. Enzymatic deficiency in primaquine-sensitive erythrocytes. Science 124 (3220), 484–485.

Anderson, P.L., Kakuda, T.N., Lichtenstein, K.A., 2004. The cellular pharmacology of nucleoside- and nucleotide-analogue reverse-transcriptase inhibitors and its relationship to clinical toxicities. Clin. Infect. Dis. 38 (5), 743–753.

Anderson, P.L., Lamba, J., Aquilante, C.L., Schuetz, E., Fletcher, C.V., 2006. Pharmacogenetic characteristics of indinavir, zidovudine, and lamivudine therapy in HIV-infected adults: a pilot study. J. Acquir. Immune. Defic. Syndr. 42 (4), 441–449.

Arnedo, M., Taffe, P., Sahli, R., Furrer, H., Hirschel, B., Elzi, L., et al., 2007. Contribution of 20 single nucleotide polymorphisms of 13 genes to dyslipidemia associated with antiretroviral therapy. Pharmacogenet. Genomics 17 (9), 755–764.

Bates, D.W., Spell, N., Cullen, D.J., Burdick, E., Laird, N., Petersen, L.A., et al., 1997. The costs of adverse drug events in hospitalized patients. Adverse Drug Events Prevention Study Group. JAMA 277 (4), 307–311.

Blum, M., Demierre, A., Grant, D.M., Heim, M., Meyer, U.A., 1991. Molecular mechanism of slow acetylation of drugs and carcinogens in humans. Proc. Natl. Acad. Sci. U.S.A. 88 (12), 5237–5241.

Boffito, M., Back, D.J., Blaschke, T.F., Rowland, M., Bertz, R.J., Gerber, J.G., et al., 2003. Protein binding in antiretroviral therapies. AIDS Res. Hum. Retroviruses 19 (9), 825–835.

Brinkman, K., Smeitink, J.A., Romijn, J.A., Reiss, P., 1999. Mitochondrial toxicity induced by nucleoside-analogue reverse-transcriptase inhibitors is a key factor in the pathogenesis of antiretroviral-therapy-related lipodystrophy. Lancet 354 (9184), 1112–1115.

Burger, D.M., Schwietert, H.R., Colbers, E.P., Becker, M., 2006. The effect of the CYP2C19*2 heterozygote genotype on the pharmacokinetics of nelfinavir. Br. J. Clin. Pharmacol. 62 (2), 250–252.

Burk, O., Arnold, K.A., Nussler, A.K., Schaeffeler, E., Efimova, E., Avery, B.A., et al., 2005. Antimalarial artemisinin drugs induce cytochrome P450 and MDR1 expression by activation of xenosensors pregnane X receptor and constitutive androstane receptor. Mol. Pharmacol. 67 (6), 1954–1965.

Calmy, A., Hirschel, B., Cooper, D.A., Carr, A., 2007. Clinical update: adverse effects of antiretroviral therapy. Lancet 370 (9581), 12–14.

Canter, J.A., Robbins, G.K., Selph, D., Clifford, D.B., Kallianpur, A.R., Shafer, R., et al., 2010. African mitochondrial DNA subhaplogroups and peripheral neuropathy during antiretroviral therapy. J. Infect. Dis. 201 (11), 1703–1707.

Carr, A., Samaras, K., Burton, S., Law, M., Freund, J., Chisholm, D.J., et al., 1998. A syndrome of peripheral lipodystrophy, hyperlipidaemia and insulin resistance in patients receiving HIV protease inhibitors. AIDS 12 (7), F51–F58.

Carr, D.F., la Porte, C.J., Pirmohamed, M., Owen, A., Cortes, C.P., 2010. Haplotype structure of CYP2B6 and association with plasma efavirenz concentrations in a Chilean HIV cohort. J. Antimicrob. Chemother. 65 (9), 1889–1893.

Chang, T.K., Bandiera, S.M., Chen, J., 2003. Constitutive androstane receptor and pregnane X receptor gene expression in human liver: interindividual variability and correlation with CYP2B6 mRNA levels. Drug Metab. Dispos. 31 (1), 7–10.

Clayman, C.B., Arnold, J., Hockwald, R.S., Yount Jr., E.H., Edgcomb, J.H., Alving, A.S., 1952. Toxicity of primaquine in Caucasians. J. Am. Med. Assoc. 149 (17), 1563–1568.

Clumeck, N., Pozniak, A., Raffi, F., 2008. European AIDS Clinical Society (EACS) guidelines for the clinical management and treatment of HIV-infected adults. HIV Med. 9 (2), 65–71.

Colombo, S., Buclin, T., Decosterd, L.A., Telenti, A., Furrer, H., Lee, B.L., et al., 2006. Orosomucoid (alpha1-acid glycoprotein) plasma concentration and genetic variants: effects on human immunodeficiency virus protease inhibitor clearance and cellular accumulation. Clin. Pharmacol. Ther. 80 (4), 307–318.

Cooper, C.L., van Heeswijk, R.P., Gallicano, K., Cameron, D.W., 2003. A review of low-dose ritonavir in protease inhibitor combination therapy. Clin. Infect. Dis. 36 (12), 1585–1592.

Cortes C, Chaikan A, Owen A, Zhang G, Siccardi M, La Porte C. Demographic and genetic correlates of efavirenz plasma concentrations in HIV+ Chilean patients, in press.

Csajka, C., Marzolini, C., Fattinger, K., Decosterd, L.A., Fellay, J., Telenti, A., et al., 2003. Population pharmacokinetics and effects of efavirenz in patients with human immunodeficiency virus infection. Clin. Pharmacol. Ther. 73 (1), 20–30.

Cutrell, A.G., Hernandez, J.E., Fleming, J.W., Edwards, M.T., Moore, M.A., Brothers, C.H., et al., 2004. Updated clinical risk factor analysis of suspected hypersensitivity reactions to abacavir. Ann. Pharmacother. 38 (12), 2171–2172.

Davies, E.C., Green, C.F., Taylor, S., Williamson, P.R., Mottram, D.R., Pirmohamed, M., 2009. Adverse drug reactions in hospital in-patients: a prospective analysis of 3695 patient-episodes. PLoS One 4 (2), e4439.
Desta, Z., Saussele, T., Ward, B., Blievernicht, J., Li, L., Klein, K., et al., 2007. Impact of CYP2B6 polymorphism on hepatic efavirenz metabolism in vitro. Pharmacogenomics 8 (6), 547–558.
di Iulio, J., Fayet, A., Arab-Alameddine, M., Rotger, M., Lubomirov, R., Cavassini, M., et al., 2009. In vivo analysis of efavirenz metabolism in individuals with impaired CYP2A6 function. Pharmacogenet. Genomics 19 (4), 300–309.
Dieterich, D.T., Robinson, P.A., Love, J., Stern, J.O., 2004. Drug-induced liver injury associated with the use of nonnucleoside reverse-transcriptase inhibitors. Clin. Infect. Dis. 38 (Suppl 2), S80–S89.
Edstein, M.D., Yeo, A.E., Kyle, D.E., Looareesuwan, S., Wilairatana, P., Rieckmann, K.H., 1996. Proguanil polymorphism does not affect the antimalarial activity of proguanil combined with atovaquone in vitro. Trans. R Soc. Trop. Med. Hyg. 90 (4), 418–421.
Erickson, D.A., Mather, G., Trager, W.F., Levy, R.H., Keirns, J.J., 1999. Characterization of the in vitro biotransformation of the HIV-1 reverse transcriptase inhibitor nevirapine by human hepatic cytochromes P-450. Drug Metab. Dispos. 27 (12), 1488–1495.
Ernest 2nd, C.S., Hall, S.D., Jones, D.R., 2005. Mechanism-based inactivation of CYP3A by HIV protease inhibitors. J. Pharmacol. Exp. Ther. 312 (2), 583–591.
Fletcher, K.A., Evans, D.A., Gilles, H.M., Greaves, J., Bunnag, D., Harinasuta, T., 1981. Studies on the pharmacokinetics of primaquine. Bull. World Health Organ. 59 (3), 407–412.
Fontaine, F., de Sousa, G., Burcham, P.C., Duchene, P., Rahmani, R., 2000. Role of cytochrome P450 3A in the metabolism of mefloquine in human and animal hepatocytes. Life Sci. 66 (22), 2193–2212.
Friis-Moller, N., Reiss, P., Sabin, C.A., Weber, R., Monforte, A., El-Sadr, W., et al., 2007. Class of antiretroviral drugs and the risk of myocardial infarction. N. Engl. J. Med. 356 (17), 1723–1735.
Friis-Moller, N., Sabin, C.A., Weber, R., d'Arminio Monforte, A., El-Sadr, W.M., Reiss, P., et al., 2003. Combination antiretroviral therapy and the risk of myocardial infarction. N Engl. J. Med. 349 (21), 1993–2003.
Frohlich, M., Hoffmann, M.M., Burhenne, J., Mikus, G., Weiss, J., Haefeli, W.E., 2004. Association of the CYP3A5 A6986G (CYP3A5*3) polymorphism with saquinavir pharmacokinetics. Br. J. Clin. Pharmacol. 58 (4), 443–444.
Gallego, L., Barreiro, P., del Rio, R., Gonzalez de Requena, D., Rodriguez-Albarino, A., Gonzalez-Lahoz, J., et al., 2004. Analyzing sleep abnormalities in HIV-infected patients treated with Efavirenz. Clin Infect Dis. 38 (3), 430–432.
Gatanaga, H., Hayashida, T., Tsuchiya, K., Yoshino, M., Kuwahara, T., Tsukada, H., et al., 2007a. Successful efavirenz dose reduction in HIV type 1-infected individuals with cytochrome P450 2B6 *6 and *26. Clin. Infect. Dis. 45 (9), 1230–1237.
Gatanaga, H., Yazaki, H., Tanuma, J., Honda, M., Genka, I., Teruya, K., et al., 2007b. HLA-Cw8 primarily associated with hypersensitivity to nevirapine. AIDS 21 (2), 264–265.
Gazzard, B.G., 1999. Efavirenz in the management of HIV infection. Int. J. Clin. Pract. 53 (1), 60–64.
Gazzard, B.G., Anderson, J., Babiker, A., Boffito, M., Brook, G., Brough, G., et al., 2008. British HIV Association Guidelines for the treatment of HIV-1-infected adults with antiretroviral therapy 2008. HIV Med. 9 (8), 563–608.

Ge, D., Fellay, J., Thompson, A.J., Simon, J.S., Shianna, K.V., Urban, T.J., et al., 2009. Genetic variation in IL28B predicts hepatitis C treatment-induced viral clearance. Nature 461 (7262), 399–401.

Gil, J.P., Gil Berglund, E., 2007. CYP2C8 and antimalaria drug efficacy. Pharmacogenomics 8 (2), 187–198.

Gonzalez de Requena, D., Bonora, S., Garazzino, S., Sciandra, M., D'Avolio, A., Raiteri, R., et al., 2005. Nevirapine plasma exposure affects both durability of viral suppression and selection of nevirapine primary resistance mutations in a clinical setting. Antimicrob. Agents Chemother. 49 (9), 3966–3969.

Haas, D.W., Ribaudo, H.J., Kim, R.B., Tierney, C., Wilkinson, G.R., Gulick, R.M., et al., 2004. Pharmacogenetics of efavirenz and central nervous system side effects: an Adult AIDS Clinical Trials Group study. AIDS 18 (18), 2391–2400.

Haas, D.W., Smeaton, L.M., Shafer, R.W., Robbins, G.K., Morse, G.D., Labbe, L., et al., 2005. Pharmacogenetics of long-term responses to antiretroviral regimens containing Efavirenz and/or Nelfinavir: an Adult Aids Clinical Trials Group Study. J. Infect. Dis. 192 (11), 1931–1942.

Haas, D.W., Bartlett, J.A., Andersen, J.W., Sanne, I., Wilkinson, G.R., Hinkle, J., et al., 2006. Pharmacogenetics of nevirapine-associated hepatotoxicity: an Adult AIDS Clinical Trials Group collaboration. Clin. Infect. Dis. 43 (6), 783–786.

Hartkoorn, R.C., Kwan, W.S., Shallcross, V., Chaikan, A., Liptrott, N., Egan, D., et al., 2010. HIV protease inhibitors are substrates for OATP1A2, OATP1B1 and OATP1B3 and lopinavir plasma concentrations are influenced by SLCO1B1 polymorphisms. Pharmacogenet. Genomics 20 (2), 112–120.

Herrlin, K., Massele, A.Y., Rimoy, G., Alm, C., Rais, M., Ericsson, O., et al., 2000. Slow chloroguanide metabolism in Tanzanians compared with white subjects and Asian subjects confirms a decreased CYP2C19 activity in relation to genotype. Clin. Pharmacol. Ther. 68 (2), 189–198.

Hill, D.R., Baird, J.K., Parise, M.E., Lewis, L.S., Ryan, E.T., Magill, A.J., 2006. Primaquine: report from CDC expert meeting on malaria chemoprophylaxis I. Am. J. Trop. Med. Hyg. 75 (3), 402–415.

Hoehe, M.R., Timmermann, B., Lehrach, H., 2003. Human inter-individual DNA sequence variation in candidate genes, drug targets, the importance of haplotypes and pharmacogenomics. Curr. Pharm. Biotechnol. 4 (6), 351–378.

Huang, Y., Paxton, W.A., Wolinsky, S.M., Neumann, A.U., Zhang, L., He, T., et al., 1996. The role of a mutant CCR5 allele in HIV-1 transmission and disease progression. Nat. Med. 2 (11), 1240–1243.

Hughes, A.R., Mosteller, M., Bansal, A.T., Davies, K., Haneline, S.A., Lai, E.H., et al., 2004. Association of genetic variations in HLA-B region with hypersensitivity to abacavir in some, but not all, populations. Pharmacogenomics 5 (2), 203–211.

Hughes, C.A., Robinson, L., Tseng, A., MacArthur, R.D., 2009. New antiretroviral drugs: a review of the efficacy, safety, pharmacokinetics, and resistance profile of tipranavir, darunavir, etravirine, rilpivirine, maraviroc, and raltegravir. Expert. Opin. Pharmacother. 10(15), 2445–2466.

Hughes, H.B., 1953. On the metabolic fate of isoniazid. J. Pharmacol. Exp. Ther. 109 (4), 444–452.

Hughes, H.B., Biehl, J.P., Jones, A.P., Schmidt, L.H., 1954. Metabolism of isoniazid in man as related to the occurrence of peripheral neuritis. Am. Rev. Tuberc. 70 (2), 266–273.

Hulgan, T., Tebas, P., Canter, J.A., Mulligan, K., Haas, D.W., Dube, M., et al., 2008. Hemochromatosis gene polymorphisms, mitochondrial haplogroups, and peripheral lipoatrophy during antiretroviral therapy. J. Infect. Dis. 197 (6), 858–866.

Ilett, K.F., Ethell, B.T., Maggs, J.L., Davis, T.M., Batty, K.T., Burchell, B., et al., 2002. Glucuronidation of dihydroartemisinin in vivo and by human liver microsomes and expressed UDP-glucuronosyltransferases. Drug Metab. Dispos. 30 (9), 1005−1012.

Izzedine, H., Hulot, J.S., Villard, E., Goyenvalle, C., Dominguez, S., Ghosn, J., et al., 2006. Association between ABCC2 gene haplotypes and tenofovir-induced proximal tubulopathy. J. Infect. Dis. 194 (11), 1481−1491.

Josephson, F., Allqvist, A., Janabi, M., Sayi, J., Aklillu, E., Jande, M., et al., 2007. CYP3A5 genotype has an impact on the metabolism of the HIV protease inhibitor saquinavir. Clin. Pharmacol. Ther. 81 (5), 708−712.

Kalow, W., Tang, B.K., Endrenyi, L., 1998. Hypothesis: comparisons of inter- and intra-individual variations can substitute for twin studies in drug research. Pharmacogenetics 8 (4), 283−289.

Kaneko, A., Kaneko, O., Taleo, G., Bjorkman, A., Kobayakawa, T., 1997. High frequencies of CYP2C19 mutations and poor metabolism of proguanil in Vanuatu. Lancet 349 (9056), 921−922.

Kaneko, A., Bergqvist, Y., Taleo, G., Kobayakawa, T., Ishizaki, T., Bjorkman, A., 1999. Proguanil disposition and toxicity in malaria patients from Vanuatu with high frequencies of CYP2C19 mutations. Pharmacogenetics 9 (3), 317−326.

Kappelhoff, B.S., van Leth, F., Robinson, P.A., MacGregor, T.R., Baraldi, E., Montella, F., et al., 2005. Are adverse events of nevirapine and efavirenz related to plasma concentrations? Antivir. Ther. 10 (4), 489−498.

Kassahun, K., McIntosh, I., Cui, D., Hreniuk, D., Merschman, S., Lasseter, K., et al., 2007. Metabolism and disposition in humans of raltegravir (MK-0518), an anti-AIDS drug targeting the human immunodeficiency virus 1 integrase enzyme. Drug. Metab. Dispos. 35 (9), 1657−1663.

Kim, Y.R., Kuh, H.J., Kim, M.Y., Kim, Y.S., Chung, W.C., Kim, S.I., et al., 2004. Pharmacokinetics of primaquine and carboxyprimaquine in Korean patients with vivax malaria. Arch. Pharm. Res. 27 (5), 576−580.

Kimchi-Sarfaty, C., Oh, J.M., Kim, I.W., Sauna, Z.E., Calcagno, A.M., Ambudkar, S.V., et al., 2007. A "silent" polymorphism in the MDR1 gene changes substrate specificity. Science 315 (5811), 525−528.

Klein, K., Lang, T., Saussele, T., Barbosa-Sicard, E., Schunck, W.H., Eichelbaum, M., et al., 2005. Genetic variability of CYP2B6 in populations of African and Asian origin: allele frequencies, novel functional variants, and possible implications for anti-HIV therapy with efavirenz. Pharmacogenet. Genomics 15 (12), 861−873.

Kwara, A., Lartey, M., Sagoe, K.W., Kenu, E., CYP2B6, Court MH., 2009. CYP2A6 and UGT2B7 genetic polymorphisms are predictors of efavirenz mid-dose concentration in HIV-infected patients. AIDS 23 (16), 2101−2106.

Li, W.W., Dammerman, M.M., Smith, J.D., Metzger, S., Breslow, J.L., Leff, T., 1995. Common genetic variation in the promoter of the human apo CIII gene abolishes regulation by insulin and may contribute to hypertriglyceridemia. J. Clin. Invest. 96 (6), 2601−2605.

Li, X.Q., Bjorkman, A., Andersson, T.B., Ridderstrom, M., Masimirembwa, C.M., 2002. Amodiaquine clearance and its metabolism to N-desethylamodiaquine is mediated by CYP2C8: a new high affinity and turnover enzyme-specific probe substrate. J. Pharmacol. Exp. Ther. 300 (2), 399−407.

Li, X.Q., Bjorkman, A., Andersson, T.B., Gustafsson, L.L., Masimirembwa, C.M., 2003. Identification of human cytochrome P(450)s that metabolise anti-parasitic drugs and predictions of in vivo drug hepatic clearance from in vitro data. Eur. J. Clin. Pharmacol. 59 (5-6), 429−442.

Littera, R., Carcassi, C., Masala, A., Piano, P., Serra, P., Ortu, F., et al., 2006. HLA-dependent hypersensitivity to nevirapine in Sardinian HIV patients. AIDS 20 (12), 1621–1626.

Lochet, P., Peyriere, H., Lotthe, A., Mauboussin, J.M., Delmas, B., Reynes, J., 2003. Long-term assessment of neuropsychiatric adverse reactions associated with efavirenz. HIV Med. 4 (1), 62–66.

Lubomirov, R., di Iulio, J., Fayet, A., Colombo, S., Martinez, R., Marzolini, C., et al., 2010. ADME pharmacogenetics: investigation of the pharmacokinetics of the antiretroviral agent lopinavir coformulated with ritonavir. Pharmacogenet. Genomics 20 (4), 217–230.

MacArthur, R.D., Novak, R.M., 2008. Reviews of anti-infective agents: maraviroc: the first of a new class of antiretroviral agents. Clin. Infect. Dis. 47 (2), 236–241.

Mahley, R.W., Rall Jr., S.C., 2000. Apolipoprotein E far more than a lipid transport protein. Annu. Rev. Genomics Hum. Genet. 1, 507–537.

Mahungu, T.W., Rodger, A.J., Johnson, M.A., 2009a. HIV as a chronic disease. Clin. Med. 9 (2), 125–128.

Mahungu, T.W., Smith, C., Turner, F., Egan, D., Youle, M., Johnson, M., et al., 2009b. Cytochrome P450 2B6 516G>T is associated with plasma concentrations of nevirapine at both 200 mg twice daily and 400 mg once daily in an ethnically diverse population. HIV Med. 10 (5), 310–317.

Mallal, S., Nolan, D., Witt, C., Masel, G., Martin, A.M., Moore, C., et al., 2002. Association between presence of HLA-B*5701, HLA-DR7, and HLA-DQ3 and hypersensitivity to HIV-1 reverse-transcriptase inhibitor abacavir. Lancet 359 (9308), 727–732.

Mallal, S., Phillips, E., Carosi, G., Molina, J.M., Workman, C., Tomazic, J., et al., 2008. HLA-B*5701 screening for hypersensitivity to abacavir. N. Engl. J. Med. 358 (6), 568–579.

Mallants, R., Van Oosterwyck, K., Van Vaeck, L., Mols, R., De Clercq, E., Augustijns, P., 2005. Multidrug resistance-associated protein 2 (MRP2) affects hepatobiliary elimination but not the intestinal disposition of tenofovir disoproxil fumarate and its metabolites. Xenobiotica 35 (10–11), 1055–1066.

Martin, A.M., Nolan, D., James, I., Cameron, P., Keller, J., Moore, C., et al., 2005. Predisposition to nevirapine hypersensitivity associated with HLA-DRB1*0101 and abrogated by low CD4 T-cell counts. AIDS 19 (1), 97–99.

Marzolini, C., Paus, E., Buclin, T., Kim, R.B., 2004. Polymorphisms in human MDR1 (P-glycoprotein): recent advances and clinical relevance. Clin. Pharmacol. Ther. 75 (1), 13–33.

Marzolini, C., Telenti, A., Decosterd, L.A., Greub, G., Biollaz, J., Buclin, T., 2001. Efavirenz plasma levels can predict treatment failure and central nervous system side effects in HIV-1-infected patients. AIDS 15 (1), 71–75.

Meyer, U.A., 2004. Pharmacogenetics—five decades of therapeutic lessons from genetic diversity. Nat. Rev. Genet. 5 (9), 669–676.

Mirghani, R.A., Sayi, J., Aklillu, E., Allqvist, A., Jande, M., Wennerholm, A., et al., 2006. CYP3A5 genotype has significant effect on quinine 3-hydroxylation in Tanzanians, who have lower total CYP3A activity than a Swedish population. Pharmacogenet. Genomics 16 (9), 637–645.

Moore, N., Lecointre, D., Noblet, C., Mabille, M., 1998. Frequency and cost of serious adverse drug reactions in a department of general medicine. Br. J. Clin. Pharmacol. 45 (3), 301–308.

Mouly, S.J., Matheny, C., Paine, M.F., Smith, G., Lamba, J., Lamba, V., et al., 2005. Variation in oral clearance of saquinavir is predicted by CYP3A5*1 genotype but not

by enterocyte content of cytochrome P450 3A5. Clin. Pharmacol. Ther. 78 (6), 605−618.
Navaratnam, V., Mansor, S.M., Sit, N.W., Grace, J., Li, Q., Olliaro, P., 2000. Pharmacokinetics of artemisinin-type compounds. Clin. Pharmacokinet. 39 (4), 255−270.
Nunez, M., Gonzalez de Requena, D., Gallego, L., Jimenez-Nacher, I., Gonzalez-Lahoz, J., Soriano, V., 2001. Higher efavirenz plasma levels correlate with development of insomnia. J. Acquir Immune. Defic. Syndr. 28 (4), 399−400.
Owen, A., Chandler, B., Back, D.J., Khoo, S.H., 2004. Expression of pregnane-X-receptor transcript in peripheral blood mononuclear cells and correlation with MDR1 mRNA. Antivir. Ther. 9 (5), 819−821.
Parikh, S., Ouedraogo, J.B., Goldstein, J.A., Rosenthal, P.J., Kroetz, D.L., 2007. Amodiaquine metabolism is impaired by common polymorphisms in CYP2C8: implications for malaria treatment in Africa. Clin. Pharmacol. Ther. 82 (2), 197−203.
Penzak, S.R., Kabuye, G., Mugyenyi, P., Mbamanya, F., Natarajan, V., Alfaro, R.M., et al., 2007. Cytochrome P450 2B6 (CYP2B6) G516T influences nevirapine plasma concentrations in HIV-infected patients in Uganda. HIV Med. 8 (2), 86−91.
Pham, Y.T., Regina, A., Farinotti, R., Couraud, P., Wainer, I.W., Roux, F., et al., 2000. Interactions of racemic mefloquine and its enantiomers with P-glycoprotein in an immortalised rat brain capillary endothelial cell line, GPNT. Biochim. Biophys. Acta 1524 (2-3), 212−219.
Ray, A.S., Cihlar, T., Robinson, K.L., Tong, L., Vela, J.E., Fuller, M.D., et al., 2006. Mechanism of active renal tubular efflux of tenofovir. Antimicrob. Agents Chemother. 50 (10), 3297−3304.
Ribaudo, H.J., Liu, H., Schwab, M., Schaeffeler, E., Eichelbaum, M., Motsinger-Reif, A.A., et al., 2010. Effect of CYP2B6, ABCB1, and CYP3A5 polymorphisms on efavirenz pharmacokinetics and treatment response: an AIDS Clinical Trials Group study. J. Infect. Dis. 202 (5), 717−722.
Ritchie, M.D., Haas, D.W., Motsinger, A.A., Donahue, J.P., Erdem, H., Raffanti, S., et al., 2006. Drug transporter and metabolizing enzyme gene variants and nonnucleoside reverse-transcriptase inhibitor hepatotoxicity. Clin. Infect. Dis. 43 (6), 779−782.
Rodriguez-Novoa, S., Barreiro, P., Rendon, A., Jimenez-Nacher, I., Gonzalez-Lahoz, J., Soriano, V., 2005. Influence of 516G>T polymorphisms at the gene encoding the CYP450-2B6 isoenzyme on efavirenz plasma concentrations in HIV-infected subjects. Clin. Infect. Dis. 40 (9), 1358−1361.
Rodriguez-Novoa, S., Labarga, P., Soriano, V., Egan, D., Albalater, M., Morello, J., et al., 2009. Predictors of kidney tubular dysfunction in HIV-infected patients treated with tenofovir: a pharmacogenetic study. Clin. Infect. Dis. 48 (11), e108−e116.
Rosenthal, P.J., 2008. Artesunate for the treatment of severe falciparum malaria. N. Engl. J. Med. 358 (17), 1829−1836.
Rotger, M., Telenti, A., 2008. Optimizing efavirenz treatment: CYP2B6 genotyping or therapeutic drug monitoring? Eur. J. Clin. Pharmacol. 64 (4), 335−336.
Rotger, M., Colombo, S., Furrer, H., Bleiber, G., Buclin, T., Lee, B.L., et al., 2005a. Influence of CYP2B6 polymorphism on plasma and intracellular concentrations and toxicity of efavirenz and nevirapine in HIV-infected patients. Pharmacogenet. Genomics 15 (1), 1−5.
Rotger, M., Taffe, P., Bleiber, G., Gunthard, H.F., Furrer, H., Vernazza, P., et al., 2005b. Gilbert syndrome and the development of antiretroviral therapy-associated hyperbilirubinemia. J. Infect. Dis. 192 (8), 1381−1386.

Saag, M., Balu, R., Phillips, E., Brachman, P., Martorell, C., Burman, W., et al., 2008. High sensitivity of human leukocyte antigen-b*5701 as a marker for immunologically confirmed abacavir hypersensitivity in white and black patients. Clin. Infect. Dis. 46 (7), 1111–1118.

Sabin, C.A., Worm, S.W., Weber, R., Reiss, P., El-Sadr, W., Dabis, F., et al., 2008. Use of nucleoside reverse transcriptase inhibitors and risk of myocardial infarction in HIV-infected patients enrolled in the D:A:D study: a multi-cohort collaboration. Lancet 371 (9622), 1417–1426.

Saitoh, A., Sarles, E., Capparelli, E., Aweeka, F., Kovacs, A., Burchett, S.K., et al., 2007. CYP2B6 genetic variants are associated with nevirapine pharmacokinetics and clinical response in HIV-1-infected children. AIDS 21 (16), 2191–2199.

Schuetz, J.D., Connelly, M.C., Sun, D., Paibir, S.G., Flynn, P.M., Srinivas, R.V., et al., 1999. MRP4: a previously unidentified factor in resistance to nucleoside-based antiviral drugs. Nat. Med. 5 (9), 1048–1051.

Setzer, B., Schlesier, M., Thomas, A.K., Walker, U.A., 2005. Mitochondrial toxicity of nucleoside analogues in primary human lymphocytes. Antivir. Ther. 10 (2), 327–334.

Shelby, M.K., Klaassen, C.D., 2006. Induction of rat UDP-glucuronosyltransferases in liver and duodenum by microsomal enzyme inducers that activate various transcriptional pathways. Drug Metab. Dispos. 34 (10), 1772–1778.

Siccardi, M., D'Avolio, A., Baietto, L., Gibbons, S., Sciandra, M., Colucci, D., et al., 2008. Association of a single-nucleotide polymorphism in the pregnane X receptor (PXR 63396C->T) with reduced concentrations of unboosted atazanavir. Clin. Infect. Dis. 47 (9), 1222–1225.

Simonsson, U.S., Lindell, M., Raffalli-Mathieu, F., Lannerbro, A., Honkakoski, P., Lang, M.A., 2006. In vivo and mechanistic evidence of nuclear receptor CAR induction by artemisinin. Eur. J. Clin. Invest. 36 (9), 647–653.

Stern, J.O., Robinson, P.A., Love, J., Lanes, S., Imperiale, M.S., Mayers, D.L., 2003. A comprehensive hepatic safety analysis of nevirapine in different populations of HIV infected patients. J. Acquir Immune Defic. Syndr. 34 (Suppl 1), S21–S33.

Storch, C.H., Theile, D., Lindenmaier, H., Haefeli, W.E., Weiss, J., 2007. Comparison of the inhibitory activity of anti-HIV drugs on P-glycoprotein. Biochem. Pharmacol. 73 (10), 1573–1581.

Stormer, E., von Moltke, L.L., Perloff, M.D., Greenblatt, D.J., 2002. Differential modulation of P-glycoprotein expression and activity by non-nucleoside HIV-1 reverse transcriptase inhibitors in cell culture. Pharm. Res. 19 (7), 1038–1045.

Sun, F., Chen, Y., Xiang, Y., Zhan, S., 2008. Drug-metabolising enzyme polymorphisms and predisposition to anti-tuberculosis drug-induced liver injury: a meta-analysis. Int. J. Tuberc. Lung. Dis. 12 (9), 994–1002.

Svensson, U.S., Ashton, M., 1999. Identification of the human cytochrome P450 enzymes involved in the in vitro metabolism of artemisinin. Br. J. Clin. Pharmacol. 48 (4), (Oct), 528–535.

Takenaka, K., Morgan, J.A., Scheffer, G.L., Adachi, M., Stewart, C.F., Sun, D., et al., 2007. Substrate overlap between Mrp4 and Abcg2/Bcrp affects purine analogue drug cytotoxicity and tissue distribution. Cancer Res. 67 (14), 6965–6972.

Tarr, P.E., Taffe, P., Bleiber, G., Furrer, H., Rotger, M., Martinez, R., et al., 2005. Modeling the influence of APOC3, APOE, and TNF polymorphisms on the risk of antiretroviral therapy-associated lipid disorders. J. Infect. Dis. 191 (9), 1419–1426.

Taylor, W.R., White, N.J., 2004. Antimalarial drug toxicity: a review. Drug Saf. 27 (1), 25–61.

Torno, M.S., Witt, M.D., Saitoh, A., Fletcher, C.V., 2008. Successful use of reduced-dose efavirenz in a patient with human immunodeficiency virus infection: case report and review of the literature. Pharmacotherapy 28 (6), 782−787.

Uwai, Y., Ida, H., Tsuji, Y., Katsura, T., Inui, K., 2007. Renal transport of adefovir, cidofovir, and tenofovir by SLC22A family members (hOAT1, hOAT3, and hOCT2). Pharm. Res. 24 (4), 811−815.

Veal, G.J., Back, D.J., 1995. Metabolism of Zidovudine. Gen. Pharmacol. 26 (7), 1469−1475.

Veldkamp, A.I., Weverling, G.J., Lange, J.M., Montaner, J.S., Reiss, P., Cooper, D.A., et al., 2001. High exposure to nevirapine in plasma is associated with an improved virological response in HIV-1-infected individuals. AIDS 15 (9), 1089−1095.

Vitezica, Z.G., Milpied, B., Lonjou, C., Borot, N., Ledger, T.N., Lefebvre, A., et al., 2008. HLA-DRB1*01 associated with cutaneous hypersensitivity induced by nevirapine and efavirenz. AIDS 22 (4), 540−541.

Wainberg, M.A., 2003. HIV resistance to nevirapine and other non-nucleoside reverse transcriptase inhibitors. J. Acquir. Immune. Defic. Syndr. 34 (Suppl 1), S2−S7.

Wang, J., Sonnerborg, A., Rane, A., Josephson, F., Lundgren, S., Stahle, L., et al., 2006. Identification of a novel specific CYP2B6 allele in Africans causing impaired metabolism of the HIV drug efavirenz. Pharmacogenet. Genomics 16 (3), 191−198.

Ward, B.A., Gorski, J.C., Jones, D.R., Hall, S.D., Flockhart, D.A., Desta, Z., 2003. The cytochrome P450 2B6 (CYP2B6) is the main catalyst of efavirenz primary and secondary metabolism: implication for HIV/AIDS therapy and utility of efavirenz as a substrate marker of CYP2B6 catalytic activity. J. Pharmacol. Exp. Ther. 306 (1), 287−300.

Weiner M, Peloquin C, Burman W, Luo CC, Engle M, Prihoda TJ, et al., 2010. The effects of tuberculosis, race and human gene SLCO1B1 polymorphisms on rifampin concentrations. Antimicrob. Agents Chemother. 54 (10), 4192−4200.

Weiss, J., Rose, J., Storch, C.H., Ketabi-Kiyanvash, N., Sauer, A., Haefeli, W.E., et al., 2007a. Modulation of human BCRP (ABCG2) activity by anti-HIV drugs. J. Antimicrob. Chemother. 59 (2), 238−245.

Weiss, J., Theile, D., Ketabi-Kiyanvash, N., Lindenmaier, H., Haefeli, W.E., 2007b. Inhibition of MRP1/ABCC1, MRP2/ABCC2, and MRP3/ABCC3 by nucleoside, nucleotide, and non-nucleoside reverse transcriptase inhibitors. Drug Metab. Dispos. 35 (3), 340−344.

Wenning, L.A., Petry, A.S., Kost, J.T., Jin, B., Breidinger, S.A., DeLepeleire, I., et al., 2009. Pharmacokinetics of raltegravir in individuals with UGT1A1 polymorphisms. Clin. Pharmacol. Ther. 85 (6), 623−627.

White, N.J., Looareesuwan, S., Edwards, G., Phillips, R.E., Karbwang, J., Nicholl, D.D., et al., 1987. Pharmacokinetics of intravenous amodiaquine. Br. J. Clin. Pharmacol. 23 (2), 127−135.

Wijnholds, J., Mol, C.A., van Deemter, L., de Haas, M., Scheffer, G.L., Baas, F., et al., 2000. Multidrug-resistance protein 5 is a multispecific organic anion transporter able to transport nucleotide analogs. Proc. Natl. Acad. Sci. U.S.A. 97 (13), 7476−7481.

Winstanley, P., Edwards, G., Orme, M., Breckenridge, A., 1987. The disposition of amodiaquine in man after oral administration. Br. J. Clin. Pharmacol. 23 (1), 1−7.

Wood, A., Armour, D., 2005. The discovery of the CCR5 receptor antagonist, UK-427,857, a new agent for the treatment of HIV infection and AIDS. Prog. Med. Chem. 43, 239−271.

Woodrow, C.J., Haynes, R.K., Krishna, S., 2005. Artemisinins. Postgrad Med J. 81 (952), 71−78.

Wortham, M., Czerwinski, M., He, L., Parkinson, A., Wan, Y.J., 2007. Expression of constitutive androstane receptor, hepatic nuclear factor 4 alpha, and P450 oxidoreductase genes determines interindividual variability in basal expression and activity of a broad scope of xenobiotic metabolism genes in the human liver. Drug Metab. Dispos. 35 (9), (Sep)1700–1710.

Wyen C, Hendra H, Siccardi M, Vogel M, Hoffmann C, Knechten H, et al. Cytochrome P450 2B6 (CYP2B6) and constitutive androstane receptor (CAR) polymorphisms are associated with early discontinuation of efavirenz-containing regimens, in press.

Wyen, C., Hendra, H., Vogel, M., Hoffmann, C., Knechten, H., Brockmeyer, N.H., et al., 2008. Impact of CYP2B6 983T>C polymorphism on non-nucleoside reverse transcriptase inhibitor plasma concentrations in HIV-infected patients. J. Antimicrob. Chemother. 61 (4), 914–918.

Zhang, H., Coville, P.F., Walker, R.J., Miners, J.O., Birkett, D.J., Wanwimolruk, S., 1997. Evidence for involvement of human CYP3A in the 3-hydroxylation of quinine. Br. J. Clin. Pharmacol. 43 (3), 245–252.

Zhang, K.E., Wu, E., Patick, A.K., Kerr, B., Zorbas, M., Lankford, A., et al., 2001. Circulating metabolites of the human immunodeficiency virus protease inhibitor nelfinavir in humans: structural identification, levels in plasma, and antiviral activities. Antimicrob. Agents Chemother. 45 (4), 1086–1093Apr 45 (4), 1086–1093.

Zucker, S.D., Qin, X., Rouster, S.D., Yu, F., Green, R.M., Keshavan, P., et al., 2001. Mechanism of indinavir-induced hyperbilirubinemia. Proc. Natl. Acad. Sci. U.S.A. 98 (22), 12671–12676.

Part 2

Specialized Chapters

21 Genetic Exchange in Trypanosomatids and Its Relevance to Epidemiology

Wendy Gibson[1,*], Michael D. Lewis[2] and Michael Miles[2]

[1]School of Biological Sciences, University of Bristol, Bristol, UK,
[2]Department of Pathogen Molecular Biology, Faculty of Infectious and Tropical Diseases, London School of Hygiene and Tropical Medicine (LSHTM), London, UK

21.1 Introduction

Genetic exchange has now been demonstrated in all three of the so-called TriTryps, the three trypanosomatids for which genome sequences have been published (Berriman et al., 2005; El-Sayed et al., 2005; Ivens et al., 2005): *Trypanosoma brucei*, *T. cruzi*, and *Leishmania major*. For *T. brucei* and *L. major* genetic exchange occurs in the insect vector and appears to follow Mendelian rules of inheritance (Jenni et al., 1986; MacLeod et al., 2005a; Akopyants et al., 2009). In contrast, for *T. cruzi* genetic exchange has been demonstrated experimentally in infected mammalian cells in vitro and hybrid formation appears to result from fusion of diploid cells followed by genome erosion (Gaunt et al., 2003).

Before the experimental confirmation of genetic exchange in these parasites, evidence for the natural occurrence of hybrids had accumulated from molecular epidemiological analysis of isolates collected from the field. Tait (1980) showed that isoenzyme data from *T. brucei* conformed to Hardy–Weinberg equilibrium and concluded that the population was undergoing random mating. Later analyses threw doubt on the extent of panmictic mating in *T. brucei* with the widespread acceptance of the concept of clonality in parasitic protozoa (Tibayrenc et al., 1990). However, rare occurrences of genetic exchange were thought to have led to the demonstrably hybrid lineages of *T. cruzi* recovered from the field (Carrasco et al., 1996; Machado and Ayala, 2001; Sturm et al., 2003; Westenberger et al., 2005) and natural hybrids were also demonstrated between several *Leishmania* species and subspecies (Kelly et al., 1991; Belli et al., 1994; Ravel et al., 2006).

*E-mail: W.Gibson@bristol.ac.uk

Here we review results from these two complementary avenues of study—analysis of naturally occurring hybrids among field isolates and experimental genetic exchange in the laboratory—for the pathogenic trypanosomatids *T. brucei*, *T. cruzi*, and *Leishmania major*.

21.2 *Trypanosoma brucei*

21.2.1 Genetic Crosses

Compelling evidence for mating in *Trypanosoma brucei* s.l. came from population genetics studies based on isoenzyme data. Gibson et al. (1980) described isoenzyme patterns consistent with those expected from homo- and heterozygotes in an extensive analysis of *T. brucei* isolates from East and West Africa, and Tait (1980) showed that similar data from 17 Ugandan isolates of *T. brucei* conformed to Hardy–Weinberg equilibrium, indicating that the population was undergoing random mating. The first successful laboratory cross was reported in 1986 when Jenni and colleagues co-transmitted two genetically distinct clones of *T. brucei* s.l. through tsetse flies and demonstrated hybrid progeny that had inherited a mixture of genetic markers from both parents (Jenni et al., 1986). The hybrids were found among metacyclics recovered from the salivary glands showing that genetic exchange had occurred some time during the trypanosome's developmental cycle in the tsetse fly. However, the metacyclic population contained a mixture of parental and hybrid genotypes, indicating that mating is not an obligatory event in the life cycle. This contrasts with the situation for the malaria parasite, *Plasmodium* spp., where gamete formation and production of a zygote is a normal part of the transmission cycle.

Subsequent trypanosome crosses have followed the same general plan: two genetically distinct parental trypanosome clones are fed to groups of newly emerged tsetse flies in their first bloodmeal (Figure 21.1). Tsetse are typically refractory to trypanosome infection, but are at their most susceptible as very young flies before they have fed (Maudlin, 1991). Not all flies become infected after the infected feed and only some infected flies produce hybrids. Thus large numbers of flies and trypanosome populations need to be screened to identify those containing hybrids. To avoid the laborious job of identifying hybrid-producing flies by screening every fly, selectable markers were incorporated into the experimental design. This became feasible following the development of methods for the stable transformation of trypanosomes with exogenous DNA in the early 1990s (Lee and Van der Ploeg, 1990; Ten Asbroek et al., 1990; Eid and Sollner-Webb, 1991). In the cross described by Gibson and Whittington (1993), each of the parental clones was transformed with a different construct designed to integrate a gene for drug resistance into the tubulin locus by homologous recombination. In this way, parental clones resistant to the antibiotics hygromycin or G418 were created. After co-transmission through the fly, hybrid progeny were selected by resistance to both drugs. This strategy has obvious advantages over the previous "finding a

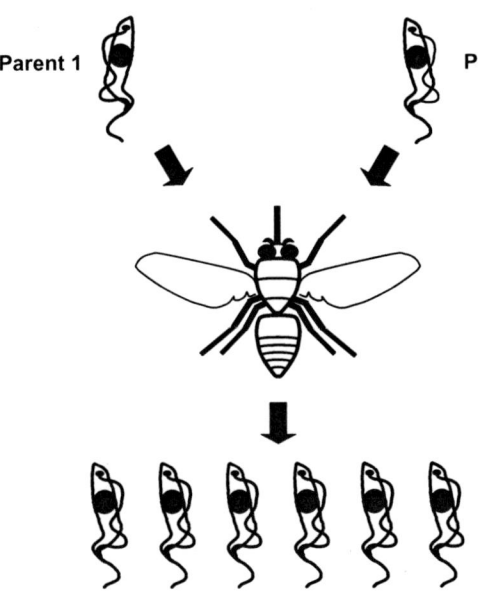

Figure 21.1 Design of an experimental cross for *Trypanosoma brucei*. The two parental trypanosomes are co-transmitted via tsetse flies and hybrid trypanosomes are found among the infective metacyclics from the salivary glands.

needle in a haystack" approach and was that used for the recent discovery of hybrids in *Leishmania major* (Akopyants et al., 2009).

The discovery of green fluorescent protein (GFP) and the development of methods to image the protein in living cells have opened exciting new approaches for studying genetic exchange in trypanosomes. Incorporation of genes for GFP in one parental line and red fluorescent protein (RFP) in the other leads to the production of hybrids with both genes, which appear yellow, making them immediately distinct from the parental cells by simple fluorescence microscopy of live cell preparations. This approach was a boon for the analysis of *T. brucei* hybrids within the tsetse fly and allowed the location and timing of genetic exchange to be determined (Gibson et al., 2008).

It has proved possible to cross all subspecies of *T. brucei* except the human pathogen *T. b. gambiense* Group 1. The difficulty in setting up crosses of *T. b. gambiense* Group 1 is that it transmits poorly or not at all through *Glossina morsitans* ssp., the standard laboratory tsetse fly (Gibson, 1986; Dukes et al., 1989). The more virulent Group 2 *T. b. gambiense* is readily transmissible via *G. morsitans* ssp. and has featured in several crosses, including the original cross of Jenni et al. (1986), where the parents were *T. b. brucei* STIB 247 and *T. b. gambiense* Group 2 STIB 386 (TH114). *T. b. gambiense* Group 2 (TH2) was also mated with both *T. b. brucei* and *T. b. rhodesiense* (Gibson et al., 1997b; Bingle et al., 2001). Several crosses have involved parental lines of *T. b. brucei* (Turner et al., 1990; Gibson and Garside, 1991; Schweizer et al., 1994; Degen et al., 1995; Gibson et al., 2008), or *T. b. brucei* and *T. b. rhodesiense* (Gibson, 1989; Gibson and Whittington, 1993). Inheritance of the trait for human infectivity has been analyzed in crosses of *T. b. brucei* and *T. b. rhodesiense* (Gibson and Mizen, 1997), and *T. b. brucei* and Group 2 *T. b. gambiense* (Turner et al., 2004).

21.2.2 Location of Genetic Exchange

Jenni and colleagues originally showed that genetic exchange took place in the fly, since cloned metacyclics from the saliva of infected flies were hybrid (Jenni et al., 1986); however, the exact location and life cycle stage remained undefined. While there were reports of hybrid formation in mixed midgut procyclic populations both in vitro and in vivo (Evans and Ellis, 1983; Schweizer and Jenni, 1991; Schweizer et al., 1991), the bulk of evidence pointed to the salivary glands as the site of genetic exchange. First the timing of hybrid appearance: hybrids were most likely to be found in flies infected for at least 28 days, after sufficient time for salivary gland invasion and colonization, despite there being a large population of procyclic trypanosomes continuously present in the midgut throughout this time (Gibson and Whittington, 1993; Gibson et al., 1997b; Schweizer et al., 1988). Selection by double drug resistance revealed that hybrids were present in populations derived from the salivary glands but not from the midgut (Gibson and Whittington, 1993; Gibson and Bailey, 1994; Gibson et al., 1997b). The direct visualization of trypanosome hybrids using fluorescent reporter proteins unequivocally established that hybrids are formed in the salivary glands and are not present in the midgut or among the migratory stages in the proventriculus and foregut (Bingle et al., 2001; Gibson et al., 2008).

The use of red and green fluorescent reporters also explained why hybrids are found infrequently, with less than a quarter of infected flies producing hybrids (Gibson and Stevens, 1999). In order to mate, both parental trypanosomes need to reach and colonize the same salivary gland, and it became evident that this is not always the case. While most flies co-infected with red and green fluorescent trypanosomes developed a mixed midgut infection, only about a third of these flies also had a mixed infection in the salivary glands (Peacock et al., 2007; Gibson et al., 2008). In a number of cases, the composition of the trypanosome population in the two salivary glands of the pair differed, perhaps with only one salivary gland infected or a mixed infection in one gland but not the other. Interestingly, all progeny from the red/green cross were hybrid and no trypanosomes with parental genotypes were recovered as in previous crosses (Gibson et al., 2008). The design of this cross may have increased the probability of finding hybrids, as analysis could be focused on salivary glands containing both parental genotypes, not possible in previous crosses.

These observations highlight the fact that few trypanosomes complete the journey from the midgut to the salivary duct, and only some of these then succeed in establishing an infection in either of the salivary glands. When together as epimastigotes in the same salivary gland, the trypanosomes readily mate, as demonstrated by the fact that most salivary glands with a mixed infection of red and green trypanosomes also contained yellow fluorescent hybrids (Gibson et al., 2008). Compatibility of different trypanosome strains may then depend on them reaching the salivary glands simultaneously. In some early crosses, the two parental clones differed substantially in their speed of colonizing the salivary glands (Schweizer et al., 1988; Gibson and Whittington, 1993; Gibson et al., 1997b). The fact that mating in *T. brucei* occurs among epimastigotes in the salivary glands makes the prospect of producing hybrids in vitro more remote, as reliable culture systems do not exist for the life cycle stages that occur in the salivary glands.

21.2.3 Mendelian Inheritance and Meiosis

It is generally accepted that *T. brucei* is diploid with respect to the 11 pairs of large chromosomes that contain the housekeeping genes, although this arrangement probably does not apply to the intermediate and mini-chromosomes (Gottesdiener et al., 1990; Melville et al., 1998; Berriman et al., 2005). Analysis of the inheritance of genetic markers in hybrid progeny from crosses of *T. brucei* is consistent with Mendelian genetics for the most part (Gibson, 1989; Sternberg et al., 1989; Turner et al., 1990; Schweizer et al., 1994; Gibson and Bailey, 1994; MacLeod et al., 2005a), leading to the assumption that meiosis occurs. Indeed, genetic linkage maps for *T. b. brucei* and Group 2 *T. b. gambiense* have been constructed from detailed analysis of microsatellite inheritance and frequency of crossing-over (MacLeod et al., 2005b; Cooper et al., 2008).

It has proved more difficult to directly visualize trypanosomes undergoing meiosis, but a potential solution to this problem is now at hand. While hybrids were easily detected in a cross of red and green fluorescent trypanosomes from day 13 onwards (Gibson et al., 2008), putative intermediate stages were neither abundant nor obvious, necessitating an alternative approach to detect trypanosomes undergoing meiosis. Phylogenomic studies had identified the presence in *T. brucei* of homologues of several genes crucial for meiosis in yeast and other eukaryotes such as *Spo11*, *Hop1*, *Dmc1*, and *Mnd1* (Ramesh et al., 2005). By constructing fusions of the *T. brucei* homologs with the gene for yellow fluorescent protein (YFP), expression of these putative meiosis genes could be monitored through the developmental cycle of *T. brucei* in the fly. Expression was observed only among trypanosomes during the early stage of colonization of the salivary glands, consistent with the first appearance of hybrids and was localized to the nucleus (Peacock et al., 2010). Surprisingly, the meiotic trypanosomes occurred with similar frequency in both single and mixed strain transmissions and it seems probable that meiosis is a normal part of the trypanosome developmental cycle in the fly. The identification of trypanosomes undergoing meiosis is the first step in elucidating the mechanism of genetic exchange in *T. brucei* at the cell biology level.

While the overall agreement with a Mendelian pattern of inheritance supported the hypothesis that a meiotic division was central for genetic exchange in *T. brucei*, a haploid life cycle stage has not been substantiated (Shapiro et al., 1984; Zampetti-Bosseler et al., 1986; Kooy et al., 1989). Indirect evidence for haploid nuclei came from the observation that many *T. brucei* spp. crosses produced triploid hybrid progeny, which most probably arise from fusion of a haploid nucleus with one that is diploid (Gibson et al., 1992, 1995, 1997b; Gibson and Bailey, 1994; Gibson and Whittington, 1993). Hybrids with high DNA contents relative to the parents were found even in the first experimental cross (Jenni et al., 1986), which created some initial confusion about the mechanism of genetic exchange

(Paindavoine et al., 1986; Wells et al., 1987). Analysis of progeny clones with high DNA contents from several further crosses demonstrated that these hybrids were triploid with DNA contents that clustered at the 3N value, and, in addition, trisomy was confirmed for several chromosomes (Gibson and Bailey, 1994; Gibson et al., 1992, 1995, 1997b; Gibson and Whittington, 1993).

As well as triploid hybrids, several tetraploid hybrids were recovered from a recent cross of red and green fluorescent trypanosomes (Gibson et al., 2008). Whereas the presence of triploid hybrids was obvious from the demonstration of three alleles at some loci, only two microsatellite alleles were present in each of the tetraploid hybrids. They were therefore not formed by fusion of the two diploid parental genomes, but more likely are the products of genome endoreplication (Gibson et al., 2008). While triploids appear to be stable during growth and fly transmission (Wells et al., 1987; Gibson et al., 1992), the tetraploids may be unstable, as flow cytometry analysis of the DNA contents of tetraploid clones frequently revealed an extra G1 peak at the 2N position (Gibson et al., 2008).

Intraclonal mating was initially thought not to occur in *T. brucei* except in the presence of outcrossing trypanosomes, leading to the hypothesis that some kind of diffusible factor produced by nonself recognition induced mating (Tait et al., 1996; Gibson et al., 1997b). Intraclonal mating explained the occasional anomalies where hybrid progeny were homozygous instead of heterozygous as expected if the parents were different homozygotes (Gibson and Bailey, 1994; Sternberg et al., 1988). Recent experiments using red and green fluorescent clones of a single *T. brucei* strain have shown that intraclonal mating occurs with some frequency in the absence of a second trypanosome strain (Peacock et al., 2009). The assumption that a *T. brucei* clone can be tsetse-transmitted without change is therefore open to question.

21.2.4 Inheritance of Kinetoplast DNA

Kinetoplast DNA (kDNA) is the mitochondrial DNA of trypanosomatids and consists of an interlocked network of about 50 (20–25 kb) maxicircles and 5,000 (1 kb) minicircles (reviewed by Shapiro and Englund, 1995). The kDNA is contained within an organelle, the kinetoplast, which is inside the mitochondrion. Initial results supported the hypothesis that inheritance of kDNA was uniparental, with the kDNA of either parent being passed on to the hybrid progeny (Gibson, 1989; Sternberg et al., 1988, 1989). However, detailed analysis of both maxi- and minicircles showed that although maxicircles were of a single parental type, the minicircles had been inherited from both parents (Gibson and Garside, 1990; Gibson et al., 1997a). To explain this result, it was assumed that the hybrid kDNA network initially consists of both maxi- and minicircles from both parents in an equal proportion; the small number of maxicircles resolves to a single type by random segregation during subsequent mitotic divisions, while the much greater number of minicircles endures as a hybrid network. This idea is supported by the observation that hybrids at an early stage of growth have mixed maxicircle networks (Turner et al., 1995; Gibson et al., 2008).

The initial formation of the hybrid kDNA network remains an intriguing problem. The first requirement is fusion of the mitochondria from both parental trypanosomes to allow the kDNA to mix. Then mini- and maxicircles from both parental kDNA networks would need to combine into a single hybrid kDNA network, which would involve detachment and reattachment of all the individual DNA circles. An alternative hypothesis is that only some minicircles are swapped while the kDNA networks of the two parents are adjacent, leaving the core maxicircle network intact; the partially hybrid kDNA networks would then be inherited by individual progeny trypanosomes (Shapiro and Englund, 1995). However, this hypothesis does not fit with the observation of mixed maxicircle networks above.

21.2.5 Implications for Epidemiology

Although genetic exchange in *T. brucei* has been amply demonstrated in the laboratory, its importance in natural trypanosome populations remains controversial. One problem has been sampling bias, with collection of trypanosome isolates usually focused on epidemics of human disease (Cibulskis, 1992; Stevens and Welburn, 1993; Hide et al., 1994; Stevens and Tibayrenc, 1996). In epidemics of human trypanosomiasis caused by *T. b. rhodesiense*, transmission may well be direct from human to human rather than from an animal reservoir, allowing the clonal expansion of particular trypanosome genotypes. This is very different from the endemic scenario where humans and their livestock are occasional hosts in the natural circulation of *T. brucei* ssp. strains in wild mammals and tsetse. Through the analysis of trypanosome mating in the laboratory, we can now define the biological circumstances in which genetic exchange will be found: at least two trypanosome strains must be present in the salivary glands of a tsetse fly for mating to occur; since flies are most readily infected on their first bloodmeal (Wijers, 1958; Maudlin, 1991), a mixed infection is likely to be acquired from one infected mammal carrying multiple trypanosome strains. Few mixed infections have been reported from humans (Truc et al., 1998), but mixed infections of more than one trypanosome strain or species are frequently encountered in livestock (Godfrey and Killick-Kendrick, 1961; Nyeko et al., 1990), tsetse (Masiga et al., 1996; MacLeod et al., 1999; Lehane et al., 2000; Adams et al., 2006), and presumably also occur frequently in the large wild mammals that sustain many tsetse populations. In-depth analysis of trypanosome samples from these transmission cycles would be informative.

Even though the frequency of genetic exchange in natural populations may be low, there is potential for significant epidemiological consequences. For example, the trait of human infectivity in *T. b. rhodesiense* is conferred by a single gene, the serum resistance associated (SRA) gene (De Greef et al., 1989; Xong et al., 1998). Any cross between *T. b. rhodesiense* and *T. b. brucei* would place this key virulence gene in new genetic backgrounds, thus creating new genotypes of *T. b. rhodesiense*. There is abundant evidence of strain heterogeneity in *T. b. rhodesiense* from several foci of human trypanosomiasis in East Africa (Gibson and Gashumba, 1983; Gibson and Wellde, 1985; Enyaru et al., 1993; Hide et al., 1994; Komba et al., 1997), and quite different *T. b. rhodesiense* genotypes have been found in neighboring foci in Uganda and Kenya (Gibson and Gashumba, 1983; Gibson and Wellde, 1985). We know little about the genetic basis of most phenotypic characteristics of *T. brucei s.l.*, but surely genetic exchange provides the opportunity for more virulent, pathogenic or fly transmissible strains of *T. b. rhodesiense* to arise.

21.3 Trypanosoma cruzi

21.3.1 T. cruzi Diversity

Trypanosoma cruzi is considered to be a single species but comprises six distinct genetic lineages or discrete typing units (DTUs). The definition of these six *T. cruzi* subgroups was originally on the basis of phenotyping using multilocus enzyme electrophoresis (MLEE). Subsequently the same six DTUs were supported by comparative analyses of a wide range of nuclear DNA targets (see Miles et al., 2009, for review). The current international consensus nomenclature for the six *T. cruzi* lineages is TcI to TcVI (Zingales et al., 2009).

TcI to TcVI show broadly distinctive but not entirely exclusive geographical, ecological, transmission cycle and disease associations; overlaps occur, and mixed infections are reported from humans, reservoir mammal hosts, and arthropod vectors (Bosseno et al., 1996; Yeo et al., 2005, 2007; Burgos et al., 2008; Cardinal et al., 2008). TcI is the principal agent of Chagas disease in Latin America north of the Amazon basin, whereas TcII, TcV, and TcVI are the main causes of Chagas disease in the Southern Cone countries of South America (Miles et al., 1978, 1981; Fernandes et al., 1998; Zingales et al., 1998; Barnabe et al., 2000). The sylvatic TcI transmission cycles are widespread throughout Latin America and are largely arboreal; the common opossums (*Didelphis* species) are obvious and abundant reservoir hosts but many other mammal species may be infected, with some transmission among rodent species and triatomines with terrestrial habitats (Yeo et al., 2005). The natural transmission cycles of TcII, TcV, and TcVI are either rare or have yet to be fully characterized, although TcII is sporadically found among mammals of the Atlantic forest region of Brazil (Lisboa et al., 2004). TcIII and TcIV are predominantly sylvatic and seldom infect humans. However, TcIV is a secondary cause of Chagas disease in Venezuela, after TcI. TcIII appears to have the most clearly defined and exclusive natural ecological and host association, with the burrowing armadillo, *Dasypus novemcinctus* (Yeo et al., 2005).

The various *T. cruzi* isolates that were characterized to establish this intraspecific genetic diversity were collected from disparate geographical locations and usually with small numbers of isolates gathered from each site. Although this striking genetic heterogeneity of *T. cruzi* was a landmark discovery, which has transformed subsequent research on the epidemiology of Chagas disease, this sampling strategy was not ideally suited for determining the structure of *T. cruzi* populations. Nevertheless, repeated analyses of the molecular diversity of *T. cruzi* based on such samples have supported the discrete nature of the six DTUs and demonstrated strong linkage disequilibrium among them (Tibayrenc et al., 1990). These observations fostered the tenet that *T. cruzi* was propagated clonally, both between and within the DTUs, and that genetic exchange, if it occurred, was rare and of little epidemiological consequence (Tibayrenc and Ayala, 2002). On the other hand, despite this clonal theory, even the early MLEE analyses indicated that TcV and TcVI resembled natural interlineage hybrids between TcII and TcIII (Miles, 1985). The application of multilocus DNA sequencing eventually confirmed that TcV and TcVI were hybrid lineages (Machado and Ayala, 2001; Brisse et al., 2003), but not before the TcVI reference strain "CL Brener" had been selected for the TriTryp genome sequencing project. As discussed later, TcV and TcVI are endemic agents of Chagas disease across much of the Southern Cone region, making the study of recombination in *T. cruzi* of profound epidemiological relevance.

21.3.2 Genome Sequence of a Natural Hybrid

The hybrid nature of the CL Brener strain made the task of sequencing its genome particularly challenging since most genes were represented by two divergent copies (El-Sayed et al., 2005). This complicated the assembly of the genome and eventually required additional sequence data from a representative of one of the parental groups, for which the TcII strain "Esmeraldo" was chosen. This allowed the putative parental TcII ("Esmeraldo-like") and TcIII ("non-Esmeraldo-like") sequences to be partially deduced from the single hybrid genome sequence. The initial assembly of the genome was further complicated by the large number of repetitive surface protein gene families that were found throughout the *T. cruzi* genome. The CL Brener haploid genome contained $\sim$12,000 genes and was approximately 55 Mb in size, considerably larger than either *T. brucei* (26 Mb) (Berriman et al., 2005), or *L. major* (33 Mb) (Ivens et al., 2005). The reassembled CL Brener sequence has produced fewer, larger contigs compared to the original assembly (Weatherly et al., 2009). Other *T. cruzi* genome sequencing projects are now in progress, with the advantages of new sequencing technologies and focus on the remaining DTUs with less complex genomes. Of special interest is the genome of TcI, because of its degree of divergence and epidemiological importance as an agent of Chagas disease, particularly north of the Amazon.

21.3.3 Genetic Crosses

In contrast with *T. brucei*, only a single successful genetic crossing experiment has so far been reported for *T. cruzi*. The earliest, and unsuccessful, attempts at genetic crosses involved passaging the TcI and TcII DTUs together in vitro, through triatomine bugs, or through mice, followed by phenotypic analysis of the resultant populations (Miles, 1985). As with *T. brucei*, such experiments were revolutionized by the ability to genetically transform *T. cruzi* strains to carry different drug resistance markers, permitting selection of any double-drug-resistant populations emerging from genetic crossing experiments. Perhaps the best experimental strategy for application of this new technology would be an attempt to cross *T. cruzi* strains from within the same DTU, an approach encouraged by the discovery of both putative hybrid and parental TcI phosphoglucomutase isoenzyme phenotypes among clones of *T. cruzi* isolates from a single undisturbed locality in Amazonian forest (Carrasco et al., 1996). Accordingly, a pair of these putative TcI parental isolates was selected, biological clones were prepared and they were transfected to carry different episomal drug resistance markers for hygromycin and neomycin (G418). The transgenic parental isolates were passaged together through triatomine bugs, mice, and mammalian cell cultures and the recovered populations were cultured in media containing both drugs to select possible hybrids. The only double-drug-resistant clones were derived from the mammalian cell cultures (Stothard et al., 1999). MLEE, karyotyping, gene sequencing, and microsatellite profile analysis demonstrated that the six double-drug-resistant clones were hybrids of the two parental strains (Gaunt et al., 2003). Thus, it was demonstrated experimentally for the first time that *T. cruzi* has an extant capacity to undergo genetic exchange, at least within TcI.

The availability of multiple fluorescent protein markers with distinct excitation and emission spectra provides a means to study genetic exchange in greater detail. As described earlier, crosses between *T. brucei* cell lines carrying either a GFP or an RFP gene led to readily identifiable hybrid organisms expressing both reporters such that they appear yellow in composite images of tsetse salivary glands (Gibson et al., 2008). Red and green fluorescent lines of *T. cruzi*, and the closely related species *T. rangeli*, with strong and stable expression of fluorescence and drug resistance markers have been generated by integrating reporter constructs into either the ribosomal RNA gene array (Guevara et al., 2005) or the tubulin locus (Pires et al., 2008). Just as *T. brucei* mating only occurs between strains residing in the same tsetse fly salivary gland, so it is expected that experimental crosses in *T. cruzi* will only be successful if two strains can be brought into close physical proximity at the appropriate point of their life cycles. The use of red and green fluorescently labeled *T. cruzi* cell lines has already enabled tracking of co-infections of mammalian cell cultures in vitro and in mice and triatomine bugs in vivo indicating there should be few technical barriers to identifying co-expressing "yellow" hybrids (Guevara et al., 2005; Pires et al., 2008) (Figure 21.2). Indeed, mammalian cell cultures simultaneously infected with several different pairs of red and green transgenic strains commonly display co-infections of individual cells.

21.3.4 Location of Genetic Exchange

In the successful experimental *T. cruzi* cross described by Gaunt et al. (2003), non-confluent mammalian cell cultures were infected with *T. cruzi* cultures containing both infective metacyclic trypomastigotes and noninfective epimastigotes. Once the metacyclic trypomastigotes had invaded the mammalian cells, the epimastigotes were removed by washing the cell monolayers. The infected mammalian cell cultures were maintained for more than 20 days, allowing completion of several rounds of intracellular replication, each of which may take as little as 5 days. The double-drug-resistant hybrids were recovered from the culture supernatant, which contained trypomastigotes released from lysed mammalian cells. Thus, the

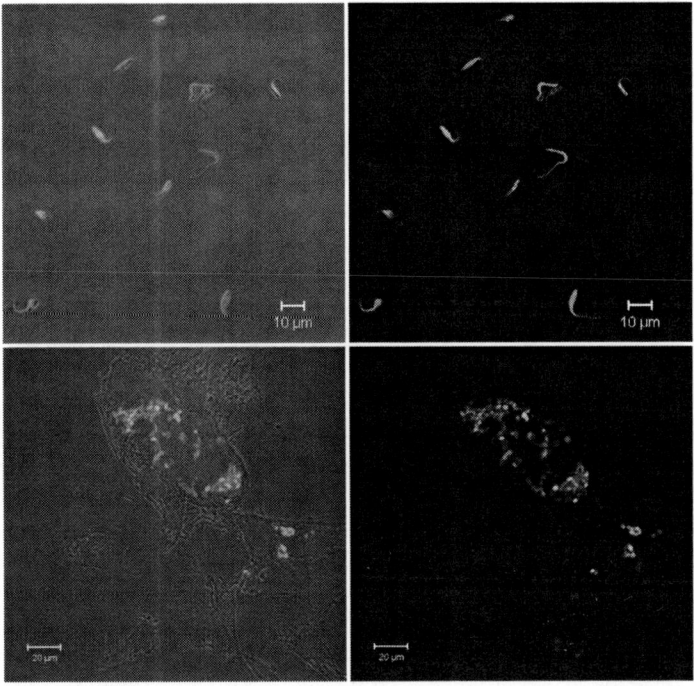

Figure 21.2 Transgenic *Trypanosoma cruzi* cell lines expressing RFP or GFP markers. Top panels show mixed samples of extracellular flagellate forms; bottom panels show mammalian cells co-infected with amastigote forms. Panels on the left show phase microscopy, with red fluorescence, green fluorescence, and DNA staining; panels on the right show red and green fluorescence only. DNA is stained with Hoechst 33342, a blue fluorescent dye. (For interpretation of the references to color in this figure legend, the reader is referred to the web version of this book.)

most obvious interpretation is that hybridization took place intracellularly in a co-infected host cell, and that it therefore involved either amastigotes or trypomastigotes. However, it is also conceivable that genetic exchange could have taken place between extracellular forms prior to host cell invasion (i.e., epimastigotes or metacyclic trypomastigotes), or after infected cells had ruptured (i.e., new trypomastigotes, including short or slender forms).

In terms of opportunity in endemic areas, where natural recombinants do exist, mixed interlineage *T. cruzi* infections and multiclonal intralineage infections are by no means rare in humans or other mammals (Bosseno et al., 1996; Brenière et al., 1998; Fernandes et al., 1999; Vago et al., 2000; Gaunt et al., 2003). Furthermore, sylvatic mammals are likely to be subject to multiple challenge infections, some of which may be orally acquired by consuming triatomine bugs. *Didelphis* can have anal gland infections that include morphological stages typically restricted to the insect vector. Nevertheless, by analogy with *T. brucei* and *Leishmania* where genetic exchange takes place in the tsetse or sand fly vectors, respectively, it would be surprising if genetic exchange in *T. cruzi* did not occur in its insect vector. There would be abundant opportunity for *T. cruzi* to undergo genetic exchange in triatomine bugs. Whereas tsetse flies have limited susceptibility to *T. brucei* and most readily acquire infection during their first feed, triatomine bugs may acquire *T. cruzi* infection at any of the five nymphal stages or as adults. Each bug takes many feeds, any one of which may be infective; infection rates rise with instar examined, overall infection prevalence rates are often 50% and may be much higher; as for mammals, mixed infections in bugs are common (Bosseno et al., 1996; Yeo et al., 2007; Cardinal et al., 2008). Although infection of the triatomine salivary glands is not described for *T. cruzi*, there is analogous intense infection of the rectum with abundant epimastigotes attached to the epithelium and free in the lumen, as well as metacyclic trypomastigotes. Salivary gland infections cannot be considered a prerequisite for genetic exchange to occur in triatomines—they are not a feature of the *Leishmania* life cycle in sand flies and many examples exist of genetic exchange in other protozoan flagellates in the gut lumen of insect species (Sleigh, 1973). There are also prominent candidate physiological triggers in triatomines, for example, hormonal changes that govern moulting, known to precipitate genetic exchange in flagellates of other insects, or starvation/nutritional depletion, which appears to encourage dispersive flight in male triatomine bugs.

Clearly, the occurrence of genetic exchange in *T. cruzi* during the vector stage of the life cycle deserves further investigation. With approximately 127 of the 140 triatomine species native to the Americas and at least six *T. cruzi* DTUs, there are multiple scenarios to be explored and new discoveries to be made, aided by the latest technological advances.

21.3.5 Behavior of Experimental Hybrids

The phenomenon of hybrid vigor (heterosis), in which hybrids tolerate unusually severe conditions or thrive and outcompete nonhybrids, is well-known for a variety of organisms (Mallet, 2007). On the other hand, hybrids may have significantly reduced viability and be difficult to recover from experimental or natural populations. Hybrid vigor was explored theoretically by Widmer et al. (1987) for a naturally occurring hybrid *T. cruzi* by comparing the catalytic efficiency of individual glucose phosphate isomerase isoenzymes: the isoenzymes differed in temperature stabilities but not in catalytic efficiencies. Comprehensive phenotypic comparisons of natural TcV and TcVI hybrid strains with TcII and TcIII representative parental strains will be required to establish whether heterosis contributed to the success of TcV and TcVI in becoming established in domestic transmission cycles. In this context it is clearly of interest to know whether experimentally derived hybrid *T. cruzi* clones have decreased or increased vigor. The experimental hybrid clones described by Gaunt et al. (2003) grew vigorously in vitro, and in preliminary in vivo comparisons in immunocompromised SCID mice, the hybrids readily established infections and produced abundant pseudocysts in heart and skeletal muscle; pseudocysts were also seen in smooth muscle of the alimentary tract (Lewis et al., unpublished data). The hybrid clones therefore appear to be at least as virulent as their parents. Genotypes of the infecting strain or mixture of strains may have an important impact upon disease pathology in humans, for instance, as a result of differences in tissue tropism (Vago et al., 1996, 2000). Results from experimental investigations of mixed infection dynamics also suggest that dual-clone mixtures of *T. cruzi* behave differently than would be expected simply based on the behavior of each clone individually, both in bugs (Pinto et al., 1998, 2000), and in animal models (de Lana et al., 2000; Martins et al., 2006, 2007).

21.3.6 Mechanism of Genetic Exchange in Experimental and Natural Hybrids

The mechanism of genetic exchange that produced the six experimental hybrid *T. cruzi* clones is at least partially understood. MLEE, karyotype analysis, random amplification of polymorphic DNA (RAPD) and multilocus microsatellite typing (MLMT) clearly showed that the hybrids had not inherited alleles in typical Mendelian ratios, at one allele per locus from each parent (Gaunt et al., 2003). In fact, for most loci tested the hybrids had inherited at least two alleles from both parents, although a minority of the parental alleles were absent. *T. cruzi* hybrid clones each displayed one of the parental mitochondrial maxicircle genotypes but not both, as is reported (21.2.4) for *T. brucei*; the dynamics for inheritance of minicircles remains unexplored in *T. cruzi*. Flow cytometric analysis of DNA content demonstrated that the hybrid clones were subtetraploid (approximately intermediate between 3n and 4n), and that DNA content was relatively stable after passage through mice (Lewis et al., 2009) (Figure 21.3). However, following long-term in vitro culture, progressive and gradual decline in DNA content has been observed without any evidence of a meiotic reductional division, which would be expected to result in a halving of ploidy in a single step (Lewis et al., unpublished data). This situation is in contrast with the general consensus for the typical program of genetic exchange in *T. brucei*, which evidently does involve meiosis and Mendelian inheritance. However, hybridization and genome erosion does have a biological precedent in the parasexual mechanism of genetic exchange in the pathogenic yeast *Candida albicans* (Bennett and Johnson, 2003). In *C. albicans* the genome erosion following fusion of diploids is by random, concerted loss of whole chromosomes over the course of repeated mitotic replication and results in diploid or near-diploid recombinants.

The mechanism of genome erosion in *T. cruzi* requires more extensive comparative genotyping of hybrids and parents. The full extent of recombination and mosaic formation that took place, at chromosomal and intragenic levels, during the experimental *T. cruzi* cross is also not clear and requires further investigation. In the context of these observations on *T. cruzi*, it is interesting to note that a proportion of polyploid progeny may also occur as products of genetic crossing experiments with *T. brucei* and *Leishmania*, and in all three species haploid gametes have never been detected.

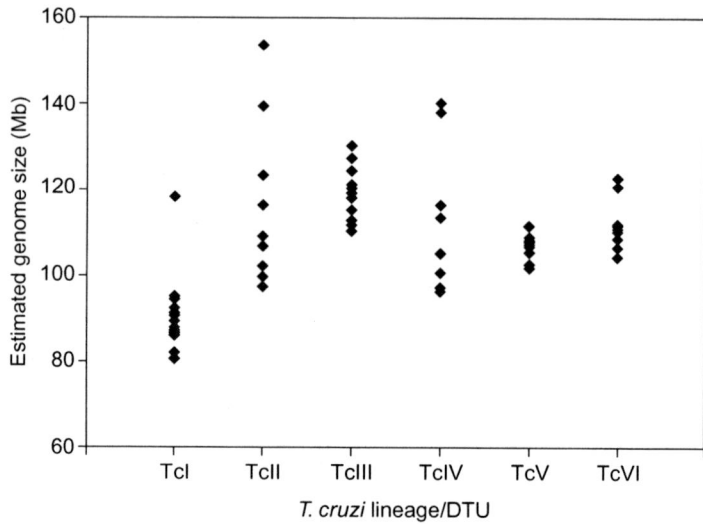

Figure 21.3 Variation in DNA content within and between *Trypanosoma cruzi* genetic lineages or DTUs. Each diamond indicates a different *T. cruzi* clone. Flow cytometry was used to measure the relative DNA content of parasites labeled with the fluorescent dye propidium iodide. Genome sizes were estimated based on the predicted size of the CL Brener genome (El-Sayed et al., 2005).
Source: Adapted from Lewis et al. (2009).

The interesting results of the DNA content analysis of the experimental cross, in particular the elevated subtetraploid DNA content of the hybrids, led us to undertake a much wider study of DNA content among field isolates representing all the known *T. cruzi* DTUs. As described by Lewis et al. (2009) and in agreement with the pioneering observations by Dvorak et al. (1982), extraordinarily wide diversity in DNA content was observed (Figure 21.3). Some trends were observed relating to the DNA content of cloned strains from the different DTUs, for example, TcI on average clearly had the lowest DNA content, with a few outlying isolates, while DTUs TcII and TcIV presented a wide range of DNA contents. The key question, however, was whether naturally occurring hybrid TcV and TcVI strains had elevated DNA contents, compatible with the fusion of diploids, as deduced from the experimental cross. This proved not to be the case. TcV and TcVI had DNA contents that are equivalent to those of their parental TcII and TcIII lineages, implying diploidy. This was confirmed by genome-scale MLMT and DNA sequencing of nuclear and mitochondrial (maxicircle) genes in TcII, TcIII, TcV, and TcVI isolates, which indicated TcV and TcVI carried one TcII allele and one TcIII allele per locus, with only rare instances of allelic aneuploidy (Lewis et al., 2009). TcV and TcVI present very little intralineage diversity but the extensive genotyping clearly shows they are distinct from each other, confirming the original MLEE-based distinction. TcV and TcVI thus resemble normal F1 meiotic progeny of hybridization events between TcII and TcIII that have undergone clonal expansion within the domestic niche—they are essentially diploid with fixed heterozygous genotypes comprising equal proportions of TcII and TcIII alleles. There are therefore contrasts with the aneuploid experimental hybrids. These could stem from mechanistic differences in interlineage, as opposed to intralineage, recombination. Alternatively the physiological cues that might be required for meiotic reductive division may well have been absent under the conditions used for experimental crosses. It is not known how the diploid state was reached by TcV/VI, and so operation of the fusion-erosion mechanism or other genetic exchange mechanisms in natural recombination events should not be ruled out at this stage.

The existence of TcV and TcVI clearly shows that natural recombination events have been a feature of the evolution of *T. cruzi*. Whether it is a contemporary phenomenon in active transmission cycles remains an open question. Some evidence does suggest that recombination may be more frequent than previously thought. For example, comparisons between the nuclear and mitochondrial genotypes of TcI, TcIII, and TcIV have revealed several striking instances of mitochondrial introgression (Machado and Ayala, 2001). Multilocus sequence typing (MLST) of reference strains has now shown that TcIII and TcIV are probably descended from ancient hybridization between TcI and TcII (Westenberger et al., 2005). Also, unexpectedly high levels of homozygosity within sylvatic populations of TcI and TcIII might be compatible with occurrence of some natural intralineage recombination (Llewellyn et al., 2009a,b). More intensive population genetics studies at the microscale of individual undisturbed transmission cycles, and in communities where active transmission to humans is occurring, are required to investigate this hypothesis. In summary, recent research suggests that genetic exchange is more frequent among natural populations than so far described, with consequent epidemiological implications.

21.3.7 Implications for Epidemiology

The successful experimental *T. cruzi* genetic cross was a significant research milestone in that it proved this organism has an extant capacity for genetic exchange, with consequent epidemiological implications, and the unusual genetic mechanism continues to be of considerable fundamental interest. However, the main implications of genetic exchange in *T. cruzi* come not from examination of the experimental hybrids but from genetic analysis of natural populations. The hybrid DTUs, TcV, and TcVI have thrived and spread dramatically among human populations in Bolivia, Argentina, Chile, Paraguay, and the extreme south of Brazil, where severe clinical manifestations of Chagas disease are common, including chagasic cardiomyopathy, megaoesophagus, megacolon, and congenital transmission (WHO, 2002; Miles et al., 2009). TcV and TcVI have probably been propagated in conjunction with the dispersion of *Triatoma infestans*, the principal domestic triatomine vector in the Southern Cone region, which was itself spread in association with the activities of humans (Bargues et al., 2006; Cortez et al., 2010). There is therefore potentially a great epidemiological risk attached to the emergence of new recombinant lineages, which would have unpredictable phenotypes, for example, with respect to pathogenicity or transmission potential. Furthermore, new high impact hybridization events cannot be excluded as the dynamics of sylvatic and domestic transmission cycles alter with development, environmental change and migration of human populations.

21.4 Leishmania

21.4.1 Genetic Crosses

As with *T. cruzi* the occurrence of genetic exchange in *Leishmania* has been controversial for many years. Recently, Akopyants et al. (2009) finally demonstrated recombination in *Leishmania* experimentally. They co-transmitted pairs of transgenic *L. major* strains resistant to different selective drugs through the natural sand fly vector (*Phlebotomus duboscqi*) and recovered parasites that were resistant to both drugs. Nine double-drug-resistant populations were recovered directly from co-infected sand flies and two from mice bitten by such flies; these yielded a total of 18 progeny clones. Genetic analysis of homozygous parental markers, assigned to several chromosomes, showed that all clones carried both parental alleles, proving they were hybrid. Maxicircle kDNA genotypes were uniparentally inherited. Phenotypic analysis also demonstrated clear segregation of dominant traits in different hybrid progeny. The genotypes were consistent with canonical meiosis generating heterozygous F1 progeny. However, haploid forms were not observed and 7 of the 18 clones were triploid. Therefore, alternative genetic mechanisms should also be considered for *Leishmania*. Moreover, it is not yet clear which *Leishmania* life cycle stages are involved in genetic exchange and whether there is any molecular mechanism that determines strain compatibility. As illustrated in *T. brucei* and *T. cruzi*, the research will be greatly aided by use of transgenic fluorescent *Leishmania* to delimit genetic exchange events.

21.4.2 Implications for Epidemiology

As with *T. cruzi*, understanding of the epidemiological implications of genetic exchange in *Leishmania* is largely derived from population genetics studies in endemic regions. Natural populations of *Leishmania* have been split into two subgenera, *Leishmania* and *Viannia*, and numerous species, partly according to the clinical presentation of infection. Molecular markers have provided a more objective means of analyzing genetic diversity and the structure of natural *Leishmania* populations. Results have been striking: several recognized species clearly do not have validity and the taxonomy deserves reassessment (Mauricio et al., 2006; Fraga et al., 2010;). More fundamentally, evidence for genetic exchange in natural populations of *Leishmania* has repeatedly emerged in the form of both inter- and intraspecific hybrids bearing recombinant genotypes. For example, natural hybrids have been described between *L. braziliensis* and *L. panamensis* in Nicaragua (Belli et al., 1994), between *L. braziliensis* and *L. guyanensis* in Venezuela (Delgado et al., 1997), and between *L. braziliensis* and *L. peruviana* in Peru (Dujardin et al., 1995). In the latter case, the *L. braziliensis/peruviana* hybrids were highly prevalent among patients, and occurred in some patients with mucosal disease. Furthermore, nine hybrid genotypes were discovered in a single Peruvian endemic region (Nolder et al., 2007). In the Old World *L. major/arabica* hybrids have been described (Evans et al., 1987; Kelly et al., 1991), and *L. major/donovani* hybrids (originally designated *L. major/infantum* hybrids) from patients with HIV infection (Ravel et al., 2006). Putative parental and hybrid phenotypes of the *L. donovani* complex (*L. donovani;* "*L. archibaldi*") occur sympatrically in East Africa and sequencing of housekeeping genes encoding enzymes shows mosaic characters across such strains (Mauricio et al., 2006; Zemanova et al., 2007). A widespread lineage of *L. tropica* appears to be disseminated from a recent recombination event (Schwenkenbecher et al., 2006). In addition, Volf et al. (2007) have shown that *L. major/donovani* hybrids have increased transmission potential since they were transmitted efficiently by the otherwise *L. major*-specific vector, *Phlebotomus papatasi*. These results highlight the potential epidemiological impact of genetic exchange in *Leishmania*: hybrids may have the disturbing potential to expand transmission of visceral disease and invade new geographical regions.

List of Abbreviations

MLEE multilocus enzyme electrophoresis
MLMT multilocus microsatellite typing
MLST multilocus sequence typing

References

Adams, E.R., Malele, I.I., Msangi, A.R., Gibson, W.C., 2006. Trypanosome identification in wild tsetse populations in Tanzania using generic primers to amplify the ribosomal RNA ITS-1 region. Acta Trop. 100, 103–109.

Akopyants, N.S., Kimblin, N., Secundino, N., Patrick, R., Peters, N., Lxawyer, P., et al., 2009. Demonstration of genetic exchange during cyclical development of *Leishmania* in the sand fly vector. Science 324, 265–268.

Bargues, M.D., Klisiowicz, D.R., Panzera, F., Noireau, F., Marcilla, A., Perez, R., et al., 2006. Origin and phylogeography of the Chagas disease main vector *Triatoma infestans* based on nuclear rDNA sequences and genome size. Infect. Genet. Evol. 6, 46–62.

Barnabe, C., Brisse, S., Tibayrenc, M., 2000. Population structure and genetic typing of *Trypanosoma cruzi*, the agent of Chagas disease: a multilocus enzyme electrophoresis approach. Parasitology 120, 513–526.

Belli, A.A., Miles, M.A., Kelly, J.M., 1994. A putative *Leishmania panamensis—Leishmania braziliensis* hybrid is a causative agent of human cutaneous leishmaniaisis in Nicaragua. Parasitology 109, 435–442.

Bennett, R.J., Johnson, A.D., 2003. Completion of a parasexual cycle in *Candida albicans* by induced chromosome loss in tetraploid strains. EMBO J. 22, 2505−2515.
Berriman, M., Ghedin, E., Hertz-Fowler, C., Blandin, G., Renauld, H., Bartholomeu, D.C., et al., 2005. The genome of the African trypanosome *Trypanosoma brucei*. Science 309, 416−422.
Bingle, L.E.H., Eastlake, J.L., Bailey, M., Gibson, W.C., 2001. A novel GFP approach for the analysis of genetic exchange in trypanosomes allowing the in situ detection of mating events. Microbiology 147, 3231−3240.
Bosseno, M.-F., Telleria, J., Vargas, F., Yaksic, N., Noireau, F., Morin, A., et al., 1996. *Trypanosoma cruzi*: study of the distribution of two widespread clonal genotypes in Bolivian *Triatoma infestans* vectors shows a high frequency of mixed infections. Exp. Parasitol. 83, 275−282.
Brenière, S.F., Morochi, W., Bosseno, M.-F., Ordoñez, J., Gutierrez, T., Vargas, F., et al., 1998. *Trypanosoma cruzi* genotypes associated with domestic *Triatoma sordida* in Bolivia. Acta Trop. 71, 269−283.
Brisse, S., Henriksson, J., Barnabe, C., Douzery, E.J.P., Berkvens, D., Serrano, M., et al., 2003. Evidence for genetic exchange and hybridization in *Trypanosoma cruzi* based on nucleotide sequences and molecular karyotype. Infect. Genet. Evol. 2, 173−183.
Burgos, J.M., Begher, S., Silva, H.M.V., Bisio, M., Duffy, T., Levin, M.J., et al., 2008. Molecular identification of *Trypanosoma cruzi* I tropism for central nervous system in Chagas reactivation due to AIDS. Am. J. Trop. Med. Hyg. 78, 294−297.
Cardinal, M.V., Lauricella, M.A., Ceballos, L.A., Lanati, L., Marcet, P.L., Levin, M.J., et al., 2008. Molecular epidemiology of domestic and sylvatic *Trypanosoma cruzi* infection in rural northwestern Argentina. Int. J. Parasitol. 38, 1533−1543.
Carrasco, H.J., Frame, I.A., Valente, S.A., Miles, M.A., 1996. Genetic exchange as a possible source of genomic diversity in sylvatic populations of *Trypanosoma cruzi*. Am. J. Trop. Med. Hyg. 54, 418−424.
Cibulskis, R.E., 1992. Genetic variation in *Trypanosoma brucei* and the epidemiology of sleeping sickness in the Lambwe Valley, Kenya. Parasitology 104, 99−109.
Cooper, A., Tait, A., Sweeney, L., Tweedie, A., Morrison, L., Turner, C.M.R., et al., 2008. Genetic analysis of the human infective trypanosome, *Trypanosoma brucei gambiense*: chromosomal segregation, crossing over and the construction of a genetic map. Genome Biol. 9.
Cortez, M.R., Monteiro, F.A., Noireau, F., 2010. New insights on the spread of *Triatoma infestans* from Bolivia—implications for Chagas disease emergence in the Southern Cone. Infect. Genet. Evol. 10, 350−353.
De Greef, C., Imberechts, H., Matthyssons, G., Van Meirvenne, N., Hamers, R., 1989. A gene expressed only in serum-resistant variants of *Trypanosoma brucei rhodesiense*. Mol. Biochem. Parasitol. 36, 169−176.
de Lana, M., Pinto, A.d.S., Bastrenta, B., Barnabe, C., Noel, S., Tibayrenc, M., 2000. *Trypanosoma cruzi*: infectivity of clonal genotype infections in acute and chronic phases in mice. Exp. Parasitol. 96, 61−66.
Degen, R., Pospichal, H., Enyaru, J., Jenni, L., 1995. Sexual compatibility among *Trypanosoma brucei* isolates from an epidemic area in southeastern Uganda. Parasitol. Res. 81, 253−257.
Delgado, O., Cupolillo, E., Bonfante-Garrido, R., Silva, S., Belfort, E., Grimaldi Jr., G., et al., 1997. Cutaneous leishmaniasis in Venezuela caused by infection with a new hybrid between *Leishmania (Viannia) braziliensis* and *L. (V.) guyanensis*. Mem. Inst. Oswaldo Cruz 92, 581−582.

Dujardin, J.-C., Banuls, A.-L., Llanos-Cuentas, A., Alvarez, E., DeDoncker, S., Jacquet, D., et al., 1995. Putative *Leishmania* hybrids in the Eastern Andean valley of Huanuco, Peru. Acta Trop. 59, 293−307.

Dukes, P., Kaukus, A., Hudson, K.M., Asonganyi, T., Gashumba, J.K., 1989. A new method for isolating *Trypanosoma brucei gambiense* from sleeping sickness patients. Trans. R. Soc. Trop. Med. Hyg. 83, 636−639.

Dvorak, J., Hall, T., Crane, M., Engel, J., McDaniel, J., Uriegas, R., 1982. *Trypanosoma cruzi*: flow cytometric analysis. I. Analysis of total DNA/organism by means of mithramycin-induced fluorescence. J. Protozool. 29, 430−437.

Eid, J., Sollner-Webb, B., 1991. Stable integrative transformation of *Trypanosoma brucei* that occurs exclusively by homologous recombination. Proc. Natl. Acad. Sci. U.S.A. 88, 2118−2121.

El-Sayed, N.M., Myler, P.J., Bartholomeu, D.C., Nilsson, D., Aggarwal, G., Tran, A.N., et al., 2005. The genome sequence of *Trypanosoma cruzi*, etiologic agent of Chagas disease. Science 309, 409−415.

Enyaru, J.C., Stevens, J.R., Odiit, M., Okuna, N.M., Carasco, J.F., 1993. Isoenzyme comparison of *Trypanozoon* isolates from two sleeping sickness areas of south-eastern Uganda. Acta Trop. 55, 97−115.

Evans, D.A., Ellis, D.E., 1983. Recent observations on the behaviour of certain trypanosomes within their insect hosts. Adv. Parasitol. 22, 1−42.

Evans, D.A., Kennedy, W.P., Elbihari, S., Chapman, C.J., Smith, V., Peters, W., 1987. Hybrid formation within the genus *Leishmania*? Parasitologia 29, 165−173.

Fernandes, O., Souto, R., Castro, J., Pereira, J., Fernandes, N., Junqueira, A., et al., 1998. Brazilian isolates of *Trypanosoma cruzi* from humans and triatomines classified into two lineages using mini-exon and ribosomal RNA sequences. Am. J. Trop. Med. Hyg. 58, 807−811.

Fernandes, O., Mangia, R.H., Lisboa, C.V., Pinho, A.P., Morel, C.M., Zingales, B., et al., 1999. The complexity of the sylvatic cycle of *Trypanosoma cruzi* in Rio de Janeiro state (Brazil) revealed by the non-transcribed spacer of the mini-exon gene. Parasitology 118, 161−166.

Fraga, J., Montalvo, A.M., De Doncker, S., Dujardin, J.-C., Van der Auwera, G., 2010. Phylogeny of *Leishmania* species based on the heat-shock protein 70 gene. Infect. Genet. Evol. 10, 238−245.

Gaunt, M.W., Yeo, M., Frame, I.A., Stothard, J.R., Carrasco, H.J., Taylor, M.C., et al., 2003. Mechanism of genetic exchange in American trypanosomes. Nature 421, 936−939.

Gibson, W.C., 1986. Will the real *Trypanosoma brucei gambiense* please stand up? Parasitol. Today 2, 255−257.

Gibson, W.C., 1989. Analysis of a genetic cross between *Trypanosoma brucei rhodesiense* and *T. b. brucei*. Parasitology 99, 391−402.

Gibson, W., Bailey, M., 1994. Genetic exchange in *Trypanosoma brucei*: evidence for meiosis from analysis of a cross between drug resistant transformants. Mol. Biochem. Parasitol. 64, 241−252.

Gibson, W., Garside, L., 1990. Kinetoplast DNA mini-circles are inherited from both parents in genetic hybrids of *Trypanosoma brucei*. Mol. Biochem. Parasitol. 42, 45−54.

Gibson, W.C., Garside, L.H., 1991. Genetic exchange in *Trypanosoma brucei brucei*: variable location of housekeeping genes in different trypanosome stocks. Mol. Biochem. Parasitol. 45, 77−90.

Gibson, W.C., Gashumba, J.K., 1983. Isoenzyme characterisation of some *Trypanozoon* stocks from a recent trypanosomiasis epidemic in Uganda. Trans. R. Soc. Trop. Med. Hyg. 77, 114–118.
Gibson, W.C., Mizen, V.H., 1997. Heritability of the trait for human infectivity in genetic crosses of *Trypanosoma brucei* ssp. Trans. R. Soc. Trop. Med. Hyg. 91, 236–237.
Gibson, W., Stevens, J., 1999. Genetic exchange in the trypanosomatidae. Adv. Parasitol. 43, 1–46.
Gibson, W.C., Wellde, B.T., 1985. Characterisation of *Trypanozoon* stocks from the South Nyanza sleeping sickness focus in Western Kenya. Trans. R. Soc. Trop. Med. Hyg. 79, 671–676.
Gibson, W., Whittington, H., 1993. Genetic exchange in *Trypanosoma brucei*: selection of hybrid trypanosomes by introduction of genes conferring drug resistance. Mol. Biochem. Parasitol. 60, 19–26.
Gibson, W., Garside, L., Bailey, M., 1992. Trisomy and chromosome size changes in hybrid trypanosomes from a genetic cross between *Trypanosoma brucei rhodesiense* and *T. b. brucei*. Mol. Biochem. Parasitol. 51, 189–199.
Gibson, W., Kanmogne, G., Bailey, M., 1995. A successful backcross in *Trypanosoma brucei*. Mol. Biochem. Parasitol. 69, 101–110.
Gibson, W., Crow, M., Kearns, J., 1997a. Kinetoplast DNA minicircles are inherited from both parents in genetic crosses of *Trypanosoma brucei*. Parasitol. Res. 83, 483–488.
Gibson, W., Winters, K., Mizen, G., Kearns, J., Bailey, M., 1997b. Intraclonal mating in *Trypanosoma brucei* is associated with out crossing. Microbiology 143, 909–920.
Gibson, W.C., Marshall, T.F.D.C., Godfrey, D.G., 1980. Numerical analysis of enzyme polymorphism: a new approach to the epidemiology and taxonomy of trypanosomes of the subgenus *Trypanozoon*. Adv. Parasitol. 18, 175–246.
Gibson, W., Peacock, L., Ferris, V., Williams, K., Bailey, M., 2008. The use of yellow fluorescent hybrids to indicate mating in *Trypanosoma brucei*. Parasit. Vector. 1, 4.
Godfrey, D.G., Killick-Kendrick, R., 1961. Bovine trypanosomiasis in Nigeria. I. The inoculation of blood into rats as a method of survey in the Donga Valley, Benue province. Ann. Trop. Med. Parasitol. 55, 287–292.
Gottesdiener, K., Garcia-Anoveros, J., Lee, G.-S.M., Van der Ploeg, L.H.T., 1990. Chromosome organization of the protozoan *Trypanosoma brucei*. Mol. Cell. Biol. 10, 6079–6083.
Guevara, P., Dias, M., Rojas, A., Crisante, G., Abreu-Blanco, M.T., Umezawa, E., et al., 2005. Expression of fluorescent genes in *Trypanosoma cruzi* and *Trypanosoma rangeli* (Kinetoplastida: Trypanosomatidae): its application to parasite–vector biology. J. Med. Entomol. 48–56.
Hide, G., Welburn, S.C., Tait, A., Maudlin, I., 1994. Epidemiological relationships of *Trypanosoma brucei* stocks from South East Uganda: evidence for different population structures in human infective and non-infective isolates. Parasitology 109, 95–111.
Ivens, A.C., Peacock, C.S., Worthey, E.A., Murphy, L., Aggarwal, G., Berriman, M., et al., 2005. The genome of the kinetoplastid parasite, *Leishmania major*. Science 309, 436–442.
Jenni, L., Marti, S., Schweizer, J., Betschart, B., Lepage, R.W.F., Wells, J.M., et al., 1986. Hybrid formation between African trypanosomes during cyclical transmission. Nature 322, 173–175.
Kelly, J.M., Law, J.M., Chapman, C.J., Van Eyes, G.J.J.M., Evans, D.A., 1991. Evidence of genetic recombination in *Leishmania*. Mol. Biochem. Parasitol. 46, 253–264.

Komba, E.K., Kibona, S.N., Ambwene, A.K., Stevens, J.R., Gibson, W.C., 1997. Genetic diversity among *Trypanosoma brucei rhodesiense* isolates from Tanzania. Parasitology 115, 571–579.

Kooy, R.F., Hirumi, H., Moloo, S.K., Nantuyla, V.M., Dukes, P., Van der Linden, P.M., et al., 1989. Evidence for diploidy in metacyclic forms of African trypanosomes. Proc. Natl. Acad. Sci. U.S.A. 5469–5472.

Lee, G.-S.M., Van der Ploeg, L.H.T., 1990. Homologous recombination and stable transfection in the parasitic protozoan *Trypanosoma brucei*. Science 250, 1583–1587.

Lehane, M.J., Msangi, A.R., Whitaker, C.J., Lehane, S.M., 2000. Grouping of trypanosome species in mixed infections in *Glossina pallidipes*. Parasitology 120, 583–592.

Lewis, M.D., Llewellyn, M.S., Gaunt, M.W., Yeo, M., Carrasco, H.J., Miles, M.A., 2009. Flow cytometric analysis and microsatellite genotyping reveal extensive DNA content variation in *Trypanosoma cruzi* populations and expose contrasts between natural and experimental hybrids. Int. J. Parasitol 39, 1305–1317.

Lisboa, C.V., Mangia, R.H., De Lima, N.R.C., Martins, A., Dietz, J., Baker, A.J., et al., 2004. Distinct patterns of *Trypanosoma cruzi* infection in *Leontopithecus rosalia* in distinct Atlantic Coastal Rainforest fragments in Rio de Janeiro—Brazil. Parasitology 129, 703–711.

Llewellyn, M.S., Lewis, M.D., Acosta, N., Yeo, M., Carrasco, H.J., Segovia, M., et al., 2009a. *Trypanosoma cruzi* IIc: phylogenetic and phylogeographic insights from sequence and microsatellite analysis and potential impact on emergent Chagas disease. PLoS Negl. Trop. Dis. 3, e510.

Llewellyn, M.S., Miles, M.A., Carrasco, H.J., Lewis, M.D., Yeo, M., Vargas, J., et al., 2009b. Genome-scale multilocus microsatellite typing of *Trypanosoma cruzi* discrete typing unit I reveals phylogeographic structure and specific genotypes linked to human infection. PLoS Pathogens 5, e1000410.

Machado, C.A., Ayala, F.J., 2001. Nucleotide sequences provide evidence of genetic exchange among distantly related lineages of *Trypanosoma cruzi*. PNAS 98, 7396–7401.

MacLeod, A., Turner, C.M.R., Tait, A., 1999. A high level of mixed *Trypanosoma brucei* infections in tsetse flies detected by three hypervariable minisatellites. Mol. Biochem. Parasitol. 102, 237–248.

MacLeod, A., Tweedie, A., McLellan, S., Taylor, S., Cooper, A., Sweeney, L., et al., 2005a. Allelic segregation and independent assortment in *Trypanosoma brucei* crosses: proof that the genetic system is Mendelian and involves meiosis. Mol. Biochem. Parasitol. 143, 12–19.

MacLeod, A., Tweedie, A., McLellan, S., Taylor, S., Hall, N., Berriman, M., et al., 2005b. The genetic map and comparative analysis with the physical map of *Trypanosoma brucei*. Nucleic Acids Res. 33, 6688–6693.

Mallet, J., 2007. Hybrid speciation. Nature 446, 279–283.

Martins, H.R., Toledo, M.J.O., Veloso, V.M., Carneiro, C.M., Machado-Coelho, G.L.L., Tafuri, W.L., et al., 2006. *Trypanosoma cruzi*: impact of dual-clone infections on parasite biological properties in BALB/c mice. Exp. Parasitol. 112, 237–246.

Martins, H.R., Silva, R.M., Valadares, H.M.S., Toledo, M.J.O., Veloso, V.M., Vitelli-Avelar, D.M., et al., 2007. Impact of dual infections on chemotherapeutic efficacy in BALB/c mice infected with major genotypes of *Trypanosoma cruzi*. Antimicrob. Agents Chemother. 51, 3282–3289.

Masiga, D.K., McNamara, J.J., Laveissiere, C., Truc, P., Gibson, W.C., 1996. A high prevalence of mixed trypanosome infections in tsetse flies in Sinfra, Cote d'Ivoire, detected by DNA amplification. Parasitology 112, 75–80.

Maudlin, I., 1991. Transmission of African Trypanosomiasis: interactions among tsetse immune system, symbionts and parasites. Adv. Disease Vector Res. 7, 117–148.

Mauricio, I.L., Yeo, M., Baghaei, M., Doto, D., Pratlong, F., Zemanova, E., et al., 2006. Towards multilocus sequence typing of the *Leishmania donovani* complex: resolving genotypes and haplotypes for five polymorphic metabolic enzymes (ASAT, GPI, NH1, NH2, PGD). Int. J. Parasitol. 36, 757–769.

Melville, S.E., Leech, V., Gerrard, C.S., Tait, A., Blackwell, J.M., 1998. The molecular karyotype of the megabase chromosomes of *Trypanosoma brucei* and the assignment of chromosome markers. Mol. Biochem. Parasitol. 94, 155–173.

Miles, M., Cedillos, R., Povoa, M., de Souza, A., Prata, A., Macedo, V., 1981. Do radically dissimilar *Trypanosoma cruzi* strains (zymodemes) cause Venezuelan and Brazilian forms of Chagas disease?. Lancet 317, 1338–1340.

Miles, M.A., 1985. Ploidy, "heterozygosity" and antigenic expression of South American trypanosomes. Parassitologia 27, 87–104.

Miles, M.A., Souza, A., Povoa, M., Shaw, J.J., Lainson, R., Toye, P.J., 1978. Isozymic heterogeneity of *Trypanosoma cruzi* in the first autochthonous patients with Chagas disease in Amazonian Brazil. Nature 272, 819–821.

Miles, M.A., Llewellyn, M.S., Lewis, M.D., Yeo, M., Baleela, R., Fitzpatrick, S., et al., 2009. The molecular epidemiology and phylogeography of *Trypanosoma cruzi* and parallel research on *Leishmania*: looking back and to the future. Parasitology 136, 1509–1528.

Nolder, D., Roncal, N., Davies, C.R., Llanos-Cuentas, A., Miles, M.A., 2007. Multiple hybrid genotypes of *Leishmania (Viannia)* in a focus of mucocutaneous leishmaniasis. Am. J. Trop. Med. Hyg. 76, 573–578.

Nyeko, J.H.P., Ole-Moiyoi, O.K., Majiwa, P., Otieno, L.H., Ociba, P.M., 1990. Characterisation of trypanosome isolates from cattle in Uganda using species-specific DNA probes reveals predominance of mixed infections. Insect Sci. Appl. 11, 271–280.

Paindavoine, P., Zampetti-Bosseler, F., Pays, E., Schweizer, J., Guyaux, M., Jenni, L., et al., 1986. Trypanosome hybrids generated in tsetse flies by nuclear fusion. EMBO J. 5, 3631–3636.

Peacock, L., Ferris, V., Bailey, M., Gibson, W., 2007. Dynamics of infection and competition between two strains of *Trypanosoma brucei brucei* in the tsetse fly observed using fluorescent markers. Kinetoplastid Biol. Dis. 6, 4.

Peacock, L., Ferris, V., Bailey, M., Gibson, W., 2009. Intraclonal mating occurs during tsetse transmission of *Trypanosoma brucei*. Parasit. Vector. 2, 43.

Peacock, L., Ferris, V., Sharma, R., Sunter, J., Bailey, M., Carrington, M., Gibson, W., 2010. Meiosis in Trypanosomes. In preparation.

Pinto, A.S., de Lana, M., Bastrenta, B., BarnabÃ©, C., Quesney, V., NoÃ«l, S., et al., 1998. Compared vectorial transmissibility of pure and mixed clonal genotypes of *Trypanosoma cruzi* in *Triatoma infestans*. Parasitol. Res. 84, 348–353.

Pinto, A.d.S., de Lana, M., Britto, C., Bastrenta, B., Tibayrenc, M., 2000. Experimental *Trypanosoma cruzi* biclonal infection in *Triatoma infestans*: detection of distinct clonal genotypes using kinetoplast DNA probes. Int. J. Parasitol. 30, 843–848.

Pires, S.F., DaRocha, W.D., Freitas, J.M., Oliveira, L.A., Kitten, G.T., Machado, C.R., et al., 2008. Cell culture and animal infection with distinct *Trypanosoma cruzi* strains expressing red and green fluorescent proteins. Int. J. Parasitol. 38, 289–297.

Ramesh, M.A., Malik, S.B., Logsdon, J.M., 2005. A phylogenomic inventory of meiotic genes: evidence for sex in *Giardia* and an early eukaryotic origin of meiosis. Curr. Biol. 15, 185–191.
Ravel, C., Cortes, S., Pratlong, F., Morio, F., Dedet, J.P., Campino, L., 2006. First report of genetic hybrids between two very divergent *Leishmania* species: *Leishmania infantum* and *Leishmania major*. Int. J. Parasitol. 36, 1383–1388.
Schweizer, J., Jenni, L., 1991. Hybrid formation in the lifecycle of *Trypanosoma (T.) brucei*: detection of hybrid trypanosomes in a midgut-derived isolate. Acta Trop. 48, 319–321.
Schweizer, J., Tait, A., Jenni, L., 1988. The timing and frequency of hybrid formation in African trypanosomes during cyclical transmission. Parasitol. Res. 75, 98–101.
Schweizer, J., Pospichal, H., Jenni, L., 1991. Hybrid formation between African trypanosomes in vitro. Acta Trop. 49, 237–240.
Schweizer, J., Pospichal, H., Hide, G., Buchanan, N., Tait, A., Jenni, L., 1994. Analysis of a new genetic cross between 2 East African *Trypanosoma brucei* clones. Parasitology 109, 83–93.
Schwenkenbecher, J.M., Wirth, T., Schnur, L.F., Jaffe, C.L., Schallig, H., Al-Jawabreh, A., et al., 2006. Microsatellite analysis reveals genetic structure of *Leishmania tropica*. Int. J. Parasitol. 36, 237–246.
Shapiro, S.Z., Naessens, J., Liesegang, B., Moloo, S.K., Magondu, J., 1984. Analysis by flow cytometry of DNA synthesis during the life cycle of African trypanosomes. Acta Trop. 41, 313–323.
Shapiro, T.A., Englund, P.T., 1995. The structure and replication of kinetoplast DNA. Annu. Rev. Microbiol. 49, 117–143.
Sleigh, M.A., 1973. The Biology of Protozoa. Edward Arnold, London.
Sternberg, J., Tait, A., Haley, S., Wells, J.M., Lepage, R.W.F., Schweizer, J., et al., 1988. Gene exchange in African trypanosomes: characterisation of a new hybrid genotype. Mol. Biochem. Parasitol. 27, 191–200.
Sternberg, J., Turner, C.M.R., Wells, J.M., Ranford-Cartwright, L.C., Le Page, R.W.F., Tait, A., 1989. Gene exchange in African trypanosomes: frequency and allelic segregation. Mol. Biochem. Parasitol. 34, 269–279.
Stevens, J.R., Tibayrenc, M., 1996. *Trypanosoma brucei* s.l.: evolution, linkage and the clonality debate. Parasitology 112, 481–488.
Stevens, J.R., Welburn, S.C., 1993. Genetic processes within an epidemic of sleeping sickness in Uganda. Parasitol. Res. 79, 421–427.
Stothard, J., Frame, I., Miles, M., 1999. Genetic diversity and genetic exchange in *Trypanosoma cruzi*: dual drug-resistant "progeny" from episomal transformants. Mem. Inst. Oswaldo Cruz 94, 189–193.
Sturm, N.R., Vargas, N.S., Westenberger, S.J., Zingales, B., Campbell, D.A., 2003. Evidence for multiple hybrid groups in *Trypanosoma cruzi*. Int. J. Parasitol. 33, 269–279.
Tait, A., 1980. Evidence for diploidy and mating in trypanosomes. Nature 287, 536–538.
Tait, A., Buchanan, N., Hide, G., Turner, M., 1996. Self-fertilisation in *Trypanosoma brucei*. Mol. Biochem. Parasitol. 76, 31–42.
Ten Asbroek, A.L.M.A., Ouellette, M., Borst, P., 1990. Targeted insertion of the neomycin phosphotransferase gene into the tubulin gene cluster of *Trypanosoma brucei*. Nature 348, 174–175.
Tibayrenc, M., Ayala, F.J., 2002. The clonal theory of parasitic protozoa: 12 years on. Trends Parasitol. 18, 405–410.
Tibayrenc, M., Kjellberg, F., Ayala, F.J., 1990. A clonal theory of parasitic protozoa: the population structures of *Entamoeba, Giardia, Leishmania, Naegleria, Plasmodium,*

Trichomonas and *Trypanosoma* and their medical and taxonomical consequences. Proc. Natl. Acad. Sci. U.S.A. 87, 2414−2418.

Truc, P., Jamonneau, V., NGuessan, P., NDri, L., Diallo, P.B., Cuny, G., 1998. *Trypanosoma brucei* ssp. and *T. congolense*: mixed human infection in Cote d'Ivoire. Trans. R. Soc. Trop. Med. Hyg. 92, 537−538.

Turner, C.M.R., Sternberg, J., Buchanan, N., Smith, E., Hide, G., Tait, A., 1990. Evidence that the mechanism of gene exchange in *Trypanosoma brucei* involves meiosis and syngamy. Parasitology 101, 377−386.

Turner, C.M.R., Hide, G., Buchanan, N., Tait, A., 1995. *Trypanosoma brucei*—inheritance of kinetoplast DNA maxicircles in a genetic cross and their segregation during vegetative growth. Exp. Parasitol. 80, 234−241.

Turner, C.M.R., McLellan, S., Lindergard, L.A.G., Bisoni, L., Tait, A., MacLeod, A., 2004. Human infectivity trait in *Trypanosoma brucei*: stability, heritability and relationship to sra expression. Parasitology 129, 445−454.

Vago, A.R., Andrade, L.O., Leite, A.A., d'Avila Reis, D., Macedo, A.M., Adad, S.J., et al., 2000. Genetic characterization of *Trypanosoma cruzi* directly from tissues of patients with chronic Chagas disease : differential distribution of genetic types into diverse organs. Am. J. Pathol. 156, 1805−1809.

Vago, A.R., Macedo, A.M., Oliveira, R.P., Andrade, L.O., Chiari, E., Galvão, L.M., et al., 1996. Kinetoplast DNA signatures of *Trypanosoma cruzi* strains obtained directly from infected tissues. Am. J. Pathol. 149, 2153−2159.

Volf, P., Benkova, I., Myskova, J., Sadlova, J., Campino, L., Ravel, C., 2007. Increased transmission potential of *Leishmania major/Leishmania infantum* hybrids. Int. J. Parasitol. 37, 589−593.

Wells, J.M., Prospero, T.D., Jenni, L., Le Page, R.W.F., 1987. DNA contents and molecular karyotypes of hybrid *Trypanosoma brucei*. Mol. Biochem. Parasitol. 24, 103−116.

Westenberger, S.J., Barnabe, C., Campbell, D.A., Sturm, N.R., 2005. Two hybridization events define the population structure of *Trypanosoma cruzi*. Genetics 171, 527−543.

Weatherly, D.B., Boehike, C., Tarleton, R.L., 2009. Chromosome assembly of the hybrid *Trypanosoma cruzi* genome. BMC Genomics 10, 255.

WHO, 2002. Control of Chagas Disease. 2nd ed. WHO, Geneva.

Widmer, G., Dvorak, J., Miles, M., 1987. Temperature modulation of growth rates and glucosephosphate isomerase isozyme activity in *Trypanosoma cruzi*. Mol. Biochem. Parasitol. 23, 55−62.

Wijers, D.J.B., 1958. Factors that may influence the infection rate of *Glossina palpalis* with *Trypanosoma gambiense*. I. The age of the fly at the time of the infected feed. Ann. Trop. Med. Parasitol. 52, 385−390.

Xong, V.H., Vanhamme, L., Chamekh, M., Chimfwembe, C.E., Van den Abbeele, J., Pays, A., et al., 1998. A VSG expression site-associated gene confers resistance to human serum in *Trypanosoma rhodesiense*. Cell 95, 839−846.

Yeo, M., Acosta, N., Llewellyn, M., Sanchez, H., Adamson, S., Miles, G.A.J., et al., 2005. Origins of Chagas disease: *Didelphis* species are natural hosts of *Trypanosoma cruzi* I and armadillos hosts of *Trypanosoma cruzi* II, including hybrids. Int. J. Parasitol. 35, 225−233.

Yeo, M., Lewis, M.D., Carrasco, H.J., Acosta, N., Llewellyn, M., de Silva Valente, S.A., et al., 2007. Resolution of multiclonal infections of *Trypanosoma cruzi* from naturally infected triatomine bugs and from experimentally infected mice by direct plating on a sensitive solid medium. Int. J. Parasitol. 37, 111−120.

Zampetti-Bosseler, F., Schweizer, J., Pays, E., Jenni, L., Steinert, M., 1986. Evidence for haploidy in metacyclic forms of *Trypanosoma brucei*. Proc. Natl. Acad. Sci. U.S.A. 83, 6063−6064.

Zemanova, E., Jirku, M., Mauricio, I.L., Horak, A., Miles, M.A., Lukes, J., 2007. The Leishmania donovani complex: genotypes of five metabolic enzymes (ICD, ME, MPI, G6PDH and FH), new targets for multilocus sequence typing. Int. J. Parasitol. 37, 149−160.

Zingales, B., Andrade, S.G., Briones, M.R., Campbell, D.A., Chiari, E., Fernandes, O., et al., 2009. A new consensus for *Trypanosoma cruzi* intraspecific nomenclature: second revision meeting recommends TCI to TCVI. Mem. Inst. Oswaldo Cruz 104, 1051−1054.

Zingales, B., Souto, R.P., Mangia, R.H., Lisboa, C.V., Campbell, D.A., Coura, J.R., et al., 1998. Molecular epidemiology of American trypanosomiasis in Brazil based on dimorphisms of rRNA and mini-exon gene sequences. Int. J. Parasitol. 28, 105−112.

22 Genomic Insights into the Past, Current and Future Evolution of Human Parasites of the Genus *Plasmodium*

Colin J. Sutherland[1,*] and Spencer D. Polley[2]

[1]HPA Malaria Reference Laboratory, London School of Hygiene and Tropical Medicine, London, UK, [2]Department of Clinical Parasitology, Hospital for Tropical Diseases, London, UK

22.1 Introduction

22.1.1 *Overview of* Plasmodium *Phylogeny and Description of Species Infecting* Homo sapiens

The protozoan genus *Plasmodium* comprises chromalveolate protists of the phylum Apicomplexa, order Haemosporida and the family Plasmodidae. Members of the genus are obligate parasites of vertebrate hosts including lizards, snakes, birds, rodents, and primates. Amphibians, marsupials, carnivores, and ungulates are major vertebrate groups not known to host *Plasmodium* sp. parasites. Natural infections of *Homo sapiens* are caused by six species: *Plasmodium falciparum*, *P. knowlesi*, *P. malariae*, *P. vivax*, and the two closely related species *P. ovale curtisi* and *P. ovale wallikeri*, only recently recognized as genetically distinct (Sutherland et al., 2010). In naïve human hosts each of these six parasites cause an acute febrile illness of varying severity and duration, known as malaria.

The evolution of the genus *Plasmodium* is punctuated by a series of host transitions, as the radiation into more than 270 current species occurred through a variety of vertebrate hosts. The primate malarias are probably the best studied group of species, and have been well described both in natural infections of simian, ape and human hosts (Garnham, 1966; Duval et al., 2009, 2010; Krief et al., 2010; Prugnolle et al., 2010), and in experimental infections in chimpanzees, baboons, and rhesus and *Aotus* monkeys (Coatney et al., 2003). Investigations of the biochemistry and cell biology of *P. falciparum* have been possible in vitro since a system for continuous culture was devised by Trager and Jensen (1976), and this is therefore by far the

[*]E-mail: Colin.Sutherland@lshtm.ac.uk

Genetics and Evolution of Infectious Diseases. DOI: 10.1016/B978-0-12-384890-1.00022-4
© 2011 Elsevier Inc. All rights reserved.

best characterized of the primate malaria parasites. However, as no other human or ape parasite species have been successfully adapted to continuous culture, with the possible exceptions of *P. knowlesi* and *P. fragile* (Wickham et al., 1980; Schuster, 2002), analysis of parasite biology in vitro has been of only indirect use in understanding the comparative evolutionary history of different members of the genus. In fact, it can be argued that the level of fundamental knowledge of each of the other species has probably suffered as a result of the focus on *in vitro* studies of *P. falciparum*.

Several species of rodent malaria, including *P. berghei*, *P. chabaudi*, *P. vinckei*, and *P. yoelii*, have been isolated from forest-dwelling thicket rats of the genus *Thamnomys*, found only in central Africa. These species are readily propagated in *Mus musculus*, and have proved excellent model systems for studies of host−parasite interactions, pathology, intra-host−parasite dynamics, and the evolution of parasite resistance to antimalarial drugs. However, as relatively few isolates were collected of each species, and many of these before 1960, relatively little is known of natural populations of these parasites in *Thamnomys*. Thus questions remain as to the extent of intra- and inter-species variation in this group of plasmodia, but nevertheless some evolutionary insights into the relationships among these parasite species can be drawn (Carter, 1978; Perkins et al., 2007). One of the avian malaria parasites, *P. gallinaceum*, a parasite of wild Galliformes in Southeast Asia (Shortt et al., 1941), is widely studied in experimental infections of domestic chickens, and frequently included as an outgroup to both primate and rodent malarias in DNA sequence-based phylogenetic studies of the genus.

Due to the significant number of severe and fatal cases of malaria caused by *P. falciparum*, and its ability to be propagated as blood-stage parasites in vitro, Laveran's parasite *P. falciparum* is the most closely studied of the human-infecting plasmodia. Geographically widespread, the parasite occurs well north of the Tropic of Cancer (e.g., Afghanistan) and also south of the Tropic of Capricorn (e.g., South Africa, Namibia). Human blood-stage infections display a 48-h cycle in experimental infections in volunteers, such as those studied in detail in the Georgia State Penitentiary in the mid-twentieth century (Coatney et al., 2003), hence the term "malignant tertian" malaria, referring to the periodicity of fevers in *P. falciparum*. Two particular biological features of this parasite distinguish it from the other human malaria infections. The first is that parasite-encoded adhesins, capable of binding to host endothelium, are expressed on the surface of the infected erythrocyte from the mid-trophozoite stage right through to schizogony and release of merozoites. This enables the mature forms of the parasite to sequester in small blood vessels in a variety of host tissues, for a period of 30−36 h, and thus only the young trophozoites ("ring forms") are observed in smears of peripheral blood, whereas for other human malaria species intra-erythrocytic parasites at all stages of maturity circulate in the periphery. The second distinguishing feature of *P. falciparum* is the production of distinctive crescent-shaped gametocytes (transmissible stages) which do not develop in synchrony with the asexual parasite stages, but appear late in infection after an extended period of development (typically 6−10 days) sequestered in endothelial beds in various tissues (reviewed by Drakeley et al., 2006). Once mature, gametocytes of *P. falciparum* are released into the peripheral circulation in order to have access to biting mosquitoes. This is the opposite of the pattern seen with mature asexual intra-erythrocytic forms, which are sequestered for the latter half of the schizogonic cycle.

P. vivax infection occurs across the broadest geographical range of all the human malaria parasites, aided partly by an important survival strategy which this parasite shares with *P. ovale*: the ability of liver schizonts to arrest development and remain dormant for weeks, months, or years as hypnozoites. Reactivation of hypnozoites some time after the primary infection, which directly follows an infective mosquito bite, can thus initiate a fresh blood-stage malaria infection in a subsequent season favorable for transmission to mosquitoes. This mechanism is thought particularly important for the continued transmission of *P. vivax* in temperate areas with a long winter in which anophelines are scarce or absent (Garnham, 1966; Coatney et al., 2003) or in areas with extreme or extended dry, hot seasons, such as Mauritania (Lekweiry et al., 2009). Malaria caused by *P. vivax* shares the 48-h periodicity of falciparum malaria, but is responsible for many fewer severe or fatal cases, hence the label "benign tertian" malaria. There is no appreciable stage-specific sequestration of parasitized erythrocytes, and all asexual and sexual forms of *P. vivax* are seen in peripheral blood smears. The lack of a reliable continuous in vitro culture system has hampered research on this parasite, but transient 48-h schizont maturation cultures have been useful in studying antimalarial drug response phenotypes in *P. vivax* (Suwanarusk et al., 2007). Vaccine development for *P. vivax* has relied on the ability of this strain to infect American monkeys, in particular the genera *Aotus* and *Saimiri*, and it is parasite material from this source that has provided the first full genome sequence of *P. vivax* (Salvador I strain) (Collins et al., 2009).

Plasmodium malariae occurs in humans throughout malaria endemic regions of the world, and has earned the name "quartan malaria" for its 72-h fever periodicity (reviewed in Collins and Jeffery, 2007; Mueller et al., 2007). This species occurs at low parasite density in the peripheral blood, frequently occurs with *P. vivax* or *P. falciparum* as mixed infections, causes generally a relatively mild malaise compared to other species (Garnham, 1966), and is likely to be substantially underreported due to misdiagnosis as *P. vivax* or *P. falciparum* (Lindo et al., 2007). *P. malariae* also occurs as a well-described zoonosis in American monkeys, under the species designation *P. brasilianum*. Although generally considered a minor species, *P. malariae* is very common in some locations in PNG, Indonesia and Africa, contributing substantially to overall malaria morbidity (Mueller et al., 2007, 2009; Bruce et al., 2008). One of the most puzzling aspects of the biology of *P. malariae* is the ability to recur years, and even decades, after the last possible exposure of the infected individual to an infected mosquito bite (Tsuchida et al., 1982). Despite the wide distribution of this parasite, and its association with severe and occasionally fatal nephritic complications, *P. malariae* is a poorly understood pathogen, and existing animal models that would enable its propagation are largely neglected, despite their potential to support detailed study of its biology (Collins and Jeffery, 2007). Of particular interest would be studies of the mechanism allowing revival of active infection after decades-long dormancy, despite no evidence of either hypnozoisis or tissue sequestration in this species.

Ovale malaria is now understood to be caused by two closely related but distinct parasite species, *P. ovale curtisi* and *P. ovale wallikeri* (Sutherland et al., 2010). We will consider aspects of this interesting speciation later in this chapter (Section 22.2.3). First described by Stephens in 1922 (reviewed in Collins and Jeffery, 2005; Mueller et al., 2007), *P. ovale* sp. causes acute febrile malaria with a tertian periodicity, but associations with severe or life-threatening complications are rarely reported. Ovale malaria does not occur outside of the tropics, and is not known in the Americas, but benefits from the ability to form hypnozoites by being able to persist in highly seasonal and arid settings (Faye et al., 1998; Lekweiry et al., 2009). The absence of *P. ovale* transmission in temperate zones despite its ability to form hypnozoites strongly suggests that the mosquito stages of this parasite are not able to complete development in the insect host at lower temperatures. Relapse episodes of ovale malaria can occur months or years apart (Davis et al., 2001; Coldren et al., 2007), and therefore this species may pose a particular challenge for the eradication of malaria from sub-Saharan Africa as relapses occur following clearance of blood-stage *P. falciparum* by combination therapy in mixed infections of the two species, which occur commonly across the continent (Sutherland et al., 2010).

Dubbed by some "the fifth human malaria," it became clear at the beginning of the twenty-first century that the well-described parasite of Asian macaques, *Plasmodium knowlesi*, was a relatively common agent of clinical malaria in humans in Malaysian Borneo in particular (Singh et al., 2004; White, 2008). This finding was soon duplicated in a number of countries in Southeast Asia where the macaques occur (Luchavez et al., 2008; Ng et al., 2008; Putaporntip et al., 2009; Van den Eede et al., 2009). Whereas it remains unlikely on evidence gathered so far that true human-centered transmission of *P. knowlesi* is occurring, the infection has caused a number of well-documented malaria deaths (Cox-Singh et al., 2008) and must be considered at the very least a zoonosis of substantial public health importance (see Section Emergence of *P. knowlesi* Zoonosis).

A radial phylogeny of the human-infecting malaria parasites, together with selected other taxa within the genus *Plasmodium*, is given in Figure 22.1. This is based on our previously published tree which establishes the likely position of *P. ovale curtisi* and *P. ovale wallikeri* within the genus, using mitochondrial (cytochrome *b*) sequence data (Sutherland et al., 2010).

22.1.2 Population Genetics and Design of Public Health Interventions

Tools for discerning heterogeneity in genetic loci within malaria species became available in the 1990s, and revealed the common occurrence of multiple clone infections in *P. falciparum*, and the dynamic turnover of these distinct genotypes in a single individual over time (Daubersies et al., 1996; Snounou et al., 1999). These studies were based on detecting variants of the merozoite antigen genes *pfmsp1* and *pfmsp2*, both known to be under strong diversifying selection from the human immune system, along with a number of other merozoite stage parasite

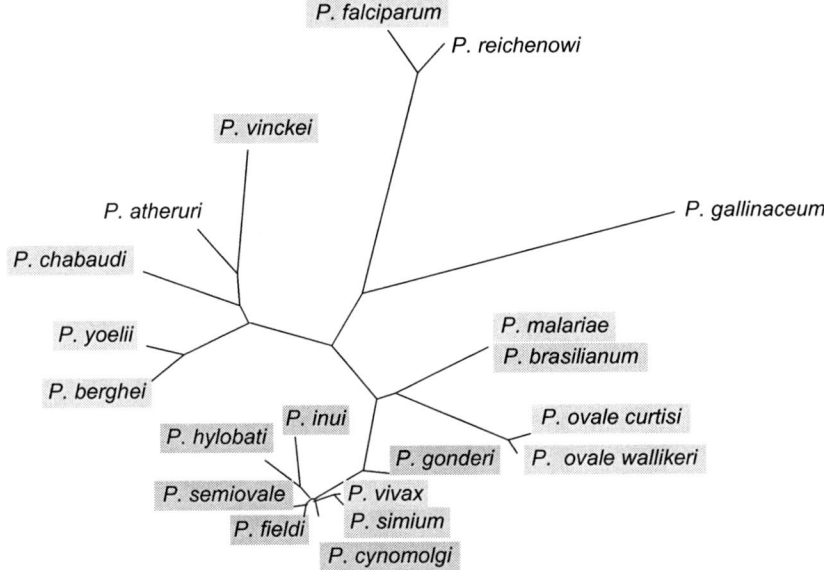

Figure 22.1 Radial phylogenetic tree showing cytochrome b—based phylogeny of selected *Plasmodium* species. Methodologies for phylogenetic reconstruction are the same as Sutherland et al. (2010). Species in yellow have rodents as their primary hosts, although *Plasmodium atheruri*, parasite of porcupines, retains the ability to infect the large vesper mouse (*Caolmys collosus*) and the Mongolian gerbil (*Meriones unguiculatus*). The five species in light blue are those known to infect humans, whilst species in light red infect Old World monkeys. Those in light green infect New World monkeys. *P. gallinaceum* is an avian malaria parasite, whilst *P. reichenowi* is a parasite of chimpanzees. (For interpretation of the references to color in this figure legend, the reader is referred to the web version of this book.)

proteins (Conway and Polley, 2002). Whereas much attention has been paid to population-level studies of polymorphic protein-coding genes as a means to empirical identification of potential candidate vaccine antigens in both *P. falciparum* and *P. vivax* (Rayner et al., 2002; Ord et al., 2005; Polley et al., 2007), these studies have also provided invaluable insights into the nature and complexity of historical selective forces in shaping populations of *Plasmodium* parasites (Polley and Conway, 2001; Ord et al., 2005), which may be significantly different even when examining homologous loci in sympatric populations of different species (Ord et al., 2008). Such understanding provides a framework for predicting the possible impact of vaccines directed at one or a very few parasite epitopes (Section 22.3.3; Takala et al., 2007). Knowledge of population-level signals of immune selection also provides tools for monitoring allele-specific selective impact of a malaria vaccine on polymorphic variants of a target antigen, or indirect changes in the multiplicity of blood-stage infections, such as those reported for the RTS,S/AS02A vaccine which targets the *pfcsp* gene of *P. falciparum* (Alloueche et al., 2003;

Enosse et al., 2006; Waitumbi et al., 2009). Thus these insights from population genetics will have a far-reaching effect on not only the design of future vaccines and vaccine development trials, but also the processes used to monitor the efficacy over time of vaccines put into widespread use.

Population genetic tools have also been powerfully deployed to assist understanding of the spread over recent decades of gene variants encoding parasite resistance to antimalarial drugs, particularly chloroquine (CQ) and sulfadoxine-pyramethamine (SP) (Roper et al., 2004; Anderson and Roper, 2005). Further, with the determination by shotgun capillary sequencing of the genome sequences of *P. falciparum*, *P. vivax*, and *P. knowlesi* (Gardner et al., 2002; Carlton et al., 2008a; Pain et al., 2008), and the deposition of these data into public databases (http://www.Plasmodb.org), genome-wide variation can now be studied among three of the malaria parasite species infecting humans. Significantly, a substantial effort is now being put into obtaining full genome sequence of multiple isolates of the human malaria species *P. falciparum* and *P. vivax* (Carlton et al., 2008b). Initial efforts have been based on Sanger di-deoxy sequencing reactions fractionated on capillary sequencing machines, employing both cultured lab-adapted *P. falciparum* lines, and recent patient isolates. Early results have demonstrated the power of multi-genome data to generate densely distributed informative polymorphic sites across the 14 parasite chromosomes, and thus provide understanding of global population structure, and patterns of linkage and selection across different genomic regions (Volkman et al., 2007). Complementary mapping approaches are based on identification of single-nucleotide polymorphisms (SNP) across the *P. falciparum* genome in a variety of parasite isolates cultured in vitro, and this has generated important clues as to where recent selection by antimlalarial drugs has left detectable signals in the parasite genome (Mu et al., 2010). A new impetus is now focused on the direct sequencing of full parasite genomes of isolates taken directly from malaria patients without ex vivo culture, using so-called "next-generation" or "massively parallel" sequencing strategies (Mardis, 2009).

22.1.3 Genomic Signals of Selection due to Host Immunity

Coevolution of *Plasmodium* species and their vertebrate hosts has left significant signals of selection on the host genome, well-known examples being the hemoglobin structural variants of *Homo sapiens* (such as sickle-cell anemia, the thalassemias, and G6PD deficiency), which provide a measure of protection against malaria. These genotypes have therefore become established in populations with ancient or recent history of *Plasmodium* infection risk (reviewed by Weatherall et al., 2002). Conversely, both mammalian and insect hosts have imposed selection upon the *Plasmodium* parasite genome as their respective immune systems adapt to minimize the harmful effects of parasitization. The mammalian immune system has left particularly strong signals of selection on the parasite genome, and perhaps one of the most spectacular examples of this is the genus-wide expansion of genes encoding large families of proteins involved in immune evasion. The *sicaVAR* family of *P. knowlesi* were the first such proteins to be discovered, followed by a number of families in other members of the genus, notably including the PfEMP1 proteins of *P. falciparum*, encoded by the ~60-strong *var* gene family, the *surf* family (N = 10) and the *Pfmc-2tm* family (N = 11) (Sam-Yellowe et al., 2004). Genome analysis of a number of *Plasmodium* species genomes has identified members of the *pir* (*Plasmodium* interspersed repeat) family of small variant proteins (reviewed by Cunningham et al., 2010). Thus sub-telomeric gene families have expanded to encode hundreds of protein variants per genome, suggesting that the benefit of avoiding immune responses from the vertebrate host has outweighed the cost of amplifying up such large gene families, and developing the complex gene regulatory pathways required to express them effectively.

Single-copy genes throughout *Plasmodium* genomes also bear signals of immune selection, and these are exemplified by diversification of the well-characterized nucleotide sequences in both *P. falciparum* and *P. vivax* that encode the immunogenic merozoite adhesion/invasion proteins such as AMA1, MSP2, MSP3, EBA-175, and the duffy-binding proteins (Polley et al., 2001, 2007; Ord et al., 2005, 2008; Tetteh et al., 2009; Osier et al., 2010). Other single-copy genes displaying evidence of selection for diversification include those coding for the sporozoite protein thrombospondin-related protein (TRAP), and SURFIN$_{4.2}$, a protein expressed in late stage intra-erythrocytic asexual parasites (Winter et al., 2005; Ochola et al., 2010). In contrast, evidence for balancing selection is not seen in population genetic analyses of the locus encoding circumsporozoite protein (CSP), despite its high degree of polymorphism (Weedall et al., 2007).

22.1.4 Summary of Genomic Studies of the Genus

Genomic studies per se of the genus *Plasmodium* began in the 1990s with a remarkably forward-thinking, multicenter project to map extended sequences of cultured *P. falciparum* parasites cloned in yeast artificial chromosomes (Dame et al., 1996). Supported by the Wellcome Trust, this early initiative smoothly dove-tailed with the full genome sequencing later carried out at the Wellcome Trust Sanger Institute, Hinxton, UK (WTSI), Stanford University, USA, and The Institute for Genomic Research (TIGR). The *P. falciparum* genome sequencing project divided up the work among the institutes chromosome-by-chromosome, deriving the data using different technologies in the three sites (Gardner et al., 2002, and accompanying articles in the same issue of *Nature*). Subsequent *Plasmodium* sp. genome sequence projects have been based at WTSI, and full capillary sequence of genomic DNA shotgun-cloned into *Escherichia coli* plasmids has been generated for *P. vivax* and *P. knowlesi* (Carlton et al., 2008a; Pain et al., 2008). Currently, annotation and assembly is ongoing from capillary di-deoxy sequence data from the parasite DNA of a single isolate of *P. ovale curtisi* from a US resident returning from Nigeria in 1977, and subsequently propagated in chimpanzees (Collins et al., 1987; Sutherland et al., 2010). Other *Plasmodium* genomes currently being sequenced at WTSI include *P. berghei*, *P. chabaudi*, *P. gallinaceum*, and *P. reichenowi* (http://www.sanger.ac.uk/Projects/Protozoa; accessed 10/07/2010).

22.2 Evolution of *Plasmodium*: The Past 10 Million Years

22.2.1 Role of Host Transitions in Speciation Events Within the Genus

The most basic definition of a species is a group of organisms which can mate to produce fertile offspring. This can be proved directly by mating experiments or, where such experiments are impossible, inferred by a lack of genetic recombination between the genomes of genetically related species. For speciation to occur at least two populations from an ancestral species must become reproductively isolated, thereby preventing gene flow between them. The result of reproductive isolation is that over time the two populations will accumulate a number of unique mutations (through either positive or neutral selection), which result in the inability to form viable offspring when inter-population breeding is allowed to occur.

There are a number of theoretical models in which reproductive isolation can occur. In allopatric speciation reproductive isolation is a direct result of the geographical separation of two (or more) populations through a physical barrier such as a river. In parapatric and peripatric speciation, the initial step of speciation occurs through a given population of organisms entering a new environmental niche. Over time the two populations become specialized for or isolated by their given niches, resulting in reproductive isolation, allowing the formation of bona fide species. In contrast to the above models of speciation, the underlying basis of sympatric speciation is genetic in origin, preventing admixture within two populations found in the same location. Putative mechanisms for sympatric speciation may include temporal isolation through a shift in the timing of gamete release, behavior isolation through different courtship routines, physical isolation through noncompatible genitalia, and gametic incompatibility mediated by mating incompatibility loci in plant, fungi, and marine organisms such as the sea urchin *Echinometra* (Palumbi and Metz, 1991) and marine diatoms (Amato et al., 2007). For an obligate parasite, an obvious mechanism for sympatric speciation is host switching (although the term niche could conceivably be used in this context). In the case of *Plasmodium*, this will be especially efficient at imposing genetic isolation where the two hosts are targeted by different species of vector, given that genetic recombination between different parasites genotypes occurs exclusively in the mosquito midgut.

Experimental infections, the comparison of parasite and host phylogenies, and most recently transgenic manipulation of parasites have all shed light on the role of host switching in the evolution of the plasmodium genus. Phylogenetic studies have allowed the evolution of mammalian and bird malarias to be elucidated and the common origin of these species to be calculated at between 120 and 160 million years ago (mya) (Escalante and Ayala, 1994; Perkins and Schall, 2002). Such a time point is much later than the divergence of sauropsid (ancestral linage of mammals) and synapsid (ancestoral lineage of birds, lizards, and snakes) lineages some 315 mya, suggesting that host switching has occurred during this time period. Human malaria species themselves appear not to be monophyletic but instead have multiple independent evolutionary origins. At least three separate simian to human host transitions are apparent from the phylogenetic reconstruction of mammalian malarial mitochondrial sequences, when only *P. vivax*, *P. knowlesi*, and *P. malariae* are considered (Mu et al., 2005). Recent studies on both bird and bat malaria species also show evidence for both host plasticity and host switching (Ricklefs and Fallon, 2002; Iezhova et al., 2005; Duval et al., 2007).

22.2.2 P. falciparum *and* P. reichenowi—*Divergent Host Specificities?*

The closest relatives of the human malaria parasite *P. falciparum* are the chimpanzee parasite *P. reichenowi* and other recently discovered members of the Laverania sub-group of *Plasmodium* found in African great apes (Liu et al., 2010). The ancestors of *Homo sapiens sapiens* (modern day humans) and *Pan troglodytes* (chimpanzees) are thought to have diverged around 4–8 mya according to phylogenetic and fossil evidence (Hacia, 2001; Brunet et al., 2002). This time point has often been used as a date for the divergence of *P. falciparum* and *P. reichenowi*, a time point which has in turn been used to date other nodes within phylogenetic studies. When *P. reichenowi* was discovered in 1917 in the blood of wild chimpanzees and gorillas it was morphologically indistinguishable from *P. falciparum*. However, in vitro experiments showed that humans could not be infected with *P. reichenowi*, and nor could chimpanzees be infected with *P. falciparum* parasitized patient blood (citations in Martin et al., 2005). The two infections were therefore deemed to be separate species, a fact which was subsequently confirmed by genetic analysis of the *csp* (CSP) locus (Lal and Goldman, 1991).

As early as 1977, differences in the erythrocyte receptors of humans and chimpanzees were postulated as the driving force behind the apparent species specificity of *P. reichenowi* and *P. falciparum* (Miller, 1977). Recently it has been suggested that changes in the sialic acid structure of Glycophorin A (a major receptor on the erythrocyte surface) are responsible for host specificity (Martin et al., 2005). Human Glycophorin A is known to be the ligand for *P. falciparum* EBA-175, and is essential for red cell invasion for some wild-type and cultured parasites (including W2-mef). The sialic acid side chains of glycophorin A can be cleaved by neuraminidase treatment leading to a significant loss of erythrocyte invasion by these parasites. Whereas the sialic acids of chimpanzee glycophorin A contain both *N*-glycolylneuraminic acid (Neu5Gc) and *N*-acetylneuraminic acid (Neu5Ac), in humans the *CMAH* gene function is disrupted, preventing the conversion of Neu5Gc to Neu5Ac (in chimpanzees Neu5Ac is the dominant sialic acid). Recombinant PfEBA-175 and PrEBA-175 show preferential binding to both human and chimpanzee and Neu5Gc and Neu5Ac, respectively. It has been postulated that the loss of *CMAH* may have rendered ancient hominids resistant to malaria, until a separate host switch from chimpanzees to humans some time later (Krief et al., 2010). The eba-175 loci of *P. falciparum* and *P. reichenowi* show a high level of positive selection driving interspecific divergence (Baum et al., 2003a; Wang et al., 2003), suggesting that it has evolved to counter the changes in host receptors.

It is of interest to note that erythrocyte invasion by a number of different *P. falciparum* isolates (including 3D7) is resistant to neuraminidase treatment (Baum et al.,

2003b). This suggests that these parasites have an alternative Glycophorin A independent invasion pathway. Laboratory clones of *P. falciparum* have shown invasion pathway switching in vitro following disruption of the *eba-175* locus (Duraisingh et al., 2003). Whilst Martin et al. state that even neuraminidase-resistant lines of *P. falciparum* exhibit reduced invasion efficiency, it is unknown how dependent *P. reichenowi* is on glycophorin A. Alternative mechanisms may therefore be the key to host species specificity including CSP affinity for human liver cells (Rathore et al., 2003) or EBL-1 specificity for glycophorin B (Mayer et al., 2009).

While early experiments showed that chimpanzees were refractory to *P. falciparum*, experimental infections can be established in splenetcomized chimpanzees. Indeed several isolates of *P. falciparum* including Vietnam Oak Knoll (FVO) (Siddiqui et al., 1972) and Uganda Palo Alto (Geiman and Meagher, 1967) have been adapted by serial passage through New World monkey hosts for use as animal models. The past few years have also seen the discovery that *P. knowlesi* infections are common in humans (Singh et al., 2004) and the presence of *P. falciparum* infection in both great apes and monkeys (Duval et al.,. 2009; Krief et al., 2010) and *P. malariae* in chimpanzees (Hayakawa et al., 2009) suggesting that the host specificity of certain Plasmodium species may be more fluid than was once thought. However, whether any of these novel infections produce viable gametocytes capable of transmitting into mosquitoes and then to fresh primate hosts, of any species, remains to be determined. The experimental adaptation of a Ghanian isolate of *P. falciparum* to *Aotus nancymai* allowed the production of viable gametocytes in splenectomized monkeys (Sullivan et al., 2003). Such infections, at the very least, suggest that mechanisms such as changes in host receptors may be insufficient to fully explain host switching and potential sympatric speciation events.

What alternatives are there to host switching that would facilitate the genetic isolation of sympatric populations? Mutations within mating incompatibility loci are known to produce genetic isolation within sympatric populations of animals and plants. The surface gamete proteins Pfs48/45 are known to be critical for the fertility of male microgametes (Van Dijk et al., 2001) and gene distributions at this locus show significant skewing at the population level (Conway et al., 2001). Following fertilization, a diploid ookinete is formed which buries into the mosquito midgut and undergoes meiotic reduction to form a sporozoite-filled oocyte. Analysis of Pfs48/45 allele frequencies in oocysts gathered from Tanzanian mosquitos shows a significantly higher inbreeding coefficients ($F(IS)$) compared with 11 unlinked microsatellite loci. Thus it would appear that assortative mating is occurring in these populations (Anthony et al., 2007), raising the possibility that genetic isolation may occur through mutations in gametic surface receptors following or possibly even preceding host switching by the novel host species.

22.2.3 Speciation Between P. ovale curtisi *and* P. ovale wallikeri: Separate Host Transitions?

It has recently been shown that parasites previously defined as *Plasmodium ovale* on the basis of their morphology when viewed on Giemsa-stained blood films comprise two genetically distinct, nonrecombining species which diverged around 2 mya. Through studies of mainly imported human malaria cases, these two newly recognized species were shown to be sympatric at country-level in their distribution, thus ruling out allopatric barriers as an explanation for the current pattern of endemicity, and suggesting that reproductive isolation between the two parasites is due to other mechanisms such as host partitioning (Sutherland et al., 2010). Nevertheless, as only country-level distribution data was available for this study, it could not be ruled out that continuing physical separation between the two species is maintained due to more subtle mechanisms such as micro-geographic discontinuities or specific habitat or seasonal requirements preventing admixture of the two types in the present. Subsequent examination of parasites from clinics in Congo-Brazzaville, and community parasitological surveys in Equatorial Guinea and two sites in Uganda have demonstrated conclusively that both *P. ovale curtisi* and *P. ovale wallikeri* occur in different people in the same village at the same time, providing stronger support for the existence of a robust biological barrier (or barriers) between the two parasite species (Oguike et al., submitted).

It is prudent to recognize that whatever the current barriers to mating and recombination might be, it is not necessarily true that these were also the *cause* of the original speciation event. In our view, a reasonably parsimonious explanation for the existence of these two related but distinct parasite lineages in the same primate host, *Homo*, is that their common ancestor in the most recent nonhuman primate host (dare we call this proto-ovale species *Plasmodium ovale ovale*?) underwent successful transit to *Homo* on two separate occasions. The two lineages antecedent to the recently identified *P. ovale curtisi* and *P. ovale wallikeri* species were thus prevented from recombination because each occupied a different primate host, and so the two lineages never occurred simultaneously in a single mosquito blood-meal, the absolute requirement for hybridization. During this time apart, differences arose which meant that recombination was no longer possible, once both lineages had become parasites of humans. It is also valid to argue an alternative to our favored "two transition hypothesis," namely that separation was due to a geographic separation of two populations after host transition to *Homo* (or the ancestors of *Homo*). This could be dubbed the "transient allopatry hypothesis" in that it requires sympatric, early *Homo*-parasitizing (hypothetical) *P. ovale ovale* populations to have developed into two geographically isolated lineages that came back into sympatry due to migration of the hominid hosts or expansion of the separated parasite populations. During the period of separation, the pair of lineages underwent sufficient divergent evolution for differences to arise, and these now keep the two lineages apart, despite their current close co-habitation across a global range (Sutherland et al., 2010; Oguike et al., submitted).

What could be the nature of these differences that maintain the species barrier between *P. ovale curtisi* and *P. ovale wallikeri* now they exist in sympatry? This is a matter of great interest, and a number of explanations can be hypothesized, and eventually tested. Here, we will briefly consider three: (1) special requirements of each species for different subsets of human or mosquito hosts; (2) divergence of specific molecules required during the mating process, or gross changes at chromosomal level that prevent viable chromosome pairing at meiosis; (3) *P. ovale curtisi* and *P. ovale wallikeri* essentially propagate clonally due to very frequent self-fertilization, and this explains why both exhibit very little intra-species polymorphism.

Hypothesis 1: Host Restriction

Meiosis and sexual recombination in malaria parasites both occur in the midgut of a blood-fed mosquito immediately after fertilization, as the transient diploid zygote stage generates 4 haploid progeny nuclei which will go on to form the ookinete, oocyst, and infective sporozoites in the mosquito host. Thus recombination will only occur between the genomes of male and female gametocytes taken up in the same blood-meal. For this reason, recombination between *P. ovale curtisi* and *P. ovale wallikeri* would require that both species be present in a single human individual, fed upon by an *Anopheles* mosquito, which is able to sustain the developmental cycle of both species. A restriction of either parasite species to mutually exclusive subsets of the human population, or to different mosquito species, would therefore prevent heterologous meiotic recombination. One caveat to this human partitioning argument is the theoretical possibility that heterologous fertilization could occur due to interruption of mosquito feeding, leading to partial blood-meals taken on two individual people in close proximity to each other, as has been shown occurs for *A. gambiae*, as implied by the distribution of *P. falciparum* genotypes among children sleeping in the same house in The Gambia (Conway and McBride, 1991). To permit mixing of gametocytes of both ovale species in a single mosquito, this would have to occur with at least two different individuals, infected with circulating gametocytes of *P. ovale curtisi* and *P. ovale wallikeri*, respectively, in the same place at the same time. This seems unlikely, but by no means impossible. However to date no evidence of such an event has been found.

Species-specific restriction to different subsets of the human population may be a plausible mechanism for the current maintenance of the species barrier between *P. ovale curtisi* and *P. ovale wallikeri*. A precedent exists in the form of the well-described reliance of *P. vivax* on the Duffy blood group antigen for invasion of host reticulocytes. We have postulated that human blood groups may delineate mutually exclusive subsets of human hosts, and are currently seeking evidence for this in field studies (Sutherland et al., 2010). Weak support for this notion comes from the observation that the two ovale homologues of the *P. vivax* gene encoding the reticulocyte-binding protein *pvrb2*, implicated in blood-stage invasion and thus potentially host cell selection, have accumulated a large number of non-synonymous mutations. Thus *P. ovale curtisi* and *P. ovale wallikeri* might be expected to exhibit phenotypic differences in erythrocyte/reticulocyte invasion, including different patterns of blood group restriction in selecting host red blood cells for invasion (Sutherland et al., 2010). The availability of genome sequence from both species in the very near future (see Section Hypothesis 2: Divergent Mating Factors) will permit direct comparison of the full repertoire of erythrocyte invasion molecules between the two parasites, and this could provide further evidence for the blood group restriction hypothesis.

On the other hand, species-specific restriction of *P. ovale curtisi* and *P. ovale wallikeri* to different species of *Anopheles* mosquito is not considered a likely explanation for the lack of recombination between the two parasites. There are a large variety of competent malaria vector species within the genus *Anopheles*, and the malaria vectors of medical importance differ within and between nations, and within and between continents. Thus the distribution of both *P. ovale curtisi* and *P. ovale wallikeri* in Asia, Africa, and the Pacific demonstrates unequivocally that both parasites are transmitted by a variety of mosquito species across this broad range (Sutherland et al., 2010). Mosquito host restriction, although possibly important in some endemic localities, therefore cannot explain the observed lack of recombination between the two ovale species.

Hypothesis 2: Divergent Mating Factors

Allopatric speciation events can generate two related taxa, physically separated, which can become secondarily sympatric due to migration or changes in the extent of suitable habitat (Section 22.2.1). If during the period of separation substantial genetic drift occurs in the sequences of genes determining mating compatibility, or in other genes such that hybrid offspring are unlikely to be viable, then the species barrier will remain intact, and recombination between the two forms will not occur. As argued above, the most likely reason for the evolution of two forms of ovale malaria parasite is that at least two independent host transitions into ancestors of modern humans occurred, separated by a lengthy period of time. This scenario is of course analogous to allopatry, in that the two lineages would have been "physically" separated by occupying different hosts. Thus the lack of recombination between twenty-first century human-dwelling populations of *P. ovale curtisi* and *P. ovale wallikeri* may be due to substantial changes in key genes encoding molecules essential for mate recognition, fertilization, or meiosis. It is also possible that gross chromosomal rearrangements have occurred in one or both lineages, thus rendering meiotic chromosome pairing impossible, and hybrid zygotes unviable.

Happily, genomic sequencing of a 1977 isolate of *P. ovale curtisi* is well advanced (Section 22.1.4; Sutherland et al., 2010), and we have now prepared two patient isolates each of *P. ovale curtisi* and *P. ovale wallikeri*, collected in 2009–2010, for next-generation highly parallel sequencing on the Solexa platform at WTSI. When these data are compared with the capillary genome sequence of the 1977 *P. ovale curtisi*, it should be possible to address directly the question of whether the two ovale parasite species have accumulated either crucial mutations (particularly in mating, fertilization, or meiotic function genes), or chromosomal structural changes sufficient to prevent viable hybridization. However, these data can also answer some very important questions about genome-wide inter-species polymorphisms: which loci have diverged the most between *P. ovale curtisi* and *P. ovale wallikeri*? Are erythrocyte invasion molecules prominent among them? Finally, we will also be able to compare contemporary twenty-first century isolates of *P. ovale curtisi* with the capillary sequenced isolate collected in Nigeria over

30 years previously, and thus also gain some interesting new insights into intra-species polymorphism, in both time and space, at the genome level.

Hypothesis 3: Clonal Propagation

It is certainly true that, to date, there is little evidence that either ovale parasite species is highly diverse genetically, and our recent study of inter-species genetic variation found only a very low level of diversity at the seven loci examined (Sutherland et al., 2010). However, the genes studied were mostly identified due to their high degree of sequence conservation with homologous loci in other better characterized *Plasmodium* species; such genes are likely to be under purifying selection and thus poor candidates with which to estimate levels of linkage disequilibrium, and coefficients of inbreeding. Some intra-specific polymorphism was found in the *potra* locus (Oguike et al., submitted) and when *P. ovale curtisi* and *P. ovale wallikeri* homologues of polymorphic loci well characterized in other species are identified, evidence of extensive intra-species polymorphism may be found. Nevertheless, on current knowledge it remains possible that frequent clonal propagation has exacerbated the isolation of these two genomes across time and space; however, detailed studies of other members of the genus would not support this as an explanation for the existence of a species barrier. Particularly relevant is the demonstration that African populations of *P. malariae*, a parasite species that occurs with a broadly similar population prevalence to ovale malaria across sub-Saharan Africa, exhibits a level of heterozygosity (H) as high as 0.8 (Bruce et al., 2007). This level of diversity is similar to that found in studies of *P. vivax* in the Amazon region of Venezuela where, despite a low intensity of transmission, population H of 0.8 was measured at the *pvmsp1*α locus in 1996, and similar estimates of H of 0.8 and 0.9 were observed at the *Pvama1* locus in 1996 and 1997, respectively (Ord et al., 2005, 2008). Detailed comparison among full, or at least extended, genome sequences of both ovale parasite species should provide the identity of several suitable polymorphic loci for studies of heterozygosity in *P. ovale curtisi* and *P. ovale wallikeri* across different endemic areas. In the meantime, a project to develop a panel of microsatellites in *P. ovale* sp. is currently underway (A. Barry, pers. comm.).

22.2.4 Importance of Host Specificity

Understanding the mechanisms of host specificity will allow a greater understanding of the pathology of infection, identifying key genes required for both asexual and sexual reproduction. Such targets may be utilized to disrupt the reproduction and transmission of the parasites through the production of novel vaccines and chemotherapeutics. Host switching may also be of both current and future importance in the pathogenicity of malaria species. The establishment of SARS and HIV in human populations shows how host switching can lead to the emergence of a highly virulent pathogen, and a relatively recent switch to human hosts might explain the increased pathogenicity of *P. falciparum* compared to *P. ovale*, *P. vivax*, and *P. malariae*. Indeed *P. knowlesi* infections are known to be highly virulent in humans (albeit less virulent than *P. falciparum* infections), with a 24-h asexual replication time leading rapidly to severe disease and death in some documented cases (Cox-Singh et al., 2010). Could a future host switch or adaptation of *P. knowlesi* to allow transmission among humans result in an even more deadly pathogen (Section 22.3.1)?

Ultimately, detailed understanding of the parasite genes contributing to the vertebrate host specificity of each *Plasmodium* parasite species may come from careful comparisons among the genomes now being sequenced. Given current understanding of the importance of erythrocyte-binding proteins and there specificity for certain host receptors (Section 22.2.2), elucidation of the reticulocyte-binding proteins, Duffy-binding proteins and erythrocyte-binding protein families in each species are a good place to start. However, an uncomfortably large proportion of each *Plasmodium* genome thus far sequenced encodes for "hypothetical proteins" of unknown function; it may be that comparative empirical approaches are required to identify many of the key molecules involved in parasite host specificity (Hall et al., 2005).

22.3 Evolution of *Plasmodium*: The Twenty-First Century in Three Courses

22.3.1 Entree: Habitat Destruction and Ecological Transitions

Host transitions in malaria parasites require contact between the novel vertebrate host species and insect vectors that have bitten infected individuals of the primary vertebrate host species. Thus human (and prehuman) migration, settlement, and the resulting encroachment of human activity into the habitats of different nonhuman primates have been the probable driving force behind the prehistorical transitions discussed in Section 22.2. In the twenty-first century, human migrations still occur, and encroachment into the habitats of wild simian and ape populations continue. Coupled with a recent improvement in our ability to discern the presence of unusual *Plasmodium* species in humans with malaria, these continued close encounters between humans and the parasites of beasts will be more commonly recognized. Further, the destruction of habitat and the resulting decline in numbers of the great apes means that the parasite species dependent on them as hosts are under selective pressure to expand their host range. This may increase the likelihood of new transitions into human-centered parasitism.

The recent description of additional wild isolates of *P. reichenowi* (Prugnolle et al., 2010), and additional falciparum-like members of *Plasmodium* in African great apes (Liu et al., 2010), raises the possibility that human infections with these parasites may be relatively common in some parts of Africa, but have failed to be recognized. This is not least because remnant populations of the nonhuman hominids exist in relatively remote locations, where people suffering from malaria who do come in contact with health services are likely to be treated (if at all) without any diagnosis, and certainly without a species-specific one (Sutherland et al., 2010). The prospect of such previously unrecognized zoonoses occurring is an interesting one from a scientific point of view, but may also have important evolutionary and public health consequences. As long as such infections occur, the evolutionary change to permit efficient human-to-human transmission of a previously ape-confined species may occur. From a public health point of view, this could lead to rapid expansion of a "new" malaria pathogen in human populations with no species-specific immunity; this threat is even more relevant where malaria elimination/eradication is expected to occur in these regions in the near future. Removal of human parasite species through vector control, immunization, and effective drug deployment will lead to a population with no naturally acquired immunity, and perhaps with greatly reduced access to effective malaria treatment as the number of cases dwindle. Thus understanding of these zoonotic pathogens should be vigorously pursued not only because of its great scientific interest, which would be enough for most of us, but also through human self-interest, which may be needed to persuade science funders.

Emergence of P. knowlesi Zoonosis

As mentioned earlier (Section 22.1.1), the story of *P. knowlesi* in Southeast Asia is one particularly important example, becoming well understood only in the past decade, of a zoonotic malaria parasite being recognized as a significant public health problem for humans. The parasite was thought to only naturally infect long-tailed and pig-tailed macaques, which are widely distributed across Southeast Asia and in which infection is benign (Coatney et al., 2003), despite one or two sporadic reports of human cases in Singapore and peninsular Malaysia (Garnham, 1966). It had been demonstrated during the use of induced malaria infections for syphilis therapy in the early twentieth century that *P. knowlesi* would infect humans with effective blood-stage multiplication, and cause full-blown clinical malaria (Garnham, 1966), but the true extent of naturally occurring human cases had not been understood, and was assumed to be negligible. However, molecular investigations during a malaria epidemic in Malaysian Borneo in the late 1990s confirmed *P. knowlesi* as the causative agent of a significant number of human cases (Singh et al., 2004). Molecular studies have since identified human cases of *P. knowlesi* in several Asian countries in addition to Malaysia, including Burma, The Philippines, Singapore, Vietnam, and Thailand (Zheng et al., 2006; Luchavez et al., 2008; Ng et al., 2008; Putaporntip et al., 2009; Van den Eede et al., 2009). *P. knowlesi* infections in humans can cause severe and fatal disease (Cox-Singh et al., 2008). To date, very limited surveillance has been carried out and the tools to define the distribution of the parasite and level of exposure are lacking.

It is probably the case that infections of humans in areas where macaques abound have always occurred, and that human *P. knowlesi* infections may have been either ignored, or extensively misdiagnosed in the past (Lee et al., 2009). However, as there is yet no evidence of human-to-mosquito transmission, the disease is best described as a well-established zoonosis with a measurable public health impact in some areas. Should efficient human-to-vector transmission arise de novo, then *P. knowlesi* could easily spread via human movements beyond the current range, which for now is absolutely restricted by the distribution of suitable host macaque species. Such a "break-out" of this species would certainly lead to substantial mortality, as this parasite has a very rapid replication cycle and quickly reaches life-threatening parasite densities in the absence of prompt antimalarial therapy (Cox-Singh et al., 2008).

Have Other Plasmodium Zoonoses Gone Undetected?

It has been recently confirmed that *Plasmodium falciparum* is present in wild African apes, with identification of this "human" parasite in six bonobos (*Pan paniscus*) from the Democratic Republic of Congo (Krief et al., 2010), in *Gorilla gorilla dielhi* and in *G. g. gorilla* (Prugnolle et al., 2010; Liu et al., 2010). Genetic analyses of these recent isolates strongly suggest continued cycling of the Laverania grouping of malaria parasites among different hominids, including *Homo*, in Africa. Conversely, as these studies also identified additional examples of *P. reichenowi* in three sub-species of *Pan* in Ivory Coast and Cameroon, and a closely related parasite in a bonobo from DRC, as well as two new species, *P. billbrayi* and *P. billcollinsi*, the possibility now occurs that a variety of hitherto overlooked parasite species, with *Pan* and *Gorilla* as primary hosts, may be zoonotic in humans. This poses two interesting questions. First, have we seriously underestimated the number of malaria parasite species that naturally infect humans in central Africa by ignoring the possibility of frequent zoonotic infections? Parasitological surveys of human communities living in proximity to ape populations would assist in answering this question. Second, will the great apes provide an eradication-proof reservoir of human malaria parasites, particularly *P. falciparum*, *P. ovale curtisi*, and *P. ovale wallikeri* (Section 22.4; Duval et al., 2009, 2010)?

22.3.2 Plat du Jour: Chemotherapy and the Evolution of Drug-Resistant Parasites

Lessons Learned from the Evolution of Parasite Resistance to CQ

A well-studied example of recent evolution in the genus *Plasmodium* is the worldwide development of resistance to the antimalarial drug CQ during the latter half of the twentieth century. The evolutionary aspects of this drug selection have been well examined (Wootton et al., 2002; Escalante et al., 2009), but some important principles can be drawn out which are relevant to understanding the likely impact of antimalarial drugs in the present century.

First, although there is evidence that CQ-resistant alleles of the transporter gene *pfcrt* evolved in different lineages of *P. falciparum* on multiple occasions, one or two of these alleles were particularly successful, spreading through contiguous parasite populations, and moving intercontinental distances, presumably in human travelers, to dominate parasite populations worldwide (Mehlotra et al., 2008). Interestingly, in the case of at least one of these alleles, that encoding the amino acid haplotype SVMNT at codons 72–76 of *pfcrt*, new results from our own laboratory cast doubt that CQ selective pressure was responsible for its expansion through Asian and American parasite populations (Beshir et al., 2010). It now seems likely that heavy use of the related amino-quinoline drug amodiaquine (AQ) in some countries in the mid-twentieth century provides a plausible alternative explanation for the spread of this parasite type, suggesting that CQ resistance per se is a secondary characteristic of the CRT protein variant encoded by this allele (Sa et al., 2009).

Second, antimalarial selection is far from uniform in endemic areas, so there are both temporal and geographic variations in the intensity of drug pressure, broadly meaning the proportion of infected people likely to use a particular drug type. This modulates the selective advantage enjoyed by parasites carrying drug-resistant alleles of key genes, particularly if there is a fitness cost to the development of resistance. Thus poor access to treatment, high levels of anti-parasite immunity leading to asymptomatic parasite carriage (particularly in African adults), and extended periods in the dry season with no transmission of new infections have contributed to the successful survival and continual transmission of wild-type, CQ-sensitive *P. falciparum* in many parts of Africa throughout the period that CQ was the main treatment option for malaria (Diallo et al., 2007; Ord et al., 2007). In the extreme case of complete removal of CQ from an entire health system, as was achieved in Malawi, the fitness advantage of wild-type CQ-sensitive parasites has ensured their relatively rapid resurgence to replace CQ-resistant genotypes that no longer have a survival advantage (Laufer et al., 2006).

Third, different antimalarial regimens generate different, sometimes opposite, selective forces on the parasite genome (Duraisingh et al., 2000; Humphreys et al., 2007). Thus as drug use diversifies in the current post-CQ era, the parasite genome appears to be showing evidence of "balancing selection" at drug resistance loci, a concept in direct opposition to the theoretical notion that drug pressure always drives advantageous alleles to "fixation" (see Section The Impact of RTS,S Vaccination on Parasite Diversity).

Finally, although genetically determined resistance to CQ has been described in several species of *Plasmodium*, only in *P. falciparum* are mutations in the *crt* locus directly linked to altered response to the drug. Stable resistance to CQ in both *P. vivax* and in *P. chabaudi* has been described, but the *pvcrt* and *pccrt* loci, respectively, remain unchanged in resistant parasites (Hunt et al., 2007; Suwanarusk et al., 2007). This is in contrast to resistance to the anti-folate drugs pyrimethamine and sulfadoxine, which involve non-synonymous mutations in the *dhfr* and *dhps* loci of each *Plasmodium* species so far investigated. Thus the use of model host–parasite systems, and even in vitro studies of drug sensitivity, may not provide adequate understanding of drug resistance of *Plasmodium* parasites in vivo.

These observations and lessons from the recent past on the evolutionary impact of CQ upon the parasite genome need to be borne in mind as we contemplate the future impact of drug selection on *Plasmodium* species.

Evolution of Parasite Resistance in the Artemisinin Combination Therapy Era

As CQ use across malaria endemic regions becomes less common, there is in effect a rapid diversification of the selective drug pressure upon malaria parasites, after decades of intense directional selection on the *Plasmodium* genomes for resistance

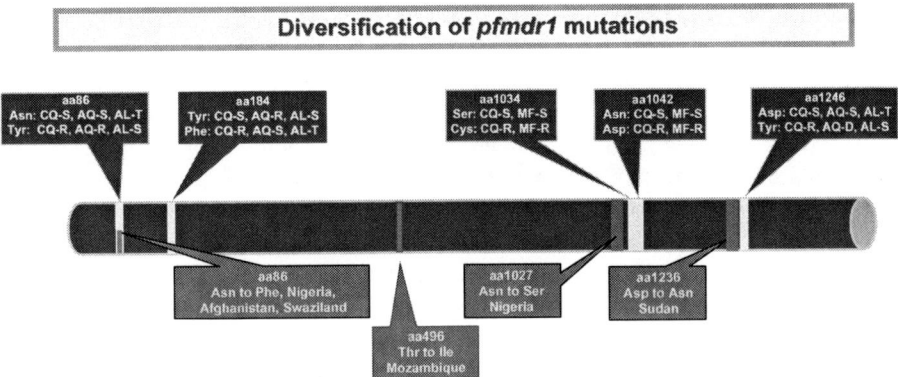

Figure 22.2 Schematic representation of the *pfmdr1*-encoded PgH-1 protein of *P. falciparum*, with five well-established mutations flagged above the figure and novel mutations recently described in our laboratory below the line. There are no studies linking any of these novel substitutions to treatment response for any drug as yet (Beshir et al., 2010; Dlamini et al. 2010; Gadalla et al., unpublished data).

to CQ and to the anti-folate fixed combination drug sulfadoxine-pyrimethamine (SP). A major contributor to this diversification is the introduction of ACT by a majority of governments as the main replacement for CQ and SP in the formal health service of most endemic countries. The most frequently used ACT in Africa are artemether-lumefantrine (AL) and artesunate-amodiaquine (AsAq) and, importantly, there is good evidence that the partner drugs lumefantrine and AQ exert strong selection on the *pfmdr1* locus in opposite directions. This gene encodes the ABC transporter PgH1, which is homologous to the multidrug resistant proteins of mammalian cells, and involved in modulating the effects of multiple antimalarial drugs. Thus, whereas recurrence of parasitemia after AL treatment is associated with the *pfmdr1* haplotype NFD at codons 86, 184, and 1246, AQ selects for the haplotype YYY (Figure 22.2; Humphreys et al., 2007). At the same time, in the private sector both artemisinin monotherapies and a variety of ACT are available in shops and pharmacies, many of whom continue to sell CQ, SP, AQ, oral quinine, and other antimalarials. Newer ACT formulations, including dihydroartemisinin-piperaquine and artesunate-pyronaridine, are also entering the marketplace, and may also soon be adopted by government health services. The selective pressure exerted by these regimens on *pfmdr1*, *pfcrt*, and other loci of importance in determining drug response in *P. falciparum* is unknown. We suggest that the variety of partner drugs used in ACT, and the slackening in CQ pressure is already leading to a pattern of diversity at the *pfmdr1* locus such that drugs are now exerting diversifying (balancing) selection pressure on this locus, rather than the intense directional selection that characterized the CQ monotherapy era (Figure 22.2). The propensity for this locus to undergo copy number amplification in both *P. falciparum* and *P. vivax* under drug selection adds additional complexity to this diversification (Price et al., 2004; Suwanarusk et al., 2008).

The diversification argument set out in the previous paragraph is largely based on consideration of the variety of partner drugs used in different ACT formulations. However, what of the selective effect of the use of the artemisinin compounds themselves? The emergence of *P. falciparum* parasites in Cambodia with markedly increased survival time in vivo following artesunate monotherapy has caused great alarm, and efforts are underway to systematically monitor parasite clearance times across the global range of *P. falciparum*, to both artesunate alone and to ACT (Noedl et al., 2008; Dondorp et al., 2009). Despite the coverage afforded by partner drugs, the almost universal adoption of ACT in endemic countries does effectively guarantee that selection on parasite genomes by artemisinins is essentially worldwide (Imwong et al., 2010). Thus it remains likely that parasites with extended post-artemisinin survival times will evolve, and could spread rapidly over intercontinental distances, as was seen for CQ resistance in the twentieth century (see Section Lessons Learnt from the Evolution of Parasite Resistance to CQ). Several strategies could minimize the impact of such newly evolving ART resistance, and these largely rely on the expectation that evolution of resistance involves a fitness cost for the parasite.

- A variety of partner drugs could be deployed in ACT, so that ART resistant parasites that also enjoy resistance to, say, AL are very likely to be contained if treated with different combinations such as AsAq.
- The use of triple combination regimens, deploying two partners (with opposite or independent selective imprints on the *Plasmodium* genome) with the artemisinins simultaneously should be strongly encouraged.
- Strategies for deployment of non-artemisinin combinations need to be devised that take into account geographic variations in the patterns of resistance to older drugs such as CQ, SP, and AQ. Such regimens may prove effective against ACT-resistant parasites.
- Continued development of new drugs must take place for additional partners to combine with the artemisinins, and for potent new compounds to replace the artemisinins should their failure be catastrophic at some point in the future.

In this way, the evolution of resistance in malaria parasites can be partially managed, in order to maximize the duration of the public health benefit currently enjoyed due to the effectiveness of ACT.

22.3.3 Dessert: Selection for Immunological Escape Variants by Malaria Vaccines—a Real or Imagined Threat?

At this time, there are no registered vaccines available that offer potential protection through the stimulation of an effective immune response against malaria parasites. However, increased investment in malaria vaccine development from private, governmental, and intergovernmental funding agencies during the first decade of the twenty-first century have pushed vaccines forward and greatly improved the chances of having a product to deploy during the next decade. A theoretical obstacle to effective vaccination against malaria is that the parasite will exploit the existing polymorphism of target antigens against which vaccines are currently being developed to escape vaccine-elicited immunity, and there is concern that this problem may not be easily overcome (Sutherland, 2007; Takala et al., 2007). Evaluation studies of two developmental vaccines have explicitly examined this threat in PNG (Combination B) and in The Gambia, Mozambique, and Kenya (RTS,S).

Combination B and Other Blood-Stage Malaria Vaccines

A program of *P. falciparum* blood-stage vaccine development based in Australia generated a promising multivalent product known as Combination B, incorporating polypeptide elements of the merozoite proteins MSP1 and MSP2, the early intra-erythrocytic stage antigen RESA and a T-cell epitope from CSP (Genton et al., 2002). The vaccine was not found to be protective against malaria in children in PNG, but had a significant effect in reducing parasite density. Analysis of parasite genotypes at each of the target antigen loci demonstrated that the MSP2 component of the vaccine in particular elicited strain-specific antibody responses that selectively protected against parasites with similar variants of the antigen (Genton et al., 2002; Flück et al., 2004, 2007). Thus widespread deployment, at high coverage, of vaccines designed to elicit protective antibody response against polymorphic parasite antigens would be expected to have a direct evolutionary impact on the parasite population, affecting the allele frequencies of variants of the target antigen(s).

The Impact of RTS,S Vaccination on Parasite Diversity

The most successful vaccine against malaria to date is the RTS,S vaccine, which at the time of writing is in Phase III trials at multiple centers across Africa. Although efficacy estimates (against clinical episodes of malaria) for this vaccine have varied between 27% and 60% (Casares et al., 2010), particularly promising evidence of protection in African children, and a likely effect in reducing the number of severe malaria cases in vaccinated infants (Alonso et al., 2004) have helped drive the continued development of the vaccine. Remarkably, given the greater efficacy demonstrated with RTS,S compared to Combination B, there has been no evidence in any study thus far of strain selection for parasites encoding particular variants of the CSP antigen, the target of the vaccine (Allouèche et al., 2003; Enosse et al., 2006; Waitumbi et al., 2009). Further, as the mechanism of protection of RTS,S is unclear (Casares et al., 2010), it remains possible that the vaccine-elicited immunity is indirect, imparting multivalent blood-stage immunity due to the "leaky" partial protection against liver-stage development achieved (Sutherland et al., 2007). Thus it remains uncertain whether this particular pre-erythrocytic vaccination strategy would provide a sufficiently focused selective force to have a direct evolutionary effect on parasite populations in vaccinated individuals.

22.4 Evolution of *Plasmodium* and the Eradication Agenda

As malaria eradication programs are envisaged for implementation around the malaria endemic world, evolutionary biologists are pondering the potential of uniform, large-scale interventions to force the parasite genome through selection bottlenecks. An example of such a bottleneck is the impact of the antimalarial CQ, which for decades in the twentieth century exerted monolithic selection on the *P. falciparum* genome for a handful of advantageous alleles of *pfcrt* and *pfmdr1*. Will the eradication agenda pose a straightforward challenge that the parasite is more than able to meet? To avoid this outcome, it is essential that multiple tools are deployed, including not just anti-parasite but anti-vector methods, and interventions to alter human behavior that enhances risk of exposure to malaria.

Is it likely that malaria will continue to exert selection for new mutations on the human genome, as it has in the past? This seems unlikely, as relatively few people are prevented from reproduction by malaria, and hopes are high that efforts toward eradication will rapidly bring this number down even further. Nevertheless, there is ample evidence that already existing variants such the thalassemias, G6PD deficiency, and sickle-cell Hb are maintained at stable frequencies in human populations by offering protection against malaria.

Will zoonotic primate malaria infections lead to a reservoir of hit-and-run malaria cases on the fringes of forests in Africa and Asia, and might these species adapt to anthropocentric transmission? This fascinating and potentially dangerous scenario cannot be ruled out, and, should human malaria be eradicated at some time in the future, will require careful vigilance in human populations that frequent the forest fringes of Southeast Asia and central Africa.

References

Alloueche, A., Milligan, P., Conway, D.J., Pinder, M., Bojang, K., Doherty, T., et al., 2003. Protective efficacy of the RTS,S/AS02 *Plasmodium falciparum* malaria vaccine is not strain specific. Am. J. Trop. Med. Hyg. 68, 97–101.

Alonso, P.L., Sacarlal, J., Aponte, J.J., Leach, A., Macete, E., Milman, J., et al., 2004. Efficacy of the RTS,S/AS02A vaccine against *Plasmodium falciparum* infection and disease in young African children: randomised controlled trial. Lancet 364, 1411–1420.

Amato, A., Kooistra, W.H., Ghiron, J.H., Mann, D.G., Proschold, T., Montresor, M., 2007. Reproductive isolation among sympatric cryptic species in marine diatoms. Protist 158, 193–207.

Anderson, T.J., Roper, C., 2005. The origins and spread of antimalarial drug resistance: lessons for policy makers. Acta Trop. 94, 269–280, Review.

Anthony, T.G., Polley, S.D., Vogler, A.P., Conway, D.J., 2007. Evidence of non-neutral polymorphism in *Plasmodium falciparum* gamete surface protein genes Pfs47 and Pfs48/45. Mol. Biochem. Parasitol. 156, 117–123.

Baum, J., Thomas, A.W., Conway, D.J., 2003a. Evidence for diversifying selection on erythrocyte-binding antigens of *Plasmodium falciparum* and *P. vivax*. Genetics 163, 1327–1336.

Baum, J., Pinder, M., Conway, D.J., 2003b. Erythrocyte invasion phenotypes of *Plasmodium falciparum* in The Gambia. Infect. Immun. 71, 1856–1863.

Beshir, K., Sutherland, C.J., Merinopoulos, I., Durrani, N., Leslie, T., Rowland, M., et al., 2010. Amodiaquine resistance in Plasmodium falciparum malaria is associated with

the pfcrt 72-76 SVMNT allele in Afghanistan. Antimicrob. Agents Chemother. 54, 3714−3716.
Bruce, M.C., Macheso, A., Galinski, M.R., Barnwell, J.W., 2007. Characterization and application of multiple genetic markers for *Plasmodium malariae*. Parasitology 134, 637−650.
Bruce, M.C., Macheso, A., Kelly-Hope, L.A., Nkhoma, S., McConnachie, A., Molyneux, M. E., 2008. Effect of transmission setting and mixed species infections on clinical measures of malaria in Malawi. *PLoS One* 3, e2775.
Brunet, M., Guy, F., Pilbeam, D., Mackaye, H.T, Likius, A., Ahounta, D., et al., 2002. A new hominid from the Upper Miocene of Chad Central Africa. Nature 418, 145−151.
Carlton, J.M., Adams, J.H., Silva, J.C., Bidwell, S.L., Lorenzi, H., Caler, E., et al., 2008a. Comparative genomics of the neglected human malaria parasite *Plasmodium vivax*. Nature 455, 757−763.
Carlton, J.M., Escalante, A.A., Neafsey, D., Volkman, S.K., 2008b. Comparative evolutionary genomics of human malaria parasites. Trends Parasitol. 24, 545−550, Review.
Carter, R., 1978. Studies on enzyme variation in the murine malaria parasites *Plasmodium berghei*, *P. yoelii*, *P. vinckei* and *P. chabaudi* by starch gel electrophoresis. Parasitology 76, 241−267.
Casares, S., Brumeanu, T.-D., Ritchie, T.L., 2010. The RTS,S malaria vaccine. *Vaccine* 28, 4880−4894.
Coatney, G.R., Collins, W.E., Warren, M., Contacos, P.G., 2003. Division of Parasitic Disease, producers. Version 1.0. Atlanta, GA, USA: CDC; 2003. CD-ROM. The primate malarias [original book published 1971].
Coldren, R.L., Jongsakul, K., Vayakornvichit, S., Noedl, H., Fukuda, M.M., 2007. Apparent relapse of imported *Plasmodium ovale* malaria in a pregnant woman. Am. J. Trop. Med. Hyg. 77, 992−994.
Collins, W.E., Jeffery, G.M., 2005. *Plasmodium ovale:* parasite and disease. Clin. Micro. Rev. 18, 570−581.
Collins, W.E., Jeffery, G.M., 2007. *Plasmodium malariae*: parasite and disease. Clin. Microbiol. Rev. 20, 579−592, Review.
Collins, W.E., Pappaioanou, M., McClure, H.M., Swenson, R.B., Strobert, E., Skinner, J.C., et al., 1987. Infection of chimpanzees with Nigerian I/CDC strain of *Plasmodium ovale*. Am. J. Trop. Med. Hyg. 37, 455−459.
Collins, W.E., Sullivan, J.S., Strobert, E., Galland, G.G., Williams, A., Nace, D., et al., 2009. Studies on the Salvador I strain of *Plasmodium vivax* in non-human primates and anopheline mosquitoes. Am. J. Trop. Med. Hyg. 80, 228−235.
Conway, D.J., McBride, J.S., 1991. Genetic evidence for the importance of interrupted feeding by mosquitoes in the transmission of malaria. Trans. R. Soc. Trop. Med. Hyg. 85, 454−456.
Conway, D.J., Polley, S.D., 2002. Measuring immune selection. Parasitology 125 (Suppl.), S3−S16.
Conway, D.J., Machado, R.L., Singh, B, Dessert, P, Mikes, Z.S., Povoa, M.M., et al., 2001. Extreme geographical fixation of variation in the *Plasmodium falciparum* gamete surface protein gene Pfs48/45 compared with microsatellite loci. Mol. Biochem. Parasitol. 115, 145−156.
Cox-Singh, J., Davis, T.M., Lee, K.S., Shamsul, S.S., Matusop, A., Ratnam, S., et al., 2008. *Plasmodium knowlesi* malaria in humans is widely distributed and potentially life threatening. Clin. Infect. Dis. 46, 165−171.

Cox-Singh, J., Hiu, J., Lucas, S.B., Divia, P.C., Zulkarnaen, M., Chandran, P., et al., 2010. Severe malaria—a case of fatal *Plasmodium knowlesi* infection with post-mortem findings: a case report. Malar. J. 9, 10.

Cunningham, D., Lawton, J., Jarra, W., Preiser, P., Langhorne, J., 2010. The *pir* multigene family of *Plasmodium*: antigenic variation and beyond. Mol. Biochem. Parasitol. 170, 65−73, Review.

Dame, J.B., Arnot, D.E., Bourke, P.F., Chakrabarti, D., Christodoulou, Z., Coppel, R.L., et al., 1996. Current status of the *Plasmodium falciparum* genome project. Mol. Biochem. Parasitol. 79, 1−12.

Daubersies, P., Sallenave-Salles, S., Magne, S., Trape, J.-F., Contamin, H., Fandeur, T., et al., 1996. Rapid turnover of *Plasmodium falciparum* populations in asymptomatic individuals living in a high transmission area. Am. J. Trop. Med. Hyg. 54, 18−26.

Davis, T.M., Singh, B., Sheridan, G., 2001. Parasitic procrastination: late-presenting ovale malaria and schistosomiasis. Med. J. Aust. 175, 146−148.

Diallo, D., Sutherland, C., Nebié, I., Konaté, A., Ord, R., Ilboudo-Sanogo, E., et al., 2007a. Sustained use of insecticide-treated curtains is not associated with greater circulation of drug-resistant malaria parasites, nor with higher risk of treatment failure among children with uncomplicated malaria in Burkina Faso. Am. J. Trop. Med. Hyg. 76, 237−244.

Dlamini, S.V., Beshir, K., Sutherland, C.J., 2010. Markers of anti-malarial drug resistance in *Plasmodium falciparum* isolates from Swaziland: identification of pfmdr1-86F in natural parasite isolates. Malar. J. 9, 68.

Dondorp, A.M., Nosten, F., Yi, P., Das, D., Phyo, A.P., Tarning, J., et al., 2009. Artemisinin resistance in *Plasmodium falciparum* malaria. N. Engl. J. Med. 361, 455−467.

Drakeley, C.J., Sutherland, C.J., Bousema, J.T., Sauerwein, R.W., Targett, G.A.T., 2006. The epidemiology of *Plasmodium falciparum* gametocytes: weapons of mass dispersion. Trends Parasitol 22, 424−430.

Duraisingh, M.T., Jones, P., Sambou, I., von Seidlein, L., Pinder, M., Warhurst, D.C., 2000. The tyrosine-86 allele of the pfmdr1 gene of *Plasmodium falciparum* is associated with increased sensitivity to the anti-malarials mefloquine and artemisinin. Mol. Biochem. Parasitol. 108, 13−23.

Duraisingh, M.T., Maier, A.G., Triglia, T., Cowman, A.F., 2003. Erythrocyte-binding antigen 175 mediates invasion in *Plasmodium falciparum* utilizing sialic acid-dependent and -independent pathways. Proc. Natl. Acad. Sci. U.S.A. 100, 4796−4801.

Duval, L., Robert, V., Csorba, G., Hassanin, A., Randrianarivelojosia, M., Walston, J., et al., 2007. Multiple host-switching of Haemosporidia parasites in bats. Malar. J. 6, 157.

Duval, L., Nerrienet, E., Rousset, D., Sadeuh Mba, S.A., Houze, S., Fourment, M., et al., 2009. Chimpanzee malaria parasites related to *Plasmodium ovale* in Africa. PLoS One 4, e5520.

Duval, L., Fourment, M., Nerrienet, E., Rousset, D., Sadeuh, S.A., Goodman, S.M., et al., 2010. African apes as reservoirs of *Plasmodium falciparum* and the origin and diversification of the Laverania subgenus. Proc. Natl. Acad. Sci. U.S.A. 107, 10561−10566.

Enosse, S., Dobano, C., Quelhas, D., Aponte, J.J., Lievens, M., Leach, A., et al., 2006. RTS, S/AS02A malaria vaccine does not induce parasite CSP T cell epitope selection and reduces multiplicity of infection. PLoS Clin. Trials 1, e5.

Escalante, A.A., Ayala, F.J., 1994. Phylogeny of the malarial genus *Plasmodium*, derived from rRNA gene sequences. Proc. Natl. Acad. Sci. U.S.A. 91, 11373−11377.

Escalante, A.A., Smith, D.L., Kim, Y., 2009. The dynamics of mutations associated with anti-malarial drug resistance in *Plasmodium falciparum*. Trends Parasitol. 25, 557−563.

Faye, F.B.K., Konate, L., Rogier, C., Trape, J.-F., 1998. *Plasmodium ovale* in a highly malaria endemic area of Senegal. Trans. Roy. Soc. Trop. Med. Hyg. 92, 522−525.

Flück, C., Smith, T., Beck, H.-P., Irion, A., Betuela, I., Alpers, M.P., et al., 2004. Strain-specific humoral response to a polymorphic malaria vaccine. Infect. Immun. 72, 6300−6305.
Flück, C., Schopflin, S., Smith, T., Genton, B., Alpers, M.P., Beck, H.-P., et al., 2007. Effect of the malaria vaccine Combination B on merozoite surface antigen 2 diversity. Infect. Genet. Evol. 7, 44−51.
Gardner, M.J., Hall, N., Fung, E., White, O., Berriman, M., Hyman, R.W., et al., 2002. Genome sequence of the human malaria parasite *Plasmodium falciparum*. Nature 419, 498−511.
Garnham, P.C.C., 1966. Malaria Parasites and Other Haemosporidia. Blackwell, Oxford, UK.
Geiman, Q.M., Meagher, M.J., 1967. Susceptibility of a new world monkey to *Plasmodium falciparum* from man. Nature 215, 437−439.
Genton, B., Betuela, I., Felger, I., Al-Yaman, F., Anders, R.F., Saul, A., et al., 2002. A recombinant blood-stage malaria vaccine reduces *Plasmodium falciparum* density and exerts selective pressure on parasite populations in a phase 1-2b trial in Papua New Guinea. J. Infect. Dis. 185, 820−827.
Hacia, J.G., 2001. Genome of the Apes. Trends Genet. 17, 637−645.
Hall, N., Karras, M., Raine, J.D., Carlton, J.M., Kooij, T.W., Berriman, M., et al., 2005. A comprehensive survey of the *Plasmodium* life cycle by genomic, transcriptomic, and proteomic analyses. Science 307, 82−86.
Hayakawa, T., Arisue, N., Udono, T., Hirai, H., Sattabongkot, J., Toyama, T., et al., 2009. Identification of *Plasmodium malariae*, a human malaria parasite, in imported chimpanzees. PLoS One 4, e7412.
Humphreys, G.A., Merinopoulos, I., Ahmed, J., Whitty, C.J.M., Mutabingwa, T.K., Sutherland, C.J., et al., 2007. Amodiaquine and artemether-lumefantrine select distinct alleles of the *Plasmodium falciparum pfmdr1* gene in Tanzanian children treated for uncomplicated malaria. Antimicrob. Agents Chemother. 51, 991−997.
Hunt, P., Afonso, A., Creasey, A., Culleton, R., Sidhu, A.B., Logan, J., et al., 2007. Gene encoding a deubiquinating enzyme is mutated in artesunate- and chloroquine-resistant rodent malaria parasites. Mol. Microbiol. 65, 27−40.
Iezhova, T.A., Valkiunas, G., Bairlein, F., 2005. Vertebrate host specificity of two avian malaria parasites of the subgenus Novyella: *Plasmodium nucleophilum* and *Plasmodium vaughani*. J. Parasitol. 91, 472−474.
Imwong, M., Dondorp, A.M., Nosten, F., Yi, P., Mungthin, M., Hanchana, S., et al., 2010. Exploring the contribution of candidate genes to artemisinin resistance in *Plasmodium falciparum*. Antimicrob. Agents Chemother. 54, 2886−2892.
Krief, S., Escalante, A.A., Pacheco, M.A., Mugisha, L., André, C., et al., 2010. On the diversity of malaria parasites in African apes and the origin of *Plasmodium falciparum* from Bonobos. PLoS Pathog. 6 (2), e1000765.
Lal, A.A., Goldman, I.F., 1991. Circumsporozoite protein gene from *Plasmodium reichenowi* a chimpanzee malaria parasite evolutionarily related to the human malaria parasite *Plasmodium falciparum*. J. Biol. Chem. 266, 6686−6689.
Laufer, M.K., Thesing, P.C., Eddington, N.D., Masonga, R., Dzinjalamala, F.K., Takala, S. L., et al., 2006. Return of chloroquine antimalarial efficacy in Malawi. N. Engl. J. Med. 355, 1959−1966.
Lee, K.S., Cox-Singh, J., Brooke, G., Matusop, A., Singh, B., 2009. *Plasmodium knowlesi* from archival blood films: further evidence that human infections are widely distributed and not newly emergent in Malaysian Borneo. Int. J. Parasitol. 39, 1125−1128.

Lekweiry, K.M., Abdallahi, M.O., Ba, H., Arnathau, C., Durand, P., Trape, J.F., et al., 2009. Preliminary study of malaria incidence in Nouakchott, Mauritania. Malar. J. 8, 92.

Lindo, J.F., Bryce, J.H., Ducasse, M.B., Howitt, C., Barrett, D.M., Morales, J.L., et al., 2007. *Plasmodium malariae* in Haitian refugees, Jamaica. Emerg. Infect. Dis. 13, 931–933.

Liu, W., Li, Y., Learn, G.H., Rudicell, R.S., Robertson, J.D., Keele, B.F., et al., 2010. Origin of the human malaria parasite *Plasmodium falciparum* in gorillas. Nature. 467, 420–425.

Luchavez, J., Espino, F., Curameng, P., Espina, R., Bell, D., Chiodini, P., et al., 2008. Human Infections with *Plasmodium knowlesi*, the Philippines. Emerg. Infect. Dis. 14, 811–813.

Mardis, E.R., 2009. New strategies and emerging technologies for massively parallel sequencing: applications in medical research. Genome Med. 1, 40.

Martin, M.J., Rayner, J.C., Gagneux, P., Barnwell, J.W., Varki, A., 2005. Evolution of human-chimpanzee differences in malaria susceptibility: relationship to human genetic loss of N-glycolylneuraminic acid. Proc. Natl. Acad. Sci. U.S.A. 102, 12819–12824.

Mayer, D.C., Cofie, J., Jiang, L., Hartl, D.L., Tracy, E., Kabat, J., et al., 2009. Glycophorin B is the erythrocyte receptor of *Plasmodium falciparum* erythrocyte-binding ligand EBL-1. Proc. Natl. Acad. Sci. U.S.A. 106, 5348–5352.

Mehlotra, R.K., Mattera, G., Bockarie, M.J., Maguire, J.D., Baird, J.K., Sharma, Y.D., et al., 2008. Discordant patterns of genetic variation at two chloroquine resistance loci in worldwide populations of the malaria parasite *Plasmodium falciparum*. Antimicrob. Agents Chemother. 52, 2212–2222.

Miller, L.H., 1977. Hypothesis on the mechanism of erythrocyte invasion by malaria merozoites. Bull. World Health Organ. 55, 157–162.

Mu, J., Joy, D.A., Duan, J., Huang, Y., Carlton, J., Walker, J., et al., 2005. Host switch leads to emergence of *Plasmodium vivax* malaria in humans. Mol. Biol. Evol. 22, 1686–1693.

Mu, J., Myers, R.A., Jiang, H., Liu, S., Ricklefs, S., Waisberg, M., et al., 2010. *Plasmodium falciparum* genome-wide scans for positive selection, recombination hot spots and resistance to antimalarial drugs. Nat. Genet. 42, 268–271.

Mueller, I., Widmer, S., Michel, D., Maraga, S., McNamara, D.T., Kiniboro, B., et al., 2009. High sensitivity detection of *Plasmodium* species reveals positive correlations between infections of different species, shifts in age distribution and reduced local variation in Papua New Guinea. Malar. J. 8, 41.

Mueller, I., Zimmerman, P.A., Reeder, J.C., 2007. *Plasmodium malariae* and *Plasmodium ovale*—the "bashful" malaria parasites. Trends Parasitol. 23, 278–283.

Ng, O.T., Ooi, E.E., Lee, C.C., Lee, P.J., Ng, L.C., Pei, S.W., et al., 2008. Naturally acquired human *Plasmodium knowlesi* infection, Singapore. Emerg. Infect. Dis. 14, 814–816.

Noedl, H., Se, Y., Schaecher, K., Smith, B.L., Socheat, D., Fukuda, M.M., et al., 2008. Evidence of artemisinin-resistant malaria in western Cambodia. N. Engl. J. Med. 359, 2619–2620.

Ochola, L.I., Tetteh, K.K., Stewart, L.B., Riitho, V., Marsh, K., Conway, D.J., 2010. Allele frequency-based and polymorphism-versus-divergence indices of balancing selection in a new filtered set of polymorphic genes in *Plasmodium falciparum*. Mol. Biol. Evol. 27, 2344–2351.

Ord, R., Polley, S., Tami, A., Sutherland, C., 2005. High sequence diversity and evidence of balancing selection in the $Pvmsp3\alpha$ gene of *Plasmodium vivax* in the Venezuelan Amazon. Mol. Biochem. Parasitol. 144, 89–93.

Ord, R., Alexander, N, Dunyo, S., Hallett, R.L., Jawara, M., Targett, G.A.T., et al., 2007. Seasonal carriage of *pfcrt* and *pfmdr1* alleles in Gambian *Plasmodium falciparum* imply reduced fitness of chloroquine-resistant parasites. J. Infect. Dis. 196, 1613–1619.

Ord, R.L., Tami, A., Sutherland, C.J., 2008. *ama1* genes of sympatric *Plasmodium vivax* and *P. falciparum* from Venezuela differ significantly in genetic diversity and recombination frequency. PLoS One 3, e3366.
Osier, F.H., Murungi, L.M., Fegan, G., Tuju, J., Tetteh, K.K., Bull, P.C., et al., 2010. Allele-specific antibodies to *Plasmodium falciparum* merozoite surface protein-2 and protection against clinical malaria. Parasite Immunol. 32, 193−201.
Pain, A., Böhme, U., Berry, A.E., Mungall, K., Finn, R.D., Jackson, A.P., et al., 2008. The genome of the simian and human malaria parasite *Plasmodium knowlesi*. Nature 455, 799−803.
Palumbi, S.R., Metz, E.C., 1991. Strong reproductive isolation between closely related tropical sea urchins (genus *Echinometra*). Mol. Biol. Evol. 8, 227−239.
Perkins, S.L., Schall, J.J., 2002. A molecular phylogeny of malarial parasites recovered from cytochrome b gene sequences. J. Parasitol. 88, 972−978.
Perkins, S.L., Sarkar, I.N., Carter, R., 2007. The phylogeny of rodent malaria parasites: simultaneous analysis across three genomes. Infect. Genet. Evol. 7, 74−83.
Polley, S.D., Conway, D.J., 2001. Strong diversifying selection on domains of the *Plasmodium falciparum* apical membrane antigen 1 gene. Genetics 158, 1505−1512.
Polley, S.D., Tetteh, K.K., Lloyd, J.M., Akpogheneta, O.J., Greenwood, B.M., Bojang, K.A., et al., 2007. *Plasmodium falciparum* merozoite surface protein 3 is a target of allele-specific immunity and alleles are maintained by natural selection. J. Infect. Dis. 195, 279−287.
Price, R.N., Uhlemann, A.C., Brockman, A., McGready, R., Ashley, E., Phaipun, L., et al., 2004. Mefloquine resistance in *Plasmodium falciparum* and increased *pfmdr1* gene copy number. Lancet 364, 438−447.
Prugnolle, F., Durand, P., Neel, C., Ollomo, B., Ayala, F.J., Arnathau, C., et al., 2010. African great apes are natural hosts of multiple related malaria species, including *Plasmodium falciparum*. Proc. Natl. Acad. Sci. U.S.A. 107, 1458−1463.
Putaporntip, C., Hongsrimuang, T., Seethamchai, S., Kobasa, T., Limkittikul, K., Cui, L., et al., 2009. Differential prevalence of *Plasmodium* infections and cryptic *Plasmodium knowlesi* malaria in humans in Thailand. J. Infect. Dis. 199, 1143−1150.
Rathore, D., Hrstka, S.C., Sacci Jr, J.B., De La Vega, P., Linhardt, R.J., Kumar, S., et al., 2003. Molecular mechanism of host specificity in *Plasmodium falciparum* infection: role of circumsporozoite protein. J. Biol. Chem. 278, 40905−40910.
Rayner, J.C., Corredor, V., Feldman, D., et al., 2002. Extensive polymorphism in the *Plasmodium vivax* merozoite surface coat protein MSP-3α is limited to specific domains. Parasitology 125, 393−405.
Ricklefs, R.E., Fallon, S.M., 2002. Diversification and host switching in avian malaria parasites. *Proc. Biol. Sci.* 269, 885−892.
Roper, C., Pearce, R., Nair, S., Sharp, B., Nosten, F., Anderson, T., 2004. Intercontinental spread of pyrimethamine-resistant malaria. Science 305, 1124.
Sá, J.M., Twu, O., Hayton, K., Reyes, S., Fay, M.P., Ringwald, P., et al., 2009. Geographic patterns of *Plasmodium falciparum* drug resistance distinguished by differential responses to amodiaquine and chloroquine. Proc. Natl. Acad. Sci. USA. 106, 18883−18889.
Sam-Yellowe, T.Y., Florens, L., Johnson, J.R., Wang, T., Drazba, J.A., Le Roch, K.G., et al., 2004. A *Plasmodium* gene family encoding Maurer's cleft membrane proteins: structural properties and expression profiling. Genome Res. 14, 1052−1059.
Schuster, F.L., 2002. Cultivation of *Plasmodium* spp. Clin. Microbiol. Rev. 15, 355−364, Review.
Shortt, H.E., Menon, K.P., Seetharama Iyer, P.V., 1941. The natural host of *Plasmodium gallinaceum* (Brumpt, 1935). J. Malar. Inst. India 4, 175 (cited in Garnham, 1966).

Siddiqui, W.A., Schnell, J.V., Geiman, Q.M., 1972. A model in vitro system to test the susceptibility of human malarial parasites to antimalarial drugs. Am. J. Trop. Med. Hyg. 21, 393–399.

Singh, B., Kim Sung, L., Matusop, A., Radhakrishnan, A., Shamsul, S.S., Cox-Singh, J., et al., 2004. A large focus of naturally acquired *Plasmodium knowlesi* infections in human beings. Lancet 363, 1017–1024.

Snounou, G., Zhu, X., Siripoon, N., Jarra, W., Thaithong, S., Brown, K.N., et al., 1999. Biased distribution of *msp1* and *msp2* allelic variants in *Plasmodium falciparum* populations in Thailand. Trans. Roy. Soc. Trop. Med. Hyg. 93, 1–6.

Sullivan, J.S., Sullivan, J.J., Williams, A., Grady, K.K., Bounngaseng, A., Huber, C.S., et al., 2003. Adaptation of a strain of *Plasmodium falciparum* from Ghana to *Aotus lemurinus griseimembra*, *A. nancymai* and *A. vociferans* monkeys. Am. J. Trop. Med. Hyg. 69, 593–600.

Sutherland, C., 2007. A challenge for the development of malaria vaccines: polymorphic target antigens. PLoS Med. 4 (3), e116.

Sutherland, C.J., Drakeley, C.J., Schellenberg, D., 2007. How is childhood development of immunity to *Plasmodium falciparum* enhanced by certain antimalarial interventions? Malar. J. 6, 161.

Sutherland, C.J., Tanomsing, N., Nolder, D., Oguike, M., Jennison, C., Pukrittayakamee, S., et al., 2010. Two non-recombining sympatric forms of the human malaria parasite *Plasmodium ovale* occur globally. J. Infect. Dis. 201, 1544–1550.

Suwanarusk, R., Russell, B., Chavchich, M., Chalfein, F., Kenangalem, E., 2007. Chloroquine resistant *Plasmodium vivax*: in vitro characterisation and association with molecular polymorphisms. PLoS One 2, e1089.

Suwanarusk, R., Chavchich, M., Russell, B., Jaidee, A., Chalfein, F., Barends, M., et al., 2008. Amplification of *pvmdr1* associated with multidrug-resistant *Plasmodium vivax*. J. Infect. Dis. 198, 1558–1564.

Takala, S.L., Coulibaly, D., Thera, M.A., Dicko, A., Smith, D.L., Guindo, A.B., et al., 2007. Dynamics of polymorphism in a malaria vaccine antigen at a vaccine-testing site in Mali. PLoS Med. 4, e93.

Tetteh, K.K., Stewart, L.B., Ochola, L.I., Amambua-Ngwa, A., Thomas, A.W., Marsh, K., et al., 2009. Prospective identification of malaria parasite genes under balancing selection. PLoS One 4, e5568.

Trager, W., Jensen, J.B., 1976. Human malaria parasites in continuous culture. Science 193, 673–675.

Tsuchida, H., Yamaguchi, K., Yamamoto, S., Ebisawa, I., 1982. Quartan malaria following splenectomy 36 years after infection. Am. J. Trop. Med. Hyg. 31, 163–165.

Van den Eede, P., Van, H.N., Van Overmeir, C., Vythilingam, I., Duc, T.N., Hung le, X., et al., 2009. Human *Plasmodium knowlesi* infections in young children in central Vietnam. Malar. J. 8, 249.

Van Dijk, M.R., Janse, C.J., Thompson, J., Waters, A.P., Braks, J.A., Dodemont, H.J., et al., 2001. A central role for P48/45 in malaria parasite male gamete fertility. Cell 104, 153–164.

Volkman, S.K., Sabeti, P.C., DeCaprio, D., Neafsey, D.E., Schaffner, S.F., Milner Jr, D.A., et al., 2007. A genome-wide map of diversity in *Plasmodium falciparum*. Nat. Genet. 39, 113–119.

Waitumbi, J.N., Anyona, S.B., Hunja, C.W., Kifude, C.M., Polhemus, M.E., Walsh, D.S., et al., 2009. Impact of RTS,S/AS02(A) and RTS,S/AS01(B) on genotypes of *P. falciparum* in adults participating in a malaria vaccine clinical trial. PLoS One 4, e7849.

Wang, H.Y., Tang, H., Shen, C.K., Wu, C.I., 2003. Rapidly evolving genes in human. I. The glycophorins and their possible role in evading malaria parasites. Mol. Biol. Evol. 20, 1795–1804.
Weatherall, D.J., Miller, L.H., Baruch, D.I., Marsh, K., Doumbo, O.K., Casals-Pascual, C., et al., 2002. Malaria and the red cell. Hematology Am. Soc. Hematol. Educ. Program 2002, 35–57, Review.
Weedall, G.D., Preston, B.M., Thomas, A.W., Sutherland, C.J., Conway, D.J., 2007. Differential evidence of natural selection on two leading sporozoite stage malaria vaccine candidate antigens. Int. J. Parasitol. 37, 77–85.
White, N.J., 2008. *Plasmodium knowlesi*: the fifth human malaria parasite. Clin. Infect. Dis. 46, 172–173.
Wickham, J.M., Dennis, E.D., Mitchell, G.H., 1980. Long term cultivation of a simian malaria parasite (*Plasmodium knowlesi*) in a semi-automated apparatus. Trans. R. Soc. Trop. Med. Hyg. 74, 789–792.
Winter, G., Kawai, S., Haeggström, M., Kaneko, O., von Euler, A., Kawazu, S., et al., 2005. SURFIN is a polymorphic antigen expressed on *Plasmodium falciparum* merozoites and infected erythrocytes. J. Exp. Med. 201, 1853–1863.
Wootton, J.C., Feng, X, Ferdig, M.T., Cooper, R.A., Mu, J., Baruch, D.I., et al., 2002. Genetic diversity and chloroquine selective sweeps in *Plasmodium falciparum*. Nature 418, 320–323.
Zheng, H., Zhu, H.M., Ning, B.F., Li, X.Y., 2006. [Molecular identification of naturally acquired *Plasmodium knowlesi* infection in a human case.] Zhongguo Ji Sheng Chong Xue Yu Ji Sheng Chong Bing Za Zhi 24, 273–276 (in Chinese).

23 Integrated Genetic Epidemiology of Chagas Disease

Michel Tibayrenc[1,]*, Jenny Telleria[1], Patricio Diosque[2], Juan Carlos Dib[3] and Christian Barnabé[1]

[1]Institut de Recherche pour le Développement (IRD)/Centre National de la Recherche Scientifique (CNRS), Montpellier, France, [2]CONICET, Argentina, University of Salta, Salta, Argentina, [3]University of Antioquia, Antioquia, Colombia

23.1 What Is Integrated Genetic Epidemiology?

As authoritatively illustrated by this book, the impressive progress of molecular megatechnologies (high-throughput sequencing, microarrays, postgenomics) and the concomitant development of bioinformatics have considerably improved our knowledge on infectious diseases. However, there is a strong tendency toward compartmentalization in the research effort: scientists working on human (and other hosts) genetic susceptibility to infectious diseases are generally not aware of research on the role played by pathogens, and vectors in the case of vector-borne diseases. This results in each community of scientists tending to overemphasize the role of its study material. This compartmentalization is all the more distressing since coevolution between hosts, pathogens, and vectors should be considered a unique biological phenomenon. The term "integrated genetic epidemiology" (Tibayrenc, 1998a) has been coined to designate the approach consisting in simultaneously analyzing the impact of the host's, the pathogen's, and the vector's genetic diversity on the transmission and severity of infectious diseases as well as the coevolution processes between the three. The present chapter aims to show that Chagas disease is an excellent model to develop this approach. It briefly summarizes what is presently known about: (i) human genetic susceptibility to Chagas disease, (ii) the vectors' species and population diversity, and (iii) the parasite's genetics and evolution. Then it demonstrates how these three components could be merged in a unique approach.

*E-mail: michel.tibayrenc@ird.fr

23.2 Chagas Disease: A Major Health Problem in Latin America and Other Countries

Chagas disease remains by far the most serious health problem in Latin America. Control of the disease has been improved, but several million people remain at risk or are stricken by the disease.

From a clinical point of view, Chagas disease is a very serious illness. After infection by the parasite (see Section 23.3), patients develop an acute phase, which actually corresponds to parasitic septicemia. Mortality at this stage is approximately 5%. After a few weeks, patients who survive enter the indeterminate phase, with no symptoms. About 70% of patients will never exhibit any symptoms again. However, 30% of them will develop symptomatic Chagas disease. The most worrisome symptom is Chagasic cardiopathy, which leads to a severe cardiac insufficiency. Other clinical forms involve the digestive system (megacolon, megaesophagus) and cause severe functional abnormalities.

The health problem of Chagas disease is worsened by its being a "neglected disease" according to official classifications. As for the infectious diseases predominant in the southern world, malaria, AIDS, and tuberculosis receive special attention from WHO and other international health authorities, while other diseases tend to be underprioritized. However, the dispersion of Chagas disease from Latin America to nonendemic countries by population movements has started to create new epidemiological, economic, social, and political challenges as *T. cruzi* has spread throughout the world (Rodrigues Coura and Albajar Viñas, 2010). In the domain of scientific research, it is notable that the scientific community involved in Chagas disease research, although very productive, is tiny, but hopefully will expand.

23.3 The Chagas Disease Cycle

The Chagas disease cycle will be only briefly summarized here, since this chapter is not intended to be an exhaustive review of what Chagas disease is, but rather attempts to explain why this disease is a valuable model for integrated genetic epidemiology.

The causative agent of Chagas disease is a parasitic protozoan of the family Kinetoplastidae, which also includes *Trypanosoma brucei*, the agent of sleeping sickness (African trypanosomiasis) and the *Leishmania*, agents of the various forms of leishmaniosis.

Trypanosoma cruzi is transmitted by "true" bugs, heteropterous insects of the family Reduviidae, subfamily Triatominae. This subfamily has specialized in obligatory blood-feeding. It is worth noting that Chagas vectors include many different species and three principal genera, namely *Triatoma*, *Rhodnius*, and *Panstrongylus*. Vectors are infected by ingesting blood that contains the parasite. They transmit the parasite, not through their biting, but by their feces, which contains the infecting forms. Most vector species present the particularity of depositing feces while they feed on their host. The parasite enters the host by excoriations, through the mucosa, even through intact skin.

Hosts comprise virtually all mammalian species, either domestic or selvatic, including of course humans.

23.4 Host Genetic Susceptibility to Chagas Disease

This section emphasizes what is known about the human species. Since Chagas disease strikes virtually all mammalian species, a general view of the role played by all these hosts' genetic diversity in the disease is difficult.

Human genetic susceptibility to Chagas disease is less well known than for other transmissible diseases such as Hepatitis C (Alric et al., 1999), tuberculosis (Bellamy et al., 2000), malaria (Garcia et al., 1998), AIDS (Dean et al., 1996), leprosy (Abel et al., 1998), schistosomiasis (Dessein et al., 1999), or visceral leishmaniosis (Bucheton et al., 2002).

Chagas disease could be a fine target for studying human genetic susceptibility. As for leprosy, the chagasic **phenotypes** are well defined: asymptomatic (ASY) versus symptomatic in the acute and chronic phases, and for chronic symptomatic Chagas, cardiac, digestive, and cardiodigestive manifestations. One can also distinguish between negative serology (Brenière et al., 1984) and positive serology in the chronic phase. Therefore, one can analyze not only genetic susceptibility to Chagas, in general, but also to its different clinical manifestations.

Another favorable situation is that Chagas disease strikes populations that are ethnically very diverse. Latin Americans have European, African, Amerindian, and mixed ancestries. Ethnic diversity is a parameter to take into account when exploring human genetic susceptibility to infectious diseases (Tibayrenc, 2007).

It is distressing that few people have considered these favorable features (clearly defined clinical phenotypes and ethnic diversity) to analyze human genetic susceptibility to Chagas disease. When looking for genes of susceptibility to transmissible diseases, two main approaches are available: (i) linkage studies and (ii) the candidate gene approach (Tibayrenc, 2007). Linkage studies are based on the analysis of pedigrees and families where some individuals have the studied disease while other individuals are used as controls. The genomes are screened with a high number of genetic markers (microsatellites or single-nucleotide polymorphisms, SNPs). If some markers happen to be statistically linked to the clinical phenotype surveyed (linkage disequilibrium), the genomic region where they are located is more finely dissected to look for the putative susceptibility genes. Candidate gene studies analyze genes that are putatively associated with the pathology surveyed. The reasons for inferring that a gene can be a candidate are either that it pertains to a genetic system frequently involved in susceptibility to infectious diseases (cytokine genes, HLA genes) or that it has proved to have such a role in animal experiments.

Genome-wide association studies (GWAS; Pennisi, 2007) are a more recent approach that screens the whole genome of thousands of individuals.

Research on human genetic susceptibility to Chagas disease has most often been based on the candidate gene approach and has chiefly, but not exclusively, explored the HLA system.

23.4.1 Genetic Heritability of Some Chagasic Characteristics

Pedigree analyses have shown an apparent heritability of some serological parameters. Chagasic seropositivity heritability in Brazil is estimated to be 0.556, which is high (Williams-Blangero et al., 1997). In the same country, the levels of IgA and IgG show a heritability of 0.33 (Barbossa et al., 1981). Zicker et al. (1990) have postulated a putative familial component in chagasic cardiopathy. This result has been challenged by Morini et al. (1994).

23.4.2 Role of the HLA System

As noted above, a great deal of the results on human genetic susceptibility to Chagas disease, especially the most recent ones, deal with the putative role of the HLA system. It has now been determined that the HLA supergene complex of several genes having a related role plays an important role in the transmission, severity, and clinical diversity of Chagas disease.

Apt (1988) surveyed the distribution of HLA antigens in 124 Chilean seropositive patients, divided into patients with chronic Chagas cardiomyopathy (CCC) and ASY patients.

Fernandes-Mestre et al. (1998) evidenced a lowered frequency of DQB1*0303 and DRB1*14 in chagasic patients compared with controls, suggesting independent protective effects to the chronic infection in this population. There was also a higher frequency of DRB1*01, DRB1*08, and DQB1*0501 and a lower frequency of DRB1*1501 in the CCC patients.

Layrisse et al. (2000) surveyed 113 seropositive patients (CCC versus ASY). They postulated that the HLA-C*03 allele constituted a risk factor for CCC.

Nieto et al. (2000) surveyed 172 Peruvian patients (85 seropositive with ASY = 52, CCC = 33; 87 seropositive controls) for the variability of the HAL-DRB1 and DQB1 genes. They recorded no allelic frequency differences between CCC and ASY patients. On the other hand, the DRB1*14-DQB1*0301 haplotype was statistically linked to seronegativity, which suggests that this haplotype has a protective role against Chagas disease.

Faé et al. (2000) were unable to find any significant relation between HLA and Chagas disease. This negative result has been challenged by many later studies.

Visentainer et al. (2002) surveyed 35 CCC patients with 72 control patients in Brazil. They found a statistically significant relation between CCC and HLA-DR2.

Moreno et al. (2004) surveyed 104 seropositive patients and 60 seronegative controls. They observed significant allelic frequency differences between seropositive patients and controls at the loci HLA D6S291 and IL-10. These results suggested epistasis between the HLA and IL loci that could be linked to susceptibility to Chagas disease.

Cruz-Robles et al. (2004) conducted a survey of 193 Mexican patients, 66 of whom were seropositive (either CCC or ASY) and 127 were seronegative controls. The results suggested that HLA alleles are associated with chronic Chagas disease and CCC. HLA-DR4 and HLA-B39 alleles could be associated directly with the infection by *Trypanosoma cruzi*, whereas HLA-DR16 might be a marker of susceptibility to CCC and HLA-A68 could be protective against CCC.

Ramasawmy et al. (2006a) analyzed the variants of BAT1, a putative anti-inflammatory gene (situated in the HLA class III region) in 76 ASY and 154 CCC patients, all seropositive. They found that some BAT1 could be used to predict the occurrence of CCC.

The same authors (Ramasawmy et al., 2008) analyzed the variants in the promoter region of the IKBL/NFKBIL-1 gene, which pertains to the MHC class I region. A total of 245 patients (76 ASY and 169 CCC) were surveyed. Subjects that were homozygous for the −62A allele, had threefold risk of having CCC compared with those having the TT genotype. Moreover, the haplotype −262A-62A was prevalent in CCC patients.

Borrs et al. (2009) surveyed 152 Argentinean subjects (71 seropositive individuals and 81 controls) for the variability of the second exon of HLA-DRB1. The DRB1*1103 allele was predominant in the control patients, which suggests that this allele plays a protective role. On the other hand, DRB1*0409 and DRB1*1503 had a significantly higher frequency in seropositive patients, while CCC subjects had a higher DRB1*1503 frequency.

23.4.3 Other Genetic Systems Involved in Susceptibility to Chagas Disease

Calzada et al. (2001a) surveyed 85 seropositive patients (53 ASY and 32 CCC) and 87 seronegative controls. A relation was found between the CCR5 59029 promoter polymorphism and susceptibility to CCC.

The same authors (Calzada et al., 2001b) surveyed 168 Peruvian patients (83 seropositive with 51 ASY and 32 CCC; 85 seronegative controls) for the variability of the natural resistance-associated macrophage protein-1 (NRAMP1) gene. No differences were observed between: (i) seropositive and seronegative subjects and (ii) CCC versus ASY patients.

Messias-Reason et al. (2003) surveyed 100 seropositive individuals (43 ASY and 57 CCC) and 100 seronegative control patients. They observed a positive relation between CCC and complement C3 and BF allotypes and a negative link between CCC and seropositive patients and the BFS haplotype. No significant associations were observed for the C3, BF, CAA, CAB, and C2 haplotypes.

Flórez et al. (2006) analyzed the relations between the IL-1A, IL-1B, and IL-1RN gene variability and Chagas disease in 260 seropositive Colombian patients (130 ASY and 130 CCC). They evidenced that the presence of the IL1-B + 5810G allele was associated with a higher CCC risk.

Ramasawmy et al. (2006b) analyzed the monocyte chemoattractant protein-1 (CCL2/MCP-1) gene variability in 245 seropositive patients (76 ASY and 169

CCC). They observed that patients harboring the CCL2-2518AA genotype had a fourfold higher CCC risk.

Drigo et al. (2007) surveyed 246 patients (80 ASY and 166 CCC). They found no link between pathology and tumor necrosis factor-α polymorphisms. This result contradicted a previous study from the same authors (Drigo et al., 2006), which showed that CCC patients having the TNF2 or TNFa2 alleles have a significantly shorter survival time compared to patients who have other alleles.

Similarly, Campelo et al. (2007) evidenced that the TNFa2, TNFa7, TNFa8, TNFb2, TNFb4, TNFd5, TNFd7, and TNFe2 alleles were overrepresented, whereas the TNFb7 and TNFd3 alleles were underrepresented in 162 Chagasic patients compared with 221 control individuals.

Zafra et al. (2007) analyzed the putative role of a 3′ untranslated region (3′ UTR) polymorphism of the interleukin (IL)12B gene on Chagas disease in 460 Colombian individuals (seronegative: 200; seropositive: 260, with 130 ASY and 130 CCC). They observed a significantly higher frequency of the IL1-2B 3′ UTR CC genotype and of the IL1-2B 3′ UTR C allele in CCC patients. The same authors (Zafra et al., 2008) analyzed the polymorphism of toll-like receptor 2 and 4 genes in 475 Colombian patients (132 ASY, 143 CCC, and 200 seronegative controls). They recorded no frequency differences between chagasic patients and controls.

Ramasawmy et al. (2007) analyzed the polymorphism of the gene coding for lymphotoxin-α in 76 ASY and 169 CCC patients. Homozygosity for the LTA +80C and LTA +252G alleles was significantly more frequent in CCC patients than in ASY patients. Haplotype LTA +80A-252A appeared to have a protective effect against CCC, whereas haplotype LTA +80C-252G was associated with CCC susceptibility.

Robledo et al. (2007) observed no link between Chagas disease and the variants of the protein tyrosine phosphatase nonreceptor 22 (PTPN22) gene in 316 chagasic patients versus 520 healthy controls in Colombia and Peru.

Costa et al. (2009) found that the IL-10 gene polymorphism and IL-10 expression seem to be strongly involved in CCC susceptibility.

Cruz-Robles et al. (2009) surveyed 86 seropositive patients (28 ASY and 58 CCC), 50 seronegative individuals with idiopathic dilated cardiomyopathy (IDC), and 109 control individuals for the distribution of IL-1B and IL-1 receptor antagonist (IL-1RN) variants. Seropositive individuals showed a higher frequency of the CC genotype of the IL-1RN4 polymorphism. CCC patients exhibited an increased frequency of the C allele and of the CC genotype of this polymorphism.

Ramasawmy et al. (2009) analyzed 76 ASY and 169 CCC patients for their variability in the MAL/TIRAP gene, which expresses an adaptor protein in the toll-like receptor pathway. Contrary to Zafra et al. (2008), who found no links between Chagas disease and the genes of the toll-like receptor pathway, these authors observed a protective role against CCC of heterozygosity in the S180L variant of the gene under survey.

Calzada et al. (2009) explored the transforming growth factor beta 1 (TGFβ1) gene polymorphisms in 626 individuals from Colombia and Peru (ASY = 175; CCC = 172; seronegative controls = 279). The frequency of the high TGFβ1 producer genotype 10 C/C was significantly increased in chagasic patients by comparison with seronegative control individuals.

This summarizes the present state of knowledge on human genetic susceptibility to Chagas disease. Although significant progress has been made in the last 15 years, many aspects could be fruitfully explored, including: (i) the possible role of major genomic rearrangements (Check, 2005; Conrad et al., 2010); (ii) the ethnic diversity parameter; and (iii) the linkage and GWAS approaches.

23.5 Vector Genetic Diversity

Chagas disease exhibits a specific epidemiological feature, namely, that the parasite can be transmitted by an impressive range of different vectors. They all pertain to the category of "true bugs" (order Hemiptera, suborder Heteroptera). They are all included in the subfamily Triatominae, family Reduviidae. While other Reduviidae are predators, the Triatominae have specialized in obligatory blood-feeding, including adults of both sexes and larvae. Within the subfamily Triatominae, three main genera of unequal ecogeographic distribution can transmit Chagas disease, namely *Triatoma*, *Rhodnius*, and *Panstrongylus*. Each of these genera includes various species that are able to transmit the disease.

The genetic diversity of the vectors at both the genera and the species levels is therefore considerable.

At the subspecific level, many studies have explored the diversity of many species, both by population genetic markers (see Chapter 15) and by computer-assisted morphometric analysis (see Chapter 16); therefore, the diversity of Chagas disease vectors at the subspecific and population levels is fairly well known.

However, little is known about the differential vectorial capacity of the various triatomine species and of different populations within species. The null hypothesis that all species and all populations are equally able to transmit *T. cruzi* and its various genotypes (see Section 23.6) can be ruled out. It is highly conceivable that refined coevolution phenomena have occurred, meaning that local vectors are better able to transmit local parasite genotypes. This remains to be explored. It is worth noting, however, that a North American vector (*Triatoma protracta*) is fully able to transmit a Latin American strain of *T. cruzi* (Theis et al., 1987).

23.6 Parasite Genetic Diversity

It is interesting to note that, although the scientific community working on Chagas disease is small compared to the numbers working on AIDS, malaria, or tuberculosis, this pathogen has long been among the pioneer species explored by advanced approaches such as molecular typing and **population genetics**. Therefore, this parasite is probably the pathogen whose evolutionary biology is the best known, together with *Escherichia coli*. It can therefore be suggested as a paradigmatic biological model, as has been done with *E. coli*, *Drosophila melanogaster*, *Mus musculus*, and *Caenorhabditis elegans* (Tibayrenc, 2009).

Pioneering molecular studies on *T. cruzi* explored **isoenzyme** variability as early as the beginning of the 1970s (Toyé, 1974). Although a now out-of-fashion technique, **multilocus enzyme electrophoresis** has clearly discriminated three principal variants or zymodemes within *T. cruzi* (Miles et al., 1978). It is interesting to note that this observation remains current, since these three zymodemes continue to be recorded today in *T. cruzi* natural populations, although their denomination and evolutionary status has changed substantially. This permanency of **multilocus genotypes** over space and time is one of the strongest arguments in favor of predominant clonal evolution (see later).

The interpretation of isoenzyme diversity in terms of population genetics and evolutionary biology has made it possible to clarify the evolutionary status of the zymodemes. The model of predominant clonal evolution has been proposed for *T. cruzi* (Tibayrenc et al., 1986) as well as for other parasitic protozoa (Tibayrenc et al., 1990). The evidence was mainly based on the observation of a considerable linkage disequilibrium (nonrandom association of genotypes occurring at different loci). Linkage disequilibrium is the very manifestation of very limited or absent genetic recombination. The model stipulates that offspring multilocus genotypes are virtually identical to the parental genotypes and are stable in space and time, whatever the precise cytological mechanism of propagation. The model therefore includes not only mitotic propagation but also various forms of parthenogenesis, extreme homogamy, and self-fertilization in haploid organisms (Tibayrenc et al., 1990). Extreme inbreeding is not an alternative model to clonal evolution (Rougeron et al., 2009), but rather a particular case of it.

The main relevance of the model concerns molecular epidemiology (tracing multilocus genotypes [strains] with molecular tools for epidemiological follow-up). If predominant clonal evolution inhibits recombination, as stated earlier, the multilocus genotypes are stable in space and time, even at an evolutionary scale, and therefore constitute convenient targets for molecular epidemiology.

Since its inception, the clonal model has stated that it was compatible with occasional bouts of genetic recombination. Recombination has long been suspected in natural populations of *T. cruzi* (Machado and Ayala, 2001) and has been experimentally evidenced (Gaunt et al., 2003). However, it is clear that such hybridization events interfere only at an evolutionary scale. The stability of *T. cruzi* multilocus genotypes in the long run, with its extreme manifestation of strong parity between **phylogenetic** trees designed from different genetic markers (Tibayrenc et al., 1993) is incompatible with frequent genetic recombination.

It has been suggested that *T. cruzi* genotypes are distributed into six different clusters (Barnabé et al., 2000; Brisse et al., 2000), which cannot be equated with real **clades** because some of them clearly originate from former hybridization events (Brisse et al., 2003; Sturm and Campbell, 2010), further stabilized by clonal propagation. The term "discrete typing unit" (DTU; Tibayrenc, 1998a) has been coined to designate sets of stocks that are genetically closer to each other than to any other stock and are identifiable by common molecular, genetic, biochemical, or immunological markers called tags. The six *T. cruzi* clusters match this definition.

Their validity has been confirmed at a recent meeting of Chagas disease experts (Zingales et al., 2009).

From the points of view of molecular epidemiology and integrated genetic epidemiology, the population structure of *T. cruzi* summarized earlier can be illustrated by two key words: stability and discreteness. *T. cruzi* natural clones, and the DTUs into which they are distributed, are genetic entities that are both stable in space and time (up to the evolutionary scale) and strictly separated from each other, with rare occasional bouts of genetic exchange.

23.7 Concluding Remarks

The data described herein was not intended to be a comprehensive review of our present knowledge on the genetic diversity of Chagas disease hosts, vectors, and parasites. Instead, the goal was to briefly highlight why these data make Chagas disease a good model for the integrated genetic epidemiology of infectious diseases, as already proposed long ago (Tibayrenc, 1998b).

The key word here again is discreteness: discreteness of the clinical phenotypes of Chagas disease in humans, discreteness of *T. cruzi* clonal genotypes and DTUs, discreteness of the many different species that are hosts (mammals) and vectors (triatomine bugs) of Chagas disease. All these discrete entities can be used as units of analysis, keys on the keyboard to be played in many different situations that can be analyzed, both in surveying natural Chagas cycles and in designing experimental evolution protocols.

There are several possible examples.

When natural cycles are considered, possible protocols could be to compare *T. cruzi* genotypes isolated from (i) cardiac versus digestive versus ASY patients, (ii) different mammal species, and (iii) different triatomine bug species.

Experimental evolution protocols are easy because: (i) *T. cruzi* is easy to culture; (ii) many triatomine bug species are easy to raise; and (iii) several experimental animal models are available, and one can compare, for example, different breeds of mice, whose genetic distinctness results in differing susceptibility to Chagas disease.

All this makes the integrated genetic epidemiology of Chagas disease an extremely promising field of research that has until now been underexplored. It could constitute a paradigmatic example to develop similar approaches in other infectious models.

Glossary

Clade evolutionary lineage defined by cladistic analysis. A clade is monophyletic (it has only a single ancestor) and is genetically isolated (which means that it evolves independently) from other clades.

Isoenzymes, multilocus enzyme electrophoresis protein extracts of given biological samples are separated by electrophoresis. The gel is then processed with a biochemical reaction involving the specific substrate of a given enzyme. This enzyme's zone of activity is then specifically stained on the gel. From one sample to another, migration differences can appear for this same enzyme. These different electrophoretic forms of the same enzyme are referred to as isoenzymes or isozymes. These differences reflect sequence differences in the genes coding for the involved enzymes.

Multilocus genotype the combined genotype of a given strain or a given individual established with several genetic loci.

Phenotype all observable properties of a given individual or a given population apart from the genotype. The phenotype is not limited to morphological characteristics and can include, for example, physiological or biochemical parameters. The pathogenicity of a microorganism is a phenotypic property, as are the different clinical forms of a given disease. The phenotype is produced by the interaction between genotype and the environment.

Phylogeny, phylogenetic evolutionary relationships between taxa, species, organisms, genes, or molecules.

Population genetics analysis of allele and genotype frequency distribution and modifications under the influence of natural selection, mutation, genetic drift, and gene flow.

References

Abel, L., Sanchez, F.O., Oberti, J., Thuc, N.V., Hoa, L.V., Lap, V.D., et al., 1998. Susceptibility to leprosy is linked to the human NRAMP1 gene. J. Infect. Dis. 177, 133−145.

Alric, L., Fort, M., Izopet, J., Vinel, J.P., Duffaut, M., Abbal, M., 1999. Association between genes of the major histocompatibility complex class II and the outcome of hepatitis C virus infection. J. Infect. Dis. 179, 1309−1310.

Apt, W., 1988. HLA Antigens in cardiomyopathic Chilean chagasics. Am. J. Hum. Genet. 43, 770−773.

Barbossa, C.A.A., Morton, N.E., Pao, D.C., Krieger, H., 1981. Biological and cultural determinants of immunoglobulin levels in a Brazilian population with Chagas' disease. Hum. Genet. 59, 161−163.

Barnabé, C., Brisse, S., Tibayrenc, M., 2000. Population structure and genetic typing of *Trypanosoma cruzi*, the agent of Chagas' disease: a multilocus enzyme electrophoresis approach. Parasitology 150, 513−526.

Bellamy, R., Beyers, N., McAdam, K.P., Ruwende, C., Gie, R., Samaai, P., et al., 2000. Genetic susceptibility to tuberculosis in Africans: a genome-wide scan. Proc. Natl. Acad. Sci. U.S.A. 97, 8005−8009.

Borrs, S.G., Racca, L., Cotorruelo, C., Biondi, C., Beloscar, J., Racca, A., 2009. Distribution of HLA-DRB1 alleles in Argentinean patients with chagas' disease cardiomyopathy. Immunol. Invest. 38, 268−275.

Brenière, S.F., Poch, O., Selaes, H., Tibayrenc, M., Lemesre, J.L., Antezana, G., et al., 1984. Specific humoral depression in chronic patients infected by *Trypanosoma cruzi*. Rev. Inst. Med. Trop. Sao Paulo 26, 254−258.

Brisse, S., Barnabé, C., Tibayrenc, M., 2000. Identification of six *Trypanosoma cruzi* phylogenetic lineages by random amplified polymorphic DNA and multilocus enzyme electrophoresis. Int. J. Parasitol. 30, 35−44.
Brisse, S., Henriksson, J., Barnabé, C., Douzery, E.J.P., Berkvens, D., Serrano, M., et al., 2003. Evidence for genetic exchange and hybridization in *Trypanosoma cruzi* based on nucleotide sequences and molecular karyotype. Infect. Genet. Evol. 2, 173−183.
Bucheton, B., Kheir, M.M., El-Safi, S.H., Hammad, A., Mergani, A., Mary, C., et al., 2002. The interplay between environmental and host factors during an outbreak of visceral leishmaniasis in eastern Sudan. Microbes Infect. 4, 1449−1457.
Calzada, J.E., Nieto, A., Beraún, Y., Martín, J., 2001a. Chemokine receptor CCR5 polymorphisms and Chagas' disease cardiomyopathy. Tissue Antigens 58, 154−158.
Calzada, J.E., Nieto, A., López-Nevot, M.A., Martín, J., 2001b. Lack of association between NRAMP1 gene polymorphisms and *Trypanosoma cruzi* infection. Tissue Antigens 57, 353−357.
Calzada, J.E., Beraún, Y., González, C.I., Martín, J., 2009. Transforming growth factor beta 1 (TGFβ1) gene polymorphisms and Chagas disease susceptibility in Peruvian and Colombian patients. Cytokine 45, 149−153.
Campelo, V., Dantas, R.O., Simões, R.T., Mendes-Junior, C.T., Sousa, S.M.B., Simões, A.L., et al., 2007. TNF microsatellite alleles in Brazilian chagasic patients. Dig. Dis. Sci. 52, 3334−3339.
Check, E., 2005. Patchwork people. Nature 437, 1084−1086.
Conrad, D.F., Pinto, D., Redon, R., Feuk, L., Gokcumen, O., Zhang, Y., et al., 2010. Origins and functional impact of copy number variation in the human genome. Nature 464, 704−712.
Costa, G.C., Rocha, M.O.D.C., Moreira, P.R., Menezes, C.A.S., Silva, M.R., Gollob, K.J., et al., 2009. Functional IL-10 gene polymorphism is associated with Chagas disease cardiomyopathy. J. Infect. Dis. 199, 451−454.
Cruz-Robles, D., Reyes, P.A., Monteón-Padilla, V.M., Ortiz-Muñiz, A.R., Vargas-Alarcón, G., 2004. MHC class I and class II genes in Mexican patients with Chagas disease. Hum. Immunol. 65, 60−65.
Cruz-Robles, D., Chvez-Gonzlez, J.P., Cavazos-Quero, M.M., Prez-Mndez, O., Reyes, P.A., Vargas-Alarcn, G., 2009. Association between IL-1B and IL-1RN gene polymorphisms and chagas' disease development susceptibility. Immunol. Invest. 38, 231−239.
Dean, M., Carrington, M., Winckler, C., Huttley, G.A., Smith, M.W., Allikmets, R., et al., 1996. Genetic restriction of HIV-1 infection and progression to AIDS by a deletion allele of the CKR5 structural gene. Science 273, 1856−1862.
Dessein, A.J., Marquet, S., Henri, S., El Wali, N.E., Hillaire, D., Rodrigues, V., et al., 1999. Infection and disease in human schistosomiasis mansoni are under distinct major gene control. Microbes Infect. 1, 561−567.
Drigo, S.A., Cunha-Neto, E., Ianni, B., Cardoso, M.R.A., Braga, P.E., Faé, K.C., et al., 2006. TNF gene polymorphisms are associated with reduced survival in severe Chagas' disease cardiomyopathy patients. Microbes Infect. 8, 598−603.
Drigo, S.A., Cunha-Neto, E., Ianni, B., Mady, C., Faé, K.C., Buck, P., et al., 2007. Lack of association of tumor necrosis factor-α polymorphisms with Chagas disease in Brazilian patients. Immunol. Lett. 108, 109−111.
Faé, K.C., Drigo, S.A., Cuha-Neto, E., Ianni, B., Mady, C., Kalil, J., et al., 2000. HLA and beta-myosin heavy chain do not influence susceptibility to Chagas disease cardiomyopathy. Microbes Infect. 2, 745−751.

Fernandes-Mestre, M.T., Layrisse, Z., Montagnani, S., Acquatella, H., Catalioti, F., Matos, M., et al., 1998. Influence of the HLA class II polymorphism in chronic Chagas disease. Parasite Immunol. 20, 197–203.

Flórez, O., Zafra, G., Morillo, C., Martín, J., González, C.I., 2006. Interleukin-1 gene cluster polymorphism in Chagas disease in a Colombian case–control study. Hum. Immunol. 67, 741–748.

Garcia, A., Marquet., S., Bucheton, B., Hillaire, D., Cot, M., Fievet, N., et al., 1998. Linkage analysis of blood *Plasmodium falciparum* levels: interest of the 5q31-q33 chromosome region. Am. J. Trop. Med. Hyg. 58, 705–709.

Gaunt, M.W., Yeo, M., Frame, I.A., Tothard, J.R., Carrasco, H.J., Taylor, M.C., et al., 2003. Mechanism of genetic exchange in American trypanosomes. Nature 421, 936–939.

Layrisse, Z., Fernandez, M.T., Montagnani, S., Matos, M., Balbas, O., Herrera, F., et al., 2000. HLA-C*03 is a risk factor for cardiomyopathy in Chagas disease. Hum. Immunol. 61, 925–929.

Machado, C.A., Ayala, F.J., 2001. Nucleotide sequences provide evidence of genetic exchange among distantly related lineages of *Trypanosoma cruzi*. Proc. Nat. Acad. Sci. U.S.A. 98, 7396–7401.

Messias-Reason, I.J., Urbanetz, L., Pereira Da Cunha, C., 2003. Complement C3 F and BFS allotypes are risk factors for Chagas disease cardiomyopathy. Tissue Antigens 62, 308–312.

Miles, M.A., Souza, A., Povoa, M., Shaw, J.J., Lainson, R., Toyé, P.J., 1978. Isozymic heterogeneity of *Trypanosoma cruzi* in the first autochtonous patients with Chagas' disease in Amazonian Brazil. Nature 272, 819–821.

Moreno, M., Silva, E.L., Ramírez, L.E., Palacio, L.G., Rivera, D., Arcos-Burgos, M., 2004. Chagas' disease susceptibility/resistance: linkage disequilibrium analysis suggest epistasis between major histocompatibility complex and interleukin-10. Tissue Antigens 64, 18–24.

Morini, J.C., Berra, H., Davila, H.O., Pividori, J.F., Bottasso, O.A., 1994. Electrocardiographic alteration among first degree relatives with serological evidence of *Trypanosoma cruzi* infection. A sibship study. Mem. Inst. Oswaldo Cruz 89, 371–375.

Nieto, A., Beraún, Y., Callado, M.D., Caballero, A., Alonso, A., González, A., et al., 2000. HLA haplotypes are associated with differential susceptibility to *Trypanosoma cruzi* infection. Tissue Antigens 55, 195–198.

Pennisi, E., 2007. Human genetic variation. Science 318, 1842–1843.

Ramasawmy, R., Cunha-Neto, E., Faé, K.C., Müller, N.G., Cavalcanti, V.L., Drigo, S.A., et al., 2006a. BAT1, a putative anti-inflammatory gene, is associated with chronic chagas cardiomyopathy. J. Infect. Dis. 193, 1394–1399.

Ramasawmy, R., Cunha-Neto, E., Faé, K.C., Martello, F.G., Müller, N.G., Cavalcanti, V.L., et al., 2006b. The monocyte chemoattractant protein-1 gene polymorphism is associated with cardiomyopathy in human Chagas disease. Clin. Infect. Dis. 43, 305–311.

Ramasawmy, R., Fae, K.C., Cunha-Neto, E., Müller, N.G., Cavalcanti, V.L., Ferreira, R.C., et al., 2007. Polymorphisms in the gene for lymphotoxin-α predispose to chronic chagas cardiomyopathy. J. Infect. Dis. 196, 1836–1843.

Ramasawmy, R., Faé, K.C., Cunha-Neto, E., Borba, S.C.P., Ianni, B., Mady, C., et al., 2008. Variants in the promoter region of IKBL/NFKBIL1 gene may mark susceptibility to the development of chronic Chagas' cardiomyopathy among *Trypanosoma cruzi*-infected individuals. Mol. Immunol. 45, 283–288.

Ramasawmy, R., Cunha-Neto, E., Faé, K.C., Borba, S.C.P., Teixeira, P.C., Ferreira, S.C.P., et al., 2009. Heterozygosity for the S180L variant of MAL/TIRAP, a gene expressing

an adaptor protein in the toll-like receptor pathway, is associated with lower risk of developing chronic chagas cardiomyopathy. J. Infect. Dis. 199, 1838−1845.
Robledo, G., González, C.I., Morillo, C., Martín, J., González, A., 2007. Association study of PTPN22 C1858T polymorphism in *Trypanosoma cruzi* infection. Tissue Antigens 69, 261−264.
Rodrigues Coura, J., Albajar Viñas, P., 2010. Chagas disease: a new world challenge. Nature 465, S6−S7.
Rougeron, V., De Meeûs, T., Hide, M., Waleckx, E., Bermudez, H., Arevalo, J., et al., 2009. Extreme inbreeding in *Leishmania braziliensis*. Proc. Nat. Acad. Sci. U.S.A. 25, 10224−10229.
Sturm, N., Campbell, D.A., 2010. Alternative lifestyles: the population structure of *Trypanosoma cruzi*. Acta Trop. 115, 35−43.
Theis, J.H., Tibayrenc, M., Mason, D.T., Ault, S.K., 1987. Exotic stock of *Trypanosoma cruzi* (*Schizotrypanum*) capable of development in and transmission by *Triatoma protracta protracta* from California. Public health implications. Am. J. Trop. Med. Hyg. 36, 523−528.
Tibayrenc, M., 1998a. Genetic epidemiology of parasitic protozoa and other infectious agents: the need for an integrated approach. Int. J. Parasitol. 28, 85−104.
Tibayrenc, M., 1998b. Integrated genetic epidemiology of infectious diseases: the Chagas model. Mem. Inst. Oswaldo Cruz 93, 577−580.
Tibayrenc, M., 2007. Human genetic diversity and epidemiology of parasitic and other transmissible diseases. Adv. Parasitol. 64, 378−428.
Tibayrenc, M., 2009. Modeling the transmission of *Trypanosoma cruzi*: the need for an integrated genetic epidemiological and population genomics approach. In: Michael, E. (Ed.), Infectious Disease Transmission Modeling and Management of Parasite Control. Landes Bioscience/Eurekah, Austin, TX.
Tibayrenc, M., Ward, P., Moya, A., Ayala, F.J., 1986. Natural populations of *Trypanosoma cruzi*, the agent of Chagas' disease, have a complex multiclonal structure. Proc. Nat. Acad. Sci. U.S.A. 83, 115−119.
Tibayrenc, M., Kjellberg, F., Ayala, F.J., 1990. A clonal theory of parasitic protozoa: the population structure of *Entamoeba, Giardia, Leishmania, Naegleria, Plasmodium, Trichomonas* and *Trypanosoma*, and its medical and taxonomical consequences. Proc. Nat. Acad. Sci. U.S.A. 87, 2414−2418.
Tibayrenc, M., Neubauer, K., Barnabé, C., Guerrini, F., Sarkeski, D., Ayala, F.J., 1993. Genetic characterization of six parasitic protozoa: parity of random-primer DNA typing and multilocus isoenzyme electrophoresis. Proc. Natl. Acad. Sci. U.S.A. 90, 1335−1339.
Toyé, P.J., 1974. Isoenzymic differences between culture forms of *Trypanosoma rangeli, T. cruzi*, and *T. lewisi*. Trans. R. Soc. Trop. Med. Hyg. 68, 266.
Visentainer, J.E., Moliterno, R.A., Sell, A.M., Petzel-Erler, M.L., 2002. Association of HLA-DR2 with chronic chagasic cardiopathy in a population at Paraná Northeast region, Brazil. Acta Sci. Biol. Health Sci. 24, 727−730.
Williams-Blangero, S., VandeBerg, J.L., Blangero, J., Teixera, A.R.L., 1997. Genetic epidemiology of seropositivity for *T. cruzi* infection in rural Goiás, Brazil. Am. J. Trop. Med. Hyg. 57, 538−543.
Zafra, G., Morillo, C., Martín, J., González, A., González, C.I., 2007. Polymorphism in the 3′ UTR of the IL12B gene is associated with Chagas' disease cardiomyopathy. Microbes Infect. 9, 1049−1052.

Zafra, G., Flórez, O., Morillo, C.A., Echeverría, L.E., Martín, J., González, C.I., 2008. Polymorphisms of toll-like receptor 2 and 4 genes in Chagas disease. Mem. Inst. Oswaldo Cruz 103, 27–30.

Zicker, F., Slith, P.G., Netto, J.C.A., Oliveira, R.M., Zicke, E.M.S., 1990. Physical activity, opportunity for reinfection, and sibling history of heart diseases as risk factors for Chagas' cardiopathy. Am. J. Trop. Med. Hyg. 43, 498–505.

Zingales, B., Andrade, S.G., Briones, M.R.S., Campbell, D.A., Chiari, E., Fernandes, O., et al., 2009. A new consensus for *Trypanosoma cruzi* intraspecific nomenclature: second revision meeting recommends TcI to TcVI. Mem. Inst. Oswaldo Cruz 104, 1051–1054.

24 The Rise and Fall of the *Mycobacterium tuberculosis* Complex

Marcel A. Behr[1],* and Sébastien Gagneux[2]

[1]McGill University, Montreal, Canada, [2]Swiss Tropical and Public Health Institute, Basel, Switzerland University of Basel, Switzerland MRC National Institute for Medical Research, London, UK

24.1 Overview

The publication in 1998 of the genome sequence for *Mycobacterium tuberculosis* strain H37Rv marked the beginning of the genomic era for the tubercle bacillus (Cole et al., 1998). At that time, there were no other mycobacterial species for comparison, so this sequence became a *de facto* benchmark for mycobacterial genomics (Brosch et al., 2001). As such, predictions about the role of genome composition in the biology of this organism were largely based on conjecture. For instance, while the large number of genes coding for lipid metabolism may be related to the pathogenicity of this organism, the larger number of such genes in the relatively avirulent organism *M. avium hominissuis* argues that lipid metabolism may be a shared feature across mycobacteria that is uncoupled with virulence. Therefore, the key challenges for the subsequent decade were to uncover the genomic basis of pathogenicity and to understand the extent to which bacterial genetic variability influenced the natural epidemiology and control of tuberculosis (TB). To these ends, the introduction of a variety of different technologies, from microarrays to massively parallel sequencing, have enabled the TB research community to move beyond this classic point of reference to pose anew a set of questions that had been largely defined in the early twentieth century, but eluded the scientific capacity at the time. As a result of these new research modalities and their associated datasets, we now have a greater breadth of information about the genetics of *M. tuberculosis*, both in terms of variability within the species and the distinctions between this epidemic pathogen of mankind and related nontuberculous mycobacteria. At the time of writing, over a dozen mycobacterial species have been subject to genome sequencing, and likewise, several dozen strains

*E-mail: marcel.behr@mcgill.ca

of *M. tuberculosis* complex organisms are underway. In this chapter, we first define the organisms which will be the subject of consideration and outline their role in human and veterinary disease. Next, we will discuss the current understanding of the genus mycobacterium, and the situation of *M. tuberculosis* within this genus. In the third section, we will consider the results of genetic studies within the *M. tuberculosis* complex and the inferences of these findings on the population biology of this organism. Finally, we will conclude with our own reflections about subjects in need of further research and the links we envision might bridge the largely theoretical world of bacterial genetics with the very real ongoing TB epidemic.

24.2 *Mycobacterium tuberculosis* Complex: Definitions and Epidemiology

Soon after the discovery of the tubercle bacillus by Koch, it was recognized that similar but distinct organisms could be cultivated from infected cattle (Smith, 1898). With time, these organisms came to be known as the species *M. tuberculosis* and *M. bovis* resulting in a convenient but simplistic view that there were two organisms that caused TB, each primarily in its namesake host. Over time, however, other related organisms were isolated from a variety of different hosts or ecosystems such that the species name became *M. tuberculosis* complex, of which named variants that shared identical 16s rRNA sequences were considered subspecies. At the time of writing, a number of these have earned formal latin names, such as *M. microti, M. caprae,* and *M. pinnipedii*; others are simply named according to their host, such as the Dassie bacillus or the Oryx bacillus (also called the Antelope bacillus). Additionally, human isolates from Western Africa had a variety of phenotypic differences that led to these being named *M. africanum*. As a result of this somewhat confusing nomenclature, at present one can consider the *M. tuberculosis* complex to be the collective term for a number of different tubercle bacilli of different mammalian hosts, some of which appear to spread efficiently between man (notably *M. tuberculosis*), some of which have never been isolated from man (e.g., the Dassie bacillus) and others which occasionally cause disease in man but typically are inefficient at propagating through human hosts (e.g., *M. bovis*).

Two classic members of the *M. tuberculosis* complex merit specific mention: *M. canetii* and *M. bovis* BCG. *M. canetii*, originally described as an atypical organism based on colony morphology (van Soolingen et al., 1997), has been associated with a few dozen cases of human TB in persons who had lived in the Horn of Africa (Gutierrez et al., 2005). While conventionally classified as a member of the *M. tuberculosis* complex, recent sequence information indicates that the 16s rRNA of this organism does not share perfect identity with that of other *M. tuberculosis* complex organisms. Furthermore, this organism has unique biochemical attributes, as compared to *M. tuberculosis*, and likely has a unique epidemiologic profile, given the small number of cases ever described in man. For these reasons, in this chapter, we will refer to *M. canetii* in the section of the mycobacterium genus, highlighting the features of this organism that provide clues into the most recent evolution of

M. tuberculosis from a nontuberculous mycobacterium ancestor. The other outlier member of the *M. tuberculosis* complex is the vaccine strain, *M. bovis* BCG. BCG Pasteur was derived through the in vitro passage of *M. bovis* from 1908 to 1961, during which time other BCG strains were derived by passaging lots of BCG in a variety of labs worldwide (Behr and Small, 1999). Because BCG strains have been cultivated in vaccine labs for about half of the twentieth century, the evolution of these organisms provides clues into in vitro adaptation, which may mirror the elements required for in vivo infection and pathogenesis. Therefore, when discussing organisms with the name *M. bovis*, we will use the term "virulent *M. bovis*" when discussing genetic attributes of the bovine pathogen in contrast to "**BCG vaccines**" when outlining lessons from the genomic studies of *M. bovis* BCG strains.

In terms of the epidemiology of human TB, it is estimated that about one-third of the world's population has latent tuberculous infection (i.e., infected with the organism but lacking any clinical manifestations) (Dye et al., 2005). As about 10% are expected to progress to active TB during their lifetime, this suggests that there are 200 million incipient cases of active TB to be expected from this large reservoir of infected individuals. Indeed, it is estimated by the World Health Organization that there are over 9 million new cases of active TB per year and 1.5 million deaths due to TB each year (http://www.who.int/tb/publications/global_report/2009/en/). As TB is spread by sick patients via contagious aerosols, the organisms that cause disease in patients with active TB are expected to be those that transmit to their contacts, thereby propagating the epidemic. The vast majority of human cases are due to *M. tuberculosis* s.s.. As noted above, only a handful of cases due to *M. canetii* have been reported, and likewise, human disease due to *M. caprae*, virulent *M. bovis* and *M. pinnipedii* is generally uncommon, even in parts of Africa where it was thought that these cases might be overlooked due to inadequate laboratory capacity. In West Africa, some series observe up to half of human TB cases due to *M. africanum*, with an epicentre around Guinea Bissau (de Jong et al., 2009). While it is clear that zoonotic reservoirs exist for virulent *M. bovis*, *M. caprae*, etc., it is not clear at this point whether *M. africanum* is restricted to humans or whether it represents a spill-over infection from a yet to be determined zoonotic source (Demers et al., 2010). Given that *M. africanum* is associated with a slower time from to disease than *M. tuberculosis* (de Jong et al., 2008) and that cases with *M. africanum* are more likely to suffer HIV infection (de Jong et al., 2005), there are features of this organism compatible with an opportunistic pathogen, without any direct evidence that this is the case. In the case of *M. canetii*, with so few cases of human TB detected due to this organism, one also wonders whether there is a nonhuman reservoir, but this is presently not known.

24.3 The Mycobacterium Genus

The availability of sequence-based modalities to define bacterial variants has led to an explosion in mycobacterial species; at last count, there are nearly 150 species in this genus (http://www.bacterio.cict.fr/m/mycobacterium.html). To some extent, this list is growing because of the discovery of new obscure organisms in habitats that had not been subject to mycobacteriologic study. However, in many cases, these new proposed species represent organisms previously considered to be variants of better known species. Classically, the genus can be divided between rapid-growing organisms (colonies seen within a week) and slow-growing organisms (Figure 24.1 for a phylogenetic representation of mycobacterial species subject to complete genome sequencing; Veyrier et al., 2009). While there is no strict association between slow growth and pathogenicity, it is noted that the group of rapid-growing organisms has only one important human pathogen (*M. abscessus*) whereas the slow-growing group of mycobacteria include the following species associated with disease in humans: *M. avium*, *M. intracellulare*, *M. leprae*, *M. marinum*, *M. ulcerans*, *M. kansasii*, and organisms of the *M. tuberculosis* complex. Because

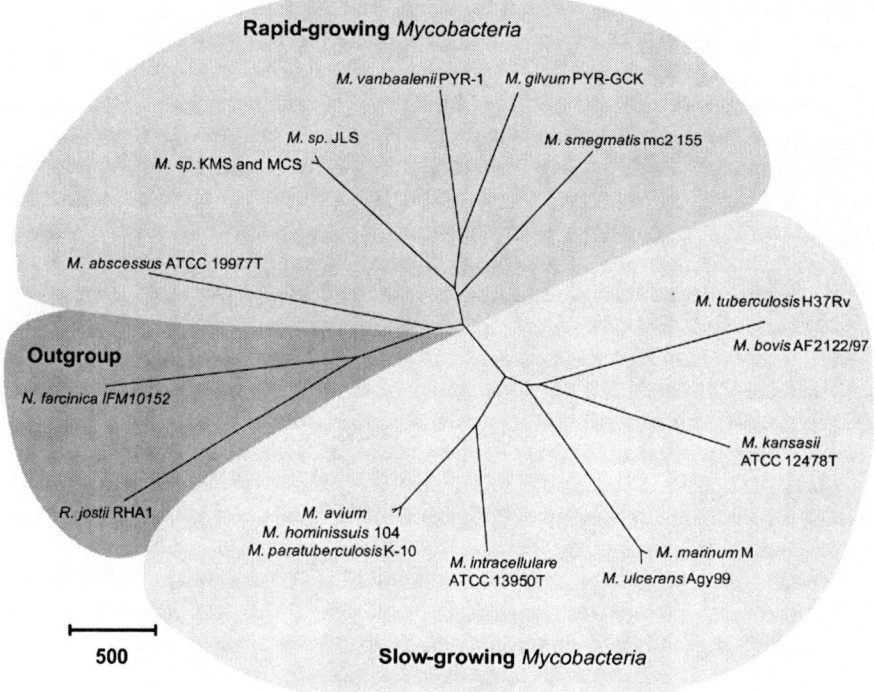

Figure 24.1 A phylogenetic representation of the mycobacterium genus, based on concatenates from 20 complete genes in all mycobacterial genomes that have been sequenced to date. *Nocardia* and *Rhodococcus* are presented as outgroups. The distinction between rapid-growing mycobacteria and slow-growing mycobacteria is highlighted through coloring. (For interpretation of the references to color in this figure legend, the reader is referred to the web version of this book.)
Source: Adapted from Veyrier et al. (2009).

research efforts are often focused on addressing threats to human health, this list is naturally biased towards human pathogens, presenting the optic that slow-growing mycobacteria are all pathogenic to humans (which is not the case). Nonetheless, of the mycobacteria that cause human disease, the vast majority are slow-growing organisms, suggesting that a study of the similarities and differences across these species can help illuminate common trends in mycobacterial pathogenicity.

A first key observation is that while the slow-growing mycobacteria can be isolated from patients with clinical disease, with the exception of *M. leprae* these are organisms that are not thought to spread from person-to-person, therefore, these are considered environmental organisms that can cause human disease, in the appropriate host. In some cases, the host defect associated with disease is well documented (e.g., disseminated *M. avium* disease in a person with AIDS, pulmonary *M. intracellulare* in a person with bronchiectasis), whereas in other cases, disease is uncommon despite a presumably common exposure, suggesting an unknown host resistance defect. *M. marinum* is the cause of TB in amphibians and fish, but is only associated with a lympho-cutaneous process in humans called fish tank granuloma. *M. ulcerans* is the cause of Buruli ulcer, a disease thought to be emerging in parts of Africa, but for which the precise epidemiology remains to be worked out. *M. kansasii* is the cause of a TB-like disease in persons with pre-existing lung disease, such as coal miners. Finally, *M. leprae* is the cause of leprosy, and this organism was the second mycobacterium to have its genome determined.

Because the *M. leprae* genome is considerably smaller than that of *M. tuberculosis*, with many pseudogenes and a drastically reduced coding capacity (Cole et al., 2001), the view about 10 years ago was that pathogenic mycobacteria are marked by a process of reductive evolution. This corresponded with microarray-based interrogation of members of the *M. tuberculosis* complex, and was consistent with the notion that pathogenic mycobacteria have limited opportunity for gene acquisition within their intracellular niches. Importantly, these themes were arguably developed based on study of terminally differentiated professional pathogens, and therefore the view of mycobacterial evolution was decidedly skewed towards organisms that cannot proliferate outside of a eukaryotic host. Consequently, while the observations were correct, the inference that mycobacteria were somehow exempt from many of the prime drivers of bacterial evolution were presumably overstated (Gal-Mor and Finlay, 2006), biased as they were on the data from outliers used to generalize to the more typical members of the genus. With further genome sequence data emerging from other mycobacterial species, it became apparent that mycobacterial genomes are shaped by more than just reductive genomics, and that horizontal gene transfer (HGT), in the form of plasmids or chromosomal DNA, was an important contributor to mycobacterial diversity.

After *M. tuberculosis* and *M. leprae*, the next mycobacteria studied by genomic approaches were *M. marinum*/*M. ulcerans* and the *M. avium*−*intracellulare* complex. Curiously, while *M. marinum* and *M. ulcerans* are classically named as stand-alone organisms, they share considerably greater genetic similarity than

M. avium versus *M. intracellulare*, suggesting that in taxonomic terms they could be called the *M. marinum—ulcerans* complex, the latter representing a clonal variant of *M. marinum*. The key lessons from genomic studies of these organisms is that *M. marinum* has a much larger genome than *M. tuberculosis* (6.6 Mb versus 4.4 Mb) (Stinear et al., 2008), that virulence determinants of *M. tuberculosis* such as the ESX-1 locus are conserved in *M. marinum* (Volkman et al., 2004) and that compared with *M. marinum*, *M. ulcerans* has acquired a giant plasmid and lost its ESX-1 locus (Stinear et al., 2007). A similar pattern has now been documented in postgenomic study of *M. avium*, where pathogenic subspecies (*M. avium* subsp. *paratuberculosis*; *M. avium* subsp. *avium*) have distinct genomic inserts that are not present in the environmental organism *M. avium* subsp. *hominissuis*, but also lost segments of the *M. avium* subsp. *hominissuis* genome (Semret et al., 2005, 2006). Through in-depth assessment of each of the *M. avium* subsp. *paratuberculosis*-specific elements, Alexander and colleagues were able to provide evidence for HGT in 6 of these genomic islands, through the detection of aberrant GC%, the presence of phage integration sites and the documentation that these elements share no homologs within the genus (Alexander et al., 2009). These authors proposed that the acquisition of novel genetic material followed by deletion of nonessential elements might represent a common process in mycobacterial evolution, stimulating a conceptually similar study seeking phylogenetic evidence of HGT in *M. tuberculosis*.

By generating a phylogeny based on multilocus sequence analysis of 20 conserved genes, Veyrier and colleagues were able to conduct homology searches of genes apparently restricted to certain mycobacterial lineages. A number of *M. tuberculosis* genes were described that lack orthologues in other mycobacterial species, some of which presented further evidence for HGT, such as aberrant GC% and presence of a vector of HGT. This study found that *M. tuberculosis* has at least 55 genes that are not present in any other member of the genus, with *M. kansasii* representing the closest comparator (Veyrier et al., 2009). Notably, this study used a different but complementary approach to looking for compositional evidence of HGT in *M. tuberculosis* where a number of the same genes were uncovered (Becq et al., 2007). Furthermore, a study of *M. canetii*-like organisms provided further evidence for HGT, documenting that *M. tuberculosis* s.s. organisms contain a set of genes absent from the closely related *M. canetii* species (Rosas-Magallanes et al., 2006). Together, these studies have shaped an emerging view that mycobacteria, like other genera, evolve through both horizontal and vertical gene flow, and that gene acquisition occurred coincident with the emergence of pathogenic mycobacterial species. A key research priority now is to define the extent to which these added genes contributed to the pathogenicity of *M. tuberculosis*, either through targeted gene disruption studies in *M. tuberculosis* or via heterologous complementation of putative virulence genes into less virulent organisms such as *M. kansasii*. As an example, it has been shown for Rv0986 and Rv0987, specific to *M. tuberculosis*, that their disruption leads to impaired infection of macrophages (Rosas-Magallanes et al., 2007). Further studies are required to elucidate the importance of HGT in the emergence of *M. tuberculosis* as an epidemic pathogen.

24.4 *Mycobacterium tuberculosis* Complex Evolution

After the completion of the *M. tuberculosis* H37Rv genome, the most readily available tools for comparative genomics were BAC libraries and DNA microarrays (Behr et al., 1999; Gordon et al., 1999). Relying on these tools, comparative genomic studies were able to detect large sequence polymorphisms that distinguished the interrogated organisms from the referent; since the comparisons were unidirectional, there was a bias towards finding genomic deletions in a particular subspecies or strains. This methodology led to the determination of genomic deletions that were then studied across selected collections of clinical isolates, revealing that some genomic regions (known as "regions of difference" or RDs) were deleted across large groups of organisms (e.g., RD9 is absent from *M. africanum* and the veterinary variants; Parsons et al., 2002), while others were specific to distinct lineages (e.g., RD1 is absent from BCG vaccines; Mahairas et al., 1996). Because *M. tuberculosis* complex exhibits essentially no ongoing HGT (Supply et al., 2000), these deletions were predicted to represent unique evolutionary events that were highly unlikely to revert. Hence, these deletions were used to derive first-generation phylogenies of *M. tuberculosis* complex organisms that largely agreed in content: *M. tuberculosis* was situated in a more ancestral position, with a larger genome, while *M. bovis* was situated in a more derived position, with a smaller genome, and served as the parental strain for BCG vaccines, which themselves were marked by a succession of further genomic deletions (Brosch et al., 2002; Mostowy et al., 2002). At the same time, the availability of spacer oligotyping, or spoligotyping, as a DNA fingerprinting tool allowed investigators to interrogate collections of organisms and determine that the number of spacers at the direct repeat locus also presented a pattern of deletions, again placing *M. tuberculosis* before *M. bovis* before BCG (Smith et al., 2006). Together, these modalities suggested that the traditional subspecies names were robust and that different strains and lineages of the *M. tuberculosis* complex had distinct genomic identities (Smith et al., 2005).

In recent years, these kinds of genomic comparisons have expanded from *ad hoc* study of the strain of interest (Mostowy et al., 2004; Nguyen et al., 2004; Tsolaki et al., 2005) towards a more comprehensive assessment of genetic variability through study of large, representative collections of strains from around the world (Gagneux et al., 2006). These studies have employed GeneChip-based interrogation for genomic deletions (Kato-Maeda et al., 2001; Hirsh et al., 2004; Tsolaki et al., 2004), and determined that the human-adapted members of *M. tuberculosis* complex consist of six main lineages that are associated with different regions of the world (Gagneux et al., 2006). The phylogeographic population structure of *M. tuberculosis* has been confirmed using a variety of SNP-typing and other genotyping approaches (Baker et al., 2004; Brudey et al., 2006; Filliol et al., 2006; Gutacker et al., 2006; Gagneux and Small, 2007).

A pertinent question about genomic deletions is whether these are merely markers of mycobacterial evolution or whether these are important drivers of functional differences between related strains (Tsolaki et al., 2004). Since 1996, an ever-increasing list of genomic deletions has been catalogued, serving to distinguish subspecies and strains, and offering molecular targets for simple and efficient branding of mycobacterial lineages (Behr and Mostowy, 2007). On first glance, these appear to be suboptimal markers of evolution, because the loss of part of a gene or a set of genes should logically translate into some sort of evolutionary selective pressure. However, somewhat remarkably, there has been very little evidence that these genomic deletions contribute to the full virulence of *M. tuberculosis*. The RD1 region, encoding the ESX-1 locus, was deleted during the derivation of BCG vaccines, which is consistent with selection in the laboratory for in vitro growth at the expense of producing a bacterial secretion system needed to interact with the host. Aside from RD1, the vast majority of genomic deletions have not been assigned a functional role and studies of genes essential for full virulence in vivo have typically not detected many of the genes in these deleted regions (Sassetti and Rubin, 2003). One interpretation of these data is genomic deletions identified in clinical isolates present a list of dispensible genes, that are not required for production of successful infection and disease in the natural host. In contrast, genes that are present in all clinical isolates, especially those that are invariant, suggest elements that are critical to the full virulence of this organism.

Despite their usefulness as phylogenic markers, large deletions are uncommon and unpredictable single events, and the data derived by deletion-based study did not permit a quantitative assessment of *M. tuberculosis* genetic variability (Gagneux and Small, 2007). Moreover, since deletions were being catalogued by comparison to a single reference strain, there was concern about the possibility that unidirectional comparisons had resulted in a biased and potentially skewed understanding of the diversity and evolution of *M. tuberculosis* and related organisms. By contrast, *de novo* DNA sequence data has the advantage of providing an unbiased assessment of genetic variability and a quantification of the nature of such variability. Initial comparative DNA sequencing efforts based on a limited set of genes yielded a low number of genetic polymorphisms (Sreevatsan et al., 1997; Musser et al., 2000), reinforcing the notion that *M. tuberculosis* complex represented the classical example of a genetically monomorphic bacterial pathogen (Achtman, 2008). However, following the publication of the genome sequence of the laboratory strains H37Rv (Cole et al., 1998), the determination of the complete genome sequence for a *M. tuberculosis* clinical strain (Fleischmann et al., 2002) and a virulent isolate of *M. bovis* (Garnier et al., 2003) permitted a first unbiased comparison of diversity across the entire genome. These genome-wide data uncovered larger amounts of genomic diversity than had been predicted from previous reports. For example, *M. bovis* was found to differ from the *M. tuberculosis* by 2,400 SNPs, suggesting that *M. tuberculosis* complex stands at the higher end from what can be considered genetically monomorphic (Achtman, 2008).

More recently, 89 complete genes were sequenced in a collection of 108 globally representative strains (Hershberg et al., 2008). Phylogenetic analyses of these multilocus sequence data revealed a tree topology that was largely congruent with the previous results based on genomic deletion and SNP-based analyses (Comas and Gagneux, 2009). However, the unbiased nature of the multilocus sequence analysis by Hershberg and colleagues yielded some important new insights (Figure 24.2). First, despite the fact that the different animal-adapted members of *M. tuberculosis* complex represent distinct ecotypes adapted to different host species (Smith et al., 2005), these animal strains all clustered together and formed an ingroup with respect to the overall diversity represented by the human-adapted lineages. This further indicates that human-adapted lineages of *M. tuberculosis* complex are more genetically diverse than previously believed. Second, the ancestral position of the two African lineages (commonly known as *M. africanum*) support the notion that human-adapted *M. tuberculosis* complex originated in Africa. Based on this observation, together with the fact that Africa is the only continent which harbors all six main lineages of human-adapted *M. tuberculosis* complex, Hershberg and colleagues proposed an "Out-of-and-back-to-Africa" scenario for the evolution and global spread of human TB; this scenario was supported by their multilocus sequence data (Hershberg et al., 2008). According to this scenario, *M. tuberculosis* originated in Africa and accompanied the ancient Out-of-Africa migrations of modern humans that occurred about 50,000 years ago. Over the last few hundred years, three particularly successful lineages of *M. tuberculosis* expanded as a consequence of human population increases in Europe, India, and China, and spread globally through recent waves of travel, trade, and conquest.

In addition to offering quantitative assessments of genetic diversity, DNA sequence data allow for detailed population genetic analyses of genetic variation by comparing the ratio of nonsynonymous to synonymous mutations (a measure referred to as dN/dS). Measuring dN/dS values across genes and genomes can be used to infer the evolutionary forces shaping the genetic diversity of populations. *M. tuberculosis* complex exhibits an unusually high dN/dS compared to most other bacteria (Hershberg et al., 2008). While several possible explanations could account for this phenomenon (Rocha et al., 2006), the study by Hershberg et al. found that the most likely explanation for the high dN/dS in *M. tuberculosis* is a strong reduction in purifying selection against nonsynonymous changes, which generally tend to be slightly deleterious (Hershberg et al., 2008). The authors hypothesized this reduction in purifying selection was likely a consequence of the serial population bottlenecks during patient-to-patient transmission which increase the impact of random genetic drift relative to selection.

Even though selective constraint appears to be strongly reduced in *M. tuberculosis*, natural selection is clearly acting in this organism, as observed in a recent study based on 21 whole-genome sequences representative of the global diversity of human-adapted *M. tuberculosis* complex (Comas et al., 2010). In this study, the authors used Illumina next-generation DNA sequencing and in-depth population genomic analyses to address the question as to whether *M. tuberculosis* evolves to evade host immune responses by varying its antigenic genes. Strategies of immune escape through accumulation of antigenic variation have been documented in various human pathogenic viruses, protozoa, and bacteria. Comas and colleagues used their comparative whole-genome dataset to study the evolutionary conservation of three experimentally confirmed groups of genes, including

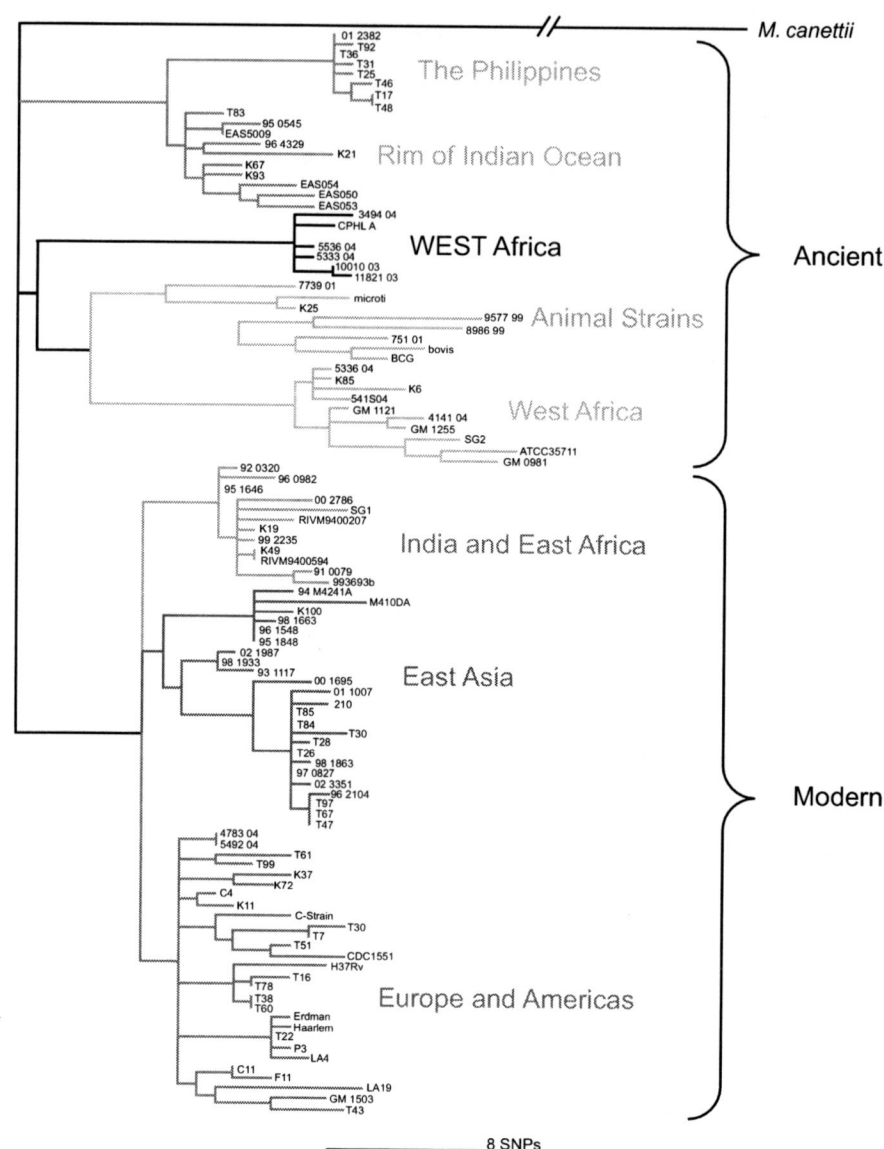

Figure 24.2 The global phylogeny of *Mycobacterium tuberculosis* complex based on concatenates of 89 complete gene sequences in 108 strains. Maximum parsimony analysis resulted in a single tree with negligible homoplasy. Analysis by the neighbor-joining method resulted in an identical tree topology with strong bootstrap support. Major lineages are indicated in different colors. (For interpretation of the references to color in this figure legend, the reader is referred to the web version of this book.)
Source: Adapted from Hershberg et al. (2008).

essential genes, nonessential genes, and human T cell antigens. As expected, the authors found that essential genes were less diverse and more evolutionarily conserved than nonessential genes. Surprisingly however, the T cell antigens were not more diverse than essential genes and equally conserved. Moreover, when the authors analyzed the specific antigen regions known to interact with human T cells (i.e., the T cell epitopes) they found that these T cell epitopes were the most conserved of all genomic regions analyzed. The study concluded that, in contrast to many other pathogens which evade host immunity by varying their antigenic genes, *M. tuberculosis* wants to be recognized, perhaps because the cellular immune responses elicited by T cell epitopes offer a net benefit to the bacteria. In support of this notion, the characteristic immuno-pathological process in human TB known as lung cavitation greatly increases the transmission potential of *M. tuberculosis*. Cavitation depends in part on an intact adaptive immune response as HIV co-infected TB patients with low CD4 T cell counts rarely develop cavities in their lungs, and are thus often less infectious than immune-competent TB patients.

24.5 Pending Questions and Concluding Thoughts

The availability of genome-level data has permitted a number of fields of research to revisit old dogma, challenge assumptions about the biology of the disease, and uncover new targets for postgenomic approaches to addressing that particular process. This is clearly the case for TB research. Over 100 years ago it was known that *M. tuberculosis* is less virulent than *M. bovis* for rabbits and cows, but more likely to cause human disease. For a number of reasons it was subsequently assumed that the human pathogen had evolved from the bovine pathogen, which set up a model whereby *M. bovis* may have had a larger genome, with greater flexibility to cause disease in a wider range of hosts, including the ability to produce antigenic proteins such as MPB70 and MPB83 that are not detected in culture filtrates of *M. tuberculosis*. Now we know that this scenario is wrong. *M. bovis* has a smaller genome, the reason for its ability to cause disease in a large number of hosts is still unknown, and the production of copious amounts of these antigenic proteins is not because *M. tuberculosis* lacks the necessary genes, but rather because *M. bovis* has an anti-sigma factor mutation resulting in constitutive expression of genes in the SigK regulon (Said-Salim et al., 2006). This one anecdote serves to remind us of the greatest value of genome-level interrogation, which eliminates many of the biases of directed research and instead lays out a large dataset that is ripe for new interpretation of old questions. So what are the key questions for TB in 2010?

At the fundamental level, we still do not know why *M. tuberculosis* spreads so efficiently between people, unlike nontuberculous mycobacteria, and unlike other members of the *M. tuberculosis* complex. Arguably, an understanding of transmission, coupled with an intervention to reduce transmission, could serve to convert TB from a public health problem (epidemic disease) into a clinical problem (infectious disease). Currently, all patients with TB are isolated and treated in a similar manner, without a clear understanding of which of these is likely to propagate the epidemic and which is merely suffering from a private malady that is unlikely to affect their family and friends. One can envision two approaches to resolving this question: starting close and working outwards, or starting at the extremes and working inwards. In the former approach, one can analyze sequence variability among closely related strains to determine the nature of genetic change (e.g., among substrains of the Beijing lineage of *M. tuberculosis*). Such an analysis will reveal the smallest number of SNPs but may point to key regions in the genome tolerating variability, such as drug-resistance genes for which variability in such comparisons is now well documented (Niemann et al., 2009). The contrasting latter approach starts with the premise that genes present in *M. tuberculosis* that are absent from *M. kansasii*, *M. marinum*, and other nontuberculous mycobacteria may hold the key to the capacity of *M. tuberculosis* to efficiently cause disease and spread between humans. Here, the challenge is not so much to detect differences, but rather to sort through the long list of differences to generate targets that lend themselves to experimental assessment in the laboratory. Notably, among the genes conserved between *M. tuberculosis* and *M. kansasii* are a number of well-established virulence genes, including the ESX-1 secretion system; this argues that the difference between a professional pathogen and an opportunistic pathogen is not necessarily whether a virulence system is present, but whether the organism is optimally tuned to fully exploit an extant virulence determinant.

Ultimately, the driving goal of TB research is to develop the tools needed to combat the epidemic. Conceptually this occurs in clusters of researchers interested in better diagnosis, better treatment, or a better vaccine against TB; likely, synergy will be achieved if these advances are applied in concert. Key questions in coming years include the importance of pathogen variability in each of these programs of research, from conception to product development to validation in the field (Gagneux and Small, 2007). Will new immunediagnostic tests based on *M. tuberculosis*-specific antigens detect all cases of TB infection, or are there strains that will elude detection? Do drug treatments produce a predictable rate of antibiotic resistance, across strains, or is there reason to believe that certain lineages have a genetically programmed propensity for antibiotic tolerance and/or selection of antibiotic resistance (Borrell and Gagneux, 2009)? Do new vaccines provide protection against all strains of *M. tuberculosis*, or just laboratory strains that are used for convenience because they lend themselves to reproducible experimental infections? Prior to the genomic era, one of the classic dogmas of TB research was that the organism is relatively invariant and that the pathogenesis of disease is largely driven by host immune processes (Comas and Gagneux, 2009). The genome era has now taught us that the first view is largely incorrect and it is clear that differences in the outcome of infection occur as a function of different mycobacterial species or strains (Nicol and Wilkinson, 2008). The challenge at hand is to harness this new data, translate it into new insights, and use this information to help inform the next generation of TB research.

Acknowledgments

Work in our laboratories is supported by the Canadian Institutes for Health Research (MAB) and by the UK Medical Research Council, the Royal Society, the National Institutes of Health (HHSN266200700022C and AI034238), and the Swiss National Science Foundation (SG).

Glossary

Bacille de Calmette et Guérin (BCG) vaccine a family of attenuated strains of *M. bovis*, originally derived by serial passage of virulent *M. bovis* in the laboratory between 1908 and 1921. Named for the scientists who developed this vaccine.
***Mycobacterium tuberculosis* complex** a bacterial species that comprises *M. tuberculosis* s.s., *M. bovis*, and other related agents of tuberculosis in their respective hosts.

List of Abbreviations

TB tuberculosis
BCG Bacille de Calmette et Guérin
RD Region of difference, representing a locus of the bacterial genome that is absent from a closely related species, as in RD1, RD2, etc.

References

Achtman, M., 2008. Evolution, population structure, and phylogeography of genetically monomorphic bacterial pathogens. Annu. Rev. Microbiol. 62, 53–70.
Alexander, D.C., Turenne, C.Y., Behr, M.A., 2009. Insertion and deletion events that define the pathogen *Mycobacterium avium* subsp. *paratuberculosis*. J. Bacteriol. 191, 1018–1025.
Baker, L., Brown, T., Maiden, M.C., Drobniewski, F., 2004. Silent nucleotide polymorphisms and a phylogeny for *Mycobacterium tuberculosis*. Emerg. Infect. Dis. 10, 1568–1577.
Becq, J., Gutierrez, M.C., Rosas-Magallanes, V., Rauzier, J., Gicquel, B., Neyrolles, O., et al., 2007. Contribution of horizontally acquired genomic islands to the evolution of the tubercle bacilli. Mol. Biol. Evol. 24, 1861–1871.
Behr, M.A., Mostowy, S., 2007. Molecular tools for typing and branding the tubercle bacillus. Curr. Mol. Med. 7, 309–317.
Behr, M.A., Small, P.M., 1999. A historical and molecular phylogeny of BCG strains. Vaccine 17, 915–922.

Behr, M.A., Wilson, M.A., Gill, W.P., Salamon, H., Schoolnik, G.K., Rane, S., et al., 1999. Comparative genomics of BCG vaccines by whole-genome DNA microarray. Science 284, 1520–1523.
Borrell, S., Gagneux, S., 2009. Infectiousness, reproductive fitness and evolution of drug-resistant *Mycobacterium tuberculosis*. Int. J. Tuberc. Lung Dis. 13, 1456–1466.
Brosch, R., Pym, A.S., Gordon, S.V., Cole, S.T., 2001. The evolution of mycobacterial pathogenicity: clues from comparative genomics. Trends Microbiol. 9, 452–458.
Brosch, R., Gordon, S.V., Marmiesse, M., Brodin, P., Buchrieser, C., Eiglmeier, K., et al., 2002. A new evolutionary scenario for the *Mycobacterium tuberculosis* complex. Proc. Natl. Acad. Sci. U.S.A. 99, 3684–3689.
Brudey, K., Driscoll, J.R., Rigouts, L., Prodinger, W.M., Gori, A., Al-Hajoj, S.A., et al., 2006. *Mycobacterium tuberculosis* complex genetic diversity : mining the fourth international spoligotyping database (SpolDB4) for classification, population genetics and epidemiology. BMC Microbiol. 6, 23.
Cole, S.T., Brosch, R., Parkhill, J., Garnier, T., Churcher, C., Harris, D., et al., 1998. Deciphering the biology of *Mycobacterium tuberculosis* from the complete genome sequence. Nature 393, 537–544.
Cole, S.T., Eiglmeier, K., Parkhill, J., James, K.D., Thomson, N.R., Wheeler, P.R., et al., 2001. Massive gene decay in the leprosy bacillus. Nature 409, 1007–1011.
Comas, I., Gagneux, S., 2009. The past and future of tuberculosis research. PLoS Pathog. 5, e1000600.
Comas, I., Chakravartti, J., Small, P.M., Galagan, J., Niemann, S., Kremer, K., et al., 2010. Human T cell epitopes of *Mycobacterium tuberculosis* are evolutionarily hyperconserved. Nat. Genet. 42, 498–503.
de Jong, B.C., Hill, P.C., Brookes, R.H., Otu, J.K., Peterson, K.L., Small, P.M., et al., 2005. *Mycobacterium africanum*: a new opportunistic pathogen in HIV infection? Aids 19, 1714–1715.
de Jong, B.C., Hill, P.C., Aiken, A., Awine, T., Antonio, M., Adetifa, I.M., et al., 2008. Progression to active tuberculosis, but not transmission, varies by *Mycobacterium tuberculosis* lineage in The Gambia. J. Infect. Dis. 198, 1037–1043.
de Jong, B.C., Antonio, M., Awine, T., Ogungbemi, K., de Jong, Y.P., Gagneux, S., et al., 2009. Use of spoligotyping and large-sequence polymorphisms to study the population structure of the *Mycobacterium tuberculosis* complex in a cohort study of consecutive smear positive tuberculosis cases in the Gambia. J. Clin. Microbiol. 47, 994–1001.
Demers, A.M., Mostowy, S., Coetzee, D., Warren, R., van Helden, P., Behr, M.A., 2010. *Mycobacterium africanum* is not a major cause of human tuberculosis in Cape Town, South Africa. Tuberculosis (Edinb). 90, 143–144.
Dye, C., Watt, C.J., Bleed, D.M., Hosseini, S.M., Raviglione, M.C., 2005. Evolution of tuberculosis control and prospects for reducing tuberculosis incidence, prevalence, and deaths globally. JAMA 293, 2767–2775.
Filliol, I., Motiwala, A.S., Cavatore, M., Qi, W., Hernando Hazbon, M., Valle, Bobadilla Del, et al., 2006. Global phylogeny of *Mycobacterium tuberculosis* based on single nucleotide polymorphism (SNP) analysis: insights into tuberculosis evolution, phylogenetic accuracy of other DNA fingerprinting systems, and recommendations for a minimal standard SNP set. J. Bacteriol. 188, 759–772.
Fleischmann, R.D., Alland, D., Eisen, J.A., Carpenter, L., White, O., Peterson, J., et al., 2002. Whole-genome comparison of *Mycobacterium tuberculosis* clinical and laboratory strains. J. Bacteriol. 184, 5479–5490.

Gagneux, S., Small, P.M., 2007. Global phylogeography of *Mycobacterium tuberculosis* and implications for tuberculosis product development. Lancet. Infect. Dis. 7 (3), 337.

Gagneux, S., Deriemer, K., Van, T., Kato-Maeda, M., de Jong, B.C., Narayanan, S., et al., 2006. Variable host–pathogen compatibility in *Mycobacterium tuberculosis*. Proc. Natl. Acad. Sci. U.S.A. 103, 2869–2873.

Gal-Mor, O., Finlay, B.B., 2006. Pathogenicity islands: a molecular toolbox for bacterial virulence. Cell Microbiol. 8, 1707–1719.

Garnier, T., Eiglmeier, K., Camus, J.C., Medina, N., Mansoor, H., Pryor, M., et al., 2003. The complete genome sequence of *Mycobacterium bovis*. Proc. Natl. Acad. Sci. U.S.A. 100, 7877–7882.

Gordon, S.V., Brosch, R., Billault, A., Garnier, T., Eiglmeier, K., Cole, S.T., 1999. Identification of variable regions in the genomes of tubercle bacilli using bacterial artificial chromosome arrays. Mol. Microbiol. 32, 643–655.

Gutacker, M.M., Mathema, B., Soini, H., Shashkina, E., Kreiswirth, B.N., Graviss, E.A., et al., 2006. Single-nucleotide polymorphism-based population genetic analysis of *Mycobacterium tuberculosis* strains from 4 geographic sites. J. Infect. Dis. 193, 121–128.

Gutierrez, C., Brisse, S., Brosch, R., Fabre, M., Omais, B., Marmiesse, M., et al., 2005. Ancient origin and gene mosaicism of the progenitor of *Mycobacterium tuberculosis*. PLoS Pathogens 1, 1–7.

Hershberg, R., Lipatov, M., Small, P.M., Sheffer, H., Niemann, S., Homolka, S., et al., 2008. High functional diversity in *Mycobacterium tuberculosis* driven by genetic drift and human demography. PLoS Biol. 6, e311.

Hirsh, A.E., Tsolaki, A.G., DeRiemer, K., Feldman, M.W., Small, P.M., 2004. Stable association between strains of *Mycobacterium tuberculosis* and their human host populations. Proc. Natl. Acad. Sci. U.S.A. 101, 4871–4876.

Kato-Maeda, M., Rhee, J.T., Gingeras, T.R., Salamon, H., Drenkow, J., Smittipat, N., et al., 2001. Comparing genomes within the species *Mycobacterium tuberculosis*. Genome Res. 11, 547–554.

Mahairas, G.G., Sabo, P.J., Hickey, M.J., Singh, D.C., Stover, C.K., 1996. Molecular analysis of genetic differences between *Mycobacterium bovis* BCG and virulent *M. bovis*. J. Bacteriol. 178, 1274–1282.

Mostowy, S., Cousins, D., Brinkman, J., Aranaz, A., Behr, M.A., 2002. Genomic deletions suggest a phylogeny for the *Mycobacterium tuberculosis* complex. J. Infect. Dis. 186, 74–80.

Mostowy, S., Onipede, A., Gagneux, S., Niemann, S., Kremer, K., Desmond, E.P., et al., 2004. Genomic analysis distinguishes *Mycobacterium africanum*. J. Clin. Microbiol. 42, 3594–3599.

Musser, J.M., Amin, A., Ramaswamy, S., 2000. Negligible genetic diversity of *Mycobacterium tuberculosis* host immune system protein targets: evidence of limited selective pressure. Genetics 155, 7–16.

Nguyen, D., Brassard, P., Menzies, D., Thibert, L., Warren, R., Mostowy, S., et al., 2004. Genomic characterization of an endemic *Mycobacterium tuberculosis* strain: evolutionary and epidemiologic implications. J. Clin. Microbiol. 42, 2573–2580.

Nicol, M.P., Wilkinson, R.J., 2008. The clinical consequences of strain diversity in *Mycobacterium tuberculosis*. Trans. R Soc. Trop. Med. Hyg. 102, 955–965.

Niemann, S., Koser, C.U., Gagneux, S., Plinke, C., Homolka, S., Bignell, H., et al., 2009. Genomic diversity among drug sensitive and multidrug resistant isolates of *Mycobacterium tuberculosis* with identical DNA fingerprints. PLoS ONE 4, e7407.

Parsons, L.M., Brosch, R., Cole, S.T., Somoskovi, A., Loder, A., Bretzel, G., et al., 2002. Rapid and simple approach for identification of *Mycobacterium tuberculosis* complex isolates by PCR-based genomic deletion analysis. J. Clin. Microbiol. 40, 2339−2345.

Rocha, E.P., Smith, J.M., Hurst, L.D., Holden, M.T., Cooper, J.E., Smith, N.H., et al., 2006. Comparisons of dN/dS are time dependent for closely related bacterial genomes. J. Theor. Biol. 239, 226−235.

Rosas-Magallanes, V., Deschavanne, P., Quintana-Murci, L., Brosch, R., Gicquel, B., Neyrolles, O., 2006. Horizontal transfer of a virulence operon to the ancestor of *Mycobacterium tuberculosis*. Mol. Biol. Evol. 23, 1129−1135.

Rosas-Magallanes, V., Stadthagen-Gomez, G., Rauzier, J., Barreiro, L.B., Tailleux, L., Boudou, F., et al., 2007. Signature-tagged transposon mutagenesis identifies novel *Mycobacterium tuberculosis* genes involved in the parasitism of human macrophages. Infect. Immun. 75, 504−507.

Said-Salim, B., Mostowy, S., Kristof, A.S., Behr, M.A., 2006. Mutations in *Mycobacterium tuberculosis* Rv0444c, the gene encoding anti-SigK, explain high level expression of MPB70 and MPB83 in *Mycobacterium bovis*. Mol. Microbiol. 62, 1251−1263.

Sassetti, C.M., Rubin, E.J., 2003. Genetic requirements for mycobacterial survival during infection. Proc. Natl. Acad. Sci. U.S.A. 100, 12989−12994.

Semret, M., Alexander, D.C., Turenne, C.Y., de Haas, P., Overduin, P., van Soolingen, D., et al., 2005. Genomic polymorphisms for *Mycobacterium avium* subsp. *paratuberculosis* diagnostics. J. Clin. Microbiol. 43, 3704−3712.

Semret, M., Turenne, C.Y., de Haas, P., Collins, D.M., Behr, M.A., 2006. Differentiating host-associated variants of *Mycobacterium avium* by PCR for detection of large sequence polymorphisms. J. Clin. Microbiol. 44, 881−887.

Smith, T., 1898. A comparative study of bovine tubercle bacilli and of human bacilli from sputum. J. Exp. Med. 3, 451−511.

Smith, N.H., Kremer, K., Inwald, J., Dale, J., Driscoll, J.R., Gordon, S.V., et al., 2005. Ecotypes of the *Mycobacterium tuberculosis* complex. J. Theor. Biol. 239, 220−225.

Smith, N.H., Gordon, S.V., de la Rua-Domenech, R., Clifton-Hadley, R.S., Hewinson, R.G., 2006. Bottlenecks and broomsticks: the molecular evolution of *Mycobacterium bovis*. Nat. Rev. Microbiol. 4, 670−681.

Sreevatsan, S., Pan, X., Stockbauer, K.E., Connell, N.D., Kreiswirth, B.N., Whittam, T.S., et al., 1997. Restricted structural gene polymorphism in the *Mycobacterium tuberculosis* complex indicates evolutionarily recent global dissemination. Proc. Natl. Acad. Sci. U.S.A. 94, 9869−9874.

Stinear, T.P., Seemann, T., Pidot, S., Frigui, W., Reysset, G., Garnier, T., et al., 2007. Reductive evolution and niche adaptation inferred from the genome of *Mycobacterium ulcerans*, the causative agent of Buruli ulcer. Genome Res. 17, 192−200.

Stinear, T.P., Seemann, T., Harrison, P.F., Jenkin, G.A., Davies, J.K., Johnson, P.D., et al., 2008. Insights from the complete genome sequence of *Mycobacterium marinum* on the evolution of *Mycobacterium tuberculosis*. Genome Res. 18, 729−741.

Supply, P., Mazars, E., Lesjean, S., Vincent, V., Gicquel, B., et al., 2000. Variable human minisatellite-like regions in the *Mycobacterium tuberculosis* genome. Mol. Microbiol. 36, 762−771.

Tsolaki, A.G., Hirsh, A.E., DeRiemer, K., Enciso, J.A., Wong, M.Z., Hannan, M., et al., 2004. Functional and evolutionary genomics of *Mycobacterium tuberculosis*: insights from genomic deletions in 100 strains. Proc. Natl. Acad. Sci. U.S.A. 101, 4865−4870.

Tsolaki, A.G., Gagneux, S., Pym, A.S., Goguet de la Salmoniere, Y.O., Kreiswirth, B.N., Van Soolingen, D., et al., 2005. Genomic deletions classify the Beijing/W strains as a distinct genetic lineage of *Mycobacterium tuberculosis*. J. Clin. Microbiol. 43, 3185–3191.

van Soolingen, D., Hoogenboezem, T., de Haas, P.E., Hermans, P.W., Koedam, M.A., Teppema, K.S., et al., 1997. A novel pathogenic taxon of the *Mycobacterium tuberculosis* complex, Canetti: characterization of an exceptional isolate from Africa. Int. J. Syst. Bacteriol. 47, 1236–1245.

Veyrier, F., Pletzer, D., Turenne, C., Behr, M.A., 2009. Phylogenetic detection of horizontal gene transfer during the step-wise genesis of *Mycobacterium tuberculosis*. BMC Evol. Biol. 9, 196.

Volkman, H.E., Clay, H., Beery, D., Chang, J.C., Sherman, D.R., Ramakrishnan, L., 2004. Tuberculous granuloma formation is enhanced by a mycobacterium virulence determinant. PLoS Biol. 2, e367.

25 The Evolution and Dynamics of Methicillin-Resistant *Staphylococcus aureus*

Patrick Basset[1], Edward J. Feil[2], Giorgio Zanetti[1] and Dominique S. Blanc[1,*]

[1]Hospital Preventive Medicine, Centre Hospitalier Universitaire Vaudois and University Hospital of Lausanne, Lausanne, Switzerland,
[2]Department of Biology and Biochemistry, University of Bath, Bath, UK

25.1 Introduction

Staphylococcus aureus is a Gram-positive bacterium which typically resides asymptomatically on the skin and in the nose of animals, and in particular of mammals. Since its discovery in the 1880s, it has been recognized as a major opportunistic pathogen in humans responsible for various diseases, ranging from minor skin infections to severe bacteraemia and necrotizing pneumonia. Before the era of antibiotics, the mortality rate of patients infected with *S. aureus* exceeded 80% (Skinner and Keefer, 1941). The introduction of penicillin in the early 1940s saved the lives of tens of thousands of wounded allied troops in the Second World War and dramatically improved the prognosis of patients with staphylococcal infections. However, as early as 1942, penicillin-resistant staphylococci were recognized, and these strains arose via the acquisition of a plasmid carrying a gene encoding a penicillinase (β-lactamase). Although the spread of penicillin-resistant *S. aureus* was initially confined to hospital settings, this was quickly followed by the wider dissemination of resistance in the community, and by the late 1960s more than 80% of both community- and hospital-acquired isolates were resistant to penicillin (Rammelkamp and Maxon, 1942). This pattern is being repeated for methicillin, an alternative β-lactam antibiotic designed to resist β-lactamase. In the 50 years since the introduction of this antibiotic, various hospital-associated methicillin-resistant *S. aureus* (HA-MRSA) clones disseminated worldwide, and virulent

*E-mail: Dominique.Blanc@chuv.ch

community-associated MRSA (CA-MRSA) have continued to emerge and spread from the mid-1990s onwards.

25.2 The Staphylococcal Cassette Chromosome *mec*

Staphylococcus aureus is naturally susceptible to most antibiotics, and resistance is often acquired by the horizontal transfer of genes from intrinsically resistant coagulase-negative staphylococci. These genes are generally located on mobile genetic elements (MGEs) such as plasmids or cassettes.

Resistance to methicillin and all other β-lactam antibiotics is conferred by the acquisition of the methicillin resistance gene *mecA* (Matsuhashi et al., 1986). This gene is carried on a MGE called the staphylococcal chromosome cassette *mec* (SCC*mec*) (Katayama et al., 2000). This MGE is likely to have been introduced into the *S. aureus* population on multiple occasions from related staphylococcal species (Couto et al., 2003; Ibrahem et al., 2009). Several structural variants of SCC*mec* have been described which differ in their gene content and size (21–67 kb) but which share four characteristics. First, they carry the *mec* gene complex (*mec*) consisting of the methicillin resistance determinants *mecA* and its regulatory genes (*mecR1* and *mecI*). Second, they carry the cassette chromosome recombinase gene complex (*ccr*) consisting of genes that are responsible for the mobility of the element. Third, they have characteristic repeated sequences at both ends. Fourth, they integrate into the *S. aureus* chromosome at a site-specific location (attBscc), located within *orfX* near the origin of replication (Hiramatsu et al., 2001; Kuroda et al., 2001; Chongtrakool et al., 2006; de Lencastre et al., 2007). Despite these common characteristics, the detailed structure of SCC*mec* elements is highly divergent. In particular, several allotypic differences have been identified in *ccr* and *mec* complexes (Ito et al., 2009), as described here.

So far, three distinct *ccr* genes have been described (*ccrA*, *ccrB*, and *ccrC*) in *S. aureus*. Whereas *ccrC* is usually found alone, *ccrA* and *ccrB* are generally found adjacently on the same element. In addition, several allotypes of *ccrA* and *ccrB* have been identified. The presence of these genes and allotypes has been used to distinguish among the five different *ccr* types, which are currently observed (Table 25.1).

The region of the *mec* gene complex differs among SCC*mec* elements in its composition of regulatory genes (*mecI* and *mecR1*) and/or insertion sequences (IS431 and IS1272). So far, three classes of *mec* gene complexes have been described (A, B, and C) in *S. aureus* (Table 25.1).

These differences of *ccr* and *mec* gene complexes have been used to classify the SCC*mec* elements into different types by combining the class of the *mec* gene complex with the *ccr* allotype. To date, eight major types of SCC*mec* elements (I–VIII) have been reported in MRSA strains (Ito et al., 2009) (Table 25.1). In addition, the major elements have been further classified into subtypes by differences in regions other than *ccr* and *mec*, which are designated junction or junkyard

Table 25.1 Major SCC*mec* Elements Identified in *S. aureus*

ccr Gene Complex		*mec* Gene Complex		SCC*mec* Type
ccr Genes	*ccr* Type	*mec* Genes	*mec* Class	
ccrA1 and *ccrB1*	1	IS1272-Δ*mecR1*-*mecA*-IS431	B	I
ccrA2 and *ccrB2*	2	*mecI*-*mecR1*-*mecA*-IS431	A	II
ccrA3 and *ccrB3*	3	*mecI*-*mecR1*-*mecA*-IS431	A	III
ccrA2 and *ccrB2*	2	IS1272-Δ*mecR1*-*mecA*-IS431	B	IV
ccrC	5	IS431-Δ*mecR1*-*mecA*-IS431	C1[a]	V
ccrA4 and *ccrB4*	4	IS1272-Δ*mecR1*-*mecA*-IS431	B	VI
ccrC	5	IS431-Δ*mecR1*-*mecA*-IS431	C2[a]	VII
ccrA4 and *ccrB4*	4	*mecI*-*mecR1*-*mecA*-IS431	A	VIII

[a]*mec* class C1 and C2 differ in the orientation of IS431 upstream of *mecA*.
Source: From Ito et al. (2009).

(J) regions. It is likely that many other variants of SCC*mec* elements will be discovered with increasing typing, especially of isolates from poorly sampled geographical regions (e.g., Africa; Okon et al., 2009). In addition, coagulase-negative staphylococci (e.g., *S. haemoliticus*, *S. epidermidis*, *S. hominis*) contain a high diversity of SCC*mec* elements, which might serve as a potential reservoir for *S. aureus* (Miragaia et al., 2005; Hanssen and Sollid, 2007; Ibrahem et al., 2009).

The typing of *SCCmec* elements has become essential for several reasons. First, the SCCmec type is an important characteristic, in combination with the genotype of the *S. aureus* chromosome, for defining MRSA clones in epidemiological studies and to understand the evolution of these clones (Kondo et al., 2007). Second, the various SCC*mec* elements also differ in their patterns of antibiotic susceptibility, which could have important clinical implications. SCC*mec* type I as well as type IV−VIII cause only resistance to β-lactam antibiotics. In contrast, the largest SCC*mec* types II and III cause resistance to multiple classes of antibiotics due to the integration within these elements of plasmids or transposons carrying multiple resistance genes.

Several SCC*mec* typing methods have been developed, among which the most widely used are based on multiplex PCR assays that identify the different *ccr* types and *mec* classes (Oliveira and de Lencastre, 2002; Oliveira et al., 2006; Kondo et al., 2007; Milheirico et al., 2007). These have a limited number of targets, which may restrict their resolution but can be combined according to the level of discrimination required by the study. Two additional sequence-based typing methods based on the *ccr* gene complex have also been proposed (Lina et al., 2006; Oliveira et al., 2006), and these are likely to provide further useful data. Although SCC*mec* typing is essential for the characterization of MRSA clones in epidemiological studies, it is only recently that a rationalized nomenclature for the SCC*mec* has been proposed (Chongtrakool et al., 2006; Ito et al., 2009).

25.3 Molecular Epidemiology of MRSA

Epidemiological surveillance of MRSA has been greatly facilitated in recent years by the development of molecular typing procedures. The grouping of isolates into clones depends on the typing method used (e.g., PFGE, MLST, *spa*-typing) (Box 25.1). MLST provides probably the most robust subtyping system for *S. aureus* and MRSA

Box 25.1 Common Typing Methods for *S. aureus*

The epidemiology of *S. aureus* has been analyzed by an array of genotypic and phenotypic typing methods. Here, we review the methods that are currently the most widely used.

Pulsed-field gel electrophoresis: Pulsed-field gel electrophoresis (PFGE) is considered the gold standard for *S. aureus* typing because it shows the highest discriminatory power. This method is based on the restriction of whole DNA with an enzyme that cuts only rarely. The enzyme *SmaI* is generally used for *S. aureus*. Digestion with this enzyme gives between 20 and 50 large fragments (between 10 and 700 kb) that can only be separated using a pulsed gel electrophoresis. Although this method is reproducible within a laboratory, the data can be ambiguous (Blanc, 2004) and inter-laboratory studies have highlighted the problem of standardization (Murchan et al., 2003). PFGE standardization can only be obtained with a strict control of all parameters. For example, standardized protocols have been developed for PFGE typing by the American and Canadian CDCs to build nationwide databases (Mulvey et al., 2001; McDougal et al., 2003).

Multilocus sequence typing: Multilocus sequence typing (MLST) is a typing method that combines the sequence of several housekeeping genes, and is essentially a sequenced-based version of multilocus enzyme electrophoresis (MLEE) (Enright and Spratt, 1999). MLST has been designed to analyze and compare genetic variation in worldwide collections of bacterial pathogens. It gives important information about the nucleotide divergence of the core genome, the clonal origin of one group of strains, the recombination rate and the phylogenetic relationship among strains. The main advantage of this method is that it gives unambiguous data that are reproducible among laboratories. Its limitations are its cost and its relatively low discriminatory power that prevent its use for local epidemiology. For *S. aureus*, the amplification and the sequencing of 450–500 bp of the seven genes *arcC, aroE, glpF, gmk, pta, tpi* and *yqiL* have been retained (Enright et al., 2000). Alleles at each locus are assigned according to differences in nucleotide sequences. The allelic profile of the seven loci defines the ST. For example, isolates with the profile 2-3-1-1-4-4-3 belong to ST 239, of which the Brazilian clone is an example. An international database containing more than 3000 isolates and 1600 STs is available at http://www.mlst.net.

spa-typing: spa-typing is based on polymorphism of the *spa* locus of *S. aureus,* which codes for the protein A. This locus is highly polymorphic due to an internal variable region of short tandem repeats. It varies not only in numbers but also because of nucleotide substitutions within individual repeats. A *spa* profile is identified by a succession of number representing each individual repetition of the X region. An international database has been created to standardize the nomenclature of the *spa* types (http://spaserver.ridom.de). Several studies have shown the value of this method for *S. aureus* typing (Shopsin et al., 1999; Koreen et al., 2004; Hallin et al., 2007b). However, this method might reflect homoplasy (Nübel et al., 2008), its discriminatory power is below PFGE (Tang et al., 2000; Kuhn et al., 2007; Basset et al., 2009) and the analysis of *spa* data is not simple.

Double Locus Sequence Typing: Recently, we developed a new typing method called Double Locus Sequence Typing (DLST) based on the analysis of partial sequences (ca. 500 bp) of the highly variable *clfB* and *spa* genes (Kuhn et al., 2007). This method was shown to be far more discriminatory than *spa*-typing and matched the high resolution of PFGE. In addition, the combination of high typeability and reproducibility with low cost, ease of use and unambiguous definition of types makes this method promising for epidemiological analyses. It is important to note that although *spa*-typing and DLST investigate polymorphisms in the *spa* gene, these methods do not analyze the same regions of the gene. Therefore the *spa* alleles determined by these two methods are not identical.

clones are usually defined as all isolates belonging to the same ST and having the same SCC*mec* type (Enright et al., 2002). However, this method might be not discriminatory enough to differentiate among recognized clones. For instance, ST 239 includes EMRSA 1, 4, 11 and the Brazilian, Portuguese, Viennese, and Hungarian clones. ST 5 includes the New York/Japan and the Paediatric clones. These clones were differentiated by other typing methods such as PFGE, *spa*- and/or SCC*mec* typing (Aires de Sousa and de Lencastre, 2004). Similarly, two Swiss clones (clone D and G) showed identical STs, SCC*mec* elements, and virulence gene contents whereas they differed by 16 bands by PFGE (Blanc et al., 2007). However, we also note that in some cases PFGE appears to underestimate the level of relatedness between strains, as evidenced by other typing data. Because of these occasional inconsistencies, the pluralist approach (combining data from different techniques) will be the most powerful, and there is a need to address this issue in the design of future databases.

Many studies have demonstrated that high frequencies of MRSA within a given location tend to reflect the clonal spread of only one or two clones (e.g., Aparicio et al., 1992; de Lencastre et al., 1997; Leski et al., 1998; Oliveira et al., 1998; Melter et al., 1999; Deplano et al., 2000; Gomes et al., 2001; Aires de Sousa and de Lencastre, 2004; Blanc et al., 2007). The domain of dominance of specific clones can range in size from single hospitals, single countries, or even neighboring countries (Ayliffe, 1997; Aires de Sousa et al., 1998; Mato et al., 1998). Analysis of more than 3000 isolates from Southern Europe, the USA, and South America showed that nearly 70% of them belong to five major pandemic clones, namely the Iberian (ST247-SCC*mec* I), Brazilian (ST239-SCC*mec* III), Hungarian (ST239-SCC*mec* III), New York/Japan (ST5-SCC*mec* II), and Pediatric (ST5-SCC*mec* IV) clones (Oliveira et al., 2001, 2002; Aires de Sousa and de Lencastre, 2004). The addition of three more clones would essentially encompass Northern Europe: the EMRSA-15 (ST22-SCC*mec* IV), EMRSA-16 (ST36-SCC*mec* II), and Berlin (ST45-SCC*mec* IV) clones (Enright et al., 2002) (Figure 25.1). It is hypothesized that these clones are particularly transmissible and/or well adapted to the hospital environment (Cookson and Phillips, 1988; Blanc et al., 1999).

The epidemiology of MRSA is highly dynamic, and clonal replacement of predominant clones within a given locale has been widely documented. Whereas cross-sectional studies showed the predominance of one or two clones in a defined setting in the 1990s, several longitudinal studies showed the replacement of the predominant clones by others within a decade (Aires de Sousa and de Lencastre, 2004; Perez-Roth et al., 2004). A very early example was the replacement in England of EMRSA-1 (ST239) by EMRSA-15 and -16 (Cookson and Phillips, 1988). Other ST239 variants (e.g., in particular the Brazillian clone and the Hungarian clone) have subsequently become very widespread throughout South America, Eastern Europe, and mainland Asia (including both China and the Middle East), where this genotype may account for at least 90% of all cases of HA-MRSA. Additionally, the Iberian clone was replaced by another pandemic clone on at least two occasions. It was first replaced by EMRSA-16 in one Spanish hospital while the rate of MRSA among *S. aureus* remained constant (Perez-Roth et al., 2004), and by the Brazilian clone in one Portuguese hospital (Amorim et al., 2002). The fact that on both occasions the Iberian clone was replaced might suggest that it lost its epidemic potential during the last decade. Other examples are the complete replacement in a 2-year period of a local clone (ST5-SCC*mec* IV) by the New York/Japan clone (ST5-SCC*mec* II) in a Mexico City hospital (Velazquez-Meza et al., 2004) and the replacement of the the Berlin clone by a variant from the New York/Japan clone (ST105-SCC*mec* II) and by the South Germany clone (ST228-SCC*mec* I) in an area of low MRSA incidence in Western Switzerland (Blanc et al., 2007). Although the reasons why some clones replace others are typically unclear, the emergence and replacement of clones might have significant medical consequences as different clones possess differing resistance and virulence attributes (Blanc et al., 2001; Amorim et al., 2002; Rossney and Keane, 2002; Pantazatou et al., 2003; Denis et al., 2004; Perez-Roth et al., 2004; Velazquez-Meza et al., 2004). For instance, during the 1990s in France, the replacement of the Iberian clone (ST 247- SCC*mec* I) by the Lyon clone (ST8- SCC*mec* IV) resulted in a change of the susceptibility profile to antibiotics, the Iberian clone being less susceptible than the Lyon clone (e.g., to gentamicin and co-trimoxasole) (Blanc et al., 2001).

Perhaps the most interesting, and also worrying, example is the emergence of MRSA clones specific within the community. Up until the 1990s, MRSA was

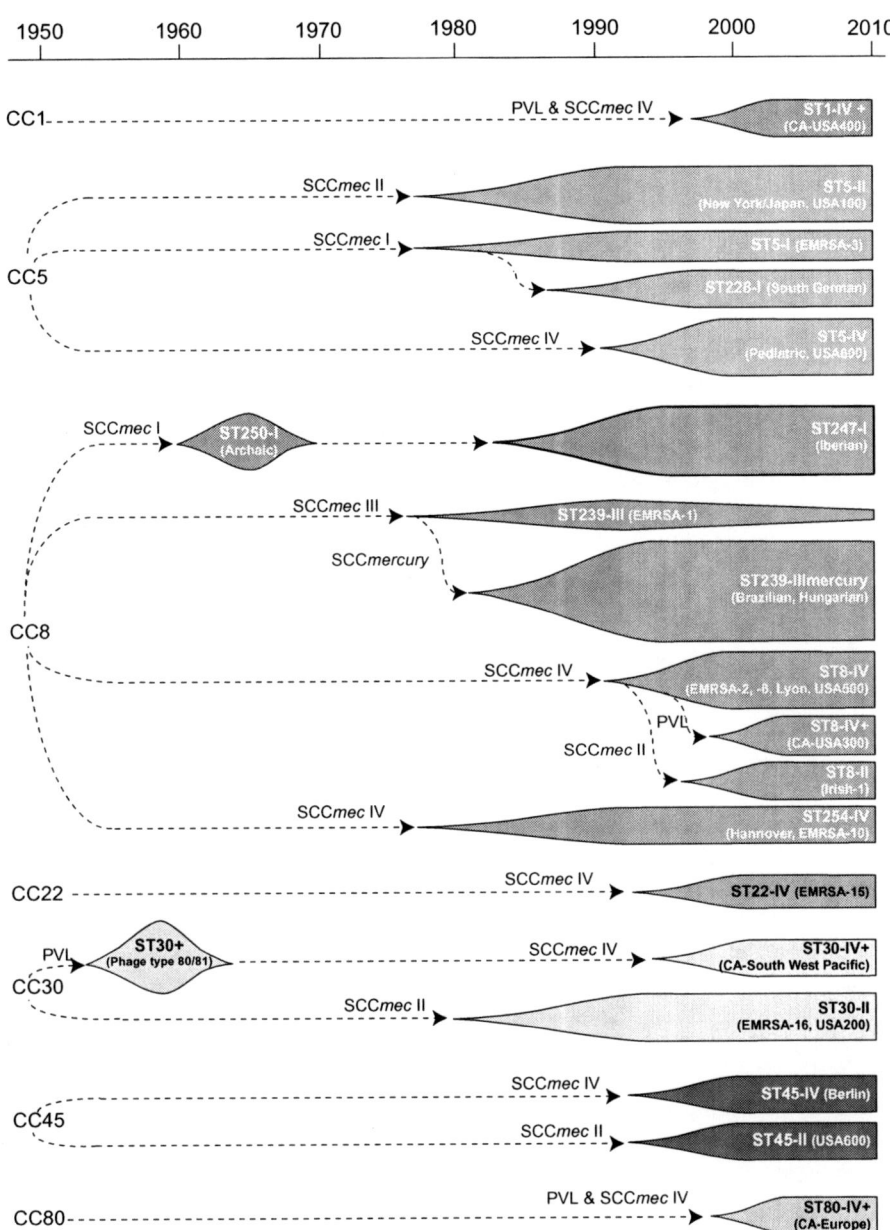

Figure 25.1 Tentative representation of the evolution in time of the major MRSA clones with the acquisition of different SCC*mec* elements and PVL genes.

found to be restricted to hospitals, but the 2000s have witnessed a dramatic increase in virulent MRSA clones in the community (CA-MRSA) (Dufour et al., 2002). These clones are generally characterized by the presence of a SCC*mec* type IV or V and the phage-borne genes encoding the Panton-Valentine leukocidin (PVL) toxin. This toxin is widely considered to be an important virulence factor, particularly for pediatric infection. Molecular typing has revealed that CA-MRSA clones are distinct from those noted in hospital settings (Aires de Sousa et al., 1998; Oliveira et al., 1998). ST80-SCC*mec* IV provides a notable example of an emerging CA-MRSA clone, which is currently restricted to several European communities with low social status (e.g., homeless people). Although the widespread HA-MRSA do not appear to have adapted to the community, it does appear that clones which emerge in the community may be able to spread in hospitals. For example, ST8-SCC*mec* IV (generally called the USA300 clone) was spread mainly in the USA, initially in the community but now also causing a major burden in hospital settings (Denis et al., 2005). In countries with low incidence of hospital MRSA such as northern European countries, CA-MRSA has become a major concern (Skov and Jensen, 2009).

The success of community-acquired MRSA clones is possibly related to the small size of the SCC*mec* types IV and V. The acquisition of antibiotic resistance is likely to confer a fitness cost to the cell which renders it uncompetitive against susceptible strains in the absence of the antibiotic. This trade-off is thought to be the reason why infection and carriage of HA-MRSA clones have remained largely confined to health-care settings. The smaller type IV and V SCC*mec* cassettes do not confer multiple resistances, but may also result in a smaller fitness cost. A useful analogy is to consider the large (type I, II, and III) SCC*mec* cassettes present within hospitals akin to a suit of armor (providing high protection but at a high cost), and the small type IV and V cassettes as a bullet-proof vest (possibly less protective but certainly less costly).

Although the epidemiological distinctions between CA-MRSA and HA-MRSA can be largely explained in terms of the fitness cost of resistance, the more general question of why a single MRSA clone can predominate in a given area, or the forces underlying clonal replacement, are far less well understood. It is probable that genetic differences underlie increased or decreased fitness (transmissibility) (Aparicio et al., 1992; Witte et al., 1994; Cookson and Phillips, 1988; Marples and Cooke, 1988; Blanc et al., 1999), and some general traits have been identified which may account for epidemic spread. These include the ability to survive in the environment, to colonize the host, to multiply on epithelial and mucosal surfaces, to "detach" from the host, and to resist to various antimicrobials. However, stochastic effects and extrinsic factors, such as local compliance to infection control measures and local use of antibiotics, may also have unpredictable consequences for the local composition of circulating MRSA clones. Furthermore, the specific genetic differences corresponding to fitness effects are very difficult to identify due to extensive gene redundancy and the possibility of subtle epistatic or regulatory effects playing a major role. The precise relationship between the "spread" (epidemicity) of a clone, and its virulence potential, is also unclear.

These complications can perhaps explain why a number of studies drawing comparisons between epidemic and sporadic MRSA have not generated clear experimental evidence consistent with the different epidemiological patterns (Peacock et al., 1981; Roberts and Gaston, 1987; Jordens et al., 1989; Duckworth and Jordens, 1990; Farrington et al., 1992; van Wamel et al., 1995; Wagenvoort and Penders, 1997). An exception is a recent study demonstrating differences in biofilm production and adhesion to epithelial cells within epidemic variants of the Brazilian clone (ST239-III) (Amaral et al., 2005). Although these laboratory comparisons were carried out on a small sample of strains, an epidemiological study also found evidence for increased virulence of an ST239 variant (TW20) which caused a recent outbreak in a London hospital (Edgeworth et al., 2007).

Molecular approaches have also not provided a clear understanding of epidemiological differences between clones. Population genetic analyses based on nucleotide sequence data of both housekeeping (MLST) genes and cell surface adhesion genes (which play a key role in host invasion) have also largely failed to detect robust links between genotype and epidemic phenotype (Kuhn et al., 2006). Comparative genome hybridization has also been used to compare epidemic and sporadic strains but this approach also failed to identify any genes likely to play a major role in increased transmission (Kuhn et al., 2010). These findings are strong evidence against the presence or absence of a single common specific factor differentiating epidemic from sporadic *S. aureus* clones.

Although the evidence linking genotype and epidemiological phenotype is in many cases weak, there are tantalizing clues. For example, the CA-MRSA strain USA300 has disseminated widely throughout the USA within the last 5 years. Genome sequencing of this strain revealed a novel genetic element, the arginine catabolic mobile element (ACME), which contained the gene for the arginine deiminase which may play a crucial role in the growth and survival of the bacterial cells (Diep et al., 2006; Highlander et al., 2007). In addition, genome sequencing of 10 other isolates from the same disseminating clone confirmed its recent expansion (Kennedy et al., 2008).

25.4 Evolution of *Staphylococcus aureus* and MRSA

Most detailed studies on the population genetics of *S. aureus* have been performed using MLST (Box 25.1). Based on MLST data, the population of *S. aureus* was classified into related groups of strains defined as clonal complexes (CCs) and isolated sequence types (STs) (Enright et al., 2002). These CCs are considered as different genetic lineages within the *S. aureus* population and only few differences are detected within groups although the characteristics of MGEs (e.g., SCC*mec*) may vary substantially (Nübel et al., 2008).

Extensive typing showed that the *S. aureus* population associated with humans consists of ten major lineages (i.e., CC1, CC5, CC8, CC12, CC15, CC22, CC25, CC30, CC45, and CC51) as well as several other minor lineages (Figure 25.2)

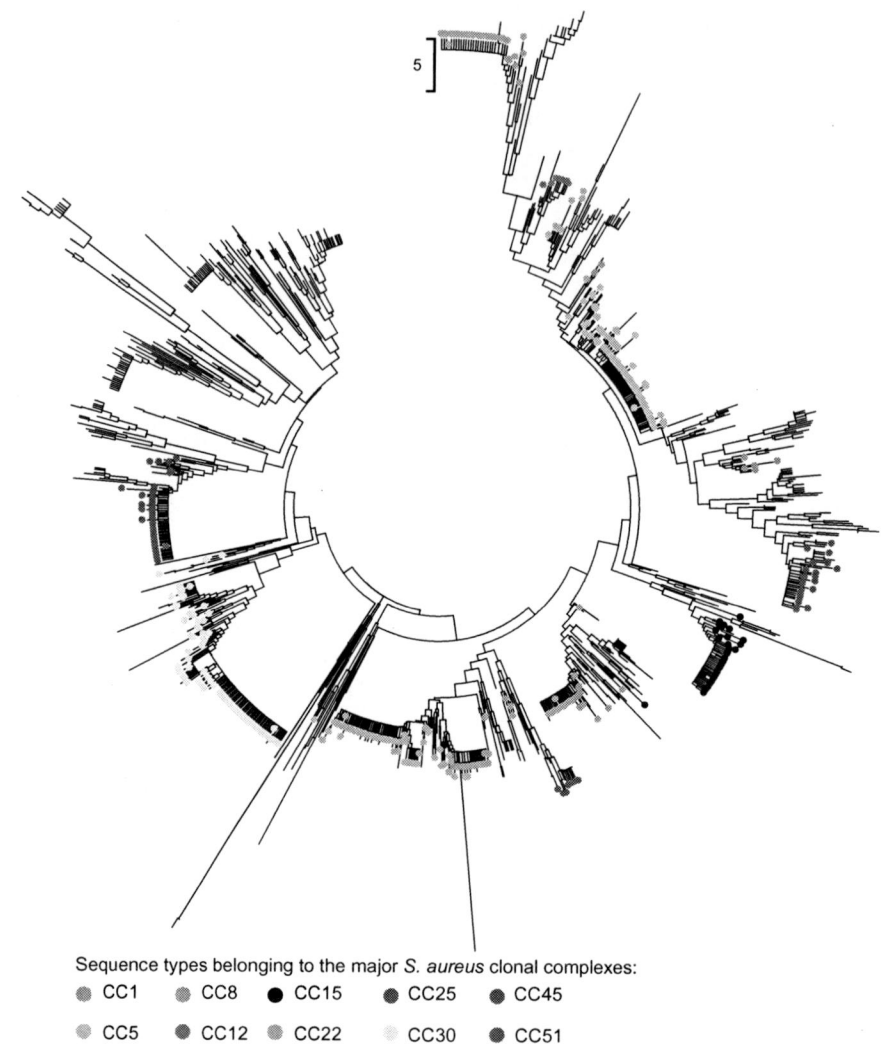

Figure 25.2 Neighbor-joining tree of the concatenated sequence of all the STs available in the *S. aureus* MLST database (http://saureus.mlst.net/). The position of the STs belonging to major *S. aureus* CCs is indicated by a colored dot. (For interpretation of the references to color in this figure legend, the reader is referred to the web version of this book.)

(Enright et al., 2002; Oliveira et al., 2002; Feil et al., 2003). These lineages have not only been identified using MLST but also using other categories of genes (Robinson and Enright, 2003) confirming the biological reality of the CCs. These CCs generally have a radial genetic structure with a founder ST surrounded by numerous single locus variants of the founder. This observation highlights that with the exception of MGEs the genetic diversity within each lineage is remarkably low

(Nübel et al., 2008; Harris et al., 2010). For example, the nonmobile genome of two strains belonging to CC1 (MW2 and MSSA476) differ at only 285 SNPs despite one was a PVL positive MRSA isolated in the USA and the other a PVL negative MSSA isolated in the UK (Lindsay, 2010). The low variability observed within CCs might be explained by recent expansion and/or strong purifying selection. Although the relative contribution of each of these factors is difficult to disentangle, purifying selection was described for several categories of genes such as the seven housekeeping genes used for MLST typing, core and accessory adhesion genes (Kuhn et al., 2006) as well as many others (Hughes and Friedman, 2005; Cooper and Feil, 2006; Sabat et al., 2008). This suggests that purifying selection is an important factor that acts on the chromosome of *S. aureus* and it probably affects the diversity observed within CCs.

Another important factor is the low rate of homologous recombination within the core genome of *S. aureus*. Although several chromosomal replacement events were described for *S. aureus* (Robinson and Enright, 2004; Narra and Ochman, 2006), this species has been shown to be highly clonal using a variety of genes: MLST (Feil et al., 2003), cell surface *sas* genes (Robinson and Enright, 2003), cell surface core and accessory (i.e., not present in all the strains) adhesion genes (Kuhn et al., 2006), accessory exotoxin-like genes (Fitzgerald et al., 2003). For example, using MLST data, it was shown that genetic differences between a single locus variant and its ancestral strains were created 15 times more frequently by a point mutation than by a recombination event (Feil et al., 2003). This low rate of recombination can help to explain why the CCs have remained discrete and coherent in the *S. aureus* population, and why the same basic groups tend to be defined regardless of the genes used for typing (with the notable exception of *agr*; Robinson et al., 2005).

The genetic diversity of MRSA is also known to be much smaller than MSSA (Grundmann et al., 2010) and the most common MRSA isolates belong to only six CCs (i.e., CC1, CC5, CC8, CC22, CC30, and CC45). In contrast to MSSA, the genetic diversity of MRSA differs considerably among countries and dominant MRSA lineages form distinctive geographical clusters, at least in Europe (Grundmann et al., 2010). This largely reflects the recent origin of many MRSA clones, that is, since the first administration of methicillin in 1961. This means that there has been insufficient time for the MRSA clones to fully homogenize geographically. The fitness cost associated with resistance is also likely to have limited the transmission of these clones outside of the hospital environment.

For regions outside of Europe, North America, and Australia the picture may be different. For example, ST239, which probably emerged in the mid-1960s (Harris et al., 2010) is a probable major cause of HA-MRSA infection throughout mainland Asia and South America, a geographical region that holds more than 50% of the world's human population (Feil et al., 2008). This ST always exhibits a variant of the large SCC*mec* type III and to our knowledge there are no cases of MSSA ST239 in the literature. ST239 is rarely found outside of the hospital environment which makes its rapid global dissemination, which must have occurred largely through very short transmission chains between hospitals, even more remarkable.

25.4.1 Mobile Genetic Elements

Bacterial genomes can be viewed as two compartments of genes, one comprising "core" genes that are ubiquitously present in all clones of a given species, and the other comprising "accessory" genes that are not necessarily present in all clones and that have the possibility to be transferred between strains (Lan and Reeves, 2000). Recently, comparison of whole genome sequences as well as comparative genome hybridization experiments showed that about 75% of the *S. aureus* genome is present in more than 95% of the strains (i.e., core genome) (Fitzgerald et al., 2001; Lindsay and Holden, 2004; El Garch et al., 2009). As expected, the majority of genes comprising the core genome are composed of species specific genes and genes associated with central metabolism and other housekeeping function. In contrast, the gene content of the remaining 25% of the genome varies significantly among strains (i.e., accessory genome).

The accessory genome mostly consists of MGEs, such as bacteriophages, pathogenicity islands, genomic islands, SCCs, plasmids, or transposons. Many of these genetic elements carry genes encoding for protein associated with virulence (e.g., *sea*, *tst*, and *eta*) (Betley and Mekalanos, 1985; Lindsay et al., 1998; Moore and Lindsay, 2001) and resistance to antibiotics (e.g., *mecA*) (Enright et al., 2002). The gene content of a particular *S. aureus* strain is thus a combination of (i) vertical inheritance of its core genome, and (ii) horizontal transfer of MGEs, allowing rapid adaptation by loss or gain of virulence and/or resistance genes (Lindsay and Holden, 2004). Thus, there is a considerable proportion of the genome that is not essential for survival and that contributes to genetic differences between strains. The distribution and horizontal spread of these elements can have important clinical implications and the characterization of these elements is providing insights into how *S. aureus* is evolving and cause diseases.

Whole genome comparisons indicated important variation in the distribution of genomic islands. This suggests that MGEs are readily exchanged in the *S. aureus* population. For example, genome comparison of one MRSA strain with one MSSA strain showed at least 5 different acquisition/loss events involved in differences in virulence factors and drug resistance (Holden et al., 2004). Horizontal transfer of MGEs is also suggested by the phylogenetic distribution of these elements, which does not correlate with the phylogenetic genetic relatedness inferred by MLST. This lack of correlation suggests that mobile elements facilitate the exchange of virulence and antibiotic resistance determinants between *S. aureus* lineages and may lead to rapid changes in the pathogenic potential or drug resistance of strains. In contrast, the sequence of the core genome is remarkably constant (Kuroda et al., 2001; Holden et al., 2004; Ohta et al., 2004; Highlander et al., 2007; Sivaraman et al., 2008).

Several studies suggested that some toxin genes (e.g., toxic shock syndrome toxin 1 (*tst*), leukocidin DE (*lukDE*) and superantigens [*sea*, *seg*, and *sei*]) are associated with particular lineages (MLST CCs) (Moore and Lindsay, 2001; Peacock et al., 2002), and there is evidence of frequent acquisition and loss of particular elements that is restricted to particular CCs. Variability of accessory genes such as resistance, toxin or virulence genes, has recently been described within two STs (ST5 and ST228) of CC5 (Monecke et al., 2009). However, the biological significance and modalities of this intra-strain variability still need to be clarified.

The importance of understanding the patterns of evolution of MGEs is illustrated by the evolution of the SCC*mec* element. The evolution from MSSA to MRSA involves the acquisition of a SCC*mec* element by an MSSA strain. The exact mechanisms explaining how the SCC*mec* elements enter the *S. aureus* cell are not clearly known. However, the transduction by phages is often postulated (de Lencastre et al., 2007). The frequency of transfer of SCC*mec* elements as well as their geographical history is also poorly known. Identical clones have been sampled in different countries suggesting a single *SCCmec* acquisition, followed by clonal spread. Yet, the presence of multiple SCC*mec* types in MRSA suggests multiple introductions into *S. aureus*. Moreover, the occurrence of isolates with identical ST but with different *SCCmec* types indicates that horizontal transfer of SCC*mec* elements is relatively frequent within *S. aureus* (Enright et al., 2002). Using MLST, it has been shown that the SCC*mec* element must have been acquired on multiple occasions (at least 20 times) during *S. aureus* evolution (Enright et al., 2002; Robinson and Enright, 2003; Lina et al., 2006). Recently, a study based on SNPs discovery on a worldwide collection of ST 5 showed a close association between phylogenetic lineages and geography (Nübel et al., 2008). These data suggest that geographical spread of MRSA over long distance is a rare event compared with the frequency with which the *SCCmec* is imported locally. Moreover, MSSA strains genetically identical to the predominant MRSA strains have been observed at a local level (Enright et al., 2000; Hallin et al., 2007a), confirming the possibility of local acquisitions of the SCC*mec*.

25.5 Conclusion

The widespread occurrence of MRSA in hospitals is recognized as a major challenge, especially with the recent emergence of strains with intermediate susceptibility to glycopeptides and of community-acquired MRSA. Given the difficulties to control MRSA, a thorough understanding of the processes underlying the emergence and spread of MRSA may help designing new strategies to counteract this evolution. Several major pandemic clones have been identified and their epidemiology may change rapidly at a regional scale. Changes in clones have significant medical consequences, since the new clones often display different antibiotic susceptibility and/or virulence patterns.

With the advance of recent sequencing technologies and development of associated bioinformatics tools, it will be possible to analyze and compare several genomes in a short period of time. These data will allow addressing many important questions about the evolution and epidemiology of MRSA and will bridge the gap left by the low discriminatory power of MLST.

References

Aires de Sousa, M., de Lencastre, H., 2004. Bridges from hospitals to the laboratory: genetic portraits of methicillin-resistant *Staphylococcus aureus* clones. FEMS Immunol. Med. Microbiol. 40, 101–111.

Aires de Sousa, M., Santos Sanches, I., Ferro, M.L., Vaz, M.J., Saraiva, Z., Tendeiro, T., et al., 1998. International spread of multidrug-resistant methicillin-resistant *Staphylococcus aureus* clone. J. Clin. Microbiol. 36, 2590–2596.

Amaral, M.M., Coelho, L.R., Flores, R.P., Souza, R.R., Silva-Carvalho, M.C., Teixeira, L.A., et al., 2005. The predominant variant of the Brazilian epidemic clonal complex of methicillin-resistant *Staphylococcus aureus* has an enhanced ability to produce biofilm and to adhere to and invade airway epithelial cells. J. Infect. Dis. 192, 801–810.

Amorim, M.L., Ires de, S.M., Sanches, I.S., Sa-Leao, R., Cabeda, J.M., Amorim, J.M., et al., 2002. Clonal and antibiotic resistance profiles of methicillin-resistant *Staphylococcus aureus* (MRSA) from a Portuguese hospital over time. Microb. Drug Resist. 8, 301–309.

Aparicio, P., Richardson, J., Martin, S., Vindel, A., Marples, R.R., Cookson, B.D., 1992. An epidemic methicillin-resistant strain of *Staphylococcus aureus* in Spain. Epidemiol. Infect. 108, 287–298.

Ayliffe, G.A., 1997. The progressive intercontinental spread of methicillin-resistant *Staphylococcus aureus*. Clin. Infect. Dis. 24 (Suppl 1), S74–S79.

Basset, P., Hammer, N.B., Kuhn, G., Vogel, V., Sakwinska, O., Blanc, D.S., 2009. *Staphylococcus aureus clfB* and *spa* alleles of the repeat regions are segregated into major phylogenetic lineages. Infect. Genet. Evol. 9, 941–947.

Betley, M.J., Mekalanos, J.J., 1985. Staphylococcal enterotoxin A is encoded by phage. Science 229, 185–187.

Blanc, D.S., 2004. The use of molecular typing for the epidemiological surveillance and investigation of endemic nosocomial infections. Infect. Genet. Evol. 4, 193–197.

Blanc, D.S., Petignat, C., Moreillon, P., Entenza, J., Eisenring, M.C., Kleiber, H., et al., 1999. Unusual spread of a penicillin-susceptible methicillin-resistant *Staphylococcus aureus* clone in a geographic area of low incidence. Clin. Infect. Dis. 29, 1512–1518.

Blanc, D.S., Francioli, P., Le Coustumier, A., Gazagne., L., Lecaillon, E., Gueudet, P., et al., 2001. Reemergence of gentamicin-susceptible strains of methicillin-resistant *Staphylococcus aureus* in France: a phylogenetic approach. J. Clin. Microbiol. 39, 2287–2290.

Blanc, D.S., Petignat, C., Wenger, A., Kuhn, G., Vallet, Y., Fracheboud, D., et al., 2007. Changing molecular epidemiology of methicillin-resistant *Staphylococcus aureus* in a small geographic area over an eight-year period. J. Clin. Microbiol. 45, 3729–3736.

Chongtrakool, P., Ito, T., Ma, X.X., Kondo, Y., Trakulsomboon, S., Tiensasitorn, C., et al., 2006. Staphylococcal cassette chromosome mec (*SCCmec*) typing of methicillin-resistant *Staphylococcus aureus* strains isolated in 11 Asian countries: a proposal for a new nomenclature for *SCCmec* elements. Antimicrob. Agents Chemother. 50, 1001–1012.

Cookson, B.D., Phillips, I., 1988. Epidemic methicillin-resistant *Staphylococcus aureus*. J. Antimicrob. Chemother. 21 (Suppl C), 57–65.

Cooper, J.E., Feil, E.J., 2006. The phylogeny of *Staphylococcus aureus*—which genes make the best intra-species markers? Microbiology-Sgm 152, 1297–1305.

Couto, I., Wu, S.W., Tomasz, A., de Lencastre, H., 2003. Development of methicillin resistance in clinical isolates of *Staphylococcus sciuri* by transcriptional activation of the mecA homologue native to the species. J. Bacteriol. 185, 645–653.

de Lencastre, H., Severina, E.P., Milch, H., Konkoly Thege, M., Tomasz, A., 1997. Wide geographic discribution of a unique methicillin-resistant *Staphylococcus aureus* clone in Hungarian hospitals. Clin. Microbiol. Infect. 3, 289–296.

de Lencastre, H., Oliveira, D., Tomasz, A., 2007. Antibiotic resistant *Staphylococcus aureus*: a paradigm of adaptive power. Curr. Opin. Microbiol. 10, 428–435.

Denis, O., Deplano, A., Nonhoff, C., de, R.R., de, M.R., Rottiers, S., et al., 2004. National surveillance of methicillin-resistant *Staphylococcus aureus* in Belgian hospitals indicates rapid diversification of epidemic clones. Antimicrob. Agents Chemother. 48, 3625–3629.

Denis, O., Deplano, A., De, B.H., Hallin, M., Huysmans, G., Garrino, M.G., et al., 2005. Polyclonal emergence and importation of community-acquired methicillin-resistant *Staphylococcus aureus* strains harbouring Panton-Valentine leucocidin genes in Belgium. J. Antimicrob. Chemother. 56, 1103–1106.

Deplano, A., Witte, W., van Leeuwen, W.J., Brun, Y., Struelens, M.J., 2000. Clonal dissemination of epidemic methicillin-resistant *Staphylococcus areus* in Belgium and neighbouring countries. Clin. Microbiol. Infect. 6, 239–245.

Diep, B.A., Gill, S.R., Chang, R.F., Phan, T.H., Chen, J.H., Davidson, M.G., et al., 2006. Complete genome sequence of USA300, an epidemic clone of community-acquired methicillin-resistant *Staphylococcus aureus*. The Lancet 367, 731–739.

Duckworth, G.J., Jordens, J.Z., 1990. Adherence and survival properties of an epidemic methicillin-resistant strains of *Staphylococcus aureus* compared with those of methicillin-sensitive strains. J. Med. Microbiol. 32, 195–200.

Dufour, P., Gillet, Y., Bes, M., Lina, G., Vandenesch, F., Floret, D., et al., 2002. Community-acquired methicillin-resistant *Staphylococcus aureus* infections in France: emergence of a single clone that produces Panton-Valentine leukocidin. Clin. Infect. Dis. 35, 819–824.

Edgeworth, J.D., Yadegarfar, G., Pathak, S., Batra, R., Cockfield, J.D., Wyncoll, D., et al., 2007. An outbreak in an intensive care unit of a strain of methicillin-resistant *Staphylococcus aureus* sequence type 239 associated with an increased rate of vascular access device-related bacteremia. Clin. Infect. Dis. 44, 493–501.

El Garch, F., Hallin, M., De Mendonca, R., Denis, O., Lefort, A., Struelens, M.J., 2009. StaphVar-DNA microarray analysis of accessory genome elements of community-acquired methicillin-resistant *Staphylococcus aureus*. J. Antimicrob. Chemother. 63, 877–885.

Enright, M.C., Spratt, B.G., 1999. Multilocus sequence typing. Trends Microbiol. 7, 482–487.

Enright, M.C., Day, N.P., Davies, C.E., Peacock, S.J., Spratt, B.G., 2000. Multilocus sequence typing for characterization of methicillin-resistant and methicillin-susceptible clones of *Staphylococcus aureus*. J. Clin. Microbiol. 38, 1008–1015.

Enright, M.C., Robinson, D.A., Randle, G., Feil, E.J., Grundmann, H., Spratt, B.G., 2002. The evolutionary history of methicillin-resistant *Staphylococcus aureus* (MRSA). Proc. Natl. Acad. Sci. U.S.A. 99, 7687–7692.

Farrington, M., Brenwald, N., Haines, D., Walpole, E., 1992. Resistance to desiccation and skin fatty acids in outbreak strains of methicillin-resistant *Staphylococcus aureus*. J. Med. Microbiol. 36, 56–60.

Feil, E.J., Cooper, J.E., Grundmann, H., Robinson, D.A., Enright, M.C., Berendt, T., et al., 2003. How clonal is *Staphylococcus aureus*? J. Bacteriol. 185, 3307–3316.

Feil, E.J., Nickerson, E.K., Chantratita, N., Wuthiekanun, V., Srisomang, P., Cousins, R., et al., 2008. Rapid detection of the pandemic methicillin-resistant *Staphylococcus aureus* clone ST 239, a dominant strain in Asian hospitals. J. Clin. Microbiol. 46, 1520–1522.

Fitzgerald, J.R., Sturdevant, D.E., Mackie, S.M., Gill, S.R., Musser, J.M., 2001. Evolutionary genomics of *Staphylococcus aureus*: insights into the origin of methicillin-resistant strains and the toxic shock syndrome epidemic. Proc. Natl. Acad. Sci. U.S.A. 98, 8821−8826.

Fitzgerald, J.R., Reid, S.D., Ruotsalainen, E., Tripp, T.J., Liu, M., Cole, R., et al., 2003. Genome diversification in *Staphylococcus aureus*: molecular evolution of a highly variable chromosomal region encoding the staphylococcal exotoxin-like family of proteins. Infect. Immun. 71, 2827−2838.

Gomes, A.R., Sanches, I.S., ires de, S.M., Castaneda, E., de, L.H., 2001. Molecular epidemiology of methicillin-resistant *Staphylococcus aureus* in Colombian hospitals: dominance of a single unique multidrug-resistant clone. Microb. Drug Resist. 7, 23−32.

Grundmann, H., Aanensen, D.M., van den Wijngaard, C.C., Spratt, B.G., Harmsen, D., Friedrich, A.W., 2010. Geographic distribution of *Staphylococcus aureus* causing invasive infections in Europe: a molecular-epidemiological analysis. Plos Medicine 7(1):e1000215.

Hallin, M., Denis, O., Deplano, A., De Mendonca, R., De Ryck, R., Rottiers, S., et al., 2007a. Genetic relatedness between methicillin-susceptible and methicillin-resistant *Staphylococcus aureus*: results of a national survey. J. Antimicrob. Chemother. 59, 465−472.

Hallin, M., Deplano, A., Denis, O., de, M.R., de, R.R., Struelens, M.J., 2007b. Validation of pulsed-field gel electrophoresis and spa typing for long-term, nationwide epidemiological surveillance studies of *Staphylococcus aureus* infections. J. Clin. Microbiol. 45, 127−133.

Hanssen, A.M., Sollid, J.U.E., 2007. Multiple staphylococcal cassette chromosomes and allelic variants of cassette chromosome recombinases in *Staphylococcus aureus* and coaaulase-neizative staphylococci from Norway. Antimicrob. Agents Chemother. 51, 1671−1677.

Harris, S.R., Feil, E.J., Holden, M.T.G., Quail, M.A., Nickerson, E.K., Chantratita, N., et al., 2010. Evolution of MRSA during hospital transmission and intercontinental spread. Science 327, 469−474.

Highlander, S.K., Hulten, K.G., Qin, X., Jiang, H., Yerrapragada, S., et al., 2007. Subtle genetic changes enhance virulence of methicillin resistant and sensitive *Staphylococcus aureus*. BMC Microbiol. 7, 99.

Hiramatsu, K., Cui, L., Kuroda, M., Ito, T., 2001. The emergence and evolution of methicillin-resistant *Staphylococcus aureus*. Trends Microbiol. 9, 486−493.

Holden, M.T., Feil, E.J., Lindsay, J.A., Peacock, S.J., Day, N.P., et al., 2004. Complete genomes of two clinical *Staphylococcus aureus* strains: evidence for the rapid evolution of virulence and drug resistance. Proc. Natl. Acad. Sci. U.S.A. 101, 9786−9791.

Hughes, A.L., Friedman, R., 2005. Nucleotide substitution and recombination at orthologous loci in *Staphylococcus aureus*. J. Bacteriol. 187, 2698−2704.

Ibrahem, S., Salmenlinna, S., Virolainen, A., Kerttula, A.M., Lyytikainen, O., Jagerroos, H., et al., 2009. Carriage of methicillin-resistant Staphylococci and their SCCmec types in a long-term-care facility. J. Clin. Microbiol. 47, 32−37.

Ito, T., Hiramatsu, K., Oliveira, D.C., Lencastre, H., Zhang, K.Y., Westh, H., et al., 2009. Classification of staphylococcal cassette chromosome mec (SCCmec): guidelines for reporting novel SCCmec elements. Antimicrob. Agents Chemother. 53, 4961−4967.

Jordens, J.Z., Duckworth, G.J., Williams, R.J., 1989. Production of "virulent factors" by "epidemic" methicillin-resistant *Staphylococcus aureus in vitro*. J. Med. Microbiol. 30, 245−252.

Katayama, Y., Ito, T., Hiramatsu, K., 2000. A new class of genetic element, Staphylococcus cassette chromosome mec, encodes methicillin resistance in *Staphylococcus aureus*. Antimicrob. Agents Chemother. 44, 1549−1555.

Kennedy, A.D., Otto, M., Braughton, K.R., Whitney, A.R., Chen, L., Mathema, B., et al., 2008. Epidemic community-associated methicillin-resistant *Staphylococcus aureus*: recent clonal expansion and diversification. Proc. Natl. Acad. Sci. U.S.A. 105, 1327−1332.

Kondo, Y., Ito, T., Ma, X.X., Watanabe, S., Kreiswirth, B.N., Etienne, J., et al., 2007. Combination of multiplex PCRs for staphylococcal cassette chromosome mec type assignment: rapid identification system for *mec*, *ccr*, and major differences in junkyard regions. Antimicrob. Agents Chemother. 51, 264−274.

Koreen, L., Ramaswamy, S.V., Graviss, E.A., Naidich, S., Musser, J.M., Kreiswirth, B.N., 2004. Spa typing method for discriminating among *Staphylococcus aureus isolates*: implications for use of a single marker to detect genetic micro- and macrovariation. J. Clin. Microbiol. 42, 792−799.

Kuhn, G., Francioli, P., Blanc, D.S., 2006. Evidence for clonal evolution among highly polymorphic genes in methicillin-resistant *Staphylococcus aureus*. J. Bacteriol. 188, 169−178.

Kuhn, G., Francioli, P., Blanc, D.S., 2007. Double-locus sequence typing using clfB and spa, a fast and simple method for epidemiological typing of methicillin-resistant *Staphylococcus aureus*. J. Clin. Microbiol. 45, 54−62.

Kuhn, G., Koessler, T., Melles, D.C., Francois, P., Huyghe, A., Dunman, P., et al., 2010. Comparative genomics of epidemic versus sporadic *Staphylococcus aureus* strains does not reveal molecular markers for epidemicity. Infect. Genet. Evol. 10, 89−96.

Kuroda, M., Ohta, T., Uchiyama, I., Baba, T., Yuzawa, H., et al., 2001. Whole genome sequencing of methicillin-resistant *Staphylococcus aureus*. Lancet 357, 1225−1240.

Lan, R.T., Reeves, P.R., 2000. Intraspecies variation in bacterial genomes: the need for a species genome concept. Trends Microbiol. 8, 396−401.

Leski, T., Oliveira, D., Trzcinski, K., Sanches, I.S., de Sousa, M.A., Hryniewicz, W., et al., 1998. Clonal distribution of methicillin-resistant *Staphylococcus aureus* in Poland. J. Clin. Microbiol. 36, 3532−3539.

Lina, G., Durand, G., Berchich, C., Short, B., Meugnier, H., Vandenesch, F., et al., 2006. Staphylococcal chromosome cassette evolution in *Staphylococcus aureus* inferred from ccr gene complex sequence typing analysis. Clin. Microbiol. Infect. 12, 1175−1184.

Lindsay, J.A., 2010. Genomic variation and evolution of *Staphylococcus aureus*. Int. J. Med. Microbiol. 300, 98−103.

Lindsay, J.A., Holden, M.T., 2004. *Staphylococcus aureus:* superbug, super genome? Trends Microbiol. 12, 378−385.

Lindsay, J.A., Ruzin, A., Ross, H.F., Kurepina, N., Novick, R.P., 1998. The gene for toxic shock toxin is carried by a family of mobile pathogenicity islands in *Staphylococcus aureus*. Mol. Microbiol. 29, 527−543.

Marples, R.R., Cooke, E.M., 1988. Current problems with methicillin-resistant *Staphylococcus aureus*. J. Hospital Infect. 11, 381−392.

Mato, R., Sanches, S., Venditti, M., Platt, D.J., Brown, A., Chung, M., et al., 1998. Spread of the multiresistant Iberian clone of methicillin-resistant *Staphylococcus aureus* (MRSA) to Italy and Scotland. Microb. Drug Resist. 4, 107−112.

Matsuhashi, M., Song, M.D., Ishino, F., Wachi, M., Doi, M., Inoue, M., et al., 1986. Molecular cloning of the gene of a penicillin-binding protein supposed to cause high resistance to beta-lactam antibiotics in *Staphylococcus aureus*. J. Bacteriol. 167, 975−980.

McDougal, L.K., Steward, C.D., Killgore, G.E., Chaitram, J.M., McAllister, S.K., Tenover, F.C., 2003. Pulsed-field gel electrophoresis typing of oxacillin-resistant *Staphylococcus aureus* isolates from the United States: establishing a national database. J. Clin. Microbiol. 41, 5113–5120.

Melter, O., Santos, S.I., Schindler, J., Aires, D.S., Mato, R., Kovarova, V., et al., 1999. Methicillin-resistant *Staphylococcus aureus* clonal types in the Czech Republic. J. Clin. Microbiol. 37, 2798–2803.

Milheirico, C., Oliveira, D.C., de Lencastre, H., 2007. Update to the multiplex PCR strategy for assignment of mec element types in *Staphylococcus aureus*. Antimicrob. Agents Chemother. 51, 4537.

Miragaia, M., Couto, I., De Lancastre, H., 2005. Genetic diversity among methicillin-resistant Staphylococcus epidermidis (MRSE). Microb. Drug Resist. Mech. Epidemiol. Dis. 11, 83–93.

Monecke, S., Ehricht, R., Slickers, P., Wiese, N., Jonas, D., 2009. Intra-strain variability of methicillin-resistant *Staphylococcus aureus* strains ST228-MRSA-I and ST5-MRSA-II. Eur. J. Clin. Microbiol. Infect. Dis. 28, 1383–1390.

Moore, P.C., Lindsay, J.A., 2001. Genetic variation among hospital isolates of methicillin-sensitive *Staphylococcus aureus*: evidence for horizontal transfer of virulence genes. J. Clin. Microbiol. 39, 2760–2767.

Mulvey, M.R., Chui, L., Ismail, J., Louie, L., Murphy, C., Chang, N., et al., 2001. Development of a Canadian standardized protocol for subtyping methicillin-resistant *Staphylococcus aureus* using pulsed-field gel electrophoresis. J. Clin. Microbiol. 39, 3481–3485.

Murchan, S., Kaufmann, M.E., Deplano, A., de, R.R., Struelens, M., Zinn, C.E., et al., 2003. Harmonization of pulsed-field gel electrophoresis protocols for epidemiological typing of strains of methicillin-resistant *Staphylococcus aureus*: a single approach developed by consensus in 10 European laboratories and its application for tracing the spread of related strains. J. Clin. Microbiol. 41, 1574–1585.

Narra, H.P., Ochman, H., 2006. Of what use is sex to bacteria? Curr. Biol. 16, R705–R710.

Nübel, U., Roumagnac, P., Feldkamp, M., Song, J.H., Ko, K.S., Huang, Y.C., et al., 2008. Frequent emergence and limited geographic dispersal of methicillin-resistant *Staphylococcus aureus*. Proc. Natl. Acad. Sci. U.S.A. 105, 14130–14135.

Ohta, T., Hirakawa, H., Morikawa, K., Maruyama, A., Inose, Y., Yamashita, A., et al., 2004. Nucleotide substitutions in *Staphylococcus aureus* strains, Mu50, Mu3, and N315. DNA Res. 11, 51–56.

Okon, K.O., Basset, P., Uba, A., Lin, J., Oyawoye, B., Shittu, A.O., et al., 2009. Cooccurrence of predominant Panton-Valentine leukocidin-positive sequence type (ST) 152 and multidrug-resistant ST 241 *Staphylococcus aureus* clones in nigerian hospitals. J. Clin. Microbiol. 47, 3000–3003.

Oliveira, D.C., de Lencastre, H., 2002. Multiplex PCR strategy for rapid identification of structural types and variants of the mec element in methicillin-resistant *Staphylococcus aureus*. Antimicrob. Agents Chemother. 46, 2155–2161.

Oliveira, D., Santos Sanches, I., Mato, R., Tamayo, M., Ribeiro, G., Costa, D., et al., 1998. Virtuallay all methicillin-resistant *Staphylococcus aureus* (MRSA) infections in the largest Portuguese teaching hospital are caused by two internationally spread multiresistant strains: the "Iberian" and the "Brazilian" clones of MRSA. Clin. Microbiol. Infect. 4, 373–384.

Oliveira, D.C., Tomasz, A., de Lencastre, H., 2001. The evolution of pandemic clones of methicillin-resistant *Staphylococcus aureus*: identification of two ancestral genetic backgrounds and the associated mec elements. Microb. Drug Resist 7, 349–361.

Oliveira, D.C., Tomasz, A., de Lencastre, H., 2002. Secrets of success of a human pathogen: molecular evolution of pandemic clones of meticillin-resistant *Staphylococcus aureus*. Lancet Infect. Dis. 2, 180–189.

Oliveira, D.C., Milheirico, C., Vinga, S., de Lencastre, H., 2006. Assessment of allelic variation in the ccrAB locus in methicillin-resistant *Staphylococcus aureus* clones. J. Antimicrob. Chemother. 58, 23–30.

Pantazatou, A., Papaparaskevas, J., Stefanou, I., Papanicolas, J., Demertzi, E., Avlamis, A., 2003. Changes in the epidemiology of methicillin-resistant *Staphylococcus aureus* in a Greek tertiary care hospital, over an 8-year-period. Int. J. Antimicrob. Agents 21, 542–546.

Peacock, J.E., Moorman, D.R., Wenzel, R.P., MANDELL, G.L., 1981. Methicillin-resistant *Staphylococcus aureus*: microbiologic characteristics, antimicrobial susceptibilities, and assessment of virulence of an epidemic strain. J. Infect. Dis. 144, 575–582.

Peacock, S.J., Moore, C.E., Justice, A., Kantzanou, M., Story, L., Mackie, K., et al., 2002. Virulent combinations of adhesin and toxin genes in natural populations of *Staphylococcus aureus*. Infect. Immun. 70, 4987–4996.

Perez-Roth, E., Lorenzo-Diaz, F., Batista, N., Moreno, A., Mendez-Alvarez, S., 2004. Tracking methicillin-resistant *Staphylococcus aureus* clones during a 5-year period (1998 to 2002) in a Spanish hospital. J. Clin. Microbiol. 42, 4649–4656.

Rammelkamp, C.H., Maxon, T., 1942. Resistance of *Staphylococcus aureus* to the action of penicillin. Proc. Soc. Exp. Biol. Med. 51, 386–389.

Roberts, J.I.S., Gaston, M.A., 1987. Protein A and coagulase expression in epidemic and non-epidemic *Staphylococcus aureus*. J. Clin. Pathol. 40, 837–840.

Robinson, D.A., Enright, M.C., 2003. Evolutionary models of the emergence of methicillin-resistant *Staphylococcus aureus*. Antimicrob. Agents Chemother. 47, 3926–3934.

Robinson, D.A., Enright, M.C., 2004. Evolution of *Staphylococcus aureus* by large chromosomal replacements. J. Bacteriol. 186, 1060–1064.

Robinson, D.A., Monk, A.B., Cooper, J.E., Feil, E.J., Enright, M.C., 2005. Evolutionary genetics of the accessory gene regulator (agr) locus in *Staphylococcus aureus*. J. Bacteriol. 187, 8312–8321.

Rossney, A.S., Keane, C.T., 2002. Strain variation in the MRSA population over a 10-year period in one Dublin hospital. Eur. J. Clin. Microbiol. Infect. Dis. 21, 123–126.

Sabat, A., Wladyka, B., Kosowska-Shick, K., Grundmann, H., Dijl, J., Kowal, J., et al., 2008. Polymorphism, genetic exchange and intragenic recombination of the aureolysin gene among *Staphylococcus aureus* strains. BMC Microbiol. 8, 129.

Shopsin, B., Gomez, M., Montgomery, S.O., Smith, D.H., Waddington, M., Dodge, D.E., et al., 1999. Evaluation of protein A gene polymorphic region DNA sequencing for typing of *Staphylococcus aureus* strains. J. Clin. Microbiol. 37, 3556–3563.

Sivaraman, K., Venkataraman, N., Tsai, J., Dewell, S., Cole, A.M., 2008. Genome sequencing and analysis reveals possible determinants of *Staphylococcus aureus* nasal carriage. BMC Genomics 9, 433.

Skinner, D., Keefer, C.S., 1941. Significance of bacteremia caused by *Staphylococcus aureus*—a study of one hundred and twenty-two cases and a review of the literature concerned wiih experimental infection in animals. Arch. Intern. Med. 68, 851–875.

Skov, R.L., Jensen, K.S., 2009. Community-associated meticillin-resistant *Staphylococcus aureus* as a cause of hospital-acquired infections. J. Hosp. Infect. 73, 364–370.

Tang, Y.W., Waddington, M.G., Smith, D.H., Manahan, J.M., Kohner, P.C., Highsmith, L.M., et al., 2000. Comparison of protein A gene sequencing with pulsed-field gel electrophoresis and epidemiologic data for molecular typing of methicillin-resistant *Staphylococcus aureus*. J. Clin. Microbiol. 38, 1347–1351.

van Wamel, W.J.B., Fluit, A.C., Wadström, T., van Dijk, H., Verhoef, J., Vandenbroucke-Grauls, C.M., 1995. Phenotypic characterization of epidemic versus sporadic strains of methicillin-resistant *Staphylococcus aureus*. J. Clin. Microbiol. 33, 1769–1774.

Velazquez-Meza, M.E., Ires de, S.M., Echaniz-Aviles, G., Solorzano-Santos, F., Miranda-Novales, G., Silva-Sanchez, J., et al., 2004. Surveillance of methicillin-resistant *Staphylococcus aureus* in a pediatric hospital in Mexico City during a 7-year period (1997 to 2003): clonal evolution and impact of infection control. J. Clin. Microbiol. 42, 3877–3880.

Wagenvoort, J.H.T., Penders, R.J.R., 1997. Long-term in-vitro survival of an epidemic MRSA phage-group III-29 strain (letter to the Editor). J. Hosp. Infect. 35, 322–325.

Witte, W., Cuny, C., Braulke, C., Heuck, D., 1994. Clonal dissemination of two MRSA strains in Germany. Epidemiol. Infect. 113, 67–73.

26 Origin and Emergence of HIV/AIDS

*Lucie Etienne, Eric Delaporte and Martine Peeters**

UMR145, Institut de Recherche pour le Développement (IRD) and University of Montpellier, Montpellier, France

26.1 History of AIDS

Acquired immunodeficiency syndrome (AIDS) was first recognized between 1979 and 1981 among men having sex with men (MSM) who presented with pneumonia caused by *Pneumocystiis carinii* and/or with symptoms of Kaposi sarcoma in New York, Los Angeles, or San Francisco (CDC 1981) (Figure 26.1). Subsequently patients with similar symptoms were seen among intravenous drug users (IDUs), hemophiliacs, Haitians, and Africans in Europe. In May 1983, the etiologic agent of AIDS, the human immunodeficiency virus (HIV), was identified (Barre-Sinoussi et al., 1983). In 1984, several authors reported AIDS cases in women and men in hospitals from sub-Saharan Africa, suggesting also a heterosexual epidemic (Ellrodt et al., 1984; Piot et al., 1984; Van de Perre et al., 1984). Sero-epidemiological studies showed subsequently that a significant proportion of the population in certain regions of Africa was infected with HIV. In the early 1990s, the epidemic exploded in south and eastern Africa, where in certain urban areas 25% of pregnant women were HIV positive (Buve et al., 2002) (Figure 26.1).

Molecular epidemiological studies revealed that the epicenter of the HIV pandemic is situated in Central Africa, and more precisely the area of Kinshasa, the capital city of the Democratic Republic of Congo (DRC) (Vidal et al., 2000; Worobey et al., 2008). The virus has been introduced from Africa in Haiti in the 1960s (most recent common ancestor MRCA 1966) before it started to circulate in North America (MRCA, 1969) about 12 years before the discovery and description of the first AIDS cases (Low-Beer, 2001; Gilbert et al., 2007). Today, more than 33 million people all over the world are infected with HIV (Figure 26.1) and about 70% of HIV-infected persons live in sub-Saharan Africa. With more than 25 million people already deceased, HIV/AIDS continues to be one of the most serious public health threats in the twenty-first century (http://www.unaids.org). It is thus important to identify where this virus came from, to understand how it has been

---

*E-mail: martine.peeters@ird.fr

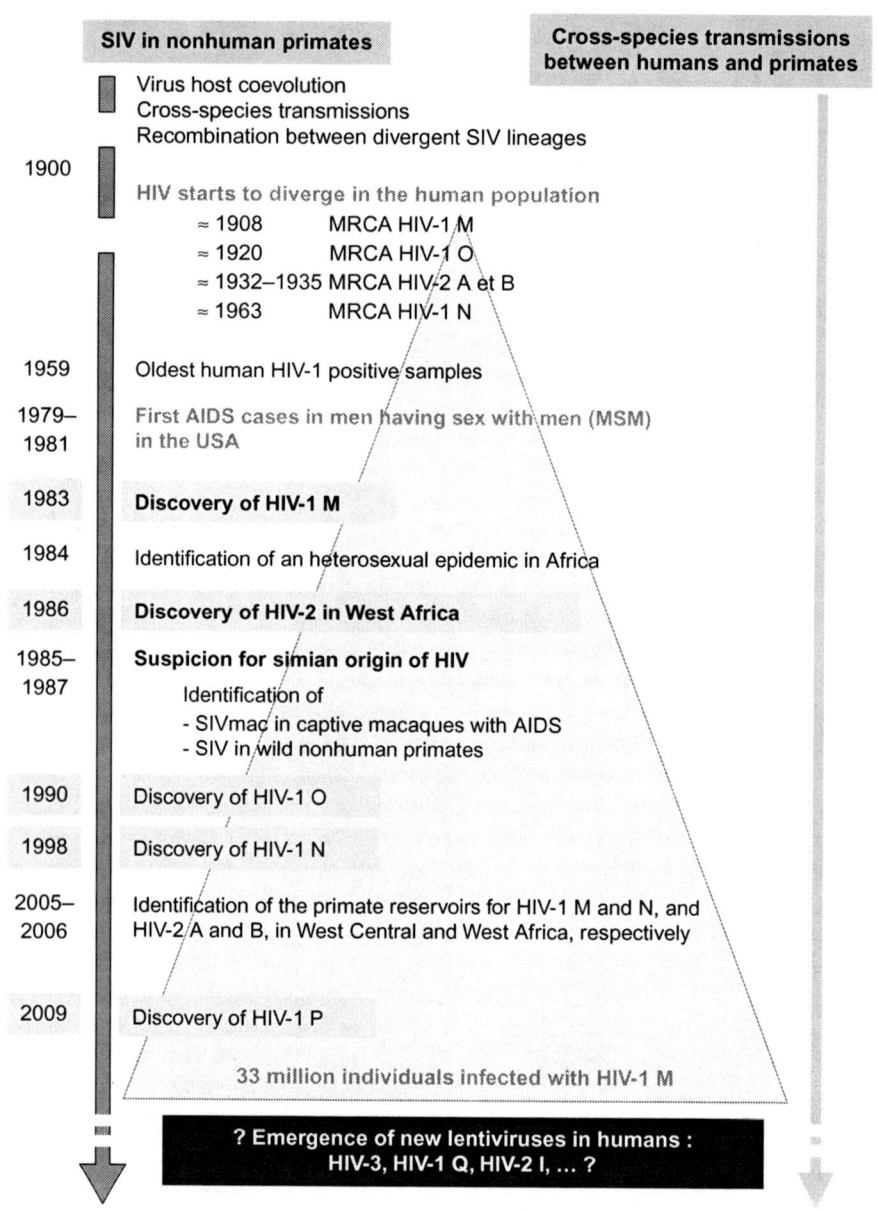

Figure 26.1 History of the AIDS epidemic: past and future events. Dates referring to events in the history of the HIV epidemic in humans are shown at the left, major events are highlighted in pink, boxes indicate subsequent discoveries of the different HIV-1 groups. The number of persons living with HIV increases overtime as illustrated by the pink triangle. Gray arrows represent a schematic timescale of the different events. (For interpretation of the references to color in this figure legend, the reader is referred to the web version of this book.)

introduced into the human population and to determine which factors are associated with host adaptation and epidemic spread.

26.2 HIV Is Closely Related to Simian Immunodeficiency Viruses (SIV) from Nonhuman Primates

Although HIV-1 has been identified in 1983, another closely related virus, HIV-2, has been described in 1986 in France among west Africans with AIDS (Clavel et al., 1986). HIV belongs to the Lentivirus genus of the Retroviridae family where five serogroups are recognized, each reflecting the vertebrate hosts with which they are associated (primates, sheep and goats, horses, cats, and cattle). Both, HIV-1 and HIV-2, are most closely related to the lentiviruses from primates, called simian immunodeficiency viruses (SIVs) and are thus most likely the result of cross-species transmissions of SIVs from African primates.

26.2.1 Discovery of the First SIV

Shortly after the identification of HIV-1 as the cause of AIDS in 1983, the first SIV, SIVmac, was isolated from rhesus macaques (*Macaca mulatta*) with immune deficiency and clinical symptoms similar to AIDS at the New England Regional Primate Research Center (NERPRC) (Henrickson et al., 1983; Daniel et al., 1985). Retrospective studies revealed that SIVmac was introduced at NERPRC by other rhesus monkeys, previously housed at the California National Primate Research Center (CNPRC), where they survived an earlier disease outbreak (late 1960s), characterized by immune suppression and opportunistic infections (Mansfield et al., 1995). A decade after the first outbreak, stump-tailed macaques (*Macaca arctoides*) developed a similar disease in the same settings and a lentivirus called SIVstm was isolated from frozen tissue from one of these monkeys (Lowenstine et al., 1986). In both cases, the infected macaques had been in contact with healthy, but retrospectively shown, SIVsmm seropositive sooty mangabeys at the CNPRC (Lowenstine et al., 1986). The close phylogenetic relationship between SIVmac, SIVstm, and SIVsmm identified mangabeys as the plausible source of SIV in macaques.

Since SIVmac induced a disease in rhesus macaques with remarkable similarity to human AIDS, a simian origin of HIV was soon suspected. The discovery in 1986 of HIV-2, the agent of AIDS in West Africa, and the remarkable high relatedness of HIV-2 with SIVsmm, naturally occurring in sooty mangabeys in West Africa, reinforced this hypothesis.

26.2.2 SIVs in African Nonhuman Primates

Currently, serological evidence of SIV infection has been shown for more than 40 different primate species and SIV infection has been confirmed by viral sequence analysis in the majority (Table 26.1 and Figure 26.2). A high genetic diversity is

Table 26.1 SIV Infections in Old World Monkeys and Apes in Africa

Genus	Species Subspecies	Common Name	SIV
Pan	*troglodytes troglodytes*	**Central chimpanzee**	**SIVcpz*Ptt***
	troglodytes schweinfurthii	East African chimpanzee	SIVcpz*Pts*
Gorilla	*gorilla gorilla*	**Western gorilla**	**SIVgor**
Colobus	*guereza*	Mantled guereza	SIVcol
Piliocolobus	*badius badius*	Western red colobus	SIVwrc*Pbb*
	badius temminckii	Temminck's Red Colobus	SIVwrc*Pbt*
	tholloni	Thollon's Red Colobus	SIVtrc*
	rufomitratus tephrosceles	Ugandan Red Colobus	SIVkrc*
Procolobus	*verus*	Olive colobus	SIVolc
Lophocebus	*albigena*	Gray-cheeked mangabey	
	aterrimus	Black crested mangabey	SIVbkm*
Papio	*anubis*	Olive baboon	
	cynocephalus	Yellow baboon	SIVagm-Ver*
	ursinus	Chacma baboon	SIVagm-Ver*
Cercocebus	*atys*	**Sooty mangabey**	**SIVsmm**
	torquatus	Red capped mangabey	SIVrcm
	agilis	agile mangabey	SIVagi
Mandrillus	*sphinx*	Mandrill	SIVmnd-1,-2
	leucophaeus	Drill	SIVdrl
Allenopithecus	*nigroviridis*	Allen's swamp monkey	
Miopithecus	*talapoin*	Angolan or southern talapoin	SIVtal*
	ogouensis	Gabon or northern talapoin	SIVtal
Erythrocebus	*patas*	Patas monkey	SIVagm-sab*
Chlorocebus	*sabaeus*	Green monkey	SIVagm-Sab
	aethiops	Grivet monkey	SIVagm-Gri
	tantalus	tantalus monkey	SIVagm-Ver
Cercopithecus	*diana*	Diana monkey	
	nictitans	Greater spot nosed monkey	SIVgsn
	mitis	Bleu monkey	SIVblu
	albogularis	Sykes' monkey	SIVsyk
	mona	Mona monkey	SIVmon
	lowei	Lowe's mona monkey	
	campbelli	Campbells mona monkey	
	pogonias	Crested mona monkey	
	denti	Dent's mona monkey	SIVden
	wolfi	Wolf's mona monkey	SIVwol*
	cephus	mustached monkey	SIVmus1,-2
	erythrotis	Red-eared monkey	SIVery
	ascanius	Red-tailed monkey	SIVasc*
	lhoest	l'Hoest monkey	SIVlho
	solatus	Sun-tailed monkey	SIVsun
	preussi	Preuss's monkey	SIVpre*
	hamlyni	Owl-faced monkey	
	neglectus	de Brazza's monkey	SIVdeb

For each species the genus, species, and subspecies (if applicable) are given. Species representing a reservoir for HIV-1 and 2 are highlighted in bold, species with only serological evidence for SIV infection are shown in gray, species with SIV infection confirmed by sequence analysis are shown in black and an asterisks indicates that only partial sequences are currently available. The following references were used for the table: Bibollet-Ruche et al. (2004), Takemura et al. (2005), Van Heuverswyn et al. (2006), VandeWoude and Apetrei (2006), Goldberg et al. (2009), Liégeois et al. (2009), Ahuka-Mundeke et al. (2010).

Origin and Emergence of HIV/AIDS

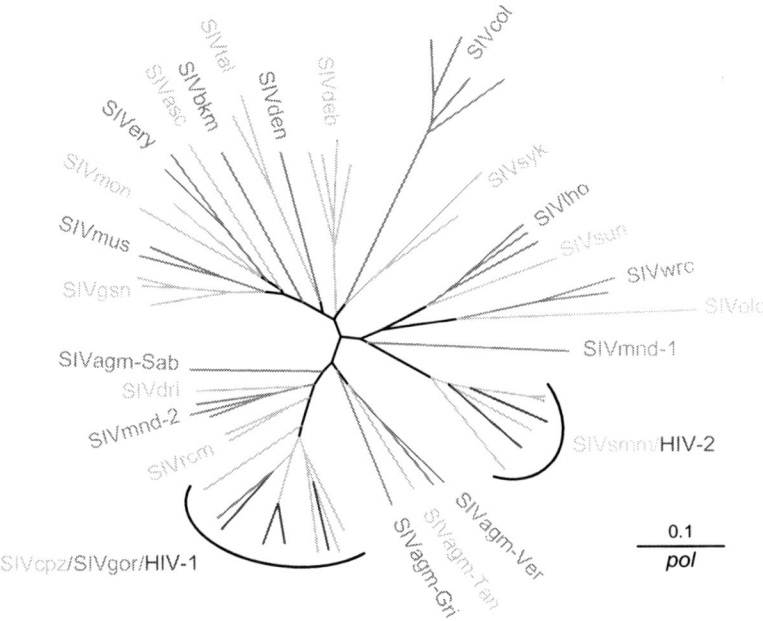

Figure 26.2 Genetic diversity and evolutionary history of the different HIV/SIV lineages. Phylogenetic tree analysis using the neighbor-joining method on a 512 bp fragment from the *pol* gene of different SIVs infecting nonhuman primates and HIVs infecting humans. Branch lengths are drawn to scale (the scale bar indicates 0.1 substitutions per site). The different gray colors are used for clarity to discriminate the different SIV lineages. The different HIV-1 and HIV-2 lineages, which are interspersed with the SIVcpz/SIVgor and SIVsmm lineages respectively, are indicated in pink. The correspondence between the SIV lineages and their natural hosts are shown in Table 26.1. (For interpretation of the references to color in this figure legend, the reader is referred to the web version of this book.)

observed among the different SIVs, but generally each primate species is infected with a species-specific virus, which forms monophyletic lineages in phylogenetic trees. These species-specific SIVs are generally identified by a lower case three-letter code, which corresponds to the initial letters of the common species name, such as for African green monkeys, SIVcpz for chimpanzees. When different subspecies of the same species are infected, an abbreviation referring to the name of the subspecies is added to the virus designation, such as SIVcpz*Ptt* and SIVcpz*Pts* to differentiate between the two chimpanzee subspecies, *Pan troglodytes troglodytes* and *P. t. schweinfurthii*, respectively, in Table 26.1 and Figure 26.2.

Importantly, the number of African primates infected with SIV is most probably underestimated, since 30 species of the 73 recognized Old World monkey and ape species in sub-Saharan Africa have not been tested yet or only very few individuals. Knowing that the vast majority (90%) of the primate species tested is SIV infected, many of the remaining species can be expected to harbor SIV infections.

In addition, all major SIV lineages known to date were discovered as a consequence of primate hosts' antibodies, which cross-reacted with HIV-1 or HIV-2 antigens. Because the extent of this cross-reactivity is not known, SIV infection can be underestimated.

Interestingly, only old world primates are infected with SIVs, and only those from the African continent. No SIVs have been identified in Asian primate species.

26.2.3 Pathogenicity of SIVs in their Natural Hosts

Although SIVs are called immune deficiency viruses, these viruses generally do not induce an AIDS-like disease in their natural hosts, suggesting that they have been associated and evolved with their hosts over an extended period of time (Silvestri et al., 2007). This absence of pathogenicity has been extensively studied among naturally infected captive sooty mangabeys and African green monkeys. Their life span as well as their immunological system do not seem to be affected by SIV (Silvestri et al., 2003; Keele et al., 2009a; Liovat et al., 2009). However, some cases of AIDS have been described among captive monkeys, but mainly at an age which is not reached in their natural habitats (Ling et al., 2004). Nevertheless, the paradigm of nonpathogenicity has been challenged recently by observations on wild chimpanzees (*P. t. schweinfurthii*) in Tanzania where SIVcpz*Pts* infection seems to have a negative impact on life span and reproduction (Keele et al., 2009b). Moreover, retrospective analysis on tissues conserved from dead SIVcpz*Pts* infected animals showed that they have also symptoms of immune deficiency similar to what is observed in humans with AIDS. More studies are needed to confirm these observations, especially in other chimpanzee communities and in the central African subspecies, *P. t. troglodytes*, infected with SIVcpz*Ptt*.

26.2.4 Evolution and Phylogeny of SIVs

The genetic diversity among nonhuman primate lentiviruses is extremely complex. There are many examples of coevolution between viruses and their hosts, but recombination between distant SIVs does not seem exceptional and one species can even harbor two different SIVs. Although it seems now clear that a simple co-divergence between viruses and their hosts is not common, phylogenies for some SIVs and their hosts suggest coevolution over long time periods, like SIVagm in the different African green monkey species (*Chlorocebus* sp.) or SIVs in arboreal *Cercopithecus* species (Charleston et al., 2002; Wertheim et al., 2007). However, cross-species transmissions could give erroneous impressions of coevolution, especially when chances for efficient host switch are higher among genetically closely related species (Charleston et al., 2002). There are numerous examples of cross-species transmissions of SIVs between primates living in the same habitats or in polyspecific associations. For example, SIVagm from African green monkeys has been transmitted to Patas monkeys in West Africa (Bibollet-Ruche et al., 1996) and to yellow and chacma baboons in South Africa (Jin et al., 1994; van Rensburg et al., 1998). There are also more complex examples of cross-species transmissions of SIVs between greater spot nosed monkeys (SIVgsn) and mustached monkeys (SIVmus), followed by recombination as seen for SIVmus-2 in mustached monkeys in Cameroon (Aghokeng et al., 2007) (Table 26.1). One of the most striking examples of cross-species transmission, followed by recombination is SIVcpz in chimpanzees. The 5′ region of SIVcpz is most similar to SIVrcm from red capped mangabeys, and the 3′ region is found to be closely related to SIVgsn from greater spot nosed monkeys (Bailes et al., 2003). Chimpanzees are known to hunt monkeys for food. Most probably, the recombination of these monkey viruses occurred within chimpanzees and gave rise to the common ancestor of today's SIVcpz lineages, which on their turn were subsequently transmitted to gorillas (Van Heuverswyn et al., 2006; Takehisa et al., 2009).

As numerous cross-species transmissions among different primate species occurred, both HIV-1 and HIV-2 in humans are also the result of cross-species transmissions of SIVs from African primates (Hahn et al., 2000). The closest simian relatives of HIV-1 are SIVcpz and SIVgor, in chimpanzees (*Pan troglodytes troglodytes*) and gorillas (*Gorilla gorilla*), respectively, from west-central Africa, and SIVsmm in sooty mangabeys (*Cercocebus atys*) from West Africa are the closest relatives of HIV-2.

26.3 HIV-1 Is Derived from SIVs Circulating among African Apes

Based on phylogenetic analyses of numerous isolates obtained from diverse geographic origins, HIV-1 is classified into four groups, M, N, O, and P (Figure 26.1). Group M (for Major), discovered in 1983 (Barre-Sinoussi et al., 1983), represents the vast majority of HIV-1 strains found worldwide and is responsible for the global pandemic. Group O (for Outlier), described in 1990 (De Leys et al., 1990), remained restricted to west-central Africa, and represents currently 1% of HIV-1 infections in Cameroon. Groups N and P, described in 1998 and 2009, respectively (Simon et al., 1998; Plantier et al., 2009), have only been observed in a handful of Cameroonian patients. Each HIV-1 group corresponds to an independent cross-species transmission from SIVs from apes to humans.

26.3.1 SIVs from Chimpanzees and Gorillas Are the Ancestors of HIV-1 in Humans

The first SIVcpz strains have been isolated from two captive wild-born chimpanzees in Gabon more than 20 years ago, SIVcpzGab1 and SIVcpzGab2 (Peeters et al., 1989). Genetic analysis of the SIVcpzGab1 genome revealed the presence of the accessory gene, *vpu*, also identified in HIV-1. Furthermore, phylogenetic analysis showed that SIVcpzGab1 was more closely related to HIV-1 than to any other SIV (Huet et al., 1990). Characterization of a third SIVcpz, SIVcpzANT, showed an unexpected high degree of divergence among the chimpanzee viruses (Peeters et al., 1992; Vanden Haesevelde et al., 1996). Subsequent subspecies identification of the chimpanzee hosts revealed that the SIVcpzANT strain was isolated from a member of the *P. t. schweinfurthii* subspecies, whereas the other chimpanzees belonged to the *P. t. troglodytes* subspecies (Gao et al., 1999). These findings suggested two distinct SIVcpz lineages according to the host subspecies: SIVcpz*Ptt* and SIVcpz*Pts* from central (*P. t. troglodytes*) and eastern chimpanzees (*P. t. schweinfurthii*), respectively. All HIV-1 groups were more closely related to SIVcpz*Ptt* than to SIVcpz*Pts* (Figure 26.3). Although these data pointed already to the west-central African chimpanzees (*P. t. troglodytes*) as the natural reservoir of the ancestors of HIV-1, the SIVcpz reservoirs in wild living apes that are at the origin of HIV in humans still needed to be identified. The major issue in studying SIVcpz infection in wild chimpanzees is their endangered status and the fact that they live in isolated forest regions. In addition, all previously studied chimpanzees were wild-caught, but they were capture as infants, which do not reflect true prevalences among wild living adult animals. The recent development of noninvasive methods to detect and characterize SIVcpz in fecal and urine samples boosted the search for new SIVcpz strains in wild ape populations in the vast tropical forest of central Africa (Santiago et al., 2002; Santiago et al., 2003a). The first full-length SIVcpz sequence from a wild infected chimpanzee, SIVcpzTan1, was obtained from a fecal sample from a *P. t. schweinfurthii* chimpanzee in Gombe National Park, Tanzania. Subsequently, additional cases of SIVcpz*Pts* infections were documented in Tanzania and the DRC (Santiago et al., 2003b; Worobey et al., 2004; Li et al., 2010). All the new SIVcpz*Pts* viruses formed phylogeographic clusters according to their geographic origin, but formed a separate lineage with the initially described SIVcpzANT strain indicating thus that the SIVcpz*Pts* strains are not at the origin of HIV-1 (Figure 26.3).

Recent molecular epidemiological studies on SIVcpz*Ptt* in central chimpanzees (*P. t. troglodytes*) allowed to trace the ancestors of the HIV-1 group M and N strains to distinct chimpanzee communities in south-east and south-central Cameroon (Keele et al., 2006) (Figure 26.3). Overall, SIVcpz*Ptt* viruses exhibited also a significant geographic clustering and SIVcpz*Ptt* is widely present in Central chimpanzees with an overall prevalence of 5.9%, although not all chimpanzee communities are infected and prevalences can range from 0% to 32% (Van Heuverswyn et al., 2007).

Despite testing of numerous samples, no SIV infection has been detected yet in the two other chimpanzee subspecies, *P. t. elioti* (previously *P. t. vellerosus*) and *P. t. verus* (Switzer et al., 2005; Van Heuverswyn et al., 2007).

While the reservoirs of the ancestors of HIV-1 M and N were identified in 2006, the reservoirs of group O and P remained unknown. In 2006, SIV infection was described for the first time in western gorillas (*Gorilla gorilla gorilla*) in Cameroon. Surprisingly, the newly characterized gorilla viruses, termed SIVgor, formed a monophyletic group within the HIV-1/SIVcpz*Ptt* radiation, but in contrast to SIVcpz*Ptt*, they were most closely related to the HIV-1 group O and P lineages (Figure 26.3) (Van Heuverswyn et al., 2006; Takehisa et al., 2009). However, the phylogenetic relationships between SIVcpz, SIVgor, and HIV-1 indicate that chimpanzees represent the original reservoir of SIVs now found in chimpanzees, gorillas

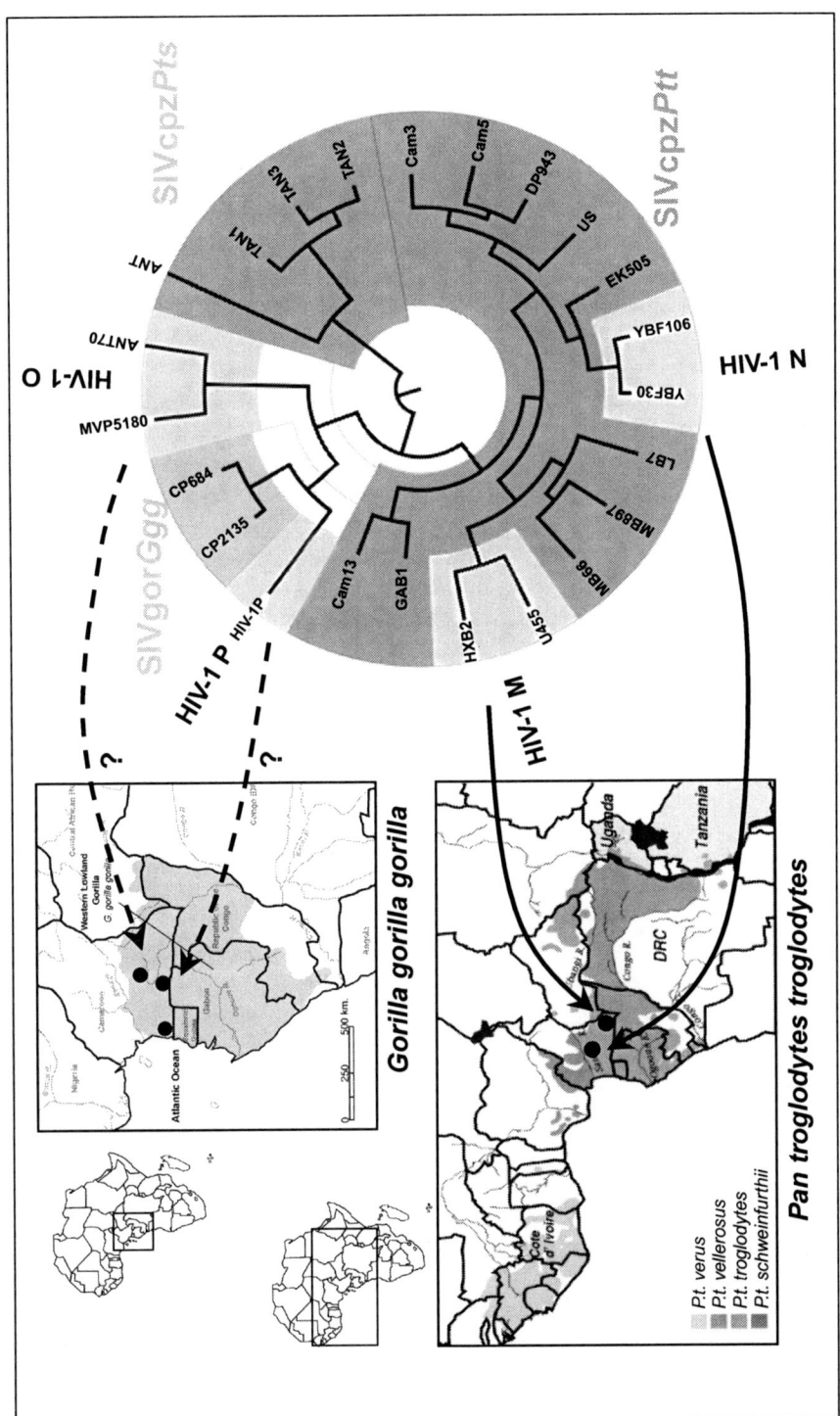

Figure 26.3 *(Continued)*

and humans (Figure 26.3). Given the herbivorous diet of gorillas and their peaceful coexistence with other primate species, especially chimpanzees, it remains a mystery by which route gorillas acquired SIVgor. A large molecular epidemiological survey among more than 2200 fecal samples from wild gorillas collected in 21 sites across southern Cameroon showed a global prevalence of 1.6% ranging from 0% to 4.6% in gorillas, which is three times less than the overall prevalence observed in chimpanzees in the same areas (Neel et al., 2010).

The close phylogenetic relationship of the recently discovered HIV-1 group P and SIVgor, suggested that group P is derived from a gorilla lentivirus (Plantier et al., 2009). However, no SIVgor strain sufficiently close to HIV-1 P has been identified yet to be the direct ancestor of HIV-1 group P or O found in humans (Figure 26.3).

It is now clear that central chimpanzees (*P. t. troglodytes*) are the reservoirs for the pandemic HIV-1 group M strain and also for HIV-1 N. However, more studies are still needed to identify the direct ancestors of HIV-1 O and P in gorillas and/or chimpanzees in other areas in Cameroon and in neighboring countries. These studies will also help to clarify the origin of SIVgor in gorillas, whether gorillas and/or chimpanzees transmitted group O and P to humans, and whether the ancestors of SIVgor, HIV-1 O and P still circulate among chimpanzees.

Importantly, the finding that SIVcpz strains from east African chimpanzees, including those from Kisangani in DRC, are more distantly related to HIV-1 provides also evidence that the oral polio vaccine (OPV), which was largely distributed in this part of Africa at the end of the 50s, is not at the origin of the HIV-1 epidemic. It has been suggested that tissues derived from SIVcpz-infected chimpanzees, captured in the northeastern part of DRC were used for the OPV production. However, this geographical region is situated in the middle of the *P. t. schweinfurthii* range and the characterization of SIVcpz*Pts* from wild chimpanzees in DRC proved once more the inconsistency of the OPV theory (Worobey et al., 2004) (Figure 26.3).

26.3.2 The Cross-species Transmissions Resulting in HIV-1 Viruses in Humans Occurred in West-Central Africa

Since the four groups of HIV-1 fall within the HIV-1/SIVcpz*Ptt*/SIVgor radiation, the cross-species transmissions giving rise to HIV-1 most likely occurred in

---

Figure 26.3 *(Cont.)* HIV-1 is derived from SIVs circulating in chimpanzees and/or gorillas from West Central Africa. Evolutionary relationship of SIVcpz*Pts* (blue), SIVcpz*Ptt* (red), SIVgor (yellow), and HIV-1 group M, N, O, and P (gray) strains based on maximum likelihood phylogenetic analysis of partial Env (gp41) sequences. Horizontal branch lengths are drawn to scale (the scale bar indicates 0.2 substitutions per site). Maps represent the geographical range of *G. g. gorilla* (upper map) and the four chimpanzee subspecies (lower map). Arrows between the phylogenetic tree and maps indicate the ape reservoirs with the ancestors or most closely related strains to the different HIV groups. Dotted arrows indicate that the direct reservoirs for HIV-1 groups O and P are not yet identified. (For interpretation of the references to color in this figure legend, the reader is referred to the web version of this book.)

western equatorial Africa, the home of *P. t. troglodytes* chimpanzees and western gorillas. Furthermore, no human counterpart is found for SIVcpz*Pts* from *P. t. schweinfurthii*. The recent studies in wild chimpanzee communities in Cameroon, not only strengthen the evidence of the west-central African origin of HIV-1, but also indicate that HIV-1 groups M and N arose from geographically distinct chimpanzee populations in south Cameroon (Figure 26.3). This coincides with the geographical area of group N infections, which remain actually restricted to Cameroon. HV-1 groups O and P are also mainly restricted to Cameroon (Brennan et al., 2008; Vessiere et al., 2010), coinciding with the geographical ranges of central chimpanzees and western gorillas.

The four HIV-1 groups have thus their seeds in west-central Africa, but only one, HIV-1 group M, has spread across Africa and all the other continents. Moreover, the reservoir of the ancestors of HIV-1 M has been identified at almost 1000 km distance from the epicenter of the HIV-1 epidemic in DRC (Vidal et al., 2000; Worobey et al., 2008). A combination of several factors (viral, host, socioeconomic, demographic, etc.) are thus most likely involved in the subsequent efficient spread of HIV-1 M. It has also to be noted that nowadays no data are available on SIV infection in wild chimpanzee and gorilla populations living between southern Cameroon and Kinshasa, DRC, and it cannot be excluded that in this area other chimpanzee populations exist that are infected with viruses also closely related to HIV-1 M.

26.3.3 HIV-1 Started to Diverge in the Human Population at the Beginning of the Twentieth Century

As mentioned above, the SIVcpz*Ptt* ancestors of group M have been identified in Cameroon, but the highest genetic diversity of HIV-1 M, in number of co circulating subtypes and intrasubtype diversity, has been observed in the western part of DRC, suggesting that this region may be the epicenter of HIV-1 group M (Vidal et al., 2000). Retrospective studies showed that 20 years before the AIDS epidemic was recognized in the United States, HIV-1 M (subtypes A and D) infection was already circulating in humans in Kinshasa (i.e., HIV-1 was identified in a serum from 1959) (Zhu et al., 1998) and a biopsy from 1960 (Worobey et al., 2008). Molecular clock analyses estimated the date of the most recent common ancestor of HIV-1 group M around 1908 with a confidence interval of 1884−1924 (Wertheim et al., 2009). A similar time frame is estimated for the origin of the HIV-1 group O radiation; 1920 with a range from 1890−1940 (Wertheim et al., 2009). Since the first identification of HIV-1 group N in 1998, less than 10 group N infections have been described, and all were from Cameroonian patients. The intragroup genetic diversity is significantly lower for group N than for group M or O, which suggest a more recent introduction of the HIV-1 N lineage into the human population and the most recent common ancestor of HIV-1 group N is estimated around 1963 with a confidence interval of 1948−1977 (Wertheim et al., 2009).

26.3.4 SIVs Are Transmitted to Humans by Exposure to Infected Primates

Although the conditions and circumstances of cross-species from SIVs from primates to humans remain unknown, human exposure to blood or other secretions of infected primates, through hunting and butchering of primate bushmeat, represents the most plausible source for human infection. Also bites and other injuries caused by primates, kept as pet animals can favor a possible viral transmission. However, factors associated with single cross-species transmission have to be differentiated from those associated with epidemic spread, the latter being a combination of multiple factors.

26.4 Origin of HIV-2: An Other Emergence, an Other Epidemic

Two independent studies in 1989 and 1992 confirmed the homologies between HIV-2 and SIVsmm infecting sooty mangabeys in West Africa (Hirsch et al., 1989; Gao et al., 1992). Sooty mangabeys are home to West Africa, from Senegal to Ivory Coast, coinciding with the endemic center of HIV-2 (Figure 26.4). In contrast to HIV-1, HIV-2 remained restricted to West Africa and HIV-2 prevalences are even decreasing, since HIV-1 M is now also predominating in West Africa (de Silva et al., 2008; van Tienen et al., 2009). The highest HIV-2 prevalences have been observed in Guinea-Bissau and southern Senegal (Casamance area). Overall, HIV-2 is less pathogenic, less transmissible with almost absence of mother to child transmission, and less-efficient sexual transmission most likely related to lower viral loads (Gottlieb et al., 2006; Hawes et al., 2008). However, a high genetic diversity is seen among HIV-2 strains, eight groups (A−H) of HIV-2 have been described so far (Damond et al., 2004) (Figure 26.4), each corresponding to a cross-species transmission. Only groups A and B are largely represented in the HIV-2 epidemic, with group A circulating in the western part of West Africa (Senegal, Guinea-Bissau) and group B being predominant in Ivory Coast. The ancestors of the HIV-2 group A and B viruses have been identified in wild sooty mangabey populations in the Tai forest in Ivory Coast, close to the border with Liberia (Santiago et al., 2005). The other HIV-2 groups have been documented in one or few individuals only. Except for groups G and H, groups C, D, E, and F were isolated in rural areas in Sierra Leone and Liberia, and these viruses are more closely related to the SIVsmm strains obtained from sooty mangabeys found in the same area than to any other HIV-2 strain. Molecular clock analyses traced the origin of the epidemic HIV-2 groups A and B to be around 1932 (1906−1955) and 1935 (1907−1961), respectively (Wertheim et al., 2009). These dates seem to coincide with a political unstable period in Guinea-Bissau, and it has been suggested that civil wars at that time amplified the rapid spread of HIV-2 into the human population (Lemey et al., 2003) (Figure 26.4).

Origin and Emergence of HIV/AIDS

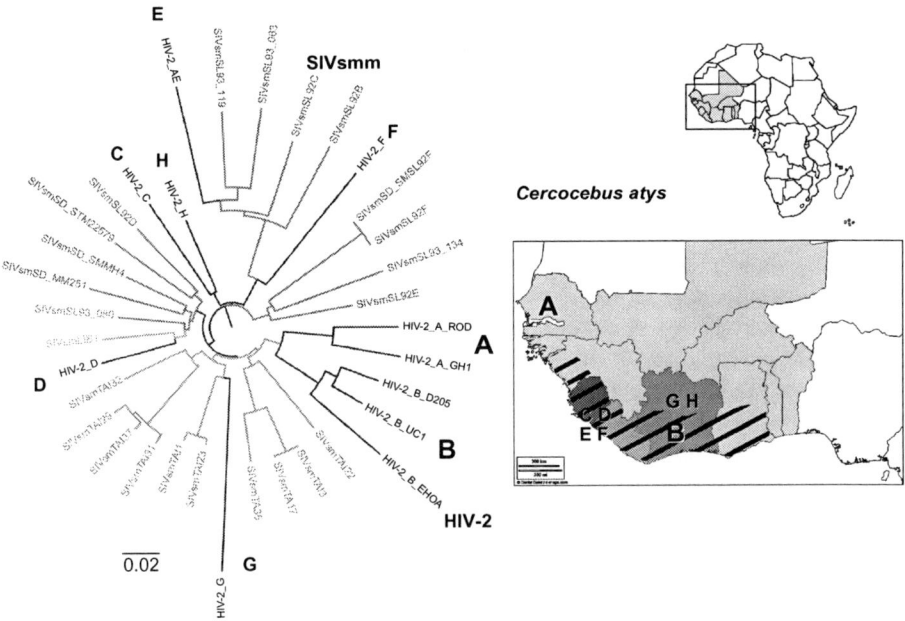

Figure 26.4 HIV-2 is derived from SIVs circulating in sooty mangabeys from West Central Africa. Evolutionary relationship of HIV-2 groups A to H (black) and SIVsmm (blue, green, and pink according to the country of origin) using the neighbor-joining method on partial *env* (741 bp) sequences. Horizontal branch lengths are drawn to scale (the scale bar indicates 0.02 substitutions per site). The map indicates the distribution of HIV-2 in the human population (gray) with letters referring to the HIV-2 groups in the different areas, capital letters refer to epidemic strains, lower cases refer to geographically isolated HIV-2 variants with no or limited spread in humans. The geographic range of sooty mangabeys (*Cercocebus atys*) is highlighted by black lines, color codes for the different SIVsmm strains in the phylogenetic tree are green for strains from Ivory Coast, pink for Liberia and blue for Sierra Leone. (For interpretation of the references to color in this figure legend, the reader is referred to the web version of this book.)

26.5 Ongoing Exposure of Humans to a Large Diversity of SIVs: Risk for a Novel HIV?

26.5.1 Exposure to SIV-Infected Primates Is Still Ongoing

Bushmeat hunting, as a source of animal proteins and income for the family, is a longstanding common component of rural households in the Congo Basin, and more generally throughout sub-Saharan Africa. However, the bushmeat trade has increased significantly during the last decades, due to the expanding logging industry and the increasing demand of bushmeat in cities (Laporte et al., 2007). Indeed, a recent study showed that in Northern Congo, the bushmeat trade increased together with the increasing presence of logging concessions in the same area (Poulsen et al., 2009). Commercial logging has led to road constructions penetrating remote forest areas, to human migration, and the development of social and economic networks (including those of sex workers), which support this industry. Furthermore, villages around logging concessions have grown from a few hundred to several thousand inhabitants in just a few years, and the number of people entering previously inaccessible forest areas increased significantly over the last decades. Importantly, the HIV prevalences in these previous rural settings are also increasing. A high number of individuals with immune deficiency are thus potentially exposed to new viruses, and recombination between newly introduced SIVs and circulating HIVs can pose an additional risk for the outbreak of a novel HIV epidemic (Laurent et al., 2004; LeBreton et al., 2007).

Our prior studies on primate bushmeat in Cameroon showed that bushmeat hunting is not limited to chimpanzees and mangabeys, but the majority of hunted primates are represented by multiple *Cercopithecus* species, colobus monkeys, mandrills, drills, etc. Moreover, these data revealed ongoing exposure of humans to a plethora of different SIVs (Peeters et al., 2002; Aghokeng et al., 2009). Cross-species transmissions with SIV strains from other primates to humans should thus be considered. It has also to be noted that apes are not only hunted for bushmeat but also medicinal uses (Meder, 1999). The socioeconomic changes, which go together with the presence of logging or other industries in remote areas, combined with the SIV prevalence and genetic complexity in wild living primates, suggest that the magnitude of human exposure to SIV has increased, as have the social and environmental conditions that support the emergence and spread of new zoonotic infections.

26.5.2 SIV Prevalences and Cross-species Transmissions

The chances for cross-species transmissions most likely increase when frequency of exposure and SIV prevalences are high. A recent study in pet monkeys in Cameroon revealed that SIV prevalence in mandrill pets could reach 23% (Ndembi et al., 2009). We recently showed among more than 2,500 samples in Cameroon and by using SIV lineage specific Elisas that about 3% of primate bushmeat is SIV infected, but prevalences can vary from 0% to 10% depending on geographic localities, or from 0% to 40% according to species (Aghokeng et al., 2009). Interestingly, the lowest prevalences (0% to 1%) were observed among the most frequently hunted species, reducing thus the risk for cross-species transmissions. However, this situation can be different in other geographic areas, for example in the DRC a pilot study showed that 20% of the primate bushmeat are infected with the highest prevalences among the most frequently hunted monkeys (Ahuka-Mundeke et al., 2010). Our studies on wild chimpanzees in Cameroon showed also that SIVcpz prevalences in chimpanzee communities infected with the ancestors of HIV-1 M and N are among the highest (i.e., around 30%) (Keele et al., 2006; Van Heuverswyn et al., 2007).

Moreover, in West Africa, the SIVsmm prevalences of wild mangabeys are around 50%, and at least eight cross-species from mangabeys to humans occurred.

In this same region, 50% of western red colobus monkeys are also infected with SIVs and are together with mangabeys highly represented among primate bushmeat, but no SIVwrc cross-species transmission to humans has been documented yet. Western red colobus monkeys represent also 80% of animal proteins among the chimpanzee from the *P. t. verus* subspecies in the Tai forest in Ivory Coast, and again no SIVwrc infection could be identified in this chimpanzee subspecies. These data suggest that in addition to SIV prevalences, other factors like viral adaptation to the new host play also an important role.

26.5.3 Viral Adaptation

The majority of the viruses infecting wild animals are not able to cross the species barrier, adapt to a new host, and spread into the new species. The viral and molecular characteristics that allowed the ancestors of HIV-1 and HIV-2 to cross and adapt to humans are not yet identified, and more studies are needed to find out why, for example, SIVs from mangabeys and chimpanzees or gorillas have been transmitted at multiple occasions and not those of western red colobus or other monkeys. Moreover, cross-species transmissions are not always followed by efficient spread into the new populations, as illustrated by HIV-1 versus HIV-2 and among the different HIV-1 groups, and depend thus not only on the virus but also on the host and the environment (Chastel, 2000; Lloyd-Smith et al., 2009).

Actually, only limited studies showed some viral adaptation, for example at the Gag-30 position in the p17 region of the *gag* gene, methionine or leucine is present among SIVcpz/SIVgor, but in humans all HIV-1 strains harbor an arginine at this position (Wain et al., 2007) suggesting an adaptation of the virus to its new host. Sauter and colleagues showed recently that the Vpu protein of the pandemic HIV-1 M strain is able to block the human restriction factor tetherin, whereas HIV-1 O viruses do not (Sauter et al., 2009). This could have lead to a higher replication of HIV-1 M in humans, better interhuman transmission, and thus allowing a better epidemic spread of the virus in the human population (Gupta et al., 2009). Nef allows viral persistence in the host but controls also for superactivation of the immune system. However, this latter function is lost in certain SIV lineages and more precisely in the ancestors of the HIV-1/SIVcpz lineage, which could thus have resulted in higher pathogenicity in humans, in contrast to SIVsmm/HIV-2 where this adaptation is not observed and which are less pathogenic (Schindler et al., 2006). Another study showed a lower viral fitness for HIV-2 compared to HIV-1, and for HIV-1 O versus M, which could also partially explain the lower prevalence and limited spread of HIV-2 and HIV-1 O (Arien et al., 2005). However, these observations need to be confirmed by additional studies on viral factors.

26.5.4 Human Factors

Human factors also play a major role in the epidemic spread of a new virus, especially among viruses that are transmitted by blood or sexual contacts, as it is the case for HIV. The difference in the localizations of the origin of HIV-1 M (South

Cameroon) and the origin of the epidemic (DRC for HIV-1 M) illustrates the importance of human and environmental factors in the epidemic spread. The main factors involved in human-to-human spread are sexual risk behavior, high prevalences of sexually transmitted infections, absence of circumcision, and transmission through unsafe blood transfusions and nonsterile needles. These factors associated with sociodemographic factors like human density in forest areas, increasing transport between urban and rural areas, human migration, urbanization, increase in commercial sex were in favor of an epidemic spread of the virus.

26.5.5 Ongoing Cross-species Transmissions from Other Retroviruses from Primates to Humans

SFVs (simian foamy viruses) are infecting primates at high levels; 70% of primates in captivity (Hussain et al., 2003), 97% of wild western red colobus monkeys (Goldberg et al., 2009), between 44% and 100% of wild chimpanzees (Liu et al., 2008). Epidemiological studies among primate care workers in the United States showed a SFV prevalence of 25% in persons who reported ape bites (Switzer et al., 2004; Calattini et al., 2007). However, human-to-human transmission or a SFV epidemic has not been documented yet, and no pathogenicity associated with SFV infection has been observed in humans. In this example, cross-species transmissions were thus possible, but viral adaptation was insufficient to allow a spread of the virus into the new host.

Exposure to nonhuman primates allowed also the emergence of HTLV (Human T-lymphotropic virus) in humans (Wolfe et al., 2005; Switzer et al., 2009) and recent studies among hunters in central Africa showed ongoing cross-species transmissions of STLVs (Sintasath et al., 2009; Zheng et al., 2010). These T-lymphotropic viruses are thus able to cross the species barrier, and certain variants were able to spread into the human population leading to HTLV-1 and HTLV-2, which are endemic in certain parts of the world.

The ongoing transmissions of SFV and STLV highlight the risk for potential emergence of a new SIV into the human population. The recent discovery in 2009 of a new HIV-1, group P, is an additional demonstration of a zoonotic transmission.

26.6 Conclusion

Actually, we have a clear picture of the origin of HIV and the seeds of the AIDS pandemic. The current HIV-1 epidemic provides evidence for the extraordinary impact that can result from a single primate lentiviral zoonotic transmission event. Despite the fact that the first AIDS cases have been observed around 1980 in the United States, the virus circulated already early in the twentieth century in the human population in central Africa (Figure 26.1). Currently, at least 12 SIV cross-species transmissions occurred, 8 for HIV-2 and 4 for HIV-1, but most likely several others occurred in the past, which remained unrecognized since some viruses were not able to adapt to the new host or the environment was not suitable for epidemic spread. Because humans are still exposed to a plethora of primate lentiviruses through hunting and handling of primate bushmeat, the possibility of additional cross-species transfers of primate lentiviruses from other primate species in addition to chimpanzees, gorillas, or sooty mangabeys has to be considered. One major public health implication is that these SIV strains are not always recognized by commercial HIV-1/HIV-2 screening assays. As a consequence, due to the long incubation period, human infection with such variants can go unrecognized for several years and lead to another epidemic. Identification of SIVs in wild primates can serve as sentinels by signaling which pathogens may be a risk for humans and allow the development of serological and molecular assays to detect transmissions with other SIVs in humans. While more insight is gained about the origin of the HIV-1 and HIV-2 viruses, some important questions concerning pathogenicity and epidemic spread of certain HIV/SIV variants need to be further elucidated.

References

Ahuka Mundeke, S., Mbala, P., Liégeois, F., Lunguya, O., Ayouba, A., Bilulu, G., et al., 2010. Infection à rétrovirus chez les primates chassés pour la viande de brousse en RD du Congo. 5e Conférence Francophone VIH/SIDA, Casablanca, Marocco. Abstract 597.

Aghokeng, A.F., Bailes, E., Loul, S., Courgnaud, V., Mpoudi-Ngolle, E., Sharp, P.M., et al., 2007. Full-length sequence analysis of SIVmus in wild populations of mustached monkeys (Cercopithecus cephus) from Cameroon provides evidence for two co-circulating SIVmus lineages. Virology 360, 407–418.

Aghokeng, A.F., Ayouba, A., Mpoudi-Ngole, E., Loul, S., Liegeois, F., Delaporte, E., et al., 2009. Extensive survey on the prevalence and genetic diversity of SIVs in primate bushmeat provides insights into risks for potential new cross-species transmissions. Infect. Genet. Evol. 10, 386–396.

Arien, K.K., Abraha, A., Quinones-Mateu, M.E., Kestens, L., Vanham, G., Arts, E.J., 2005. The replicative fitness of primary human immunodeficiency virus type 1 (HIV-1) group M, HIV-1 group O, and HIV-2 isolates. J. Virol. 79, 8979–8990.

Bailes, E., Gao, F., Bibollet-Ruche, F., Courgnaud, V., Peeters, M., Marx, P.A., et al., 2003. Hybrid origin of SIV in chimpanzees. Science 300, 1713.

Barre-Sinoussi, F., Chermann, J.C., Rey, F., Nugeyre, M.T., Chamaret, S., Gruest, J., et al., 1983. Isolation of a T-lymphotropic retrovirus from a patient at risk for acquired immune deficiency syndrome (AIDS). Science 220, 868–871.

Bibollet-Ruche, F., Galat-Luong, A., Cuny, G., Sarni-Manchado, P., Galat, G., Durand, J.P., et al., 1996. Simian immunodeficiency virus infection in a patas monkey (Erythrocebus patas): evidence for cross-species transmission from African green monkeys (Cercopithecus aethiops sabaeus) in the wild. J. Gen. Virol. 77, 773–781, Part 4.

Bibollet-Ruche, F., Bailes, E., Gao, F., Pourrut, X., Barlow, K.L., Clewley, J.P., et al., 2004. New simian immunodeficiency virus infecting De Brazza's monkeys (Cercopithecus neglectus): evidence for a cercopithecus monkey virus clade. J. Virol. 78, 6864–6874.

Brennan, C.A., Bodelle, P., Coffey, R., Devare, S.G., Golden, A., Hackett Jr., J., et al., 2008. The prevalence of diverse HIV-1 strains was stable in Cameroonian blood donors from 1996 to 2004. J. Acquir. Immune Defic. Syndr. 49, 432–439.

Buve, A., Bishikwabo-Nsarhaza, K., Mutangadura, G., 2002. The spread and effect of HIV-1 infection in sub-Saharan Africa. Lancet 359, 2011–2017.

Calattini, S., Betsem, E.B., Froment, A., Mauclere, P., Tortevoye, P., Schmitt, C., et al., 2007. Simian foamy virus transmission from apes to humans, rural Cameroon. Emerg. Infect. Dis. 13, 1314–1320.

Charleston, M.A., Robertson, D.L., 2002. Preferential host switching by primate lentiviruses can account for phylogenetic similarity with the primate phylogeny. Syst. Biol. 51, 528–535.

Chastel, C., 2000. Emergential success: a new concept for a better appraisal of viral emergences and reemergences. Acta. Virol. 44, 375–376.

Clavel, F., Guyader, M., Guetard, D., Salle, M., Montagnier, L., Alizon, M., 1986. Molecular cloning and polymorphism of the human immune deficiency virus type 2. Nature 324, 691–695.

Damond, F., Worobey, M., Campa, P., Farfara, I., Colin, G., Matheron, S., et al., 2004. Identification of a highly divergent HIV type 2 and proposal for a change in HIV type 2 classification. AIDS Res. Hum. Retroviruses 20, 666–672.

Daniel, M.D., Letvin, N.L., King, N.W., Kannagi, M., Sehgal, P.K., Hunt, R.D., et al., 1985. Isolation of T-cell tropic HTLV-III-like retrovirus from macaques. Science 228, 1201–1204.

De Leys, R., Vanderborght, B., Vanden Haesevelde, M., Heyndrickx, L., van Geel, A., Wauters, C., et al., 1990. Isolation and partial characterization of an unusual human immunodeficiency retrovirus from two persons of west-central African origin. J. Virol. 64, 1207–1216.

de Silva, T.I., Cotten, M., Rowland-Jones, S.L., 2008. HIV-2: the forgotten AIDS virus. Trends Microbiol. 16, 588–595.

Ellrodt, A., Barre-Sinoussi, F., Le Bras, P., Nugeyre, M.T., Palazzo, L., Rey, F., et al., 1984. Isolation of human T-lymphotropic retrovirus (LAV) from Zairian married couple, one with AIDS, one with prodromes. Lancet 1, 1383–1385.

Gao, F., Yue, L., White, A.T., Pappas, P.G., Barchue, J., Hanson, A.P., et al., 1992. Human infection by genetically diverse SIVSM-related HIV-2 in west Africa. Nature 358, 495–499.

Gao, F., Bailes, E., Robertson, D.L., Chen, Y., Rodenburg, C.M., Michael, S.F., et al., 1999. Origin of HIV-1 in the chimpanzee *Pan troglodytes troglodytes*. Nature 397, 436–441.

Gilbert, M.T., Rambaut, A., Wlasiuk, G., Spira, T.J., Pitchenik, A.E., Worobey, M., 2007. The emergence of HIV/AIDS in the Americas and beyond. Proc. Natl. Acad. Sci. U.S.A. 104, 18566–18570.

Goldberg, T.L., Sintasath, D.M., Chapman, C.A., Cameron, K.M., Karesh, W.B., Tang, S., et al., 2009. Coinfection of Ugandan red colobus (*Procolobus [Piliocolobus] rufomitratus tephrosceles*) with novel, divergent delta-, lenti-, and spumaretroviruses. J. Virol. 83, 11318–11329.

Gottlieb, G.S., Hawes, S.E., Agne, H.D., Stern, J.E., Critchlow, C.W., Kiviat, N.B., et al., 2006. Lower levels of HIV RNA in semen in HIV-2 compared with HIV-1 infection: implications for differences in transmission. AIDS 20, 895–900.

Gupta, R.K., Towers, G.J., 2009. A tail of Tetherin: how pandemic HIV-1 conquered the world. Cell Host Microbe 6, 393–395.

Hahn, B.H., Shaw, G.M., De Cock, K.M., Sharp, P.M., 2000. AIDS as a zoonosis: scientific and public health implications. Science 287, 607–614.

Hawes, S.E., Sow, P.S., Stern, J.E., Critchlow, C.W., Gottlieb, G.S., Kiviat, N.B., 2008. Lower levels of HIV-2 than HIV-1 in the female genital tract: correlates and longitudinal assessment of viral shedding. AIDS 22, 2517–2525.

Henrickson, R.V., Maul, D.H., Osborn, K.G., Sever, J.L., Madden, D.L., Ellingsworth, L.R., et al., 1983. Epidemic of acquired immunodeficiency in rhesus monkeys. Lancet 1, 388–390.

Hirsch, V.M., Olmsted, R.A., Murphey-Corb, M., Purcell, R.H., Johnson, P.R., 1989. An African primate lentivirus (SIVsm) closely related to HIV-2. Nature 339, 389–392.

Huet, T., Cheynier, R., Meyerhans, A., Roelants, G., Wain-Hobson, S., 1990. Genetic organization of a chimpanzee lentivirus related to HIV-1. Nature 345, 356–359.

Hussain, A.I., Shanmugam, V., Bhullar, V.B., Beer, B.E., Vallet, D., Gautier-Hion, A., et al., 2003. Screening for simian foamy virus infection by using a combined antigen Western blot assay: evidence for a wide distribution among Old World primates and identification of four new divergent viruses. Virology 309, 248–257.

Jin, M.J., Rogers, J., Phillips-Conroy, J.E., Allan, J.S., Desrosiers, R.C., Shaw, G.M., et al., 1994. Infection of a yellow baboon with simian immunodeficiency virus from African green monkeys: evidence for cross-species transmission in the wild. J. Virol. 68, 8454–8460.

Keele, B.F., Van Heuverswyn, F., Li, Y., Bailes, E., Takehisa, J., Santiago, M.L., et al., 2006. Chimpanzee reservoirs of pandemic and nonpandemic HIV-1. Science 313, 523–526.

Keele, B.F., Jones, J.H., Terio, K.A., Estes, J.D., Rudicell, R.S., Wilson, M.L., et al., 2009a. Increased mortality and AIDS-like immunopathology in wild chimpanzees infected with SIVcpz. Nature 460, 515–519.

Keele, B.F., Li, H., Learn, G.H., Hraber, P., Giorgi, E.E., Grayson, T., et al., 2009b. Low-dose rectal inoculation of rhesus macaques by SIVsmE660 or SIVmac251 recapitulates human mucosal infection by HIV-1. J. Exp. Med. 206, 1117–1134.

Laporte, N.T., Stabach, J.A., Grosch, R., Lin, T.S., Goetz, S.J., 2007. Expansion of industrial logging in Central Africa. Science 316, 1451.

Laurent, C., Bourgeois, A., Mpoudi, M., Butel, C., Peeters, M., Mpoudi-Ngolé, E., et al., 2004. Commercial logging and HIV epidemic, rural Equatorial Africa. Emerg. Infect. Dis. 10, 1953–1956.

LeBreton, M., Yang, O., Tamoufe, U., Mpoudi-Ngole, E., Torimiro, J.N., Djoko, C.F., et al., 2007. Exposure to wild primates among HIV-infected persons. Emerg. Infect. Dis. 13, 1579–1582.

Lemey, P., Pybus, O.G., Wang, B., Saksena, N.K., Salemi, M., Vandamme, A.M., 2003. Tracing the origin and history of the HIV-2 epidemic. Proc. Natl. Acad. Sci. U.S.A. 100, 6588–6592.

Li, Y, Ndjango, J.B., Learn, G., Robertson, J., Takehisa, J., Bibollet-Ruche, F., et al., 2010 Molecular epidemiology of SIV in eastern chimpanzees and gorillas. In: 17th Conference on Retroviruses and Opportunistic Infections, San Francisco, USA. Abstract 440.

Liégeois, F., Lafay, B., Formenty, P., Locatelli, S., Courgnaud, V., Delaporte, E., Peeters, M., 2009. Full-length genome characterization of a novel simian immunodeficiency virus lineage (SIVolc) from olive Colobus (*Procolobus versus*) and new SIVwrcPbb strains from Western Red Colobus (*Piliocolobus badius badius*) from the Tai Forest in Ivory Coast. J. Virol. 83, 428–439.

Ling, B., Apetrei, C., Pandrea, I., Veazey, R.S., Lackner, A.A., Gormus, B., et al., 2004. Classic AIDS in a sooty mangabey after an 18-year natural infection. J. Virol. 78, 8902–8908.
Liovat, A.S., Jacquelin, B., Ploquin, M.J., Barre-Sinoussi, F., Muller-Trutwin, M.C., 2009. African non human primates infected by SIV—why don't they get sick? Lessons from studies on the early phase of non-pathogenic SIV infection. Curr. HIV Res. 7, 39–50.
Liu, W., Worobey, M., Li, Y., Keele, B.F., Bibollet-Ruche, F., Guo, Y., et al., 2008. Molecular ecology and natural history of simian foamy virus infection in wild-living chimpanzees, PLoS Pathog., 4, e1000097.
Lloyd-Smith, J.O., George, D., Pepin, K.M., Pitzer, V.E., Pulliam, J.R., Dobson, A.P., et al., 2009. Epidemic dynamics at the human-animal interface. Science 326, 1362–1367.
Low-Beer, D., 2001. The distribution of early acquired immune deficiency syndrome cases and conditions for the establishment of new epidemics. Philos. Trans. R. Soc. Lond B. Biol. Sci. 356, 927–931.
Lowenstine, L.J., Pedersen, N.C., Higgins, J., Pallis, K.C., Uyeda, A., Marx, P., et al., 1986. Seroepidemiologic survey of captive Old-World primates for antibodies to human and simian retroviruses, and isolation of a lentivirus from sooty mangabeys (*Cercocebus atys*). Int. J. Cancer 38, 563–574.
Mansfield, K.G., Lerch, N.W., Gardner, M.B., Lackner, A.A., 1995. Origins of simian immunodeficiency virus infection in macaques at the New England Regional Primate Research Center. J. Med. Primatol. 24, 116–122.
Meder, A., 1999. Gorillas in African culture and medicine. Gorilla J. 18, 11–15.
Ndembi, N., Kaptue, L., Ido, E., 2009. Exposure to SIVmnd-2 in southern Cameroon: public health implications. AIDS Rev. 11, 135–139.
Neel, C., Etienne, L., Li, Y., Takehisa, J., Rudicell, R.S., Bass, I.N., et al., 2010. Molecular epidemiology of simian immunodeficiency virus infection in wild-living gorillas. J. Virol. 84, 1464–1476.
Peeters, M., Honore, C., Huet, T., Bedjabaga, L., Ossari, S., Bussi, P., et al., 1989. Isolation and partial characterization of an HIV-related virus occurring naturally in chimpanzees in Gabon. AIDS 3, 625–630.
Peeters, M., Fransen, K., Delaporte, E., Van den Haesevelde, M., Gershy-Damet, G.M., Kestens, L., et al., 1992. Isolation and characterization of a new chimpanzee lentivirus (simian immunodeficiency virus isolate cpz-ant) from a wild-captured chimpanzee. AIDS 6, 447–451.
Peeters, M., Courgnaud, V., Abela, B., Auzel, P., Pourrut, X., Bibollet-Ruche, F., et al., 2002. Risk to human health from a plethora of simian immunodeficiency viruses in primate bushmeat. Emerg. Infect. Dis. 8, 451–457.
Piot, P., Quinn, T.C., Taelman, H., Feinsod, F.M., Minlangu, K.B., Wobin, O., et al., 1984. Acquired immunodeficiency syndrome in a heterosexual population in Zaire. Lancet 2, 65–69.
Plantier, J.C., Leoz, M., Dickerson, J.E., De Oliveira, F., Cordonnier, F., Lemee, V., et al., 2009. A new human immunodeficiency virus derived from gorillas. Nat. Med. 15, 871–872.
Poulsen, J.R., Clark, C.J., Mavah, G., Elkan, P.W., 2009. Bushmeat supply and consumption in a tropical logging concession in northern Congo. Conserv. Biol. 23, 1597–1608.
Santiago, M.L., Rodenburg, C.M., Kamenya, S., Bibollet-Ruche, F., Gao, F., Bailes, E., et al., 2002. SIVcpz in wild chimpanzees. Science 295, 465.
Santiago, M.L., Bibollet-Ruche, F., Gross-Camp, N., Majewski, A.C., Masozera, M., Munanura, I., et al., 2003a. Noninvasive detection of Simian immunodeficiency virus

infection in a wild-living L'Hoest's monkey (*Cercopithecus lhoesti*). AIDS Res. Hum. Retroviruses 19, 1163−1166.
Santiago, M.L., Lukasik, M., Kamenya, S., Li, Y., Bibollet-Ruche, F., Bailes, E., et al., 2003b. Foci of endemic simian immunodeficiency virus infection in wild-living eastern chimpanzees (*Pan troglodytes schweinfurthii*). J. Virol. 77, 7545−7562.
Santiago, M.L., Range, F., Keele, B.F., Li, Y., Bailes, E., Bibollet-Ruche, F., et al., 2005. Simian immunodeficiency virus infection in free-ranging sooty mangabeys (*Cercocebus atys atys*) from the Tai Forest, Cote d'Ivoire: implications for the origin of epidemic human immunodeficiency virus type 2. J. Virol. 79, 12515−12527.
Sauter, D., Schindler, M., Specht, A., Landford, W.N., Munch, J., Kim, K.A., et al., 2009. Tetherin-driven adaptation of Vpu and Nef function and the evolution of pandemic and nonpandemic HIV-1 strains. Cell Host Microbe 6, 409−421.
Schindler, M., Munch, J., Kutsch, O., Li, H., Santiago, M.L., Bibollet-Ruche, F., et al., 2006. Nef-mediated suppression of T cell activation was lost in a lentiviral lineage that gave rise to HIV-1. Cell 125, 1055−1067.
Silvestri, G., Sodora, D.L., Koup, R.A., Paiardini, M., O'Neil, S.P., McClure, H.M., et al., 2003. Nonpathogenic SIV infection of sooty mangabeys is characterized by limited bystander immunopathology despite chronic high-level viremia. Immunity 18, 441−452.
Silvestri, G., Paiardini, M., Pandrea, I., Lederman, M.M., Sodora, D.L., 2007. Understanding the benign nature of SIV infection in natural hosts. J. Clin. Invest. 117, 3148−3154.
Simon, F., Mauclere, P., Roques, P., Loussert-Ajaka, I., Muller-Trutwin, M.C., Saragosti, S., et al., 1998. Identification of a new human immunodeficiency virus type 1 distinct from group M and group O. Nat. Med. 4, 1032−1037.
Sintasath, D.M., Wolfe, N.D., Zheng, H.Q., LeBreton, M., Peeters, M., Tamoufe, U., et al., 2009. Genetic characterization of the complete genome of a highly divergent simian T-lymphotropic virus (STLV) type 3 from a wild *Cercopithecus mona* monkey. Retrovirology 6, 97.
Switzer, W.M., Bhullar, V., Shanmugam, V., Cong, M.E., Parekh, B., Lerche, N.W., et al., 2004. Frequent simian foamy virus infection in persons occupationally exposed to nonhuman primates. J. Virol. 78, 2780−2789.
Switzer, W.M., Parekh, B., Shanmugam, V., Bhullar, V., Phillips, S., Ely, J.J., et al., 2005. The epidemiology of simian immunodeficiency virus infection in a large number of wild- and captive-born chimpanzees: evidence for a recent introduction following chimpanzee divergence. AIDS Res. Hum. Retroviruses 21, 335−342.
Switzer, W.M., Salemi, M., Qari, S.H., Jia, H., Gray, R.R., Katzourakis, A., et al., 2009. Ancient, independent evolution and distinct molecular features of the novel human T-lymphotropic virus type 4. Retrovirology 6, 9.
Takehisa, J., Kraus, M.H., Ayouba, A., Bailes, E., Van Heuverswyn, F., Decker, J.M., et al., 2009. Origin and biology of simian immunodeficiency virus in wild-living western gorillas. J. Virol. 83, 1635−1648.
Takemura, T., Ekwalanga, M., Bikandou, B., Ido, E., Yamaguchi-Kabata, Y., Ohkura, S., et al., 2005. A novel simian immunodeficiency virus from black mangabey (*Lophocebus aterrimus*) in the Democratic Republic of Congo. J. Gen. Virol. 86, 1967−1971.
Vanden Haesevelde, M., Peeters, M., Jannes, G., Janssens, W., van der Groen, G., Sharp, P.M., et al., 1996. Sequence analysis of a highly divergent HIV-1-related lentivirus isolated from a wild captured chimpanzee. Virology 221, 346−350.
Van de Perre, P., Rouvroy, D., Lepage, P., Bogaerts, J., Kestelyn, P., Kayihigi, J., et al., 1984. Acquired immunodeficiency syndrome in Rwanda. Lancet 2, 62−65.

VandeWoude, S., Apetrei, C., 2006. Going wild: lessons from naturally occurring T-lymphotropic lentiviruses. Clin. Microbiol. Rev. 19, 728–762.
Van Heuverswyn, F., Li, Y., Neel, C., Bailes, E., Keele, B.F., Liu, W., et al., 2006. Human immunodeficiency viruses: SIV infection in wild gorillas. Nature 444, 164.
Van Heuverswyn, F., Li, Y., Bailes, E., Neel, C., Lafay, B., Keele, B.F., et al., 2007. Genetic diversity and phylogeographic clustering of SIVcpzPtt in wild chimpanzees in Cameroon. Virology 368, 155–171.
van Rensburg, E.J., Engelbrecht, S., Mwenda, J., Laten, J.D., Robson, B.A., Stander, T., et al., 1998. Simian immunodeficiency viruses (SIVs) from eastern and southern Africa: detection of a SIVagm variant from a chacma baboon. J. Gen. Virol. 79 (Part 7), 1809–1814.
van Tienen, C., van der Loeff, M.S., Zaman, S.M., Vincent, T., Sarge-Njie, R., Peterson, I., et al., 2010. Two distinct epidemics: the rise of HIV-1 and decline of HIV-2 infection between 1990 and 2007 in rural Guinea-Bissau. J. Acquir. Immune Defic. Syndr. 53, 640–647.
Vessiere, A., Rousset, D., Kfutwah, A., Leoz, M., Depatureaux, A., Simon, F., et al., 2010. Diagnosis and monitoring of HIV-1 group O-infected patients in Cameroun. J. Acquir. Immune. Defic. Syndr. 53, 107–110.
Vidal, N., Peeters, M., Mulanga-Kabeya, C., Nzilambi, N., Robertson, D., Ilunga, W., et al., 2000. Unprecedented degree of human immunodeficiency virus type 1 (HIV-1) group M genetic diversity in the democratic republic of Congo suggests that the HIV-1 pandemic originated in Central Africa. J. Virol. 74, 10498–10507.
Wain, L.V., Bailes, E., Bibollet-Ruche, F., Decker, J.M., Keele, B.F., Van Heuverswyn, F., et al., 2007. Adaptation of HIV-1 to its human host. Mol. Biol. Evol. 24, 1853–1860.
Wertheim, J.O., Worobey, M., 2007. A challenge to the ancient origin of SIVagm based on African green monkey mitochondrial genomes. PLoS Pathog 3, e95.
Wertheim, J.O., Worobey, M., 2009. Dating the age of the SIV lineages that gave rise to HIV-1 and HIV-2. PLoS Comput. Biol. 5, e1000377.
Wolfe, N.D., Heneine, W., Carr, J.K., Garcia, A.D., Shanmugam, V., Tamoufe, U., et al., 2005. Emergence of unique primate T-lymphotropic viruses among central African bushmeat hunters. Proc. Natl. Acad. Sci. U.S.A. 102, 7994–7999.
Worobey, M., Santiago, M.L., Keele, B.F., Ndjango, J.B., Joy, J.B., Labama, B.L., et al., 2004. Origin of AIDS: contaminated polio vaccine theory refuted. Nature 428, 820.
Worobey, M., Gemmel, M., Teuwen, D.E., Haselkorn, T., Kunstman, K., Bunce, M., et al., 2008. Direct evidence of extensive diversity of HIV-1 in Kinshasa by 1960. Nature 455, 661–664.
Zheng, H., Wolfe, N.D., Sintasath, D.M., Tamoufe, U., Lebreton, M., Djoko, C.F., et al., 2010. Emergence of a novel and highly divergent HTLV-3 in a primate hunter in Cameroon. Virology 401, 137–145.
Zhu, T., Korber, B.T., Nahmias, A.J., Hooper, E., Sharp, P.M., Ho, D.D., 1998. An African HIV-1 sequence from 1959 and implications for the origin of the epidemic. Nature 391, 594–597.

27 Evolution of SARS Coronavirus and the Relevance of Modern Molecular Epidemiology

Zhengli Shi[1], and Lin-Fa Wang[2]*

[1]State Key Laboratory of Virology, Wuhan Institute of Virology, Chinese Academy of Sciences (CAS), Wuhan, China, [2]CSIRO Livestock Industries, Australian Animal Health Laboratory, Geelong, Australia

27.1 A Brief History of SARS

As outlined in Table 27.1, the first reported case of "atypical pneumonia," now known as severe acute respiratory syndrome (SARS), occurred in Guangzhou, Guangdong province, China, on November 16, 2002. Before the end of February 2003, a total of 11 index cases occurred independently in 9 cities of Guangdong Province, which was the early phase of the SARS epidemic (Chinese, 2004). These index cases spread the virus to their close relatives and hospital staffs and provided the early demonstration of the respiratory transmission mode of the disease. The clinical symptoms of SARS are nonspecific. The index cases all began to have fever higher than 38°C and displayed common respiratory symptoms such as cough, headache, and shortness of breath.

The dynamics of the outbreak was largely shaped by the presents of the so-called super spread event (SSE), in which a single patient was shown to spread the virus to a large number of contacts (Chinese, 2004). It was the SSEs that triggered the large-scale SARS pandemic in China. The first SSE patient is a businessman specialized in fishery wholesale. He was treated in three hospitals from January 30, 2003 to February 10, 2003 and along the way infected at least 78 other individuals including hospital staffs, patients, and close relatives and friends (Chinese, 2004). The second SSE individual, who caused the major spread of the disease out of Guangdong, was a native of Shanxi province. She went to Guangdong for business in late February and become sick while traveling. She went back to her home province and infected eight family members as well as five hospital staff members. The spread continued to Beijing when she decided to seek better treatment in Beijing (Chinese, 2004; Zhao, 2007).

*E-mail: zlshi@wh.iov.cn

Table 27.1 Chronological Events of the SARS Outbreaks

Date	Event
Nov 16, 2002	The first recognized SARS patient, in Foshan, Guangdong province, China
Nov 16, 2002–Mar 10, 2003	Eleven independent index cases in Foshan, Heyuan, Jiangmen, Zhongshan, Shunde, Guanzhou, Zhaoqing, Shenzhen, Dongguan, China, resulting in more than 50 secondary infections
Jan 22, 2003	SARS spreading in Guangdong province
Mar 22, 2003	SARS spreading to Shanxi and Beijing
Feb 21, 2003	SARS spreading to Hong Kong, marking the beginning of the global pandemic
Feb 28, 2003	SARS spreading to Vietnam
Mar 12, 2003	WHO Global travel alert for the SARS pandemic
Mar 14, 2003	SARS spreading to Canada
Mar 6, 2003	SARS spreading to Singapore
Mar 17, 2003	WHO established a 9-nation/11-institute international laboratory network
Mar 24, 2003	Coronavirus was isolated from SARS patient
Apr 4, 2003	SARS spreading to Philippines
Apr 12, 2003	Full-length genome of SARS-CoV determined
Apr 17, 2003	The international laboratory network announced conclusive identification of SARS-CoV as the causative agent
May 23, 2003	Detected SARS coronavirus in market animals
July 5, 2003	WHO removed the last region from the effected list, effectively marking the end of the outbreak
Aug 7, 2003	WHO reported a total of 8,096 cases and 774 deaths covering the major 2002–2003 outbreaks
Sep 2003–Apr 2004	Outbreaks caused by laboratory incidents in Singapore, Taiwan, and Beijing
Dec 16, 2003–Jan 8, 2004	Four independent SARS cases in Guangdong, causing mild disease with no death

The beginning of the global transmission occurred in Metropole Hotel of Hong Kong where a visiting urologist from a Guangdong hospital stayed during a private visit. Without his knowledge, the urologist was infected with SARS coronavirus (SARS-CoV) a few days before he traveled to Hong Kong. It is later found that he spread the virus to at least 15 other persons in the hotel and in the hospital where he was treated. Among them, five of the hotel contacts continued to their international journey and further transmitted the disease to Vietnam and Singapore. This marks the true beginning of a disastrous worldwide pandemic (http://www.who.int/csr/sars/en/).

WHO played a key role in the investigation and control of the SARS outbreak from the very beginning. For the first time in history, WHO issued a global travel alert on March 12, 2003, which greatly reduced the rate of long-distance transmission of the disease. On March 17, 2003, WHO established a 9-nation/11-institute SARS network, which played a major role in the rapid identification of the causative agent and development of diagnostic tests. Thanks to the international effort coordinated by WHO, the SARS outbreaks were effectively under control by July 5, 2003. This was the first powerful demonstration of the kind of devastation a new infectious disease can cause worldwide and the effectiveness of an international organization when it is running at its peak.

Following the major SARS outbreaks of 2003–2004, there were several minor outbreaks with much smaller impacts. In December 2003 and January 2004, four independent SARS cases were reported in Guangdong, and none of them led to fetal infection or widespread transmission. Subsequent epidemiological tracing revealed that all cases could be linked to civet trading activities (Wang et al., 2005). In addition, there were laboratory outbreaks reported in September 2003, December 2003, and April 2004 in Singapore, Taiwan, and Beijing, respectively. The most severe outbreak was associated with the incident in Beijing, which resulted in a total of nine infection cases with one death. None of the other two laboratory infections resulted in further spread of the virus (Lim et al., 2006).

27.2 SARS Coronavirus

Rapid identification of causative agent in an outbreak caused by unknown pathogen is the key for an effective response. However, in the case of SARS outbreak, this was not the case. Due to the association of nonspecific clinical symptoms associated with SARS patients, several pathogens were initially "identified" as the potential causes of SARS, which included Chlamydia, influenza virus, and paramyxovirus (WHO, 2003). The confusion continued until March 2003, when three laboratories independently confirmed that a previously unknown coronavirus was the most likely etiological agent of SARS (Drosten et al., 2003; Ksiazek et al., 2003; Peiris et al., 2003).

Coronaviruses are enveloped viruses with the largest single-stranded, positive-sense RNA genomes currently known, ranging in size from 27 to nearly 32 kb in length. Coronaviruses can infect and cause disease in a broad array of avian and mammal species, including humans. The name "coronavirus" is derived from the Greek word for crown, as the virus envelope appears under electron microscopy to be crowned by a characteristic ring of small bulbous structures. Within the virion, the ssRNA genome is encased in a helical nucleocapsid composed of many copies of the nucleocapsid (N) protein. The lipid bilayer envelope contains three proteins: envelope (E) and membrane (M), which coordinate virion assembly and release, and the large spike protein (S), which confers the virus's characteristic corona shape and serves as the principle mediator of host cell attachment and entry via virus- and host-specific cell receptors. The size of the SARS-CoV viral particle is approximately 80–90 nm and its genomic size is around 29.7 kb (Marra et al., 2003; Rota et al., 2003). The SARS-CoV genome contains 14 open reading frames (ORFs) flanked by 5'- and 3'-untranslated regions of 265 and 342 nucleotides (nt), respectively. While all CoVs carry strain-specific accessory genes in their downstream ORFs, the order of essential genes—the replicase/transcriptase gene, S gene, E gene, M gene, and N are highly conserved (Graham and Baric, 2010).

Similar to other known coronaviruses, the SARS-CoV genome expression starts with two long ORFs, ORF1a and ORF1b, which account for two thirds of the genomic capacity, followed by ORFs encoding S, E, M, and N proteins (Figure 27.1). In addition to these conserved core genes in coronaviruses, the SARS-CoV genome contains several accessory genes that are specific to SARS-CoV and have no homolog to known proteins. Phylogenetic analysis based on the most conserved gene ORF1b indicated that SARS-CoV is distantly related to the group 2 coronaviruses in the family *Coronaviridae* and represents a distinct cluster, named group 2b (Figure 27.2) (Snijder et al., 2003).

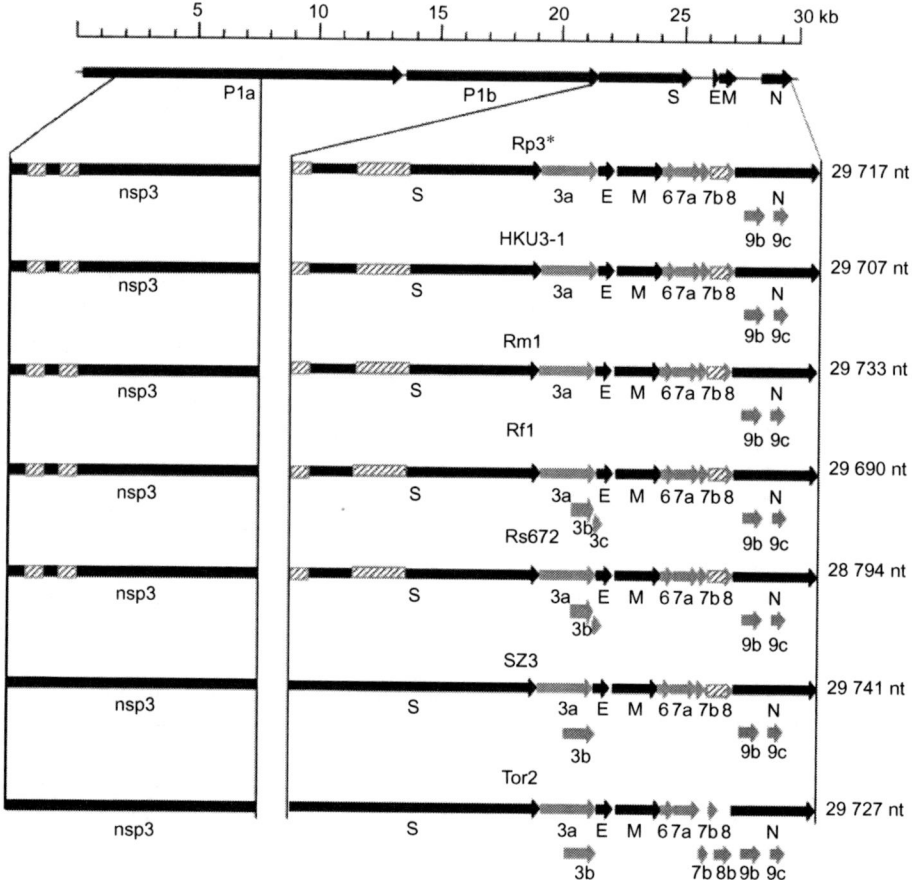

Figure 27.1 Genomic structure of SARS-CoV and bat SL-CoV. The highly conserved genes present in all coronaviruses are shown in dark-colored arrows and the group 2b-specific ORFs in light-colored arrows. The most variable regions were marked with shaded boxes. The asterisk (*) indicates the host of Rp3 was previously identified as *Rhinolophus pearsoni* and later corrected to be *R. sinicus*.
Source: Yuan et al. (2010).

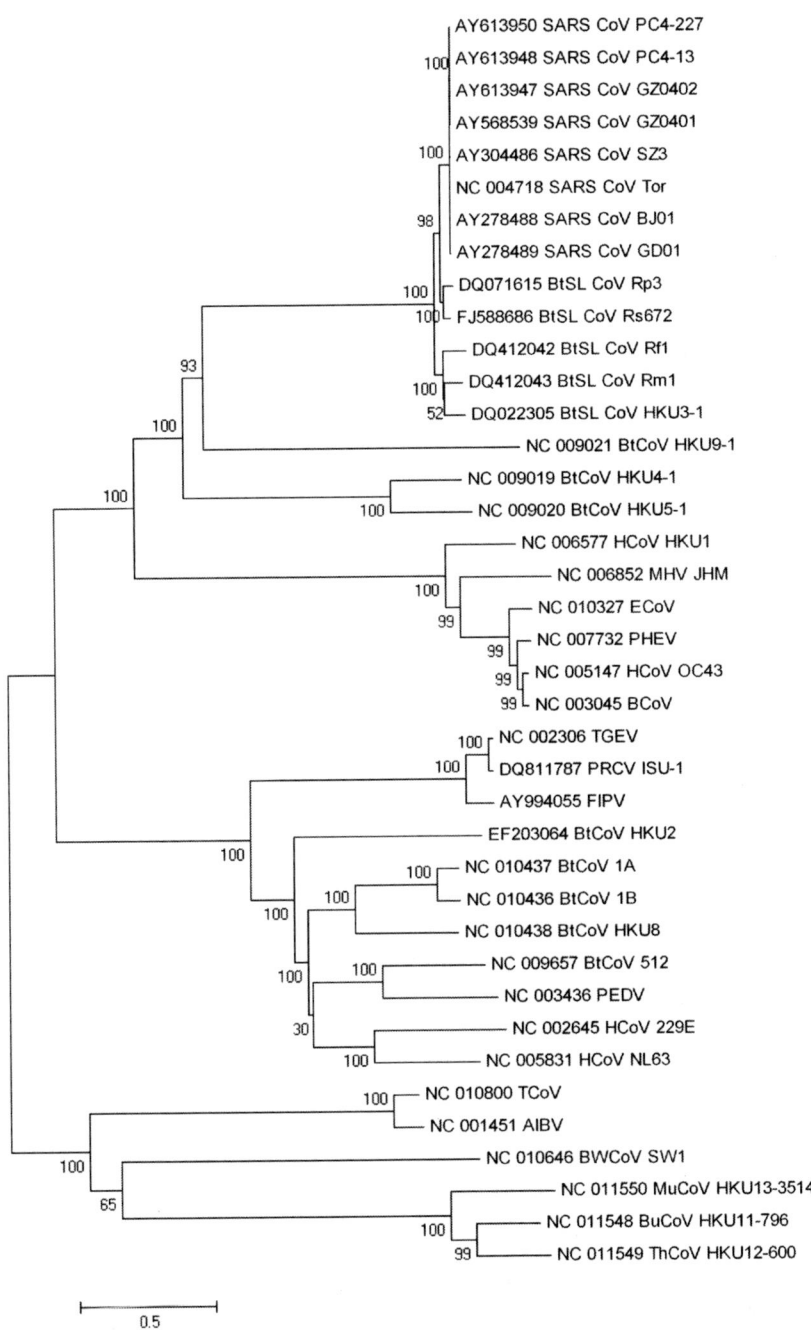

Figure 27.2 *(Continued)*

27.3 The Animal Link

Due to the rapid spread of the disease and the delay in the identification of the causative agent, there was no detailed epidemiological tracing done at the beginning of the outbreaks, and it was therefore impossible to trace the origin of the virus. However, through retrospective investigation, it emerged that the majority of the early index cases were limited in several cities of the Guangdong province and most of them have a history of contact directly or indirectly with wild animals, including handling, killing and selling them, as well as preparing and serving animal meat in restaurants (He et al., 2003; Xu et al., 2004a,b).

As these epidemic regions have a unique dietary tradition favoring freshly slaughtered game meat, there is a huge trafficking and trading industry dedicated to live animal trading in specialized market, the "wet market." So naturally, immediately after SARS-CoV was identified as the etiological agent of SARS, studies were conducted in those markets for evidence of SARS-CoV. One of the earliest and most important studies was conducted by a joint team from Hong Kong and Shenzhen in mainland China (Guan et al., 2003). In this investigation, out of 25 samples collected from market animals, SARS-CoV like viruses were isolated from 4 out of 6 masked palm civets (*Paguma larvata*) and one raccoon dog (*Nyctereutes procyonoides*). Antibodies against SARS-CoV were detected in masked palm civets, raccoon dogs, and Chinese ferret-badgers (*Melogale moschata*). Genome sequencing indicated that the viruses isolated from civets were almost identical to those from human, suggesting a highly possible zoonotic transmission of SARS-CoV from animal(s) to human (Guan et al., 2003). These data indicated that at least three different animal species were infected by a coronavirus that is closely related to SARS-CoV. This important study provided the first direct evidence that SARS-CoV existed in animals, pointing to an animal link of the SARS outbreaks.

Figure 27.2 *(Cont.)* **Phylogenetic tree of representative coronaviruses.** The phylogenetic tree is generated based on full-length genome sequences of selected coronaviruses using the Neighbor-Joining algorithm in the MEGA4 program (Tamura et al., 2007) with a bootstrap of 1000 replicates. Numbers above branches indicate bootstrap values from 1000 replicates. Scale bar, 0.5 substitutions per site. GD01: SARS-CoV isolate from early phase patient during 2002–2003 SARS ourbreak; Tor2, BJ01: SARS-CoV isolate from late phase patient during 2002–2003 SARS ourbreak; SZ: SARS-CoV isolate from civet during 2002–2003 SARS ourbreak; GZ0401/02: SARS-CoV isolate from patient during 2003–2004 SARS ourbreak; PC4-13, PC4-227: SARS-CoV isolate from civet during 2003–2004 SARS ourbreak. HCoV, human coronavirus; PEDV, porcine epidemic diarrhea virus; TGEV, porcine transmissible gastroenteritis virus; PRCV, porcine respiratory coronavirus; BtSL-CoV, bat SARS-like CoV; BtCoV, bat coronavirus; MHV, mouse hepatitis virus; BCoV, bovine coronavirus; FIPV, feline infectious peritonitis virus; PHEV, porcine hemagglutinating encephalomyelitis virus; ECoV, equine coronavirus; AIBV, avian infectious bronchitis virus; TCoV, turkey coronavirus; BWCoV, beluga whale coronavirus; BuCoV, bulbul coronavirus; ThCoV, thrush coronavirus; MuCoV, munia coronavirus.

Although three animals were identified as susceptible to SARS-CoV infection, the larger sale volume of civets in comparison to other animals in the market led to them being the focus of subsequent surveillance studies. The role of civets as a major carrier of SARS-CoV in the markets was further confirmed by serological studies involving much large samples (Tu et al., 2004; Kan et al., 2005).

The most detailed epidemiological data proving a direct civet to human transmission of SARS-CoV was obtained during the investigation of the second wave of SARS outbreaks during December 2003 and January 2004. There were two lines of evidences suggesting a direct transmission. First, all four independent cases had the history of direct or indirect contact with civets. Second, sequencing analysis indicated that sequences derived from human samples were more closely related to those in the civets during that period than those from human samples obtained in the major 2002−2003 outbreaks (Wang et al., 2005).

In summary, there is little doubt now that the civet to human transmission is a major source of SARS-CoV introduction into the human population (Wang et al., 2006; Wang and Eaton, 2007; Shi and Hu, 2008).

27.4 Natural Reservoirs of SARS-CoV

A natural reservoir is a long-term host of the pathogen of an infectious disease. It is often the case that hosts do not get the disease carried by the pathogen or the infection in the reservoir host is subclinical, asymptomatic, and nonlethal. Once discovered, natural reservoirs elucidate the complete life cycle of infectious diseases, which in turn will help to provide effective prevention and control strategies.

As stated earlier, it is clear that civets played a pivotal role in the 2002−2004 outbreaks of SARS in southern China. Culling of civets seemed to be effective in controlling further outbreaks in the region. However, the role of civets as a potential natural reservoir host was less evident and eventually ruled out by several studies. Serological studies indicated that only civets in the markets were infected with SARS-CoV, whereas the populations of civets in the wild or on farms are free of major infections (Tu et al., 2004; Lan et al., 2005; Poon et al., 2005). Civets produced overt clinical syndromes when experimentally infected with SARS-CoV (Wu et al., 2005). Comparative genome sequence analysis indicated that SARS-CoVs civets experience rapid ongoing mutation, suggesting that the viruses were still adapting to the host rather than persisting in equilibrium expected for viruses in their natural reservoir species (Kan et al., 2005; Song et al., 2005).

Continuing searching for potential reservoir host of SARS-CoV resulted in the simultaneous discovery of SARS-like coronaviruses (SL-CoVs) in bats by two independent teams in 2005. Using serological and PCR surveillance, both groups discovered that SL-CoVs were present in different horseshoe bats in the genus *Rhinolophus* (Lau et al., 2005; Li et al., 2005c). While neither team was able to isolate live virus from any bat samples, the high level of viral RNA materials enabled them to determine the whole length genome sequence from several different

Table 27.2 Comparison of Gene Products Between SARS-CoV and Bat SL-CoV

Gene/ORF	Gene Product Size (aa)							Amino Acid Sequence Identity with Tor2/SZ3 (%)[a]				
	Tor2	SZ3	Rf1	Rp3	Rm1	HKU3-1	Rs1	Rf1	Rp3	Rm1	HKU3-1	Rs672
P1a	4382	4382	4377	4380	4388	4376	4189	94	96	93	94	94
P1b	2628	2628	2628	2628	2628	2628	2628	98	99	98	98	99
nsp3[b]	1922	1922	1917	1920	1928	1916	1729	92	95	90	92	87
S	1255	1255	1241	1241	1241	1242	1241	76	78	78	78	79
S1	680	680	666	666	666	667	666	63	63	64	6	64
S2	575	575	575	575	575	575	575	92	96	96	94	96
ORF3a	274	274	274	274	274	274	274	86	83	83	82	90
ORF3b	154	154	113	56	56	39	114	89	NA	NA	NA	97
ORF3c	NP	NP	32	NP	NP	NP	NP	NA	NA	NA	NA	NA
E	76	76	76	76	76	76	76	96	100	98	100	100
M	221	221	221	221	221	221	221	97	97	97	99	99
ORF6	63	63	63	63	63	63	63	93	92	92	94	98
ORF7a	122	122	122	122	122	122	122	91	95	93	94	96
ORF7b	44	44	44	44	44	44	44	90	93	93	93	93
ORF8a	39	NP	NP	NP	NP	NP	NP	NA	NA	NA	NA	NA
ORF8b	84	NP	NP	NP	NP	NP	NP	NA	NA	NA	NA	NA
ORF8	NP	122	122	121	121	121	121	80	35	35	34	36
N	422	422	421	421	420	421	422	95	97	97	96	99
ORF9a	98	98	96	97	97	97	98	81	85	90	88	92
ORF9b	70	70	70	70	70	70	70	80	91	91	88	94

NP, not present; NA, not applicable.
[a]Tor2 was used for all homology calculations with the exception of ORF8, which is absent in Tor2, the SZ3 was used instead.
[b]The region of nsp3 is high variable and was calculated alone.

samples. Complete genome sequence analysis revealed that bat SL-CoVs have an identical genome organization and a nucleotide sequence identity of 88–92% to SARS-CoV (Figure 27.1, Table 27.2). Except for the S, ORF3, and ORF8 gene products, all deduced amino acid (aa) sequences of the other gene products have a sequence identity above 93% with those of SARS-CoV. The variable regions between SARS-CoV and bat SL-CoV are mainly located in the coding regions for the nonstructural protein 3 (Nsp 3), S protein, ORF3, and ORF8, the products of these genes have aa sequence identity of 87–95%, 76–78%, 82–90%, 34–80%, respectively. Among the different bat SL-CoVs, the coding regions for these proteins also represent the most variable regions (Ren et al., 2006; Lau et al., 2010; Yuan et al., 2010).

The phylogenetic analysis indicated that bat SL-CoVs were grouped in the same cluster of SARS-CoV and only distantly related to other previously known coronaviruses (Figure 27.2). To date, these bat SL-CoVs represent naturally occurring CoVs, which are most closely related to the SARS-CoVs isolated from humans and civets.

Analysis of nonsynonymous and synonymous substitution rates in bat SL-CoVs suggests that these viruses are not experiencing a positive selection pressure that would be expected if horseshoe bats are new hosts to these viruses. Instead, these data would argue that these viruses have been associated with the bat hosts for a long time (Ren et al., 2006; Tang et al., 2009; Lau et al., 2010). These observations would support the notion that bats in the genus *Rhinolophus* are the likely natural reservoir hosts of bat SL-CoVs. It can be further postulated that similar bat species may serve as natural reservoirs of viruses with closer evolutionary relationship to the viruses that were responsible for the 2002–2004 SARS outbreaks.

27.5 Molecular Evolution of SARS-CoV in Humans and Animals

Analysis of the large number of SARS-CoV and SL-CoV sequence datasets accumulated during the last few years has clearly demonstrated the importance of virus evolution in cross-species transmission and in pathogenesis. The following is a review of the major evolutionary findings in host switching, recombination, and virus–receptor interactions.

27.5.1 Rapid Adaptation of SARS-CoVs in Humans

On the basis of the epidemiological data, the Chinese SARS molecular epidemiology consortium divided the course of the 2002–2004 outbreaks into three stages: early, middle, and late (Chinese, 2004). The early phase is defined as the period from the first emergence of SARS to the first documented SSE. The middle phase refers to the ensuing events up to the first cluster of SARS cases in a hotel (Hotel M) in Hong Kong, while cases following this cluster fall into the late phase.

Analysis of all the viral sequences available from human patients and animals revealed two major hallmarks of rapid virus evolution during the initial stages of the 2002–2003 outbreaks: (1) All isolates from early patients and market animals contained a 29-nt sequence in ORF8 that is absent in most of the publicly available human SARS-CoV sequences derived from later phases of the outbreaks; (2) a characteristic motif of single-nucleotide variations (SNVs) were identified in SARS-CoV of different phases, and all these SNVs were located in the S gene that codes for the spike protein responsible for attachment to host cellular receptor (Song et al., 2005). All SARS-CoV isolates from epidemic countries and regions outside the mainland China could be traced to Guangdong or Hong Kong based on the S-gene SNV motif (Lan et al., 2005; Tang et al., 2007).

During the second sporadic outbreaks of 2003–2004, it was shown that the SARS-CoV sequences from index patients were almost identical to that from civets collected in the same period and all retained the 29-nt sequence in the ORF8 gene. The mild disease symptoms associated with these viruses and the lack of rapid human-to-human transmission provided further evidence that the rapid adaptation of the SARS-CoV in the first major outbreak of 2002–2003 was essential for its establishment and pathogenesis in the humans.

With the available genomic variation data and the sampling time, it is now possible to calculate the neutral mutation rate and to estimate the date for the most recent common ancestors (MRCAs) of SARS-CoV. The estimate obtained is around 8.00×10^{-6}/nt/day, suggesting that SARS-CoV evolves at a relatively constant neutral rate both in human and palm civet. From these calculations, it was estimated that the MRCAs for palm civet and human of different transmission lineages lie in mid-November 2002. This estimate was consistent with the first observed SARS case around November 16, 2002 in Foshan, Guangdong (Chinese, 2004; Song et al., 2005; Zhao, 2007).

27.5.2 Generation of Viral Genetic Diversity by Recombination

At the present time, a total of 18 full-length genome sequences of bat SL-CoVs are determined (Lau et al., 2005, 2010; Li et al., 2005c; Ren et al., 2006; Yuan et al., 2010). Figure 27.1 shows a comparison of the genome structures for five selected bat SL-CoVs and one each of civet and human SARS-CoV isolates. All bat SL-CoVs with the exception of HKU3-8 (Lau et al., 2010) contain the 29-nt sequence in ORF8, which is present in SARS-CoV from early phase patients and civets, indicating the common ancestor between civet SARS-CoV and bat SL-CoV. The SL-CoV HKU3-8 contained a 26-nt deletion, which is located 14-nt downstream from the commonly observed 29-nt deletion, indicating that the ORF8 coding region is a "hotspot" for deletions (Lau et al., 2010).

SL-CoVs from different bat species share 88–92% nt identity among themselves, indicating that the genetic diversity of SL-CoVs in bats is much greater than that observed among civet or human isolates, which provides further support that bats are likely the natural reservoir of this group of coronaviruses. The most dramatic sequence difference between human SARS-CoV and bat SL-CoV is in the S protein, which has only 76–78% aa identity for the whole S protein and 64% aa identity if the N-terminal region (or the S1 region) was compared (Table 27.2). This observed great genetic diversity among bat SL-CoVs and the major difference between the S1 regions of SL-CoV and SARS-CoV S proteins imply that the currently identified SL-CoVs are not the direct progenitor of human SARS-CoV and continued search is required to find a bat SL-CoV that is much closely related to SARS-CoV. It can also be concluded from the earlier observations that genetic recombination may be required to bridge the gap between SL-CoV and SARS-CoV.

It is well documented that the positive-sense ssRNA genomes of coronaviruses are prone to homologous recombination during co infection of two different coronaviruses and that recombination plays an important role in generating new coronavirus species, in facilitating cross-species transmission and in modulating virus virulence (Brian and Baric, 2005; Woo et al., 2005a; Decaro et al., 2009; Woo et al., 2009a; Graham and Baric, 2010).

The first line of evidence for co infection and recombination came from analysis of SL-CoVs in bats (Tang et al., 2006; Cui et al., 2007; Vijaykrishna et al., 2007;

Hon et al., 2008; Lau et al., 2010). Several studies have also confirmed that recombination can occur at multiple sites along the SL-CoV genome (Hon et al., 2008; Graham and Baric, 2010; Lau et al., 2010; Yuan et al., 2010). For example, detailed sequence analysis of two genotypes of bat SL-CoV, Rp3, and Rs672 (both were identified from *R. sinicus*) suggested that they may represent a recombinant of two bat SL-CoVs and one of them is more closely related to the human SARS-CoVs (Hon et al., 2008; Yuan et al., 2010).

Although the exact origin of SARS-CoV remains elusive, it appears reasonable to hypothesize that the virus that successfully infected civets and humans may have evolved from multiple progenitor viruses through mutation and recombination events in one or more reservoir and intermediate hosts.

27.5.3 Receptor Usage and Evolutionary Selection

The S protein of coronavirus is responsible for attachment to cellular receptor to initiate the first step of virus infection. The angiotensin-converting enzyme 2 (ACE2) was identified as a main functional receptor for SARS-CoV (Li et al., 2003). Further analysis demonstrated that the region covering aa 318–520 of S protein is the key receptor binding domain (RBD), which is both essential and sufficient to bind the human ACE2 molecule (Wong et al., 2004). Detailed analysis of the crystal structure of the RBD-ACE2 complex revealed that 19 key residues have close contact with the receptor molecules, which are located from aa 424 to 474. This region is termed the receptor binding motif (RBM) (Li et al., 2005a).

When the existing epidemiological data were analyzed in combination with the data on infectivity of SARS-CoV isolated in humans at the different phases of the outbreaks and SARS-CoV isolates in civets, a clear correlation could be established between the evolution of the S proteins and virus infectivity. It was observed that the S protein is the fastest-evolving protein of SARS-CoV during interspecies transmission from animal to human and in the following phases of human-to-human transmission. The majority of the mutations are located in the S1 domain (31 out of a total of 48 SNVs), particularly in the RBD (Chinese, 2004; Wong et al., 2004). The interaction analysis between the S proteins of different isolates and the ACE2 molecules demonstrated that two aa residues in the S protein, aa 479 and aa 487, played a key role in virus infectivity (Li et al., 2005b; Qu et al., 2005). For aa residue 479, all 2002–2003 human isolates and some 2003–2004 palm civet isolates have a codon for asparagine (N), all 2002–2003 and some 2003–2004 civet isolates have a codon for lysine (K), while some 2003–2004 civet isolates have a codon for arginine (R). For aa residue 487, all isolates including those from early and middle phase patient, civets of 2002–2003 and 2003–2004 have a codon for serine (S), whereas all isolates from late phase patients have a codon for threonine (T) (Figure 27.3). When examined using an HIV-based pseudovirus infection assay, S proteins with all combinations of residues 487/479 could efficiently use the civet ACE2 as an entry receptor but showed different infectivity in human ACE2-mediated infection (Li et al., 2005b; Qu et al., 2005). The combination of N479/T487 had the highest infectivity, N479/S487 medium infectivity, and K479/S487

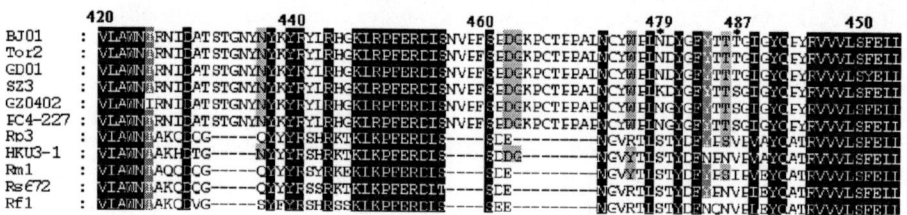

Figure 27.3 Alignment of amino acid sequences covering the RBM from viruses of different species origin. GD01, SARS-CoV isolate from early phase patient during 2002–2003 SARS ourbreak; Tor2, BJ01, SARS-CoV isolate from late phase patient during 2002–2003 SARS ourbreak; SZ, SARS-CoV isolate from civet during 2002–2003 SARS ourbreak; GZ0402, SARS-CoV isolate from patient during 2003–2004 SARS ourbreak; PC4-227, SARS-CoV isolate from civet during 2003–2004 SARS ourbreak. The asterisks (*) indicate two key residues 479 and 487.

the lowest, which almost abolished the infection. These results demonstrated elegantly at the molecular interface that the rapid evolution of the S protein, especially at the aa positions important for host receptor engagement, was essential for the adaptation to and establishment of an effective and productive human infection.

When the genome sequences of SL-CoVs were analyzed, it immediately became evident that the N-terminal region of their S proteins is substantially different from that of the SARS-CoV S proteins. As shown in Figure 27.3, there is a major sequence difference involving deletions of 17–18 aa right in the middle of RBM. We have since demonstrated experimentally that SL-CoV S proteins are unable to use ACE2 molecule as a functional receptor, regardless of the origin. More specifically, ACE2 from bat, civet, or human all failed to function as an entry receptor for a pseudovirus containing the SL-CoV S protein (Ren et al., 2008). It can therefore be concluded that the SL-CoVs identified in bats to date are unlikely to be the direct progenitor virus of human SARS-CoV (Ren et al., 2008).

This raises two questions: (1) What is the functional receptor for SL-CoVs and do these viruses have the potential to spill over into other animals and humans to cause disease like SARs-CoV? (2) Can the SARS-CoV use any bat ACE2 as a functional receptor to satisfy the precondition for bats acting as the natural reservoir of SARS-CoV or a closely related progenitor virus? In the absence of a live SL-CoV, addressing the first question is difficult, and there has been no real progress made in this area. However, recent studies in our group have provided some useful insight into the second question.

Sequence analysis revealed that ACE2 molecules from human and different animals share a significant level of sequence identities, including the key contact points involved in S-ACE2 interaction for SARS-CoV entry (Li et al., 2003, 2005a,b, 2006; Li, 2008). This is also true for the bat ACE2 molecule derived from *R. pearsonii* (Ren et al., 2008). Recently, we have demonstrated that a 3-aa change from SHE to FYQ at aa 40–42 is sufficient for the *R. pearsonii* ACE2 to function as an entry receptor for the human SARS-CoV (Yu et al., 2010). Furthermore, we have also demonstrated that the native ACE2 molecules from other bat species, including the microbat *Myotis daubentonii* and the megabat *Rousettus leschenaultii*, were fully functional as an entry receptor for the human SARS-CoV (Yu et al., 2010; Hou et al., unpublished results). Taking together, these studies demonstrated that a subtle change in sequence was sufficient to convert a nonsusceptible horseshoe bat ACE2 into a functional receptor for SARS-CoV. Considering that there are more than 60 different horseshoe species around the world (Rossiter et al., 2007; Flanders et al., 2009), it is highly conceivable that one or more of them may serve as the natural reservoir of SARS-CoV and/or its progenitor virus(es). Moreover, the existence of functional ACE2 receptors in other bat species would suggest that the host range for SARS-CoV or SL-CoVs in bats may be much wider than originally thought.

27.6 Virus Surveillance in Wild Animals

Zoonosis contributes to the majority of emerging diseases in the last 30 years, many of them originated from wild animals (Bengis et al., 2004; Woolhouse and Gowtage-Sequeria, 2005; Chomel et al., 2007; Jones et al., 2008). The story of SARS is just one of such examples that spectacularly demonstrated the seamless evolution of an animal (probably a bat) virus into a human pathogen responsible for one of the most severe global pandemic outbreaks in the modern history of mankind. In general, pathogens carried by wildlife reservoir animals usually do not cause clinical symptoms, and they lie dormant until they spill over into and cause diseases in domestic animals or humans. Classical outbreak response measures, such as those deployed during the SARS outbreaks, are still useful, but no longer sufficient for early detection and prevention of major infectious disease outbreaks in the twenty-first century.

With the recent demonstration of an increasing number of spillover events that led to severe disease outbreaks in human and domestic animals, we believe it is paramount that from now we include active surveillance of wild animals as part of an integrated infectious disease prevention and control strategy. Surveillance of wildlife animals has also been made more feasible and productive thanks to the advance in modern molecular techniques including PCR with virus group-specific primers, virus discovery using next generation high-throughput sequencing technologies, and high-density virus microarrays (Breitbart et al., 2003; Gaynor et al., 2007; Ng et al., 2009; Yanai-Balser et al., 2010). Since the SARS outbreaks, especially after the discovery of SL-CoVs in bats, there is a significant surge in international effort for surveillance of coronaviruses in wildlife animals. Before the SARS outbreak, there were only 10 coronaviruses with complete genomes sequenced. This number has increased more than threefold as a result of the active surveillance works conducted around the world (Vijgen et al., 2005; Woo et al., 2005b, 2006, 2009a,b; Ren et al., 2006; Vijgen et al., 2006; Alekseev et al., 2008; Lau et al., 2010; Yuan et al., 2010). Although this only marks the beginning of our understanding coronaviruses in wildlife animals, it is fair to say that we have learned a lot more about coronaviruses in general than the past 50 years or so during that period studying of viruses was only possible and called for in response to disease outbreaks. Based on phylogenetic analysis of the large number of bat coronavirus sequences available presently, it is postulated that all known disease-causing coronaviruses previously identified in human or animals originated from bat strains (Vijaykrishna et al., 2007). It is therefore likely that another outbreak could occur on a similar scale as that of the SARS-CoV outbreaks. It is our strong belief that our response to a future outbreak caused by any bat-borne coronavirus will be much more effective than what was done for the SARS outbreaks.

Facilitation of new disease outbreak by prior knowledge accumulated through wildlife surveillance played a major role in the discovery of the new zoonotic reovirus. Melaka virus was a novel bat virus that jumped species and caused direct bat-to-human transmission, followed by human-to-human transmission in a small cluster of patients in Malaysia (Chua et al., 2007). It caused severe respiratory or enteric infections in affected humans and represented a new class of orthoreoviruses, which are capable of infecting and causing disease in humans. The rapid identification and characterization of Melaka virus was made possible by using exiting reagents (primers and antibodies) and knowledge (sequence and reservoir species distribution) gained from a previous surveillance study in bats, which resulted in the discovery of a bat orthoreovirus called Pulau virus (Pritchard et al., 2006). It turned out that the Melaka virus was very closely related to the Pulau virus in genetic organization of the genome segments and in antigenic cross reactivity (Chua et al., 2007). Since then, at least two other related bat have undergone cross-species transmission and caused diseases in humans (Chua et al., 2008; Cheng et al., 2009).

27.7 Concluding Remarks

The emergence of SARS-CoV has had a huge impact on the global health and economy. It served as a warning to what may come out of a seemingly harmless virus-reservoir equilibrium in bats or any other wildlife species. At the same time, the experience gained from the SARS outbreaks and the following in-depth studies on SARS-like coronaviruses has provided and will continue to provide invaluable knowledge and guideline to our future fight against new and emerging infectious diseases. One of the major lessons is that we need to pay much more attention to the reservoir species in understanding the genetic diversity of different viruses, the intricate interplay at the virus−host interface and the major factors responsible for the disturbance of virus−host equilibrium, which in turn trigger spillover events leading to disease outbreaks.

References

Alekseev, K.P., Vlasova, A.N., Jung, K., Hasoksuz, M., Zhang, X., Halpin, R., et al., 2008. Bovine-like coronaviruses isolated from four species of captive wild ruminants are

homologous to bovine coronaviruses, based on complete genomic sequences. J. Virol. 82, 12422−12431.
Bengis, R.G., Leighton, F.A., Fischer, J.R., Artois, M., Morner, T., Tate, C.M., 2004. The role of wildlife in emerging and re-emerging zoonoses. Rev. Sci. Tech. 23, 497−511.
Breitbart, M., Hewson, I., Felts, B., Mahaffy, J.M., Nulton, J., Salamon, P., et al., 2003. Metagenomic analyses of an uncultured viral community from human feces. J. Bacteriol. 185, 6220−6223.
Brian, D.A., Baric, R.S., 2005. Coronavirus genome structure and replication. Curr. Top. Microbiol. Immunol. 287, 1−30.
Cheng, P., Lau, C.S., Lai, A., Ho, E., Leung, P., Chan, F., et al., 2009. A novel reovirus isolated from a patient with acute respiratory disease. J. Clin. Virol. 45, 79−80.
Chinese, S.M.E.C., 2004. Molecular evolution of the SARS coronavirus during the course of the SARS epidemic in China. Science 303, 1666−1669.
Chomel, B.B., Belotto, A., Meslin, F.X., 2007. Wildlife, exotic pets, and emerging zoonoses. Emerg. Infect. Dis. 13, 6−11.
Chua, K.B., Crameri, G., Hyatt, A., Yu, M., Tompang, M.R., Rosli, J., et al., 2007. A previously unknown reovirus of bat origin is associated with an acute respiratory disease in humans. Proc. Natl. Acad. Sci. U.S.A. 104, 11424−11429.
Chua, K.B., Voon, K., Crameri, G., Tan, H.S., Rosli, J., McEachern, J.A., et al., 2008. Identification and characterization of a new orthoreovirus from patients with acute respiratory infections. PLoS One 3, e3803.
Cui, J., Han, N., Streicker, D., Li, G., Tang, X., Shi, Z., et al., 2007. Evolutionary relationships between bat coronaviruses and their hosts. Emerg. Infect. Dis. 13, 1526−1532.
Decaro, N., Mari, V., Campolo, M., Lorusso, A., Camero, M., Elia, G., et al., 2009. Recombinant canine coronaviruses related to transmissible gastroenteritis virus of swine are circulating in dogs. J. Virol. 83, 1532−1537.
Drosten, C., Gunther, S., Preiser, W., van der Werf, S., Brodt, H.R., Becker, S., et al., 2003. Identification of a novel coronavirus in patients with severe acute respiratory syndrome. N. Engl. J. Med. 348, 1967−1976.
Flanders, J., Jones, G., Benda, P., Dietz, C., Zhang, S., Li, G., et al., 2009. Phylogeography of the greater horseshoe bat, *Rhinolophus ferrumequinum*: contrasting results from mitochondrial and microsatellite data. Mol. Ecol. 18, 306−318.
Gaynor, A.M., Nissen, M.D., Whiley, D.M., Mackay, I.M., Lambert, S.B., Wu, G., et al., 2007. Identification of a novel polyomavirus from patients with acute respiratory tract infections. PLoS Pathog. 3, e64.
Graham, R.L., Baric, R.S., 2010. Recombination, reservoirs, and the modular spike: mechanisms of coronavirus cross-species transmission. J. Virol. 84, 3134−3146.
Guan, Y., Zheng, B.J., He, Y.Q., Liu, X.L., Zhuang, Z.X., Cheung, C.L., et al., 2003. Isolation and characterization of viruses related to the SARS coronavirus from animals in southern China. Science (New York, NY) 302, 276−278.
He, J.F., Xu, R.H., Yu, D.W., Peng, G.W., Liu, Y.Y., Liang, W.J., et al., 2003. [Severe acute respiratory syndrome in Guangdong Province of China: epidemiology and control measures]. Zhonghua Yu Fang Yi Xue Za Zhi 37, 227−232.
Hon, C.C., Lam, T.Y., Shi, Z.L., Drummond, A.J., Yip, C.W., Zeng, F., et al., 2008. Evidence of the recombinant origin of a bat severe acute respiratory syndrome (SARS)-like coronavirus and its implications on the direct ancestor of SARS coronavirus. J. Virol. 82, 1819−1826.
Jones, K.E., Patel, N.G., Levy, M.A., Storeygard, A., Balk, D., Gittleman, J.L., et al., 2008. Global trends in emerging infectious diseases. Nature 451, 990−993.

Kan, B., Wang, M., Jing, H., Xu, H., Jiang, X., Yan, M., et al., 2005. Molecular evolution analysis and geographic investigation of severe acute respiratory syndrome coronavirus-like virus in palm civets at an animal market and on farms. J. Virol. 79, 11892–11900.

Ksiazek, T.G., Erdman, D., Goldsmith, C.S., Zaki, S.R., Peret, T., Emery, S., et al., 2003. A novel coronavirus associated with severe acute respiratory syndrome. N. Engl. J. Med. 348, 1953–1966.

Lan, Y.C., Liu, T.T., Yang, J.Y., Lee, C.M., Chen, Y.J., Chan, Y.J., et al., 2005. Molecular epidemiology of severe acute respiratory syndrome-associated coronavirus infections in Taiwan. J. Infect. Dis. 191, 1478–1489.

Lau, S.K., Woo, P.C., Li, K.S., Huang, Y., Tsoi, H.W., Wong, B.H., et al., 2005. Severe acute respiratory syndrome coronavirus-like virus in Chinese horseshoe bats. Proc. Natl. Acad. Sci. U.S.A. 102, 14040–14045.

Lau, S.K., Li, K.S., Huang, Y., Shek, C.T., Tse, H., Wang, M., et al., 2010. Ecoepidemiology and complete genome comparison of different strains of severe acute respiratory syndrome-related *Rhinolophus* bat coronavirus in China reveal bats as a reservoir for acute, self-limiting infection that allows recombination events. J. Virol. 84, 2808–2819.

Li, F., 2008. Structural analysis of major species barriers between humans and palm civets for severe acute respiratory syndrome coronavirus infections. J. Virol. 82, 6984–6991.

Li, W., Moore, M.J., Vasilieva, N., Sui, J., Wong, S.K., Berne, M.A., et al., 2003. Angiotensin-converting enzyme 2 is a functional receptor for the SARS coronavirus. Nature 426, 450–454.

Li, F., Li, W., Farzan, M., Harrison, S.C., 2005a. Structure of SARS coronavirus spike receptor-binding domain complexed with receptor. Science (New York, NY) 309, 1864–1868.

Li, W., Zhang, C., Sui, J., Kuhn, J.H., Moore, M.J., Luo, S., et al., 2005b. Receptor and viral determinants of SARS-coronavirus adaptation to human ACE2. EMBO J. 24, 1634–1643.

Li, W.D., Shi, Z.L., Yu, M., Ren, W.Z., Smith, C., Epstein, J.H., et al., 2005c. Bats are natural reservoirs of SARS-like coronaviruses. Science (New York, NY) 310, 676–679.

Li, W., Wong, S.K., Li, F., Kuhn, J.H., Huang, I.C., Choe, H., et al., 2006. Animal origins of the severe acute respiratory syndrome coronavirus: insight from ACE2-S-protein interactions. J. Virol. 80, 4211–4219.

Lim, W., Ng, K.C., Tsang, D.N., 2006. Laboratory containment of SARS virus. Ann. Acad. Med. Singapore 35, 354–360.

Marra, M.A., Jones, S.J.M., Astell, C.R., Holt, R.A., Brooks-Wilson, A., Butterfield, Y.S.N., et al., 2003. The genome sequence of the SARS-associated coronavirus. Science 300, 1399–1404.

Ng, T.F., Manire, C., Borrowman, K., Langer, T., Ehrhart, L., Breitbart, M., 2009. Discovery of a novel single-stranded DNA virus from a sea turtle fibropapilloma by using viral metagenomics. J. Virol. 83, 2500–2509.

Peiris, J.S., Lai, S.T., Poon, L.L., Guan, Y., Yam, L.Y., Lim, W., et al., 2003. Coronavirus as a possible cause of severe acute respiratory syndrome. Lancet 361, 1319–1325.

Poon, L.L., Chu, D.K., Chan, K.H., Wong, O.K., Ellis, T.M., Leung, Y.H., et al., 2005. Identification of a novel coronavirus in bats. J. Virol. 79, 2001–2009.

Pritchard, L.I., Chua, K.B., Cummins, D., Hyatt, A., Crameri, G., Eaton, B.T., et al., 2006. Pulau virus; a new member of the Nelson Bay orthoreovirus species isolated from fruit bats in Malaysia. Arch. Virol. 151, 229–239.

Qu, X.X., Hao, P., Song, X.J., Jiang, S.M., Liu, Y.X., Wang, P.G., et al., 2005. Identification of two critical amino acid residues of the severe acute respiratory syndrome coronavirus spike protein for its variation in zoonotic tropism transition via a double substitution strategy. J. Biol. Chem. 280, 29588–29595.

Ren, W., Li, W., Yu, M., Hao, P., Zhang, Y., Zhou, P., et al., 2006. Full-length genome sequences of two SARS-like coronaviruses in horseshoe bats and genetic variation analysis. J. Gen. Virol. 87, 3355–3359.

Ren, W., Qu, X., Li, W., Han, Z., Yu, M., Zhou, P., et al., 2008. Difference in receptor usage between severe acute respiratory syndrome (SARS) coronavirus and SARS-like coronavirus of bat origin. J. Virol. 82, 1899–1907.

Rossiter, S.J., Benda, P., Dietz, C., Zhang, S., Jones, G., 2007. Rangewide phylogeography in the greater horseshoe bat inferred from microsatellites: implications for population history, taxonomy and conservation. Mol. Ecol. 16, 4699–4714.

Rota, P.A., Oberste, M.S., Monroe, S.S., Nix, W.A., Campagnoli, R., Icenogle, J.P., et al., 2003. Characterization of a novel coronavirus associated with severe acute respiratory syndrome. Science 300, 1394–1399.

Shi, Z., Hu, Z., 2008. A review of studies on animal reservoirs of the SARS coronavirus. Virus Res. 133, 74–87.

Snijder, E.J., Bredenbeek, P.J., Dobbe, J.C., Thiel, V., Ziebuhr, J., Poon, L.L.M., et al., 2003. Unique and conserved features of genome and proteome of SARS-coronavirus, an early split-off from the coronavirus group 2 lineage. J. Mol. Biol. 331, 991–1004.

Song, H.D., Tu, C.C., Zhang, G.W., Wang, S.Y., Zheng, K., Lei, L.C., et al., 2005. Cross-host evolution of severe acute respiratory syndrome coronavirus in palm civet and human. Proc. Natl. Acad. Sci. U.S.A. 102, 2430–2435.

Tamura, K., Dudley, J., Nei, M., Kumar, S., 2007. MEGA4: Molecular Evolutionary Genetics Analysis (MEGA) software version 4.0. Mol. Biol. Evol. 24, 1596–1599.

Tang, X.C., Zhang, J.X., Zhang, S.Y., Wang, P., Fan, X.H., Li, L.F., et al., 2006. Prevalence and genetic diversity of coronaviruses in bats from China. J. Virol. 80, 7481–7490.

Tang, J.W., Cheung, J.L., Chu, I.M., Ip, M., Hui, M., Peiris, M., et al., 2007. Characterizing 56 complete SARS-CoV S-gene sequences from Hong Kong. J. Clin. Virol. 38, 19–26.

Tang, X., Li, G., Vasilakis, N., Zhang, Y., Shi, Z., Zhong, Y., et al., 2009. Differential stepwise evolution of SARS coronavirus functional proteins in different host species. BMC Evol. Biol. 9, 52.

Tu, C., Crameri, G., Kong, X., Chen, J., Sun, Y., Yu, M., et al., 2004. Antibodies to SARS coronavirus in civets. Emerg. Infect. Dis. 10, 2244–2248.

Vijaykrishna, D., Smith, G.J., Zhang, J.X., Peiris, J.S., Chen, H., Guan, Y., 2007. Evolutionary insights into the ecology of coronaviruses. J. Virol. 81, 4012–4020.

Vijgen, L., Keyaerts, E., Moes, E., Thoelen, I., Wollants, E., Lemey, P., et al., 2005. Complete genomic sequence of human coronavirus OC43: molecular clock analysis suggests a relatively recent zoonotic coronavirus transmission event. J. Virol. 79, 1595–1604.

Vijgen, L., Keyaerts, E., Lemey, P., Maes, P., Van Reeth, K., Nauwynck, H., et al., 2006. Evolutionary history of the closely related group 2 coronaviruses: porcine hemagglutinating encephalomyelitis virus, bovine coronavirus, and human coronavirus OC43. J. Virol. 80, 7270–7274.

Wang, L.F., Eaton, B.T., 2007. Bats, civets and the emergence of SARS. Curr. Top. Microbiol. Immunol. 315, 325–344.

Wang, M., Yan, M., Xu, H., Liang, W., Kan, B., Zheng, B., et al., 2005. SARS-CoV infection in a restaurant from palm civet. Emerg. Infect. Dis. 11, 1860–1865.

Wang, L.F., Shi, Z., Zhang, S., Field, H., Daszak, P., Eaton, B.T., 2006. Review of bats and SARS. Emerg. Infect. Dis. 12, 1834–1840.
WHO, 2003. Severe acute respiratory syndrome (SARS). Wkly. Epidemiol. Rec. 78, 81–88.
Wong, S.K., Li, W., Moore, M.J., Choe, H., Farzan, M., 2004. A 193-amino acid fragment of the SARS coronavirus S protein efficiently binds angiotensin-converting enzyme 2. J. Biol. Chem. 279, 3197–3201.
Woo, P.C., Lau, S.K., Huang, Y., Tsoi, H.W., Chan, K.H., Yuen, K.Y., 2005a. Phylogenetic and recombination analysis of coronavirus HKU1, a novel coronavirus from patients with pneumonia. Arch. Virol. 150, 2299–2311.
Woo, P.C., Lau, S.K., Chu, C.M., Chan, K.H., Tsoi, H.W., Huang, Y., et al., 2005b. Characterization and complete genome sequence of a novel coronavirus, coronavirus HKU1, from patients with pneumonia. J. Virol. 79, 884–895.
Woo, P.C., Lau, S.K., Yip, C.C., Huang, Y., Tsoi, H.W., Chan, K.H., et al., 2006. Comparative analysis of 22 coronavirus HKU1 genomes reveals a novel genotype and evidence of natural recombination in coronavirus HKU1. J. Virol. 80, 7136–7145.
Woo, P.C., Lau, S.K., Huang, Y., Yuen, K.Y., 2009a. Coronavirus diversity, phylogeny and interspecies jumping. Exp. Biol. Med. (Maywood) 234, 1117–1127.
Woo, P.C., Lau, S.K., Lam, C.S., Lai, K.K., Huang, Y., Lee, P., et al., 2009b. Comparative analysis of complete genome sequences of three avian coronaviruses reveals a novel group 3c coronavirus. J. Virol. 83, 908–917.
Woolhouse, M.E., Gowtage-Sequeria, S., 2005. Host range and emerging and reemerging pathogens. Emerg. Infect. Dis. 11, 1842–1847.
Wu, D.L., Tu, C.C., Xin, C., Xuan, H., Meng, Q.W., Liu, Y.G., et al., 2005. Civets are equally susceptible to experimental infection by two different severe acute respiratory syndrome coronavirus isolates. J. Virol. 79, 2620–2625.
Xu, H.F., Wang, M., Zhang, Z.B., Zou, X.Z., Gao, Y., Liu, X.N., et al., 2004a. [An epidemiologic investigation on infection with severe acute respiratory syndrome coronavirus in wild animals traders in Guangzhou]. Zhonghua Yu Fang Yi Xue Za Zhi 38, 81–83.
Xu, R.H., He, J.F., Evans, M.R., Peng, G.W., Field, H.E., Yu, D.W., et al., 2004b. Epidemiologic clues to SARS origin in china. Emerg. Infect. Dis. 10, 1030–1037.
Yanai-Balser, G.M., Duncan, G.A., Eudy, J.D., Wang, D., Li, X., Agarkova, I.V., et al., 2010. Microarray analysis of *Paramecium bursaria* chlorella virus 1 transcription. J. Virol. 84, 532–542.
Yu, M., Tachedijian, M., Crameri, G., Shi, Z., Wang, L.F., 2010. Identification of key amino acid residues required for horseshoe bat ACE2 to function as a receptor for the SARS coronavirus. J. Gen. Virol. 91, 1708.
Yuan, J., Hon, C.C., Li, Y., Wang, D., Xu, G., Zhang, H., et al., 2010. Intraspecies diversity of SARS-like coronaviruses in *Rhinolophus sinicus* and its implications for the origin of SARS coronaviruses in humans. J. Gen. Virol. 91, 1058–1062.
Zhao, G.P., 2007. SARS molecular epidemiology: a Chinese fairy tale of controlling an emerging zoonotic disease in the genomics era. Philos. Trans. R. Soc. Lond. B Biol. Sci. 362, 1063–1081.

28 Ecology and Evolution of Avian Influenza Viruses

Josanne H. Verhagen[1], Vincent J. Munster[2] and Ron A.M. Fouchier[1],*

[1]Department of Virology and National Influenza Center, Erasmus Medical Center, Rotterdam, The Netherlands, [2]Laboratory of Virology, Rocky Mountain Laboratories, National Institute of Allergy and Infectious Diseases, National Institutes of Health, Hamilton, MT, USA

28.1 Introduction to Influenza A Virus

28.1.1 Taxonomy and Host Range

Influenza A viruses belong to the family *Orthomyxoviridae*, a family of viruses with a negative sense, single-stranded, segmented RNA genome (Palese and Shaw, 2007; Wright et al., 2007). Other members of the family include the human pathogens influenza B virus and influenza C virus, the tick-borne viruses in the *Thogotovirus* genus, and the *Isavirus* infectious salmon anemia virus (International Committee on Taxonomy of Viruses, 2000). Influenza A viruses have been isolated from many host species including humans, pigs, horses, mink, cats, dogs, marine mammals, and a wide range of domestic birds, but wild birds in the orders *Anseriformes* (ducks, geese, and swans) and *Charadriiformes* (gulls, terns, and waders) are thought to form the virus reservoir in nature. From this virus reservoir, viruses may be transmitted occasionally to other animals and humans, the latter generally via intermediate hosts such as domestic birds and pigs (Figure 28.1; Webster et al., 1992).

28.1.2 Influenza A Virus Structure and Genome Organization

The influenza A virus genome consists of 8 gene segments (Figure 28.2; Palese and Shaw, 2007; Wright et al., 2007). Segments 1—3 encode the polymerase proteins basic polymerase 2 (PB2), basic polymerase 1 (PB1), and acidic polymerase (PA) respectively. Segment 2 also encodes a second open reading frame, PB1-F2, which has been implicated in the induction of cell death (Chen et al., 2001; Conenello and Palese, 2007). Segments 4 and 6 encode the viral surface

*E-mail: r.fouchier@erasmusmc.nl

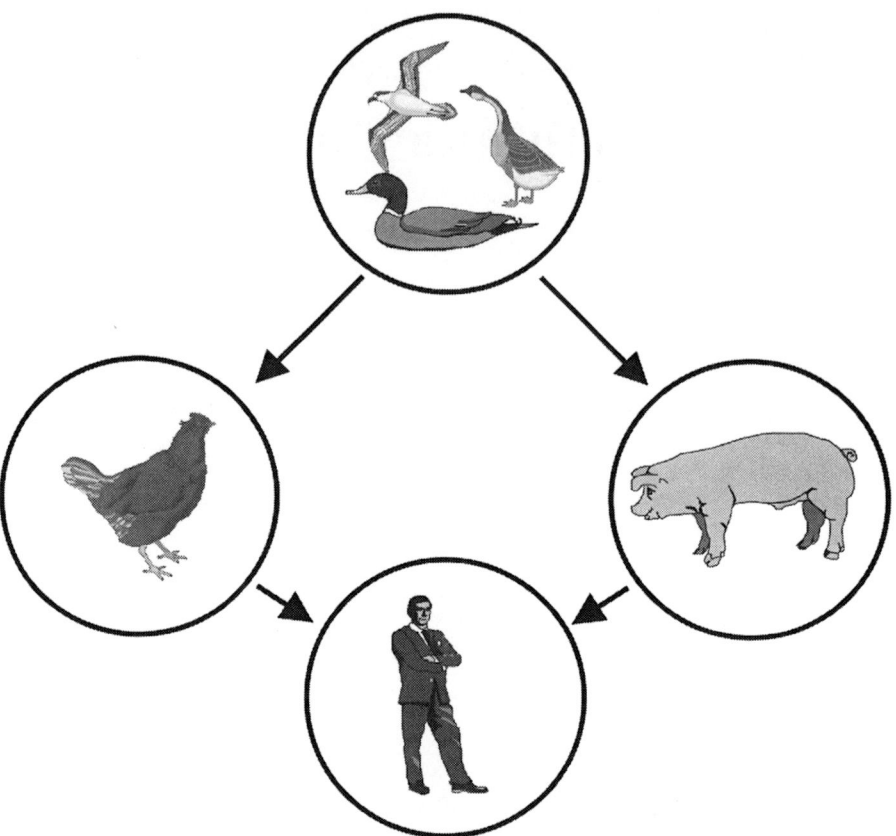

Figure 28.1 Wild birds form the reservoir for influenza A viruses in nature. Avian viruses are occasionally transmitted to other hosts, in which they may cause serious disease. Only major routes of transmission are indicated with arrows, and in this chapter, only influenza viruses of wild and domestic birds are discussed.

glycoproteins hemagglutinin (HA) and neuraminidase (NA). HA is responsible for binding to sialic acids, the viral receptors on host cells, and for fusion of the viral and host cell membranes upon endocytosis. NA is a sialidase, responsible for cleaving sialic acids from virus-producing host cells and virus particles, thus facilitating virus release. Segment 5 codes for the nucleocapsid protein (NP) that binds to viral RNA and together with the polymerase proteins forms the ribonucleoprotein (RNP) complexes. Segment 7 codes for the classical viral matrix structural protein M1, and the ion channel protein M2 that is incorporated in the viral membrane. Segment 8 encodes the nonstructural protein NS1 and the nucleic export protein (NEP) previously known as NS2. NS1 is an antagonist of host innate immune responses and interferes with host gene expression, while NEP is involved in the nuclear export of RNPs into the cytoplasm prior to virion assembly (Palese and Shaw, 2007; Wright et al., 2007).

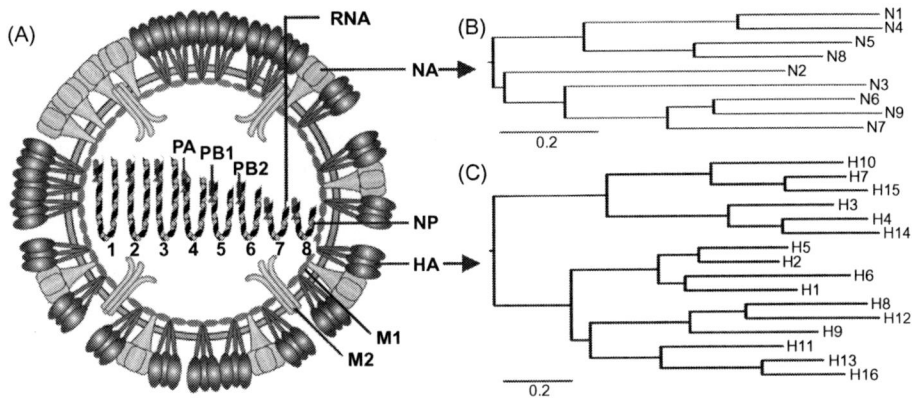

Figure 28.2 Influenza A virus and genetic variation of the surface glycoproteins HA and NA. A schematic presentation of an influenza A virus particle with its eight RNA gene segments (numbered) and virus-associated proteins (named) is shown in (A). (B and C) represent DNA-ML phylogenetic trees to visualize the genetic variation of the surface glycoprotein genes NA (B) and HA (C) in the wild bird reservoir. Contemporary representative Eurasian wild bird virus sequences were used to construct both trees. *Source*: Panel (A) adapted from Karlsson Hedestam et al. (2008).

28.1.3 Influenza A Virus Classification

Influenza A viruses are highly variable. Most pronounced is the genetic and antigenic variation of the surface glycoproteins HA and NA (Webster et al., 1992). Influenza A viruses can be classified based on the antigenic variation of these HA and NA proteins. To date, 16 major antigenic variants of HA and 9 antigenic variants of NA have been recognized which are found in numerous combinations, and these so-called virus subtypes (e.g., H1N1, H7N7, H16N3) are used in influenza A virus classification and nomenclature (Webster et al., 1992; Fouchier et al., 2005). The classification system is biologically relevant, as host antibodies that recognize one HA or NA subtype will generally not cross-react with other HA and NA subtypes. The antigenic variation of the HA and NA proteins is in agreement with the major genetic variation of the respective genes of avian influenza A viruses (Figure 28.2). For instance, the maximum amino acid sequence identity between HA of any two different HA subtypes is 79%, and within a subtype can be as low as 86% (Fouchier et al., 2005). The genetic variation of the HA and NA genes in the avian reservoir is of the same order of magnitude as the genetic variation of the surface glycoproteins of primate lentiviruses, a notoriously variable group of viruses (Karlsson Hedestam et al., 2008).

Besides classification based on the antigenic properties of HA and NA, avian influenza A viruses can be classified based on their pathogenic phenotype in chickens. Highly pathogenic avian influenza (HPAI) is an acute generalized disease of poultry, for which mortality may be as high as 100%. HPAI is restricted to infections with influenza A virus of the H5 and H7 subtypes. All other avian influenza

A virus subtypes are low pathogenic avian influenza (LPAI) viruses, which cause a much milder disease. The vast majority of H5 and H7 influenza A viruses are also of the LPAI phenotype. HPAI viruses are generally thought to arise in poultry upon introduction of LPAI H5 and H7 viruses from the wild bird reservoir (Alexander, 2000, 2007a). The HA protein of influenza A viruses is initially synthesized as a single polypeptide precursor (HA_0), which is cleaved into HA_1 and HA_2 subunits by trypsin-like proteases. The switch from the LPAI to the HPAI virus phenotype for H5 and H7 viruses is generally achieved by the introduction of basic amino acid residues into the HA_0 cleavage site. In contrast to the HA of LPAI viruses, a cleavage site in HA that is composed of multiple basic amino acid residues can be cleaved by ubiquitously expressed proteases and it is generally believed that this property facilitates systemic virus replication and a mortality of up to 100% in poultry (Webster and Rott, 1987; Alexander, 2007a).

28.2 Influenza Viruses in Birds

28.2.1 LPAI Virus Subtypes in Wild Birds

The first recorded isolation of influenza A virus from wild birds was from common terns (*Sterna hirundo*) in 1961 in South Africa, when at least 1300 of these birds died in an outbreak (Becker, 1966). This outbreak was unusual because it is the only recorded outbreak caused by an HPAI virus in wild birds with no direct evidence for association with poultry. The viruses most frequently detected in gulls are LPAI viruses of subtypes H13 and H16 (Hinshaw et al., 1982; Kawaoka et al., 1988; Fouchier et al., 2005; Munster et al., 2007). Gulls and terns appear to represent the H13 and H16 reservoir hosts in nature, as these subtypes are frequently found in gulls and terns but only very rarely in other bird species.

Ducks, in particular dabbling ducks such as mallards (*Anas platyrhynchos*), have been most frequently reported to be infected with avian influenza viruses in surveillance studies (Hinshaw et al., 1980a,b; Hanson et al., 2003; Krauss et al., 2004; Olsen et al., 2006; Wallensten et al., 2006; Krauss et al., 2007; Munster et al., 2007; Wallensten et al., 2007; Ellstrom et al., 2008; Parmley et al., 2008). In contrast to H13 and H16 influenza viruses, influenza virus subtypes H1–H12 have been detected predominantly in wild ducks and in a wide variety of HA/NA subtype combinations.

Extensive surveillance studies in wader species within the *Charadriidae* and *Scolopacidae* families have primarily been reported from North America. LPAI viruses of subtypes H1–12 have been isolated from waders in the eastern USA (Kawaoka et al., 1988; Krauss et al., 2004). LPAI viruses have also been detected in waders in Europe, Africa, and Australia, but at relatively low prevalence (Hurt et al., 2006; Munster et al., 2007; Gaidet et al., 2008).

LPAI viruses have been detected in numerous other bird species as well (Olsen et al., 2006), but it is unclear whether the viruses are truly enzootic in these species or whether these birds are "transient" host species. Such transient host species may share the same habitat for part of the year with species in which LPAI virus prevalence is relatively high, such as ducks. For instance, in geese, swans, rails, petrels, cormorants, passerines, and other bird species, LPAI viruses are detected occasionally but prevalence is generally much lower than in dabbling ducks. Due to limitations of wild bird surveillance studies, it cannot be excluded that LPAI virus reservoir species may exist beyond the orders *Anseriformes* and *Charadriiformes*.

While viruses of subtypes H1 through H12, H13, and H16 thus have consistently been detected in global surveillance studies in aquatic birds throughout the years and in a wide range of geographical locations, H14 and H15 subtype viruses have only been detected in a few years (H14; 1982, H15; 1979–1983) and in a few locations (H14; former USSR, H15; Australia). It is unclear whether these virus subtypes are still circulating today, and if so, in which species (Kawaoka et al., 1990; Rohm et al., 1996).

28.2.2 LPAI Virus Transmission and Epidemiology in Wild Birds

The prevalence of LPAI viruses in wild birds varies with geographical location, time of the year, and bird species (Olsen et al., 2006). Although general patterns of virus prevalence have been described, large variations within these patterns may exist from year to year and between different surveillance studies. In mallards in North America and Europe, seasonal prevalence varies from very low (<1%) in spring and summer to very high ($\sim$30%) during fall migration and winter (Halvorson et al., 1985; Webster et al., 1992; Munster et al., 2007; Wallensten et al., 2007). The peak prevalence in fall migration is believed to be related to the large numbers of young, immunologically naive birds of the breeding season that aggregate prior to and during southbound migration (Hinshaw et al., 1978; Webster et al., 1992). Indeed, age-related differences in LPAI virus prevalence were detected in mallards and Eurasian wigeons (*Anas penelope*) in Europe, with virus prevalence of 6.8% in juvenile ducks and 2.8% in adults (Munster et al., 2007). The peak virus prevalence likely gradually declines when migration proceeds, forming a gradient in LPAI virus prevalence from the breeding grounds of the birds in the north to the wintering areas in the south (Munster et al., 2007). It is likely that with increasing age, ducks mount an immune response that limits subsequent infections with LPAI viruses. Upon experimental inoculation of mallards, it has been shown that birds could become re-infected with a heterologous LPAI virus subtype, but that the duration of virus shedding was markedly reduced. This reduction was more pronounced upon re-infection with an LPAI virus of homologous HA subtype (Kida et al., 1980; Fereidouni et al., 2009; Jourdain et al., 2010). Also under field conditions, consecutive or simultaneous infections with different LPAI virus subtypes are common in dabbling ducks, suggesting that heterosubtypic immunity induced by LPAI viruses indeed is only partial (Latorre-Margalef et al., 2009). In wild-caught mallards, it was further estimated that virus shedding occurred for $\sim$10 days (Latorre-Margalef et al., 2009), in agreement with the duration of virus shedding under experimental conditions of 7–17 days (Kida et al., 1980; Fereidouni et al., 2009; Jourdain et al., 2010).

Also in gulls, there is evidence that juveniles play a key role in starting local outbreaks of LPAI. The highest virus prevalence reported in gulls is late summer, shortly after the introduction of naïve juveniles into the population (Fouchier et al., 2005; Munster et al., 2007). Most gull species breed in dense colonies, with adults and juveniles crowded in a small space, potentially creating good opportunities for virus spread. Breeding of gulls in dense colonies contrasts with dabbling ducks which do not breed in dense colonies (Del Hoyo et al., 1996). This could provide an explanation why epizootics in gulls start within the breeding colony, while epizootics in ducks are more likely initiated when birds congregate in large numbers during molt, migration, or wintering.

In waders migrating along the Atlantic America flyway, which runs along the Atlantic Coast of North America, the patterns of LPAI virus presence appeared to be reversed compared to the prevalence in ducks. Peak LPAI virus prevalence of ~14% was observed for waders, most notably ruddy turnstones (*Arenaria interpres*), during their spring migration in Delaware Bay rather than fall migration (Krauss et al., 2004). This has led Krauss et al. to hypothesize that waders and ducks could both be important reservoir hosts to maintain the annual cycle of LPAI virus epidemics, in which ducks carry viruses southbound in the fall and waders bring the viruses northbound in spring. While this may be a valid hypothesis based on data from North America, data supporting such an annual cycle elsewhere is not available. In Europe, Australia, and Alaska, LPAI virus prevalence in waders was shown to be low, and hot spots for LPAI virus detection equivalent to Delaware Bay have not been found elsewhere in the world (Hurt et al., 2006; Munster et al., 2007; Hanson et al., 2008; Winker et al., 2008; Haynes et al., 2009).

As opposed to the endemic behavior of LPAI viruses in dabbling ducks, the transient virus prevalence in several other *Anseriformes* species suggests a more epidemic behavior in those species. For instance, in white-fronted geese in Northern Europe, LPAI viruses were only detected upon their arrival at the wintering grounds in The Netherlands, suggesting the birds only contracted LPAI viruses after contact with other reservoir species, such as mallards (Kleijn et al., 2010). It is possible that many virus hosts identified in surveillance studies act as "transient hosts," while a more limited number of species act as "reservoir hosts," in which LPAI viruses are truly endemic.

Many LPAI virus host species migrate regularly over long distances. During migration, birds have the potential to distribute viruses between countries or continents. Within the large continents and along the major flyways, migration connects bird populations in time and space, either at common breeding areas, common foraging areas during migration, or at shared nonbreeding areas. As a result, bird populations may transmit their pathogens to new migratory and nonmigratory populations and to new areas. Virus transmission and geographical spread is thus dependent on the ecology of the migrating hosts. Migrating birds rarely fly the full distance between breeding and nonbreeding areas without stopping over and "refueling" along the way. Rather, birds make frequent stopovers during migration, and spend more time foraging and preparing for migration than actively performing flights. Many species aggregate at favorable stopover or wintering sites, resulting in high local densities. Such sites may be important for transmission of viruses between different species of wild birds (Olsen et al., 2006).

The maintenance and circulation of LPAI viruses within the wild bird host populations relies on effective transmission of the virus between susceptible hosts or host populations. LPAI viruses generally infect cells lining the intestinal tract and are believed to be transmitted primarily via the fecal−oral route (Webster et al., 1992). LPAI viruses can stay infectious for prolonged periods of time in surface waters, potentially allowing viruses to infect different bird populations (Stallknecht et al., 1990a,b; Brown et al., 2009; Stallknecht and Brown, 2009). Dabbling ducks feed mainly on water surfaces, allowing effective fecal−oral transmission. Diving ducks forage at deeper depths and more often in marine habitats. Dabbling ducks display a propensity for migration and the switching of breeding grounds between years, in part due to mate choice (Del Hoyo et al., 1996). These behavioral differences between ecological guilds of ducks could provide an explanation for differences in LPAI virus prevalence in different species.

Given the relatively short duration of LPAI virus shedding by individual infected birds, the spatial and temporal dynamics of LPAI virus circulation may be explained by the continuous circulation within and between bird flocks, or by viral persistence in abiotic reservoirs such as lakes. The data presently available is insufficient to exclude either of these two possibilities. It has been shown that LPAI viruses remain infectious in surface waters at cold conditions for considerable periods of time, and hence a role of abiotic reservoirs is plausible (Stallknecht and Brown, 2009). On the other hand, the continuous LPAI virus prevalence throughout the annual cycle in dabbling ducks may be sufficient for the year-round perpetuation of viruses in these species without the need for environmental persistence (Wallensten et al., 2006).

28.2.3 LPAI and HPAI Viruses in Domestic Birds

Influenza viruses may infect virtually all species of domestic birds, depending mostly on their direct contact with wild birds and their excretions or indirect contact via human activities. In general, influenza viruses originating from wild birds do not cause serious disease in domestic birds. Infection of domestic birds with most influenza virus subtypes may result in decreased egg production, mild respiratory disease, and other mild clinical symptoms, referred to as LPAI. Most LPAI outbreaks have a limited duration and geographical scale, although large-scale and long-term outbreaks have been reported, for instance, the outbreaks caused by viruses of the H9N2 subtype in the eastern hemisphere (Alexander, 2000, 2007a). HPAI viruses can emerge when LPAI viruses of the H5 or H7 subtype are introduced from wild birds into poultry, through a change in the HA cleavage site (Webster and Rott, 1987). These HPAI viruses can have a devastating impact on chickens and turkeys, with mortality rates of $\sim 100\%$ (Alexander, 2000, 2007a). In the past two decades, HPAI outbreaks have occurred frequently, caused by influenza viruses of subtype H5N1 in Asia, Russia, the Middle East, Europe, and Africa (ongoing since 1997, see later), H5N2 in Mexico (1994), Italy (1997), and Texas

(2004), South Africa (2004), H7N1 in Italy (1999), H7N3 in Australia (1994), Pakistan (1994), Chile (2002) and Canada (2003), H7N4 in Australia (1997), and H7N7 in the Netherlands (2003), North Korea (2005) and England (2008) (Alexander, 2000, 2007b; Alexander and Brown, 2009).

While most HPAI outbreaks have been controlled relatively quickly, HPAI H5N1 virus has been circulating in poultry continuously since 1997. In 1997, the HPAI H5N1 virus was detected in chicken farms and the live bird markets of Hong Kong, and caused the first reported case of human influenza and fatality attributable directly to avian influenza virus (de Jong et al., 1997). The H5N1 HPAI virus reappeared in 2002 in waterfowl in two parks in Hong Kong, and was detected in other captive and wild birds as well (Ellis et al., 2004). In 2003 the virus resurfaced again, and devastated the poultry industry in large parts of Southeastern Asia since 2004. In 2005, the virus was isolated during an outbreak among migratory birds in Qinghai Lake, China, where it affected large numbers of wild birds (Liu et al., 2005). After this event, the virus has appeared across Asia, Europe, the Middle East, and in several African countries. The large-scale spread is thought to be caused by both human activities and wild bird movements. The numbers of poultry involved in these outbreaks is unknown, but is likely in order of hundreds of millions (Alexander and Brown, 2009). In addition, the virus has continued to cause disease and fatalities in humans, due to zoonotic transmissions (transmissions from birds to humans) in numerous countries.

28.2.4 HPAI H5N1 Virus in Wild Birds

Compared to all other HPAI virus outbreaks, the current epizootic of HPAI H5N1 virus is highly unusual in many regards, such as the spread of HPAI H5N1 virus throughout Asia and into Europe and Africa, the large number of countries affected, the loss of hundreds of millions of poultry, the zoonotic transmission to humans and other mammals, the continuously changing genotypes and the spillback of the virus into wild birds, leading to outbreaks and circulation of HPAI H5N1 virus in those birds. The ancestral HPAI H5N1 virus likely originated from a virus circulating in domestic geese in Guandong province, China in 1996 and was introduced in Hong Kong poultry markets in 1997 (Chen et al., 2004). The 1997 outbreak in poultry in Hong Kong led to another paradigm shift, as direct transmission of a purely avian influenza virus from poultry caused respiratory disease and death in humans, something that was previously deemed impossible (de Jong et al., 1997). After the local containment of the HPAI H5N1 virus outbreak in Hong Kong, the virus reappeared in 2002 when it caused an outbreak in waterfowl and various other bird species in two waterfowl parks in Hong Kong (Sims et al., 2003; Ellis et al., 2004; Sturm-Ramirez et al., 2004). In 2003, the HPAI H5N1 virus was again transmitted to humans, leading to at least one fatal case. There is little information on the circulation of HPAI H5N1 virus from 1997 to 2002, but it is believed that the virus continued to circulate in China during that period. HPAI H5N1 virus resurfaced again in 2004 to spread across a large part of Southeast Asia, including Cambodia, China, Hong Kong, Indonesia, Japan, Laos, Malaysia, South Korea,

Thailand, and Vietnam. Until 2005, HPAI H5N1 viruses had been isolated only sporadically from wild birds. In 2005, the first reported outbreak in wild migratory birds occurred in April−June at Lake Qinghai, China. This HPAI H5N1 virus outbreak in wild birds affected large numbers of birds such as bar-headed geese (*Anser indicus*), brown-headed gulls (*Larus brunnicephallus*), great black-headed gulls (*Larus ichthyaetus*), and great cormorants (*Phalacrocorax carbo*) (Chen et al., 2005; Liu et al., 2005). After the HPAI H5N1 virus outbreak in wild birds in 2005, the virus rapidly spread westwards across Asia, Europe, the Middle East, and Africa. Affected wild birds have been reported in several countries, predominantly in mute swans (*Cygnus olor*), whooper swans (*Cygnus cygnus*), and tufted ducks (*Aythya fuligula*), although small numbers of cases in many other species have been reported as well (raptors, gulls, and herons) (Olsen et al., 2006; Hesterberg et al., 2009).

While in the years prior to 2005 the transmission of the HPAI H5N1 virus to new areas is thought to have primarily occurred via movement of poultry and poultry products (Alexander and Brown, 2000; Sims et al., 2003; Gilbert et al., 2006), dispersal via wild migratory birds seems to be a likely route for several of the reported outbreaks. Some have argued that infected birds would be too severely affected to continue migration and would thus be unlikely to spread the HPAI H5N1 virus (Feare and Yasue, 2006). However, it has been shown that the pathogenesis of the HPAI H5N1 virus infection and the susceptibility of wild bird species to this infection may vary considerably, depending on bird species and previous exposure to viruses of the same or other avian influenza virus subtypes. Recent experimental infections suggest that pre-exposure to LPAI viruses of homologous or heterologous subtypes may result in partial immunity to HPAI H5N1 virus infection (Fereidouni et al., 2009). Such preexisting immunity might protect birds from developing severe disease upon infection but may still allow replication and thus shedding and spreading of the virus. Upon experimental HPAI H5N1 virus infection, some duck species proved to develop minor, if any, disease signs while still excreting the virus, predominantly from the respiratory tract, whereas other species developed a largely fatal infection that would not allow them to spread the virus efficiently over a considerable distance (Brown et al., 2006, 2008; Kalthoff et al., 2008; Keawcharoen et al., 2008). The outcome of HPAI H5N1 virus infections in wild bird species generally ranges from high morbidity and mortality (geese, swan, and certain duck species) to minimal morbidity without mortality (dabbling duck species). The situation in Europe from 2005 to 2009, when infected wild birds have been found in several countries that have not reported outbreaks in poultry (Globig et al., 2009; Hesterberg et al., 2009), suggests that wild birds can indeed carry the virus to previously unaffected areas. In addition, analysis of the spread of HPAI H5N1 virus indicated that it was likely that most of the introductions into several parts of Asia were likely poultry related, whereas in Europe most introductions were probably caused by migratory birds (Starick et al., 2008; Si et al., 2009). Although swan deaths have been the first indicator for the presence of the HPAI H5N1 virus in several European countries, this does not necessarily imply a role as predominant vectors; they could merely have functioned as sentinel birds infected via other migrating bird species.

These recent introductions of HPAI H5N1 virus in wild birds and the subsequent spread of the virus through large parts of the eastern hemisphere has put a focus on the role of wild birds in the geographical spread of HPAI H5N1 virus. Wild bird surveillance programs have been implemented on an unprecedented scale in many parts of the world to determine the role of wild birds in the spread of the virus and potentially to serve as a warning system for virus incursions into new geographical regions (Munster et al., 2006; Gaidet et al., 2007; Dusek et al., 2009; Haynes et al., 2009; Kou et al., 2009; Fereidouni et al., 2010; Sharshov et al., 2010; Siembieda et al., 2010). Whereas the initial influenza A virus surveillance studies in the 1960s and 1970s relied on classical virological tools such as virus isolation in embryonated chicken eggs and were able to process, at most, a couple of thousand samples on a yearly basis, the implementation of high-throughput molecular diagnostic methods has allowed the intensity to increase to an astonishing 400,000 samples a year (Munster et al., 2009), making avian influenza viruses one of the most extensively studied wildlife diseases of our time.

Despite intensive surveillance programs, HPAI H5N1 virus has predominantly been found in dead wild birds (Artois et al., 2009; Hesterberg et al., 2009). Only in a limited number of cases was HPAI H5N1 virus detected in apparently healthy birds (Chen et al., 2006a,b). Many national surveillance programs aimed at the early detection of HPAI H5N1 virus have therefore focused on collecting samples from birds exhibiting morbidity or mortality. The intrinsic problem associated with establishing a clear picture of the prevalence of HPAI H5N1 virus in wild bird populations is the number of birds that have to be caught and sampled for this purpose. The more prevalent a virus is in the respective bird population, the fewer individuals need to be sampled to actually detect the virus. However, the number of birds that would need to be caught and sampled to detect viruses with a very low prevalence with a 95% probability of detection will rapidly become unfeasible, as may currently be the case with the lack of detection of HPAI H5N1 virus in wild bird populations (Artois et al., 2009).

This raises the question whether HPAI viruses have indeed become endemic in wild bird populations, or whether the HPAI H5N1 virus is being re-introduced repeatedly by poultry or human activities. A recent study reported high prevalence of HPAI H5N1 in wild birds in China, suggesting that HPAI H5N1 circulates endemically in China (Kou et al., 2009). Whether HPAI H5N1 viruses will eventually also cross the Atlantic or the Pacific Oceans to reach the Americas, remains a matter of speculation.

28.3 Evolutionary Genetics of Avian Influenza Viruses

28.3.1 LPAI Viruses

LPAI viruses can be divided into two main phylogenetic lineages: the Eurasian and American lineages (Chen and Holmes, 2006; Webster et al., 1992; Widjaja et al., 2004; Olsen et al., 2006). Apparently, the long-term ecological and geographical

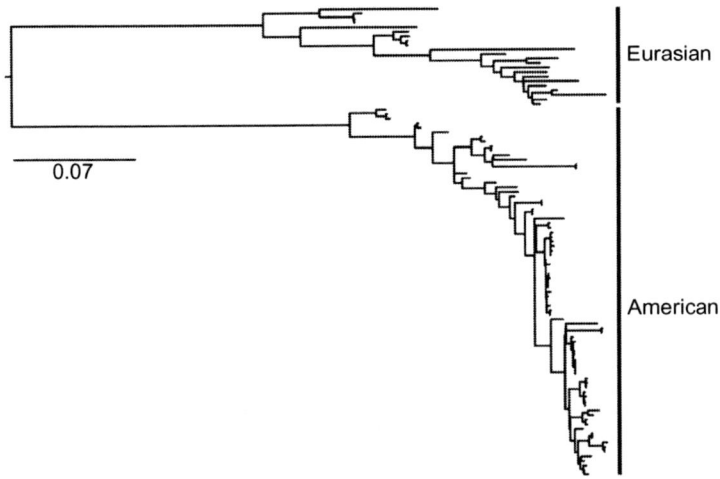

Figure 28.3 DNA maximum likelihood tree for the HA gene of subtype H10 influenza A viruses. All nearly full-length H10 HA genes available from GenBank were used to construct the tree. The major division into the Eurasian and American lineages of influenza viruses is clearly visible.

separation of the viruses isolated from bird species that utilize the migratory flyways of the Americas and Eurasia/Africa/Australia have led to allopatric speciation, resulting in a major phylogenetic split between the Eurasian and American genetic lineages of LPAI viruses. An example of the geographical differences between the genes within an HA subtype (H10 in this example) is given in Figure 28.3, but similar phylogenies can be observed for other viral genes and subtypes. Despite these phylogenetic splits, the separation of the American and Eurasian wild bird and virus populations is not absolute. Some ducks and shorebirds cross the Bering Strait during migration or have breeding ranges that include both the Russian Far East and northwestern America (1996). The majority of tundra shorebirds from the Russian Far East winter in Southeast Asia and Australia, but some species winter along the West coast of the Americas. The overlap in distribution of ducks is not as profound as that of shorebirds, but a few species (e.g., Northern pintail, *Anas acuta*) are common in both North America and Eurasia and could also provide an intercontinental bridge for influenza A virus. As a result, LPAI viruses carrying a mix of genes from the American and Eurasian lineages have been isolated occasionally, indicating that allopatric speciation is only partial and that gene segments may be exchanged between the two virus populations (Makarova et al., 1999; Wallensten et al., 2005; Krauss et al., 2007; Dugan et al., 2008; Koehler et al., 2008; Wahlgren et al., 2008; Ramey et al., 2010). Whole-genome analyses of LPAI virus isolates obtained from Northern pintails in Alaska, a species that migrates between North America and Asia, suggested that intercontinental virus exchange can occur at a relatively high frequency (Ramey et al., 2010). Of the viruses analyzed in this study, a large proportion had at least one gene segment originating from the Eurasian lineage. In addition, analyses of H6 LPAI viruses pointed to introductions of the Eurasian H6 gene segment in North America on several occasions (zu Dohna et al., 2009). So far, the cross-hemisphere transmission of LPAI viruses has not resulted in the detection of entire virus genomes but only introduction of single gene segments that reassorted with other segments found in the new hemisphere (Dugan et al., 2008). The partial geographic isolation of influenza virus hosts therefore seems to be sufficient to facilitate divergent evolution and the continuous existence of separate gene pools.

Besides the influence of geographical separation on the evolutionary genetics of LPAI viruses, differences in host species have also resulted in clearly distinguishable virus populations. Good examples are the LPAI viruses of the H13 and H16 subtypes that are predominantly isolated from gulls and terns (Hinshaw et al., 1982; Fouchier et al., 2005). These viruses belong to a group of distinct LPAI viruses based on genetic, functional, and ecological properties and have evolved into separate genetic lineages from the viruses isolated from other *Charadriiformes* and *Anseriformes* (H1−H12 subtypes). Gene segments of gull viruses are genetically distinct from those circulating in other wild birds, suggesting that they have been separated for a sufficient amount of time to allow genetic differentiation by sympatric speciation (Fouchier et al., 2005; Olsen et al., 2006). An example of the diversification of the gull lineage from other lineages of avian influenza virus genes is shown for gene segment 5 (NP) in Figure 28.4 (Gorman et al., 1990). Gull influenza viruses do not readily infect ducks upon experimental inoculation (Hinshaw

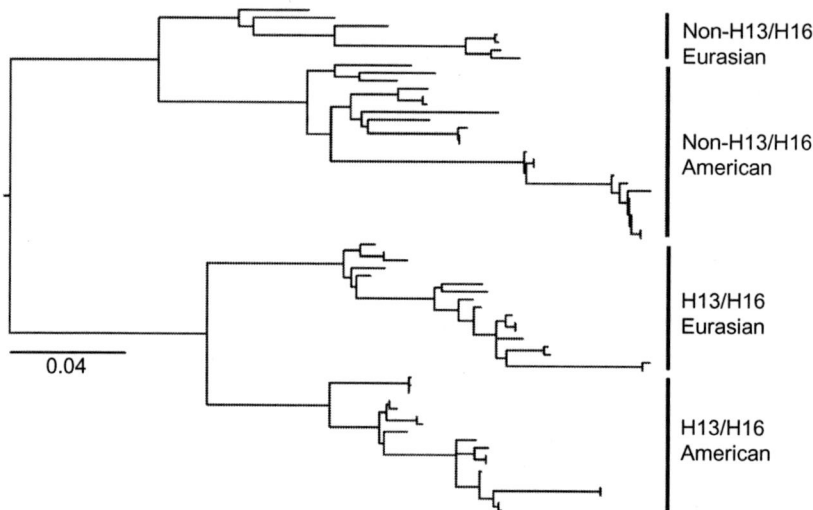

Figure 28.4 DNA maximum likelihood tree for the NP gene of gull influenza A viruses. All nearly full-length gull virus NP genes from GenBank were used to construct the tree. Virus of the H13 and H16 subtypes form a gull-specific genetic lineage (bottom), which is distinct from the lineage of influenza viruses isolated from other avian hosts. In both the gull lineage and non-gull lineage, a separation is observed for American and Eurasian strains.

et al., 1982), providing a biological explanation for the limited detection of these viruses in other avian influenza host species, although a limited number of gull viruses has been isolated from ducks and vice versa (Kawaoka et al., 1988; Krauss et al., 2004; Munster et al., 2007). Genetic data from duck and shorebird influenza A virus isolates from the Americas suggested an active interplay exists between these host populations, as recent genetic analyses did not reveal striking differences between influenza viruses from ducks and waders in the Americas, suggesting that the viruses are not separated and that the duck and shorebird populations might function as one influenza virus host population (Widjaja et al., 2004; Spackman et al., 2005). Although certain HA subtypes have been reported to be more prevalent in either shorebirds or ducks in North America, this also does not seem to have resulted in differences in the genetic composition of influenza viruses obtained from these two reservoirs (Obenauer et al., 2006; Krauss et al., 2007), in contrast to what is observed for gulls.

Evolution of the genes of avian influenza viruses in their natural hosts is often considered slow (Webster et al., 1992), but it certainly is not (Chen and Holmes, 2006; Dugan et al., 2008). Recent estimates indicated that avian influenza viruses evolve at a rate of $>10^{-3}$ substitutions per site per year (Chen and Holmes, 2006; Webster et al., 1992). For each gene segment, and within both the Eurasian and American genetic lineages, multiple sublineages of viral genes seem to co-circulate, but without apparently consistent temporal or spatial correlations. The only additional noticeable peculiarity is the evolution of gene segment 8 (NS) into two highly divergent genetic lineages, referred to as alleles A and B (Figure 28.5; Spackman et al., 2005; Dugan et al., 2008). The biological significance of these two alleles remains unknown, but the alleles have been detected on both hemispheres, in all virus subtypes irrespective of wild bird host species.

The segmented genome of influenza viruses enables evolution by a process known as genetic reassortment, the mixing of genes from two (or more) viruses (Webster et al., 1992). Reassortment is one of the driving forces of the genetic variation of LPAI and HPAI viruses and contributes greatly to their phenotypic

Figure 28.5 DNA maximum likelihood tree for the NS gene of avian influenza A viruses. All nearly full-length NS genes of Dutch avian viruses available from GenBank were used to construct the tree. These sequences are phylogenetically divided in two groups, representing two different "alleles," with no obvious correlation with time, location, or host species.

variability. A study of LPAI viruses obtained from Canadian ducks showed that genetic "sublineages" do not persist in wild birds, but frequently reassort (Hatchette et al., 2004). Analysis of the genome constellation of five H4N6 LPAI viruses isolated from mallards on one day and one location revealed four different genome constellations with only one pair of viruses having a identical genome composition (Dugan et al., 2008). The high LPAI virus prevalence in some wild bird species and the detection of concomitant infections in single birds supports the notion that reassortment occurs at a relatively high rate in wild birds (Macken et al., 2006; Dugan et al., 2008). Thus, LPAI viruses do not present as "fixed" genome constellations but reassortment leads to new "transient" genome constellations continuously.

28.3.2 HPAI H5N1 Virus

The evolution of the Eurasian HPAI H5N1 viruses since their first detection in 1997 has been relatively well recorded (Neumann et al., 2010). Unique for HPAI viruses, the evolutionary path of H5N1 virus has been characterized by very frequent reassortment events, creating numerous novel so-called "genotypes" in time (Neumann et al., 2010). Thus, starting with viruses similar to A/Goose/Guangdong/1/1996, a series of reassortment events between HPAI H5N1 viruses and LPAI viruses from wild birds or poultry, have led to the diverse H5N1 virus lineages circulating today (Neumann et al., 2010). After the depopulation of the poultry markets of Hong Kong in 1997, the virus is thought to have remained present in Southern China. Sporadically, viruses entered the markets in the Hong Kong area and South Korea (Neumann et al., 2010). From December 2003 to February 2004, eight countries in East and Southeast Asia reported H5N1 outbreaks. During these and later outbreaks, a wide range of domestic birds, including free-ranging commercial ducks, village poultry, birds in live bird markets, fighting cocks, etc., were found to be infected, in addition to regular commercial poultry. The wide range of domestic birds infected, and the contact of these birds with wild bird species may not only have provided opportunities for rapid spread of the virus to new areas, but also may have caused multiple cycles of periods of positive selection in new host species. Indeed, as compared to LPAI viruses in wild birds, the Asian H5N1 HPAI virus lineage displays a relatively high evolutionary rate and selection pressures, as measured by dN/dS ratios (Chen and Holmes, 2006).

As a consequence, the HA gene of the Eurasian HPAI H5N1 virus lineage has also evolved rapidly, diverging into at least 10 antigenically and genetically distinct "clades" (WHO/OIE/FAO H5N1 evolution working group, 2009; Figure 28.6). While some of the "clades" have predominantly been detected in particular regions (e.g., clade 2.1 in Indonesia, clade 2.2 in Europe and Africa), there is no clear correlation between virus genetics and geography, and different clades appear to continue to be introduced to "new" areas (e.g., the recent spread of clade 2.3 in Southeast Asia). This evolutionary pattern is unprecedented in the recent history of HPAI viruses, and is likely due to multiple factors, including the enormous geographical spread of the outbreak throughout the eastern hemisphere, the

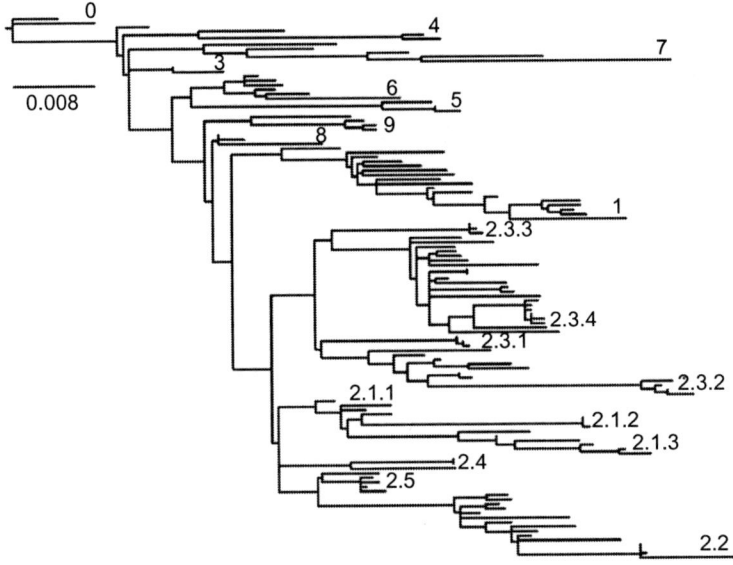

Figure 28.6 DNA maximum likelihood tree for the HA gene of representative HPAI H5N1 viruses. Nearly full-length HA genes available from GenBank were used to construct the tree. The phylogenetic tree shows the known "clades" of viruses, as identified by numbers (Ramey et al., 2010). These clades were defined using criteria based on sharing of a common ancestor and monophyletic grouping with high bootstrap support.

involvement of large number of hosts—both numbers of individual birds and numbers of different species—and the duration of the outbreak. Each of these factors is unprecedented in HPAI outbreaks of the recent history, and likely contributed to the rapid evolution of the HPAI H5N1 virus lineage.

The result of this diversification of the virus—genetically and antigenically—is that preparation for a new potential influenza pandemic caused by the H5N1 virus is very difficult. Currently, vaccination is the method of choice to protect humans against influenza. However, due to the antigenic variation of the H5N1 viruses that currently cause outbreaks in poultry and morbidity and mortality in humans, it has been shown to be impossible to prepare vaccines protective against all circulating "clades." In the interest of both animal and human health, continued monitoring of emerging virus variants during new outbreaks is warranted.

28.4 Future Perspective

Wild bird surveillance programs have been implemented at an unprecedented scale in many parts of the world to determine the role of wild birds in the spread of avian influenza viruses and potentially to serve as a warning system for HPAI H5N1 virus incursions into new geographical regions. At the same time, the increased capacity of current sequencing technology has facilitated the genetic analyses of large numbers of influenza A virus genomes. Together, these developments provide a unique opportunity to advance our understanding of the ecology and evolution of avian influenza in wild birds, and in the longer term will help to limit the impact of influenza on human and animal health. To this end, we hope that all specialists in the field, including virologists, epidemiologists, ornithologists, geneticists, ecologists, veterinarians, clinicians, mathematicians, bio-informaticists, will work together closely to make optimal use of the wealth of data available, ultimately leading to a better understanding of the ecology of avian influenza—LPAI and HPAI—in wild birds.

Acknowledgments

We would like to thank all of our collaborators who have supported our work through many years and with whom we have had numerous stimulating discussions to help us with our research. The research of the authors is sponsored by NIAID-NIH contract HHSN266200700010C.

References

Alexander, D.J., 2000. A review of avian influenza in different bird species. Vet. Microbiol. 74, 3−13.
Alexander, D.J., 2007a. An overview of the epidemiology of avian influenza. Vaccine 25, 5637−5644.
Alexander, D.J., 2007b. Summary of avian influenza activity in Europe, Asia, Africa, and Australasia, 2002−2006. Avian Dis. 51, 161−166.
Alexander, D.J., Brown, I.H., 2000. Recent zoonoses caused by influenza A viruses. Rev. Sci. Tech. 19, 197−225.
Alexander, D.J., Brown, I.H., 2009. History of highly pathogenic avian influenza. Rev. Sci. Tech. 28, 19−38.
Artois, M., Bicout, D., Doctrinal, D., Fouchier, R., Gavier-Widen, D., Globig, A., et al., 2009. Outbreaks of highly pathogenic avian influenza in Europe: the risks associated with wild birds. Rev. Sci. Tech. 28, 69−92.
Becker, W.B., 1966. The isolation and classification of Tern virus: influenza A-Tern South Africa—1961. J. Hyg. (Lond.) 64, 309−320.
Brown, J.D., Stallknecht, D.E., Beck, J.R., Suarez, D.L., Swayne, D.E., 2006. Susceptibility of North American ducks and gulls to H5N1 highly pathogenic avian influenza viruses. Emerg. Infect. Dis. 12, 1663−1670.
Brown, J.D., Stallknecht, D.E., Swayne, D.E., 2008. Experimental infection of swans and geese with highly pathogenic avian influenza virus (H5N1) of Asian lineage. Emerg. Infect. Dis. 14, 136−142.

Brown, J.D., Goekjian, G., Poulson, R., Valeika, S., Stallknecht, D.E., 2009. Avian influenza virus in water: infectivity is dependent on pH, salinity and temperature. Vet. Microbiol. 136, 20–26.

Chen, R., Holmes, E.C., 2006. Avian influenza virus exhibits rapid evolutionary dynamics. Mol. Biol. Evol. 23, 2336–2341.

Chen, W., Calvo, P.A., Malide, D., Gibbs, J., Schubert, U., Bacik, I., et al., 2001. A novel influenza A virus mitochondrial protein that induces cell death. Nat. Med. 7, 1306–1312.

Chen, H., Deng, G., Li, Z., Tian, G., Li, Y., Jiao, P., et al., 2004. The evolution of H5N1 influenza viruses in ducks in southern China. Proc. Natl. Acad. Sci. U.S.A. 101, 10452–10457.

Chen, H., Smith, G.J., Zhang, S.Y., Qin, K., Wang, J., Li, K.S., et al., 2005. Avian flu: H5N1 virus outbreak in migratory waterfowl. Nature 436, 191–192.

Chen, H., Li, Y., Li, Z., Shi, J., Shinya, K., Deng, G., et al., 2006a. Properties and dissemination of H5N1 viruses isolated during an influenza outbreak in migratory waterfowl in western China. J. Virol. 80, 5976–5983.

Chen, H.X., Shen, H.G., Li, X.L., Zhou, J.Y., Hou, Y.Q., Guo, J.Q., et al., 2006b. Seroprevalence and identification of influenza A virus infection from migratory wild waterfowl in China (2004–2005). J. Vet. Med. B Infect. Dis. Vet. Public Health 53, 166–170.

Conenello, G.M., Palese, P., 2007. Influenza A virus PB1-F2: a small protein with a big punch. Cell Host Microbe 2, 207–209.

de Jong, J.C., Claas, E.C., Osterhaus, A.D., Webster, R.G., Lim, W.L., 1997. A pandemic warning? Nature 389, 554.

Del Hoyo, J., Elliot, A., Sargatal, J., 1996. Handbook of the Birds of the World. Lynx Edicions, Barcelona.

Dugan, V.G., Chen, R., Spiro, D.J., Sengamalay, N., Zaborsky, J., Ghedin, E., et al., 2008. The evolutionary genetics and emergence of avian influenza viruses in wild birds. PLoS Pathog. 4, e1000076.

Dusek, R.J., Bortner, J.B., DeLiberto, T.J., Hoskins, J., Franson, J.C., Bales, B.D., et al., 2009. Surveillance for high pathogenicity avian influenza virus in wild birds in the Pacific Flyway of the United States, 2006–2007. Avian Dis. 53, 222–230.

Ellis, T.M., Bousfield, R.B., Bissett, L.A., Dyrting, K.C., Luk, G.S., Tsim, S.T., et al., 2004. Investigation of outbreaks of highly pathogenic H5N1 avian influenza in waterfowl and wild birds in Hong Kong in late 2002. Avian Pathol. 33, 492–505.

Ellstrom, P., Latorre-Margalef, N., Griekspoor, P., Waldenstrom, J., Olofsson, J., Wahlgren, J., et al., 2008. Sampling for low-pathogenic avian influenza A virus in wild Mallard ducks: oropharyngeal versus cloacal swabbing. Vaccine 26, 4414–4416.

Feare, C.J., Yasue, M., 2006. Asymptomatic infection with highly pathogenic avian influenza H5N1 in wild birds: how sound is the evidence? Virol. J. 3, 96.

Fereidouni, S.R., Starick, E., Beer, M., Wilking, H., Kalthoff, D., Grund, C., et al., 2009. Highly pathogenic avian influenza virus infection of mallards with homo- and heterosubtypic immunity induced by low pathogenic avian influenza viruses. PLoS One 4, e6706.

Fereidouni, S.R., Werner, O., Starick, E., Beer, M., Harder, T.C., Aghakhan, M., et al., 2010. Avian influenza virus monitoring in wintering waterbirds in Iran, 2003–2007. Virol. J. 7, 43.

Fouchier, R.A., Munster, V., Wallensten, A., Bestebroer, T.M., Herfst, S., Smith, D., et al., 2005. Characterization of a novel influenza A virus hemagglutinin subtype (H16) obtained from black-headed gulls. J. Virol. 79, 2814–2822.

Gaidet, N., Dodman, T., Caron, A., Balanca, G., Desvaux, S., Goutard, F., et al., 2007. Avian influenza viruses in water birds, Africa. Emerg. Infect. Dis. 13, 626–629.

Gaidet, N., Newman, S.H., Hagemeijer, W., Dodman, T., Cappelle, J., Hammoumi, S., et al., 2008. Duck migration and past influenza A (H5N1) outbreak areas. Emerg. Infect. Dis. 14, 1164–1166.

Gilbert, M., Chaitaweesub, P., Parakamawongsa, T., Premashthira, S., Tiensin, T., Kalpravidh, W., et al., 2006. Free-grazing ducks and highly pathogenic avian influenza, Thailand. Emerg. Infect. Dis. 12, 227–234.

Globig, A., Staubach, C., Beer, M., Koppen, U., Fiedler, W., Nieburg, M., et al., 2009. Epidemiological and ornithological aspects of outbreaks of highly pathogenic avian influenza virus H5N1 of Asian lineage in wild birds in Germany, 2006 and 2007. Transbound. Emerg. Dis. 56, 57–72.

Gorman, O.T., Bean, W.J., Kawaoka, Y., Webster, R.G., 1990. Evolution of the nucleoprotein gene of influenza A virus. J. Virol. 64, 1487–1497.

Halvorson, D.A., Kelleher, C.J., Senne, D.A., 1985. Epizootiology of avian influenza: effect of season on incidence in sentinel ducks and domestic turkeys in Minnesota. Appl. Environ. Microbiol. 49, 914–919.

Hanson, B.A., Stallknecht, D.E., Swayne, D.E., Lewis, L.A., Senne, D.A., 2003. Avian influenza viruses in Minnesota ducks during 1998–2000. Avian Dis. 47, 867–871.

Hanson, B.A., Luttrell, M.P., Goekjian, V.H., Niles, L., Swayne, D.E., Senne, D.A., et al., 2008. Is the occurrence of avian influenza virus in Charadriiformes species and location dependent? J. Wildl. Dis. 44, 351–361.

Hatchette, T.F., Walker, D., Johnson, C., Baker, A., Pryor, S.P., Webster, R.G., 2004. Influenza A viruses in feral Canadian ducks: extensive reassortment in nature. J. Gen. Virol. 85, 2327–2337.

Haynes, L., Arzey, E., Bell, C., Buchanan, N., Burgess, G., Cronan, V., et al., 2009. Australian surveillance for avian influenza viruses in wild birds between July 2005 and June 2007. Aust. Vet. J. 87, 266–272.

Hesterberg, U., Harris, K., Stroud, D., Guberti, V., Busani, L., Pittman, M., et al., 2009. Avian influenza surveillance in wild birds in the European Union in 2006. Influenza Other Respi. Viruses 3, 1–14.

Hinshaw, V.S., Webster, R.G., Turner, B., 1978. Novel influenza A viruses isolated from Canadian feral ducks: including strains antigenically related to swine influenza (Hsw1N1) viruses. J. Gen. Virol. 41, 115–127.

Hinshaw, V.S., Webster, R.G., Bean, W.J., Sriram, G., 1980a. The ecology of influenza viruses in ducks and analysis of influenza viruses with monoclonal antibodies. Comp. Immunol. Microbiol. Infect. Dis. 3, 155–164.

Hinshaw, V.S., Webster, R.G., Turner, B., 1980b. The perpetuation of orthomyxoviruses and paramyxoviruses in Canadian waterfowl. Can. J. Microbiol. 26, 622–629.

Hinshaw, V.S., Air, G.M., Gibbs, A.J., Graves, L., Prescott, B., Karunakaran, D., 1982. Antigenic and genetic characterization of a novel hemagglutinin subtype of influenza A viruses from gulls. J. Virol. 42, 865–872.

Hurt, A.C., Hansbro, P.M., Selleck, P., Olsen, B., Minton, C., Hampson, A.W., et al., 2006. Isolation of avian influenza viruses from two different transhemispheric migratory shorebird species in Australia. Arch. Virol. 151, 2301–2309.

International Committee on Taxonomy of Viruses, 2000. Virus Taxonomy, Seventh Report of the International Committee on Taxonomy of Viruses. Academic Press, San Diego, CA.

Jourdain, E., Gunnarsson, G., Wahlgren, J., Latorre-Margalef, N., Brojer, C., Sahlin, S., et al., 2010. Influenza virus in a natural host, the mallard: experimental infection data. PLoS One 5, e8935.

Kalthoff, D., Breithaupt, A., Teifke, J.P., Globig, A., Harder, T., Mettenleiter, T.C., et al., 2008. Highly pathogenic avian influenza virus (H5N1) in experimentally infected adult mute swans. Emerg. Infect. Dis. 14, 1267−1270.

Karlsson Hedestam, G.B., Fouchier, R.A., Phogat, S., Burton, D.R., Sodroski, J., Wyatt, R.T., 2008. The challenges of eliciting neutralizing antibodies to HIV-1 and to influenza virus. Nat. Rev. Microbiol. 6, 143−155.

Kawaoka, Y., Chambers, T.M., Sladen, W.L., Webster, R.G., 1988. Is the gene pool of influenza viruses in shorebirds and gulls different from that in wild ducks? Virology 163, 247−250.

Kawaoka, Y., Yamnikova, S., Chambers, T.M., Lvov, D.K., Webster, R.G., 1990. Molecular characterization of a new hemagglutinin, subtype H14, of influenza A virus. Virology 179, 759−767.

Keawcharoen, J., van Riel, D., van Amerongen, G., Bestebroer, T., Beyer, W.E., van Lavieren, R., et al., 2008. Wild ducks as long-distance vectors of highly pathogenic avian influenza virus (H5N1). Emerg. Infect. Dis. 14, 600−607.

Kida, H., Yanagawa, R., Matsuoka, Y., 1980. Duck influenza lacking evidence of disease signs and immune response. Infect. Immun. 30, 547−553.

Kleijn, D., Munster, V.J., Ebbinge, B.S., Jonkers, D.A., Muskens, G.J., Van Randen, Y., et al., 2010. Dynamics and ecological consequences of avian influenza virus infection in greater white-fronted geese in their winter staging areas. Proc. Biol. Sci. 277, 2041−2048.

Koehler, A.V., Pearce, J.M., Flint, P.L., Franson, J.C., Ip, H.S., 2008. Genetic evidence of intercontinental movement of avian influenza in a migratory bird: the northern pintail (*Anas acuta*). Mol. Ecol. 17, 4754−4762.

Kou, Z., Li, Y., Yin, Z., Guo, S., Wang, M., Gao, X., et al., 2009. The survey of H5N1 flu virus in wild birds in 14 provinces of China from 2004 to 2007. PLoS One 4, e6926.

Krauss, S., Walker, D., Pryor, S.P., Niles, L., Chenghong, L., Hinshaw, V.S., et al., 2004. Influenza A viruses of migrating wild aquatic birds in North America. Vector Borne Zoonotic Dis. 4, 177−189.

Krauss, S., Obert, C.A., Franks, J., Walker, D., Jones, K., Seiler, P., et al., 2007. Influenza in migratory birds and evidence of limited intercontinental virus exchange. PLoS Pathog. 3, e167.

Latorre-Margalef, N., Gunnarsson, G., Munster, V.J., Fouchier, R.A., Osterhaus, A.D., Elmberg, J., et al., 2009. Effects of influenza a virus infection on migrating mallard ducks. Proc. Biol. Sci. 276, 1029−1036.

Liu, J., Xiao, H., Lei, F., Zhu, Q., Qin, K., Zhang, X.W., et al., 2005. Highly pathogenic H5N1 influenza virus infection in migratory birds. Science 309, 1206.

Macken, C.A., Webby, R.J., Bruno, W.J., 2006. Genotype turnover by reassortment of replication complex genes from avian influenza A virus. J. Gen. Virol. 87, 2803−2815.

Makarova, N.V., Kaverin, N.V., Krauss, S., Senne, D., Webster, R.G., 1999. Transmission of Eurasian avian H2 influenza virus to shorebirds in North America. J. Gen. Virol. 80 (Pt 12), 3167−3171.

Munster, V.J., Veen, J., Olsen, B., Vogel, R., Osterhaus, A.D., Fouchier, R.A., 2006. Towards improved influenza A virus surveillance in migrating birds. Vaccine 24, 6729−6733.

Munster, V.J., Baas, C., Lexmond, P., Waldenstrom, J., Wallensten, A., Fransson, T., et al., 2007. Spatial, temporal, and species variation in prevalence of influenza A viruses in wild migratory birds. PLoS Pathog. 3, e61.

Munster, V.J., Baas, C., Lexmond, P., Bestebroer, T.M., Guldemeester, J., Beyer, W.E., et al., 2009. Practical considerations for high-throughput influenza A virus surveillance studies of wild birds by use of molecular diagnostic tests. J. Clin. Microbiol. 47, 666−673.

Neumann, G., Green, M.A., Macken, C.A., 2010. Evolution of highly pathogenic avian H5N1 influenza viruses and the emergence of dominant variants. J. Gen. Virol. 91 (Pt 8), 1984−1995.

Obenauer, J.C., Denson, J., Mehta, P.K., Su, X., Mukatira, S., Finkelstein, D.B., et al., 2006. Large-scale sequence analysis of avian influenza isolates. Science 311, 1576−1580.

Olsen, B., Munster, V.J., Wallensten, A., Waldenstrom, J., Osterhaus, A.D., Fouchier, R.A., 2006. Global patterns of influenza a virus in wild birds. Science 312, 384−388.

Palese, P., Shaw, M.L., 2007. Orthomyxoviridae: the viruses and their replication. In: Knipe, D.M., Howley, P.M., Griffin, D.E., Lamb, R.A., Martin, M.A., Roizman, B., Strauss, S. (Eds.), Fields Virology, fifth ed. Lippincott Williams & Wilkins, Philadelphia, PA, pp. 1647−1690.

Parmley, E.J., Bastien, N., Booth, T.F., Bowes, V., Buck, P.A., Breault, A., et al., 2008. Wild bird influenza survey, Canada, 2005. Emerg. Infect. Dis. 14, 84−87.

Ramey, A.M., Pearce, J.M., Flint, P.L., Ip, H.S., Derksen, D.V., Franson, J.C., et al., 2010. Intercontinental reassortment and genomic variation of low pathogenic avian influenza viruses isolated from northern pintails (*Anas acuta*) in Alaska: examining the evidence through space and time. Virology 401, 179−189.

Rohm, C., Zhou, N., Suss, J., Mackenzie, J., Webster, R.G., 1996. Characterization of a novel influenza hemagglutinin, H15: criteria for determination of influenza A subtypes. Virology 217, 508−516.

Sharshov, K., Silko, N., Sousloparov, I., Zaykovskaya, A., Shestopalov, A., Drozdov, I., 2010. Avian influenza (H5N1) outbreak among wild birds, Russia, 2009. Emerg. Infect. Dis. 16, 349−351.

Si, Y., Skidmore, A.K., Wang, T., de Boer, W.F., Debba, P., Toxopeus, A.G., et al., 2009. Spatio-temporal dynamics of global H5N1 outbreaks match bird migration patterns. Geospat. Health 4, 65−78.

Siembieda, J.L., Johnson, C.K., Cardona, C., Anchell, N., Dao, N., Reisen, W., et al., 2010. Influenza A viruses in wild birds of the Pacific flyway. Vector Borne Zoonotic Dis. 2005−2008.

Sims, L.D., Ellis, T.M., Liu, K.K., Dyrting, K., Wong, H., Peiris, M., et al., 2003. Avian influenza in Hong Kong 1997−2002. Avian Dis. 47, 832−838.

Spackman, E., Stallknecht, D.E., Slemons, R.D., Winker, K., Suarez, D.L., Scott, M., et al., 2005. Phylogenetic analyses of type A influenza genes in natural reservoir species in North America reveals genetic variation. Virus Res. 114, 89−100.

Stallknecht, D.E., Brown, J.D., 2009. Tenacity of avian influenza viruses. Rev. Sci. Tech. 28, 59−67.

Stallknecht, D.E., Kearney, M.T., Shane, S.M., Zwank, P.J., 1990a. Effects of pH, temperature, and salinity on persistence of avian influenza viruses in water. Avian Dis. 34, 412−418.

Stallknecht, D.E., Shane, S.M., Kearney, M.T., Zwank, P.J., 1990b. Persistence of avian influenza viruses in water. Avian Dis. 34, 406−411.

Starick, E., Beer, M., Hoffmann, B., Staubach, C., Werner, O., Globig, A., et al., 2008. Phylogenetic analyses of highly pathogenic avian influenza virus isolates from Germany in 2006 and 2007 suggest at least three separate introductions of H5N1 virus. Vet. Microbiol. 128, 243–252.

Sturm-Ramirez, K.M., Ellis, T., Bousfield, B., Bissett, L., Dyrting, K., Rehg, J.E., et al., 2004. Reemerging H5N1 influenza viruses in Hong Kong in 2002 are highly pathogenic to ducks. J. Virol. 78, 4892–4901.

Wahlgren, J., Waldenstrom, J., Sahlin, S., Haemig, P.D., Fouchier, R.A., Osterhaus, A.D., et al., 2008. Gene segment reassortment between American and Asian lineages of avian influenza virus from waterfowl in the Beringia area. Vector Borne Zoonotic Dis. 8, 783–790.

Wallensten, A., Munster, V.J., Elmberg, J., Osterhaus, A.D., Fouchier, R.A., Olsen, B., 2005. Multiple gene segment reassortment between Eurasian and American lineages of influenza A virus (H6N2) in Guillemot (*Uria aalge*). Arch. Virol. 150, 1685–1692.

Wallensten, A., Munster, V.J., Karlsson, M., Lundkvist, A., Brytting, M., Stervander, M., et al., 2006. High prevalence of influenza A virus in ducks caught during spring migration through Sweden. Vaccine 24, 6734–6735.

Wallensten, A., Munster, V.J., Latorre-Margalef, N., Brytting, M., Elmberg, J., Fouchier, R. A., et al., 2007. Surveillance of influenza a virus in migratory waterfowl in northern Europe. Emerg. Infect. Dis. 13, 404–411.

Webster, R.G., Rott, R., 1987. Influenza virus A pathogenicity: the pivotal role of hemagglutinin. Cell 50, 665–666.

Webster, R.G., Bean, W.J., Gorman, O.T., Chambers, T.M., Kawaoka, Y., 1992. Evolution and ecology of influenza A viruses. Microbiol. Rev. 56, 152–179.

WHO/OIE/FAO H5N1 evolution working group, 2009. Continuing progress towards a unified nomenclature for the highly pathogenic H5N1 avian influenza viruses: divergence of clade 2.2 viruses. Influenza Other Respi. Viruses 3, 59–62.

Widjaja, L., Krauss, S.L., Webby, R.J., Xie, T., Webster, R.G., 2004. Matrix gene of influenza a viruses isolated from wild aquatic birds: ecology and emergence of influenza a viruses. J. Virol. 78, 8771–8779.

Winker, K., Spackman, E., Swayne, D.E., 2008. Rarity of influenza A virus in spring shorebirds, southern Alaska. Emerg. Infect. Dis. 14, 1314–1316.

Wright, P.F., Neumann, G., Kawaoka, Y., 2007. Orthomyxoviruses. In: Knipe, D.M., Howley, P.M., Griffin, D.E., Lamb, R.A., Martin, M.A., Roizman, B., Strauss, S. (Eds.), Fields Virology, fifth ed. Lippincott Williams & Wilkins, Philadelphia, PA, pp. 1691–1740.

zu Dohna, H., Li, J., Cardona, C.J., Miller, J., Carpenter, T.E., 2009. Invasions by Eurasian avian influenza virus H6 genes and replacement of the virus' North American clade. Emerg. Infect. Dis. 15, 1040–1045.